W9-BIP-509

Glencoe
Algebra 2

Integration
Applications
Connections

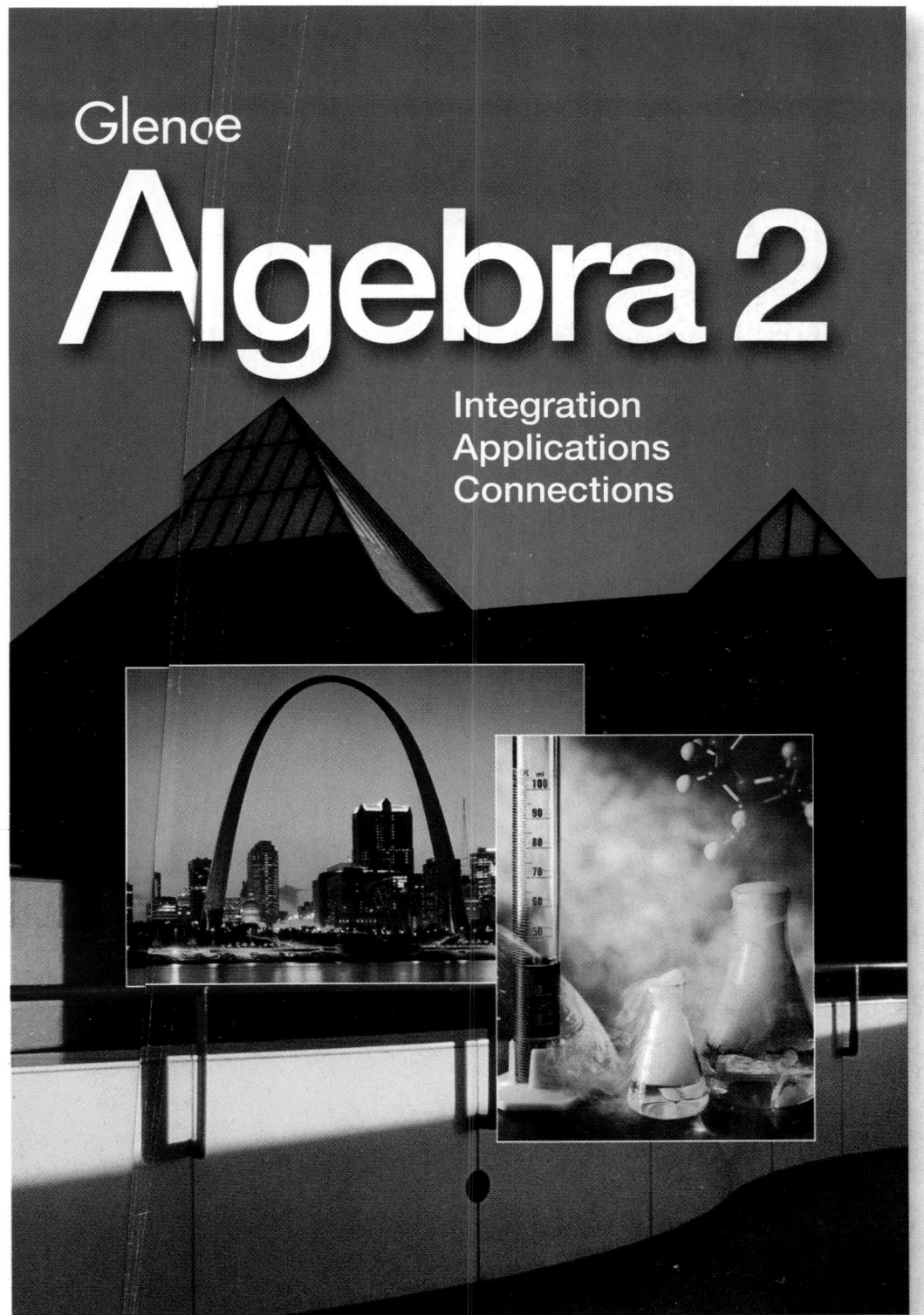

GLENCOE
McGraw-Hill

New York, New York   Columbus, Ohio   Woodland Hills, California   Peoria, Illinois

Glencoe/McGraw-Hill

*A Division of The **McGraw·Hill** Companies*

Send all inquiries to:
Glencoe/McGraw-Hill
8787 Orion Place
Columbus, OH 43240

ISBN: 0-02-825178-4

10 11 12 13 14 071/043 04 03 02 01 00

WHY IS ALGEBRA IMPORTANT?

Algebra is a tool that you'll be able to use throughout your life.

Algebra 2 is designed to illustrate how you'll be using algebra in the real world. This goal is accomplished through **integration, applications,** and **connections.**

Geometry

In addition to algebra, you'll study many other math topics. You'll learn how different branches of mathematics, such as geometry and statistics, are interrelated.

You'll use matrices to perform geometric transformations such as dilations. (Lesson 4-1, page 190)

Space Science

Mathematics can usually be related to real-life events, even the making of a hit movie. Relevant, real-world uses of mathematics are featured throughout this book.

You'll learn how quadratic equations and parabolas are related to the making of the hit movie *Apollo 13.* (Lesson 6-1, page 334)

Literature

Many authors incorporate their interest in math in their writing. Mathematics topics are connected to other subjects that you study, even literature.

You'll use Edgar Allen Poe's *The Pit and the Pendulum* to evaluate expressions with rational exponents. (Lesson 5-7, page 296)

Authors

WILLIAM COLLINS teaches mathematics at James Lick High School in San Jose, California. He has served as the mathematics department chairperson at James Lick and Andrew Hill High Schools. Mr. Collins received his B.A. in mathematics and philosophy from Herbert H. Lehman College in Bronx, New York, and his M.S. in mathematics education from California State University, Hayward. Mr. Collins is a member of the Association of Supervision and Curriculum Development and the National Council of Teachers of Mathematics, and is active in several professional mathematics organizations at the state level. He is also currently serving on the Teacher Advisory Panel of the *Mathematics Teacher*.

"In this era of educational reform and change, it is good to be part of a program that will set the pace for others to follow. This program integrates the ideas of the NCTM Standards with real tools for the classroom, so that algebra teachers and students can expect success every day."

GILBERT CUEVAS is a professor of mathematics education at the University of Miami in Miami, Florida. Dr. Cuevas received his B.A. in mathematics and M.Ed. and Ph.D., both in educational research, from the University of Miami. He also holds a M.A.T. in mathematics from Tulane University. Dr. Cuevas is a member of many mathematics, science, and research associations on the local, state, and national levels and has been an author and editor of several National Council of Teachers of Mathematics (NCTM) publications. He is also a frequent speaker at NCTM conferences, particularly on the topics of equity and mathematics for all students.

ALAN G. FOSTER is a former mathematics teacher and department chairperson at Addison Trail High School in Addison, Illinois. He obtained his B.S. from Illinois State University and his M.A. in mathematics from the University of Illinois. Mr. Foster is a past president of the Illinois Council of Teachers of Mathematics (ICTM) and was a recipient of the ICTM's T.E. Rine Award for Excellence in the Teaching of Mathematics. He also was a recipient of the 1987 Presidential Award for Excellence in the Teaching of Mathematics for Illinois. Mr. Foster was the chairperson of the MATHCOUNTS question writing committee in 1990 and 1991. He frequently speaks and conducts workshops on the topic of cooperative learning.

BERCHIE GORDON is the mathematics/science coordinator for the Northwest Local School District in Cincinnati, Ohio. Dr. Gordon has taught mathematics at every level from junior high school to college. She received her B.S. in mathematics from Emory University in Atlanta, Georgia, her M.A.T. in education from Northwestern University in Evanston, Illinois, and her Ph.D. in curriculum and instruction at the University of Cincinnati. Dr. Gordon has developed and conducted numerous inservice workshops in mathematics and computer applications. She has also served as a consultant for IBM, and has traveled throughout the country making presentations on graphing calculators to teacher groups.

"Using this textbook, you will learn to think mathematically for the 21st century, solve a variety of problems based on real-world applications, and learn the appropriate use of technological devices so you can use them as tools for problem solving."

BEATRICE MOORE-HARRIS is an educational specialist at the Region IV Education Service Center in Houston, Texas. She is also the Southwest Regional Director of the Benjamin Banneker Association. Ms. Moore-Harris received her B.A. from Prairie View A&M University in Prairie View, Texas. She has also done graduate work there, at Texas Southern University in Houston, Texas, and at Tarleton State University in Stephenville, Texas. Ms. Moore-Harris is a consultant for the National Council of Teachers of Mathematics (NCTM) and serves on the Editorial Board of the NCTM's *Mathematics Teaching in the Middle School.*

"This program will bring algebra to life by engaging you in motivating, challenging, and worthwhile mathematical tasks that mirror real-life situations. Opportunities to use technology, manipulatives, language, and a variety of other tools are an integral part of this program, which allows all students full access to the algebra curriculum."

JAMES RATH has 30 years of classroom experience in teaching mathematics at every level of the high school curriculum. He is a former mathematics teacher and department chairperson at Darien High School in Darien, Connecticut. Mr. Rath earned his B.A. in philosophy from The Catholic University of America and his M.Ed. and M.A. in mathematics from Boston College. He has also been a Visiting Fellow in the mathematics department at Yale University in New Haven, Connecticut.

DORA SWART is a mathematics teacher and department chairperson at W.F. West High School in Chehalis, Washington. She received her B.A. in mathematics education at Eastern Washington University in Cheney, Washington, and has done graduate work at Central Washington University in Ellensburg, Washington, and Seattle Pacific University in Seattle, Washington. Ms. Swart is a member of the National Council of Teachers of Mathematics, the Western Washington Mathematics Curriculum Leaders, and the Association of Supervision and Curriculum Development. She has developed and conducted numerous inservices and presentations to teachers in the Pacific Northwest.

"Glencoe's algebra series provides the best opportunity for you to learn algebra well. It explores mathematics through hands-on learning, technology, applications, and connections to the world around us. Mathematics can unlock the door to your success— this series is the key."

LESLIE J. WINTERS is the former secondary mathematics specialist for the Los Angeles Unified School District and is currently supervising student teachers at California State University, Northridge. Mr. Winters received bachelor's degrees in mathematics and secondary education from Pepperdine University and the University of Dayton, and master's degrees from the University of Southern California and Boston College. He is a past president of the California Mathematics Council-Southern Section, and received the 1983 Presidential Award for Excellence in the Teaching of Mathematics and the 1988 George Polya Award for being the Outstanding Mathematics Teacher in the state of California.

Consultants, Writers, and Reviewers

Consultants

Cindy J. Boyd
Mathematics Teacher
Abilene High School
Abilene, Texas

Eva Gates
Independent Mathematics
    Consultant
Pearland, Texas

Melissa McClure
Consultant, Tech Prep
Mathematics Consultant
Teaching for Tomorrow
Fort Worth, Texas

Gail Burrill
National Center for Research/
    Mathematical & Science Education
University of Wisconsin
Madison, Wisconsin

Joan Gell
Mathematics Department Chairman
Palos Verdes High School
Palos Verdes Estates, California

Dr. Luis Ortiz-Franco
Consultant, Diversity
Associate Professor of Mathematics
Chapman University
Orange, California

David Foster
Glencoe Author and Mathematics
    Consultant
Morgan Hill, California

Daniel Marks
Consultant, Real-World Applications
Associate Professor of Mathematics
Auburn University at Montgomery
Montgomery, Alabama

Writers

Gail Burrill
Writer, Statistics
National Center for
    Research/Mathematical & Science
    Education
University of Wisconsin
Madison, Wisconsin

Jeri Nichols-Riffle
Writer, Graphing Technology
Assistant Professor
Teacher Education/Mathematics and
    Statistics
Wright State University
Dayton, Ohio

Carol Damian
Writer, Investigations
Physics Teacher
Dublin Scioto High School
Dublin, Ohio

David Foster
Writer, Investigations
Glencoe Author and Mathematics
    Consultant
Morgan Hill, California

Reviewers

Jacqueline L. Austin
Mathematics Teacher
Henry Clay High School
Lexington, Kentucky

Bill Bernhardt
Mathematics Teacher
Moses Lake High School
Moses Lake, Washington

Melody L. Boring
Mathematics Department Chairperson
Lafayette High School
St. Joseph, Missouri

Judy L. Buchholtz
Mathematics Department Chairperson
Dublin Scioto High School
Dublin, Ohio

Linda R. Baker
Mathematics Teacher
New Hanover High School
Wilmington, North Carolina

Gary Boe
Mathematics Teacher
Platteview High School
Springfield, Nebraska

Sabina Wade Bradberry
Mathematics Teacher
Hueytown High School
Hueytown, Alabama

Judith A. Fish
Mathematics Department
    Chairperson/Teacher
West Deptford High School
West Deptford, New Jersey

Nancy Barr
Mathematics Teacher
Monterey High School
Lubbock, Texas

Paul J. Bohney
Supervisor of Mathematics/Science
Gary Community School Corporation
Gary, Indiana

Teresa Call Brown
Mathematics Teacher
Western Guilford High School
Greensboro, North Carolina

Robert Gillham
Mathematics Department Chairperson
Huffman High School
Birmingham, Alabama

Table of Contents

Technology
Explorations **8, 45**
Mathematics and Society
What, No Census? **26**

INTEGRATION

Content Integration

What does geometry have to do with algebra? Believe it or not, you can study most math topics from more than one point of view. Here are some examples.

Geometry You'll use your skills with matrices to find the coordinates of the vertices of a parallelogram.
(Lesson 3–3, pages 141 and 143)

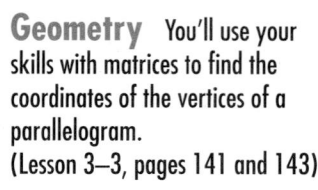

Look Back features refer you to skills and concepts that have been taught earlier in the book.

LOOK BACK

You can refer to Lesson 1-5 for information about absolute value equations.

(Lesson 2–6, page 104)

▲ Problem Solving
You'll solve real-world problems by using matrix logic.
(Lesson 4–1, page 187)

Discrete Mathematics ▲
You'll use permutations to find out why certain movies are shown at some theaters but not others. (Lesson 12–2, page 718)

▼ Number Theory You'll learn how complex numbers are used in building houses.
(Lesson 5–9, pages 310 and 313)

T.V. Channels

Number

30
20
17
27
10
0
1984 1986 1988 1990
Year

◄ Statistics You'll learn how to draw scatter plots and from the scatter plots to determine prediction equations.
(Lesson 2–5, pages 95–100)

Probability You'll model the behavior of mice by using probability and transition matrices. (Lesson 4–3, page 200)

APPLICATIONS

Real-Life Applications

Have you ever wondered if you'll ever actually use math? Every lesson in this book is designed to help show you where and when math is used in the real world. Since you'll explore many interesting topics, we believe you'll discover that math is relevant and exciting. Here are some examples.

Top Five List, FYI, and **Fabulous Firsts** contain interesting facts that enhance the applications.

Singles of All Time in the U.S.

1. *White Christmas,* Bing Crosby, 1942

2. *I Want to Hold Your Hand,* Beatles, 1964

3. *Hound Dog/Don't Be Cruel,* Elvis Presley, 1956

4. *It's Now or Never,* Elvis Presley, 1960

5. *I Will Always Love You,* Whitney Houston, 1992

(Lesson 12–2, page 721)

Auto Racing You'll use racing data from the Indianapolis 500 in your study of the slope of lines. (Lesson 2–3, pages 80–81)

Cryptography You'll learn how matrices can be used to encode secret messages. (Lesson 4–5, pages 212–213)

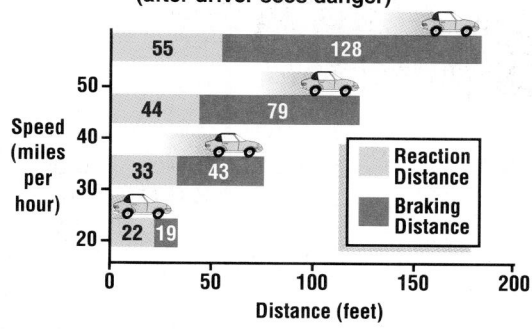

◄ **Solar Energy** You'll learn how to write an equation of a parabola that models the shape of a mirror used to harness solar energy. (Lesson 7–2, page 418)

BRAKING DISTANCES
(after driver sees danger)

Speed (miles per hour) | Distance (feet)

55 / 128
44 / 79
33 / 43
22 / 19

Reaction Distance
Braking Distance

▲ **Law Enforcement** You'll evaluate a radical expression that police officers often use at the scene of an auto accident. (Lesson 5–6, page 288)

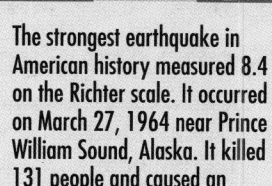

F Y I

The strongest earthquake in American history measured 8.4 on the Richter scale. It occurred on March 27, 1964 near Prince William Sound, Alaska. It killed 131 people and caused an estimated $750 million in property damage.

(Lesson 7–3, page 423)

World Cultures You'll relate the solution of a number puzzle to the study of functions and their inverses. (Lesson 8–8, page 528)

Mae Carol Jemison (1956–)

Mae Carol Jemison was the first African-American woman astronaut. She was the computer engineer and physician aboard the space shuttle Endeavor, launched September 12, 1992.

(Lesson 3–4, page 148)

Mathematics and SOCIETY

Divine Mathematics

Did you know that the great American poet Henry Wadsworth Longfellow enjoyed math? Reprints of actual articles or works of literature illustrate how mathematics is a part of our society. (Lesson 5–6, page 295)

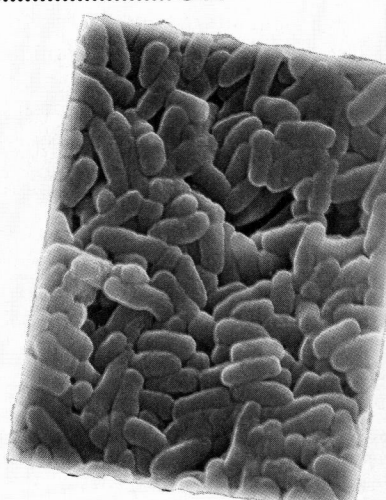

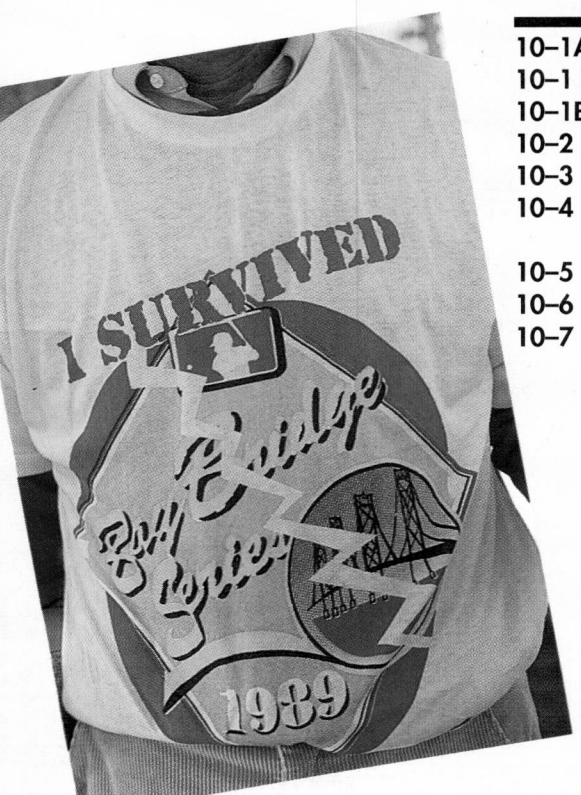

CONNECTIONS

Interdisciplinary Connections

Global Connections features introduce you to a variety of world cultures. ▼

GLOBAL CONNECTIONS

Malaysian foot tennis is a very popular sport in Malaysia and other countries in the Far East. It is played by teams of two using a volleyball-type net. A ball woven from rattan must be kept in the air using only feet, knees, or thighs.

(Lesson 3–3, page 145)

CAREER CHOICES

Engineers apply the principles of science and mathematics to solve practical technical problems. **Electrical engineers** design, develop, test, and supervise the manufacturing of electrical and electronic equipment. They comprise more than one fourth of all engineers. A bachelor's degree in engineering is the minimum requirement.

For more information, contact:

Institute of Electrical and Electronic Engineers
1828 L St. NW, Suite 1202
Washington, DC 20036

(Lesson 5–10, page 321)

S ome people believe that mathematics has very little use in other subjects such as chemistry or music. Of course, this isn't true. In this textbook, mathematics is frequently connected to other subjects that you are studying.

▲ **Biology** You'll model the shape of a deadly frog's leap by using an equation of a parabola. (Lesson 6–6, page 370)

Physics You'll model speed of sound data by using a linear equation. (Lesson 2–2, page 73)

Chemistry You'll solve a problem involving Avogadro's constant. (Lesson 5–1, page 258)

Health You'll evaluate a polynomial that is a model for the number of people who are likely to get the flu. (Lesson 5–3, page 272)

◀ **Career Choices** features include information on interesting careers.

◀ **Literature** You'll solve a Sherlock Holmes mystery. (Lesson 13–1, page 772)

Geography The growth of the population of Arizona is used to introduce mathematical relations. (Lesson 2–1, page 64)

Math Journal exercises give you the opportunity to assess yourself and write about your understanding of key math concepts. (Lesson 3–1, Exercise 6)

6. **Write** a paragraph explaining how to identify whether the graph of a system of linear equations would be two intersecting lines, two distinct parallel lines, or two coincident lines.

TECHNOLOGY

Do you know how to use computers and graphing calculators? If you do, you'll have a much better chance of being successful in today's society and workplace.

GRAPHING CALCULATORS

There are several ways in which graphing calculators are integrated.

- **Introduction to Graphing Calculators**   On pages 2–3, you'll get acquainted with the basic features and functions of a graphing calculator.

- **Graphing Technology Lessons**   In Lesson 3–1A, you'll learn how to solve a system of linear equations using a graphing calculator.

- **Graphing Calculator Explorations**   You'll learn how to use a graphing calculator to model sun intensity using a quadratic equation in Lesson 6–6.

- **Graphing Calculator Programs**   In Lesson 5–3, Exercise 46 includes a graphing calculator program that performs synthetic division.

- **Graphing Calculator Exercises**   Many exercises are designed to be solved using a graphing calculator. For example, see Exercise 45 in Lesson 2–3.

COMPUTER SOFTWARE

- **Spreadsheets**   On page 188 of Lesson 4–1, a spreadsheet is used to help manage a small business.

- **BASIC Programs**   The program on page 143 of Lesson 3–3 is designed to solve systems of equations using Cramer's Rule.

- **Graphing Software**   The graphing software Exploration on page 571 of Lesson 9–4 involves simplifying rational expressions.

Technology Tips, such as this one on page 207 of Lesson 4–4, are designed to help you make more efficient use of technology through practical hints and suggestions.

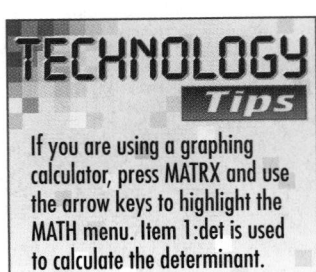

TECHNOLOGY *Tips*

If you are using a graphing calculator, press MATRX and use the arrow keys to highlight the MATH menu. Item 1:det is used to calculate the determinant.

Student Handbook 875

SYMBOLS AND MEASURES

Symbols

$=$	is equal to		$!$	factorial
$\neq$	is not equal to		$\log_b x$	logarithm, base b, of x
$>$	is greater than		$P(A)$	probability of A
$<$	is less than		$f(x)$	f of x, the value of f at x
$\geq$	is greater than or equal to		$f^{-1}(x)$	f inverse of x, inverse of $f(x)$
$\leq$	is less than or equal to		(a, b)	ordered pair a, b
$\approx$	is approximately equal to		$\overline{AB}$	line segment AB
$\sim$	is similar to		$\overset{\frown}{AB}$	arc AB
$\times$ or $\cdot$	times		$\overrightarrow{AB}$	ray AB
$\div$	divided by		$\overleftrightarrow{AB}$	line AB
$-$	negative or minus		AB	measure of $\overline{AB}$
$+$	positive or plus		$\angle$	angle
$\pm$	positive or negative		$\triangle$	triangle
$-a$	opposite or additive inverse of a		$\cos A$	cosine of A
$\lvert a \rvert$	absolute value of a		$\sin A$	sine of A
$a \overset{?}{=} b$	Does a equal b?		$\tan A$	tangent of A
$a : b$	ratio of a to b		$\sec A$	secant of A
$\sqrt{a}$	square root of a		$\csc A$	cosecant of A
$[a]$	greatest integer not greater than a		$\cot A$	cotangent of A
O	origin		$(\ )$	parentheses; *also* ordered pairs
$\varnothing$	empty set		$[\ ]$	brackets; *also* matrices
i	imaginary unit		$\{\ \}$	braces; *also* sets
x	mean of a set of data		$\circ$	degree
$\sigma_{\bar{x}}$	standard deviation of a set of data		$'$	minute

Measures

mm	millimeter		in.	inch
cm	centimeter		ft	foot
m	meter		yd	yard
km	kilometer		mi	mile
g	gram		in^2 or sq in.	square inch
kg	kilogram		s	second
mL	milliliter		min	minute
L	liter		h	hour

Greek Letters

α	alpha		θ	theta
β	beta		σ	sigma
ϕ	phi		ω	omega
π	pi		Σ	capital sigma

GETTING ACQUAINTED WITH THE GRAPHING CALCULATOR

What is it?
What does it do?
How is it going to help me learn math?

These are just a few of the questions many students ask themselves when they first see a graphing calculator. Some students may think, "Oh, no! Do we *have* to use one?", while others may think, "All right! We get to use these neat calculators!" There are as many thoughts and feelings about graphing calculators as there are students, but one thing is for sure: a graphing calculator *can* help you learn mathematics.

So what is a graphing calculator? Very simply, it is a calculator that draws graphs. This means that it will do all of the things that a "regular" calculator will do, *plus* it will draw graphs of simple or very complex equations. In algebra, this capability is nice to have because the graphs of some complex equations take a lot of time to sketch by hand. Some are even considered impossible to draw by hand. This is where a graphing calculator can be very useful.

But a graphing calculator can do more than just calculate and draw graphs. You can program it, work with matrices, and make statistical graphs and computations, just to name a few things. If you need to generate random numbers, you can do that on the graphing calculator. If you need to find the absolute value of numbers, you can do that, too. It's really a very powerful tool—so powerful that it is often called a pocket computer. But don't let that intimidate you. A graphing calculator can save you time and make doing mathematics easier.

As you may have noticed, graphing calculators have some keys that other calculators do not. The Texas Instruments TI-83 will be used throughout this text. The keys located on the bottom half of the calculator are probably familiar to you as they are the keys found on basic scientific calculators. The keys located just below the screen are the graphing keys. You will also notice the up, down, left, and right arrow keys. These allow you to move the cursor around on the screen and to "trace" graphs that have been plotted. The other keys located on the top half of the calculator access the special features such as statistical and matrix computations.

There are some keystrokes that can save you time when using the graphing calculator. A few of them are listed below.

- Any yellow commands written above the calculator keys are accessed with the 2nd key. Similarly, any green characters above the keys are accessed with the ALPHA key.

- 2nd ENTRY copies the previous calculation so you can edit and use it again.

- Pressing ON while the calculator is graphing stops the calculator from completing the graph.

- 2nd QUIT will return you to the home (or text) screen.

- 2nd A-LOCK locks the ALPHA key, which is like pressing "shift lock" or "caps lock" on a typewriter or computer. The result is that all caps will be typed and you do not have to hold the shift key down. (This is handy for programming.)

- 2nd OFF turns the calculator off.

Graphing on the TI–83

Before graphing, we must instruct the calculator how to set up the axes in the coordinate plane. To do this, we define a **viewing window**. The viewing window for a graph is the portion of the coordinate grid that is displayed on the **graphics screen** of the calculator. The viewing window is written as [left, right] by [bottom, top] or [Xmin, Xmax] by [Ymin, Ymax]. A viewing window of $[-10, 10]$ by $[-10, 10]$ is called the **standard viewing window** and is a good viewing window to start with to graph an equation. The standard viewing window can be easily obtained by pressing ZOOM 6. Try this. Move the arrow keys around and observe what happens. You are seeing a portion of the coordinate plane that includes the region from -10 to 10 on the x-axis and from -10 to 10 on the y-axis. Move the cursor, and you can see the coordinates of the points for the current position of the cursor.

Any viewing window can be set manually by pressing the WINDOW key. The window screen will appear and display the current settings for your viewing window. First press ENTER. Then, using the arrow and ENTER keys, move the cursor to edit the window settings. Xscl and Yscl refer to the x-scale and y-scale. This is the number of tick marks placed on the x- and y-axes. Xscl=1 means that there will be a tick mark for every unit of one along the x-axis. In the standard viewing window both the Xscl and Yscl are 1.

Graphing equations is as simple as defining a viewing window, entering the equations in the Y= list, and pressing GRAPH. It is often important to view enough of a graph so you can see all of the important characteristics of the graph and understand its behavior. The term **complete graph** refers to a graph that shows all of the important characteristics such as intercepts or maximum and minimum values.

Example: Graph $y = x + 20$ in the standard viewing window.

Enter: $\boxed{Y=}$

$\boxed{X,T,\theta,n}$ $\boxed{+}$

20 $\boxed{GRAPH}$

It appears that nothing happens. Why?

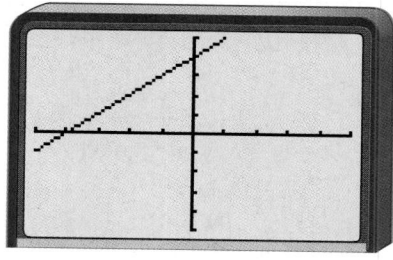

The graph of $y = x + 20$ is a line whose y-intercept is 20 and whose x-intercept is -20. The graph is plotted off the screen and is not complete. A better viewing window for this graph would be $[-25, 25]$ by $[-25, 25]$, which includes both intercepts.

Graphing calculators allow you to analyze graphs by approximating the coordinates of individual points. Suppose we wanted to determine the x-intercept of the graph of $y = x^2 - 2$. Begin by graphing the equation in the standard viewing window.

Enter: $\boxed{Y=}$

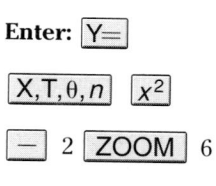 $\boxed{X,T,\theta,n}$ $\boxed{x^2}$

$\boxed{-}$ 2 $\boxed{ZOOM}$ 6

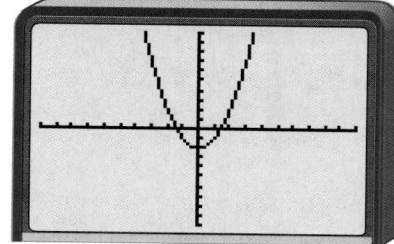

We will use a process called zoom-in to approximate the intercepts. The ZOOM IN feature of the calculator is very useful for determining the coordinates of any point, such as an intercept, with greater accuracy. Use the TRACE function and the arrow keys to determine an approximation for the x-coordinate of the left-most intercept. Begin by placing the cursor on the left-most intercept and observing the x-coordinate. Now, ZOOM IN by entering $\boxed{ZOOM}$ 2 $\boxed{ENTER}$.

TRACE to position the cursor at the intercept again. Observe the coordinates. Now move the cursor one time to the left or right. Any digits that remain unchanged as you move are accurate. Repeat the process of zooming and checking digits until you have the number of accurate digits that you desire.

$x \approx -1.414214$

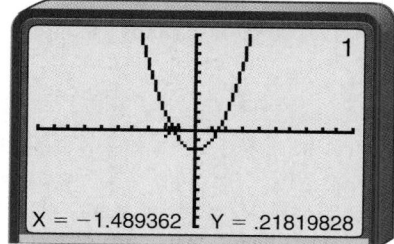

Programming on the TI–83

The TI–83 has programming features that allow us to write and execute a series of commands to perform tasks that may be too complex or cumbersome to perform otherwise. Each program is given a name. Commands begin with a colon (:) followed by an expression or an instruction. Most of the features of the calculator are accessible from program mode.

When you press $\boxed{PRGM}$, you see three menus: EXEC, EDIT, and NEW. EXEC allows you to execute a stored program by selecting the name of the program from the menu. EDIT allows you to edit or change an existing program and NEW allows you to create a new program. To break during program execution, press $\boxed{ON}$. The following example illustrates how to create and execute a new program that stores an expression as Y and evaluates the expression for a designated value of X.

1. Enter $\boxed{PRGM}$ $\boxed{▶}$ $\boxed{▶}$ $\boxed{ENTER}$ to create a new program.

2. Type EVAL $\boxed{ENTER}$ to name the program. (Make sure that the caps lock is on.) You are now in the program editor, which allows you to enter commands. The colon (:) in the first column of the line indicates that this is the beginning of the command line.

3. The first command lines will ask the user to designate a value for x. Enter $\boxed{PRGM}$ $\boxed{▶}$ 3 $\boxed{2nd}$ $\boxed{A-LOCK}$ $\boxed{"}$ ENTER THE VALUE FOR X $\boxed{"}$ $\boxed{ALPHA}$ $\boxed{ENTER}$ $\boxed{PRGM}$ $\boxed{▶}$ 1 $\boxed{X,T,\theta,n}$ $\boxed{ENTER}$.

4. The expression to be evaluated for the value of x is $3x^2 - \sqrt{x}$. To store the expression as Y, enter 3 $\boxed{X,T,\theta,n}$ $\boxed{x^2}$ $\boxed{-}$ $\boxed{2nd}$ $\boxed{\sqrt{\ }}$ $\boxed{X,T,\theta,n}$ $\boxed{STO▶}$ $\boxed{ALPHA}$ $\boxed{Y}$ $\boxed{ENTER}$.

5. Finally, we want to display the value for the expression. Enter $\boxed{PRGM}$ $\boxed{▶}$ 3 $\boxed{ALPHA}$ $\boxed{Y}$ $\boxed{ENTER}$.

6. Now press $\boxed{2nd}$ $\boxed{QUIT}$ to return to the home screen.

7. To execute the program, press $\boxed{PRGM}$. Then press the down arrow to locate the program name and press $\boxed{ENTER}$, or press the number or letter next to the program name. The program asks for a value for x. You will input any value for which the expression is defined and press $\boxed{ENTER}$. To immediately re-execute the program, simply press $\boxed{ENTER}$ when Done appears on the screen.

While a graphing calculator cannot do everything, it can make some things easier. To prepare for whatever lies ahead, you should try to learn as much as you can. The future will definitely involve technology, and using a graphing calculator is a good start toward becoming familiar with technology. Who knows? Maybe one day you will be designing the next satellite, building the next skyscraper, or helping students learn mathematics with the aid of a graphing calculator!

Analyzing Equations and Inequalities

Objectives

In this chapter, you will:

- evaluate and simplify expressions,
- display and interpret data using line plots and stem-and-leaf plots,
- solve equations, and
- solve and graph inequalities.

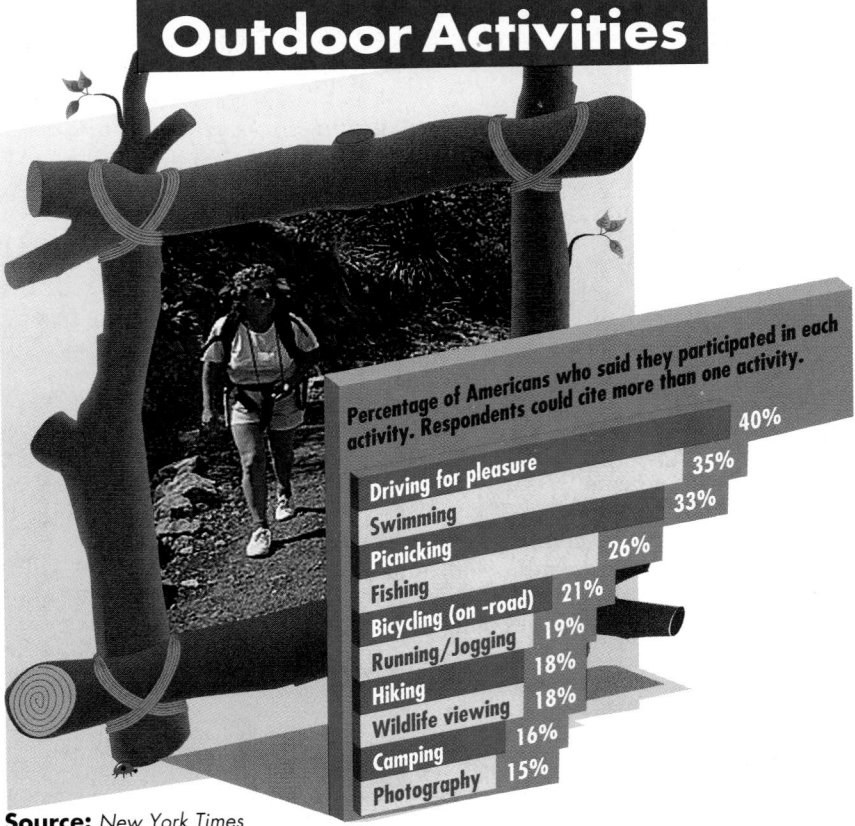

Outdoor Activities

Percentage of Americans who said they participated in each activity. Respondents could cite more than one activity.

Activity	Percentage
	40%
Driving for pleasure	35%
Swimming	33%
Picnicking	26%
Fishing	21%
Bicycling (on -road)	19%
Running/Jogging	18%
Hiking	18%
Wildlife viewing	16%
Camping	15%
Photography	

Source: *New York Times*

In American culture, sports are very important. Our sports heroes have fame and fortune. Why does American society place so much emphasis on sports? Why is so much money spent on spectator sports such as football and baseball? Should more emphasis be placed on participation than on passive viewing? What benefits do each provide an individual? What are your views on these issues?

TIME Line

A.D. 100 Nichomachus of Gerasa's *Introductio Arithmetica* lists the only four known perfect numbers: 6, 28, 496, and 8128.

1718 Lady Mary Wortley Montagu starts smallpox inoculation in England.

0 A.D. 100 200 1650 1675 1700 1725 1750 1775 1800 1825 1850 1875

1869 First college football game between Rutgers and College of New Jersey.

Chapter Project

Sports statistics involve data that are often easy to display and interpret. Look at the topics below. Research to find the data for each. Then display the data in a graphical format such as a line plot, stem-and-leaf plot, bar graph, circle graph, or line graph that best represents the data. Present the data to the class.

- Find the winning times for the gold medal in the women's 100-meter freestyle in the last eight Olympic Games.

- Find the dollar amounts for the top twenty money winners on the men's PGA tour last year.

- Pick a professional baseball team and find the final batting averages at the end of last season. Show this data for each player and also show the averages listed by positions.

Eldrick "Tiger" Woods of Cypress, California, has been playing golf for as long as he can remember. At the age of 9 months, he was introduced to golf, and at age 6, he made his first hole-in-one.

Today Tiger is busy winning tournaments. In August of 1996, he won the U.S. Amateur Golf Championship for the third time at the age of 20. His first U.S. amateur title in 1994 made him the youngest winner ever, the first black player to win, and the only player to have won both the Amateur and the U.S. Junior Amateur Championships, a title he won in 1991, 1992, and 1993.

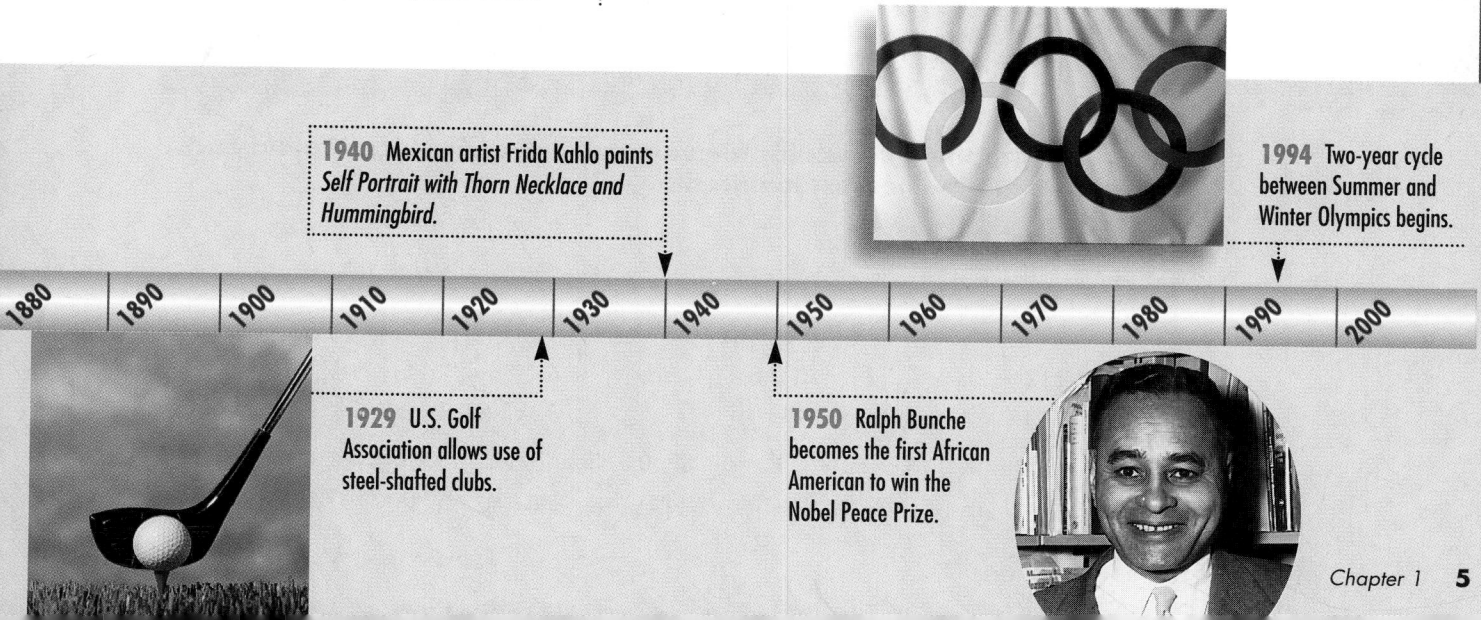

1940 Mexican artist Frida Kahlo paints *Self Portrait with Thorn Necklace and Hummingbird.*

1994 Two-year cycle between Summer and Winter Olympics begins.

1880 | 1890 | 1900 | 1910 | 1920 | 1930 | 1940 | 1950 | 1960 | 1970 | 1980 | 1990 | 2000

1929 U.S. Golf Association allows use of steel-shafted clubs.

1950 Ralph Bunche becomes the first African American to win the Nobel Peace Prize.

1–1A Graphing Technology
Expressions

A Preview of Lesson 1–1

The Graphing Technology lessons in this text will use the TI-83 graphing calculator. In these lessons, you will be introduced to keying sequences that will allow you to perform mathematical computations, graph equations, and utilize other features of the calculator.

Remember that, as with any scientific calculator, graphing calculators follow the order of operations.

Example ❶ Evaluate $39 - 3^3 \cdot 2 + \dfrac{5^2 + 9}{4}$.

Enter: 39 ⊟ 3 ⌃ 3 ⊠ 2 ⊞ ⦅ 5 x^2 ⊞ 9 ⦆ ÷ 4 ENTER   −6.5

You may enter 5^2 as 5 ⌃ 2 or as 5 x^2 .

Use parentheses for all grouping symbols on a graphing calculator.

Occasionally, you will need to evaluate an algebraic expression for several different values of a variable. This can be done efficiently using the STO▸ and ENTRY features.

Example ❷ Evaluate $6x(14 - x)^2$ for $x = -6$, 11, and 1.25.

Enter: (−) 6 STO▸ X,T,θ,*n*   *Stores −6 as x.*

2nd : 6 X,T,θ,*n* ⦅ 14   *Evaluates the expression.*

⊟ X,T,θ,*n* ⦆ x^2 ENTER   −14400

Now press 2nd ENTRY and use the editing features of the calculator to change the value for x. Move the cursor with the arrow keys to the −6 and change it to 11. The new value is 594 and is computed when you press ENTER . Repeat for $x = 1.25$.

EXERCISES

Use a graphing calculator to evaluate each expression. Round each answer to the nearest hundredth.

1. $5 \times \dfrac{4 + 7}{6}$

2. 231.5^6

3. $(52 \times 4)^3 + \dfrac{76}{6}$

4. $3\left[\dfrac{4 + \frac{3(4 + 6)}{5}}{4}\right] + 11$

5. $112 - 2\{16 + 2[14 - 2(7 + 1)]\} - 4^2$

6. $[(-4 + 11)^2 \div 4]3$

7. $(1.12 \times 10^4)(3.65 \times 10^{-3})$

8. $(5.24 \times 10^{-6})(3.24 \times 10^4)$

9. Evaluate $\dfrac{5}{9}(F - 32)$ for $F = 50$, 32, 0, −5, and 80.

10. Evaluate $x^2 - y^2$ for $x = 15$, $y = 13$ and $x = 0.8$, $y = 0.4$.

Expressions and Formulas

What YOU'LL LEARN

- To use the order of operations to evaluate expressions, and
- to use formulas.

Why IT'S IMPORTANT

You can use expressions to solve problems involving exercise, baseball, and geometry.

F Y I

In 1993, there were more than 12.5 million in-line skaters in the United States. This was a third more than there were in 1992 and double the number in 1991.

APPLICATION
Exercise

Chris Edwards of Escondido, California, earned his nickname "Airman" by soaring 13 feet off of a ramp while on in-line skates. He has been in-line skating since he was 13 years old and is currently a member of Team Rollerblade, a group that performs shows around the country.

According to a recent study by the University of Massachusetts, you burn as many Calories in-line skating at moderate speeds or faster as you do running. The chart below shows the Calories burned per minute for various body weights and skating speeds.

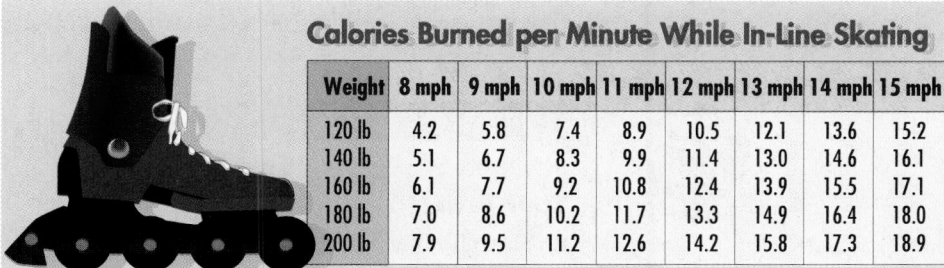

Calories Burned per Minute While In-Line Skating

Weight	8 mph	9 mph	10 mph	11 mph	12 mph	13 mph	14 mph	15 mph
120 lb	4.2	5.8	7.4	8.9	10.5	12.1	13.6	15.2
140 lb	5.1	6.7	8.3	9.9	11.4	13.0	14.6	16.1
160 lb	6.1	7.7	9.2	10.8	12.4	13.9	15.5	17.1
180 lb	7.0	8.6	10.2	11.7	13.3	14.9	16.4	18.0
200 lb	7.9	9.5	11.2	12.6	14.2	15.8	17.3	18.9

To determine your skating speed in miles per hour, divide 60 (the number of minutes in an hour) by the number of minutes you skated. Then multiply that number by the number of miles you skated. You can represent this by the expression $(60 \div t) \times d$, where t represents the time skated in minutes and d represents the distance traveled in miles.

Maya works out each day by in-line skating. She skates about three miles in 20 minutes, and she weighs 140 pounds. About how many Calories does she burn during a workout?

First, find how fast Maya was skating.

$$(60 \div t) \times d = (60 \div 20) \times 3 \quad \textit{t = 20 and d = 3}$$

How do you evaluate $(60 \div 20) \times 3$? What do you do first? Do you multiply or divide? A numerical expression must have exactly one value. In order to find that value, you must follow an **order of operations.**

Order of Operations	1. **Simplify the expressions inside grouping symbols, such as parentheses, brackets, braces, and fraction bars.** 2. **Evaluate all powers.** 3. **Do all multiplications and divisions from left to right.** 4. **Do all additions and subtractions from left to right.**

Using the rules above, evaluate the expression.

$$(60 \div 20) \times 3 = 3 \times 3 \quad \textit{Perform the operation inside the parentheses first.}$$
$$= 9 \quad \textit{Multiply.}$$

Maya skated at an average speed of 9 mph during her 20-minute workout.

To determine how many Calories Maya burns, find her weight on the chart. Then go across to the column with her average speed of 9 mph. Maya burns 6.7 Calories per minute during her 20-minute workout. So, she burns a total of 20×6.7 or 134 Calories.

As in the expression $(60 \div 20) \times 3$, grouping symbols can be used to change or clarify the order of operations. Frequently-used grouping symbols are parentheses, ( ), brackets, [ ], braces, { }, and fraction bars, as in $\frac{2+4}{3}$. When calculating the value of an expression, begin with the operation in the innermost set of grouping symbols.

Example **Find the value of $[(3 + 6)^2 \div 3] \cdot 4$.**

$$
\begin{aligned}
[(3 + 6)^2 \div 3] \cdot 4 &= [(9)^2 \div 3] \cdot 4 && \textit{First add 3 and 6.} \\
&= (81 \div 3) \cdot 4 && \textit{Then find } 9^2. \\
&= 27 \cdot 4 && \textit{Divide 81 by 3.} \\
&= 108 && \textit{Multiply 27 by 4.}
\end{aligned}
$$

The value is 108.

Scientific calculators follow the order of operations.

EXPLORATION

SCIENTIFIC CALCULATORS

Your Turn
a. Simplify $3 + 5 \times 2 - 1$ using a scientific calculator.
b. Explain how your calculator arrived at the answer.
c. Suppose someone concluded that $3 + 5 \times 2 - 1 = 15$. How is this possible?
d. Evaluate $2045 - (18^2 + 711)$ using your calculator. Explain how the answer was calculated.
e. If you remove the parentheses in part d above, would the solution remain the same? Explain.
f. Write a set of directions for someone to find the value of $16^2 - (135 + 10 \times 8)$ without a scientific calculator.

Algebraic expressions contain at least one variable. You can evaluate an algebraic expression by replacing each variable with a value and then applying the rules for the order of operations.

Example **a. Evaluate $a[b^2(b + a)]$ if $a = 12$ and $b = 0.5$.**

$$
\begin{aligned}
a[b^2(b+a)] &= 12[(0.5)^2 (0.5 + 12)] && \textit{Replace a with 12 and b with 0.5.} \\
&= 12[0.25(0.5 + 12)] && \textit{Find } (0.5)^2. \\
&= 12[0.25(12.5)] && \textit{Add 0.5 and 12.} \\
&= 12[3.125] && \textit{Multiply 0.25 by 12.5.} \\
&= 37.5 && \textit{Multiply 12 by 3.125.}
\end{aligned}
$$

The value is 37.5.

b. Evaluate $\dfrac{y^3}{3ab + 2}$ **if** $y = 4$, $a = -2$, **and** $b = -5$.

The fraction bar, which indicates division, is a grouping symbol. Evaluate the expressions in the numerator and denominator separately before dividing.

$$\frac{y^3}{3ab + 2} = \frac{4^3}{3(-2)(-5) + 2} \quad \textit{Replace y with 4, a with } -2\textit{, and b with } -5.$$

$$= \frac{64}{3(10) + 2} \quad \textit{Evaluate the numerator and denominator separately.}$$

$$= \frac{64}{32} \quad \textit{Divide.}$$

$$= 2$$

The value is 2.

Pythagoras

A **formula** is a mathematical sentence that expresses the relationship between certain quantities. If you know a value for every variable in the formula except one, you can find the value of the remaining variable.

One formula you may use is the **Pythagorean theorem,** which was developed by the Greek mathematician Pythagoras. In a right triangle, the side opposite the right angle is called the **hypotenuse.** This side is always the longest side of a right triangle. The other two sides are called the **legs** of the right triangle. The Pythagorean theorem states that in a right triangle, if a and b are the measures of the legs and c is the measure of the hypotenuse, then $c^2 = a^2 + b^2$.

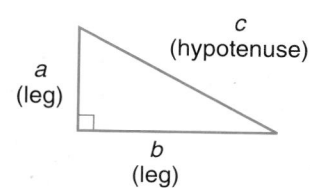

Example ③

Geometry

Suppose the distance from a person flying a kite to the ground directly below the kite is 100 meters, and the length of the string from the person to the kite is 125 meters. Use the Pythagorean theorem to find the height of the kite.

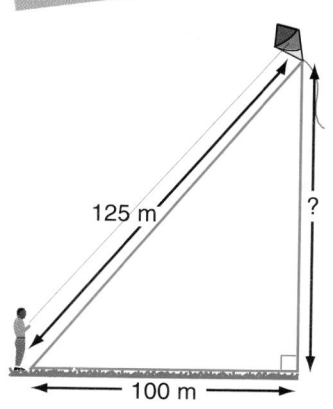

You can think of the kite and the ground as a right triangle as shown at the right.

$$c^2 = a^2 + b^2$$

$$125^2 = 100^2 + b^2 \quad \textit{Substitute.}$$

$$15{,}625 = 10{,}000 + b^2 \quad \textit{Evaluate all powers.}$$

$$b^2 = 5625 \quad \textit{Subtract.}$$

$$b = \sqrt{5625} \quad \textit{Take the square root of}$$

$$b = 75 \quad \textit{each side.}$$

The kite is 75 meters high.

Communicating Mathematics

Study the lesson. Then complete the following.

1. **Explain** which operation to perform first when evaluating an expression that has both brackets and parentheses.

2. **Describe** how you would evaluate the expression $\frac{1}{3} - \frac{12(77-11)}{2}$.

3. **List** three formulas that you have used before.

4. **Write** three examples of algebraic expressions.

5. **You Decide**   To find the volume of a cone, you multiply the area of the base by the height, then divide by 3. Andrea developed the formula $V = \frac{1}{3}\pi r^2 h$ to find the volume of a cone while Lorenzo developed the formula $V = \frac{\pi r^2 h}{3}$. The cone has a height of 16.3 centimeters, and its base has a radius of 14.7 centimeters. Explain which formula will work, and why.

Guided Practice

Find the value of each expression.

6. $7 - 6 \div 3$

7. $9(4 + 2)$

8. $18 \cdot 6 + 12$

9. $25 \cdot 2 - 3$

10. $\frac{45(4 + 32)}{10}$

11. $(4 - 3)^5 \cdot 9$

Evaluate each expression if $a = 3$, $b = -4$, and $c = 5$.

12. $a + b - c$

13. $a + c^2$

14. $a(b + c)$

The relationship between Celsius temperature C and Fahrenheit temperature F is given by $C = \frac{5(F - 32)}{9}$. Find the Celsius temperature for each Fahrenheit temperature.

15. normal body temperature, $98.6°$

16. freezing point of water, $32°$

17. **Geometry**   Write a formula to represent the area A of the rectangle shown at the right.

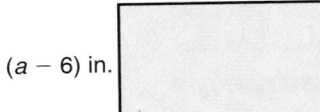
$(a + 6)$ in.

$(a - 6)$ in.

Practice

Find the value of each expression.

18. $4(3^2 + 3)$

19. $2(9 + 2) - 3$

20. $(7 + 5)3 - 3$

21. $4 + 2^2 - 15 + 4$

22. $10 + 16 \div 4 + 8$

23. $5 + 9(3) \div 3 - 8$

24. $7 - [4 + (6 \cdot 5)]$

25. $[21 - (9 - 2)] \div 2$

26. $4 + [9 \div (9 - 2(4))]$

27. $[(-7 + 4) \times 5 - 2] \div 6$

28. $\frac{1}{2}(5^2 + 3)$

29. $\frac{14(8 - 15)}{2}$

30. $-3(2^2 + 3)$

31. $4 + (49 \div 7) \times 8 \div 2$

32. $0.4(0.6 + 3.2) \div 2$

33. $0.5(2.3 + 25) \div 1.5$

34. $3 + [8 \div (9 + 2(-4))]$

35. $\frac{1}{3} - \frac{12(77 \div 11)}{9}$

Evaluate each expression if $a = -5$, $b = 0.25$, $c = \frac{1}{2}$, and $d = 4$.

36. $d(3 + c)$

37. $a + b + d$

38. $a + 2b - c$

39. $a + 10 \div c$

40. $2^d + a$

41. $\frac{3ab}{cd}$

42. $\frac{3a + 4c}{2c}$

43. $(a + c)^2 - bd$

44. Geometry The formula for the area A of a triangle is $A = \frac{1}{2}bh$ where b is the measure of the base and h is the measure of the height. Write an expression to represent the area of the triangle at the right.

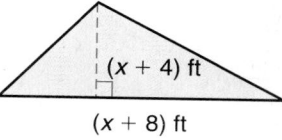

$(x + 4)$ ft

$(x + 8)$ ft

The formula for the area of a trapezoid is $A = \frac{h}{2}(b_1 + b_2)$. A represents the area, h represents the measure of the altitude, and b_1 and b_2 represent the measures of the bases. Find the area of each trapezoid given the following values.

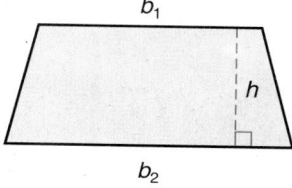

b_1

h

b_2

45. $h = 6, b_1 = 22, b_2 = 17$

46. $h = 10, b_2 = 17, b_1 = 52$

47. $b_2 = 6.25, b_1 = 12.50, h = 4$

48. $h = \frac{5}{8}, b_1 = \frac{3}{4}, b_2 = \frac{1}{2}$

Simple interest is the amount paid or earned for the use of money for a unit of time. It is calculated using the formula $I = prt$ where p represents the principal in dollars, r represents the annual interest rate, and t represents the time in years. Find the simple interest I given each of the following values.

49. $p = \$1500, r = 6.5\%, t = 3$ years

50. $p = \$2500, r = 7.25\%, t = 2$ years

51. $p = \$20,005, r = 7.9\%, t = 2$ years, 3 months

52. $p = \$65,283.21, r = 9.32\%, t = 78$ months

Programming

53. The graphing calculator program at the right finds the area of a trapezoid once the height and lengths of the two bases are entered.

Run the program to find the area of each trapezoid.

a. $h = 8, b_1 = 112, b_2 = 20$

b. $h = 7, b_1 = 4, b_2 = 11$

c. $h = 4.8, b_1 = 5.6, b_2 = 6.4$

```
PROGRAM:AREA
: Disp "HEIGHT =?"
: Input H
: Disp "B1 ="
: Input B
: Disp "B2 = "
: Input C
: Disp "AREA =",
  H(B+C)/2
```

(continued on the next page)

Alter the program to find the area of a triangle if the formula is $A = \frac{1}{2}bh$, where b is the measure of the base (in cm) and h is the measure of the height (in cm). Find the area of each triangle.

    d. $b = 6.7, h = 13.8$ **e.** $b = 127.2, h = 82.6$

Critical Thinking

54. Insert grouping symbols as needed to make each statement true.

    a. $1 + 3 \cdot 2^2 = 16$

    b. $1 + 3 \cdot 2^2 = 49$

    c. $1 + 3 \cdot 2^2 = 13$

    d. $1 + 3 \cdot 2^2 = 37$

55. Geometry Study the rectangular prism at the right.

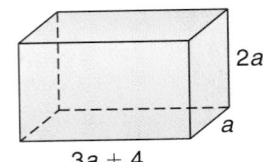

    a. Create a formula that will determine the total surface area of this rectangular prism.

    b. What is the surface area if $a = 4$?

    c. What is the surface area if $a = 6.2$?

Applications and Problem Solving

56. Medicine Dosages of medicine are based upon the weight and/or age of a patient. Suppose an asthma patient must take a medicine that is dispensed in 100-milligram tablets. The dosage is 5 milligrams per kilogram of body weight and is given every six hours.

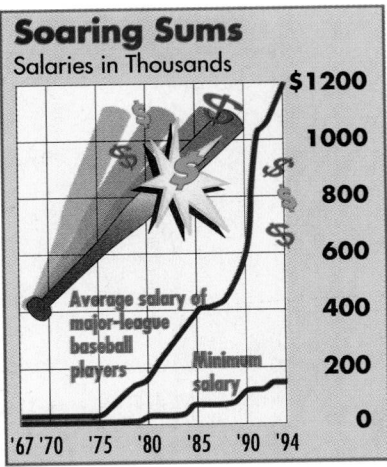

    a. If the patient weighs 20 kilograms, what dosage should he receive?

    b. How many tablets would be needed for a 30-day supply?

57. Number Theory The sum of the factors of 6 that are less than 6 is $1 + 2 + 3$, or 6. Thus, 6 is called a **perfect number** because the number 6 equals the sum of its factors, excluding itself. 28 is also a perfect number because the sum of the factors, $1 + 2 + 4 + 7 + 14$, is equal to 28. The formula for generating perfect numbers is $p = (2^{n-1})(2^n - 1)$. In order for the formula to work, both n and $(2^n - 1)$ must be prime numbers.

    a. List the first 25 prime numbers. Recall that a prime number is an integer greater than 1 whose only positive factors are 1 and itself.

    b. Find four prime numbers n where $2^n - 1$ is also prime.

    c. Find the next two perfect numbers after 28.

58. Baseball The graph at the right gives the average salary of a major-league baseball player. The expression $2.9t^2 + 5.8t + 50$, where t is the time in years since 1975, closely models the graph. Evaluating the expression gives the average salary in thousands of dollars.

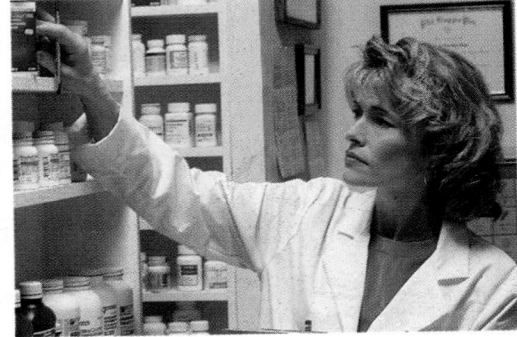

Soaring Sums
Salaries in Thousands

Source: Major League Baseball Players Assoc.

    a. What was the average salary in 1975?

    b. What was the average salary in the strike-shortened 1994 season?

    c. Explain how to find the salary for 1970 and 1965.

Properties of Real Numbers

What YOU'LL LEARN

• To determine the sets of numbers to which a given number belongs, and

• to use the properties of real numbers to simplify expressions.

Why IT'S IMPORTANT

You can use the properties of real numbers to evaluate expressions and solve equations.

INTEGRATION
Statistics

Tables, charts, and graphs are used to represent data in a manner that is easy to read and understand.

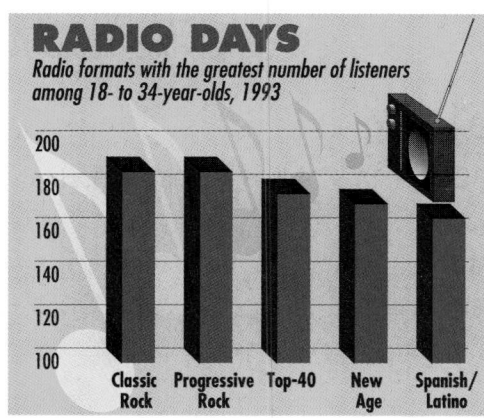

RADIO DAYS
Radio formats with the greatest number of listeners among 18- to 34-year-olds, 1993

Source: American Demographics

POPULAR SPORTS ACTIVITIES
Top sports of Americans 7 years and older, 1994
Participation (in millions):

70.8 — Exercising/Walking
60.3 — Swimming
49.8 — Bicycling
45.7 — Fishing
43.8 — Exercising with Equipment

Source: National Sporting Goods Association

Both of the figures above use numbers to convey their message to the reader. All of the numbers that you use in everyday life are **real numbers.** Each real number corresponds to exactly one point on the number line, and every point on the number line represents exactly one real number.

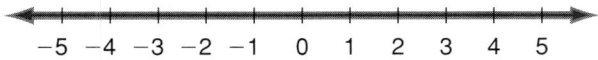

$$-5 \quad -4 \quad -3 \quad -2 \quad -1 \quad 0 \quad 1 \quad 2 \quad 3 \quad 4 \quad 5$$

Every real number can be classified as either **rational** or **irrational.** A rational number can be expressed as a ratio $\frac{m}{n}$, where m and n are integers and n is not zero. The decimal form of a rational number is either a terminating or repeating decimal. Some examples of rational numbers are $\frac{1}{3}$, $1.\overline{34}$, 5.8, -6, and 0. Any real number that is not rational is irrational. $\sqrt{2}$, π, and $\sqrt{7}$ are irrational numbers.

The sets of natural numbers, $\{1, 2, 3, 4, 5, \ldots\}$, whole numbers, $\{0, 1, 2, 3, 4, \ldots\}$, and integers, $\{\ldots, -3, -2, -1, 0, 1, 2, \ldots\}$ are all subsets of the rational numbers.

Reals, R

Q
Z
W
N
I

The Venn diagram at the left shows the relationship among these sets of numbers.

R = reals   Q = rationals

I = irrationals   Z = integers

W = wholes   N = naturals

Example **1** Find the value of each expression. Then name the sets of numbers to which each value belongs.

a. $\sqrt{17}$

$\sqrt{17} = 4.1231056\ldots$ reals (R), irrationals (I)

b. $8 \div 4$

$8 \div 4 = 2$ reals (R), rationals (Q), integers (Z), whole numbers (W), natural numbers (N)

c. 0.25×0

$0.25 \times 0 = 0$ reals (R), rationals (Q), integers (Z), whole numbers (W)

d. $10 - 25$

$10 - 25 = -15$ reals (R), rationals (Q), integers (Z)

e. $6 \div 10$

$6 \div 10 = 0.6 \text{ or } \frac{3}{5}$ reals (R), rationals (Q)

Operations with real numbers have several important properties. The chart below summarizes the properties of real numbers for addition and multiplication.

For any real numbers *a*, *b*, and *c*:		
	Addition	**Multiplication**
Commutative	$a + b = b + a$	$a \cdot b = b \cdot a$
Associative	$(a + b) + c = a + (b + c)$	$(a \cdot b) \cdot c = a \cdot (b \cdot c)$
Identity	$a + 0 = a = 0 + a$	$a \cdot 1 = a = 1 \cdot a$
Inverse	$a + (-a) = 0 = (-a) + a$	If $a \neq 0$, then $a \cdot \frac{1}{a} = 1 = \frac{1}{a} \cdot a.$
Distributive	$a(b + c) = ab + ac$ and $(b + c)a = ba + ca$	

−a is read "the opposite of a".

Example **2** Name the property illustrated by each equation.

a. $(3 + 4a)2 = 2(3 + 4a)$

commutative property of multiplication

The commutative property says that the order in which you multiply does not change the product.

b. $62 + (38 + 75) = (62 + 38) + 75$

associative property of addition

The associative property says that the way you group three numbers when adding does not change their sum.

Example 3

Name the additive inverse and multiplicative inverse for each number.

a. $\dfrac{3}{8}$

Since $\dfrac{3}{8} + \left(-\dfrac{3}{8}\right) = 0$, the additive inverse of $\dfrac{3}{8}$ is $-\dfrac{3}{8}$.

Since $\left(\dfrac{3}{8}\right)\left(\dfrac{8}{3}\right) = 1$, the multiplicative inverse of $\dfrac{3}{8}$ is $\dfrac{8}{3}$.

b. -2.5

Since $-2.5 + 2.5 = 0$, the additive inverse of -2.5 is 2.5.

To find the multiplicative inverse of -2.5, find $\dfrac{1}{-2.5}$.

Enter: 1 [÷] 2.5 [+/−] [=]   −0.4

The multiplicative inverse of -2.5 is -0.4.

You can use properties to simplify algebraic expressions.

Example 4

Simplify $4(2b - 6c) + 2(3b + c)$.

$4(2b - 6c) + 2(3b + c)$

$= 4(2b) - 4(6c) + 2(3b) + 2(c)$ *Use the distributive property.*

$= 8b - 24c + 6b + 2c$ *Multiply.*

$= 14b - 22c$ *Combine like terms.*

The distributive property is often used in real-world applications.

Example 5

APPLICATION
Consumerism

Taco Bell® is offering its new Border Lights™ Light Taco Supreme for a special introductory price of 99¢ to attract customers who would like a healthier choice. The number of Light Taco Supremes sold each day for 5 days in one restaurant are given in the table below. What was the average income from the sale of the tacos during each day?

Monday	Tuesday	Wednesday	Thursday	Friday
182	341	246	303	378

An average is calculated by adding to find the total amount, then dividing the total amount by the number of items. In this case, it would be the total dollar amount divided by the number of days, 5.

There are two ways to find the total dollar amount.

Method 1

Multiply each daily amount by 0.99 and then add.

$T = 0.99(182) + 0.99(341) + 0.99(246) + 0.99(303) + 0.99(378)$

$= 180.18 + 337.59 + 243.54 + 299.97 + 374.22$

$= 1435.50$

(continued on the next page)

Method 2

Add the daily amounts and then multiply the total by 0.99.

$T = 0.99(182 + 341 + 246 + 303 + 378)$

$\quad = 0.99(1450)$

$\quad = 1435.50$

Now find the mean or average by dividing the total by 5.

$1435.50 \div 5 = 287.10$

The average daily income from Light Taco Supremes was $287.10.

The two methods of calculating the average income illustrate the distributive property of multiplication over addition.

CHECK FOR UNDERSTANDING

Communicating Mathematics

Study the lesson. Then complete the following.

1. **Explain** how you know π is an irrational number given that the digits of π never end and never repeat a pattern.

2. **Examine** the Venn diagram on page 13 that shows the relationships among all real numbers. Explain why natural numbers are enclosed by whole numbers, integers, and rationals.

3. **List** five rational numbers and five irrational numbers.

4. **Explain** the difference between the commutative and associative properties.

MODELING MATHEMATICS

5. You can model algebraic expressions with algebra tiles. The unit tile $\boxed{1}$ represents 1. The x tile $\boxed{}$ represents x.
   a. Model $x + 1$ with algebra tiles.
   b. Model $2(x + 1)$ with algebra tiles.
   c. Use algebra tiles to show that $2(x + 1) = 2x + 2$.

Guided Practice

Find the value of each expression. Then name the sets of numbers to which each value belongs.

6. $7 - 6$
7. $6 - 7$
8. $6 \div 2^2$
9. $\sqrt{49 + 8}$

Determine whether each statement is _true_ or _false_. If _false_, give an example of a number that shows the statement is false.

10. Every integer is a whole number.

11. Every whole number is an integer.

Name the property illustrated by each equation.

12. $8(4) = 4(8)$
13. $2 + (-2) = 0$
14. $6 + (0 + 4) = 6 + (4 + 0)$

Name the additive inverse and the multiplicative inverse of each number.

15. 7
16. $-\dfrac{2}{3}$

Simplify each expression.

17. $2(4c + 5d) + 6(2c - d)$
18. $3x + 5y + 7x - 3y$

19. Baby-sitting Maria baby-sat for 5 hours on Friday night and 7 hours on Saturday night to earn money for band camp. She charges $3 per hour. How much money did Maria earn for band camp?

EXERCISES

Practice

Find the value of each expression. Then name the sets of numbers to which each value belongs.

20. $2.9 + 3.7$ **21.** $-56 \div 8$ **22.** $58 \div 100$ **23.** -4.2×10

24. $1 - 5$ **25.** $\sqrt{25} - 6$ **26.** $3^3 + 2^2$ **27.** $1\frac{1}{2} + \frac{3}{4}$

28. $10 \times (-3.9)$ **29.** $4 \div 2^3$ **30.** $-81 \div (-9)$ **31.** $\sqrt{64 + 3}$

Determine whether each statement is *true* or *false*. If *false*, give an example of a number that shows the statement is false.

32. Every real number is irrational.

33. Every integer is a rational number.

34. Every rational number is an integer.

35. Every irrational number is a real number.

36. Every natural number is an integer.

37. Every real number is either a rational number or an irrational number.

Name the property illustrated by each equation.

38. $m(4 - 3) = m \cdot 4 - m \cdot 3$ **39.** $(5 + 9) + 13 = 13 + (5 + 9)$

40. $s + t + 0 = s + t$ **41.** $(a + b) + [-(a + b)] = 0$

42. $(3 + 9) \cdot 5 = 3(5) + 9(5)$ **43.** $(48)3 = 3(48)$

44. $\frac{4}{5}(1) = \frac{4}{5}$ **45.** $\left(\frac{1}{4}\right)4 = 1$

Name the additive inverse and the multiplicative inverse of each number.

46. 8 **47.** 0.2 **48.** -1.25

49. -1 **50.** $\frac{5}{6}$ **51.** $-3\frac{5}{7}$

Simplify each expression.

52. $6x - 2y - 3x + 2y$ **53.** $4(14c - 10d) - 6(d + 4c)$

54. $\frac{1}{2}(17 - 4x) - \frac{3}{4}(6 - 16x)$ **55.** $\frac{3}{4}(2x - 5y) + \frac{1}{2}\left(\frac{2}{3}x + 4y\right)$

56. $\frac{1}{4}(12 + 20a) + \frac{3}{4}(12 + 20a)$ **57.** $7(0.2m + 0.3n) + 5(0.6m - n)$

58. Use the properties of real numbers to answer these questions.
  a. If $a + b = a$, what is the value of b?
  b. If $ab = 1$, what is the value of b? What is b called?
  c. If $ab = a$, what is the value of b?

59. Accounting A number is divisible by 9 if the sum of its digits is divisible by 9. This fact is used by accountants to check figures in double entry books. If the totals of the credit and debit columns do not match, and the difference between the totals is divisible by 9, then the error was probably made when two digits were reversed in one of the entries. Explain whether the errors in the following might have come from reversing the digits of an entry.
  a. credit = $638, debit = $577
  b. credit = $1050, debit = $1095
  c. Why does this error check work?

60. Geometry Find the area of the tennis court at the right in two different ways.

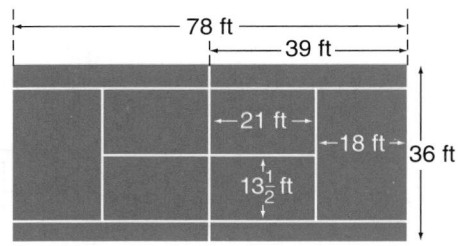

Find the value of each expression. (Lesson 1–1)

61. $0.2(0.5 + 2.2) \div 6$

62. $8 - [21 - (3 \cdot 5)]$

63. $3(13 - 7) + (7 - 5)2$

64. $17 - [22 \div (21 - 2(5))]$

65. Evaluate the expression $\dfrac{3a + 4c}{b}$ if $a = -3$, $b = 2$, and $c = 0.5$.
(Lesson 1–1)

66. Evaluate the expression $d - [c \div (a - b(a))]$ if $a = -1$, $b = 5$, $c = 12$, and $d = 17$. (Lesson 1–1)

67. Human Body Humans blink their eyes about once every 5 seconds. About how many times do humans blink their eyes in two hours?
(Lesson 1–1)

68. Geometry The formula for the area of a trapezoid is $A = \dfrac{h}{2}(b_1 + b_2)$. Find the area of a trapezoid with height 3 and bases 4 and 8. (Lesson 1–1)

69. Use the formula $I = prt$ to find the simple interest if $p = \$2500$, $r = 7.37\%$, and $t = 4$ years.
(Lesson 1–1)

70. Buildings The Sears Tower, located in Chicago, Illinois, is the tallest building in the United States. It is 1454 feet tall, only 22 feet shorter than the Petronas Towers in Kuala Lumpur, Malaysia, the tallest buildings in the world. Use the formula $Y = \dfrac{F}{3}$ to find the height of both buildings in yards if Y is the height in yards and F is the height in feet. (Lesson 1–1)

71. Evaluate $12 + 18 \div 6 + 7$. (Lesson 1–1)

Petronas Towers

Integration: Statistics
Graphs and Measures of Central Tendency

APPLICATION
Education

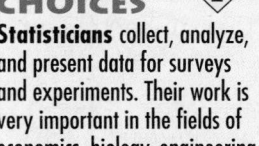

The average salaries for teachers in 1993–94 are listed in the table below. How do teacher salaries in your state compare with other states?

Average Teacher Salaries (1993–94)							
State	Salary	State	Salary	State	Salary	State	Salary
AL	$28,705	IN	$35,711	NE	$29,564	SC	$29,566
AK	$46,581	IA	$30,760	NV	$33,955	SD	$25,059
AZ	$31,800	KS	$33,919	NH	$34,121	TN	$30,514
AR	$27,873	KY	$31,640	NJ	$44,693	TX	$30,519
CA	$40,289	LA	$26,285	NM	$27,922	UT	$28,056
CO	$33,826	ME	$30,996	NY	$45,772	VT	$34,517
CT	$49,910	MD	$39,463	NC	$29,727	VA	$33,063
DE	$37,469	MA	$40,852	ND	$25,506	WA	$35,855
FL	$31,944	MI	$42,500	OH	$35,684	WV	$30,549
GA	$30,527	MN	$36,146	OK	$27,009	WI	$35,990
HI	$36,564	MS	$25,153	OR	$37,590	WY	$30,952
ID	$27,756	MO	$30,324	PA	$42,411		
IL	$39,387	MT	$28,200	RI	$39,261		

Source: NEA Research

The answer to this question is not immediately obvious from the table. You are familiar with statistical graphs such as *bar graphs, line graphs,* and *circle graphs* that are used to organize and illustrate data. Another way to display statistical data such as that provided in the table above is on a number line called a **line plot.** Like other statistical graphs, line plots can help you see patterns and variability in data.

To make a line plot, determine a scale that includes all of the data and appropriate intervals. Then plot each number using a symbol to represent the data. The teacher salaries data ranges from 25,059 to 49,910. Round each salary to the nearest thousand. Let's use a scale of 25,000 to 50,000 with intervals of 2500 and denote each salary with an "×."

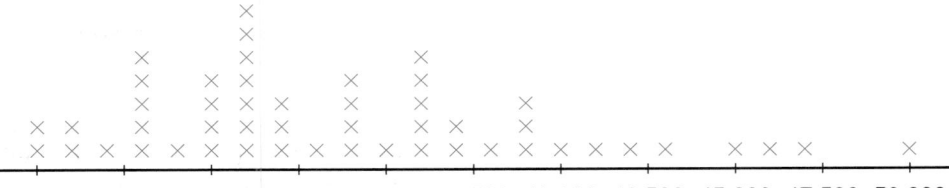

Locate your state's data on the line plot. Now it should be easy to see how teachers' salaries in your state compare with those in other states.

A **stem-and-leaf plot** can also be used to display data in a compact way. When making a stem-and-leaf plot, each item of data is separated into two parts. The *stems* usually consist of the digits in the greatest common place value of each item of data. The *leaves* contain the other digits of each item of data. For example, if the greatest common place value is thousands, then the stem for 1573 is 1, and the leaf is 573.

If data values have many digits, you may want to round each item of data before you plot it in a stem-and-leaf plot. This way each leaf will have only one digit. If 1573 is rounded to 1600, then the stem is 1 and the leaf is 6.

1 Refer to the application at the beginning of the lesson. Make a stem-and-leaf plot of the rounded salaries.

Round each item of data to the nearest thousand.
So, using rounded data, $3 \mid 0 = 29,500$ to $30,499$, inclusive.

Stem	Leaf
2	
·	5 5 6 6 7 8 8 8 8 9
3	0 0 0 0 1 1 1 1 1 1 2 2 2 3 4 4 4 4
·	5 6 6 6 6 6 7 7 8 9 9 9
4	0 1 2 3
·	5 6 7
5	0

To make the plot easier to interpret, the stems have been broken into two parts. $4 \mid = \$40,000$ to $\$44,000$ and $· \mid = \$45,000$ to $\$49,000$

A **back-to-back stem-and-leaf plot** is sometimes used to compare two sets of data or rounded values of the same set of data. In a back-to-back plot, the same stem is used for the leaves of both plots.

2 The number of Republicans and Democrats in the U.S. Senate from 1965–1995 is given in the table below.
a. Make a back-to-back stem-and-leaf plot of the data.
b. Compare the number of Republican and Democratic senators since 1965.

Political Divisions of the U.S. Senate (1965–1995)

Years	Number of Democrats	Number of Republicans	Years	Number of Democrats	Number of Republicans
1965–67	68	32	1981–83	46	53
1967–69	64	36	1983–85	46	54
1969–71	58	42	1985–87	47	53
1971–73	54	44	1987–89	54	46
1973–75	56	42	1989–91	57	43
1975–77	61	37	1991–93	57	43
1977–79	61	38	1993–95	56	44
1979–81	58	41	1995–97	47	53

Source: *World Almanac, 1995*

a.

Democrats	Stem	Republicans
	3	2 6 7 8
6 6 7 7	4	1 2 2 3 3 4 4 6
4 4 6 6 7 7 8 8	5	3 3 3 4
1 1 4 8	6	

Nydia M. Velazquez (1953–)

In 1992, Nydia Velazquez became the first Puerto Rican woman elected to the U.S. House of Representatives. She earned a master's degree from New York University and served on the City Council of New York City.

b. The stem-and-leaf plot shows that there have usually been more Democratic senators than Republican senators. The number of Democratic senators ranged from 46 to 68, while the number of Republican senators ranged from 32 to 54.

Sometimes it is convenient to have one number that describes a set of data. This number is called a **measure of central tendency** because it represents the center or middle of the data. The most commonly-used measures of central tendency are the **median, mode,** and **mean.**

Definition of Median, Mode, and Mean	**Median:**	The median of a set of data is the middle value. If there are two middle values, it is the mean of the two middle values.
	Mode:	The mode of a set of data is the most frequent value. Some sets of data have multiple modes and others have no mode.
	Mean:	The mean of a set of data is the sum of all the values divided by the number of values.

Example ❸

APPLICATION
Consumerism

Over the years, jeans have become very popular with consumers. Some prices for these jeans are listed below. Find the median, mode, and mean prices.

$44.99	$39.99	$39.99	$27.99	$32.00	$34.99
$29.99	$32.99	$44.99	$29.99	$24.99	$44.99

Median: To find the median of the jean data, arrange the dollar values in order.

44.99 44.99 44.99 39.99 39.99 34.99 32.99 32.00 29.99
29.99 27.99 24.99

If there were an odd number of dollar values, the middle one would be the median. However, since there is an even number of dollar values, the median is the average of the two middle values, 34.99 and 32.99.

$$\text{median} = \frac{34.99 + 32.99}{2}$$
$$= \frac{67.98}{2}$$
$$= 33.99$$

The median is $33.99. Notice that the number of values that are greater than the median is the same as the number of values that are less than the median.

Mode: To find the mode of the jean data, look for the number that occurs most often. In this set, $44.99 appears three times. Thus, the mode of the data is $44.99.

Mean: To find the mean of the jean data, find the sum of the dollar values and divide by 12, the number of values in the set.

$$\text{mean} = \frac{44.99 + 44.99 + 44.99 + \ldots + 24.99}{12}$$
$$= \frac{427.89}{12}$$
$$= 35.6575$$

The mean of the data is approximately $35.66.

As you can see, the median, mode, and mean are not always the same number. In Example 3, the mode is greater than both the median and the mean. This means that more of the jeans cost $44.99 than any other price. In this example, the mean is the most representative average of the prices. If you buy any pair of jeans in the set of data, you will pay an average of $35.66.

Extreme values are those data values that vary greatly from the central group of data values. Every data value affects the value of the mean, so when extreme values are included in a set of data, the mean may become less representative of the set. However, the values of the median and the mode are not affected by extreme values in the set.

Example ④

APPLICATION

Business

The amounts of money spent per student in 1992 in two regions of the U.S. are listed below. Determine the mean of the values in each column. To what extent is each mean representative of the data?

Pacific States		Southwest Central States	
State	Expenditures Per Student	State	Expenditures Per Student
Alaska	$8450	Texas	$4632
California	4746	Arkansas	4031
Washington	5271	Louisiana	4354
Oregon	5913	Oklahoma	4078

Source: *World Almanac, 1995*

$$\text{mean} = \frac{8450 + 4746 + 5271 + 5913}{4}$$

$$= \frac{24{,}380}{4} \text{ or } 6095$$

$8450 is an extreme value. In this case, the mean is not representative of the data.

$$\text{mean} = \frac{4632 + 4031 + 4354 + 4078}{4}$$

$$= \frac{17{,}095}{4} \text{ or } 4273.75$$

There are no extreme values in this set. In this case, the mean is representative of the data.

CHECK FOR UNDERSTANDING

Communicating Mathematics

Study the lesson. Then complete the following.

1. **Describe** what the line plot below tells you about the overall scores for the floor exercises of the Buckeye Gymnastics team members.

```
                                    ×
                              ×     ×        ×
                ××     ×      ×     ×        ××
        +----+----+----+----+----+----+
        5    6    7    8    9    10
```

2. **Explain** how the number 98.6 would be plotted on a stem-and-leaf plot if 54.6 is plotted using stem 5 and leaf 4.

3. **Explain** what an extreme value does to the median.

4. **Tell** which measure, the median, mode, or mean, must be a member of the set of data.

5. Assess Yourself List the main advantages and disadvantages for using the median, mode, and mean to describe a set of data.

Guided Practice

Find the median, mode, and mean for each set of data.

6. 0, 2, 2, 3, 4   **7.** 4, 5, 8, 10, 12   **8.** 7, 7, 7, 7, 7, 7

9. The table at the right shows the salary offers to graduating college students with bachelor's degrees in mathematics- and science-related fields of study in 1992.

   a. Make a stem-and-leaf plot of the data rounded to the nearest thousand.

   b. What is the mean salary?

   c. What is the median salary?

   d. Which is the mode salary?

   e. Write three questions that can be answered using this data.

Field of Study	Bachelor's Degree
Accounting	$27,179
Business, general	24,305
Marketing	23,914
Engineering:	
Civil	29,376
Chemical	39,203
Computer	32,848
Electrical	33,754
Mechanical	34,462
Nuclear	34,447
Petroleum	40,679
Engineering tech.	31,051
Chemistry	27,557
Mathematics	28,434
Physics	29,019
Humanities	22,941
Social sciences	21,623
Computer science	30,523

EXERCISES

Practice

Find the median, mode, and mean for each set of data.

10. 298, 256, 399, 388, 276   **11.** 3, 75, 58, 7, 34

12. 4.8, 5.7, 2.1, 2.1, 4.8, 2.1   **13.** 80, 50, 65, 55, 70, 65, 75, 50

14. 61, 89, 93, 102, 45, 89   **15.** 13.3, 15.4, 12.5, 10.7

16. 101, 192, 121, 153, 101   **17.** 43, 43, 55, 43, 54, 42, 51

18. 2301, 2324, 2000, 1999, 2738, 1947, 1989, 2004, 2938

19. Find the median, mode, and mean for the stem-and-leaf plot at the right. Round to the nearest whole number if necessary.

Stem	Leaf
5	5 5
6	0 4 5 6 7 8
7	0 1 1 1 2 2 5 6 6
8	0 5 6 8 8 $8\|5 = 85$

20. In Mrs. Elizondo's algebra class, report card averages are based on 100 points. Tests and quizzes account for $\frac{2}{3}$ of the final average. Homework, classwork, and journals account for $\frac{1}{3}$ of the final average. Jennifer's test and quiz scores are 100, 100, 88, 76, 95, 88, and 93. What is Jennifer's average on tests and quizzes?

21. The Millersburg school board is negotiating a pay raise with the teacher's union. Three of the administrators have salaries of $80,000 each. However, a majority of the teachers have salaries of about $35,000 per year.

   a. You are a member of the school board and would like to show that the current salaries are reasonable. Would you quote the median, mode, or mean as the "average" salary to justify your claim? Explain.

   b. You are the head of the teacher's union and maintain that a pay raise is in order. Which of the median, mode, or mean would you quote to justify your claim? Explain your reasoning.

22. Use a graphing calculator to find the mean and median for each set of data below. Access the LIST MATH function by pressing [2nd] [LIST] [▶]. Then choose 3, for mean or 4, for median. Enter the numbers by first pressing [2nd] [{] and ending with [2nd] [}], a right parenthesis, and [ENTER].

a. 3, 5, 7, 5, 3, 8, 2

b. 45.7, 64.8, 33.2, 66.1, 54.4, 64.5

Critical Thinking

23. Statistics can sometimes be misleading if an incorrect representation of the data leads to wrong conclusions. Study the two graphs below.

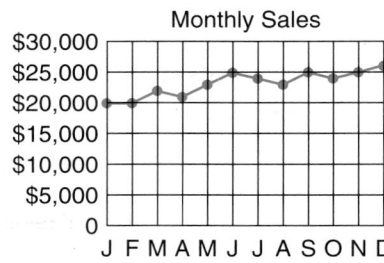
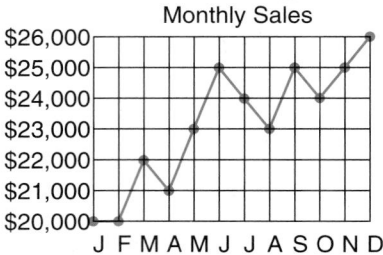

a. Explain why the graphs made from the same data look different.

b. Explain a situation where graph A might be used.

c. Explain a situation where graph B might be used.

24. Choose five whole numbers from 1 to 10 for each situation below. Numbers may be used more than once.

a. Construct a set of data in which the median, mode, and mean are the same number.

b. Find the median of a set of data with the greatest possible mean.

c. Construct a set of data in which the median and mean are the same number and there is no mode.

Applications and Problem Solving

25. Employment The table below lists the employment figures for 11 states during October, 1994.

Employment, October 1994		
State	**Employed**	**Unemployed**
California	14,411,000	1,197,000
Florida	6,384,000	445,000
Illinois	5,672,000	378,000
Massachusetts	2,979,000	205,000
Michigan	4,570,000	247,000
New Jersey	3,830,000	277,000
New York	8,048,000	561,000
North Carolina	3,443,000	180,000
Ohio	5,282,000	274,000
Pennsylvania	5,428,000	344,000
Texas	8,842,000	555,000

Source: *USA Today*

a. Round each number to the nearest hundred thousand. Make a back-to-back stem-and-leaf plot of the data.

b. What observations can you make from the data?

c. How could you calculate the unemployment rate for each state?

d. Calculate the approximate unemployment rate for each state.

26. Advertising Cordless Camera placed an ad in the newspaper showing five videocameras for sale. The ad says, "Our videocameras average $695." The prices of the videocameras are $1200, $999, $1499, $895, $695, $1100, $1300, and $695.

a. Find the median, mode, and mean of the prices.

b. Which measure was Cordless Camera using in its ad? Why did they choose this measure?

c. As a consumer, which measure would you want to see advertised? Explain your reasoning.

27. Business The minimum start-up costs for the fastest-growing franchises in 1993 are listed below. Find the median, mode, and mean for the costs. Which average is the most representative of the data? Explain.

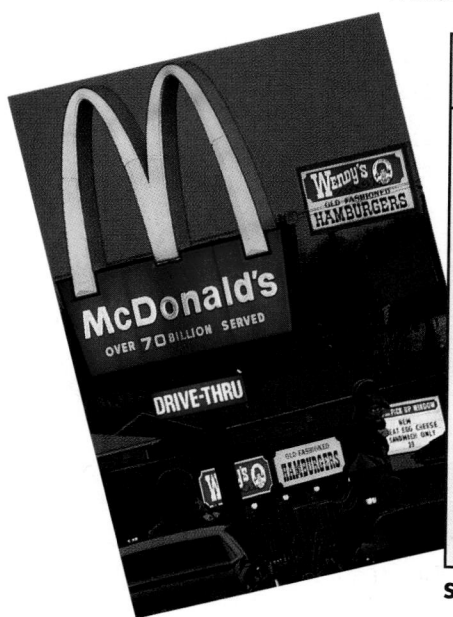

Company	Business	Minimum Start-up Cost
7-Eleven Convenience Stores	convenience stores	$12,500
Subway	submarine sandwiches	38,900
Snap-On Tools	retail hardware	12,600
Maico Tools	retail hardware	42,500
McDonald's	hamburgers	varies
Chem-Dry Carpet Drapery & Upholstery Cleaning	carpet, upholstery, and drapery services	3,550
Little Caesar's Pizza	pizza	170,000
Burger King Corp.	hamburgers	73,000
Coverall North America Inc.	commercial cleaning services	3,250
Mail Boxes Etc.	postal and business services	28,180
CleanNet USA Inc.	commercial cleaning services	425
Jani-King	commercial cleaning services	2,500
Dunkin' Donuts	donuts	175,000

Source: Reprinted with permission from *Entrepreneur* Magazine, January 1994

28. Shopping Malls The table below shows the largest U.S. shopping malls.

	Mall	Gross Leasable Area (sq ft)
1	Del Amo Fashion Center, Torrance, California	3,000,000
2	South Coast Plaza/Crystal Court, Costa Mesa, California	2,918,236
3	Mall of America, Bloomington, Minnesota	2,472,500
4	Lakewood Center Mall, Lakewood, California	2,390,000
5	Roosevelt Field Mall, Garden City, New York	2,300,000
6	Gurnee Mills, Gurnee, Illinois	2,200,000
7	The Galleria, Houston, Texas	2,100,000
8	Randall Park Mall, North Randall, Ohio	2,097,416
9	Oakbrook Shopping Center, Oak Brook, Illinois	2,006,688
10	Sawgrass Mills, Sunrise, Florida	2,000,000
10	The Woodlands Mall, The Woodlands, Texas	2,000,000
10	Woodfield, Schaumburg, Illinois	2,000,000

Source: Blackburn Marketing Service

a. Find the median, mode, and mean of the gross leasable area.

b. You are a realtor who is trying to lease mall space in different areas of the country to a large retailer. Which measure would you talk about if the customer felt that the price per square foot was expensive to lease? Explain.

c. Which measure would you talk about if the customer had lots of inventory to display? Explain.

29. Weather The average monthly temperatures for Mobile, Alabama, are listed below.

January, 59.7°	May, 84.6°	September, 86.9°
February, 63.6°	June, 90.0°	October, 79.5°
March, 70.9°	July, 91.3°	November, 70.3°
April, 78.5°	August, 90.5°	December, 62.9°

a. Make a line plot. Round each temperature to the nearest degree.

b. Use the line plot to find the median, mode, and mean.

c. Suppose each of the average monthly temperatures rose 5°. Make a line plot of these new temperatures.

d. How do the two line plots compare? Are the median, mode, and mean the same or different? Explain.

e. How could you make the new line plot from the original line plot?

Mixed Review

30. Name the property illustrated by $5 + (2 + x) = (2 + x) + 5$. (Lesson 1–2)

31. Simplify $a(3 + 5) - 6(3a - 1)$. (Lesson 1–2)

32. Simplify $\frac{2}{3}(6a - 18) + 3(2a - 9)$. (Lesson 1–2)

33. Evaluate $\frac{7(3 + 2)}{4 - 9}$. (Lesson 1–1)

34. Evaluate $(r + 7) \div s$ if $r = 20$ and $s = 3$. (Lesson 1–1)

35. Chemistry The boiling point of the metal zinc is 787.1°F. Use the formula $C = \frac{5(F - 32)}{9}$ to find the equivalent Celsius temperature. (Lesson 1–1)

Mathematics and SOCIETY

What, No Census?

The article below appeared in *Business Week* on November 28, 1994.

SINCE 1790, THE GOVERNMENT HAS FAITH-fully tried to count every resident once a decade. Now, a panel of statisticians has a blunt message for the Census Bureau: Give up. For 2000, a National Academy of Sciences committee says a "fundamentally redesigned" census should rely heavily on statistical sampling and estimation. The committee says that these methods can produce a more accurate—and far less costly—count of residents who don't respond to the census' mail-in questionnaire, especially minority groups. Congress, angered by the inaccuracy and expense of the 1990 census, is likely to embrace the committee's proposal. ■

1. Each census compiles large amounts of data about us, including age, sex, race, education, home address, household size and composition, and income level. What uses could the government make of the census data? Who else might be interested in the data?

2. Why would a census using statistical sampling and estimation be less expensive than the past method of trying to count everyone by using questionnaires?

3. Describe how you might use statistical sampling to study a group of people that you want to learn more about. What factors would you want to consider in planning your procedures?

Solving Equations

1-4

What YOU'LL LEARN

- To translate verbal expressions and sentences into algebraic expressions and equations,
- to solve equations by using the properties of equality, and
- to solve equations for a specific variable.

Why IT'S IMPORTANT

You can use equations to solve problems involving history, astronomy, and travel.

CONNECTION
History

Have you ever thought it would have been great to live 4500 years ago in Babylon because you wouldn't have had algebra homework? Those teenagers may not have had shopping malls, CD-ROMS, or TVs, but they did have algebra homework!

Babylonian students wrote their assignments on clay tablets with little sticks used to make wedge-shaped marks. They didn't use letters for unknown values in equations because today's alphabet hadn't been invented. Instead, they drew pictures to stand for the unknowns.

A Greek mathematician named Diophantus introduced algebraic symbols to write equations. The chart below shows examples of his equations along with today's algebraic form.

Diophantine Equation	Modern Meaning
ζισβ	$x = 2$
ζγισθ	$x + 3 = 9$
ζγβισθ	$3x + 2 = 9$
ζθΛγισβ	$9x - 3 = 2$
ζβΛθισζγ	$2x - 9 = x + 3$

Diophantus of Alexandria, who lived about A.D. 250, was famous for his work in algebra. His main work was titled *Arithmetica* and introduced symbolism to Greek algebra as well as propositions in number theory and polygonal numbers.

The language of today's algebra provides a powerful way to translate word expressions into algebraic or mathematical expressions. **Variables** are used to represent numbers that are not known. Any letter can be used as a variable.

Verbal Expression	Algebraic Expression
• *a number* increased by 4	$x + 4$
• twice the cube of *a number*	$2n^3$
• the square of *a number* decreased by the cube of the *same number*	$c^2 - c^3$
• three times the sum of *a number* and 6	$3(b + 6)$

Sentences with variables to be replaced, such as $4x - 8 = 32$ and $2x + 4 > 9$, are called **open sentences.** An open sentence that states that two mathematical expressions are equal is called an **equation.** Equations can be used to represent verbal mathematical sentences.

Verbal Sentence	Equation
• Nine is equal to five plus four.	$9 = 5 + 4$
• *A number* decreased by 6 is -3.	$m - 6 = -3$
• *A number* divided by 3 is equal to $\frac{3}{4}$.	$\frac{x}{3} = \frac{3}{4}$

To solve an equation, you find replacements for the variables that make the equation true. Each of these replacements is called a **solution** of the equation.

We can use certain properties of equality to solve equations or open sentences. Some of those properties are listed below.

Reflexive Property of Equality	For any real number a, $a = a$.
Symmetric Property of Equality	For all real numbers a and b, if $a = b$, then $b = a$.
Transitive Property of Equality	For all real numbers a, b, and c, if $a = b$ and $b = c$, then $a = c$.
Substitution Property of Equality	If $a = b$, then a may be replaced by b.

Example ❶ Name the property illustrated by each statement.
a. If $1.5(7.5) = 11.25$, then $11.25 = 1.5(7.5)$.

symmetric property of equality

b. If $7 = 1 + 2 + 4$ and $1 + 2 + 4 = 4 + 3$, then $7 = 4 + 3$.

transitive property of equality

Sometimes an equation can be solved by adding or subtracting the same number on each side.

Addition and Subtraction Properties of Equality	For any numbers a, b, and c, if $a = b$, then $a + c = b + c$ and $a - c = b - c$.

Example ❷ **Solve $x + 54.57 = 78$.**   *Estimate:* $80 - 55 = 25$

$$x + 54.57 = 78$$
$$x + 54.57 - 54.57 = 78 - 54.57 \quad \textit{Subtract 54.57 from each side.}$$
$$x = 23.43$$

Check:
$$x + 54.57 = 78$$
$$23.43 + 54.57 \stackrel{?}{=} 78 \quad \textit{Replace x with 23.43.}$$
$$78 = 78 \quad \checkmark$$

The solution is 23.43.

Some equations may be solved by multiplying or dividing each side by the same number.

Multiplication and Division Properties of Equality	For any real numbers a, b, and c, if $a = b$, then $a \cdot c = b \cdot c$ and, if $c \neq 0$, $\frac{a}{c} = \frac{b}{c}$.

Example **3** **Solve each equation.**

a. $4x = -12$

$$4x = -12$$

$$\frac{1}{4}(4x) = \frac{1}{4}(-12) \quad \textit{Multiply each side by } \frac{1}{4},$$
$$x = -3 \qquad \qquad \textit{the reciprocal of 4.}$$

The solution is -3.

This equation could also be solved by dividing each side by 4.

Check:

$$4x = -12$$
$$4(-3) \stackrel{?}{=} -12$$
$$-12 = -12 \checkmark$$

b. $-\frac{3}{4}t = 15$

$$-\frac{3}{4}t = 15$$

$$-\frac{4}{3}\left(-\frac{3}{4}\right)t = \left(-\frac{4}{3}\right)(15) \quad \textit{Multiply each side by } -\frac{4}{3},$$
$$t = -20 \qquad \qquad \textit{the reciprocal of } -\frac{3}{4}.$$

The solution is -20.

Check:

$$-\frac{3}{4}t = 15$$
$$-\frac{3}{4}(-20) \stackrel{?}{=} 15$$
$$15 = 15 \checkmark$$

In order to solve some equations, it may be necessary to apply more than one property.

Example **4** **Solve $3(2a + 25) - 2(a - 1) = 78$.**

$$3(2a + 25) - 2(a - 1) = 78$$
$$6a + 75 - 2a + 2 = 78 \quad \textit{Distributive and substitution properties}$$
$$4a + 77 = 78 \quad \textit{Commutative, distributive, and substitution properties}$$
$$4a = 1 \quad \textit{Subtraction and substitution properties}$$
$$a = \frac{1}{4} \quad \textit{Division and substitution properties}$$

The solution is $\frac{1}{4}$. *Check this result.*

Sometimes you need to solve an equation or formula for a variable.

Example **5**

Geometry

The formula for the volume of a cylinder is $V = \pi r^2 h$, where V is the volume, r represents the radius of the circular base and top and h represents the height of the cylinder. Solve the formula for h.

$$V = \pi r^2 h$$

$$\frac{V}{\pi r^2} = \frac{\pi r^2 h}{\pi r^2} \quad \textit{Divide each side by } \pi r^2.$$

$$\frac{V}{\pi r^2} = h$$

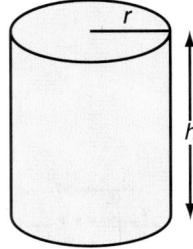

You can use a four-step **problem-solving plan** to help you solve problems.

Problem-Solving Plan

1. **Explore the problem.**
2. **Plan the solution.**
3. **Solve the problem.**
4. **Examine the solution.**

Example **6**

Geometry

The perimeter of a parallelogram is 48 inches. What is the length of the longer side if the shorter side measures 9 inches?

Explore   Draw a diagram and let ℓ represent the measure of the longer side.

Plan   The perimeter equals the sum of the lengths of the sides. So, we can write the following equation.

$$2(9) + 2(\ell) = 48$$

Solve $2(9) + 2(\ell) = 48$

$$18 + 2\ell = 48$$

$$\ell = 15$$

The length of the longer side is 15 inches.

Examine   If one of the two longer sides has length 15 inches and one of the shorter sides has length 9 inches, the perimeter is $15 + 15 + 9 + 9 = 48$ inches. Thus, the answer is correct.

CHECK FOR UNDERSTANDING

Communicating Mathematics

Study the lesson. Then complete the following.

1. **Write** two verbal expressions and two verbal sentences containing unknown quantities. Write each as an algebraic expression or equation.

2. **Summarize** the properties you studied in this lesson. Exchange summaries with another student. Discuss your summaries and make any necessary revisions.

3. **Explain** the difference between an equation and an expression.

4. **Write** an equation to find the length of a side of a regular pentagon if its perimeter is 250 inches.

5. **You Decide**   Was this equation solved correctly? If not, explain what error was made. Then show how to solve the equation correctly.

$$4(a + 5) - 2(a + 6) = a + 16$$

$$4a + 20 - 2a - 12 = a + 16$$

$$2a + 8 = a + 16$$

$$a = 8$$

Write an algebraic expression to represent each verbal expression.

6. three decreased by twice a number

7. five times a number decreased by three

Name the property illustrated by each statement.

8. If $r + 2 = 8$, then $r = 6$.

9. If $4x = 16$, then $12x = 48$.

Solve each equation.

10. $10 + 5x = 110$

11. $-2(a + 4) = 2$

12. $3b + 4b + 5b = 30$

13. $7 + 5n = -58$

14. $-1.4t + 3 = -7.5$

15. $-\frac{2}{3}k = 14$

Solve each equation or formula for the variable specified.

16. $2x - 3m = 6$, for x

17. $V = \frac{1}{3}\pi r^2 h$, for h

Define a variable, write an equation, and solve the problem. Then check your solution.

18. Marisa is 16 years old. Her parents are both the same age. The three of them have lived a total of 100 years. How old are Marisa's parents?

EXERCISES

Write an algebraic expression to represent each verbal expression.

19. fourteen decreased by the square of a number

20. twice the sum of a number and 11

21. four times the sum of a number and its square

22. the product of the square of a number and five

23. the sum of 7 and three times a number

24. the square of the sum of a number and 13

Name the property illustrated by each statement.

25. $(5 + 6) + 7 = (5 + 6) + 7$

26. If $4 + 8 = 12$, then $12 = 4 + 8$.

27. If $3x = 10$, then $3x + 6 = 10 + 6$.

28. If $3m = 5n$ and $5n = 10p$, then $3m = 10p$.

29. If $7 + s = 21$, then $s = 14$.

30. If $q + (8 + 5) = 32$, then $q + 13 = 32$.

Solve each equation.

31. $3y + 16 = 22$

32. $14 - x = -7$

33. $34 - 10w = 6w + 2$

34. $t + 2t + 3t + 4t + 5t = 45$

35. $\frac{1}{8} - \frac{3}{4}x = \frac{1}{16}$

36. $\frac{3}{4} - \frac{3}{5}x = \frac{2}{5}x + \frac{2}{4}$

37. $5 = -5(y + 3)$

38. $2d + 5 = 8d + 2$

39. $280 - 26f = 1098$

40. $3(4 - 5k) = 2k - 4$

41. $4m - 9 = 5m + 7$

42. $32g + 245 = 3829$

43. $12x - 24 = -14x + 28$

44. $18 = -6(p + 5)$

45. $4.5(b + 1) - 2 = 4(b + 3)$

46. $2.3n + 1 = 1.3n + 7$

47. $\frac{5}{7}x - 4 = \frac{3}{7}x + 1$

48. $\frac{3}{4}n - 2 = \frac{1}{2}n + 7$

Solve each equation or formula for the variable specified.

49. $x(y + 2) = z$, for y

50. $I = prt$, for t

51. $5a - 6b = 9$, for b

52. $de - 4f = 5g$, for e

53. $F = G\frac{Mm}{r^2}$, for M

54. $qr + s = t$, for q

Define a variable, write an equation, and solve the problem. Then check your solution.

55. The high school principal has given Mrs. Diaz $420 to buy tickets to *Phantom of the Opera* for her English class and chaperones. The school requires that there be one adult chaperone for every five students on such a trip. How many $15.00 student tickets and $30.00 adult tickets can she order?

56. Hearthstone Toyota's price for a Toyota Camry is $18,999. Sheila Rayburn offered the dealership a price of $17,099. Sheila's price is what percent of the dealership's price?

57. You have $32 to spend on supplies for your science fair project. If you buy two plants for experiments, you will have $18 left for other supplies. How much is each plant?

58. Geometry The perimeter of an isosceles triangle is 116 centimeters. The length of the base is 36 cm. What is the length of one of the equal sides?

Scene from Phantom of the Opera

Critical Thinking

59. Write a verbal expression to represent the algebraic expression $2x(x + 4) + 2(x + 6)$.

Applications and Problem Solving

60. Packaging Two designs for a tuna can are shown at the right. If each can holds the same amount of tuna, what is the height of can A?

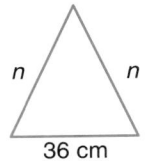
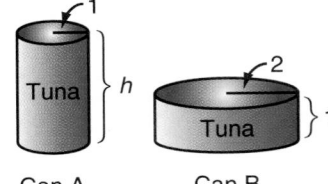

Can A   Can B

61. Transportation The U.S. Department of Transportation regulations prohibit a truck driver from driving more than 70 hours in any 8-day period. Over the past 7 days, a driver has accumulated 54.75 hours of driving time. Does he have enough hours available on the eighth day to deliver freight if he must drive 510 miles at a speed of 50 mph? Explain.

62. Astronomy Earth is about 93,000,000 miles from the sun. When Venus is on the opposite side of the sun from Earth, it is about 69,000,000 from the sun. What is the distance from Earth to Venus?

63. Travel The distance by water from New York City to San Francisco by way of Cape Horn is about 13,200 miles. By going through the Panama Canal, the distance is only 5280 miles. How many miles does a ship save by going through the Panama Canal?

64. Animals The table at the right lists major U.S. zoological parks and the number of species at each park. (Lesson 1–3)

Zoo	Species
Bronx	670
Cleveland	563
Dallas	329
Detroit	413
Houston	596
Memphis	445
Minnesota	311
Philadelphia	560
Phoenix	324
San Antonio	700
San Francisco	270

a. Make a stem-and-leaf plot of the data. Round each number to the nearest ten.

b. Find the median, mode, and mean of the data.

65. Statistics Find the median, mode, and mean for the following set of data. (Lesson 1–3)

216, 399, 219, 179, 180, 399

66. Hockey Use the line plot below to answer each question. (Lesson 1–3)

Most NHL Goals in a Season

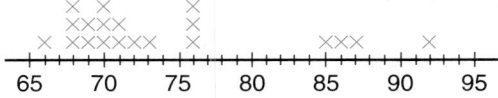

```
                X  X          X
                XXXX          X
      X  XXXXXXX     X              XXX       X
    +--+--+--+--+--+--+--+--+--+--+--+--+--+--+--+
   65    70    75    80    85    90    95
```

Mario Lemieux

a. What is the greatest number of goals scored in a season?

b. In the 1992–93 season, Mario Lemieux scored 69 goals. In the 1988–89 season, he scored 85 goals. How many scores were greater than Mario's in 1992–93 but less than Mario's in 1988–89?

67. Weather The average monthly temperatures in selected cities for January and July are given in the table below. Make a back-to-back stem-and-leaf plot of the temperatures of the cities rounded to the nearest degree. (Lesson 1–3)

City	January Temperature	July Temperature
Baton Rouge, LA	50.8	82.1
Caribou, ME	10.7	65.1
Charlotte, NC	40.5	78.5
Chicago, IL	21.4	73.0
Dallas, TX	44.0	86.3
Denver, CO	29.5	73.4
Indianapolis, IN	26.0	75.1
Jacksonville, FL	53.2	81.3
Juneau, AK	21.8	55.7
Roswell, NM	41.4	77.7
San Diego, CA	56.8	70.3
Tulsa, OK	35.2	83.2

Source: U.S. National Oceanic and Atmospheric Administration

68. Statistics Find the median, mode, and mean for the following set of data. (Lesson 1–3)

$$11, 10, 13, 12, 12, 13, 15$$

69. Simplify $2(9a - 2) - 3(5 + a)$. (Lesson 1–2)

Name the property illustrated by each equation. (Lesson 1–2)

70. $11(3a + 2b) = 11(2b + 3a)$ **71.** $a + b + 0 = a + b$

72. Evaluate $\sqrt{9} \div 3$ and name the sets of numbers to which it belongs. (Lesson 1–2)

73. Banking Find the interest earned in 6 years on a savings account containing $20,000 if the interest rate is 14.5%. (Lesson 1–1)

74. Evaluate $3^b - a + c$ if $a = 13$, $b = 2$ and $c = 9$. (Lesson 1–1)

75. Evaluate $4 - [32 \div (16 - 12)]$. (Lesson 1–1)

76. Evaluate $[18 - (5 + 22)] \times 2$. (Lesson 1–1)

SELF TEST

Evaluate each expression if $m = 2$, $n = -3$, and $p = 4$. (Lesson 1–1)

1. $m + n - p$ **2.** $m(n + p)$

Find the value of each expression. Then name the sets of numbers to which each value belongs. (Lesson 1–2)

3. $7 - 8$ **4.** $3.9 + 2.6$ **5.** $\sqrt{36} + 5$

6. Consumerism The Super Shoes catalog contains 29 pairs of shoes that can be ordered through the mail. The prices are $53, $42, $49, $38, $39, $48, $37, $48, $37, $39, $58, $59, $32, $50, $59, $37, $36, $30, $40, $33, $30, $45, $40, $30, $35, $48, $37, $48, and $50. (Lesson 1–3)
   a. Make a stem-and-leaf plot of the shoe prices.
   b. Find the median, mode, and mean of the prices.

Solve each equation. (Lesson 1–4)

7. $4.5 - 3.9m = 20.1$ **8.** $9 = 16d + 51$

9. $2y - 8 = 14 - 9y$ **10.** $285 - 38x = 2033$

1-5A Graphing Technology
Using Tables to Estimate Solutions

A Preview of Lesson 1–5

You can use a graphing calculator to estimate solutions to equations by building tables of values.

Example ❶ **Estimate the solution of $12x - 3 = 5$ to the nearest hundredth.**

Rewrite the equation in an equivalent form to get 0 on one side.

$$12x - 3 = 5 \quad \rightarrow \quad 12x - 8 = 0$$

To estimate the solution means to find a value for x so that $12x - 8$ is very close to 0. Let $y = 12x - 8$ and make a table of values for x and y. First, enter $y = 12x - 8$.

Enter: [Y=] 12 [X,T,θ,n] [−] 8

Then, set up a table of values for x and y. You must enter a starting value for x and an increment for successive values. Let's start with $x = 0$ and use increments of 1.

Enter: [2nd] [TblSet] 0 [ENTER]

　　　 1 [2nd] [Table]

We need to find the value of x when $y = 0$. From the table, we can see that x is a value between 0 and 1. To get a better approximation for x, create a new table that starts at $x = 0$ and use increments of 0.1.

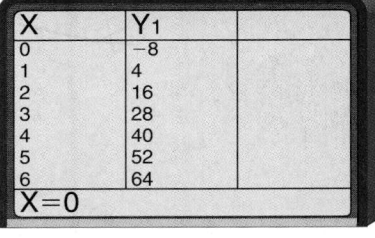

X	Y₁
0	−8
1	4
2	16
3	28
4	40
5	52
6	64
X=0	

Enter: [2nd] [TblSet] 0 [ENTER]

　　　 .1 [2nd] [Table]

Use the arrow keys to scroll down the table to determine where $y = 0$. The solution is between 0.6 and 0.7. Continue to adjust the estimated value for x.

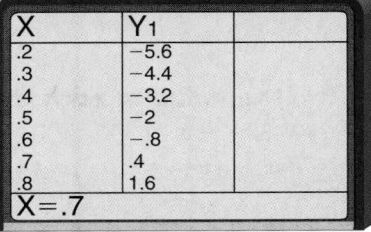

X	Y₁
.2	−5.6
.3	−4.4
.4	−3.2
.5	−2
.6	−.8
.7	.4
.8	1.6
X=.7	

Enter: [2nd] [TblSet] .6 [ENTER]

　　　 .01 [2nd] [Table]

Repeating this process one more time, we see that the solution is between 0.666 and 0.667. Thus, the solution is about 0.67.

X	Y₁
.62	−.56
.63	−.44
.64	−.32
.65	−.2
.66	−.08
.67	.04
.68	.16
X=.66	

Example 2 involves an absolute value equation.

Example ② **Estimate the solutions of $|1.5x - 3| = 0.7$ to the nearest hundredth.**

Rewrite the equation as $|1.5x - 3| - 0.7 = 0$. Then, enter the equation and set up the table. Let's start with $x = 0$ and then use increments of 1 for the x values.

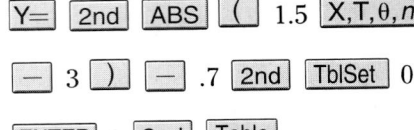

Enter:

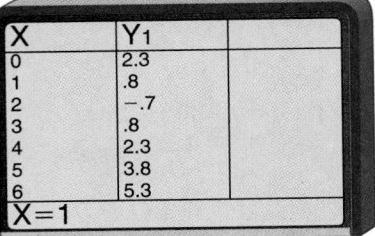

We need to find the value of x when $y = 0$. From the table, we can see that x is a value between 1 and 2 and also between 2 and 3. Begin to adjust your estimate by repeating the process.

Enter: 2nd TblSet 1 ENTER .1 2nd
Table

The solution is between 1.5 and 1.6. Continue the process to estimate the first solution. Then adjust your estimate for the second solution.

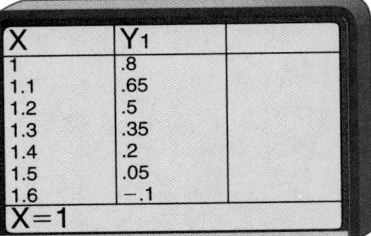

Enter: 2nd TblSet 2 ENTER .1
2nd Table

The solution is between 2.4 and 2.5. Continue the process to estimate the second solution.

The solutions are about 1.53 and 2.47.

EXERCISES

Use the TABLE feature to estimate the solution(s) of each equation to the nearest hundredth.

1. $4x + 6 = 9$

2. $3.5x + 7 = 11$

3. $-1.25 - 0.3x = 8$

4. $5(x - 3) = -2$

5. $2x + 1 = 12 - x$

6. $|6x + 4| - 7 = 2$

7. $|3 - x| = 5$

8. $|2.21 + 0.55x| = 1.75$

9. $\left|\frac{1}{2}x - 5\right| = 17$

10. $40 = \frac{5}{9}(x - 32)$

Solving Absolute Value Equations

What YOU'LL LEARN

- To solve equations containing absolute value, and
- to solve problems by making lists.

Why IT'S IMPORTANT

You can use absolute value equations to solve problems involving travel and manufacturing.

APPLICATION
Time Zones

Joan Haghiri works as a consultant. Her job consists of providing or developing training for businesses or organizations to help their employees perform better. Joan travels frequently as a part of her job and often goes from one time zone to another.

Joan lives in El Paso, Texas. One week she flew to New York City to give a presentation. She called home the first night to talk to her 6-year-old son. She called at 10:00 P.M. to be in time for his 8:00 P.M. bedtime. Since New York is in the Eastern time zone and El Paso is in the Mountain time zone, it is two hours later in New York City than in El Paso.

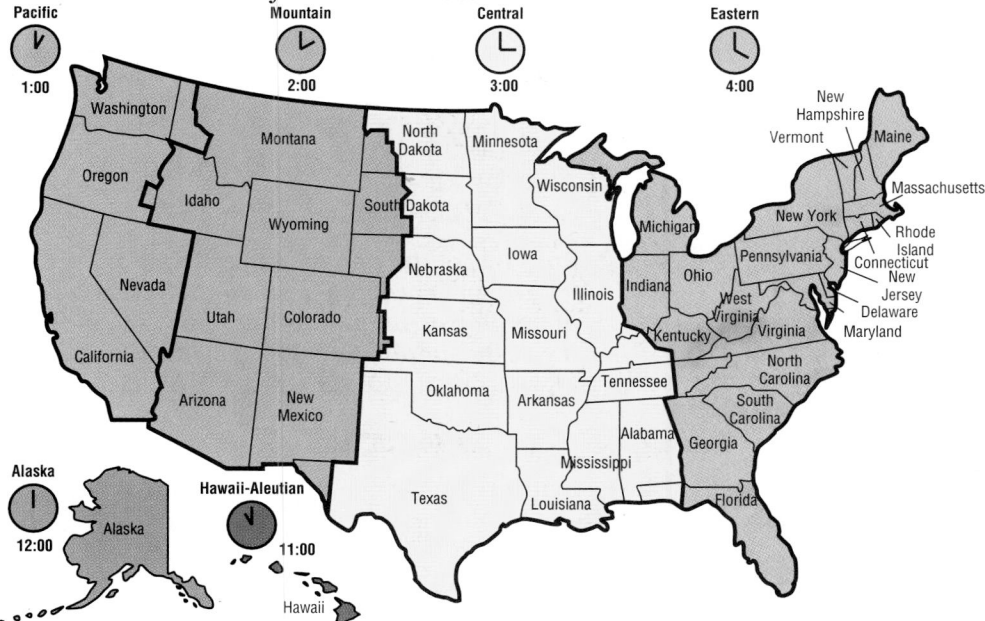

While in Fairbanks, Alaska, the next week, Joan had to remember to call her son at 6:00 P.M. Alaska time. This is because it is two hours later in El Paso.

On a number line, the time zone for El Paso is the starting point and corresponds to 0. New York City is at $+2$, while Fairbanks is at -2.

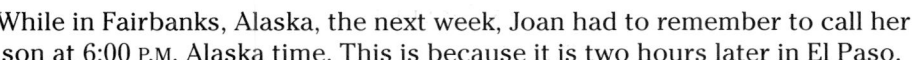

Certainly -2 and 2 are quite different, but they do have something in common. They are the same distance from 0 on the number line.

We say that -2 and 2 have the same **absolute value.** The absolute value of a number is the number of units it is from 0 on the number line. We use the symbol $|x|$ to represent the absolute value of a number x.

The absolute value of -2 is 2.   The absolute value of 2 is 2.

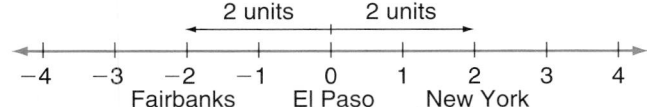

$$|-2| = 2 \qquad\qquad |2| = 2$$

We can also define absolute value in the following way.

Absolute Value	For any real number *a*:
	if $a \geq 0$, then $|a| = a$;
	if $a < 0$, then $|a| = -a$.

The symbol $\geq$ means "is greater than or equal to."

Example **Find each absolute value.**

a. $|9|$

$|9| = 9$

b. $|-14|$

$|-14| = -(-14)$ or 14

You will solve certain kinds of problems by using the strategy **list possibilities**. List possibilities is one of many *problem-solving strategies* that you can use to solve problems. Here are some other problem-solving strategies.

Problem-Solving Strategies	
draw a diagram	solve a simpler (or a similar) problem
make a table or chart	eliminate the possibilities
make a model	look for a pattern
guess and check	act it out
check for hidden assumptions	work backward
use a graph	identify subgoals

Example **2** **Rosa forgot the personal identification number (PIN) for her automatic teller bank card. She remembered that her PIN is the rearranged digits of her house number. If her house number is 1256, what PINs should she try in the automatic teller machine?**

List the possible PINs.

Possible PINs starting with 1:

1265   1526   1562   *Why isn't*

1625   1652   *1256 listed?*

Possible PINs starting with 2:

2156   2165   2516

2561   2615   2651

Possible PINs starting with 5:

5126   5162   5216

5261   5612   5621

Possible PINs starting with 6:

6125   6152   6215

6251   6512   6521

These are the 23 possible PINs that Rosa should try in the automatic teller machine.

You can list the possibilities to help you find absolute values.

Example **Find the absolute value of $x - 15$.**

Make a list of the possible cases.

Case 1: If *x* is 15 or greater, then $x - 15 \geq 0$.

So, $|x - 15| = x - 15$.

Case 2: If *x* is less than 15, then $x - 15 < 0$.

So, $|x - 15| = -(x - 15)$ or $15 - x$.

You can evaluate expressions that contain absolute values. The absolute value bars can be grouping symbols.

Example **4** **Evaluate** $|3x - 6| + 3.2$ **if** $x = -2$.

$$|3x - 6| + 3.2 = |3(-2) - 6| + 3.2 \quad \textit{Replace x with } -2.$$
$$= |-6 - 6| + 3.2 \quad \textit{Simplify within absolute value bars first.}$$
$$= |-12| + 3.2$$
$$= 12 + 3.2$$
$$= 15.2$$

The value is 15.2.

Some equations contain absolute value expressions. The definition of absolute value is used in solving the equations. When an equation has more than one solution, the solutions are often written as a set, {a, b}.

Example **5** **Solve** $|x - 25| = 17$. **Check each solution.**

$|x - 25| = 17$ means $x - 25 = 17$ or $-(x - 25) = 17$.

$$x - 25 = 17 \qquad \text{or} \qquad -(x - 25) = 17$$
$$x - 25 + 25 = 17 + 25 \qquad\qquad x - 25 = -17$$
$$x = 42 \qquad\qquad x - 25 + 25 = -17 + 25$$
$$x = 8$$

Check:
$$|x - 25| = 17 \qquad\qquad |x - 25| = 17$$
$$|42 - 25| \stackrel{?}{=} 17 \quad \text{or} \quad |8 - 25| \stackrel{?}{=} 17$$
$$|17| \stackrel{?}{=} 17 \qquad\qquad |-17| \stackrel{?}{=} 17$$
$$17 = 17 \checkmark \qquad\qquad 17 = 17 \checkmark$$

The solutions are 42 and 8. Thus, the solution set is {42, 8}.

Example **6**

Health

Hypothermia **and** *hyperthermia* **are similar words but have opposite meanings. Hypothermia is defined as a lowered body temperature. Hyperthermia means an extremely high body temperature. Both are potentially dangerous conditions and occur when a person's body temperature is more than 8° above or below the normal body temperature of 98.6°F. At what temperatures do these conditions begin to occur?**

Let b = normal body temperature. Then solve $|b - 98.6| = 8$.

$$|b - 98.6| = 8$$

$$b - 98.6 = 8 \qquad \text{or} \qquad b - 98.6 = -8$$
$$b = 106.6 \qquad\qquad b = 90.6$$

The solutions are 106.6 and 90.6. Hypothermia occurs when the body temperature is below 90.6°F. Hyperthermia occurs when the body temperature is above 106.6°F.

Sometimes an equation has no solution. For example, $|x| = -6$ is never true. Since the absolute value of a number is always positive or zero, there is no replacement for x that will make that sentence true. The solution set has no members. It is called the **empty set** and is symbolized by $\{\ \}$ or $\varnothing$.

Example **7** Solve $|2x + 7| + 5 = 0$.

$$|2x + 7| + 5 = 0$$
$$|2x + 7| = -5$$

This sentence is *never* true, so the solution set is $\varnothing$.

It is important to check your answers when solving absolute value equations. Even if the correct procedure for solving the equations is used, the answers may not be actual solutions to the original equation.

Example **8** Solve $|x - 2| = 2x - 10$.

$|x - 2| = 2x - 10$ means $x - 2 = 2x - 10$ or $x - 2 = -(2x - 10)$.

$$x - 2 = 2x - 10 \quad \text{or} \quad x - 2 = -(2x - 10)$$
$$-2 = x - 10 \qquad\qquad x - 2 = -2x + 10$$
$$8 = x \qquad\qquad\qquad 3x = 12$$
$$x = 4$$

Check: $\quad |x - 2| = 2x - 10 \qquad\qquad |x - 2| = 2x - 10$

$$|8 - 2| \stackrel{?}{=} 2(8) - 10 \quad \text{or} \quad |4 - 2| \stackrel{?}{=} 2(4) - 10$$
$$|6| \stackrel{?}{=} 16 - 10 \qquad\qquad |2| \stackrel{?}{=} 8 - 10$$
$$6 = 6 \checkmark \qquad\qquad\qquad 2 \neq -2$$

The only solution is 8.

CHECK FOR UNDERSTANDING

Communicating Mathematics

Study the lesson. Then complete the following.

1. **Explain** why an absolute value equation can have two solutions.

2. **Write** a convincing argument for why $|x - 7| + 4 = 0$ has no solution.

3. **You Decide** Janelle says that $-a$ is always negative in the definition of absolute value. (If $a < 0$, then $|a| = -a$.) Bobby says that even $-a$ can be positive. Who is correct, and why?

4. **Determine** whether $-x < x$ is always true, sometimes true, or never true. Explain your reasoning.

5. **Write** an absolute value equation that has no solution.

Guided Practice

Evaluate each expression if $x = 2.5$.

6. $|x + 6|$
7. $|-2x|$
8. $-|x + 10|$

Solve each equation.

9. $|x + 5| = 18$
10. $|x + 9| = 25$

11. $|x - 6| = 12$
12. $|x - 3| = 15$

13. $|3 + x| = 45$
14. $6|5x + 2| = 312$

15. **Manufacturing** A machine is to fill each of several boxes with 16 ounces of sugar. After the boxes are filled, another machine weighs the boxes. If the box is more than 0.2 ounces above or below the desired weight, the box is rejected.

    a. Write an absolute value equation to find the heaviest and lightest box the machine will approve.

    b. Solve the equation.

EXERCISES

Practice

Evaluate each expression if $x = -4$, $y = 5$, and $z = 1.2$.

16. $|-4x|$ 17. $|2y - 5|$ 18. $|3z|$

19. $|x + 5|$ 20. $|-2y|$ 21. $-|2z - 4|$

22. $6 - |4y + 10|$ 23. $7 - |3z + 10|$ 24. $3|x + 4| + |3x|$

Solve each equation.

25. $|x - 3| = 17$ 26. $|x + 6| = 18$

27. $|x + 11| = 42$ 28. $3|x + 6| = 36$

29. $11|x - 9| = 121$ 30. $|2x + 9| = 30$

31. $8|x - 3| = 88$ 32. $|2x + 7| = 0$

33. $|4x - 3| = -27$ 34. $8|4x - 3| = 64$

35. $3|3x + 2| = 51$ 36. $5|x + 4| = 45$

37. $4|6x - 1| = 29$ 38. $|3t - 5| = 2t$

39. $|2a + 7| = a - 4$ 40. $|x - 3| + 7 = 2$

41. $3|x + 6| = 9x - 6$ 42. $5|3x - 4| = x + 1$

Programming

43. The graphing calculator program at the right tests decimal values to estimate the solution for $|x^2 - 2| = 0$. Enter a possible solution for x. The program will test it, give you the value of $|x^2 - 2|$, and tell you if you need to try again. If you guess correctly, the program will give you both solutions.

```
PROGRAM:ABSVALUE
:Prompt X
:Disp "abs(X²-2)=",abs(X²-2)
:If abs(X²-2)=0
:Then
:Goto 1
:End
:Disp "TRY AGAIN"
:Disp "PRESS ENTER"
:Stop
:Lbl 1
:Disp "SOLUTIONS ARE",X
:Disp "AND",-X
```

Use the program to approximate to the nearest tenth the solutions for each equation. You will need to enter the equation into Y₁ on the Y= list for each exercise. All solutions are between −10 and 10.

    a. $x^2 - 2x - 4 = 0$; 2 solutions

    b. $x^3 - 3x = 0$; 3 solutions

    c. $|3x - 2| - 4 = 0$; 2 solutions

    d. $5x^3 + 3x^2 - 25x - 15 = 0$; 3 solutions

44. Solve $|x + 2| = |2x - 4|$ and explain your method of solution.

45. Contests Foust Honda is having a contest to win a new Honda Civic. To win a chance at the car, you must guess the number of keys in a jar within 5 of the actual number. The people who are within this range get to try a key in the ignition of the Civic. Suppose there are 697 keys in the jar.

  a. Write an equation to determine the highest and lowest guesses that will win a chance at the car.

  b. Solve the equation.

46. Chemistry For hydrogen to be a liquid, its temperature must be within 2°C of −257°C.

  a. Write an equation to determine the least and greatest temperatures for this substance to remain a liquid.

  b. Solve the equation.

47. List Possibilities The telephone number of a local business is 555-1829. They are trying to make a word from the last three digits of their number so that customers will remember it easily. The digit 8 can be T, U, or V; 2 can be A, B, or C; 9 can be W, X, or Y. List the possible combinations of letters that their number can represent.

48. Write an algebraic expression to represent *four plus three times a number*. (Lesson 1–4)

49. Solve $3 - 2x = 18$. (Lesson 1–4)

50. Geometry The perimeter of a square is 42 inches. Find the length of one side of the square. (Lesson 1–4)

51. Zoology The population of gorillas at a zoo was decreased when five of them were moved to another zoo. Write an expression to represent the original number of gorillas at the zoo if there are p gorillas there now. (Lesson 1–4)

52. Statistics Find the median, mode, and mean for the following set of data. (Lesson 1–3)
$-29, -35, -31, -27, -29, -32, -28$

53. Statistics Find the median, mode, and mean for the stem-and-leaf plot at the right. (Lesson 1–3)

Stem	Leaf
13	0 2 3 3 5 7
14	1 1 1 5 7 8 9 9
15	5 6 7 7 $14 \mid 5 = 145$

54. Name the property illustrated by $\frac{3}{8}(1) = \frac{3}{8}$. (Lesson 1–2)

55. Evaluate -2.4×10. Then name the sets of numbers to which it belongs. (Lesson 1–2)

56. Find the value of $12a^2 + bc$ if $a = 3$, $b = 7$, and $c = -2$. (Lesson 1–1)

Solving Inequalities

What YOU'LL LEARN

- To solve inequalities and graph the solution sets.

Why IT'S IMPORTANT

You can use inequalities to solve problems involving health, school, and shopping.

CONNECTION

Health

Kristin read in a magazine about an effective way for women to calculate the optimal daily Calorie intake, which is the number of Calories used while at rest. The steps are listed below.

1. Multiply your weight in pounds w by 4.3.
2. Multiply your height in inches h by 4.7.
3. Add the numbers.
4. Add 655 to your result from step 3.
5. Multiply your age a times 4.7.
6. Subtract the product in step 5 from the expression in step 4.

The expression for optimal Calorie intake is $4.3w + 4.7h + 655 - 4.7a$. To find the number of Calories you burn with moderate activity, multiply the expression by 1.3. This gives you the number of Calories per day to maintain your weight.

$$1.3(4.3w + 4.7h + 655 - 4.7a)$$

If you keep your daily Calorie intake between the two numbers, you should lose weight. If you take in fewer Calories than you burn at rest, your metabolism will slow down and you may lose less weight.

A 30-year old woman weighing 130 pounds and 65 inches tall burns about 1400 Calories a day at rest and about 1800 Calories with moderate activity. If she eats 1800 Calories per day, she should maintain her weight. Her actual intake will either be greater than, less than, or equal to her optimal intake of 1800 Calories.

Let x represent her optimal intake and a represent her actual intake. You can compare the optimal and actual intakes using an inequality or an equation.

$$x > a \qquad\qquad x < a \qquad\qquad x = a$$

This is an illustration of the **trichotomy property.**

Trichotomy Property	For any two real numbers, a and b, exactly one of the following statements is true. $a < b \qquad a = b \qquad a > b$

Adding the same number to each side of an inequality does not change the truth of the inequality.

Addition and Subtraction Properties for Inequalities	For any real numbers, a, b, and c: 1. if $a > b$, then $a + c > b + c$ and $a - c > b - c$; 2. if $a < b$, then $a + c < b + c$ and $a - c < b - c$.

These properties can be used to solve inequalities. The solution sets of inequalities can then be graphed on number lines.

Example **1** **Solve $6x + 3 > 5x - 2$. Graph the solution set.**

$$6x + 3 > 5x - 2$$

$$-5x + 6x + 3 > -5x + 5x - 2 \quad \textit{Add } -5x \textit{ to each side.}$$

$$x + 3 > -2$$

$$x + 3 + (-3) > -2 + (-3) \quad \textit{Add } -3 \textit{ to each side.}$$

$$x > -5$$

A circle means that this point is <u>not</u> included.

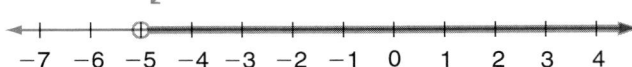

$$-7 \quad -6 \quad -5 \quad -4 \quad -3 \quad -2 \quad -1 \quad 0 \quad 1 \quad 2 \quad 3 \quad 4$$

Any real number greater than -5 is a solution.

Check: Substitute -5 for x in $6x + 3 > 5x - 2$. The two sides should be equal. Then substitute a number greater than -5. The inequality should be true.

You know that $12 > -3$ is a true inequality. What happens if you multiply the numbers on each side by a positive number or by a negative number? Is it still true?

Multiply by 6.

$$12 > -3$$

$$6(12) \overset{?}{>} 6(-3)$$

$$72 > -18 \quad \text{true}$$

Multiply the inequality by other positive numbers. Do you think that the inequality will always remain true?

Multiply by $-\frac{1}{3}$.

$$12 > -3$$

$$-\frac{1}{3}(12) \overset{?}{>} -\frac{1}{3}(-3)$$

$$-4 > 1 \quad \text{false}$$

If you reverse the inequality, the statement is true.

$$-4 < 1 \quad \text{true}$$

Try other negative numbers as multipliers.

This suggests that when you multiply each side of an inequality by a negative number, the order of the inequality must be reversed.

These and other examples suggest the following properties.

Multiplication and Division Properties for Inequalities	**For any real numbers a, b, and c:** 1. if c is positive and $a < b$, then $ac < bc$ and $\frac{a}{c} < \frac{b}{c}$; 2. if c is positive and $a > b$, then $ac > bc$ and $\frac{a}{c} > \frac{b}{c}$; 3. if c is negative and $a < b$, then $ac > bc$ and $\frac{a}{c} > \frac{b}{c}$; 4. if c is negative and $a > b$, then $ac < bc$ and $\frac{a}{c} < \frac{b}{c}$.

Examples 2 and 3 show how to use these properties when solving inequalities.

Example **Solve $-0.4p > 10$. Graph the solution set.**

$$-0.4p > 10$$

$$\frac{-0.4p}{-0.4} < \frac{10}{-0.4} \qquad \text{\textit{Reverse the inequality sign because each}}$$
$$p < -25 \qquad \text{\textit{side is divided by a negative number.}}$$

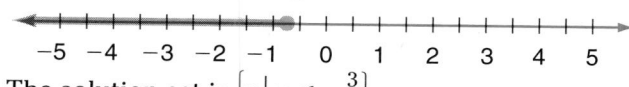

Any real number less than -25 is a solution.

The solution in Example 2 can be written using set-builder notation. This solution set can be written as $\{p \mid p < -25\}$. This is read as *the set of all numbers p such that p is less than -25.*

Example ③ **Solve $-y \geq \dfrac{y+6}{7}$.**

$$-y \geq \frac{y+6}{7}$$

$$-7y \geq y + 6 \qquad \text{\textit{Multiply each side by 7.}}$$

$$-8y \geq 6 \qquad \text{\textit{Add $-y$ to each side.}}$$

$$y \leq -\frac{3}{4} \qquad \text{\textit{Divide each side by -8, reversing the inequality sign.}}$$

A dot means that this point is included.

The solution set is $\left\{y \mid y \leq -\dfrac{3}{4}\right\}$.

You can use a graphing calculator to find the solution of an inequality graphically.

You can use the inequality symbols in the TEST menu on the TI-83 graphing calculator to find the solution to an inequality in one variable. Use the standard viewing window.

Your Turn

a. Clear the Y= list. Enter $8x + 5 < 6x - 3$ as Y1. Press GRAPH . Describe what you see. (The symbol $<$ is item 5 on the TEST menu.)

b. Use the TRACE function to scan the values along the graph. What values of x are on the graph? What do you notice about the values of Y on the graph?

c. Solve the inequality algebraically. How does your solution compare to the pattern you noticed in part b?

Inequalities can be used to solve many verbal problems. You can solve problems with inequalities in the same way you solve problems with equations.

Example **4** **Ron's scores on the first three of four 100-point chemistry tests were 90, 96, and 86. What score must he receive on the fourth test to have an average of at least 92 for all the tests?**

Example

APPLICATION

School

Explore   Let *s* represent the score needed on the fourth test. *The phrase at least 92 means greater than or equal to 92.*

Plan   The average of Ron's test scores is their sum divided by 4. This number must be greater than or equal to 92. Write an inequality. Let *s* represent the score on the fourth test.

Ron's average	*is greater than or equal to*	*92.*
$\dfrac{90 + 96 + 86 + s}{4}$	$\geq$	92

Solve
$$\frac{90 + 96 + 86 + s}{4} \geq 92$$
$$90 + 96 + 86 + s \geq 368 \quad \text{\textit{Multiply each side by 4.}}$$
$$s \geq 96$$

Examine   Ron must score at least 96 on the fourth test to average at least 92 for all the tests.

CHECK FOR UNDERSTANDING

Communicating Mathematics

Study the lesson. Then complete the following.

1. **Draw** a graph that shows the solution set $\{x \,|\, x > -2\}$.

2. **Write** *half of five times a number is less than or equal to 10* as an inequality.

3. **Solve** the following problems.

   a. Choose the correct symbol ($<$, $>$, or $=$) to make each statement true.

   $6 \underline{\ ?\ } 3 \qquad 6 \underline{\ ?\ } -3 \qquad -6 \underline{\ ?\ } 3 \qquad -6 \underline{\ ?\ } -3$

   b. Divide each side of each inequality above by 2. Record each result. Are the inequalities still true?

   c. Divide each side of the original inequalities by -2. Are the inequalities still true? Explain your results.

Solve each inequality. Graph the solution set.

4. $x > 4.5$

5. $7 \le 4a$

6. $7 - b \ge 5$

7. $3x + 4 \ge 19$

8. $2c + 15 \ge 3$

9. $\frac{d}{10} - 2 \le 0$

10. $-0.5y < 6$

11. $\frac{7x + 1}{8} > \frac{7x}{8} + 1$

Define a variable and write an inequality for each problem. Then solve.

12. Four times a number is less than 32.

13. A number plus fifteen is greater than or equal to 27.

EXERCISES

Solve each inequality. Graph the solution set.

14. $6x < 30$

15. $-5r > 25$

16. $11 - 5y < -77$

17. $0.06 + x < 2$

18. $15 - 5t \ge 55$

19. $6x + 4 \ge 34$

20. $3(4x + 7) < 21$

21. $8x + 5 \ge 10$

22. $40 \le -6(5r - 7)$

23. $7x - 5 > 3x + 4$

24. $9(2x + 3) > 10$

25. $5(3z - 3) \le 60$

26. $7 - 2m \ge 0$

27. $2(m - 5) - 3(2m - 5) < 5m + 1$

28. $0.01x - 4.23 \ge 0$

29. $3b - 2(b - 5) < 2(b + 4)$

30. $0.75x - 0.5 < 0$

31. $2.55x - 4.24 \le 0$

32. $\frac{2x + 3}{5} \le 0.03$

33. $\frac{3x - 3}{5} < \frac{6(x - 1)}{10}$

34. $\frac{4x + 2}{5} \ge -0.04$

35. $-x \ge \frac{x + 4}{7}$

36. $\frac{x + 8}{4} - 1 > \frac{x}{3}$

37. $20\left(\frac{1}{5} - \frac{w}{4}\right) \ge -2w$

Define a variable and write an inequality for each problem. Then solve.

38. The product of 11 and a number is less than 53.

39. Three fourths of a number decreased by 25 is at least 8.

40. The opposite of five times a number is less than 321.

41. Fifty-seven is greater than one-half a number.

42. Ninety decreased by 5 is greater than or equal to the product of a number and 10.

43. Sixty-two is less than the opposite of 6 times a number.

Use a graphing calculator to solve each inequality.

44. $-49 > 7(2x + 3)$

45. $8r + 3(2 + 7.5) < 25$

46. $2(3k + 4) - 3 \le 2(-1)$

47. $-5s + 4(3 - 5) < 7$

Critical Thinking

48. Find the set of all numbers that satisfy $3x - 2 \ge 0$ and $5x - 1 \le 0$.

Applications and Problem Solving

49. Health The National Heart Association recommends that less than 30% of a person's total daily caloric intake come from fat. One gram of fat yields nine Calories. Jason is a healthy 21-year old male whose average daily caloric intake is between 2500 and 3300 Calories.

  a. Write an inequality that represents the suggested fat intake for Jason.

  b. What is the greatest suggested fat intake for Jason?

50. Consumerism Tala is buying holiday gifts for her family early this year to take advantage of sales. Best Buy has a select group of CDs on sale for $4.99 each. Tala finds 2 jazz CDs for her father, who loves jazz. If she has $75 to spend on her family, write an inequality that tells how much money Tala has to spend on her other family members.

Mixed Review

Solve each equation. (Lesson 1–5)

51. $|x - 4| = 11$

52. $|x - 8| = 3x - 4$

53. Evaluate $|2x - 4| + 1.2$ if $x = -3$. (Lesson 1–5)

54. Geometry A piece of wire was cut into two pieces. One was bent into a square and the other was bent into an equilateral triangle. The side of the equilateral triangle has the same whole-number length (in cm) as the side of the square. If the length of the piece of wire is less than 50 cm, find all possible measurements for the sides of the figures. (Lesson 1–4)

55. Use a calculator to solve $68x + 373 = 802$. (Lesson 1–4)

56. Statistics Find the median, mode, and mean for the following set of data. 2, 56, 8, 43, 44 (Lesson 1–3)

57. Name the property illustrated by $x(7 - 5) = x \cdot 7 - x \cdot 5$. (Lesson 1–2)

58. Simplify $5(3m - 7n) + 3(4m + n)$. (Lesson 1–2)

59. Electricity Find the amount of current I (in amperes) produced if the electromotive force E is 1.5 volts, the circuit resistance R is 2.35 ohms, and the resistance r within a battery is 0.15 ohms, using the formula $I = \dfrac{E}{R + r}$.

(Lesson 1–1)

Solving Absolute Value Inequalities

▲ APPLICATION
Postal Service

In January 1995, the postage for a first-class stamp rose from $0.29 to $0.32. The table below shows the past and present postage rates.

Postal Rates		
Item	Before January 1995	After January 1995
Postcard	$0.19	$0.20
Birthday card (1 oz)	$0.29	$0.32
Heavy letter (2 oz)	$0.52	$0.55
Bank statement (3 oz)	$0.75	$0.78
Insured mail ($50)	$0.75	$0.75
Parcel post (2 lb)	$1.74	$2.10
Priority mail (1 lb)	$2.90	$3.00
Registered mail ($500)	$4.85	$5.40
Express mail (8 oz)	$9.95	$10.75

Source: U.S. Postal Service

Most popular types of greeting cards mailed in the U.S.

1. Christmas
2. Valentine's Day
3. Easter
4. Mother's Day
5. Father's Day

If you were to mail an oversized birthday card, you would expect to pay at least $0.32 but no more than $0.55. Let c stand for the cost of mailing the card. The two inequalities, $c \geq 0.32$ and $c \leq 0.55$, describe the cost of mailing the card. A sentence like this is called a **compound inequality.** A compound inequality containing *and* is true only if both parts of it are true.

Another way of writing $c \geq 0.32$ and $c \leq 0.55$ is $0.32 \leq c \leq 0.55$. This inequality is read *c is greater than or equal to 0.32 and is less than or equal to 0.55.*

To solve a compound inequality, you must solve each part of the inequality. Thus, the graph of a compound inequality containing *and* is the **intersection** of the graphs of the two inequalities. The intersection can be found by graphing the two inequalities and then determining where these graphs overlap or intersect.

Example ❶ **Solve $9 < 3x + 6 < 15$. Then graph the solution set.**

Method 1

Write the compound inequality using the word *and*. Then solve each part.

$9 < 3x + 6$ and $3x + 6 < 15$

$3 < 3x$ $\qquad\qquad$ $3x < 9$

$1 < x$ $\qquad\qquad$ $x < 3$

Method 2

Solve both parts at the same time by adding -6 to each part of the inequality. Then divide each part by 3.

$$9 < \qquad 3x + 6 \qquad < 15$$
$$9 + (-6) < 3x + 6 + (-6) < 15 + (-6)$$
$$3 \div 3 < \qquad 3x \div 3 \qquad < 9 \div 3$$
$$1 < \qquad\quad x \qquad\quad < 3$$

Graph each inequality and find the intersection.

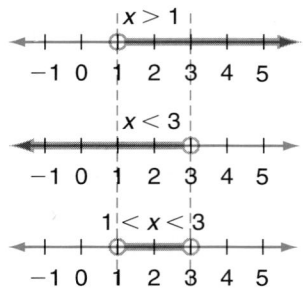

The solution set is $\{x \mid 1 < x < 3\}$.

Another type of compound inequality contains the word *or* instead of *and*. A compound inequality containing *or* is true if one or more of the inequalities is true. The graph of a compound inequality containing *or* is the **union** of the graphs of the two inequalities.

Example ❷ **Solve $x - 3 > 1$ or $x + 2 < 1$. Then graph the solution set.**

Solve each part separately.

$x - 3 > 1$ $\qquad$ or $\qquad$ $x + 2 < 1$

$\quad x > 4$ $\qquad\qquad\qquad\quad x < -1$

The last graph shows the solution set, $\{x \mid x > 4 \text{ or } x < -1\}$.

There is no short way to write an inequality containing "or."

Recall that the absolute value of a number is its distance from 0 on the number line. You can use this idea to solve inequalities involving absolute value.

Example ③ **Solve** $|y| < 7$.

$|y| < 7$ means that the distance between y and 0 is less than 7 units. To make $|y| < 7$ true, you must substitute values for y that are less than 7 units from 0. *Note that $|y| < 7$ is the same as $y < 7$ and $y > -7$.*

$$\begin{array}{c}\xleftarrow{\hspace{0.3cm}+\;+\;\oplus\;+\;+\;+\;+\;+\;+\;+\;+\;+\;+\;+\;+\;\oplus\;+\;+\;}\\ -9\text{-}8\text{-}7\text{-}6\text{-}5\text{-}4\text{-}3\text{-}2\text{-}1\;\;0\;\;1\;\;2\;\;3\;\;4\;\;5\;\;6\;\;7\;\;8\;\;9\end{array}$$

All of the numbers between -7 and 7 are less than 7 units from 0. The solution set is $\{y \mid -7 < y < 7\}$.

Example ④ **Solve** $|2x + 4| \geq 12$. **Graph the solution set.**

This inequality says that $2x + 4$ is greater than or equal to 12 units from 0.

$$\begin{array}{ccc} 2x + 4 \geq 12 & \text{or} & 2x + 4 \leq -12 \\ 2x \geq 8 & & 2x \leq -16 \\ x \geq 4 & & x \leq -8 \end{array}$$

$$\begin{array}{c}\xleftarrow{\hspace{0.3cm}+\;+\;+\;\bullet\;+\;+\;+\;+\;+\;+\;+\;+\;+\;+\;+\;\bullet\;+\;+\;}\\ -11\text{-}10\text{-}9\text{-}8\text{-}7\text{-}6\text{-}5\text{-}4\text{-}3\text{-}2\text{-}1\;\;0\;\;1\;\;2\;\;3\;\;4\;\;5\;\;6\;\;7\end{array}$$

The solution set is $\{x \mid x \geq 4 \text{ or } x \leq -8\}$.

Example ⑤

APPLICATION
Entertainment

Steven Spielberg is the most successful filmmaker in history. The ten top-grossing movies he has directed are listed below.

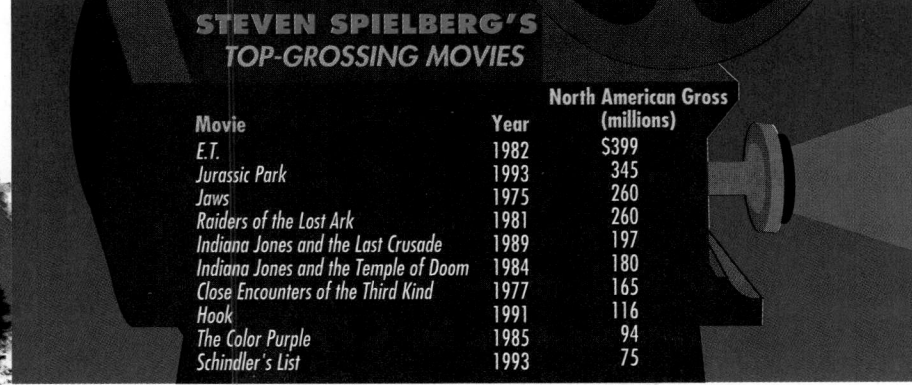

STEVEN SPIELBERG'S
TOP-GROSSING MOVIES

Movie	Year	North American Gross (millions)
E.T.	1982	$399
Jurassic Park	1993	345
Jaws	1975	260
Raiders of the Lost Ark	1981	260
Indiana Jones and the Last Crusade	1989	197
Indiana Jones and the Temple of Doom	1984	180
Close Encounters of the Third Kind	1977	165
Hook	1991	116
The Color Purple	1985	94
Schindler's List	1993	75

Source: *The Guinness Book of Records, 1995*

a. **Write a compound inequality that expresses the range of the grosses from Steven Spielberg's movies.**

b. **Suppose Mr. Spielberg wanted his next film to gross within $3 million of the average gross of his movies, $210.3 million. Write and solve an absolute value inequality for the difference between the gross of the next film and the average.**

(continued on the next page)

a. Let *r* represent the range of the films' grosses.

$$75 \le r \le 399$$

b. $|x - 210.3| \le 3$

$$x - 210.3 \le 3 \qquad \text{and} \qquad x - 210.3 \ge -3$$
$$x \le 3 + 210.3 \qquad\qquad x \ge -3 + 210.3$$
$$x \le 213.3 \qquad\qquad x \ge 207.3$$

The solution set is $\{x \mid x \ge 207.3 \text{ and } x \le 213.3\}$, which may be written as $\{x \mid 207.3 \le x \le 213.3\}$. Spielberg wanted his next movie to gross between \$207.3 and \$213.3 million.

Some absolute value inequalities have no solution. For example, $|4x - 3| < -6$ is never true. Since the absolute value of a number is never negative, there is no replacement for *x* that will make this sentence true. So, the solution set to this inequality is the empty set.

Other absolute value inequalities are always true. One such inequality is $|x + 5| > -10$. The solution set of this inequality is all real numbers. Can you see why? *Think of the definition of absolute value.*

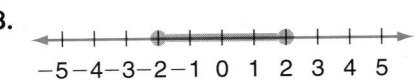

CHECK FOR UNDERSTANDING

Communicating Mathematics

Study the lesson. Then complete the following.

1. **Give** an example of an absolute value inequality whose solution set is the empty set.

2. **State** an absolute value inequality for all numbers less than 5 and greater than −5.

3. **Explain** why $|x + 2| \ge -4$ has all real numbers as its solution set.

4. Explain the difference between a compound inequality containing the word *or* and the word *and.*

Guided Practice

State an absolute value inequality for each of the following. Then graph each solution set.

5. all numbers less than 18 and greater than −18

6. all numbers between −3 and 3

State an absolute value inequality for each graph.

7.
$$-5\ -4\ -3\ -2\ -1\ \ 0\ \ 1\ \ 2\ \ 3\ \ 4\ \ 5$$

8.
$$-5\ -4\ -3\ -2\ -1\ \ 0\ \ 1\ \ 2\ \ 3\ \ 4\ \ 5$$

Solve each inequality. Graph the solution set.

9. $3x + 1 < 7$ or $7 < 2x - 9$

10. $|x| \ge 4$

11. $|x + 2| > 3$

12. $1 \le x - 2 \le 7$

13. $|3x + 12| > 42$

14. $|x| \ge x$

Practice

State an absolute value inequality for each of the following. Then graph each solution set.

15. all numbers less than 7 and greater than -7

16. all numbers less than or equal to 15, and greater than or equal to -15

17. all numbers greater than 11 or less than -11

18. all numbers less than or equal to 5, and greater than or equal to -5

19. all numbers between -8 and 8

20. all numbers greater than or equal to -10, and less than or equal to 10

State an absolute value inequality for each graph.

21.
```
<-+--+--+-○═══════○--+--+->
 -5-4-3-2-1 0 1 2 3 4 5
```

22.
```
<-+--+--+--+--●═══●--+--+--+->
 -5-4-3-2-1 0 1 2 3 4 5
```

23.
```
◄═══●--+--+--+--+--+--●═══►
 -5-4-3-2-1 0 1 2 3 4 5
```

24.
```
<-+--+--○--+--+--+--○--+->
 -5.6-4.2-2.8-1.4 0 1.4 2.8 4.2
```

25.
```
◄═══○--+--+--+--○═══►
 -6-5-4-3-2-1 0 1 2 3 4
```

26.
```
<-+--+--+--+--○═══○--+--+--+->
 -4-3-2-1 0 1 2 3 4 5 6
```

Solve each inequality. Graph the solution set.

27. $|8x| \leq 10$

28. $|2x| < 6$

29. $|x| > 5$

30. $|2x - 9| \leq 27$

31. $|3x| \geq 7$

32. $|5x| < -25$

33. $|2x| > 1$

34. $x - 4 < 1$ or $x + 2 > 1$

35. $|x - 6| \leq -12$

36. $-1 < 3x + 2 < 14$

37. $|3x + 11| > 1$

38. $|x| \leq x$

39. $x + 6 \geq -1$ or $x - 2 \leq 4$

40. $-4 \leq 4x + 24 \leq 4$

41. $|3x| + 3 \leq 0$

42. $|5x - 7| < 81$

43. $2x - 1 < -5$ or $3x + 2 \geq 5$

44. $|x + 2| - x \geq 0$

Critical Thinking

45. Solve $|x + 1| + |x - 1| \leq 2$.

Applications and Problem Solving

46. **Transportation** On some interstate highways, the maximum speed a car may drive is 65 miles per hour. A tractor-trailer may not drive more than 55 miles per hour. The minimum speed for all vehicles is 45 miles per hour.

  a. Write an inequality to represent the allowable speed for a car on an interstate highway.

  b. Write an inequality to represent the speed at which a tractor-trailer may travel on an interstate highway.

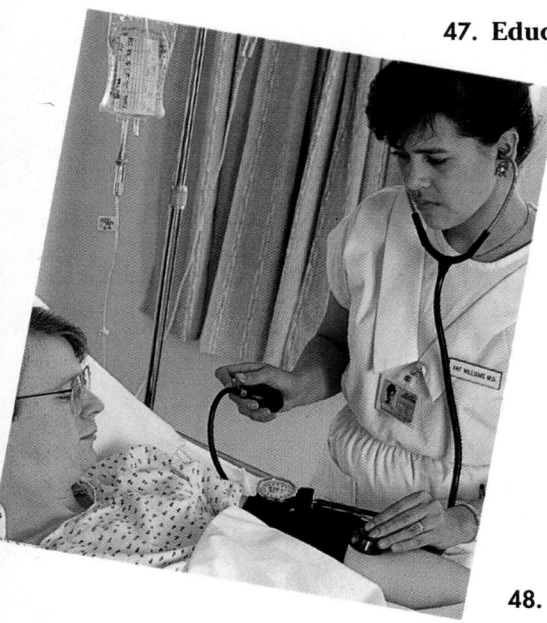

47. Education Use the chart below to answer the following questions.

Average Nursing Salaries	
Type	Salary
Nurse anesthetist	$76,000
Nurse midwife	57,000
Corporate nurse	47,000
Nurse practitioner	43,600
Nurse employed by federal government	43,200
Legal nurse consultant at law firm	40,000

Source: Bureau of Labor Statistics

  a. Write an absolute value inequality that describes the range in overall salaries.

  b. Write two questions that can be answered using the mean and mode of the data.

48. Consumerism Marcus is buying his first tank of gasoline since he got his driver's license. Where he lives, gasoline is selling for between $1.20 and $1.40 per gallon. If he has $10.50 to spend on gas, how many gallons can he buy?

Mixed Review

Solve each inequality. Then graph the solution set. (Lesson 1–6)

49. $9(x + 2) < 72$ **50.** $3(3x + 2) > 7x - 2$

51. Solve $8x + 5 < 7x - 3$. (Lesson 1–6)

52. Solve $-4(3m - 7) - (3 - m) < 13$. (Lesson 1–6)

53. Solve $|3x - 4| = -1$. (Lesson 1–5)

54. Find the value of $|7(-3) + 10|$. (Lesson 1–5)

55. Consumerism Beto has gone to a doughnut shop with $10 his father gave him. He needs to buy 6 glazed doughnuts at $0.50 each for his father. He can then buy some frosted cake doughnuts for himself at $0.35 each. (Lesson 1–4)

  a. Let x represent the number of frosted cake doughnut Beto buys. Translate "Beto bought 6 glazed doughnuts and some frosted cake doughnuts for $10.00" into an equation.

  b. Solve the equation to find out how many frosted cake doughnuts Beto can buy.

56. Solve $y = 8(0.3) + 1.2$. (Lesson 1–4)

57. Olympics The time in seconds of 15 Olympic Games Champions' scores for the 200-meter run are listed below. Find the median, mode, and mean for these scores. 22.2, 21.6, 22.6, 21.7, 22, 21.6, 21.8, 20.7, 20.7, 20.5, 20.3, 20.01, 19.80, 20.19, 19.75 (Lesson 1–3)

VOCABULARY

After completing this chapter, you should be able to define each term, property, or phrase and give an example or two of each.

Algebra

absolute value   (p. 37)
addition property of equality   (p. 28)
addition property of inequality   (p. 43)
algebraic expressions (p. 8)
associative properties (p. 14)
commutative properties (p. 14)
compound inequality (p. 49)
distributive properties (p. 14)
division property of equality   (p. 28)
division property of inequality   (p. 44)
empty set   (p. 40)
equation   (p. 27)
formula   (p. 9)
identity properties   (p. 14)
intersection   (p. 49)

inverse properties   (p. 14)
irrational number   (p. 13)
multiplication property of equality   (p. 28)
multiplication property of inequality   (p. 44)
open sentences   (p. 27)
order of operations   (p. 7)
perfect number   (p. 12)
rational number   (p. 13)
real numbers   (p. 13)
reflexive property of equality (p. 28)
solution   (p. 28)
substitution property of equality (p. 28)
subtraction property of equality (p. 28)
subtraction property of inequality (p. 43)
symmetric property of equality (p. 28)
transitive property of equality (p. 28)
trichotomy property   (p. 43)

union   (p. 50)
variables   (p. 27)

Geometry
hypotenuse   (p. 9)
legs   (p. 9)
Pythagorean theorem   (p. 9)

Problem Solving
list possibilities   (p. 38)
problem-solving plan   (p. 30)
problem-solving strategies (p. 38)

Statistics
back-to-back stem-and-leaf plot (p. 20)
extreme values   (p. 22)
line plot   (p. 19)
mean   (p. 21)
measure of central tendency (p. 21)
median   (p. 21)
mode   (p. 21)
stem-and-leaf plot   (p. 20)

UNDERSTANDING AND USING THE VOCABULARY

Choose the letter that best matches each description.

1. $-b = -b$
2. $(4x) \cdot 1 = 4x$
3. $8(4x - 1) = 32x - 8$
4. $x = 4$, then $4 = x$
5. $4x + 6(y - 5x) - 3z$
6. $10 > t > 4.5$
7. $x = 4y, 4y = 12$, then $x = 12$
8. $5a - 7(a - 6) = 12$
9. $9 + (4 + 7) = (9 + 4) + 7$
10. $|-3m|$
11. $-5 + 5 = 0$
12. $xy = yx$

a. absolute value
b. algebraic expression
c. associative property
d. commutative property
e. compound inequality
f. distributive property
g. equation
h. identity property
i. inverse property
j. reflexive property
k. symmetric property
l. transitive property

SKILLS AND CONCEPTS

OBJECTIVES AND EXAMPLES	REVIEW EXERCISES

Upon completing this chapter, you should be able to:

Use these exercises to review and prepare for the chapter test.

- use the order of operations to evaluate expressions (Lesson 1–1)

$$5 - 8(3 - 6) \div 2^2 + 10 = 5 - 8(-3) \div 4 + 10$$
$$= 5 - (-24) \div 4 + 10$$
$$= 5 - (-6) + 10$$
$$= 5 + 6 + 10$$
$$= 21$$

Find the value of each expression.

13. $6(5 - 8) \div 9 + 4$

14. $(3 + 7)^2 - 16 \div 2$

15. $(6 + 5)4 - 3$

16. $-7 + [28 \div (18 - 7(2))]$

Evaluate each expression if $a = 4$, $b = 5$, $c = -0.5$, and $d = -3$.

17. $\frac{8c + ab}{a}$

18. $(a - d + b) \div c$

- determine the sets of numbers to which a given number belongs (Lesson 1–2)

Find the value of $3(-4.5)$. Then name the sets of numbers to which this value belongs.

$$3(-4.5) = -13.5$$

rationals, reals

Find the value of each expression. Then name the sets of numbers to which each value belongs.

19. $4 - 12$

20. $42 \div 8$

21. $\sqrt{2 + 3}$

22. $2^3 + 10$

23. $2\pi(8.75)$

24. $-20 \div 2^2$

- use the properties of real numbers to simplify expressions (Lesson 1–2)

$$2a(5.4 - 4b) - 4a(3.1 + 8b)$$
$$= 10.8a - 8ab - 12.4a - 32ab$$
$$= -1.6a - 40ab$$

Name the property illustrated by each equation.

25. $7 \cdot \frac{1}{7} = 1$

26. $4(3 \cdot 5.5) = (4 \cdot 3) 5.5$

Simplify each expression.

27. $7a + 2b - 5a - 6b$

28. $3(a + 4b) - 2(4a + 2b)$

- find and use the median, mode, and mean to interpret data (Lesson 1–3)

Find the median, mode, and mean of 78, 67, 73, 69, 84, 68, 74, 76, 66, 78, 70.

66, 67, 68, 69, 70, 73, 74, 76, 78, 78, 84

median: 6th value = 73

mode: 78

mean: $\frac{66 + 67 + 68 + \ldots + 78 + 84}{11} = 73$

Find the median, mode, and mean for each set of data.

29. 5, 92, 64, 18, 25

30. 66, 48, 35, 52, 48, 41, 59, 61

31. 4.6, 6.1, 8.9, 3.6, 6.1, 10, 2.9

32. 3, 6, 7, 10, 7, 14, 19, 21, 10, 10, 31, 17, 16, 9, 7, 17, 20, 14, 7, 10, 13, 10

OBJECTIVES AND EXAMPLES	REVIEW EXERCISES

OBJECTIVES AND EXAMPLES

- solve equations by using the properties of equality   (Lesson 1–4)

  Solve $2(a - 1) = 8a - 6$.

  $2(a - 1) = 8a - 6$

  $2a - 2 = 8a - 6$ *Distributive property*

  $-2 = 6a - 6$ *Subtract 2a from each side.*

  $4 = 6a$ *Add 6 to each side.*

  $\frac{2}{3} = a$ *Divide each side by 6.*

- solve equations for a specific variable (Lesson 1–4)

  Solve $3(x - 2) = y$ for x.

  $3(x - 2) = y$

  $3x - 6 = y$ *Distributive property*

  $3x = y + 6$ *Add 6 to each side.*

  $x = \frac{y + 6}{3}$ *Divide each side by 3.*

- solve equations containing absolute value (Lesson 1–5)

  Solve $|r + 14| = 23$.

  $|r + 14| = 23$

  $r + 14 = 23$ or $r + 14 = -23$

  $r = 9$ $r = -37$

- solve inequalities and graph the solution sets. (Lesson 1–6)

  Solve $3 - 4x \le 6x - 2$.

  $3 - 4x \le 6x - 2$

  $3 \le 10x - 2$ *Add 4x to each side.*

  $5 \le 10x$ *Add 2 to each side.*

  $\frac{1}{2} \le x$ *Divide each side by 10.*

REVIEW EXERCISES

Solve each equation.

33. $12z + 36 = 8z - 48$
34. $4.2x + 6.4 = 40$
35. $14y - 3 = 25$
36. $7w + 2 = 3w + 94$
37. $4 - 2(1 - w) = -38$
38. $4y - \frac{1}{10} = 3y + \frac{4}{5}$
39. $48 + 5y = 96 - 3y$
40. $\frac{x}{3} + \frac{x}{2} = \frac{3}{4}$

Solve each equation or formula for the variable specified.

41. $A = p + prt$ for t
42. $df - 3g = 4h$ for f $\dfrac{4h + 3G}{D}$
43. $\frac{3a^2 - 1}{2b} = c$ for b
44. $s = \frac{1}{2}gt^2$ for g

Solve each equation.

45. $|y - 5| - 2 = 10$
46. $|5y - 8| = 12$
47. $|2x - 36| = 14$
48. $|x + 4| + 3 = 17$
49. $|q - 3| + 7 = 2$
50. $4|3x + 4| = 4x + 8$
51. $2|w + 6| = 10$
52. $5|6 - 5x| = 15x - 35$

$\dfrac{4|3x + 4|}{4} = \dfrac{4x + 8}{4}$

$3x + 4 = x + 2$

$3x + 4 = x + 2$

$3x + 4 = x + 2$

Solve each inequality. Graph the solution set.

53. $5z - 6 > 14$
54. $5(x - 2) < 75$
55. $57 - 4t \ge 13$
56. $-3(2x + 5) > 13x - 4$
57. $18 - 2(y + 6) < 76$
58. $3(6 - 5x) \le 12x - 36$
59. $2 - 3z \ge 7(8 - 2z) + 12$
60. $8(2x - 1) > 11x - 17$

OBJECTIVES AND EXAMPLES

• solve compound inequalities using *and* and *or*
(Lesson 1–7)

$$-2 \leq x - 4 < 3$$

$$-2 \leq x - 4 \text{ and } x - 4 < 3$$

$$2 \leq x \qquad\qquad x < 7$$

$$2 \leq x < 7$$

• solve inequalities involving absolute value and graph the solutions   (Lesson 1–7)

$$|3x + 7| \geq 26$$

$$3x + 7 \geq 26 \text{ or } 3x + 7 \leq -26$$

$$3x \geq 19 \qquad\qquad 3x \leq -33$$

$$x \geq \frac{19}{3} \qquad\qquad x \leq -11$$

$$x \geq \frac{19}{3} \text{ or } x \leq -11$$

REVIEW EXERCISES

Solve each inequality. Graph the solution set.

61. $11 < 3x + 2 < 20$

62. $4x - 10 < -10 \text{ or } 6x + 4 \geq 10$

63. $-1 < 3(y - 2) \leq 9$

64. $5y - 4 > 16 \text{ or } 3y + 2 < 1$

Solve each inequality. Graph the solution set.

65. $|2x + 6| \leq 4$

66. $7 + |9 - 5x| > 1$

67. $|4x| + 3 \leq 0$

68. $|x| + 1 < 12$

69. $|3x| < 27$

70. $|2x + 3| - 6 \geq 7$

APPLICATIONS AND PROBLEM SOLVING

71. Geometry   The perimeter of a rectangle is 150 centimeters. The length is 15 centimeters greater than the width. Find the dimensions of the rectangle.   (Lesson 1–1)

72. Car Expenses   Rafael spent $2011 to operate his car last year. He drove 7400 miles. He also paid $972 for insurance and $114 for the registration fee. Rafael's only other expense was for gasoline. How much did the gasoline cost per mile?   (Lesson 1–4)

73. Test Scores   Your quiz scores are 73, 75, 89, and 91. What is the lowest score you can obtain on the last quiz and still achieve an average of at least 85?   (Lesson 1–6)

74. Bowling   Bowling at Sunset Lanes cost Danny and Zorina $9. This included shoe rental of $0.75 a pair. How much did each game cost if Danny bowled 3 games and Zorina bowled 2 games?   (Lesson 1–4)

75. Oceans   The depths, in meters, of certain points of the oceans and seas of the world are: 10918, 9219, 7455, 5625, 4632, 5016, 4773, 3787, 3658, 2782, 3742, 3777, 660, 22, 421, 6946, and 183. Find the median, mode, and mean for this set.   (Lesson 1–3)

A practice test for Chapter 1 is provided on page 912.

ALTERNATIVE ASSESSMENT

COOPERATIVE LEARNING PROJECT

Chicken Farming

In this project, you will complete a chicken and egg math problem. Janice flew to Iowa to visit her grandparents on their chicken farm for a week in the summer. She was raised in the city so it was quite a treat for her to be on a farm. One night as she and her grandfather were sitting at the table enjoying lemonade, her grandfather pulled out a piece of paper that he had been working on to give her.

The following is what was on the paper. Answer all the questions.

> If a chicken and a half lays an egg and a half in a day and a half, then
>
> **a.** how many eggs will nine chickens lay in ten days?
>
> **b.** how many eggs will c chickens lay in twelve days?
>
> **c.** how many eggs will twelve chickens lay in d days?
>
> **d.** how many days will it take c chickens to lay twelve eggs?
>
> **e.** how many days will it take twelve chickens to lay e eggs?
>
> **f.** how many chickens will it take to lay e eggs in twelve days?
>
> **g.** how many chickens will it take to lay twelve eggs in d days?

Suppose you want all of the answers in parts d–g to be whole numbers. Then what must be true about c, d, and e?

Follow these steps to accomplish your task.

- Construct a pattern for this situation.
- Read and reread each of the verbal sentences in order to understand them.
- Determine what each of the variables stands for and how they should be used.
- Write a paragraph describing the problem and your solution.

THINKING CRITICALLY

- Translate each of the following into at least two different word phrases that mean the same thing.

$$x + 8, \; n - 3, \text{ and } 2w$$

- When, if ever, is $|a + b| = |a| + |b|$ true?

PORTFOLIO

Translating word expressions into algebraic expressions requires reading the words and determining their meaning mathematically. Another way to enhance this skill is to do the opposite—translate mathematical symbols and numbers into words.

Write a mathematical expression or equation and create a story about it. Be creative and make it fun to read. Then exchange your story with someone else's story and translate the story into a mathematical expression or equation. Place your story in your portfolio.

SELF EVALUATION

There are five steps to solid reasoning.

1. *Clarify*—Determine precisely what must be decided.

2. *Evaluate*—Determine the facts and assumptions.

3. *Decide*—Determine what information should be used and what is not relevant.

4. *Implement*—Develop a plan of how to implement the decision.

5. *Monitor and Modify*—Watch the conclusions of the decision and be prepared to revise the plan or take a different course of action based on new information.

Assess yourself. Do you use these five steps in your reasoning? Determine a problem or situation that you have had recently, or think you will have in the future, and use these steps to reason your way through it.

In·ves·ti·ga·tion

Through the Looking Glass

MATERIALS NEEDED

four cardboard tubes of different sizes

tape measure

grid paper

pens or pencils in two colors

You are a naturalist. You have heard that other naturalists have been using view tubes called *calibration scopes* to estimate the sizes of animals in the wild. You have been told that by using the scopes, along with mathematics, you can determine the actual heights of objects.

In this Investigation, you will experiment with different sizes of calibration scopes. You will answer the following questions.

- How does the size of a particular scope affect the animal's image?
- Is the size of the image you see affected by how far away you are from an object?
- Is there a mathematical relationship between the size of the scope and the distance from the object?

Work in groups of three. You will need four different sizes of cardboard tubes.

A: a tube of any length or diameter

B: a tube longer than Tube A, but of the same diameter

C: a tube the same length as Tube B but wider in diameter

D: a tube that has the same diameter as Tube C and the same length as Tube A

Label each tube as Scope A, B, C, or D.

SCOPE:	
DIAMETER:	
LENGTH:	
Distance	**Height**

EXPERIMENT—FIRST SCOPE

1 Begin by copying the chart above onto a sheet of paper.

2 Choose one of the four scopes. Record the label of the tube, its diameter, and its length in your chart.

3 Measure a set distance from the wall. Mark this distance with masking tape and record it in your chart. Move to another distance closer to the wall, mark it with masking tape, and record the distance in your chart. Do the same for two more distances farther than the first distance from the wall.

4 Have one of the group members hold a measuring tape vertically against the wall. Have another group member stand directly behind the first mark and view the tape measure through the scope. Record the height of the image he or she sees through the tube. Repeat this process at each of the other three marks.

EXPERIMENT—SECOND SCOPE

5 Select another scope. Make another chart like the one you made for the first scope. Repeat the experiment with the new scope using the same distances you marked in the first experiment.

6 Explain any relationship that you notice in the data. Discuss how naturalists might use these scopes to find the actual heights of animals.

You will continue working on this Investigation throughout Chapters 2 and 3.

Be sure to keep your chart and materials in your Investigation Folder.

Through the Looking Glass Investigation

Working on the Investigation
Lesson 2–5, p.100

Working on the Investigation
Lesson 2–6, p.108

Working on the Investigation
Lesson 3–2, p.140

Working on the Investigation
Lesson 3–5, p.159

Closing the Investigation
End of Chapter 3, p.174

Graphing Linear Relations and Functions

Objectives

In this chapter, you will:

- identify different types of relations and functions,
- graph relations and functions on the coordinate plane,
- look for patterns to solve problems,
- model real-world data using scatter plots, and
- graph inequalities on the coordinate plane.

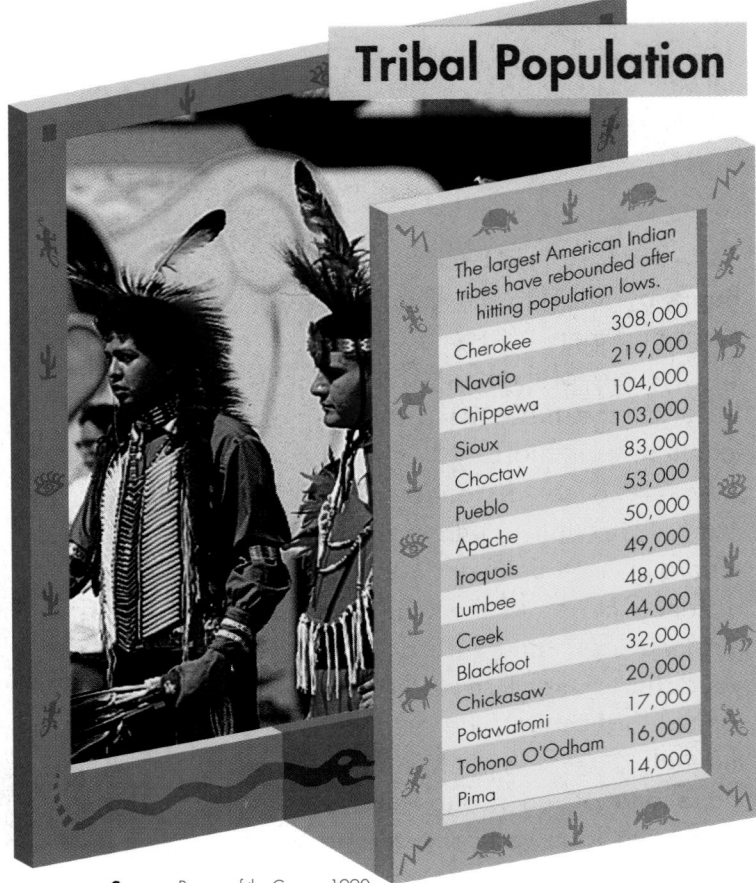

Tribal Population

The largest American Indian tribes have rebounded after hitting population lows.

Cherokee	308,000
Navajo	219,000
Chippewa	104,000
Sioux	103,000
Choctaw	83,000
Pueblo	53,000
Apache	50,000
Iroquois	49,000
Lumbee	48,000
Creek	44,000
Blackfoot	32,000
Chickasaw	20,000
Potawatomi	17,000
Tohono O'Odham	16,000
Pima	14,000

Source: Bureau of the Census, 1990

TIME *Line*

The 1990 U.S. census report indicates that the number of people who call themselves American Indians has tripled since 1960. Are you aware of the rich, Native American diversity that weaves its thread through the history of our continent?

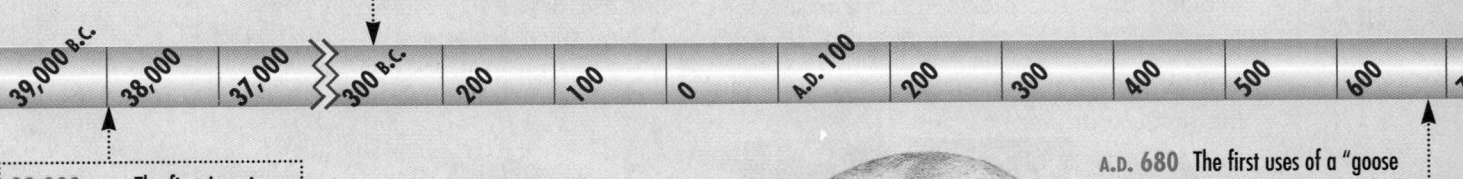

260 The Olmecs of Mesoamerica develop their system of numeration, a place-value system based on 20. It is in use during the next 1000 years.

| 39,000 B.C. | 38,000 | 37,000 | 300 B.C. | 200 | 100 | 0 | A.D. 100 | 200 | 300 | 400 | 500 | 600 | 70 |

38,000 B.C. The first Americans cross the temporarily dry Bering Straits from Asia and begin to colonize the new continent.

A.D. 680 The first uses of a "goose egg" sign for zero appear in Cambodia and Sumatra.

Cheryl Beno, of Gallup, New Mexico, keeps her Navajo heritage alive by weaving traditional Navajo rugs—blankets of heavy yarn. When she was 8, Cheryl's mother taught her to weave. "My mom makes a big rug, and I do a miniature version," she says. Because she enjoys weaving so much, when school begins, her mother disassembles Cheryl's loom to prevent her from weaving instead of completing her homework.

One of Cheryl's rugs, a type called a chief's blanket, won first prize in the youth division of the arts and crafts contest at the 1991 Intertribal Ceremonial in Gallup. It also took second prize in the young artists' division of the 1992 Navajo Nation Window Rock Fair.

Do research to learn about traditional Native American weaving patterns. Draw several designs and describe the patterns in each, paying particular attention to the use of parallel and perpendicular lines in the design. Then try to design a pattern of your own on paper. Be creative in your use of color. Then actually create the design using fabric, construction paper, or any other materials that will best show your design.

1600 Shakespeare's *Hamlet* is first performed.

1976 Patricia Roberts Harris becomes the first African-American female cabinet member, under President Carter.

1643 Evangelista Torricelli, an Italian mathematician, physicist, and a pupil of Galileo, was the first to measure air pressure.

1700 The clarinet was invented in Nuremburg, Germany.

1993 *Rollerblade* was added to the *World Book Dictionary* because it had become an important part of our everyday language.

Relations and Functions

What YOU'LL LEARN

- To graph a relation, state its domain and range, and determine if it is a function, and
- to find values of functions for given elements of the domain.

Why IT'S IMPORTANT

You can use relations to solve problems involving geography, forestry, and sports.

CONNECTION
Geography

The growth in the population of the state of Arizona over the last several decades can be shown by using a table.

Year	1930	1940	1950	1960	1970	1980	1990
Population (millions)	0.4	0.5	0.7	1.3	1.8	2.7	3.7

Source: U.S. Census

Another way to represent the population growth is to use **ordered pairs**. The ordered pairs for the data above are: (1930, 0.4), (1940, 0.5), (1950, 0.7), (1960, 1.3), (1970, 1.8), (1980, 2.7), and (1990, 3.7). The first number in the ordered pair is the year, and the second number is the population in millions.

You can *graph* these ordered pairs by creating a **coordinate system** with two axes. The horizontal axis represents the year, and the vertical axis represents the population. Each point represents an ordered pair shown in the table above. Remember that each point in the coordinate plane can be named by exactly one ordered pair and that every ordered pair names exactly one point in the coordinate plane.

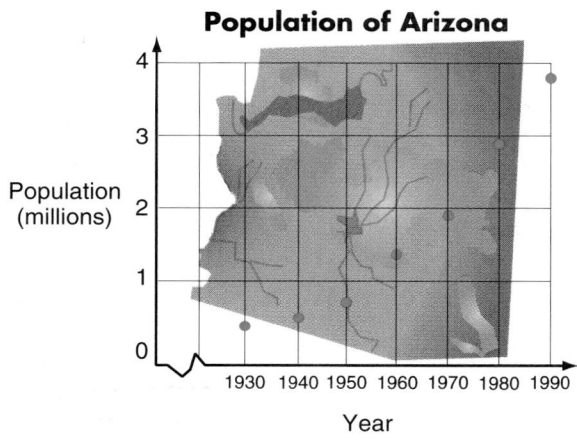

Population of Arizona

In the graph above, only one part of the **Cartesian coordinate plane** was shown—the one with all positive numbers. The Cartesian coordinate plane is composed of the *x*-**axis** (horizontal) and the *y*-**axis** (vertical), which meet at the **origin** (0, 0) and divide the plane into four **quadrants.** *The points on the two axes do not lie in any quadrant.*

The ordered pairs graphed on this plane can be represented by (*x*, *y*).

In this book, assume that each square on a graph represents 1 unit unless otherwise labeled.

A set of ordered pairs, such as the one for the population of Arizona, forms a **relation.** The **domain** of a relation is the set of all first coordinates (*x*-coordinates) from the ordered pairs, and the **range** is the set of all second coordinates (*y*-coordinates) from the ordered pairs.

A **mapping** shows how each member of the domain is paired with each member of the range.

$\{(4, 5), (-2, 3), (5, 6)\}$ $\{(1, 3), (4, -9) (6, 3)\}$ $\{(2, 3), (-4, 8), (2, 6), (7, -3)\}$

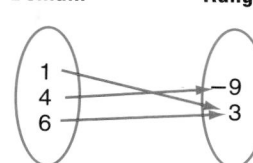   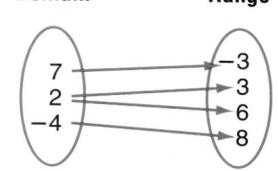

Domain	Range
-2	3
4	5
5	6

Domain	Range
1	-9
4	3
6	

Domain	Range
7	-3
2	3
-4	6
	8

A **function** is a special type of relation in which each element of the domain is paired with *exactly one* element from the range. The first two relations above are functions. The third relation is *not* a function because the 2 in the domain is paired with both 3 and 6 in the range.

Example **1** **State the domain and range of the relation shown in the graph. Is the relation a function?**

The relation is $\{(-4, 5), (-3, -4), (-2, 0), (1, 1), (2, -4)\}$.
The domain is $\{-4, -3, -2, 1, 2\}$.
The range is $\{-4, 0, 1, 5\}$

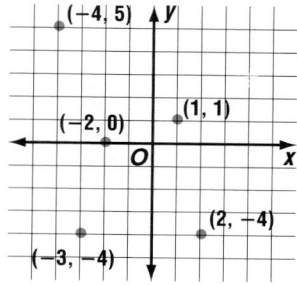

Each member of the domain is paired with exactly one member of the range, so this relation is a function.

Since the domain of the function in Example 1 is a set of individual points, it is called a **discrete function.** Notice that its graph consists of points that are not connected.

You can use the **vertical line test** to determine if a relation is a function. Using the graph in Example 1, place your pencil at the left of the graph to represent a vertical line. Slowly move the pencil to the right across the graph. At each point of the domain, the vertical line intersects the graph of the relation at only one point. Therefore, the relation is a function. If the vertical line intersects the graph at more than one point, the relation is *not* a function.

Is the population growth in Arizona a function? Why or why not?

2 The table below shows the number of fires and the number of acres burned over six years on the lands owned by the Bureau of Indian Affairs in the state of Washington. Graph this information and determine if it is a function.

APPLICATION
Forestry

Year	Number of Fires	Acreage
1987	198	10,288
1988	152	17,842
1989	250	3320
1990	227	1726
1991	250	6218
1992	310	24,499

Source: Washington State Department of Natural Resources, 1993

Forest Fires

Acres (thousands) vs Number of Fires

Using the vertical line test, you can see that 250 in the domain is mapped to two different range values, 3320 and 6218. Therefore, the relation is *not* a function.

Functions and their graphs can help you discover many relationships in mathematics.

MODELING MATHEMATICS

Volume

Materials: centimeter grid paper scissors

In this activity, you will make open boxes from identical square pieces of centimeter paper and investigate their volumes.

Your Turn

a. Cut identical squares from the corners of one square piece of paper and fold along the edges to form an open box.

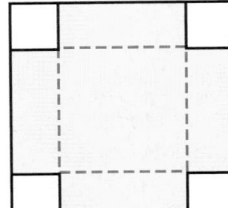

b. Find the volume of the open box.

c. Repeat steps a and b for other cutout squares with whole-number lengths.

d. Organize the data into ordered pairs (length of the side of cutout square, volume of open box). Is this relation a function?

e. Graph the ordered pairs.

f. Which ordered pair results in the greatest volume?

g. Why might some businesses be interested in data like these?

An equation is another way to represent a relation. The solutions of an equation in x and y are the set of ordered pairs (x, y) that make the equation true. Consider the equation $y = 5x - 4$. Since x can be any real number, the domain has an infinite number of elements. To determine whether an equation represents a function, it is often simpler to look at the graph of the relation.

One way to graph an equation is to make a table of solutions, graph enough ordered pairs to see a pattern, and then connect the points with a line or smooth curve. You can then use the vertical line test to determine whether the equation represents a function.

When the domain of the function has an infinite number of elements and can be graphed with a line or smooth curve, the function is a **continuous function**.

Example **3** **Determine whether each equation represents a function.**
a. $y = 3x - 4$

Prepare a table of values to find ordered pairs that satisfy the equation. Choose values for x and find the corresponding values for y. Then graph the ordered pairs.

x	y
−1	
0	
1	
2	

$\rightarrow$

x	y
−1	−7
0	−4
1	−1
2	2

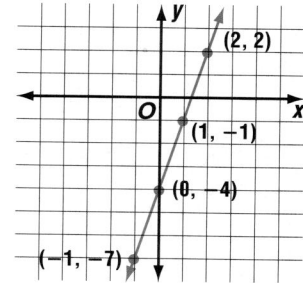

Since x can be any real number, there is an infinite number of ordered pairs that can be graphed. If all of them were graphed, they would form a line. For each x value, there is exactly one y value. Thus, this set of ordered pairs passes the vertical line test and is a function.

b. $x = y^2 + 1$

Complete the table. In this case, it is easier to choose y values and then find the corresponding values for x. Then sketch the graph, connecting the points with a curved line.

x	y
	−2
	−1
	0
	1
	2

$\rightarrow$

x	y
5	−2
2	−1
1	0
2	1
5	2

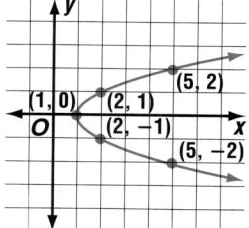

This graph is called a *parabola*. You can see from the table as well as the vertical line test that there are two y values for all but one of the x values. Therefore, this set of ordered pairs is *not* a function.

Letters other than f can be used to represent a function. For example, the equation $y = 4x + 3$ can also be written as $g(x) = 4x + 3$.

Equations that represent functions are often written in *functional notation*. The equation $y = 2x + 3$ can be written as $f(x) = 2x + 3$. The symbol $f(x)$ replaces the y and is read "f of x." The f is just the name of the function, not a variable. Suppose you want to find the value in the range that corresponds to the element 6 in the domain. This is written $f(6)$ and is read "f of 6." The value $f(6)$ is found by substituting 6 for each x in the equation. Therefore, $f(6) = 2(6) + 3$ or 15.

Example **4** Given the function $g(x) = x^2 - 8$, find each value.
  a. $g(-2)$ b. $g(7a)$

$$g(x) = x^2 - 8 \qquad\qquad\qquad g(x) = x^2 - 8$$
$$g(-2) = (-2)^2 - 8 \quad \textit{Substitute.} \qquad g(7a) = (7a)^2 - 8 \quad \textit{Substitute.}$$
$$= 4 - 8 \text{ or } -4 \qquad\qquad\qquad = 49a^2 - 8 \quad (ab)^2 = a^2b^2$$

Example **5** Use a calculator to find $f(4.6)$ if $f(x) = 0.5x^2 + 4x - 2.5$.

$$f(4.6) = 0.5(4.6)^2 + 4(4.6) - 2.5$$
Estimate: $0.5(5)^2 + 4(5) - 2.5 = 12.5 + 20 - 2.5 \text{ or } 30$

Enter: 0.5 $\boxed{\times}$ 4.6 $\boxed{x^2}$ $\boxed{+}$ 4 $\boxed{\times}$ 4.6 $\boxed{-}$ 2.5 $\boxed{=}$ *26.48*

Therefore, $f(4.6) = 26.48$.   *Compare with the estimate.*

CHECK FOR UNDERSTANDING

Communicating Mathematics

Study the lesson. Then complete the following.

1. **Graph** a function that has a domain of $-1 \le x \le 5$ and a range of $-3 \le y \le 2$. Be sure that your function passes the vertical line test.

2. **Explain** the difference between a discrete function and a continuous function. Give an example of a graph of each type of function.

3. **Find a counterexample** for the statement "Every straight line is a function."

4. **Draw** a Cartesian coordinate plane. Name seven possible locations on the plane where a point might be graphed. Then graph and label a point in each of these locations.

5. **List** four ways to show how the relationship between members of a domain and range can be represented. Give an example of each.

MODELING MATHEMATICS

6. Suppose you are going to construct a graph that shows the relationship between the number of hours worked per week by a student at a fast-food restaurant and the amount of money earned. Identify a reasonable domain and range for the graph. Justify your answer. Then draw a sample graph.

Guided Practice

State whether each relation is a function or not.

7.
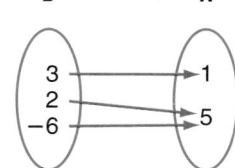

8.
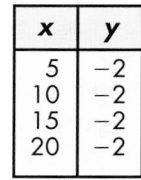

x	y
5	-2
10	-2
15	-2
20	-2

9.
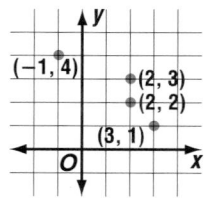

State the domain and range of each relation. Then graph the relation and identify whether it is a function or not. For each function, state whether it is discrete or continuous.

10. $\{(7, 8), (7, 5), (7, 2), (7, -1)\}$ 11. $\{(6, 2.5), (3, 2.5), (4, 2.5)\}$

12. $y = -2x + 1$ 13. $x = y^2$

14. Find $f(5)$ if $f(x) = x^2 - 3x$.   15. Find $h(-2)$ if $h(x) = x^3 + 1$.

Practice

State whether each relation is a function or not.

16.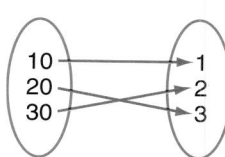

17.

18.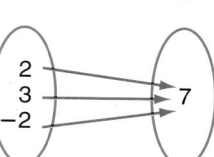

19.

x	y
0.5	−3
2	0.8
0.5	8

20.

Year	Expenses
1994	$4000
1995	$4300
1996	$4000
1997	$4500

21.

x	y
3	5
3	10
3	15
3	20

22.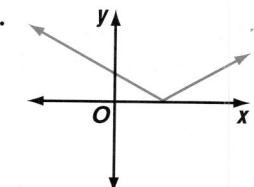

23.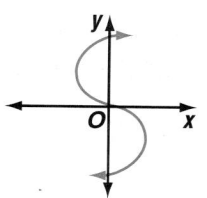

24.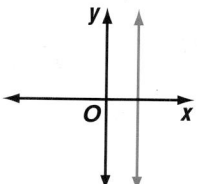

State the domain and range of each relation. Then graph the relation and identify whether it is a function or not. For each function, state whether it is discrete or continuous.

25. $\{(4, 5), (6, 5), (3, 5)\}$

26. $\{(−2, 5), (3, 7), (−2, 8)\}$

27. $\{(3, 4), (4, 3), (6, 5), (5, 6)\}$

28. $y = 7x − 6$

29. $y = −5x$

30. $x = y^2 + 1$

A function h includes the ordered pairs $(−2, 1)$, $(1, 2)$, and $(3.5, −0.3)$. State whether h will still be a function if each ordered pair given below is also included in h.

31. $(−2, 2)$

32. $(0, 0)$

33. $(2, 1)$

Find each value if $f(x) = 3x − 5$ and $g(x) = x^2 − x$.

34. $f(−3)$

35. $g(3)$

36. $g\left(\dfrac{1}{3}\right)$

37. $f\left(\dfrac{2}{3}\right)$

38. $f(a)$

39. $g(5n)$

Find each value if $h(x) = \dfrac{x^2 + 5x − 6}{x + 3}$.

40. $h(3)$

41. $h(−2)$

42. $h(a − 1)$

43. Write an example of a function that has a value of 5 when $x = 2$.

44. The function $f(x) = 3x$ can be represented by the notation $f : x \rightarrow 3x$, where x "maps" to $3x$. If a function g is defined as $g : x \rightarrow x^2 + x$, find the number that $−2$ maps to in function g.

The graph of each figure described below can be a function or a relation that is *not* a function depending on how it appears on a coordinate plane. Graph each figure as both a function and as a relation that is *not* a function.

45. a set of ordered pairs

46. a wavy line

47. an angle

48. a semicircle

Critical Thinking

49. When a fraction contains a variable in the denominator, there are some values of the variable for which the fraction is undefined. Find the domain of $f(x) = \frac{15}{x^2 - 9}$.

50. If $f(x) = x^2 + 2x + 1$, for what value(s) of x would $f(x) = 0$? On a graph of $f(x)$, describe the graph at which $f(x) = 0$.

Applications and Problem Solving

51. Finance On January 1, 1984, the Bell Telephone System gave up its local telephone monopolies and became AT&T. Despite Bell's break-up, the stock price nearly tripled over the next ten years.

Year	'83	'84	'85	'86	'87	'88	'89	'90	'91	'92	'93
Stock Price	19	$20	$25	$25	$28	$29	$45	$30	$39	$50	$55

a. Identify the domain and range.

b. Write the ordered pairs and graph.

c. Is this a function?

d. Would buying AT&T stock in 1983 have been a sound investment? Explain.

52. Sports Sketch a graph of the flight of a baseball, in which the vertical coordinate at each point measures the height of the baseball above the ground and the horizontal coordinate measures the time since the ball was hit. Does the graph represent a function? Explain.

53. Geometry Identify the domain and range of the circle shown below. Is this relation a function?

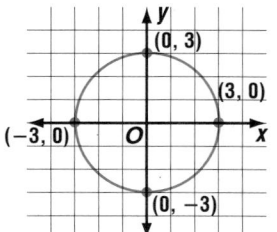

54. Government The table below shows the number of Latino representatives in the U.S. Congress for 1981–1995.

Year	1981	1983	1985	1987	1989	1991	1993	1995
Latino Representatives	6	8	10	11	10	11	17	15

Source: National Association of Latino Elected and Appointed Officials

a. Identify the domain and range.

b. Write the ordered pairs and draw the graph.

c. Is this relation a function? If it is a function, determine whether it is discrete or continuous.

d. If the domain and range were switched, would the relation be a function? Explain.

Mixed Review

Solve each equation or inequality.

55. $|y + 1| < 7$ (Lesson 1–7)

56. $|5 - m| < 1$ (Lesson 1–7)

57. $x - 5 < 0.1$ (Lesson 1–6)

58. $3|2x - 5| = -\frac{1}{3}$ (Lesson 1–5)

59. Consumerism Ryan had $25.04 with him when he went to the mall. His friend, Tim, had $32.67. Ryan wanted to buy a golf shirt for $27.89. (Lesson 1–4)

a. How much money did he have to borrow from Tim in order to buy the shirt?

b. How much money did that leave Tim?

60. State the property used in $9 + (2 + 10) = 9 + 12$. (Lesson 1–4)

61. Write an algebraic expression to represent *seven less than the sum of a number and two times its square.* (Lesson 1–4)

62. Statistics At a bowling party, the members of the junior class decided to separate into two teams and the team with the lower mean would have to make dinner for the other team. The list at the right shows the two teams and their scores.

Team A	Score	Team B	Score
Ed	281	Laurel	101
Kelley	212	Tammy	236
Paul	72	Smitty	143
Mandi	147	Jeff	154
Maria	110	Renee	111
Bryce	212	Yolanda	88
Ryan	69	Percy	69
Sue	28	Debra	205

a. What are the mean, median, and mode for Team A? for Team B? (Lesson 1–3)

b. Make a back-to-back stem-and-leaf plot of the data. (Lesson 1–3)

c. Which team had to make dinner?

Simplify each expression.

63. $3(5a + 6b) + 8(2a - b)$ (Lesson 1–2)

64. $3^2(2^2 - 1^2) + 4^2$ (Lesson 1–1)

2-2A Graphing Technology
Linear Equations

A Preview of Lesson 2-2

Graphing calculators are powerful tools for studying a wide variety of graphs. The examples below show graphs of linear equations.

Example ● **Graph each equation in the standard viewing window.**
  a. $y - 2x = 3$

First, rewrite the equation so y is isolated on one side. Enter the equation for the line and graph.

Enter: [Y=] 2 [X,T,θ,n] [+]
        3 [ZOOM] 6

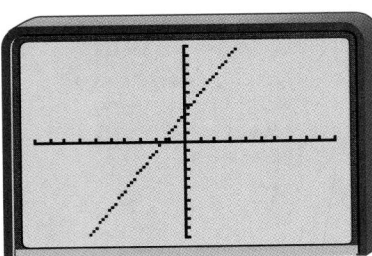

A graph that appears on the graphics screen showing all important characteristics of the graph is called a **complete graph**. The graph of $y = 2x + 3$ is complete because we can see both the x- and y-intercepts.

  b. $y = -x + 14$

Enter: [Y=] [(-)] [X,T,θ,n] [+]
        14 [ZOOM] 6

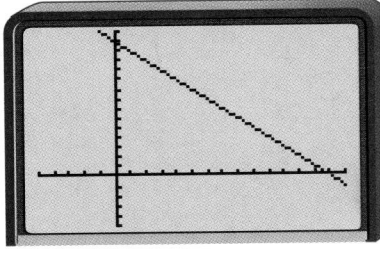

What happened? Only a small portion of the graph is shown in the $[-10, 10]$ by $[-10, 10]$ window. This graph is not complete. Adjust the viewing window to include more of the graph. Try $[-5, 15]$ by $[-5, 15]$. This graph is shown at the right.

EXERCISES

Use a graphing calculator to graph each equation. Describe the viewing window that you used to view a complete graph for each equation.

1. $y = 3x - 3$ **2.** $y = -2x + 5$ **3.** $y = 4 - x$

4. $y = 5x - 35$ **5.** $y = -12x$ **6.** $y = 0.1x - 1$

7. $y = -0.3x + 15$ **8.** $y = 0.01x$ **9.** $y = 100x + 5$

Linear Equations

CONNECTION

Physics

You might guess that sound would travel fastest through air since air is less dense than other mediums like water, glass, or steel. However, just the opposite is true. Sound travels through air most slowly of all. Through air it travels 1129 feet per second, through water about 4760 feet per second, and through glass and steel about 16,000 feet per second. Distances through air for various numbers of seconds are given in the table below.

Time (seconds)	0	1	2	3	4
Distance (feet)	0	1129	2258	3387	4516

The open sentence that describes this relationship is $y = 1129x$, where x represents the number of seconds, and y represents the distance in feet. Since the value of y *depends* on the value of x, y is called the **dependent variable**, and x is called the **independent variable**.

When the relation is graphed on a coordinate plane, the independent variable is graphed on the horizontal axis, and the dependent variable is graphed on the vertical axis. In this case, the points appear to lie on a line. This graph is a function.

Suppose we connect the points with a line. The line would contain an infinite number of points, whose ordered pairs are solutions of the equation $y = 1129x$. An equation whose graph is a line is called a **linear equation**. A linear equation is an equation that can be written in **standard form**, $Ax + By = C$.

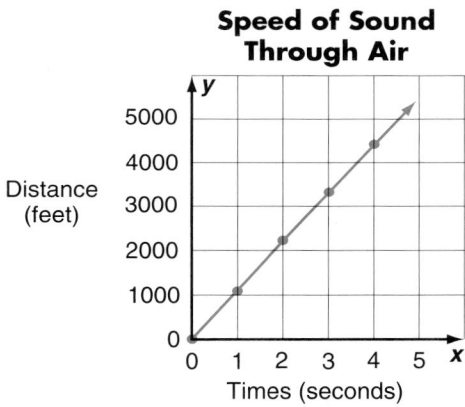

Speed of Sound Through Air

| | Standard Form of a Linear Equation | The standard form of a linear equation is $$Ax + By = C,$$ where *A*, *B*, and *C* are real numbers and *A* and *B* are not both zero. |

Usually A, B, and C are given as integers whose greatest common factor is 1.

When variables other than x are used, assume that the letter coming first in the alphabet represents the domain or horizontal coordinate.

Linear equations contain one or two variables, with no variable having an exponent other than 1.

Linear equations
$5x - 3y = 7$

$x = 9$

$6s = -3t - 15$

Not linear equations
$7a + 4b^2 = -8$

$y = \sqrt{x + 5}$

$x + xy = 1$

Example **1** Write each equation in standard form where A, B, and C are integers whose greatest common factor is 1. Identify A, B, and C.

a. $y = -5x + 6$

$$y = -5x + 6$$
$$5x + y = 6 \qquad \text{\textit{Add 5x to each side.}}$$

So $A = 5$, $B = 1$, and $C = 6$.

b. $\frac{2}{7}x = 2y + 5$

$$\frac{2}{7}x = 2y + 5$$
$$\frac{2}{7}x - 2y = 5 \qquad \text{\textit{The equation is in standard form.}}$$
$$2x - 14y = 35 \qquad \text{\textit{Multiply each side by 7.}}$$

So $A = 2$, $B = -14$, and $C = 35$.

c. $5x - 10y = 25$

$$5x - 10y = 25 \qquad \text{\textit{The GCF of 5, 10, and 25 is 5.}}$$
$$x - 2y = 5 \qquad \text{\textit{Divide each side by 5.}}$$

So $A = 1$, $B = -2$, and $C = 5$.

Any function whose ordered pairs satisfy a linear equation is called a **linear function.**

Definition of Linear Function	A function is linear if it can be defined by $f(x) = mx + b$, where m and b are real numbers.

In the definition of a linear function, m or b may be zero. If $m = 0$, then $f(x) = b$. The graph is a horizontal line. This function is called a **constant function.** If $f(x) = 0$, the function is called the *zero function.*

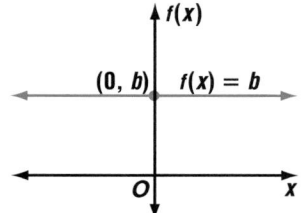

Example **2** State whether each function is a linear function.

a. $f(x) = 4x + 5$ This is a linear function because it is in the form $f(x) = mx + b$, with $m = 4$ and $b = 5$.

b. $g(x) = x^3 + 2$ This is *not* a linear function because x has an exponent other than 1.

c. $f(x) = 9 - 6x$ This is a linear function because it can be written as $f(x) = -6x + 9$, with $m = -6$ and $b = 9$.

In Lesson 2–1, you graphed an equation or function by making a table of values, graphing enough ordered pairs to see a pattern, and connecting the points with a line or smooth curve. However, there are quicker ways to graph a linear equation or function. One way is to find the points at which the graph intersects each axis and connect them with a line. The y-coordinate of the point at which a graph crosses the y-axis is called the **y-intercept.** Likewise, the x-coordinate of the point at which it crosses the x-axis is the **x-intercept.**

Example ③ Graph $5x - 3y = 15$ using the x- and y-intercepts.

The x-intercept is the value of x when $y = 0$.

$$5x - 3y = 15$$
$$5x - 3(0) = 15 \quad \textit{Substitute 0 for y.}$$
$$x = 3$$

The x-intercept is 3. The graph crosses the x-axis at $(3, 0)$.

Likewise, the y-intercept is the value of y when $x = 0$.

$$5x - 3y = 15$$
$$5(0) - 3y = 15 \quad \textit{Substitute 0 for x.}$$
$$y = -5$$

The y-intercept is -5. The graph crosses the y-axis at $(0, -5)$.

Use these ordered pairs to graph the equation.

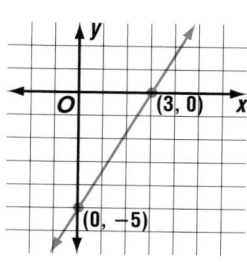

Linear equations, functions, and their graphs can be used to model situations that occur in real life.

Example ④

APPLICATION

Education

The total score of 1300 is only an illustration. Different colleges have different entrance requirements.

Each year, more than two million high school juniors and seniors across the nation tackle one or more college admission tests. One of them is the SAT (Scholastic Assessment Test), which is divided into two sections, verbal and mathematical. For each section, the possible scores range from 200 to 800. Suppose your counselor advises you that, as one factor for admission, some colleges expect a combined score of 1300. This situation can be represented by the equation $x + y = 1300$, where x is the verbal score and y is the mathematical score.

a. Graph the linear equation.
b. Name an ordered pair that satisfies the equation and explain what it represents.

a. Find the x- and y-intercepts.
$$x + y = 1300$$
$$x + 0 = 1300 \quad \textit{Substitute 0 for y.}$$
$$x = 1300$$

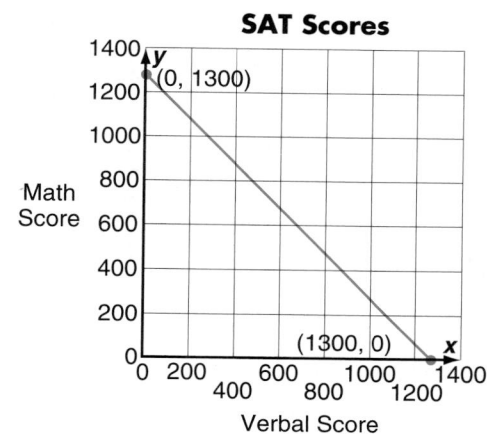

SAT Scores

The x-intercept is 1300. Similarly, the y-intercept is 1300.

(continued on the next page)

b. There are many ordered pairs that satisfy the equation. However, not all of them are solutions of the problem. Since the scores on each section range from 200 to 800, it is necessary to restrict the domain to $200 \leq x \leq 800$ and the range to $200 \leq y \leq 800$.

Therefore, the ordered pairs that are solutions to the problem are shown at the right. One ordered pair is (500, 800). It represents having a verbal score of 500 and a math score of 800 for a total of 1300.

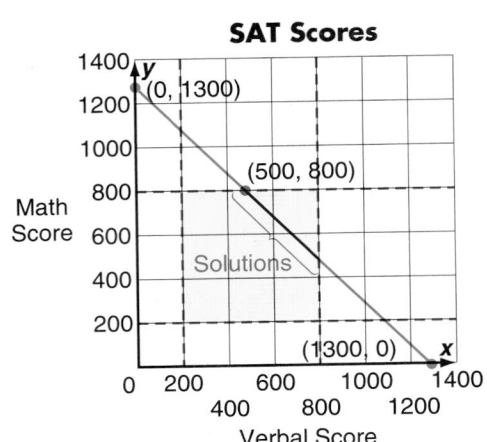

SAT Scores

CHECK FOR UNDERSTANDING

Communicating Mathematics

Study the lesson. Then complete the following.

1. **Write** the equation $4y = 3x + 5$ in standard form. Identify A, B, and C.

2. **Name** the x- and y-intercepts of the graph shown at the right.

3. **Explain** how to find the x- and y-intercepts of the graph of $2x + y = 7$.

4. **List** at least two ways to graph a linear equation.

5. **Write** a paragraph explaining why both graphs at the right are graphs of linear equations, but only one is a linear function. Use drawings with your explanation.

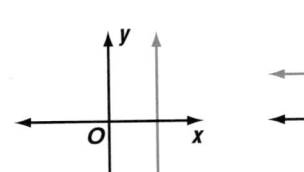 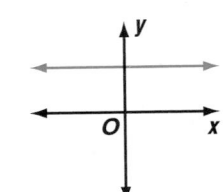

Guided Practice

State whether each equation is linear. Write *yes* or *no* and explain your answer.

6. $x^2 + y^2 = 4$

7. $h(x) = 1.1 - 2x$

Write each equation in standard form where *A*, *B*, and *C* are integers whose greatest common factor is 1. Identify *A*, *B*, and *C*.

8. $y = 3x - 5$

9. $4x = 10y + 6$

10. $y = \frac{2}{3}x + 1$

Find the *x*-intercept and the *y*-intercept of the graph of each equation.

11. $6x + y = 9$

12. $y = -3x - 5$

13. $f(x) = x - 2$

Graph each equation.

14. $3x + 2y = 6$

15. $y = -2x$

16. $4x + 8y = 12$

17. **Economics** On June 20, 1995, the function $f(x) = 0.718x$ was used to convert German marks, x, to U.S. dollars, $f(x)$. For that date, find the value in U.S. dollars of 100 German marks.

Practice

State whether each equation is linear. Write *yes* or *no* and explain your answer.

18. $x + y = 5$

19. $\frac{1}{x} + 3y = -5$

20. $x + xy = 4$

21. $g(x) = 10$

Write each equation in standard form where *A*, *B*, and *C* are integers whose greatest common factor is 1. Identify *A*, *B*, and *C*.

22. $y = -3x + 4$

23. $y = 12x$

24. $x = 4y - 5$

25. $5y = 10x - 25$

26. $\frac{1}{2}x + \frac{1}{2}y = 6$

27. $0.25y = 10$

Find the *x*-intercept and *y*-intercept of the graph of each equation.

28. $y + 6 = 5x$

29. $3x = y$

30. $y = -2$

31. $x = 8$

32. $g(x) = 4x - 1$

33. $5x + 3y = 15$

Graph each equation.

34. $y = x$

35. $y = 4x + 2$

36. $x + y = 7$

37. $2x - y = 5$

38. $2x + 5y = 10$

39. $y = 0.5x - 3$

40. $b = 2a - 3$

41. $x - y = 6$

42. $2a + 3b = 6$

43. $3 = 3x$

44. $x + 2y = 7$

45. $4x + 3y = 12$

46. $\frac{1}{3}x + \frac{1}{2}y = 1$

47. $\frac{x}{4} - \frac{y}{3} = 2$

48. $\frac{x}{3} + \frac{y}{2} = \frac{15}{2}$

Critical Thinking

49. Graph $x + y = 0$, $x + y = 5$, and $x + y = -5$ on the same coordinate plane.
    a. Compare and contrast the graphs.
    b. Write a linear equation whose graph is between the graphs of $x + y = 0$ and $x + y = 5$.

Applications and Problem Solving

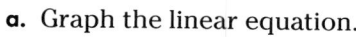

50. **Geology** Geothermal energy is generated wherever water comes into contact with heated underground rocks. The underground temperature of rocks varies with their depth below the surface. The temperature *t* in degrees Celsius is estimated by the function $t(d) = 35d + 20$, where *d* is the depth in kilometers of the rocks.
    a. Graph the linear equation.
    b. Find the temperature of the rocks at a depth of 3 kilometers.
    c. Is this function discrete or continuous? Explain your reasoning.
    d. Find the depth if the temperature of the rocks is 195°C.

51. **Commercial Fishing** Fishing boats are usually equipped with sonar—a device used to locate schools of fish by the reflection of sound waves.
    a. Refer to the application at the beginning of the lesson. Write a function that is a model for the relationship between the number of seconds it takes the sound signal to return to the boat and the depth of the school of fish.
    b. Suppose the sound signal returned to the boat in 0.05 seconds. Estimate the depth of the school of fish.

52. Entertainment In the movie *Crimson Tide*, the crew of the submarine *Alabama* tried desperately to restore power to the ship before it sunk to a depth of 1850 feet. Use the function $f(x) = 1.15x$, where x is the depth in miles and $f(x)$ is the pressure in tons per square inch, to estimate the water pressure on the outside of the hull at that depth.

53. Fundraising The Central High School Band Boosters have a concessions stand for home football games. They sell beverages for $1.50 and candy for $1.25.
Their goal is to sell a total of $375 for each game.
  a. Write an equation that is a model for the different numbers of beverages and candy that can be sold to meet the goal.
  b. Graph the equation.
  c. Is this equation also a function? If so, is it discrete or continuous? Explain your reasoning.
  d. If they sell 100 beverages and 200 pieces of candy, will the Band Boosters meet their goal?

Mixed Review

54. Which of the following graphs represent functions? (Lesson 2–1)

  a. 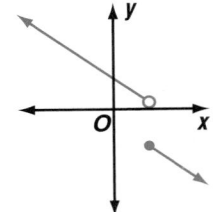   **b.** 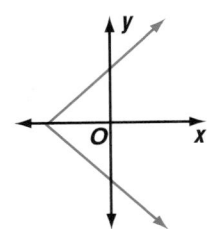   **c.**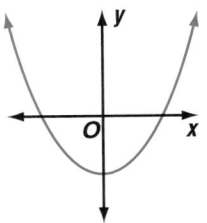

Solve each equation.

55. $|x + 7| > -2$ (Lesson 1–7)   **56.** $5(2x - 7) > 10$ (Lesson 1–6)

57. $7|3x + 5| = 35$ (Lesson 1–5)   **58.** $x + 28.3 = 56.0$ (Lesson 1–4)

59. Statistics Suppose four tests had been given in your Algebra 2 class this quarter. On the first three, you scored 87, 92, and 81. If you must have at least 350 points to earn an A, what must you score on the fourth test to earn an A? (Lesson 1–6)

60. The heights in feet of 20 mountains are given below.

26,504	26,041	26,400	26,750	26,810
29,108	26,470	26,360	26,090	26,000
29,064	26,291	25,910	25,895	28,208
25,925	27,890	27,790	26,760	26,660

  a. Find the mean, median, and mode for the heights. (Lesson 1–3)
  b. Make a stem-and-leaf plot of the heights. (Lesson 1–3)

Simplify each expression.

61. $(9s - 4) - 3(2s - 6)$ (Lesson 1–2)   **62.** $[19 - (8 - 1)] \div 3$ (Lesson 1–1)

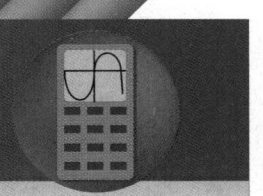

2–2B Graphing Technology
Using Graphs to Estimate Solutions

An Extension of Lesson 2–2

The graphing features of a graphing calculator allow you to approximate solutions to an equation in one variable from a graph. First set each side of the equation equal to y. Then graph the two equations on the same screen. The solution of the original equation is the x-coordinate of the point of intersection of the two graphs.

Example **Solve $5x - 5 = 2x + 1$ graphically.**

Set each side of the equation equal to y.
$y = 5x - 5$
$y = 2x + 1$

Then graph both equations in the standard viewing window.

Enter: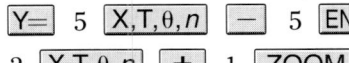

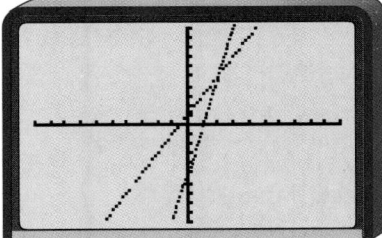

You are interested in the point at which the values of x are equal—namely, the point of intersection. Use TRACE with the left and right arrow keys to approximate the point of intersection. The display gives an approximation for the coordinates of the point. Or press

[2nd] [CALC] 5 [ENTER] [ENTER]

[ENTER] and read the x-coordinate at the bottom of the screen.

You can see that the point of intersection occurs when $x = 2$.

Therefore, the solution of $5x - 5 = 2x + 1$ is 2.

Verify the solution by substituting 2 into the original equation.

EXERCISES

Solve each equation graphically to the nearest tenth. Check your solutions algebraically.

1. $2x - 1 = 2$
2. $3x + 9 = 25$
3. $2x + 1 = 16 - x$
4. $4x + 3 = 5x + 7$
5. $7x + 9 = 3(x + 3)$
6. $5(8 - 2x) = 4x - 2$
7. $-1.5x = -2x + 5.75$
8. $16x - 3.8 = 12x - 3.8$
9. $5.2x + 0.7 = 2.8 + 2.2x$
10. $-2(3x - 5) + 3x = 2 - x$

Slope

What YOU'LL LEARN
- To determine the slope of a line,
- to use slope and a point to graph an equation,
- to determine if two lines are parallel, perpendicular, or neither, and
- to solve problems by identifying and using a pattern.

Why IT'S IMPORTANT
You can use slope to describe lines and solve problems involving auto racing and entertainment.

APPLICATION
Auto Racing

Lady and gentlemen, start your engines! At 11:00 A.M. on May 28, 1995, 32 men and one woman began the Indianapolis 500. Three hours and fifteen minutes later, Canadian racer Jacques Villeneuve crossed the finish line as the winner. Meanwhile, the Diaz family left Indianapolis, Indiana, in the family car, heading for Atlanta, Georgia. They took turns driving, stopped only for gasoline, and completed their 500 mile trip in 9.5 hours. Both of these situations can be modeled graphically.

The graph representing Villeneuve's race is much steeper than the graph representing the Diaz family's trip because, on average, Villeneuve traveled a greater distance for each unit of time. The **slope** of each line is the ratio of the change in the vertical units (distance) to the change in the horizontal units (time).

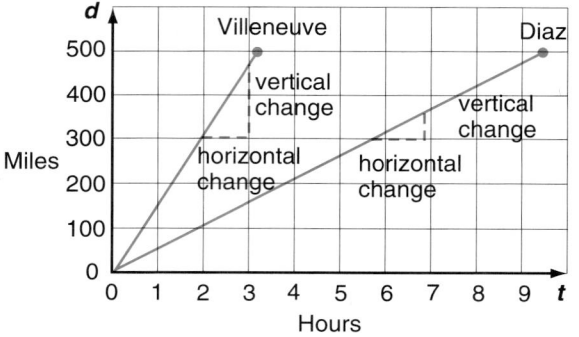

$$\text{slope} = \frac{\text{vertical change}}{\text{horizontal change}}$$

The slope of the line representing Villeneuve's race is $\frac{500}{3.25}$ or about 153.8.

The slope of the line representing the Diaz' trip is $\frac{500}{9.5}$ or about 52.6.

The slope of each line indicates its steepness, and in this case, it also indicates the average speed in miles per hour.

The problem-solving strategy **look for a pattern** is one of the most-used strategies in mathematics. When using this strategy, you will often need to make a table to organize the information.

Example

PROBLEM SOLVING

Look for a Pattern

1 **The symbol used for a cent is a lowercase c with a vertical line through it. The line separates the c into 3 parts, as shown at the right. How many parts would there be if the c had 101 lines through it?**

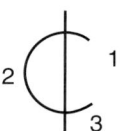

Make a table to show the pattern.

Number of Lines	1	2	3	4	5	x
Number of Parts	3	5	7	9	11	2x + 1

If the c had x lines through it, there would be $2x + 1$ parts. Thus, if the c had 101 lines through it, there would be $2(101) + 1$ or 203 parts.

You can look for a pattern to help solve many problems involving slope and graphs.

Example ② **The winner of the first Indianapolis 500 was Ray Harroun. In 1911, he completed the race in a Marmon Wasp at an average speed of 74.6 miles per hour. Harroun's race can be modeled by the linear function $d(t) = 74.6t$, where t is the time in hours and $d(t)$ is the distance in miles.**
a. Make a table of ordered pairs and graph the linear function.
b. Use the graph to predict the value of t when $d(t) = 500$.
c. The winning speed for the Indianapolis 500 seems to increase each year. Suppose next year's winning speed is 160 mph. Predict where the line representing this race will be graphed.

Auto Racing

Ray Harroun winning the Indianapolis 500

a. Choose ordered pairs for $1 \le t \le 5$.

Time t (hours)	Distance $d(t)$ (miles)
1	74.6
2	149.2
3	223.8
4	298.4
5	373.0

+1 ⟍ +74.6
+1 ⟍ +74.6
+1 ⟍ +74.6
+1 ⟍ +74.6

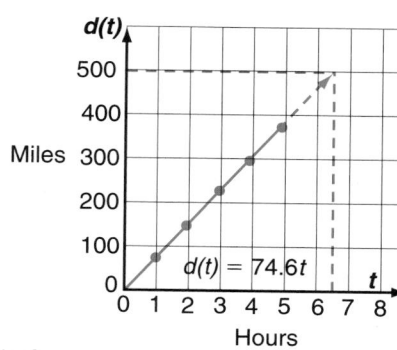

Notice that the ratio of the change in the vertical units to the change in horizontal units is $\frac{74.6}{1}$ or 74.6. Therefore, the slope is 74.6.

b. To predict the value of t when $d(t) = 500$, extend the graph to $d(t) = 500$. The corresponding value of t is between 6 and 7 hours.

c. To predict where the line representing a winning speed of 160 mph will be graphed, look for a pattern in the graphs of Harroun's race and the graphs in the application at the beginning of the lesson. Notice that as the average speed increases, the slope of the line also increases.

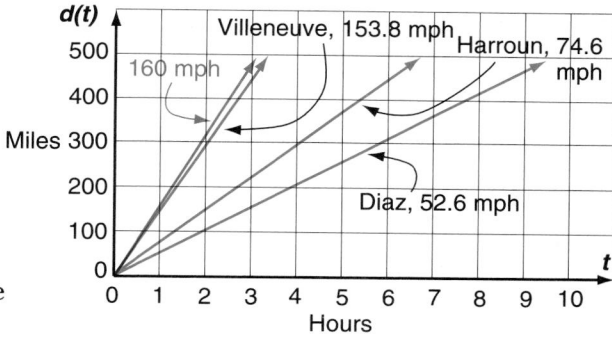

The graph representing a winning speed of 160 mph should have the steepest slope and, following the pattern in the graphs, should be to the left of the graph representing Villeneuve's race.

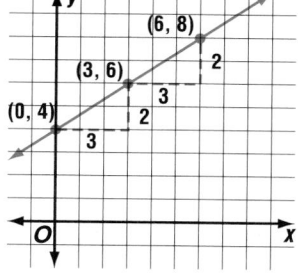

The slopes of linear functions can be defined by looking for a pattern. Consider the equation $y = \frac{2}{3}x + 4$.

In the table of values shown at the right and the graph shown at the left, look for a pattern in the relationship of the change in the y-coordinates to the change in the x-coordinates for the points on the graph.

x	y
0	4
3	6
6	8

+3 ⟍ +2
+3 ⟍ +2

Notice that the y-coordinates increase 2 units for each 3-unit increase in the x-coordinates. Thus, the slope of the line whose equation is $y = \frac{2}{3}x + 4$ is $\frac{2}{3}$.

These examples suggest that the slope of a line can be determined from the coordinates of two points on the line.

Definition of Slope	The slope m of the line passing through points (x_1, y_1) and (x_2, y_2) is given by $m = \dfrac{y_2 - y_1}{x_2 - x_1}$, where $x_1 \neq x_2$.

x_2 is read "x sub 2." The 2 is called a <u>subscript</u>.

Example 3 Determine the slope of the line that passes through the points at $(3, 4)$ and $(6, -8)$. Then graph the line.

$$m = \frac{y_2 - y_1}{x_2 - x_1}$$

$$= \frac{-8 - 4}{6 - 3} \quad (x_1, y_1) = (3, 4), (x_2, y_2) = (6, -8)$$

$$= \frac{-12}{3} \text{ or } -4$$

The slope of the line is -4.

Graph the two ordered pairs and draw the line. Use the slope to check your graph by selecting any point on the line. Then go down 4 units and right 1 unit or go up 4 units and left 1 unit. This point should also be on the line.

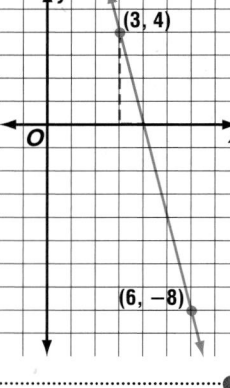

Example 4 Graph the line passing through the point $(-2, -5)$ with a slope of $\frac{3}{5}$.

Graph the ordered pair $(-2, -5)$. Then, using the definition of slope, go up 3 units and 5 units to the right. Plot the point. This new point is $(3, -2)$. *You can also go 5 units right and 3 units up to plot the new point.*

Connect the points to draw the line.

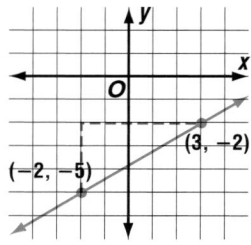

The slope of a line tells the direction in which it rises or falls.

If the line rises to the right, then the slope is positive.	*If the line is horizontal, then the slope is zero.*	*If the line falls to the right, then the slope is negative.*	*If the line is vertical, then the slope is <u>undefined</u>.*

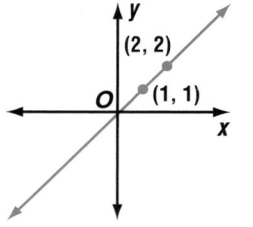

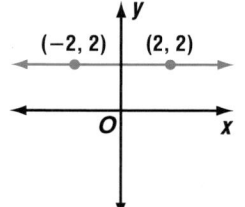

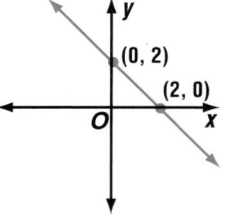

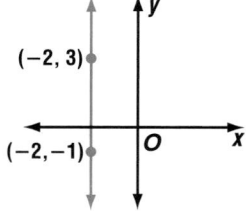

$m = \dfrac{2 - 1}{2 - 1} \text{ or } 1$ $\qquad$ $m = \dfrac{2 - 2}{2 - (-2)} \text{ or } 0$ $\qquad$ $m = \dfrac{0 - 2}{2 - 0} \text{ or } -1$ $\qquad$ $m = \dfrac{3 - (-1)}{-2 - (-2)} \text{ or } \dfrac{4}{0}$

A **family of graphs** is a group of graphs that displays one or more similar characteristics. The **parent graph** is the simplest of the graphs in a family. You can graph families on the same screen and observe their common traits.

Your Turn

a. Graph $y = 3x$, $y = 3x + 2$, $y = 3x - 2$, and $y = 3x - 5$ on the same screen.

b. Identify the parent function and describe the family of graphs.

c. Find the slope of each line.

d. Write a function that has the same characteristics as this family of graphs. Check by graphing.

In the Exploration, you saw that lines that have the same slope are parallel.

Definition of Parallel Lines	**In a plane, nonvertical lines with the same slope are parallel.**

All vertical lines are parallel.

If you know the slope of a line and the coordinates of a point not on the line, you can graph a line through the point that is parallel to the first line.

Example **5** Graph the line that goes through the point at (6, −2) and is parallel to the line whose equation is $-4x + y = -2$.

The y-intercept is −2, and the x-intercept is $\frac{1}{2}$.

Use the x- and y-intercepts to graph $-4x + y = -2$. The slope of the line is 4.

Now use the slope and the point at (6, −2) to graph the line parallel to the graph of $-4x + y = -2$.

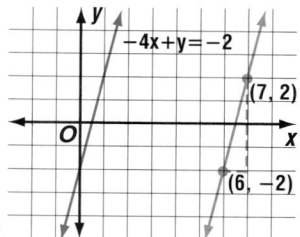

The figure at the left shows the graphs of two lines that are perpendicular. We found that parallel lines have the same slope. Is there a special relationship between the slopes of two perpendicular lines?

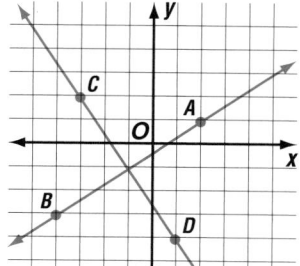

slope of line AB	*slope of line CD*
$\dfrac{1 - (-3)}{2 - (-4)} = \dfrac{4}{6}$ or $\dfrac{2}{3}$	$\dfrac{2 - (-4)}{-3 - 1} = \dfrac{6}{-4}$ or $-\dfrac{3}{2}$

The slopes are negative reciprocals of each other. This and other examples suggest that when you multiply the slopes of two perpendicular lines, the product is always −1.

Definition of Perpendicular Lines	**In a plane, two oblique lines are perpendicular if and only if the product of their slopes is −1.**

Lines that are not vertical or horizontal are called <u>oblique</u>.
Any vertical line is perpendicular to any horizontal line.

You can use the fact that the slopes of perpendicular lines are negative reciprocals of each other to solve problems involving figures with right angles.

Example ⑥ **The consecutive sides of a rectangle are perpendicular. In rectangle *ABCD*, the coordinates of point *B* are (2, 0), and the coordinates of point *C* are (5, 1). Find the slope of the line containing side $\overline{CD}$ of the rectangle.**

INTEGRATION

Geometry

In rectangle *ABCD*, $\overline{BC}$ is perpendicular to $\overline{CD}$. First find the slope of side $\overline{BC}$.

slope of $\overline{BC} = \dfrac{1-0}{5-2}$ or $\dfrac{1}{3}$

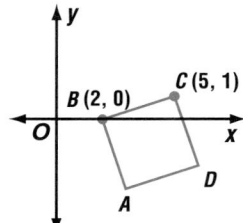

Let *m* represent the slope of $\overline{CD}$. Since the product of the slopes must be -1, you can use this equation.

$$m\left(\frac{1}{3}\right) = -1$$

$$m = -1\left(\frac{3}{1}\right) \text{ or } -3$$

The slope of the line containing $\overline{CD}$ is -3.

CHECK FOR UNDERSTANDING

Communicating Mathematics

Study the lesson. Then complete the following.

1. **Explain** how to find the slope of the line at the right.

2. **Graph** a line with a slope of 2 and a *y*-intercept of 3.

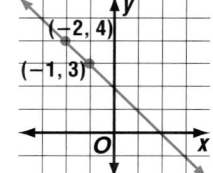

3. **Choose** the line that has a negative slope.

a. 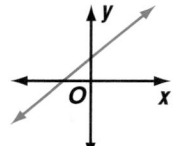 b. 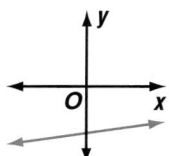 c. 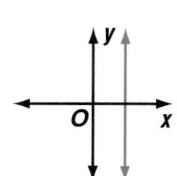 d.

4. **Describe** the relationship between the slopes of two parallel lines.

MATH **J**OURNAL 5. Use a dictionary to find other definitions of slope.

Guided Practice

State the slope of each line.

6.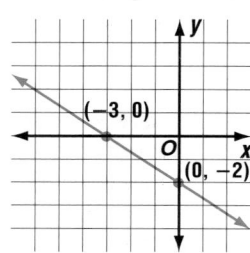

7.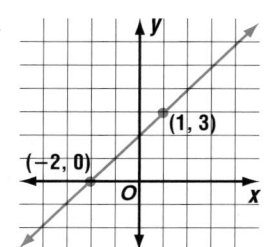

8.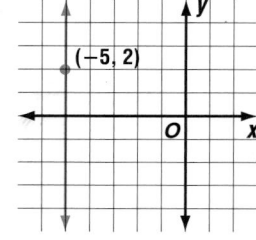

Find the slope of the line that passes through each pair of points. Then determine whether the line rises to the right, falls to the right, is horizontal, or is vertical.

9. $(1, 1), (3, 1)$ **10.** $(-1, 0), (3, -2)$ **11.** $(3, 4), (1, 2)$

Find the slope of the graph of each equation.

12. $2x - y = 4$ **13.** $x + y = 3$ **14.** $2x + 3y = 6$

15. State the slope of a line parallel to the line passing through $(-1, -1)$ and $(3, 2)$. Then state the slope of a line perpendicular to it.

16. Graph a line that passes through $(0, 0)$ and has a slope of 3.

17. Graph the line that passes through $(0, 3)$ and is parallel to the line whose equation is $6y - 10x = 30$.

EXERCISES

Practice

Find the slope of the line that passes through each pair of points. Then determine whether the line rises to the right, falls to the right, is horizontal, or is vertical.

18. $(6, 1), (8, -4)$ **19.** $(6, 8), (5, -5)$ **20.** $(-6, -5), (4, 1)$

21. $(7, 8), (1, 8)$ **22.** $(2.5, 3), (1, -9)$ **23.** $(a, 2), (a, -2)$

Find the slope of the graph of each equation.

24. $x + y = 5$ **25.** $3x + 9 = 0$ **26.** $2x - y = 8$

27. $2x + 3y + 32 = 0$ **28.** $3x - 4y = 0$ **29.** $y = 5$

Determine the value of r so that a line through the points with the given coordinates has the given slope. Draw a sketch of each situation.

30. $(r, 2), (4, -6)$; slope $= -\dfrac{8}{3}$ **31.** $(5, r), (2, 3)$; slope $= 2$

32. $(r, 6), (8, 4)$; slope $= \dfrac{1}{2}$ **33.** $(6, r), (9, 2)$; slope $= \dfrac{1}{3}$

34. Graph a line through $(2, 6)$ that has a slope of $\dfrac{2}{3}$.

35. Graph a line through $(-2, 2)$ that is parallel to a line whose slope is -1.

36. Graph a line through $(-4, 1)$ that is perpendicular to a line whose slope is $-\dfrac{3}{2}$.

37. Graph a line through $(3, 3)$ that is perpendicular to the graph of $y = 3$.

38. Graph a line through the origin that is parallel to the graph of $x + y = 10$.

39. Graph a line through $(-4, -2)$ that has an undefined slope.

40. One line has a slope of 0 and another line has an undefined slope, but they both pass through $(-3, -3)$. Graph the lines.

41. Graph the line perpendicular to the graph of $3x - 2y = 24$ that intersects it at its x-intercept.

42. Geometry   In the ordered pairs $(3, 0), (4, 2), (5, 5), (6, 9)$, and $(7, \underline{\ ?\ })$, the first coordinate is the number of sides in a polygon, and the second coordinate is the number of diagonals that can be drawn in each polygon. Find the pattern to complete the last ordered pair.

43. Determine whether the diagonals of parallelogram *ABCD* at the right are perpendicular. Explain your answer.

44. Are the diagonals of a rectangle perpendicular? Use analytic methods to explain your answer.

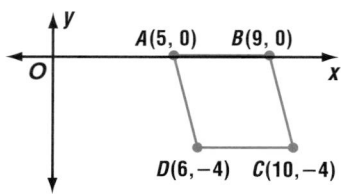

A(5, 0)   *B*(9, 0)

O

x

D(6, −4)   *C*(10, −4)

Graphing Calculator

45. Use a graphing calculator to investigate each family of graphs. Explain how changing the slope affects the graph of the line.

    a. $y = 2x + 3$, $y = 4x + 3$, $y = 8x + 3$, $y = x + 3$

    b. $y = -3x + 1$, $y = -x + 1$, $y = -5x + 1$, $y = -7x + 1$

Programming

46. Points that lie on the same line are called *collinear* points. The graphing calculator program at the right will help you determine whether three points are collinear.

Draw $\overline{AB}$ and $\overline{BC}$ for each set of points on a graphing calculator. Then use the program at the right to find the slopes of $\overline{AB}$ and $\overline{BC}$ and determine which points are collinear.

    a. $A(3, 6)$, $B(5, 7)$, $C(7, 8)$

    b. $A(5, 9)$, $B(7, 12)$, $C(11, 17)$

    c. $A(0, 1.5)$, $B(5, 8.2)$, $C(11, 15.4)$

    d. $A(2.2, -2.1)$, $B(0.6, -1.3)$, $C(-3.8, 0.9)$

```
PROGRAM: Slope
: Disp "ENTER X AND Y FOR POINT
  1"
: Input A
: Input B
: Disp "ENTER X AND Y FOR POINT
  2"
: Input C
: Input D
: If (C−A)=0
: Goto 1
: (D−B)/(C−A) → M
: Disp "THE SLOPE IS  "
: Disp M
: Stop
: Lbl 1
: Disp "THE SLOPE IS UNDEFINED"
: End
```

Critical Thinking

47. If the graph of the equation $ax + 2y = 8$ is perpendicular to the graph of the equation $2x + y = -3$, find the value of a.

Applications and Problem Solving

48. Ancient Cultures   Probably the most famous use of pyramids occurred 4500 years ago as tombs for Egyptian pharaohs and their relatives. But the Western Hemisphere has also had its share of pyramids. Mayan Indians of Mexico and Central America built pyramids that were used as their temples.

    a. The Pyramid of the Sun in Teotihuacán, Mexico, measures about 700 feet on each side of its square base and is about 210 feet high. Estimate the slope that one of its faces makes with the base.

    b. The Great Pyramid in Egypt measures 756 feet on each side of its square base and was originally 481 feet high. Estimate the slope that one of its faces makes with the base.

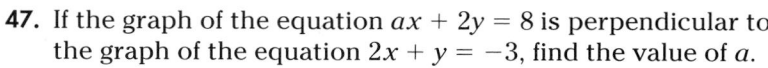

49. Look for a Pattern   Find the next number in each sequence.

    a. 6, 10, 15, 21, 28, . . .   **b.** 1, 4, 9, 16, 25, . . .

50. Entertainment In 1992, CDs passed cassette tapes as the most popular form for pre-recorded music. The graph at the right shows that most of the growth in sales from 1991 to 1994 has been in CDs. If x represents the year and y represents the number of cassettes and CDs sold in millions, find the rate of increase (slope) for both cassette tapes and CDs.

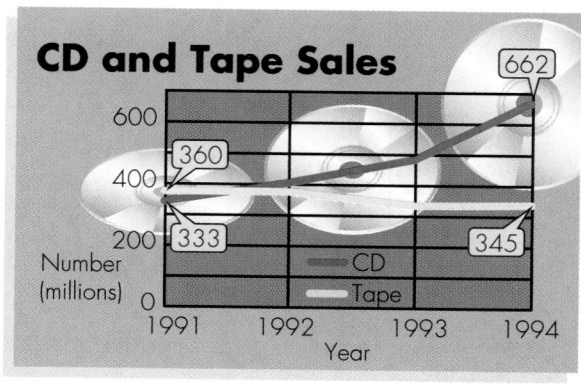

CD and Tape Sales

Number (millions)

Source: Recording Industry Association of America

Mixed Review

51. Write $y = -2x + 4$ in standard form. (Lesson 2–2)

52. Aviation The air pressure in the cabin of a fighter jet decreases as the plane ascends. (Lesson 2–1)

Altitude (feet)	10,000	20,000	30,000	40,000	50,000
Air Pressure (lb/in²)	10.2	6.4	4.3	2.7	1.6

a. Graph the data above.

b. Predict what you think the air pressure would be at 60,000 feet.

53. Solve $5 < 2x + 7 < 13$. (Lesson 1–7)

54. One number is twice another. Twice the lesser number increased by the greater number is at least 85. Find the least possible value for the lesser number. (Lesson 1–6)

55. Solve $|7 + 3a| = 11 - a$. (Lesson 1–5)

56. Solve $0.75(8a + 20) - 2(a - 1) = 3$. (Lesson 1–4)

57. Statistics Find the mean, median, and mode for 100, 45, 105, 98, 97, and 101. (Lesson 1–3)

58. Simplify $\frac{1}{3}(15a + 9b) - \frac{1}{7}(28b - 84a)$. (Lesson 1–2)

59. Simplify $3 + (21 \div 7) \times 8 \div 4$. (Lesson 1–1)

SELF TEST

1. Meteorology When the temperature is 30° F, the speed of the wind makes the temperature feel colder. This is called the windchill factor. The chart below shows how the wind affects your perception of how cold it is when the temperature is 30°. (Lesson 2–1)

a. State the domain and range of the relation shown in the table below.

Wind Speed (mph)	0	5	10	15	20	25	30	35	40
Windchill Factor (°F)	30	27	16	9	4	1	-2	-4	-5

b. Graph the relation. Is it a function?

2. Find the value of $f(15)$ if $f(x) = 100x - 5x^2$. (Lesson 2–1)

3. Write $y = -6x + 4$ in standard form. (Lesson 2–2)

4. Graph $3x + 5y = 30$ using the x- and y-intercepts. (Lesson 2–2)

5. Graph the line that goes through $(4, -3)$ and is parallel to the line whose equation is $2x + 3y = 6$. (Lesson 2–3)

Writing Linear Equations

What YOU'LL LEARN

- To write an equation of a line in slope-intercept form given the slope and one or two points, and
- to write an equation of a line that is parallel or perpendicular to the graph of a given equation.

Why IT'S IMPORTANT

You can use equations to explore relations in telecommunications and business.

APPLICATION
Telecommunications

Did you ever wonder why there are so many advertisements for long-distance telephone companies on television? Well, the long-distance market is huge! In 1989, there were 4.6 *billion* hours of long-distance calls made in the United States. Since 1989, the number of hours has increased by about 0.4 billion each year.

The equation $y = 0.4x + 4.6$ can be used to find y, the number of hours of long-distance calls (in billions) for any number of years, x, after 1989. This equation is graphed at the right. The y-intercept, 4.6, represents the number of billions of hours in 1989. The slope, 0.4, represents the yearly increase.

In Lesson 2–2, you learned if a function can be written in the form $y = mx + b$, then it is a linear function. In the example above, m is 0.4, which is the slope, and b is 4.6, which is the y-intercept. Is it always true that m is the slope and b is the y-intercept?

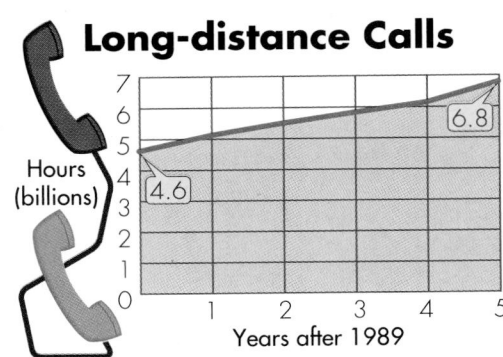

Long-distance Calls

Hours (billions)

Years after 1989

Source: Commonwealth Associates

Consider the graph below. The line passes through points $A(0, b)$ and $C(x, y)$. Notice that b is the y-intercept of $\overline{AC}$. Suppose you need to find the slope of $\overline{AC}$. Substitute the coordinates of points A and C into the slope equation.

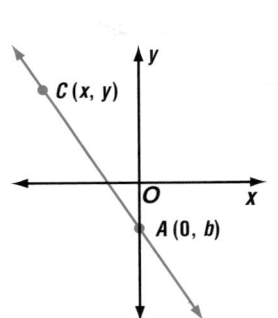

$$m = \frac{y_2 - y_1}{x_2 - x_1}$$

$$m = \frac{y - b}{x - 0} \quad \text{Substitute (0, b) for } (x_1, y_1) \text{ and } (x, y) \text{ for } (x_2, y_2).$$

$$m = \frac{y - b}{x}$$

Now solve the equation for y.

$$mx = y - b \quad \text{Multiply each side by x.}$$

$$mx + b = y \quad \text{Add b to each side.}$$

$$y = mx + b \quad \text{Symmetric property of equality}$$

When an equation is written in this form, it is in **slope-intercept form**.

Slope-Intercept Form of a Linear Equation	**The slope-intercept form of the equation of a line is $y = mx + b$, where m is the slope and b is the y-intercept.**

If you are given the slope and the y-intercept of a line, you can find an equation of the line by substituting the values of m and b into the slope-intercept form. For example, if you know that the slope of a line is -4 and the y-intercept is 5, the equation of the line is $y = -4x + 5$, or, in standard form, $4x + y = 5$.

You can also use the slope-intercept form to find an equation of a line if you know the slope and the coordinates of a point on the line.

Example ① Find the slope-intercept form of an equation of the line that has a slope of $-\frac{2}{3}$ and passes through $(-6, 1)$.

You know the slope and the x and y values of one point on the graph. Substitute for m, x, and y in the slope-intercept form.

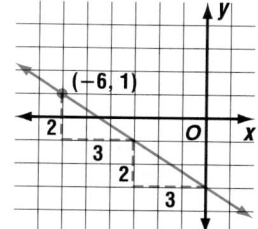

$$y = mx + b$$
$$1 = \left(-\frac{2}{3}\right)(-6) + b$$
$$1 = 4 + b$$
$$-3 = b$$

The y-intercept is -3. So, the equation in slope-intercept form is $y = -\frac{2}{3}x - 3$.

If you are given the coordinates of two points on a line, you can use the **point-slope form** to find the equation of the line that passes through them.

Point-Slope Form of a Linear Equation	The point-slope form of the equation of a line is $y - y_1 = m(x - x_1)$, where (x_1, y_1) are the coordinates of a point on the line and m is the slope of the line.

Example ② Find an equation of the line that passes through $(3, 2)$ and $(5, 3)$.

First, use the two points given to find the slope of the line.

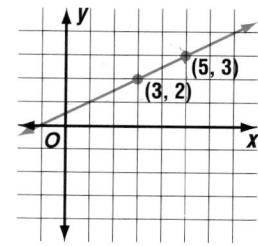

$$m = \frac{3 - 2}{5 - 3} \text{ or } \frac{1}{2}$$

Then use the point-slope form to write the linear equation.

$$y - y_1 = m(x - x_1)$$
$$y - 2 = \frac{1}{2}(x - 3) \quad \textit{Replace } m \textit{ with } \tfrac{1}{2} \textit{ and } (x_1, y_1) \textit{ with the}$$
$$y - 2 = \frac{1}{2}x - \frac{3}{2} \quad \textit{coordinates of either point. We chose } (3, 2).$$
$$y = \frac{1}{2}x + \frac{1}{2}$$

The slope-intercept form of the equation of the line is $y = \frac{1}{2}x + \frac{1}{2}$.

In standard form, the equation is $x - 2y = -1$.

When changes in real-life situations occur at a linear rate, a linear equation can be used as a model for describing the situation.

Example ③ With a certain long-distance company, the price of a 4-minute long-distance call is $1.70. An 11-minute call with the same company costs $4.15.

APPLICATION
Business

a. Write a linear equation that describes the cost of these telephone calls. Assume that the changes increase linearly.

b. How much would a 20-minute telephone call cost?

a. The line passes through the points (4, 1.7) and (11, 4.15). Find the slope of the line. *Use $(x_1, y_1) = (4, 1.7)$ and $(x_2, y_2) = (11, 4.15)$.*

$$m = \frac{y_2 - y_1}{x_2 - x_1}$$

$$= \frac{4.15 - 1.7}{11 - 4}$$

$$= \frac{2.45}{7} \text{ or } 0.35$$

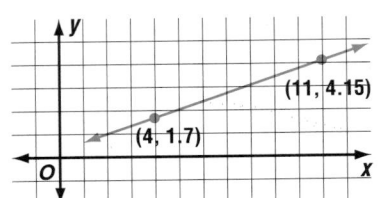

(11, 4.15)

(4, 1.7)

Now use the point-slope form to write the linear equation.

$$y - y_1 = m(x - x_1)$$
$$y - 1.7 = 0.35(x - 4) \quad \text{\textit{Replace m with 0.35 and } } (x_1, y_1) \text{ \textit{with} } (4, 1.7).$$
$$y - 1.7 = 0.35x - 1.4$$
$$y = 0.35x + 0.3$$

The slope-intercept form of the equation of the line is $y = 0.35x + 0.3$.

b. Use the equation to find the cost of a 20-minute call.

$$y = 0.35x + 0.3$$
$$y = 0.35(20) + 0.3 \quad \text{\textit{Replace x with 20.}}$$
$$y = 7 + 0.3 \text{ or } 7.3$$

A 20-minute call would cost $7.30.

Check this by estimating the cost from the graph.

Top Five
List

Countries with the most telephones per 100 people

1. Sweden, 68.4
2. Switzerland, 60.8
3. Canada, 59.2
4. Denmark, 58.3
5. United States, 56.1

The slope-intercept form can also be used to find equations of lines that are parallel or perpendicular.

Example **4** Write an equation of the line that passes through $(-9, 5)$ and is perpendicular to the line whose equation is $y = -3x + 2$.

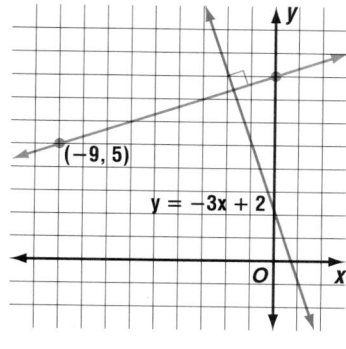

Geometry

The slope of the given line is -3. Since the product of this slope and the slope of the perpendicular line is -1, the slope of the perpendicular line is $\frac{1}{3}$.

Use the slope-intercept form and the ordered pair $(-9, 5)$ to write the equation.

$y = mx + b$

$5 = \left(\frac{1}{3}\right)(-9) + b$ *Replace m with $\frac{1}{3}$ and (x, y) with $(-9, 5)$.*

$5 = -3 + b$

$8 = b$

Use a graphing calculator to verify that the equation is correct.

An equation of the line is $y = \frac{1}{3}x + 8$.

CHECK FOR UNDERSTANDING

Communicating Mathematics

Study the lesson. Then complete the following.

1. **Explain** how to write an equation in slope-intercept form and tell what each variable represents.

2. **Write** an equation of a line with slope of 5 and y-intercept of -4.

3. **Explain** how to write the equation of a line if you know the ordered pairs for two points on the line.

4. **Choose** the graph that shows $y = mx + b$, $m < 0$, $b > 0$.

   a.   b.   c.   d.

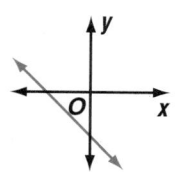

5. **You Decide**   Maribela thinks that the graphs of $y = 4x + 2$ and $4x - y = -2$ are different lines. Karen thinks that they are the same. Who is correct? Explain your reasoning.

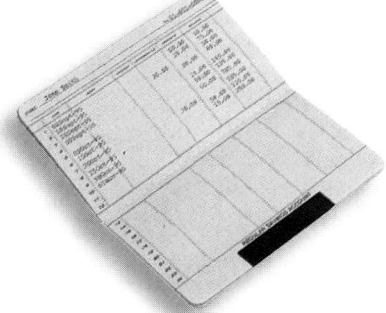 

6. **Assess Yourself**   Suppose you have $250 in a savings account and decide to save an additional $10 each month. Write an equation to find the total amount, y, in your savings account after x months. Graph the equation. Then explain what the slope and y-intercept represent.

Guided Practice

The slope and y-intercept of a line are given. Write the slope-intercept form of the equation for each line described.

7. $m = 7, b = -3$ 8. $m = 1.5, b = 0$ 9. $m = -\frac{1}{3}, b = 4$

State the slope and y-intercept of the graph of each equation.

10. $y = 2x - 5$ 11. $2y = 4x + 6$ 12. $3x + 2y = 10$

Write an equation in slope-intercept form for each graph.

13.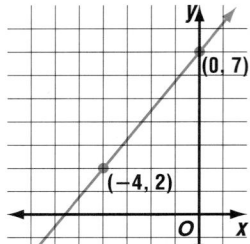
(0, 7)
(−4, 2)

14.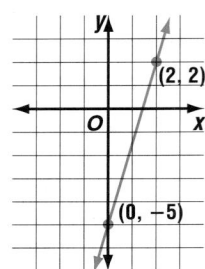
(2, 2)
(0, −5)

15.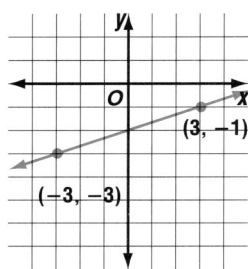
(3, −1)
(−3, −3)

Write an equation in slope-intercept form that satisfies each condition.

16. slope = 0.5, passes through (6, 4)

17. passes through (6, 1) and (8, −4)

18. passes through (0, 5) and is parallel to the graph of $y = 4x + 12$

19. passes through (0, −2) and is perpendicular to the graph of $y = x - 2$

20. Ecology A nature-preserve worker estimates there are 6000 deer in Sharon Woods Park. She also estimates that the population will increase by 75 deer each year thereafter. Write an equation that represents how many deer will be in the park in x years.

EXERCISES

Practice

State the slope and y-intercept of the graph of each equation.

21. $y = -\frac{2}{3}x - 4$

22. $y = \frac{3}{4}x$

23. $-y = 0.3x + 6$

24. $4y = 2x - 10$

25. $-5y = 3x - 30$

26. $y = cx + d$

Write an equation in slope-intercept form for each graph.

$Y = mx + b$

27.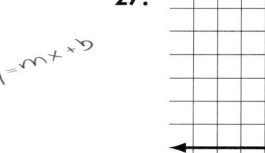
(2.5, 2)

28.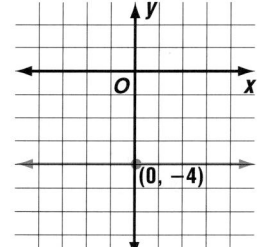
(0, −4)

29.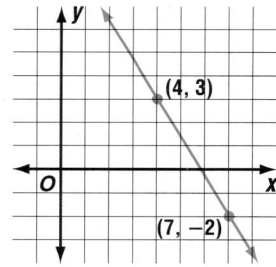
(4, 3)
(7, −2)

Write an equation in slope-intercept form that satisfies each condition.

$Y = mx + b$

30. slope = 0.25, passes through (0, 4)

31. slope = −0.5, passes through (2, −3)

32. slope = 4, passes through the origin

33. passes through (−2, 5) and (3, 1)

34. passes through (7, 1) and (7, 8)

35. passes through (−2, −3) and (0, 0)

36. x-intercept = −4, y-intercept = 4

37. *x*-intercept = $\frac{1}{3}$, *y*-intercept = $-\frac{1}{4}$

38. *x*-intercept = 0, *y*-intercept = 4

39. passes through (4, 6), parallel to the graph of $y = \frac{2}{3}x + 5$

40. passes through (−2, 0), perpendicular to the graph of $y = -3x + 7$

41. passes through (−3, −1), parallel to the line that passes through (3, 3) and (0, 6)

42. passes through (6, −5), perpendicular to the line whose equation is $3x - \frac{1}{5}y = 3$

Find the value of *k* in each equation if the given ordered pair is a solution of the equation.

43. $5x + ky = 8$, (3, −1) **44.** $4x - ky = 7$, (4, 3)

45. $3x + 8y = k$, (0, 0.5) **46.** $kx + 3y = 11$, (7, 2)

Graphing Calculator

Compare and contrast the graphs of each pair of equations. Use a graphing calculator to check your answers.

47. $2y = 6x + 14$
$3x - y = 6$

48. $2y - 4 = x$
$y = -2x + 2$

49. $3x + 5y = 15$
$y = -\frac{3}{5}x + 3$

Critical Thinking

50. Geometry Given $\triangle ABC$ with $A(-6, -8)$, $B(6, 4)$, and $C(-6, 10)$, write the equation of the line containing the altitude from A. Remember that the altitude from A is perpendicular to $\overline{BC}$.

Applications and Problem Solving

51. Telecommunications Refer to the application at the beginning of the lesson. Estimate the number of hours of long-distance calls that will be made in 1999.

52. Geometry The equation $d = 180(c - 2)$ can be used to find the number of degrees, d, in any convex polygon with c sides.
   a. Write this equation in slope-intercept form.
   b. Identify the slope and *y*-intercept.
   c. Find the number of degrees in a pentagon.

53. Science At 20°F, 1 gallon of water weighs approximately 8.33 pounds. Write a linear equation to represent this situation.

54. Business Benny's Floral Shop charges $3 per mile for delivery of a $20 floral arrangement. Carmelita's Floral Shop charges $2 per mile for delivery of a $30 arrangement. When is it less expensive to buy from Carmelita?

55. Technology The graph shows the increase in the number of subscribers to on-line computer services from 1991 until 1995.
   a. Assuming that the increase is linear, write an equation to represent this situation.
   b. Predict the number of subscribers in the year 2000.
   c. What is the meaning of the slope and *y*-intercept?

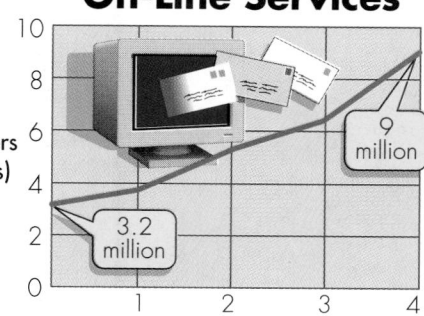

On-Line Services

Customers (millions)

3.2 million

9 million

Years after 1991

Source: Information and Interactive Services Report, Dataquest

Mixed Review

56. Zoology According to the *World Almanac*, a zebra can run at speeds up to 40 mph. Suppose a zebra could run at 40 mph for a long period of time. If at 1:30 P.M, a zebra had already traveled 32 miles, use the definition of slope to find how many miles the zebra could have traveled by 3:00 P.M. if it was running at top speed. (Lesson 2–3)

57. Determine if $g(x) = x(2 - x)$ is a linear function. (Lesson 2–2)

58. Find the value of $h(a + 1)$ if $h(x) = 3x - 1$ and $a = 4$. (Lesson 2–1)

Solve each inequality.

59. $|x - 2| \leq -99$ (Lesson 1–7)

60. $2(r - 4) + 5 \geq 9$ (Lesson 1–6)

61. Solve $|x - 3| = 2x$. (Lesson 1–5)

62. Write an algebraic expression to represent *one-fifth the sum of four and a number*. (Lesson 1–4)

63. Statistics Find the mean, median, and mode of the hourly wages of 200 employees. One hundred earn $5.00 per hour, ten earn $6.25 per hour, ten earn $7.75 per hour, twenty earn $4.50 per hour, and sixty earn $5.90 per hour. (Lesson 1–3)

64. Name the property illustrated by $11 + a = a + 11$. (Lesson 1–2)

65. Evaluate $(5a + 3d)^2 - e^2$ if $a = 3$, $d = 0.5$, and $e = 0.3$. (Lesson 1–1)

Honeybee Counting

The excerpts below appeared in an article in *New Scientist* on March 4, 1995.

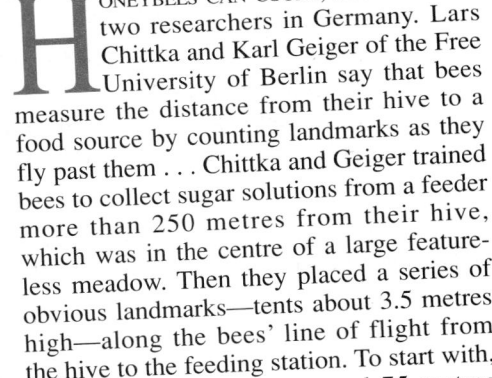

HONEYBEES CAN COUNT, ACCORDING TO two researchers in Germany. Lars Chittka and Karl Geiger of the Free University of Berlin say that bees measure the distance from their hive to a food source by counting landmarks as they fly past them . . . Chittka and Geiger trained bees to collect sugar solutions from a feeder more than 250 metres from their hive, which was in the centre of a large feature-less meadow. Then they placed a series of obvious landmarks—tents about 3.5 metres high—along the bees' line of flight from the hive to the feeding station. To start with, there were four tents, spaced 75 metres apart, so the feeder was between the third and fourth landmarks. Even if sugar was also available between the second and third tents, most of the bees flew to the original feeder. The researchers then changed the number or position of the landmarks . . . Many of the bees simply stopped at the first feeder they encountered after the third land-mark, even though this meant that they were nowhere near the feeder on which they had been trained . . . Chittka and Geiger say that bees clearly react to the number of landmarks they have passed, rather than to the distance they have flown, which means they must have the begin-nings of an ability to count. ■

1. Under natural conditions, what types of landmarks might bees use to find their way to food sources?

2. If bees do in fact count landmarks, compare their behavior with the methods you use to find your way when traveling.

3. How is the bees' behavior similar to your use of graphs or measuring lines when illustrating or solving problems?

Integration: Statistics
Modeling Real-World Data Using Scatter Plots

CONNECTION
Science

What YOU'LL LEARN

- To draw scatter plots, and
- to find and use prediction equations.

Why IT'S IMPORTANT

You can use scatter plots to display data, examine trends, and make predictions.

When you ride the express elevator to the top floor of a skyscraper, you experience the effects of changing air pressure. Because of the rapid increase in altitude, the air pressure outside your eardrum is lower than the air pressure inside your eardrum. This difference in air pressure causes your eardrums to push out until some air finally forces its way out of your ears and goes pop!

The weight of the air pressing down around us produces air pressure. Generally, air pressure is greatest near Earth's surface, where it averages 14.7 pounds per square inch (psi). It decreases as you move out toward space, simply because there is less air pressing down. The chart at the right shows average air pressure measured at different altitudes.

Altitude (thousand feet)	Air Pressure (psi)
0	14.7
5	12.3
10	10.2
15	7.0
20	6.4
25	5.2
30	4.3
35	3.5
40	2.7
45	2.0
50	1.6

To determine the relationship between altitude and air pressure, graph the data points in a **scatter plot.** When real-life data is collected, the points graphed usually do not form a straight line, but may *approximate* a linear relationship. When this is the case, a **best-fit line** can be drawn, as shown at the left.

A **prediction equation** can be determined by employing a process similar to that used to determine an equation of a line when you know two points. You can use the prediction equation to estimate, or predict, one of the variables given the other.

Use two points from the line, (45, 2.0) and (25, 5.2), to find the slope.

$$m = \frac{2.0 - 5.2}{45 - 25} \text{ or } -0.16$$

Let x represent the altitude, and let y represent the air pressure. Use the slope and one of the points in the slope-intercept form to find a prediction equation.

$y - y_1 = m(x - x_1)$ *Use point-slope form.*

$y - 2 = -0.16(x - 45)$ *Let $(x_1, y_1) = (45, 2)$.*

$y = -0.16x + 9.2$

A prediction equation is $y = -0.16x + 9.2$.

Air Pressure

Altitude (thousand feet)

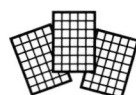

MODELING MATHEMATICS

Head Versus Height

Materials: tape measure   grid paper

In this activity, you will collect data to determine whether there is a relationship between the horizontal circumference of a person's head and his or her height.

Your Turn

a. Collect data from several of your classmates. Measure the circumference of each person's head and his or her height. Record the data as ordered pairs (circumference, height).

b. Graph the data in a scatter plot.

c. Choose two ordered pairs and write a prediction equation.

d. Explain the meaning of the slope in the prediction equation.

e. Predict the head size of a person who is 66 inches tall.

f. Predict the height of an individual whose head size is 18 inches.

Example

APPLICATION
Keyboarding

1 Draw a scatter plot and find two prediction equations to show how keyboarding speed and experience are related. Predict the keyboarding speed in words per minute (wpm) of a student who has 11 weeks of experience.

Experience (weeks)	4	7	8	1	6	3	5	2	9	6	7	10
Speed (wpm)	33	45	49	20	40	30	38	22	52	44	42	55

The best-fit line does not necessarily contain any points from the data.

Explore   The problem asks for two prediction equations.

Plan   Make a scatter plot to determine the relationship between experience and speed. Since keyboarding speed is dependent on experience, the independent variable is the number of weeks of experience.

Solve   The pattern of points suggests a possible line that passes through (5, 36) and (8, 49).

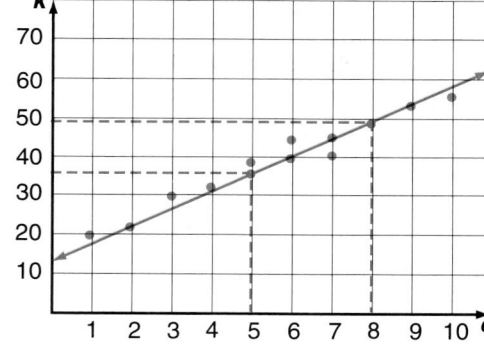

$$m = \frac{49 - 36}{8 - 5}$$

$$= \frac{13}{3} \text{ or about } 4.3$$

Let e stand for experience in weeks. Let k stand for keyboarding speed in words per minute.

$$y = mx + b$$
$$k = 4.3e + b$$
$$36 = 4.3(5) + b$$
$$14.5 = b$$

One prediction equation is $k = 4.3e + 14.5$.

Another line can be suggested by using (2, 22) and (9, 52). Using these points results in a prediction equation of $t = 4.3e + 13.4$.

If a student had 11 weeks of experience, the first equation would predict that the student could type approximately 62 words per minute.

Examine   Locate the ordered pair (11, 62) on the scatter plot. The point lies close to the graph of the prediction equation. Therefore, the solution is reasonable.

Rita Moreno
(1931–   )

Rita Moreno, a Hispanic-American actress, singer, and dancer, was the first and only artist to win an Oscar®, a Tony, an Emmy, and a Grammy award.

The procedure for determining a prediction equation is dependent upon your judgment. You decide where to draw the best-fit line. You decide which two points on the line are used to find the slope and intercept. Your prediction equation may be different from someone else's. The prediction equation is used when a rough estimate is sufficient.

Example 2

APPLICATION

Entertainment

Each year, the entertainment industry publishes a list of the top earners. The table at the right shows the top ten earners for 1993–1994.

a. Draw a scatter plot for the data.

b. Predict the earnings of the fifteenth person on the list.

Rank	Person or Group	Earnings (millions)
1	Steven Spielberg	$335
2	Oprah Winfrey	105
3	Barney	84
4	Pink Floyd	62
5	Bill Cosby	60
6	Barbra Streisand	57
7	Eagles	56
8	David Copperfield	55
9	Rolling Stones	53
10	Harrison Ford	44

a.

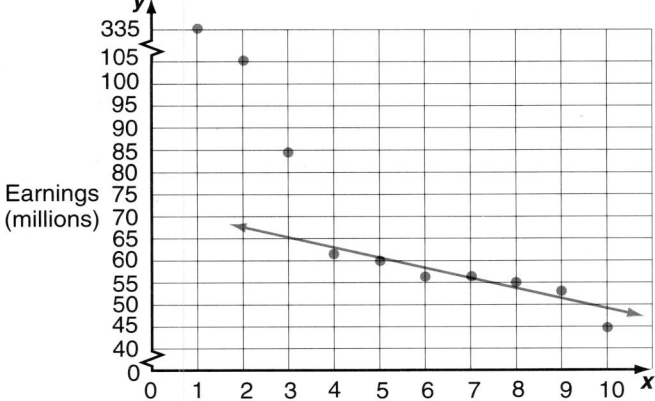

b. The pattern of dots suggests a possible line that passes through (5, 60) and (7, 56), as shown above.

Find the slope.

$$m = \frac{56 - 60}{7 - 5}$$

$$= \frac{-4}{2} \ or -2$$

Then find a prediction equation.

$$y = mx + b$$

$$60 = -2(5) + b$$

$$70 = b$$

A prediction equation is $y = -2x + 70$. If an entertainer was fifteenth on the list, the equation would predict earnings of $-2(15) + 70$ or $40 million. *Do you think a line is a good predictor in this case?*

Communicating Mathematics

Study the lesson. Then complete the following.

1. **Explain** why best-fit lines are helpful.

2. **Make** a scatter plot that shows the relationship between your test scores and the number of hours you do homework for each class you take.

3. **Choose** the scatter plot that has a prediction equation with a positive slope.

a. 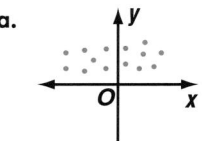 b. 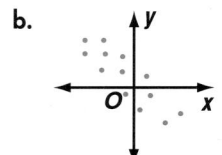 c. 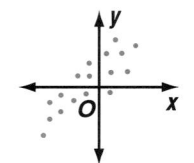 d.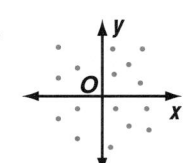

4. **You Decide**   Refer to Example 1. Yoki thinks a prediction equation is accurate for values of *e* greater than 25. Juanita thinks a prediction equation probably won't be accurate at those values. Who is correct? Explain your reasoning.

5. Collect data to determine whether there is any relationship between the circumference of a person's wrist and neck. If so, write a prediction equation for the relationship.

Guided Practice

In a study of the relationship between the height (*h*), in inches, and the ideal weight (*w*), in pounds, of adult men, a prediction equation is *w* = 5*h* − 187. Predict the ideal weight for each height.

6. 66 inches   7. 72 inches   8. 78 inches

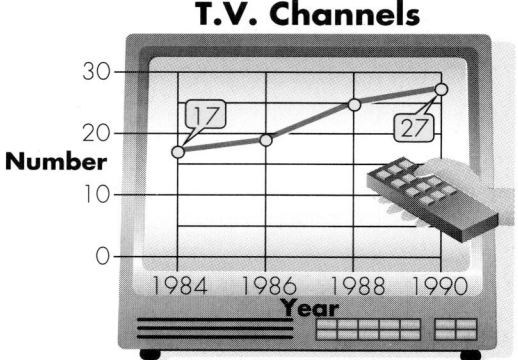

T.V. Channels

Number

Year

Source: Nielsen Media Research

9. **Television**   The graph at the left shows the average number of television stations that were received in U.S. households from 1984 until 1990.

a. Copy the graph and draw a best-fit line.

b. Find the slope and *y*-intercept of a prediction equation.

c. Predict the average number of television stations that a household will receive in 1999.

Applications and Problem Solving

10. **Safety**   The graph at the right shows how the percentage of traffic fatalities that were in alcohol-related crashes has decreased from 1982 through 1994. Predict the percentage of fatalities for the year 2000 if this trend continues.

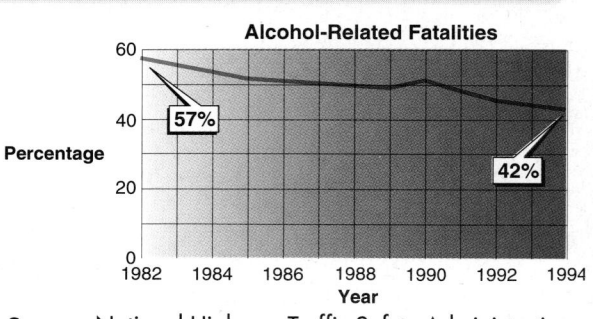

Alcohol-Related Fatalities

Percentage

57%

42%

Year

Source: National Highway Traffic Safety Administration

11. **Health** The table below shows the age and systolic blood pressure for a group of people who recently donated blood.

Age	35	24	48	50	34	55	30	26	41	37
Blood Pressure	128	108	140	135	119	146	132	104	132	121

a. Draw a scatter plot to show how age x and systolic blood pressure y are related.

b. Write a prediction equation that relates a person's age to their approximate systolic blood pressure.

c. Find the approximate systolic blood pressure of a person 54 years old.

12. **Geography** The table below shows elevation and average precipitation for selected cities.

City	Elevation (feet)	Average Precipitation (inches)
Beirut, Lebanon	111	35
London, England	149	23
Paris, France	164	22
Montreal, Canada	187	41
Algiers, Algeria	194	30
Bucharest, Romania	269	23
Warsaw, Poland	294	22
Oslo, Norway	308	27
Rome, Italy	377	30
Toronto, Canada	379	32
Budapest, Hungary	394	24
Moscow, Russia	505	25

a. Draw a scatter plot to show how elevation e and precipitation p are related.

b. Write a prediction equation.

c. Check your equation by using the elevation of Dublin, Ireland, which is 155 feet with an average precipitation of 30 inches.

13. **Agriculture** Farmers will sometimes hold their crops from market until the price goes up to a level they think is satisfactory. The table below records the price per bushel and how many thousand bushels of wheat were sold at that price during a 10-day selling period in Iowa.

Price ($ / bushel)	3.84	3.66	3.87	3.96	3.60	4.05	3.63	3.60	3.72	3.87
Bushels Sold (thousands)	50	47	38	28	49	23	47	46	39	42

a. Draw a scatter plot and find a prediction equation for the data.

b. If the market price of wheat is $3.90/bushel next week, how many bushels of wheat can you predict will be sold?

c. Estimate what the price of wheat was when 25,500 bushels were sold.

Seed Germination Rate

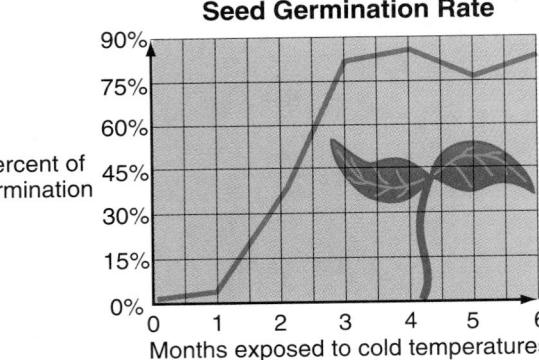

14. **Botany** The graph at the right shows the germination rate of batches of bristlecone pine seeds that have been exposed to cold temperatures. Explain why one best-fit line may not be the best solution to finding a prediction equation.

15. Suppose there are three lines in a plane. Line *a* passes through Quadrants I, II, and IV. Line *b* passes through Quadrants I, III, and IV. Line *c* only passes through Quadrants I and II. Line *a* is perpendicular to Line *b*. All three lines pass through (3, 3). Given that the slope of Line *b* is 3, write the equations of the three lines in slope-intercept form.   (Lesson 2–4)

16. Find $g(3)$ if $g(x) = -\dfrac{4x}{2} + 7$. (Lesson 2–1)

17. Solve $|x + 4| > 3$. (Lesson 1–7)

18. Write an algebraic expression to represent *the sum of a number and its square*. (Lesson 1–4)

19. Simplify $3(2x + 2) - 2(x - 1)$. (Lesson 1–2)

WORKING ON THE

In·ves·ti·ga·tion

Refer to the Investigation on pages 60–61.

Through the Looking Glass

To determine if their data are characteristic of a species or just random observations, naturalists often plots their observances on a graph such as a scatter plot. Then they look for patterns and draw conclusions from the patterns they observe.

1 Graph the data in your two charts in a single scatter plot. Use different colors for the data from each chart. Let one axis represent the height of the image and let the other axis represent the distance the viewer stood from the wall. Look for a pattern or relationship in the number pairs. How does the distance from the wall compare to the height of the image? Find

a mathematical relationship between those two measures. Then explain your findings.

2 Write a function for each scope you used that relates distance from an image and its height. Explain the similarities and differences between the functions. Include a description of a reasonable domain and range in your discussion.

3 How does the size of the scope compare with the other data? Find a ratio between the width and the length of each scope. Explain how the equations relate to the sizes of the scopes.

4 Find two best-fit lines for the points on the scatter plot. How does the ratio for the size of the scope compare to the slope of the corresponding best-fit line? Explain the relationships that exist.

Add the results of your work to your Investigation Folder.

2–5B Graphing Technology
Lines of Regression

An Extension of Lesson 2–5

You can use a graphing calculator to draw scatter plots and a line that best fits the points in the scatter plot. This line is called a **regression line.** Once you have drawn the regression line, you can use the TRACE feature on the graphing calculator to make predictions about the data.

Example **Some scientists believe that global warming is a result of carbon dioxide emissions from fuel consumption. The table below shows world carbon dioxide emissions for 1950–1989. Draw a scatter plot and regression line to show how the year is related to the level of carbon dioxide emissions.**

Year	Emissions (millions of metric tons)
1950	6002
1955	7511
1960	9475
1965	11,556
1970	14,989
1975	16,961
1980	19,287
1985	19,672
1990	22,588

Source: Carbon Dioxide Information Analysis Center, 1992

Let the independent variable be the years since 1940, and let the dependent variable be the emissions. First, set the window parameters. The values of the data suggest a viewing window of [0, 55] by [5000, 25000] with Xscl = 5 and Yscl = 1000.

Next, enter the data. Press STAT 1 to display lists for storing data.

If old data has previously been stored, enter STAT 1 ▲ CLEAR ENTER ▶ ▲ CLEAR ENTER to clear the lists. The years will be entered into column L1.

Enter: 10 ENTER 15 ENTER 20 ENTER ...50 ENTER

Use ▶ to move the cursor to column L2 to enter carbon dioxide emissions.

Enter: 6002 ENTER 7511 ENTER 9475 ENTER ... 22588 ENTER

We are now ready to draw the scatter plot.

Enter: 2nd STAT PLOT 1 ENTER ▼ ENTER ▼ ENTER ▼ ▶ ENTER ▼ ENTER GRAPH

(continued on the next page)

Next, graph the equation for the regression line.

Enter: STAT ▶ 5 2nd L1
, 2nd L2 ENTER Y=
VARS 5 ▶ ▶ 7
GRAPH

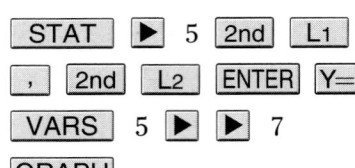

The TRACE feature allows you to move a cursor along the graph or scatter plot and read the coordinates of the points. Press TRACE and any of the arrow keys to observe what happens. Approximately what would you expect the carbon dioxide emissions to have been in 1971?

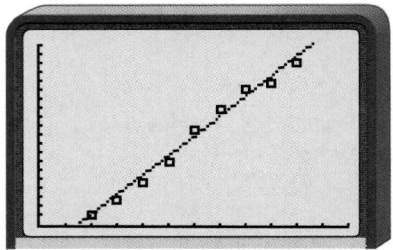

EXERCISES

Use a graphing calculator to draw a scatter plot and a regression line for the data in each table.

1.
x	−2	1	4	10	23	25
y	1	−0.5	2	−4.5	−10	−11.5

2.
x	0.2	1	3	4	5	8
y	0.11	0.31	0.9	1.25	1.75	2.3

3. **Employment**   The table below shows the average yearly incomes (in dollars) of men and women in the United States.

Year	1960	1970	1980	1985	1990
Women	3257	5323	11,197	15,624	19,822
Men	5368	8966	18,612	24,195	27,678

Source: *The 1993 Information Please Almanac*

a. Use a graphing calculator to draw a scatter plot and regression line to show how the year is related to women's salaries for 1960 to 1990. Let the independent variable be the years since 1950, and let the dependent variables be the incomes. Predict the average salary for women in the year 2000.

b. Repeat part a with the men's salaries.

c. Compare and contrast the data in the two scatter plots.

Special Functions

What YOU'LL LEARN

- To identify and graph special functions.

Why IT'S IMPORTANT

You can use functions to explore relations in technology and finance.

Of course, you know that the formula $C = \pi d$ describes the relationship between the diameter and circumference of a circle. Notice that the slope, 3.32, is close to π.

INTEGRATION
Geometry

Rebeca's little brother, Jorge, was doing a project for his mathematics class in which he measured the diameter and circumference of several circular objects. The data are shown below.

Diameter (centimeters)	4.8	10.5	17.3	23.8	25.0
Circumference (centimeters)	15.4	31.5	54.1	75.2	79.1

Rebeca helped Jorge graph the data, and it appeared to be a linear function. Using the points (10.5, 31.5) and (17.3, 54.1), the slope of the prediction equation is about 3.32. It appears that the *y*-intercept is the origin.

Whenever a linear function in the form $y = mx + b$ has $b = 0$ and $m \neq 0$, the function is called a **direct variation.** In this situation, the circumference varies directly as the diameter. In other words, as the diameter gets larger, the circumference also gets larger.

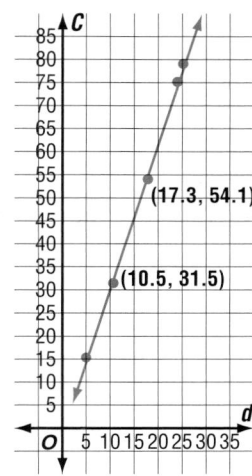

There are other special cases of linear functions. Two of these, the **constant function** and the **identity function** are shown below.

constant function
$m = 0$

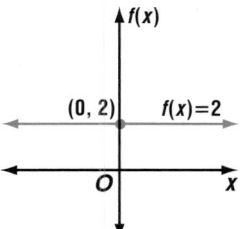

identity function
$m = 1, b = 0$

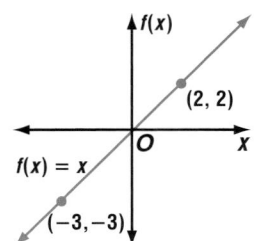

Step functions like the ones shown below are also related to linear functions. The open circle means that the point is *not* included in the graph.

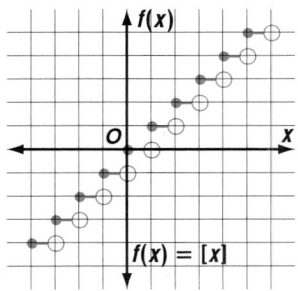

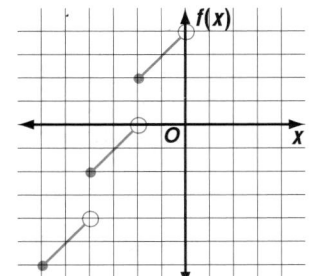

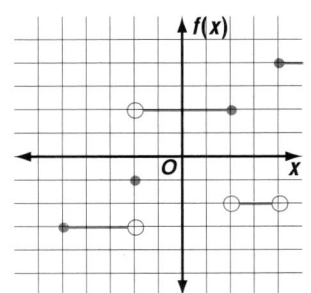

One type of step function is the **greatest integer function.** The symbol [x] means *the greatest integer not greater than x.* For example, [7.3] = 7 and [−1.5] = −2 because −1 > −1.5. The greatest integer function is given by f(x) = [x]. Its graph is the first step function graph shown on the previous page.

The graphs of step functions are often used to model real-world problems.

Example **1**

When you go to the post office to mail a first-class letter, you may need to ask the clerk how much it will cost to mail it—that is, how much postage is required. In 1995, first-class mail cost 32¢ for the first ounce and 23¢ for each additional ounce. So if your letter weighed one ounce or less, it cost you 32¢ to mail. A letter weighing 1.1 ounces cost 32 + 23 or 55¢ to mail. Graph the function that describes the relationship between the number of ounces and the cost of postage.

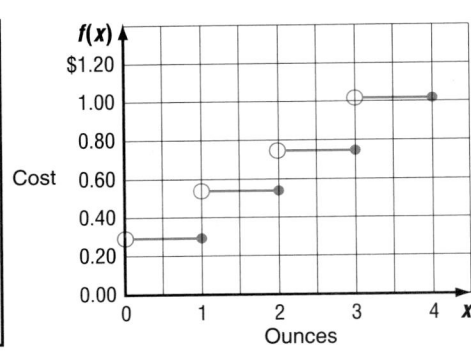

The equation that describes this function is f(x) = 0.32 + 0.23[x − 0.1], where x is the number of ounces.

Make a table of values to help you draw the graph.

x	0.32 + 0.23[x − 0.1]	f(x)
0.1	0.32 + 0	0.32
0.7	0.32 + 0	0.32
1.0	0.32 + 0	0.32
1.5	0.32 + 0.23(1)	0.55
1.9	0.32 + 0.23(1)	0.55
2.0	0.32 + 0.23(1)	0.55
2.3	0.32 + 0.23(2)	0.78
2.8	0.32 + 0.23(2)	0.78
3.0	0.32 + 0.23(2)	0.78
3.2	0.32 + 0.23(3)	1.01

Another function that is closely related to linear functions is the **absolute value function.** Consider f(x) = |x| or y = |x|. Look for a pattern when studying the values in the chart.

LOOK BACK

You can refer to Lesson 1-5 for information about absolute value equations.

x	−3	−2	−1	0	1	2	3
f(x)	3	2	1	0	1	2	3

You can see that when x is positive or zero, the absolute value function looks like the graph of y = x. When x is negative, the absolute value function looks like the graph of y = −x.

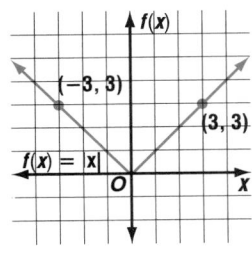

Example **2** Graph $f(x) = |x| + 2$ and $g(x) = 3|x| + 2$ on the same coordinate plane. Determine the similarities and differences in the two graphs.

Find several ordered pairs for each function.

x	|x| + 2
−2	4
−1	3
0	2
1	3
2	4

→

x	3|x| + 2
−2	8
−1	5
0	2
1	5
2	8

Graph the points and connect them. Both graphs have the same *y*-intercept. The graph of $g(x) = 3|x| + 2$ is narrower than the graph of $f(x) = |x| + 2$.

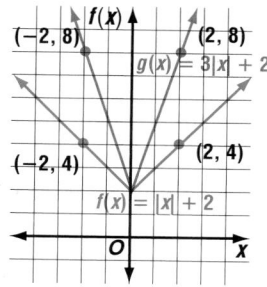

Recall that families of graphs are groups of graphs that display one or more similar characteristics. Graphs of absolute value functions may also display similar characteristics.

EXPLORATION

GRAPHING CALCULATORS

When a linear equation is written in the form $y = mx + b$, *m* is the slope, and *b* is the *y*-intercept. In this Exploration, you will use a graphing calculator to investigate absolute value functions of the form $y = a|x|$.
The parent graph of most families of absolute value functions is the graph of $y = |x|$.

Your Turn

a. Graph $y = |x|$, $y = 2|x|$, $y = 3|x|$, and $y = 5|x|$ on the same screen.

b. Describe this family of graphs. What pattern do you see?

c. In $y = mx + b$, as *m* increases, the slope of the line also increases. Describe how the graph of $y = a|x|$ changes as *a* increases.

d. Write an absolute value function whose graph is between the graphs of $y = 2|x|$ and $y = 3|x|$.

e. Graph $y = |x|$ and $y = -|x|$ on the same screen. Then graph $y = 2|x|$ and $y = -2|x|$ on the same screen.

f. Describe this family of graphs. What pattern do you see?

g. In a linear equation $y = mx + b$ with $m < 0$, the line falls to the right. Describe how the graph of $y = a|x|$ changes when $a < 0$.

h. Write an absolute value function whose graph opens down.

Communicating Mathematics

Study the lesson. Then complete the following.

1. **Explain** why the slope was not 3.14 in the application at the beginning of the lesson.

2. **Choose** which function is a direct variation.
   a. $f(x) = 4$ **b.** $y = -3$ **c.** $y = |x| + 2$ **d.** $f(x) = 5x$

3. **Explain** why the value of $[4.3]$ is 4, but the value of $[-4.3]$ is -5.

4. **Describe** the pattern for graphing an absolute value function.

5. **Compare and contrast** the graphs of $f(x) = |x|$ and $f(x) = |x| - 2$.

Guided Practice

Identify each function as C for constant, D for direct variation, A for absolute value, or G for greatest integer function. Then graph each function.

6. $f(x) = |3x - 2|$ 7. $g(x) = -[x]$

8. $f(x) = 2x$ 9. $h(x) = 0.5$

10. If $g(x) = |x - 5|$, find $g(4)$.

11. If $f(x) = 2[x - 1]$, find $f(6.9)$.

12. **Business** The Fix-It Auto Repair Shop has a sign in the Service Department that states that the labor costs are $35 per hour or any fraction thereof. What type of function does this relationship represent?

Practice

If $h(x) = [2x + 3]$, find each value.

13. $h(2)$ 14. $h(-2)$ 15. $h(-2.3)$ 16. $h\left(\frac{1}{4}\right)$

Identify each function as C for constant, D for direct variation, A for absolute value, or G for greatest integer function. Then graph each function.

17. $h(x) = x$ 18. $f(x) = |2x|$ 19. $g(x) = -3$

20. $f(x) = [2x + 1]$ 21. $f(x) = \left|x - \frac{1}{4}\right|$ 22. $f(x) = -\frac{2}{3}x$

23. $f(x) = x + 3$ 24. $f(x) = |x + 3|$ 25. $f(x) = [x + 3]$

26. $g(x) = |x| + 3$ 27. $g(x) = [x] + 3$ 28. $g(x) = 3|x|$

Graph each pair of equations on the same coordinate plane. Discuss the similarities and differences in the two graphs.

29. $y = |x + 2|, y = |x - 2|$ 30. $y = |x| + 4, y = |x| - 4$

31. $y = |x + 2|, y = |x + 2| - 1$ 32. $y = 2[x], y = [2x]$

33. $y = [x + 5], y = [x] + 5$ 34. $y = |2x|, y = 2|x|$

35. $y = -2|4x|, y = 4|-2x|$ 36. $y = -3[x], y = [-3x]$

Graph each equation.

37. $y = [\,|x|\,]$ 38. $y = |\,[x]\,|$ 39. $y = x - [x]$ 40. $y = x + |x|$

Use a graphing calculator to solve each equation graphically.

41. $|2x - 4| = 6$ 42. $|x + 2| = 4$

43. Compare and contrast the graphs of $y = |x|$ and $|y| = x$. Identify the domain and range of each.

44. **Technology** The graph at the right shows the number of CD-ROM multimedia software packages shipped since 1991.

   a. Write an equation to represent a best-fit line for this graph.

   b. What type of function does this relationship represent?

   c. Predict the number of software packages that will be shipped in 1999.

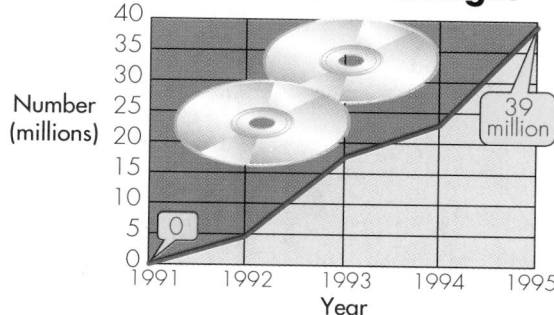

CD-Rom Software Packages

Number (millions)

Year

Source: Information and Interactive Services Report, Dataquest

45. **Transportation** When the Westerville North boy's basketball team went to the Ohio state tournament, the school chartered buses so that the student body could attend the games. Each bus held a maximum of 60 students.

   a. Make a graph that shows the relationship between the number of students that went to the game by bus x and the number of buses that were needed y.

   b. What type of function does this relationship represent?

46. **Finance** Lupe earns $5.25 per hour working at a video store after school. About 25% of her earnings are taken out for taxes and other deductions.

   a. Write an equation that shows the relationship between the number of hours worked per week and Lupe's take-home pay.

   b. About how many hours will Lupe need to work to take home $100?

   c. What type of function does this relationship represent?

47. **Smoking** The chart at the right shows the percent of people ages 20 to 24 who smoked for selected years. (Lesson 2–5)

   a. Draw a scatter plot and find a prediction equation for the data.

   b. Estimate what percent of the population in this age group will be smoking by the year 2000, if current trends continue.

Year	People Ages 20–24 Who Smoke
1965	47.8%
1970	41.5
1974	39.6
1976	39.6
1978	35.3
1979	35.9
1980	36.1
1983	36.9
1985	31.8
1987	29.5
1988	27.6
1993	25.0

48. Write an equation in slope-intercept form for the line with slope $\frac{2}{3}$ that passes through $(6, -5)$.   (Lesson 2–4)

49. Graph a line that passes through $(0, 3)$ and is perpendicular to the line $y - 2x = 4$.   (Lesson 2–3)

50. Graph $b = 3a - 2$.   (Lesson 2–2)

51. Find $h(-1)$ if $h(x) = \frac{x^2 + 2x - 5}{x^2 - 2}$.   (Lesson 2–1)

52. Solve $28 - 6y < 23$.   (Lesson 1–6)

53. Statistics   Use the line plot below to answer each question.   (Lesson 1–3)

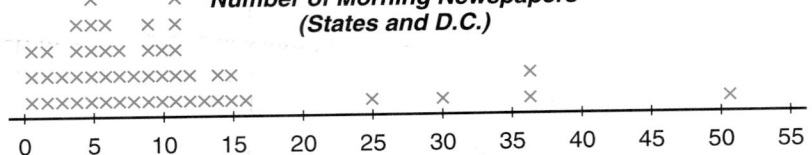

a. What is the greatest number of newspapers in one state?
b. What is the least number of newspapers in one state?
c. How many states have 10 to 20 newspapers?
d. How many newspapers do most states have?

54. Simplify $3 + \{8 \div [9 - 2(4)]\}$.   (Lesson 1–1)

WORKING ON THE

In·ves·ti·ga·tion

Refer to the Investigation on pages 60–61.

Through the Looking Glass

Many species react when an invader to their territory comes too close. They often become defensive and may attack. So, it is often handy for naturalists to have equipment, such as a scope, to observe animals from a distance without creating a threat to the animal or to the naturalist's safety.

1 Select one of the scopes that you used in your experiment. Suppose you were 60 feet from a giraffe and the animal's image exactly filled the viewer. How tall would the giraffe be?

2 Using a second scope that was a different size than your first scope, how would the image of the giraffe change when looking through the second tube? Where would you need to stand to see the entire image of the giraffe fill the view of the second scope? Can you see the same image if you are standing in the same location and looking through two view tubes of different sizes?

3 Explain the methods you discovered for determining heights using the view tubes. Analyze those methods and discuss your findings in writing.

Add the results of your work to your Investigation Folder.

2–7A Graphing Technology
Linear Inequalities

A Preview of Lesson 2–7

You can graph inequalities with a graphing calculator by using the Shade(command located in the DRAW menu. You must enter *two* functions to activate the shading. The first function defines the lower boundary of the region to be shaded. The second function defines the upper boundary of the region. The calculator will graph both functions and shade between the two. If the inequality is "$y \leq$", you can use the Ymin window value as the lower boundary since the points that satisfy the inequality are below the graph of the related equation. If the inequality is "$y \geq$", you can use the Ymax window value as the upper boundary since the points that satisfy the inequality are above the graph of the related function.

Before using the Shade(option, be sure to clear equations stored in the Y = list.

Example

Graph $y \geq 3x - 2$ in the standard viewing window.

The inequality asks for points where y is greater than or equal to $3x - 2$, so we will use Ymax, or 10, as the upper boundary and $3x - 2$ as the lower boundary. This will shade all points on the graphics screen between $y = 3x - 2$ and $y = 10$.

Enter: ZOOM 6 2nd DRAW 7
3 X,T,θ,*n* − 2 , 10
) ENTER

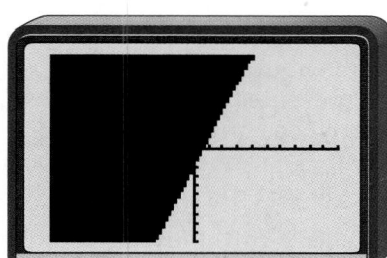

Since both the *x*- and *y*-intercepts of the line $y = 3x - 2$ are within the current viewing window, the graph of the inequality is complete.

EXERCISES

Use a graphing calculator to graph each inequality. Then sketch your graph on a sheet of paper.

1. $y \geq 3$ **2.** $y \geq x + 2$ **3.** $y \leq -2x - 4$

4. $y > \frac{1}{3}x + 7$ **5.** $y \geq \frac{1}{2}x - 3$ **6.** $x - 7 \leq y$

7. $y + 1 \leq 0.5x$ **8.** $y - 3 > -2x$ **9.** $2 \geq x - 2y$

Linear Inequalities

What YOU'LL LEARN

- To draw graphs of inequalities in two variables.

Why IT'S IMPORTANT

You can use inequalities to solve problems involving manufacturing and shopping.

F Y I

In 1994, over 600 million CDs and 345 million cassettes were sold!

APPLICATION
Shopping

In 1978, the Sony Corporation revolutionized the music industry when it introduced its portable cassette player, the Walkman. In effect, this marked the end of records as a means to enjoy music. Today, CDs are the most popular format for recorded music.

Suppose you have $35 to spend and want to purchase some cassettes that cost $7 and some CDs that cost $14. If c represents the number of cassettes, and d represents the number of CDs, the equation $7c + 14d = 35$ describes the different ways for you to spend *exactly* $35. The graph at the right shows all of the ordered pairs that are solutions of the equation $7c + 14d = 35$. One solution is (3, 1).

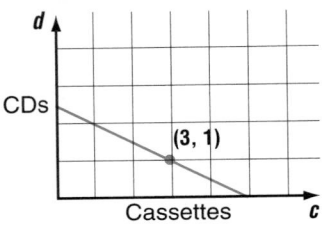

However, when you're shopping, it's not necessary to spend exactly $35. You want to spend *at most* $35. You know that if you spend *more than* $35, you won't have enough money left to go out with your friends. These situations can be represented by the linear inequalities $7c + 14d \leq 35$ and $7c + 14d > 35$, respectively. The graph of $7c + 14d = 35$ separates the coordinate plane into two regions. The line is called the *boundary* of the regions. To graph an inequality, first graph the boundary and then determine which region to shade.

The graph of $7c + 14d \leq 35$ contains points for which the cost of the cassettes and CDs is less than or equal to 35. For example, the ordered pair (2, 1) is in the shaded region and therefore, is one of the solutions. Note that the boundary is a solid line. If the inequality uses the symbols $\leq$ or $\geq$, the boundary is solid to show that it is included.

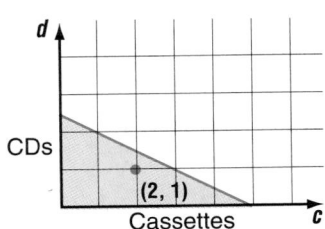

The graph of $7c + 14d > 35$ contains points for which the cost of the cassettes and CDs is greater than $35. For example, the ordered pair (3, 2) is in the shaded region and therefore, is one of the solutions. Note that the boundary is a dashed line. If the inequality uses the symbols $<$ or $>$, the boundary is dashed to show that it is *not* included.

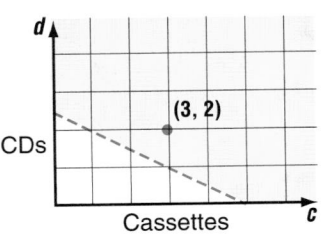

You can graph an inequality by following these steps.

1. Graph the boundary. Determine whether it should be solid or dashed.
2. Test a point in each region.
3. Shade the region whose ordered pair results in a true inequality.

Example **Graph $2y - 3x \le 6$.**

The boundary will be the graph of $2y - 3x = 6$. Let's use intercepts to graph the boundary more easily.

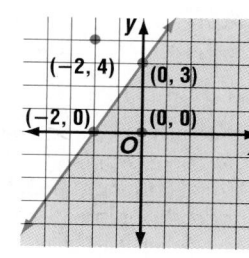

x-intercept	*y-intercept*
$2(0) - 3x = 6$	$2y - 3(0) = 6$
$-3x = 6$	$2y = 6$
$x = -2$	$y = 3$

Since the inequality is "less than or equal to," draw a solid line connecting the two intercepts. This is the boundary.

Try to test the origin because it is easy to substitute 0 for x and y.

Now test a point in each region.

Try $(0, 0)$.

$2(0) - 3(0) \le 6$

$0 \le 6$ true

Try $(-2, 4)$.

$2(4) - 3(-2) \le 6$

$8 - (-6) \le 6$

$14 \le 6$ false

The region that contains $(0, 0)$ should be shaded.

Sometimes inequalities contain absolute value. When this occurs, you must consider that the expression within the absolute value bars may be positive *or* negative.

Example **Graph $y > |x| - 2$.**

This absolute value function has two conditions to consider.

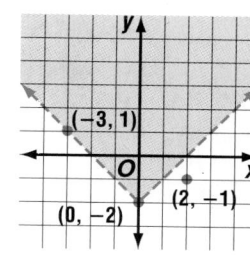

when x < 0   *when x ≥ 0*

$y > -x - 2$ and $y > x - 2$

Graph each inequality for the specified values of x. The lines will be dashed.

When one ordered pair results in a true inequality, it is not necessary to test a point in the other region.

Test $(0, 0)$.

$y > |x| - 2$

$0 > |0| - 2$ *Note that the boundary is __not__ included.*

$0 > -2$ true

The shaded region should include $(0, 0)$.

Inequalities can sometimes be used to analyze a situation and determine the trends in business and profitability.

Example **③**

APPLICATION

Business

Amanda Harris wants to rent a car for a business trip of about 100 miles. Reasonable Car Rental advertises that their daily rental rate is $30 plus $0.25 a mile. She knows that Executive Car Rental charges $70 for a car rental with 100 free miles. Which company has the better rate?

Explore   One way to solve the problem is to make a graph for one company that shows the relationship between the number of miles driven d and the total cost of the rental r.

Plan   The initial cost of a car from Reasonable Car Rental is $30. Since this is the point at which no miles are driven, it is the y-intercept of the graph. The slope would be the rate of change in the total cost. In this case, the rate is $0.25 per mile. Thus, the slope is 0.25, and an equation of the line is $r = 0.25d + 30$.

Solve   Graph the equation $r = 0.25d + 30$. Then graph the point (100, 70), which represents Executive Car Rental's charge of $70 for a rental of 100 miles.

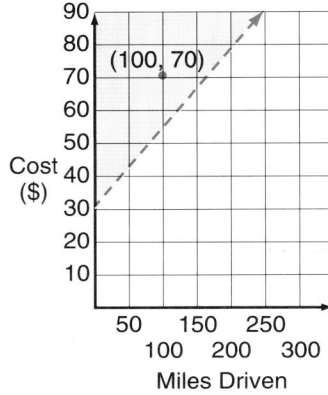

Since the point (100, 70) lies above the boundary, it is in the graph of $r > 0.25d + 30$. For all points in this region, the cost of renting a car is greater than the cost of renting from Reasonable Car Rental. Thus, Reasonable Car Rental has a better rate than Executive Car Rental for a car rental of 100 miles.

Examine   For a rental of 100 miles, Reasonable Car Rental charges $0.25(100) + 30$ or $55. This is a better rate than $70. Thus, the solution seems reasonable.

CHECK FOR UNDERSTANDING

Communicating Mathematics

Study the lesson. Then complete the following.

1. **Explain** how you decide if the boundary of an inequality should be solid or dashed.

2. **Explain** why (0, 2) is a solution of $y \geq -8x + 2$.

3. **Choose** the graph of $y < 3x + 2$.

   a.   b.   c.   d.

4. **Compare** the graphs of $y = |x| + 1$ and $y < |x| + 1$.

State which points, (0, 0), (3, −4), or (−1, 3), satisfy each inequality.

5. $x + 2y < 5$ **6.** $4x + 3y \leq 0$ **7.** $5x - y \geq 6$

Graph each inequality.

8. $y > 2$ **9.** $x - y \geq 0$ **10.** $y > 2x$

11. Graph $y > |2x|$.

12. Graph all the points on the coordinate plane to the right of $x = 4$. Write an inequality to describe these points.

EXERCISES

Graph each inequality.

13. $x + y > -5$ **14.** $y + 1 < 4$ **15.** $y > 6x - 2$

16. $x - 5 \leq y$ **17.** $y \geq -4x + 3$ **18.** $y - 2 < 3x$

19. $y > \frac{1}{3}x + 5$ **20.** $y \geq \frac{1}{2}x - 5$ **21.** $3 \geq x - 3y$

22. $y \leq |x|$ **23.** $y + |x| < 3$ **24.** $y > |4x|$

25. Graph all the points on the coordinate plane to the left of $x = -2$. Write an inequality to describe these points.

26. Graph all the points in the first quadrant bounded by the two axes and the line $x + 2y = 5$.

27. Graph all second quadrant points bounded by the lines $x = -3$, $x = -6$, and $y = 4$.

28. Graph all points in the fourth quadrant bounded by the two axes and the lines $3x - y = 4$ and $x - y = 5$.

Graph each inequality.

29. $|x| \leq |y|$ **30.** $|x| - |y| = 1$

31. $|x| + |y| \geq 1$ **32.** $|x + y| > 1$

33. Describe the graph of $|y| < x$.

34. Business The No-Drip Sponge Company must produce a certain number of sponges each day to keep the assembly-line staff busy. If production falls, then layoffs may be necessary. The equation that describes this relationship is $s > 50e + 25$, where e is the number of employees and s is the number of sponges. Graph this inequality.

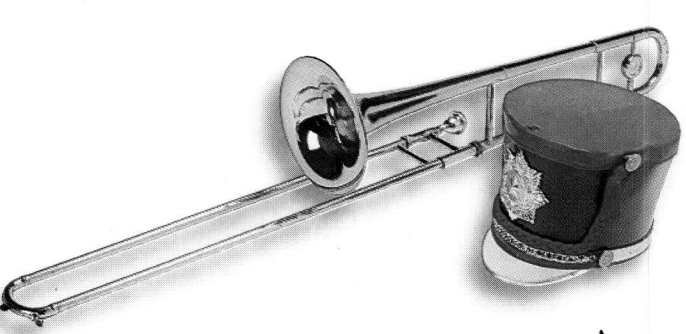

35. Drama Tickets for the Kingwood High School Drama Club's production of *The Music Man* cost $5 for adults and $4 for students. In order to cover expenses, at least $2500 worth of tickets must be sold.

     a. Write an inequality that describes this situation.

     b. Graph the inequality.

     c. If 175 adult and 435 student tickets are sold, will the Drama Club cover its expenses?

36. Manufacturing   The Lone Star Auto Company has a daily production quota of $100,000 worth of cars. They produce two types of cars. Their compact model C is valued at $10,000 and their luxury car L is valued at $20,000. The equation $10,000C + 20,000L = 100,000$ describes the production quota.

   a. Make a graph of the quota equation.

   b. On December 17, the factory produced 5 compacts and 2 luxury cars. Was the company above, below, or on target with their quota? Write the equation or inequality that contains this point.

   c. On February 24, the factory produced 6 compacts and 2 luxury cars. Write the equation or inequality that contains this point.

   d. On March 5, the factory produced 9 compacts and 2 luxury cars. Write the equation or inequality that contains this point.

37. Education   Your school counselor advises you that to be admitted to the college of your choice, you need to score at least 1200 for the combined verbal and mathematics parts of the SAT.

   a. Write an inequality that describes this situation.

   b. Graph the inequality.

   c. What restrictions, if any, are necessary to impose on the domain and range?

Mixed Review

38. Graph $y = [x] - 4$.   (Lesson 2–6)

39. Statistics   The table below shows the years of experience for eight encyclopedia sales representatives and the amount of sales during a given period of time.   (Lesson 2–5)

Sales	$9000	$6000	$4000	$3000	$3000	$5000	$8000	$2000
Years	6	5	3	1	4	3	6	2

   a. Draw a scatter plot to show how the years of experience and the amount of sales are related.

   b. Write a prediction equation from these data.

   c. Predict the amount of sales for a representative with 8 years of experience.

40. Write an equation in slope-intercept form for the line with slope $\frac{2}{3}$ that passes through the point at $(6, -5)$.   (Lesson 2–4)

41. Graph a line that passes through the point at $(-2, 7)$ and is perpendicular to the graph of $x - 2y = 3$.   (Lesson 2–3)

42. Find the x-intercept and y-intercept of the graph of $4x + 5y = 10$.   (Lesson 2–2)

43. Geometry   The points at $(-2, 2)$, $(7, 1)$, and $(-2, 1)$ form three of the vertices of a rectangle. Find the coordinates of the fourth vertex. (Lesson 2–1)

VOCABULARY

After completing this chapter, you should be able to define each term, property, or phrase and give an example or two of each.

Algebra

absolute value function (p. 104)

Cartesian coordinate plane (p. 64)

coordinate system (p. 64)

constant function (pp. 74, 103)

continuous function (p. 67)

dependent variable (p. 73)

direct variation (p. 103)

domain (p. 65)

family of graphs (p. 83)

function (p. 65)

greatest integer function (p. 104)

identity function (p. 103)

independent variable (p. 73)

linear equation (p. 73)

linear function (p. 74)

mapping (p. 65)

ordered pairs (p. 64)

origin (p. 64)

parent graph (p. 83)

point-slope form (p. 89)

quadrants (p. 64)

range (p. 65)

relation (p. 65)

slope (p. 80)

slope-intercept form (p. 88)

standard form (p. 73)

step functions (p. 103)

vertical line test (p. 65)

x-axis (p. 64)

x-intercept (p. 75)

y-axis (p. 64)

y-intercept (p. 75)

Discrete Mathematics

discrete function (p. 65)

Geometry

parallel lines (p. 83)

perpendicular lines (p. 83)

Problem Solving

look for a pattern (p. 80)

Statistics

best-fit line (p. 95)

prediction equation (p. 95)

regression line (p. 101)

scatter plot (p. 95)

UNDERSTANDING AND USING THE VOCABULARY

Choose the term from the list above that best completes each statement or phrase.

1. The equation of the line suggested by the points on a scatter plot is a _____.

2. An _____ is a linear function described by $f(x) = x$.

3. A _____ is a function of the form $f(x) = b$, and the slope of the function is zero.

4. A linear function described by $f(x) = mx$, where $m \neq 0$, is known as a _____.

5. An _____ is closely related to a linear function, except that the graph of this function always forms a V-shape and is described by $f(x) = |x|$.

6. The _____ of a linear function is $Ax + By = C$, where A, B, and C are real numbers and A and B are not both zero.

7. Two or more lines in the same plane having the same slope are _____.

8. The _____ is the set of all x-coordinates of the ordered pairs of a relation.

9. Two lines with slopes that are negative reciprocals of each other form right angles and therefore are _____.

10. The set of all y-coordinates of the ordered pairs of a relation are known as the _____.

11. The _____ of a line may be described in two ways, the first as the vertical change divided by the horizontal change and the second as $(y_2 - y_1)$ divided by $(x_2 - x_1)$.

STUDY GUIDE AND ASSESSMENT

SKILLS AND CONCEPTS

OBJECTIVES AND EXAMPLES

Upon completing this chapter, you should be able to:

• graph a relation, state its domain and range, and determine if it is a function   (Lesson 2–1)

$\{(-5, 6), (-3, -4), (-1, -6), (2, 6)\}$

The domain is $\{-5, -3, -1, 2\}$.

The range is $\{6, -4, -6\}$.

It is a function because each element of the domain is paired with exactly one element of the range.

• find values of functions for given elements of the domain   (Lesson 2–1)

If $f(x) = x^3 - 5$, find $f(-3)$.

$f(-3) = (-3)^3 - 5$
$\qquad = -27 - 5 \text{ or } -32$

• identify equations that are linear and graph them
(Lesson 2–2)

$4x + y = -3$ is a linear equation.

$4x + y^2 = -3$ is *not* a linear equation.

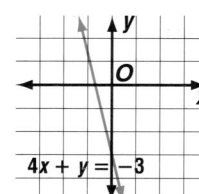

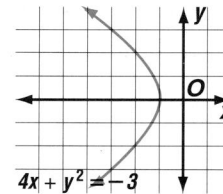

• write linear equations in standard form
(Lesson 2–2)

Write $2x - 6 = y + 8$ in standard form where A, B, and C are integers whose greatest common factor is 1.

$\qquad 2x - 6 = y + 8$

$2x - y - 6 = 8$ *Subtract y from each side.*

$\qquad 2x - y = 14$ *Add 6 to each side.*

REVIEW EXERCISES

Use these exercises to review and prepare for the chapter test.

State the domain and range of each relation. Then graph the relation and identify whether it is a function or not.

12. $\{(4, -7), (-4, -7), (-4, 7), (4, 7)\}$

13. $\{(9, 1), (7, 2), (5, 3), (3, 4), (1, 5)\}$

14. $3x - 4y = -7$

15. $-2y + 4 = 1 - 5x$

Find each value if $f(x) = 4x^2 + 5x - 9$.

16. $f(6)$

17. $f(-2)$

18. $f(3y)$

19. $f(-2v)$

State whether each equation is linear. Write *yes* or *no*. If it is a linear equation, graph it.

20. $3x^2 - y = 6$

21. $2x + y = 11$

22. $y = -7$

23. $x^2 + y^2 = 25$

Write each equation in standard form where A, B, and C are integers whose greatest common factor is 1.

24. $y = 7x + 15$

25. $0.5x = -0.2y - 0.4$

26. $\frac{2}{3}x - \frac{3}{4}y = 6$

27. $-\frac{1}{5}y = x + 4$

28. $6x = -12y + 48$

29. $y - x = -9$

OBJECTIVES AND EXAMPLES

• determine the slope of a line   (Lesson 2–3)

Determine the slope of the line that passes through $(-5, 3)$ and $(7, 9)$.

$$m = \frac{9 - 3}{7 - (-5)}$$

$$= \frac{6}{12} \text{ or } \frac{1}{2}$$

• write an equation of a line in slope-intercept form given the slope and one or two points (Lesson 2–4)

The slope-intercept form of the line that has a slope of $\frac{2}{3}$ and a y-intercept of 3 is $y = \frac{2}{3}x + 3$.

The standard form of this equation is $-2x + 3y = 9$.

• write an equation of a line that is parallel or perpendicular to the graph of a given equation (Lesson 2–4)

The equation of a line parallel to $y = 2x - 2$ is $y = 2x + 1$.

The equation of a line perpendicular to $y = 2x - 2$ is $y = -\frac{1}{2}x + 1$.

• draw a scatter plot and find a prediction equation   (Lesson 2–5)

Construct a scatter plot for the given data. Then sketch the line that appears to best fit the points and find an equation of the line that best fits the data.

x	−1	0	0.5	1	2	3.5	4
y	3	3.5	1.5	2	0	−3.5	−2

The slope-intercept form of the equation of the best-fit line is $y = -\frac{5}{4}x + \frac{17}{8}$.

REVIEW EXERCISES

Determine the slope of the line that passes through each pair of points.

30. $(-6, -3)$ and $(6, 7)$

31. $(5.5, -5.5)$ and $(11, -7)$

32. $(-3, 24)$ and $(10, -41)$

Write an equation in slope-intercept form that satisfies each condition.

33. slope $= \frac{3}{4}$, passes through $(-6, 9)$

34. slope $= 2$, x-intercept $\frac{3}{2}$

35. passes through $(3, -8)$ and $(-3, 2)$

36. passes through $(0.35, 0.7)$ and $(0.7, 0.35)$

Write an equation in slope-intercept form for each given situation.

37. passes through $(1, 2)$ and is parallel to the graph of $y = -3x + 7$

38. passes through $(-1, 2)$ and is parallel to the graph of $x - 3y = 14$

39. passes through $(3, 2)$ and is perpendicular to the graph of $4x - 3y = 12$

40. passes through $(1, 3)$ and is perpendicular to the graph of $y = -\frac{2}{3}x + \frac{11}{3}$

41. On average, a person that is 70 inches tall (5′10″) weighs about 167 pounds. A person who is 62 inches tall (5′2″) weighs about 125 pounds. Let h represent the height, and let w represent the weight.

   a. Draw a scatter plot to show how height and weight are related.

   b. Find a prediction equation for the data.

   c. Predict the weight of a person who is 77 inches tall.

   d. Predict the height of a person who weighs 155 pounds.

OBJECTIVES AND EXAMPLES

- identify and graph special functions
  (Lesson 2–6)

  $y = [0.3x] + 5$ is a greatest integer function.

  $y = |0.5x + 9| - 21$ is an absolute value function.

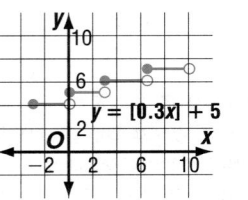

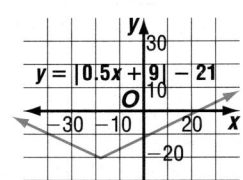

REVIEW EXERCISES

Identify each function as C for constant, D for direct variation, A for absolute value, or G for greatest integer function. Then graph each function.

42. $f(x) = |x| + 4$
43. $f(x) = [x] - 2$
44. $g(x) = \frac{2}{3}$
45. $h(x) = 3x$
46. $h(x) = [2x + 1]$
47. $g(x) = |x - 1| + 7$

- draw graphs of inequalities in two variables
  (Lesson 2–7)

  Replace the inequality symbol with an equal sign to find the equation for the boundary line. Graph the boundary line, using a solid line for ≤ or ≥, or a dashed line for < or >. Test a point to determine which region to shade.

Graph each inequality.

48. $y \leq 3x - 5$
49. $x > y - 1$
50. $y \geq |x| + 2$
51. $y + 0.5x < 4$

APPLICATIONS AND PROBLEM SOLVING

52. **Employment**   Ley works at a clothing store for men. He earns $7.00 an hour plus 50¢ for every item over 25 items that he sells. He works 40 hours a week. How much money will he make if he sells x items?   (Lesson 2–6)

53. **Consumer Awareness**   Monique's Fine Fashions allows its customers to charge their purchases on a delayed payment plan. When the customer receives a bill from Monique's,

there is a $5 charge for using the delayed payment plan, plus 2% interest on the purchase. Calculate the bill for a purchase of $110.   (Lesson 2–2)

54. **Geometry**   What are the equations of the five lines that make up the sides of the star shown below?   (Lesson 2–4)

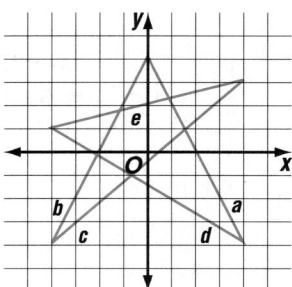

A practice test for Chapter 2 is provided on page 913.

ALTERNATIVE ASSESSMENT

COOPERATIVE LEARNING PROJECT

Predicting the Future   The better we understand nature, the more accurately we are able to predict its behavior.

Ancient peoples had no understanding of the movements of the sun, the moon, or the stars. Predicting such events with accuracy was impossible. As a result, eclipses were believed to be a sign that the gods were angry. When astronomers discovered the cause of an eclipse and calculated the orbits of the sun and the other planets in our solar system, they were able to predict the timing of an eclipse with great accuracy.

In an effort to make predictions about natural occurrences that are not fully understood, scientists search for predictors or events which, for no apparent reason, seem to correlate with the occurrences. For example, geologist Ruth Simon discovered that cockroaches become more active a few hours preceding a major earthquake. Political analysts have noted that the candidate with the longer name wins nearly 75% of the presidential elections.

In this project, you will attempt to find a predictor that can be used to predict the behavior of some natural phenomenon. You will analyze several possible predictors to find the one with the best promise. Then you will use that one to make a prediction about the future.

Follow these steps to carry out your project.

- Choose a phenomenon that interests your group, whose behavior cannot be predicted with complete accuracy. Be sure there is plenty of data available on the phenomenon you choose and that the phenomenon is not completely arbitrary in its behavior.

- Collect data on the phenomenon. Study it, looking for patterns, and discuss your findings with the rest of your group.

- Brainstorm with your fellow group members to find at least three possible predictors of the phenomenon you chose.

- Collect data on your predictors.

- Analyze your predictors. You may wish to draw scatter plots and find a prediction equation. You can use a graphing calculator to help you draw lines of regression. After you complete your analysis, choose the predictor that shows the best promise.

- Write a report summarizing your work as a group. Explain how you chose your predictors and how they were analyzed. Describe the results of your prediction. List any discrepancies found between predicted and actual results.

THINKING CRITICALLY

You have learned several ways to find the slope of a line: by using the vertical change divided by the horizontal change, by using the formula $m = \frac{y_2 - y_1}{x_2 - x_1}$, and by simply viewing the graph of a line. Is one way better than another? Write a one-page paper on the method you prefer, citing examples, and explaining why you prefer it.

PORTFOLIO

Select one of the assignments from this chapter that you found especially challenging and place it in your portfolio.

SELF EVALUATION

In this chapter, you examined many types of graphs. They show how quantities relate to each other. Some graphs are used to show change. Jonathan Swift said, "There is nothing in this world constant, but inconstancy."

Assess yourself. Do you like change? How do you react when your plans get changed at the last minute? At school, home or the workplace, having the flexibility to change and adapt is necessary to solve problems and to work together.

COLLEGE ENTRANCE EXAM PRACTICE

CHAPTERS 1–2

There are eight multiple-choice questions in this section. After working each problem, write the letter of the correct answer on your paper.

1. Determine which graph represents a function.

A.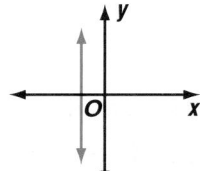

B.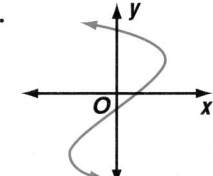

C.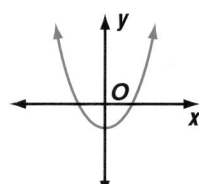

D.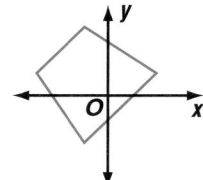

2. Choose the equation that is equivalent to $3(a + 2b) - c = 0$.

A. $a + 2b = c$

B. $b = \frac{c - 3a}{6}$

C. $a = \frac{2b - c}{3}$

D. $3a + 6b - 3c = 0$

3. What is the mean age of those in attendance at the Cruz family reunion if their ages are represented in the stem-and-leaf plot below?

Stem	Leaf
0	3 7
1	2 2 5 8 8
2	3 5 5 6 7
3	6
4	0 3 3 3
5	0 3
6	2 6 8 *4	3 = 43 years*

A. 32.5 years

B. 26.5 years

C. 43 years

D. 102 years

4. Evaluate the expression $\left(\frac{a}{x} + b\right)^2 - gh$ if $a = 9$, $b = 2$, $g = 4$, $h = 6$, and $x = 3$.

A. 19

B. −11

C. −3

D. 1

5. Determine which description of the graph of the equation $2x - 4y = 8$ is true.

A. It is parallel to the graph of the equation $2x - y = 8$.

B. Its slope is 2.

C. The point $(6, 1)$ lies on it.

D. It is perpendicular to the equation $4x - 8y = 16$.

6. Choose the statement that is false for all real numbers a, b, and c.

A. If $a = b$, then $a + c = b + c$ and $a - c = b - c$.

B. $a \cdot \frac{1}{a} = 1 \cdot a$

C. If $a = b$ and $b = c$, then $a = c$.

D. $(a + b) + c = a + (b + c)$

7. Choose the inequality that describes the graph below.

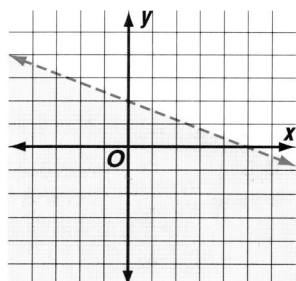

A. $2x + 5y < 10$

B. $y > 2 - \frac{2}{5}x$

C. $y \geq -2x - 5$

D. $5x + 2y \leq 10$

8. Choose the type of function that represents the following relationship. For every yard Melissa mows, she earns $7.00.

A. absolute value

B. step

C. greatest integer

D. direct variation

SECTION TWO: SHORT ANSWER

This section contains seven questions for which you will provide short answers. Write your answer on your paper.

9. Solve $8|2b - 3| = 64$.

10. The width of a rectangular rug is four feet more than one third its length. The perimeter is 64 feet. What are the length and width of the rug?

11. Intercity Car Rental offers a mid-size car that costs 20¢ per mile plus an initial fee of $20. To rent the same car from Big Wheels Car Rental, each mile costs 23¢ with an initial fee of $10. Barry plans to drive about 875 miles. Which company provides the cheaper offer, and how much will Barry save with this offer?

12. Solve $\left\{\frac{x+8}{4} - 1 > \frac{x}{3} \text{ and } 7 < 2x - 11\right\}$. Then graph the solution set.

13. Determine the slope of a line perpendicular to the line that passes through $(-3, -1)$ and $(4, -7)$.

14. The marching band at Scoring High School is raising money for new uniforms by selling candy. They need to earn at least $1025 to receive a 10% discount on the price of uniforms. During the first week of the sale, the amounts turned in by each music class were $125, $86, $98, $72, $63, and $135. What is the minimum amount the band must raise to receive the 10% discount?

15. Describe the steps you would take to graph the equation $3y - 2x = 15$.

SECTION THREE: COMPARISON

This section contains five comparison problems that involve comparing two quantities, one in column A and one in column B. In certain questions, information related to one or both quantities is centered above them. All variables used represent real numbers.

Compare quantities A and B below.

- Write A if quantity A is greater.
- Write B if quantity B is greater.
- Write C if the two quantities are equal.
- Write D if there is not enough information to determine the relationship.

Column A	Column B
16. the slope of a line perpendicular to the line that passes through $(-3, 2)$ and $(5, -1)$	$\frac{8}{3}$

$$g = 3, k = -2, m = 5, p = 4$$

17. $g^2 + (k - p) \div g$	$\left(\frac{p}{k} + m\right)^2 \div g + k$

18. $y - 3 \geq 4x$	$3x + 5 < 10$

19. the median of $\{41, 33, 12, 38, 27, 19\}$	the mean of $\{14, 23, 35, 18, 32, 27, 39\}$

$$a > b, b = c$$

20. $(b + a) - d$	$-d + a + c$

Solving Systems of Linear Equations and Inequalities

Objectives

In this chapter, you will:

- solve systems of equations in two or three variables,
- solve systems of inequalities,
- use linear programming to find maximum and minimum values of functions, and
- solve problems by solving a simpler problem.

What Will You Do for a Living?

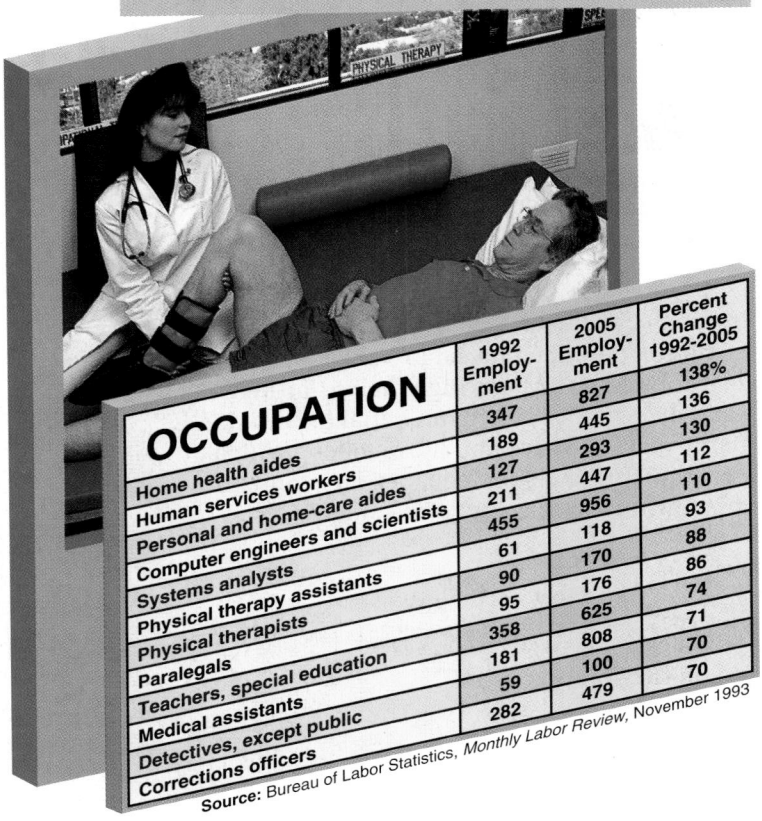

OCCUPATION	1992 Employ-ment	2005 Employ-ment	Percent Change 1992-2005
Home health aides	347	827	138%
Human services workers	189	445	136
Personal and home-care aides	127	293	130
Computer engineers and scientists	211	447	112
Systems analysts	455	956	110
Physical therapy assistants	61	118	93
Physical therapists	90	170	88
Paralegals	95	176	86
Teachers, special education	358	625	74
Medical assistants	181	808	71
Detectives, except public	59	100	70
Corrections officers	282	479	70

Source: Bureau of Labor Statistics, *Monthly Labor Review*, November 1993

The 21st century will truly be the age of science and technology, the age of information. How will the age of information affect the job market of the future? The table above lists the occupations that will be most in demand in 2005.

TIME Line

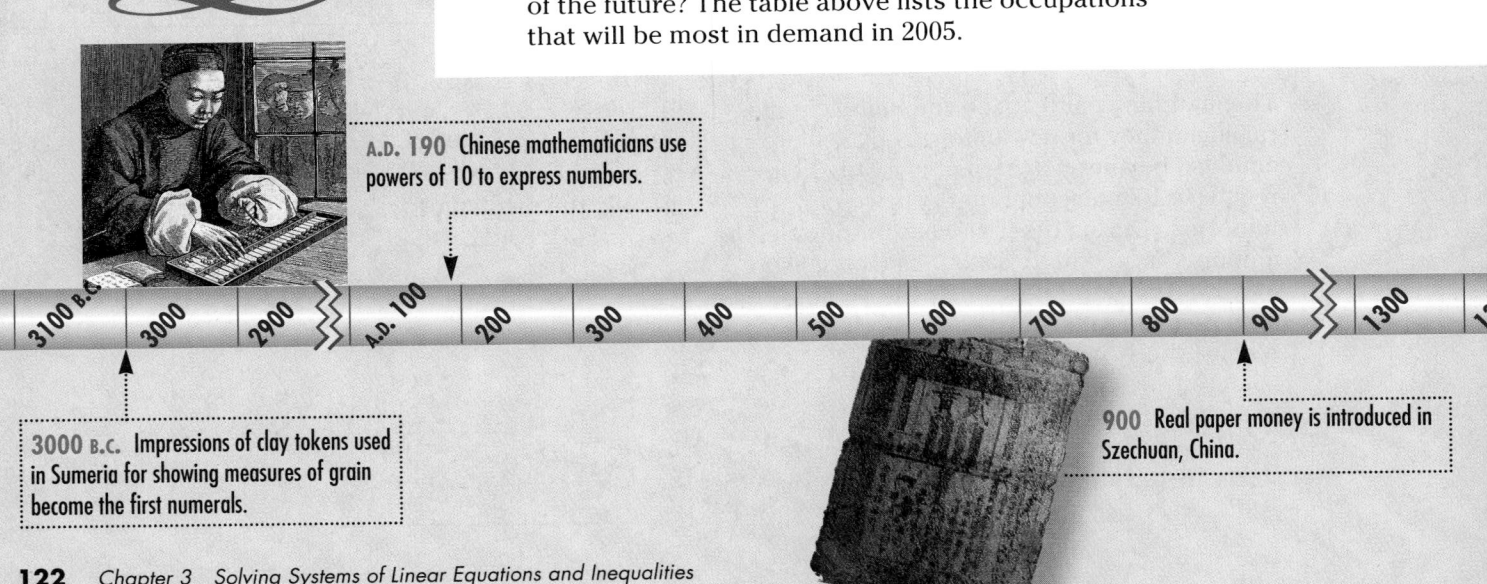

A.D. 190 Chinese mathematicians use powers of 10 to express numbers.

3100 B.C. 3000 2900 A.D. 100 200 300 400 500 600 700 800 900 1300 13

3000 B.C. Impressions of clay tokens used in Sumeria for showing measures of grain become the first numerals.

900 Real paper money is introduced in Szechuan, China.

Choose five stocks on the New York Stock Exchange that sell for about the same amount of money. Suppose you purchase 100 shares of each of these stocks.

- Create a graph to track your stock purchases showing their ups and downs over the course of a few weeks.

- Use your graph to compare the values of the stocks. Are there times when two or more of the stocks are worth the same amount of money? Which stock was worth the most at the beginning of the project? Which stock was worth the most at the end of the project?

- Determine the percent of gain or loss on your original capital investment.

Matt Seto, a 17-year-old from Troy, Michigan, is a teen financial wizard who earns respect as he outperforms professional money managers. In 1994, Matt invested in 300 of Best Buy's shares at $21 each and then unloaded all of them for $41 a share. The transaction brought in a $6000 profit, which grew to an incredible $40,700 by the end of 1994 because of his additional investments. Compare that with his $27,900 balance at the beginning of 1994. "It was the best year of my life, definitely," Matt says. "It was the year when everyone started to respect me."

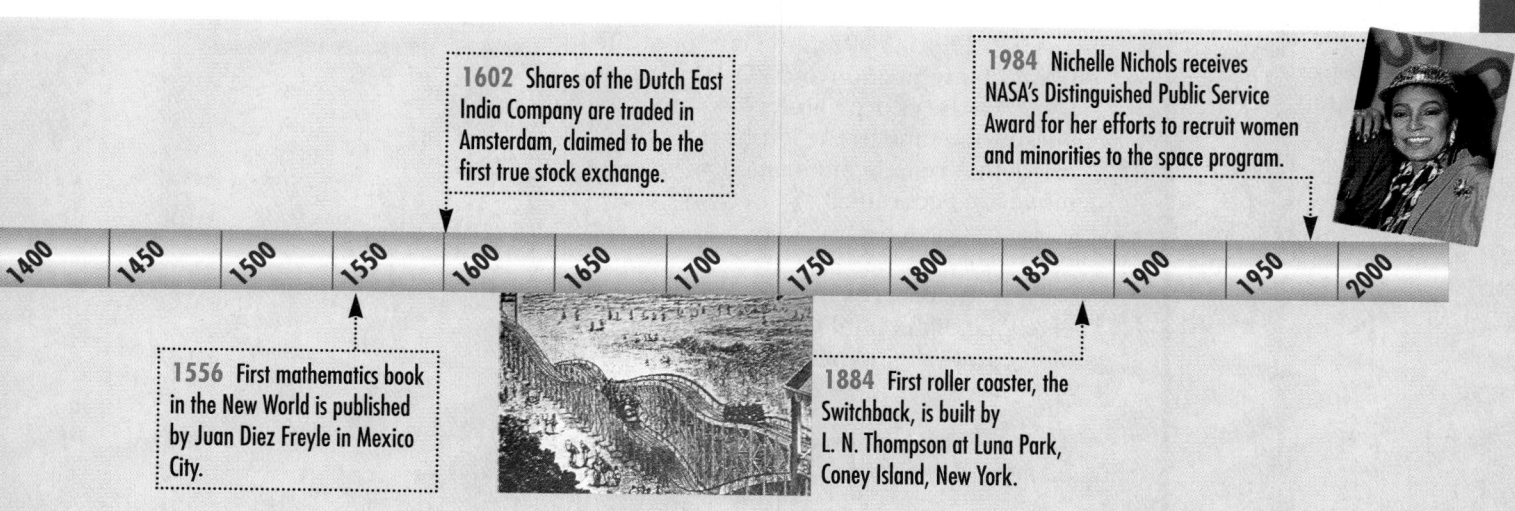

1602 Shares of the Dutch East India Company are traded in Amsterdam, claimed to be the first true stock exchange.

1984 Nichelle Nichols receives NASA's Distinguished Public Service Award for her efforts to recruit women and minorities to the space program.

1400　1450　1500　1550　1600　1650　1700　1750　1800　1850　1900　1950　2000

1556 First mathematics book in the New World is published by Juan Diez Freyle in Mexico City.

1884 First roller coaster, the Switchback, is built by L. N. Thompson at Luna Park, Coney Island, New York.

3–1A Graphing Technology
Systems of Equations

A Preview of Lesson 3–1

You can use a graphing calculator to graph systems of equations because several equations can be graphed on the screen at the same time. If a system of equations has a solution, it is located where the graphs intersect. The coordinates of the intersection point, (x, y), can be determined by using the TRACE function.

Example **1** **Solve the system of equations to the nearest hundredth.**

$y = -x + 4.35$

$y = 3x - 3.25$

Begin by graphing both equations in the standard viewing window.

Enter: [Y=] [(−)] [X,T,θ,n] [+]

4.35 [ENTER] 3 [X,T,θ,n]

[−] 3.25 [ZOOM] 6

Use the TRACE function to determine the coordinates of the point of intersection. Press [TRACE] and use the arrow keys to place the cursor as close as possible to the intersection point. Observe the coordinates at the bottom of the screen. Then ZOOM IN to determine the coordinates of the intersection point with greater accuracy.

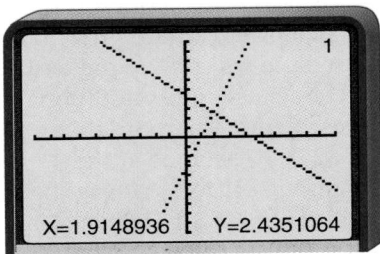

Enter: [ZOOM] 2 [ENTER]

Use TRACE to position the cursor at the point of intersection and ZOOM IN again. Observe the coordinates. Now move the cursor one time to the left or right. Any digits that remain unchanged as you move are accurate.

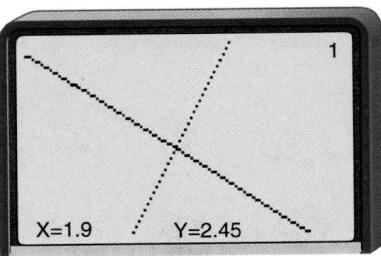

Repeat the process of zooming and checking digits until you have the number of accurate digits that you desire.

The solution is approximately (1.90, 2.45). *Verify this result.*

The graphing calculator also has an INTERSECT feature that automatically finds the coordinates of the point of intersection.

Enter: 2nd CALC 5 ENTER ENTER ENTER

You can use the TABLE function on the TI-83 to solve systems of equations.

Example **2** **Solve the system of equations to the nearest hundredth.**
$$x + y = 4$$
$$2x + 3y = 9$$

First solve each equation for y.
$$y = -x + 4$$
$$y = -\frac{2}{3}x + 3$$

It is not necessary to graph the equations first.

Graph both equations in the standard viewing window.

Enter: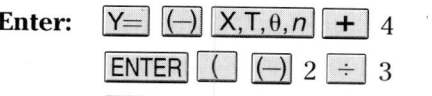

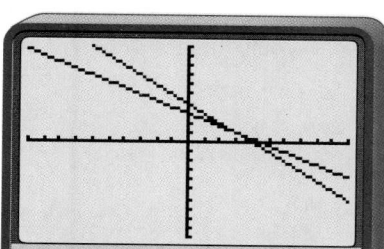

Then press 2nd TABLE . On the screen, you will see the coordinates of points on both lines. Use the arrow keys to scroll up or down and watch the trend of the coordinates. When you find a row at which Y1 = Y2, you have found the solution.

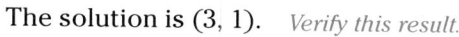

The solution is (3, 1).   *Verify this result.*

EXERCISES

Use a graphing calculator to solve each system of equations to the nearest hundredth.

1. $y = 3x - 2$
$y = -0.5x + 5$

2. $y = \frac{1}{4}x + 3$
$y = -2x + 21$

3. $2x + 3y = 8$
$3x - 8y = -13$

4. $y = -x + 7$
$8 = 2x - y$

5. $y = 0.125x - 3.005$
$y = -2.58$

6. $\frac{1}{2}x + \frac{1}{3}y = 1$
$3x + 2y = 6$

7. $3.14x + 2.03y = 1.99$
$9.32x - 3.77y = -4.21$

8. $12y = 4x - 16$
$9y - 3x = 3$

Graphing Systems of Equations

What YOU'LL LEARN

- To solve systems of equations by graphing.

Why IT'S IMPORTANT

You can graph systems of equations to solve problems involving business and nutrition.

APPLICATION

Business

Terri Shackelford is the chief executive officer (CEO) of a commuter airline company. One day, the business section of a newspaper reported that a competing airline had purchased eight new aircraft. They purchased two different sizes of planes for a total of $13.5 million. Ms. Shackelford calls the manufacturer of the planes and learns that the smaller planes cost $1.5 million and the larger planes cost $2 million. She would like to know how many of each size plane her competitor purchased.

She could find the answer by guess and check, but she did not become a CEO by guessing. She decides that using equations might be more straightforward. Let a represent the number of smaller planes purchased, and let b represent the number of larger planes purchased. We can then write the following equations.

$$a + b = 8 \qquad \text{\textit{A total of 8 planes was purchased.}}$$

$$1.5a + 2b = 13.5 \qquad \text{\textit{The total cost of the planes was \$13.5 million.}}$$

By graphing these two equations, we can find the number of each type of plane. Each point of each line has coordinates that satisfy the equation of that line. The one point that satisfies both equations is the point at which the two lines intersect, namely, (5, 3).

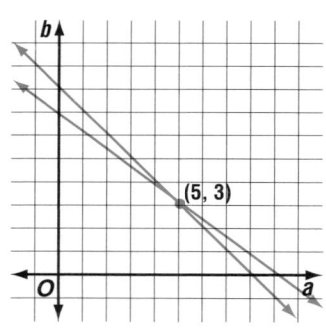

Therefore, the competing company purchased 5 of the smaller planes and 3 of the larger planes.

Together the equations $a + b = 8$ and $1.5a + 2b = 13.5$ are called a **system of equations.** The *solution* of this system is (5, 3). To check this solution, replace a with 5 and b with 3 in each equation.

$$
\begin{array}{ll}
a + b = 8 & 1.5a + 2b = 13.5 \\
5 + 3 \stackrel{?}{=} 8 & 1.5(5) + 2(3) \stackrel{?}{=} 13.5 \\
8 = 8 \ \checkmark & 7.5 + 6 \stackrel{?}{=} 13.5 \\
& 13.5 = 13.5 \ \checkmark
\end{array}
$$

The solution (5, 3) is correct.

Use a graphing calculator to graph each system of equations. You must first solve each equation for y.

1. $\frac{1}{2}x + \frac{1}{3}y = 2$ 2. $2x + 3y = 5$ 3. $y = \frac{x}{2}$

     $x - y = -1$ $-6x - 9y = -15$ $2y = x + 4$

a. Describe the graphs of each system of equations.

b. How are the graphs of the systems of equations similar?

c. How are the graphs of the systems of equations different?

When two lines have different slopes, the graphs of the equations are intersecting lines.

Example ① **Solve the system of equations by graphing.**
$x + y = 6$
$3x - 4y = 4$

The slope-intercept form of $x + y = 6$ is $y = -x + 6$.

The slope-intercept form of $3x - 4y = 4$ is $y = \frac{3}{4}x - 1$.

In this case, the lines have different slopes and intersect at $(4, 2)$. The solution of the system is $(4, 2)$.

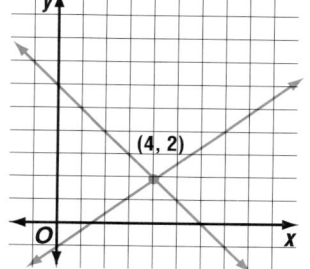

$(4, 2)$

Check: $x + y = 6$ $3x - 4y = 4$
         $4 + 2 \overset{?}{=} 6$ $3(4) - 4(2) \overset{?}{=} 4$
            $6 = 6$ ✓ $12 - 8 \overset{?}{=} 4$
                             $4 = 4$ ✓

A system of equations that has at least one solution is called a **consistent system** of equations. If a system has exactly one solution, it is an **independent system**. So, the system in Example 1 is consistent and independent.

Example ② **Solve the system of equations by graphing.**
$12x - 9y = 27$
$8x - 6y = 18$

The slope-intercept form of $12x - 9y = 27$ is $y = \frac{4}{3}x - 3$.

The slope-intercept form of $8x - 6y = 18$ is $y = \frac{4}{3}x - 3$.

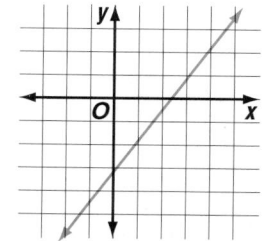

Since the equations are equivalent, their graphs are the same line. Any ordered pair representing a point on that line will satisfy both equations. So, there are *infinitely many* solutions to this system.

A system is **dependent** if it has an infinite number of solutions. So, the system in Example 2 is consistent and dependent.

Example ❸ **Solve the system of equations by graphing.**
$$4x + 6y = 18$$
$$6x + 9y = 18$$

The slope-intercept form of $4x + 6y = 18$ is
$$y = -\frac{2}{3}x + 3.$$

The slope-intercept form of $6x + 9y = 18$ is
$$y = -\frac{2}{3}x + 2.$$

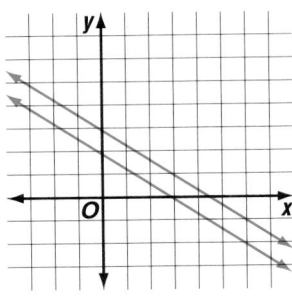

The empty set is also called the null set. The empty or null set can be represented as ∅ or {}.

The lines have the same slope, but different y-intercepts. Their graphs are parallel lines. Since they never intersect, there are no solutions to this system. *The solution set is the empty set, ∅.*

A system with no solutions, like the one in Example 3, is called an **inconsistent system.**

The chart below summarizes the possibilities for the graphs and solutions of two linear equations in two variables, which are illustrated in Examples 1–3.

Example	Graphs of Equations	Slopes and Intercepts	Name of System of Equations	Number of Solutions
1	lines intersect	different slopes	consistent, independent	one
2	lines coincide	same slope, same intercepts	consistent, dependent	infinite
3	lines parallel	same slope, different intercepts	inconsistent	zero

A business can use equations to represent both its costs and its income. A graph of the resulting system of equations clearly illustrates when the business is making a profit and when it is not.

Example ❹ **Lina Sanchez is starting a business in Vail, Colorado. She plans to make souvenir sweatshirts to sell to skiers. She has initial start-up costs of $900. Each sweatshirt costs $18 to produce, and she plans to sell the sweatshirts for $30. How many sweatshirts must Lina sell before she starts to make a profit?**

APPLICATION
Business

The cost of making x sweatshirts is represented by $y = 900 + 18x$.

The income from x sweatshirts is represented by $y = 30x$.

Graph this system of equations. The graphs of these two equations intersect at (75, 2250). This point is called the *break-even point*. If Lina sells fewer than 75 sweatshirts, she loses money. If she sells more than 75 sweatshirts, she makes a profit.

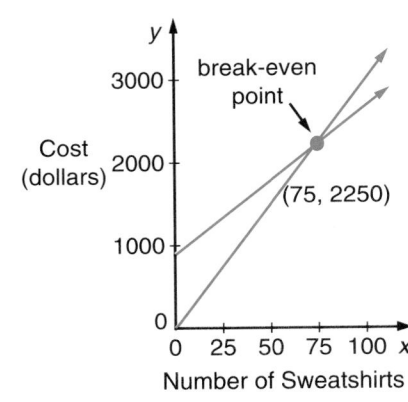

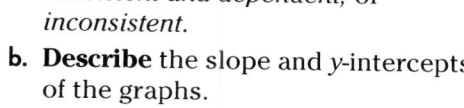
Communicating Mathematics

Study the lesson. Then complete the following.

1. Refer to the graph at the right.
   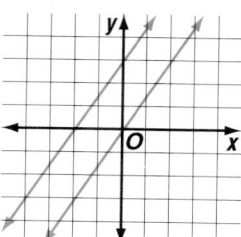
   a. **Explain** whether the graph represents a system of equations that is *consistent and independent, consistent and dependent,* or *inconsistent.*
   b. **Describe** the slope and *y*-intercepts of the graphs.

2. **Sketch** and describe the graphs of $y = -2$ and $x = 4$. What is the solution to this system of linear equations?

3. **Explain** why a system of linear equations cannot have exactly two solutions.

4. **Write** a system of equations that is consistent.

5. Refer to the graph in Example 4.
   a. Determine if Lina Sanchez would have a profit or a loss if she sold 100 sweatshirts. Estimate the amount of the profit or loss.
   b. How many sweatshirts must Lina sell in order to make a profit of $600?

6. **Write** a paragraph explaining how to identify whether the graph of a system of linear equations would be two intersecting lines, two distinct parallel lines, or two coincident lines.

Guided Practice

State the number of solutions to each system of equations. State whether the system is *consistent and independent, consistent and dependent,* or *inconsistent.* If the system is consistent and independent, estimate the solution.

7.
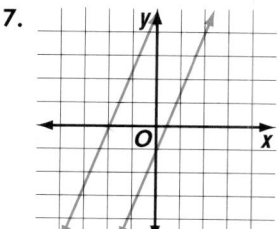

8.
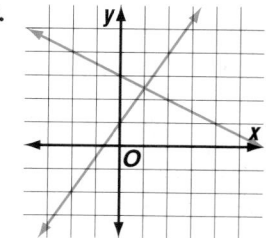

Graph each system of equations and state its solution. Also, state whether the system is *consistent and independent, consistent and dependent,* or *inconsistent.*

9. $5x - y = 3$
   $y = 5x - 3$

10. $2x - 3y = 7$
    $2x + 3y = 7$

11. $x + y = 6$
    $3x + 3y = 3$

12. $\frac{1}{2}x - y = 0$
    $\frac{1}{4}x + \frac{1}{2}y = -2$

13. Which graph illustrates the solution to the system of equations $x + y = 4$ and $y = x - 2$?

a.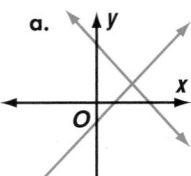

b.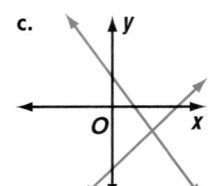

c.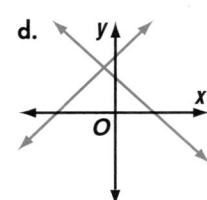

d.

EXERCISES

Practice

State the number of solutions to each system of equations. State whether the system is *consistent and independent, consistent and dependent,* or *inconsistent.* If the system is consistent and independent, estimate the solution.

14.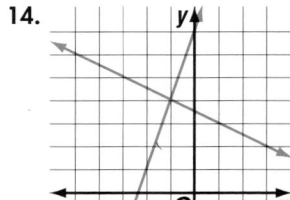

15.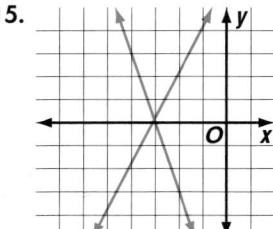

16.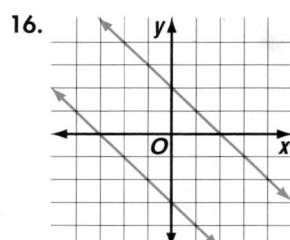

17.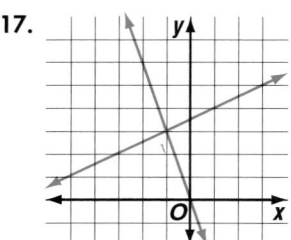

Graph each system of equations and state its solution. Also, state whether the system is *consistent and independent, consistent and dependent,* or *inconsistent.*

18. $2x + 3y = 12$
$2x - y = 4$

19. $3x - y = 8$
$x - y = 8$

20. $4x - 6y = 5$
$2x - 3y = 5$

21. $x + 2y = 6$
$2x + y = 9$

22. $x + 1 = y$
$2x - 2y = 8$

23. $2x + 4y = 8$
$x + 2y = 4$

24. $3x - 8y = 4$
$6x - 42 = 16y$

25. $3x + 6 = 7y$
$x + 2y = 11$

26. $\frac{3}{4}x + \frac{1}{6}y = \frac{2}{3}$
$9x + 2y = 8$

27. $\frac{2}{3}x + y = -3$
$y - \frac{1}{3}x = 6$

28. $\frac{4}{3}x + \frac{1}{5}y = 3$
$\frac{2}{3}x - \frac{3}{5}y = 5$

29. $9x + 8y = 8$
$\frac{3}{4}x + \frac{2}{3}y = 8$

30. $3x + 4y = 7$
$1.5x + 2y = 3.5$

31. $1.2x + 2.5y = 4$
$0.8x - 1.5y = -10$

32. $5x - 7y = 70$
$-10x + 14y = 120$

Find values of _m_ and _n_ that satisfy the condition given for each system.

33. $5x - 3y = 8$
    $mx + ny = 4$
    consistent and dependent

34. $4x - 5y = 10$
    $mx - y = n$
    consistent and independent

35. $3x + 4y = 8$
    $mx + 2y = n$
    inconsistent

36. $2x + 7y = 5$
    $x + my = n$
    consistent and dependent

37. **Geometry**   The sides of an angle are parts of two lines whose equations are $y = -\frac{3}{2}x - 6$ and $y = \frac{2}{3}x + 7$. Find the coordinates of the vertex of the angle.

38. **Geometry**   The length of the base of an isosceles triangle is 2 centimeters shorter than the length of either of the other sides. If the perimeter of the triangle is 16 centimeters, find the length of each side of the triangle.

Graphing Calculator

Use a graphing calculator to solve each system of equations to the nearest hundredth.

39. $3.6x + 4.8y = -7.2$
    $5.8x - 7.1y = 32.9$

40. $-14x + 18y = 75$
    $9.1x - 11.7y = 36$

41. $3.6x - 2y = 4$
    $-2.7x + y = 3$

42. $7x + 13.5y = 31$
    $9.8x + 18.9y = 43.4$

Critical Thinking

43. State the conditions for which the system below is: (a) consistent and dependent, (b) consistent and independent, and (c) inconsistent.
    $ax + by = c$
    $dx + ey = f$

Applications and Problem Solving

44. **Consumer Awareness**   During the week of February 27 to March 5, 1995, the Kroger and Meijer grocery stores in Columbus, Ohio, had sales on General Mills cereals. At Kroger, all General Mills cereals were $33\frac{1}{3}\%$ off, while at Meijer, the cereals were $1 off. Assume that the regular prices of the cereals were the same at each store.
    a. Graph a system of equations representing the sales at the two stores.
    b. What does the point of intersection represent?
    c. For what regular prices would Kroger's sale be a better deal?
    d. For what regular prices would Meijer's sale be a better deal?

45. **Football**   Mani Peters is the kicker for the Winston College football team. In one game, Mani kicked the ball 9 times for a total of 13 points. If 2 of the 9 kicks were no good, how many field goals (worth 3 points each) and how many points after touchdowns (worth 1 point each) did Mani make?

46. **Consumer Awareness**   The Photo Shop charges $1.60 to develop a roll of film plus 11¢ for each print, while Photos R Us charges $1.20 to develop a roll plus 11¢ a print. Under what conditions is it best to use the Photo Shop and when is it best to use Photos R Us?

47. Nutrition The graph at the right shows how the consumption of butter and margarine has changed over the years.

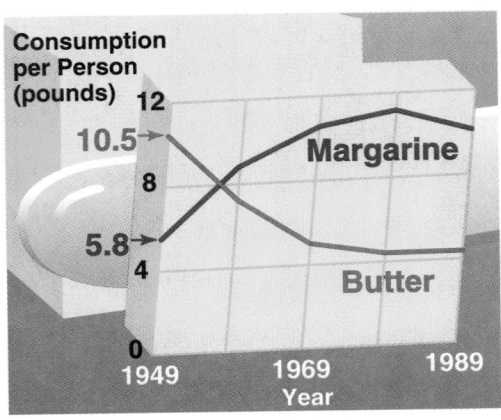

a. During the early 1950s, which was more popular, butter or margarine?

b. Estimate the year that the consumption of butter and margarine was equal.

c. A recent study has shown that the consumption of margarine raises cholesterol levels. Do you think that this information has changed the consumption of margarine? Explain.

d. Do you think that the consumption of butter and margarine will be the same again sometime in the future? Explain.

48. Guess and Check Mr. and Mrs. Leshin have fewer than ten children. The sum of the squares of the number of boys and the number of girls in the family equals 25. How many children do Mr. and Mrs. Leshin have?

Mixed Review

49. Graph $y = |x| + 1$. (Lesson 2–6)

50. Speed Skating In the 1988 Winter Olympics, Bonnie Blair set a world record for women's speed skating by skating approximately 12.79 meters per second in the 500-meter race. Suppose she could maintain that speed. Write an equation that represents how far she could travel in t seconds. What type of function does the equation represent? (Lesson 2–6)

51. Economics A developer surveyed families in a suburb of Raleigh, North Carolina, to determine their monthly household income and the percent of their income spent on housing. The table below shows the data from eight families. (Lesson 2–5)

Monthly Income ($)	870	1430	1920	2460	2850	3240	3790	4510
Percent Spent on Housing	44	39	40	35	43	38	37	33

a. Write a prediction equation for this relationship.

b. Predict the percent of income spent on housing by a family with a monthly income of $3000.

52. Write an equation of the line that passes through $(-16, 7)$ and is perpendicular to the line whose equation is $y = 8x - 4$. (Lesson 2–4)

53. Graph a line that passes through $(4, -2)$ and is perpendicular to a line whose slope is -3. (Lesson 2–3)

54. Find the slope of a line passing through $(5, 4)$ and $(2, 2)$. (Lesson 2–3)

55. Write $x = \frac{1}{3}y + 3$ in standard form. (Lesson 2–2)

56. Evaluate $[25 - (5 - 2)^2 + 5] \div 7$. (Lesson 1–1)

Solving Systems of Equations Algebraically

APPLICATION
Consumerism

Austin is moving to a new condominium. He is sure that he can complete the move in one day, but he does not know how many trips he will need to make between his old and new residences. He plans to rent a 16-foot moving van for a day. When he checks the cost, he finds the following.

Rent-A-Truck: $59.95 a day plus 49¢ per mile
Sam's U-Drive: $81 per day plus 38¢ per mile

Austin needs to determine the mileage at which it is better to rent from Rent-A-Truck and the mileage at which it is better to rent from Sam's U-Drive.

To solve the problem, let m represent the miles driven and let c represent the total cost. Then we can write and graph the following equations.

$$c = 59.95 + 0.49m \quad \textit{Cost for Rent-A-Truck}$$
$$c = 81 + 0.38m \quad \textit{Cost for Sam's U-Drive}$$

It is very difficult to determine an exact solution from the graph. However, we can use the graph to estimate the solution. An estimate of the solution of this system of equations is (190, 150).

The cost will be the same if the mileage is somewhere around 190 miles. For lesser distances, Rent-A-Truck is cheaper, and for greater distances, Sam's U-Drive is cheaper.

For systems of equations like this one, it may be easier to solve the system by using algebraic methods rather than by graphing. Two algebraic methods are the **substitution method** and the **elimination method.** *This system of equations will be solved algebraically in Example 1.*

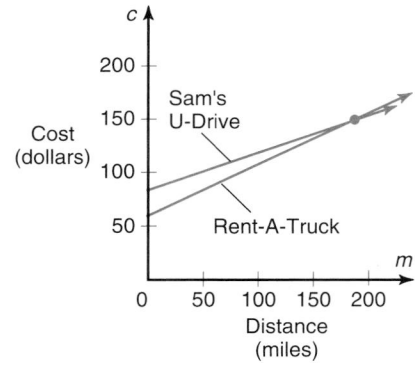

The Modeling Mathematics activity below suggests how to solve a system of equations using the substitution method.

MODELING MATHEMATICS

Systems of Equations

Materials: ▭ equation mat   ⬗ cups and counters

Use modeling to solve the system of equations.

$3x + 2y = 9$
$y = x + 2$

Your Turn

a. Let one cup represent x. If $y = x + 2$, how can you represent y?

b. Represent $3x + 2y = 9$ on the equation mat. On one side of the mat, place three cups to represent $3x$ and two representations of y from step a. On the other side of the mat, place nine positive counters.

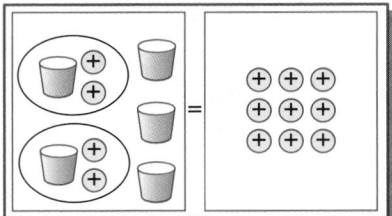

c. Use what you know about equation mats and zero pairs to solve the equation. What value of x solves the system of equations?

d. Use the value of x from step c and the equation $y = x + 2$ to find the value of y.

e. What is the solution of the system of equations?

In the substitution method, one equation is solved for one variable in terms of the other. Then, this expression for the variable is substituted in the other equation.

Example ① Refer to the application at the beginning of the lesson.

Consumerism

a. Determine the mileage for which the cost for Rent-A-Truck and Sam's U-Drive would be the same. Find the cost.

b. Under what conditions should Austin rent from Rent-A-Truck and under what conditions should Austin rent from Sam's U-Drive?

a. The two equations are $c = 59.95 + 0.49m$ and $c = 81 + 0.38m$. To solve this system of equations, use the substitution method.

$$c = 59.95 + 0.49m$$

$$81 + 0.38m = 59.95 + 0.49m \quad \text{Substitute } (81 + 0.38m) \text{ for } c.$$

$$8100 + 38m = 5995 + 49m \quad \text{Multiply by 100. Why?}$$

$$2105 = 11m$$

$$191.36 \approx m \quad \text{Solve for } m.$$

Substitute 191.36 for m in $c = 59.95 + 0.49m$.

$c = 59.95 + 0.49m$
$c \approx 59.95 + 0.49(191.36)$ *Substitute 191.36 for m.*
$c \approx 153.72$ *Solve for c.*

The solution is about (191.36, 153.72). Therefore, at about 191.36 miles, the cost of $153.72 is equal for both Rent-A-Truck and Sam's U-Drive. *Compare this result to the estimate we obtained from the graph.*

b. Austin will need to decide how many trips he thinks he will need to make. If he thinks he will be driving fewer than 191.36 miles, he should rent from Rent-A-Truck. If he thinks he will be driving more than 191.36 miles, he should rent from Sam's U-Drive.

The second algebraic method is the elimination method. To use the elimination method effectively, first compare the coefficients of the variables.

Example ❷ Use the elimination method to solve the system of equations.
$3a - 2b = -3$
$3a + b = 3$

In each equation, the coefficient of a is 3. If one equation is subtracted from the other, the variable a will be eliminated.

$$
\begin{array}{r}
3a - 2b = -3 \\
(-)\ 3a + b = 3 \\
\hline
-3b = -6 \\
b = 2
\end{array}
$$
   Subtract.
   The variable a is eliminated.
   Solve for b.

Now, find a by substituting 2 for b in either original equation.

First Equation   or   **Second Equation**

First Equation	Second Equation
$3a - 2b = -3$	$3a + b = 3$
$3a - 2(2) = -3$ *Substitute 2 for b.*	$3a + 2 = 3$
$3a - 4 = -3$	$3a = 1$
$3a = 1$	$a = \dfrac{1}{3}$
$a = \dfrac{1}{3}$	

The solution is $\left(\dfrac{1}{3}, 2\right)$.

Check by replacing (a, b) with $\left(\dfrac{1}{3}, 2\right)$ in each equation.

First Equation: $3\left(\dfrac{1}{3}\right) - 2(2) \stackrel{?}{=} -3$ *Second Equation:* $3\left(\dfrac{1}{3}\right) + 2 \stackrel{?}{=} 3$

                          $-3 = -3$ ✓ $3 = 3$ ✓

In the following example, adding or subtracting the two equations will not eliminate either variable.

Example ③ **Use the elimination method to solve the system of equations.**
$$3x + 5y = -4$$
$$2x - 3y = 29$$

To use the elimination method we must write equivalent equations containing the same coefficient for either x or y.

Method 1
Multiply the first equation by 3 and the second by 5. Then the variable y can be eliminated by addition.

$3x + 5y = -4$ | Multiply by 3. ➡ | $9x + 15y = -12$

$2x - 3y = 29$ | Multiply by 5. ➡ | $10x - 15y = 145$

Now, add to eliminate y.

$$\begin{array}{rl} 9x + 15y = -12 & \\ (+)\ 10x - 15y = 145 & \textit{Add.} \\ \hline 19x = 133 & \textit{The variable x is eliminated.} \\ x = 7 & \end{array}$$

Find y by substituting 7 for x in $3x + 5y = -4$.

$$\begin{array}{rl} 3(7) + 5y = -4 & \textit{Substitute 7 for x.} \\ 21 + 5y = -4 & \\ 5y = -25 & \\ y = -5 & \textit{Solve for y.} \end{array}$$

The solution is $(7, -5)$.   *Check this solution.*

Method 2
We could also solve the system by eliminating x first. Multiply the first equation by 2 and the second by -3. Then add.

$3x + 5y = -4$ | Multiply by 2. ➡ | $6x + 10y = -8$

$2x - 3y = 29$ | Multiply by -3. ➡ | $\begin{array}{r} (+)\ -6x + 9y = -87 \\ \hline 19y = -95 \\ y = -5 \end{array}$

Finally, solve for x.

$$\begin{array}{rl} 3x + 5(-5) = -4 & \textit{Substitute -5 for y.} \\ 3x - 25 = -4 & \\ 3x = 21 & \\ x = 7 & \textit{Solve for x.} \end{array}$$

The solution is $(7, -5)$.

Communicating Mathematics

Study the lesson. Then complete the following.

1. **Describe** how we could have solved Example 1 by using elimination.

2. **Explain** when you might use substitution rather than elimination.

3. Consider the system of equations $y = 4x + 3$ and $y = 2x - 5$.
   a. Use the symmetric and transitive properties of equality to write a statement about $4x + 3$ and $2x - 5$.
   b. Explain how you could solve the system of equations now.
   c. If you graph the system of equations, where would the lines intersect?

4. **You Decide** When Helen solves the system of equations $4y - 8x = 28$ and $y = 2x - 5$, the result is $-20 = 28$. Helen decides that the graphs of the equations are the same line. Juanita says that the graphs of the equations are parallel lines. Who is correct, and why?

MODELING MATHEMATICS

5. Use cups and counters to model and solve the system of equations.
$$x + 2y = -5$$
$$y = x + 2$$

Guided Practice

Solve each system of equations by using substitution.

6. $x - 2y = 1$
   $x + y = 4$

7. $2p + 3q = 2$
   $p - 3q = -17$

Solve each system of equations by using elimination.

8. $m + n = 6$
   $m - n = 5$

9. $5x + 3y = 0$
   $4x + 5y = 13$

Solve each system of equations. Use either algebraic method.

10. $4a - b = 26$
    $8a - b = 54$

11. $6x + 9y = -45$
    $2x + 3y = -15$

12. $3s - 2t = 10$
    $4s + t = 6$

13. $\frac{1}{4}x + y = \frac{7}{2}$
    $2x - y = 4$

14. **Geometry** If the measure of one angle is five less than two-thirds of the measure of its supplementary angle, find the measure of the angles.

Practice

Solve each system of equations by using substitution.

15. $4x - 3y = 18$
    $3x + y = 7$

16. $x + 3y = 13$
    $-3x + 2y = 27$

17. $3r + 9s = 36$
    $r = 8s - 10$

18. $n = 3m + 7$
    $4m + 9n = 1$

19. $4x + 6y = -9$
    $2x - 10y = -11$

20. $4x + 3y = 4$
    $6x - 6y = -1$

Solve each system of equations by using elimination.

21. $x - 3y = -12$
    $2x + 11y = -7$

22. $3x - 4y = 1$
    $5x + 2y = 45$

23. $4p + 5q = 7$
    $3p - 2q = 34$

24. $5c - 6d = -27$
    $7c + 3d = -15$

25. $\frac{1}{3}x + \frac{1}{2}y = 7$
    $\frac{2}{3}x - y = -2$

26. $\frac{2}{5}x - \frac{1}{2}y = 6$
    $\frac{4}{5}x + \frac{3}{2}y = -8$

Solve each system of equations. Use either substitution or elimination.

27. $3x - 7y = 5$
$x + 3y = -1$

28. $2p - 5q = -53$
$6p + 7q = 39$

29. $2a - b = 8$
$6a - 3b = -9$

30. $3u + 5v = -12$
$2u - 3v = -8$

31. $y = 5x + 37$
$2x - 3y = -20$

32. $5s - t = 2$
$t = 4s + 3$

33. $y = 3x - 27$

$y = \frac{1}{2}x - 7$

34. $2.5m - 1.3n = 0.9$

$10m - 5.2n = 3.6$

35. $\frac{1}{4}x + \frac{3}{5}y = -3$

$\frac{3}{4}x - \frac{2}{5}y = 13$

36. $\frac{3}{5}s - \frac{1}{6}t = 1$

$\frac{1}{5}s + \frac{5}{6}t = 11$

37. $1.5a - 0.2b = -8.3$

$0.4a + 0.4b = -0.4$

38. $4m + 9n = 3$

$8m - 3n = -8$

39. Geometry Find the coordinates of the vertices of the triangle whose sides are contained in the lines whose equations are $5x - 3y = -7$, $x + 2y = 9$, and $3x - 7y = 1$.

40. Geometry Find the coordinates of the vertices of the parallelogram whose sides are contained in the lines whose equations are $2x + y = -12$, $2x - y = -8$, $2x - y - 4 = 0$, and $4x + 2y = 24$.

Programming

41. The graphing calculator program at the right will help you solve systems of equations of the form $Ax + By = C$ and $Dx + Ey = F$.

Run the program to find the solution for each system of equations.

a. $x - 3y = 6$
$2x + 6y = 24$

b. $x + 4y = 2$
$-x + y = -7$

c. $2x - y = 36$
$3x - 0.5y = 26$

d. $2x + y = 45$
$3x - y = 5$

```
PROGRAM:SLVSYSTM
: Disp "ENTER COEFFICIENTS"
: Prompt A,B,C,D,E,F
: If AE-BD=0
: Then
: Goto 1
: End
: (CE-BF)/(AE-BD) → X
: (AF-CD)/(AE-BD) → Y
: Disp "THE SOLUTION IS"
: Disp "X=",X
: Disp "Y=",Y
: Stop
: Lbl 1
: If CE-BF=0 or AF-CD=0
: Then
: Disp "INFINITE","SOLUTIONS"
: Else
: Disp "NO SOLUTION"
```

Critical Thinking

42. Solve the system of equations. (*Hint:* Let $n = \frac{1}{x}$ and $m = \frac{1}{y}$.)

$\frac{1}{x} + \frac{3}{y} = \frac{3}{4}$

$\frac{3}{x} - \frac{2}{y} = \frac{5}{12}$

Applications and Problem Solving

43. Sales The drama club at Lincoln High School sells hot chocolate and coffee at the school's football games to make money for a special trip. At one game, they sold $200 worth of hot drinks. They need to report how many of each type of drink they sold for their club records. Macha knows that they used 295 cups that night. If hot chocolate sells for 75¢ and coffee sells for 50¢, how many of each type of hot drinks did they sell?

44. Literature   Lewis Carroll was a mathematician and an author. He used mathematical logic when writing his stories. In his book, *Through the Looking Glass*, there is a conversation between Tweedledum and Tweedledee. Tweedledum says, "The sum of your weight and twice mine is 361 pounds." Tweedledee answers, "Contrariwise, the sum of your weight and twice mine is 362 pounds." What are the weights of Tweedledum and Tweedledee?

45. Population Growth   The chart below shows the states with the greatest percentage of population growth during the 1980s. In 1990, the population of New York was 17.99 million, which represented a growth of 0.43 million over its 1980 population. Assume that each state continues to gain the same number of residents every ten years.

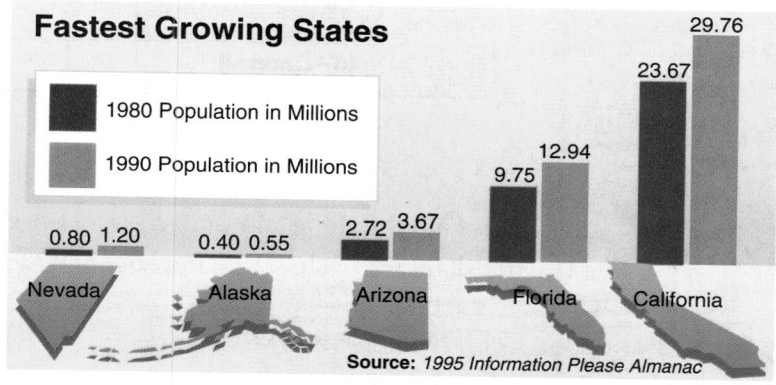

Fastest Growing States

1980 Population in Millions
1990 Population in Millions

Nevada 0.80 1.20
Alaska 0.40 0.55
Arizona 2.72 3.67
Florida 9.75 12.94
California 23.67 29.76

Source: *1995 Information Please Almanac*

a. Write an equation that represents the population of New York *d* decades after 1990.

b. Write an equation that represents the population of Arizona *d* decades after 1990.

c. When will the populations of Arizona and New York be equal?

46. City Planning   A portion of the subway in Washington, D.C., heads out of the main part of town in a northwesterly direction. It goes under New Hampshire Avenue as shown at the right. If distances are measured in kilometers, the path of the subway can be represented by the equation $y = -2.5x + 2.5$, and the path of New Hampshire Avenue can be represented by the equation $y = x$. What are the coordinates of the point at which the subway goes under New Hampshire Avenue?

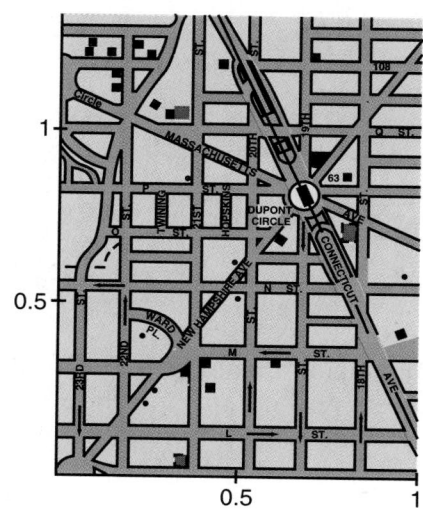

Top Five
List

Longest Subway Networks
1. Washington, D.C., 612 km
2. London, 430 km
3. New York, 370 km
4. Paris, 301 km
5. Moscow, 225 km

47. Consumer Awareness   A new car rental company wants to offer two rental options similar to the truck rental plans offered in the application at the beginning of the lesson. Design two rate structure options that will offer different rates per mile, but will be equal for customers driving 150 miles.

48. Which graph illustrates the solution to the system of equations $y = 2x - 5$ and $x + y = -3$? (Lesson 3–1)

a. 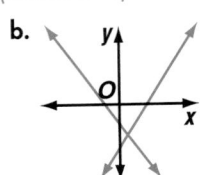   b. 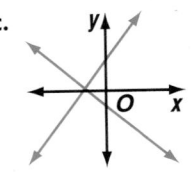   c. 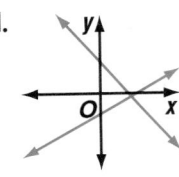   d.

49. Business   The Fix-It Auto Repair Shop has a sign that states that the labor costs are $40 per hour or any fraction thereof. What type of function does this relationship represent? (Lesson 2–6)

50. If $g(x) = |4x + 17|$, find the value of $g(-2)$. (Lesson 2–6)

51. Postal Service   The table shows the price for first-class stamps since the U.S. Postal Service was created on July 1, 1971. (Lesson 2–5)

Age of U.S. Postal Service (years)	0	3	4	7	10	10	14	17	20	25
Price of Stamp (¢)	8	10	13	15	18	20	22	25	29	32

   a. Write a prediction equation for this relationship.

   b. Predict the price for a first-class stamp issued in the year 2010.

52. Find the slope and y-intercept of the graph of $-5x = 6 - 2y$. (Lesson 2–4)

53. Find the value of k in the equation $kx - 3y = 12$ if $(2, 2)$ is a solution of the equation. (Lesson 2–4)

54. Find the x- and y-intercepts of the graph of $x = 9$. (Lesson 2–3)

55. Find the mean, median, and mode of the prime numbers between 0 and 35. (Lesson 1–3)

WORKING ON THE
In·ves·ti·ga·tion

Refer to the Investigation on pages 60–61.

Through the Looking Glass

By making observations with different instruments and comparing the results, naturalists can make more reliable conclusions than by using just one instrument.

1 You have four different sizes of calibration scopes. Record the dimensions of each of the four tubes.

2 Suppose you were standing 50 feet from an animal. Looking through Tube A, the animal's image fills your view exactly. Explain what you need to do to be able to look through Tube B so that the animal's image fills your view exactly. Explain what you need to do to be able to look through Tube D so that the animal's image fills your view exactly.

3 Suppose you had a Tube E that was 3 times as long and twice as wide as Tube A. Exactly where would you need to stand so that the animal's image fills your view exactly?

Add the results of your work to your Investigation Folder.

Cramer's Rule

3-3

INTEGRATION
Geometry

WHAT YOU'LL LEARN

- To find the values of second-order determinants, and
- to solve systems of equations by using Cramer's rule.

WHY IT'S IMPORTANT

You can use Cramer's rule to solve problems involving sports and politics.

Two sides of a parallelogram are contained in the lines whose equations are $2.3x + 1.2y = 2.1$ and $4.1x - 0.5y = 14.3$. To find the coordinates of a vertex of the parallelogram, we must solve the system of equations. However, solving this system by using substitution or elimination would require many calculations. *This problem will be solved in Example 3.*

Another method for solving systems of equations is **Cramer's rule.** The rule gives us a quick way to find the solution to a system of two equations with two variables. It is especially useful when the coefficients are large or involve fractions or decimals. Cramer's rule makes use of **determinants**. A determinant is a square array of numbers or variables enclosed between two parallel vertical bars. The numbers or variables written within a determinant are called **elements**. The determinant below has two rows and two columns and is called a **second-order determinant.**

$$rows \Longleftarrow \begin{vmatrix} a & b \\ c & d \end{vmatrix}$$

$$columns$$

A second-order determinant is evaluated as follows.

Value of a Second-Order Determinant	$\begin{vmatrix} a & b \\ c & d \end{vmatrix} = ad - bc$

Notice that the value of the determinant is found by calculating the difference of the products of the two diagonals.

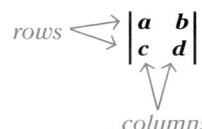

$$ad - bc$$

Example **1** **Find the value of each determinant.**

a. $\begin{vmatrix} 3 & 5 \\ 2 & 6 \end{vmatrix}$

b. $\begin{vmatrix} -2 & 7 \\ 5 & 8 \end{vmatrix}$

$\begin{vmatrix} 3 & 5 \\ 2 & 6 \end{vmatrix} = 3(6) - 5(2)$

$\begin{vmatrix} -2 & 7 \\ 5 & 8 \end{vmatrix} = -2(8) - 7(5)$

$= 8$

$= -51$

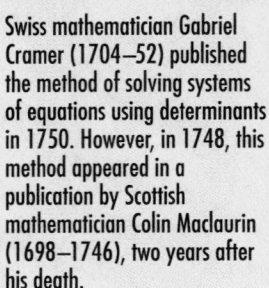
To discover how Cramer's rule uses determinants to solve a system of linear equations, consider the following system.

$$ax + by = e$$
$$cx + dy = f$$ *a, b, c, d, e, and f represent constants, not variables.*

Solve for x by using elimination.

$$
\begin{array}{ll}
adx + bdy = de & \text{\textit{Multiply the first equation by d.}} \\
(-)\ bcx + bdy = bf & \text{\textit{Multiply the second equation by b.}} \\
\hline
adx - bcx = de - bf & \text{\textit{Subtract.}} \\
(ad - bc)x = de - bf & \text{\textit{Factor.}} \\
x = \dfrac{de - bf}{ad - bc} & \text{\textit{Notice that ad − bc must not be zero.}}
\end{array}
$$

Solving for y in the same way produces the following expression.

$$y = \frac{af - ce}{ad - bc}$$

So, the solution to the system of equations $\begin{cases} ax + by = e \\ cx + dy = f \end{cases}$ is $\left(\dfrac{de - bf}{ad - bc}, \dfrac{af - ce}{ad - bc} \right)$.

Notice that the two fractions have the same denominator. It can be written as a determinant. The numerators can also be written as determinants.

$$ad - bc = \begin{vmatrix} a & b \\ c & d \end{vmatrix} \qquad de - bf = \begin{vmatrix} e & b \\ f & d \end{vmatrix} \qquad af - ce = \begin{vmatrix} a & e \\ c & f \end{vmatrix}$$

So, now we can find the solution to a system of two linear equations in two variables by using determinants. This method is called Cramer's rule.

Cramer's Rule

The solution to the system $\begin{cases} ax + by = e \\ cx + dy = f \end{cases}$ is (x, y),

where $x = \dfrac{\begin{vmatrix} e & b \\ f & d \end{vmatrix}}{\begin{vmatrix} a & b \\ c & d \end{vmatrix}}$, $y = \dfrac{\begin{vmatrix} a & e \\ c & f \end{vmatrix}}{\begin{vmatrix} a & b \\ c & d \end{vmatrix}}$, and $\begin{vmatrix} a & b \\ c & d \end{vmatrix} \neq 0$.

Example **2** Use Cramer's rule to solve the system of equations.

$$2x - 3y = 9$$
$$x + 5y = -2$$

$$x = \frac{\begin{vmatrix} 9 & -3 \\ -2 & 5 \end{vmatrix}}{\begin{vmatrix} 2 & -3 \\ 1 & 5 \end{vmatrix}} \qquad\qquad y = \frac{\begin{vmatrix} 2 & 9 \\ 1 & -2 \end{vmatrix}}{\begin{vmatrix} 2 & -3 \\ 1 & 5 \end{vmatrix}}$$

$$= \frac{9(5) - (-3)(-2)}{2(5) - (-3)(1)} \qquad\qquad = \frac{2(-2) - 9(1)}{2(5) - (-3)(1)}$$

$$= \frac{39}{13} \qquad\qquad\qquad\qquad = \frac{-13}{13}$$

$$= 3 \qquad\qquad\qquad\qquad\quad = -1$$

The solution is $(3, -1)$.

Check by graphing.

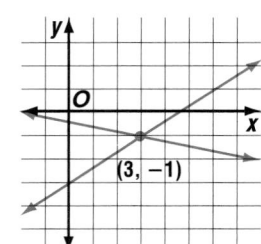

Example **3**

Geometry

Refer to the application at the beginning of the lesson. **Find the coordinates of a vertex of the parallelogram.**

To find the coordinates of a vertex, you need to solve the system of equations.

$2.3x + 1.2y = 2.1$
$4.1x - 0.5y = 14.3$

Since the numbers in the equations would make solving the system by substitution or elimination difficult, use Cramer's rule.

$$x = \frac{\begin{vmatrix} 2.1 & 1.2 \\ 14.3 & -0.5 \end{vmatrix}}{\begin{vmatrix} 2.3 & 1.2 \\ 4.1 & -0.5 \end{vmatrix}} \qquad y = \frac{\begin{vmatrix} 2.3 & 2.1 \\ 4.1 & 14.3 \end{vmatrix}}{\begin{vmatrix} 2.3 & 1.2 \\ 4.1 & -0.5 \end{vmatrix}}$$

$$= \frac{2.1(-0.5) - 14.3(1.2)}{2.3(-0.5) - 4.1(1.2)} \qquad = \frac{2.3(14.3) - 4.1(2.1)}{2.3(-0.5) - (4.1)(1.2)}$$

$$= \frac{-18.21}{-6.07} \qquad\qquad = \frac{24.28}{-6.07}$$

$$= 3 \qquad\qquad\qquad = -4$$

The coordinates of a vertex of the parallelogram are $(3, -4)$.

Check by using substitution.

First Equation: $2.3(3) + 1.2(-4) = 2.1$ ✓
Second Equation: $4.1(3) - 0.5(-4) = 14.3$ ✓

Systems of equations can also be solved by using a BASIC program.

EXPLORATION

B A S I C

The BASIC program below finds the solution of the system of equations of the form $ax + by = c$ and $dx + ey = f$.

```
10   PRINT "ENTER THE COEFFICIENTS."
20   INPUT A, B, C, D, E, F
30   IF A*E-B*D = 0 THEN 80
40   LET X = (C*E-B*F)/(A*E-B*D)
50   LET Y= (A*F-C*D)/(A*E-B*D)
60   PRINT "(";X;",";Y;") IS A SOLUTION"
70   GOTO 10
80   IF C*E-B*F = 0 OR A*F-C*D=0 THEN 110
90   PRINT "NO SOLUTION."
100   GOTO 10
110   PRINT "INFINITE NUMBER OF SOLUTIONS."
120   GOTO 10
130   END
```

Your Turn

a. Use the program to solve the system of equations in Example 2.

b. Use the program to solve the system of equations in Example 3.

c. Study Steps 40 and 50. How does this program relate to Cramer's rule?

Communicating Mathematics

Study the lesson. Then complete the following.

1. **Evaluate** a determinant if both elements of a row or column are 0.

2. **Describe** the elements of a determinant when the value of the determinant is 0 and none of the elements is 0.

3. In Cramer's rule, if the value of the determinant in the denominator is 0, what must be true of the graph of the system of equations represented by the determinant?

4. Carmen used Cramer's rule and solved for x as follows.

$$x = \frac{\begin{vmatrix} 18 & -5 \\ -4 & 8 \end{vmatrix}}{\begin{vmatrix} 2 & -5 \\ 3 & 8 \end{vmatrix}}$$

Write the system of equations that she was solving.

Guided Practice

Find the value of each determinant.

5. $\begin{vmatrix} 5 & 2 \\ 4 & 1 \end{vmatrix}$

6. $\begin{vmatrix} 6 & -2 \\ 7 & 3 \end{vmatrix}$

7. $\begin{vmatrix} \frac{2}{5} & 6 \\ \frac{1}{3} & 2 \end{vmatrix}$

Use Cramer's rule to solve each system of equations.

8. $2x - y = 1$
   $3x + 2y = 19$

9. $5x + 2y = 8$
   $2x - 3y = 7$

10. $\frac{1}{6}x - \frac{1}{9}y = 0$
    $x + y = 15$

11. $2m - 5n = 2$
    $3m + 4n = -5$

12. **Geometry** The two sides of an angle are contained in lines whose equations are $4x + y = -4$ and $2x - 3y = -9$. Find the coordinates of the vertex of the angle.

Find the value of each determinant.

13. $\begin{vmatrix} 8 & 5 \\ 6 & -2 \end{vmatrix}$

14. $\begin{vmatrix} -2 & 4 \\ 8 & -7 \end{vmatrix}$

15. $\begin{vmatrix} -8 & 3 \\ -9 & 7 \end{vmatrix}$

16. $\begin{vmatrix} -6 & -2 \\ 8 & 5 \end{vmatrix}$

17. $\begin{vmatrix} 2 & -7 \\ -5 & 3 \end{vmatrix}$

18. $\begin{vmatrix} 21 & 43 \\ 17 & -29 \end{vmatrix}$

19. $\begin{vmatrix} -54 & 39 \\ 18 & -13 \end{vmatrix}$

20. $\begin{vmatrix} -3.2 & -5.8 \\ 4.1 & 3.9 \end{vmatrix}$

21. $\begin{vmatrix} 7 & -5.2 \\ 1.3 & 2.29 \end{vmatrix}$

Use Cramer's rule to solve each system of equations.

22. $5x + 7y = 13$
    $2x - 5y = 13$

23. $3a + 5b = 33$
    $5a + 7b = 51$

24. $2m + 7n = 4$
    $m - 2n = -20$

25. $3x - 2y = 4$
    $\frac{1}{2}x - \frac{2}{3}y = 1$

26. $1.5x + 0.7y = 0.5$
    $2.2x - 0.6y = -7.4$

27. $4u + 3v = 6$
    $8u - v = -9$

28. $\frac{1}{2}r + \frac{3}{4}s = -4$

$\frac{3}{4}r - \frac{7}{8}s = 10$

29. $\frac{1}{3}x + \frac{2}{5}y = 5$

$\frac{2}{3}x - \frac{1}{2}y = -3$

30. $\frac{1}{2}x - y = -1$

$\frac{3}{4}x + \frac{1}{2}y = -\frac{1}{4}$

31. $\frac{3}{4}a + \frac{1}{2}b = \frac{11}{12}$

$\frac{1}{2}a - \frac{1}{4}b = \frac{1}{8}$

32. $0.2a = 0.3b$

$0.4a - 0.2b = 0.2$

33. $3.5x + 4y = -5$

$2(x - y) = 10$

34. Geometry The sides of a triangle are contained in the lines represented by $x - 4y = -6$, $5x + y = 33$, and $2x - y = 2$. Find the coordinates of the vertices of the triangle.

Critical Thinking

35. When using Cramer's rule, how can you tell whether there is no solution or an infinite number of solutions?

Applications and Problem Solving

36. Sports The winning times for various speed-skating events in the 1994 Winter Olympics are given in the chart below. Suppose the men and women continue to decrease their times for the 1000-meter event by the same amount each Winter Olympics. In what Olympics will the women's times be faster than the men's in this event?

GLOBAL CONNECTIONS

Malaysian foot tennis is a very popular sport in Malaysia and other countries in the Far East. It is played by teams of two using a volleyball-type net. A ball woven from rattan must be kept in the air using only feet, knees, or thighs.

Winning Times for Speed Skaters in the 1994 Winter Olympics				
Event	Men's Time	Change from 1992 Winner	Women's Time	Change from 1992 Winner
500 m	36.33 s	−0.81 s	39.25 s	−1.08 s
1000 m	72.43 s	−2.42 s	78.74 s	−3.16 s
1500 m	111.29 s	−3.52 s	122.19 s	−3.68 s

Source: *The World Almanac and Book of Facts 1995*

37. Exercise Rosa rides the bus to work. Usually she rides 35 minutes and then walks 6 minutes to get to her place of employment. Since the bus travels 30 miles per hour and her walking speed is 5 miles per hour, she can easily determine that the total distance is 18 miles. One day when the weather is nice, she decides to get off the bus earlier and walk the rest of the way for exercise. She only has a total of 75 minutes to get to work, so she can't afford to get off the bus too early. How much time should she spend on the bus before getting off to walk?

38. Politics The three states with the most electoral votes are California (54), New York (33), and Texas (32). The graph at the right shows the results of the popular votes in the 1992 presidential election. Bill Clinton received about 5,700,000 popular votes from the voters in New York and Texas, and George Bush received about 4,800,000 from the same voters. About how many people voted in the 1992 presidential election in New York? in Texas?

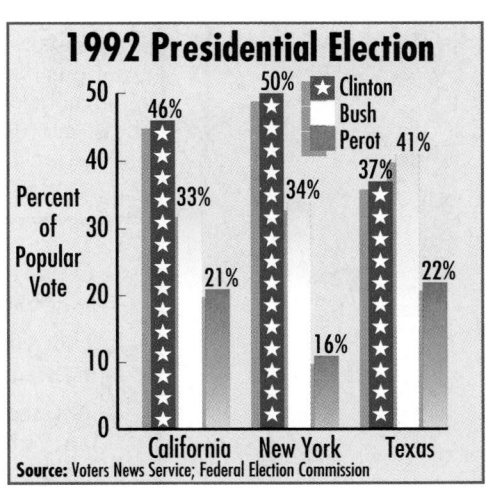

39. Solve the system of equations $2x + y = 0$ and $5x + 3y = 1$ by using either substitution or elimination. (Lesson 3–2)

40. Consumer Awareness Mrs. Katz is planning a family vacation. She bought 8 rolls of film and 2 camera batteries for $23.00. The next day, her daughter went back and bought 6 more rolls of film and 2 batteries for her camera. This bill was $18.00. What is the price of a roll of film and a camera battery? (Lesson 3–2)

41. Graph the system of equations $y = -3x$ and $y - x = 4$. State the solution. Is the system of equations consistent and independent, consistent and dependent, or inconsistent? (Lesson 3–1)

42. If $f(x) = [5x - 3]$, find $f\left(-\frac{1}{2}\right)$. (Lesson 2–6)

43. Write a prediction equation for the data in the following table. (Lesson 2–5)

x	-3	0	4	6	8
y	5	3	0	-3	-5

44. Solve $\frac{w - 6}{3} \leq 6 - w$. (Lesson 1–6)

45. Geometry The formula for the area of a trapezoid is $A = \frac{1}{2} h(b_1 + b_2)$, where h is the height of the trapezoid and b_1 and b_2 are the lengths of the bases. Find the area of the trapezoid at the right. (Lesson 1–1)

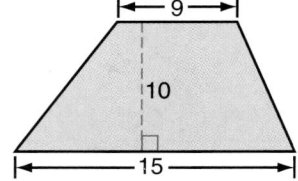

Quantum Computing

The article below appeared in *Science News* on May 14, 1994.

IN THE QUANTUM WORLD, A PARTICLE—undisturbed by any attempt to observe it—can be in myriad places at the same time. Thus, a single photon traveling through a crystal simultaneously follows all possible optical paths through the material. . . . Computer scientists have speculated that computers operating according to the rules of quantum mechanics can potentially take advantage of a similar multiplicity of paths to solve certain types of mathematical problems much more quickly than conventional computers can. Now, mathematician Peter W. Shor . . . has proved that, in principle, quantum computation can provide the shortcut needed to convert the factoring of large numbers from a time-consuming chore into an amazingly quick operation. . . . Quantum computers don't exist yet, and building them involves surmounting significant technological barriers. Nonetheless, some researchers are starting to produce designs . . . that may lead eventually to working models. ■

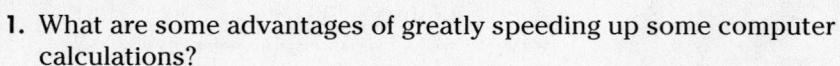

1. What are some advantages of greatly speeding up some computer calculations?

2. Can you think of an example in everyday life where moving numerous items along many paths is faster than using only one path?

3. If a quantum computer is built, we will learn a lot if it works successfully, and we will learn a lot if it doesn't work successfully. Explain.

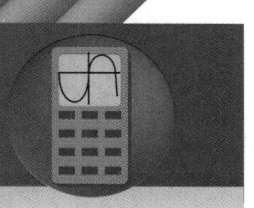

3–4A Graphing Technology
Systems of Linear Inequalities

A Preview of Lesson 3–4

You can graph systems of linear inequalities with a graphing calculator by using the SHADE feature described in Lesson 2–7A. It shades above the first function entered and below the second function entered. Be sure to clear any equations currently stored in the Y= list of the calculator before you begin.

Example **Graph the system of inequalities in the standard viewing window.**

$$y \geq x + 2$$
$$y \leq -2x + 5$$

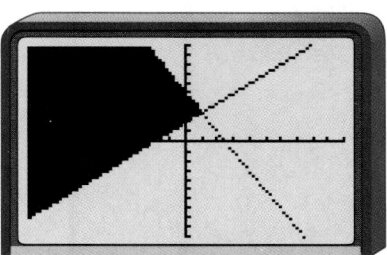

The greater than or equal to symbol in $y \geq x + 2$ indicates the values on or above the line $y = x + 2$. Similarly, the less than or equal to symbol in $y \leq -2x + 5$ indicates values on or below the line $y = -2x + 5$. Therefore, the function $y = x + 2$ will be entered first, and $y = -2x + 5$ will be entered second.

Enter: $\boxed{\text{ZOOM}}$ 6 $\boxed{\text{2nd}}$ $\boxed{\text{DRAW}}$ 7 $\boxed{\text{X,T,}\theta\text{,}n}$ $\boxed{+}$ 2 $\boxed{,}$ $\boxed{(-)}$
2 $\boxed{\text{X,T,}\theta\text{,}n}$ $\boxed{+}$ 5 $\boxed{)}$ $\boxed{\text{ENTER}}$

The shaded area indicated points that satisfy the system of inequalities $y \geq x + 2$ and $y \geq -2x + 5$.

Before graphing again, you must clear the graphics screen.

Enter: $\boxed{\text{2nd}}$ $\boxed{\text{DRAW}}$ 1 $\boxed{\text{ENTER}}$

EXERCISES

Use a graphing calculator to solve each system of inequalities. Sketch each graph on a sheet of paper.

1. $y \geq x$
$y \leq 5$

2. $y \geq -2x + 4$
$y \leq x - 1$

3. $y \leq 6x - 3$
$y \geq 0.5x$

4. $y \leq 0.1x + 1$
$y \geq -0.5x - 3$

5. $3x - 4y \leq 12$
$2x + y \leq 10$

6. $y - 2 \leq x$
$y \geq -3x + 5$

7. $y \geq 0$
$y \leq -2x + 12$

8. $y \leq \frac{2}{3}x - 1$
$y \geq -\frac{1}{5}x + 3$

9. $-5y \leq -2x$
$2y \leq 3x - 8$

Graphing Systems of Inequalities

What YOU'LL LEARN

• To solve systems of inequalities by graphing.

Why IT'S IMPORTANT

You can use systems of inequalities to solve problems involving space science and higher education.

APPLICATION
Space Science

When the National Aeronautics and Space Administration (NASA) chose the first astronauts in 1959, size was important since the space available in the Mercury capsule was very limited. NASA called for men who were between 5 feet 4 inches and 5 feet 11 inches tall, inclusively, and who were between 21 and 40 years of age.

This information can be represented by a **system of inequalities**. To solve such a system of inequalities, we need to find the ordered pairs that satisfy all the inequalities involved. One way to do this is to graph the inequalities on the same coordinate plane. The solution set is then represented by the intersection, or overlap, of the graphs.

Mae Carol Jemison (1956–)

Mae Carol Jemison was the first African-American woman astronaut. She was the computer engineer and physician aboard the space shuttle Endeavor, launched September 12, 1992.

Since the heights of the astronauts are between 5 feet 4 inches (or 64 inches) and 5 feet 11 inches (or 71 inches), inclusively, we can write this information as the following two inequalities using h as the height in inches.

$$h \geq 64 \text{ and } h \leq 71$$

This could also be expressed as $64 \leq h \leq 71$.

The acceptable ages, a, can be expressed as the following inequalities.

$$a > 21 \text{ and } a < 40$$

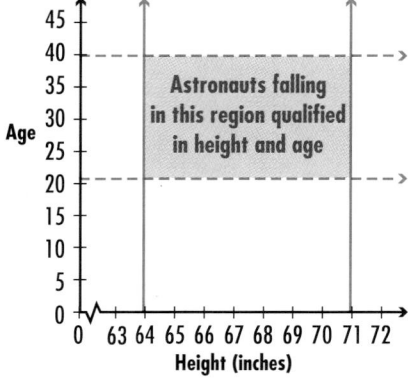

The broken lines indicate that the boundaries are not part of the graphs.

This could also be expressed as $21 < a < 40$.

Graph both inequalities. Any point in the intersection of the two graphs is a solution to the system.

Example **Solve the system of inequalities by graphing.**

$$x - 2y \geq -2$$
$$x + y \leq -1$$

$x - 2y \geq -2$ represents Regions 2 and 3.
$x + y \leq -1$ represents Regions 1 and 2.

The intersection of these regions is Region 2, which is the solution of the system of inequalities.

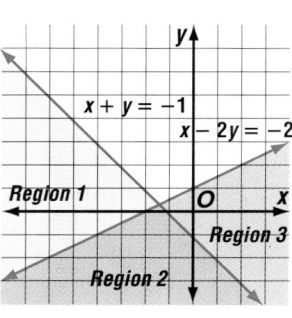

It is possible that two regions do *not* intersect. In such cases, we say the solution is the empty set, shown as ∅, and no solution exists.

Example ❷ **Solve the system of inequalities by graphing.**

$$y > \frac{2}{3}x + 2$$

$$y < \frac{2}{3}x - 1$$

The two solutions have no points in common. The solution set is ∅.

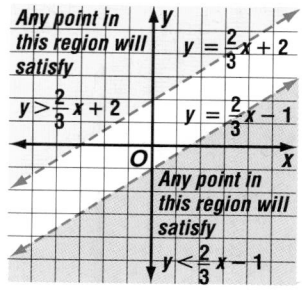

As you recall, an absolute value inequality can be restated as two inequalities using an *and* or an *or*. So, an absolute value inequality can be graphed like a system of inequalities.

Example ❸ **Solve the system of inequalities by graphing.**

$$|x| < 3$$
$$y \geq x - 2$$
$$y \leq -\frac{1}{2}x + 4$$

The inequality $|x| < 3$ can be rewritten as $x > -3$ and $x < 3$.

Graph all of the inequalities on the same grid and look for the region or regions that are common to all.

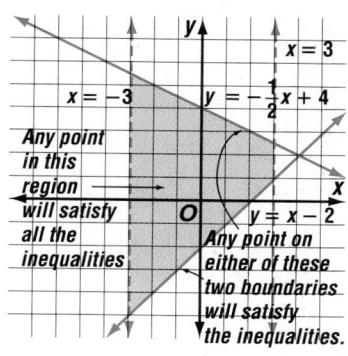

Check to be sure the proper region is shaded.

Example ❹ **The system of inequalities $x - y \leq 7$, $3x - 11y \geq -11$, and $x + y + 1 \geq 0$ form a triangle and its interior.**
a. Graph the system and describe the triangle.
b. Name the coordinates of the vertices of the triangle.

Geometry

a. The lines are graphed at the right. The triangle is a right triangle.

b. The coordinates (11, 4) and (3, −4) can be determined from the graph. To find the coordinates of the third vertex, solve the system of equations $3x - 11y = -11$ and $x + y + 1 = 0$.

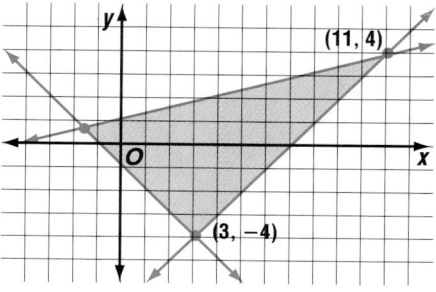

$$x + y + 1 = 0 \qquad \textit{Solve for x.}$$
$$x = -y - 1$$

(continued on the next page)

Substitute $-y - 1$ for x.

$$3x - 11y = -11$$
$$3(-y - 1) - 11y = -11$$
$$-3y - 3 - 11y = -11$$
$$-14y = -8$$
$$y = \frac{8}{14} \text{ or } \frac{4}{7}$$

Evaluate when $y = \frac{4}{7}$.

$$x = -y - 1$$
$$= -\left(\frac{4}{7}\right) - 1$$
$$= -\frac{11}{7}$$

The coordinates of the vertices of the triangle are $(11, 4)$, $(3, -4)$, and $\left(-\frac{11}{7}, \frac{4}{7}\right)$.

CHECK FOR UNDERSTANDING

Communicating Mathematics

Study the lesson. Then complete the following.

1. How do you determine whether a point is a solution to a system of inequalities?

2. **Describe** the difference between the graphs of the following two systems of inequalities.

   a. $2x + 3y > 6$ or $2x + 3y < 3$ **b.** $2x + 3y > 6$ and $2x + 3y < 3$

3. **Describe** how the graphs of $|x| \geq 2$ and $|x| \leq 2$ differ.

4. **Research** the names, ages, and heights of the original seven astronauts. Graph this information on a graph similar to the one on page 148. Describe your results in a paragraph.

5. **Write** a system of two inequalities that will have no intersection.

MATH JOURNAL

6. **Assess Yourself**   Write a paragraph describing what you like and/or dislike about graphing systems of inequalities.

Guided Practice

7. Which graph illustrates the solution to the system of inequalities $x + y \geq 1$ and $x - y \leq 0$.

   a. 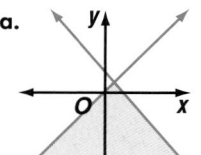   **b.** 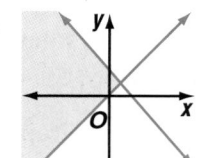   **c.** 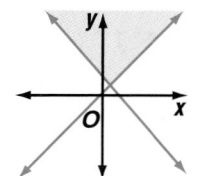   **d.**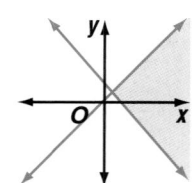

Solve each system of inequalities by graphing.

8. $x \leq 1$
   $y > 3$

9. $y \geq 2x - 2$
   $y \leq -x + 2$

10. $x - 3y \leq -3$
    $2x + 3y < 12$

11. $x - 3y \geq -9$
    $4x - y \leq 4$
    $x + 2y \geq -2$

12. **Geometry**   Graph the system of inequalities $x \geq 0$, $y \geq 0$, and $x + y \leq 6$. Find the area of the region defined by the system of inequalities.

Practice

Solve each system of inequalities by graphing.

13. $x < 2$
$y \geq 1$

14. $x - y \leq 3$
$x + y \geq 2$

15. $x + y > 2$
$y > 3$

16. $y - x \leq 3$
$y \geq x + 2$

17. $y \geq x - 3$
$y \geq -x + 1$

18. $y < -x - 3$
$x > y - 2$

19. $y \leq 2x - 3$
$y \leq \frac{1}{2}x + 1$

20. $y > \frac{2}{3}x - 1$
$y \leq -\frac{3}{4}x + 2$

21. $|x| < 3$
$|y| > 2$

22. $|x + 1| \leq 2$
$x - 2y \geq 1$

23. $3x + 2y \leq 6$
$4x - y < 2$

24. $4x - 3y \geq 7$
$y < 2$

25. $x - 3y > 2$
$2x - y < 4$
$3x + 4y > 0$

26. $5x - y < 0$
$4x + 3y > 6$
$x - 3y > 3$

27. $y < 2x + 1$
$y > 2x - 2$
$3x + y > 8$

28. Geometry Study the graph at the right.
  a. Write the system of inequalities whose solution forms the triangular region shown in the graph.
  b. Replace $y \geq -1$ with an inequality that would result in a triangular region with four times the area of the first region.

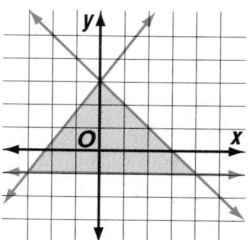

29. Geometry Write a system of inequalities that would form a polygonal region having at least three sides.

30. Geometry Study the region at the right.
  a. What is the most descriptive name for the quadrilateral?
  b. Write the system of inequalities whose solution forms the region.
  c. Find the area of the region.

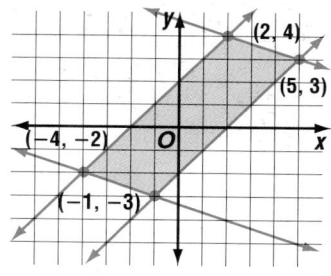

Critical Thinking

31. Geometry Find the area of the region defined by $|x| + |y| \leq 5$ and $|x| + |y| \geq 2$.

Applications and Problem Solving

32. Higher Education According to *Lovejoy's College Guide,* the middle 50% of the freshmen entering the University of Notre Dame have a score between 540 and 650, inclusively, on the verbal portion of the SAT. The middle 50% of these same freshmen have between 620 and 720, inclusively, on the math portion of the SAT. Write a system of inequalities that describes the SAT scores of the middle 50% of Notre Dame freshmen. Then, graph the system of inequalities.

33. Time Management Paloma attends Jones High School. Each school day, she spends 7 hours in school and usually some time after school studying and doing homework. She tries to sleep exactly 8 hours out of every 24-hour period. Let s represent the time Paloma spends on school and studying, and let a represent the time she spends on activities other than school, studying, and sleep. Write a system of inequalities that describes the way Paloma manages her time during a typical school day. Then graph the system of inequalities.

University of Notre Dame

34. Use Cramer's rule to solve the system of equations $x - 4y = 1$ and $2x + 3y = 13$.   (Lesson 3–3)

Solve each system of equations. Use either substitution or elimination.   (Lesson 3–2)

35. $4a - 3b = -4$
$3a - 2b = -4$

36. $2r + s = 1$
$r - s = 8$

37. Graph the system of equations $x + 5y = 10$ and $x + 5y = 15$. State its solution.   (Lesson 3–1)

38. Which equation is represented by the graph at the right?   (Lesson 2–6)

a. $g(x) = |x| + 3$
b. $g(x) = |x + 3|$
c. $g(x) = |x| - 3$
d. $g(x) = |x - 3|$

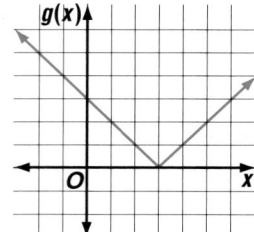

39. Salina earned $8 more than Tyler selling newspaper subscriptions. At the end of the day, Salina was given $1 as a sales bonus, and Tyler spent $3 on snacks. When they returned home, they had a total of $12. How much money did Tyler earn selling newspaper subscriptions?   (Lesson 1–4)

SELF TEST

Solve the system of equations by graphing.   (Lesson 3–1)

1. $2x - 5y = 14$
$x - y = 1$

2. $4x - 3y = -9$
$x + 2y = -5$

Solve each system of equations by using substitution or elimination.   (Lesson 3–2)

3. $y = 3x$
$x + 21 = -2y$

4. $4a + 3b = -2$
$5a + 7b = 17$

5. Find the value of the determinant $\begin{vmatrix} -5 & -2 \\ -3 & 11 \end{vmatrix}$.   (Lesson 3–3)

Use Cramer's rule to solve each system of equations.   (Lesson 3–3)

6. $3x - 7y = 2$
$6x - 13y = 4$

7. $6x + 5y = -7$
$2x - 3y = 7$

8. Travel   A hotel at the Downhill Ski Resort advertises two package deals. One package offers three nights at the hotel and a two-day lift ticket for $245. The other offers five nights at the hotel and a three-day lift ticket for $400. For these specials, what is the cost of staying at the hotel one night and what is the cost of a one-day lift ticket?   (Lesson 3–3)

Solve each system of inequalities by graphing.   (Lesson 3–4)

9. $x + y > 2$
$x - 2y \leq -1$

10. $y < 2$
$y \geq 2x$
$y \geq x + 1$

Linear Programming

What YOU'LL LEARN

- To find the maximum and minimum values of a function over a region by using linear programming techniques, and
- to solve problems by solving a simpler problem.

Why IT'S IMPORTANT

You can use linear programming to solve problems involving agriculture and manufacturing.

APPLICATION

Agriculture

Harris Grunden has 20 days in which to plant corn and soybeans. The corn can be planted at a rate of 60 acres per day and the soybeans at a rate of 70 acres per day. He has 1300 acres available for planting these two crops. Write inequalities to show the possible ways he can plant the available acres.

Let c represent the acres planted in corn and let s represent the acres planted in soybeans. Since Harris cannot plant negative acres of corn or soybeans, c and s must be nonnegative numbers.

$$c \geq 0 \text{ and } s \geq 0$$

Since corn can be planted at a rate of 60 acres per day, Harris can plant no more than 20×60 or 1200 acres of corn.

$$c \leq 1200 \quad \textit{Thus, } 0 \leq c \leq 1200.$$

Since soybeans can be planted at a rate of 70 acres per day, Harris can plant no more than 20×70 or 1400 acres of soybeans.

$$s \leq 1400 \quad \textit{Thus, } 0 \leq s \leq 1400.$$

Harris cannot plant more than 1300 acres of corn and soybeans.

$$c + s \leq 1300$$

Since Harris has 20 days to plant the corn and soybeans, the number of days spent planting these crops must be less than or equal to 20.

$$\frac{c}{60} + \frac{s}{70} \leq 20 \qquad \frac{c}{60} \textit{ represents the time needed to plant the corn.}$$

$$\frac{s}{70} \textit{ represents the time needed to plant the soybeans.}$$

$$7c + 6s \leq 8400 \quad \textit{Multiply each side by 420 to eliminate fractions.}$$

If we graph these inequalities, all of the points in their intersection are possible ways Harris can plant the available acres. The inequalities are called the **constraints**. The area of intersection of the graphs, in which every constraint is met, is called the **feasible region** of the planting.

Let's graph the constraints.

$0 \le c \le 1200$
$0 \le s \le 1400$
$c + s \le 1300$
$7c + 6s \le 8400$

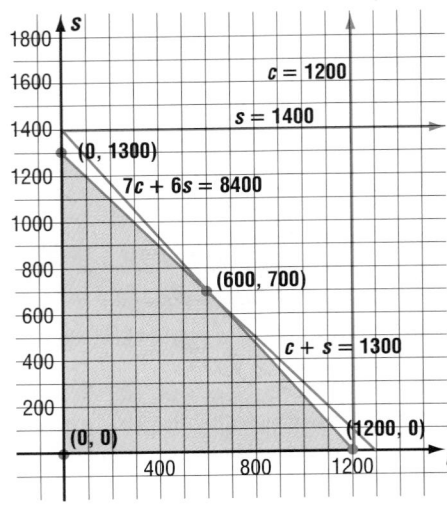

The first two constraints indicate that the graph is in the first quadrant.

The feasible region contains all possible solutions. If we choose any point within the region, its ordered pair should be a solution to each inequality. Let's try (600, 400).

Check: $0 \le 600 \le 1200$
$0 \le 400 \le 1400$
$600 + 400 \le 1400$
$7 \cdot 600 + 6 \cdot 400 \le 8400$ *All are true.*

Choose a point outside the region and test it in each inequality. Is its ordered pair a solution to all of the inequalities?

Harris needs to decide how many acres of each crop to plant. There are many options, as we have seen by graphing the constraints. Of course Harris would like to make as much of a profit as possible. If the profit on corn is $30 per acre and the profit on soybeans is $26 per acre, how much of each should he plant to maximize his profit? Harris uses a spreadsheet to organize his results.

The notation f(c, s) is used to represent a function with two variables c and s.

The profit can be defined by the function $f(c, s) = 30c + 26s$. Mathematicians have shown that the maximum and minimum values of a function, $f(c, s)$, occur at the vertices of the feasible region. The coordinates of the vertices of a feasible region can be found by reading a graph or by solving systems of equations. The points we need to try have coordinates (0, 1300), (600, 700), (1200, 0), and (0, 0).

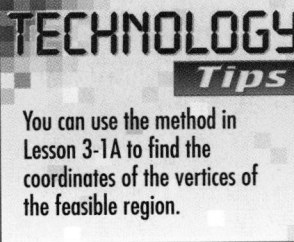

You can use the method in Lesson 3-1A to find the coordinates of the vertices of the feasible region.

(c, s)	$30c + 26s$	Profit $f(c, s)$
(0, 1300)	30(0) + 26(1300)	$33,800
(600, 700)	30(600) + 26(700)	36,200
(1200, 0)	30(1200) + 26(0)	36,000
(0, 0)	30(0) + 26(0)	0

Check some other values within the feasible region to convince yourself that we have found the maximum profit for the 1300 acres. According to our results, Harris should plant 600 acres of corn and 700 acres of soybeans for a profit of $36,200.

The process we have just used is called **linear programming**. This procedure is used to find the maximum or minimum value of a linear function subject to given conditions on the variables, called constraints. When a system of inequalities produces a convex polygonal region as a solution, the maximum or minimum value of a related function will occur at a vertex of the region.

Example ❶ Find the maximum and minimum values of $f(x, y) = 2x - 3y$ for the polygonal region determined by the system of inequalities.
$x \geq 1$
$y \geq 2$
$x + 2y \leq 9$

First we must find the vertices of the feasible region. Graph the inequalities.

The polygon formed is a triangle with vertices at $(1, 2)$, $(5, 2)$, and $(1, 4)$.

Use a chart to find the maximum and minimum values of the function.

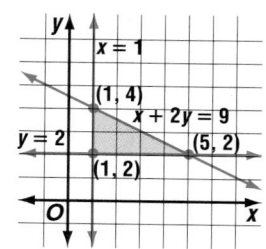

(x, y)	$2x - 3y$	$f(x, y)$
$(1, 2)$	$2(1) - 3(2)$	-4
$(5, 2)$	$2(5) - 3(2)$	4
$(1, 4)$	$2(1) - 3(4)$	-10

The maximum value is 4 at $(5, 2)$. The minimum value is -10 at $(1, 4)$.

Sometimes a polygonal region is not formed. In this case, the function is said to be **unbounded**.

Example ❷ Graph the following constraints. Then find the maximum and minimum values of the function $f(x, y) = 5x + 2y$.
$x - 3y \leq 0$
$x - 3y \geq -15$
$4x + 3y \geq 15$

Graph the system of inequalities. There are only two points of intersection, $(0, 5)$ and $(3, 1)$.

(x, y)	$5x + 2y$	$f(x, y)$
$(0, 5)$	$5(0) + 2(5)$	10
$(3, 1)$	$5(3) + 2(1)$	17

The minimum is 10 at $(0, 5)$.

Although $f(3, 1)$ is 17, it is not the maximum value since there are other points in the feasible region that produce greater values. For example, $f(4, 2) = 24$ and $f(300, 101) = 1702$. It appears that $f(x, y) = 5x + 2y$ has no maximum value when using the given constraints. Thus, the region is unbounded.

When using linear programming, we do not check every ordered pair in the feasible region to find the maximum and minimum values of the function. Instead we solve a simpler problem by checking only the values at the vertices. **Solve a simpler problem** is one of many problem-solving strategies that you can use to solve problems.

Example **3** **Find the sum of the whole numbers 1 to 1000, inclusive.**

We could add all of the numbers directly, but even with a calculator that would be time-consuming and tedious. Let's look at the sum of the whole numbers 1 to 10 to see if we can find a faster way.

$$
\begin{aligned}
S &= 1 + 2 + 3 + \ldots + 10 \\
(+)\,S &= 10 + 9 + 8 + \ldots + 1 \\
\hline
2S &= 11 + 11 + 11 + \ldots + 11 \\
2S &= 10 \cdot 11 \\
2S &= 110 \\
S &= 55
\end{aligned}
$$

S represents the sum of the whole numbers.

Now, extend this concept to the original problem.

$$
\begin{aligned}
S &= 1 + 2 + 3 + \ldots + 1000 \\
(+)\,S &= 1000 + 999 + 998 + \ldots + 1 \\
\hline
2S &= 1001 + 1001 + 1001 + \ldots + 1001 \\
2S &= 1000 \cdot 1001 \\
2S &= 1,001,000 \\
S &= 500,500
\end{aligned}
$$

The sum of the numbers of the right side of the equals sign has 1000 addends of 1001.

Therefore, the product of 1000 and 1001 equals twice the sum.

The sum of the whole numbers from 1 to 1000 is 500,500.

CHECK FOR UNDERSTANDING

Communicating Mathematics

Study the lesson. Then complete the following.

1. A feasible region is defined by the graph at the right.

   a. **Name** the points at which a maximum or minimum value of a function could occur for the feasible region.

   b. **Find** the maximum and minimum value of the function $f(x, y) = 2x + 3y$.

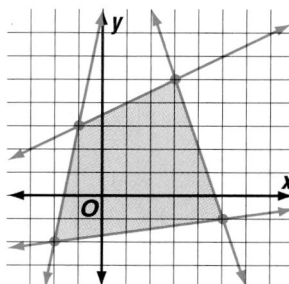

2. **Define** linear programming in your own words.

3. **Describe** a problem that might be solved by the strategy of solving a simpler problem.

A feasible region has vertices at $(-3, 2)$, $(4, 1)$, $(2, 6)$, and $(1, -2)$. Find the maximum and minimum values of each function.

4. $f(x, y) = x + 2y$

5. $f(x, y) = 4x - y$

Graph each system of inequalities. Name the coordinates of the vertices of the feasible region. Find the maximum and minimum values of the given function for this region.

6. $y \geq 1$
$x \leq 6$
$y \leq 2x + 1$
$f(x, y) = x + y$

7. $4y \leq x + 8$
$x + y \geq 2$
$y \geq 2x - 5$
$f(x, y) = 4x + 3y$

8. $y \geq 2$
$1 \leq x \leq 5$
$y \leq x + 3$
$f(x, y) = 3x - 2y$

9. Solve a Simpler Problem   Carl Friedrich Gauss (1777–1855) of Germany was the greatest mathematician of his time. When he was in elementary school, his teacher wanted to keep the students busy by asking them to add the numbers from 1 to 100, inclusive. Within seconds, Gauss declared that the answer was 5050.

  a. How did Gauss add the numbers so quickly?

  b. What is the sum of the whole numbers from 1 to 2000, inclusive?

EXERCISES

A feasible region has vertices at $(-1, 3)$, $(3, 5)$, $(4, -1)$, and $(-1, -2)$. Find the maximum and minimum values of each function.

10. $f(x, y) = x - y$

11. $f(x, y) = 3x + 2y$

12. $f(x, y) = 3y - x$

13. $f(x, y) = -2x - y$

Graph each system of inequalities. Name the coordinates of the vertices of the feasible region. Find the maximum and minimum values of the given function for this region.

14. $2x + 3y \geq 6$
$3x - 2y \geq -4$
$5x + y \leq 15$
$f(x, y) = x + 3y$

15. $x \geq 1$
$y \geq 0$
$2x + y \leq 6$
$f(x, y) = 3x + y$

16. $x + y \geq 4$
$3x - 2y \leq 12$
$x - 4y \geq -16$
$f(x, y) = x - 2y$

17. $y \leq 2x + 1$
$1 \leq y \leq 3$
$y \leq -0.5x + 6$
$f(x, y) = 3x + y$

18. $y \leq x + 6$
$y + 2x \geq 6$
$2 \leq x \leq 6$
$f(x, y) = -x + 3y$

19. $x + y \geq 2$
$2y \geq 3x - 6$
$4y \leq x + 8$
$f(x, y) = 3y + x$

20. $x - 3y \geq -7$
$5x + y \leq 13$
$x + 6y \geq -9$
$3x - 2y \geq -7$
$f(x, y) = x - y$

21. $x \geq 2$
$x \leq 4$
$y \geq 1$
$x - 2y \geq -4$
$f(x, y) = x - 3y$

22. $x \geq 0$
$y \geq 0$
$x + 2y \leq 6$
$2y - x \leq 2$
$x + y \leq 5$
$f(x, y) = 3x - 5y$

Use a graphing calculator to find the coordinates of the vertices of the feasible region. Find the maximum and minimum values of the given function for this region.

23. $0 \leq x \leq 5$
$y \geq 0$
$-x + y \leq 2$
$x + y \leq 6$
$f(x, y) = 5x - 3y$

24. $x \leq 3$
$y \leq 5$
$x + y \geq 1$
$x \geq 0$
$y \geq 0$
$f(x, y) = 2x + 8y + 10$

25. $y \leq 1$
$y \geq -2$
$5x \leq -2$
$1.2x - y \geq -2.9$
$f(x, y) = 4x + 2y$

**Critical
Thinking**

26. The vertices of a feasible region are $A(1, 2)$, $B(5, 2)$, and $C(1, 4)$. Write a function that satisfies each condition.
  a. A is the maximum and B is the minimum.
  b. C is the maximum and B is the minimum.
  c. B is the maximum and A is the minimum.
  d. A is the maximum and C is the minimum.
  e. A is the minimum and both B and C are maximums.

**Applications and
Problem Solving**

27. Employment Rosalyn works no more than 20 hours a week during the school year. She is paid $10 an hour for tutoring geometry students and $7 an hour for delivering pizzas for Pizza King. She wants to spend at least 3 hours but no more than 8 hours a week tutoring. Find Rosalyn's maximum earnings.

28. Manufacturing The Northern Wisconsin Paper Mill can convert wood pulp to either notebook paper or newsprint. The mill can produce at most 200 units of paper a day. At least 10 units of notebook paper and 80 units of newspaper are required daily by regular customers. If the profit on a unit of notebook paper is $500 and the profit on a unit of newsprint is $350, how many units of each type of paper should the manager have the mill produce each day to maximize profits?

29. Agriculture Refer to the application at the beginning of the lesson. Suppose Harris has an opportunity to plant 50 acres of his father's land in corn and soybeans. Do you think Harris should farm the land? Explain.

30. Solve a Simpler Problem A team is eliminated from the Central Indiana Women's Basketball Tournament when it loses a game. If there are 30 teams playing in the tournament, how many games will need to be played to determine a champion?

31. **Solve a Simpler Problem** Opa Azul is a marketing research executive for a soft drink company. The research team has arranged to perform a survey in 16 different local shopping malls. To ensure that competing soft drink companies will not learn of the survey results, Opa has arranged for telephone lines to be set up so that each survey station has a direct line to each of the other stations. How many telephone lines does Opa need to have installed?

Mixed Review

32. **Geometry** Consider the system of inequalities $x - 2y \leq -1$, $5x - 2y \geq -21$, $x - 2y \geq -9$, and $3x + 2y \leq 13$. (Lesson 3–4)

   a. What is the most descriptive name for the region formed by this system?

   b. Name the coordinates of the vertices.

33. Use Cramer's rule to solve the system of equations. (Lesson 3–3)

   $3x + 2y = 9$
   $2x - 3y = 19$

34. Write an equation for the line that passes through $(-7, 9)$ and is perpendicular to the y-axis. (Lesson 2–4)

35. Is the relation at the right a function? (Lesson 2–1)

36. **Business** Reliable Rentals rents cars for $12.95 per day plus 15¢ per mile. Luis Romero works for a company that limits expenses for car rentals to $90 per day. What is the maximum number of miles that Mr. Romero can drive each day? (Lesson 1–6)

37. Simplify $4(12 - 5b) - 3(4b - 2)$. (Lesson 1–2)

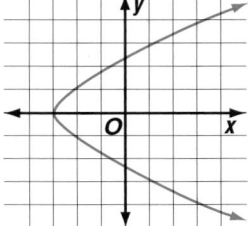

WORKING ON THE

In·ves·ti·ga·tion

Refer to the Investigation on pages 60–61.

Through the Looking Glass

A proportion is an equality in which $\frac{a}{b} = \frac{c}{d}$. A proportion exists that relates the dimensions of the scope, the height of the animal's image through the tube, and the distance you are from the animal.

1 Look for this relationship and compare your results from the different scopes as worked on the Investigation in Lesson 3–2.

2 Use your data to write a proportion for each scope.

3 Graph these equations on the same coordinate plane. Write any observations you make about these graphs.

Add the results of your work to your Investigation Folder.

Applications of Linear Programming

3-6

What YOU'LL LEARN

- To solve problems involving maximum and minimum values by using linear programming techniques.

Why IT'S IMPORTANT

You can use linear programming to solve problems involving manufacturing and business.

APPLICATION
Business

Mrs. Fernandez's Cookie Factory makes different-sized packages of cookies that contain a combination of chocolate chip and peanut butter cookies. Mrs. Fernandez places at least three of each type of cookie in each of the combination packages. The largest package, the "Baker's Dozen," contains thirteen cookies. It costs Mrs. Fernandez 19¢ to make a chocolate chip cookie and 13¢ to make a peanut butter cookie. She sells them for 44¢ and 39¢, respectively. How should Mrs. Fernandez package the cookies to maximize profit? Which packaging will yield the least profit?

Linear programming can be used to solve many types of problems like the one above. These problems have certain restrictions placed on the variables, and some function of the variables must be maximized or minimized. The steps used to solve a problem using linear programming are listed below.

Linear Programming Procedure	1. **Define the variables.** 2. **Write a system of inequalities.** 3. **Graph the system of inequalities.** 4. **Find the coordinates of the vertices of the feasible region.** 5. **Write an expression to be maximized or minimized.** 6. **Substitute the coordinates of the vertices into the expression.** 7. **Select the greatest or least result. Answer the problem.**

Use this procedure to solve the problem given above.

F Y I

The largest cookie ever made was a chocolate chip cookie made in Arcadia, California, on October 15, 1993. It was a rectangle 35 feet by 28 feet 7 inches, and it contained more than 3 million chocolate chips.

Let c represent the number of chocolate chip cookies, and let p represent the number of peanut butter cookies in a package. Write the system of inequalities and graph.

$$c \geq 3$$
$$p \geq 3$$
$$c + p \leq 13$$

The vertices of the feasible region are at $(3, 3)$, $(3, 10)$, and $(10, 3)$.

Then write an equation for the profit.

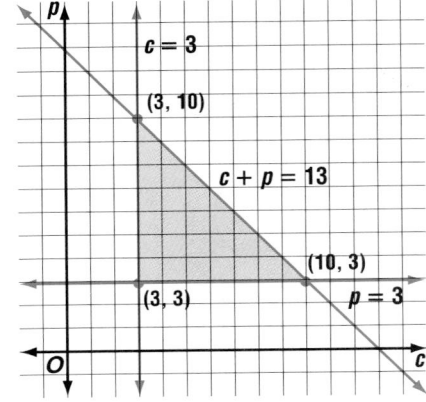

	Selling Price		Baking Cost		Profit
Chocolate Chip:	44¢	−	19¢	=	25¢
Peanut Butter:	39¢	−	13¢	=	26¢

The profit function is $f(c, p) = 25c + 26p$.

Make a chart to find the maximum and minimum profit.

(c, p)	25c + 26p	f(c, p)
(3, 3)	25(3) + 26(3)	153
(3, 10)	25(3) + 26(10)	335
(10, 3)	25(10) + 26(3)	328

Using the chart, we can see that packaging 3 chocolate chip cookies with 10 peanut butter cookies yields a maximum profit, $3.35. Packaging 3 chocolate chip cookies with 3 peanut butter cookies yields a minimum profit, $1.53.

Students who use test-taking strategies can improve their scores. Making the best use of the time available is one important strategy.

Example

APPLICATION

Education

Dolores Acosta arrives at school late because her car broke down, and therefore, has only 45 minutes to complete a history exam. The exam has 2 open-ended questions and 30 multiple-choice questions. Each correct open-ended question is worth 20 points, and each multiple-choice question is worth 2 points. She knows that it usually takes her 15 minutes to answer an open-ended question and only one minute to answer a multiple-choice question. Assume that for each question Dolores answers, she receives full credit. How many of each type of question should she answer to receive the maximum possible points?

Let e represent the number of open-ended questions answered, and let m represent the number of multiple-choice questions answered.

The number of open-ended questions must be between 0 and 2, inclusive.

$0 \le e \le 2$

The number of multiple-choice questions must be between 0 and 30, inclusive.

$0 \le m \le 30$

The total time spent on the test must be less than or equal to 45 minutes.

$15e + 1m \le 45$

Graph the system. The vertices of the feasible region are at (0, 0), (2, 0), (2, 15), (1, 30), and (0, 30).

The function that describes the number of points earned is $f(e, m) = 20e + 2m$.

(e, m)	20e + 2m	f(e, m)
(0, 0)	20(0) + 2(0)	0
(2, 0)	20(2) + 2(0)	40
(2, 15)	20(2) + 2(15)	70
(1, 30)	20(1) + 2(30)	80
(0, 30)	20(0) + 2(30)	60

Dolores should answer 1 open-ended question and 30 multiple-choice questions to get a maximum score of 80.

Communicating Mathematics

Study the lesson. Then complete the following.

1. **You Decide** Pan says that a function always has a minimum value for a given region. Lula says that sometimes a function does not have a minimum value. Who is correct and why?

2. **Explain** why coordinates of the vertices of the feasible region produce the maximum and minimum values in a linear programming situation.

Math Journal

3. **Write** a paragraph explaining in your own words how to solve a linear programming problem.

Guided Practice

4. NaKisha Heyman has just finished writing a research paper. She has hired a typist who will type the paper using a word processor. The typist charges $3.50 per page if no charts or graphs are used and $8.00 per page if a chart or graph appears on the page. NaKisha knows there will be at most 40 pages having no charts or graphs. There will be no more than 16 pages with charts or graphs, and the paper will be 50 pages or less.

   a. Write inequalities that limit the number of plain pages to be typed.

   b. Write inequalities that limit the number of pages with charts or graphs to be typed.

   c. Write an equality that expresses the total number of pages to be prepared.

   d. Draw the graph showing the feasible region.

   e. List the coordinates of all the vertices of the feasible region.

   f. Write an expression for the cost to have the paper typed.

   g. Which vertex produces the greatest cost?

   h. What is the greatest possible cost to have the paper typed?

Applications and Problem Solving

5. **Manufacturing** Superbats, Inc., manufactures two different quality wood baseball bats, the Wallbanger and the Dingbat. The Wallbanger takes 8 hours to trim and turn on a lathe and 2 hours to finish it. It has a profit of $17. The Dingbat takes 5 hours to trim and turn on a lathe and 5 hours to finish, but its profit is $29. The total time per day available for trimming and lathing is 80 hours and for finishing is 50 hours.

   a. If w represents the number of Wallbangers produced per day and d represents the number of Dingbats produced per day, write a system of inequalities to represent the number of Wallbanger and Dingbats that can be produced per day.

   b. Draw the graph showing the feasible region.

   c. Write an expression for the profit per day.

   d. How many of each type of bat should be produced to have the maximum profit? What is the maximum profit?

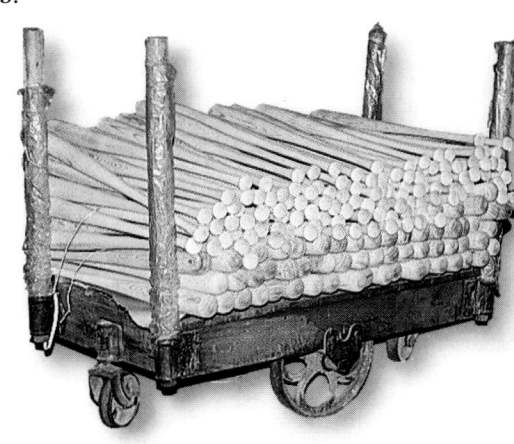

6. **Business** The available parking area of a parking lot is 600 square meters. A car requires 6 square meters of space, and a bus requires 30 square meters of space. The attendant can handle no more than 60 vehicles.

   a. Let c represent the number of cars, and let b represent the number of buses. Write a system of inequalities to represent the number of cars and buses that can be parked on the lot.

   b. If the parking fees are $2.50 for cars and $7.50 for buses, how many of each type of vehicle should the attendant accept to maximize income? What is the maximum income?

   c. The parking fees for special events are $4.00 for cars and $8.00 for buses. How many of each vehicle should the attendant accept during a special event to maximize income? What is the maximum income?

7. **Veterinary Medicine** The table below shows the amounts of nutrient A and nutrient B in two types of dog food, X and Y.

Food Type	Amount of Ingredient A	Amount of Ingredient B
X	1 unit per pound	$\frac{1}{2}$ unit per pound
Y	$\frac{1}{3}$ unit per pound	1 unit per pound

   The dogs in Kay's K-9 Kennel must get at least 40 pounds of food per day. The food may be a mixture of foods X and Y. The daily diet must include at least 20 units of nutrient A and at least 30 units of nutrient B. The dogs must not get more than 100 pounds of food per day.

   a. Food X costs $0.80 per pound and food Y costs $0.40 per pound. What is the least possible cost per day for feeding the dogs?

   b. If the price of food X is raised to $1.00 per pound, and the price of food Y stays the same, should Kay change the combination of foods she is using? Explain why or why not.

8. **Manufacturing** One of the dolls that Dolls R Us manufactures is Talking Tommy. Another doll without the talking mechanism is called Silent Sally. In one hour, the company can produce 8 Talking Tommy dolls or 20 Silent Sally dolls. Because of the demand, the company knows that it must produce at least twice as many Talking Tommy dolls as Silent Sally dolls. The company spends no more than 48 hours per week making these two dolls. The profit on each Talking Tommy is $3.00, and the profit on each Silent Sally is $7.50.

   a. How many of each doll should be produced to maximize profit each week?

   b. What is the profit?

9. **Retail** A sales associate at a paint store plans to mix color A and color B. The sales associate has exactly 32 units of blue dye and 54 units of red dye. Each gallon of color A requires 4 units of blue dye and one unit of red dye. Each gallon of color B requires one unit of blue dye and 6 units of red dye.

   a. Let a represent the number of gallons of color A, and let b represent the number of gallons of color B. Write the inequalities that represent the number of gallons of paint that can be mixed.

   b. Find the maximum number of gallons, $a + b$, that can be mixed.

10. **Manufacturing** TeeVee Electronics, Inc., makes console and wide-screen televisions. The equipment in the factory allows for making at most 450 console televisions and 200 wide-screen televisions in one month. The chart below shows the cost of making each type of television, as well as the profit for each type.

Television	Cost per Unit	Profit per Unit
Console	$600	$125
Wide Screen	$900	$200

During the month of November, the company can spend $360,000 to make these televisions. To maximize profits, how many of each type should they make?

11. **Education** Carol Sommers has 50 minutes to take an English test that has 20 multiple-choice questions and 20 short-answer questions. She knows she can answer a multiple-choice question in $1\frac{1}{2}$ minutes and a short-answer question in 2 minutes. Each correct multiple-choice answer receives 2 points, and each correct short-answer receives 3 points. Assume that any question Carol answers, she gets correct.
   a. What is the maximum possible score she can receive?
   b. What advice might you give Carol to improve her score?

Critical Thinking

12. Consider the feasible region defined by the system of inequalities below.

$0 \leq x \leq 5$
$0 \leq y \leq 6$
$x + 2y \leq 13$
$2x + y \leq 11$

   a. Suppose the profit function for the feasible region is $f(x, y) = 3x + 4y$. Graph the feasible region. On the same coordinate plane, graph the profit function for $f(x, y) = 32, 28, 24, 20$, and 16. What does the graph tell you about the maximum point?
   b. Suppose the profit function for the feasible region is $g(x, y) = 3x + 6y$. Graph the feasible region on another coordinate plane. Then add the graph of the profit function for $g(x, y) = 42, 36, 30$, and 24. What does the graph tell you about the maximum point?

Mixed Review

13. Graph the system of inequalities $x \geq 0, y \geq 3, y \geq 2x + 1$ and $y \leq -0.5x + 6$. Name the coordinates of the vertices of the feasible region. Find the maximum and minimum values of the function $f(x, y) = 3x - 2y$ for this region. (Lesson 3–5)

14. **Geometry** Write a system of inequalities that will form a region shaped like an isosceles right triangle. (Lesson 3–4)

15. At 10:00 A.M., Monsa had traveled 195 miles across the plains states on his way to California. At 2:00 P.M., he had traveled 415 miles. Use slope to calculate his rate of travel. (Lesson 2–3)

16. Graph $5y - 25x = -10$. (Lesson 2–2)

17. State an absolute value inequality for the graph at the right. (Lesson 1–7)

$-5\ -4\ -3\ -2\ -1\ \ 0\ \ 1\ \ 2\ \ 3\ \ 4\ \ 5$

Solving Systems of Equations in Three Variables

What YOU'LL LEARN

• To solve a system of three equations in three variables.

Why IT'S IMPORTANT

You can use systems of equations to solve problems involving banking and consumer awareness.

APPLICATION

Business

The Nutty Food Company sells trail mixes and other snacks by the pound. A clerk is filling a barrel with peanuts, raisins, and carob-coated pretzels. The manager wants the associate to make 80 pounds of this mixture and to sell it for $3.35 per pound. The peanuts sell for $3.20 per pound, the raisins sell for $2.40 per pound, and the carob-coated pretzels sell for $4.00 per pound. If the mixture has twice as many pounds of carob-coated pretzels as raisins, how many pounds of each ingredient should the clerk use?

Problems like this can be expressed using a system of equations in three variables. Let p represent the number of pounds of peanuts, let r represent the number of pounds of raisins, and let c represent the number of pounds of carob-coated pretzels. Then write a system of equations using the information given.

$$p + r + c = 80 \qquad \textit{The clerk makes 80 pounds of the mixture.}$$
$$3.20p + 2.40r + 4.00c = 268 \qquad \textit{The total selling price is } 80 \times \$3.35, \textit{ or } \$268.$$
$$c = 2r \qquad \textit{There are twice as many pounds of carob-coated pretzels as raisins.}$$

Solving systems such as this one is similar to solving systems of equations in two variables.

Since $c = 2r$, substitute $2r$ for c in each of the first two equations.

$$p + r + (2r) = 80 \quad \rightarrow \quad p + 3r = 80$$
$$3.20p + 2.40r + 4.00(2r) = 268 \quad \rightarrow \quad 3.20p + 10.40r = 268$$

The result is two equations with the same two variables. Use elimination to solve for r.

$$\begin{array}{l} p + 3r = 80 \\ 3.20p + 10.40r = 268 \end{array} \quad \boxed{\textbf{Multiply by 3.20.}} \rightarrow$$

$$3.20p + 9.60r = 256$$
$$(-)\ 3.20p + 10.40r = 268$$

Subtract to eliminate p. $\quad -0.8r = -12$

Solve for r. $\qquad r = 15$

Since $r = 15$, substitute 15 for r in the equation $p + 3r = 80$.

$$p + 3(15) = 80$$
$$p + 45 = 80$$
$$p = 35 \quad \textit{Solve for p.}$$

Finally, substitute 15 for r in the original equation $c = 2r$.

$$c = 2(15)$$
$$= 30$$

The associate should use 35 pounds of peanuts, 15 pounds of raisins, and 30 pounds of carob-coated pretzels.

The solution of a system of equations in three variables x, y, and z, is called an **ordered triple** (x, y, z).

Example **Solve the system of equations.**

$x + 2y - 3z = 50$
$2x + y + 2z = 3$
$2x - 5y + 4z = -79$

Use elimination to make a system of two equations in two variables.

$x + 2y - 3z = 50$ ► Multiply by 2. ► $2x + 4y - 6z = 100$

$2x + y + 2z = 3$ ► Multiply by -1. ► $(+) -2x - y - 2z = -3$
Add to eliminate x. $\quad 3y - 8z = 97$

$2x + y + 2z = 3$
$2x - 5y + 4z = -79$ ► Multiply by -1. ► $2x + y + 2z = 3$
$(+) -2x + 5y - 4z = 79$
Add to eliminate x. $\quad 6y - 2z = 82$

The result is two equations with the same two variables. Use elimination to solve for z.

$3y - 8z = 97$
$6y - 2z = 82$ ► Multiply by -2. ► $-6y + 16z = -194$
$(+) 6y - 2z = \quad 82$
Add to eliminate y. $\quad 14z = -112$
Solve for z. $\quad z = -8$

Substitute -8 for z in the equation $3y - 8z = 97$.

$3y - 8(-8) = 97$
$3y + 64 = 97$
$3y = 33$
$y = 11$ *Solve for y.*

Substitute 11 for y and -8 for z in the original equation $x + 2y - 3z = 50$.

$x + 2(11) - 3(-8) = 50$
$x + 22 + 24 = 50$
$x + 46 = 50$
$x = 4$ *Solve for x.*

The solution is $(4, 11, -8)$.

LOOK BACK

You can refer to Lesson 3-1 for information on the number of solutions for systems of two linear equations in two variables.

You know that a system of two linear equations in two variables does not always have a unique solution that is an ordered pair. Similarly, a system of three linear equations in three variables does not always have a unique solution that is an ordered triple. The graph of each equation in a system of three linear equations in three variables is a plane. The three planes can appear in various configurations.

The three planes intersect at one point, so the system has a unique solution, an ordered triple (x, y, z).

The three planes intersect in a line. There is an infinite number of solutions to the system.

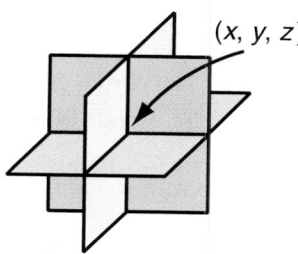
(x, y, z)

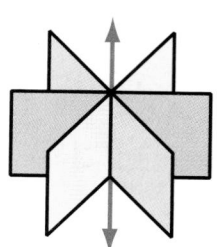

Each of the figures below shows three planes that have *no* points in common. These systems of equations have *no* solutions.

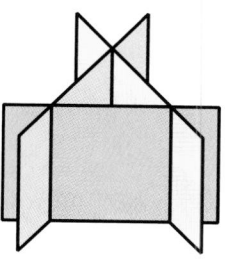 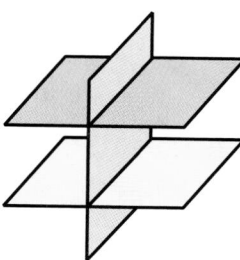 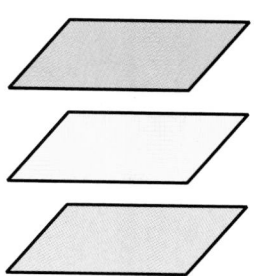

If all three planes coincide, there are infinite number of solutions. Also if two planes coincide and intersect the third plane in a line, there are infinite number of solutions.

When graphing two equations in two variables, it is obvious when the system is inconsistent or dependent. However, it is usually *not* obvious when solving a system of three equations algebraically whether there is a unique solution, no solutions, or many solutions.

Example **Solve the system of equations.**
$3x - 6y + 3z = 33$
$2x - 4y + 2z = 22$
$4x + 2y - z = -6$

Eliminate x in the first two equations.

$3x - 6y + 3z = 33$ **Multiply by 2.** ➡ $6x - 12y + 6z = 66$

$2x - 4y + 2z = 22$ **Multiply by −3.** ➡ $\underline{(+) -6x + 12y - 6z = -66}$
$$0 = 0$$

The equation $0 = 0$ is always true. This indicates that there are infinite number of solutions. In this case, the first two equations represent the same plane. This plane intersects the plane represented by the third equation. The coordinates of any point on the line of intersection are a solution to the system of equations.

3 The three American universities with the greatest endowments are Harvard, Yale, and Princeton. Their combined endowments are $12.09 billion. Together Yale and Princeton have $0.53 billion more in endowments than Harvard. Princeton's endowments trail Harvard's by $2.70 billion. What are the endowments of each of these universities?

APPLICATION

Education

F Y I

Funds held by a university, hospital, or other institution are called *endowments*. Usually these funds are invested, and only the income from these investments is spent. In this way, the endowment can last forever.

Explore Read the problem and define the variables.
Let h represent Harvard's endowments, let y represent Yale's endowments, and let p represent Princeton's endowments.

Plan Write three equations.

$h + y + p = 12.09$ *The combined endowments are $12.09 billion.*
$y + p = h + 0.53$ *Together Yale and Princeton have $0.53 billion more than Harvard.*

$p = h - 2.70$ *Princeton's endowments trail Harvard's by $2.70 billion.*

Solve Use the elimination method to solve for h.

$$\begin{array}{lll} h + y + p = 12.09 & \rightarrow & h + y + p = 12.09 \\ y + p = h + 0.53 & \rightarrow & (-) \; -h + y + p = 0.53 \\ \hline & & 2h = 11.56 \\ & & h = 5.78 \end{array}$$

Substitute 5.78 for h in the equation $p = h - 2.70$.

$$p = 5.78 - 2.70$$
$$= 3.08$$

Substitute 5.78 for h and 3.08 for p in the equation $h + y + p = 12.09$.

$$5.78 + y + 3.08 = 12.09$$
$$y + 8.86 = 12.09$$
$$y = 3.23 \quad \text{Solve for } y.$$

Harvard has $5.78 billion, Yale has $3.23 billion, and Princeton has $3.08 billion.

Examine Check to see if all of the criteria are met.

The combined endowments are $12.09 billion.
$$5.78 + 3.23 + 3.08 = 12.09 \;\checkmark$$

Yale and Princeton have $0.53 billion more than Harvard.
$$3.23 + 3.08 = 5.78 + 0.53 \;\checkmark$$

Princeton's endowments trail Harvard's by $2.70 billion.
$$3.08 = 5.78 - 2.70 \;\checkmark$$

Communicating Mathematics

Study the lesson. Then complete the following.

1. Refer to the application at the beginning of the lesson. Explain why substitution is the best first step in solving the system of equations.

2. **Describe** a situation that might occur when you solve a system of three equations in three variables that has each of the following number of solutions.

   a. none **b.** an infinite number

Guided Practice

For each system of equations, an ordered triple is given. Determine whether it is a solution of the system.

3. $3x - 7y + 2z = 43$
   $5x + 2y - 3z = -1$
   $2x + 5y - z = -12$; $(4, -3, 5)$

4. $5a - 3b + c = -3$
   $7a + 2b - 3c = -35$
   $a - 6b + 7c = 51$; $(-2, 0, 7)$

Solve each system of equations.

5. $4x - 3y + 5z = 43$
   $2x + y = 9$
   $3y - 2z = -9$

6. $6a - 2b = 18$
   $3b + 5c = -34$
   $a + 6c = -28$

7. $4x + 3y + 2z = 34$
   $2x + 4y + 3z = 45$
   $3x + 2y + 4z = 47$

8. $x + y + z = -1$
   $3x - 2y - 4z = 16$
   $2x - y + z = 19$

9. Write a system of three equations in three variables that has $(-6, 2, -5)$ as a solution.

10. Write a system of three equations in three variables in which $(-2, -3, 6)$ satisfies only two of the three equations.

Practice

For each system of equations, an ordered triple is given. Determine whether it is a solution of the system.

11. $3a + 7b - 4c = -17$
    $2a - 8b - c = 8$
    $6a - b + 3c = 23$; $(2, -1, 4)$

12. $x + 3z = -5$
    $5x - 2y = -22$
    $5y - 6z = 36$; $(-2, 6, -1)$

Solve each system of equations.

13. $5r + 2s = 0$
    $-3t = 12$
    $6s + 5t = 10$

14. $2b - c = -13$
    $2a = 12$
    $3a + b = 13$

15. $x + y - z = -1$
    $x + y + z = 3$
    $3x - 2y - z = -4$

16. $b + c = 4$
    $2a + 4b - c = -3$
    $3b = -3$

17. $5x + 7y = -1$
    $-2y + 3z = 9$
    $7x - z = 27$

18. $r - s + 3t = -8$
    $2s - t = 15$
    $3r + 2t = -7$

19. $5a - b + 3c = 5$
    $2a + 7b - 2c = 5$
    $4a - 5b - 7c = -65$

20. $6x + 2y - 3z = -17$
    $7x - 5y + z = 72$
    $2x + 8y + 3z = -21$

Solve each system of equations.

21. $3x + 4y - 3z = 5$
$x + 6y + 2z = 3$
$6x + 2y + 3z = 4$

22. $4x + 7y - z = -10$
$6x - 3y + 6z = 3$
$2x + y + 8z = 9$

23. $2r + 3s + 4t = 3$
$5r - 9s + 6t = 1$
$\frac{1}{3}r - \frac{1}{2}s + \frac{1}{3}t = \frac{1}{12}$

24. $2x + y + z = 7$
$12x - 2y - 2z = 2$
$\frac{2x}{3} - y + \frac{z}{3} = -\frac{1}{3}$

25. The sum of three numbers is 12. The first is five times the second and the sum of the first and third is 9. Find the numbers.

26. The sum of three numbers is 20. The first number is the sum of the second and the third. The third number is three times the first. Find the numbers.

27. The sum of three numbers is 18. The first is eight times the sum of the second and third. The sum of the first number and the last number is 11. Find the numbers.

Critical Thinking

28. Now that you know how to solve a system of three equations in three variables, use what you know to solve the system of equations below.

$w + x + y + z = 2$
$2w - x - y + 2z = 7$
$2w + 3x + 2y - z = -2$
$3w - 2x - y - 3z = -2$

Applications and Problem Solving

29. Banking Maria Hernandez has $15,000 that she would like to invest in certificates of deposit. The bank has the following rates.

Number of Years	1	2	3
Rate	3.4%	5.0%	6.0%

She does not want to have all her money committed for more than one year, so she plans to invest some money at each rate. She wants her total interest for one year to be $800, so she will not be in a higher tax bracket. She decides to put $1000 more in a 2-year certificate than in a 1-year certificate and invest the rest in a 3-year certificate. How much should she invest in each type of certificate?

30. Consumer Awareness Jack-in-the-Box offers three different types of hamburgers at three different prices. The types are the Hamburger, the Double Cheeseburger, and the Jumbo Jack. The decathlon team at Kennedy High School went to the local Jack-in-the-Box on three different occasions and ordered different burgers. The first time, they ordered 3 Hamburgers, 5 Double Cheeseburgers, and 6 Jumbo Jacks and paid $25.24. The second time, they ordered 2 Hamburgers, 7 Double Cheeseburgers, and 5 Jumbo Jacks and paid $25.68. The last time, they ordered 4 Hamburgers, 4 Double Cheeseburgers, and 7 Jumbo Jacks and paid $26.59. The coach, who did not go along on the burger-buying trips, needs to know the price of each type of burger. What is the price of each kind of burger?

31. **Basketball** One night Glen Rice of the NBA's Miami Heat scored a total of 35 points against the Los Angeles Clippers. In basketball, it is possible to make a 3-point field goal, a 2-point field goal, or a 1-point free throw. He made as many 2-pointers as 3-pointers and free throws combined. He scored one point more with 2-pointers than he did with 3-pointers and free throws combined. How many of each did he score?

Mixed Review

32. **Manufacturing** Stitches Inc. can make at most 30 jean jackets and 20 leather jackets in a week. It takes a worker 10 hours to make a jean jacket and 20 hours to make a leather jacket. The total number of hours by all of the employees can be no more than 500 hours per week. (Lesson 3–6)

   a. If the profit on a jean jacket is the same as the profit on a leather jacket, how many of each should be made to maximize profit?

   b. How many of each should be made if the profit on a leather jacket is three times the profit on a jean jacket?

33. Which system of inequalities is represented by the graph at the right? (Lesson 3–4)

   a. $x + y \geq 4$ and $x \leq 2y$
   b. $x + y \geq 4$ and $x \geq 2y$
   c. $x + y \leq 4$ and $x \leq 2y$
   d. $x + y \leq 4$ and $x \geq 2y$

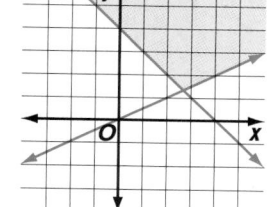

34. **Telecommunications** The formula relating the cost of a long-distance phone call in which the initial cost for the call is $0.50 for the first minute and each additional minute costs $0.95 is $c = 0.95(t - 1) + 0.50$, where t is the length of the call in minutes. The length of a call is always rounded up to the next minute if it includes part of a minute. For example, a 2.3 minute call is charged at a rate of 3 minutes. (Lesson 2–6)

   a. Determine the cost of calls lasting 2 minutes, 2.25 minutes, 2.5 minutes, 3 minutes, 3.75 minutes, and 4 minutes.

   b. Write the minute amounts and charges as ordered pairs and graph the function.

   c. Use your graph to predict the cost of a call lasting 8 minutes.

   d. Describe how the complete graph of the formula for calculating the cost of phone calls would look. What type of function is this?

35. **Economics** The Serves-You-Best Rental Car Company has two rental offers. The first offer charges the customer $20 plus 25¢ a mile for a compact car. The second offer charges the customer $35 plus 25¢ a mile for a luxury car. (Lesson 2–4)

   a. Write an equation to represent each offer, if the cars are rented for one day.

   b. What is the relationship between the graphs of these offers?

   c. If both cars are driven 750 miles, what is the cost difference between renting the compact car and the luxury car?

36. If $f(x) = x^2 + 3x$, find $f(5)$. (Lesson 2–1)

37. Solve $2 \left| -x - 6 \right| = -3x$. (Lesson 1–5)

3–7B Graphing Equations in Three Variables

Materials: isometric dot paper

An Extension of Lesson 3–7

To draw the graph of an equation in three variables, it is necessary to add a third dimension to our coordinate system. The graph of an equation of the form $Ax + By + Cz = D$, where either A, B, C, or D can be equal to zero, is a plane.

When graphing in space (three dimensions), space is separated into eight regions, called **octants.** Think of three coordinate planes intersecting at right angles as shown at the right. Any point lying on a coordinate plane is not in an octant.

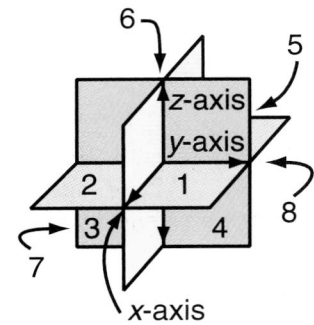

The octants are numbered as shown.

Activity 1 **Use isometric dot paper to draw and label a three-dimensional axis system. Then graph the ordered triple (3, 6, 1).**

Step 1 Draw the x-, y-, and z-axes as shown below.

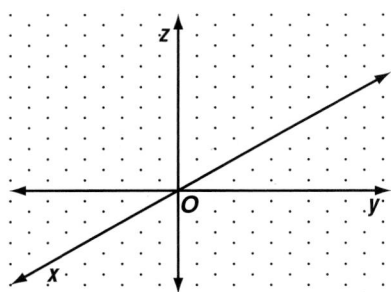

Step 2 Locate 3 on the positive x-axis, 6 on the positive y-axis, and 1 on the positive z-axis. Complete a "box" by drawing lines parallel to the axes through each intercept. Draw the graph, which is a point, at (3, 6, 1).

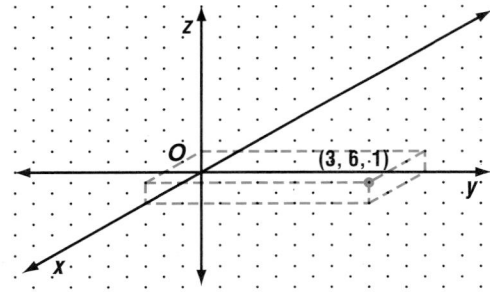

This point is in octant 1, which is also called the <u>first octant</u>.

It is not necessary to show the entire "box" when you graph an ordered triple. The desired point will always be the corner farthest from the point of origin.

To graph a linear equation in three variables, first find the intercepts of the graph. Connect the intercepts on each axis. This forms a portion of a plane that lies in a single octant.

Activity 2   Graph $2x + 4y + 3z = 12$.

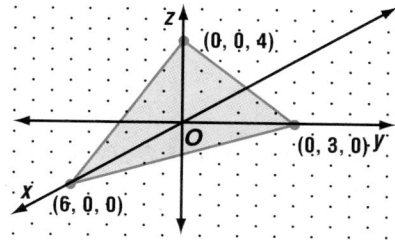

To find the x-intercept, let $y = 0$ and $z = 0$.
 $2x = 12$
 $x = 6$
To find the y-intercept, let $x = 0$ and $z = 0$.
 $4y = 12$
 $y = 3$
To find the z-intercept, let $x = 0$ and $y = 0$.
 $3z = 12$
 $z = 4$
To indicate the plane, graph the intercepts, which have coordinates $(6, 0, 0)$, $(0, 3, 0)$, and $(0, 0, 4)$, respectively, and then connect the points. Remember that the plane extends indefinitely.

....

Draw **Graph each ordered triple using isometric dot paper. Name the octant in which each point lies.**

    1. $(5, 2, 3)$ **2.** $(7, 5, -6)$ **3.** $(3, 0, 1)$ **4.** $(3, -7, 2)$

    Graph each equation using isometric dot paper. Name the coordinates for the x-, y-, and z-intercepts.

    5. $4x + y + 2z = 4$ **6.** $3x - 2y + 2z = 6$

    7. $3x - y + 6z = 3$ **8.** $4x + 5y - 10z = 20$

    9. $3z - 2x = 6$ **10.** $3x - 4y = -12$

Write **Write an equation of the plane given its x-, y-, and z-intercepts, respectively.**

    11. $2, -2, 5$ **12.** $\frac{1}{2}, 3, -2$

    13. Describe the relationship between quadrants and octants.

    14. Consider the graph of $x = 2$ in one, two, and three dimensions.

        a. Describe the graph on a number line.
        b. Describe the graph on a coordinate plane.
        c. Describe the graph in a three-dimensional coordinate axis.
        d. Compare the graphs in parts a, b, and c.

In·ves·ti·ga·tion

Through the Looking Glass

Refer to the Investigation on pages 60–61.

Through observations of species in their habitat, naturalists have been able to identify characteristics of the species, their migratory and breeding patterns, and the size of their population in any given area. These efforts have helped in preventing extinction of some species and better control for those species that are over-populating some areas of Earth. Continued observations with more advanced equipment will make the naturalist's job easier in the future and offer an abundance of information that cannot be acquired from other long-distance observations.

Analyze

You have conducted experiments and organized your data in various ways. It is now time to analyze your findings and state your conclusions.

PORTFOLIO ASSESSMENT

You may want to keep your work on this Investigation in your portfolio.

1 Write an expression that relates the size of the image through your scope with the distance you are from the image for each scope. How does this relate to the dimensions of each scope?

2 Study the various graphs you made for each of the scopes. Write an explanation of how the scope can help you determine the actual size of an animal you are observing.

3 If you were trying to observe an object from 100 yards away, describe the tube that would enable you to see the entire object in your view.

4 Make a general statement about the dimensions of the tube in relation to the distance you are from an object when trying to view it in its entirety.

Write

Imagine that you are a tourist on safari on an African plain. There is an elephant standing a distance away across the river. The current of the river is swift and you cannot cross it. You have in your possession a tape measure, a pen, a calculator, and a piece of paper that can be rolled into various sizes of view tubes.

5 You would like to determine the height of the elephant. Write how you would explain to another tourist how you could accurately determine the elephant's height.

VOCABULARY

After completing this chapter, you should be able to define each term, property, or phrase and give an example or two of each.

Algebra

consistent system   (p. 127)
constraints   (p. 153)
Cramer's rule   (p. 141)
dependent system   (p. 128)
determinant   (p. 141)
element   (p. 141)
elimination method   (p. 133)
inconsistent system   (p. 128)
independent system   (p. 127)
linear programming   (p. 155)

octants   (p. 172)
ordered triple   (p. 166)
second-order determinant   (p. 141)
substitution method   (p. 133)
system of equations   (p. 126)
system of inequalities   (p. 148)
unbounded   (p. 155)

Problem Solving

solve a simpler problem   (p. 156)

UNDERSTANDING AND USING THE VOCABULARY

Choose the letter of the term that best matches each phrase.

1. a square array of numbers or variables having a numerical value

2. a system of equations that has an infinite number of solutions

3. the region of intersection of graphs of inequalities, where every constraint is met

4. a method of solving equations in which one equation is solved for one variable in terms of the other variable

5. a system of equations that has at least one solution

6. a method of solving equations in which one variable is eliminated when the two equations are combined

7. the solution of a system of equations in three variables (x, y, z)

8. a method for finding the maximum or the minimum value of a function with two variables.

9. the numbers or variables written within a determinant

10. a system of equations that has exactly one solution

11. a function in which no maximum value exists

a. consistent system
b. dependent system
c. determinant
d. elements
e. elimination method
f. feasible region
g. independent system
h. linear programming
i. ordered triple
j. substitution method
k. unbounded

OBJECTIVES AND EXAMPLES

Upon completing this chapter, you should be able to:

- solve systems of equations by graphing
  (Lesson 3–1)

Solve the system of equations.
$$y + x = 3$$
$$3x - y = 1$$

Graph each equation. The intersection of the graphs is the solution.

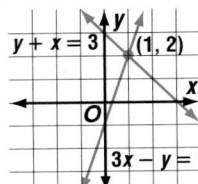

The solution is (1, 2).

REVIEW EXERCISES

Use these exercises to review and prepare for the chapter test.

Graph each system of equations and state its solution. Also, state whether the system is *consistent and independent, consistent and dependent*, or *inconsistent*.

12. $3x + 2y = 12$
$x - 2y = 4$

13. $8x - 10y = 7$
$4x - 5y = 7$

14. $y - 2x = 8$
$y = \frac{1}{2}x - 4$

15. $20y + 13x = 10$
$0.65x + y = 0.5$

- use the substitution and elimination methods to solve systems of equations   (Lesson 3–2)

Solve the system of equations.
$$x = 4y + 7$$
$$x = -y - 3$$

$$-y - 3 = 4y + 7 \qquad \Big| \qquad x = 4(-2) + 7$$
$$-10 = 5y \qquad \qquad = -8 + 7$$
$$-2 = y \qquad \qquad = -1$$

The solution is (−1, −2).

Solve each system of equations. Use either substitution or elimination.

16. $x + y = 4$
$x - y = 8.5$

17. $2x + 3y = -6$
$3x + 2y = 25$

18. $7y - 2x = 10$
$-3y + x = -3$

19. $-6y - 2x = 0$
$11y + 3x = 4$

20. $3x - 5y = -13$
$4x + 2y = 0$

21. $c + d = 5$
$2c - d = 4$

- solve systems of equations by using Cramer's rule   (Lesson 3–3)

Solve the system of equations.
$$4x + 7y = -1$$
$$2x + y = 7$$

$$x = \frac{\begin{vmatrix} -1 & 7 \\ 7 & 1 \end{vmatrix}}{\begin{vmatrix} 4 & 7 \\ 2 & 1 \end{vmatrix}} = \frac{-1(1) - 7(7)}{4(1) - 7(2)} \text{ or } 5$$

$$y = \frac{\begin{vmatrix} 4 & -1 \\ 2 & 7 \end{vmatrix}}{\begin{vmatrix} 4 & 7 \\ 2 & 1 \end{vmatrix}} = \frac{4(7) - (-1)(2)}{4(1) - 7(2)} \text{ or } -3$$

The solution is (5, −3).

Use Cramer's rule to solve each system of equations.

22. $2x - 3y = 4$
$x + 5y = 2$

23. $7x + 3y = 5$
$2x + 4y = 3$

24. $2x - y = 7$
$x + 3y = 7$

25. $u + 11 = 8v$
$8(u - v) = 3$

26. $f - 2g = -1$
$2f + 3g = -16$

27. $m - n = 0$
$4m + 10n = -6$

OBJECTIVES AND EXAMPLES

- solve systems of inequalities by graphing (Lesson 3–4)

  Solve the system of inequalities by graphing.
  $y - x \leq 2$
  $0.5x + y \geq -4$

  $y - x \leq 2$ represents Regions 1 and 2.
  $0.5x + y \geq -4$ represents Regions 2 and 3.

  The intersection is Region 2, which is the solution of this system of inequalities.

  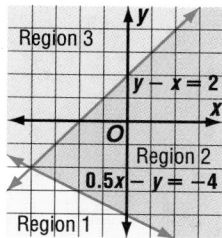

- find the maximum and minimum values of a function over a region using linear programming techniques   (Lesson 3–5)

  $x \geq 0$
  $y \geq 0$
  $3x + y \leq 15$
  $y \leq 6$
  $f(x, y) = 3x + y$

  vertices: (0, 0), (5, 0),
  (3, 6), (0, 6)

(x, y)	3x + y	f(x, y)
(0, 0)	3(0) + 0	0
(5, 0)	3(5) + 0	15
(3, 6)	3(3) + 6	15
(0, 6)	3(0) + 6	6

  maximum value = 15, minimum value = 0

- solve problems involving maximum and minimum values by using linear programming techniques   (Lesson 3–6)

  The available parking area of a parking lot is 600 m². A car requires 6 m² of space, and a bus requires 30 m² of space. The attendant can handle no more than 60 vehicles. If a car is charged $3.00 to park and a bus is charged $8.00, how many of each should the attendant accept to maximize income?   50 cars, 10 buses

REVIEW EXERCISES

Solve each system of inequalities by graphing.

28. $y \leq 4$
    $y > -3$

29. $y > 3$
    $x \leq 1$

30. $y < x + 1$
    $x > 5$

31. $x + y \geq 3$
    $x \leq 0$

32. $y \leq x + 4$
    $2y \geq x - 3$

33. $y < 2$
    $y \geq -7$
    $y \geq 2x$
    $y \leq x + 1$

Graph each system of inequalities. Name the coordinates of the vertices of the feasible region. Find the maximum and minimum values of the given function.

34. $f(x, y) = -2x + y$
    $x \geq -5$
    $x \leq 4$
    $y \geq -1$
    $y \leq 3$

35. $f(x, y) = 3x + 2y$
    $x \geq 0$
    $y \geq 0$
    $x + 3y \leq 15$
    $4x + y \leq 16$

36. **Community Service**   A theater at which a drug abuse program is being presented seats 150 people. The proceeds will be donated to a local drug information center. Admission is $2 for adults and $1 for students. Every two adults must bring at least one student. How many adults and students should attend in order to raise the maximum amount of money?

OBJECTIVES AND EXAMPLES

• solve a system of three equations in three variables (Lesson 3–7)

Solve the system of equations.
$2x + y - z = 2$
$x + 3y + 2z = 1$
$x + y + z = 2$

$$\begin{array}{r} -2x - y + z = -2 \\ (+)\ 2x + 6y + 4z = 2 \\ \hline 5y + 5z = 0 \\ y + z = 0 \\ y = -z \end{array}$$

$$\begin{array}{r} 2(2) + (-z) - z = 2 \\ 4 - 2z = 2 \\ -2z = -2 \\ z = 1 \end{array}$$

$$\begin{array}{r} x + (-z) + z = 2 \\ x = 2 \end{array}$$

$$\begin{array}{r} 2 + y + 1 = 2 \\ y = -1 \end{array}$$

The solution is $(2, -1, 1)$.

REVIEW EXERCISES

Solve each system of equations.

37. $x + 4y - z = 6$
$3x + 2y + 3z = 16$
$2x - y + z = 3$

38. $2a + b - c = 5$
$a - b + 3c = 9$
$3a - 6c = 6$

39. $e + f = 4$
$2d + 4e - f = -3$
$3e = -3$

APPLICATIONS AND PROBLEM SOLVING

40. **Donkey Basketball** Your school has contracted with a professional animal trainer to host a donkey basketball game at the school. The school has guaranteed an attendance of at least 1000 people and $4800 in total ticket sales. The tickets are $4 for students $6 for nonstudents, of which the animal trainer receives $3 from students and $4 from nonstudents. What is the minimum amount of money the animal trainer could receive? What is the maximum amount of money the animal trainer could receive? (Lesson 3–6)

41. **Lunch Costs** Melissa, Wes, and Daryl went to Fred's Burgers to get food for their friends at school. Melissa spent $6.35 on two burgers, one order of french fries, and two colas. Wes ordered 1 burger, 2 orders of french fries, and 2 colas. His bill was $5.45. Daryl's order of 3 burgers, 3 orders of french fries, and 3 colas totaled $11.01. Find the price of each item. (Lesson 3–7)

42. **State Fair** A dairy makes three types of cheese—cheddar, Monterey Jack and Swiss—and sells the cheese in three booths at the state fair. At the beginning of one day, the first booth received x pounds of each type of cheese. The second booth received y pounds of each type of cheese, and the third booth received z pounds of each type of cheese. By the end of the day, the dairy had sold 131 pounds of cheddar, 291 pounds of Monterey Jack, and 232 pounds of Swiss. The table below shows the percent of the cheese delivered in the morning that was sold at each booth. How many pounds of cheddar cheese did each booth receive in the morning? (Lesson 3–7)

Type	Booth 1	Booth 2	Booth 3
Cheddar	40%	30%	10%
Monterey Jack	40%	90%	80%
Swiss	30%	70%	70%

A practice test for Chapter 3 is provided on page 914.

ALTERNATIVE ASSESSMENT

COOPERATIVE LEARNING PROJECT

You have just been promoted to the position of national purchasing agent for a large rental car company. Having worked as a manager for several years in one of their local offices, you have become familiar with many different makes and models of automobiles. The executive officers of the company have narrowed the field to two models having the following characteristics.

Characteristics	Model A	Model B
Purchase price per unit	$10,000	$15,000
Number of units needed	5000	5000
Total cash outlay	$50,000,000	$75,000,000
Projected revenue per year	$36,500,000	$36,500,000
Projected expense per year	$21,900,000	$18,250,000
Projected utilization	85%	85%
Useful life per unit	3 years	4 years
Resale value per unit	$1,500	$2,000

The executive officers have asked you to create a graph of the information on the chart and make a recommendation. Here are some things you might want to show on your graph.

- total cash outlay
- purchase price of each unit
- profit line
- when each model has paid for itself
- when the two models generate the same profit

You have three recommendations to choose from: Model A, Model B or a combination of the two. Tell which recommendation you would choose and why.

What other factors might influence your decision?

THINKING CRITICALLY

- Write a system of equations in three variables that has one unique solution. Explain why there is only one solution.
- Write a system of equations in three variables that has no solution. Explain why there is no solution.
- Write a system of equations in three variables that has an infinite number of solutions. Explain why this is the case.

PORTFOLIO SUGGESTIONS

Select an item from this chapter that you feel shows your best work and place it in your portfolio. Explain why you selected it.

SELF EVALUATION

Systems of linear equations and linear inequalities occur often as models of real-life situations. As you grow older, these real-life situations will become more evident in your own life. How will you solve these problems?

Assess yourself. How well am I able to solve the problems I encounter? Write a paragraph that describes what you know about systems of equations and inequalities. Describe the kinds of problems you have solved using systems of equations. Write down parts that are difficult for you to understand and parts that are easy. This will help you to identify sections that need further study to better understand these topics.

In·ves·ti·ga·tion

3-2-1-Blast-Off!

MATERIALS NEEDED

- shoe box
- paper cups
- wooden craft sticks
- ruler with groove down the middle
- Ping-Pong™ ball
- plastic spoon
- golf tee
- masking tape
- string
- rubber bands (various sizes)
- paper clips
- measuring tape

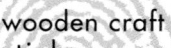

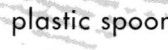
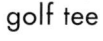

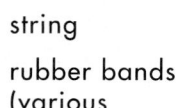

You work for a scientific research company that has just received a government contract. This contract involves designing, building, and testing a prototype launching system. Currently, the government is using inefficient designs supplied by previous companies. The government has requested rush status for this operation and is requiring a shoot-off to demonstrate the accuracy and features of the system.

Your research company must design, build, and test the launch system. It must be calibrated by shooting a number of test shots. To calibrate the launcher, shoot several test shots, mark the results, and adjust the launcher until it will hit a designated target. The test data will show the accuracy of the

system. Then a detailed individual report must be written.

The government has furnished the following raw materials to be used to construct the launching system.

- launcher component kit (shoe box, paper cups, and craft sticks)
- linear scale (ruler with groove down the middle)
- projectile (Ping-Pong™ ball)
- cradles (plastic spoon and golf tee)
- adhesive lamination (masking tape)
- rope (string)
- power supply (rubber bands of various size)
- bars (paper clips)
- test range calibration device (measuring tape)

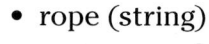
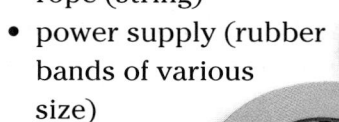

Your team of four researchers must design and build a launcher out of the raw materials provided. The launcher must be capable of hitting the government's specified target at a distance somewhere between 50 cm and 250 cm from the launcher. Since you will need to calibrate your launcher to hit a specified target at a given distance within that range, you will need to conduct several tests to perfect your design.

In this Investigation, you will examine ways in which the launcher can be designed. Once the design has been determined, the launcher will need to be built and then tested for accuracy.

Make an Investigation Folder in which you can store all of your work on this Investigation for future use.

PLAN THE DESIGN

1 Think about several different launcher designs. Look in books, magazines, and science literature to find pictures of various launchers.

2 Discuss these designs with your teammates. Narrow the list of options down to one launcher design on which you all agree.

3 Write a design proposal in which you include a blueprint of your design and launching instructions.

BUILD THE LAUNCHER

4 Use the materials that the government has furnished to build the launcher.

5 Be sure that the blueprint in your proposal matches the actual model that you built. Modify the blueprint as necessary to match your model.

6 Review the launching instructions that you wrote in your proposal for clarity and understandability. Revise your proposal as necessary.

You will continue working on this Investigation throughout Chapters 4 and 5.

Be sure to keep your chart and materials in your Investigation Folder.

3–2–1–Blast-Off! Investigation

Working on the Investigation
Lesson 4–1, p. 193
....
Working on the Investigation
Lesson 4–7, p. 231
....
Working on the Investigation
Lesson 5–3, p. 273
....
Working on the Investigation
Lesson 5–7, p. 302
....
Closing the Investigation
End of Chapter 5, p. 322
....

Using Matrices

Objectives

In this chapter, you will:

- create matrices and box-and-whisker plots to represent data,
- solve problems by using matrix logic,
- perform operations with matrices,
- use matrices to achieve transformations of geometric figures, and
- use matrices to solve systems of equations.

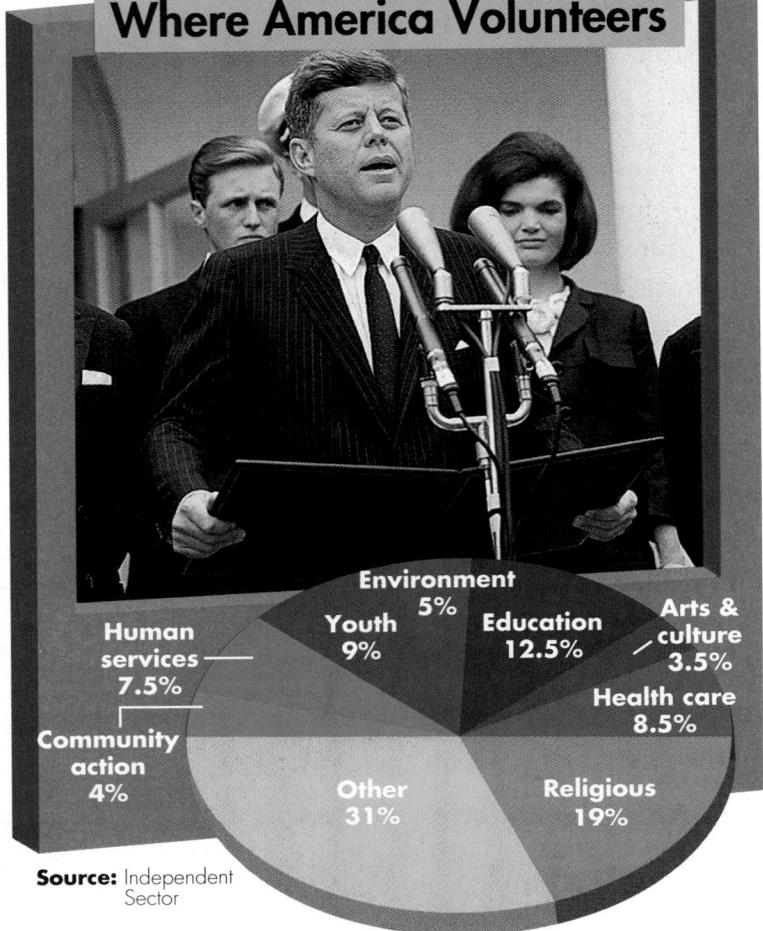

Where America Volunteers

- Environment 5%
- Youth 9%
- Education 12.5%
- Arts & culture 3.5%
- Human services — 7.5%
- Health care 8.5%
- Community action 4%
- Other 31%
- Religious 19%

Source: Independent Sector

"**A**sk not what your country can do for you—ask what you can do for your country." These famous words from President John Kennedy's 1961 inaugural address are meaningful and significant today. It may seem that the actions of one individual high school student can't make a major difference in the course of America's history. However, energy put into volunteering in your local community is an excellent way to make a difference.

TIME Line

A.D. 690 The Maya create the Great Plaza at Tikal in present-day Guatemala.

1760 The jigsaw puzzle is created as an educational toy in England and in France.

A.D. 600 | 700 | 800 | 900 | 1000 | 1100 | 1200 | 1300 | ⚡ 1720 | 1730 | 1740 | 1750 | 1760 | 17

1202 Leonardo Fibonacci, born in Pisa, Italy, writes *Liber abaci* (Book of the Abacus), which introduces the famous Fibonacci sequence 1, 1, 2, 3, 5, 8, 13, ... in a problem about the reproduction of rabbits.

PEOPLE IN THE
NEWS

Amy Banna, a Rochester Hills, Michigan, teenager who likes to help others, won an award for her volunteer efforts. She was one of only five students in southeastern Michigan who received a Young Metro Volunteer Award in 1995, recognizing her service to the community, her leadership, commitment, and character. Amy tutors elementary students after school, packs and delivers food monthly to a Detroit housing project, helps with child care at a shelter for mothers and their children, and is a teacher's aide for a Head Start program.

Investigate volunteering in your community. Find some of the interesting ways you can help your city and country in your spare time. Do research by contacting as many organizations as possible who use volunteers. Also, conduct a survey of your classmates to find the number of hours per week they volunteer in the community.

Report to the class on the different opportunities for volunteering, especially groups that can use teen volunteers. In your report, organize your data into graphs and box-and-whisker plots. Explain the advantages and disadvantages of each method of data presentation.

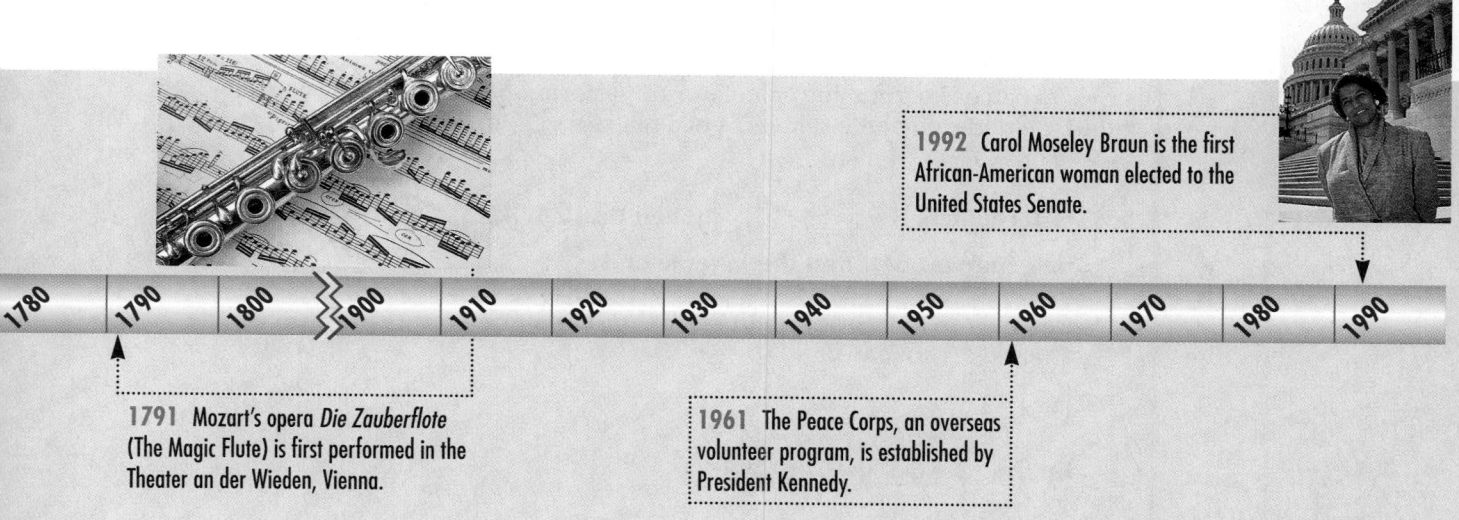

1992 Carol Moseley Braun is the first African-American woman elected to the United States Senate.

1791 Mozart's opera *Die Zauberflote* (The Magic Flute) is first performed in the Theater an der Wieden, Vienna.

1961 The Peace Corps, an overseas volunteer program, is established by President Kennedy.

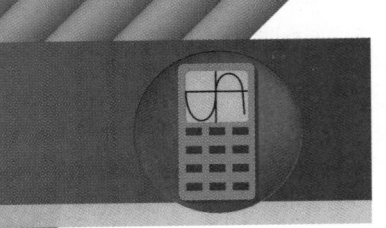

4–1A Graphing Technology Matrices

A Preview of Lesson 4–1

Most graphing calculators can perform operations with matrices, as well as find determinants and inverse matrices. On a TI-82, the MATRX key accesses the matrix operation menus. The EDIT menu allows you to define matrices. When the EDIT menu is accessed, the dimensions of matrices *A–E* are listed. A matrix dimension of 2 × 3 indicates a matrix has 2 rows and 3 columns.

To enter a matrix into the calculator, choose the EDIT menu and select matrix *A*. Then enter the dimensions and elements of the matrix.

Example ① Define matrix $A = \begin{bmatrix} 6 & 7 \\ 1 & 2 \end{bmatrix}$ with a graphing calculator.

This is a 2 × 2 matrix. Enter the matrix dimensions. Then enter the matrix elements.

Enter: MATRX ▶ ▶ ENTER 2 ENTER 2 ENTER

6 ENTER 7 ENTER 1 ENTER 2 ENTER

The ENTER *key fills the row, not the column.*

You can display the matrix by quitting the matrix menus and then requesting matrix *A* by name.

Enter: 2nd QUIT MATRX 1 ENTER

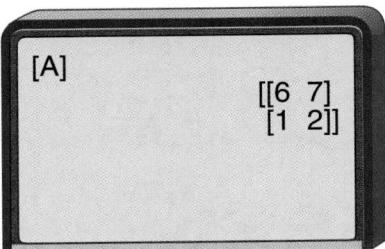

[A]

[[6 7]
 [1 2]]

You can use a graphing calculator to perform operations on matrices and to find the determinant and inverse of a matrix.

Example ② Enter matrix $B = \begin{bmatrix} 1 & 0 & 8 \\ -2 & -3 & 6 \end{bmatrix}$. Then find $2A$, AB, A^2, $B + AB$, the determinant of A and the inverse of A.

Use the procedure shown in Example 1 to enter matrix *B*.
This is a 2 × 3 matrix.

Find $2A$.

Enter: 2 MATRX 1 ENTER

[[12 14]
 [2 4]]

Find AB.

Enter: [MATRX] 1 [MATRX] 2 [ENTER]

$[[-8\ -21\ 90]$
$[-3\ -6\ 20]]$

Find A^2.

Enter: [MATRX] 1 [x^2] [ENTER]

$[[43\ 56]$
$[8\ 11]]$

Find $B + AB$.

Enter: [MATRX] 2 [+] [MATRX] 1 [MATRX] 2 [ENTER]

$[[-7\ -21\ 98]$
$[-5\ -9\ 26]]$

Find the determinant of A.   *The determinant of A is denoted det A.*

Enter: [MATRX] [▶] 1 [MATRX] 1 [ENTER]

5

Find the inverse of A.   *The inverse of A is denoted A^{-1}.*

Enter: [MATRX] 1 [x^{-1}] [ENTER]

$[[.4\ -1.4]$
$[-.2\ 1.2]]$

EXERCISES

Enter the matrices below into a graphing calculator. Then find each of the following.

$$A = \begin{bmatrix} 2 & 1 & 4 \\ 0 & 1 & -1 \\ 4 & 2 & 3 \end{bmatrix} \qquad B = \begin{bmatrix} 6 & -2 & 5 \\ 0 & 7 & -1 \end{bmatrix} \qquad C = \begin{bmatrix} 1 & 4 \\ -3 & 6 \\ 7 & -2 \end{bmatrix}$$

1. $-C$

2. $4B$

3. det A

4. $-2A$

5. A^{-1}

6. CB

7. BC

8. det BC

9. BA

10. $CB - A$

11. det CB

12. $A + CB$

13. $(BC)^{-1}$

14. A^2

15. $(BC)^2$

16. $B + BA$

17. BAC

18. CBA

An Introduction to Matrices

What YOU'LL LEARN

- To perform scalar multiplication on a matrix,
- to solve matrices for variables, and
- to solve problems using matrix logic.

Why IT'S IMPORTANT

You can use matrices to make decisions and solve many types of problems.

The plural of matrix is *matrices.*

GLOBAL CONNECTIONS

The term *matrix* was first used by the British-born mathematician James Joseph Sylvester in 1850 to designate a rectangular array of numbers from which determinants may be formed.

APPLICATION
Decision Making

Emilio has been accepted at three colleges in Ohio: Denison University, Marietta College, and Muskingum College. He and his parents are trying to make a final decision based on cost, distance from home, campus life, and educational quality. Emilio rates each criteria on a scale from 1 (least favorable) to 10 (most favorable) and organizes the information in a **matrix** like the one shown below. A matrix is a rectangular array of variables or constants in horizontal rows and vertical columns, usually enclosed in brackets.

Matrices are often used as problem-solving tools.

$$
\begin{array}{c}
\\
\text{Denison} \\
\text{Marietta} \\
\text{Muskingum}
\end{array}
\begin{array}{cccc}
\text{cost} & \text{distance} & \text{campus} & \text{quality} \\
\end{array}
\begin{bmatrix}
5 & 6 & 6 & 8 \\
6 & 6 & 8 & 9 \\
5 & 6 & 7 & 7
\end{bmatrix}
$$

When the information is shown in a matrix, it is easy to see that distances from home are not a useful criteria, because each college received the same score. You can also see that all of the entries in the second row are greater than the entries in either the first or third row. Based on these criteria, Emilio should attend Marietta College.

In a matrix, numbers or data are organized so that each position in the matrix has a purpose. Each value in the matrix is called an **element**.

$$
C = \begin{bmatrix}
5 & 6 & 6 & 8 \\
6 & 6 & 8 & 9 \\
5 & 6 & 7 & 7
\end{bmatrix} \Big\} \; 3 \; rows
$$

4 columns

The element 9 is in row 2, column 4.

A matrix that has only one row is called a <u>row matrix</u>. A matrix that has only one column is called a <u>column matrix</u>.

A matrix is usually named using an uppercase letter, as in matrix C on the previous page. A matrix can also be named by using the matrix **dimensions** with the letter name. The dimensions tell how many rows and columns, in that order, are in the matrix. The matrix above would be named $C_{3 \times 4}$ since it has 3 rows and 4 columns.

Many problems can be solved using a method sometimes referred to as **matrix logic**. When you use matrix logic, you create a matrix that helps you organize all the information in the problem. By using the matrix, you can eliminate one possibility after another until you eventually arrive at a solution.

Example **1**

PROBLEM SOLVING
Use Matrix Logic

Miko, Amanda, Latisha, and Tara are friends, and each has one of these pets: dog, cat, parrot, and gerbil. Use these clues to match each girl with her pet.

- Latisha likes to visit the friend with the gerbil.
- Tara and Amanda frequently help their friend walk her dog.
- Miko cannot have a dog or a cat because she is allergic to them.
- Tara plans to teach her pet how to talk.

Explore   There are 4 girls and 4 pets. You must match each girl with her pet by using the information from the statements above.

Plan   Make a 4 × 4 matrix to organize the information. Through the process of elimination, each girl can be matched with her pet.

A matrix that has the same number of rows and columns is called a <u>square matrix</u>.

Solve   Put an × in the first row under gerbil to show that Latisha does not have the gerbil. Put two ×s to show that Tara and Amanda do not own the dog. Put two ×s to show that Miko cannot have a cat or dog. By the process of elimination, Latisha owns the dog. Put a circle in this box. Since only one girl owns the dog, put ×s in the rest of the boxes in that row. Continue to eliminate possibilities in this manner.

	dog	cat	parrot	gerbil
Latisha	O	×	×	×
Tara	×	×	O	×
Amanda	×	O	×	×
Miko	×	×	×	O

Miko has the gerbil, Amanda has the cat, Latisha has the dog, and Tara has the parrot.

Examine   Check the result against the statements. The first statement says that Latisha likes to visit the girl with the gerbil, and the answer says that Miko has the gerbil. There is no conflict here. Using the same method for each sentence, you can see that there are no conflicts.

LOOK BACK

You can refer to Lesson 2-1 for information on the graphs of continuous and discrete functions.

Although matrices are sometimes used as a problem-solving tool, their importance extends to another branch of mathematics called **discrete mathematics**. Discrete mathematics deals with finite or discontinuous quantities. The distinction between continuous and discrete quantities is one that you have encountered before. Think of a staircase. You can slide your hand up the banister, but you have to climb the steps one by one. The banister represents a continuous quantity, like a linear function. However, each step represents a discrete quantity, like a point on a scatter plot or an element of a matrix.

Just as algebraic rules exist for functions, matrices have special algebraic rules. For example, you can multiply any matrix by a constant called a **scalar**. This is called **scalar multiplication**. When scalar multiplication is performed, each element is multiplied by that constant, and a new matrix is formed.

Scalar Multiplication of a Matrix	$k\begin{bmatrix} a & b & c \\ d & e & f \end{bmatrix} = \begin{bmatrix} ka & kb & kc \\ kd & ke & kf \end{bmatrix}$

Example ②

APPLICATION
Business

The manager of Just Sports keeps track of monthly sales on a spreadsheet. The spreadsheet below shows the number of baseball and softball bats, balls, shoes, and gloves sold last May. This May, the store is going to have a promotion and hopes to increase sales by 8%. Write a matrix that shows the store's sales goals for this May.

	A	B	C	D	E
1		**bats**	**balls**	**shoes**	**gloves**
2	**baseball**	38	29	18	43
3	**softball**	42	25	16	51

First, write the matrix for last May.

$$\begin{bmatrix} 38 & 29 & 18 & 43 \\ 42 & 25 & 16 & 51 \end{bmatrix}$$

Multiply the matrix by 1.08 to show an increase of 8% for this May.

$$1.08 \begin{bmatrix} 38 & 29 & 18 & 43 \\ 42 & 25 & 16 & 51 \end{bmatrix} = \begin{bmatrix} 41.04 & 31.32 & 19.44 & 46.44 \\ 45.36 & 27.00 & 17.28 & 55.08 \end{bmatrix}$$

TECHNOLOGY Tips

If you are using a graphing calculator, enter the sales matrix into matrix A. Then find 1.08A. You may need to use the arrow keys to see the entire matrix.

Two matrices are considered to be *equal* if they have the same dimensions and if each element of one matrix is equal to the corresponding element of the other matrix.

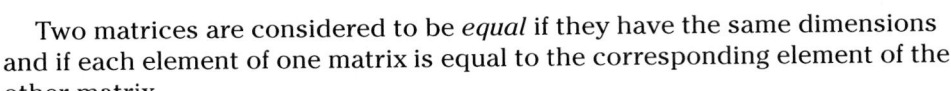

$$\begin{bmatrix} 2 & 5 & 4 \\ 8 & 6 & 1 \end{bmatrix} \neq \begin{bmatrix} 2 & 8 \\ 5 & 6 \\ 4 & 1 \end{bmatrix} \qquad \begin{bmatrix} 4 & 16 \\ 5 & 3 \end{bmatrix} \neq \begin{bmatrix} 4 & 16 \\ 5 & 3 \\ 0 & 0 \end{bmatrix}$$

$$\begin{bmatrix} 8 & 3 \\ 9 & 1 \end{bmatrix} \neq \begin{bmatrix} 8 & 9 \\ 3 & 1 \end{bmatrix} \qquad \begin{bmatrix} 3 & 12 \\ 20 & 8 \end{bmatrix} = \begin{bmatrix} 3 & 12 \\ 20 & 8 \end{bmatrix}$$

The definition of equal matrices can be used to find values when elements of the matrices are algebraic expressions.

Example ❸ Solve $\begin{bmatrix} 2x \\ 2x + 3y \end{bmatrix} = \begin{bmatrix} y \\ 12 \end{bmatrix}$ for x and y.

Since the matrices are equal, the corresponding elements are equal. When you write the sentences that show this equality, two linear equations are formed.

$$2x = y$$
$$2x + 3y = 12$$

The first equation gives you a value for y that can be substituted into the second equation. Then you can find a value for x.

$$2x + 3y = 12$$
$$2x + 3(2x) = 12 \quad \text{Replace } y \text{ with } 2x.$$
$$2x + 6x = 12 \quad \text{Simplify.}$$
$$8x = 12 \quad \text{Combine like terms.}$$
$$x = 1.5 \quad \text{Divide each side by 8.}$$

To find a value for y, you can substitute 1.5 into either equation.

$$2x = y$$
$$2(1.5) = y \quad \text{Replace } x \text{ with } 1.5.$$
$$3 = y$$

The solution is (1.5, 3).

Check your solution by substituting the values into the equation you did *not* use to find y.

$$2x + 3y = 12$$
$$2(1.5) + 3(3) \stackrel{?}{=} 12 \quad \text{Replace } x \text{ with } 1.5 \text{ and } y \text{ with } 3.$$
$$12 = 12 \quad \checkmark$$

LOOK BACK

You can refer to Lesson 3-2 for information on using substitution to solve systems of equations.

A matrix containing coordinates of a geometric figure is often called a coordinate matrix.

Matrices are an important tool for integrating algebra and geometry because points and polygons can be represented by matrices. The ordered pair (x, y) is usually represented by the column matrix $\begin{bmatrix} x \\ y \end{bmatrix}$, where the x-coordinate is in row 1, and the y-coordinate is in row 2. Similarly, polygons can be represented by grouping all of the column matrices of the coordinates of the vertices into one matrix.

coordinates of vertices

$$\triangle ABC = \begin{bmatrix} 3 & -2 & 1 \\ 2 & 1 & -4 \end{bmatrix} \begin{array}{l} \leftarrow x\text{-coordinate} \\ \leftarrow y\text{-coordinate} \end{array}$$

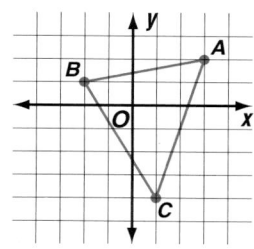

One of the ways that matrices help connect algebra and geometry is through **transformations.** Transformations are functions that map points of a shape onto its image. When a geometric figure is enlarged or reduced, this transformation is called a **dilation.** When the size of a figure changes, all linear measures of its image change in the same ratio. For example, if the perimeter of a figure triples, the length of each side of the figure also triples.

Example ④ △ABC has vertices $A(1, -4)$, $B(2, 3)$, and $C(-2, -1)$. Enlarge △ABC so that its perimeter is twice the original perimeter. What are the coordinates of the vertices of △A'B'C'?

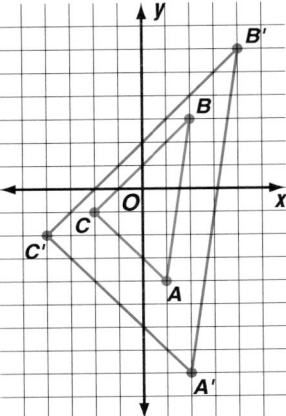

Graph △ABC. Since perimeter is a linear measurement, multiply the coordinate matrix by the scalar 2.

$$2\begin{bmatrix} 1 & 2 & -2 \\ -4 & 3 & -1 \end{bmatrix} = \begin{bmatrix} 2 & 4 & -4 \\ -8 & 6 & -2 \end{bmatrix}$$

The coordinates of the vertices of △A'B'C' are $A'(2, -8)$, $B'(4, 6)$, and $C'(-4, -2)$. Graph △A'B'C'.

You can measure to verify that the perimeter of △A'B'C' is twice the original perimeter.

CHECK FOR UNDERSTANDING

Communicating Mathematics

Study the lesson. Then complete the following.

1. **Define** a matrix in your own words.

2. **Find** an example of a matrix in a newspaper and name it using its dimensions.

3. **Choose** the matrix that represents the ordered pair $(-1, 3)$.

   a. $[-1, 3]$ 　　　　　　　　**b.** $\begin{bmatrix} -1 \\ 3 \end{bmatrix}$

   c. $[3, -1]$ 　　　　　　　　**d.** $\begin{bmatrix} 3 \\ -1 \end{bmatrix}$

4. **Write** a coordinate matrix for the triangle shown at the right.

5. **Explain** the meaning of *dilation*.

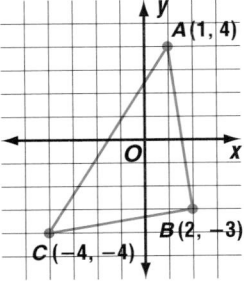

MATH JOURNAL

6. **Draw** a figure on a coordinate plane and write a coordinate matrix for its vertices. Explain what happens to the figure when the matrix is multiplied by a number greater than 1. Explain what happens when the matrix is multiplied by a number between 0 and 1. Use drawings to justify your answers.

Perform the indicated operation.

7. $-2[7 \ 3 \ -1]$

8. $4\begin{bmatrix} -1 & 0 \\ 3 & -2 \end{bmatrix}$

Solve for the variables.

9. $[2x \ 3 \ 3z] = [5 \ 3y \ 9]$

10. $\begin{bmatrix} 6x \\ y \end{bmatrix} = \begin{bmatrix} 62 + 8y \\ 6 - 2x \end{bmatrix}$

11. **Business**   On Monday, the Main Street Deli sold the following number of sandwiches: 15 turkey, 12 turkey and cheese, 8 ham, 10 ham and cheese, 8 roast beef, 11 roast beef and cheese. Organize the information into a 3×2 matrix.

12. **Geometry**   Triangle ABC with $A(4, 5)$, $B(-3, -2)$, and $C(1, -4)$ is reduced so that its perimeter is one-half the original perimeter.

   a. Write the coordinate matrix for $\triangle ABC$.

   b. Write the coordinates of $\triangle A'B'C'$ in matrix form.

   c. Graph this situation.

EXERCISES

Perform the indicated operation.

13. $3\begin{bmatrix} 5 & -2 & 7 \\ -3 & 8 & 4 \end{bmatrix}$

14. $-2\begin{bmatrix} 6 & -4 \\ -2 & 4 \end{bmatrix}$

15. $\frac{1}{3}[6 \ -5]$

16. $0.2\begin{bmatrix} 10.50 \\ 8.75 \end{bmatrix}$

17. $-5\begin{bmatrix} 1.3 & 0 & 5.1 \\ 0.4 & 1.0 & 2.5 \end{bmatrix}$

18. $-0.3[8.95 \ 7.50]$

Solve for the variables.

19. $[4x \ 3y] = [12 \ -1]$

20. $\begin{bmatrix} 2x + y \\ x - 3y \end{bmatrix} = \begin{bmatrix} 5 \\ 13 \end{bmatrix}$

21. $x\begin{bmatrix} 4 & y \\ 7 & 2 \end{bmatrix} = \begin{bmatrix} 12 & -15 \\ 21 & z \end{bmatrix}$

22. $4\begin{bmatrix} x & y - 1 \\ 3 & z \end{bmatrix} = \begin{bmatrix} 20 & 8 \\ 6z & x + y \end{bmatrix}$

23. $\begin{bmatrix} x^2 & 7 & 9 \\ 5 & 12 & 6 \end{bmatrix} = \begin{bmatrix} 25 & 7 & y \\ 5 & 2z & 6 \end{bmatrix}$

24. $\begin{bmatrix} x + 3y \\ 3x + y \end{bmatrix} = \begin{bmatrix} -13 \\ 1 \end{bmatrix}$

25. **Geometry**   The vertex of the right angle of a right triangle is located at the origin with its other vertices at $(0, 12)$ and $(5, 0)$. Find the coordinates of the vertices of a similar triangle whose perimeter is one-fourth that of the original triangle.

26. **Geometry**   Enlarge $\triangle ABC$ shown at the right so that the resulting perimeter is three times the original perimeter.

   a. Graph $\triangle ABC$ and $\triangle A'B'C'$.

   b. Write the coordinates of $\triangle A'B'C'$ in matrix form.

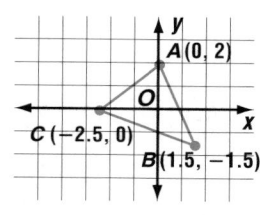

27. Geometry The coordinate matrix for $\triangle XYZ$ is $\begin{bmatrix} -2 & 4 & -1 \\ -1 & 2 & 3 \end{bmatrix}$. Explain what happens to the triangle when the matrix is multiplied by -0.5. Make a drawing to justify your answer.

Solve for the variables.

28. $\begin{bmatrix} r^2 - 24 & 17 \\ 7 & t^3 \end{bmatrix} = \begin{bmatrix} 1 & 2y + 3 \\ z^2 - 12 & 27 \end{bmatrix}$

29. $\begin{bmatrix} 5x - 7 & 11 \\ 5 & 23 \end{bmatrix} = \begin{bmatrix} 8 & 21 - m \\ r^3 - 3 & 4y + x \end{bmatrix}$

30. $\begin{bmatrix} 13 - 7y & a \\ 1 & 2b - 38 \end{bmatrix} = \begin{bmatrix} 5x & 2 - 6b \\ 2x + 3y & 5a \end{bmatrix}$

Critical Thinking

31. When the size of a figure changes, all linear measures of its image, such as the perimeter, change in the same ratio. Is it also true that the area of the figure changes in the same ratio? Justify your answer with matrices and a graph.

Applications and Problem Solving

32. Use Matrix Logic Fred, Ted, and Ed are taking Mary, Carrie, and Terri to the homecoming dance. Use these clues to find which couples will be attending the dance.

Mary is Ed's sister and lives on Fifth Avenue.

Ted drives a car to school each day.

Ed is taller than Terri's date.

Carrie and her date ride their bicycles to school every day.

Fred's date lives on State Street.

33. Sports The graph below shows the percent of the U.S. population that participates in the ten most popular sporting activities. Estimate the percents from the graph and organize the information in a matrix.

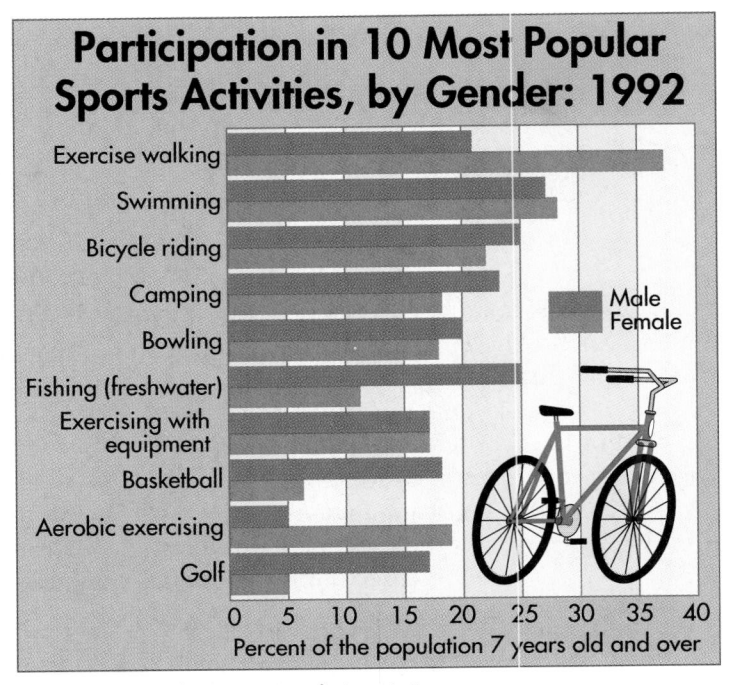

Participation in 10 Most Popular Sports Activities, by Gender: 1992

Source: National Sporting Goods Association

34. Find the x-, y-, and z-intercepts for $3x + 6y - 8z = 24$. (Lesson 3–7)

35. Theater   The Woodward Park High School auditorium seats 150 people. Admission to the spring play is $2.00 for adults and $1.00 for students. The Drama Club has already sold fifty student tickets and the rest are to be sold at the door. How many of each type of ticket should be sold for the Drama Club to earn the maximum amount of money? (Lesson 3–6)

36. Solve the system of inequalities by graphing. (Lesson 3–4)

$x + y < 8$

$x + y > 5$

37. Find the value of $f(-2)$, if $f(x) = x^2 - 4$. (Lesson 2–1)

38. Solve $|9 - 3t| > 5$. (Lesson 1–7)

39. Statistics   The table at the right lists the nine cities in the United States with the fewest average rainy days per year. (Lesson 1–3)

   a. Find the median, mode, and mean of the average number of rainy days.

   b. What do you notice about the locations of these cities?

City	Days
San Diego, CA	42
Bakersfield, CA	37
Los Angeles, CA	35
Long Beach, CA	32
Santa Barbara, CA	30
Bishop, CA	29
Las Vegas, NV	26
Phoenix, AZ	26
Yuma, AZ	17

WORKING ON THE In·ves·ti·ga·tion

Refer to the Investigation on pages 180–181.

Using the design of your launcher, conduct tests by shooting the Ping-Pong™ ball various distances. During these tests, the launcher should be operated by one to four people in the group, the launcher should be shot from the floor, and the distance should be measured from the launcher's location to the spot where the Ping-Pong ball hits the floor.

Also, use this testing time to calibrate your launcher to hit different distances in your firing range. In other words, determine the launch settings for several distances.

1 List your test results.

2 Describe your calibration system.

3 What is the process you used to hit a target at 50 cm, 100 cm, 150 cm, 200 cm, or 250 cm?

4 How accurate have your tests been? Explain.

5 List your launch settings for at least six target distances between 50 cm and 250 cm.

6 Create a table of launch settings and target distances. How can this data be put into a matrix? What would be the dimensions of the matrix? Create this matrix.

7 Draw a scatter plot of the relationship between the launch settings and the target distances. Describe the graph.

8 Determine a mathematical relationship between the two values.

Add the results of your work to your Investigation Folder.

Adding and Subtracting Matrices

What YOU'LL LEARN

- To add and subtract matrices.

Why IT'S IMPORTANT

You can use matrices to solve problems involving meteorology, geography, and recreation.

APPLICATION
Meteorology

At the turn of the century, weather forecasters were unable to give residents of coastal areas much warning of an approaching hurricane. On September 8, 1900, Isaac Cline, the head of the Galveston Weather Bureau in Texas, rode a horse along the beach front, urging people to evacuate as an unnamed hurricane approached. Even so, nearly 6000 people lost their lives later that day due to the flooding caused by the storm surge. In contrast, when hurricane Andrew hit the Florida and Louisiana coasts in 1992, only 23 deaths were attributed to it—most likely because of better forecasting and evacuation planning than in 1900.

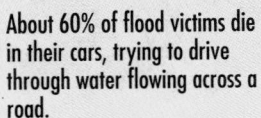

F Y I

About 60% of flood victims die in their cars, trying to drive through water flowing across a road.

Other types of severe weather give little notice. Between 1940 and 1990, 21,447 people have died in the United States due to lightning, tornadoes, floods, and hurricanes. The data for each decade since the 1940s are shown in the spreadsheet below.

	A	B	C	D	E
1		lightning	tornadoes	floods	hurricanes
2	1940s	3293	1788	619	216
3	1950s	1841	1409	791	877
4	1960s	1332	935	1297	587
5	1970s	978	986	1819	217
6	1980s	726	521	1097	118

Source: National Weather Service

The information in the spreadsheet can also be represented by column matrices, with one for each type of severe weather. To find the total number of deaths due to severe weather, add the corresponding elements.

$$\begin{bmatrix} 3293 \\ 1841 \\ 1332 \\ 978 \\ 726 \end{bmatrix} + \begin{bmatrix} 1788 \\ 1409 \\ 935 \\ 986 \\ 521 \end{bmatrix} + \begin{bmatrix} 619 \\ 791 \\ 1297 \\ 1819 \\ 1097 \end{bmatrix} + \begin{bmatrix} 216 \\ 877 \\ 587 \\ 217 \\ 118 \end{bmatrix} = \begin{bmatrix} 5916 \\ 4918 \\ 4151 \\ 4000 \\ 2462 \end{bmatrix} \begin{matrix} \leftarrow 1940s \\ \leftarrow 1950s \\ \leftarrow 1960s \\ \leftarrow 1970s \\ \leftarrow 1980s \end{matrix}$$

Since 1940, the following number of deaths occurred due to lightning, tornadoes, floods, or hurricanes: 1940s, 5916; 1950s, 4918; 1960s, 4151; 1970s, 4000; 1980s, 2462.

This example illustrates that in order to add matrices, they must have the same dimensions.

Addition of Matrices

If **A** and **B** are two $m \times n$ matrices, then **A** + **B** is an $m \times n$ matrix in which each element is the sum of the corresponding elements of **A** and **B**.

$$\begin{bmatrix} a & b & c \\ d & e & f \\ g & h & i \end{bmatrix} + \begin{bmatrix} j & k & l \\ m & n & o \\ p & q & r \end{bmatrix} = \begin{bmatrix} a+j & b+k & c+l \\ d+m & e+n & f+o \\ g+p & h+q & i+r \end{bmatrix}$$

Similarly, it is possible to subtract matrices.

Example ❶

APPLICATION

Recreation

Top Five List

Sports activities with greatest participation, in millions
1. walking, 67.8
2. swimming, 63.1
3. bicycle riding, 54.6
4. camping, 47.3
5. bowling, 42.4

The matrices below show sporting goods sales in the United States in millions of dollars for 1992 and 1993. By how many dollars did each category change from 1992 to 1993?

	1992		**1993**

$$A = \begin{matrix} \text{athletic clothing} \\ \text{athletic shoes} \\ \text{athletic equipment} \\ \text{recreational vehicles} \end{matrix} \begin{bmatrix} 12{,}057 \\ 6{,}300 \\ 12{,}063 \\ 12{,}524 \end{bmatrix} \qquad B = \begin{matrix} \text{athletic clothing} \\ \text{athletic shoes} \\ \text{athletic equipment} \\ \text{recreational vehicles} \end{matrix} \begin{bmatrix} 10{,}101 \\ 6{,}242 \\ 12{,}816 \\ 13{,}275 \end{bmatrix}$$

To determine the change in each category, find $B - A$.

$$B - A = \begin{bmatrix} 10{,}101 \\ 6{,}242 \\ 12{,}816 \\ 13{,}275 \end{bmatrix} - \begin{bmatrix} 12{,}057 \\ 6{,}300 \\ 12{,}063 \\ 12{,}524 \end{bmatrix} \text{ or } \begin{bmatrix} -1956 \\ -58 \\ 753 \\ 751 \end{bmatrix}$$

From 1992 until 1993, sales of athletic clothing decreased $1956 million, athletic shoes decreased $58 million, athletic equipment increased $753 million, and recreational vehicles increased $751 million.

In Lesson 4–1, you used matrices to represent polygons and their dilation images. Another type of transformation is a **translation**. A translation occurs when a figure is moved from one location to another on the coordinate plane without changing its size, shape, or orientation. You can use matrix addition to find the coordinates of translated figures.

INTEGRATION
Geometry

2 **Find the coordinates of the vertices of quadrilateral *QUAD* with** *Q*(−2, −3), *U*(−1, 2), *A*(3, 4), and *D*(1, −2) if it is moved 3 units to the right and 1 unit down.**

Write the coordinates of quadrilateral *QUAD* as a coordinate matrix.

$$\begin{bmatrix} -2 & -1 & 3 & 1 \\ -3 & 2 & 4 & -2 \end{bmatrix}$$

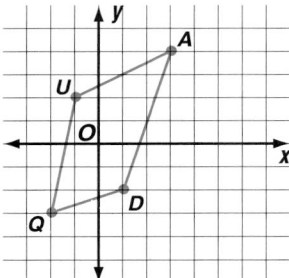

To translate the quadrilateral 3 units to the right means that each *x*-coordinate increases by 3. Translating 1 unit down means that each *y*-coordinate decreases by 1.

The matrix, called the *translation matrix*, that will increase each *x*-coordinate by 3 and decrease each *y*-coordinate by 1 is
$$\begin{bmatrix} 3 & 3 & 3 & 3 \\ -1 & -1 & -1 & -1 \end{bmatrix}.$$

To find the coordinates of the vertices of the translated quadrilateral *Q'U'A'D'*, add the translation matrix to the coordinate matrix of *QUAD*.

$$\begin{bmatrix} -2 & -1 & 3 & 1 \\ -3 & 2 & 4 & -2 \end{bmatrix} + \begin{bmatrix} 3 & 3 & 3 & 3 \\ -1 & -1 & -1 & -1 \end{bmatrix} = \begin{bmatrix} 1 & 2 & 6 & 4 \\ -4 & 1 & 3 & -3 \end{bmatrix}$$

The coordinates of the vertices of *Q'U'A'D'* are *Q'*(1, −4), *U'*(2,1), *A'*(6, 3), and *D'*(4, −3).

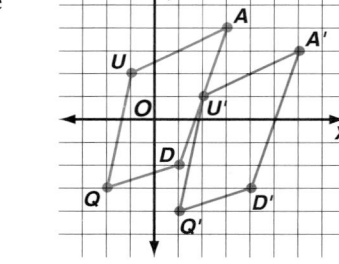

Graph the coordinates of *Q'U'A'D'* to check the accuracy of your coordinates. The two quadrilaterals have the same size and shape. *Q'U'A'D'* has moved to the right 3 units and 1 unit down from *QUAD*.

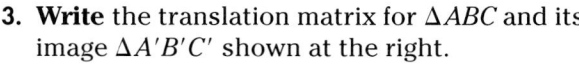

CHECK FOR UNDERSTANDING

Communicating Mathematics

Study the lesson. Then complete the following.

1. **Explain** the conditions under which matrices can be added.

2. **Illustrate** the difference between a dilation and a translation.

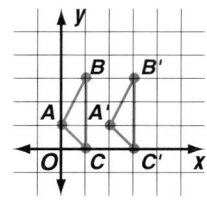

3. **Write** the translation matrix for △*ABC* and its image △*A'B'C'* shown at the right.

MATH **J**OURNAL

4. **Write** a convincing argument for the statement *Matrix addition is commutative and associative.* If the statement is not true, find a counterexample.

Guided Practice

Perform the indicated operations.

5. $\begin{bmatrix} 3 & 7 \\ -2 & 1 \end{bmatrix} - \begin{bmatrix} 2 & -3 \\ 5 & -4 \end{bmatrix}$ 6. $\begin{bmatrix} 4 \\ 1 \\ -3 \end{bmatrix} + \begin{bmatrix} 6 \\ -5 \\ 8 \end{bmatrix}$

7. $2[3 \ -1] + 3[5 \ 0]$

8. **Meteorology** Refer to the application at the beginning of the lesson. For each decade since 1940, how many more people died as a result of lightning than hurricanes?

9. **Geometry** Triangle ABC with vertices $A(-2, 2)$, $B(3, 5)$, and $C(5, -2)$ is translated so that A' is at $(1, -5)$.
   a. Draw a graph of this situation.
   b. Find the translation matrix.
   c. Write the coordinates of $A'B'C'$ in matrix form.

EXERCISES

Practice

Perform the indicated operations.

10. $\begin{bmatrix} 3 & -9 \\ 4 & 2 \end{bmatrix} + \begin{bmatrix} -8 & -4 \\ 3 & 10 \end{bmatrix}$

11. $[5 \ 8 \ -4] + [-1 \ 12 \ 5]$

12. $4\begin{bmatrix} 2 & 7 \\ -3 & 6 \end{bmatrix} + 5\begin{bmatrix} -6 & -4 \\ 3 & 0 \end{bmatrix}$

13. $\frac{1}{2}\begin{bmatrix} 4 & 6 \\ 3 & 0 \end{bmatrix} - \frac{2}{3}\begin{bmatrix} 9 & 27 \\ 0 & 3 \end{bmatrix}$

14. $5\begin{bmatrix} 1 \\ -1 \\ -3 \end{bmatrix} + 6\begin{bmatrix} -4 \\ 3 \\ 5 \end{bmatrix} - 2\begin{bmatrix} -3 \\ 8 \\ -4 \end{bmatrix}$

15. $2\begin{bmatrix} -2 & 4 \\ 1 & -1 \\ 3 & 0 \end{bmatrix} - 3\begin{bmatrix} 5 & 3 \\ -3 & 2 \\ 8 & -9 \end{bmatrix} + \begin{bmatrix} 0 & -5 \\ 9 & -3 \\ -2 & 7 \end{bmatrix}$

Geometry

16. Translate $\triangle ABC$ shown at the right so that A' is at $(3, 4)$.
   a. Graph $\triangle ABC$ and $\triangle A'B'C'$.
   b. Write the coordinates of $\triangle A'B'C'$ in matrix form.

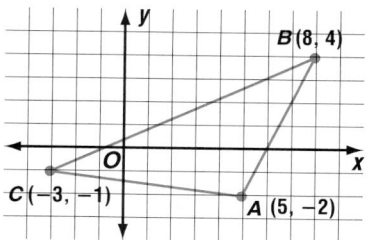

17. Quadrilateral $BURT$ has vertices $B(6, 1)$, $U(3, 5)$, $R(-1, 4)$, and $T(-3, -5)$.
   a. What translation matrix would you need to use to translate $BURT$ so that R' has coordinates $(3, 2)$?
   b. Use your translation matrix to find the coordinates of B', T', and U'.

18. Dilate and then translate $\triangle XYZ$ with vertices $X(-6, 2)$, $Y(-2, 8)$, and $Z(4, -5)$ so that X' has coordinates $(2, 2)$ and the perimeter of $\triangle X'Y'Z'$ is one-half the perimeter of $\triangle XYZ$. State the coordinates of Y' and Z'.

19. Solve for the variables.
$$\begin{bmatrix} x \\ 7z \\ 2y \end{bmatrix} - \begin{bmatrix} 4z \\ -3y \\ 3x \end{bmatrix} + \begin{bmatrix} -2y \\ 2x \\ -5z \end{bmatrix} = \begin{bmatrix} -4 \\ 11 \\ 18 \end{bmatrix}$$

20. Geometry Find the coordinates of the vertices of quadrilateral *MNPQ* that is a translation of quadrilateral *XYZW* whose vertices are *X*(5, −3), *Y*(2, 7), *Z*(−3, 3), and *W*(−5, 1), if *M* is located at the origin.

21. Geography The matrices below represent the number of births and deaths in seven Atlantic seaboard states in 1992.

<table>
<tr><th></th><th>Births</th><th></th><th>Deaths</th></tr>
<tr><td>Delaware</td><td>10,902</td><td>Delaware</td><td>5937</td></tr>
<tr><td>Maryland</td><td>76,173</td><td>Maryland</td><td>37,806</td></tr>
<tr><td>Virginia</td><td>97,600</td><td>Virginia</td><td>49,541</td></tr>
<tr><td>B = North Carolina</td><td>103,047</td><td>D = North Carolina</td><td>59,478</td></tr>
<tr><td>South Carolina</td><td>56,635</td><td>South Carolina</td><td>30,609</td></tr>
<tr><td>Georgia</td><td>111,397</td><td>Georgia</td><td>53,288</td></tr>
<tr><td>Florida</td><td>192,291</td><td>Florida</td><td>140,401</td></tr>
</table>

a. Does it make sense to find the sum of the matrices? Why or why not? If so, explain the meaning of the sum.

b. Does it make sense to find the difference of the matrices? Why or why not? If so, explain the meaning of the difference.

c. Suppose the Census Bureau predicts a 1% decrease in the number of deaths for this region. How would you show this in a matrix?

22. Business The Cookie Cutter Bakery keeps a log of each type of cookie sold in a spreadsheet at three of their branch stores so that they can monitor their purchases of supplies. Two days of sales are shown below.

A	B	C	D	E	
1	**Friday**	chocolate chip	peanut butter	sugar	cut-out
2	Store 1	120	97	64	75
3	Store 2	80	59	36	60
4	Store 3	72	84	29	48

A	B	C	D	E	
1	**Saturday**	chocolate chip	peanut butter	sugar	cut-out
2	Store 1	112	87	56	74
3	Store 2	84	65	39	70
4	Store 3	88	98	43	60

a. Write a matrix for each day's sales. Then find the sum of the two days' sales expressed as a matrix.

b. Each cookie takes approximately one-fourth cup of flour. If there are four cups of flour in one pound, how many pounds of flour were needed for these two days of baking?

23. Find $4\begin{bmatrix} -7 & 5 & -11 \\ 2 & -4 & 9 \end{bmatrix}$. (Lesson 4–1)

24. Geometry In which octant does the point (5, −1, 9) lie? (Lesson 3–7)

25. Graph $y > x + 4$. (Lesson 2–7)

26. State whether $y = x^2 - 4$ is a linear equation. (Lesson 2–2)

27. Solve $\frac{3}{4}t + 1 = 10$. (Lesson 1–4)

Multiplying Matrices

What YOU'LL LEARN

• To multiply matrices.

Why IT'S IMPORTANT

You can use matrices to solve problems involving probability and track and field.

APPLICATION

Sales

The manager of DK's Donuts makes a daily report to the owner that summarizes the cost of each kind of donut and the number of donuts sold for that day. The sales for one day are summarized in the cost matrix C and sales matrix S shown below.

cost ($)

$$C = \begin{array}{cccc} plain & jelly & glazed & specialty \\ [0.45 & 0.55 & 0.50 & 0.85] \end{array}$$

number

$$S = \begin{array}{c} plain \\ jelly \\ glazed \\ specialty \end{array} \begin{bmatrix} 191 \\ 122 \\ 98 \\ 69 \end{bmatrix}$$

You can use matrix multiplication to find the income for the day. In this case, multiply each element in the cost matrix by its corresponding element in the sales matrix and find the total.

$$CS = [0.45 \; 0.55 \; 0.50 \; 0.85] \cdot \begin{bmatrix} 191 \\ 122 \\ 98 \\ 69 \end{bmatrix}$$

$$= [0.45(191) + 0.55(122) + 0.50(98) + 0.85(69)]$$

$$= [85.95 + 67.10 + 49.00 + 58.65]$$

$$= [260.70]$$

Notice that each element in the row matrix is multiplied by an element in the column matrix.

The income for the day was $260.70.

In general, the product of the two matrices is found by multiplying rows and columns.

Multiplying Matrices	The product of $A_{m \times n}$ and $B_{n \times r}$ is $(AB)_{m \times r}$. The element in the *i*th row and the *j*th column of AB is the sum of the products of the corresponding elements in the *i*th row of A and the *j*th column of B.

Notice that you can multiply two matrices only if the number of columns in the first matrix is equal to the number of rows in the second matrix.

$$\underbrace{\begin{bmatrix} 2 & -1 \\ 3 & 4 \end{bmatrix}}_{2 \times 2} \cdot \underbrace{\begin{bmatrix} 3 & -9 & 2 \\ 5 & 7 & -6 \end{bmatrix}}_{2 \times 3}$$

possible

$$\underbrace{\begin{bmatrix} 3 & -9 & 2 \\ 5 & 7 & -6 \end{bmatrix}}_{2 \times 3} \cdot \underbrace{\begin{bmatrix} 2 & -1 \\ 3 & 4 \end{bmatrix}}_{2 \times 2}$$

not possible

The product on the left is defined, but the product on the right is not.

Example **1** If $A = \begin{bmatrix} 3 & -5 \\ 2 & 7 \end{bmatrix}$ and $B = \begin{bmatrix} 5 & 1 & -3 \\ 8 & -4 & 9 \end{bmatrix}$, find AB.

$$AB = \begin{bmatrix} 3(5) + (-5)(8) & 3(1) + (-5)(-4) & 3(-3) + (-5)(9) \\ 2(5) + 7(8) & 2(1) + 7(-4) & 2(-3) + 7(9) \end{bmatrix}$$

$$= \begin{bmatrix} 15 - 40 & 3 + 20 & -9 - 45 \\ 10 + 56 & 2 - 28 & -6 + 63 \end{bmatrix} \quad \textit{Note that } A_{2 \times 2} \cdot B_{2 \times 3} = (AB)_{2 \times 3}.$$

$$= \begin{bmatrix} -25 & 23 & -54 \\ 66 & -26 & 57 \end{bmatrix}$$

In many situations involving chance, matrices can be used to represent probabilities. Matrix multiplication can be used to predict future events.

Example **2**

Probability

A transition matrix contains information about the transition from one event to another.

A psychologist notes the behavior of mice at a certain point in a maze. For any particular trial, 70% of the mice that went right on the previous trial will go right on this trial, and 60% of those that went left on the previous trial will go right on this trial. This information can be represented by the following *transition matrix*.

$$T = \begin{matrix} & R & L & \leftarrow second\ trial \\ R & \begin{bmatrix} 0.7 & 0.3 \\ L & 0.6 & 0.4 \end{bmatrix} \end{matrix}$$

first trial $\longrightarrow$

Suppose 50% of the mice went right on the first trial. This is represented by the following *probability matrix*.

$$\begin{matrix} R & L \\ P = [0.5 & 0.5] \end{matrix}$$

a. **Make a prediction for the second trial.**

b. **Make a prediction for the third trial.**

a. To make a prediction about what will happen on the second trial, find PT.

$$PT = [0.5 \quad 0.5] \cdot \begin{bmatrix} 0.7 & 0.3 \\ 0.6 & 0.4 \end{bmatrix} \text{ or } [0.65 \quad 0.35]$$

On the second trial, 65% of the mice should go right, and 35% should go left.

b. For the third trial, the new probability matrix is $P = [0.65 \quad 0.35]$. Find PT.

$$PT = [0.65 \quad 0.35] \cdot \begin{bmatrix} 0.7 & 0.3 \\ 0.6 & 0.4 \end{bmatrix} \text{ or } [0.665 \quad 0.335]$$

On the third trial, 66.5% of the mice should go right, and 33.5% should go left.

Another use of matrix multiplication is in transformational geometry. You have already learned how to translate a geometric figure and change its size by using matrices. Another type of transformation is a **rotation**. A rotation occurs when a figure is moved around a center point. To move a figure by rotation, you can use a rotation matrix.

 MODELING MATHEMATICS

Rotations

Materials:  grid paper, ☐ tracing paper, protractor

The matrix $\begin{bmatrix} 0 & -1 \\ 1 & 0 \end{bmatrix}$ will rotate a figure on the coordinate plane about the origin. In this activity, you will determine the direction and degrees of rotation.

Your Turn

a. Draw a triangle on a coordinate plane and label it $\triangle ABC$. Write the coordinates of the vertices as a coordinate matrix.

b. Multiply the rotation matrix shown above by your coordinate matrix. Graph the resulting triangle on the same coordinate plane and label it $\triangle A'B'C'$. *Note that the rotation matrix should be on the <u>left</u> when multiplying.*

c. Place a piece of tracing paper over $\triangle ABC$ and trace it. With your pencil at the origin as a pivot point, slowly turn the tracing paper until the drawing of $\triangle ABC$ matches $\triangle A'B'C'$. Describe the motion of the triangle.

d. On the coordinate plane, draw $\overline{OA}$ and $\overline{OA'}$. Find the measure of $\angle A'OA$. Repeat for the remaining vertices.

e. Write a sentence that describes the effect of multiplying a coordinate matrix by the rotation matrix $\begin{bmatrix} 0 & -1 \\ 1 & 0 \end{bmatrix}$.

Example ③

 **INTEGRATION**

Geometry

Line AB passes through points $A(4, -2)$ and $B(-3, 5)$. Find the coordinates of two points on line $A'B'$ that has been rotated 90° counterclockwise about the origin. Draw its graph and describe the relationship between lines AB and $A'B'$.

Write the ordered pairs in a coordinate matrix. Then multiply the coordinate matrix by the rotation matrix.

$$\begin{bmatrix} 0 & -1 \\ 1 & 0 \end{bmatrix} \cdot \begin{bmatrix} 4 & -3 \\ -2 & 5 \end{bmatrix} = \begin{bmatrix} 2 & -5 \\ 4 & -3 \end{bmatrix}$$

Coordinates of two points on the line are $A'(2, 4)$ and $B'(-5, -3)$. The two lines appear to be perpendicular.

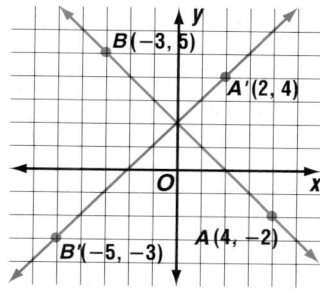

You can check that the lines are perpendicular by finding the slope of each line.

Communicating Mathematics

Study the lesson. Then complete the following.

1. **Name** the conditions under which two matrices can be multiplied.

2. **Find** the dimensions of matrix M if $M = A_{3 \times 2} \cdot B_{2 \times 4}$.

3. **Write** a convincing argument for the statement *Matrix multiplication is commutative*. If the statement is not true, find a counterexample.

4. **Give an example** of two matrices M and N for which the products MN and NM are both defined.

5. **You Decide** Brandon thinks two matrices can always be multiplied if they can be added. Dolores thinks that isn't necessarily true. Who is correct? Explain your reasoning.

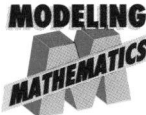

6. Apply a 90° counterclockwise rotation about the origin twice to a triangle on the coordinate plane. Compare the new coordinates to the original ones. Make a conjecture about what effect this rotation has on any figure.

Guided Practice

Find the dimensions of each matrix product.

7. $A_{3 \times 5} \cdot B_{5 \times 2}$ 8. $P_{2 \times 2} \cdot Q_{2 \times 4}$

Perform the indicated operations, if possible.

9. $\begin{bmatrix} 4 & -2 & -7 \\ 6 & 3 & 5 \end{bmatrix} \cdot \begin{bmatrix} -2 \\ 5 \\ 3 \end{bmatrix}$ 10. $\begin{bmatrix} 4 & -1 & 6 \\ 1 & 5 & -8 \end{bmatrix} \cdot \begin{bmatrix} 1 & 3 \\ 9 & -6 \end{bmatrix}$

11. **Geometry** Find the new coordinates of the vertices of triangle ABC with vertices $A(-5, 2)$, $B(3, 4)$, and $C(1, -4)$, when the triangle is rotated 90° counterclockwise about the origin. Graph the original triangle and its rotation $A'B'C'$.

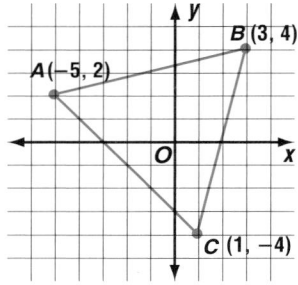

$2 \times 2 \quad 2 \times 1$
2×1

Practice

Find the dimensions of each matrix product.

12. $A_{5 \times 2} \cdot B_{2 \times 5}$ 13. $M_{4 \times 2} \cdot N_{1 \times 3}$

14. $R_{2 \times 3} \cdot S_{3 \times 4}$ 15. $X_{3 \times 4} \cdot Y_{4 \times 1}$

16. $P_{1 \times 5} \cdot Q_{5 \times 1}$ 17. $A_{3 \times 2} \cdot B_{3 \times 2}$

$1 \times 2 \quad 2 \times 1$
1×1
$2 \times 2 \cdot 2 \times 3$
2×3

Perform the indicated operations, if possible.

18. $[2 \ -1] \cdot \begin{bmatrix} 5 \\ 3 \end{bmatrix}$ 19. $\begin{bmatrix} 2 & -1 \\ 3 & 4 \end{bmatrix} \cdot \begin{bmatrix} 3 & -9 & -2 \\ 5 & 7 & -6 \end{bmatrix}$

20. $\begin{bmatrix} 4 & -1 \\ 3 & 5 \end{bmatrix} \cdot \begin{bmatrix} 7 \\ 4 \end{bmatrix}$ 21. $\begin{bmatrix} 5 & -2 & -1 \\ 8 & 0 & 3 \end{bmatrix} \cdot \begin{bmatrix} -4 & 2 \\ 1 & 0 \end{bmatrix}$

Perform the indicated operations, if possible.

22. $3\begin{bmatrix} 5 & 7 \\ 1 & -2 \end{bmatrix} + 2\begin{bmatrix} -3 & 0 \\ -4 & 2 \end{bmatrix}$

23. $\begin{bmatrix} 0 & 8 \\ 3 & 1 \\ -1 & 5 \end{bmatrix} \cdot \begin{bmatrix} 3 & 1 & -2 \\ 0 & 8 & -5 \end{bmatrix}$

24. **Geometry** Find the new coordinates of the vertices of $\triangle ABC$ shown at the right after it has been rotated 90° counterclockwise about the origin.

25. **Geometry** Given that A is any 2×2 matrix, and R is the rotation matrix, is $RA = AR$? Explain.

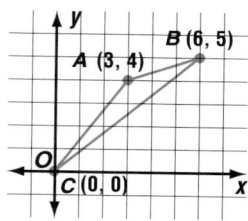

Use the matrices A, B, C, and D to evaluate each expression.

$$A = \begin{bmatrix} 3 & -1 \\ 2 & 4 \end{bmatrix} \qquad B = \begin{bmatrix} 4 & 0 & -3 \\ 7 & -5 & 9 \end{bmatrix} \qquad C = \begin{bmatrix} -6 & 4 \\ -2 & 8 \\ 3 & 0 \end{bmatrix} \qquad D = \begin{bmatrix} -1 & 0 \\ 3 & 7 \end{bmatrix}$$

26. $AB + B$

27. $CB + B$

28. $AD + CB$

29. $AD + BC$

30. **Geometry** After a triangle was rotated 90° counterclockwise about the origin, the coordinates of the vertices are $(-3, -5)$, $(-2, 7)$, and $(1, 4)$. What were the coordinates of the vertices of the triangle in its original position?

Graphing Calculator

Use a graphing calculator to determine the effect of multiplying the unit square matrix, $S = \begin{bmatrix} 0 & 1 & 1 & 0 \\ 0 & 0 & 1 & 1 \end{bmatrix}$, illustrated in the graph at the right, by each matrix below.

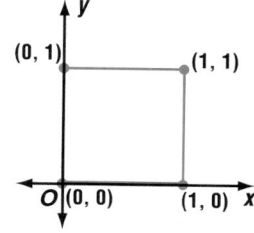

31. $A = \begin{bmatrix} -1 & 0 \\ 0 & -1 \end{bmatrix}$

32. $B = \begin{bmatrix} 1 & 0 \\ 0 & -1 \end{bmatrix}$

33. $C = \begin{bmatrix} -1 & 0 \\ 0 & 1 \end{bmatrix}$

34. $D = \begin{bmatrix} 0 & 1 \\ 1 & 0 \end{bmatrix}$

Critical Thinking

35. Find the values of w, x, y, and z to make the statement $\begin{bmatrix} 1 & 2 \\ 3 & 4 \end{bmatrix} \cdot \begin{bmatrix} w & x \\ y & z \end{bmatrix} = \begin{bmatrix} 1 & 2 \\ 3 & 4 \end{bmatrix}$ true. If the matrix containing w, x, y, and z were multiplied by any other matrix containing two columns, what do you think the result would be?

Applications and Problem Solving

36. **Track and Field** In a three-team track meet, the following numbers of first-, second-, and third-place finishes were recorded.

School	First Place	Second Place	Third Place
Blendon	4	10	6
Walnut Springs	7	6	9
Heritage	8	3	4

If 5 points are awarded for first, 3 for second, and 1 for third, use matrix multiplication to find the final scores for each school.

37. Probability A weather station in a certain area gathers data about the chance of precipitation. It predicts that, if it rains on a given day, 50% of the time it will rain on the next day. If it is not raining, it will rain on the next day only 30% of the time. The weather forecast for Monday predicts the chance of rain is 80%. Find the chance of rain on Wednesday using this pattern.

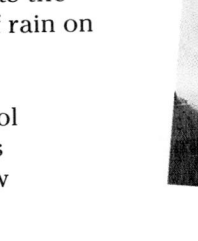

38. Health Due to a flu epidemic, the school nurse estimates that 30% of the students who are well today will be sick tomorrow and 50% of the students who are sick today will be well tomorrow.

 a. Write a transition matrix to show this situation.

 b. If 80% of the student population is well today, predict what percent will be sick tomorrow.

39. Finance Isabel "bought" shares of stock in three U.S. companies for a project in her economics class. She bought 150 shares of General Electric, 100 shares of General Mills, and 200 shares of General Motors. At the time she purchased the stocks, General Electric was $54 per share, General Mills was $60 per share, and General Motors was $43.50 per share.

 a. Organize the data into two matrices and use matrix multiplication to find the total amount she spent for the shares of stock.

 b. At the end of the project, she "sold" all of her stock. General Electric was $55.25 per share, General Mills was $61 per share, and General Motors was $41.75 per share. Use matrix operations to determine how much money Isabel "made" or "lost" in her project.

Mixed Review

40. Find $[4 \; 1 \; -3] + [6 \; -5 \; 8]$. (Lesson 4–2)

41. Find $-6\begin{bmatrix} 2 & -1 \\ -5 & 7 \end{bmatrix}$. (Lesson 4–1)

42. School Your semester test in English consists of short answers and essay questions. Each short answer question is worth 5 points, and each essay question is worth 15 points. You may choose up to 20 questions of any type to answer. It takes 2 minutes to answer each short answer question and 12 minutes to answer each essay question. If you have one hour to complete the test, and assuming you answer all of the questions that you attempt correctly, how many of each type of question should you answer to earn the highest score? (Lesson 3–6)

43. Solve the system of equations by using the elimination method. (Lesson 3–2)
$$3x - 6y = 15$$
$$-3x + 5y = -8$$

44. Write the slope-intercept form of the equation of the line with slope of -2 that passes through the point $(3, 1)$. (Lesson 2–4)

45. State the x- and y-intercepts of the graph of the line with equation $3x - 12y = 24$. (Lesson 2–2)

46. Telecommunications A call to a sports hotline costs $3.38 for the first three minutes and $0.96 for each minute thereafter. What is the cost of a 12-minute phone call? (Lesson 1–4)

Matrices and Determinants

What YOU'LL LEARN

- To evaluate the determinant of a 3 × 3 matrix, and
- to find the area of a triangle, given the coordinates of its vertices.

Why IT'S IMPORTANT

You can use matrices and determinants to solve problems involving geometry and geography.

CONNECTION
Math History

Although matrices are a relatively new notation in mathematics, the *idea* of matrices can be traced back to the Chinese book *Nine Chapters on the Mathematical Art*, which was published about 250 B.C. This book contained many problems that are solved by using matrices. The Chinese matrix was a large counting board resembling a checkerboard, and bamboo rods were placed on the squares to represent equations.

When the Japanese mathematician Seki Kowa (1683) investigated the Chinese system of solving systems of equations, his calculations were similar to those used today to simplify a **determinant**.

Every square matrix has a number associated with it, called its determinant. The notation for the determinant of $\begin{bmatrix} -5 & -7 \\ 11 & 8 \end{bmatrix}$ is $\begin{vmatrix} -5 & -7 \\ 11 & 8 \end{vmatrix}$. To evaluate the determinant, use the rule for second-order determinants.

$$\begin{vmatrix} -5 & -7 \\ 11 & 8 \end{vmatrix} = -5(8) - (-7)(11) \quad \textit{Recall that} \begin{vmatrix} a & b \\ c & d \end{vmatrix} = ad - bc.$$

$$= -40 + 77$$

$$= 37$$

LOOK BACK

You can refer to Lesson 3-3 for information about second-order determinants.

Determinants of 3 × 3 matrices are called **third-order determinants**. One method of evaluating third-order determinants is called **expansion by minors**. The **minor** of an element is the determinant formed when the row and column containing that element are deleted. For the determinant $\begin{vmatrix} -2 & 3 & 8 \\ 6 & 7 & -1 \\ -4 & 5 & 9 \end{vmatrix}$, the minor of 5 is $\begin{vmatrix} -2 & 3 & 8 \\ 6 & 7 & -1 \\ -4 & ⑤ & 9 \end{vmatrix}$ or $\begin{vmatrix} -2 & 8 \\ 6 & -1 \end{vmatrix}$.

To use expansion by minors with third-order determinants, each member of one row is multiplied by its minor. The signs of the products alternate, beginning with a positive sign in the first and third row and a negative sign in the second row. The following definition shows an expansion using the elements in the first row of the determinant. However, any row can be used.

$$\begin{vmatrix} a & b & c \\ d & e & f \\ g & h & i \end{vmatrix} = a\begin{vmatrix} e & f \\ h & i \end{vmatrix} - b\begin{vmatrix} d & f \\ g & i \end{vmatrix} + c\begin{vmatrix} d & e \\ g & h \end{vmatrix}$$

Example **Evaluate** $\begin{vmatrix} 2 & 3 & 4 \\ 6 & 5 & 7 \\ -1 & 9 & 8 \end{vmatrix}$ **using expansion by minors.**

Decide which row of elements you will use for the expansion. Let's use the first row.

$$\begin{vmatrix} 2 & 3 & 4 \\ 6 & 5 & 7 \\ -1 & 9 & 8 \end{vmatrix} = 2\begin{vmatrix} 5 & 7 \\ 9 & 8 \end{vmatrix} - 3\begin{vmatrix} 6 & 7 \\ -1 & 8 \end{vmatrix} + 4\begin{vmatrix} 6 & 5 \\ -1 & 9 \end{vmatrix}$$

$$= 2(40 - 63) - 3(48 + 7) + 4(54 + 5)$$
$$= -46 - 165 + 236$$
$$= 25$$

You can check your work by evaluating the determinant again using a different row of elements.

Another method for evaluating a third-order determinant is using diagonals. In this method, you begin by writing the first two columns on the right side of the determinant.

$$\begin{vmatrix} a & b & c \\ d & e & f \\ g & h & i \end{vmatrix} \rightarrow \begin{vmatrix} a & b & c \\ d & e & f \\ g & h & i \end{vmatrix}\begin{matrix} a & b \\ d & e \\ g & h \end{matrix}$$

Next, draw diagonals from each element of the top row of the determinant downward to the right. Find the product of the elements on each diagonal.

$$\begin{vmatrix} a & b & c \\ d & e & f \\ g & h & i \end{vmatrix} \rightarrow \begin{vmatrix} a & b & c \\ d & e & f \\ g & h & i \end{vmatrix}\begin{matrix} a & b \\ d & e \\ g & h \end{matrix}$$

aei   bfg   cdh

Then, draw diagonals from the elements in the third row of the determinant upward to the right. Find the product of the elements on each diagonal.

gec   hfa   idb

$$\begin{vmatrix} a & b & c \\ d & e & f \\ g & h & i \end{vmatrix} \rightarrow \begin{vmatrix} a & b & c \\ d & e & f \\ g & h & i \end{vmatrix}\begin{matrix} a & b \\ d & e \\ g & h \end{matrix}$$

To find the value of the determinant, add the products of the first set of diagonals and then subtract the products of the second set of diagonals. The value is *aei + bfg + cdh − gec − hfa − idb*.

Example **2** Evaluate $\begin{vmatrix} -1 & 0 & 8 \\ 7 & 3 & 4 \\ 2 & 2 & 5 \end{vmatrix}$ using diagonals.

First, rewrite the first two columns to the right of the determinant.

$$\begin{vmatrix} -1 & 0 & 8 \\ 7 & 3 & 4 \\ 2 & 2 & 5 \end{vmatrix}\begin{matrix} -1 & 0 \\ 7 & 3 \\ 2 & 2 \end{matrix}$$

Next, find the products of the elements of the diagonals.

$$\begin{matrix} & 48 & -8 & 0 \\ \end{matrix}$$
$$\begin{vmatrix} -1 & 0 & 8 \\ 7 & 3 & 4 \\ 2 & 2 & 5 \end{vmatrix}\begin{matrix} -1 & 0 \\ 7 & 3 \\ 2 & 2 \end{matrix}$$
$$\begin{matrix} & -15 & 0 & 112 \\ \end{matrix}$$

Then, add the bottom products and subtract the top products.

$$-15 + 0 + 112 - 48 - (-8) - 0 = 57$$

The value of the determinant is 57.

One very powerful application of determinants is finding the areas of polygons. The formula below shows how determinants serve as a mathematical tool to find the area of a triangle when the coordinates of the three vertices are given.

Area of Triangles

The area of a triangle having vertices at (a, b), (c, d), and (e, f) is $|A|$, where

$$A = \frac{1}{2}\begin{vmatrix} a & b & 1 \\ c & d & 1 \\ e & f & 1 \end{vmatrix}.$$

Notice that it is necessary to use the absolute value of A to guarantee a nonnegative value for area.

Example **3** Find the area of the triangle whose vertices are located at $(3, -4)$, $(5, 4)$, and $(-3, 2)$.

Geometry

Assign values to $a, b, c, d, e,$ and f and substitute them into the area formula and evaluate.

$$A = \frac{1}{2}\begin{vmatrix} a & b & 1 \\ c & d & 1 \\ e & f & 1 \end{vmatrix} \quad \begin{matrix} (a, b) = (3, -4) \\ (c, d) = (5, 4) \\ (e, f) = (-3, 2) \end{matrix}$$

$$= \frac{1}{2}\begin{vmatrix} 3 & -4 & 1 \\ 5 & 4 & 1 \\ -3 & 2 & 1 \end{vmatrix} \quad \textit{Substitute.}$$

$$= \frac{1}{2}[(3)4 + (-4)(-3) + (5)(2) - (-3)(4) - (2)(3) - (5)(-4)]$$

$$= \frac{1}{2}[12 + 12 + 10 - (-12) - 6 - (-20)]$$

$$= \frac{1}{2}(60) \text{ or } 30$$

The area of the triangle is 30 square units.

Maps usually have grids, similar to a coordinate system, that make it easier for you to locate cities, states, or landmarks. A coordinate system can also be used to find the area of large regions.

Example **4**

The figure at the right shows a map of the state of Nevada that has been placed on a coordinate plane in which 1 unit = 1 mile. Estimate the area of Nevada from the coordinates of the vertices.

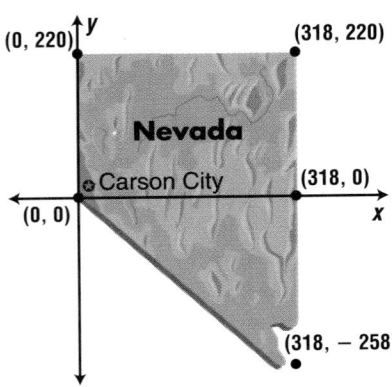

The *x*-axis separates the map into a triangular region and a rectangular region. Find the area of the rectangular region.

$$A = \ell w$$
$$= 318 \cdot 220 \text{ or } 69{,}960$$

Now find the area of the triangular region using the coordinates $(0, 0)$, $(318, 0)$, and $(318, -258)$. Use expansion by minors.

$$A = \frac{1}{2} \begin{vmatrix} a & b & 1 \\ c & d & 1 \\ e & f & 1 \end{vmatrix} \quad \begin{array}{l} (a, b) = (0, 0) \\ (c, d) = (318, 0) \\ (e, f) = (318, -258) \end{array}$$

$$= \frac{1}{2} \begin{vmatrix} 0 & 0 & 1 \\ 318 & 0 & 1 \\ 318 & -258 & 1 \end{vmatrix} = \frac{1}{2} \cdot 1 \begin{vmatrix} 318 & 0 \\ 318 & -258 \end{vmatrix} \text{ or } -41{,}022$$

Finally, add the areas of the two regions. The area of Nevada is $69{,}960 + 41{,}022$ or about 111,000 square miles. *Compare this to the actual area.*

CHECK FOR UNDERSTANDING

Communicating Mathematics

Study the lesson. Then complete the following.

1. **Explain** how $\begin{bmatrix} 7 & 8 \\ 3 & -2 \end{bmatrix}$ and $\begin{vmatrix} 7 & 8 \\ 3 & -2 \end{vmatrix}$ are different.

2. **Describe** how to find the minor of 8 in $\begin{bmatrix} 1 & 2 & 3 \\ 4 & 5 & 6 \\ 7 & 8 & 9 \end{bmatrix}$

3. **State** the condition(s) under which a matrix has a determinant.

4. **Write** a matrix that will help you use a determinant to find the area of the triangle shown at the right.

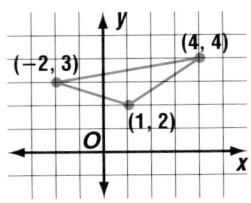

Guided Practice

Determine whether each matrix has a determinant. Write *yes* or *no*. If *yes*, find the value of the determinant.

5. $\begin{bmatrix} 6 \\ 2 \end{bmatrix}$

6. $\begin{bmatrix} -8 & 0 \\ 5 & -4 \end{bmatrix}$

7. $\begin{bmatrix} 1 & 0 & 0 \\ 0 & 1 & 0 \\ 0 & 0 & 1 \end{bmatrix}$

8. Evaluate $\begin{vmatrix} 2 & 3 & 4 \\ 6 & 5 & 7 \\ 1 & 2 & 8 \end{vmatrix}$ using expansion by minors.

9. Evaluate $\begin{vmatrix} -1 & 4 & 0 \\ 3 & -2 & -5 \\ -3 & -1 & 2 \end{vmatrix}$ using diagonals.

10. Geometry Use a determinant to find the area of the triangle whose vertices have coordinates $(-2, 3)$, $(5, 8)$, and $(1, 2)$.

EXERCISES

Practice

Determine whether each matrix has a determinant. Write *yes* or *no*. If *yes*, find the value of the determinant.

11. $\begin{bmatrix} -3 & 5 \\ 6 & -10 \end{bmatrix}$

12. $\begin{bmatrix} 3 \\ -2 \\ 6 \end{bmatrix}$

13. $\begin{bmatrix} 4 & 3 \\ 8 & -1 \\ 7 & 2 \end{bmatrix}$

14. $\begin{bmatrix} -5 & 8 \\ 3 & 0 \end{bmatrix}$

15. $\begin{bmatrix} -2 & 0 & 1 \\ 1 & 2 & 0 \\ 4 & -1 & 1 \end{bmatrix}$

16. $\begin{bmatrix} 5 & 7 & -2 \\ 3 & -2 & 6 \\ 1 & -4 & 3 \end{bmatrix}$

17. Geometry Use a determinant to find the area of $\triangle ABC$ shown at the right. Check your answer by using the formula $A = \frac{1}{2}bh$.

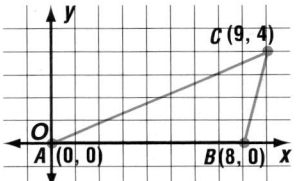

Evaluate each determinant using expansion by minors.

18. $\begin{vmatrix} -3 & 0 & 6 \\ 6 & 5 & -2 \\ 1 & 4 & 2 \end{vmatrix}$

19. $\begin{vmatrix} 0 & -4 & 0 \\ 3 & -2 & 5 \\ 2 & -1 & 1 \end{vmatrix}$

20. $\begin{vmatrix} -2 & 7 & -2 \\ 4 & 6 & 2 \\ 1 & 0 & -1 \end{vmatrix}$

Evaluate each determinant using diagonals.

21. $\begin{vmatrix} 1 & 6 & 4 \\ -2 & 3 & 1 \\ 1 & 6 & 4 \end{vmatrix}$

22. $\begin{vmatrix} 1 & -1 & 1 \\ 3 & 3 & 1 \\ 0 & 5 & 2 \end{vmatrix}$

23. $\begin{vmatrix} 2 & -3 & 4 \\ -2 & 1 & 5 \\ 5 & 3 & -2 \end{vmatrix}$

Use a determinant to find the area of each triangle below.

24.

25.

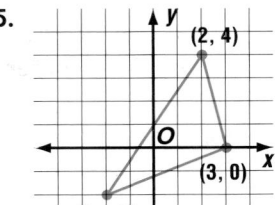

26.

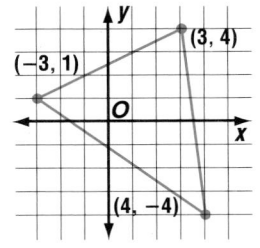

Solve for the variable.

27. $\begin{vmatrix} 2 & x \\ 5 & -3 \end{vmatrix} = 24$

28. $\begin{vmatrix} 4 & x & -2 \\ -x & -3 & 1 \\ -6 & 2 & 3 \end{vmatrix} = -3$

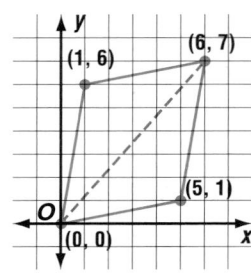

Use determinants to find the area of each polygon below.

29.

(1, 6) (6, 7)
(5, 1)
(0, 0) O x

30.

(4, 5)
(−2, 2) (2, 2)
O x
(5, −2)

31. Find the value of x such that the area of a triangle whose vertices have coordinates $(6, 5)$, $(8, 2)$, and $(x, 11)$ is 30.

Graphing Calculator

32. Use a graphing calculator to find two third-order matrices that are not equal, but have equal determinants.

Critical Thinking

33. Find a third-order matrix in which no element is 0, but for which the determinant is 0.

Applications and Problem Solving

34. **World Cultures** Easter Island is the easternmost island of Polynesia, which is a triangular area containing thousands of islands in the South Pacific. Its northern vertex is Midway Island and the southern boundary runs from New Zealand to Easter Island. A map of Polynesia is shown below. Explain how to use a coordinate grid to estimate the area of Polynesia.

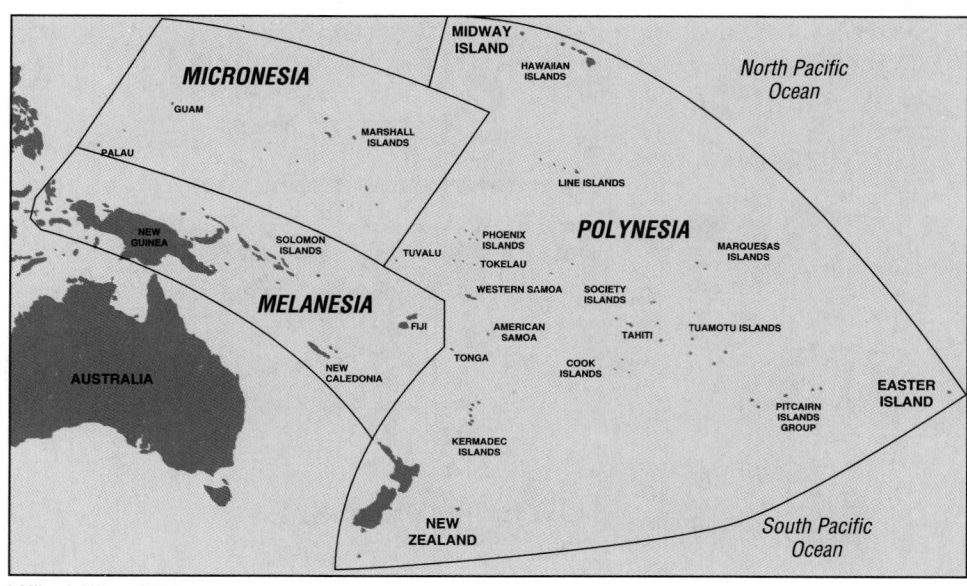

35. Geography The region in North Carolina bounded by Chapel Hill, Durham, and Raleigh is known as the Research Triangle Park. If a coordinate grid in which 1 unit = 1 mile is placed over the map of North Carolina with Chapel Hill at the origin, the coordinates of these three cities are $(0, 0)$, $(6, 3.5)$ and $(21, -10.5)$. Estimate the area of Research Triangle Park.

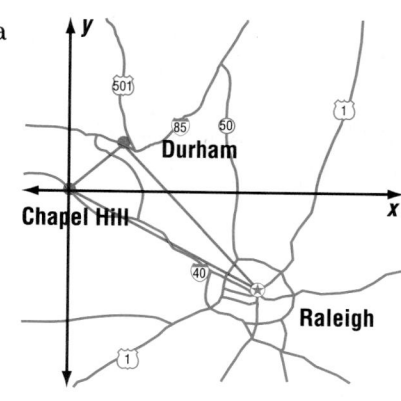

Mixed Review

36. Find $\begin{bmatrix} 4 & 0 & -8 \\ 7 & -2 & 10 \end{bmatrix} \cdot \begin{bmatrix} -1 & 3 \\ 6 & 0 \end{bmatrix}$. (Lesson 4–3)

37. Solve $\begin{bmatrix} 2 & x \\ y & 5 \end{bmatrix} = \begin{bmatrix} 2 & 1 \\ 3 & z \end{bmatrix}$ for the variables. (Lesson 4–1)

38. Given $f(x, y) = 12x - 8y$, find the value of $f(-2, -4)$. (Lesson 3–5)

39. Use Cramer's rule to solve the system of equations. (Lesson 3–3)
$$2x - y = 7$$
$$x + 3y = 7$$

40. Graph $f(x) = |x - 3|$. (Lesson 2–6)

41. Ecology If you recycle a $3\frac{1}{2}$-foot stack of newspapers, you can save one 20-foot loblolly pine tree. Use the formula $f(x) = \frac{x}{3.5}$, where x is the height of the newspaper in feet, to determine how many 20-foot loblolly pine trees you can save when you recycle a pile of newspapers 20 feet tall. (Lesson 2–1)

42. Name the property illustrated by $x(a + b) = xa + xb$. (Lesson 1–2)

State Capitol, Raleigh, North Carolina

SELF TEST

1. Investments Three women and their husbands invested a total of $5400 in a sandwich shop. The women invested $2400 in all; Sue invested $200 more than Tamara, and Elisa invested $200 more than Sue. Lou invested half as much as his wife, Bob invested the same as his wife, and Mateo invested twice as much as his wife. Who is married to whom? (Lesson 4–1)

2. Find the dimensions of the matrix product $A_{3 \times 4} \cdot B_{4 \times 3}$. (Lesson 4–3)

Perform the indicated operations, if possible. (Lessons 4–1, 4–2, and 4–3)

3. $\begin{bmatrix} -2 & 1.5 \\ 3 & -0.25 \end{bmatrix} - \begin{bmatrix} -6 & 2 \\ 3 & 1.25 \end{bmatrix}$

4. $\begin{bmatrix} 5.3 & -1.2 \\ 1.6 & 2 \end{bmatrix} + \begin{bmatrix} 0.3 \\ 0.2 \end{bmatrix}$

5. $2\begin{bmatrix} 5 & 4 \\ -1 & 6 \end{bmatrix} - 3\begin{bmatrix} -3 & 0 \\ -2 & 0 \end{bmatrix}$

6. $-4\begin{bmatrix} -1 & -0.25 \\ 0 & 2 \\ 0.5 & 4 \end{bmatrix}$

7. $\begin{bmatrix} -2 & 3 \\ 1 & 10 \\ 0 & -6 \end{bmatrix} \cdot \begin{bmatrix} 9 & 3 \\ 1 & 4 \end{bmatrix}$

8. $\begin{bmatrix} 1 & 0 & 2 \\ 0 & 4 & 2 \\ 3 & 5 & 0 \end{bmatrix} \cdot \begin{bmatrix} 1 & 0 \\ 0 & 1 \end{bmatrix}$

9. Evaluate the determinant of $\begin{bmatrix} -1 & 3 & 4 \\ 0 & 5 & 1 \\ 6 & -2 & 3 \end{bmatrix}$. (Lesson 4–4)

10. Geometry Use determinants to find the area of a triangle with vertices having coordinates $(-1, -2)$, $(5, 3)$, and $(2, 6)$. (Lesson 4–4)

Identity and Inverse Matrices

APPLICATION
Cryptology

You're sitting in math class reading a note that has been passed to you by your friend. Suddenly, the silence is broken by the teacher's voice saying, "If that note is so interesting, why don't you read it to the whole class?" As you slowly walk to the front of the room you think, "If I had written this note in code, I wouldn't be so embarrassed to read it!"

Cryptology deals with coding messages so that only people with the key can decipher them. In ancient Greece, Spartans wound a belt in a spiral around a stick and wrote messages along the length of the stick. When they unwound the belt, only those people who had a stick exactly the same size as the first could read the message. Since then, cryptology has been important in military communications, particularly in time of war. Today, programmers use cryptology to protect secret data stored in computers.

An important advancement in cryptology occurred in the 1930s when American mathematician Lester Hill used matrices to encode messages. Here's a simplified version of how it works.

Step 1

Suppose the first word in a message is MEET. Assign each letter a number based on its position in the alphabet (A = 1, B = 2, ..., Y = 25, Z = 26). Thus, M = 13, E = 5, and T = 20. Write the numbers in a matrix.

$$\begin{bmatrix} M & E \\ E & T \end{bmatrix} = \begin{bmatrix} 13 & 5 \\ 5 & 20 \end{bmatrix}$$

Step 2

Multiply the matrix by a *coding matrix*. Let's use $\begin{bmatrix} 0 & 1 \\ 1 & 1 \end{bmatrix}$.

$$\begin{bmatrix} 0 & 1 \\ 1 & 1 \end{bmatrix} \cdot \begin{bmatrix} 13 & 5 \\ 5 & 20 \end{bmatrix} = \begin{bmatrix} 5 & 20 \\ 18 & 25 \end{bmatrix}$$

Step 3

Assign a letter to each number of the matrix based on Step 1.

$$\begin{bmatrix} 5 & 20 \\ 18 & 25 \end{bmatrix} = \begin{bmatrix} E & T \\ R & Y \end{bmatrix}$$

Two U.S. Marines use their Navajo code during battle, 1943

Therefore, MEET would encode as ETRY. *Notice that each E in the uncoded message is assigned to a different letter in the coded message.*

When the person receives the message, he or she needs to decipher it by *undoing* the multiplication to get back to the original matrix. Mathematically, this means finding the **inverse** of the coding matrix. *You will use this code in Example 4.*

Recall from your work with real numbers that the inverse and identity of a number are related. In real numbers, 1 is the identity for multiplication because $a \cdot 1 = 1 \cdot a = a$. Similarly, the inverse of a matrix is related to the **identity matrix**. The identity matrix is a square matrix that, when multiplied by another matrix, equals that same matrix.

With 2×2 matrices, $\begin{bmatrix} 1 & 0 \\ 0 & 1 \end{bmatrix}$ is the identity matrix because $\begin{bmatrix} a & b \\ c & d \end{bmatrix} \cdot \begin{bmatrix} 1 & 0 \\ 0 & 1 \end{bmatrix} = \begin{bmatrix} a & b \\ c & d \end{bmatrix}$ and $\begin{bmatrix} 1 & 0 \\ 0 & 1 \end{bmatrix} \cdot \begin{bmatrix} a & b \\ c & d \end{bmatrix} = \begin{bmatrix} a & b \\ c & d \end{bmatrix}$. The identity matrix is symbolized by I.

Since one of the properties of the identity matrix is that it is commutative, only a square matrix can have an identity. In an identity matrix, the principal diagonal goes from upper left to lower right and consists only of ones.

Identity Matrix for Multiplication	The identity matrix for multiplication, *I*, is a square matrix with 1 for every element of the principal diagonal and 0 in all other positions. For any square matrix **A** of the same order as *I*, $$\mathbf{A} \cdot \mathbf{I} = \mathbf{I} \cdot \mathbf{A} = \mathbf{A}.$$

Example **1** If $N = \begin{bmatrix} -2 & 4 & 7 \\ 5 & -3 & 6 \\ -8 & 2 & -1 \end{bmatrix}$, find *I* so that $N \cdot I = N$.

The dimensions of N are 3×3. So, *I* must also be 3×3. The principal diagonal contains only 1s. Complete the matrix with 0s.

$$\begin{bmatrix} 1 & & \\ & 1 & \\ & & 1 \end{bmatrix} \rightarrow \begin{bmatrix} 1 & 0 & 0 \\ 0 & 1 & 0 \\ 0 & 0 & 1 \end{bmatrix}$$ The 3×3 identity matrix is $\begin{bmatrix} 1 & 0 & 0 \\ 0 & 1 & 0 \\ 0 & 0 & 1 \end{bmatrix}$.

$$\begin{bmatrix} -2 & 4 & 7 \\ 5 & -3 & 6 \\ -8 & 2 & -1 \end{bmatrix} \cdot \begin{bmatrix} 1 & 0 & 0 \\ 0 & 1 & 0 \\ 0 & 0 & 1 \end{bmatrix} = \begin{bmatrix} -2 & 4 & 7 \\ 5 & -3 & 6 \\ -8 & 2 & -1 \end{bmatrix}$$

Therefore, $N \cdot I = N$.

When we refer to the inverse of a matrix, it implies the multiplicative inverse unless otherwise stated.

Another property of real numbers is that every real number, except 0, has a multiplicative inverse. That is, $\frac{1}{a}$ is the multiplicative inverse of a because $a \cdot \frac{1}{a} = \frac{1}{a} \cdot a = 1$. Likewise, if matrix A has an inverse named A^{-1}, then $A \cdot A^{-1} = A^{-1} \cdot A = I$. The following example shows how the inverse of a 2×2 matrix can be found.

Example **2** If $M = \begin{bmatrix} 7 & 4 \\ 2 & 3 \end{bmatrix}$, find M^{-1}. Check your result.

Let M^{-1} be $\begin{bmatrix} w & x \\ y & z \end{bmatrix}$. By the definition of an inverse, $M \cdot M^{-1} = I$.

$$\begin{bmatrix} 7 & 4 \\ 2 & 3 \end{bmatrix} \cdot \begin{bmatrix} w & x \\ y & z \end{bmatrix} = \begin{bmatrix} 1 & 0 \\ 0 & 1 \end{bmatrix}$$

$$\begin{bmatrix} 7w + 4y & 7x + 4z \\ 2w + 3y & 2x + 3z \end{bmatrix} = \begin{bmatrix} 1 & 0 \\ 0 & 1 \end{bmatrix} \quad \textit{Multiply.}$$

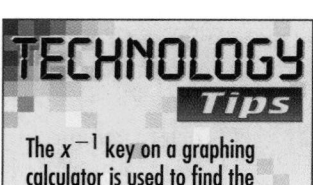
When two matrices are equal, their corresponding elements are equal. So the following equations can be generated from the two equal matrices.

(1) $7w + 4y = 1$ (2) $7x + 4z = 0$ (3) $2w + 3y = 0$ (4) $2x + 3z = 1$

Use equations (1) and (3) to find values for w and y.

First solve for w. Then substitute the w value into one of the equations to find y.

$$\begin{aligned} 7w + 4y &= 1 \\ 2w + 3y &= 0 \end{aligned} \rightarrow \begin{aligned} 21w + 12y &= 3 \\ (-)\ 8w + 12y &= 0 \\ \hline 13w &= 3 \\ w &= \frac{3}{13} \end{aligned} \longrightarrow \begin{aligned} 7w + 4y &= 1 \\ 7\left(\frac{3}{13}\right) + 4y &= 1 \\ 4y &= -\frac{8}{13} \\ y &= -\frac{2}{13} \end{aligned}$$

Use equations (2) and (4) to find values for x and z.

First solve for z. Then substitute the z value into one of the equations to find x.

$$\begin{aligned} 7x + 4z &= 0 \\ 2x + 3z &= 1 \end{aligned} \rightarrow \begin{aligned} 14x + 8z &= 0 \\ (-)\ 14x + 21z &= 7 \\ \hline -13z &= -7 \\ z &= \frac{7}{13} \end{aligned} \longrightarrow \begin{aligned} 7x + 4z &= 0 \\ 7x + 4\left(\frac{7}{13}\right) &= 0 \\ 7x &= -\frac{28}{13} \\ x &= -\frac{4}{13} \end{aligned}$$

Therefore, $M^{-1} = \begin{bmatrix} \dfrac{3}{13} & -\dfrac{4}{13} \\ -\dfrac{2}{13} & \dfrac{7}{13} \end{bmatrix}$.

Check: $\begin{bmatrix} 7 & 4 \\ 2 & 3 \end{bmatrix} \cdot \begin{bmatrix} \dfrac{3}{13} & -\dfrac{4}{13} \\ -\dfrac{2}{13} & \dfrac{7}{13} \end{bmatrix} = \begin{bmatrix} \dfrac{21}{13} - \dfrac{8}{13} & -\dfrac{28}{13} + \dfrac{28}{13} \\ \dfrac{6}{13} - \dfrac{6}{13} & -\dfrac{8}{13} + \dfrac{21}{13} \end{bmatrix}$ or $\begin{bmatrix} 1 & 0 \\ 0 & 1 \end{bmatrix}$ ✓

You should also check to be sure that $M^{-1} \cdot M = I$.

The same method used in Example 2 can be used to develop the general form of the inverse of a 2×2 matrix.

The inverse of $\begin{bmatrix} a & b \\ c & d \end{bmatrix}$ is $\begin{bmatrix} \dfrac{d}{ad-bc} & \dfrac{-b}{ad-bc} \\ \dfrac{-c}{ad-bc} & \dfrac{a}{ad-bc} \end{bmatrix}$ or $\dfrac{1}{ad-bc}\begin{bmatrix} d & -b \\ -c & a \end{bmatrix}$.

Notice that $ad - bc$ is the value of the determinant of the matrix. Remember that $\dfrac{1}{ad-bc}$ is not defined when $ad - bc = 0$. Therefore, if the value of the determinant of a matrix is 0, the matrix cannot have an inverse.

Inverse of a 2×2 Matrix	Any matrix $M = \begin{bmatrix} a & b \\ c & d \end{bmatrix}$ will have an inverse M^{-1} if and only if $\begin{vmatrix} a & b \\ c & d \end{vmatrix} \neq 0$. Then $M^{-1} = \dfrac{1}{ad-bc}\begin{bmatrix} d & -b \\ -c & a \end{bmatrix}$.

Example 3 If $Q = \begin{bmatrix} 2 & -1 \\ 1 & -3 \end{bmatrix}$, find Q^{-1}. Check your result.

Find the value of the determinant.

$$\begin{vmatrix} 2 & -1 \\ 1 & -3 \end{vmatrix} = -6 - (-1) \text{ or } -5$$

Since the determinant does not equal 0, Q^{-1} exists.

$$Q^{-1} = \frac{1}{ad-bc}\begin{bmatrix} d & -b \\ -c & a \end{bmatrix}$$

$$= -\frac{1}{5}\begin{bmatrix} -3 & 1 \\ -1 & 2 \end{bmatrix}$$

Check: $-\dfrac{1}{5}\begin{bmatrix} -3 & 1 \\ -1 & 2 \end{bmatrix} \cdot \begin{bmatrix} 2 & -1 \\ 1 & -3 \end{bmatrix} = -\dfrac{1}{5}\begin{bmatrix} -6+1 & 3-3 \\ -2+2 & 1-6 \end{bmatrix} = \begin{bmatrix} 1 & 0 \\ 0 & 1 \end{bmatrix}$ ✓

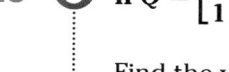

TECHNOLOGY Tips

If you get a SINGULAR MATRIX error on a graphing calculator when trying to find an inverse, it means that the matrix has no inverse.

In the application at the beginning of the lesson, the coding matrix $\begin{bmatrix} 0 & 1 \\ 1 & 1 \end{bmatrix}$ was used to encode the word MEET as ETRY. In the following Example, you will *decode* part of the message by finding the inverse of the coding matrix.

Example 4

**APPLICATION
Cryptology**

Suppose a person receives the message ETRYATNYEMYULAMMIXCQ that has been encoded using the matrix $C = \begin{bmatrix} 0 & 1 \\ 1 & 1 \end{bmatrix}$. You already know that the first four letters, ETRY, correspond to MEET. Decode the next four letters in the message.

First, write the letters in a 2×2 matrix and assign each letter a number based on its position in the alphabet.

$$\begin{bmatrix} A & T \\ N & Y \end{bmatrix} = \begin{bmatrix} 1 & 20 \\ 14 & 25 \end{bmatrix}$$

(continued on the next page)

Now, find the inverse of the coding matrix $C = \begin{bmatrix} 0 & 1 \\ 1 & 1 \end{bmatrix}$. The determinant is $0 - 1$ or -1.

$$C^{-1} = \frac{1}{-1} \begin{bmatrix} 1 & -1 \\ -1 & 0 \end{bmatrix} \text{ or } \begin{bmatrix} -1 & 1 \\ 1 & 0 \end{bmatrix}$$

Finally, multiply the inverse matrix by the first matrix and assign letters to the elements in the product.

$$\begin{bmatrix} -1 & 1 \\ 1 & 0 \end{bmatrix} \cdot \begin{bmatrix} 1 & 20 \\ 14 & 25 \end{bmatrix} = \begin{bmatrix} 13 & 5 \\ 1 & 20 \end{bmatrix} \text{ or } \begin{bmatrix} M & E \\ A & T \end{bmatrix}$$

Therefore, the next four letters of the message are MEAT.

The first eight letters of the message are MEETMEAT. You will decode the rest of the message in Exercises 10 and 33.

CHECK FOR UNDERSTANDING

Communicating Mathematics

Study the lesson. Then complete the following.

1. **Explain** how the multiplicative inverse and identity for real numbers are similar to the matrix inverse and identity.

2. **Write** the 4×4 identity matrix.

3. **Choose** the inverse of $\begin{bmatrix} 2 & 2 \\ 3 & 4 \end{bmatrix}$.

   a. $\begin{bmatrix} \frac{1}{2} & \frac{1}{2} \\ \frac{1}{3} & \frac{1}{4} \end{bmatrix}$ b. $\begin{bmatrix} 1 & 0 \\ 0 & 1 \end{bmatrix}$ c. $\begin{bmatrix} 2 & -1 \\ -\frac{3}{2} & 1 \end{bmatrix}$ d. $\begin{bmatrix} 4 & -2 \\ -3 & 2 \end{bmatrix}$

4. **Create** a square matrix that does not have an inverse.

5. **You Decide**   Miyoki says that the matrix $\begin{bmatrix} 1 \\ 4 \end{bmatrix}$ does not have a multiplicative identity. Hector says the identity is $\begin{bmatrix} 1 & 0 \\ 0 & 1 \end{bmatrix}$ because $\begin{bmatrix} 1 & 0 \\ 0 & 1 \end{bmatrix} \cdot \begin{bmatrix} 1 \\ 4 \end{bmatrix} = \begin{bmatrix} 1 \\ 4 \end{bmatrix}$. Who is correct? Explain your reasoning.

Guided Practice

Find the inverse of each matrix, if it exists. If it does not exist, explain why not.

6. $\begin{bmatrix} 3 & 1 \\ 5 & 2 \end{bmatrix}$ 7. $\begin{bmatrix} 1 \\ 2 \end{bmatrix}$ 8. $\begin{bmatrix} 6 & 3 \\ 8 & 4 \end{bmatrix}$ 9. $\begin{bmatrix} -5 & 1 \\ 7 & 4 \end{bmatrix}$

10. **Cryptology**   Decode the next four letters, EMYU, of the message in Example 4.

Practice

Find the inverse of each matrix, if it exists. If it does not exist, explain why not.

11. $\begin{bmatrix} 5 & 0 \\ 0 & 1 \end{bmatrix}$

12. $\begin{bmatrix} 8° & -5 \\ -3 & 2 \end{bmatrix}$

13. $\begin{bmatrix} 4 & -8 \\ -1 & 2 \end{bmatrix}$

14. $\begin{bmatrix} 6 \\ 4 \end{bmatrix}$

15. $\begin{bmatrix} 4 & -3 \\ 2 & 7 \end{bmatrix}$

16. $\begin{bmatrix} -1 & 5 & -2 \\ 4 & 2 & -3 \end{bmatrix}$

17. $\begin{bmatrix} 2 & -5 \\ 6 & 1 \end{bmatrix}$

18. $\begin{bmatrix} 4 & 4 \\ 4 & 4 \end{bmatrix}$

19. $\begin{bmatrix} -2 & 0 \\ 5 & 6 \end{bmatrix}$

Determine whether each statement is *true* or *false*.

20. $\begin{bmatrix} 0 & 1 \\ 1 & 1 \end{bmatrix} \cdot \begin{bmatrix} -1 & 1 \\ 1 & 0 \end{bmatrix} = I$

21. $\begin{bmatrix} 2 & 1 & -4 \\ -3 & 6 & 5 \\ 1 & 2 & -2 \end{bmatrix} \cdot I = \begin{bmatrix} 2 & 1 & -4 \\ -3 & 6 & 5 \\ 1 & 2 & -2 \end{bmatrix}$

22. $\begin{bmatrix} \frac{1}{3} & -\frac{2}{3} \\ \frac{2}{3} & -\frac{1}{3} \end{bmatrix} \cdot \begin{bmatrix} 1 & 2 \\ 2 & 1 \end{bmatrix} = I$

23. $\begin{bmatrix} 1 & 5 \\ 1 & -2 \end{bmatrix} \cdot \begin{bmatrix} \frac{2}{7} & \frac{5}{7} \\ \frac{1}{7} & -\frac{1}{7} \end{bmatrix} = I$

24. The inverse of $\begin{bmatrix} 3 & 1 & 2 \\ -2 & 0 & 4 \\ 3 & 5 & 2 \end{bmatrix}$ is $-\frac{1}{64}\begin{bmatrix} -20 & 8 & 4 \\ 16 & 0 & -16 \\ -10 & -12 & 2 \end{bmatrix}$.

25. All square matrices have multiplicative identities.

26. Only square matrices have multiplicative inverses.

27. Some square matrices do not have multiplicative inverses.

28. Some square matrices do not have multiplicative identities.

29. All multiplicative identities are square matrices.

Graphing Calculator

30. When Mike used a graphing calculator to find the inverse of $\begin{bmatrix} -3 & -2 \\ 6 & 4 \end{bmatrix}$, there was an ERROR statement. Explain why.

Critical Thinking

31. Prove that $A \cdot I = I \cdot A = A$ for all second-order matrices.

Applications and Problem Solving

32. **Geometry** Recall that the matrix $\begin{bmatrix} 0 & -1 \\ 1 & 0 \end{bmatrix}$ will rotate a figure on a coordinate plane 90° counterclockwise about the origin.
  a. Find the inverse of this rotation matrix.
  b. Make a conjecture about what movement the inverse describes on a coordinate plane.
  c. Test your conjecture on the triangle shown at the right. Make a drawing to verify your conjecture.

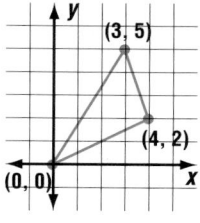

33. Cryptology Refer to Example 4 and Exercise 10.

  a. Decode the last eight letters, LAMMIXCQ, of the coded message. (*Hint:* Negative integers and zero are assigned letters as follows: $0 = Z$, $-1 = Y$, $-2 = X$, $-3 = W$, and so on.)

  b. Write the entire decoded message from Example 4.

  c. Write a message and code it using your own coding matrix. (*Hint:* Use a coding matrix whose determinant is 1 or -1.) Trade messages with a partner and decode the messages.

Mixed Review

34. Solve $\begin{vmatrix} 4 & -a \\ 7 & 3a \end{vmatrix} = 57.$ (Lesson 4–4)

35. Geometry The perimeter of a right triangle is 24 centimeters. Three times the length of the longer leg minus two times the length of the shorter leg exceeds the hypotenuse by 2 centimeters. The length of the shorter leg is one centimeter more than half the hypotenuse. What are the lengths of all three sides? (Lesson 3–7)

36. Solve the system of equations by using substitution. (Lesson 3–2)

$$3x - 2y = -3$$
$$3x + y = 3$$

37. Find the value of $g(12)$ when $g(x) = \dfrac{26 - x}{2}$. (Lesson 2–1)

38. Solve $5 < 2x - 9 < 11$. (Lesson 1–6)

39. Solve $|a + 5| + 5 = 3.$ (Lesson 1–5)

Mathematics and SOCIETY

DSS Code

The excerpt below appeared in an article in *Popular Science* in December, 1994.

THE GROWING AMOUNT OF BUSINESS being conducted electronically these days raises thorny questions: How does your broker, lawyer, or accountant know that your e-mail message really came from you? And just how valid is a computerized contract when there are no signatures on the bottom line? These are among the questions being addressed by the National Institute of Standards and Technology, which has proposed a solution called the Digital Signal Standard (DSS). Essentially, the DSS is a method for creating a mathematical "signature" on your documents . . . . The DSS relies on a well-known cryptology concept called public and private keys. You use a private key—a long number you keep to yourself—to generate an encoded, "signed" version of your message. A recipient verifies this signature using your public key, another long number. The public key is tied to the private key by a mathematical equation that makes it easy to compute the public key from the private key, but nearly impossible to perform the reverse calculation. If the math indicates a match, your signature has been verified. ■

1. You could call the DSS code a "one-way" function because it is relatively easy to do in one direction, but much more difficult to do in the reverse direction. Can you think of another example of a one-way function?

2. If the federal government used the DSS, who might want to keep a "master key" so they could decode all messages flowing through the government's systems?

3. List some of the advantages and disadvantages of having a "master key."

Using Matrices to Solve Systems of Equations

What YOU'LL LEARN

* To solve systems of linear equations by using inverse matrices.

Why IT'S IMPORTANT

You can use matrices and systems of equations to solve problems involving chemistry and business.

APPLICATION

Business

Katie earns extra money by making stuffed teddy bears and rabbits and then selling them to a local craft store. She has arranged to sell a total of 15 stuffed animals to the owner of the craft store each week. Her profit on each rabbit is $5 and each bear is $9. Her goal is to earn at least $120 each week. Of course, Katie could make 15 bears and easily meet her goals. But it takes her longer to make a bear than a rabbit, and she doesn't have time to make 15 bears. Katie wants to know what combination of stuffed animals she should make each week to guarantee a profit of $120. *This problem will be solved in Example 3.*

The problem above can be solved by using a system of equations. You have already learned several methods for solving a system of equations, including graphing, substitution, addition and subtraction, and Cramer's rule. Matrices can also be used to solve systems of equations. In this lesson, you will be using inverses of matrices.

Remember, two matrices are equal if their corresponding elements are equal.

Consider the system of equations below. You can write this system with matrices by using the left and right sides of the equations.

$$\begin{aligned} 7x + 5y &= 3 \\ 3x - 2y &= 22 \end{aligned} \quad \rightarrow \quad \begin{bmatrix} 7x + 5y \\ 3x - 2y \end{bmatrix} = \begin{bmatrix} 3 \\ 22 \end{bmatrix}$$

Write the matrix on the left as the product of the coefficients and variables.

$$\begin{bmatrix} 7 & 5 \\ 3 & -2 \end{bmatrix} \cdot \begin{bmatrix} x \\ y \end{bmatrix} = \begin{bmatrix} 3 \\ 22 \end{bmatrix}$$

coefficient matrix *variable matrix* *constant matrix*

LOOK BACK

You can refer to Lessons 3-1, 3-2, and 3-3 for information about solving systems of equations.

The system of equations is now expressed as a **matrix equation**. *This system will be solved in Example 2.*

Example **1** Write each system of equations as a matrix equation.

a. $3x - 2y = 7$
   $4x + y = 8$

b. $3a - 5b + 2c = 9$
   $4a + 7b + c = 3$
   $2a - c = 12$

The matrix equation is

$$\begin{bmatrix} 3 & -2 \\ 4 & 1 \end{bmatrix} \cdot \begin{bmatrix} x \\ y \end{bmatrix} = \begin{bmatrix} 7 \\ 8 \end{bmatrix}.$$

The matrix equation is

$$\begin{bmatrix} 3 & -5 & 2 \\ 4 & 7 & 1 \\ 2 & 0 & -1 \end{bmatrix} \cdot \begin{bmatrix} a \\ b \\ c \end{bmatrix} = \begin{bmatrix} 9 \\ 3 \\ 12 \end{bmatrix}.$$

A linear equation in the form $ax = b$ and a matrix equation in the form $AX = B$, where A is the coefficient matrix, X is the variable matrix, and B is the constant matrix, can be solved in a similar manner.

$ax = b$		$AX = B$
$\left(\dfrac{1}{a}\right)ax = \left(\dfrac{1}{a}\right)b$	*Multiply by the inverse if it exists.*	$A^{-1}AX = A^{-1}B$
$1x = \left(\dfrac{1}{a}\right)b$	*1 and I are identities.*	$IX = A^{-1}B$
$x = \left(\dfrac{1}{a}\right)b$		$X = A^{-1}B$

Since matrix multiplication is not commutative, the inverse matrix should be at the left on each side of the matrix equation.

The solution of the linear equation is the product of the inverse of the coefficient and the constant term. In the matrix equation, the solution is the product of the inverse of the coefficient matrix and the constant matrix.

Example 2 **Use a matrix equation to solve the system of equations.**
$$7x + 5y = 3$$
$$3x - 2y = 22$$

The matrix equation is $\begin{bmatrix} 7 & 5 \\ 3 & -2 \end{bmatrix} \cdot \begin{bmatrix} x \\ y \end{bmatrix} = \begin{bmatrix} 3 \\ 22 \end{bmatrix}$.

First, find the inverse of the coefficient matrix. The inverse of $\begin{bmatrix} 7 & 5 \\ 3 & -2 \end{bmatrix}$ is
$$\frac{1}{-14 - (15)} \begin{bmatrix} -2 & -5 \\ -3 & 7 \end{bmatrix} \text{ or } -\frac{1}{29} \begin{bmatrix} -2 & -5 \\ -3 & 7 \end{bmatrix}.$$

Next, multiply each side of the matrix equation by the inverse matrix.

The identity matrix on the left verifies that the inverse matrix has been calculated correctly.

$$-\frac{1}{29} \begin{bmatrix} -2 & -5 \\ -3 & 7 \end{bmatrix} \cdot \begin{bmatrix} 7 & 5 \\ 3 & -2 \end{bmatrix} \cdot \begin{bmatrix} x \\ y \end{bmatrix} = -\frac{1}{29} \begin{bmatrix} -2 & -5 \\ -3 & 7 \end{bmatrix} \cdot \begin{bmatrix} 3 \\ 22 \end{bmatrix}$$

$$-\frac{1}{29} \begin{bmatrix} -29 & 0 \\ 0 & -29 \end{bmatrix} \cdot \begin{bmatrix} x \\ y \end{bmatrix} = -\frac{1}{29} \begin{bmatrix} -116 \\ 145 \end{bmatrix}$$

$$\begin{bmatrix} 1 & 0 \\ 0 & 1 \end{bmatrix} \cdot \begin{bmatrix} x \\ y \end{bmatrix} = \begin{bmatrix} 4 \\ -5 \end{bmatrix}$$

$$\begin{bmatrix} x \\ y \end{bmatrix} = \begin{bmatrix} 4 \\ -5 \end{bmatrix}$$

The solution is $(4, -5)$.

The graph at the right confirms the solution.

How might you check the solution?

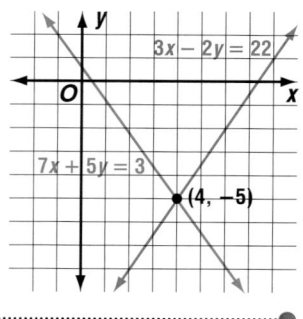

To solve a system of equations with three variables, you will use the 3×3 identity matrix. However, as you may imagine, finding the inverse of a 3×3 matrix can be tedious. Graphing calculators and computer programs offer fast and accurate methods for performing the necessary calculations.

You will not be asked to find the inverse of a 3×3 matrix in this chapter.

You can use a graphing calculator and a matrix equation to solve systems of equations. Consider the system of equations below.

$3x - 2y + z = 0$
$2x + 3y = 12$
$y + 4z = -18$

- Write the system so that each equation is in standard form and contains all three variables. Then write the coefficient matrix.

$$\begin{matrix} 3x - 2y + 1z = 0 \\ 2x + 3y + 0z = 12 \\ 0x + 1y + 4z = -18 \end{matrix} \rightarrow \begin{bmatrix} 3 & -2 & 1 \\ 2 & 3 & 0 \\ 0 & 1 & 4 \end{bmatrix}$$

- Write the matrix equation in the form $AX = B$.

$$\begin{bmatrix} 3 & -2 & 1 \\ 2 & 3 & 0 \\ 0 & 1 & 4 \end{bmatrix} \cdot \begin{bmatrix} x \\ y \\ z \end{bmatrix} = \begin{bmatrix} 0 \\ 12 \\ -18 \end{bmatrix}$$

TECHNOLOGY
Tips

You may need to use the arrow keys to see the entire matrix on the graphing calculator screen.

- Use a graphing calculator to find A^{-1}. The inverse is shown below.

[A]⁻¹
[[.2222222222...
[−.1481481481...
[.037037037...
■

- Now, find the product of A^{-1} and B.

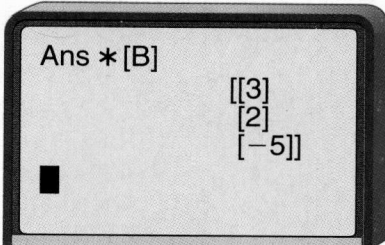

Ans * [B]

[[3]
[2]
[−5]]
■

Therefore, $\begin{bmatrix} x \\ y \\ z \end{bmatrix} = \begin{bmatrix} 3 \\ 2 \\ -5 \end{bmatrix}$, and $x = 3$, $y = 2$, and $z = -5$.

Your Turn

Solve the matrix equation $AX = B$ for each value of A and B.

a. $A = \begin{bmatrix} 1 & -2 & 1 \\ 3 & 1 & -1 \\ 2 & 3 & 2 \end{bmatrix}$, $B = \begin{bmatrix} 7 \\ 2 \\ 7 \end{bmatrix}$

b. $A = \begin{bmatrix} 2 & 6 & 8 \\ -2 & 9 & -12 \\ 4 & 6 & -4 \end{bmatrix}$, $B = \begin{bmatrix} 5 \\ -1 \\ 3 \end{bmatrix}$

You can use matrices to help solve problems that involve systems of equations. Matrices often simplify the process of solving these systems.

Example ③

APPLICATION
Business

Refer to the application at the beginning of the lesson. What combination of stuffed animals should Katie make each week to guarantee her a profit of $120?

Explore   Let b represent the number of bears, and let r represent the amount of rabbits. Therefore, $b + r = 15$.
The total number of animals must be 15.

Now, write an equation that represents Katie's profit.

$$9b + 5r = 120$$

Plan   Write a system of equations. Then write the system as a matrix equation.

$$\begin{aligned} b + r &= 15 \\ 9b + 5r &= 120 \end{aligned} \quad \rightarrow \quad \begin{bmatrix} 1 & 1 \\ 9 & 5 \end{bmatrix} \cdot \begin{bmatrix} b \\ r \end{bmatrix} = \begin{bmatrix} 15 \\ 120 \end{bmatrix}$$

Solve   To solve the equation, first find the inverse of the coefficient matrix.

$$\frac{1}{ad - bc}\begin{bmatrix} d & -b \\ -c & a \end{bmatrix} \quad \rightarrow \quad -\frac{1}{4}\begin{bmatrix} 5 & -1 \\ -9 & 1 \end{bmatrix} \text{ or } \begin{bmatrix} -1.25 & 0.25 \\ 2.25 & -0.25 \end{bmatrix}$$

Now multiply each side of the matrix equation by the inverse and solve.

$$\begin{bmatrix} -1.25 & 0.25 \\ 2.25 & -0.25 \end{bmatrix} \cdot \begin{bmatrix} 1 & 1 \\ 9 & 5 \end{bmatrix} \cdot \begin{bmatrix} b \\ r \end{bmatrix} = \begin{bmatrix} -1.25 & 0.25 \\ 2.25 & -0.25 \end{bmatrix} \cdot \begin{bmatrix} 15 \\ 120 \end{bmatrix}$$

$$\begin{bmatrix} b \\ r \end{bmatrix} = \begin{bmatrix} 11.25 \\ 3.75 \end{bmatrix}$$

Examine   Since Katie cannot make a fraction of an animal, the solution (11.25, 3.75) is unreasonable. However, try $b = 11$, $r = 4$ and $b = 12$, $r = 3$ as possible solutions.

Bears	Rabbits	Profit
11	4	9(11) + 5(4) or $119
12	3	9(12) + 5(3) or $123 ✓

Therefore, one solution to the problem is 12 bears and 3 rabbits.

What other solutions are possible?

LOOK BACK

You can refer to Lesson 3-1 for information about consistent and inconsistent systems.

When the determinant of the coefficient matrix is 0, the system of equations has no unique solution. You can graph the equations to determine whether the system is consistent and has infinitely many solutions or is inconsistent and has no solutions.

Example **4** **Solve the matrix equation** $\begin{bmatrix} 3 & 1 \\ 6 & 2 \end{bmatrix} \cdot \begin{bmatrix} x \\ y \end{bmatrix} = \begin{bmatrix} -2 \\ 10 \end{bmatrix}.$

The determinant of the coefficient matrix is 0, so there is no unique solution. Graph the system of equations.

The matrix equation represents the system of equations below.

$3x + y = -2$

$6x + 2y = 10$

Since the lines are parallel, this system has no solution. Therefore, the system is inconsistent.

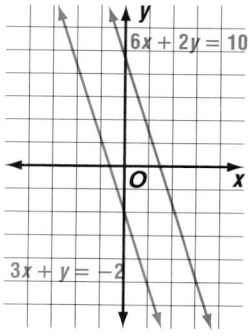

CHECK FOR UNDERSTANDING

Communicating Mathematics

Study the lesson. Then complete the following.

1. **Write** a matrix equation for the system.

   $2x - 8y = 3$

   $7x - 2y = 5$

2. **Write** the matrix equation $\begin{bmatrix} 2 & -5 \\ 3 & 8 \end{bmatrix} \cdot \begin{bmatrix} r \\ s \end{bmatrix} = \begin{bmatrix} 2 \\ 9 \end{bmatrix}$ as a system of linear equations.

3. **List** the steps you would use to solve a system of linear equations with inverse matrices.

4. **Write** an example of a system of equations that does not have a unique solution.

Guided Practice

Write a matrix equation for each system.

5. $4x - 7y = 2$
   $3x + 5y = 9$

6. $2a + 3b - 5c = 1$
   $7a + 3c = 7$
   $3a - 6b + c = -5$

7. $y = 3x$
   $x + 2y = -21$

8. Given that the inverse of the coefficient matrix is $-\dfrac{1}{28}\begin{bmatrix} -3 & -8 \\ -2 & 4 \end{bmatrix}$, solve the matrix equation $\begin{bmatrix} 4 & 8 \\ 2 & -3 \end{bmatrix} \cdot \begin{bmatrix} x \\ y \end{bmatrix} = \begin{bmatrix} 7 \\ 0 \end{bmatrix}.$

Use a matrix equation to solve each system of equations.

9. $x + 2y = 8$
   $3x + 2y = 6$

10. $5s + 4t = 12$
    $3s = -4 + 4t$

11. Determine whether the system of equations has *one* solution, *no* solution, or *infinitely many* solutions.

    $x + 2y = 5$

    $3x - 15 = -6y$

Practice

Write a matrix equation for each system.

12. $5a - 6b = -47$
$3a + 2b = -17$

13. $3m - 7n = -43$
$6m + 5n = -10$

14. $2r + 3s = -17$
$s = r - 4$

15. $y = -x$
$y = 2x$

16. $y = x + 3$
$3y + x = 5$

17. $2x = 3y$
$y = 4x - 5$

Matrix M^{-1} is the inverse of the coefficient matrix. Use M^{-1} to solve each matrix equation.

18. $\begin{bmatrix} 3 & 1 \\ 4 & -2 \end{bmatrix} \cdot \begin{bmatrix} x \\ y \end{bmatrix} = \begin{bmatrix} 13 \\ 24 \end{bmatrix}$ $M^{-1} = -\frac{1}{10} \begin{bmatrix} -2 & -1 \\ -4 & 3 \end{bmatrix}$

19. $\begin{bmatrix} 3 & 1 & 1 \\ -6 & 5 & 3 \\ 9 & -2 & -1 \end{bmatrix} \cdot \begin{bmatrix} x \\ y \\ z \end{bmatrix} = \begin{bmatrix} -1 \\ -9 \\ 5 \end{bmatrix}$ $M^{-1} = -\frac{1}{9} \begin{bmatrix} 1 & -1 & -2 \\ 21 & -12 & -15 \\ -33 & 15 & 21 \end{bmatrix}$

20. $\begin{bmatrix} 1 & 2 & 2 \\ 2 & -1 & 1 \\ 3 & -2 & 3 \end{bmatrix} \cdot \begin{bmatrix} a \\ b \\ c \end{bmatrix} = \begin{bmatrix} 0 \\ -1 \\ -4 \end{bmatrix}$ $M^{-1} = -\frac{1}{9} \begin{bmatrix} -1 & -10 & 4 \\ -3 & -3 & 3 \\ -1 & 8 & -5 \end{bmatrix}$

Solve each matrix equation or system of equations by using inverse matrices.

21. $\begin{bmatrix} 2 & 6 \\ 4 & -3 \end{bmatrix} \cdot \begin{bmatrix} x \\ y \end{bmatrix} = \begin{bmatrix} 3 \\ 1 \end{bmatrix}$

22. $\begin{bmatrix} 8 & -1 \\ 2 & 3 \end{bmatrix} \cdot \begin{bmatrix} a \\ b \end{bmatrix} = \begin{bmatrix} 16 \\ -9 \end{bmatrix}$

23. $\begin{bmatrix} 5 & -3 \\ 8 & 5 \end{bmatrix} \cdot \begin{bmatrix} a \\ b \end{bmatrix} = \begin{bmatrix} -30 \\ 1 \end{bmatrix}$

24. $3x = 13 - y$
$2x - y = 2$

25. $6a + 2b = 11$
$3a = 8b + 1$

26. $4x = -3y + 5$
$8x = 9y$

Determine whether each system has *one* unique solution, *no* solution, or *infinitely many* solutions. If it has one unique solution, name it. If it has no solution or many solutions, graph the system.

27. $3x - y = 4$
$6x + 2y = -8$

28. $x + 3y = 6$
$3x - 18 = -9y$

29. $3x - 8y = 4$
$6x - 42 = 16y$

Graphing Calculator

Use a graphing calculator to solve each system of equations.

30. $5x + y = 1$
$9x + 3y = 1$

31. $1.8x + 5y = 19.5$
$5.2x - 2.9y = 4.3$

Critical Thinking

32. According to Cramer's rule, the solution of the system $\begin{cases} ax + by = e \\ cx + dy = f \end{cases}$ is

(x, y), where $x = \dfrac{\begin{vmatrix} e & b \\ f & d \end{vmatrix}}{\begin{vmatrix} a & b \\ c & d \end{vmatrix}}$, $y = \dfrac{\begin{vmatrix} a & e \\ c & f \end{vmatrix}}{\begin{vmatrix} a & b \\ c & d \end{vmatrix}}$, and $\begin{vmatrix} a & b \\ c & d \end{vmatrix} \neq 0$. Generalize Cramer's rule

to be used with a system of equations in three variables.

33. Chemistry   Sonia Ramos is
a chemist who is preparing
an acid solution to be used
as a cleaner for machine
parts. The machine shop
needs several 200-mL
batches of solution at a 48%
concentration. Sonia only has
60% and 40% concentration
solutions. The two solutions
can be combined in some
ratio to make the 48%
solution. How much of each
solution should Sonia use to
make 200 mL of solution?

34. Food Service   The manager of the Snack Shack is gathering information
about the time it takes to make and serve hamburgers and chicken
sandwiches. It takes 5 minutes to prepare a hamburger and 2 additional
minutes to serve it with cheese, lettuce, and ketchup. It takes 7 minutes
to prepare a chicken sandwich and 1 additional minute to serve it with
lettuce, tomato, and mayonnaise. How many sandwiches can be prepared
and served by one employee if 42 minutes is spent on preparation and
15 minutes is spent on serving?

35. *True or false:* $\begin{bmatrix} \frac{9}{2} & \frac{1}{2} \\ 4 & \frac{2}{3} \end{bmatrix} \cdot \begin{bmatrix} \frac{2}{3} & -\frac{1}{2} \\ -4 & \frac{9}{2} \end{bmatrix} = I.$ (Lesson 4–5)

36. Find $\begin{bmatrix} 2 \\ 9 \\ 0 \end{bmatrix} + 4\begin{bmatrix} -1 \\ 3 \\ 5 \end{bmatrix} + 3\begin{bmatrix} -6 \\ -3 \\ -1 \end{bmatrix}.$ (Lesson 4–2)

37. Baseball   In a baseball game, a ball that lands to the right of the right field
baseline or to the left of the left field baseline is a foul ball. If the ball lands
on the line, it is a fair ball. Suppose a baseball diamond could be placed on
a coordinate plane with home plate at the origin, first base on the x-axis,
and third base on the y-axis. Write a system of inequalities that would
describe foul territory.   (Lesson 3–4)

38. Solve the system of equations by graphing.   (Lesson 3–1)
$3x + 4y = 8$
$6y - 8x = 12$

39. Oceanography   The Mariana Trench is the deepest point in any of the
oceans. It is located in the western Pacific Ocean north of Australia. The
deepest point in the trench is 36,198 feet, or about 6.8 miles below sea
level. Water pressure in the ocean is represented by the function $f(x) =$
$1.15x$, where x is the depth in miles and $f(x)$ is the pressure in tons per
square inch. Find the approximate water pressure at the deepest point in
the Mariana Trench.   (Lesson 2–2)

40. Retail Sales   Leon bought a 10-speed bicycle on sale for 75% of its original
price. The sale price was $41 less than the original price. Find the original
price and the sale price.   (Lesson 1–4)

Using Augmented Matrices

What YOU'LL LEARN

- To solve systems of linear equations by using augmented matrices.

Why IT'S IMPORTANT

You can use augmented matrices to solve problems involving finance and business.

APPLICATION
Investing

When Cleveland Jackson inherited $5000, he went to a financial planner for help in investing the money. The financial planner suggested that he invest the money in a stock fund that earns an average of 6%, a bond fund that earns an average of 7%, and a term fund that earns an average of 4%. The planner told Mr. Jackson that he shouldn't invest all the money in the highest paying fund, because although it would make more money, higher paying funds are more risky. Since the earnings from the stock fund will be available sooner, Mr. Jackson wants to earn three times as much from the stock fund as he will from the term fund. If he wants to earn a total of $300, how much should he invest in each fund? *This problem will be solved in Example 2.*

In the last lesson, you solved systems of equations using inverse matrices. A system of equations may also be solved using a matrix called an **augmented matrix**. The augmented matrix of a system contains the coefficient matrix with an extra column containing the constant terms. Study how the system below is written as an augmented matrix.

$$\begin{array}{c} x + 5y + 6z = -8 \\ 3x - 2y - 2z = 17 \\ 2x + 3y + 4z = 1 \end{array} \quad \rightarrow \quad \left[\begin{array}{ccc|c} 1 & 5 & 6 & -8 \\ 3 & -2 & -2 & 17 \\ 2 & 3 & 4 & 1 \end{array} \right]$$

The system of equations can be solved by manipulating the rows of the matrix rather than the equations themselves. For example, you could multiply each side of the first equation by -2. The result would be $-2x - 10y - 12z = 16$. The corresponding change in the matrix is that the first row becomes $[-2 \quad -10 \quad -12 \quad 16]$.

When you use an augmented matrix, you perform the same operations as you would in working with the equations, but you do not have to bother writing the variables or worrying about the order in which the terms are written—the organization of the matrix keeps all of this in its proper place. Here is a summary of the **row operations** that can be performed on an augmented matrix.

Notice that the row operations can only be performed on rows, not columns.

- Any two rows can be interchanged.
- Any row can be replaced with a nonzero multiple of that row.
- Any row can be replaced with the sum of that row and a multiple of another row.

The solution of the system of equations above is $(4, -3, 0.5)$; that is, $x = 4$, $y = -3$, and $z = 0.5$.

Suppose we write these three equations in the form of an augmented matrix.

$$x = 4$$
$$y = -3 \quad \rightarrow \quad \begin{bmatrix} 1 & 0 & 0 & | & 4 \\ 0 & 1 & 0 & | & -3 \\ 0 & 0 & 1 & | & 0.5 \end{bmatrix}$$
$$z = 0.5$$

Notice that the first three columns are the same as a 3×3 identity matrix. When doing row operations, your goal should be to find an augmented identity matrix.

Just as there is no single order of steps to solve a system of equations, there is also no one single group of row operations that arrives at the correct solution. The order in which you solve a system may be different from the way a classmate solves it, but you may both be correct.

Example 1

Use an augmented matrix to solve the system of equations.
$$a + 2b + c = 0$$
$$2a + 5b + 4c = -1$$
$$a - b - 9c = -5$$

Write the augmented matrix.
$$\begin{bmatrix} 1 & 2 & 1 & | & 0 \\ 2 & 5 & 4 & | & -1 \\ 1 & -1 & -9 & | & -5 \end{bmatrix}$$
The first element in row 1 is already 1.

Multiply row 1 by -1 and add to row 3.
$$\begin{bmatrix} 1 & 2 & 1 & | & 0 \\ 2 & 5 & 4 & | & -1 \\ 0 & -3 & -10 & | & -5 \end{bmatrix}$$
The first element in row 3 is now 0.

Multiply row 1 by -2 and add to row 2.
$$\begin{bmatrix} 1 & 2 & 1 & | & 0 \\ 0 & 1 & 2 & | & -1 \\ 0 & -3 & -10 & | & -5 \end{bmatrix}$$
The first element in row 2 is now 0, and the second element in row 2 is now 1.

Multiply row 2 by -2 and add to row 1.
$$\begin{bmatrix} 1 & 0 & -3 & | & 2 \\ 0 & 1 & 2 & | & -1 \\ 0 & -3 & -10 & | & -5 \end{bmatrix}$$
The second element in row 1 is now 0.

Multiply row 2 by 3 and add to row 3.
$$\begin{bmatrix} 1 & 0 & -3 & | & 2 \\ 0 & 1 & 2 & | & -1 \\ 0 & 0 & -4 & | & -8 \end{bmatrix}$$
The second element in row 3 is now 0.

Multiply row 3 by $-\frac{1}{4}$.
$$\begin{bmatrix} 1 & 0 & -3 & | & 2 \\ 0 & 1 & 2 & | & -1 \\ 0 & 0 & 1 & | & 2 \end{bmatrix}$$
The third element in row 3 is now 1.

Multiply row 3 by 3 and add to row 1.
$$\begin{bmatrix} 1 & 0 & 0 & | & 8 \\ 0 & 1 & 2 & | & -1 \\ 0 & 0 & 1 & | & 2 \end{bmatrix}$$
The third element in row 1 is now 0.

Multiply row 3 by -2 and add to row 2.
$$\begin{bmatrix} 1 & 0 & 0 & | & 8 \\ 0 & 1 & 0 & | & -5 \\ 0 & 0 & 1 & | & 2 \end{bmatrix}$$
This matrix contains an augmented identity matrix. Now you can read the solution.

The solution is $(8, -5, 2)$.

Notice that the matrix has all zeros in the "bottom triangle." The system can now be solved by setting $c = 2$ and substituting into the other equations.

The process of performing row operations to get the desired matrix is called **reducing a matrix**. The resulting matrix is called a **reduced matrix**. Reducing a matrix is the method used by computers to solve systems of equations with many variables and equations.

Example **2**

APPLICATION

Investing

Refer to the application at the beginning of the lesson. How much money should Mr. Jackson invest in each fund so that the total earnings will be $300?

Explore   Let *s* represent the amount in stocks, let *b* represent the amount in bonds, and let *t* represent the amount in the term fund.

Plan   Write a system of equations.

$s + b + t = 5000$ *The total amount invested is $5000.*

$0.06s + 0.07b + 0.04t = 300$ *The total amount earned is $300.*

$0.06s = 3(0.04)t$ *The amount earned from the stock fund is three times the amount earned from the term fund.*

Solve   Then write an augmented matrix.

$s + b + t = 5000$

$0.06s + 0.07b + 0.04t = 300$ $\rightarrow$

$0.06s = 3(0.04)t$

$$\begin{bmatrix} 1 & 1 & 1 & | & 5000 \\ 0.06 & 0.07 & 0.04 & | & 300 \\ 0.06 & 0 & -0.12 & | & 0 \end{bmatrix}$$

After applying row operations on the matrix, we get the following.

$$\begin{bmatrix} 1 & 0 & 0 & | & 2000 \\ 0 & 1 & 0 & | & 2000 \\ 0 & 0 & 1 & | & 1000 \end{bmatrix}$$

The solution is (2000, 2000, 1000), which means that Mr. Jackson should invest $2000 in the stock fund, $2000 in the bond fund, and $1000 in the term fund.

Examine   Examine this solution to see if it makes sense.

Stocks: 6% of $2000 = $120   *Three times the amount earned in*
Bonds: 7% of $2000 = $140   *the term fund is 3($40) or $120.*
Term: 4% of $1000 = $ 40
Total: $300 ✓

As with other methods of solving systems of equations, there is not always a unique solution. In systems where no solution or multiple solutions exist, solving by augmented matrices can identify these solutions. Study the solution of the system shown below.

$\begin{array}{l} x + y + 2z = -5 \\ 3x + y + 12z = -19 \\ 2x + y + 7z = -12 \end{array}$ $\rightarrow$ $\begin{bmatrix} 1 & 1 & 2 & | & -5 \\ 3 & 1 & 12 & | & -19 \\ 2 & 1 & 7 & | & -12 \end{bmatrix}$ $\rightarrow$ $\begin{bmatrix} 1 & 0 & 5 & | & -7 \\ 0 & 1 & -3 & | & 2 \\ 0 & 0 & 0 & | & 0 \end{bmatrix}$

The row of zeros indicates that the last equation is some combination of the other two equations, meaning that this is a dependent system and there is no unique solution. However, there is a solution. In fact, there is an infinite number of solutions.

Let's write the equations represented by the matrix and solve each of them in terms of z.

$$x + 5z = -7 \qquad\qquad y - 3z = 2$$
$$x = -5z - 7 \qquad\qquad y = 3z + 2$$

The solution is the ordered triple $(-5z - 7, 3z + 2, z)$. By choosing any value for z, you can find coordinates of points on a line that are solutions to the system.

Let's look at another system of equations.

$$\begin{aligned} 3x - 5y + 2z &= -7 \\ x + 4y - z &= 10 \\ 6x + 7y - z &= -18 \end{aligned} \rightarrow \begin{bmatrix} 3 & -5 & 2 & \vdots & -7 \\ 1 & 4 & -1 & \vdots & 10 \\ 6 & 7 & -1 & \vdots & -18 \end{bmatrix} \rightarrow \begin{bmatrix} 1 & 4 & -1 & \vdots & 10 \\ 0 & -17 & 5 & \vdots & -37 \\ 0 & 0 & 0 & \vdots & 5 \end{bmatrix}$$

Notice the last row of the final matrix. This represents the equation $0 = 5$. Since this cannot be true, the system is inconsistent, and no solution exists.

CHECK FOR UNDERSTANDING

Communicating Mathematics

Study the lesson. Then complete the following.

1. **Discuss** the advantages and disadvantages of using an augmented matrix to solve a system of equations.

2. **Write** an augmented matrix for the system of equations.

   $x + 3z = 5$

   $2x + y = 5$

   $-2x + 3y - z = 8$

3. **State** the row operations you would use to change $\begin{bmatrix} 4 & -1 & \vdots & -19 \\ 3 & 4 & \vdots & 19 \end{bmatrix}$ to $\begin{bmatrix} 1 & 0 & \vdots & -3 \\ 0 & 1 & \vdots & 7 \end{bmatrix}$.

4. **Choose** the operation that cannot be used with the augmented matrix below.

   a. Multiply row 1 by -3 and add it to row 2.

   b. Switch row 2 and row 3.

   c. Multiply column 3 by $\frac{1}{2}$.

   d. Multiply row 3 by 6.

   $$\begin{bmatrix} 1 & 5 & 6 & \vdots & -8 \\ 3 & -2 & -2 & \vdots & 17 \\ 2 & 3 & 4 & \vdots & 1 \end{bmatrix}$$

5. **Describe** the solution for the system of equations represented by each reduced augmented matrix.

   a. $\begin{bmatrix} 1 & 0 & 0 & \vdots & 5 \\ 0 & 1 & 0 & \vdots & -2 \\ 0 & 0 & 1 & \vdots & 8 \end{bmatrix}$

   b. $\begin{bmatrix} 1 & 0 & 4 & \vdots & -3 \\ 0 & 0 & 0 & \vdots & 7 \\ 0 & 2 & -3 & \vdots & 8 \end{bmatrix}$

   c. $\begin{bmatrix} 1 & 0 & -2 & \vdots & 5 \\ 0 & 2 & 1 & \vdots & 6 \\ 0 & 0 & 0 & \vdots & 0 \end{bmatrix}$

*M*ATH *J*OURNAL

6. **Assess Yourself** List all of the algebraic methods that can be used to solve systems of equations. Which method do you prefer, and why?

Guided Practice

Write a system of equations represented by each augmented matrix.

7. $\begin{bmatrix} 3 & -5 & \vdots & 25 \\ 2 & 4 & \vdots & 24 \end{bmatrix}$

8. $\begin{bmatrix} 3 & -5 & 2 & \vdots & 9 \\ 1 & -7 & 3 & \vdots & 11 \\ 4 & 0 & -3 & \vdots & -1 \end{bmatrix}$

Write an augmented matrix for each system. Then solve each system.

9. $4m - 7n = -19$
   $3m + 2n = 22$

10. $3a - b + 5c = -1$
    $a + 3b - c = 25$
    $2a + 4c = 2$

11. Describe the solution for the system of equations represented by
$$\begin{bmatrix} 1 & 0 & -2 & | & 3 \\ 0 & 0 & 3 & | & 1 \\ 0 & 0 & 0 & | & 3 \end{bmatrix}.$$

EXERCISES

Practice

Write an augmented matrix for each system of equations. Then solve each system.

12. $3x + 2y = 7$
    $x - 3y = 17$

13. $4x - 3y = 5$
    $2x + 9y = 6$

14. $7m - 3n = 41$
    $2m = -5n$

15. $2a - b + 4c = 6$
    $a + 5b - 2c = -6$
    $3a - 2b + 6c = 8$

16. $3x - 5y + 2z = 22$
    $2x + 3y - z = -9$
    $4x + 3y + 3z = 1$

17. $2q + r + s = 2$
    $-q - r + 2s = 7$
    $-3q + 2r + 3s = 7$

Describe the solution for the system of equations represented by each reduced augmented matrix.

18. $\begin{bmatrix} 3 & 0 & | & 6 \\ 0 & 2 & | & -8 \end{bmatrix}$

19. $\begin{bmatrix} 4 & 0 & 1 & | & 4 \\ 0 & 1 & -2 & | & 5 \\ 0 & 0 & 0 & | & 0 \end{bmatrix}$

20. $\begin{bmatrix} 4 & 0 & 0 & | & -8 \\ 0 & 0 & 1 & | & 3 \\ 0 & 2 & 0 & | & 5 \end{bmatrix}$

21. $\begin{bmatrix} 1 & 0 & -7 & | & 2 \\ 0 & 0 & 5 & | & 0 \\ 0 & 0 & 0 & | & 2 \end{bmatrix}$

Solve each system of equations by using augmented matrices.

22. $6a + 5b = -12$
    $12a + 10b = -20$

23. $6r + s = 9$
    $3r = -2s$

24. $4m + 2n + 5p = 24$
    $3m + 5n - p = -13$
    $m + 7n + 3p = 33$

25. $6x + 2y - 6z = 4$
    $3x - 5y - 3z = -1$
    $2x + 4y + z = 1$

26. $2x - 3y + z = -2$
    $x + y - 2z = 1$
    $4x + 4y - 8z = 4$

27. $2a - b - c = 3$
    $a + b - 3c = 5$
    $4a - 2b - 2c = -2$

28. **Geometry** In triangle ABC, the measure of $\angle A$ is twice the measure of $\angle B$. The measure of $\angle C$ exceeds four times the measure of $\angle B$ by 12 degrees. Find the measure of each angle.

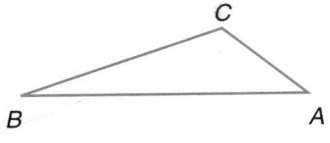

Critical Thinking

29. When one of the three rows in a reduced augmented matrix contains all zeros, the solution of the system of equations is a line. What kind of solution do you have when two rows contain all zeros? Explain your answer.

30. **Geometry** The perimeter of a triangle is 83 inches. The longest side is three times the length of the shortest side and 17 inches more than one-half the sum of the other two sides. Use augmented matrices to find the length of each side.

31. **Business** The Yogurt Shoppe sells cones in three sizes: small, $0.89; medium, $1.19; and large, $1.39. One day, Kyle Miller sold 52 cones. He sold two more than twice as many medium cones as large cones. If he sold $58.98 in cones, how many of each size did he sell?

Mixed Review

32. Solve the system of equations by using an inverse matrix. (Lesson 4–6)

$$3a + 2b = 7$$
$$-a - 7b = 23$$

33. Find the inverse of $\begin{bmatrix} 4 & -5 \\ 2 & -1 \end{bmatrix}$. (Lesson 4–5)

34. Find $\begin{vmatrix} -2 & 0 \\ 7 & -6 \end{vmatrix}$. (Lesson 3–3)

35. **Business** The parking garage at Burrough's Department Store charges $1.50 for each hour or fraction of an hour for parking. What type of function does this relationship represent? (Lesson 2–6)

36. Graph $5 = 5x$. (Lesson 2–2)

37. Evaluate $\dfrac{3ab^2 - c^3}{a + c}$ if $a = 3$, $b = 7$, and $c = -2$. (Lesson 1–1)

WORKING ON THE

In·ves·ti·ga·tion

Refer to the Investigation on pages 180–181.

3–2–1–Blast-Off!

Conduct a shoot-off day. Create a shooting area with masking tape. Label the tape with a scale, marking distances 50 cm to 250 cm from the launching area.

The teacher selects a distance between 50 cm and 250 cm and announces that distance to the class. No test shots may be taken after the announcement. Groups are randomly selected to demonstrate their launch system. During these launches, the launcher should be shot from the floor, and the distance should be measured from the launcher's location to the spot where the Ping-Pong™ ball hits the floor.

Each group should shoot ten shots at the same distance. Then each group must accurately measure and record the shot distances.

1 State the target distance and list the ten distances shot by your launcher.

2 Create a matrix of the class results from the launcher shoot-off. The matrix should consist of all ten distances shot by each of the groups in the class.

3 Find the median, quartiles, greatest and least values, and interquartile range of each set of shots by each of the launchers.

4 How do these three measures compare to the target distance of your launcher? Explain.

5 How do these three measures help to determine which launcher was most accurate? Explain.

6 Make a box-and-whisker plot for each launcher. Which launcher would you consider to be the most accurate? Explain your reasoning.

Add the results of your work to your Investigation Folder.

4–7B Graphing Technology
Matrix Row Operations

An Extension of Lesson 4–6

You can solve a system of linear equations by using a graphing calculator and the MATRX function. The row operation functions are located in the MATH menu when you press the MATRX key. Each function is listed below with instructions on the keying procedures. Suppose your augmented matrix has been entered as matrix *A*.

Row Swap(

MATRX ▶ 8

Interchange two rows.
1. Enter the name of the matrix followed by a comma.
2. Enter one of the rows you want to interchange followed by a comma.
3. Enter the other row you want to interchange followed by ▭).

Example: To interchange rows 1 and 2 in matrix A, enter rowSwap([A], 1, 2).

*Row

MATRX ▶ 0

Multiply one row by a number.
1. Enter the number you want to multiply by, followed by a comma.
2. Enter the name of the matrix followed by a comma.
3. Enter the row you want multiplied, followed by ▭).

Example: To multiply row 2 by −3 in matrix A, enter *row(−3, [A], 2).

Row+(

MATRX ▶ 9

Add two rows and store the result in the last row you entered.
1. Enter the name of the matrix followed by a comma.
2. Enter the row you want to add followed by a comma.
3. Enter the row you want it added to, followed by ▭).

Example: To add row 2 to row 1 in matrix A, enter row+([A], 2, 1).

*Row+(

MATRX ▶

ALPHA A

Multiply one row by a number and add the result to another.
1. Enter the number you want to multiply by, followed by a comma.
2. Enter the name of the matrix followed by a comma.
3. Enter the row you want multiplied, followed by a comma.
4. Enter the row you want the result added to, followed by ▭).

Example: To multiply row 1 by $\frac{1}{2}$ and add it to row 2 in matrix A, enter *row+(0.5, [A], 1, 2).

To perform one operation after another in completely reducing a matrix, let ANS be your matrix name so the operations will be done on the matrix you just finished.

Example ● Write an augmented matrix for the following system of equations. Then solve the system by reducing the matrix with a graphing calculator.

$$15x + 11y = 36$$
$$4x - 3y = -26$$

The augmented matrix $A = \begin{bmatrix} 15 & 11 & \vdots & 36 \\ 4 & -3 & \vdots & -26 \end{bmatrix}$.

Begin by entering the matrix.

Enter: [MATRX] [▶] [▶] [ENTER]

2 [ENTER] 3 [ENTER] 15 [ENTER]

11 [ENTER] 36 [ENTER] 4

[ENTER] [(−)] 3 [ENTER] [(−)] 26

[ENTER] [2nd] [QUIT]

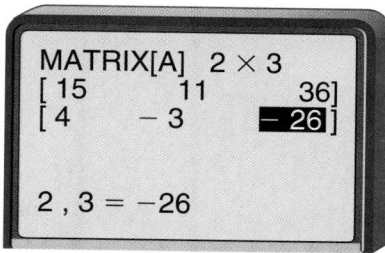

Multiply row 1 by 3.

Enter: [MATRX] [▶] 0 3 [,] [MATRX]

1 [,] 1 [)] [ENTER]

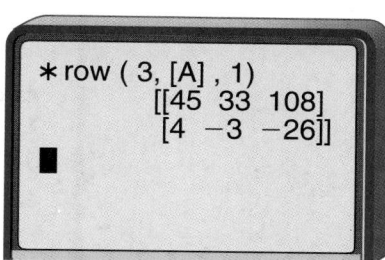

Multiply row 2 by 11 and add it to row 1.

Enter: [MATRX] [▶] [ALPHA] [A] 11

[,] [2nd] [ANS] [,] 2 [,] 1

[)] [ENTER]

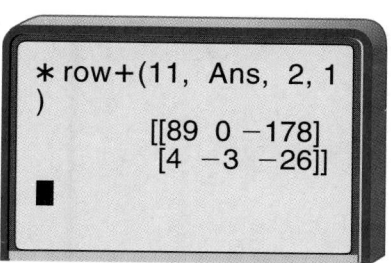

(continued on the next page)

Multiply row 1 by $\frac{1}{89}$.

Enter: MATRX ▶ 0 89 x^{-1} ,

2nd ANS , 1) ENTER

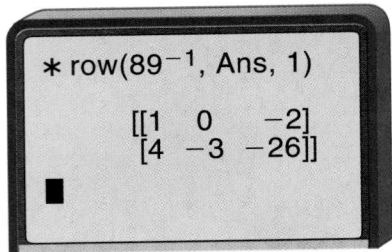

Multiply row 1 by −4 and add it to row 2.

Enter: MATRX ▶ ALPHA A (−) 4

, 2nd ANS , 1 , 2

) ENTER

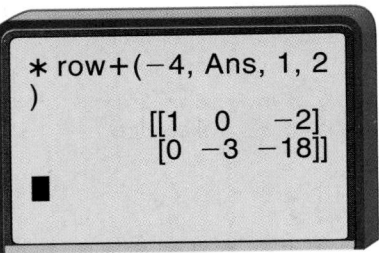

Multiply row 2 by $-\frac{1}{3}$.

Enter: MATRX ▶ 0 (−) 3 x^{-1} ,

2nd ANS , 2) ENTER

The solution is $(-2, 6)$.

EXERCISES

Write an augmented matrix for each system of equations. Then solve with a graphing calculator.

1. $3x + 2y = -2$
$2x + 3y = 7$

2. $x - 3y = 5$
$2x + y = 1$

3. $3x - y = 0$
$2x - 3y = 1$

4. $2x + y = 5$
$2x - 3y = 1$

5. $x - y + z = 2$
$x - z = 1$
$y + 2z = 0$

6. $3x - 2y + z = -2$
$x - y + 3z = 5$
$-x + y + z = -1$

7. $3x + y + 3z = 2$
$2x + y + 2z = 1$
$4x + 2y + 5z = 5$

8. $-3x + y - 2z = -7$
$2x - y - 3z = 1$
$x + 2y + z = -2$

9. $-2x + y + z = 4$
$4x - 3y - 2z = -2$
$-3x + y + z = 5$

Integration: Statistics
Box-and-Whisker Plots

What YOU'LL LEARN

• To find the range, quartiles, and interquartile range for a set of data,

• to determine if any values in a set of data are outliers, and

• to represent data using box-and-whisker plots.

Why IT'S IMPORTANT

Box-and-whisker plots are a useful way to display data. They allow you to see important characteristics of the data at a glance.

APPLICATION
News Media

The 1994 edition of the *Editor and Publisher International Yearbook* lists the total number of morning and evening newspapers that are published in each state. One way to organize these data is to show them in a 1×50 matrix.

$$N = [26\ \ 7\ \ 22\ \ 32\ \ 104\ \ ...]$$

However, this is not the best way to display the data. A more convenient way is to organize them in the table below.

Number of Morning and Evening Newspapers in the United States							
State	**Number**	**State**	**Number**	**State**	**Number**	**State**	**Number**
AL	26	IN	73	NE	18	SC	16
AK	7	IA	39	NV	8	SD	11
AZ	22	KS	47	NH	11	TN	27
AR	32	KY	23	NJ	21	TX	92
CA	104	LA	25	NM	18	UT	6
CO	28	ME	7	NY	71	VT	8
CT	19	MD	15	NC	49	VA	28
DE	3	MA	39	ND	10	WA	24
FL	40	MI	52	OH	84	WV	23
GA	34	MN	25	OK	46	WI	36
HI	6	MS	22	OR	19	WY	9
ID	12	MO	45	PA	89		
IL	68	MT	11	RI	6		

Even when the data are organized in a table, they are difficult to analyze because the values vary and there are no trends apparent in the data. However, there are several ways to measure variation in the data. The simplest is called the **range.**

Definition of Range	The range of a set of data is the difference between the greatest and least values in the set.

In this case, the greatest number of newspapers in any state is 104 in California, and the least is 3 in Delaware. So the range of the number of newspapers is $104 - 3$ or 101.

Because the range is the difference between the greatest and least values in a set of data, it is affected by extreme values. In these cases, it is not a good measure of variation.

Another way to analyze a set of data is to determine how the data are distributed between the least and greatest values. The **quartiles** are the values in a set that separate the data into four sections, each containing 25% of the data.

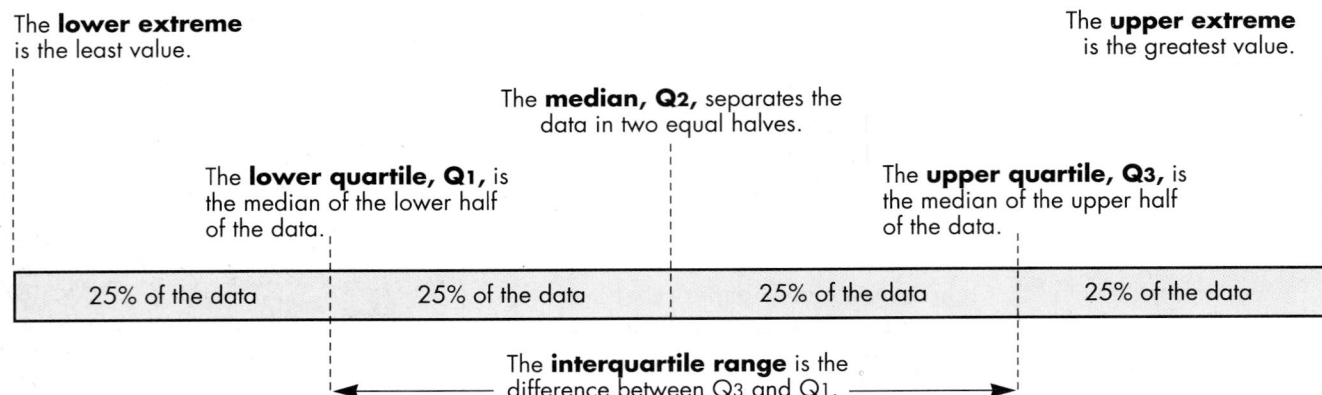

The **lower extreme** is the least value.

The **upper extreme** is the greatest value.

The **median, Q2,** separates the data in two equal halves.

The **lower quartile, Q1,** is the median of the lower half of the data.

The **upper quartile, Q3,** is the median of the upper half of the data.

| 25% of the data | 25% of the data | 25% of the data | 25% of the data |

The **interquartile range** is the difference between Q3 and Q1.

Example **1**

APPLICATION

Sports

On January 29, 1995, the San Francisco 49ers became the first team to win five Super Bowls as they defeated the San Diego Chargers 49–26. The number of points scored by the winning teams in all Super Bowls through 1995 are as follows.

35, 33, 16, 23, 16, 24, 14, 24, 16, 21, 32, 27, 35, 31, 27, 26, 27, 38, 38, 46, 39, 42, 20, 55, 20, 37, 52, 30, 49

a. Find Q1, Q2, Q3, the range, and the interquartile range.
b. Analyze how San Francisco's score in 1995 compares to the other winning scores.

F Y I

Before 1995, the record of four Super Bowl wins was shared by the San Francisco 49ers, the Dallas Cowboys, and the Pittsburgh Steelers.

a. First arrange the data in order.

14, 16, 16, 16, 20, 20, **21, 23**, 24, 24, 26, 27, 27, 27, **30**, 31, 32, 33, 35, 35, 37, **38, 38**, 39, 42, 46, 49, 52, 55

The range is 55 − 14 or 41.

There are 29 values in all. The median, Q2, is the middle, or 15th, value. Therefore, the median score is 30.

The lower quartile, Q1, is the median of the lower half of the data. Since there are 14 values in the lower half, the lower quartile falls midway between the 7th and 8th value. The lower quartile is $\frac{21 + 23}{2}$ or 22.

The upper quartile, Q3, is the median of the upper half of the data. It falls midway between the 22nd and 23rd values. Since both these values are 38, the upper quartile is 38.

The interquartile range is Q3 − Q1 = 38 − 22 or 16.
Fifty percent of the time, the number of points scored was between 22 and 38.

b. San Francisco's score of 49 is above the upper quartile. Therefore, it is in the upper 25% of the data.

Numerical data can often be represented using a **box-and-whisker plot**. In a box-and-whisker plot, the quartiles and the extreme values of a set of data are displayed using a number line.

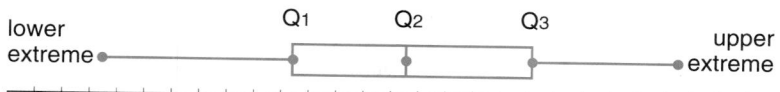

Thus, a box-and-whisker plot is a pictorial representation of the variability of the data and a way to summarize a data set with five points.

Example 2

Sports

Refer to the data in Example 1 about the Super Bowl winners. Make a box-and-whisker plot of the data.

To make a box-and-whisker plot, draw a number line and plot the quartiles, the median, and the extreme values. The lower extreme is 14, the lower quartile is 22, the median is 30, the upper quartile is 38, and the upper extreme is 55.

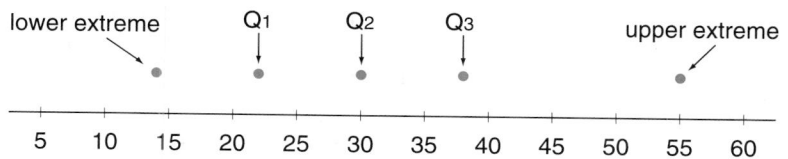

Draw a box to designate the interquartile range and mark the median by drawing a segment containing its point in the box. Draw segments (whiskers) connecting the lower quartile to the least value and the upper quartile to the greatest value.

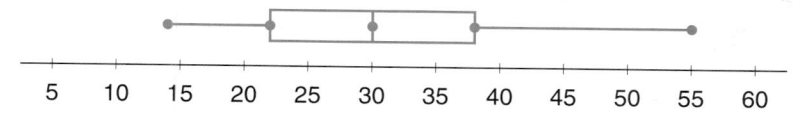

The dimensions of the box-and-whisker plot can help you characterize the data. Each whisker and each small box contains 25% of the data. If the whisker or box is short, the data are concentrated over a narrower range of values. The longer the whisker or box, the more diverse the data. Extreme values are referred to as **outliers**.

Definition of Outlier	**An outlier is any value in the set of data that is at least 1.5 interquartile ranges beyond the upper or lower quartile.**

Example **3**

Refer to the data in the application at the beginning of the lesson about the number of newspapers in circulation.

a. Find Q1, Q2, Q3, and the interquartile range.

b. List any outliers.

c. Make a box-and-whisker plot.

d. If any outliers exist, analyze them to determine possible reasons why they exist.

a. First arrange the data in order.

3, 6, 6, 6, 7, 7, 8, 8, 9, 10, 11, 11, 11, 12, 15, 16, 18, 18, 19, 19, 21, 22, 22, 23, 23, 24, 25, 25, 26, 27, 28, 28, 32, 34, 36, 39, 39, 40, 45, 46, 47, 49, 52, 68, 71, 73, 84, 89, 92, 104

There are 50 values in all.

Q1 is the 13th value, 11.

Q2 is between the 25th value, 23, and the 26th value, 24. Therefore, Q2 is 23.5.

Q3 is the 38th value, 40.

The interquartile range is 40 − 11 or 29.

b. Find the outliers.

Q1 − 1.5(29) = 11 − 43.5 or −32.5 *There are no values less than −32.5.*

Q3 + 1.5(29) = 40 + 43.5 or 83.5 *There are four values greater than 83.5.*

Therefore, 84, 89, 92, and 104 are outliers.

c. Draw a number line and plot the quartiles, the lower extreme, the upper extreme, and the outliers. Also, plot 73, since this is the last data value that is not an outlier. Extend the whiskers to the lower extreme, 14, and to 73. The outliers remain as single points.

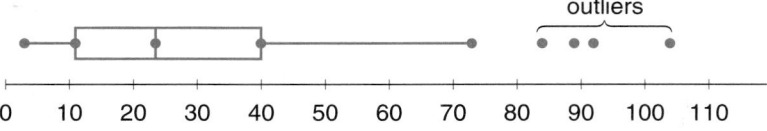

d. The outliers are from the states of Ohio, Pennsylvania, Texas, and California. All of these states have large populations that can support many daily newspapers.

In addition to showing how data within a set vary, box-and-whisker plots can be used to compare two or more sets of data.

Example 4

CONNECTION
Sociology

The table below shows the median ages of men and women at the time of their first marriage for the decades of 1890 through 1990.

Year	Men	Women	Year	Men	Women
1890s	26.1	22.0	1950s	22.8	20.3
1900s	25.9	21.9	1960s	22.8	20.3
1910s	25.1	21.6	1970s	23.2	20.8
1920s	24.6	21.2	1980s	24.7	22.0
1930s	24.3	21.3	1990s	26.2	25.1
1940s	24.3	21.5			

a. Make a box-and-whisker plot for the men's and women's ages.
b. Analyze the information as it is displayed in the two plots.

a. For the men's data, the lower extreme is 22.8, the upper extreme is 26.2, and the quartiles are 23.2, 24.6, and 25.9. There are no outliers.

For the women's data, the lower extreme is 20.3, the upper extreme is 25.1, and the quartiles are 20.8, 21.5, and 22.0. There is one outlier, 25.1. The last value of the data that is not an outlier is 22.0.

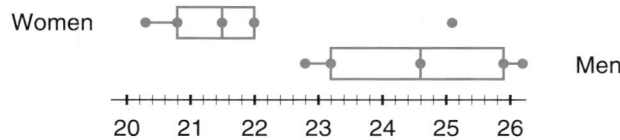

b. It appears from the data that over the years, men marry at a later age than women. It also appears that the interquartile range for the women is 1.2 years, but for the men it is 2.7 years. For 50% of the years, women married between the ages of 20.8 and 22.0. Thus, there is not much difference or spread in the ages at which women married, except in 1990. The ages at which men married varies much more.

EXPLORATION

GRAPHING CALCULATORS

The table below shows the 1990 populations of ten large cities and the predicted populations of those same cities for the year 2000.

City	1990 Population (millions)	2000 Population (millions)
Tokyo	27.0	30.0
Mexico City	20.2	27.9
Sao Paulo	18.1	25.4
Seoul	16.3	22.0
New York	14.6	14.6
Osaka	13.8	14.3
Bombay	11.8	15.4
Calcutta	11.7	14.1
Buenos Aires	11.5	12.9
Rio de Janeiro	11.4	14.2

Source: U.S. Bureau of the Census, International Data Base, 1994

(continued on the next page)

There are many ways in which these data could be displayed. You can make box-and-whisker plots using these data fairly quickly by using a graphing calculator.

Your Turn

a. Press ⎣ STAT ⎦ 1. Enter the 1990 population values into L1 and the 2000 population values into L2.

b. Press ⎣2nd⎦ ⎣STAT PLOT⎦ 1. Turn on Plot 1 and define it as a box-and-whisker plot, using L1 and Frequency 1. Press ⎣2nd⎦ ⎣STAT PLOT⎦ 2. Turn on Plot 2 and define it as a box-and-whisker plot, using L2 and Frequency 1.

c. Clear the Y= list and change the window settings to Xscl = 1, Ymin = 0, and Yscl = 0. Then press ⎣ ZOOM ⎦ 9.

d. Write a paragraph to describe the difference in the spread of the data between 1990 and 2000.

CHECK FOR UNDERSTANDING

Communicating Mathematics

Study the lesson. Then complete the following.

1. **Show** how a set of data can be separated into quartiles.

2. **Describe** what you can tell about a set of data from a box-and-whisker plot.

3. **Write** an example of two sets of data with the same lower and upper extreme but different interquartile ranges.

4. **Explain** how to find the outliers in a set of data if $Q_1 = 52.5$, $Q_2 = 60$, and $Q_3 = 72.5$.

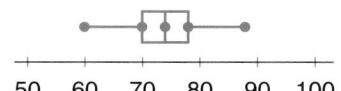

5. **Describe** how the data shown in the box-and-whisker plot at the right are distributed.

6. **You Decide**   Michelle thinks that $Q_3 - Q_2 = Q_2 - Q_1$ for any set of data. Mei thinks that this isn't necessarily true. Who is correct? Explain your reasoning.

Guided Practice

7. Use the box-and-whisker plot at the right to answer each question.

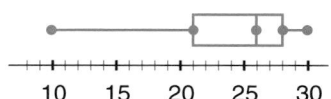

   a. What is the range of the data?

   b. What is the median of the data?

   c. What percent of the data is greater than 28?

   d. Between what two values of the data is the middle 50% of the data?

Find the range, quartiles, interquartile range, and outliers for each set of data. Then make a box-and-whisker plot for each set of data.

8. {12, 19, 20, 1, 15, 14, 19}

9. {24, 32, 38, 38, 26, 33, 37, 39, 23, 31, 40, 21}

10. **Astronomy** The table at the right lists the approximate length of a day, in Earth hours, for each of the planets in our solar system.

Planet	Length of Day (Earth hours)
Mercury	1416
Venus	5832
Earth	24
Mars	24
Jupiter	10
Saturn	11
Uranus	22
Neptune	16
Pluto	153

a. Which planet's day length separates the data into two equal halves?

b. What is the median length of a day for the planets in our solar system?

c. Fifty percent of the data lie between what two values?

d. Estimate which value(s) may be outliers. Then determine if the data set has any outliers.

e. What problems would you encounter if you tried to make a box-and-whisker plot with these data?

EXERCISES

Practice

11. Use box-and-whisker plot I to answer each question.

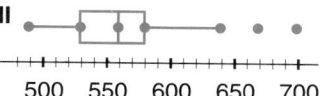
I

a. What percent of the data is less than 76?

b. What percent of the data is less than 92?

c. What percent of the data is greater than 64 and less than 92?

d. Under what conditions would a set of data have this type of box-and-whisker plot?

12. Use box-and-whisker plot II to answer each question.

II

a. What is the range of the data?

b. What values of the data are outliers?

c. What percent of the data is greater than 560?

d. What percent of the data is less than 580?

Find the range, quartiles, interquartile range, and outliers for each set of data. Then make a box-and-whisker plot for each set of data.

13. {25, 46, 31, 53, 39, 59, 48, 43, 68, 64, 29}

14. {25, 51, 29, 43, 32, 17, 21, 29, 36, 47}

15. {51, 69, 46, 27, 60, 53, 55, 39, 81, 54, 46, 23}

16. {110, 110, 330, 200, 88, 110, 88, 110, 165, 390, 150, 440, 536, 200, 110, 165, 88, 147, 110, 165}

17. {13.6, 15.1, 14.9, 15.7, 16.0, 14.1, 16.3, 14.3, 13.8}

18. {15.1, 11.5, 5.8, 6.2, 10.5, 7.6, 9.0, 8.5, 8.8, 8.5}

Graphing Calculator

19. **Meteorology** San Francisco, California, and Springfield, Missouri, are both located at 37° N latitude. However, their average monthly high temperatures are very different.

a. Use a graphing calculator to make box-and-whisker plots for each set of data.

b. Write a sentence that compares the average monthly high temperatures in each city.

c. In which city would you prefer to live? Give reasons to support your answer.

Average High Temperatures (°F)		
Month	San Francisco	Springfield
January	59	41
February	61	48
March	61	57
April	62	69
May	65	72
June	69	82
July	69	89
August	70	88
September	72	80
October	70	70
November	61	59
December	58	45

Source: *1995 Weather Almanac*

Critical Thinking

20. Suppose a set of data has a small interquartile range and a large range. What does this tell you about how the data are distributed?

21. Give an example of a set of data having ten values and containing at least one outlier at each end of the data. What conditions must be present for this to happen?

Applications and Problem Solving

22. **School** Two geometry classes took the same exam, and their scores are shown in box-and-whisker plots below.

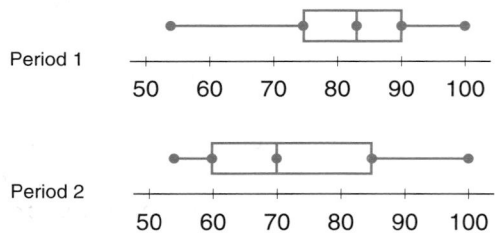

a. Which class has the higher median?
b. Which class has the greater range?
c. Which class has the greater interquartile range?
d. Which class appears to have done better?
e. Describe the spread of the scores in the two classes.

23. **History** The stem-and-leaf plot below represents the age at death of the presidents of the United States.

Stem	Leaf
4	6 9
5	3 6 6 7 8
6	0 0 3 3 4 4 5 6 7 7 7 8
7	0 1 1 2 3 4 7 8 8 9
8	0 1 3 5 8
9	0 0 4|6 = 46

a. Make a box-and-whisker plot of the data.

b. What are some advantages and disadvantages of representing these data in a stem-and-leaf plot? What are some advantages and disadvantages of representing these data in a box-and-whisker plot?

24. Food The number of Calories in a regular serving of French fries at different restaurants are listed below.

Restaurant	Calories	Restaurant	Calories
Burger Chef	250	Hardee's	239
Burger King	240	McDonald's	211
Carl's Jr.	220	Roy Rogers	240
Dairy Queen	200	Wendy's	327
Friendly's	125		

a. Make a box-and-whisker plot of the data.

b. Explain why this box-and-whisker plot has such short whiskers.

25. Literature Maya Angelou is a Reynolds Professor of African Studies at Wake Forest University in North Carolina and is well known for her poetry. She was commissioned by then President-elect Bill Clinton to write a poem for his 1993 inauguration. An excerpt from *On the Pulse of Morning* is shown below.

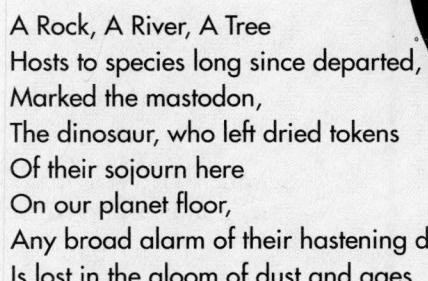

> A Rock, A River, A Tree
> Hosts to species long since departed,
> Marked the mastodon,
> The dinosaur, who left dried tokens
> Of their sojourn here
> On our planet floor,
> Any broad alarm of their hastening doom
> Is lost in the gloom of dust and ages.

a. Make a box-and-whisker plot that shows the number of letters in each word of this excerpt.

b. Find a newspaper article and make a box-and-whisker plot that shows the number of letters in each word of the article.

c. Compare and contrast the two plots.

26. Demographics The chart at the right shows the Asian and Pacific Islander population projections for 2000 and 2010 in the western United States.

a. For each year, make a box-and-whisker plot.

b. How will the median population change from 2000 until 2010?

c. How will the interquartile range change from 2000 until 2010?

d. Write a sentence that summarizes how the population of Asian and Pacific Islanders is expected to change from 2000 to 2010.

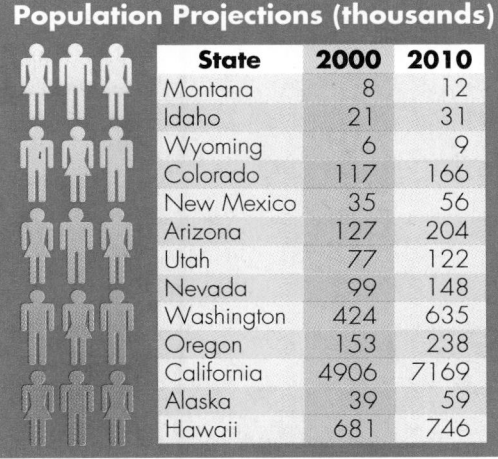

Population Projections (thousands)		
State	**2000**	**2010**
Montana	8	12
Idaho	21	31
Wyoming	6	9
Colorado	117	166
New Mexico	35	56
Arizona	127	204
Utah	77	122
Nevada	99	148
Washington	424	635
Oregon	153	238
California	4906	7169
Alaska	39	59
Hawaii	681	746

Source: U.S. Bureau of the Census

27. Snacks Who can resist the snack counter when you're at the movies? Not many people! But the Center for Science in the Public Interest reports that many snacks contain as many Calories and fat as a fast-food lunch. The chart below lists the Calories and fat content for several popular movie snacks.

Snack	Calories	Fat (grams)
KitKat Bar/4 oz	588	33
Twizzlers/5 oz	500	5
Butterfinger/4 oz	492	21
Reese's Peanut Butter Cups/3.2 oz	380	22
M&M's, peanut/2.6 oz	363	20
Junior Mints/3 oz	360	9
M&M's plain/2.6 oz	350	16
Goobers/2.2 oz	320	21
Skittles/2.6 oz	286	2
Raisinets/2.3 oz	270	10
Buttered popcorn popped in coconut oil/medium bucket	1221	97

a. Make box-and-whisker plots for the number of Calories and grams of fat.

b. Add the following values to the data.

Unbuttered popcorn popped in coconut oil/medium bucket: 901 Calories, 60 grams fat

Plain air-popped popcorn/medium bucket: 180 Calories, <1 gram fat

How do these values affect the plot in part a?

c. Find a snack that would fall in the bottom 25% of the data for both number of Calories and grams of fat.

Mixed Review

28. Solve the system of equations by using augmented matrices. (Lesson 4–7)

$2x + y + z = 0$

$3x - 2y - 3z = -21$

$4x + 5y + 3z = -2$

29. Find $[2 \ -6 \ 3] \cdot \begin{bmatrix} 3 & -3 \\ 9 & 0 \\ -2 & 4 \end{bmatrix}$.  (Lesson 4–3)

30. Running The length of a marathon was determined by the first marathon in the 1908 Olympic Games in London. The race began at Windsor Castle and ended in front of the royal box at London's Olympic Stadium, which was a distance of 26 miles 385 yards. Determine how many feet the marathon covers using the formula $f(m, y) = 5280m + 3y$, where m is the number of miles and y is the number of yards. (Lesson 3–5)

31. Find the slope of the line that passes through (5, 4) and (2, 2). (Lesson 2–3)

32. Name the sets of numbers to which $2.121221222\ldots$ belongs. (Lesson 1–2)

33. You are about to buy a new car. Your mother offers you a simple interest loan to finance it. Simple interest is calculated using the formula $I = prt$, where p represents the principal in dollars, r represents the annual interest rate, and t represents the time in years. Find the amount of interest you would pay for a two-year loan if the principal is $6000 and the rate is 12%. (Lesson 1–1)

VOCABULARY

After completing this chapter, you should be able to define each term, property, or phrase and give an example or two of each.

Discrete Mathematics
addition of matrices (p. 195)
augmented matrix (p. 226)
coding matrix (p. 212)
column matrix (p. 187)
coordinate matrix (p. 189)
determinant (p. 205)
dimensions (p. 187)
discrete mathematics
  (p. 188)
element (p. 186)
equal matrices (p. 188)
expansion by minors (p. 205)
identity matrix (p. 213)
inverse of a matrix (p. 213)
matrix (p. 186)
matrix equation (p. 219)

minor (p. 205)
multiplying matrices (p. 199)
probability matrix (p. 200)
reduced matrix (p. 227)
reducing a matrix (p. 227)
row operations (p. 226)
row matrix (p. 187)
scalar (p. 188)
scalar multiplication (p. 188)
square matrix (p. 187)
third-order determinant
  (p. 205)
transition matrix (p. 200)
translation matrix (p. 196)

Problem Solving
use matrix logic (p. 187)

Geometry
dilation (p. 190)
rotation (p. 201)
transformation (p. 190)
translation (p. 195)

Statistics
box-and-whisker plot (p. 237)
interquartile range (p. 236)
lower extreme (p. 236)
lower quartile (p. 236)
median (p. 236)
outlier (p. 237)
quartiles (p. 236)
range (p. 235)
upper extreme (p. 236)
upper quartile (p. 236)

UNDERSTANDING AND USING THE VOCABULARY

Choose the correct term to complete each sentence.

1. The matrix $\begin{bmatrix} 1 & 0 & 0 \\ 0 & 1 & 0 \\ 0 & 0 & 1 \end{bmatrix}$ is a(n) _____ for multiplication.

2. The matrix $\begin{bmatrix} 1 & 0 & 7 \\ 0 & 1 & -3 \end{bmatrix}$ is a(n) _____ .

3. _____ is the process of multiplying a matrix by a constant.

4. A(n) _____ is when a figure is moved around a center point.

5. The _____ of $\begin{bmatrix} -1 & 4 \\ 2 & -3 \end{bmatrix}$ is −5.

6. The matrix $\begin{bmatrix} 1 & 5 & 3 & 4 \\ 4 & -3 & 2 & 4 \\ 8 & -6 & 4 & 14 \end{bmatrix}$ is a(n) _____ .

7. The _____ of a matrix tell how many rows and columns are in the matrix.

8. A _____ occurs when a figure is moved from one location to another on the coordinate plane without changing its size, shape, or orientation.

9. The matrices $\begin{bmatrix} 3x \\ x + 2y \end{bmatrix}$ and $\begin{bmatrix} y \\ 7 \end{bmatrix}$ are _____ if $x = 1$ and $y = 3$.

10. A _____ is when a geometric figure is enlarged or reduced.

augmented matrix
determinant
dilation
dimensions
equal matrices
identity matrix
reduced matrix
rotation
scalar multiplication
translation

| **OBJECTIVES AND EXAMPLES** | **REVIEW EXERCISES** |

Upon completing this chapter, you should be able to:

Use these exercises to review and prepare for the chapter test.

● perform scalar multiplication on a matrix
(Lesson 4–1)

$$4\begin{bmatrix} 2 & -3 \\ 4 & 1 \\ 0 & 3 \end{bmatrix} = \begin{bmatrix} 4(2) & 4(-3) \\ 2(4) & 4(1) \\ 4(0) & 4(3) \end{bmatrix} \text{ or } \begin{bmatrix} 8 & -12 \\ 16 & 4 \\ 0 & 12 \end{bmatrix}$$

Find each product.

11. $3\begin{bmatrix} 8 & -3 & 2 \\ 4 & 1 & 7 \end{bmatrix}$

12. $-5\begin{bmatrix} -3 & 2 \\ 6 & 4 \end{bmatrix}$

13. $\frac{2}{3}\begin{bmatrix} 3 & \frac{3}{4} & -6 \end{bmatrix}$

14. $1.2\begin{bmatrix} -2 \\ 0.3 \\ 1 \end{bmatrix}$

15. $4\begin{bmatrix} 1.3 & 5.1 \\ -2 & -3.7 \\ -2.8 & 4.5 \end{bmatrix}$

16. $-\frac{1}{2}\begin{bmatrix} -2 & 4 \\ -8 & 2 \end{bmatrix}$

● solve matrices for variables (Lesson 4–1)

To find x and y in $\begin{bmatrix} 2x \\ y \end{bmatrix} = \begin{bmatrix} 32 + 6y \\ 7 - x \end{bmatrix}$, solve the system $2x = 32 + 6y$ and $y = 7 - x$.

$x = 9.25, y = -2.25$

Solve for the variables.

17. $\begin{bmatrix} 2y - x \\ x \end{bmatrix} = \begin{bmatrix} 3 \\ 4y - 1 \end{bmatrix}$

18. $\begin{bmatrix} 7x \\ x + y \end{bmatrix} = \begin{bmatrix} 5 + 2y \\ 11 \end{bmatrix}$

19. $\begin{bmatrix} 3x + y \\ x - 3y \end{bmatrix} = \begin{bmatrix} -3 \\ -1 \end{bmatrix}$

20. $\begin{bmatrix} 2x - y \\ 6x - y \end{bmatrix} = \begin{bmatrix} 2 \\ 22 \end{bmatrix}$

● add and subtract matrices (Lesson 4–2)

$$2\begin{bmatrix} 8 & -1 \\ 3 & 4 \end{bmatrix} - 3\begin{bmatrix} 1 & 6 \\ -2 & -3 \end{bmatrix} = \begin{bmatrix} 16 & -2 \\ 6 & 8 \end{bmatrix} + \begin{bmatrix} -3 & -18 \\ 6 & 9 \end{bmatrix}$$
$$= \begin{bmatrix} 13 & -20 \\ 12 & 17 \end{bmatrix}$$

Perform the indicated operations.

21. $\begin{bmatrix} -4 & 3 \\ -5 & 2 \end{bmatrix} + \begin{bmatrix} 1 & -3 \\ 3 & -8 \end{bmatrix}$

22. $[0.2 \ 1.3 \ -0.4] - [2 \ 1.7 \ 2.6]$

23. $\begin{bmatrix} 1 & -5 \\ -2 & 3 \end{bmatrix} + \frac{3}{4}\begin{bmatrix} 0 & 4 \\ -16 & 8 \end{bmatrix}$

24. $\begin{bmatrix} 1 & 0 & -3 \\ 4 & -5 & 2 \end{bmatrix} - 2\begin{bmatrix} -2 & 3 & 5 \\ -3 & -1 & 2 \end{bmatrix}$

● multiply matrices (Lesson 4–3)

$$[6 \ 4 \ 1] \cdot \begin{bmatrix} 2 & 5 \\ -3 & 0 \\ -1 & 3 \end{bmatrix} = [12 - 12 - 1 \ \ 30 + 0 + 3]$$
$$= [-1 \ 33]$$

Perform the indicated operations, if possible.

25. $[2 \ 7] \cdot \begin{bmatrix} 5 \\ -4 \end{bmatrix}$

26. $\begin{bmatrix} 8 & -3 \\ 6 & 1 \end{bmatrix} \cdot \begin{bmatrix} 2 & -3 \\ 1 & -5 \end{bmatrix}$

27. $\begin{bmatrix} 3 & 4 \\ 1 & 0 \\ 2 & -5 \end{bmatrix} \cdot \begin{bmatrix} -2 & 4 & 5 \\ 3 & 0 & -1 \\ 1 & 0 & -1 \end{bmatrix}$

• evaluate the determinant of a 3 × 3 matrix
(Lesson 4–4)

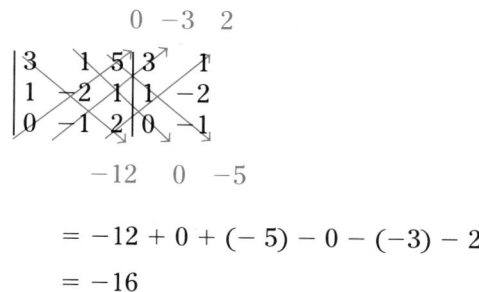

$$= -12 + 0 + (-5) - 0 - (-3) - 2$$

$$= -16$$

Determine whether each matrix has a determinant. Write *yes* or *no*. If *yes*, find the value of the determinant.

28. $\begin{bmatrix} 4 & 11 \\ -7 & 8 \end{bmatrix}$

29. $\begin{bmatrix} 7 & -4 & 5 \\ 1 & 3 & -6 \\ 5 & -1 & -2 \end{bmatrix}$

30. $\begin{bmatrix} 5 & -1 & 2 \\ -6 & -7 & 3 \\ 7 & 0 & 4 \end{bmatrix}$

31. $\begin{bmatrix} 2 & -3 & 1 \\ 0 & 7 & 8 \\ 2 & 1 & 3 \end{bmatrix}$

32. $\begin{bmatrix} -2 & 9 \\ 7 & -4 \\ -6 & 1 \end{bmatrix}$

33. $\begin{bmatrix} 6 & 3 & -2 \\ -4 & 2 & 5 \\ -3 & -1 & 0 \end{bmatrix}$

• find the inverse of a matrix   (Lesson 4–5)

Any matrix $M = \begin{bmatrix} a & b \\ c & d \end{bmatrix}$, will have an

inverse M^{-1} if and only if $\begin{vmatrix} a & b \\ c & d \end{vmatrix} \neq 0$.

Then $M^{-1} = \frac{1}{ad - bc} \begin{bmatrix} d & -b \\ -c & a \end{bmatrix}$.

Find the inverse of each matrix, if it exists. If it does not exist, explain why not.

34. $\begin{bmatrix} 3 & 2 \\ 4 & -2 \end{bmatrix}$

35. $\begin{bmatrix} 8 & 6 \\ 9 & 7 \end{bmatrix}$

36. $\begin{bmatrix} 2 & -4 \\ -3 & 6 \end{bmatrix}$

37. $\begin{bmatrix} -6 & 2 \\ 3 & 1 \end{bmatrix}$

38. $\begin{bmatrix} 0 & 2 \\ 5 & -4 \end{bmatrix}$

39. $\begin{bmatrix} 6 & -1 & 0 \\ 5 & 8 & -2 \end{bmatrix}$

• solve systems of linear equations by using inverse matrices   (Lesson 4–6)

$$\begin{bmatrix} 4 & 8 \\ 2 & -3 \end{bmatrix} \cdot \begin{bmatrix} x \\ y \end{bmatrix} = \begin{bmatrix} 7 \\ 0 \end{bmatrix}$$

$$-\frac{1}{28} \cdot \begin{bmatrix} -3 & -8 \\ -2 & 4 \end{bmatrix} \cdot \begin{bmatrix} 4 & 8 \\ 2 & -3 \end{bmatrix} \cdot \begin{bmatrix} x \\ y \end{bmatrix} = -\frac{1}{28} \cdot \begin{bmatrix} -3 & -8 \\ -2 & 4 \end{bmatrix} \cdot \begin{bmatrix} 7 \\ 0 \end{bmatrix}$$

$$-\frac{1}{28} \cdot \begin{bmatrix} -28 & 0 \\ 0 & -28 \end{bmatrix} \begin{bmatrix} x \\ y \end{bmatrix} = -\frac{1}{28} \cdot \begin{bmatrix} -21 \\ -14 \end{bmatrix}$$

$$\begin{bmatrix} 1 & 0 \\ 0 & 1 \end{bmatrix} \begin{bmatrix} x \\ y \end{bmatrix} = \begin{bmatrix} \frac{3}{4} \\ \frac{1}{2} \end{bmatrix} \text{ or } \begin{bmatrix} x \\ y \end{bmatrix} = \begin{bmatrix} \frac{3}{4} \\ \frac{1}{2} \end{bmatrix}$$

The solution is $\left(\frac{3}{4}, \frac{1}{2} \right)$.

Solve each matrix equation or system of equations by using inverse matrices.

40. $\begin{bmatrix} 5 & -2 \\ 1 & 3 \end{bmatrix} \cdot \begin{bmatrix} x \\ y \end{bmatrix} = \begin{bmatrix} 16 \\ 10 \end{bmatrix}$

41. $\begin{bmatrix} 4 & 1 \\ 3 & -2 \end{bmatrix} \cdot \begin{bmatrix} a \\ b \end{bmatrix} = \begin{bmatrix} 9 \\ 4 \end{bmatrix}$

42. $3x + 8 = -y$
    $4x - 2y = -14$

43. $3x - 5y = -13$
    $4x + 3y = 2$

OBJECTIVES AND EXAMPLES

• solve systems of linear equations by using augmented matrices   (Lesson 4–7)

$5a - 3b = 7$

$3a + 9b = -3$

$$\begin{bmatrix} 5 & -3 & \vdots & 7 \\ 3 & 9 & \vdots & -3 \end{bmatrix} \rightarrow \begin{bmatrix} 1 & 0 & \vdots & 1 \\ 0 & 1 & \vdots & -\frac{2}{3} \end{bmatrix}$$

The solution is $\left(1, -\dfrac{2}{3}\right)$.

REVIEW EXERCISES

Solve each system of equations by using augmented matrices.

44. $9a - b = 1$

  $3a + 2b = 12$

45. $x + 5y = 14$

  $-2x + 6y = 4$

46. $6x - 7z = 13$

  $8y + 2z = 14$

  $7x + z = 6$

47. $2a - b - 3c = -20$

  $4a + 2b + c = 6$

  $2a + b - c = -6$

• find the range, quartiles, and interquartile range for a set of data and represent the data using box-and-whisker plots   (Lesson 4–8)

Find the range, interquartile range, and any outliers for the set of data below.

60, 61, 62, 72, 72, 78, 80, 82, 83, 83, 99

greatest value = 99   least value = 60
range = 39   median = 6th score, 78
lower quartile = 62   upper quartile = 83
interquartile range = 83 − 62 or 21

There are no outliers.

Find the range, quartiles, interquartile range, and outliers for each set of data. Then make a box-and-whisker plot for each set of data.

48. {90, 92, 78, 93, 79, 85, 89, 88, 84, 86}

49. {10, 50, 90, 40, 60, 40, 50, 90, 0}

50. {0.4, 0.2, 0.5, 0.9, 0.3 0.4, 0.5, 1.9, 0.5, 0.7, 0.8, 0.6, 0.2, 0.1, 0.4}

51. {1055, 1075, 1095, 1125, 1005, 975, 1125, 1100, 1145, 1025, 1075}

APPLICATIONS AND PROBLEM SOLVING

52. Horticulture A rose garden is being planted as a border around two sides of a triangular shaped lawn in a city park. Two of the vertices of the triangle have coordinates $(-2, 4)$ and $(3, -5)$. The gardener wishes to locate the third vertex so that the lawn's area is 25 square feet. Find the value of f if the coordinates of the third vertex are $(3, f)$.   (Lesson 4–4)

A practice test for Chapter 4 is provided on page 915.

53. Auto Mechanics Ann Braun is inventory manager for a local repair shop. If she orders 6 batteries, 5 cases of spark plugs, and two dozen pairs of wiper blades, she will pay $830. If she orders 3 batteries, 7 cases of spark plugs, and 4 dozen pairs of wiper blades, she will pay $820. If the batteries are $22 less than twice the price of a dozen wiper blades, what is the cost of each item on her order?   (Lesson 4–7)

ALTERNATIVE ASSESSMENT

COOPERATIVE LEARNING PROJECT

Who Owns Muffin?   In this project, you will manipulate data by using logic. You will not need to use operational algebraic skills, but you will need to write things down, keep track of information, and analyze data. There are 16 pieces of information below. From this information, you are to figure out who owns Muffin and who has no children.

1. There are five houses in a row.
2. The owners of the FORD live in the RED house.
3. The VOLKSWAGEN owners have TWO children.
4. The family that lives in the GREEN house has a pet named SPOT.
5. The CHEVROLET owners' pet is called ROVER.
6. The GREEN house is just to the right of the WHITE house.
7. The family that owns a DUCK has FOUR children.
8. A PIG lives in the YELLOW house.
9. The family in the MIDDLE house has an animal named ROSIE.
10. The car at the FIRST house is a BUICK.
11. The CAT lives next door to the house with FIVE children.
12. The family that owns the PIG lives next door to the family with SEVEN children.
13. THOMAS, the GOAT, lives next door to SPOT.
14. The DODGE owners have a ZEBRA.
15. The BUICK's owners live next door to the BLUE house.
16. ROSIE's owners have a FORD.

Follow these steps to solve your problem.

- Set up a table.
- Determine what information you can be sure of and put it in the table.
- The remaining information will be retrieved by elimination and deduction.
- Continue in this manner until you have filled in the table.
- The two empty spaces point out where Muffin lives and where no children live.
- Write a paragraph describing your plan and how you attacked the problem.

THINKING CRITICALLY

- Find the new coordinates of quadrilateral *ROSE*, with vertices $R(-2, -1)$, $O(3, 0)$, $S(2, 2)$, and $E(-1, 2)$, if it is rotated 90° counterclockwise about the origin *twice*. Compare the new coordinates to the original ones. Make a conjecture about what effect this rotation has on any figure.

- If the determinant of a coefficient matrix is 0, can you use inverse matrices to solve the system of equations? Why or why not? Describe the graph of such a system of equations.

PORTFOLIO

Using the addition, subtraction, and multiplication operations with matrices, determine whether the following properties, when used with matrices, parallel these same properties under whole number operations.

- associative property
- commutative property
- distributive property

Write a convincing proof or argument for each property. Place this in your portfolio.

SELF EVALUATION

Are you a person that works independently or do you ask questions and also share your ideas with others? When you work independently, do you work for an answer or do you work to understand the problem or concept? Do you feel comfortable *learning* with others and not just *doing* with others?

Assess yourself. What type of a learner are you? Are you open to others' opinions and input or do you work and learn from your own experiences? Describe how you can become more or less dependent on yourself and more or less open to helping and/or receiving help from others both in your study of mathematics and in your daily life.

CHAPTERS 1–4

SECTION ONE: MULTIPLE CHOICE

There are eight multiple-choice questions in this section. After working each problem, write the letter of the correct answer on your paper.

1. Write the domain of $g = \{(9, 0), (3, 1), (12, 7), (1, -4), (12, 8), (-11, -3), (0, -6)\}$ and determine if g is a function.

   A. $\{9, 3, 12, 1, -11, 0\}$; g is a function.

   B. $\{0, 1, 7, -4, 8, -3, -6\}$; g is not a function.

   C. $\{9, 3, 12, 1, -11, 0\}$; g is not a function.

   D. $\{0, 1, 7, -4, 8, -3, -6\}$; g is a function.

2. The augmented matrix for a system is $\begin{bmatrix} 1 & 5 & 1 \\ 2 & -3 & 15 \end{bmatrix}$. What is the solution?

   A. $(6, -1)$ **B.** $(1, 15)$

   C. $(5, -13)$ **D.** $(13, 1)$

3. Write an algebraic expression for the verbal expression, "twice the sum of a number and 5 is at most 18."

   A. $2x + 5 \geq 18$ **B.** $2(x + 5) > 18$

   C. $2(x + 5) \leq 18$ **D.** $2x + 2(5) = 18$

4. A set of data has a mean of 86.4, a median of 87, a range of 15, an interquartile range of 6, and an upper quartile of 90. Which box-and-whisker plot represents this information?

   A.

   74 78 82 86 90 94

   B.

   74 78 82 86 90 94

   C.

   74 78 82 86 90 94

   D.

   74 78 82 86 90 94

5. What is the slope-intercept form of a line that passes through $(1, -4)$ and is perpendicular to a line whose equation is $3x - 2y = 8$?

   A. $y = -\frac{2}{3}x - \frac{10}{3}$

   B. $2x + 3y = -10$

   C. $y = \frac{3}{2}x - 4$

   D. $y = \frac{2}{3}x - 4$

6. A feasible region has vertices $(0, 0)$, $(4, 0)$, $(5, 5)$, and $(0, 8)$. Find the maximum and minimum of the function $f(x,y) = x + 3y$ over this region.

   A. max: $f(0, 8) = 24$

   min: $f(0, 0) = 0$

   B. min: $f(0, 0) = 0$

   max: $f(5, 5) = 20$

   C. max: $f(5, 5) = 20$

   min: $f(0, 8) = 8$

   D. min: $f(4, 0) = 4$

   max: $f(0, 0) = 0$

7. Name all the sets of numbers to which -12 belongs.

   A. integers

   B. integers, rationals

   C. rationals, reals

   D. integers, rationals, reals

8. Find the value for a for which the graph of $y = ax - 3$ is perpendicular to the graph of $6x + y = 4$.

   A. $\frac{11}{4}$

   B. $\frac{1}{6}$

   C. $\frac{2}{3}$

   D. 6

SECTION TWO: SHORT ANSWER

This section contains seven questions for which you will provide short answers. Write your answer on your paper.

9. Create a system of inequalities for which the graph will be a regular hexagon with its interior located in the first quadrant.

10. Solve $\left| x - \frac{7}{3} \right| = 6$.

11. Triangle ABC has vertices with coordinates $A(5, -2)$, $B(-3, 4)$, and $C(-2, -3)$. Find the coordinates of the vertices of triangle $A'B'C'$ if its perimeter is three times that of triangle ABC.

12. Sherry bought 24 cans of soda at the store. She bought r cans at \$0.19 per can and t cans at \$0.29 per can. Find r and t if she spent \$5.46 on soda.

13. During the seventh inning stretch at Fame Stadium Tuesday night, one concession stand sold 48 hot dogs and 72 soft drinks for a total of \$264. At the same time, another concession stand in the stadium sold 117 soft drinks and 54 hot dogs totaling \$387. Determine the individual price of a hot dog and a soft drink by using Cramer's Rule.

14. As an employee of Yogurt Delight, Maria is paid \$5.50 an hour less \$10 a week for cleaning her uniforms. Write an equation to describe her weekly pay and then find her pay for 32 hours of work.

15. Latasha has a blueprint, as shown below, of a metal plate that is to be used in a sculpture. She needs to know the area of the quadrilateral in order to calculate the amount of ore needed to make the plate. Find the area.

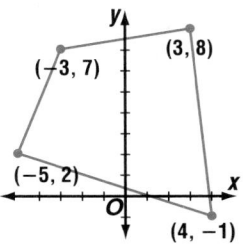

SECTION THREE: COMPARISON

This section contains five comparison problems that involve comparing two quantities, one in column A and one in column B. In certain questions, information related to one or both quantities is centered above them. All variables used represent real numbers.

Compare quantities A and B below.

- Write A if quantity A is greater.
- Write B if quantity B is greater.
- Write C if the two quantities are equal.
- Write D if there is not enough information to determine the relationship.

Column A	**Column B**

16. the value of x in
$$\begin{vmatrix} 3 & -2 & x \\ x & 1 & -5 \\ 2 & 0 & -1 \end{vmatrix} = 1$$

the value of y in the system of equations
$$x + y + 3z = 7$$
$$2x - 2y - 3z = 2$$
$$3x - y - 2z = 1$$

$$5a + 4b = -1$$
$$2a - b = 10$$

17. b

$$\begin{vmatrix} -4 & 6 \\ -13 & 24 \end{vmatrix}$$

$$t + 10 > 9$$

18. $3t - 2$

$-t + 2(t + 3)$

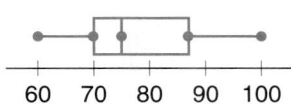

19. the mean of the data set

the median of the data set

$$2x + 3y = 7$$
$$3x - 4y = 2$$

20. x

y

Exploring Polynomials and Radical Expressions

Objectives

In this chapter, you will:

- simplify expressions containing polynomials, radicals, complex numbers, or rational exponents,
- factor polynomials,
- solve equations containing radicals, and
- solve problems by identifying and achieving subgoals.

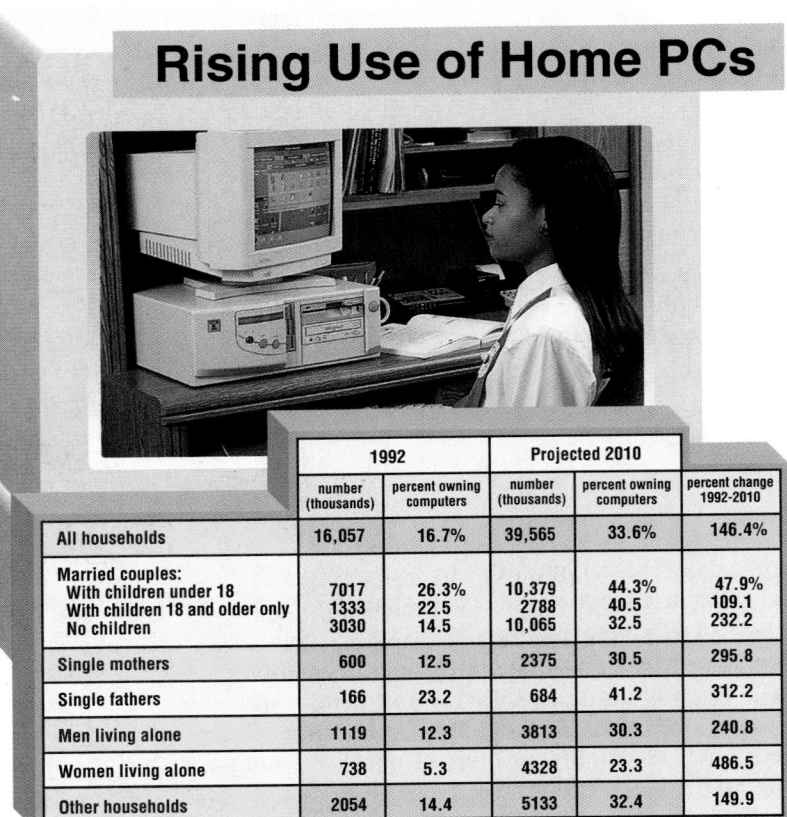

Rising Use of Home PCs

	1992		Projected 2010		
	number (thousands)	percent owning computers	number (thousands)	percent owning computers	percent change 1992-2010
All households	16,057	16.7%	39,565	33.6%	146.4%
Married couples:					
With children under 18	7017	26.3%	10,379	44.3%	47.9%
With children 18 and older only	1333	22.5	2788	40.5	109.1
No children	3030	14.5	10,065	32.5	232.2
Single mothers	600	12.5	2375	30.5	295.8
Single fathers	166	23.2	684	41.2	312.2
Men living alone	1119	12.3	3813	30.3	240.8
Women living alone	738	5.3	4328	23.3	486.5
Other households	2054	14.4	5133	32.4	149.9

Source: *American Demographics*, Feb. 1994

TIME*Line*

Home computer use is on the rise. Computer dependency has already become established in the workplace. Are you computer literate? Do you have a home computer? It may be very important to your future.

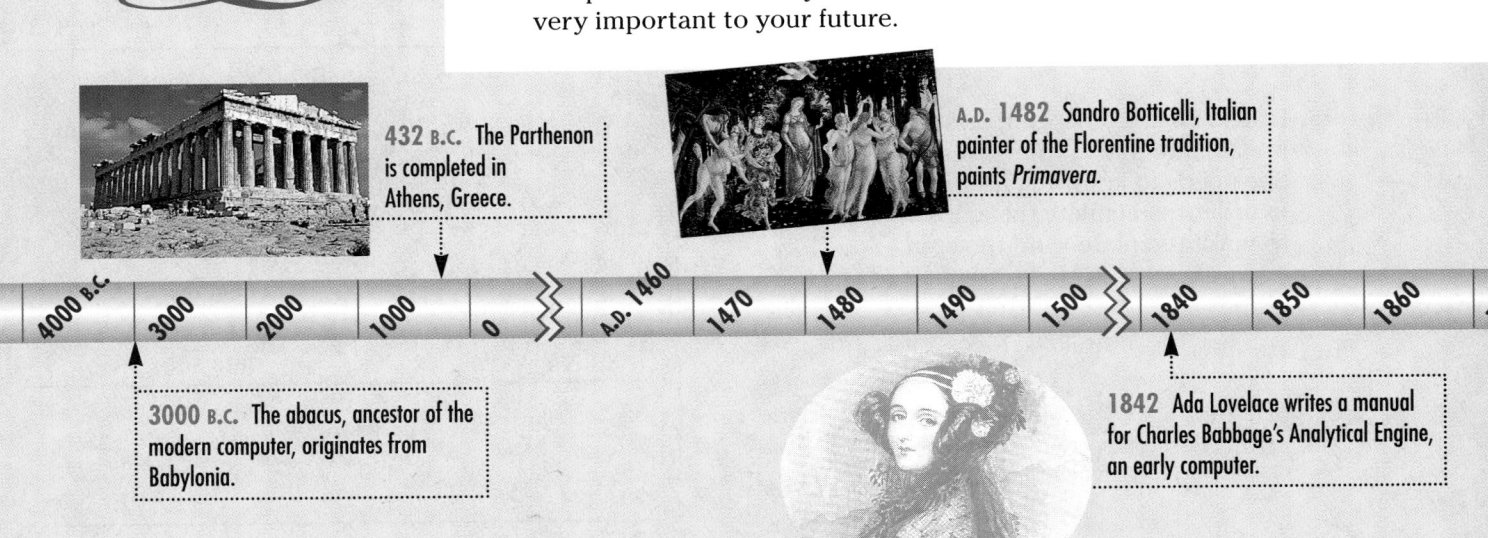

432 B.C. The Parthenon is completed in Athens, Greece.

A.D. 1482 Sandro Botticelli, Italian painter of the Florentine tradition, paints *Primavera*.

4000 B.C. 3000 2000 1000 0 A.D. 1460 1470 1480 1490 1500 1840 1850 1860

3000 B.C. The abacus, ancestor of the modern computer, originates from Babylonia.

1842 Ada Lovelace writes a manual for Charles Babbage's Analytical Engine, an early computer.

A business specialist is conducting a seminar for computer store managers so they can be more efficient in sales and customer service. The specialist tells the managers that the polynomial $2n^2 - 7n - 15$ can be used to estimate the profits on the sale of n computers. Suppose you are one of the managers attending this seminar.

- Determine how many computers you would have to sell to get a profit of $330.

- Determine the profit per computer for the sale of 10, 20, 30, 40, and 50 computers.

- Analyze the data from the sales of the different numbers of computers. Explain how this relates to computer superstores being able to sell computers at cheaper prices than smaller independent dealers.

- Include graphics and analyze the profit curve; i.e., when it is going up, going down, maximized, and so on.

When **David Bray** works on his home computer in Alexandria, Virginia, he gets things done. He recently won the grand prize in the largest science fair in the northern hemisphere for a computer program he created. His program deals with oil spills, how to predict their spread on water, and how to contain and remove them. This program also won "Best in Fair" at the International Youth Fair of Science and Technology in Argentina.

1960 Dr. Grace Hopper was first to demonstrate her company's version of COBOL on two computers.

1989 Biochemist Josephine Kong-Chan patents a chemical fat substitute for use in foods.

1880 1890 1900 1910 1920 1930 1940 1950 1960 1970 1980 1990 2000

1893 Dr. Daniel Hale Williams performs the first successful open-heart surgery.

1992 Ben Nighthorse Campbell becomes the first American Indian U.S. Senator.

Monomials

What YOU'LL LEARN

- To multiply and divide monomials,
- to represent numbers in scientific notation, and
- to multiply and divide expressions written in scientific notation.

Why IT'S IMPORTANT

You can use monomials to solve problems involving science and economics.

CONNECTION
Science

Exponents are used to express very large or very small numbers in **scientific notation.** A number is in scientific notation when it is in the form $a \times 10^n$, where $1 \leq a < 10$ and n is an integer.

The table below shows examples of numbers written in scientific notation.

	Fact	Numeral	Scientific Notation
	The temperatures produced in the center of a thermonuclear fusion bomb are as high as 400 million degrees Celsius.	400,000,000	4.0×10^8
	The most powerful laser is the Nova at the Lawrence Hall of Science in California. It generates 100 trillion watts of power.	100,000,000,000,000	1.0×10^{14}
	The most powerful microscope was invented at IBM's research labs. It is capable of focusing to one-hundredth the diameter of an atom, measured in meters.	0.0000000003	3×10^{-10}

Notice that the last number in the table is written using negative exponents. Negative exponents are another way of expressing the inverse of a number. For example, $\frac{1}{x^2}$ can be written as x^{-2}.

Negative Exponents	For any real number a, and any integer n, where $a \neq 0$, $$a^{-n} = \frac{1}{a^n} \text{ and } \frac{1}{a^{-n}} = a^n.$$

Example **Express each number in scientific notation.**
a. 2,340,000

$$2,340,000 = 2.34 \times 1,000,000$$
$$= 2.34 \times 10^6$$

b. 0.00012

$$0.00012 = 1.2 \times 0.0001$$
$$= 1.2 \times \frac{1}{10^4} \qquad 0.0001 = \frac{1}{10,000} \text{ or } \frac{1}{10^4}$$
$$= 1.2 \times 10^{-4}$$

Use the following program to explore how a TI-83 graphing calculator expresses small numbers.

```
PROGRAM: SMALLNOS
:For (N,1000,10000,1000)
:Disp "1/N IF N IS",N,"EQUALS",1/N
:Pause
:End
```

The :For statement tells the calculator to evaluate N for values from 1000 to 10,000 in increments of 1000.

The calculator will pause after each calculation. Press ENTER to continue for the next value.

Your Turn

a. How would you edit the program so that it will display N for values from 10,000 to 100,000 in increments of 1000? Rerun the program for these values.

b. How does your calculator display very small numbers? Give an example.

c. How might this differ from other scientific calculator displays?

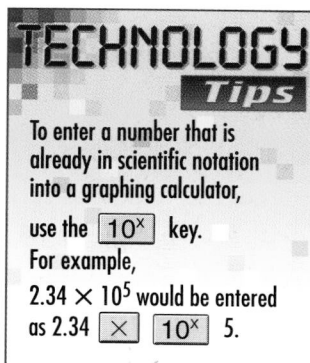

TECHNOLOGY Tips

To enter a number that is already in scientific notation into a graphing calculator,

use the $\boxed{10^X}$ key.

For example,

2.34×10^5 would be entered

as 2.34 $\boxed{\times}$ $\boxed{10^X}$ 5.

Exponents are also used in algebraic expressions called **monomials.** A monomial is an expression that is a number, a variable, or the product of a number and one or more variables. Some examples of monomials are $5c$, $-a$, 17, x^3, and $\frac{1}{2}x^4y^2$. Monomials cannot contain variables whose exponents cannot be written as whole numbers. Thus, expressions such as $\frac{1}{n^2}$ and $\sqrt{n}$ are not monomials. *Why?*

Constants are monomials that contain no variables. The numerical factor of a monomial is the **coefficient** of the variable. For example, the coefficient of m in $-6m$ is -6. The **degree** of a monomial is the sum of the exponents of its variables. For example, the degree of $12g^7h^4$ is $7 + 4$ or 11. The degree of a nonzero constant is 0.

The term power is sometimes used to refer to the exponent itself.

A **power** is an expression in the form of x^n. To **simplify** an expression containing powers means to rewrite the expression without parentheses or negative exponents.

Example ❷ Simplify $(2x^2y^3)(-5x^4y^2)$.

$$(2x^2y^3)(-5x^4y^2) = (2 \cdot x \cdot x \cdot y \cdot y \cdot y)(-5 \cdot x \cdot x \cdot x \cdot x \cdot y \cdot y) \quad \text{\textit{Definition of exponents}}$$

$$= 2(-5) \cdot x \cdot x \cdot x \cdot x \cdot x \cdot x \cdot y \cdot y \cdot y \cdot y \cdot y \quad \text{\textit{Commutative property}}$$

$$= -10x^6y^5$$

Example 2 suggests the following property of exponents.

Multiplying Powers	For any real number a and integers m and n, $$a^m \cdot a^n = a^{m+n}.$$

To multiply powers of the same variable, you add the exponents. Knowing this, it seems reasonable to expect that when dividing powers, you would subtract exponents. Consider $\dfrac{x^7}{x^4}$.

$$\dfrac{x^7}{x^4} = \dfrac{\overset{1}{\cancel{x}} \cdot \overset{1}{\cancel{x}} \cdot \overset{1}{\cancel{x}} \cdot \overset{1}{\cancel{x}} \cdot x \cdot x \cdot x}{\underset{1}{\cancel{x}} \cdot \underset{1}{\cancel{x}} \cdot \underset{1}{\cancel{x}} \cdot \underset{1}{\cancel{x}}}$$ *Remember that x cannot equal 0.*

$$= x \cdot x \cdot x$$

$$= x^3 \qquad \text{Note that } x^3 = x^{7-4}.$$

It appears that our hypothesis is true. To divide powers of the same base, you subtract exponents. This is stated formally below.

Dividing Powers	**For any real number *a*, except *a* = 0, and integers *m* and *n*,** $$\dfrac{a^m}{a^n} = a^{m-n}.$$

In the next example, the check uses the definition of exponents to verify the rule for division of powers.

Example ③   Simplify $\dfrac{p^5}{p^9}$. Assume that $p \neq 0$.

$$\dfrac{p^5}{p^9} = p^{5-9} \qquad \textit{Dividing powers}$$

$$= p^{-4} \text{ or } \dfrac{1}{p^4} \qquad \textit{Remember that a simplified expression cannot contain negative exponents.}$$

Check: $\dfrac{p^5}{p^9} = \dfrac{\overset{1}{\cancel{p}} \cdot \overset{1}{\cancel{p}} \cdot \overset{1}{\cancel{p}} \cdot \overset{1}{\cancel{p}} \cdot \overset{1}{\cancel{p}}}{\underset{1}{\cancel{p}} \cdot \underset{1}{\cancel{p}} \cdot \underset{1}{\cancel{p}} \cdot \underset{1}{\cancel{p}} \cdot \underset{1}{\cancel{p}} \cdot p \cdot p \cdot p \cdot p}$

$$= \dfrac{1}{p^4} \text{ or } p^{-4}$$

Let's use the property of division of powers and the definition of exponents to simplify $\dfrac{y^5}{y^5}$.

Method 1

$$\dfrac{y^5}{y^5} = y^{5-5} \qquad \textit{Dividing powers}$$

$$= y^0$$

Method 2

$$\dfrac{y^5}{y^5} = \dfrac{\overset{1}{\cancel{y}} \cdot \overset{1}{\cancel{y}} \cdot \overset{1}{\cancel{y}} \cdot \overset{1}{\cancel{y}} \cdot \overset{1}{\cancel{y}}}{\underset{1}{\cancel{y}} \cdot \underset{1}{\cancel{y}} \cdot \underset{1}{\cancel{y}} \cdot \underset{1}{\cancel{y}} \cdot \underset{1}{\cancel{y}}}$$

$$= 1$$

Since $\dfrac{y^5}{y^5}$ cannot have two different values, we can conclude that $y^0 = 1$, where $y \neq 0$. In general, any nonzero number raised to the zero power is equal to 1.

The properties we have presented can be used to verify other properties of powers listed below.

Properties of Powers	**Suppose m and n are integers and a and b are real numbers. Then the following properties hold.**
	Power of a Power: $(a^m)^n = a^{mn}$
	Power of a Product: $(ab)^m = a^m b^m$
	Power of a Quotient: $\left(\dfrac{a}{b}\right)^n = \dfrac{a^n}{b^n}$, $b \neq 0$ and
	$\left(\dfrac{a}{b}\right)^{-n} = \left(\dfrac{b}{a}\right)^n$ or $\dfrac{b^n}{a^n}$, $a \neq 0$, $b \neq 0$

Example **4** **Simplify each expression.**

a. $(a^4)^5$

$$(a^4)^5 = a^{4(5)} \qquad \textit{Power of}$$
$$= a^{20} \qquad \textit{a power}$$

b. $(-5p^2 s^4)^3$

$$(-5p^2 s^4)^3 = (-5)^3 \cdot (p^2)^3 \cdot (s^4)^3 \qquad \textit{Power of}$$
$$= -125p^6 s^{12} \qquad \textit{a product}$$

c. $\left(\dfrac{-2m}{n}\right)^4$

$$\left(\dfrac{-2m}{n}\right)^4 = \dfrac{(-2m)^4}{n^4} \qquad \textit{Power of a quotient}$$
$$= \dfrac{(-2)^4 m^4}{n^4}$$
$$= \dfrac{16m^4}{n^4}$$

d. $\left(\dfrac{a}{3}\right)^{-2}$

$$\left(\dfrac{a}{3}\right)^{-2} = \left(\dfrac{3}{a}\right)^2 \qquad \textit{Definition of negative exponent}$$
$$= \dfrac{3^2}{a^2} \text{ or } \dfrac{9}{a^2} \qquad \textit{Power of a quotient}$$

To simplify some expressions, you must use several of the properties of powers.

Example **5** **Simplify** $\left(\dfrac{-4x^{2n}}{x^{3n}z^2}\right)^3$.

Method 1

$$\left(\dfrac{-4x^{2n}}{x^{3n}z^2}\right)^3 = \dfrac{(-4x^{2n})^3}{(x^{3n}z^2)^3} \qquad \textit{Power}\text{/}\textit{of a quotient}$$

$$= \dfrac{(-4)^3 (x^{2n})^3}{(x^{3n})^3 (z^2)^3} \qquad \textit{Power of a product}$$

$$= \dfrac{-64x^{6n}}{x^{9n}z^6} \qquad \textit{Power of a power}$$

$$= \dfrac{-64x^{6n-9n}}{z^6} \qquad \textit{Dividing powers}$$

$$= \dfrac{-64x^{-3n}}{z^6} \text{ or } \dfrac{-64}{x^{3n}z^6}$$

Method 2

Simplify the fraction first before cubing.

$$\left(\dfrac{-4x^{2n}}{x^{3n}z^2}\right)^3 = \left(\dfrac{-4x^{2n-3n}}{z^2}\right)^3$$

$$= \left(\dfrac{-4}{x^n z^2}\right)^3$$

$$= \dfrac{(-4)^3}{(x^n)^3 (z^2)^3}$$

$$= \dfrac{-64}{x^{3n}z^6}$$

You can also multiply and divide expressions involving numbers written in scientific notation. A calculator provides the most efficient method for finding the product or quotient.

Example Use a scientific calculator to evaluate each expression.

a. $(3.69 \times 10^{-5})(4.1 \times 10^{8})$

Estimate: $(4 \times 10^{-5})(4 \times 10^{8}) = 16 \times 10^{3}$ or 16,000

3.69 [EXP] [+/−] 5 [×] 4.1 [EXP] 8 [=] 15129

b. $\dfrac{7.6 \times 10^{2}}{3.2 \times 10^{-6}}$

Estimate: $\dfrac{7 \times 10^{2}}{3.5 \times 10^{-6}} = 2 \times 10^{8}$ or 200,000,000

7.6 [EXP] 2 [÷] 3.2 [EXP] [+/−] 6 [=] 237500000

Scientists use scientific notation because they often deal with very large or very small quantities.

Example

A chemist works in a laboratory, doing research for a fertilizer company. For one batch of fertilizer, she needs 20 moles of sulfuric acid. A mole is a standard unit of measure in chemistry that contains 6.02×10^{23} molecules of a substance, Avogadro's constant. If she adds 1.08×10^{25} molecules of sulfuric acid, has she added enough to make one batch of fertilizer?

Divide the number of molecules she has by the number of molecules in a mole.

$$\frac{1.08 \times 10^{25} \text{ molecules}}{6.02 \times 10^{23} \text{ molecules/mole}} = \frac{1.08}{6.02} \cdot \frac{10^{25} \text{ molecules}}{10^{23} \text{ molecules/mole}}$$

$$\approx 0.179 \times 10^{2} \text{ or } 17.9 \text{ moles} \quad \textit{Use a calculator.}$$

The chemist did not add enough sulfuric acid.

CHECK FOR UNDERSTANDING

Communicating Mathematics

Study the lesson. Then complete the following.

1. **Explain** why $a \neq 0$ in the expression $\dfrac{a^{m}}{a^{n}}$.
2. **Determine** if $4x^{2}$ and $(4x)^{2}$ are equivalent.
3. **You Decide**   Is a negative number raised to the 777th power a negative or positive number? Explain.
4. Is $2(x^{2}y)^{3}$ in simplest form? Explain your answer.

5. Write $3wz^{-4}$ without using negative exponents.

Simplify. Assume that no variable equals 0.

6. $y^5 \cdot y^7$

7. $(3a)^4$

8. $(m^2)^2(m^{-2})^2$

9. $\frac{40x^4}{-5x^2}$

10. $\frac{-2c^3d^6}{24c^2d^2}$

11. $\frac{16s^6t^5}{(2s^2t)^2}$

12. $\left(\frac{1}{x^3y^2}\right)^4$

13. $\left(\frac{bc}{2}\right)^{-3}$

14. $\left(\frac{-6y^5}{3y^2}\right)^{-2}$

Express each number in scientific notation.

15. 386,000

16. 0.000346

Evaluate. Express each answer in both scientific and decimal notation.

17. $\frac{8 \times 10^{-1}}{16 \times 10^{-2}}$

18. $(3.42 \times 10^8)(1.1 \times 10^{-5})$

EXERCISES

Simplify. Assume that no variable equals 0.

19. $b^3 \cdot b^5$

20. $x^2 \cdot x \cdot x^3$

21. $(m^3)^3$

22. $(-3y)^3$

23. $\frac{an^6}{n^5}$

24. $\frac{-x^6y^6}{x^3y^4}$

25. $(a^3b^3)(ab)^{-2}$

26. $(4x^3y^{-4})(7xy^2)$

27. $(-3r^2s)^2(2rs^3)$

28. $(3a^3b)(-5a^2b^2)$

29. $(2mn^2)(5m^2n)$

30. $(-5x^2y)(-2x^4y^7)$

31. $\left(-\frac{3}{4}m^2n^3\right)\left(\frac{8}{9}mn^4\right)$

32. $2b^2(2ab)^3$

33. $4x(-3x)^3$

34. $4a^2(3b^3)(2a^2b)$

35. $5mn^2(m^3n)(-3p^2)$

36. $5x(6x^2y)(3xy^3)$

37. $\frac{-6x^2y^3z^3}{24x^2y^7z^3}$

38. $\frac{2x^5y^3z^3}{8x^3y^7z}$

39. $\frac{-15m^5n^8(m^3n^2)}{45m^4n}$

40. $\frac{2a^3b}{(-2ab^3)^{-2}}$

41. $\left(\frac{5a^3b}{10a^2b^2}\right)^4$

42. $\left(\frac{a}{b^{-1}}\right)^{-2}$

43. $\frac{40a^{-1}b^{-7}}{20a^{-5}b^{-9}}$

44. $\frac{5^{2x}}{5^{2x+2}}$

45. $\frac{8}{m^0 + n^0}$

Express each number in scientific notation.

46. 810,4

47. 786,500,000

48. 0.0008742

49. 0.001250

50. 901,010,000

51. 0.03331

Evaluate. Express each answer in both scientific and decimal notations.

52. $(6.23 \times 10^4)(2.0 \times 10^5)$

53. $(2 \times 10^{-3})(2.01 \times 10^{-2})$

54. $(45,000)(0.0025)$

55. $(9.5 \times 10^3)^2$

56. $(6.9 \times 10^3)(1.4 \times 10^3)^{-1}$

57. $\frac{(93,000,000)(0.005)}{0.0015}$

Find the value of r that makes each sentence true.

58. $y^{28} = y^{3r} \cdot y^7$

59. $2^{r+5} = 2^{2r-1}$

60. $2^{2r+1} = 32$

61. $(x^3 \cdot x^r)^5 = x^{30}$

62. $\frac{x^{2r}}{x^{-3r}} = x^{15}$

63. $\frac{m^r}{m^{15}} = (m^3)^{r+2}$

64. Which is greater, 100^{10} or 10^{100}? Explain your answer.

65. Express the quotient $\dfrac{x + x^2 + x^3 + x^4 + x^5 + x^6 + x^7}{x^{-3} + x^{-4} + x^{-5} + x^{-6} + x^{-7} + x^{-8} + x^{-9}}$ in simplest form.

Assume that x is not equal to zero. (*Hint:* Simplify the denominator first.)

66. Economics In 1991, the U.S. government received approximately $468,000,000,000 in personal income taxes. The population of the country at that time was about 250,000,000. If everyone paid taxes, what was the average amount paid by each man, woman, and child?

Largest Planetary
Moons (diameter)

1. Ganymede, Jupiter
(3273 miles)

2. Titan, Saturn (3200 miles)

3. Callisto, Jupiter
(2995 miles)

4. Io, Jupiter (2257 miles)

5. Moon, Earth (2159 miles)

67. Astronomy In April of 1983, the space probe *Pioneer 10* was as far from Earth as the planet Pluto. *Pioneer 10* sent radio signals that traveled at the speed of light, 3.00×10^5 kilometers per second. If *Pioneer 10* was 4.58×10^9 kilometers from Earth, how long would it take a tracking station to send a message indicating a mid-course correction in the space probe's travel course? (Use $t = \dfrac{d}{r}$, where d is distance and r is rate.)

68. Statistics The chart at the right shows the percentage of the population that is made up of people in their twenties for 18 large U.S. cities. (Lesson 4–8)

 a. Make a box-and-whisker plot of the data.

 b. List three reasons why you think someone would want this information.

69. Find M if $\begin{bmatrix} 3 & 6 & 1 \\ 2 & -1 & 0 \end{bmatrix} \cdot M = \begin{bmatrix} 3 & 6 & 1 \\ 2 & -1 & 0 \end{bmatrix}$.

(Lesson 4–5)

City	Percent of Population
Boston, MA	26%
Columbus, OH	24
San Diego, CA	22
Seattle, WA	20
Washington, DC	20
Nashville, TN	19
Anchorage, AK	18
Baltimore, MD	18
Chicago, IL	18
Indianapolis, IN	18
Jacksonville, FL	18
Milwaukee, WI	18
Philadelphia, PA	18
Phoenix, AZ	18
New York, NY	17
Detroit, MI	16
Honolulu, HI	16
U.S. Average	16

Solve for the indicated variables.

70. $\begin{vmatrix} 5a & 3 \\ a & 5 \end{vmatrix} = 7$ (Lesson 4–4)

71. $y\begin{bmatrix} 3 & -4 \\ 2 & x \end{bmatrix} = \begin{bmatrix} 15 & -20 \\ z & 5 \end{bmatrix}$ (Lesson 4–1)

Solve each system algebraically or by graphing.

72. $8a - 3b + c = 7$
$-2a + b - c = 0$
$2a - 3b + 9c = -1$ (Lesson 3–7)

73. $x + y < 9$
$x - y < 3$
$y - x > 4$ (Lesson 3–4)

74. Describe the graphs of two linear equations that are dependent. (Lesson 3–1)

75. Write an equation of the line that passes through (2, 2) and is parallel to the line $4x - y + 3 = 0$. (Lesson 2–5)

76. Find the x- and y-intercepts of $f(x) = 13x + 26$. (Lesson 2–3)

77. Graph $x - 3y = -3$. (Lesson 2–2)

78. Solve $-2(4 - 3x) > 4$. (Lesson 1–7)

79. Evaluate $15 - 3(2) \div 8 - 11$. (Lesson 1–1)

Polynomials

5-2

What YOU'LL LEARN

- To add, subtract, and multiply polynomials.

Why IT'S IMPORTANT

You can use polynomials to solve problems involving biology and genetics.

F Y I

The Punnett square is named for Reginald Crundall Punnett (1875–1967). He was a British mathematician and geneticist.

CONNECTION
Biology

Scientists can use algebraic expressions to summarize the possible outcomes in genetic breeding. Certain traits result from the pairing of two genes, one from the female parent and one from the male parent. For example, suppose a red-flowering, sweet pea plant has *genotype RR*, a white-flowering, sweet pea plant has genotype *WW*, and a pink-flowering, sweet pea plant has genotype *RW*. Each letter represents one of the two genes that make up the characteristic.

Suppose two pink-flowering plants are bred. The offspring can be expressed using algebra and a model called a *Punnett square.*

One gene from the mother pairs with one gene from the father for each possible offspring.

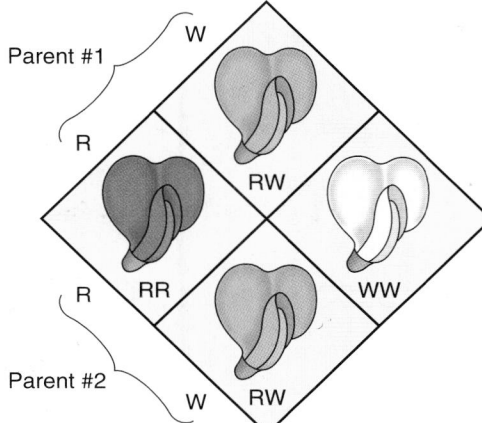

The sum of the possible results for four offspring can be written as $RR + RW + RW + WW$; that is, one red-, two pink-, and one white-flowering plants. Suppose we substitute x for R and y for W. The result would be a sum of four monomials, $xx + xy + xy + yy$, or $x^2 + 2xy + y^2$. The two monomials xy and xy can be combined because they are **like terms**. Like terms are two monomials that are the same, or differ only by their numerical coefficients.

The expression $x^2 + 2xy + y^2$ is called a **polynomial.** A polynomial is a monomial or a sum of monomials. The monomials that make up the polynomial are called the **terms** of the polynomial. An expression like $m^2 - 7mb - 12cd$ with three unlike terms is called a **trinomial.** An expression like $xy + b^3$ with two unlike terms is called a **binomial.** The *degree* of a polynomial is the degree of the monomial with the greatest degree. Thus, the degree of $x^2 + 2xy + y^2$ is 2.

Remember that a difference like $m^2 - 7mb - 12cd$ can be written as a sum, $m^2 + (-7mb) + (-12cd)$.

Example ① **Determine whether or not each expression is a polynomial. Then state the degree of each polynomial.**

a. $\frac{2}{7}x^4y^3 - 21x^3$

This expression is a polynomial. The degree of the first term is $4 + 3$ or 7, and the degree of the second term is 3. The degree of the polynomial is 7.

b. $9 + \sqrt{x} - 3$

This expression is not a polynomial because $\sqrt{x}$ is not a monomial.

To simplify a polynomial means to perform the operations indicated and combine like terms.

Example **2** **Find the perimeter of quadrilateral _ABCD_.**

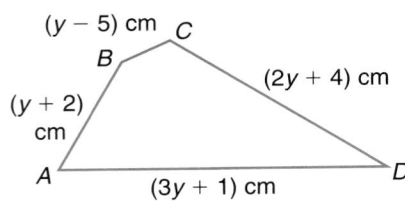

The perimeter is the sum of the measures of the sides.

$P = AB + BC + CD + DA$
$\quad = (y + 2) + (y - 5) + (2y + 4) + (3y + 1)$
$\quad = y + y + 2y + 3y + 2 - 5 + 4 + 1$
$\quad = 7y + 2$

The perimeter of quadrilateral _ABCD_ is $(7y + 2)$ cm.

Example **3** **Simplify $(4x^2 - 3x) - (x^2 + 2x - 1)$.**

$(4x^2 - 3x) - (x^2 + 2x - 1) = 4x^2 - 3x - x^2 - 2x + 1$
$\qquad\qquad\qquad\qquad\qquad = (4x^2 - x^2) + (-3x - 2x) + 1$
$\qquad\qquad\qquad\qquad\qquad = 3x^2 - 5x + 1$

You can use the distributive property to multiply polynomials.

Example **4** **Find $3x(5x^4 - x^3 + 4x)$.**

$3x(5x^4 - x^3 + 4x) = 3x(5x^4) + 3x(-x^3) + 3x(4x)$
$\qquad\qquad\qquad\quad = 15x^5 - 3x^4 + 12x^2$

You can use algebra tiles to make a geometric model of the product of two binomials.

Multiplying Binomials

Materials: algebra tiles

Use algebra tiles to find the product of $x + 4$ and $x + 3$.

Your Turn

a. Draw a 90° angle on your paper.

b. Use an _x_-tile and a 1-tile to mark off a length equal to $x + 4$ along the top.

c. Use the tiles to mark off a length equal to $x + 3$ along the side.

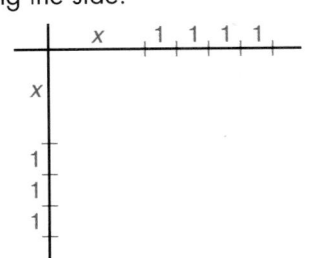

d. Draw lines to show the grid formed by these measures.

e. Fill in the lines with the appropriate tiles to show the area product. The model shows the polynomial $x^2 + 7x + 12$.

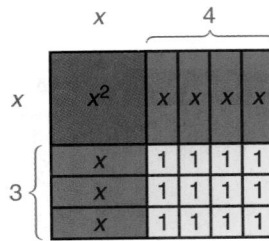

The area of the rectangle is the product of its length and width. Substituting the area, length, and width with the corresponding polynomials, we find that $x^2 + 7x + 12 = (x + 4)(x + 3)$.

In the following example, two different methods are used to multiply binomials.

Example **5** Find $(4n + 3)(3n + 1)$.

Method 1: Distributive Property

$$(4n + 3)(3n + 1) = 4n(3n + 1) + 3(3n + 1)$$
$$= 4n \cdot 3n + 4n \cdot 1 + 3 \cdot 3n + 3 \cdot 1$$
$$= 12n^2 + 4n + 9n + 3$$
$$= 12n^2 + 13n + 3$$

Method 2: FOIL Method

$$(4n + 3)(3n + 1) = \underbrace{4n \cdot 3n}_{First\ terms} + \underbrace{4n \cdot 1}_{Outside\ terms} + \underbrace{3 \cdot 3n}_{Inside\ terms} + \underbrace{3 \cdot 1}_{Last\ terms}$$
$$= 12n^2 + 13n + 3$$

The **FOIL method** is an application of the distributive property that makes the multiplication easier.

FOIL Method of Multiplying Polynomials	**The product of two binomials is the sum of the products of** **F** the *first* terms, **O** the *outer* terms, **I** the *inner* terms, and **L** the *last* terms.

Example **6** Find $(k^2 + 3k + 9)(k + 3)$.

$$(k^2 + 3k + 9)(k + 3)$$
$$= k^2(k + 3) + 3k(k + 3) + 9(k + 3) \qquad \textit{Distributive property}$$
$$= k^2 \cdot k + k^2 \cdot 3 + 3k \cdot k + 3k \cdot 3 + 9 \cdot k + 9 \cdot 3 \quad \textit{Distributive property}$$
$$= k^3 + 3k^2 + 3k^2 + 9k + 9k + 27$$
$$= k^3 + 6k^2 + 18k + 27 \qquad \textit{Combine like terms.}$$

CHECK FOR UNDERSTANDING

Communicating Mathematics

Study the lesson. Then complete the following.

1. **Demonstrate** the FOIL method by multiplying $(3a + 4b)$ and $(a - b)$.

2. **Show** another way to multiply the expression in Example 6 by distributing $(k^2 + 3k + 9)$ instead of $(k + 3)$.

3. **Write** a polynomial of degree 6 that has four terms.

4. Draw a geometric representation of $2x^2 + 6x$.

5. Write two factors and a product for the model shown at the right.

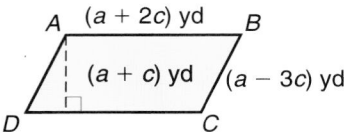

Guided Practice

Determine whether each expression is a polynomial. Write *yes* or *no* and explain your reasoning. Then state the degree of each polynomial.

6. $14x + 3y$

7. $\dfrac{y^3}{4} - 7x$

8. $\dfrac{ax^2 + 6}{by^3 + 5}$

Simplify.

9. $(5x - 7y) + (6x + 8y)$

10. $(-2y^2 - 4y + 7) - (2y^2 + 4y - 7)$

11. $3y(2x + 6)$

12. $2m^2n(5mn - 3m^3n^2 + 4mn^4)$

13. $(x + 6)(x + 3)$

14. $(y - 10)(y + 7)$

15. $(3m - 1)(3m + 1)$

16. $(2p - 3s)^2$

17. Geometry Quadrilateral *ABCD* is a parallelogram.
   a. Find the perimeter of *ABCD*.
   b. Find the area of *ABCD*.

A $(a + 2c)$ yd B
$(a + c)$ yd $(a - 3c)$ yd
D C

EXERCISES

Practice

Determine whether each expression is a polynomial. Write *yes* or *no* and explain your reasoning. Then state the degree of each polynomial.

18. $x^2 + 2x + 3$

19. $y^3 + 3$

20. $\sqrt{s - 5}$

21. $\dfrac{3k^2}{2} + \dfrac{4k^7}{5}$

22. $\dfrac{4ab}{c} - \dfrac{2d}{x}$

23. $x\sqrt{3} + 8x^2y^4$

Simplify.

24. $(3r + s) - (r - s) - (r + 3s)$

25. $(z^2 - 6z - 10) + (2z^2 + 4z - 11)$

26. $(-12y - 6y^2) + (-7y + 6y^2)$

27. $(3m^2 + 5m - 6) + (7m^2 - 9)$

28. $(10x^2 - 3xy + 4y^2) - (3x^2 + 5xy)$

29. $(8r^2 + 5r + 14) - (7r^2 + 6r + 8)$

30. $4a(3a^2b)$

31. $4f(gf - bh)$

32. $\dfrac{2}{3} x^2 (6x + 9y - 12xy^2)$

33. $-5mn^2(-3m^2n + 6m^3n - 3m^4n^4)$

34. $(c^2 - 6cd - 2d^2) + (7c^2 - cd + 8d^2) - (-c^2 + 5cd - d^2)$

35. $(4x^2 - 3y^2 + 5xy) - (8xy + 6x^2 + 3y^2)$.

INTEGRATION
Geometry

Find the perimeter of each figure.

36.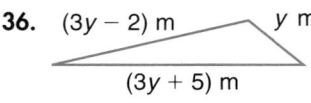
$(3y - 2)$ m y m
$(3y + 5)$ m

37.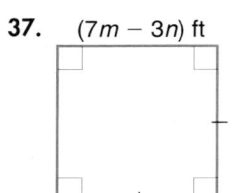
$(7m - 3n)$ ft

38.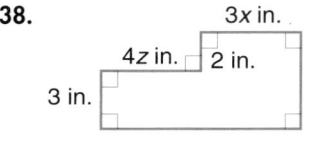
3x in.
4z in.   2 in.
3 in.

Simplify.

39. $(q - 7)(q + 5)$

40. $(m + 7)(m + 2)$

41. $(5 - r)(5 + r)$

42. $(2x + 7)(3x + 5)$

43. $(3y - 8)(2y + 7)$

44. $(x^3 - y)(x^3 + y)$

45. $g^{-3}(g^5 - 2g^3 + g^{-1})$

46. $x^{-3}y^2(yx^4 + y^{-1}x^3 + y^{-2}x^2)$

47. $(y - 3x)^2$

48. $(1 + 4m)^2$

49. $(2p + q^3)^2$

50. $(w^2 - 5)(2w^2 + 3)$

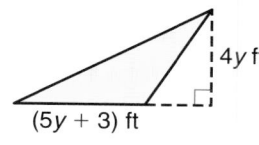 **INTEGRATION**

Geometry

Find the area of each figure.

51.
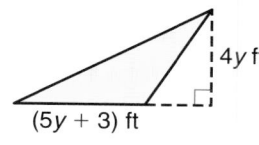
4y ft
(5y + 3) ft

52.
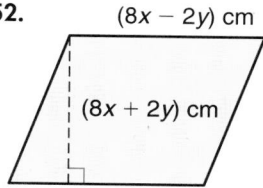
(8x − 2y) cm
(8x + 2y) cm

53.
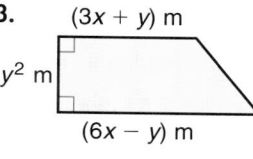
(3x + y) m
y^2 m
(6x − y) m

Simplify.

54. $(3y + 1)(3y - 1)(y + 2)$

55. $(x^2 + xy + y^2)(x - y)$

56. $(2q + 1)(q - 2)^2$

57. $(x - 2)(x + 2)(x^2 + 5)$

58. $(3b - c)^3$

59. $(y + x)^2(y - x)^2$

Critical Thinking

60. Draw a geometric representation of $(x + 3)(x - 3)$. Explain your results.

61. a. Write two different polynomial expressions that represent the area of the figure at the right.

    b. Write a polynomial for the perimeter of the figure.

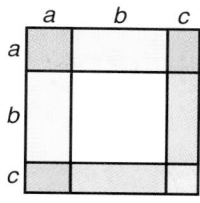

Applications and Problem Solving

62. **Personal Finance** Abey has $1500 to invest. She would like to have a return of at least $100 a year on her investment. She can invest some of the money in a mutual fund yielding 6% annually. She can also buy bonds yielding 7% annually. She decides to invest her money in both sources to diversify her investment.

    a. Express the amount of return Abey will make as a polynomial in one variable.

    b. How much money should Abey place in each of these investments to have the desired return?

63. **Geometry** Recall that the measure of an angle inscribed in a circle is half the measure of its intercepted arc. That is, $m\angle B = \frac{1}{2}m\widehat{ADC}$.

    a. Given a circle with inscribed quadrilateral $ABCD$ with the given arc measures, find the ratio of $m\angle A$ to $m\angle B$.

    b. If $m\widehat{AB} + m\widehat{BC} + m\widehat{CD} + m\widehat{AD} = 360°$, find the value of x.

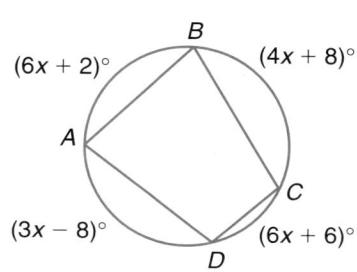

64. Make a Drawing Suppose you were trying to model the product $(2x + 3y)(x + 5)$. Make a drawing of rectangles to represent each type of monomial in the product. Then make a drawing to illustrate the product and write the result.

65. Genetics Suppose p represents the ratio of dominant gene A in a population and q represents the ratio of recessive gene a in a population. The next generation is the result of $(p + q)^2$. This results in p^2 pure dominant genotypes (AA), q^2 pure recessive genotypes (aa), and $2pq$ hybrid genotypes (Aa). Since all members of the population must have at least one recessive gene or dominant gene present in their genotypes, $p + q = 1$.

Suppose that in the population of a certain village, the recessive left-handedness gene r had a frequency of 1:4 and the dominant right-handedness gene R had a frequency of 3:4. In the next generation, what would you predict the population genotypes to be?

Mixed Review

66. Simplify $2(rk)^2(5rt^2) - k(2rk)(2rt)^2$. (Lesson 5–1)

67. Solve the system of equations by using augmented matrices. (Lesson 4–7)

$4x - y + z = 6$
$2x + y + 2z = 3$
$3x - 2y + z = 3$

68. Write a matrix equation for the system of equations. (Lesson 4–6)

$3x - y = 5$
$2x + 33y = 29$

69. If $A = \begin{bmatrix} 2 & -3 \\ 1 & 4 \end{bmatrix}$ and $B = \begin{bmatrix} -4 & 0 \\ 2 & 5 \end{bmatrix}$, find AB. (Lesson 4–3)

70. Community Service A theater where a drug abuse program is being presented seats 150 people. The proceeds will be donated to a local drug information center. Admission is $2.00 for adults and $1.00 for students. Every two adults must bring at least one student. How many adults and students should attend in order to raise the maximum amount of money? (Lesson 3–5)

71. Solve the system of equations by using substitution. (Lesson 3–2)

$6x + 4y = 80$
$x - 7y = -2$

72. Solve the system of equations by graphing. (Lesson 3–1)

$2x + 2y = -12$
$3x - 2y = -3$

73. Name the points, $(0, 0)$, $(-1, -3)$, or $(4, 0)$, that satisfy $4x - |y| \leq 12$. (Lesson 2–6)

74. Find $h(-7)$ if $h(x) = \dfrac{3 + x}{4}$. (Lesson 2–1)

75. Statistics Find the mean, median, and mode of {45, 49, 40, 39, 39, 46, 44, 41, 42}. (Lesson 1–4)

76. Banking Karen invests $7500 in a certificate of deposit at the Lombard Bank. The simple interest rate is 7.2% per year. How much will she make in 5 years? (Lesson 1–1)

Dieding Polynomials

APPLICATION
Entertainment

What YOU'LL LEARN

- To divide polynomials using long division, and
- to divide polynomials by binomials using synthetic division.

Why IT'S IMPORTANT

You can use polynomials to solve problems involving manufacturing and entertainment.

Tionna, a senior at Franklin High School, spends one free period a day as a teacher's aide at the middle school that is located next to the high school. A magician was performing at the middle school and asked a member of the audience to participate in a number game. The magician said:

- Choose any number.
- Multiply your number by 3.
- Then add the sum of your number and 8 to the number you got when you multiplied.
- Now divide by the sum of your number and 2.

Then the magician said, "Without asking you what number you chose, I can tell you that your final result was . . . 4!"

Tionna wondered how the magician's trick worked. But magicians don't tell their secrets. So, she used her algebraic skills to write expressions that modeled the steps in the trick.

Choose a number.	x
Multiply by 3.	$3x$
Add the sum of your number and 8 to the previous result.	$3x + (x + 8)$ or $4x + 8$
Divide this result by your number plus 2.	$\dfrac{4x + 8}{x + 2}$

The final expression is $\dfrac{4x + 8}{x + 2}$. But how does this help Tionna solve the secret of the magician's trick? *You will be asked to solve the mystery in Exercise 3.*

In Lesson 5–1, you learned to divide monomials. You can divide a polynomial by a monomial by using those same skills.

Example **1** Simplify $\dfrac{6r^2s^2 + 3rs^2 - 9r^2s}{3rs}$.

This expression means that each term in the numerator shares a common denominator. Rewrite the expression as a sum of quotients.

$$\frac{6r^2s^2 + 3rs^2 - 9r^2s}{3rs} = \frac{6r^2s^2}{3rs} + \frac{3rs^2}{3rs} - \frac{9r^2s}{3rs}$$

$$= \frac{6}{3} \cdot r^{2-1}s^{2-1} + \frac{3}{3} \cdot r^{1-1}s^{2-1} - \frac{9}{3} \cdot r^{2-1}s^{1-1}$$

$$= 2rs + s - 3r \quad r^{1-1} = r^0 \text{ or } 1$$

You can use a process similar to long division of whole numbers to divide a polynomial by a polynomial. When doing the division, remember that you can only add and subtract like terms.

Example **2** **Simplify** $\dfrac{c^2 - c - 30}{c - 6}$.

In this lesson, assume that the denominator never equals zero.

$$
\begin{array}{r}
c \quad\quad\quad \\
c - 6 \overline{)c^2 - c - 30} \\
\underline{c^2 - 6c} \quad\quad \\
5c - 30 \quad
\end{array}
$$
  $-c - (-6c) = -c + 6c$ or $5c$

$$
\begin{array}{r}
c + 5 \quad\quad\quad \\
c - 6 \overline{)c^2 - c - 30} \\
\underline{c^2 - 6c} \quad\quad \\
5c - 30 \\
\underline{5c - 30} \\
0
\end{array}
$$

Therefore, the quotient is $c + 5$.

Just as with the division of whole numbers, the division of two polynomials may result in a quotient with a remainder. Remember that $9 \div 4 = 2 + R1$ and is often written as $2\frac{1}{4}$. The division of polynomials with a remainder is presented in the same manner.

Example **3** **Simplify** $(s^2 + 4s - 16)(6 - s)^{-1}$.

$$(s^2 + 4s - 16)(6 - s)^{-1} = \dfrac{s^2 + 4s - 16}{6 - s}$$

$$
\begin{array}{r}
-s - 10 \quad\quad\quad \\
-s + 6 \overline{)s^2 + 4s - 16} \\
\underline{s^2 - 6s} \quad\quad \\
10s - 16 \\
\underline{10s - 60} \\
44
\end{array}
$$
For ease in dividing, rewrite $6 - s$ as $-s + 6$.

The quotient is $-s - 10 + \dfrac{44}{6 - s}$.   *The remainder is $\dfrac{44}{-s + 6}$ or $\dfrac{44}{6 - s}$.*

In Example 4, solving for a variable involves dividing polynomials.

Example **4**

Geometry

The area of triangle ABC is $(10y^2 + 6y)$ mm^2 and its base is $(5y + 3)$ mm. Find the measure of the altitude of the triangle.
An altitude of a triangle is a line segment from a vertex perpendicular to the line containing the opposite side. The measure of an altitude is the height h of the triangle.

$$A = \frac{1}{2}bh$$

$$10y^2 + 6y = \frac{1}{2}(5y + 3)(h) \quad A = 10y^2 + 6y, \; b = 5y + 3$$

$$20y^2 + 12y = (5y + 3)h \quad\quad \textit{Multiply each side by 2.}$$

$$\frac{20y^2 + 12y}{5y + 3} = h \quad\quad \textit{Divide each side by } 5y + 3.$$

$$4y = h$$

The length of the altitude is $4y$ mm.

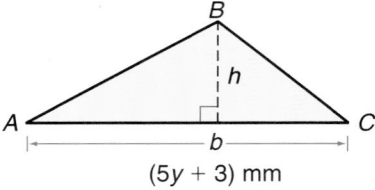

A simpler process called **synthetic division** has been devised to divide a polynomial by a binomial. Suppose we wanted to divide $(6x^3 - 19x^2 + x + 6)$ by $(x - 3)$. Long division would produce the following result.

$$
\begin{array}{r}
6x^2 - 1x - 2 \\
x - 3\overline{)6x^3 - 19x^2 + x + 6} \\
\underline{(-)6x^3 - 18x^2} \\
-1x^2 + x + 6 \\
\underline{(-)-1x^2 + 3x} \\
-2x + 6 \\
\underline{(-)-2x + 6} \\
0
\end{array}
$$

Compare the coefficients in this division with those in Example 5.

Study the next example.

Example **5** **Use synthetic division to find $(6x^3 - 19x^2 + x + 6) \div (x - 3)$.**

Write the terms of the polynomial so that the degrees of the terms are in descending order. Then write just the coefficients as shown at the right. *There must be a coefficient for every possible power of the variable.*

$$
\begin{array}{cccc}
6x^3 & -19x^2 & +x & +6 \\
\downarrow & \downarrow & \downarrow & \downarrow \\
6 & -19 & 1 & 6
\end{array}
$$

Write the constant r of the divisor $x - r$ to the left.

$$
\begin{array}{r|rrrr}
3 & 6 & -19 & 1 & 6 \\
\hline
 & 6 & & &
\end{array}
$$

In this case, $r = 3$.
Bring the first coefficient, 6, down as shown.

Multiply the first coefficient by r: $3 \cdot 6 = 18$.
Write the product under the second coefficient.
Then add the product and the second coefficient: $-19 + 18 = -1$.

$$
\begin{array}{r|rrrr}
3 & 6 & -19 & 1 & 6 \\
 & & 18 & & \\
\hline
 & 6 & -1 & &
\end{array}
$$

Multiply the sum, -1, by r: $3(-1) = -3$.
Write the product under the next coefficient and add: $1 + (-3) = -2$.

$$
\begin{array}{r|rrrr}
3 & 6 & -19 & 1 & 6 \\
 & & 18 & -3 & \\
\hline
 & 6 & -1 & -2 &
\end{array}
$$

Multiply the sum, -2, by r: $-2 \cdot 3 = -6$.
Write the product under the next coefficient and add: $6 + (-6) = 0$. The remainder is 0.

$$
\begin{array}{r|rrrr}
3 & 6 & -19 & 1 & 6 \\
 & & 18 & -3 & -6 \\
\hline
 & 6 & -1 & -2 & 0
\end{array}
$$

Writing the quotient is easy. The numbers along the bottom row are the coefficients of the powers of x in descending order. Start with the power that is one less than that of the dividend. Thus, the result of this division is $6x^2 - x - 2$.

Check this result. Does $(x - 3)(6x^2 - x - 2) = 6x^3 - 19x^2 + x + 6$?

To use synthetic division, the divisor must have a leading coefficient of 1. You can rewrite other division expressions so you can use synthetic division.

Example ⑥ **Use synthetic division to find $(4x^4 - 5x^2 + 2x + 4) \div (2x - 1)$.**

Use division to rewrite the divisor so it has a coefficient of 1.

$$\frac{4x^4 - 5x^2 + 2x + 4}{2x - 1} = \frac{(4x^4 - 5x^2 + 2x + 4) \div 2}{(2x - 1) \div 2} \text{ or } \frac{2x^4 - \frac{5}{2}x^2 + x + 2}{x - \frac{1}{2}}$$

Since the numerator does not contain all powers of x, you must include a 0 coefficient for the x^3 term.

Now use synthetic division.

$$\frac{1}{2} \begin{array}{|ccccc} 2 & 0 & -\frac{5}{2} & 1 & 2 \\ & 1 & \frac{1}{2} & -1 & 0 \\ \hline 2 & 1 & -2 & 0 & | 2 \end{array} \quad x - r = x - \frac{1}{2}$$

The quotient is $2x^3 + x^2 - 2x + \dfrac{2}{x - \frac{1}{2}}$ or $2x^3 + x^2 - 2x + \dfrac{4}{2x - 1}$.

Check: Divide using long division.

$$\begin{array}{r} 2x^3 + x^2 - 2x \\ 2x - 1 \overline{) 4x^4 + 0x^3 - 5x^2 + 2x + 4} \\ \underline{4x^4 - 2x^3} \\ 2x^3 - 5x^2 \\ \underline{2x^3 - x^2} \\ -4x^2 + 2x \\ \underline{-4x^2 + 2x} \\ 0 + 4 \end{array}$$

The quotient is $2x^3 + x^2 - 2x + \dfrac{4}{2x - 1}$. ✓

CHECK FOR UNDERSTANDING

Communicating Mathematics

Study the lesson. Then complete the following.

1. **Show** how you would set up the synthetic division for $(5y^3 + y^2 - 7) \div (y + 1)$.

2. **Illustrate** why it is necessary to include terms with zero coefficients in the row of numbers for synthetic division.

3. **You Decide** Refer to the magician's trick at the beginning of the lesson. Jocelyn says that the answer is always 4, regardless of the number chosen. Marie says that the magician knows a special way to compute numbers quickly and the answer depends on the number chosen. Who is correct, and why?

MATH JOURNAL

4. **Assess Yourself** Compare the long division method of dividing polynomials with synthetic division. Are there any advantages to either method? Which do you prefer and why?

Guided Practice

Simplify.

5. $\dfrac{5xy^2 - 4xy + 7x^2y}{xy}$

6. $(6xy^2 - 3xy + 2x^2y)(xy)^{-1}$

7. $(a^2 - 10a - 24) \div (a + 2)$

8. $(9b^2 + 9b - 10) \div (3b - 2)$

9. $(a^3 + b^3) \div (a + b)$

10. $(y^5 - 3y^2 - 20) \div (y - 2)$

Use synthetic division to find each quotient.

11. $(3x^4 - 6x^3 - 2x^2 + x - 6) \div (x + 1)$

12. $(t^4 - 2t^3 + t^2 - 3t + 2)(t - 2)^{-1}$

13. $(12x^2 + 36x + 15) \div (6x + 3)$

14. $(x^3 + 13x^2 - 12x - 8) \div (x + 2)$

EXERCISES

Practice

Simplify.

15. $\dfrac{8x^2y^3 - 28x^3y^2}{4xy^2}$

16. $\dfrac{2mn^3 + 4m^2 - 9m^3n^2}{mn}$

17. $(12rs^3 + 9r^2s^2 - 15r^2s) \div (3rs)$

18. $(28k^3p - 42kp^2 + 56kp^3) \div (14kp)$

19. $(a^3b^2 - a^2b + 2a)(-ab)^{-1}$

20. $(b^3 + 8b^2 - 20b) \div (b - 2)$

21. $(x^2 - 12x - 45) \div (x + 3)$

22. $(n^3 + 2n^2 - 5n + 12) \div (n + 4)$

23. $(g^2 + 8g + 15)(g + 3)^{-1}$

24. $(2b^3 + b^2 - 2b + 3)(b + 1)^{-1}$

25. $(6t^3 + 5t^2 + 9) \div (2t + 3)$

26. $(50a^2 - 98b^2) \div (10a + 14b)$

27. $(5y^3 + y^2 - 7) \div (y + 1)$

28. $(2h^3 - 5h^2 + 22h + 51) \div (2h + 3)$

29. $(2y^2 + y - 16) \div (y - 3)$

30. $(x^3 - 4x^2) \div (x - 4)$

31. $(x^3 - 27) \div (x - 3)$

32. $(8x^3 - 1) \div (2x - 1)$

Use synthetic division to find each quotient.

33. $\dfrac{9d^3 + 5d - 8}{3d - 2}$

34. $\dfrac{m^3 - 7m + 3m^2 - 21}{m + 3}$

35. $(2c^3 - 3c^2 + 3c - 4) \div (c - 2)$

36. $(2x^3 - x^2 + 5x - 12) \div (2x - 3)$

37. $(w^2 - w^3)(w - 1)^{-1}$

38. $(6w^5 - 18w^2 - 120) \div (w - 2)$

39. $\dfrac{2m^4 - 5m^3 - 10m + 8}{m - 3}$

40. $\dfrac{a^4 - 5a^3 - 13a^2 + 53a + 60}{a + 1}$

41. $(y^5 + 32)(y + 2)^{-1}$

42. $(t^5 - 3t^2 - 20)(t - 2)^{-1}$

43. Solve $(b + 1)y = 2b^3 + b^2 - 2b + 3$ for y.

Critical Thinking

44. Use synthetic division to find each quotient. Then evaluate the dividend for the value stated.

    a. $\dfrac{3x^3 - 5x - 2}{x - 2}, x = 2$

    b. $\dfrac{2x^4 - 3x^2 + 1}{x + 1}, x = -1$

    c. Study the results of parts a and b. If the divisor represents $x - r$ and the dividend is $f(x)$, how is the division $\dfrac{f(x)}{x - r}$ related to $f(r)$?

45. Suppose the quotient resulting from dividing one polynomial by another is $r^2 - 6r + 9 - \dfrac{1}{r - 3}$. What two polynomials were divided?

46. The graphing calculator program below uses synthetic division to compute the coefficients of the quotient and the remainder when a polynomial is divided by a linear binomial. When prompted by a question mark, input the degree of the polynomial (N), the constant R for r in the divisor $x - r$, and the coefficients (C) of the polynomial. Press ENTER after each entry.

```
PROGRAM: SYNDIV              : C →L1(X)
: ClrHome                    : End
: ClrList (L1,L2)            : L1(1)→L2(1)
: Disp "LET N = DEGREE",     : Disp "COEFFICIENTS OF",
  "OF POLYNOMIAL"              "QUOTIENT ARE"
: Disp "LET R = CONSTANT",   : For (X,1,N−1,1)
  "IN DIVISOR X-R"           : L1(X+1)+R*L2(X)→
: Prompt N,R                   L2(X+1)
: Disp "ENTER COEFFICIENTS"  : End
: For (X,1,N+1,1)            : Disp L2
: Prompt C                   : Disp "REMAINDER",
                               L1(N+1)+R*L2(N)
```

Use the program to find each quotient.

a. $(2x^5 + 3x^4 - 6x^3 + 6x^2 - 8x + 3) \div (x + 1)$
b. $(a^4 + 2a^3 - 7a^2 + 2a - 8) \div (a - 3)$
c. $(x^5 - 3x^2 - 20) \div (x - 2)$

47. Manufacturing A machinist who makes square metal pipes found a formula for the amount of metal she needed to make a pipe. She found that to make a pipe $8x$ inches long, she needed $32x^2 + x$ square inches of metal. In figuring the area needed, the machinist allowed some fixed length of metal for overlap of the seam. If the width of the finished pipe will be x inches, how much did the machinist leave for the seam?

Metal Needed

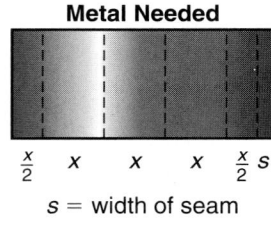

$\frac{x}{2}$ x x x $\frac{x}{2}$ s

s = width of seam

Finished Pipe

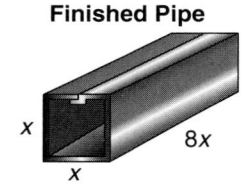

48. Health In the 1995 film *Outbreak,* a deadly disease threatens to wipe out the human race. Outbreaks of disease can become epidemics if there are no health preventatives available. The number of students and teachers at a large high school who will catch the flu during an outbreak in a certain year can be estimated by the formula $n = \dfrac{170t^2}{t^2 + 1}$, where t is the number of weeks from the beginning of the epidemic and n is the number of people who have become ill.

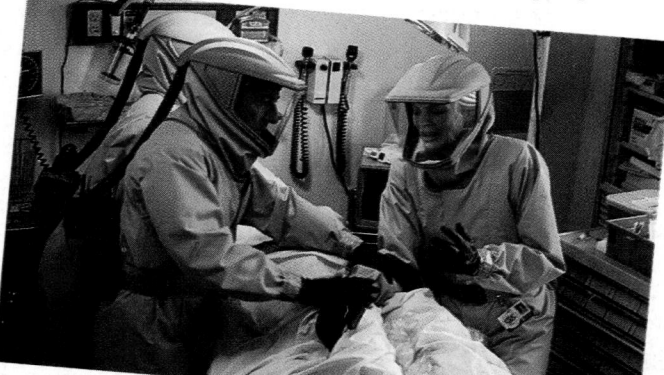

a. Perform the division indicated by $\dfrac{170t^2}{t^2 + 1}$.

b. Use the formula to estimate how many people will become ill during the first week.

c. Use your calculator to evaluate increasingly greater values of t. What happens to the value of n as t becomes very great?

Mixed Review

49. Find $(a - b)^2$. (Lesson 5–2)

50. **Astronomy** Earth is an average 1.496×10^8 kilometers from the sun. If light travels 3×10^5 kilometers per second, how long does it take sunlight to reach Earth? (Lesson 5–1)

51. Solve the system of equations by using augmented matrices. (Lesson 4–7)
$$a + b + c = -2$$
$$2a - 3b + c = -11$$
$$-a + 2b - c = 8$$

52. Find $\begin{bmatrix} -2 & 1 \\ 3 & -6 \\ 4 & 5 \end{bmatrix} \cdot \begin{bmatrix} 1 & 2 & -3 & 7 \\ -3 & 2 & 9 & -1 \end{bmatrix}$. (Lesson 4–3)

53. Given $f(x, y) = 5x - 2y$, find $f(4, 1)$. (Lesson 3–5)

54. **Photography** The perimeter of a rectangular picture is 86 inches. Twice the width exceeds the length by 2 inches. What are the dimensions of the picture? (Lesson 3–2)

55. Find $h\left(-\dfrac{1}{2}\right)$ if $h(x) = [3x + 7]$. (Lesson 2–6)

56. Write an equation in slope-intercept form for the line that has a slope of -2 and passes through the point at $(5, -3)$. (Lesson 2–4)

57. Solve $|x - 7| = 13$. (Lesson 1–6)

WORKING ON THE In·ves·ti·ga·tion

Refer to the Investigation on pages 180–181.

You are interested in using mathematics to further evaluate the performance of the launchers during the shoot-off. There are four measures you can use to help calculate accuracy around the given target.

Error Range The longest shot minus the shortest shot is the error range.

Tolerance All data must lie within a tolerance distance around the target (target distance ± tolerance), Therefore, the tolerance is the distance between the target distance and the farthest shot from the target.

Relative Error The ratio of the tolerance to the target distance. This is expressed as a percent.

Average Error Using the target distance and each of the distance shots, find the absolute value of the difference between each of these values. Calculate the mean of those differences, by adding all the absolute differences and dividing the sum by 10 shots.

1. Find these four measures for each of the launchers used in your class' shoot-off.

2. Compare these measures for each of the launchers. How do these four measures help to determine which launcher was most accurate? What do each of these measures represent when analyzing the data? Does a higher number or a lower number show more accuracy for each of the measures? Explain for each measure.

3. Consider all the data. From that information, which launcher would you consider to be the most accurate? Explain your reasoning.

Add the results of your work to your Investigation Folder.

Factoring

What YOU'LL LEARN

- To factor polynomials, and
- to use factoring to simplify polynomial quotients.

Why IT'S IMPORTANT

You can use factoring to solve problems involving history and geometry.

CONNECTION

History

You would think that someone would have found a formula for computing all the possible prime numbers, but no one has. A **prime number** is a whole number greater than 1 whose only **factors** are 1 and itself. In 1963, a computer at the University of Illinois calculated the largest prime number known at that time. In honor of this accomplishment, the postage meter at the mathematics department printed the number on their mail in 1968.

As of 1993, the largest prime number discovered was $391,581 \times 2^{216,193} - 1$. It was generated by a computer at the Amdahl Corporation in 1989 and has 65,087 digits.

Prime numbers can be used to factor whole numbers, as in the example $36 = 2 \cdot 2 \cdot 3 \cdot 3$. Many polynomials can also be factored. Their factors, however, are other polynomials. Polynomials that cannot be factored are called *prime*. The table below summarizes the most common factoring techniques used with polynomials.

Any Number of Terms	
Greatest Common Factor (GCF)	$a^3b^2 + 2a^2b - 4ab^2 = ab(a^2b + 2a - 4b)$
Two Terms	
Difference of Two Squares **Sum of Two Cubes** **Difference of Two Cubes**	$a^2 - b^2 = (a + b)(a - b)$ $a^3 + b^3 = (a + b)(a^2 - ab + b^2)$ $a^3 - b^3 = (a - b)(a^2 + ab + b^2)$
Three Terms	
Perfect Square Trinomials	$a^2 + 2ab + b^2 = (a + b)^2$ $a^2 - 2ab + b^2 = (a - b)^2$
General Trinomials	$acx^2 + (ad + bc)x + bd = (ax + b)(cx + d)$
Four or More Terms	
Grouping	$ra + rb + sa + sb = r(a + b) + s(a + b)$ $= (r + s)(a + b)$

Whenever you factor a polynomial, always look for a common factor first. Then determine if the resulting polynomial factor can be factored again using one or more of the methods listed on the previous page.

Example 1 Factor $5k^3p - 3kp^2 + k^3p^5$.

$$5k^3p - 3kp^2 + k^3p^5 = (5 \cdot k \cdot k \cdot k \cdot p) - (3 \cdot k \cdot p \cdot p) + (k \cdot k \cdot k \cdot p \cdot p \cdot p \cdot p \cdot p)$$
$$= (kp \cdot 5k^2) - (kp \cdot 3p) + (kp \cdot k^2p^4)$$
$$= kp(5k^2 - 3p + k^2p^4)$$

The GCF is kp. The remaining polynomial is not factorable.

Check this result by finding the product.

The GCF is also used in grouping to factor a polynomial of four or more terms.

Example 2 Factor $b^3 - 3b^2 + 4b - 12$.

$$b^3 - 3b^2 + 4b - 12 = (b^3 - 3b^2) + (4b - 12) \quad \textit{Group to find a GCF.}$$
$$= b^2(b - 3) + 4(b - 3) \quad \textit{Factor the GCF of each binomial.}$$
$$= (b - 3)(b^2 + 4) \quad \textit{Distributive property}$$

Check this result by finding the product.

You can use algebra tiles to model factoring a trinomial.

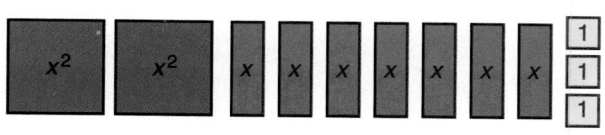

MODELING MATHEMATICS

Factoring Trinomials

Materials: algebra tiles

Use algebra tiles to factor $2x^2 + 7x + 3$.

Your Turn

a. Use algebra tiles to model $2x^2 + 7x + 3$.

b. To find the product that resulted in this polynomial, arrange the tiles to form a rectangle.

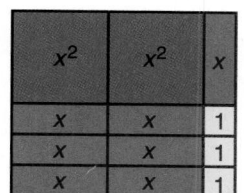

c. Determine the dimensions of the rectangle.

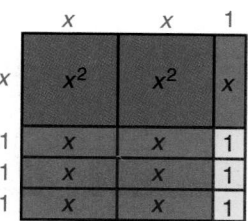

The rectangle is $2x + 1$ units long and $x + 3$ units wide. The area of the rectangle can be expressed as $(2x + 1)(x + 3)$. Since the area of the rectangle is also $2x^2 + 7x + 3$, we can say that $2x^2 + 7x + 3 = (2x + 1)(x + 3)$. *Verify by using FOIL.*

You have used FOIL to multiply two binomials. Thinking about FOIL can help you factor a polynomial into the product of two binomials. Study the following example.

$$(ax + b)(cx + d) = \overset{F}{\overbrace{ax \cdot cx}} + \overset{O}{\overbrace{ax \cdot d}} + \overset{I}{\overbrace{b \cdot cx}} + \overset{L}{\overbrace{b \cdot d}}$$
$$= acx^2 + (ad + bc)x + bd$$

Notice that the product of the *coefficient* of x^2 and the *constant* term is *abcd*. The product of the two coefficients of the *x* terms, *bc* and *ad*, is also *abcd*.

Example **3** **Factor each polynomial.**

a. $7x^2 - 16x + 4$

The product of the coefficient of the first term and the constant term is $7 \cdot 4$ or 28. So the two coefficients of the *x* terms must have a sum of -16 and a product of 28. You may need to use the guess-and-check strategy to find the two coefficients you need. The two coefficients must be -14 and -2 since $(-14)(-2) = 28$ and $-14 + (-2) = -16$.

Rewrite the expression using $-14x$ and $-2x$ in place of $-16x$ and factor by grouping.

$$
\begin{aligned}
7x^2 - 16x + 4 &= 7x^2 - 14x - 2x + 4 &&\textit{Substitute } -14x - 2x \textit{ for } -16x. \\
&= (7x^2 - 14x) + (-2x + 4) &&\textit{Associative property} \\
&= 7x(x - 2) - 2(x - 2) &&\textit{Factor out the GCF of each group.} \\
&= (7x - 2)(x - 2) &&\textit{Distributive property}
\end{aligned}
$$

Check by using FOIL.

b. $2b^2x - 50x$

$$
\begin{aligned}
2b^2x - 50x &= 2x(b^2 - 25) &&\textit{Factor out the GCF.} \\
&= 2x(b + 5)(b - 5) &&b^2 - 25 \textit{ is the difference of two squares.}
\end{aligned}
$$

c. $x^3y^3 + 64$

$x^3y^3 = (xy)^3$ and $64 = 4^3$. Thus, this is the sum of two cubes.
In this example, $a = xy$ and $b = 4$.
$$
\begin{aligned}
x^3y^3 + 64 &= (xy + 4)[(xy)^2 - 4(xy) + 4^2] \\
&= (xy + 4)(x^2y^2 - 4xy + 16)
\end{aligned}
$$

d. $a^6 - b^6$

This polynomial could be considered the difference of two squares or the difference of two cubes. The difference of two squares should always be done before the difference of two cubes.
$$
\begin{aligned}
a^6 - b^6 &= (a^3 + b^3)(a^3 - b^3) &&\textit{Difference of two squares} \\
&= (a + b)(a^2 - ab + b^2)(a - b)(a^2 + ab + b^2) &&\textit{Difference and sum} \\
& &&\textit{of two cubes}
\end{aligned}
$$

Check by using the distributive property.

You can use a graphing calculator to check that the factored form of a polynomial is correct.

EXPLORATION

GRAPHING CALCULATORS

Is the factored form of $2x^2 - 11x - 21$ equal to $(2x - 7)(x + 3)$? You can graph the polynomial and then the factored expression. If the two graphs coincide, the factored form is probably correct.

a. Press $\boxed{Y=}$. Enter $2x^2 - 11x - 21$ for Y1 and $(2x - 7)(x + 3)$ for Y2.

b. Set your viewing window for $[-10, 10]$ by $[-40, 10]$.

c. Press $\boxed{\text{GRAPH}}$. Notice that two different graphs appear. Thus, $2x^2 - 11x - 21 \neq (2x - 7)(x + 3)$.

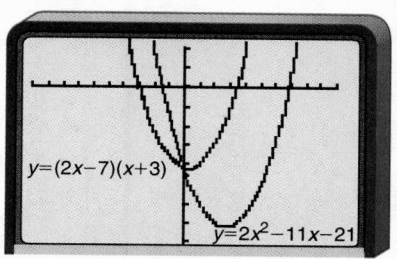

Your Turn

a. Determine if $x^2 + 5x - 6 = (x - 3)(x - 2)$ is a true statement. If it is not correct, state the correct factorization.

b. Why doesn't this method guarantee a way to check the factored form of a polynomial?

You can use this method to check the factoring you do in this lesson.

In Lesson 5–3, you learned to simplify the quotient of two polynomials by using long division or synthetic division. You can determine if one polynomial is a factor of another by using division.

Example **4**
a. Show that $4 - w$ is a factor of $w^3 - 6w^2 + 13w - 20$.
b. Find another factor of $w^3 - 6w^2 + 13w - 20$.

a.
$$
\begin{array}{r}
-w^2 + 2w - 5 \\
-w + 4 \overline{)w^3 - 6w^2 + 13w - 20} \\
\underline{w^3 - 4w^2} \\
-2w^2 + 13w \\
\underline{-2w^2 + 8w} \\
5w - 20 \\
\underline{5w - 20} \\
0
\end{array}
$$
$4 - w = -w + 4$

Since the remainder is 0, $4 - w$ is a factor of $w^3 - 6w^2 + 13w - 20$.

b. The quotient in part a, $-w^2 + 2w - 5$, is another factor since $(4 - w)(-w^2 + 2w - 5) = w^3 - 6w^2 + 13w - 20$. Since $-w^2 + 2w - 5$ is not factorable, these are the only two factors of the polynomial.

Example **5** Find the width of rectangle $ABCD$ if its area is $(3x^2 + 9xy + 6y^2)$ cm^2.

INTEGRATION
Geometry

$A = \ell w$

$w = \dfrac{A}{\ell}$ *Solve for w.*

$w = \dfrac{3x^2 + 9xy + 6y^2}{3x + 6y}$ $A = 3x^2 + 9xy + 6y^2,\ \ell = 3x + 6y$

$w = \dfrac{(3x + 6y)(x + y)}{3x + 6y}$ *Factor the numerator.*

$w = \dfrac{\overset{1}{\cancel{(3x + 6y)}}(x + y)}{\underset{1}{\cancel{3x + 6y}}}$

$w = x + y$

The width of the rectangle is $(x + y)$ cm.

The diagram shows rectangle with top side labeled $(3x + 6y)$ cm, with vertices A (top left), D (top right), B (bottom left), C (bottom right).

CHECK FOR UNDERSTANDING

Communicating Mathematics

Study the lesson. Then complete the following.

1. **Decide** which of the following is the complete factorization of $12x^2 - 8x - 15$.

    a. $(4x + 3)(3x - 5)$ **b.** $(6x - 5)(2x + 3)$

    c. $(6x + 5)(2x - 3)$ **d.** cannot be factored

2. **List** the cubes and squares of the numbers from 1 to 10. Explain how this list might help you in factoring.

3. **Explain** how to completely factor $2x^2 - 6x - 20$.

4. **Show** another way to group the terms to factor the polynomial in Example 2. Was the final result the same?

5. **Factor** the polynomial in Example 3d by using the difference of two cubes. Explain why it is better to factor a difference of squares first.

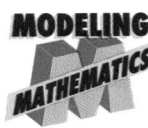
MODELING MATHEMATICS

6. Draw two different geometric models of $2x^2 + 6x$.

    a. What property is shown in each model?

    b. What is the completely-factored form of the polynomial?

Guided Practice

Factor completely. If the polynomial is not factorable, write *prime*.

7. $-15x^2 - 5x$ 8. $m^2 - 6m + 8$

9. $x^2 + xy + 3x$ 10. $16r^2 - 169$

11. $y^2 - 3y - 10$ 12. $a^2 + 5a + 6$

13. $3h^2 - 48$ 14. $2r^3 - 16s^3$

15. $g^3 + 8000$ 16. $21 - 7y + 3x - xy$

17. Determine if $3y - 2$ is a factor of $6y^3 - y^2 - 5y + 2$. If so, what is the other factor?

Practice

Factor completely. If the polynomial is not factorable, write *prime*.

18. $3a^2bx + 15cx^2y + 25ad^3y$

19. $10a^3b - 12a^2b^2$

20. $w^2 + 10w + 9$

21. $16n^2 + 25m^2$

22. $3x^2 - 3y^2$

23. $y^2 - 12y + 20$

24. $12ab^3 - 8a^2b^2 + 10a^5b^3$

25. $y^2 + 7y + 6$

26. $x^2 - 5x + 4$

27. $x^4 - y^2$

28. $6m^2 + 13m + 6$

29. $3n^2 + 21n - 24$

30. $3ay^2 + 9a$

31. $3a^2 - 27b^2$

32. $a^2 + 8ab + 16b^2$

33. $5x - 14 + x^2$

34. $2x^2 + 3x + 1$

35. $5x^2 + 15x - 10$

36. $2a^2 + 13a - 7$

37. $3a^2 + 24a + 45$

38. $12z^2 - z - 6$

39. $m^2n^2 + mn + 1$

40. $8ax - 6x - 12a + 9$

41. $4ax + 14ay - 10bx - 35by$

42. $10w^2 - 14wv - 15w + 21v$

43. $81y^2 - 49$

44. $6a^2 + 27a - 15$

45. $2x^4 + 4x^3 + 2x^2$

46. $m^4 - 1$

47. $y^4 - 16$

48. $7mx^2 + 2nx^2 - 7my^2 - 2ny^2$

49. $8a^2 + 8ab + 8ac + 3a + 3b + 3c$

50. $5a^2x + 4aby + 3acz - 5abx - 4b^2y - 3bcz$

51. $3x^3 + 2x^2 - 5x + 9x^2y + 6xy - 15y$

52. Determine the value of k so that $x - 4$ is a factor of $x^2 + 8x + k$.

53. Use factoring to simplify $\left(\dfrac{n^2 + 2n - 15}{n^2 + 3n - 10}\right)\left(\dfrac{n^2 - 9}{n^2 - 9n + 14}\right)^{-1}$.

54. One factor of $m^2 - k^2 + 6k - 9$ is $(m - k + 3)$. What is the other factor?

Graphing Calculator

Use a graphing calculator to determine if each polynomial is factored correctly. For each correct factorization, sketch the graph shown on the screen. For each incorrect factorization, write the correct factorization and sketch the graph.

55. $x^2 + 6x + 9 = (x + 3)(x + 3)$

56. $3x^2 + 5x + 2 = (3x + 2)(x + 1)$

57. $x^3 + 8 = (x + 2)(x^2 - x + 4)$

58. $3x^2 - 48 = 3(x + 4)(x - 4)$

59. $2x^2 - 5x - 3 = (x - 1)(2x + 3)$

Critical Thinking

60. Factor $49p^{2n} + 14p^n + 1$.

Applications and Problem Solving

61. Geometry The figure at the right is made of three squares.

   a. Suppose the area of square I is y^2 square units and the perimeter of square II is $4x$ units. Write a polynomial for the area of square III.

   b. Suppose the area of square I is 400 square units and the perimeter of square II is 44 units. Find the area of square III.

62. Flags The largest flag flown from a flag-pole is a Brazilian national flag measuring 100 meters by 70 meters in Brasilia, the capital of Brazil.

a. Find the area and the perimeter of the flag.

b. Suppose a company made a flag 1 meter longer and wider than the Brazilian flag. What would be the area and the perimeter of that flag?

c. Suppose a company makes a flag and increases the length and width by x feet. Find the area and perimeter of that flag. Write your answers as polynomials.

Mixed Review

63. Find $(t^3 - 3t + 2) \div (t + 2)$ by using synthetic division. (Lesson 5–3)

64. Use the FOIL method to find $(2x + 4)(7x - 1)$. (Lesson 5–2)

65. Physics Light from a laser travels about 300,000 kilometers per second. How many kilometers can it travel in a day? Write your answer in scientific notation. (Lesson 5–1)

66. Write the system of equations represented by $\begin{bmatrix} 5 & 4 \\ 3 & -5 \end{bmatrix} \cdot \begin{bmatrix} x \\ y \end{bmatrix} = \begin{bmatrix} -3 \\ -24 \end{bmatrix}$. (Lesson 4–6)

67. Find $-2 \begin{bmatrix} \frac{1}{2} & 3 \\ 5 & 7 \end{bmatrix} + \begin{bmatrix} 1 & 2 \\ 3 & 4 \end{bmatrix}$. (Lesson 4–2)

68. Manufacturing Oaken Treasures makes two different kinds of chairs—rockers and swivels. Work on machines A and B is required to make both kinds of chairs. Machine A can run no more than 20 hours per day. Machine B is limited to 15 hours per day. The chart below shows the amount of time on each machine that is required to make one chair and the profit from each chair.

Chair	Machine A	Machine B	Profit
rocker	2 h	3 h	$12
swivel	4 h	1 h	$10

How many chairs of each kind should Oaken Treasures make each day to maximize their profit? (Lesson 3–6)

69. Solve the system of equations by using Cramer's Rule. (Lesson 3–3)
$2x + 3y - 8 = 0$
$3x + 2y - 17 = 0$

70. Statistics The table below shows the ideal weight for a man for a given height. (Lesson 2–5)

Height (inches)	66	68	70	72	74	76	78
Weight (pounds)	143	153	164	171	183	198	206

a. Use the information in the chart to write a prediction equation for the relationship between a man's height and his ideal weight.

b. Predict the ideal weight for a man who is 71 inches tall.

c. Predict the height of a man who is at his ideal weight of 190 pounds.

71. Write an equation for the line that passes through the point (2, 2) that is perpendicular to the graph of $2x + 3y - 1 = 4$. (Lesson 2–4)

72. Write an algebraic expression to represent *the sum of a number and five times its square*. (Lesson 1–5)

73. State the property illustrated by $(4 + 11)6 = 4(6) + 11(6)$. (Lesson 1–2)

Roots of Real Numbers

APPLICATION
Tourism

The table below lists the five highest mountain peaks in North America, according to the *World Almanac*.

Highest Peaks in North America

Teresa and some of her friends are among a group of students visiting Mexico as a part of their study of Spanish. They are visiting Monte Albán in the Oaxaca Plateau of Mexico, which is near Orizaba. Monte Albán was the ancient religious center of the Zapotec Indians.

From an observation point at Monte Albán, Teresa and her friends can see a nearby river that is about 35 miles away. She guesses that the observation point is about 1000 feet high. Her friends guess the point to be higher than that. The formula for estimating the distance d (in miles) that can be seen from the top of a mountain of height h (in feet) is $d = 1.2\sqrt{h}$. Whose guess will be closer to the actual height of the observation point?

To determine the better estimate, we can rewrite the formula as $\sqrt{h} = \frac{d}{1.2}$. If $d = 35$ miles, then the following is true.

$$\sqrt{h} = \frac{35}{1.2} \qquad \textbf{\textit{Estimate:}} \textit{ Will } \sqrt{h} \textit{ be greater or less than 35?}$$
$$\sqrt{h} = 29.1\overline{6} \qquad \textit{Use a calculator.}$$

Now we need to determine a number h whose square root is $29.1\overline{6}$.

Finding the square root of a number and squaring a number are inverse operations. To find the square root of a number n, you must find a number whose square is n. For example, a square root of 36 is 6 since $6^2 = 36$. Since $(-6)^2 = 36$, -6 is also a square root of 36.

Definition of Square Root	For any real numbers a and b, if $a^2 = b$, then a is a square root of b.

Using the definition for the situation above, $29.1\overline{6}$ is the square root of h. So, $h = (29.1\overline{6})^2$ or about 850.7 feet. Teresa's estimate of the height is fairly close.

Since finding the square root of a number and squaring a number are inverse operations, it makes sense that the inverse of raising a number to the nth power is finding the **nth root** of the number. For example, the table below shows the relationship between raising a number to a power and taking that root of the number.

Powers	Factors	Roots
$a^3 = 64$	$4 \cdot 4 \cdot 4 = 64$	4 is a cube root of 64
$a^4 = 16$	$2 \cdot 2 \cdot 2 \cdot 2 = 16$	2 is a fourth root of 16
$a^5 = 243$	$3 \cdot 3 \cdot 3 \cdot 3 \cdot 3 = 243$	3 is a fifth root of 243
$a^n = b$	$\underbrace{a \cdot a \cdot a \cdot a \cdot \cdots \cdot a}_{n \text{ factors of } a} = b$	a is an nth root of b

The pattern suggests the following formal definition of the nth root.

Definition of nth Root	**For any real numbers a and b, and any positive integer n, if $a^n = b$, then a is an nth root of b.**

$\sqrt[n]{98}$ *is read "the nth root of 98."*

The symbol $\sqrt[n]{}$ indicates an nth root.

$$\overset{index \searrow}{\underset{\nearrow radical\ sign}{\sqrt[n]{98}}} \longleftarrow radicand$$

F Y I

The radical sign was first introduced in 1525 by Christoff Rudolff in his algebra book *Die Cross.* This symbol was probably chosen because it resembled a small *r*, the first letter in the word *radix*, which means root.

Some numbers have more than one real nth root. For example, 49 has two square roots, 7 and -7. When there is more than one real root, the nonnegative root is called the **principal root**. When no index is given, as in $\sqrt{49}$, the radical sign indicates the principal square root. The symbol $\sqrt[n]{b}$ stands for the principal nth root of b. If n is odd and b is negative, there will be no nonnegative root. In this case, the principal root is negative.

$\sqrt{64} = 8$ $\sqrt{64}$ indicates the principal square root of 64.

$-\sqrt{64} = -8$ $-\sqrt{64}$ indicates the opposite of the principal square root of 64.

$\pm\sqrt{64} = \pm 8$ $\pm\sqrt{64}$ indicates both square roots of 64.
              ± means positive or negative.

$\sqrt[3]{-27} = -3$ $\sqrt[3]{-27}$ indicates the principal cube root of -27.

$-\sqrt[4]{16} = -2$ $-\sqrt[4]{16}$ indicates the opposite of the principal fourth root of 16.

The chart below gives a summary of the real nth roots of a number b.

Real nth Roots of b, $\sqrt[n]{b}$, or $-\sqrt[n]{b}$			
n	$b > 0$	$b < 0$	$b = 0$
even	one positive root one negative root	no real roots	one real root, 0
odd	one positive root no negative roots	no positive roots one negative root	

Example ① **Find each root.**

a. $\pm\sqrt{49x^8}$

$\pm\sqrt{49x^8} = \pm\sqrt{(7x^4)^2}$

$\quad\quad\quad\quad = \pm 7x^4$

The square roots of $49x^8$ are $\pm 7x^4$.

b. $-\sqrt{(a^2+1)^4}$

$-\sqrt{(a^2+1)^4} = -\sqrt{[(a^2+1)^2]^2}$

$\quad\quad\quad\quad = -(a^2+1)^2$

The opposite of the principal square root of $(a^2+1)^4$ is $-(a^2+1)^2$.

c. $\sqrt[5]{32x^{10}y^{15}}$

$\sqrt[5]{32x^{10}y^{15}} = \sqrt[5]{(2x^2y^3)^5}$

$\quad\quad\quad\quad = 2x^2y^3$

The principal fifth root of $32x^{10}y^{15}$ is $2x^2y^3$.

d. $\sqrt{-16}$

In this case of $\sqrt[n]{b}$, n is even and b is negative. Thus, $\sqrt{-16}$ has no real root.

When you find the nth root of an even power and an odd power is the result, you must take the absolute value of the result to ensure that the value is nonnegative.

$$\sqrt{(-3)^2} = |-3| \text{ or } 3 \quad\quad \sqrt{(-2)^{10}} = |(-2)^5| \text{ or } 32$$

If the result is an even power or you find the nth root of an odd power, there is no need to take the absolute value. *Why?*

Example ② **Find each root.**

a. $\sqrt[6]{x^6}$

Since $x^6 = x \cdot x \cdot x \cdot x \cdot x \cdot x$, x is the sixth root of x^6. The index is even, so the principal root is nonnegative. However, since x could be negative, we must take the absolute value of x to identify the principal root.

$\sqrt[6]{x^6} = |x|$

b. $\sqrt[4]{16(x+3)^{12}}$

$\sqrt[4]{16(x+13)^{12}} = \sqrt[4]{2^4[(x+3)^3]^4}$

Since the index is even and the power is odd, we must use the absolute value of $(x+3)^3$.

$\sqrt[4]{16(x+3)^{12}} = 2|(x+3)^3|$

LOOK BACK

You can review real and irrational numbers in Lesson 2-8.

Recall that real numbers that cannot be expressed as terminating or repeating decimals are *irrational numbers*. $\sqrt{2}$ and $\sqrt{3}$ are examples of irrational numbers. Decimal approximations for irrational numbers, such as 3.14 for π, are often used in applications. You can use a calculator to find decimal approximations.

Example ③ **Use a calculator to find a decimal approximation for $\sqrt[5]{279}$.**

Use the root key. *It may be a second function key.*

Enter: 279 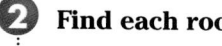 5 = *3.084045954*

Check: 3.084045954 [y^x] 5 = *279* ✓

Example ④

APPLICATION
Fishing

Carla, upon returning from a fishing trip, bragged to her friends that she caught a Pacific halibut that weighed 23 kilograms and was at least 2.5 meters long, but it got away as she was hauling it in. The length-to-weight relationship for Pacific halibut can be estimated by the formula $L = 0.46\sqrt[3]{W}$, where W is the weight in kilograms and L is the length in meters. Is there a possibility that Carla told the truth, or is her story a little "fishy?"

Suppose that Carla is correct about the weight of her fish. Calculate the approximate length of a Pacific halibut that weighs 23 kilograms.

$L = 0.46\sqrt[3]{W}$
$ = 0.46\sqrt[3]{23}$
$ = 0.46(2.84)$
$ \approx 1.3064$ *Use a calculator.*

Carla's fish probably was a little more than 1 meter long. So her story is a little "fishy."

CHECK FOR UNDERSTANDING

Communicating Mathematics

Study the lesson. Then complete the following.

1. **Explain** why it is not always necessary to take the absolute value of a result to indicate the principal root.

2. **Describe** how you would check the nth root of a number using your calculator.

3. **Explain** whether or not $\sqrt[5]{-32}$ is a real number.

4. **Determine** if each statement is true, regardless of the value of x. Explain your answers.

   a. $\sqrt[4]{(-x)^4} = x$ b. $\sqrt[5]{(-x)^5} = x$

MATH JOURNAL

5. Copy and complete the table below. If there are things you do not understand about powers and roots, reread and study the lesson as necessary.

Powers	In words	Roots	In words
$2^3 = 2\cdot2\cdot2 = 8$	2 cubed is 8.	$\sqrt[3]{8} = \sqrt[3]{2\cdot2\cdot2} = 2$	The cube root of 8 is 2.
$8^4 = 4096$	8 to the fourth power is __?__ .		A fourth root of 4096 is 8.
			The fifth root of 16,807 is 7.
		$\sqrt[6]{729} = $ __?__	
$a^n = b$			

Use a calculator to approximate each value to three decimal places.

6. $\sqrt{99}$ 7. $-\sqrt[3]{23}$ 8. $\sqrt[4]{64}$

Simplify.

9. $\sqrt{(-3)^2}$ 10. $\sqrt[3]{27}$ 11. $\sqrt[5]{-32}$

12. $\sqrt[4]{-10,000}$ 13. $\sqrt[3]{m^3}$ 14. $\sqrt[4]{x^4}$

15. $-\sqrt{25x^6}$ 16. $\sqrt{25x^2y^4}$ 17. $\sqrt{(3m-2n)^2}$

18. Find the principal fifth root of 32.

19. Find the principal third root of -125.

EXERCISES

Use a calculator to approximate each value to three decimal places.

20. $\sqrt{121}$ 21. $-\sqrt{144}$ 22. $\sqrt[3]{65}$

23. $\sqrt{0.81}$ 24. $\sqrt[3]{-670}$ 25. $\sqrt[4]{625}$

26. $\sqrt[7]{82,567}$ 27. $\sqrt[6]{(345)^3}$ 28. $\sqrt[4]{(3600)^2}$

Simplify.

29. $\sqrt{16}$ 30. $\pm\sqrt{121}$ 31. $\sqrt{196}$

32. $\sqrt{-(-6)^2}$ 33. $\sqrt[4]{81}$ 34. $-\sqrt{121}$

35. $\sqrt[4]{\left(-\frac{1}{2}\right)^4}$ 36. $\sqrt[3]{-27}$ 37. $\sqrt[3]{1000}$

38. $\sqrt{0.36}$ 39. $\sqrt[3]{-0.125}$ 40. $\sqrt[7]{-1}$

41. $\sqrt[3]{y^3}$ 42. $\sqrt[4]{t^8}$ 43. $-\sqrt[4]{x^4}$

44. $\sqrt{(5f)^4}$ 45. $\sqrt{36g^6}$ 46. $\sqrt{64h^8}$

47. $\sqrt[3]{m^6n^9z^{12}}$ 48. $\sqrt[3]{8b^3c^3}$ 49. $\sqrt[3]{-27a^9b^{12}}$

50. $\pm\sqrt{(a^3+2)^2}$ 51. $\sqrt[3]{(s+t)^3}$ 52. $\sqrt{(3x+y)^2}$

53. $-\sqrt{x^2+2x+1}$ 54. $\sqrt{x^2+6x+9}$ 55. $\pm\sqrt{s^2-2st+t^2}$

56. $\sqrt[5]{-(2m-3n)^5}$ 57. $\sqrt{-4y^2-12y-9}$

58. $\pm\sqrt{16m^2-24mn+9n^2}$

59. Under what condition does $\sqrt{x^2+y^2}=x+y$?

60. **Aerospace Engineering**   Scientists expect that in future space stations, artificial gravity will be created by rotating all or part of the space station. The formula $N=\frac{1}{2\pi}\sqrt{\frac{a}{r}}$ gives the number of rotations N required per second to maintain an acceleration of gravity of a meters per second squared on a satellite with a radius of r meters. The acceleration of gravity of Earth is 9.8 m/s^2. How many rotations per minute will produce an artificial gravity that is equal to half the acceleration of gravity on Earth in a space station with a 25-meter radius?

61. Baseball The Toronto Blue Jays defeated the Philadelphia Phillies four games to two in the 1993 World Series of baseball. During a practice session between games, the catcher for the Blue Jays practiced throwing the baseball to each of the bases. If a standard baseball diamond has sides 90 feet long, how far did the catcher have to throw the ball to reach second base? (*Hint:* Draw a diagram and use the Pythagorean formula $c = \sqrt{a^2 + b^2}$, where c is the length of the hypotenuse of a right triangle and a and b are the lengths of the legs.)

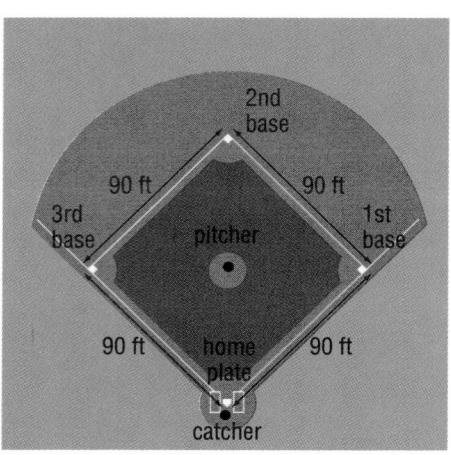

62. Physics Rashad and Alberto are part of the staff in a fun house at their school carnival that is raising money for Muscular Dystrophy. Their job is to drop water-filled balloons from a 25-foot wall on unsuspecting students as they stand on a particular spot in the fun house. The formula for determining the time t it takes a water balloon to reach the ground is $t = \sqrt{\dfrac{2d}{g}}$, where d is the height in feet from which the object is dropped and g is the acceleration due to gravity, which equals 32.2 ft/s^2. Migdalia and Lawanda have told Rashad and Alberto that they can avoid getting hit by a balloon by moving quickly out of the way. How much time do the girls have to avoid getting wet? (Assume very little air friction and the girls are about 5 feet tall.)

63. Simplify $\sqrt{(5b)^4}$. (Lesson 5–5)

64. Simplify $(2p + q^3)^2$. (Lesson 5–2)

65. Simplify $y^2x^{-3}(yx^4 + y^{-1}x^3 + y^{-2}x^2)$. (Lesson 5–2)

66. Astronomy Moonlight takes 1.28 seconds to reach Earth. If the speed of light is 3×10^5 kilometers per second, how far is the moon from Earth? (Lesson 5–1)

67. Solve $\begin{vmatrix} 5 & 7 \\ -2 & 2x \end{vmatrix} = 54$. (Lesson 4–4)

68. Find $\begin{bmatrix} 5 & -2 \\ \frac{1}{2} & -3 \end{bmatrix} \cdot \begin{bmatrix} 2 \\ 1 \end{bmatrix}$. (Lesson 4–3)

69. Graph the system below. Find the maximum and minimum values given the function for the following region. (Lesson 3–5)

$y \geq x$

$y \leq x + 5$

$x \geq -3$

$y + 2x \leq 5$

$f(x, y) = x - 2y$

70. Solve the system of equations by graphing. (Lesson 3–1)

$2x + 3y = -16$

$2y = 4x$

Graph each equation.

71. $b = 2[a] - 3$ (Lesson 2–6)   72. $b = 2a - 3$ (Lesson 2–2)

73. **Time Management**   The chart below illustrates the time in hours per week that unmarried young people between the ages of 12 and 17 and the ages of 18 and 29 spend on various activities. (Lesson 1–4)

Activity	12–17 Years	18–29 Years
Leisure activities:		
Eating	8.1	7.5
Sleeping	62.6	55.9
Attending sports events	0.8	0.3
Attending cultural events	0.3	0.4
Going to movies	0.6	0.7
Visiting, socializing	4.4	7.8
Participating in sports	3.3	1.8
Watching TV	17.7	14.2
Reading	1.3	1.9
Work/school activities:		
Attending classes	18.1	4.1
Cleaning, laundry	2.2	3.2
Doing homework	3.1	3.4
Washing, dressing	6.4	6.8
Working	2.8	29.0
Yardwork, repairs	1.9	1.2

a. Find the mean and median of the leisure activities for both age groups to the nearest tenth.

b. Find the mean and median of the work/school activities for both age groups to the nearest tenth.

c. Does the distribution of the data say anything to you about the priorities of these two groups? What can you say about the differences between the ways the two groups of people spend their time?

SELF TEST

1. **Astronomy**   The distance from Earth to the sun is approximately 93,000,000 miles. Write this number in scientific notation. (Lesson 5–1)

2. Write $8mn^{-5}$ without using negative exponents. (Lesson 5–1)

Simplify. (Lessons 5–1, 5–2, and 5–3)

3. $(-3x^2y)^3(2x)^2$ 4. $(9x + 2y) - (7x - 3y)$

5. $(n + 2)(n^2 - 3n + 1)$ 6. $(2d^4 + 2d^3 - 9d^2 - 3d + 9) \div (2d^2 - 3)$

7. Use synthetic division to find $(m^3 - 4m^2 - 3m - 7) \div (m - 4)$. (Lesson 5–3)

Factor completely. (Lesson 5–4)

8. $ax^2 + 6ax + 9a$ 9. $8r^3 - 64s^6$

Simplify. (Lesson 5–5)

10. $\sqrt[3]{-64a^6b^9}$ 11. $\sqrt{4n^2 + 12n + 9}$

Radical Expressions

5-6

What YOU'LL LEARN

- To simplify radical expressions,
- to rationalize the denominator of a fraction containing a radical expression, and
- to add, subtract, multiply, and divide radical expressions.

Why IT'S IMPORTANT

You can use radical expressions to solve problems involving sports and low enforcement.

APPLICATION

Law Enforcement

The graph at the right illustrates the distance a car travels after the driver sees danger and applies the brakes. These distances are for good weather conditions. If the road is wet or if the car has worn brakes, the car will travel farther than the distances shown.

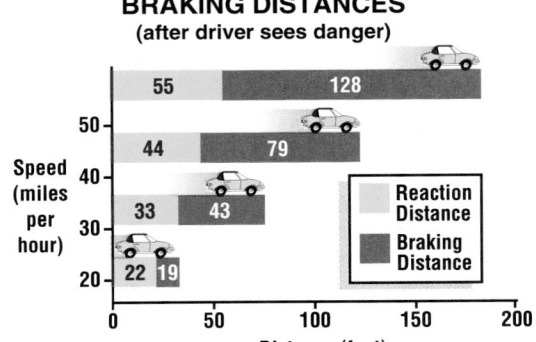

BRAKING DISTANCES
(after driver sees danger)

After an accident, police investigators use the formula $s = 2\sqrt{5\ell}$ to estimate the speed s of a car in miles per hour. The variable ℓ represents the length in feet of the tire skid marks on the pavement. On one occasion, an accident scene investigation team measured skid marks 120 feet long. How fast was the car traveling?

Let's explore two ways of applying the formula.

Method 1

$$s = 2\sqrt{5\ell}$$
$$= 2\sqrt{5 \cdot 120}$$
$$= 2\sqrt{5} \cdot \sqrt{120}$$
$$\approx 2(2.2361)(10.9545)$$
$$\approx 48.99$$

The car was going about 49 miles per hour. Using this method, we first find each of the roots and then multiply.

Method 2

$$s = 2\sqrt{5\ell}$$
$$= 2\sqrt{5 \cdot 120}$$
$$= 2\sqrt{600}$$
$$\approx 2(24.4949)$$
$$\approx 48.99$$

The car was going about 49 miles per hour. Using this method, we find the root of the product.

The result is the same using either method. These examples demonstrate the following property of radicals.

Product Property of Radicals	For any real numbers a and b, and any integer n, $n > 1$, 1. if n is even, then $\sqrt[n]{ab} = \sqrt[n]{a} \cdot \sqrt[n]{b}$ when a and b are both nonnegative, and 2. if n is odd, then $\sqrt[n]{ab} = \sqrt[n]{a} \cdot \sqrt[n]{b}$.

When you simplify a square root, first write the prime factorization of the radicand. Then use the product property to isolate the perfect squares. Then simplify each radical.

Example **Simplify $\sqrt{81p^4q^3}$.**

$$\sqrt{81p^4q^3} = \sqrt{9^2 \cdot (p^2)^2 \cdot q^2 \cdot q} \qquad \text{\textit{Factor into squares if possible.}}$$
$$= \sqrt{9^2} \cdot \sqrt{(p^2)^2} \sqrt{q^2} \cdot \sqrt{q} \quad \text{\textit{Product property of radicals}}$$
$$= 9 \cdot p^2|q|\sqrt{q}$$

However, in order for $\sqrt{q}$ to be defined, q must be positive. Therefore, the absolute value is unnecessary.

$$\sqrt{81p^4q^3} = 9p^2q\sqrt{q}$$

Simplifying nth roots is very similar to simplifying square roots. Find the factors that are nth powers and use the product property.

Example **2** **Simplify $7\sqrt[3]{27n^2} \cdot 4\sqrt[3]{8n}$.**

$$7\sqrt[3]{27n^2} \cdot 4\sqrt[3]{8n} = 7 \cdot 4 \cdot \sqrt[3]{27n^2 \cdot 8n} \qquad \text{\textit{Product property of radicals}}$$
$$= 28 \cdot \sqrt[3]{3^3 \cdot 2^3 \cdot n^3} \qquad \text{\textit{Factor into cubes where possible.}}$$
$$= 28 \cdot \sqrt[3]{3^3} \cdot \sqrt[3]{2^3} \cdot \sqrt[3]{n^3} \quad \text{\textit{Product property of radicals}}$$
$$= 28 \cdot 3 \cdot 2 \cdot n \text{ or } 168n$$

Let's look at the radicals that involve division to see if there is a quotient property of radicals similar to the product property. Consider $\sqrt{\dfrac{25}{16}}$. First simplify the radical using the product property of radicals. Then try what you think the quotient property of radicals might be.

Method 1: Product Property

$$\sqrt{\frac{25}{16}} = \sqrt{25 \cdot \frac{1}{16}}$$
$$= \sqrt{25} \cdot \sqrt{\frac{1}{16}}$$
$$= \sqrt{5^2} \cdot \sqrt{\left(\frac{1}{4}\right)^2}$$
$$= 5 \cdot \frac{1}{4} \text{ or } \frac{5}{4}$$

Method 2: Quotient Property

$$\sqrt{\frac{25}{16}} = \frac{\sqrt{25}}{\sqrt{16}}$$
$$= \frac{5}{4}$$

These results suggest that there is a quotient property of radicals.

Quotient Property of Radicals	**For real numbers a and b, $b \neq 0$, and any integer n, $n > 1$,** $\sqrt[n]{\dfrac{a}{b}} = \dfrac{\sqrt[n]{a}}{\sqrt[n]{b}}$, **if all roots are defined.**

A radical expression is simplified when the following conditions are met.

- The index, n, is as small as possible.
- The radicand contains no factors (other than 1) that are nth powers of an integer or polynomial.
- The radicand contains no fractions.
- No radicals appear in the denominator.

To eliminate radicals from a denominator or fractions from a radicand, you can use a process called **rationalizing the denominator**. To rationalize a denominator, you must multiply the numerator and denominator by a quantity so that the radicand has an exact root. Study the examples below.

Example **Simplify each expression.**

a. $\dfrac{\sqrt{b^4}}{\sqrt{a^3}}$

$$\dfrac{\sqrt{b^4}}{\sqrt{a^3}} = \dfrac{\sqrt{b^2 \cdot b^2}}{\sqrt{a^2 \cdot a}}$$

$$= \dfrac{\sqrt{b^2} \cdot \sqrt{b^2}}{\sqrt{a^2} \cdot \sqrt{a}}$$

$$= \dfrac{b \cdot b}{a \cdot \sqrt{a}}$$

$$= \dfrac{b^2}{a\sqrt{a}} \cdot \dfrac{\sqrt{a}}{\sqrt{a}} \qquad \text{\textit{Rationalize the denominator.}}$$

$$= \dfrac{b^2\sqrt{a}}{a\sqrt{a^2}} \text{ or } \dfrac{b^2\sqrt{a}}{a^2}$$

b. $\sqrt[5]{\dfrac{3}{4s^2}}$

$$\sqrt[5]{\dfrac{3}{4s^2}} = \dfrac{\sqrt[5]{3}}{\sqrt[5]{4s^2}} \cdot \dfrac{\sqrt[5]{8s^3}}{\sqrt[5]{8s^3}} \qquad \text{\textit{Why use }} \dfrac{\sqrt[5]{8s^3}}{\sqrt[5]{8s^3}} \text{ ?}$$

$$= \dfrac{\sqrt[5]{24s^3}}{\sqrt[5]{2^5 s^5}}$$

$$= \dfrac{\sqrt[5]{24s^3}}{2s} \qquad 4 = 2^2, \ 8 = 2^3$$

Can you add radicals in the same way you multiply them? In other words, if $\sqrt{a} \cdot \sqrt{a} = \sqrt{a \cdot a}$, does $\sqrt{a} + \sqrt{a} = \sqrt{a + a}$?

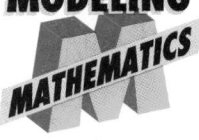

MODELING MATHEMATICS

Adding Radicals

Materials: geometric dot paper straightedge

You can use geometric dot paper to show the sum of two like radicals, such as $\sqrt{2} + \sqrt{2}$. Is $\sqrt{2} + \sqrt{2} = \sqrt{2 + 2}$ or 2?

a. First find the length $\sqrt{2}$ units by using the Pythagorean theorem with the dot paper.

$$a^2 + b^2 = c^2$$
$$1^2 + 1^2 = c^2$$
$$2 = c^2$$
$$\sqrt{2} = c$$

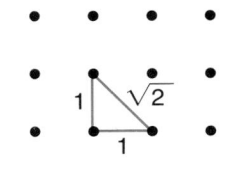

b. Extend the segment to a length twice the length of $\sqrt{2}$ to represent $\sqrt{2} + \sqrt{2}$.

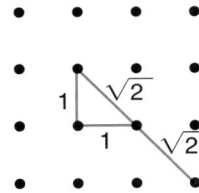

c. Is $\sqrt{2} + \sqrt{2} = \sqrt{2 + 2}$ or 2? Justify your answer.

Your Turn

Try this method to model other irrational numbers.

In the activity on the previous page, you discovered that you do not add radicals in the same manner as you multiply them. You add radicals in the same manner as adding monomials. That is, you can add only like terms or like radicals.

Two radical expressions are called **like radical expressions** if both the indices and the radicands are alike. Some examples of like and unlike radical expressions are given below.

$\sqrt[3]{2}$ and $\sqrt{2}$ are not like expressions.   *Different indices*

$\sqrt[4]{3a}$ and $\sqrt[4]{3}$ are not like expressions   *Different radicands*

$3\sqrt[4]{2x}$ and $4\sqrt[4]{2x}$ are like expressions.   *Radicands are 2x; indices are 4.*

Example **Simplify $2\sqrt{20} - 2\sqrt{45} + 3\sqrt{80}$.**

$2\sqrt{20} - 2\sqrt{45} + 3\sqrt{80}$

$= 2\sqrt{2^2 \cdot 5} - 2\sqrt{3^2 \cdot 5} + 3\sqrt{2^2 \cdot 2^2 \cdot 5}$ *Rewrite each radicand*
$= 2\sqrt{2^2} \cdot \sqrt{5} - 2\sqrt{3^2} \cdot \sqrt{5} + 3\sqrt{2^2} \cdot \sqrt{2^2} \cdot \sqrt{5}$ *using its factors.*
$= 2 \cdot 2\sqrt{5} - 2 \cdot 3\sqrt{5} + 3 \cdot 2 \cdot 2\sqrt{5}$
$= 4\sqrt{5} - 6\sqrt{5} + 12\sqrt{5}$ *These are like radicals.*
$= 10\sqrt{5}$

Example **5**

Sports

As part of the tryout for a girls' softball team, each player must hit a series of balls in a batting cage. Ms. Johnson, the coach, determines the velocity of each hit with a speed gun. She uses the formula $d = v\sqrt{\dfrac{h}{4.9}}$ to estimate the distance the ball would have traveled if the ball had been hit in an open field. In the formula, v represents the velocity (in meters per second) of the baseball, and h is the height (in meters) from which the ball is hit. Zenobia hit two balls with speeds of 45 m/s and 47 m/s.

If her bat is at a height of 0.8 meters from the ground, what is the difference between the distances these baseballs would have traveled?

Use the formula to express the differences of the two distances. Let D represent the difference of the distances, v_1 represent the velocity of the first hit, and v_2 represent the velocity of the second hit.

$D = v_2\sqrt{\dfrac{h}{4.9}} - v_1\sqrt{\dfrac{h}{4.9}}$

$= 47\sqrt{\dfrac{0.8}{4.9}} - 45\sqrt{\dfrac{0.8}{4.9}}$ *$h = 0.8$, $v_1 = 45$, and $v_2 = 47$*

$= (47 - 45)\sqrt{\dfrac{0.8}{4.9}}$ *Combine like radical expressions.*

$= 2\sqrt{\dfrac{0.8}{4.9}}$ *Estimate: Will the result be less than or greater than 2?*

Use a calculator to find an approximate value for this expression.

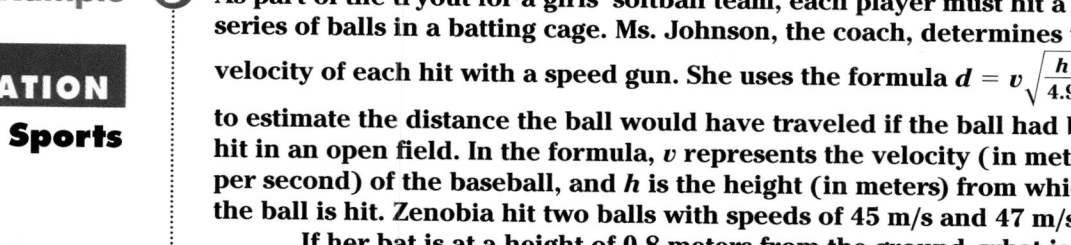

Enter: 2 ☒ ⟮ 0.8 ÷ 4.9 ⟯ √x̄ ═ *0.808122035*

There is a difference of about 0.8 meters between the two hits.

Just as you can add and subtract radicals like monomials, you can multiply radicals using FOIL like you multiply binomials.

Example 6 Simplify each expression.

a. $(2\sqrt{6} - 3\sqrt{2})(3 + \sqrt{2})$

$$\begin{array}{cccc} F & O & I & L \end{array}$$

$$(2\sqrt{6} - 3\sqrt{2})(3 + \sqrt{2}) = 2\sqrt{6} \cdot 3 + 2\sqrt{6} \cdot \sqrt{2} - 3\sqrt{2} \cdot 3 - 3\sqrt{2} \cdot \sqrt{2}$$
$$= 6\sqrt{6} + 2\sqrt{2^2 \cdot 3} - 9\sqrt{2} - 3\sqrt{2^2}$$
$$= 6\sqrt{6} + 4\sqrt{3} - 9\sqrt{2} - 6$$

b. $(3\sqrt{5} - 4)(3\sqrt{5} + 4)$

$$(3\sqrt{5} - 4)(3\sqrt{5} + 4) = 3\sqrt{5} \cdot 3\sqrt{5} + 3\sqrt{5} \cdot 4 - 4 \cdot 3\sqrt{5} - 4 \cdot 4$$
$$= 9\sqrt{5^2} + 12\sqrt{5} - 12\sqrt{5} - 16$$
$$= 45 - 16$$
$$= 29$$

Binomials like those in Example 6b, of the form $a\sqrt{b} + c\sqrt{d}$ and $a\sqrt{b} - c\sqrt{d}$ where a, b, c, and d are rational numbers, are called **conjugates** of each other. The product of conjugates is always a rational number. We can use conjugates to rationalize denominators.

Example 7 Simplify $\dfrac{\sqrt{2} - 3}{\sqrt{2} + 7}$.

$$\frac{\sqrt{2} - 3}{\sqrt{2} + 7} = \frac{(\sqrt{2} - 3)(\sqrt{2} - 7)}{(\sqrt{2} + 7)(\sqrt{2} - 7)} \quad \textit{Multiply by } \frac{\sqrt{2} - 7}{\sqrt{2} - 7} \textit{ because } \sqrt{2} - 7 \textit{ is}$$
$$= \frac{(\sqrt{2} - 3)(\sqrt{2} - 7)}{(\sqrt{2})^2 - 49} \quad \textit{the conjugate of } \sqrt{2} + 7.$$
$$= \frac{(\sqrt{2})^2 + \sqrt{2}(-7) + (-3)\sqrt{2} + (-3)(-7)}{-47}$$
$$= \frac{2 - 7\sqrt{2} - 3\sqrt{2} + 21}{-47}$$
$$= \frac{23 - 10\sqrt{2}}{-47}$$

CHECK FOR UNDERSTANDING

Communicating Mathematics

Study the lesson. Then complete the following.

1. **Explain** how you would rationalize the denominator of $\dfrac{1}{\sqrt[3]{a^2b^3c}}$.

2. **Explain** why $7\sqrt{3} - 4\sqrt[3]{3} + 2\sqrt[3]{4}$ cannot be simplified further.

3. **Write** examples for each of the following.

   a. two like radical expressions

   b. two unlike radical expressions

   c. two expressions that are conjugates of each other

4. **Explain** why the product of two conjugates is always a rational number.

5. Why is the product property of radicals for odd indices different than the product property for even indices?

6. **Replace** the $\underline{\ ?\ }$ in $\sqrt[n]{\dfrac{1}{c}}\ \underline{\ ?\ }\ \dfrac{1}{\sqrt[n]{c}}$ with an $=$ or $\neq$ to make the statement true. Write a reason for your answer.

7. Use geometric dot paper to draw segments that represent each length.
   a. $\sqrt{10}$ b. $\sqrt{13}$ c. $\sqrt{17}$ d. $\sqrt{29}$

8. Describe how you could use models to show the sum $\sqrt{5} + \sqrt{10}$. Include a drawing in your explanation.

Guided Practice

Simplify.

9. $\sqrt{80}$

10. $4\sqrt{54}$

11. $\sqrt[3]{64x^6y^3}$

12. $\sqrt[4]{81m^4n^5}$

13. $(7\sqrt{6})(-3\sqrt{10})$

14. $\sqrt{3x^2z^3} \cdot \sqrt{15x^2z}$

15. $\dfrac{\sqrt[3]{81}}{\sqrt[3]{9}}$

16. $\sqrt{\dfrac{5}{12a}}$

17. $\sqrt{\dfrac{y^2}{y-3}}$

18. $\sqrt{2} + 5\sqrt[3]{2} + 7\sqrt{2} - 4\sqrt[3]{2}$

19. $5\sqrt[3]{135} - 2\sqrt[3]{81}$

20. $(5 + \sqrt{3})(2 - \sqrt{2})$

21. $(7 + \sqrt{11y})(7 - \sqrt{11y})$

22. **Geometry**   Find the perimeter and area of the rectangle shown at the right.

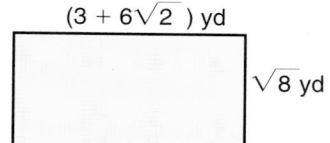

EXERCISES

Practice

Simplify.

23. $\sqrt{32}$

24. $5\sqrt{50}$

25. $\sqrt[3]{16}$

26. $\sqrt{98y^4}$

27. $\sqrt[3]{32}$

28. $\sqrt[4]{48}$

29. $\sqrt{y^3}$

30. $\sqrt{8a^2b^3}$

31. $5\sqrt{3} - 4\sqrt{3}$

32. $8\sqrt[3]{6} + 3\sqrt[3]{6}$

33. $\sqrt{90x^3y^4}$

34. $3\sqrt[3]{56a^6b^3}$

35. $(-3\sqrt{24})(5\sqrt{20})$

36. $\sqrt{26} \cdot \sqrt{39} \cdot \sqrt{14}$

37. $(4\sqrt{18})(2\sqrt{14})$

38. $8\sqrt{y^2} + 7\sqrt{y} - 4\sqrt{y}$

39. $\sqrt[3]{40} - 2\sqrt[3]{5}$

40. $(\sqrt{10} - \sqrt{6})(\sqrt{5} + \sqrt{3})$

41. $-3\sqrt{7}(2\sqrt{14} + 5\sqrt{2})$

42. $\sqrt{a}(\sqrt{b} + \sqrt{ab})$

43. $8\sqrt[3]{2x} + 3\sqrt[3]{2x} - 8\sqrt[3]{2x}$

44. $5\sqrt{20} + \sqrt{24} - \sqrt{180} + 7\sqrt{54}$

45. $(6 - \sqrt{2})(6 + \sqrt{2})$

46. $(5 + \sqrt{6})(5 - \sqrt{2})$

47. $(\sqrt{3} - \sqrt{5})^2$

48. $(x + \sqrt{y})^2$

49. $\sqrt{98} - \sqrt{72} + \sqrt{32}$

50. $\sqrt[4]{a^2} + \sqrt[4]{a^6}$

51. $\sqrt{\dfrac{a^4}{b^3}}$

52. $\sqrt[4]{\dfrac{2}{3}}$

Simplify.

53. $\sqrt{\dfrac{2}{5}} + \sqrt{40} + \sqrt{10}$

54. $\dfrac{7}{4 - \sqrt{3}}$

55. $\dfrac{\sqrt{6}}{5 + \sqrt{3}}$

56. $\dfrac{2 + \sqrt{6}}{2 - \sqrt{6}}$

57. $\dfrac{\sqrt{x+1}}{\sqrt{x-1}}$

58. $\dfrac{1}{\sqrt{x^2 - 1}}$

59. $\sqrt[4]{x^4} + \sqrt[3]{x^6} + \sqrt{x^8}$

60. $(4\sqrt{5} - 3\sqrt{2})(2\sqrt{5} + 2\sqrt{2})$

Geometry

61. Use the Pythagorean theorem to find the length of the hypotenuse of the right triangle shown at the right.

62. Find the radius, r, of a sphere whose surface area S is 2464 square inches. Use the formula $r = \dfrac{1}{2}\sqrt{\dfrac{S}{\pi}}$.

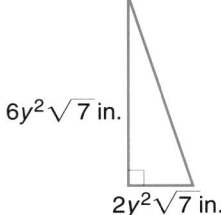

$6y^2\sqrt{7}$ in.

$2y^2\sqrt{7}$ in.

Critical Thinking

63. Under what conditions is the equation $\sqrt{x^3 y^2} = xy\sqrt{x}$ true?

64. Under what conditions is $\sqrt[n]{(-x)^n} = x$?

Applications and Problem Solving

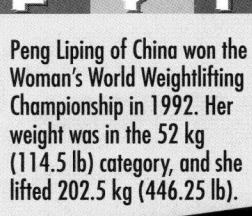

Peng Liping of China won the Woman's World Weightlifting Championship in 1992. Her weight was in the 52 kg (114.5 lb) category, and she lifted 202.5 kg (446.25 lb).

65. **Sports** Akikta and Francisco are in a weightlifting competition. They weigh 70 kg and 110 kg, respectively. Akikta's final and best lift is 190 kg, and Francisco's is 240 kg. The judges will use O'Carroll's formula for determining the superior weight lifter. The formula, $W = \dfrac{w}{\sqrt[3]{b - 35}}$, involves the weight b of the lifter and the weight w lifted. W represents the handicapped weight.

   a. Calculate each lifter's rating.

   b. Who will be the superior weight lifter, Akikta or Francisco?

66. **Physics** Find the time that it takes a pendulum to complete a swing if its length is 10 inches. Use the formula $T = 2\pi\sqrt{\dfrac{L}{384}}$, where T represents time in seconds, and L represents the length of the pendulum in inches.

67. **Automotive Engineering** An automotive engineer is trying to design a safer car. The maximum force a road can exert on the tires of the car being redesigned is 2000 pounds. What is the maximum velocity v in ft/s at which this car can safely round a turn of radius 320 ft? Use the formula $v = \sqrt{\dfrac{F_c r}{100}}$, where F_c is the force the road exerts on the car and r is the radius of the turn.

Mixed Review

68. Simplify $\sqrt{(y + 2)^2}$. (Lesson 5–5)

69. Factor $a + b + 3a^2 - 3b^2$. (Lesson 5–4)

70. Find $(n^4 - 8n^3 + 54n + 105) \div (n - 5)$ by using synthetic division. (Lesson 5–3)

71. Simplify $(x^3 - 3x^2y + 4xy^2 + y^3) - (7x^3 + x^2y - 9xy^2 + y^3)$. (Lesson 5–2)

72. Chemistry Wavelengths of light are measured in Angstroms. An Angstrom is 10^{-8} centimeter. The wavelength of cadmium's green line is 5085.8 Angstroms. How many wavelengths of cadmium's green line are in one meter? (Lesson 5–1)

73. Find $\dfrac{2}{3}\begin{bmatrix} 9 & 0 \\ 12 & 15 \end{bmatrix} + \begin{bmatrix} -2 & 3 \\ -7 & -7 \end{bmatrix}$. (Lesson 4–2)

74. Solve the system of equations. (Lesson 3–7)

$x - 2y + z = -9$

$2y + 3z = 16$

$2y = 4$

75. Use Cramer's rule to solve the system of equations. (Lesson 3–3)

$s + t = 5$

$3s - t = 3$

76. Graph $y + 3x > -1$. (Lesson 2–7)

77. Find $f(-3)$ if $f(x) = x^2 - 3x - 9$. (Lesson 2–1)

78. Transportation A San Antonio parking garage charges $1.50 for the first hour and $0.50 for each additional hour or part of an hour. For how many hours can you park your car if you only have $4.50? (Lesson 1–7)

79. Evaluate $\dfrac{3ab}{cd}$ if $a = 3$, $b = 7$, $c = -2$, and $d = 0.5$. (Lesson 1–1)

Divine Mathematics

Henry Wadsworth Longfellow (1807–1882) was one of America's most outstanding poets. He also was a lover of mathematics. Through one of the characters in his book *Kavanagh: A Tale*, published in 1965, Longfellow said the following.

HOW DULL AND PROSAIC THE STUDY OF mathematics is made in our schoolbooks; as if the grand science of numbers has been discovered and perfected merely to further the purpose of trade. There is something divine in the science of numbers . . . It holds the sea in the hollow of its hand. It measures the earth; it weighs the stars; it illumines the universe; it is law, it is order, it is beauty. And yet we imagine—that is, most of us—that its highest end and culminating point is bookkeeping by double entry. It is our way of teaching it which makes it so prosaic. ■

1. Explain in your own words what Longfellow was trying to say in this passage.

2. Longfellow enjoyed including mathematical problems in his poetry and prose. See if you can solve the following problems posed by Longfellow.

   a. "A tree one hundred cubits high is distant from a well two hundred cubits; from this tree one monkey descends and goes to the well; another monkey takes a leap upwards, and then descends by the hypotenuse; and both pass over an equal space." What is the height of the leap?

   b. "Ten times the square root of a flock of geese, seeing the clouds collect, flew to the Manus lake; one-eighth of the whole flew from the edge of the water amongst a multitude of water lilies; and three couples were observed playing in the water." How many geese were there in the flock?

Rational Exponents

CONNECTION
Literature

In *The Pit and the Pendulum,* Edgar Allan Poe describes a situation in which a man is strapped to a wooden bench-like structure. A pendulum hangs from the ceiling that is "some thirty or forty feet overhead." The pendulum begins to swing, and with each swing the pendulum gets closer and closer to the man. The pendulum is at a right angle to his body, and a blade attached to the end of the pendulum is designed to cross right over his heart. As we read the story, we wonder if this man will be able to avoid being killed by the pendulum. Here is one way we can find out.

The formula describing the relationship of the length L of a pendulum to the time t it takes to make one swing (one full cycle back and forth) is given by $t = \sqrt{\dfrac{\pi^2 L}{8}}$. Suppose you wished to know how long it would take the pendulum to swing when it is 30 feet long and swinging right above the man's body.

Like a good mystery that unfolds little by little, page by page, a good problem solver approaches a problem using a series of small steps, or **subgoals.** By doing so, you make problem solving a simpler process.

Subgoal 1

Replace L by the values given in the story to obtain the possible range of time it takes the pendulum to swing.

30 feet overhead

$t = \sqrt{\dfrac{\pi^2 (30)}{8}}$

$t \approx 6.08$ seconds

40 feet overhead

$t = \sqrt{\dfrac{\pi^2 (40)}{8}}$

$t \approx 7.02$ seconds

It would have taken about 6 or 7 seconds for the pendulum to swing one cycle.

Subgoal 2

Determine if the man has enough time to escape the blade on the pendulum if the blade is just above his body and will strike him as it lowers for the next swing.

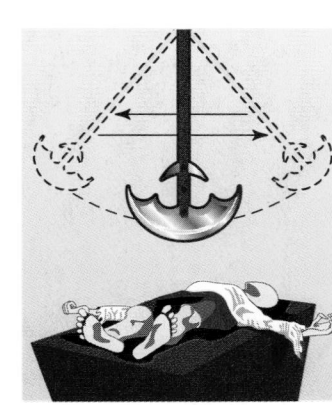

Since it takes at least 6 seconds for the pendulum to swing (that is, to move across to the other side and back to its original position), he would only have one-half the time to escape from the last safe pass of the blade over his body. One-half of 6 seconds is 3 seconds. Does this seem like enough time for him to escape?

You have learned that squaring a number and taking the square root of a number are inverse operations. But how would we evaluate an expression that contains a fractional exponent? Assume that fractional exponents behave as integral exponents.

For any number $a > 0$, $a^1 = a^{\frac{1}{2}(2)}$ or $\left(a^{\frac{1}{2}}\right)^2$.

So, $a^{\frac{1}{2}}$ is a number that when squared equals a. Since $\left(\sqrt{a}\right)^2 = a$, it follows that $a^{\frac{1}{2}} = \sqrt{a}$. This suggests the following definition.

Definition of $b^{\frac{1}{n}}$	**For any real number b and for any integer $n > 1$,** $$b^{\frac{1}{n}} = \sqrt[n]{b}$$ **except when $b < 0$ and n is even.**

Example **1**

PROBLEM SOLVING

Identify Subgoals

Economists refer to inflation as increases in the average cost of purchases. The formula $C = c(1 + r)^n$ can be used to predict the cost of consumer items at some projected time. In this formula, C represents the projected cost of the item at a given annual inflation rate, c the present cost of the item, r the rate of inflation (in decimal form), and n the number of years for the projection. Suppose a gallon of milk costs $2.69 now. How much would the price increase in 6 months with an inflation rate of 5.3%?

Subgoal 1: Identify the known values.

$c = \$2.69$, $r = 5.3\%$ or 0.053, and $n = 6$ months or $\frac{1}{2}$ year

Subgoal 2: Find the value of C.

$C = c(1 + r)^n$

$\quad = 2.69(1 + 0.053)^{\frac{1}{2}}$

Use a calculator to evaluate this expression. *Remember that $a^{\frac{1}{2}} = \sqrt{a}$.*

Enter: 2.69 $\boxed{\times}$ $\boxed{(}$ 1 $\boxed{+}$ .053 $\boxed{)}$ $\boxed{\sqrt{x}}$ $\boxed{=}$ *2.760364704*

The cost of the milk in 6 months will be $2.76.

Subgoal 3: Find the increase in cost.

$C - c = \$2.76 - \$2.69$ or $\$0.07$

The increase in cost in 6 months is predicted to be 7 cents.

From the definition of $b^{\frac{1}{n}}$, we can say that $7^{\frac{1}{4}} = \sqrt[4]{7}$ and $(-8)^{\frac{1}{3}} = \sqrt[3]{-8}$ or -2. The expression $(-16)^{\frac{1}{4}}$ is not defined since $-16 < 0$ and 4 is even. Why do we need this restriction?

In Example 2, each expression is evaluated in two ways. Method 1 uses the definition of $b^{\frac{1}{n}}$. Method 2 uses the properties of powers.

Example **Evaluate each expression.**

a. $81^{-\frac{1}{4}}$

Method 1

$$81^{-\frac{1}{4}} = \frac{1}{81^{\frac{1}{4}}} \qquad \textit{Remember that } b^{-n} = \frac{1}{b^n}.$$
$$= \frac{1}{\sqrt[4]{81}}$$
$$= \frac{1}{\sqrt[4]{3^4}}$$
$$= \frac{1}{3}$$

Method 2

$$81^{-\frac{1}{4}} = (3^4)^{-\frac{1}{4}}$$
$$= 3^{4\left(-\frac{1}{4}\right)}$$
$$= 3^{-1}$$
$$= \frac{1}{3}$$

b. $32^{\frac{3}{5}}$

Method 1

$$32^{\frac{3}{5}} = 32^{3\left(\frac{1}{5}\right)}$$
$$= (32^3)^{\frac{1}{5}}$$
$$= \sqrt[5]{32^3}$$
$$= \sqrt[5]{(2^5)^3}$$
$$= \sqrt[5]{2^5 \cdot 2^5 \cdot 2^5}$$
$$= 2 \cdot 2 \cdot 2 \text{ or } 8$$

Method 2

$$32^{\frac{3}{5}} = (2^5)^{\frac{3}{5}}$$
$$= 2^{5\left(\frac{3}{5}\right)}$$
$$= 2^3$$
$$= 8$$

In part b of Example 2, Method 1 uses a combination of the definition of $b^{\frac{1}{n}}$ and the properties of powers. This example suggests the following general definition of rational exponents.

Definition of Rational Exponents	**For any nonzero real number b, and any integers m and n, with $n > 1$,** $$b^{\frac{m}{n}} = \sqrt[n]{b^m} = \left(\sqrt[n]{b}\right)^m$$ **except when $b < 0$ and n is even.**

When simplifying expressions containing rational exponents, it is usually easier to leave the exponent in rational form rather than to write the expression as a radical. To simplify an expression, you must write the expression with all positive exponents. Furthermore, any exponents in the denominator of a fraction must be positive *integers*. That is, it may be necessary to rationalize a denominator.

All of the properties of powers you learned in Lesson 5-1 apply to rational exponents.

Example ❸ **Simplify each expression.**

a. $x^{\frac{2}{3}} \cdot x^{\frac{5}{3}}$

$$x^{\frac{2}{3}} \cdot x^{\frac{5}{3}} = x^{\left(\frac{2}{3}+\frac{5}{3}\right)}$$
$$= x^{\frac{7}{3}}$$

b. $y^{-\frac{5}{6}}$

$$y^{-\frac{5}{6}} = \frac{1}{y^{\frac{5}{6}}}$$
$$= \frac{1}{y^{\frac{5}{6}}} \cdot \frac{y^{\frac{1}{6}}}{y^{\frac{1}{6}}} \quad \textit{Why use } \frac{y^{\frac{1}{6}}}{y^{\frac{1}{6}}}?$$
$$= \frac{y^{\frac{1}{6}}}{y^{\frac{6}{6}}}$$
$$= \frac{y^{\frac{1}{6}}}{y}$$

When simplifying a radical expression, always find the smallest index possible. Using rational exponents make this process easier.

Example ❹ **Simplify each expression.**

a. $\dfrac{\sqrt[8]{16}}{\sqrt[6]{2}}$

$$\frac{\sqrt[8]{16}}{\sqrt[6]{2}} = \frac{16^{\frac{1}{8}}}{2^{\frac{1}{6}}}$$
$$= \frac{(2^4)^{\frac{1}{8}}}{2^{\frac{1}{6}}}$$
$$= \frac{2^{\frac{1}{2}}}{2^{\frac{1}{6}}}$$
$$= 2^{\left(\frac{1}{2}-\frac{1}{6}\right)}$$
$$= 2^{\frac{1}{3}} \text{ or } \sqrt[3]{2}$$

b. $\sqrt[4]{4n^2}$

$$\sqrt[4]{4n^2} = \left(4n^2\right)^{\frac{1}{4}}$$
$$= \left(2^2 \cdot n^2\right)^{\frac{1}{4}}$$
$$= 2^{2\left(\frac{1}{4}\right)} \cdot n^{2\left(\frac{1}{4}\right)}$$
$$= 2^{\frac{1}{2}} \cdot n^{\frac{1}{2}}$$
$$= \sqrt{2} \cdot \sqrt{n}$$
$$= \sqrt{2n}$$

c. $\dfrac{a^{\frac{1}{2}} + 1}{a^{\frac{1}{2}} - 1}$

$$\frac{a^{\frac{1}{2}} + 1}{a^{\frac{1}{2}} - 1} = \frac{a^{\frac{1}{2}} + 1}{a^{\frac{1}{2}} - 1} \cdot \frac{a^{\frac{1}{2}} + 1}{a^{\frac{1}{2}} + 1} \quad \textit{The conjugate of } a^{\frac{1}{2}} - 1 \textit{ is } a^{\frac{1}{2}} + 1.$$
$$= \frac{a + 2a^{\frac{1}{2}} + 1}{a - 1}$$

In summary, an expression is simplified when the following conditions are met.

- It has no negative exponents.
- It has no fractional exponents in the denominator.
- It is not a complex fraction.
- The index of any remaining radical is the least number possible.

Communicating Mathematics

Study the lesson. Then complete the following.

1. **Explain** how examining the subgoals helped to solve *The Pit and the Pendulum* application at the beginning of the lesson.

2. **Determine** if $3 \cdot 4^{\frac{1}{6}}$ is the simplest form of $2916^{\frac{1}{6}}$. If not, tell why not and write the expression in simplest form.

3. **Explain** why $(-81)^{\frac{1}{4}}$ is not defined.

4. **Write** a general rule for rationalizing expressions such as $\dfrac{1}{x^{\frac{n}{m}}}$.

5. Describe the steps you go through when solving a complicated word problem.

Guided Practice

Express using rational exponents.

6. $\sqrt{14}$　　7. $\sqrt[4]{27}$　　8. $\sqrt[6]{b^3}$　　9. $\sqrt[3]{16a^5b^7}$

Evaluate.

10. $16^{\frac{1}{4}}$　　11. $8^{-\frac{1}{3}}$　　12. $64^{\frac{2}{3}}$　　13. $\dfrac{16}{4^{\frac{3}{2}}}$

Simplify.

14. $\sqrt[6]{9x^3}$　　　15. $x^{\frac{3}{2}}y^{\frac{7}{3}}z^{\frac{9}{6}}$　　　16. $\dfrac{\sqrt[4]{125}}{\sqrt[4]{5}}$

17. $\dfrac{1}{5x^{\frac{1}{3}}}$　　18. $(x^2y)^{-\frac{1}{3}}$　　19. $c(a-4b)^{-\frac{1}{2}}$

20. $x^{\frac{1}{3}} \cdot x^{\frac{3}{4}}$　　21. $\dfrac{y^{\frac{5}{6}}}{y^{\frac{1}{6}}}$　　22. $\dfrac{x^3}{y^{\frac{1}{2}}} \cdot \dfrac{y}{x^{\frac{1}{3}}}$

23. To what power do you have to raise
   a. 4 to get 2?　　**b.** 32 to get 8?　　**c.** x^3 to get x?

EXERCISES

Practice

Evaluate.

24. $125^{\frac{1}{3}}$　　25. $100^{-\frac{1}{2}}$　　26. $16^{-\frac{3}{4}}$　　27. $81^{\frac{2}{3}} \cdot 81^{\frac{3}{2}}$

28. $9^{\frac{5}{2}} \cdot 9^{\frac{3}{2}}$　　29. $(-64)^{-\frac{2}{3}}$　　30. $\left(\dfrac{16}{81}\right)^{\frac{1}{4}}$　　31. $\left(\dfrac{1}{32}\right)^{-\frac{3}{5}}$

32. $\left(\dfrac{27}{64}\right)^{-\frac{1}{3}}$　　33. $\dfrac{24}{6^{\frac{2}{3}}}$　　34. $\dfrac{21}{7^{\frac{1}{3}}}$　　35. $\dfrac{8}{3^{\frac{1}{2}}}$

Simplify.

36. $2^{\frac{5}{3}}a^{\frac{7}{3}}$　　　37. $(2m)^{\frac{1}{2}}m^{\frac{1}{2}}$　　　38. $11^{\frac{1}{3}}p^{\frac{7}{3}}q^{\frac{2}{3}}$

39. $\dfrac{1}{w^{\frac{4}{5}}}$　　40. $\dfrac{1}{(u-v)^{\frac{1}{3}}}$　　41. $x^{-\frac{5}{6}}$

42. $\dfrac{1}{b^{\frac{1}{2}}+1}$　　43. $\dfrac{b^{-\frac{1}{2}}}{8b^{\frac{1}{3}} \cdot b^{-\frac{1}{4}}}$　　44. $\dfrac{g^{\frac{3}{2}}+3g^{-\frac{1}{2}}}{g^{\frac{1}{2}}}$

45. $\dfrac{3x^{-\frac{1}{3}}+x^{\frac{5}{3}}y}{x^{\frac{2}{3}}}$　　46. $\dfrac{2a^{\frac{1}{2}}+a^{\frac{3}{2}}}{a^{\frac{1}{2}}}$　　47. $\dfrac{pq}{\sqrt[3]{r}}$

48. $\left(m^{-\frac{2}{3}}\right)^{-\frac{1}{6}}$

49. $b^{-\frac{1}{3}} - b^{\frac{1}{3}}$

50. $\dfrac{z^{\frac{3}{2}}}{z^{\frac{1}{2}} + 2}$

51. $\left(\dfrac{m^{-2}n^{-6}}{121}\right)^{-\frac{1}{2}}$

52. $\dfrac{8^{\frac{1}{6}} - 9^{\frac{1}{4}}}{\sqrt{3} + \sqrt{2}}$

53. $\dfrac{a^{\frac{5}{3}} - a^{\frac{1}{3}}b^{\frac{4}{3}}}{a^{\frac{2}{3}} + b^{\frac{2}{3}}}$

54. $\sqrt[4]{49}$

55. $r^{\frac{1}{2}}s^{\frac{1}{3}}$

56. $\sqrt[6]{81p^4q^8}$

57. $\sqrt{13} \cdot \sqrt[3]{13^2}$

58. $a^{\frac{5}{6}}b^{\frac{7}{3}}c^{\frac{3}{2}}$

59. $\sqrt[3]{\sqrt{27}}$

Evaluate $f(x) = x^{-\frac{2}{3}} + x^{-3}$ **for each value of x.**

60. $f(-8)$

61. $f(1000)$

62. $f(0.001)$

Calculator

Evaluate each expression to the nearest hundredth by using your calculator.

63. $45^{0.33}$

64. $2.75^{\frac{2}{3}}$

65. $\left(4\frac{1}{2}\right)^{0.075}$

Critical Thinking

66. Explain how you would solve $9^x = 3^{x+\frac{1}{2}}$ for x.

Applications and Problem Solving

67. Music   On a piano, the frequency of the A note above middle C should be set at 440 vibrations per second. The frequency f_n of a note that is n notes above A should be $f_n = 440\left(\sqrt[12]{2}\right)^{n-1}$.

   a. At what frequency should a piano tuner set the A that is one octave, or 12 notes, above the A above middle C?

   b. Middle C is nine notes below A that has a frequency 440 vibrations per second. What should the frequency of middle C be?

68. Environment   When estimating the amount of pollutants contained in the hot gases released by a smokestack, EPA agents must take into account the phenomenon that the velocity of the gas varies throughout the cylindrical smokestack. In such situations, the gas near the center of a cross section of the smokestack has a greater velocity than the gas near the perimeter. This can be described algebraically by the formula $V = V_{max}\left[1 - \left(\frac{r}{r_0}\right)^2\right]$, where V_{max} is the maximum velocity of the gas, r_0 is the radius of the smokestack, and V is the velocity of the gas at a distance r from the center of the circular cross section of the smokestack. A cross section of a smokestack has a radius of 25 feet. The gas flowing from it has a measured velocity of 10 mph, and the velocity of the gas at a distance r from the center is 7 mph.

   a. Find the value of r.

   b. Write the subgoals that you followed to solve the problem.

69. Simplify $4\sqrt{(x-5)^2}$. (Lesson 5–6)

70. Simplify $\sqrt{\dfrac{9}{36}x^4}$. (Lesson 5–5)

71. Factor $a^3b^3 - 27$. (Lesson 5–4)

72. Simplify $\dfrac{-15r^5s^2}{5r^5s^{-4}}$. (Lesson 5–1)

73. Solve the system by using augmented matrices. (Lesson 4–7)

$$2x + y = 0$$
$$3x - 4y = 22$$

74. Can a 3×4 matrix have an inverse? Explain your answer. (Lesson 4–5)

75. Three numbers have a sum of 6. The first number is twice the second, and the third is three times the second. What are the three numbers? (Lesson 3–7)

76. Find a and b such that the solution of the system $2x + 3y = 7$ and $ax + by = -10$ is $(2, 1)$. (Lesson 3–1)

77. Statistics A prediction equation in a study of the relationship between minutes spent studying s and test scores t is $t = 0.36s + 61.4$. Predict the score a student would receive if she spent 1 hour studying. (Lesson 2–5)

78. Find $h(m - 2)$ if $h(x) = \dfrac{x - 5}{7}$. (Lesson 2–1)

79. Solve $4 < 2x - 2 < 10$. (Lesson 1–6)

80. Name the property illustrated by $6 + a = 6 + a$. (Lesson 1–2)

WORKING ON THE
In·ves·ti·ga·tion

Refer to the Investigation on pages 180–181.

3-2-1-Blast-Off!

Standard deviation is the most commonly used measure of variation. It is the average measure of how much each value in a set of data differs from the mean. To find the standard deviation, follow the steps below.

a. Find the mean of the set of data.

b. Find the difference between each value in the set of data and the mean.

c. Square each difference.

d. Find the mean of the squares.

e. Take the principal square root of this mean.

1 Find the standard deviation for each of the launchers used in your class shoot-off.

2 How does the standard deviation fit with the other measures you found for each of the launchers? Does its value change your analysis of which launcher is the most accurate? Explain.

3 Write a general formula for standard deviation. Also write it in a form with rational exponents.

4 Plot the data from each of the launchers used in your class shoot-off on box-and-whisker plots and relate the standard deviation to the length of the whiskers. Is there a correlation? Explain.

Add the results of your work to your Investigation Folder.

Solving Radical Equations and Inequalities

What YOU'LL LEARN

- To solve equations and inequalities containing radicals.

Why IT'S IMPORTANT

You can use radical equations to solve problems involving aviation and manufacturing.

APPLICATION

Construction

At a building construction site, carts loaded with concrete must cross a 25-foot gap between two towers that are 20 stories high. The construction manager needs to select a beam strong enough to support a worker pushing a fully loaded cart. The beam must be at least 2 feet wide to accommodate the wheels of the cart. Will a beam 6 inches thick be able to safely support the load of the cart (865 lb) and the worker (250 lb maximum)?

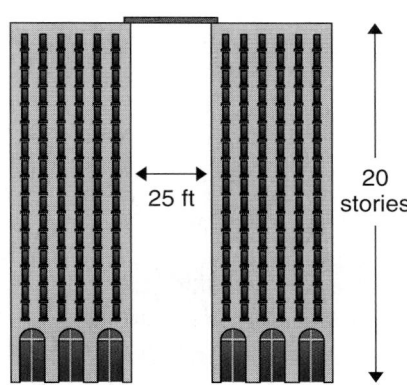

25 ft

20 stories

The formula that expresses the relationship of the safe load s of a beam to its width w in feet and depth d in inches is $s = \dfrac{kwd^2}{\ell}$, where k is a constant equal to 576 and ℓ is the distance in feet between the supports. Since we wish to know the thickness or depth of the beam, let's solve the formula for d.

$$s = \frac{kwd^2}{\ell}$$

$$\ell s = kwd^2 \quad \text{\textit{Multiply each side by }} \ell.$$

$$\frac{\ell s}{kw} = d^2 \quad \text{\textit{Divide each side by kw.}}$$

$$\pm\sqrt{\frac{\ell s}{kw}} = d \quad \text{\textit{Since d is squared, take the square root of each side.}}$$

The safe load s is the sum of the weight of the cart (865 lb) and the maximum weight of a construction worker (250 lb) or 1115 pounds.

$$d = \sqrt{\frac{25 \cdot 1115}{576 \cdot 2}} \approx 4.92 \quad \text{\textit{$\ell = 25$, $s = 1115$, $k = 576$, and $w = 2$}}$$

A beam at least 5 inches thick would safely support the load across the gap. A 6-inch beam would give an extra safety margin.

Some equations contain irrational numbers expressed as radicals. You can use the properties of radicals to solve these equations.

Example **1** Solve $x - 4 = x\sqrt{3}$.

$$x - 4 = x\sqrt{3}$$
$$x - x\sqrt{3} = 4 \qquad \text{\textit{Isolate the variable on one side of the equation}}$$
$$x(1 - \sqrt{3}) = 4 \qquad \text{\textit{Factor the GCF.}}$$
$$x = \frac{4}{1 - \sqrt{3}}$$

(continued on the next page)

$$x = \frac{4}{(1 - \sqrt{3})} \cdot \frac{(1 + \sqrt{3})}{(1 + \sqrt{3})}$$ *Rationalize the denominator.*

$$x = \frac{4(1 + \sqrt{3})}{1 - 3}$$

$$x = \frac{4 + 4\sqrt{3}}{-2} \text{ or } -2 - 2\sqrt{3}$$

Check:
$$x - 4 = x\sqrt{3}$$
$$-2 - 2\sqrt{3} - 4 \stackrel{?}{=} (-2 - 2\sqrt{3})(\sqrt{3})$$
$$-2\sqrt{3} - 6 = -2\sqrt{3} - 6 \checkmark$$

The solution is $-2 - 2\sqrt{3}$.

The Pythagorean theorem is often helpful when solving real-life problems.

Example  **2**

APPLICATION

Aviation

A student pilot leaves the Tamiami Municipal Airport at 9:00 A.M. on her first solo flight. She is traveling due west at 125 mph. At 10:00 A.M., she turns and flies in a southeasterly direction at the same speed. At noon, she notices that the fuel gauge indicates a low level of fuel. At this time, she is directly south of the airport. How far is it to the airport?

Explore   First draw a diagram of the flight path. The problem gives us the following information.

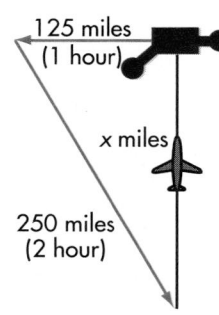

125 miles (1 hour)

x miles

250 miles (2 hour)

- She leaves the airport at 9:00 A.M.
- She flies west at 125 mph until 10:00 A.M.
- At 10:00, she turns southeast and flies until noon.
- At noon, the plane is directly south of the airport, and she turns north to return to the airport.

Plan   Since the flight path makes a right triangle, we know we can use the Pythagorean theorem to find an expression for the length *x* of one of the legs of the triangle.

Solve
$$250^2 = 125^2 + x^2$$
$$250^2 - 125^2 = x^2$$ *Subtract 125^2 from each side.*
$$62{,}500 - 15{,}625 = x^2$$
$$46{,}875 = x^2$$
$$\pm\sqrt{46{,}875} = x$$ *Take the square root of each side.*
$$\pm 216.5 \approx x$$ *Use your calculator.*

Since distance cannot be negative, we know that the pilot is approximately 216.5 miles away from the airport.

Examine   Use the Pythagorean theorem to examine the time involved.
$$1^2 + y^2 = 2^2$$ *Let y represent the unknown time flying due north.*
$$y^2 = 4 - 1 \text{ or } 3$$
$$y = \pm\sqrt{3}$$

The pilot flew $\sqrt{3}$ hours at 125 mph, or about 216.5 miles. $\checkmark$

Sometimes variables appear in the radicand. Equations with radicals like this are called **radical equations**. To solve this type of equation, you will need to raise each side of the equation to a power equal to the index to remove the variable from the radical.

Solving radical equations sometimes yields **extraneous solutions**, which are solutions that do not satisfy the original equation. You must check all the possible solutions in the *original* equation and disregard the extraneous solutions.

Example ❸ Solve $\sqrt{x-3} = \sqrt{2} - \sqrt{x}$.

$$\sqrt{x-3} = \sqrt{2} - \sqrt{x}$$
$$(\sqrt{x-3})^2 = (\sqrt{2} - \sqrt{x})^2 \quad \textit{Square each side.}$$
$$x - 3 = 2 - 2\sqrt{2x} + x$$
$$-5 = -2\sqrt{2x} \quad \textit{Isolate the square root.}$$
$$(-5)^2 = (-2\sqrt{2x})^2 \quad \textit{Square each side again.}$$
$$25 = 8x$$
$$\frac{25}{8} = x$$

Check:
$$\sqrt{x-3} = \sqrt{2} - \sqrt{x}$$
$$\sqrt{\frac{25}{8} - 3} \overset{?}{=} \sqrt{2} - \sqrt{\frac{25}{8}}$$
$$\sqrt{\frac{1}{8}} \overset{?}{=} \sqrt{2} - \sqrt{\frac{25}{8}}$$
$$\frac{1}{2}\sqrt{\frac{1}{2}} \overset{?}{=} \sqrt{2} - \frac{5}{2}\sqrt{\frac{1}{2}} \quad \frac{1}{8} = \frac{1}{2^2 \cdot 2}, \ \frac{25}{8} = \frac{5^2}{2^2 \cdot 2}$$
$$3\sqrt{\frac{1}{2}} \overset{?}{=} \sqrt{2} \quad \textit{Add } \frac{5}{2}\sqrt{\frac{1}{2}} \textit{ to each side.}$$
$$3\sqrt{\frac{1}{2} \cdot \frac{2}{2}} \overset{?}{=} \sqrt{2} \quad \textit{Rationalize the denominator.}$$
$$\frac{3\sqrt{2}}{2} \neq \sqrt{2} \quad \textit{Why?}$$

The solution does not check. The equation has no real solution.

You can apply the same methods used in solving square root equations to solving equations of nth roots. Remember to undo a square root, you squared the expression. To undo the nth root, you must raise the expression to the nth power.

Example ❹ Solve $2(7n-1)^{\frac{1}{3}} - 4 = 0$.

In order to remove the cube root, we must first isolate it and then raise each side of the equation to the third power.

$$2(7n-1)^{\frac{1}{3}} - 4 = 0 \quad \textit{Remember that } (7n-1)^{\frac{1}{3}} = \sqrt[3]{7n-1}.$$
$$2(7n-1)^{\frac{1}{3}} = 4 \quad \textit{Add 4 to each side.}$$
$$(7n-1)^{\frac{1}{3}} = 2 \quad \textit{Isolate the cube root.}$$
$$\left[(7n-1)^{\frac{1}{3}}\right]^3 = 2^3 \quad \textit{Cube each side.}$$
$$7n - 1 = 8$$
$$7n = 9 \quad \textit{Add 1 to each side.}$$
$$n = \frac{9}{7} \quad \textit{Divide each side by 7.} \quad \text{(continued on the next page)}$$

Check: $2(7n - 1)^{\frac{1}{3}} - 4 \overset{?}{=} 0$

$$2[7\left(\frac{9}{7}\right) - 1]^{\frac{1}{3}} - 4 \overset{?}{=} 0$$

$$2(8)^{\frac{1}{3}} - 4 \overset{?}{=} 0$$

$$2(2) - 4 \overset{?}{=} 0$$

$$0 = 0 \checkmark$$

The solution is $\frac{9}{7}$.

You can use what you now know about solving square root equations to solve *square root inequalities*.

Example **5** **Solve $3 + \sqrt{4x - 5} \leq 10$.**

Since the radicand of a radical expression must be greater than or equal to zero, first solve $4x - 5 \geq 0$.

$$4x - 5 \geq 0$$
$$4x \geq 5$$
$$x \geq \frac{5}{4}$$

Now solve $3 + \sqrt{4x - 5} \leq 10$.

$3 + \sqrt{4x - 5} \leq 10$

$\quad \sqrt{4x - 5} \leq 7$ *Subtract 3 from each side.*

$\quad\quad 4x - 5 \leq 49$ *Square each side.*

$\quad\quad\quad 4x \leq 54$ *Add 5 to each side.*

$\quad\quad\quad x \leq 13\frac{1}{2}$ *Divide each side by 4 and simplify.*

It appears that $\frac{5}{4} \leq x \leq 13\frac{1}{2}$. Test values on each part of the number line below to check the result.

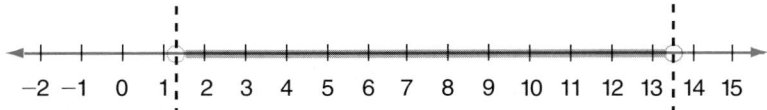

Try $x = 0$.

$3 + \sqrt{4x - 5} \leq 10$

$3 + \sqrt{-5} \leq 10$

False; negative square roots are not defined in the set of real numbers.

So, $\frac{5}{4} \leq x \leq 13\frac{1}{2}$.

Try $x = 7.5$.

$3 + \sqrt{4x - 5} \leq 10$

$3 + \sqrt{30 - 5} \leq 10$

$3 + 5 \leq 10 \checkmark$

true

Try $x = 21.5$.

$3 + \sqrt{4x - 5} \leq 10$

$3 + \sqrt{86 - 5} \leq 10$

$3 + 9 \not\leq 10$

false

To solve inequalities that involve radicals, complete the following steps.

1. Identify the excluded values.

2. Solve the inequality.

3. Test values to find the solution.

Communicating Mathematics

Study the lesson. Then complete the following.

1. **Explain** how you would solve $\sqrt{2x-3} - \sqrt{x+3} = 0$.

2. **Explain** why you need to isolate the nth root before raising it to the nth power to remove it.

3. **You Decide** Pedro and Rochelle were working in pairs to solve the equation $(x+5)^{\frac{1}{4}} = -4$. Pedro said he could tell that there was no solution without even working the problem. Rochelle said they should solve the equation and then they could test their results to see if there was a solution. Who is correct and why?

Guided Practice

Solve each equation. Be sure to check for extraneous solutions.

4. $\sqrt{3d+1} = 4$

5. $3 - (2-y)^{\frac{1}{2}} = 0$

6. $1 + x\sqrt{2} = 0$

7. $\sqrt{a-4} - 3 = 0$

8. $\frac{2}{3} \cdot (4m)^{\frac{1}{3}} = 4$

9. $\sqrt[3]{y+1} = 2$

10. $\sqrt{2x+3} - 4 \le -5$

11. $\sqrt{n+12} - \sqrt{n} > 2$

12. Solve $\sqrt{x+9} = 9 - \sqrt{x}$. Choose the best answer.

    a. 4   **b.** ± 4 **c.** 2   **d.** 16

13. Solve $y = \sqrt{r^2 + s^2}$ for r.

Practice

Solve each equation. Be sure to check for extraneous solutions.

14. $\sqrt{x} = 3$

15. $\sqrt{q} - 8 = 0$

16. $x^{\frac{1}{2}} + 4 = 0$

17. $7 + 6n\sqrt{5} = 0$

18. $\sqrt[3]{r-1} = 3$

19. $\sqrt[3]{2p+1} = 3$

20. $\sqrt[4]{7+3z} = 2$

21. $x\sqrt{x} = 8$

22. $y\sqrt{3} - y = 7$

23. $\sqrt{2x-9} = -\frac{1}{3}$

24. $9 + \sqrt{4x+8} = 11$

25. $3 + \sqrt{4n-5} = 10$

26. $6 - \sqrt{2y+1} < 3$

27. $(7-5x)^{\frac{1}{2}} \ge 8$

28. $13 - 3r = r\sqrt{5}$

29. $3x + 5 = x\sqrt{3}$

30. $\sqrt{g-4} = \sqrt{2g-3}$

31. $\sqrt{2r-6} = \sqrt{3+r}$

32. $\sqrt{2} - \sqrt{x+6} \le -\sqrt{x}$

33. $\sqrt{k+9} - \sqrt{k} > \sqrt{3}$

34. $\sqrt{x-6} = 3 + \sqrt{x}$

35. $\sqrt{h+3} + \sqrt{h-1} = 5$

36. $\sqrt{x+21} - 1 = \sqrt{x+12}$

37. $\sqrt{4x+1} = 3 + \sqrt{4x-2}$

38. $(6g-5)^{\frac{1}{3}} + 2 = -3$

39. $\sqrt{a+1} = \sqrt{a+6} - 1$

40. $\sqrt{b+5} + \sqrt{b+10} > 2$

41. $\sqrt{a-5} - \sqrt{a+7} \le 4$

42. Explain why $\{25, 36\}$ is not the solution set for $x + \sqrt{x} - 30 = 0$.

Solve for the variable indicated.

43. $T = 2\pi\sqrt{\dfrac{\ell}{g}}$ for ℓ

44. $t = \sqrt{\dfrac{2s}{g^2}}$ for s

45. $m^2 = \sqrt[3]{\dfrac{rp}{g^2}}$ for p

46. $r = \sqrt[3]{\dfrac{2mM}{c}}$ for c

47. **Geometry**   The surface area of a cone can be found by using $S = \pi r\sqrt{r^2 + h^2}$, where r is the radius of the base and h is the height of the cone.

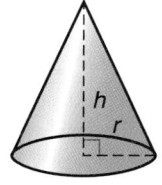

    a. Solve the equation for h.

    b. Find h if $S = 225$ and $r = 5$.

Critical Thinking

48. Is $\sqrt[n]{b^m} = (\sqrt[n]{b})^m$? Justify your answer.

49. Is $\sqrt[km]{b} = \sqrt[k]{\sqrt[m]{b}}$? Justify your answer.

Applications and Problem Solving

50. **Manufacturing**   A company that manufactures ROM (Read Only Memory) chips for computers uses the formula $c = 100\sqrt[3]{n^2} + 1200$ to determine the cost of producing the chips, where c is the cost of production in dollars and n is the number produced. If a computer company places an order totaling \$10,000, how many chips will be produced?

51. **Concerts**   The organizers of a rock concert are preparing for the arrival of 50,000 enthusiastic fans in the open field where the concert will take place. It is reasonable to allow each person 5 square feet of space, so the organizers need to rope off a circular area of 250,000 square feet. Using the formula $A = \pi r^2$, where A represents the area of the circular region and r represents the radius of the region, find the radius of this region.

52. **History**   Johann Kepler (1571–1630) is known mainly as a mathematician of the sky. He is credited with several laws of planetary motion. His third law states that the square of the time of revolution of each planet (its period) is proportional to the cube of its mean distance from the sun. This can be expressed as $\dfrac{T_a}{T_b} = \left(\dfrac{r_a}{r_b}\right)^{\frac{3}{2}}$, where T_a is the time it takes one planet to orbit the sun, r_a is that planet's average distance from the sun, and T_b and r_b are another planet's period and average distance from the sun.

    a. Solve the formula for r_a.

    b. Find the distance (to the nearest million miles) that Jupiter is from the sun, if its period is 12 years. Mercury is 36 million miles from the sun and has a period of 88 days.

308   *Chapter 5   Exploring Polynomials and Radical Expressions*

53. Aerospace Engineering The radius of the orbit of a satellite is found by $r = \sqrt[3]{\dfrac{GMt^2}{4\pi^2}}$, where t represents the time it takes for the satellite to complete one orbit, G represents the constant of universal gravitation, and M is the mass of the central object. Solve the formula for t.

54. Solar Energy The energy of direct sunlight on a solar cell with an area of one square centimeter is converted into 0.01 watt of electrical energy. Suppose that a square solar cell must deliver 15 watts of energy. What should the dimensions of the cell be?

Solar-powered car

Mixed Review

55. How would you write *the seventh root of 5 cubed* using exponents? (Lesson 5–7)

Simplify.

56. $\sqrt{3}(\sqrt{6} - 2)$ (Lesson 5–6)

57. $\sqrt{x^2 + 10x + 25}$ (Lesson 5–5)

58. $(2t^3 - 2t - 3) \div (t - 1)$ (Lesson 5–3)

59. $(y^5)^2$ (Lesson 5–1)

60. Write an augmented matrix for the system of equations. Then solve the system.   (Lesson 4–7)

$x + 5y + 2z = 10$
$3x - 3y + 2z = 2$
$2x + 4y - z = -15$

61. Evaluate $\begin{vmatrix} 4 & -2 \\ 3 & 7 \end{vmatrix}$.   (Lesson 4–4)

62. Business Mr. Whitner bought 7 drums of two different cleaning fluids for his dry-cleaning business. One of the fluids cost $30 per drum, and the other was $20 per drum. The total cost of the supplies was $160. How much of each fluid did Mr. Whitner buy? Write a system of equations and solve by graphing.   (Lesson 3–1)

63. Entertainment Under the terms of a royalty agreement, a radio station is charged $60 per month for a new release plus $0.20 for each time the song is played. The songwriter gets $\dfrac{1}{4}$ of this amount.   (Lesson 2–2)

   a. If x is the number of times the song is played, write an equation that shows how much money y the songwriter gets.

   b. How much would the songwriter get if the song was played 42 times in one month at the radio station?

64. Statistics The cost per cup of several different types of juices are given below.

| 11¢ | 9¢ | 6¢ | 9¢ | 12¢ | 4¢ | 4¢ | 8¢ | 6¢ | 4¢ | 16¢ |
| 12¢ | 11¢ | 19¢ | 7¢ | 19¢ | 7¢ | 19¢ | 4¢ | 6¢ | 5¢ | 6¢ |

Make a stem-and-leaf plot of the costs.   (Lesson 1–3)

Complex Numbers

What YOU'LL LEARN

- To simplify square roots containing negative radicands,
- to solve quadratic equations that have pure imaginary solutions, and
- to add, subtract, and multiply complex numbers.

Why IT'S IMPORTANT

You can use complex numbers to solve problems involving electricity and physics.

INTEGRATION
Number Theory

Keesha and Juanita formed a study group for their algebra class. Their teacher, Mrs. Rodriguez, gave them an assignment that included solving the equation $3x^2 + 15 = 0$. Keesha and Juanita solved the equation as follows.

$$3x^2 + 15 = 0$$
$$3x^2 = -15$$
$$x^2 = -5$$

They were puzzled at the last step because they knew that there is no real number x that when squared results in a negative number. They were stumped! After checking with a few of their classmates, they found that everyone had the same difficulty. Is there a solution to the equation?

About 400 years ago, a solution to this type of equation was proposed by French mathematician René Descartes (1596–1650). He proposed that the solution to the equation $x^2 = -1$ be represented by a number i, where i is not a real number. Thus, $i = \sqrt{-1}$.

Keesha's class could now solve the equation.

$$x^2 = -5$$
$$x = \pm\sqrt{-5}$$
$$x = \pm\sqrt{5} \cdot \sqrt{-1} \quad \textit{Extension of product property of radicals}$$
$$x = \pm\sqrt{5} \cdot i \qquad \sqrt{-1} = i$$
$$x = \pm i\sqrt{5}$$

To avoid $\sqrt{5} \cdot i$ being read as $\sqrt{5i}$, write $\sqrt{5} \cdot i$ as $i\sqrt{5}$.

Numbers such as $i\sqrt{5}$, $2i$, and $-5i$ are called **pure imaginary numbers**, and i is called the **imaginary unit**. Using i as you would any constant, you can define square roots of negative numbers.

Since $i = \sqrt{-1}$, it follows that $i^2 = -1$. Therefore,

$(3i)^2 = 3^2 \cdot i^2$ or $-9 \quad \rightarrow \quad \sqrt{-9} = \sqrt{9} \cdot \sqrt{-1}$ or $3i$

$(i\sqrt{3})^2 = i^2(\sqrt{3})^2$ or $-3 \quad \rightarrow \quad \sqrt{-3} = \sqrt{3} \cdot \sqrt{-1}$ or $i\sqrt{3}$

Definition of Pure Imaginary Numbers	For any positive real number b, $$\sqrt{-b^2} = \sqrt{b^2} \cdot \sqrt{-1} \text{ or } bi,$$ where i is the imaginary unit, and bi is called a pure imaginary number.

Example **1** **Simplify each expression.**

a. $\sqrt{-12}$

$$\sqrt{-12} = \sqrt{2^2} \cdot \sqrt{3} \cdot \sqrt{-1}$$
$$= 2 \cdot \sqrt{3} \cdot i$$
$$= 2i\sqrt{3}$$

b. $\sqrt{-27x^3}$

$$\sqrt{-27x^3} = \sqrt{3^2} \cdot \sqrt{-3} \cdot \sqrt{x^2} \cdot \sqrt{x}$$
$$= 3x\sqrt{-3x}$$
$$= 3ix\sqrt{3x}$$

The commutative and associative properties for multiplication hold true for pure imaginary numbers.

Example **2** **Simplify each expression.**

a. $-3i \cdot 8i$

$$-3i \cdot 8i = -24i^2$$
$$= -24(-1) \quad i^2 = -1$$
$$= 24$$

b. $\sqrt{-6} \cdot \sqrt{-10}$

$$\sqrt{-6} \cdot \sqrt{-10} = i \cdot \sqrt{6}\, i \cdot \sqrt{10}$$
$$= i^2 \sqrt{60}$$
$$= -1 \cdot 2\sqrt{15} \text{ or } -2\sqrt{15}$$

You can use the properties of powers to simplify powers of i.

Example **3** **Simplify i^{25}.**

LOOK BACK

You can review the properties of powers in Lesson 5-1.

$$i^{25} = i \cdot i^{24} \qquad \textit{Product of powers}$$
$$= i \cdot (i^2)^{12} \qquad \textit{Power of a power}$$
$$= i \cdot (-1)^{12} \quad i^2 = -1$$
$$= i \cdot 1 \text{ or } i$$

When you solve some equations, the answer may involve two pure imaginary numbers.

Example **4** **Solve $4x^2 + 36 = 0$.**

$$4x^2 + 36 = 0$$
$$4x^2 = -36 \qquad \textit{Subtract 36 from each side.}$$
$$x^2 = -9 \qquad \textit{Divide each side by 4.}$$
$$x = \pm\sqrt{-9} \quad \textit{Take the square root of each side.}$$
$$x = \pm 3i$$

Check:

$$4x^2 + 36 = 0$$
$$4(3i)^2 + 36 \stackrel{?}{=} 0$$
$$4 \cdot 9 \cdot i^2 + 36 \stackrel{?}{=} 0$$
$$-36 + 36 \stackrel{?}{=} 0$$
$$0 = 0 \checkmark$$

$$4x^2 + 36 = 0$$
$$4(-3i)^2 + 36 \stackrel{?}{=} 0$$
$$4 \cdot 9 \cdot i^2 + 36 \stackrel{?}{=} 0$$
$$-36 + 36 \stackrel{?}{=} 0$$
$$0 = 0 \checkmark$$

Suppose you were asked to simplify the expression $5 + 2i$. Since 5 is a real number and $2i$ is a pure imaginary number, the terms are not like terms and cannot be combined. This type of expression is called a **complex number.**

Definition of a Complex Number	A complex number is any number that can be written in the form $a + bi$, where a and b are real numbers and i is the imaginary unit; a is called the real part, and bi is called the imaginary part.

The diagram below shows the complex number system.

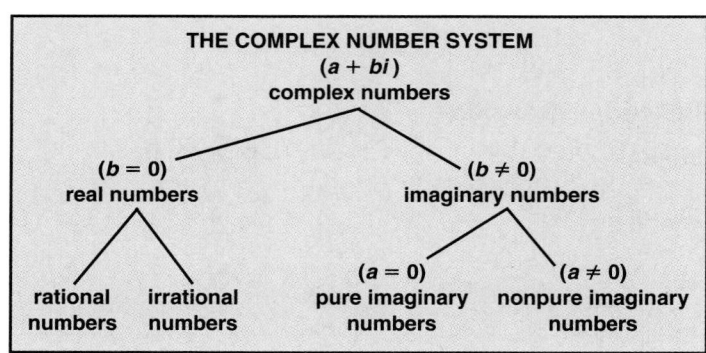

THE COMPLEX NUMBER SYSTEM
$(a + bi)$
complex numbers

$(b = 0)$
real numbers

$(b \neq 0)$
imaginary numbers

rational numbers   irrational numbers

$(a = 0)$
pure imaginary numbers

$(a \neq 0)$
nonpure imaginary numbers

- If $b = 0$, a real number results.
- If $b \neq 0$, the number is imaginary.
- If $a = 0$, the number is a pure imaginary number.

Two complex numbers are equal if and only if their real parts are equal and their imaginary parts are equal.

Definition of Equal Complex Numbers	$a + bi = c + di$ **if and only if** $a = c$ **and** $b = d$

To add or subtract complex numbers, combine like terms; that is, combine the real parts and combine the imaginary parts.

Example **5** **Simplify each expression.**

a. $(8 - 5i) + (2 + i)$ **b.** $(4 + 7i) - (2 + 3i)$

$(8 - 5i) + (2 + i)$
$= (8 + 2) + (-5i + i)$
$= 10 - 4i$

$(4 + 7i) - (2 + 3i)$
$= (4 - 2) + (7i - 3i)$
$= 2 + 4i$

You can model the addition of complex numbers geometrically.

MODELING MATHEMATICS

Adding Complex Numbers

Materials: grid paper   straightedge

You can model the addition of complex numbers on a coordinate plane. The horizontal axis represents the real part a of the complex number and the vertical axis represents the coefficient b of the imaginary part. Use a coordinate plane to find $(5 + 3i) + (-2 + 2i)$.

- Create a coordinate plane and label the axes appropriately.
- Graph $5 + 3i$ by drawing a segment from the origin to $(5, 3)$ on the coordinate plane.
- Graph $-2 + 2i$ by drawing a segment from the origin to $(-2, 2)$ on the coordinate plane.
- Draw a parallelogram that has the two segments you drew as sides.
- The diagonal of the parallelogram drawn from the origin represents the sum of the two complex

numbers. The endpoint of the diagonal is $(3, 5)$, which represents $3 + 5i$. So, $(5 + 3i) + (-2 + 2i) = (3 + 5i)$.

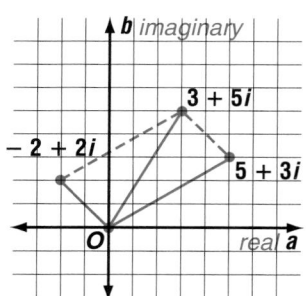

Your Turn

Model $(-2 + 3i) + (1 - 4i)$ on a coordinate plane.

You can multiply complex numbers by using the FOIL method.

Example **6** **Simplify $(4 + 2i)(3 - 5i)$.**

$$
(4 + 2i)(3 - 5i) = \overset{F}{4(3)} - \overset{O}{4(5i)} + \overset{I}{(2i)3} - \overset{L}{2i(5i)}
$$
$$
= 12 - 20i + 6i - 10i^2
$$
$$
= 12 - 14i - 10(-1)
$$
$$
= 22 - 14i
$$

One of the real-world uses of imaginary numbers is in electricity. However, electrical engineers use j instead of i to represent the imaginary unit. This avoids any confusion with the I used as a symbol for current. Imaginary numbers are used to represent the impedance of a circuit. The impedance is the resistance to the flow of electricity through the circuit.

Example **7**

Electricity

The Habitat for Humanity program utilizes volunteers to help build houses for low-income families who might otherwise not be able to afford the purchase of a home. At a recent site, Habitat workers built a small storage shed attached to the house. The electrical blueprints for the shed called for two AC circuits connected in series with a total voltage of 220 volts. One of the circuits must have an impedance of $7 - 10j$ ohms, and the other needs to have an impedance of $9 + 5j$. According to the building codes, the impedance cannot exceed $20 - 5j$ ohms. Will the circuits, as designed, meet the code?

Former President Jimmy Carter

Explore   First we need to understand the electricity terms used in the problem. In a simplified electrical circuit, there are three basic components to be considered:

* the flow of the electrical current, I
* the resistance to that flow, Z, called impedance, and
* the electromotive force, E, called voltage.

The formula $E = I \cdot Z$ illustrates the relationship among these components.

Plan   The total impedance is the sum of the individual impedances.

Solve
$$
(7 - 10j) + (9 + 5j) = 7 + 9 - 10j + 5j
$$
$$
= 16 + (-10 + 5)j
$$
$$
= 16 - 5j
$$

The total impedance is $16 - 5j$, which is less than $20 - 5j$. The circuits will meet the code.

Generally, you cannot order complex numbers that contain imaginary parts. However, if the imaginary parts of the two numbers are identical, you can compare the real number parts.

Examine   Since both numbers contain the same imaginary part, we can compare the real part. $16 < 20$, so the conclusion is correct.

Communicating Mathematics

Study the lesson. Then complete the following.

1. **Determine** if each statement is true or false.
   a. Every real number is a complex number.
   b. Every imaginary number is a complex number.

2. **Show** where each of the following lies on the complex coordinate plane.
   a. real numbers
   b. pure imaginary numbers

3. Which complex number is equivalent to $\sqrt{-50}$?
   a. $-5i$ b. $25i$ c. $5i\sqrt{2}$ d. $-5i\sqrt{2}$

4. Identify the complex numbers and their sum shown in the graph at the right.

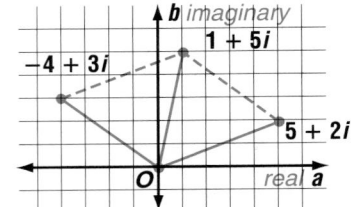

5. Graph the addends $(-2 + i)$ and $(4 + 4i)$ on the complex plane. Then find their sum geometrically.

Guided Practice

Simplify.

6. $\sqrt{-64}$

7. $\sqrt{-98m^2n^2}$

8. $(4i)(-3i)$

9. $5\sqrt{-24} \cdot 3\sqrt{-18}$

10. $\sqrt{3} \cdot \sqrt{-27}$

11. i^{16}

12. $(15 + 10i) - (4 + 6i)$

13. $(4 + 2i) + (1 + 3i)$

14. $(4 - 3i)(5 + 7i)$

15. $(3 + 2i)(3 - 2i)$

16. Find the product $(2 - 4i)(3 + 9i)$.

17. Verify that $-2i\sqrt{2}$ is a solution for $5x^2 + 40 = 0$. Find the other solution.

Solve each equation.

18. $5x^2 + 30 = 5$

19. $a^2 + 16 = 0$

Practice

Simplify.

20. $\sqrt{-169}$

21. $\sqrt{-\frac{4}{9}}$

22. $\sqrt{-49}$

23. $\sqrt{-100k^4}$

24. $\sqrt{-36m^4n^2}$

25. $\sqrt{-\frac{9x^3}{25y^8}}$

26. $(2i)^2$

27. $(-4i)(-5i)(3i)$

28. $5i(-2i)^2$

29. $(\sqrt{-11})(\sqrt{-22})$

30. $(2\sqrt{-50})(\frac{1}{8}\sqrt{-2})$

31. $\sqrt{-8} \cdot \sqrt{-18}$

32. $\sqrt{-5} \cdot \sqrt{20}$ **33.** $\sqrt{-8} \cdot \sqrt{6}$ **34.** $-2\sqrt{-x} \cdot -5\sqrt{-y}$

35. i^{17} **36.** i^{59} **37.** i^{34}

38. $2\sqrt{-18} + 3\sqrt{-2}$ **39.** $(4 - i) + (3 + 3i)$

40. $(8 - 5i) - (2 + i)$ **41.** $(7 - 6i) - (5 - 6i)$

42. $(2 - 4i) + (2 + 4i)$ **43.** $(11 - \sqrt{-3}) - (-4 + \sqrt{-5})$

44. $(4 + i)(4 - i)$ **45.** $(4 - i)(3 + 2i)$

46. $(3 - 4i)^2$ **47.** $(2 - \sqrt{-3})(2 + \sqrt{-3})$

48. $3(-5 - 2i) + 2(-3 + 2i)$ **49.** $(3 + 2i)^2 + (3 + 4i)^2$

Solve each equation.

50. $-6x^2 - 30 = 0$ **51.** $5x^2 + 40 = 0$

52. $3x^2 + 18 = 0$ **53.** $7x^2 + 84 = 0$

54. $\frac{2}{3}x^2 + 30 = 0$ **55.** $4x^2 + 5 = 0$

Simplify.

56. $(-6 + 2i)(7 - i)(4 + 3i)$ **57.** $(7 - 5i)(7 + 5i)(2 - 3i)$

58. $(7 - i)(4 + 2i)(5 + 2i)$ **59.** $(2 + i)(1 + 2i)(3 - 4i)$

Find the values of *m* and *n* that make each equation true.

60. $18 + 7i = 3m + 2ni$ **61.** $(2m + n) + (m - n)i = 7 - i$

62. $(m + 2n) + (2m - n)i = 5 + 5i$ **63.** $(2m - 3n)i + (m + 4n) = 13 + 7i$

64. Give an example to demonstrate that the product of two complex numbers in the form $a + bi$, where $a \neq 0$ and $b \neq 0$ may not result in a complex number of the same form.

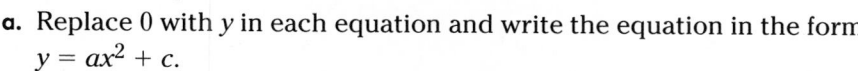

65. Refer to the equations in Exercises 50–55.
  a. Replace 0 with y in each equation and write the equation in the form $y = ax^2 + c$.
  b. Graph each equation and make a sketch of the graph.
  c. What characteristic(s) do all of the graphs have in common?
  d. How does the statement "the real roots of a function are its x-intercepts" relate to what you found in your graphs?

Critical Thinking

66. **Number Theory** Under which of the operations—addition, subtraction, or multiplication—is the set of imaginary numbers closed? Give examples to support your answer.

67. Show that $2 - 3i$ is a solution of $x^2 - 4x + 13 = 0$. Are there other solutions? Explain.

Applications and Problem Solving

68. **Electricity** Refer to the information in Example 7.
  a. A circuit has a current of $(7 + 3j)$ amps and an impedance of $(5 - j)$ ohms. What will the voltage of the circuit be?
  b. A circuit has been tested to have a current of $(10 + 5j)$ amps. The current needed to upgrade the circuit is $(50 + 20j)$ amps. How many additional amps are needed to upgrade the circuit?

69. Quantum Mechanics Wolfgang Pauli (1900–1958) won the 1945 Nobel Prize for physics for his discovery of the Pauli exclusion principle. The principle states that in an atom, no two electrons can have the same energy. In his study of electron spin, he used matrices that have become known as the Pauli spin matrices. These matrices are $A = \begin{bmatrix} 0 & 1 \\ 1 & 0 \end{bmatrix}$, $B = \begin{bmatrix} 0 & -i \\ i & 0 \end{bmatrix}$, and $C = \begin{bmatrix} 1 & 0 \\ 0 & -1 \end{bmatrix}$. Suppose matrix $D = \begin{bmatrix} -1 & 0 \\ 0 & -1 \end{bmatrix}$.

  a. Verify that AB equals the product of D and BA.
  b. Verify that CB equals the product of D and BC.

70. Look for a Pattern Find the values for all of the powers of i from i^0 to i^{12}.
  a. What pattern do you notice in the values?
  b. Describe how you would simplify i^n, where n is a positive integer.
  c. Evaluate each expression. Assume that k is an integer.
    (1) i^{4k} (2) i^{4k+1} (3) i^{4k+2} (4) i^{4k+3}
  d. Use your findings in part c to evaluate each expression.
    (1) i^{784} (2) i^{503} (3) $i^{8,413,634}$

Mixed Review

71. Solve $\sqrt{3y^2 + 11y - 5} = y\sqrt{3} + 1$. (Lesson 5–8)

72. Identify Subgoals What fraction of the perfect squares between 0 and 100 are odd? (Lesson 5–7)

73. Physics If a stone is dropped from a cliff, the equation $t = \frac{1}{4}\sqrt{d}$ represents the time t in seconds that it takes for the stone to reach the ground. If d represents the distance in feet that the stone falls, find how long it would take for a stone to fall from a 150-foot cliff. (Lesson 5–6)

74. Simplify $3p(p^2 - 2p + 3)$. (Lesson 5–2)

75. Solve the system of equations by using augmented matrices. (Lesson 4–7)
$$x + 3y - 2z = 9$$
$$-x + 5y + 2z = 31$$
$$2x - 9z = -32$$

76. Find $3\begin{bmatrix} 3 & -2 & 5 \\ 2 & 7 & -5 \end{bmatrix} + 2\begin{bmatrix} -1 & 3 & 4 \\ 2 & -3 & 0 \end{bmatrix}$. (Lesson 4–2)

77. Decorating Carol and Frank are buying some new living room furniture. A sofa, loveseat, and coffee table cost $2100. The sofa costs twice as much as the love seat. The sofa and the coffee table cost $1510. What are the prices of each piece of furniture? (Lesson 3–7)

78. Solve the system of equations by using the elimination method. (Lesson 3–2)
$$3x - 2y = 7$$
$$-3x + 9y = 14$$

79. Write an equation in standard form for the line whose x-intercept is 6 and y-intercept is -5. (Lesson 2–4)

80. Graph $2x + 3y = 12$. (Lesson 2–2)

81. A number increased by 27 is 46. Find the number. (Lesson 1–4)

82. State the property illustrated by $3 + (a + b) = (a + b) + 3$. (Lesson 1–2)

Simplifying Expressions Containing Complex Numbers

What YOU'LL LEARN

- To simplify rational expressions containing complex numbers in the denominator.

Why IT'S IMPORTANT

You can use complex numbers to solve problems involving fractals and electricity.

APPLICATION
Electricity

We take electricity for granted as a part of our everyday life. In most third-world countries, only large metropolitan areas have access to electricity. The International Foundation for the Promotion of New and Emerging Sciences and Technology (NEST) has been working to bring electricity to smaller rural communities.

In the last lesson, you used the formula $E = I \cdot Z$ to solve problems dealing with electrical circuits. Impedance Z is the resistance to the flow of electricity and is measured in ohms. Voltage E refers to the electrical potential in a circuit and is measured in volts. Current I is measured in amperes (amps). If a circuit of 110 volts were to be installed in a house, what would the impedance be for a current of $15 + 3j$ amps?

$$E = I \cdot Z$$
$$110 = (15 + 3j)Z \quad \text{\textit{E = 110, I = 15 + 3j}}$$
$$\frac{110}{15 + 3j} = Z \qquad \text{\textit{Divide each side by (15 + 3j).}}$$

Remember that in electricity j is used instead of i to represent the imaginary unit.

Since j represents a radical, this expression is not in simplest form. *You will simplify this expression in Example 3.*

Remember that two radical expressions $a\sqrt{b} + c\sqrt{d}$ and $a\sqrt{b} - c\sqrt{d}$ are conjugates. Since imaginary numbers also involve radicals, numbers of the form $a + bi$ and $a - bi$ are called **complex conjugates**. Recall that the product of two radical conjugates is a rational number. Let's investigate the product of two complex conjugates.

Example **Simplify $(15 + 3i)(15 - 3i)$.**

$$(15 + 3i)(15 - 3i) = 15^2 - 9i^2 \quad \text{\textit{Difference of two squares}}$$
$$= 225 - (-9) \quad \text{\textit{9i}}^2 = 9(-1)$$
$$= 234$$

The product in Example 1 is rational. Let's explore the general case $(a + bi)(a - bi)$.

$$(a + bi)(a - bi) = a^2 - (bi)^2 \quad \text{\textit{Difference of two squares}}$$
$$= a^2 - b^2i^2$$
$$= a^2 + b^2 \qquad \text{\textit{b}}^2i^2 = b^2(-1)$$

Since a and b are real numbers, $a^2 + b^2$ will also be real. Thus, the product of complex conjugates is a real number.

We use conjugates of radicals to rationalize the denominators of expressions with radicals in the denominator. We can also use conjugates of complex numbers to rationalize denominators of expressions with complex numbers in the denominator.

Example ② **Simplify each expression.**

LOOK BACK

You can review rationalizing the denominator in Lesson 5-6.

a. $\dfrac{8i}{1+3i}$

$$\dfrac{8i}{1+3i} = \dfrac{8i}{1+3i} \cdot \dfrac{1-3i}{1-3i} \quad \begin{array}{l} 1+3i \text{ and} \\ 1-3i \text{ are} \\ \text{conjugates.} \end{array}$$

$$= \dfrac{8i(1-3i)}{1+3^2}$$

$$= \dfrac{8i-24i^2}{10}$$

$$= \dfrac{8i+24}{10}$$

$$= \dfrac{4i+12}{5} \text{ or } \dfrac{12+4i}{5}$$

b. $\dfrac{2+i}{5i}$

$$\dfrac{2+i}{5i} = \dfrac{2+i}{5i} \cdot \dfrac{i}{i} \quad \begin{array}{l} \textit{Why multiply by } \dfrac{i}{i} \\ \textit{instead of } \dfrac{5i}{5i}? \end{array}$$

$$= \dfrac{2i+i^2}{5i^2}$$

$$= \dfrac{2i-1}{-5} \text{ or } \dfrac{-1(1-2i)}{-1(5)}$$

$$= \dfrac{1-2i}{5}$$

Example ③ **Refer to the application at the beginning of the lesson. What would the impedance Z be for a current of $15 + 3j$ amps in a 110-volt circuit?**

APPLICATION

Electricity

We found that $Z = \dfrac{110}{15+3j}$.

$$Z = \dfrac{110}{15+3j} \cdot \dfrac{15-3j}{15-3j} \quad \textit{Rationalize the denominator.}$$

$$= \dfrac{110(15-3j)}{234} \qquad (15+3j)(15-3j) = 15^2 + 3^2 \text{ or } 234$$

$$= \dfrac{1650-330j}{234}$$

$$\approx 7.05 - 1.41j$$

The impedance would be approximately $7.05 - 1.41j$ ohms.

The picture at the left, called a **fractal,** was created with the aid of a computer. Benoit Mandelbrot coined the word *fractal* as a label for computer-generated, irregular and fragmented, self-similar shapes. Fractal objects are created using functions that are **iterated,** that is, repeated over and over. The function is evaluated for some initial value of x, and then the function is evaluated again using that result. Repeating this process and plotting the points produces interesting and sometimes beautiful pictures. Some of the most exciting fractals are created using this process with a number like $4 + 3i$ as the initial value.

4 Suppose the function $f(x) = \dfrac{1}{x^2 + 1}$ is to be iterated to produce a fractal.
Find the first two points of the iteration if the initial value is $(1 + i)$.

1st iteration

Replace x with $(1 + i)$.

$$f(1 + i) = \frac{1}{(1 + i)^2 + 1}$$

$$= \frac{1}{1 + 2i - 1 + 1}$$

$$= \frac{1}{1 + 2i}$$

$$= \frac{1}{1 + 2i} \cdot \frac{1 - 2i}{1 - 2i}$$

$$= \frac{1 - 2i}{5} \text{ or } \frac{1}{5} - \frac{2i}{5}$$

2nd iteration

Replace x with $\dfrac{1 - 2i}{5}$.

$$f\left(\frac{1 - 2i}{5}\right) = \frac{1}{\left(\dfrac{1 - 2i}{5}\right)^2 + 1}$$

$$= \frac{1}{\dfrac{(1 - 4i - 4) + 25}{25}}$$

$$= \frac{25}{22 - 4i}$$

$$= \frac{25}{2(11 - 2i)} \cdot \frac{11 + 2i}{11 + 2i}$$

$$= \frac{275 + 50i}{2(121 + 4)}$$

$$= \frac{275 + 50i}{250} \text{ or } \frac{11}{10} + \frac{i}{5}$$

The first two points of the iteration are $\dfrac{1}{5} - \dfrac{2i}{5}$ and $\dfrac{11}{10} + \dfrac{i}{5}$.

CHECK FOR UNDERSTANDING

Communicating Mathematics

Study the lesson. Then complete the following.

1. **Describe** how to rationalize the denominator of $\dfrac{1}{a + bi}$.

2. **Explain** why the product of a complex number and its conjugate is always a real number.

3. **Evaluate** each expression if $z = 1 + 2i$.

   a. z^2

   b. $\dfrac{1}{z}$

   c. What would you think the product $z^2 \cdot \dfrac{1}{z}$ would be without substituting $1 + 2i$ for z? Use substitution to verify your answer.

4. Does every complex number have a multiplicative inverse and an additive inverse?

Guided Practice

Find the conjugate of each complex number.

5. $7i$

6. $3 + 5i$

Find the product of each complex number and its conjugate.

7. $-10i$

8. $12 + 5i$

Simplify.

9. $\dfrac{7}{-2i}$

10. $\dfrac{9 + 3i}{2i}$

11. $\dfrac{5}{2 + i}$

12. $\dfrac{5 + i}{1 + 2i}$

13. $\dfrac{3 - 2i}{1 - i}$

14. $\dfrac{7}{\sqrt{2} - 3i}$

15. Show that $7 + 3i$ and $\dfrac{7 - 3i}{58}$ are multiplicative inverses of each other.

Practice   **Find the conjugate of each complex number.**

16. $10i$ 17. $12 + i$ 18. $-15i$

19. $10 - 4i$ 20. $7 + i\sqrt{5}$ 21. $6 + \sqrt{-7}$

Find the product of each complex number and its conjugate.

22. $-2i$ 23. $5 - 2i$ 24. $4 + 6i$

25. $1 + i$ 26. $3 + 5i\sqrt{2}$ 27. $8 - 2i$

Simplify.

28. $\dfrac{2 + 8i}{3i}$ 29. $\dfrac{3 + 7i}{2i}$ 30. $\dfrac{11 + i}{2 - i}$

31. $\dfrac{3i}{2 + i}$ 32. $\dfrac{-3i}{5 + 4i}$ 33. $\dfrac{3}{6 + 4i}$

34. $\dfrac{3 + 5i}{1 + i}$ 35. $\dfrac{2 + i}{3 - i}$ 36. $\dfrac{1 - i}{4 - 5i}$

37. $\dfrac{2 + 3i}{3 - 2i}$ 38. $\dfrac{5 - 6i}{-3i}$ 39. $\dfrac{3 - 9i}{4 + 2i}$

40. $\dfrac{8}{\sqrt{2} + i}$ 41. $\dfrac{1}{3 - i\sqrt{2}}$ 42. $\dfrac{4}{\sqrt{3} + 2i}$

Find the multiplicative inverse of each complex number.

43. $6 - 5i$ 44. $\dfrac{-i}{3 + 5i}$ 45. $x + yi$

INTEGRATION

Fractal Geometry

Find the 1st and 2nd iteration of each function for the given initial value.

46. $f(x) = x^2 + 2$ for $x = 1 + i$

47. $f(x) = 3x^2 + 2$ for $x = 1 - i$

48. $f(x) = x^2 - x$ for $x = i + 3$

Simplify.

49. $\dfrac{3 - i\sqrt{5}}{3 + i\sqrt{5}}$ 50. $\dfrac{1 + i\sqrt{3}}{1 - i\sqrt{3}}$

51. $\dfrac{1 - i}{(1 + i)^2}$ 52. $\left(\dfrac{\sqrt{3}}{2 + 3i}\right)^2$

53. $\dfrac{(2 + 3i)^2}{(3 + i)^2}$ 54. $\dfrac{(4 + 3i)^2}{(3 - 4i)^2}$

Critical Thinking

55. If $z = a + bi$ and $w = c + di$ are complex numbers, then their conjugates are denoted by $\overline{z} = a - bi$ and $\overline{w} = c - di$, respectively. Determine if the conjugate of the product zw is equal to the product of the conjugates. That is, would $\overline{z \cdot w} = \overline{z} \cdot \overline{w}$? Why or why not?

56. Show that $\dfrac{-1 + i\sqrt{3}}{2}$ is a cube root of 1.

57. Electrical Engineering   Refer to the application at the beginning of the lesson. Copy and complete the table.

	a.	b.	c.	d.
E (volts)	$60 + 112j$	$85 + 110j$	$-50 + 100j$	$-70 + 240j$
Z (ohms)	$10 + 6j$	$3 - 4j$		
I (amps)			$-6 + 2j$	$-5 + 4j$

58. Electrical Circuitry   In a two-battery flashlight, the positive terminal of the first battery touches the negative terminal of the second. The positive terminal of the second battery touches the center terminal of the light bulb. A metal strip connects the bulb to the switch, which is connected to the negative terminal of the first battery. When you turn the flashlight on, the switch completes the circuit, and the bulb lights up.

a. Suppose each of the batteries is 2.5 volts. What is the total impedance of the circuit in the flashlight if the current is $(1 + 2j\sqrt{2})$?

b. What is the total current if two 1.5 volt batteries are used and the impedance is $(2 + 3j)$ ohms?

59. Fractal Geometry   Suppose the function $f(x) = x^2 - 1$ is to be iterated to produce a fractal. Find the first four points of the iteration if the initial value of x is $(1 + i)$. Write the points as ordered pairs.

Solve each equation.

60. $3a^2 + 24 = 0$ (Lesson 5–9)

61. $2x + 7 = -x\sqrt{2}$ (Lesson 5–8)

62. Factor $(x + y)^2 - \frac{1}{4}$.   (Lesson 5–4)

63. Retail   The store Bunches of Boxes and Bags assembles boxes to package items for mailing. The store manager found that the volume of a box made from a piece of cardboard with a square of length x inches cut from each corner is $(4x^3 - 168x^2 + 1728x)$ in^3. If the piece of cardboard is 48 inches long, how wide is it?   (Lesson 5–3)

64. Find the inverse of $\begin{bmatrix} 3 & 1 \\ -4 & 1 \end{bmatrix}$.   (Lesson 4–5)

Solve each system algebraically or by graphing.

65. $6x - 2y - 3z = -10$
$-6x + y + 9z = 3$
$8x - 3y = -16$ (Lesson 3–7)

66. $x > 1$
$y < -1$
$y < x$ (Lesson 3–4)

67. Is the relation $\{(0, 0), (1, 0)\}$ a function? Explain your answer.   (Lesson 2–1)

68. Solve $3(2m - 3) \geq 9$.   (Lesson 1–6)

69. Evaluate $[(-8 + 3) \times 4 - 2] \div 6$.   (Lesson 1–1)

3-2-1-Blast-Off!

Refer to the Investigation on pages 180–181.

When a new launch system is needed and a contract is given to a research company, the company employs engineers, scientists, physicists, drafters, and others to work together toward the final plans for the launcher. Accuracy is examined as well as the cost factor. Many of the skills you have used in this Investigation are used by professionals to design the right launcher for the job.

Analyze

You have conducted experiments and organized your data in various ways. It is now time to analyze your findings and state your conclusions.

> **PORTFOLIO ASSESSMENT**
>
> You may want to keep your work on this Investigation in your portfolio.

1 Look over the data and organize a table of all the data from the shoot-off including all the distances of each team.

2 Make a list of statistical measures and accuracy measures calculated from each team's data. These should include two sets of measures: the median, quartiles, least and greatest values, and interquartile range; the error range, tolerance, relative error, average error, and standard deviation. Did the data collected from the two sets of measures lead you to the same conclusion about the launcher's accuracy? Explain.

3 Analyze each of the launchers that were used in the shoot-off and rank them from most accurate to least accurate. Use mathematical measures to justify your rankings. Explain what measures you used and how you used them to evaluate the accuracy of the launchers.

Write

The government has asked you for a report on both the design and calibration of your launcher and an analysis of the results of the shoot-off.

4 Summarize the procedures you used to design the launcher. Describe the steps and procedures used to design your launcher, from the receipt of the instructions to the demonstration shoot-off.

5 Draw a detailed blueprint of the launch system that specifies its parts and the construction of the system.

6 Include in this report an operation manual that describes how the launcher operates. Include a description of the calibration system used for accuracy. A list or table that compares the launch system settings with actual shot distances should also be included. Describe the mathematical relationship between the settings and target distances.

7 Summarize the accuracy of the launch system. Include test data recorded from shots taken at each distance: 50 cm, 100 cm, 150 cm, 200 cm, and 250 cm. A graph depicting the location of the test shots and the target distance should also be included.

VOCABULARY

After completing this chapter, you should be able to define each
term, property, or phrase and give an example or two of each.

Algebra
binomial   (p. 261)
coefficient   (p. 255)
complex conjugates   (p. 317)
complex number   (p. 311)
conjugates   (p. 292)
constant   (p. 255)
degree   (p. 255)
extraneous solution
   (p. 305)
factors   (p. 274)
FOIL method   (p. 263)
imaginary unit   (p. 310)
iterate   (p. 318)

like radical expressions
   (p. 291)
like terms   (p. 261)
monomial   (p. 255)
nth root   (p. 282)
polynomial   (p. 261)
power   (p. 255)
prime number   (p. 274)
principal root   (p. 282)
pure imaginary number
   (p. 310)
radical equations   (p. 305)
radical inequalities   (p. 306)
rational exponent   (p. 298)

rationalizing the denominator
   (p. 290)
scientific notation   (p. 254)
simplify   (p. 255)
square root   (p. 281)
synthetic division   (p. 269)
term   (p. 261)
trinomial   (p. 261)

Geometry
fractal   (p. 318)

Problem Solving
identify subgoals   (p. 296)

UNDERSTANDING AND USING THE VOCABULARY

Choose a word or term that best completes each statement or phrase.

1. A number is expressed in _____ when it is in the form of $a \cdot 10^n$, where $1 \le a < 10$ and n is an integer.
2. Monomials that contain no variables are known as _____.
3. _____ is the process used to eliminate radicals from the denominator or fractions from a radicand.
4. A shortcut method known as _____ is used to divide polynomials by binomials.
5. Real numbers that cannot be written as terminating or repeating decimals are _____.
6. The _____ is used to multiply two binomials.
7. In an algebraic term that is the product of a number and a variable, the number is the _____ of the variable.
8. A _____ is an expression that is a number, a variable, or the product of a number and one or more variables.
9. A solution of a transformed equation that is not a solution of the original equation is an
   _____.
10. _____ are imaginary numbers of the form $a + b\boldsymbol{i}$ and $a - b\boldsymbol{i}$.
11. For any number a and b, if $a^2 = b$, then a is the _____ of b.
12. A polynomial comprised of three unlike terms is known as a _____.
13. The _____ is the degree of the monomial of the greatest degree.
14. When there is more than one root, the _____ is the nonnegative root.
15. _____ are computer-generated, irregular and fragmented, self-similar shapes created using functions that are iterated—that is, repeated over and over.

SKILLS AND CONCEPTS

Upon completing this chapter, you should be able to:

Use these exercises to review and prepare for the chapter test.

• multiply and divide monomials   (Lesson 5–1)

$(3x^4y^6)(-8x^3y)$

$\quad = (3)(-8)x^{4+3}y^{6+1}$ *Multiplying powers*

$\quad = -24x^7y^7$

Simplify. Assume that no variable equals 0.

16. $m^3 \cdot m^5$

17. $f^{-7} \cdot f^4$

18. $(3x^2)^3$

19. $(2y)(4xy^3)$

20. $\left(\dfrac{3}{5}c^2f\right)\left(\dfrac{4}{3}cd\right)^2$

21. $\dfrac{1}{x^0 + y^0} - \dfrac{x^0 + y^0}{1}$

22. $3(ab)^3(4ac^2) + c(4ab)(5a^3b^2c)$

• represent numbers in scientific notation
(Lesson 5–1)

$31{,}000 = 3.1 \times 10{,}000$

$\quad\quad\quad = 3.1 \times 10^4$

$0.007 = 7 \times 0.001$

$\quad\quad\quad = 7 \times 10^{-3}$

Evaluate. Express each answer in both scientific and decimal form.

23. $(2000)(85{,}000)$

24. $(0.0014)^2$

25. $5{,}400{,}000 \div 6000$

• add, subtract, and multiply polynomials
(Lesson 5–2)

$(5x^2 + 4x) - (3x^2 + 6x - 7)$

$\quad = 5x^2 + 4x - 3x^2 - 6x + 7$

$\quad = (5x^2 - 3x^2) + (4x - 6x) + 7$

$\quad = 2x^2 - 2x + 7$

$(9k + 4)(7k - 6)$

$\quad = (9k)(7k) + (9k)(-6) + (4)(7k) + (4)(-6)$

$\quad = 63k^2 - 54k + 28k - 24$

$\quad = 63k^2 - 26k - 24$

Simplify.

26. $(4c - 5) - (c + 11) + (-6c + 17)$

27. $(11x^2 + 13x - 15) - (7x^2 - 9x + 19)$

28. $-6m^2(3mn + 13m - 5n)$

29. $(d - 5)(d + 3)$

30. $x^{-8}y^{10}\,(x^{11}y^{-9} + x^{10}y^{-6})$

31. $(2a^2 + 6)^2$

32. $-5f^{12}(4f^3g + 2f)$

33. $(2b - 3c)^3$

• divide polynomials by binomials using synthetic
division   (Lesson 5–3)

Find $(4x^4 - x^3 - 19x^2 + 11x - 2) \div (x - 2)$.

$$4x^4 - x^3 - 19x^2 + 11x - 2$$

$$
\begin{array}{r|rrrrr}
2 & 4 & -1 & -19 & 11 & -2 \\
  &   & 8 & 14 & -10 & 2 \\
\hline
  & 4 & 7 & -5 & 1 & 0 \\
\end{array}
$$

The quotient is $4x^3 + 7x^2 - 5x + 1$.

Find each quotient.

34. $(2x^4 - 6x^3 + x^2 - 3x - 3) \div (x - 3)$

35. $(10x^4 + 5x^3 + 4x^2 - 9) \div (x + 1)$

36. $(x^2 - 5x + 4) \div (x - 1)$

37. $(5x^4 + 18x^3 + 10x^2 + 3x) \div (x^2 + 3x)$

CHAPTER 5   STUDY GUIDE AND ASSESSMENT

OBJECTIVES AND EXAMPLES

• factor polynomials   (Lesson 5–4)

$4x^3 - 6x^2 + 10x - 15$

$= (4x^3 - 6x^2) + (10x - 15)$

$= 2x^2(2x - 3) + 5(2x - 3)$

$= (2x^2 + 5)(2x - 3)$

• simplify radicals having various indices
(Lesson 5–5)

$\pm\sqrt{81x^4} = \pm\sqrt{(9x^2)^2}$ or $\pm 9x^2$

$\sqrt[7]{2187x^{14}y^{35}} = \sqrt[7]{(3x^2y^5)^7}$ or $3x^2y^5$

• add, subtract, multiply, and divide radical
expressions   (Lesson 5–6)

$6\sqrt[5]{32m^3} \cdot 5\sqrt[5]{1024m^2}$

$= 6 \cdot 5\sqrt[5]{(32m^3 \cdot 1024m^2)}$

$= 30\sqrt[5]{2^5 4^5 m^5}$

$= 30\sqrt[5]{2^5} \cdot \sqrt[5]{4^5} \cdot \sqrt[5]{m^5}$

$= 30 \cdot 2 \cdot 4 \cdot m$ or $240m$

• evaluate expressions in either exponential or
radical form   (Lesson 5–7)

$32^{\frac{4}{5}} \cdot 32^{\frac{2}{5}} = 32^{\left(\frac{4}{5}+\frac{2}{5}\right)}$

$= 32^{\frac{6}{5}}$

$= (2^5)^{\frac{6}{5}}$

$= 2^6$ or 64

$\dfrac{3x}{y^{-\frac{3}{2}} \cdot \sqrt{z}} = \dfrac{3xy^{\frac{3}{2}}}{z^{\frac{1}{3}}} \cdot \dfrac{z^{\frac{2}{3}}}{z^{\frac{2}{3}}}$

$= \dfrac{3xy^{\frac{3}{2}}z^{\frac{2}{3}}}{z}$

• solve equations containing radicals   (Lesson 5–8)

$\sqrt{3x - 8} + 1 = 3$

$\sqrt{3x - 8} = 2$

$(\sqrt{3x - 8})^2 = 2^2$

$3x - 8 = 4$

$x = 4$

REVIEW EXERCISES

Factor completely. If the polynomial is not factorable, write *prime*.

38. $200x^2 - 50$

39. $10a^3 - 20a^2 - 2a + 4$

40. $5w^3 - 20w^2 + 3w - 12$

41. $s^3 + 512$ 42. $x^2 - 7x + 5$

Simplify.

43. $\pm\sqrt{256}$ 44. $\sqrt[3]{-216}$

45. $\sqrt{-(-8)^2}$ 46. $\sqrt[5]{c^5 d^{15}}$

47. $\pm\sqrt{(x^4 - 3)^2}$ 48. $\sqrt[3]{(512 + x^2)^3}$

49. $\sqrt[4]{16m^8}$ 50. $\sqrt{a^2 - 10a + 25}$

Simplify.

51. $\sqrt[4]{64}$ 52. $\sqrt{5} + \sqrt{20}$

53. $5\sqrt{12} - 3\sqrt{75}$ 54. $6\sqrt[5]{11} - 8\sqrt[5]{11}$

55. $(\sqrt{8} + \sqrt{12})^2$ 56. $\sqrt{8} \cdot \sqrt{15} \cdot \sqrt{21}$

57. $\dfrac{1}{3 + \sqrt{5}}$ 58. $\dfrac{\sqrt{10}}{4 + \sqrt{2}}$

Evaluate.

59. $27^{-\frac{2}{3}}$ 60. $9^{\frac{1}{3}} \cdot 9^{\frac{5}{3}}$ 61. $\left(\dfrac{8}{27}\right)^{-\frac{2}{3}}$

Simplify.

62. $\dfrac{1}{y^{\frac{2}{5}}}$ 63. $\dfrac{xy}{\sqrt[3]{z}}$ 64. $\dfrac{3x + 4x^2}{x^{-\frac{2}{3}}}$

Solve each equation. Be sure to check for extraneous solutions.

65. $y^{\frac{1}{3}} - 7 = 0$ 66. $(x - 2)^{\frac{3}{2}} = -8$

67. $6 + 2x\sqrt{3} = 0$ 68. $\sqrt{3t - 5} - 3 = 4$

69. $\sqrt{1 + 8v} - 2 = v$ 70. $\sqrt[4]{2x - 1} = 2$

71. $\sqrt{y + 5} = \sqrt{2y - 3}$

72. $\sqrt{y + 1} + \sqrt{y - 3} = 5$

OBJECTIVES AND EXAMPLES

- add, subtract, and multiply complex numbers
  (Lesson 5–9)

$(15 - 2i) + (5i - 11) = (15 - 11) + (-2i + 5i)$

$$= 4 + 3i$$

$(2 + 3i)(4i - 11) = (2)(4i) + (2)(-11) +$
$$(3i)(4i) + (3i)(-11)$$

$$= 8i - 22 + 12i^2 - 33i$$

$$= 8i - 22 + 12(-1) - 33i$$

$$= 8i - 33i - 22 - 12$$

$$= -34 - 25i$$

- simplify rational expressions containing complex
  numbers in the denominator   (Lesson 5–10)

$\dfrac{7i}{2 + 4i} = \dfrac{7i}{2 + 4i} \cdot \dfrac{2 - 4i}{2 - 4i}$

$$= \dfrac{7i(2 - 4i)}{4 + 4^2}$$

$$= \dfrac{14i - 28i^2}{20}$$

$$= \dfrac{14i + 28}{20} \text{ or } \dfrac{7}{5} + \dfrac{7}{10}i$$

REVIEW EXERCISES

Simplify.

73. $\sqrt{-256}$

74. $\sqrt[6]{-64m^{12}}$

75. $(13i - 2)5i$

76. $(7 - 4i) - (-3 + 6i)$

77. $-6\sqrt{-a} \cdot 2\sqrt{-b}$

78. i^6

79. i^{85}

80. $(3 + 4i)(5 - 2i)$

81. $(\sqrt{6} + i)(\sqrt{6} - i)$

Simplify.

82. $\dfrac{1 + i}{1 - i}$

83. $\dfrac{4 - 3i}{4 + 3i}$

84. $\dfrac{4}{4 + 5i}$

85. $\dfrac{1 + i\sqrt{2}}{1 - i\sqrt{2}}$

APPLICATIONS AND PROBLEM SOLVING

86. **Aerospace Engineering**   Scientists expect that on future space stations, artificial gravity will be created by rotating all or part of the space station. The formula $N = \dfrac{1}{2\pi}\sqrt{\dfrac{a}{r}}$ gives the number of rotations N required per second to maintain an acceleration of gravity of a meters per second squared (m/s^2) on a satellite with a radius of r meters. The acceleration of gravity on Earth is 9.8 m/s^2. How many rotations per minute will produce an artificial gravity that is equal to half of the gravity on Earth in a space station 25 meters wide?   (Lesson 5–5)

87. **Law Enforcement**   A police investigator measured the skid marks left by a car to be approximately 120 feet. The driver of the car claims that she was not exceeding the 40-mph speed limit. Is she telling the truth? (Recall that $s = 2\sqrt{5\ell}$, where s represents speed, measured in miles per hour, and ℓ represents the length of the skidmarks, measured in feet.) How fast was she driving?   (Lesson 5–6)

88. **Archaeology**   Since carbon-14 is present in all living organisms and decays at a predictable rate after death, archaeologists use the amount of carbon-14 left in a fossil to extimate the age of the fossil. This is commonly called *carbon dating*. The approximate number of milligrams A of carbon-14 left in a fossil after 5000 years can be found using the formula $A = A_0(2.7)^{-\frac{3}{5}}$, where A_0 is the initial amount of carbon-14 in the organism. Find the amount of carbon-14 left in an organism that contained 500 milligrams of carbon-14.   (Lesson 5–7)

A practice test for Chapter 5 is provided on page 916.

ALTERNATIVE ASSESSMENT

PERFORMANCE ASSESSMENT TASK

Area and Volume   An open box can be made from a piece of cardboard 30 inches wide by 36 inches long by folding congruent square tabs from each corner.

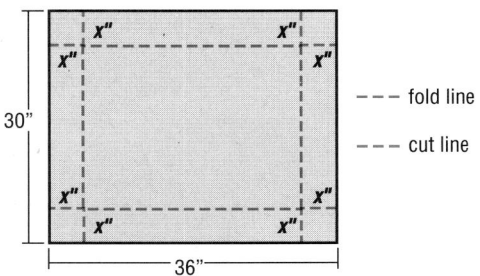

30"

--- fold line

--- cut line

36"

1. Obtain a piece of cardboard the same size as the one shown above. Mark the box as shown with 3-inch squares at each corner. Cut along the cut lines, fold in the tabs, and fold up the sides to make the box. What is the volume of the box?

2. Find the algebraic representation $V(x)$ of the volume of the box when x is the length of one side of the square tab in the corner. Make a table of values if the side length of the squares is 2, 4, 6, 8, 10, 12, and 14 inches. How does the volume of the box change when the size of the square tabs increases in size?

3. Find the algebraic representation $A(x)$ of the area of the bottom of the box. How does the area of the bottom of the box change when the size of the square tabs increases in size?

4. Is the measure of the area of the box ever equal to the measure of the volume of the box? What is the maximum volume of the box? What is the minimum volume of the box?

THINKING CRITICALLY

As you may recall, a set is closed under an operation if the result of performing the operation on any two elements of the set is an element of the set. For example, adding two whole numbers always results in a whole number.

Is the set of irrational numbers closed under any of the four basic operations? Give examples to support your answer.

PORTFOLIO

Select one of the assignments from this chapter that you found to be difficult to master. Explain why you found this to be the case. How might this assignment be presented in a different way that may help other students having the same difficulty?

SELF EVALUATION

One characteristic of a good problem solver is knowing the language and properties of mathematics. If you don't understand something, do you seek clarification from you teacher or your parents?

Assess yourself. How well do you understand the language and properties of mathematics? Without looking first, write a paragraph about the language and properties of mathematics that you have learned in this chapter. When you're done, compare it to your notes or to your textbook. Note which lessons you need to study more in order to better understand the mathematics in this chapter.

the River Canyon Bridge

MATERIALS NEEDED

- cardboard box
- scissors
- glue
- ruler
- toothpicks
- string
- pipe cleaners
- straws
- wooden craft sticks

You work in the engineering division of a company that builds bridges. Your company has just been awarded a contract to build a six-lane highway bridge across a deep river canyon. The canyon is 550 yards deep and 110 yards wide.

The walls of the canyon are very steep with an almost vertical drop. The river below the proposed bridge is very swift, with strong rapids, especially during the spring when the snow thaws. The river bed consists of soft sand.

Your team of four designers is contemplating the best design for this bridge. A traditional bridge built on stilts would be weak, especially in strong winds. The stilts would also be very expensive and difficult to construct, so you remove that design from consideration.

The company's consultant states that a bridge that includes a parabolic arch is best suited for this situation. Your task is to determine the dimensions of the bridge while ensuring that the arch is shaped like a parabola.

In this Investigation, you will examine several ways in which a bridge could be constructed over the river canyon.

1. Think about several different bridge designs. Look in books, magazines, and travel brochures to find pictures of these bridges. Research the history of spanning spaces with bridges going back to Greek and Roman designs.

2. Discuss these designs with your classmates. Narrow the list to four designs that agree with the consultant's recommendation for this project.

3. Write a proposal in which you draw and explain the possible designs. Be sure to list the pros and cons of each design in your proposal. Include any historical references that may have influenced your choices.

MODEL THE BRIDGE

4. Use a box to simulate the canyon walls. Use toothpicks, straws, wooden craft sticks, glue, string, and/or any other building materials to build a bridge to span the river canyon.

5. Make a blueprint of your bridge. Include the measurements of the span of the bridge, its height, and where it attaches to the canyon wall.

6. Test your bridge with small weights to see how durable it is.

You will continue working on this Investigation throughout Chapters 6 and 7.

Be sure to keep your chart and materials in your Investigation Folder.

The River Canyon Bridge Investigation

Working on the Investigation
Lesson 6–1, p. 340

Working on the Investigation
Lesson 6–6, p. 375

Working on the Investigation
Lesson 7–2, p. 422

Working on the Investigation
Lesson 7–6, p. 455

Closing the Investigation
End of Chapter 7, p. 468

Exploring Quadratic Functions and Inequalities

Objectives

In this chapter, you will:

- graph quadratic functions,
- solve quadratic equations
- solve problems using the guess-and-check strategy,
- analyze graphs of quadratic functions and inequalities, and
- solve problems involving standard deviation and the normal distribution.

Do you play an instrument? Many teens play an instrument, not because they have to, but because they enjoy it. Other teens have claimed that playing an instrument helps develop confidence and creativity, provides a way to make new friends, and instills a greater appreciation of other arts.

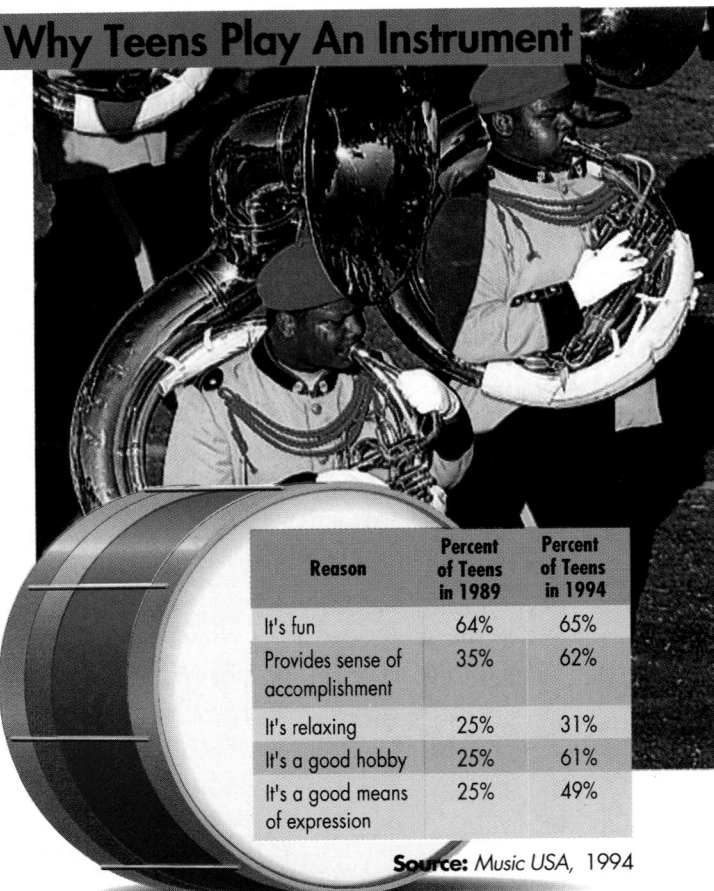

Why Teens Play An Instrument

Reason	Percent of Teens in 1989	Percent of Teens in 1994
It's fun	64%	65%
Provides sense of accomplishment	35%	62%
It's relaxing	25%	31%
It's a good hobby	25%	61%
It's a good means of expression	25%	49%

Source: *Music USA*, 1994

TIME *Line*

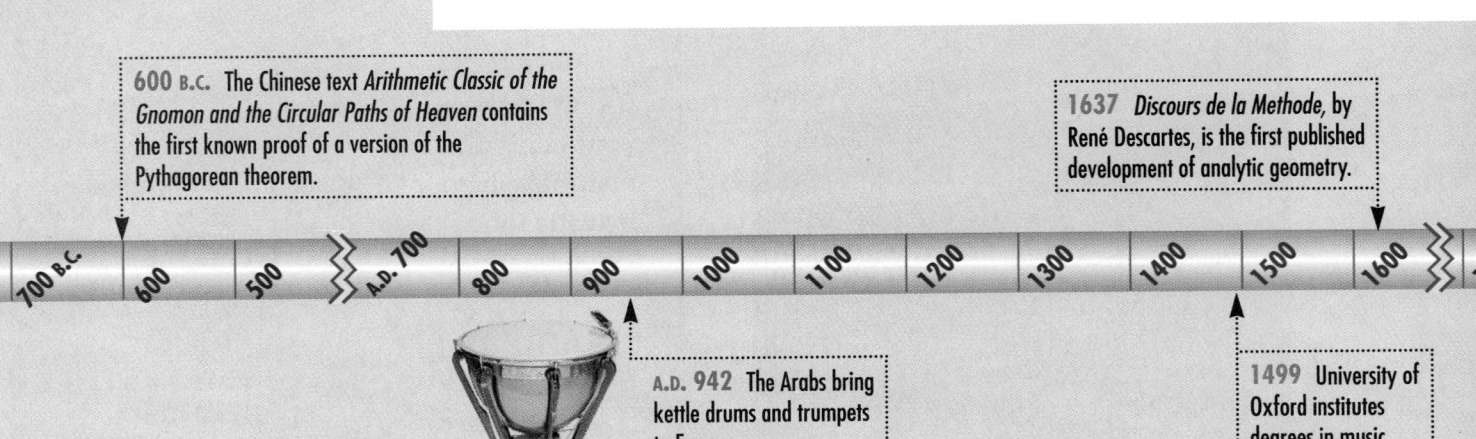

600 B.C. The Chinese text *Arithmetic Classic of the Gnomon and the Circular Paths of Heaven* contains the first known proof of a version of the Pythagorean theorem.

1637 *Discours de la Methode*, by René Descartes, is the first published development of analytic geometry.

700 B.C | 600 | 500 | A.D. 700 | 800 | 900 | 1000 | 1100 | 1200 | 1300 | 1400 | 1500 | 1600

A.D. 942 The Arabs bring kettle drums and trumpets to Europe.

1499 University of Oxford institutes degrees in music.

PEOPLE IN THE NEWS

Chapter Project

Form teams and conduct your own survey of various aspects of music and its significance in the lives of family and neighbors. Choose categories such as musical preference, CD and tape purchasing, time spent listening to or playing music, live performance attendance, instruments played, and the number of people who study music.

Keep track of the age, gender, and number of people questioned, their responses, and percentages of your sample for each question. Then see if you can make conclusions about the attitudes of people toward music.

Create charts or graphs using your data and report to the class on your findings.

Fourteen-year-old violin prodigy **Sarah Chang** began expressing herself through music in her hometown of Philadelphia at the age of 4. She has played with the New York Philharmonic, the Philadelphia Orchestra, the Chicago Symphony, and orchestras in several European cities. When she was 12, she had an album that hit the classical charts a month after it was released. Sarah, who doesn't mind practicing violin four hours a day, says, "I think performing is part of me, and I really love what I'm doing."

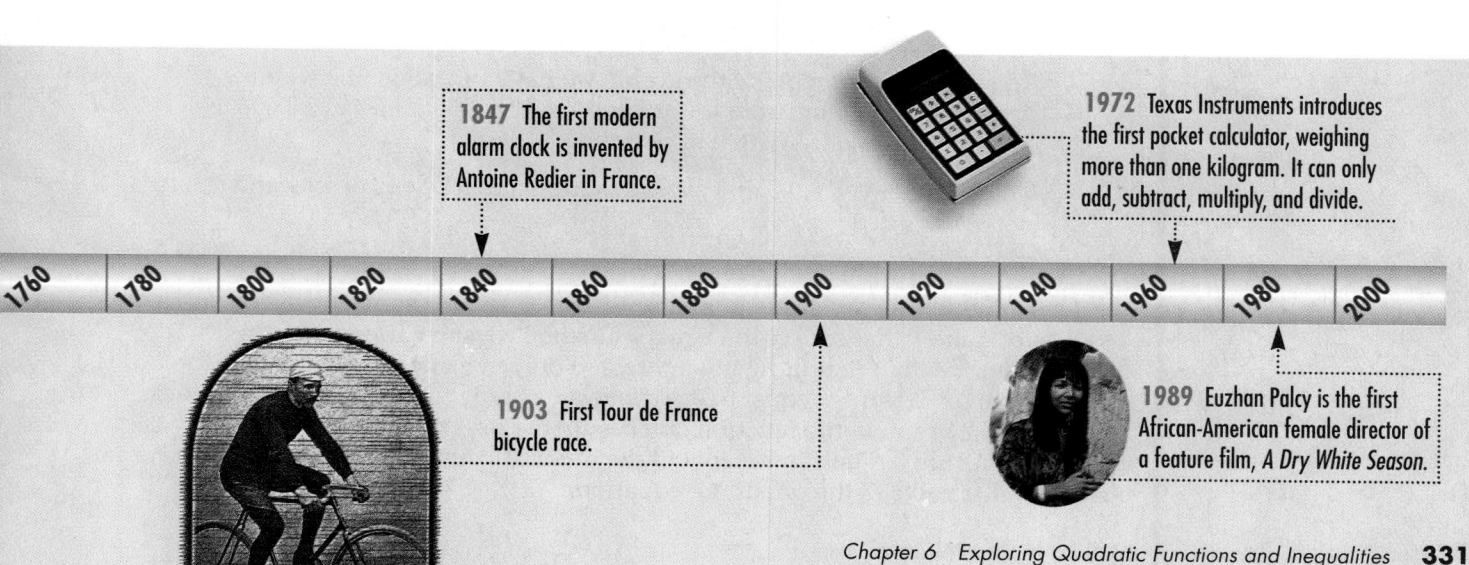

1847 The first modern alarm clock is invented by Antoine Redier in France.

1972 Texas Instruments introduces the first pocket calculator, weighing more than one kilogram. It can only add, subtract, multiply, and divide.

1760 1780 1800 1820 1840 1860 1880 1900 1920 1940 1960 1980 2000

1903 First Tour de France bicycle race.

1989 Euzhan Palcy is the first African-American female director of a feature film, *A Dry White Season*.

6-1A Graphing Technology
Quadratic Functions

A Preview of Lesson 6-1

The graphing calculator is a powerful tool for studying graphs of functions. In this lesson, we will study graphs of functions of the form $y = ax^2 + bx + c$, where $a \neq 0$. These are called **quadratic functions,** and their graphs are called **parabolas.**

Example **Graph each function in the standard viewing window.**

a. $y = x^2$

This function is of the form $y = ax^2 + bx + c$, where $a = 1$, $b = 0$, and $c = 0$.

Enter: [Y=] [X,T,θ,n] [x²]

[ZOOM] 6

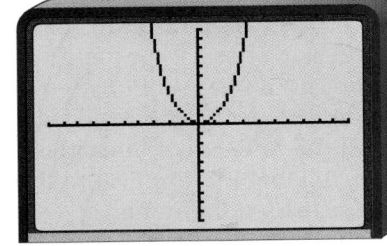

The graph of the function is shaped like a cup. This is the general shape of a parabola. *Notice that $a > 0$.*

b. $y = -0.3x^2 + 42$

The function $y = -0.3x^2 + 42$ is of the form $y = ax^2 + bx + c$, where $a = -0.3$, $b = 0$, and $c = 42$.

Enter: [Y=] [(−)] .3 [X,T,θ,n] [x²] [+] 42 [ZOOM] 6

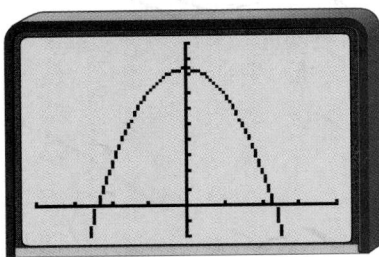

Nothing appears on the graphics screen. We must change the window to be able to view the complete graph. Press [TRACE] and notice that the coordinates of the vertex, (0, 42), are shown at the bottom of the screen. Since we cannot see the graph in the standard viewing window, change Ymax to 50. Then press [GRAPH].

To see both intercepts of the graph, we have to change the window again. Try changing Xmin to -20 and Xmax to 20. *You may want to use scale factors of 5 on both axes.*

The graph of $y = -0.3x^2 + 42$ is a parabola that opens downward. *Notice that $a < 0$.*

The graph of the quadratic function $y = ax^2 + bx + c$ represents all the values of x and y that satisfy the equation. When we solve the quadratic equation $ax^2 + bx + c = 0$, we are interested only in values of x that make the expression $ax^2 + bx + c$ equal to 0. These values are represented by the points at which the graph of the function crosses the x-axis, since the y values of these points are 0. The x-intercepts of the quadratic function are called the **solutions** or **roots** of the quadratic equation.

There are three possible outcomes when solving a quadratic equation. The equation will have two real solutions, one real solution, or no real solutions.

Example **2** **Use a graphing calculator to solve $2x^2 - 6x - 4 = 0$ to the nearest hundredth.**

Begin by graphing the related function $y = 2x^2 - 6x - 4$ in the standard viewing window.

Enter: [Y=] 2 [X,T,θ,n] [x²] [−] 6 [X,T,θ,n] [−] 4 [ZOOM] 6

TRACE to one of the x-intercepts and press [ZOOM] 2 [ENTER]. The more times you repeat this process, the more accurate your answer will be.

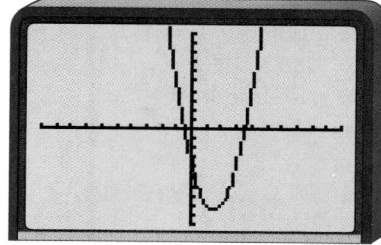

The solutions of this equation are -0.56 and 3.56, to the nearest hundredth.

You can also use the ROOT feature of the calculator to find the solutions automatically. This involves defining an interval that includes the root. A lower bound is a point on the graph just to the left of the solution and an upper bound is a point on the graph just to the right of the solution.

Enter: [2nd] [CALC] 2

Using the arrow keys, locate a lower bound for the first solution and press [ENTER]. Similarly, locate an upper bound for the root and press [ENTER].

Finally, after the Guess? prompt, press [ENTER].

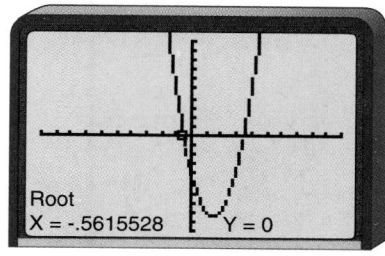

Root
X = -.5615528   Y = 0

Repeat this process for the second root.

It should be noted that solving equations graphically provides only approximate solutions. While approximate solutions are adequate for many applications, if an exact answer is required, an algebraic technique usually must be used.

EXERCISES

Determine a viewing window that gives a complete graph of each function.

1. $y = 4x^2 + 11$
2. $y = 7.5x^2 + 9.5$
3. $y = 6x^2 + 250x + 725$
4. $y = x^2 + 4x - 15$
5. $y = -2x^2 - x - 15$
6. $y = x^2 + 30x + 225$

Solve each quadratic equation to the nearest hundredth by using a graphing calculator.

7. $4x^2 + 11 = 0$
8. $7.5x^2 + 9.5 = 0$
9. $6x^2 + 250x + 725 = 0$
10. $x^2 + 4x - 15 = 0$

What YOU'LL LEARN

- To write functions in quadratic form,
- to graph quadratic functions, and
- to solve quadratic equations by graphing.

Why IT'S IMPORTANT

You can graph quadratic functions to solve problems involving space science and physics.

APPLICATION

Space Science

In the movie *Apollo 13,* actors Tom Hanks, Kevin Bacon, and Bill Paxton play real-life astronauts Jim Lovell, John Swigart, and Fred Haise in the story of the 1970 moon mission that nearly ended in disaster. During the filming of the movie, the actors trained on an airplane called the "Vomit Comet"—a stripped-down KC-135 used to help astronauts get accustomed to weightlessness. After climbing to a designated height, weightlessness begins as the jet arcs over a 36,000-foot peak and then dives toward the ground. For 23 seconds, the occupants are weightless.

The path of the Vomit Comet can be represented by the graph of a **quadratic function.** A quadratic function is a function described by an equation that can be written in the form $f(x) = ax^2 + bx + c$, where $a \neq 0$. In a quadratic function, ax^2 is called the **quadratic term**, bx is the **linear term,** and c is the **constant term.**

Example **Write each function in quadratic form. Identify the quadratic term, the linear term, and the constant term.**

a. $f(x) = 2x^2 + 7 - 9x$

In quadratic form, the function is written as $f(x) = 2x^2 - 9x + 7$. The quadratic term is $2x^2$, the linear term is $-9x$, and the constant term is 7.

b. $f(x) = (x + 4)^2 - 20$

$$
\begin{aligned}
f(x) &= (x + 4)^2 - 20 \\
&= x^2 + 8x + 16 - 20 \quad \textit{Square } x + 4. \\
&= x^2 + 8x - 4 \quad\quad\;\; \textit{Simplify.}
\end{aligned}
$$

The quadratic term is x^2, the linear term is $8x$, and the constant term is -4.

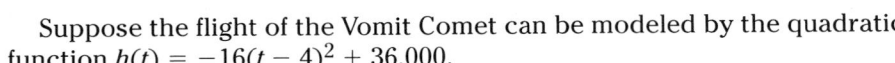

Suppose the flight of the Vomit Comet can be modeled by the quadratic function $h(t) = -16(t - 4)^2 + 36,000$.

To graph this function, we can create a table of values, as shown below.

Next, we can plot the points $(t, h(t))$, where height is in thousands of feet.

t	$-16(t - 4)^2 + 36,000$	$h(t)$
0	$-16(-4)^2 + 36,000$	35,744
4	$-16(0)^2 + 36,000$	36,000
8	$-16(4)^2 + 36,000$	35,744
12	$-16(8)^2 + 36,000$	34,976
16	$-16(12)^2 + 36,000$	33,696
20	$-16(16)^2 + 36,000$	31,904
24	$-16(20)^2 + 36,000$	29,600

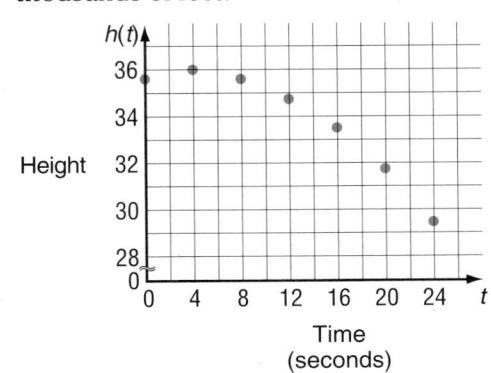

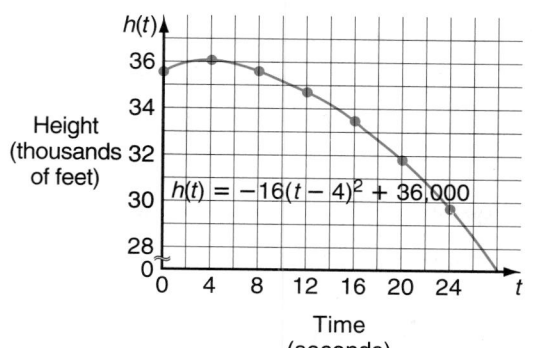

$$h(t) = -16(t - 4)^2 + 36,000$$

Height (thousands of feet)

Time (seconds)

We can now connect the points in a smooth curve. The weightlessness begins when the jet arcs at an elevation of 36,000 feet. Weightlessness continues for 23 seconds. The jet then pulls out of the dive and climbs for another weightless session.

The graph of any quadratic function is a **parabola.** The graph of the Vomit Comet is part of a parabola. All parabolas have an **axis of symmetry.** The axis of symmetry is the line about which the parabola is symmetric. That is, if you could fold the coordinate plane along the axis of symmetry, the portions of the parabola on each side of the line would match. The axis of symmetry is named by the equation of the line. All parabolas have a **vertex** as well. The vertex is the point of intersection of the parabola and the axis of symmetry. Notice that this parabola intersects the x-axis twice. The x-coordinates of these intersection points are called the **zeros** of the function.

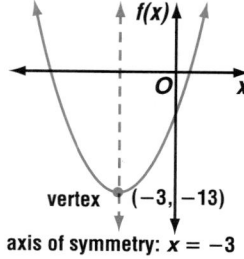

vertex $(-3, -13)$

axis of symmetry: $x = -3$

Example ❷

Physics

An arrow is shot upward with an initial velocity of 64 feet per second. The height of the arrow $h(t)$ in terms of the time t since the arrow was released is $h(t) = 64t - 16t^2$.

a. Draw the graph of the function relating the height of the arrow to the time.

b. Name the axis of symmetry and the vertex.

c. How long after the arrow is released does it reach its maximum height? What is that height?

a. First find and graph the ordered pairs that satisfy the function $h(t) = 64t - 16t^2$. Then graph the parabola suggested by the points.

This is the graph of the function that describes the height at any given time. The actual path of the arrow is an entirely different parabola.

t	$h(t)$
0	0
0.5	28
1.0	48
1.5	60
2.0	64
2.5	60
3.0	48
3.5	28
4.0	0

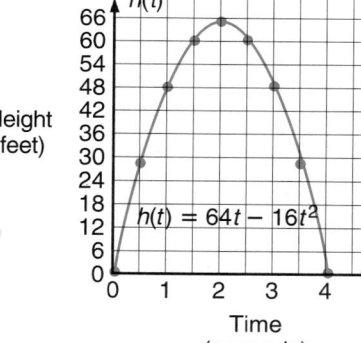

Height (feet)

$$h(t) = 64t - 16t^2$$

Time (seconds)

b. The equation of the axis of symmetry is $x = 2$. Since the vertex is the point of intersection of the parabola and the axis of symmetry, it is at $(2, 64)$.

c. You can see from the graph that the arrow reaches its maximum height at 2 seconds. The maximum height is 64 feet.   *How does the vertex relate to the maximum height and the time of the maximum height?*

You can use quadratic functions to describe different mathematical situations as well as situations occurring in everyday life.

Example **3**

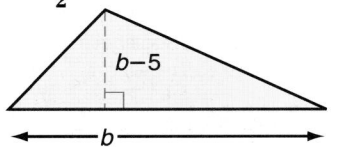

Write a function to represent the area of a triangle if the height is 5 cm less than the base. Then sketch the graph. Use $A = \frac{1}{2}bh$.

Let b represent the measure of the base of the triangle. Then $b - 5$ represents the height. Since the area is a function of the length of the base, let $f(b)$ represent the area.

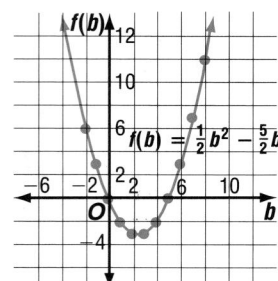

$$f(b) = \frac{1}{2}b(b - 5)$$
$$= \frac{1}{2}b^2 - \frac{5}{2}b$$

Which points on the graph could actually represent possible lengths of the base and areas of the triangles?

Find several ordered pairs that satisfy the quadratic function $f(b) = \frac{1}{2}b^2 - \frac{5}{2}b$. Then graph these ordered pairs and draw a parabola.

b	f(b)
−1	3
0	0
1	−2
2	−3
3	−3
4	−2
5	0
6	3
7	7
8	12

When a quadratic function is set equal to zero, the result is a **quadratic equation.** A quadratic equation is an equation that can be written in the form $ax^2 + bx + c = 0$, where $a \neq 0$. Since the largest exponent of the variable is 2, we say that a quadratic equation has a degree of 2. A quadratic equation contains only one variable, and all of the exponents are positive integers.

The **roots,** or solutions, of a quadratic equation are values of the variable that satisfy the equation. There are several methods you can use to find the roots of a quadratic equation. One method is to graph its related quadratic function. The related quadratic function of a quadratic equation in the form $ax^2 + bx + c = 0$ is $f(x) = ax^2 + bx + c$. The zeros of the function are the solutions of the equation, since $f(x) = 0$ at those points.

Example

Solve $x^2 + 3x - 18 = 0$ by graphing.

Graph the quadratic function $f(x) = x^2 + 3x - 18$ by finding and graphing the ordered pairs that satisfy the function. Then graph the parabola by connecting the points. Solve the equation by noting the points at which the graph intersects the x-axis.

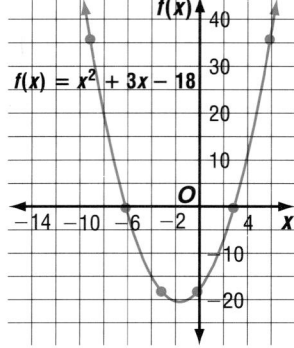

x	f(x)
−9	36
−6	0
−3	−18
0	−18
3	0
6	36

The table shows that $f(x) = 0$ when $x = -6$ and $x = 3$. The graph crosses the x-axis at −6 and 3. Thus, the solutions of the equation are −6 and 3.

Check your solutions by substituting each solution into the equation to see if it is satisfied.

$$x^2 + 3x - 18 = 0$$
$$(-6)^2 + 3(-6) - 18 \stackrel{?}{=} 0$$
$$0 = 0 \checkmark$$

$$x^2 + 3x - 18 = 0$$
$$(3)^2 + 3(3) - 18 \stackrel{?}{=} 0$$
$$0 = 0 \checkmark$$

A spreadsheet is a tool that can be used to manipulate numbers easily. Accountants, insurance underwriters, and loan officers all use spreadsheets to simplify calculations.

EXPLORATION

SPREADSHEET

To solve the quadratic equation $2x^2 + x - 3 = 0$ by using a spreadsheet, first enter the coefficients 2, 1, and −3 into the cells. Establish a starting value for x. Then choose an increment for the values of x to increase. You can change any of these values by typing a different number into the cell. The spreadsheet automatically computes and updates all other values in the cells. The bottom row of cells contains the values of the equation for the given values of x. The spreadsheet below indicates that the roots are −1.5 and 1. *You can verify these solutions by graphing.*

Quadratic Equation									
2	X^2+	1	X+	−3	= 0				
Starting Value =			−2						
Increment =			0.5						
X	−2	−1.5	−1	−0.5	0	0.5	1	1.5	2
Y	3	0	−2	−3	−3	−2	0	3	7

Your Turn

a. Work through the other examples in this lesson by using a spreadsheet.

b. Describe the difference between using a graphing calculator and using a spreadsheet to solve quadratic equations.

c. How did you determine the starting values and increments for x?

d. Find the roots of the equation $2x^2 + 2x - 12 = 0$ by using a spreadsheet.

There are three possible outcomes when solving a quadratic equation. The equation will have two real solutions, one real solution, or no real solutions. When a quadratic equation has one real solution, it really has two solutions that are the same number. A graph of each of these outcomes is shown below.

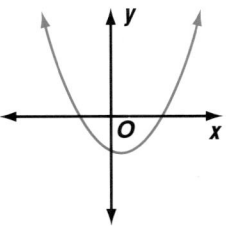

two real solutions

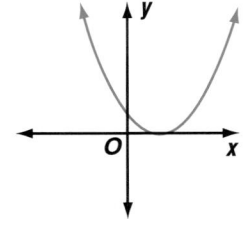

one real solution

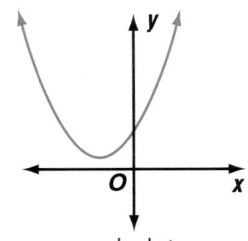

no real solutions

Example 5 illustrates a quadratic equation for which the solutions are the same number.

Example **5** **Solve $x^2 - 6x = -9$ by graphing.**

Write the equation in standard form.

$$x^2 - 6x = -9 \quad \rightarrow \quad x^2 - 6x + 9 = 0$$

Find and graph the ordered pairs that satisfy the related function $f(x) = x^2 - 6x + 9$.

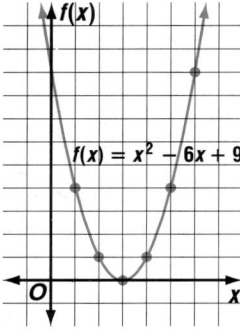

x	1	2	3	4	5	6
$f(x)$	4	1	0	1	4	9

Notice that the graph has only one x-intercept, 3. Thus, the solution is 3.

CHECK FOR UNDERSTANDING

Communicating Mathematics

Study the lesson. Then complete the following.

1. **Determine** if $3z^4 - 5z^3 + 6z - 6 = 0$ is a quadratic equation. Explain.

2. **Define** each term and explain how they are related.
   a. solution   b. root   c. zero of a function   d. x-intercept

3. Refer to Example 3.
   a. **List** three possible sets of dimensions for the triangle described.
   b. **Describe** the restrictions that must be placed on the graph of the function if you were to use it to solve the equation $\frac{1}{2}b^2 - \frac{5}{2}b = 0$.

4. **State** the difference between a quadratic equation and a quadratic function.

5. **Explain** why a quadratic equation cannot have more than two solutions.

6. **Determine** the solution of the equation $x^2 + 3x - 4 = 0$ from its related graph shown at the right. Explain how you found your answer.

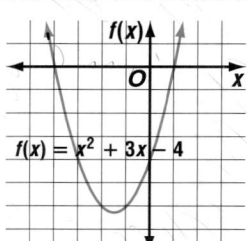

Guided Practice

Identify the quadratic term, the linear term, and the constant term in each function.

7. $f(x) = x^2 + x - 4$

8. $f(x) = -4x^2 - 8x - 9$

Graph each function. Name the vertex and the axis of symmetry.

9. $g(x) = x^2 - 4x + 4$ 10. $f(x) = x^2 + 9$ 11. $h(x) = x^2 + 6x + 9$

Solve each equation by graphing.

12. $d^2 + 5d + 6 = 0$ 13. $a^2 - 4a + 4 = 0$ 14. $2x^2 - x - 3 = 0$

15. A quadratic function $f(x)$ has values $f(3) = -4$, $f(6) = -7$, and $f(9) = 8$. Between which two x values is $f(x)$ sure to have a zero? Explain how you know.

Practice

Identify the quadratic term, the linear term, and the constant term in each function.

16. $g(x) = 5x^2 - 7x + 2$

17. $g(n) = 3n^2 - 1$

18. $f(n) = \frac{1}{3}n^2 + 4$

19. $f(z) = z^2 + 3z$

20. $f(x) = (x + 3)^2$

21. $f(t) = (3t + 1)^2 - 8$

Use the related graph of each equation to determine its solutions.

22. $2x^2 + 2x - 4 = 0$

23. $x^2 + 8x + 16 = 0$

24. $x^2 - 16x = 0$

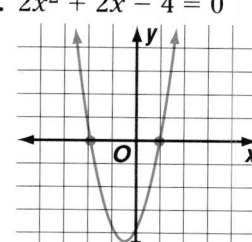 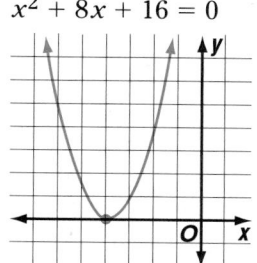 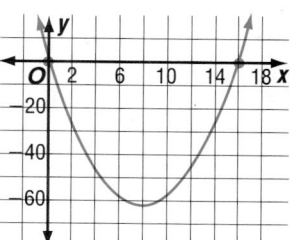

Graph each function. Name the vertex and the axis of symmetry.

25. $f(x) = x^2$

26. $f(x) = x^2 + 12x + 36$

27. $g(x) = x^2 - 9x + 9$

28. $h(x) = x^2 + 4$

29. $f(x) = x^2 - 9$

30. $h(x) = x^2 - 10x + 27$

31. $f(x) = x^2 + 20x + 93$

32. $g(x) = x^2 - \frac{2}{5}x + \frac{26}{25}$

33. $f(x) = x^2 + 3x - 0.95$

Solve each equation by graphing.

34. $m^2 + 3m = 28$

35. $p^2 - 2p - 24 = 0$

36. $4n^2 - 7n - 15 = 0$

37. $c^2 + 4c + 4 = 0$

38. $n^2 - 3n = 0$

39. $2w^2 - 3w = 9$

40. $4v^2 - 8v - 5 = 0$

41. $2c^2 + 5c - 12 = 0$

42. $(3x + 4)(2x + 7) = 0$

Critical Thinking

43. Number Theory   Although no one has found a formula that generates prime numbers, eighteenth-century Swiss mathematician Leonhard Euler discovered that $y = x^2 + x + 17$ produces prime numbers up to a certain point.

   a. Based on the graph of $f(x) = x^2 + x + 17$, what are the roots? Explain your answer.

   b. Describe how the nature of the graph relates to the equation for generating prime numbers.

Applications and Problem Solving

44. World Records   In 1940, Emanual Zacchini of Italy was fired a record distance of 175 feet from a cannon while performing in the United States. Suppose his initial upward velocity was 80 feet per second. The height y can be represented by the function $y = 80x - 16x^2$, where x represents the number of seconds that have passed.

   a. Draw the graph of the function, relating heights Mr. Zacchini reaches to the time after he was shot from the cannon.

   b. How long after he was shot out of the cannon did he reach his maximum height? What was that height?

45. Free-Falling   In 1942, I.M. Chisov of the USSR bailed out of an airplane without a parachute at 21,980 feet and survived. The function $h(t) = -16t^2 + 21,980$ describes the relationship between height $h(t)$ in feet and time t in seconds.

   a. Graph the function.

   b. About how many seconds did he fall before reaching the ground?

46. Electrical Engineering The relationship between the flow of electricity I in a circuit, the resistance to the flow Z, called impedance, and the electromotive force E, called voltage, is given by the formula $E = I \cdot Z$. Electrical engineers use j to represent the imaginary unit. An electrical engineer is designing a circuit that is to have a current of $(6 - j8)$ amps. If the impedance of the circuit is $(14 + j8)$ ohms, find the voltage. (Lesson 5–9)

47. Geometry The area of a square is $169x^2$. The length of one of its sides minus 14 is equal to 77. Solve for x. (Lesson 5–8)

48. Write an augmented matrix for $4x + 3y = 10$ and $5x - y = 3$. Then solve the system of equations. (Lesson 4–6)

49. Evaluate the determinant of $\begin{bmatrix} \frac{1}{2} & -1 & 3 \\ -3 & 5 & -2 \\ 3 & 2 & 4 \end{bmatrix}$ using expansion by minors. (Lesson 4–5)

50. Jill has three sons. The age of the oldest son is 3 times the age of the youngest son, and the age of the middle son is half the age of the oldest son. The sum of their ages is 9 years less than the age of their mother. If the age of their mother is 42, find the ages of the three sons. (Lesson 3–7)

51. Write an equation in slope-intercept form and in standard form for the line that has a slope of -3 and passes through $(-7, 5)$. (Lesson 2–4)

52. Graph a line that passes through $(3, 0)$ and is perpendicular to $y - 3x = 1$. (Lesson 2–3)

53. Sonia's aquarium holds 31.5 L of water. The container she uses to fill the aquarium holds 1.5 L. How many full containers of water does it take to fill the aquarium? (Lesson 1–4)

WORKING ON THE
In·ves·ti·ga·tion

Refer to the Investigation on pp. 328–329.

the **River Canyon Bridge**

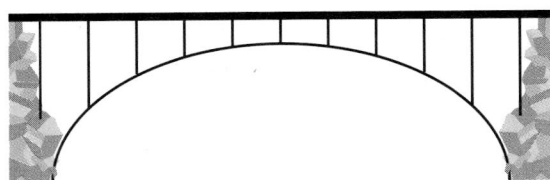

After reviewing your proposal for four types of bridges, your company's consultant suggests a bridge that is supported by an arch as shown below.

The arch would be in the shape of a parabola that opens downward with its vertex at the center of the bridge. The bridge would span the canyon and be anchored to the canyon walls.

It would be supported by a set of short struts that are anchored to the parabolic arch.

1 Draw a scaled blueprint of this bridge on grid paper. Let the road lie along the x-axis and let the vertex of the parabola lie on the y-axis.

2 Make a table of values for points that lie along the arch of the bridge.

3 Look for a pattern in your table and estimate at what points the arch of the bridge will touch the sides of the canyon.

4 How does this design compare with the bridges you considered?

Add the results of your work to your Investigation Folder.

Solving Quadratic Equations by Factoring

- To solve problems by using the guess-and-check strategy, and
- to solve quadratic equations by factoring.

Why IT'S IMPORTANT

You can use quadratic equations to solve problems involving tennis, meteorology, and air travel.

Players with the most Wimbledon titles
1. Billie Jean King, 20
2. Elizabeth Ryan, 19
3. Martina Navratilova, 18
4. Suzanne Lenglen, 15
5. William Renshaw, 14

APPLICATION

Tennis

In 1993, German tennis player Steffi Graf won three of the four biggest tournaments in tennis: Wimbledon, the French Open, and the U.S. Open. When an object, like a tennis ball, is hit straight up into the air, the height of the object is given by the function $h(t) = v_o t - 16t^2$, where $h(t)$ represents the height of the object, v_o represents the initial velocity, and t represents the time that the object has traveled. If a tennis ball is hit upward with an initial velocity of 48 feet per second, how long does it take for the ball to fall to the ground? To find the solution, substitute the known values into the formula $h(t) = v_o t - 16t^2$ to get the resulting function $h(t) = 48t - 16t^2$.

There are several methods that you can use to find the solution. In the last lesson, you learned to find the solution by graphing.

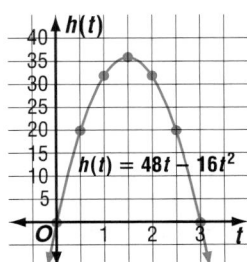

$h(t) = 48t - 16t^2$

t	0	0.5	1	1.5	2	2.5	3
$h(t)$	0	20	32	36	32	20	0

For this function, the zeros are 0 and 3. This means that the ball starts out at 0 feet and travels for 3 seconds before reaching the ground again.

Another way to solve a quadratic equation is by **factoring.** To solve by factoring, you must use the **zero product property.**

Zero Product Property	**For any real numbers a and b, if $ab = 0$, then either $a = 0$, $b = 0$, or both.**

We can solve the same equation by factoring that we solved above by graphing.

$0 = 48t - 16t^2$
$0 = 16t(3 - t)$

Since the product of $16t$ and $3 - t$ is 0, either $16t$ or $3 - t$ equals 0. So set each factor equal to 0 and solve.

$16t = 0$ or $3 - t = 0$ *Zero product property*
  $t = 0$ $t = 3$

The solutions (or roots) are 0 and 3. These are the same as the zeros obtained when you graphed the related function, $h(t) = 48 - 16t^2$. These answers seem reasonable since a tennis ball might not stay in the air for more than 3 seconds.

The **guess-and-check** strategy can be useful when factoring quadratic equations. You may have to try several combinations of factors before finding the right one.

Example **1** **Solve $x^2 + 6x - 16 = 0$ by factoring.**

Start guessing by using factors of 16: 2, 4, 8, or 16.

Guess	Check	Correct?
$(x - 4)(x + 4)$	$x^2 - 16 = 0$	*No, the linear term is 0.*
$(x - 8)(x + 2)$	$x^2 - 6x - 16 = 0$	*No, the linear term is negative.*
$(x + 8)(x - 2)$	$x^2 + 6x - 16 = 0$	*Yes*

You can now solve the equation by using the zero product property to find the values of x.

$$(x + 8)(x - 2) = 0$$
$$x + 8 = 0 \quad \text{or} \quad x - 2 = 0$$
$$x = -8 \qquad\qquad x = 2$$

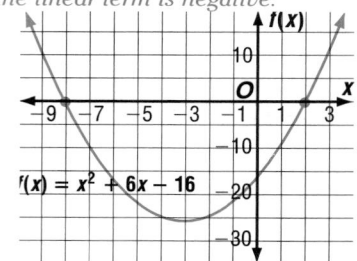

$f(x) = x^2 + 6x - 16$

The solutions are -8 and 2. You can check these solutions by graphing the related quadratic function and checking the x-intercepts.

When using factoring to solve a real-life problem, you need to examine each solution carefully to see if it is reasonable for that situation.

Example **2** **International auction houses sell hundreds of oriental rugs each year to collectors from all over the world. A 17th-century Mughal carpet was recently purchased from a museum for $253,000, despite having moth damage, corrosion, and holes. The rug has a border of uniform width depicting lilies, asters, and roses and a red center called a raspberry field. The auction brochure for the rug describes it as measuring 9 feet by 15 feet, with a raspberry field of 91 square feet. If you were interested in purchasing the rug, you might want to know how wide that beautiful border is. How wide is it?**

Explore Read the problem and make a drawing that models the Mughal carpet.

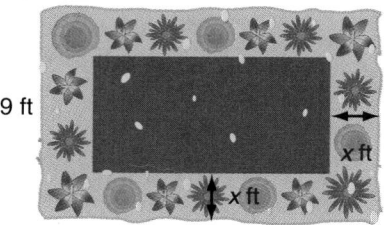

9 ft

x ft

x ft

15 ft

Let x feet be the width of the border.
The length of the red center is $15 - 2x$ feet.
The width of the red center is $9 - 2x$ feet.

17th-Century Mughal carpet, sold for $253,000

Plan Write an equation. The area of the red center can be expressed as the product of the length and width.

$$A = \ell w$$
$$= (15 - 2x)(9 - 2x) \quad \textit{Substitute for } \ell \textit{ and } w.$$
$$= 135 - 48x + 4x^2 \quad \textit{Multiply the binomials.}$$
$$= 4x^2 - 48x + 135$$

Solve Since the area of the red center is 91 square feet, replace A with 91 and solve this equation.

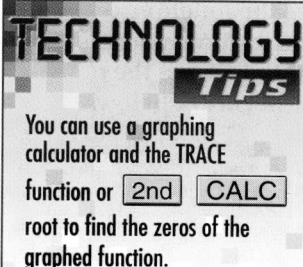

$A = 4x^2 - 48x + 135$

$91 = 4x^2 - 48x + 135$ *Replace A with 91.*

$0 = 4x^2 - 48x + 44$ *Subtract 91 from each side.*

$0 = x^2 - 12x + 11$ *Divide each side by 4.*

$0 = (x - 11)(x - 1)$ *Factor.*

$x - 11 = 0$ or $x - 1 = 0$

$x = 11$ $x = 1$

The solutions are 11 and 1. Use $x = 1$ since 11 is an unreasonable answer. Thus, the width of the floral border is 1 foot.

Examine   If the border is 1 foot wide all the way around, the width of the center is $9 - 2(1)$ or 7 and the length is $15 - 2(1)$ or 13. Since $7 \times 13 = 91$, the answer makes sense.

You have seen that a quadratic equation may have one solution. Factoring shows this is true because the two factors of the quadratic function are the same.

Example ❸ **Solve $x^2 + 10x = -25$ by factoring.**

$x^2 + 10x = -25$

$x^2 + 10x + 25 = 0$

$(x + 5)(x + 5) = 0$

$x + 5 = 0$ or $x + 5 = 0$

$x = -5$ $x = -5$

The only solution is -5.

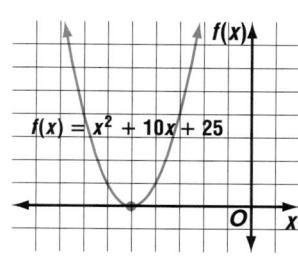

The graph of the related function intersects the x-axis in one point.

No matter which method you use, you can always check your solutions by substituting the values into the equation and simplifying.

Check: $x^2 + 10x = -25$

$(-5)^2 + 10(-5) \overset{?}{=} -25$

$-25 = -25$ ✓

CHECK FOR UNDERSTANDING

Communicating Mathematics

Study the lesson. Then complete the following.

1. **Explain** why the solution of 11 feet is unreasonable in Example 2.

2. **State** the zeros of the function graphed at the right. Verify by solving the related equation by factoring.

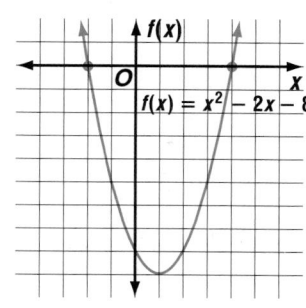

3. **You Decide** Chelsea was trying to find the zeros of the function $f(x) = x^2 - 13x + 36$. Carmen explained that she was making a mistake, but Chelsea insisted she was right. "All you have to do is make the $f(x)$ equal to zero and solve for x," Chelsea said. Look at her work below.

$$x^2 - 13x + 36 = 0$$
$$(x - 9)(x + 4) = 0$$
$$x - 9 = 0 \quad \text{or} \quad x + 4 = 0$$
$$x = 9 \qquad\qquad x = -4$$

Carmen said there was still something wrong. Who is correct? Explain.

Guided Practice

Solve each equation.

4. $(y - 8)(y + 6) = 0$

5. $(3y + 7)(y + 5) = 0$

Solve each equation by factoring.

6. $q^2 - 5q - 24 = 0$

7. $r^2 - 16r + 64 = 0$

8. $a^3 = 81a$

9. $3y^2 + y - 14 = 0$

10. **Physics** According to the the *Guinness Book of World Records,* the longest pendulum in the world is 73 feet $9\frac{3}{4}$ inches. It was installed in Tokyo, Japan, in 1983. The time in seconds t for a pendulum to swing back and forth is given by the formula $t^2 = 1.23L$, where L is the length of the pendulum in feet.

  a. What is the length of time needed for this pendulum to swing back and forth once?

  b. How long would it take for a pendulum that is 6 feet long to swing back and forth once?

  c. How long should the pendulum be if you want it to swing back and forth exactly 15 times a minute?

EXERCISES

Practice

Solve each equation.

11. $(a + 4)(a + 1) = 0$

12. $z(z - 1) = 0$

13. $2x + 6)(x - 3) = 0$

14. $(3y - 5)(2y + 7) = 0$

Solve each equation by factoring.

15. $x^2 - x = 12$

16. $d^2 - 5d = 0$

17. $z^2 - 12z + 36 = 0$

18. $y^2 + y - 30 = 0$

19. $r^2 - 3r = 4$

20. $3c^2 = 5c$

21. $18u^2 - 3u = 1$

22. $4y^2 = 25$

23. $9y^2 + 16 = -24y$

24. $4x^2 - 13x = 12$

Solve each equation by graphing or by factoring.

25. $b^2 + 3b = 40$

26. $4a^2 - 17a + 4 = 0$

27. $4s^2 - 11s = 3$

28. $6r^2 + 7r = 3$

29. $12m^2 + 25m + 12 = 0$

30. $18n^2 - 3n = 15$

31. $n^3 = 9n$

32. $x^3 = 64x$

33. $35z^3 + 16z^2 = 12z$

34. $18r^3 + 16r = 34r^2$

Critical Thinking

35. A parabola has intercepts at $x = -8$, $x = 4$, and $y = -8$. What is the equation of the parabola?

36. Air Travel A Delta Airlines route map states the formula $(VM)^2 = 1.22A$, where VM represents the number of miles to the horizon you can see from an airplane if you are flying on a clear day. The altitude, in feet, is represented by A.

  a. Suppose you are flying at an altitude of 36,000 feet on a clear day. About how many miles are there to the horizon?

  b. Suppose the distance to the horizon is 236 miles. What is your altitude in feet?

  c. The Sears Tower in Chicago is the tallest building in the world. The observation deck is 1454 feet above ground. About how far could you see on a clear day if you were standing on the observation deck?

37. Meteorology Weather forecasters can determine the approximate time that a thunderstorm will last if they know the diameter d of the storm in miles. The time t in hours can be found by using the formula $216t^2 = d^3$.

  a. Draw the graph of $y = 216t^2 - 5^3$ and use it to estimate how long a thunderstorm will last if its diameter is 5 miles.

  b. Find how long a thunderstorm will last if its diameter is 5 miles and compare this time with your estimate in part a.

38. Guess and Check Find two integers whose sum is 15 and whose product is 54.

39. Geometry The diagram below shows the relationship between the number of chords drawn through a circle and the maximum number of parts into which chords can divide a circle.

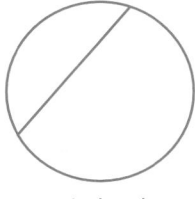 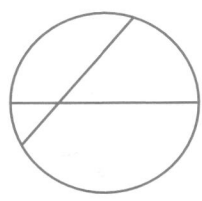 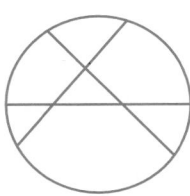

| 1 chord | 2 chords | 3 chords |
| 2 parts | 4 parts | 7 parts |

This relationship can be described by the formula $p = \frac{1}{2}x^2 + \frac{1}{2}x + 1$, where p represents the number of parts and x represents the number of chords. How many chords would you have to draw to divide a circle into 37 parts?

40. Geometry The length of a rectangle is 2 centimeters greater than its width. (Lesson 6–1)

  a. Write a function to represent the area of the rectangle. Then graph the function.

  b. Use the graph to estimate the area of a rectangle whose width is 4 cm. Compare your estimate to the actual area.

41. Solve $6y^2 = -96$. (Lesson 5–9)

42. Simplify $= \dfrac{rs}{\frac{1}{r^2} + \frac{3}{r^2}}$. (Lesson 5–7)

43. Find the inverse of $\begin{bmatrix} 4 & -2 \\ -3 & 6 \end{bmatrix}$. (Lesson 4–4)

44. Graph the system $3x - 2y = 10$ and $y - x = -1$ and state its solution. (Lesson 3–1)

45. Solve $3(2m + 9n) - (4 + 6m) = -5m$ for m. (Lesson 2–2)

46. Simplify $\frac{2}{3}\left(\frac{1}{2}a + 3b\right) + \frac{1}{2}\left(\frac{2}{3}a + b\right)$. (Lesson 1–2)

Completing the Square

APPLICATION
Architecture

An architect for Windham Homes is changing the floor plan of a house to meet the needs of a new customer. The present floor plan calls for a square dining area that measures 13 feet by 13 feet. The customer would also like for the dining area to be square, but with an area of 250 square feet. How much will this add to the dimensions of the room?

A strip must be added to the length and width of the dining room as shown below. The width of the strip that will be added is represented by x feet. The length and width of the new dining area would then be (13 + x) feet.

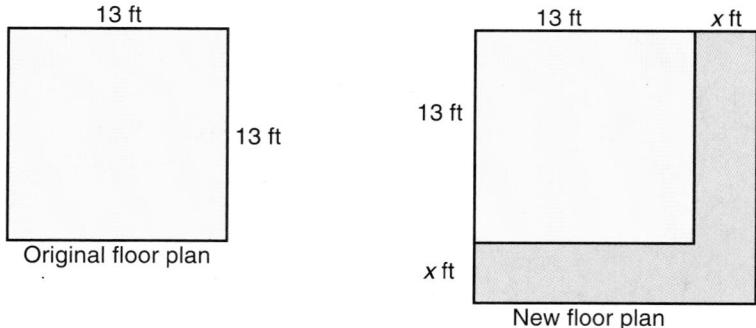

13 ft / 13 ft / Original floor plan

13 ft / x ft / 13 ft / x ft / New floor plan

We can use the formula $A = s^2$ to find the area of the new dining room. Each side measures $13 + x$, and the new area is 250 square feet.

$$A = s^2$$
$$250 = (13 + x)^2 \quad \textit{Replace A with 250 and s with 13 + x.}$$
$$\pm\sqrt{250} = \sqrt{(13 + x)^2} \quad \textit{Take the square root of each side.}$$
$$\pm\sqrt{250} = 13 + x$$
$$\pm\sqrt{250} - 13 = x$$

Use a scientific calculator to find approximate values for x.

Enter: 250 $\boxed{\sqrt{x}}$ $\boxed{-}$ 13 $\boxed{=}$ _2.8113883O1_

Enter: 250 $\boxed{\sqrt{x}}$ $\boxed{+/-}$ $\boxed{-}$ 13 $\boxed{=}$ _-28.8113883_

Since we need the increased width of the dining area, we can ignore the negative solution. The width of the increase would be approximately 2.81 feet.

You can graph the related function to show that the answer is reasonable. Rewrite $250 = (13 + x)^2$ so that one side is zero. Then graph the related function.

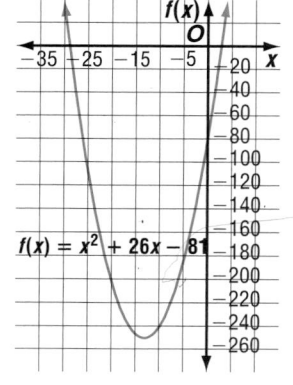

$$250 = (13 - x)^2$$
$$(13 + x)^2 - 250 = 0 \quad \rightarrow \quad f(x) = (13 + x)^2 - 250$$
$$= x^2 + 26x - 81$$

$f(x) = x^2 + 26x - 81$

The estimated zeros are about -30 and 3. Thus, the answer 2.8 is reasonable.

The quadratic equation $250 = (13 + x)^2$ contained one side that was a perfect square, $(13 + x)^2$. This allowed us to solve it by taking the square root of each side. When an equation does not contain a perfect square, you may create a perfect square by applying a process called **completing the square.**

In a perfect square, there is a relationship between the coefficient of the middle term and the constant term.

Specific Case	**General Case**
$(x + 9)^2 = x^2 + 18x + 81$	$(x + c)^2 = x^2 + 2cx + c^2$
$9 = \frac{1}{2}(18) \rightarrow 9^2 = 81$	$c = \frac{1}{2}(2c) \rightarrow c^2$

To complete the square in the expression below, you would use the same process. Using the pattern of coefficients, take half the coefficient of the linear term and square it.

$$x^2 - 8x + \underline{\ ?\ }$$

$$\left(-\frac{8}{2}\right)^2 \rightarrow (-4)^2 \text{ or } 16 \quad x^2 - 8x + 16 \text{ is a perfect square trinomial,}$$
$$\text{which can be written as } (x - 4)^2.$$

In Example 1, use the pattern of coefficients to make the trinomial a perfect square.

Example 1 **Find the value of c that makes $x^2 + 14x + c$ a perfect square.**

$c = \left(\frac{14}{2}\right)^2$ *Square half the coefficient of the linear term.*

$= 7^2$ or 49

The value of c is 49. Therefore, the trinomial $x^2 + 14x + 49$ is a perfect square. It can be written $(x + 7)^2$.

Example 2 **Solve $x^2 - 6x = 40$ by completing the square.**

First, you must find the term that completes the square on the left side of the equation. Then add that term to each side.

$x^2 - 6x + \square = 40 + \square$

$x^2 - 6x + 9 = 40 + 9$ $\left(-\frac{6}{2}\right)^2 = 9$

$(x - 3)^2 = 49$ *Factor the perfect square trinomial.*

$x - 3 = \pm 7$ *Take the square root of each side.*

$x - 3 = 7$ or $x - 3 = -7$

$x = 10$ $x = -4$

The solutions are 10 and -4.   *Check by factoring.*

A good way to help you visualize completing the square is to model the process using algebra tiles.

MODELING MATHEMATICS

Completing the Square

Materials: ⌐ algebra tiles ▢=▢ equation mat

Use algebra tiles to complete the square for the equation $x^2 + 6x + 2 = 0$.

Step 1 Subtract 2 from each side of the equation to model the equation $x^2 + 6x = -2$.

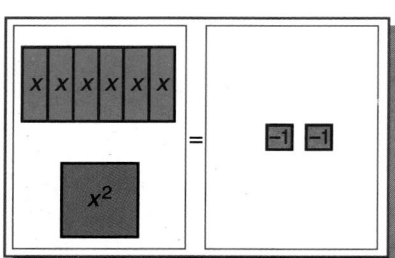

Step 2 Begin to arrange the x^2-tile and x-tiles into a square.

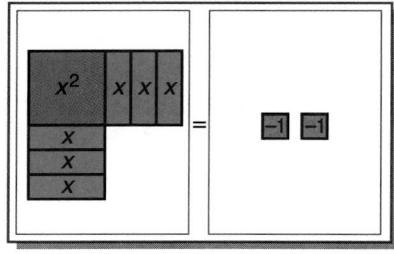

Step 3 To complete the square, add 9 1-tiles to the left side of the mat. Since it is an equation, add 9 1-tiles to the right side.

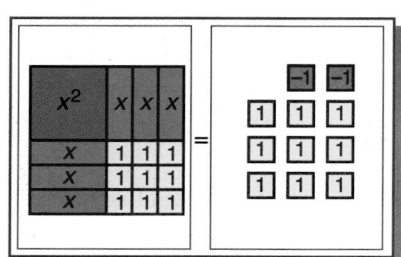

Step 4 Remove the zero pairs on the right side of the mat. After completing the square, the equation is $x^2 + 6x + 9 = 7$ or $(x + 3)^2 = 7$.

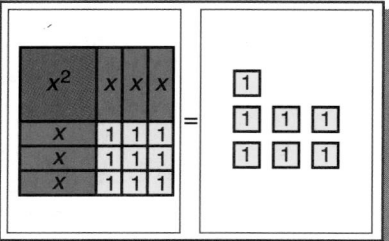

Your Turn

a. Use algebra tiles to complete the square for the equation $x^2 + 4x + 1 = 0$.

b. The equation $x^2 + 5x - 2 = 0$ has an odd number for the coefficient of x. Complete the square by using algebra tiles or by making a drawing.

c. Write a paragraph explaining how you could complete the square with models without first rewriting the equation. Include a drawing.

When the coefficient of the second-degree term is not 1, you must first divide the equation by that coefficient before completing the square.

Example ❸ **Solve $4x^2 - 5x - 21 = 0$ by completing the square.**

$$4x^2 - 5x - 21 = 0$$

$$x^2 - \frac{5}{4}x - \frac{21}{4} = 0 \qquad \textit{Divide each side by 4.}$$

$$x^2 - \frac{5}{4}x = \frac{21}{4} \qquad \textit{Isolate the constant on one side.}$$

$$x^2 - \frac{5}{4}x + \frac{25}{64} = \frac{21}{4} + \frac{25}{64} \qquad \textit{Add } \left(-\frac{5}{4} \div 2\right)^2 \textit{ or } \frac{25}{64} \textit{ to each side.}$$

$$\left(x - \frac{5}{8}\right)^2 = \frac{361}{64} \qquad \textit{Factor.}$$

$$x - \frac{5}{8} = \pm \frac{19}{8} \qquad \textit{Take the square root of each side.}$$

$$x = \frac{5}{8} \pm \frac{19}{8}$$

$$x = \frac{5}{8} + \frac{19}{8} \qquad x = \frac{5}{8} - \frac{19}{8}$$

$$= \frac{24}{8} \text{ or } 3 \qquad\qquad = -\frac{14}{8} \text{ or } -\frac{7}{4}$$

The solutions are 3 and $-\frac{7}{4}$. *Verify using the graph of the related function.*

Not all roots of quadratic equations will be rational numbers. Roots that are irrational numbers may be written as *exact* answers in radical form or as *approximate* answers in decimal form when a calculator is used.

Example 4

CONNECTION

Physics

The distance *d* that an object travels can be calculated when the initial speed v_i, elapsed time *t*, and the rate of constant acceleration *a* are known. A formula that relates these factors is $d(t) = v_i t + \frac{1}{2} a t^2$.

If a motorcycle has an initial speed of 30 m/s and a constant acceleration of 6 m/s^2, how much time will it take to travel 200 meters?

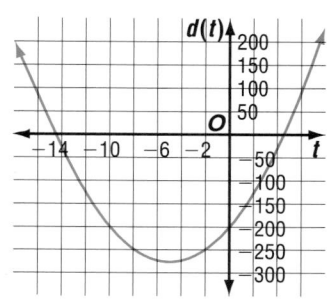

$$d(t) = v_i t + \frac{1}{2} a t^2$$

$$200 = 30t + \frac{1}{2} \cdot 6t^2 \qquad \textit{Substitute the known values into the formula.}$$

$$200 = 30t + 3t^2$$

$$\frac{200}{3} = 10t + t^2 \qquad \textit{Divide by 3.}$$

$$\frac{200}{3} + 25 = t^2 + 10t + 25 \qquad \textit{Complete the square.}$$

$$\frac{275}{3} = (t + 5)^2 \qquad \textit{Factor.}$$

$$\pm \sqrt{\frac{275}{3}} = t + 5 \qquad \textit{Take the square root of each side.}$$

$$\pm \sqrt{\frac{275}{3}} - 5 = t \qquad \textit{Subtract 5 from each side.}$$

The solutions are $\sqrt{\frac{275}{3}} - 5$ or about 4.57 and $-\sqrt{\frac{275}{3}} - 5$, or about -14.57. Verify by looking at the graph of the related function.

We can eliminate -14.57 since negative time has no meaning in this example. Thus, the motorcycle will travel 200 meters in about 4.57 seconds.

Not all solutions to quadratic equations are real. In some cases, the solutions are complex numbers of the form $a + bi$, where $b \neq 0$.

Example ⑤ **Solve $x^2 + 8x + 20 = 0$ by completing the square.**

$$x^2 + 8x + 20 = 0$$
$$x^2 + 8x = -20 \qquad \textit{Subtract 20 from each side.}$$
$$x^2 + 8x + 16 = -20 + 16 \qquad \left(\frac{8}{2}\right)^2 = 16$$
$$(x + 4)^2 = -4 \qquad \textit{Factor the left side.}$$
$$x + 4 = \pm 2i \qquad \textit{Take the square root of each side. Remember}$$
$$x = -4 \pm 2i \qquad \textit{that } \sqrt{-1} = i.$$

The roots are the complex numbers $-4 + 2i$ and $-4 - 2i$.

LOOK BACK

You can refer to Lesson 5-9 for information on complex numbers.

CHECK FOR UNDERSTANDING

Communicating Mathematics

Study the lesson. Then complete the following.

1. **Explain** how you would solve the equation $x^2 + 21x - 5 = 0$ by completing the square.

2. **Discuss** how you can tell if a quadratic equation has imaginary roots just by looking at a sketch of its graph.

3. **State** whether $b^2 + 5b + 23$ is a perfect square. Explain.

4. **Write** a paragraph explaining why we sometimes eliminate negative answers to real-world applications.

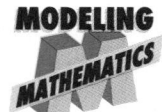

5. Use algebra tiles or make a drawing to solve the equation $x^2 + 4x - 5 = 0$ by completing the square.

Guided Practice

Find the value of c that makes each trinomial a perfect square.

6. $x^2 + 12x + c$

7. $x^2 - 7x + c$

Find the exact solutions for each equation by completing the square.

8. $x^2 + 8x = 20$

9. $12t^2 - 17t = 5$

10. $r^2 + 14 = 8r$

11. $x^2 - 7x + 4 = 0$

12. $\frac{1}{2}x^2 - 4x + 8 = 0$

13. $x^2 + 2x + 6 = 0$

14. **Safety** Juanita is driving a truck at an initial velocity of 60 ft/s. She sees a stop sign 240 feet ahead of her. If she begins to decelerate at the rate of $7\frac{1}{2}$ ft/s^2, how long will it take her before she stops at the stop sign? Use the formula $s = v_i t + \frac{1}{2}at^2$.

If acceleration is a positive number, what is deceleration?

Practice

Find the value of *c* that makes each trinomial a perfect square.

15. $x^2 + 2x + c$

16. $x^2 + 18x + c$

17. $t^2 + 40t + c$

18. $r^2 - 9r + c$

19. $a^2 - 100a + c$

20. $x^2 + 15x + c$

Find the exact solution for each equation by completing the square.

21. $x^2 + 3x - 18 = 0$

22. $x^2 + 2x - 120 = 0$

23. $x^2 - 8x + 11 = 0$

24. $x^2 + 7x - 17 = 0$

25. $x^2 + 9x + 20.25 = 0$

26. $9x^2 + 96x + 256 = 0$

27. $x^2 + 4x + 11 = 0$

28. $2x^2 - 7x + 12 = 0$

29. $3x^2 + 7x + 7 = 0$

30. $16x^2 + 9x + 20 = 0$

31. $x^2 - 3x - 20 = 0$

32. $2x^2 - x - 31 = 0$

33. $12x^2 - 13x - 35 = 0$

34. $x^2 + 19x - 12 = 0$

35. $ax^2 + bx + c = 0$

36. $px^2 + rx + m = 0$

Programming

37. The graphing calculator program at the right determines if an equation written in standard form contains a perfect square trinomial. It uses the same process that you use when completing the square to determine if the expression is a perfect square trinomial.

Use the guess-and-check strategy in the program at the right to find the value for *k* that makes each expression a perfect square trinomial.

a. $x^2 + kx + 64$
b. $x^2 - 14x + k$
c. $4x^2 - 12x + k$
d. $16x^2 + 40x + k$
e. $4x^2 + kx + 1$
f. $kx^2 - 20x + 4$

```
PROGRAM:PERFSQR
: Prompt A,B,C
: If A<0 or C<0
: Then
: Goto 2
: End
: If √(A)≠(int √(A))
: Then
: Goto 2
: End
: If √(C)≠(int √(C))
: Then
: Goto 2
: End
: If B^2=4*A*C
: Then
: Disp "CONGRATULATIONS!",
  "IT IS A","PERFECT SQUARE."
: Else
: Goto 2
: Stop
: Lbl 2
: Disp "TRY AGAIN","IT IS
  NOT A","PERFECT SQUARE."
```

Critical Thinking

38. Find the values of *m* that make $f(x) = x^2 + (m + 5)x + (5m + 1)$ a perfect square trinomial function. How many roots does it have?

Applications and Problem Solving

39. Physics When a moving car hits an object, the damage it can cause can be measured by the *collision impact*. For a certain car, the collision impact *I* can be represented by the formula $I = 2(s)^2$, where *s* represents the speed in kilometers per minute.

a. Sketch a graph of the function.

b. What is the collision impact if the speed is 1 km/min? 2 km/min? 4 km/min?

c. As the speed doubles, explain what happens to the value of the collision impact.

40. Law Enforcement The police can use the length of skid marks to help determine the speed of a vehicle before the brakes were applied. If the skid marks were on concrete, the formula $\frac{s^2}{24} = d$ can be used to approximate the speed of the vehicle. In the formula, s represents the speed in miles per hour, and d represents the length of the skid marks in feet. If the length of a car's skid marks on dry concrete were 50 feet, how fast was the car traveling when the brakes were applied?

Mixed Review

41. Physics A ball is thrown straight up with an initial velocity of 56 feet per second. The height of the ball t seconds after it is thrown is given by the formula $h(t) = 56t - 16t^2$. (Lesson 6–2)
  a. What is the height of the ball after 1 second?
  b. What is its maximum height?
  c. After how many seconds will it return to the ground?

42. Forestry A forester is often responsible for choosing which stands of trees will be harvested. The formula $V = 0.0027Ld^2 + 0.0027L\left(d + \frac{L}{A}\right)^2$ can be used to estimate how many cubic feet of wood will come from one log. In the equation, V represents the volume of the log in cubic feet, L represents the length of the log in feet, d represents the diameter of the top of the log in inches, and A represents the length of the log required for 1 inch of taper. Suppose that a 16-foot long log has a 1-inch taper over an 8-foot length. What diameter log does the forester want to find if he needs to get 150 cubic feet of wood out of one log? (Lesson 6–1)

43. Simplify $\sqrt[4]{5m^3n^5} \cdot \sqrt[4]{125m^2n^3}$. (Lesson 5–6)

44. Factor $2ab(c - d) + 10d(c - d)$. (Lesson 5–4)

45. Consumerism The prices of 14 video cameras are listed below.

$877	$819	$1100	$1450	$812	$973	$1399
$890	$1409	$949	$900	$775	$1299	$1399

Make a box-and-whisker plot of the prices. (Lesson 4–8)

46. Graph the system of inequalities. Name the vertices of the polygon formed. Find the maximum and minimum values of the given function. (Lesson 3–5)

$y \le 7$
$y \ge -x + 6$
$y \le x + 4$
$x \le 5$
$f(x, y) = 2x - 3y$

47. Use Cramer's rule to solve $3m - 4n = 0$ and $n + 7 = -m$. (Lesson 3–3)

48. Graph a line that only goes through Quadrants II and III and passes through the point $(-4, 27)$. (Lesson 2–3)

The Quadratic Formula and the Discriminant

What YOU'LL LEARN

- To solve quadratic equations by using the quadratic formula, and
- to use discriminants to determine the nature of the roots of quadratic equations.

Why IT'S IMPORTANT

You can use the quadratic formula to solve quadratic equations that involve health and probability.

F Y I

By June 30, 1992, more than 1,313,887,324 vehicles had crossed the Golden Gate Bridge.

Landmarks

The Golden Gate Bridge in San Francisco, California, is a magnificent structure that was completed on May 27, 1937, after more than four years of construction and a cost of approximately $27 million. The Golden Gate is the tallest bridge in the world, with its towers extending 746 feet above the water and the floor of the bridge extending 200 feet above the water.

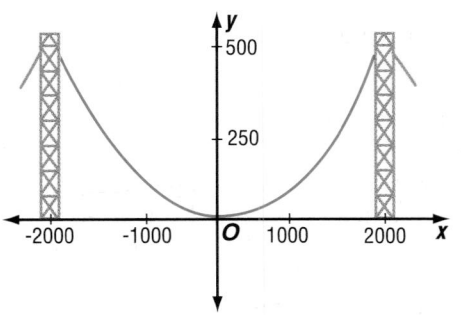

The two supporting cables that pass over the tops of the towers are each 7650 feet long and 36.5 inches in diameter. They are the largest bridge cables ever made. These supporting cables approximate the shape of a parabola with the lowest point reaching about 6 feet above the floor of the bridge. This parabola can be approximated by the quadratic function $y = 0.00012244898x^2 + 6$, where x represents the distance from the axis of symmetry and y represents the height of the cables.

How would you solve the quadratic equation $0.00012244898x^2 + 6 = 0$? In Lessons 6–1, 6–2, and 6–3, you learned several ways to solve quadratic equations. But it would be difficult to use any of those techniques to solve this equation. You might ask, "Isn't there a formula that will work for any quadratic equation?" The answer is yes! This formula can be derived by solving the general form of a quadratic equation for x.

$$ax^2 + bx + c = 0$$

$$x^2 + \frac{b}{a}x + \frac{c}{a} = 0 \qquad \textit{Divide each side by a.}$$

$$x^2 + \frac{b}{a}x = -\frac{c}{a} \qquad \textit{Subtract } \frac{c}{a} \textit{ from each side.}$$

$$x^2 + \frac{b}{a}x + \frac{b^2}{4a^2} = -\frac{c}{a} + \frac{b^2}{4a^2} \qquad \textit{Complete the square.}$$

$$\left(x + \frac{b}{2a}\right)^2 = \frac{b^2 - 4ac}{4a^2} \qquad \textit{Simplify.}$$

$$\sqrt{\left(x + \frac{b}{2a}\right)^2} = \pm\sqrt{\frac{b^2 - 4ac}{4a^2}} \qquad \textit{Take the square root of each side.}$$

$$x + \frac{b}{2a} = \frac{\pm\sqrt{b^2 - 4ac}}{2a} \qquad \textit{Simplify.}$$

$$x = \frac{-b \pm\sqrt{b^2 - 4ac}}{2a} \qquad \textit{Subtract } \frac{b}{2a} \textit{ from each side.}$$

This equation is known as the **quadratic formula.**

The Quadratic Formula	The solutions of a quadratic equation of the form $ax^2 + bx + c = 0$, where $a \neq 0$, are given by the following formula. $$x = \frac{-b \pm \sqrt{b^2 - 4ac}}{2a}$$

Example **1** **Solve $r^2 - 7r = 18$ by using the quadratic formula.**

First, write the equation in standard form. $r^2 - 7r = 18 \rightarrow r^2 - 7r - 18 = 0$

Then, substitute the values of a, b, and c directly into the quadratic formula. *$a = 1$, $b = -7$, and $c = -18$*

$$r = \frac{-b \pm \sqrt{b^2 - 4ac}}{2a}$$

$$= \frac{-(-7) \pm \sqrt{(-7)^2 - 4(1)(-18)}}{2(1)}$$

$$= \frac{7 \pm \sqrt{121}}{2} \text{ or } \frac{7 \pm 11}{2}$$

$$r = \frac{7 + 11}{2} \text{ or } 9 \quad \text{and} \quad r = \frac{7 - 11}{2} \text{ or } -2$$

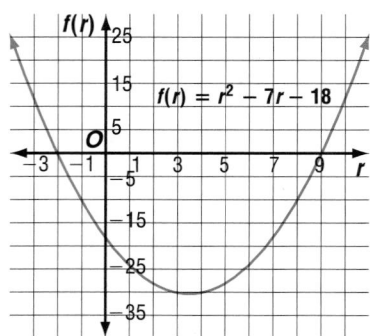

The solutions are 9 and -2.
Check these solutions.

The graph of the related function shows that there are two solutions.

Sometimes roots are irrational. You can express these roots exactly by writing them in radical form.

Example **2** **Solve $x^2 + 9x - 11 = 0$ by using the quadratic formula.**

The equation is in standard form. Substitute the values of a, b, and c directly into the quadratic formula. *$a = 1$, $b = 9$, and $c = -11$*

$$x = \frac{-b \pm \sqrt{b^2 - 4ac}}{2a}$$

$$= \frac{-9 \pm \sqrt{9^2 - (4)(1)(-11)}}{2(1)} \text{ or } \frac{-9 \pm \sqrt{125}}{2}$$

The solutions are $\dfrac{-9 + 5\sqrt{5}}{2}$ and $\dfrac{-9 - 5\sqrt{5}}{2}$.

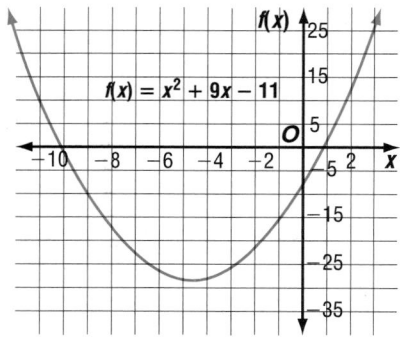

The graph of the related function shows that there are two solutions.

You can use a scientific calculator to help you find approximate solutions.

First evaluate the radical part of the expression $\dfrac{-9 \pm \sqrt{9^2 - (4)(1)(-11)}}{2(1)}$ and store the result in memory.

Enter: 9 $\boxed{x^2}$ $\boxed{-}$ 4 $\boxed{\times}$ 11 $\boxed{+/-}$ $\boxed{=}$ $\boxed{\text{2nd}}$ $\boxed{\sqrt{x}}$ $\boxed{\text{STO} \blacktriangleright}$ *11.18033989*

Now find the solutions.

Enter: 9 [+/−] [+] [RCL] [=] [÷] 2 [=] *1.090169944*

Enter: 9 [+/−] [−] [RCL] [=] [÷] 2 [=] *−10.09016994*

The two solutions are approximately 1.1 and −10.1.
How do these solutions compare with the graph?

When using the quadratic formula, if the radical contains a negative value, the solutions will be imaginary. Imaginary solutions always appear in conjugate pairs.

Example Solve $-3x^2 + 4x - 4 = 0$ by using the quadratic formula.

Substitute the values of a, b, and c directly into the quadratic formula.
$a = -3$, $b = 4$, and $c = -4$

$$x = \frac{-b \pm \sqrt{b^2 - 4ac}}{2a}$$

$$= \frac{-4 \pm \sqrt{(4)^2 - 4(-3)(-4)}}{2(-3)}$$

$$= \frac{-4 \pm \sqrt{-32}}{-6}$$

$$= -\frac{4 \pm 4i\sqrt{2}}{-6} \text{ or } \frac{2 \pm 2i\sqrt{2}}{3}$$

The solutions are the imaginary numbers
$\frac{2 + 2i\sqrt{2}}{3}$ and $\frac{2 - 2i\sqrt{2}}{3}$.

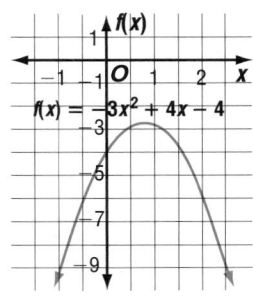

A graph of the related function shows that the solutions are imaginary.

Example  A car is traveling at 26 meters per second (m/s) and accelerating at −13 m/s². After traveling 26 m, the driver brings the car to a complete stop. The equation $26 = 26t - \frac{13}{2}t^2$, where t is the time it takes to stop, can be used to represent this situation. How long did it take the driver to stop the car?

CONNECTION
Physics

First, write the equation in standard form. $-\frac{13}{2}t^2 + 26t - 26 = 0$

Then, substitute directly into the quadratic formula.

$a = -\frac{13}{2}$, $b = 26$, and $c = -26$

$$t = \frac{-b \pm \sqrt{b^2 - 4ac}}{2a}$$

$$= \frac{-26 \pm \sqrt{(26)^2 - 4\left(-\frac{13}{2}\right)(-26)}}{2\left(-\frac{13}{2}\right)}$$

$$= \frac{-26 \pm \sqrt{0}}{-13}$$

$$= 2$$

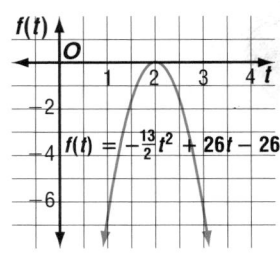

A graph of the related function shows that there is one solution.

It took 2 seconds to stop the car.

Study Examples 1, 2, 3, and 4 and observe the relationship between the expression under the radical, $b^2 - 4ac$, and the roots of the quadratic equation. This expression, $b^2 - 4ac$, is called the **discriminant.** The value of the discriminant determines the nature of the roots of a quadratic equation. The table below summarizes all the possibilities.

Example	Value of $b^2 - 4ac$	Discriminant a Perfect Square?	Nature of Root(s)	Nature of Related Graph
1	$b^2 - 4ac > 0$	yes	2 real, rational	intersects x-axis twice
2	$b^2 - 4ac > 0$	no	2 real, irrational	intersects x-axis twice
3	$b^2 - 4ac < 0$	—	2 imaginary	does not intersect x-axis
4	$b^2 - 4ac = 0$	—	1 real	intersects x-axis once

Example **Find the value of the discriminant for each quadratic equation. Then describe the nature of the roots.**

a. $x^2 - 8x + 16 = 0$

$a = 1, b = -8, c = 16$
$b^2 - 4ac = (-8)^2 - 4(1)(16)$
$\qquad = 64 - 64$
$\qquad = 0$

The value of the discriminant is 0, so there is one real root.

b. $5x^2 + 42 = 0$

$a = 5, b = 0, c = 42$
$b^2 - 4ac = (0)^2 - 4(5)(42)$
$\qquad = 0 - 840$
$\qquad = -840$

The value of the discriminant is negative, so there are two imaginary roots.

c. $x^2 - 5x - 50 = 0$

$a = 1, b = -5, c = -50$
$b^2 - 4ac = (-5)^2 - 4(1)(-50)$
$\qquad = 25 + 200$
$\qquad = 225$

The value of the discriminant is 225, which is a perfect square. There are two real, rational roots.

d. $2x^2 - 9x + 8 = 0$

$a = 2, b = -9, c = 8$
$b^2 - 4ac = (-9)^2 - 4(2)(8)$
$\qquad = 81 - 64$
$\qquad = 17$

The value of the discriminant is 17, which is not a perfect square. There are two real, irrational roots.

CHECK FOR UNDERSTANDING

Communicating Mathematics

Study the lesson. Then complete the following.

1. **Explain** why the roots of a quadratic equation are imaginary if the value of the discriminant is less than 0.

2. **Draw** graphs to illustrate the relationship between the nature of the roots determined by the discriminant and the number of times the graph of the related function intersects the x-axis.

3. **Refer** to the application at the beginning of the lesson. Calculate the value of the discriminant for the equation of the supporting cables for the Golden Gate Bridge. What does it mean?

4. **Describe** the value of the discriminant for the equation whose graph is at the right. How many real roots are there?

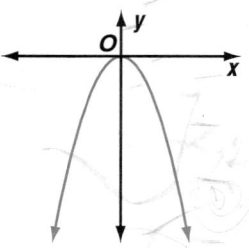

5. Assess Yourself Of the methods you have used to solve quadratic equations—graphing, factoring, completing the square, and the quadratic formula—which do you prefer? Why?

Guided Practice

6. Which equation shows how to solve $3x^2 - x + 2 = 0$ by using the quadratic formula?

a. $x = \dfrac{1 \pm \sqrt{3^2 - 4(3)(2)}}{2(3)}$

b. $x = \dfrac{-2 \pm \sqrt{(-1)^2 - 4(3)(2)}}{2}$

c. $x = \dfrac{1 \pm \sqrt{1^2 - 4(3)(2)}}{2(3)}$

d. $x = \dfrac{-1 \pm \sqrt{(-1)^2 - 4(3)(2)}}{2}$

State the values of a, b, and c for each equation. Then find the value of the discriminant.

7. $3x^2 - 5x = 2$　　　**8.** $x^2 - 24x + 144 = 0$　　**9.** $x^2 + 7 = -5x$

Find the value of the discriminant and describe the nature of the roots (real, imaginary, rational, irrational) of each quadratic equation. Then solve the equation. Express irrational roots as exact and approximate to the nearest hundredth.

10. $x^2 + 10x = -25$　　　　　　　　**11.** $2x^2 = 72$

12. $x^2 + x - 5 = 0$　　　　　　　　　**13.** $x^2 + 5x + 10 = 0$

14. If the discriminant of the equation of a parabola is 2025, how many times does the graph of the equation intersect the x-axis? Justify your answer.

15. Solve $x^2 + 2x - 5 = 0$.

　a. How many roots does the function $y = x^2 + 2x - 5$ have? What are they?

　b. Approximately where does the graph of $y = x^2 + 2x - 5$ cross the x-axis?

EXERCISES

Practice

Find the value of the discriminant and describe the nature of the roots (real, imaginary, rational, irrational) of each quadratic equation. Then solve the equation. Express irrational roots as exact and approximate to the nearest hundredth.

16. $x^2 + 12x + 32 = 0$　　　　　　**17.** $2x^2 - 12x + 18 = 0$

18. $x^2 - 4x + 1 = 0$　　　　　　　**19.** $3x^2 + 5x - 2 = 0$

20. $x^2 - 2x + 5 = 0$　　　　　　　**21.** $3x^2 + 11x + 4 = 0$

22. $x^2 - 12x + 42 = 0$　　　　　　**23.** $x^2 = 6x$

24. $2x^2 + 7x - 11 = 0$　　　　　　**25.** $5x^2 - 8x + 9 = 0$

26. $0.4x^2 + x = 0.3$　　　　　　　**27.** $2x^2 - 13x = 7$

28. $x^2 - 16x + 4 = 0$　　　　　　　**29.** $4x^2 - 9x = -7$

Critical Thinking

30. Probability Suppose you picked an integer at random from 1 through 12 as a value for c in the equation $y = x^2 + 6x + c$. What is the probability that the resulting equation will have imaginary roots?

Applications and Problem Solving

31. Health A person's blood pressure depends on his or her age. For women, normal systolic blood pressure is given by the formula $P = 0.01A^2 + 0.05A + 107$, where P is the normal blood pressure in millimeters of mercury (mm Hg) and A is the age. For men, the normal systolic blood pressure is given by the formula $P = 0.006A^2 - 0.02A + 120$.

(continued on the next page)

a. Graph both functions on the same set of axes. Describe the differences between the graphs and explain how the differences reflect the blood pressures of men and women.

b. Find the normal blood pressure of a woman who is 35 years old.

c. Find the approximate age of a man whose blood pressure is 134 mm Hg.

32. **Business**   Bryan Starr is a 17-year-old high school senior from Upper Arlington, Ohio, who started his own lawn service when he was 12. His business has grown substantially and he has been able to put away money for college, buy a truck, and invest in new lawn equipment to keep up with the growing demand for his services. Suppose his weekly revenue R can be represented by the formula $R = -p^2 + 50p - 125$, where p is the average price he charges for each lawn.

a. Sketch a graph of the related function. Explain why it behaves like it does, considering Bryan's business.

b. Explain how Bryan could earn $400 each week.

c. What price should he charge to earn the maximum revenue? What would be his revenue?

d. Use the discriminant to find if there is a price he could charge that would make his weekly revenue $600. Explain.

Mixed Review

33. **Geometry**   The area of a square plus its perimeter minus 12 is equal to 0. Find x if the area is $4x^2$.   (Lesson 6–3)

34. Solve $3t^2 + 4t = 15$ by factoring.   (Lesson 6–2)

35. Simplify $\sqrt{x^2 + 6x + 9}$.   (Lesson 5–5)

36. **Ecology**   On an average day, 958,904,100 photocopies are made. Of these, 356,164,384 are unnecessary. On an average day, how many photocopies are necessary? Round your answer to the nearest million and express it in scientific notation.   (Lesson 5–1)

37. Solve the system $x - y = 10$ and $2y - 3x = -1$ by using a matrix equation.   (Lesson 4–6)

38. Solve $|2x - 5| \leq 9$.   (Lesson 1–7)

SELF TEST

Solve each equation by graphing.   (Lesson 6–1)

1. $z^2 + 4z + 3 = 0$

2. $m^2 + 6m = 27$

Solve each equation by factoring.   (Lesson 6–2)

3. $x^2 + 5x - 36 = 0$

4. $2x^2 + 7x = -3$

Solve each equation by completing the square.   (Lesson 6–3)

5. $x^2 + 6x = 55$

6. $x^2 - 7x + 21 = 0$

7. Find the value of the discriminant for the equation $x^2 - 8x + 2 = 0$ and describe the nature of the roots.   (Lesson 6–4)

Solve each equation by using the quadratic formula.   (Lesson 6–4)

8. $3x^2 + 5x - 1 = 0$

9. $2x^2 - 25x + 72 = 0$

10. **Horticulture**   The length of a tropical garden at a local conservatory is 5 feet more than its width. A walkway 2 feet wide surrounds the outside of the garden. If the total area of the walkway and garden is 594 square feet, find the dimensions of the garden.   (Lesson 6–2)

Sum and Product of Roots

What YOU'LL LEARN

- To find the sum and product of the roots of quadratic equations, and
- to find a quadratic equation to fit a given condition.

Why IT'S IMPORTANT

You can use the sum and product of roots to write quadratic equations involving space flight and engineering.

APPLICATION

Engineering

One of the major tasks of civil engineers is to design roads that are safe and comfortable. In highway design, the quadratic function $y = ax^2 + bx + c$ is called a *transition curve* because it has properties that provide a smooth transition between peaks and valleys.

A road with an initial gradient, or slope, of 3% can be represented by the formula $y = ax^2 + 0.03x + c$, where y is the elevation and x is the distance along the curve. Suppose the elevation of the road is 1105 feet at points 200 feet and 1000 feet along the curve. You can find the equation of the transition curve. *This problem will be solved in Example 4.*

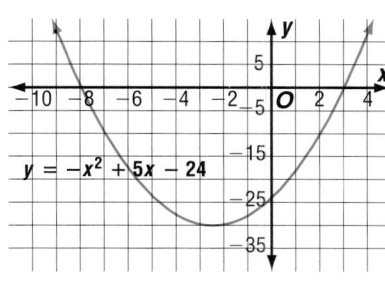

$y = -x^2 + 5x - 24$

As in the situation above, you may know the roots of a quadratic equation without knowing the equation itself. For example, suppose the roots of a quadratic equation are 3 and -8 and you want to find the equation. In a previous lesson, you used factoring to solve an equation. You applied the zero product property and set both equations equal to 0 to find the solutions. You can work backward to find the equation when you know the solutions.

$x = 3$ or	$x = -8$	*Start with the solutions.*
$x - 3 = 0$	$x + 8 = 0$	*Rewrite equations equal to 0.*

$$(x - 3)(x + 8) = 0 \quad \text{\textit{Multiplicative property of zero}}$$
$$x^2 + 5x - 24 = 0 \quad \text{\textit{Multiply.}}$$

The last equation has roots 3 and -8 and is written in standard form. The sum and product of the roots can help to develop the equation in another way.

Add the roots. $\quad 3 + (-8) = -5 \quad$ *-5 is the opposite of the coefficient of x.*

Multiply the roots. $\quad 3(-8) = -24 \quad$ *-24 is the constant term.* $\quad x^2 + 5x - 24 = 0$

This pattern can be generalized for any quadratic equation by using the roots defined by the quadratic formula. Let s_1 and s_2 represent the roots.

$$s_1 = \frac{-b + \sqrt{b^2 - 4ac}}{2a} \qquad\qquad s_2 = \frac{-b - \sqrt{b^2 - 4ac}}{2a} \qquad \text{\textit{Quadratic formula}}$$

Their sum can be represented as follows.

$$s_1 + s_2 = \frac{-b + \sqrt{b^2 - 4ac}}{2a} + \frac{-b - \sqrt{b^2 - 4ac}}{2a} \qquad \textit{Add the roots.}$$

$$= \frac{-2b + 0}{2a} \text{ or } -\frac{b}{a} \qquad \textit{Simplify.}$$

The sum of the roots is $-\frac{b}{a}$.

The product of the roots may be represented as follows.

$$s_1(s_2) = \left(\frac{-b + \sqrt{b^2 - 4ac}}{2a} \right)\left(\frac{-b - \sqrt{b^2 - 4ac}}{2a} \right)$$

$$= \frac{b^2 - (b^2 - 4ac)}{4a^2} \qquad \textit{Multiply.}$$

$$= \frac{b^2 - b^2 + 4ac}{4a^2} \qquad \textit{Use the distributive property.}$$

$$= \frac{4ac}{4a^2} \text{ or } \frac{c}{a}$$

The product of the roots is $\frac{c}{a}$.

The rule below can help you find a quadratic equation if you know the roots.

Sum and Product of Roots	If the roots of $ax^2 + bx + c = 0$ with $a \neq 0$ are s_1 and s_2, then $s_1 + s_2 = -\frac{b}{a}$ and $s_1 \cdot s_2 = \frac{c}{a}$.

Example ❶ **Write a quadratic equation that has roots $\frac{3}{4}$ and $-\frac{12}{5}$.**

Find the sum and product of the roots. Begin by expressing the sum and product of the roots with the same denominator.

$$s_1 + s_2 = \frac{3}{4} + \left(-\frac{12}{5} \right)$$

$$= \frac{15}{20} - \frac{48}{20} \text{ or } -\frac{33}{20}$$

$$s_1 \cdot s_2 = \left(\frac{3}{4} \right)\left(-\frac{12}{5} \right)$$

$$= -\frac{36}{20}$$

So, $-\frac{33}{20} = -\frac{b}{a}$ and $-\frac{36}{20} = \frac{c}{a}$.

Therefore, $a = 20$, $b = 33$, and $c = -36$. The equation is $20x^2 + 33x - 36 = 0$.

Check by solving the equation by factoring.

$$20x^2 + 33x - 36 = 0$$
$$(4x - 3)(5x + 12) = 0$$
$$4x - 3 = 0 \quad \text{or} \quad 5x + 12 = 0$$
$$4x = 3 \qquad\qquad 5x = -12$$
$$x = \frac{3}{4} \checkmark \qquad\qquad x = -\frac{12}{5} \checkmark$$

The method used in Example 1 can also be used with equations whose roots are imaginary.

Example ② **Write a quadratic equation that has roots $7 - 3i$ and $7 + 3i$.**

$$s_1 + s_2 = (7 - 3i) + (7 + 3i)$$
$$= 14 \quad -\frac{b}{a} = \frac{14}{1}$$

$$s_1(s_2) = (7 - 3i)(7 + 3i)$$
$$= 49 + 9 \text{ or } 58 \quad \frac{c}{a} = \frac{58}{1}$$

Since $-\frac{b}{a} = \frac{14}{1}$ and $\frac{c}{a} = \frac{58}{1}$, $a = 1$, $b = -14$, $c = 58$. Replace a, b, and c in

$ax^2 + bx + c = 0$ with these values. The resulting equation is
$x^2 - 14x + 58 = 0$.

You can also use the sum and product of roots to check the solutions of an equation.

Example ③ **Solve $2x^2 - 7x + 3 = 0$. Check by using the sum and product of the roots.**

$$x = \frac{-b + \sqrt{b^2 - 4ac}}{2a}$$

$$= \frac{-(-7) \pm \sqrt{(-7)^2 - 4(2)(3)}}{2(2)} \qquad a = 2, b = -7, \text{ and } c = 3$$

$$= \frac{7 \pm \sqrt{25}}{4} \text{ or } \frac{7 \pm 5}{4}$$

The solutions are $\frac{7 + 5}{4}$ and $\frac{7 - 5}{4}$ or 3 and $\frac{1}{2}$.

Check: *The sum of the roots, $s_1 + s_2$,*   *The product or the roots, $s_1 s_2$,*

       should be $-\frac{b}{a}$ or $\frac{7}{2}$.   *should be $\frac{c}{a}$ or $\frac{3}{2}$.*

       $3 + \frac{1}{2} = \frac{7}{2}$ ✓ $3\left(\frac{1}{2}\right) = \frac{3}{2}$ ✓

You can use the sum and product of roots to solve real-world problems.

Example ④

APPLICATION
Engineering

Refer to the application at the beginning of the lesson. Find the equation of the transition curve if the formula for a road with a gradient of 3% is $y = ax^2 + 0.03x + c$.

Explore   Read the problem and determine two points on the road. The points on the road are (200, 1105) and (1000, 1105). So $y = 1105$ when $x = 200$ and $x = 1000$.

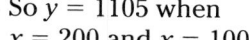

(continued on the next page)

Plan Since $y = 1105$, rewrite the equation equal to 0.

$$1105 = ax^2 + 0.03x + c \qquad \textit{Substitute 1105 for y.}$$
$$0 = ax^2 + 0.03x + (c - 1105) \qquad \textit{Rewrite equation equal to 0.}$$

Let $b = 0.03$, $s_1 = 200$, and $s_2 = 1000$.

Solve Use the sum and product of the roots to find the equation.

$$s_1 + s_2 = -\frac{b}{a} \qquad\qquad s_1 \cdot s_2 = \frac{c}{a}$$

$$200 + 1000 = -\frac{(0.03)}{a} \qquad 200 \cdot 1000 = \frac{c - 1105}{a} \qquad \textit{Why substitute}$$
$$\textit{c} -1105 \textit{ for c?}$$

$$1200 = -\frac{0.03}{a} \qquad\qquad 200{,}000 = \frac{c - 1105}{-0.000025}$$

$$a = -0.000025 \qquad\qquad -5 = c - 1105$$

$$1100 = c$$

Thus, the equation of the transition curve is $y = -0.000025x^2 + 0.03x + 1100$.

Examine Check to see if the equation makes sense. Substitute $x = 200$ and $x = 1000$ into the equation $y = -0.000025x^2 + 0.03x + 1100$.

CHECK FOR UNDERSTANDING

Communicating Mathematics

Study the lesson. Then complete the following.

1. **Write** the sum and the product of the roots of a quadratic equation expressed in terms of a, b, and c.

2. **Explain** how to use the zero product property to form a quadratic equation when its roots are known.

3. **Describe** how to create an equation when only the sum and product of the roots are known.

 MATH JOURNAL

4. When might it be difficult to find an equation using the sum and product of roots? Give some examples.

Guided Practice

State the sum and product of the roots of each quadratic equation.

5. $x^2 - 12x + 22 = 0$

6. $x^2 - 22 = 0$

7. $3x^2 + 75 = 0$

8. $2x^2 - \frac{1}{4}x = \frac{-8}{15}$

Write a quadratic equation that has the given roots.

9. $4 \pm \sqrt{5}$

10. $\sqrt{5} \pm 8i$

11. $-7, \frac{2}{3}$

12. $\frac{-3}{4}, \frac{5}{8}$

Solve each equation. Check by using the sum and product of the roots.

13. $x^2 - 49 = 0$

14. $2x^2 + 15x = 27$

15. $16x^2 - 81 = 0$

16. Suppose a quadratic equation has two real roots, r_1 and r_2. If the sum of the roots is $-\frac{19}{2}$ and the product of the roots is -30, write a quadratic equation that has roots r_1 and r_2.

Practice

Write a quadratic equation that has the given roots.

17. 6, −9

18. 5, −1

19. 2, $\frac{5}{8}$

20. 6, 6

21. $-\frac{2}{5}, \frac{2}{5}$

22. $-\frac{2}{5}, \frac{2}{7}$

23. $-4, -\frac{2}{3}$

24. $\frac{4}{3}, -\frac{1}{6}$

25. $4 \pm \sqrt{3}$

26. $\frac{3}{7} \pm 2i$

27. $\frac{-2 \pm 5i}{4}$

28. $\frac{-2 \pm 3\sqrt{5}}{7}$

Solve each equation. Check by using the sum and product of the roots.

29. $-2x^2 - 11x - 12 = 0$

30. $-3x^2 + 22x - 24 = 0$

31. $x^2 - 8x = 0$

32. $x^2 - 16 = 0$

33. $x^2 + \frac{1}{6}x - \frac{1}{3} = 0$

34. $\frac{1}{2}x^2 - \frac{13}{20}x - \frac{3}{20} = 0$

35. $x^2 - 8x - 18 = 0$

36. $2x^2 + 10x = -10$

37. $4x^2 - 31x - 45 = 0$

38. $3x^2 + 17 = 0$

39. Write a quadratic equation whose roots satisfy the following conditions.

  a. The sum of the roots is 4. The product of the roots is $\frac{13}{12}$.

  b. The sum of the roots is $\frac{1}{6}$. The product of the roots is $\frac{5}{21}$.

Critical Thinking

40. Find a value k such that -3 is a root of $2x^2 + kx - 21 = 0$.

41. Find a value k such that $\frac{1}{2}$ is a root of $2x^2 + 11x = -k$.

42. List all possible integral roots of $x^2 + bx - 24 = 0$. Assume that b represents an integer.

Applications and Problem Solving

43. Space Flight The United States is currently researching a project called the National Aerospace Plane, which would be able to regularly fly passengers directly into space. If the space plane is successful, it will be able to take off from an airport and rocket to orbit by accelerating to Mach 25 or 17,700 miles per hour. Scientists exploring the concepts of velocity and acceleration in space launch a model rocket that returns to Earth 18 seconds after takeoff. Use the formula $h = v_i t - \frac{1}{2}gt^2$, where $g = 9.8$ m/s^2, to determine the initial velocity of the rocket.   *Remember that the rocket starts on Earth when $t = 0$.*

44. Geometry Imagine that square *AEFD* is cut off the end of rectangle *ABCD* at the right. The remaining rectangle *EBCF* has the same ratio of length to width as the original rectangle *ABCD*. A rectangle with those similarities is called a "golden rectangle." Because of their pleasing shape, golden rectangles can be found in ancient and modern architecture.

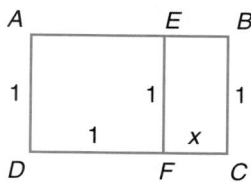

  a. Find the ratio of length to width for each rectangle.

  b. Set the ratios equal, write the equation, and solve for x.   *Remember that measures cannot be negative.*

  c. Substitute the value of x into the ratios you found in part a. What number do you get? This number is known as the *golden ratio*.

45. Solve $x^2 - 2x - 35 = 0$ by using the quadratic formula.   (Lesson 6–4)

46. Solve $m^2 + 3m - 180 = 0$ by completing the square.   (Lesson 6–3)

47. Graph the function $f(x) = x^2 + 8x - 5$. Name the vertex and axis of symmetry.   (Lesson 6–1)

48. Simplify $\sqrt{108} - \sqrt{48} + (\sqrt{3})^3$.   (Lesson 5–6)

49. Factor $4x^2 - 9$.   (Lesson 5–4)

50. Solve the system $3a - b + 2c = 7$, $-a + 4b - c = 3$, and $a + 4b - c = 1$ by using augmented matrices.   (Lesson 4–7)

51. Solve for a.   (Lesson 4–5)

$$\begin{vmatrix} a & 2 & -1 \\ 4a & -7 & -1 \\ 2 & 2 & -2 \end{vmatrix} = 45$$

52. Graph $y = -|5x - 12| + 1$.   (Lesson 2–6)

Mathematics and Juggling

The excerpt below appeared in an article in *New Scientist* on March 18, 1995.

MATHEMATICS AND MUSIC, IT IS SAID, often go together. Mathematicians are also reputed to be unusually good at chess. But there is a less well-known activity that also enjoys a time-honoured association with mathematics: juggling. Although the connection may seem tenuous at first sight, there has recently been a rush to apply mathematics to juggling, as mathematicians have come up with a clever way to invent new juggling patterns. . . . The pure mathematics of juggling concerns itself only with the patterns of throws, ignoring detail such as the precise timings, the exact trajectories and even the objects being thrown. Some theoretically minded jugglers recently introduced an idea called "site swaps" to describe the patterns in a compact form. And last year, four mathematicians—turned it into a mathematical theory. . . . (Site swap notation) is an easy way to remember a wide range of patterns that look very impressive when performed. The advantage for mathematicians is that a neat, compact notation makes it easier to count or classify patterns. ∎

1. What are the characteristics of the motions of objects being juggled?
2. If you were to construct a mathematical theory about juggling, what are some of the variables that might be involved?
3. Use your creativity to design a new juggling pattern that could actually be performed. What are some factors that limit any pattern design?

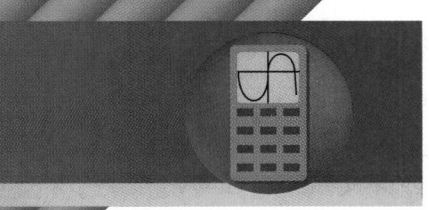

6–6A Graphing Technology
Families of Parabolas

A Preview of Lesson 6–6

The equations for the parabolas in a family are closely related. In the general form of a quadratic equation, $y = a(x - h)^2 + k$, a, h, and k may change. Changing the value of a, h, or k results in a different parabola in the family.

Example **Graph the following equations on the same screen in the standard viewing window. Describe any similarities and differences among the graphs.**

The parent graph in this example is the graph of $y = x^2$.

$y = x^2$, $y = x^2 + 4$, $y = x^2 - 2$

Enter: [Y=] [X,T,θ,n] [x²] [ENTER]
[X,T,θ,n] [x²] [+] 4 [ENTER]
[X,T,θ,n] [x²] [−] 2 [ZOOM] 6

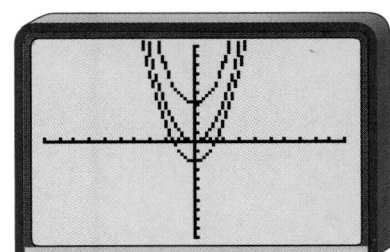

You can also graph the three equations by entering the following single equation.

Enter: [Y=] [X,T,θ,n] [x²] [+] [2nd] [{] 0 [,]
4 [,] [(−)] 2 [2nd] [}] [ZOOM] 6

The graphs have the same shape and all open upward. The vertex of each graph is on the y-axis. However, the graphs have different vertical positions.

Example 1 shows how changing the value of k in the equation $y = a(x - h)^2 + k$ translates the parabola along the y-axis. If $k > 0$, the parabola is translated k units upward, and if $k < 0$, it is translated k units downward. How do you think changing the value of h will change the graphs in a family of parabolas?

Example **Graph the following equations in the standard viewing window and describe any similarities and differences among the graphs.**

The parent graph in this example is the graph of $y = x^2$.

$y = x^2$, $y = (x + 4)^2$, $y = (x - 2)^2$

Enter: [Y=] [X,T,θ,n] [x²] [ENTER] [(]
[X,T,θ,n] [+] 4 [)] [x²] [ENTER] [(]
[X,T,θ,n] [−] 2 [)] [x²] [ZOOM] 6

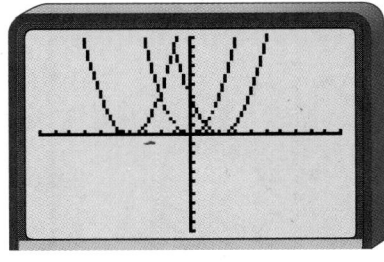

You can also graph the three equations by entering the following single equation.

Enter: [Y=] [(] [X,T,θ,n] [+] [2nd]
[{] 0 [,] 4 [,] [(−)] 2 [2nd] [}] [)] [x²] [ZOOM] 6

The graphs have the same shape and all open upward. The vertex of each graph is on the x-axis. However, the graphs have different horizontal positions.

Example 2 demonstrates that changing the value of h in $y = a(x - h)^2 + k$ translates the graph horizontally. If $h > 0$, the graph translates to the right h units. If $h < 0$, the graph translates to the left h units.

Changing the value of a in $y = a(x - h)^2 + k$ affects the direction of the opening and the shape of the graph. If $a > 0$, the graph opens upward, and if $a < 0$, the graph opens downward. If $|a| < 1$, the graph is wider than the graph of $y = x^2$, and if $|a| > 1$, then the graph is narrower than the graph of $y = x^2$. Graphs of equations with a values that have the same absolute value, such as $y = 2x^2$ and $y = -2x^2$, have the same shape.

Example **③**

The parent graph in this example is the graph of $y = x^2$.

Graph the following equations in the standard viewing window and describe any similarities and differences among the graphs.
$y = x^2, y = -x^2, y = 2x^2, y = -2x^2, y = 0.5x^2, y = -0.5x^2$

Enter: Y= X,T,θ,n x² ENTER
(−) X,T,θ,n x² ENTER
2 X,T,θ,n x² ENTER
(−) 2 X,T,θ,n x² ENTER
.5 X,T,θ,n x² ENTER
(−) .5 X,T,θ,n x² ZOOM 6

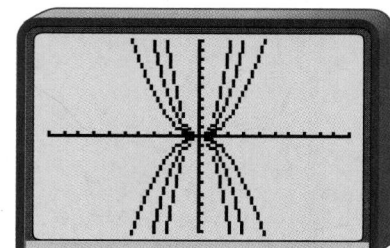

The graphs of $y = x^2$, $y = 2x^2$, and $y = 0.5x^2$ open upward, while the graphs of $y = -x^2$, $y = -2x^2$, and $y = -0.5x^2$ open downward. The graphs of $y = 2x^2$ and $y = -2x^2$ are narrower than the graph of $y = x^2$, while the graphs of $y = 0.5x^2$ and $y = -0.5x^2$ are wider than the graph of $y = x^2$.

EXERCISES

Study the lesson. Then complete the following.

1. Describe the effect that changing the value of k in an equation of the form $y = a(x - h)^2 + k$ has on the graph of the equation. Give an example.

2. Describe the effect that changing the value of h in an equation of the form $y = a(x - h)^2 + k$ has on the graph of the equation. Give an example.

3. How do the graphs of $y = a(x - h)^2 + k$ and $y = -a(x - h)^2 + k$ compare? Give an example.

Examine each pair of equations below and predict the graphs for each. Then use a graphing calculator to confirm your results. Write one sentence that compares the two graphs.

4. $y = x^2, y = (x + 6)^2$
5. $y = x^2, y = (x - 8)^2$
6. $y = x^2, y = x^2 + 1.5$
7. $y = x^2, y = x^2 - 11$
8. $y = -x^2, y = -5x^2$
9. $y = x^2, y = -2x^2$
10. $y = x^2, y = -\frac{1}{2}x^2 + 4$
11. $y = -\frac{1}{3}x^2, y = -\frac{1}{3}x^2 + 2$
12. $y = x^2, y = -6(x + 1)^2 - 11$
13. $y = (x + 2)^2 + 1, y = (x + 2)^2 - 4$
14. $y = 2(x + 3)^2 + 1,$
    $y = 4(x + 3)^2 + 1$
15. $y = 2(x - 4)^2 + 3,$
    $y = \frac{1}{2}(x - 4)^2 - 5$

Analyzing Graphs of Quadratic Functions

What YOU'LL LEARN

- To graph quadratic functions of the form $y = a(x - h)^2 + k$, and
- to determine the equation of a parabola by using points on its graph.

Why IT'S IMPORTANT

You can graph quadratic functions to solve problems involving biology and number theory.

APPLICATION
World Cultures

A recent article in *National Geographic* featured the Cherokee Indians living in Oklahoma. The Cherokee Nation is a federally recognized sovereign nation that has its own court system, legislature, and tax commission. The Cherokee are working toward self-sufficiency while trying to preserve their cultural identity.

One of the people featured in the *National Geographic* article was Lorene Drywater of Tahlequah, Oklahoma. She makes buffalo grass dolls, a craft she learned from her mother. She sells the dolls to tourists for additional income. As anyone who sells things can tell you, deciding on an appropriate price is very important. If your price is too low, you will not make much of a profit. If your price is too high, you will also probably not make much of a profit because fewer people will buy what you are selling. The best price is the price that leads to the maximum profit.

Suppose Ms. Drywater's profit $P(x)$ can be found by $P(x) = -x^2 + 24x - 60$, where x represents the price of each doll. What price should she charge to receive the maximum profit?

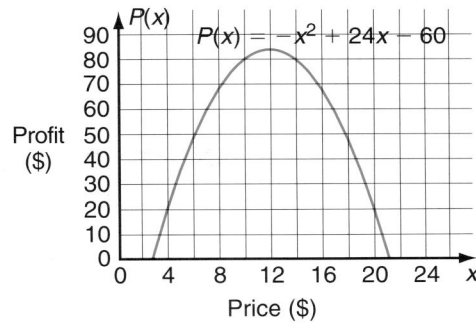

You can see from this graph that the vertex is (12, 84) and the axis of symmetry is $x = 12$. Thus, she should charge $12 for each doll to receive the maximum profit of $84. Is there a way to find this information without graphing the function?

You know that the graph of $ax^2 + bx + c = 0$ is a parabola. Quadratic functions can also be expressed in the general form $y = a(x - h)^2 + k$. We can write $P(x) = -x^2 + 24x - 60$ in the general form by completing the square.

$$
\begin{aligned}
P(x) &= -x^2 + 24x - 60 \\
&= -1(x^2 - 24x) - 60 \\
&= -1(x^2 - 24x + 144) - 60 + 144 \quad \textit{Add and subtract } \left(\tfrac{-24}{2}\right)^2 \textit{ or 144 to} \\
&= -1(x - 12)^2 + 84 \qquad\qquad\qquad\quad \textit{obtain an equivalent equation.}
\end{aligned}
$$

In this equation, $h = 12$ and $k = 84$. Compare these values with the coordinates of the vertex and axis of symmetry you found above by graphing. What pattern do you notice?

The graphs of $y = x^2$, $y = (x - 2)^2$, $y = x^2 + 3$, and $y = (x - 2)^2$ are shown at the right on the same set of axes. Study these graphs.

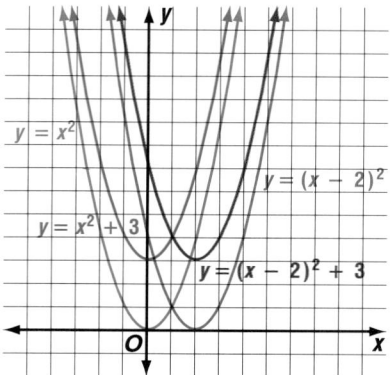

Equation	Vertex	Axis of Symmetry
$y = x^2$	(0, 0)	$x = 0$
$y = (x - 2)^2$	(2, 0)	$x = 2$
$y = x^2 + 3$	(0, 3)	$x = 0$
$y = (x - 2)^2 + 3$	(2, 3)	$x = 2$

Notice that the graphs all have the same shape. The difference is their position.

We can express the equations for these parabolas in the general form $y = (x - h)^2 + k$. When a quadratic function is written in this form, the vertex is (h, k), and the equation of the axis of symmetry is $x = h$.

Equation	General Form	h	k
$y = x^2$	$y = (x - 0)^2 + 0$	0	0
$y = (x - 2)^2$	$y = (x - 2)^2 + 0$	2	0
$y = x^2 + 3$	$y = (x - 0)^2 + 3$	0	3
$y = (x - 2)^2 + 3$	$y = (x - 2)^2 + 3$	2	3

In Chapter 4, you learned that a translation slides a figure on the coordinate plane without changing its shape or size. As the values of h and k change, the graph of $y = a(x - h)^2 + k$ is the graph of $y = x^2$ translated $|h|$ units left or right and $|k|$ units up or down. If h is positive, the parabola is translated to the right. If h is negative, it is translated to the left. Likewise, for k, the translation is up if k is positive and down if k is negative.

Example **Name the vertex and the axis of symmetry for the graph of $f(x) = (x + 6)^2 - 3$. Then graph the function. How is this graph different from the graph of $f(x) = x^2$?**

This function can be rewritten as $f(x) = [x - (-6)]^2 + (-3)$. Then $h = -6$ and $k = -3$. The vertex is at $(-6, -3)$, and the axis of symmetry is $x = -6$.

Finding several points on the graph makes graphing easier.

It is helpful to choose points that are close to h and on either side of h.

x	$(x + 6)^2 - 3$	$f(x)$
-8	$(-8 + 6)^2 - 3$	1
-7	$(-7 + 6)^2 - 3$	-2
-6	$(-6 + 6)^2 - 3$	-3
-5	$(-5 + 6)^2 - 3$	-2
-4	$(-4 + 6)^2 - 3$	1
-3	$(-3 + 6)^2 - 3$	6

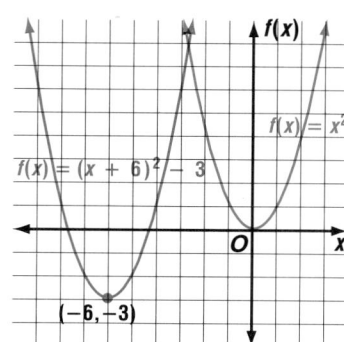

The shape of the graph is the same as the shape of the graph of $f(x) = x^2$, but it is translated 6 units left and 3 units down. *Notice that points with the same y-coordinates are the same distance from the axis of symmetry, $x = -6$.*

How does the value of a in the general form $y = a(x - h)^2 + k$ affect a parabola? Consider the equation in the application at the beginning of the lesson.

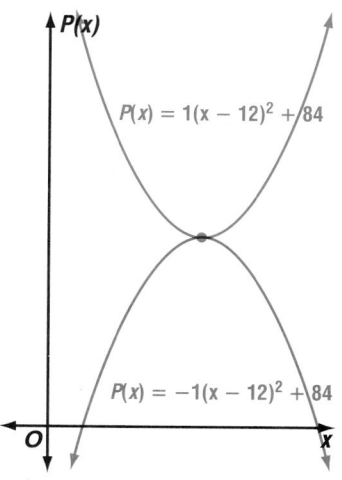

Graph $P(x) = -1(x - 12)^2 + 84$ and $P(x) = 1(x - 12)^2 + 84$ on the same coordinate plane and compare the graphs.

The graphs have the same vertex and are shaped the same. The graph of $P(x) = -1(x - 12)^2 + 84$ opens downward, and the graph of $P(x) = 1(x - 12)^2 + 84$ opens upward.

Graph the following functions to further investigate the relationships between similar functions. What do you find?

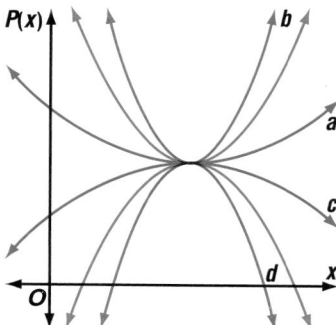

a. $P(x) = \frac{1}{4}(x - 12)^2 + 84$

b. $P(x) = 2(x - 12)^2 + 84$

c. $P(x) = -\frac{1}{4}(x - 12)^2 + 84$

d. $P(x) = -2(x - 12)^2 + 84$

All of the graphs for quadratic functions a, b, c, and d have the vertex $(12, 84)$ and the axis of symmetry $x = 12$. When a is negative, the graph opens downward. When a is positive, the graph opens upward. As the value of $|a|$ increases, the graph becomes narrower.

The chart below summarizes the characteristics of the graph of $y = a(x - h)^2 + k$.

$y = a(x - h)^2 + k$	a is positive.	a is negative.
Vertex	(h, k)	(h, k)
Axis of Symmetry	$x = h$	$x = h$
Direction of Opening	upward	downward
As the value of $|a|$ increases, the graph of $y = a(x - h)^2 + k$ narrows.		

Example ② **Graph $f(x) = -5x^2 + 80x - 319$. Name the vertex, axis of symmetry, and direction of opening for the graph.**

Write the function in the form $f(x) = a(x - h)^2 + k$ by completing the square.

$$f(x) = -5x^2 + 80x - 319$$
$$= -5(x^2 - 16x) - 319$$
$$= -5(x^2 - 16x + 64) - 319 - (-5)(64) \quad \textit{Why subtract } (-5)(64)?$$
$$= -5(x - 8)^2 + 1$$

The general form of this function is $f(x) = -5(x - 8)^2 + 1$. So, $a = -5$, $h = 8$, and $k = 1$. The vertex is at $(8, 1)$, and the axis of symmetry is $x = 8$. Since $a = -5$, the graph opens downward and is narrower than the graph of $f(x) = (x - 8)^2$.

(continued on the next page)

x	$-5(x-8)^2+1$	$f(x)$
6	$-5(6-8)^2+1$	-19
7	$-5(7-8)^2+1$	-4
8	$-5(8-8)^2+1$	1
9	$-5(9-8)^2+1$	-4
10	$-5(10-8)^2+1$	-19

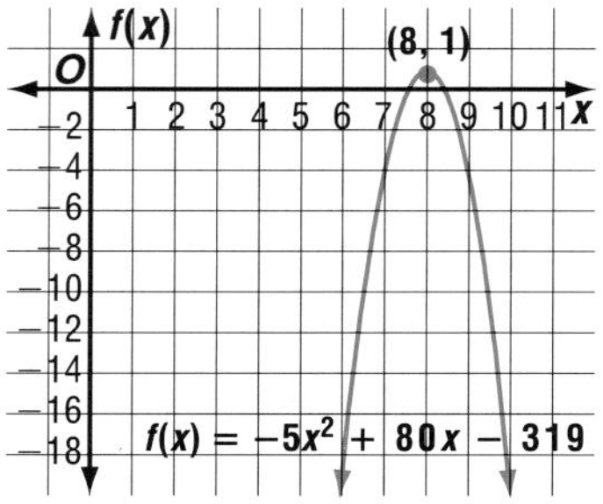
f(x)
(8, 1)
O
1 2 3 4 5 6 7 8 9 10 11 x
−2
−4
−6
−8
−10
−12
−14
−16
−18
f(x) = −5x² + 80x − 319

One way to find the equation of a parabola is by using the values of the vertex and one other point on the graph.

Example 3 CONNECTION Biology

A deadly frog found in an area of rain forest in western Colombia can be lethal even to the touch because it exudes a toxic substance. Suppose the graph below represents the path, or trajectory, that one of the frogs takes while hopping through the rain forest. Write the equation of the parabola.

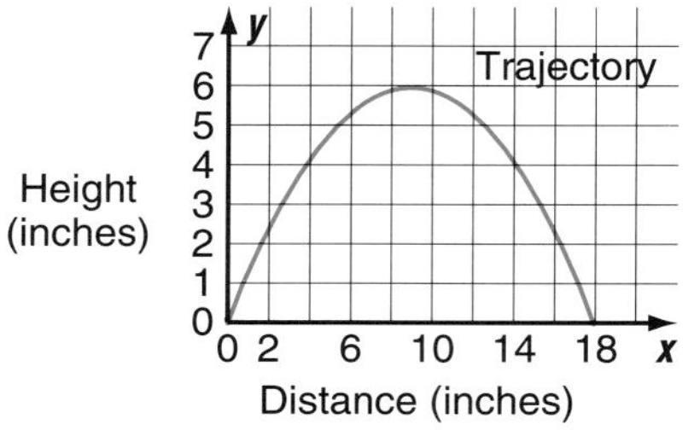
y
7
6
5
4
3
2
1
0
Trajectory
Height (inches)
0 2 6 10 14 18 x
Distance (inches)

The vertex of the parabola is at (9, 6). So $h = 9$ and $k = 6$.

Substitute the values of h and k and the coordinates of one other point on the graph into the general form of the equation and solve for a. *Use (0, 0).*

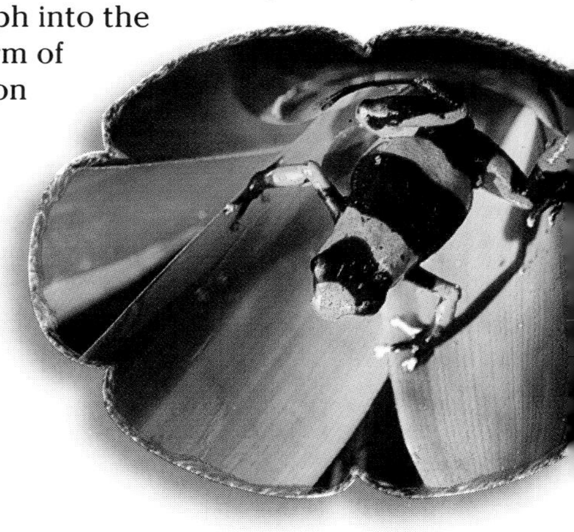

$$y = a(x-h)^2 + k$$
$$0 = a(0-9)^2 + 6 \quad \text{Substitute 9 for h, 6 for k, 0 for x, and 0 for y.}$$
$$-6 = a(-9)^2$$
$$a = -\frac{6}{81} \text{ or } -\frac{2}{27}$$

The equation of the parabola is
$y = -\frac{2}{27}(x-9)^2 + 6$ or $y = -\frac{2}{27}x^2 + \frac{4}{3}x$.

FYI

The launch angle of a frog's jump is approximately 45°. This helps the frog cover maximum distance on flat ground.

It is also possible to write the equation of a parabola if you know three of its points.

Example 4 **Write the equation of the parabola that passes through the points at (0, −3), (1, 4), and (2, 15).**

Each point should satisfy the equation of the parabola. Using the form of the equation $y = ax^2 + bx + c$, find a, b, and c by substituting the coordinates of the points into the quadratic form of the equation.

Ordered Pairs	Substitution	Simplify
(0, −3)	$-3 = a(0)^2 + b(0) + c$	$-3 = c$
(1, 4)	$4 = a(1)^2 + b(1) + c$	$4 = a + b + c$
(2, 15)	$15 = a(2)^2 + b(2) + c$	$15 = 4a + 2b + c$

LOOK BACK

Refer to Lesson 3-7 for information on solving systems of equations in three variables.

Now solve the system of equations. From the first equation, $c = -3$, so substitute -3 for c in the other two equations.

$$4 = a + b - 3 \quad \rightarrow \quad 7 = a + b$$
$$15 = 4a + 2b - 3 \quad \rightarrow \quad 18 = 4a + 2b$$

Solve by using elimination.

$$7 = a + b$$
$$18 = 4a + 2b$$

Multiply by −2.

$$-14 = -2a - 2b$$
$$\underline{18 = 4a + 2b}$$
$$4 = 2a$$
$$2 = a$$

Now substitute 2 for a and solve for b.

$$7 = 2 + b$$
$$b = 5$$

The solution to the system is $(2, 5, -3)$. So the equation of the parabola is $y = 2x^2 + 5x - 3$. *Check to see if each of the three points satisfies the equation.*

You can use a graphing calculator to model real-world data.

EXPLORATION

GRAPHING CALCULATORS

The *Mesa Tribune* (Mesa, Arizona) printed the sun intensity guide at the right on August 8, 1993. The data represent the average number of minutes of exposure to the sun required to redden untanned Caucasian skin.

Time of Day	Number of Minutes
9 A.M.	34
10 A.M.	20
11 A.M.	15
noon	13
1 P.M.	14
2 P.M.	18
3 P.M.	32
4 P.M.	60

Your Turn

a. Use the Edit option on the STAT menu feature to list the ordered pairs. Let x represent the number of hours since 8 A.M. Use the window $[-1, 9]$ with a scale factor of 1 and $[0, 65]$ with a scale factor of 5. Then press 2nd STAT PLOT 1 ENTER ▼ ENTER GRAPH .

b. To find a quadratic equation whose graph best fits the data, go to the CALC option on the STAT menu. Then choose 6, for QuadReg, and enter 2nd L1 , 2nd L2 . Then press ENTER .

c. Write the equation of the graph that best fits the data. Then press Y= VARS 5 ▶ ▶ 7 GRAPH .

d. How well do you feel the graph of the equation fits the data? Justify your answer.

Communicating Mathematics

Study the lesson. Then complete the following.

1. **Describe** the difference between the graphs of $y = 4(x + 6)^2 - 2$ and $y = -\frac{1}{2}(x + 6)^2 - 2$.

2. **List** the information you need to find the equation of a parabola.

3. **Analyze** the equation $y = -\frac{1}{6}(x + 4)^2 + 7$ and sketch its graph.

4. **Write** the equation of the function whose graph is 1 unit to the right and 4 units down from the graph shown at the right.

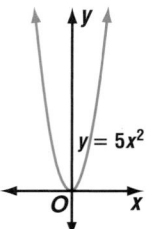
$y = 5x^2$

5. **You Decide**   Leticia and Marisel both wrote the equation $f(x) = 2x^2 - 12x - 3$ in general form by completing the square. However, they each got different answers. Look at their solutions and find which one is correct. Describe the error in the other person's solution.

Leticia	Marisel
$f(x) = 2x^2 - 12x - 3$	$f(x) = 2x^2 - 12x - 3$
$= 2(x^2 - 6x) - 3$	$= 2(x^2 - 6x) - 3$
$= 2(x^2 - 6x + 9) - 3 - 9$	$= 2(x^2 - 6x + 9) - 3 - 18$
$= 2(x - 3)^2 - 12$	$= 2(x - 3)^2 - 21$

Guided Practice

Write each equation in the form $y = a(x - h)^2 + k$ if not already in that form. Then name the vertex, axis of symmetry, and direction of opening for the graph of each quadratic function.

6. $f(x) = -2(x + 3)^2$

7. $f(x) = 5x^2 - 6$

8. $f(x) = x^2 - 4x + 5$

9. $f(x) = -3x^2 + 12x$

Write an equation for each parabola.

10.
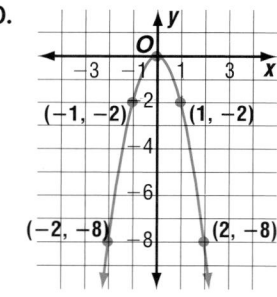

11.
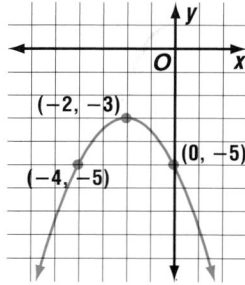

Write an equation for the parabola that passes through the given points.

12. $(0, 2), (2, 2), (3, 4)$

13. $(1, 6), (-2, 27), (2, 11)$

Graph each function.

14. $f(x) = 5(x + 3)^2 - 1$

15. $f(x) = -4x^2 + 16x - 11$

16. $f(x) = \frac{1}{3}(x - 1)^2 + 2$

17. Write the equation of a parabola with position 5 units below the parabola with equation $f(x) = 3x^2$.

Practice **Write each equation in the form $y = a(x - h)^2 + k$ if not already in that form. Then name the vertex, axis of symmetry, and direction of opening for the graph of each quadratic function.**

18. $f(x) = 4(x + 3)^2 + 1$

19. $f(x) = -(x + 11)^2 - 6$

20. $f(x) = -2(x - 2)^2 - 2$

21. $f(x) = 3(x - \frac{1}{2})^2 + \frac{1}{4}$

22. $f(x) = x^2 + 6x - 3$

23. $f(x) = -x^2 - 4x + 8$

24. $f(x) = 4x^2 + 24x$

25. $f(x) = -6x^2 + 24x$

26. $f(x) = 3x^2 - 18x + 11$

27. $f(x) = -2x^2 - 20x - 50$

28. $f(x) = -\frac{1}{2}x^2 + 5x - \frac{27}{2}$

29. $f(x) = \frac{1}{3}x^2 - 4x + 15$

Write an equation for each parabola.

30.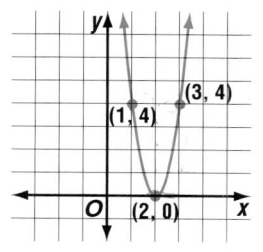

31.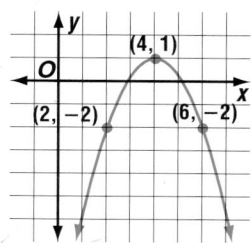

32.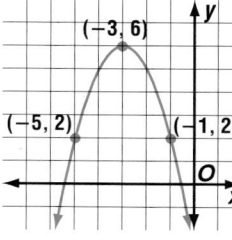

33.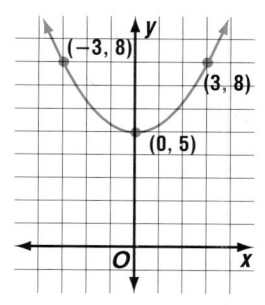

34.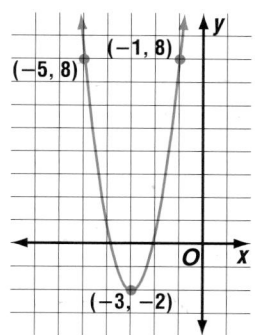

35.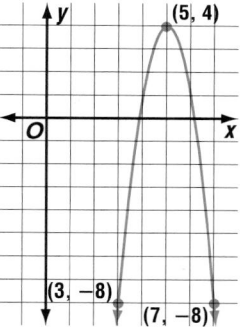

Write an equation for the parabola that passes through the given points.

36. $(0, 0)$, $(2, 6)$, $(-1, 3)$

37. $(2, -3)$, $(0, -1)$, $\left(-1, \frac{3}{2}\right)$

38. $(1, 0)$, $(3, 38)$, $(-2, 48)$

39. $(-1, -10)$, $(0, 6)$, $(2, 88)$

Graph each function.

40. $f(x) = 3(x + 3)^2$

41. $f(x) = 2(x + 3)^2 - 5$

42. $f(x) = \frac{1}{2}(x + 3)^2 - 5$

43. $f(x) = \frac{1}{3}(x - 1)^2 + 3$

44. $f(x) = x^2 + 6x + 2$

45. $f(x) = -2x^2 + 16x - 31$

46. $f(x) = -5x^2 - 40x - 80$

47. $f(x) = 2x^2 + 8x + 10$

48. $f(x) = -9x^2 - 18x - 6$

49. $f(x) = -0.25x^2 - 2.5x - 0.25$

50. Write an equation for a parabola whose vertex is at $(6, 1)$ and for which $a = 9$.

51. Write the equation of a parabola with position 2 units to the right and 9 units above the parabola with equation $f(x) = -2x^2$.

Graphing
Calculators

52. A recent article in *USA Today* listed the average high temperatures for Death Valley, California, the hottest place in the United States.

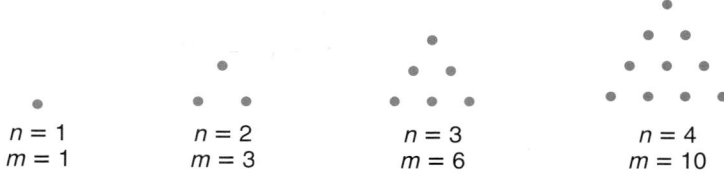

Month	Temperature (°F)
January	65
February	73
March	81
April	88
May	100
June	110
July	116
August	113
September	106
October	91
November	75
December	66

Average High Temperatures for Death Valley

a. Use a graphing calculator to model the data at the right. Find a quadratic equation whose graph best fits the data.

b. Do you think the graph of the equation fits the data? Justify your answer.

c. How would you expect a graph representing the average low temperatures to compare with the graph you found above?

**Critical
Thinking**

53. Given $f(x) = ax^2 + bx + c$ with $a \neq 0$, complete the square and rewrite the equation in the form $f(x) = a(x - h)^2 + k$. State an expression for h and k in terms of a, b, and c.

54. Sketch the family of graphs $y = (x + 2)^2 + 3$, $y = (x + 2)^2$, and $y = (x + 2)^2 - 3$. Fully describe the nature of the graphs and the roots of each equation.

**Applications and
Problem Solving**

55. Number Theory *Triangular numbers* are numbers that can be represented by a triangular array of m dots, with n dots on each side.

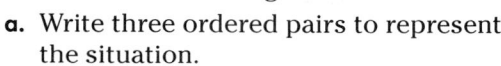

| $n = 1$ | $n = 2$ | $n = 3$ | $n = 4$ |
| $m = 1$ | $m = 3$ | $m = 6$ | $m = 10$ |

a. The relationship between the number of dots on each side and the total number of dots can be modeled by a quadratic function. Identify a reasonable domain and range for this function.

b. Write the quadratic function.

c. Graph the function. What is the least triangular number?

56. Entertainment Did you know that Tarzan has been featured in 43 movies? The story of the man who swings through the trees has fascinated us for years. Suppose Tarzan is in a tree 14 meters above the ground and he decides to use a vine to swing to another tree. One second after he begins his swing, he is 10.5 meters above the ground. Three seconds after he begins his swing, he is 6.5 meters above the ground.

a. Write three ordered pairs to represent the situation.

b. Write an equation for the parabola that passes through these points.

c. Sketch the graph of the parabola.

d. It seems that Tarzan gets pretty close to the ground as he swings through the trees. How close does he actually get?

57. Horticulture Helene Jonson has a rectangular garden 25 feet by 50 feet. She wants to increase the garden on all sides by an equal amount. If the area of the garden will be increased by 400 square feet, by how much will each dimension be increased? (Lesson 6–5)

58. Solve $4x^2 - 8x + 13 = 0$ by using the quadratic formula. (Lesson 6–4)

59. Graph $h(x) = x^2 - 2x + 5$. Name the vertex and the axis of symmetry. (Lesson 6–1)

60. Entertainment A magician asked a member of his audience to choose any number. He said, "Multiply your number by 3. Add the sum of your number and 8 to that result. Now divide by the sum of your number and 2." The magician announced the final answer without asking the original number. What was the final answer and how did he know what it was? (Lesson 5–3)

61. Simplify $(-x + 4)(-2 - 3x)$. (Lesson 5–2)

62. Find $\begin{bmatrix} -6 & 3 \\ 4 & 7 \end{bmatrix} \cdot \begin{bmatrix} 2 & -5 \\ -3 & 6 \end{bmatrix}$. (Lesson 4–3)

63. Find $-2 \begin{bmatrix} -3 & 0 & 12 \\ -7 & \frac{1}{3} & 4 \end{bmatrix}$. (Lesson 4–1)

64. Solve the system of equations. (Lesson 3–7)

$$x + 2y - z = -7$$
$$3x + y + z = -12$$
$$4z = 24$$

65. Refer to the graph at the right. The slope of $\overleftrightarrow{AB}$ is $\frac{4}{3}$. Line $\overleftrightarrow{EF}$ is perpendicular to $\overleftrightarrow{AB}$ and has a y-intercept of -2. Write the equation of $\overleftrightarrow{EF}$. (Lesson 2–4)

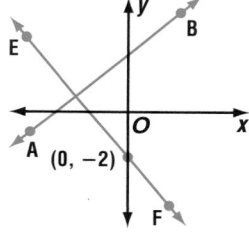

66. Solve $23x - 7 > 62$. (Lesson 1–6)

WORKING ON THE Investigation

Refer to the Investigation on pages 328–329.

the River Canyon Bridge

The formula for a parabolic curve whose vertex lies along the y-axis is $ax^2 + y = b$, where b represents the y-coordinate of the vertex.

1 Knowing the value of b, how might you determine the value of a for a particular bridge design? Describe your method for finding a.

2 Depending on the arch of the bridge, the value of a varies. For the bridges you might consider, the value of a must be within certain ranges. Describe possible values for a and justify your answer.

3 Determine the values of a and b for each of your designs and for the design proposed by the consultant in the Investigation in Lesson 6–1. With this information, write an equation for the parabolic curve of the arch in each design.

Add the results of your work to your Investigation Folder.

6-7A Graphing Technology
Quadratic Inequalities

A Preview of Lesson 6-7

To graph quadratic inequalities in two variables, we will use a procedure similar to that discussed in Lesson 2-7A on graphing linear inequalities. We will utilize the SHADE(feature found in the DRAW menu.

Example Graph $y \leq 0.5x^2 + x - 3$ in the standard viewing window.

You will recall that you must enter two functions when graphing an inequality in two variables. The calculator shades between the designated functions. The first function entered is the lower boundary of the region to be shaded and the second function is the upper boundary of the region.

Since the inequality asks for all points such that *y is less than or equal to* $0.5x^2 + x - 3$, we will shade below the parabola. The lower boundary will be $y = -10$, and the upper boundary will be $y = 0.5x^2 + x - 3$.

Before you begin, clear any functions stored in the Y = list. Then clear the DRAW menu by pressing [2nd] [DRAW] [ENTER] [ENTER] [CLEAR].

Make sure Ymin is greater than or equal to −10.

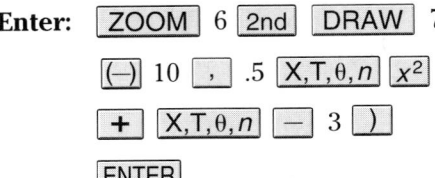

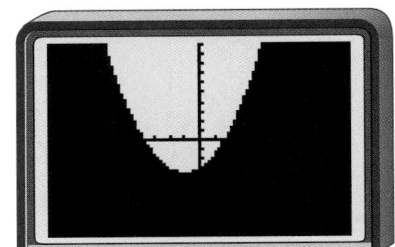

To graph an inequality that has a > or ≥ sign, the function itself will be the lower boundary.

Example Graph $y \geq x^2 - 4x + 1$ in the standard viewing window.

Since the inequality asks for all points such that *y is greater than or equal to* $x^2 - 4x + 1$, we will shade above the parabola. The lower boundary will be $x^2 - 4x + 1$, and the upper boundary will be $y = 10$. *Be sure to clear the DRAW menu first.*

Make sure Ymax is less than or equal to 10.

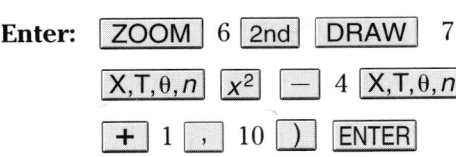

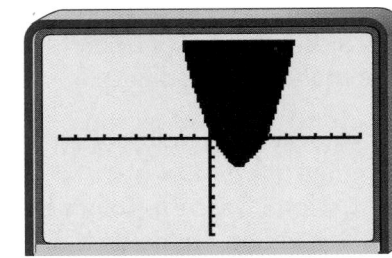

Sometimes you will want to be able to trace the graphs drawn or use other graphing features of the calculator. This often occurs when solving inequalities in one variable. In this case, equations should be entered in the Y = list.

Example ③ **Solve $2x^2 + 3x - 4 < 0$ to the nearest hundredth using a graphing calculator.**

Begin by obtaining a graph of $y = 2x^2 + 3x - 4$. We are interested in determining values for x so that $2x^2 + 3x - 4$ is less than 0. So, we look for points where the y value is less than 0, or where the graph falls below the x-axis.

Graphing calculators cannot show dashed lines for inequalities involving less than or greater than signs.

Enter: $\boxed{Y=}$ 2 $\boxed{X,T,\theta,n}$ $\boxed{x^2}$
$\boxed{+}$ 3 $\boxed{X,T,\theta,n}$ $\boxed{-}$
4 $\boxed{ZOOM}$ 6

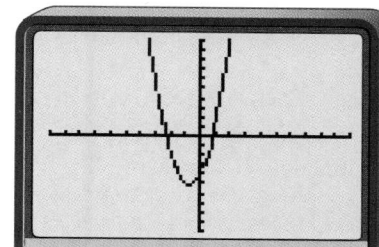

Zooming in on the x-intercepts, we find that $2x^2 + 3x - 4 < 0$ when $-2.35 < x < 0.85$.

EXERCISES

Use a graphing calculator to graph each inequality.

1. $y \geq x^2 + 11x - 3$ 2. $y \leq -0.5x^2 + 9$
3. $y \leq 1.2x^2 + 15x$ 4. $y \geq 6x^2 - 15x + 7$
5. $y > -x^2 + 6x + 8$ 6. $y > -4x^2 - 3x - 6$

Use a graphing calculator to solve each inequality to the nearest hundredth.

7. $x^2 + 4x - 21 > 0$ 8. $2x^2 - 4x + 1 \leq 0$
9. $x(2x + 1) \geq 0$ 10. $x^2 - 3 > 0$
11. $x^2 - 9x < 4$ 12. $0.5x^2 > 1.8x$

13. **Geometry**   A length of a rectangle is 4 inches longer than the width. Find the possible dimensions of the rectangle if the area must be at least 28 square inches.

Graphing and Solving Quadratic Inequalities

What YOU'LL LEARN

- To graph quadratic inequalities, and
- to solve quadratic inequalities in one variable.

Why IT'S IMPORTANT

You can use quadratic inequalities to solve problems involving sports and forensic science.

LOOK BACK

You can refer to Lesson 2-7 to review linear inequalities.

APPLICATION
Business

Athletic Advantage, Inc. makes athletic shoes for aerobics and running. According to recent sales figures, their profit $P(x)$ on x pairs of shoes can be found by the inequality $P(x) \leq -x^2 + 120x - 400$. The relation is an inequality because the wholesale price is lower for large orders.

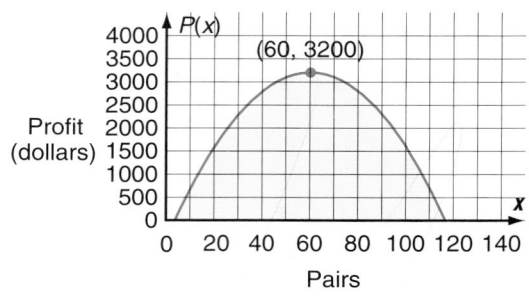

The graph of this inequality is the part of the plane enclosed by the parabola whose equation is $P(x) = -x^2 + 120x - 400$. The parabola is the **boundary** of each region. Points on the parabola represent the profit if all the shoes are sold at the list price. Points in the interior of the curve represent profits if some of the shoes are sold at a discount.

You can graph **quadratic inequalities** using the same techniques you used to graph linear inequalities.

1. Graph the boundary. Determine if it should be solid or dashed.
2. Test a point in each region.
3. Shade the region whose ordered pair results in a true inequality.

Example 1

Graph $y \leq x^2 - 6x + 2$.

The boundary will be the graph of $y = -x^2 - 6x + 2$.

$$y = -x^2 - 6x + 2$$
$$= -(x^2 + 6x) + 2$$
$$= -(x^2 + 6x + 9) + 2 + 9 \quad \text{Complete the square.}$$
$$= -(x + 3)^2 + 11$$

The boundary is a parabola that opens downward with its vertex at $(-3, 11)$. The boundary is included in the graph, so it should be solid. Test points not on the parabola to see whether points inside or outside the parabola belong to the graph.

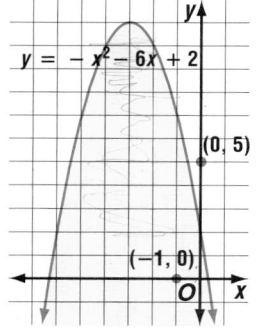

Region inside parabola

Test $(-1, 0)$: $y \leq -x^2 - 6x + 2$
$$0 \overset{?}{\leq} -(-1)^2 - 6(-1) + 2$$
$$0 \leq -1 + 6 + 2$$
$$0 \leq 7 \quad \text{true}$$

The point at $(-1, 0)$ does belong.

Region outside parabola

Test $(0, 5)$: $y \leq -x^2 - 6x + 2$
$$5 \overset{?}{\leq} -(0)^2 - 6(0) + 2$$
$$5 \leq 0 - 0 + 2$$
$$5 \leq 2 \quad \text{false}$$

The point at $(0, 5)$ does not belong.

Since $(-1, 0)$ is part of the solution and $(0, 5)$ is not, shade the region inside the parabola.

Example **2**

Geometry

A rectangle is 6 centimeters longer than it is wide. Find the possible dimensions if the area of the rectangle is more than 216 square centimeters.

Draw a diagram of the rectangle. Let w represent the width of the rectangle. Then $w + 6$ represents the length and $w(w + 6)$ represents the area.

w cm

$(w + 6)$ cm

$$w(w + 6) > 216$$
$$w^2 + 6w > 216$$
$$w^2 + 6w - 216 > 0$$
$$(w^2 + 6w + 9) - 216 - 9 > 0 \quad \text{Complete the square.}$$
$$(w + 3)^2 - 225 > 0$$

The boundary of the graph of this inequality is a parabola that opens upward with its vertex at $(-3, -225)$. Test a point not on the boundary to determine which region should be included in the graph.

Test (0, 0): $\quad w(w + 6) > 216$
$$0(0 + 6) > 216$$
$$0 > 216 \quad \text{false}$$

Since the graph of $(0, 0)$ is inside the parabola and it does not satisfy the inequality, the region outside the parabola is included in the graph.

The points on the graph outside Quadrant I should be disregarded since length cannot be negative. So the width of the rectangle should be greater than 12 cm and the length should be greater than 18 cm.

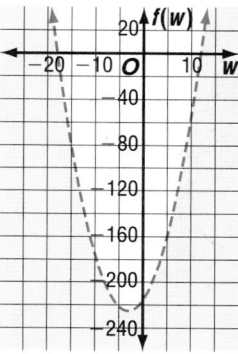

Just as you solve a quadratic equation by graphing its related quadratic function, you can solve a quadratic inequality in one variable by graphing its related quadratic inequality in two variables. For example, to solve $0 > x^2 - 6x - 7$, you can graph $y > x^2 - 6x - 7$. The solutions of the inequality are each point on the x-axis that is included in the graph.

You can also solve quadratic inequalities algebraically.

Example **3**

Solve $0 > x^2 - 6x - 7$.

Method 1: Graphing
Graph the related inequality $y > x^2 - 6x - 7$. First complete the square and rewrite the inequality as $y > (x - 3)^2 - 16$. The graph of the $y = (x - 3)^2 - 16$ is a parabola that opens upward with its vertex at $(3, -16)$.

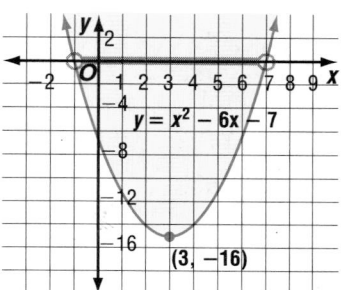

The points on the x-axis that satisfy $y > x^2 - 6x - 7$ are solutions to the inequality $0 > x^2 - 6x - 7$. Those solutions are $\{x \mid -1 < x < 7\}$.

(continued on the next page)

Method 2: Factoring

$0 < x^2 - 6x - 7$

$0 < (x - 7)(x + 1)$

Since the product of two numbers is negative if one number is negative and one is positive, we can write the following.

(x − 7) is negative and   or   *(x − 7) is positive and*
(x + 1) is positive: *(x + 1) is negative:*

$\quad x - 7 < 0 \quad$ and $\quad x + 1 > 0$ $\quad x - 7 > 0 \quad$ and $\quad x + 1 < 0$

$\qquad x < 7 \qquad\qquad x > -1$ $\qquad x > 7 \qquad\qquad x < -1$

$\qquad\qquad -1 < x < 7$

The graphs of x > 7 and x < −1
never intersect, so x > 7 and
x < −1 can never be true.

You can also solve quadratic inequalities by using three test points.

Example **4** **Solve $x^2 - x - 12 > 0$.**

First solve the equation $x^2 - x - 12 = 0$ by factoring.

$x^2 - x - 12 = 0$

$(x - 4)(x + 3) = 0$

$x - 4 = 0 \quad$ or $\quad x + 3 = 0$

$\qquad x = 4 \qquad\qquad x = -3$

The points at 4 and −3 separate the x-axis into three regions:
$x < -3$, $-3 < x < 4$, and $x > 4$.

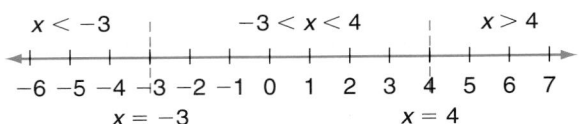

Choose a value from each part and substitute it into $x^2 - x - 12 > 0$.
Determine if the result is true or false. Organize your results in a table.

Part of x-axis	Chosen point, x	$x^2 - x - 12$	Is $x^2 - x - 12 > 0$?
$x < -3$	−5	$(-5)^2 - (-5) - 12 = 18$	yes
$-3 < x < 4$	1	$(1)^2 - 1 - 12 = -12$	no
$x > 4$	6	$(6)^2 - 6 - 12 = 18$	yes

The solution set to the inequality $x^2 - x - 12 > 0$ is $\{x \mid x < -3 \text{ or } x > 4\}$, as shown on the number line below.

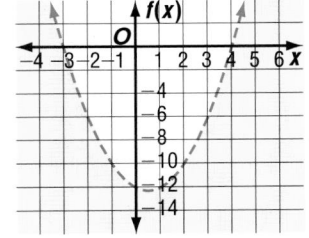

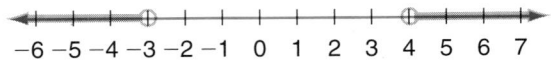

The graph of the related inequality $x^2 - x - 12 > y$, shown above, confirms the solutions.

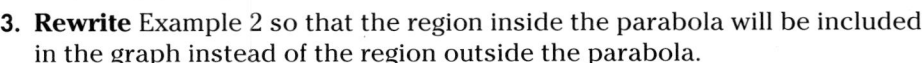

Communicating Mathematics

Study the lesson. Then complete the following.

1. The equation $y = x^2 - 5x + 4$ is graphed at the right.
   a. If you were to graph the inequality $y \geq x^2 - 5x + 4$, would you include the region inside or outside of the parabola? Explain why.
   b. What are the solutions of $0 \geq x^2 - 5x + 4$?

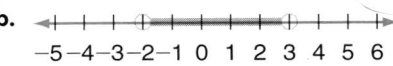

2. **State** the points you would test to find the solution to $(x - 8)(x + 2) > 0$.

3. **Rewrite** Example 2 so that the region inside the parabola will be included in the graph instead of the region outside the parabola.

4. **Explain** how factoring can help determine the solution set of a quadratic inequality.

 **MATH JOURNAL**

5. **Write** what you have learned about graphing quadratic inequalities. Explain why some boundaries are contained in the region and some are not included. How can you determine if the graph is a region inside or outside of the boundary?

Guided Practice

Determine if the ordered pair is a solution of the given inequality.

6. $y < 2x^2 + 4$, $(3, 5)$

7. $y \geq x^2 - 9$, $(0, 5)$

8. $y \leq 2x^2 - 3x + 1$, $(-1, 4)$

9. $y > 5x^2 + 2x - 3$, $(-1, -1)$

10. Which of the following is the graph of the solution set for $x^2 < x + 6$?

   a. $-5\ -4\ -3\ -2\ -1\ \ 0\ \ 1\ \ 2\ \ 3\ \ 4\ \ 5\ \ 6$
   b. $-5\ -4\ -3\ -2\ -1\ \ 0\ \ 1\ \ 2\ \ 3\ \ 4\ \ 5\ \ 6$
   c. $-6\ -5\ -4\ -3\ -2\ -1\ \ 0\ \ 1\ \ 2\ \ 3\ \ 4\ \ 5$
   d. $-6\ -5\ -4\ -3\ -2\ -1\ \ 0\ \ 1\ \ 2\ \ 3\ \ 4\ \ 5$

Graph each inequality.

11. $y \leq x^2 + 4x + 4$

12. $y > x^2 - 36$

13. $y \leq -x^2 + 7x + 8$

14. $y \leq -x^2 - 3x + 10$

Use the related graph of each inequality to write its solutions.

15. $x^2 - 4x - 12 \leq 0$

16. $x^2 - 9 > 0$

17. $-x^2 + 10x - 25 \geq 0$

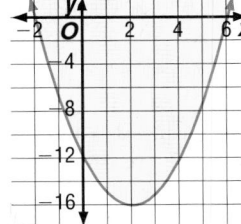
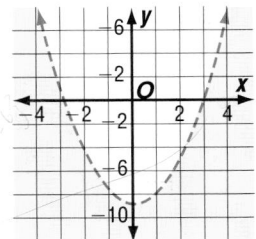
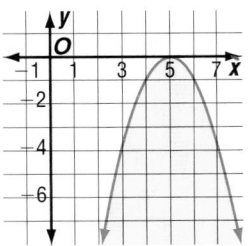

Solve each inequality.

18. $(x + 11)(x - 3) > 0$

19. $(n - 2.5)(n + 3.8) \geq 0$

20. $x^2 - 4x \leq 0$

21. $b^2 \geq 10b - 25$

22. $2x^2 > 25$

23. $2b^2 - b < 6$

24. The graph of the quadratic function $y = x^2 - 4x - 5$ is shown at the right.

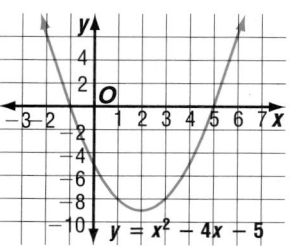

   a. What are the solutions of the equation $0 = x^2 - 4x - 5$?

   b. What are the solutions of the inequality $0 \leq x^2 - 4x - 5$?

   c. What are the solutions of the inequality $0 \geq x^2 - 4x - 5$?

25. **Recreation**   The YMCA has a 40-foot by 60-foot area in which to build a swimming pool. The pool will be surrounded by a concrete sidewalk of uniform width. What could the width of the sidewalk surrounding the pool be if organizers want the pool to be at least 1500 square feet?

EXERCISES

Practice

Graph each inequality.

26. $y \geq x^2 - 10x + 25$

27. $y < x^2 - 16$

28. $y \leq x^2 - x - 20$

29. $y \geq x^2 + 3x - 18$

30. $y \geq 2x^2 + x - 3$

31. $y \leq -x^2 + 5x + 6$

32. $y > 2x^2 + 3x - 5$

33. $y < -x^2 + 13x - 36$

34. $y \leq -x^2 + 5x + 14$

35. $y \geq -3x^2 + 5x + 2$

36. $y > 4x^2 - 8x + 3$

37. $y \leq -x^2 - 7x + 10$

Solve each inequality.

38. $(x - 4)(x + 7) < 0$

39. $x^2 - 3x - 18 > 0$

40. $m^2 + m - 6 > 0$

41. $q^2 + 2q \geq 24$

42. $p^2 - 4p \leq 5$

43. $2x^2 + 5x - 12 \leq 0$

44. $6s^2 + 5s > 4$

45. $w^2 \geq 2w$

46. $9v^2 - 6v + 1 \leq 0$

47. $2g^2 - 5g - 3 < 0$

48. $f^2 + 12f + 36 < 0$

49. $n^2 \leq 3$

50. $8d + d^2 \geq -16$

51. $4t^2 - 9 \leq -4t$

52. $(x - 1)(x + 4)(x - 3) > 0$

53. $(x + 2)(x + 4)(x - 8) \leq 0$

54. $(x + 5)(x + 1)(x - 4)(x - 6) > 0$

55. $(x - 2)(x + 2)(x - 1)(x + 3) \geq 0$

Critical Thinking

56. Find the intersection of the graphs of $y \geq x^2 - 3$ and $y \leq x^2 + 3$.

Applications and Problem Solving

57. **Sports**   The instant replay facility at the Superdome in New Orleans, Louisiana, was moved because it was hit by a high punt kicked by Oakland Raider Ray Guy. The original position of the facility was 90 feet above the playing field. Cody, a high school punter, can kick a football with an initial velocity of 65 feet per second. The height of the football t seconds after he kicks it is found by the function $h(t) = -16t^2 + 65t$.

   a. If Cody were to have kicked a football in the Superdome before the instant replay facility was moved, would he have been able to hit it? Explain.

   b. What was the speed of the ball Ray Guy punted if it hit the facility in 3 seconds?

58. Geometry A rectangle is 5 centimeters longer than it is wide. Find the possible dimensions if the area of the rectangle is more than 104 square centimeters.

59. Business Lorena drives a shuttle bus for the National Park Service. The charge is $1.00 to ride from the parking lot to one of the major attractions in the National Park. During the winter months, about 100 people ride the bus each day. It is estimated that 5 more passengers will ride per day for each $0.20 decrease in fare. The cost of operating the shuttle bus is $66 per day. How many $0.20 decreases in fare could be made and still allow the shuttle bus to make a profit?

60. Forensic Science Police are investigating the shooting of a police helicopter. They found a weapon at the scene of the crime that has a suspect's fingerprints on it. Forensic experts have deduced that the weapon is capable of firing with an initial velocity of 980 feet per second. So the height of the bullet t seconds after firing is found by the function $h(t) = -16t^2 + 980t$.

  a. Draw a graph to represent this situation.
  b. If the helicopter was flying at an altitude of 7000 feet at the time it was shot, is it possible that this weapon shot the helicopter? Explain your answer.

Mixed Review

61. Architecture The Gateway Arch of the Jefferson National Expansion Memorial in St. Louis is shaped like a parabola whose equation is $f(x) = \frac{1}{315}(-2x^2 + 1260x)$. (Lesson 6–6)

  a. Write the equation in the form $y = a(x - h)^2 + k$.
  b. If the bases of the arch are 630 feet apart, how tall is the arch?
  c. Graph $f(x)$. Compare your graph to a photo of the Gateway Arch. Describe what you observe.

62. Find a value k such that 1 is a root of $x^2 + kx - 5 = 0$. (Lesson 6–5)

63. Solve $11m^2 - 12m = 10$ by using the quadratic formula. (Lesson 6–4)

64. Solve $x^2 + bx + c = 0$ by completing the square. (Lesson 6–3)

65. Simplify $(5 + \sqrt{8})^2$. (Lesson 5–6)

66. Divide $(y^4 + 3y^3 + y - 1) \div (y + 3)$ by using synthetic division. (Lesson 5–3)

67. Simplify $(5a - 3)(1 - 3a)$. (Lesson 5–2)

68. Find the inverse of $\begin{bmatrix} 4 & 6 \\ -1 & 5 \end{bmatrix}$. (Lesson 4–4)

69. Solve the system of equations using Cramer's rule. (Lesson 3–3)
$$6x + 7y = 10$$
$$3x - 4y = 20$$

70. Graph $3x - 2y = 8$. Find the slope, the x-intercept, and the y-intercept. (Lesson 2–3)

71. Solve $4x - (2x + 8) + 3x = 5x - 8$. (Lesson 1–4)

Integration: Statistics
Standard Deviation

APPLICATION

Lifestyles

As we grow older, we spend our money in different ways. The chart below shows the money people spend on selected items each year in the United States for several age groups.

Item/Age Group	under 25	35–44	55–64	75 and up
Food at home	$1410	$3324	$2639	$1864
Food away from home	1140	2129	1634	703
Own a home	421	4549	3433	1587
Rent a home	2465	1733	751	1091
Footwear	158	315	206	153
Vehicle purchases	1916	2682	2215	528
Public transportation	181	340	378	156
Health insurance	144	583	749	1152
Electronic items	393	503	404	163
Pets and toys	139	577	442	172
Education	871	507	465	38

It appears that expenses for people in the under 25 category vary less than those in the 35–44 category. But how can you tell for sure?

Sometimes you need information about the spread, or variation, of data. A measure of variation called the **standard deviation** measures how much each value in a set of data differs from the mean. The symbol commonly used for standard deviation is SD or the lowercase Greek letter sigma, σ. The mean is usually labeled with the symbol $\overline{x}$ read "x bar."

To find the standard deviation of a set of data, follow these steps.
1. Find the mean, $\overline{x}$.
2. Find the difference between each value in the set of data and the mean.
3. Square each difference.
4. Find the mean of the squares.
5. Take the principal square root of this mean.

Standard Deviation

> From a set of data with n values, where x_1 represents the first term and x_n represents the nth term, if $\overline{x}$ represents the mean, then the standard deviation can be found as follows.
>
> $$SD \text{ or } \sigma_{\overline{x}} = \sqrt{\frac{(x_1 - \overline{x})^2 + (x_2 - \overline{x})^2 + \ldots + (x_n - \overline{x})^2}{n}}$$

To find the standard deviation of the data given for people under 25 and people aged 35-44, first find the mean of the expenses for both groups.

under 25:

$$\bar{x} = \frac{1410 + 1140 + 421 + 2465 + 158 + 1916 + 181 + 144 + 393 + 139 + 871}{11}$$

$$\approx 839.82$$

35–44:

$$\bar{x} = \frac{3324 + 2129 + 4549 + 1733 + 315 + 2682 + 340 + 583 + 503 + 577 + 507}{11}$$

$$\approx 1567.45$$

Then substitute each mean into the formula for standard deviation.

under 25:

$$SD = \sqrt{\frac{(1410 - 839.82)^2 + (1140 - 839.82)^2 + \ldots + (139 - 839.82)^2 + (871 - 839.82)^2}{11}}$$

$$= \sqrt{\frac{(570.18)^2 + (300.18)^2 + \ldots + (-700.82)^2 + (31.18)^2}{11}}$$

$$\approx 766.63$$

35–44:

$$SD = \sqrt{\frac{(3324 - 1567.45)^2 + (2129 - 1567.45)^2 + \ldots + (577 - 1567.45)^2 + (507 - 1567.45)^2}{11}}$$

$$= \sqrt{\frac{(1756.55)^2 + (561.55)^2 + \ldots + (990.45)^2 + (1060.45)^2}{11}}$$

$$\approx 1376.53$$

The standard deviation for the under–25 age group is about \$766.63. The standard deviation for the 35–44 age group is about \$1376.53. So the expenses for the under–25 age group are closer to their mean than those in the 35–44 age group, and they vary less.

Most of the members of a set of data are within 1 standard deviation from the mean. The under 25 age group expenses can be broken down as shown in the diagram below.

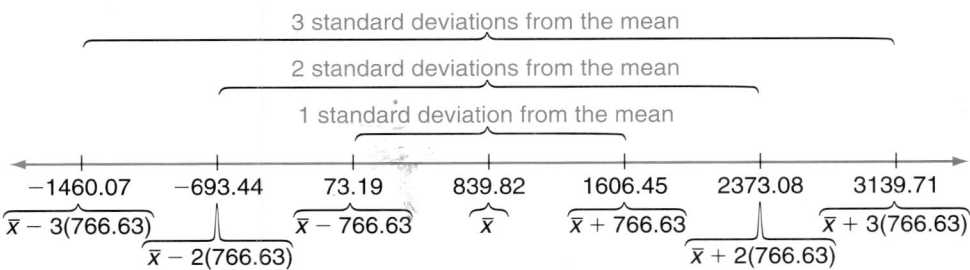

Standard deviation is often used to classify large sets of data.

Example ❶

Sports

A day at a National Football League game for a family of four represents the most expensive outing in professional sports, averaging $184.19, according to *Team Marketing Report*, a Chicago-based newsletter. The cost in 1994 represented a 6.2% increase over the previous year, when the average was $173.37. The average ticket price for each NFL team for 1994 is listed in the table below.

Team	1994 Average	Team	1994 Average
Arizona	$27.69	LA Rams	$29.13
Atlanta	27.00	Miami	29.65
Buffalo	33.73	Minnesota	29.79
Chicago	32.23	New England	34.34
Cincinnati	28.43	New Orleans	26.71
Cleveland	27.27	N.Y. Giants	35.59
Dallas	32.85	N.Y. Jets	25.00
Denver	32.34	Philadelphia	40.00
Detroit	30.04	Pittsburgh	30.99
Green Bay	26.13	San Diego	33.86
Houston	31.88	San Francisco	39.75
Indianapolis	26.48	Seattle	28.00
Kansas City	29.16	Tampa Bay	29.57
LA Raiders	31.32	Washington	35.70

a. Estimate the mean ticket price.

b. Calculate the mean ticket price.

c. Find the standard deviation of the data.

a. A quick glance at the data shows that about half of the data are above $30 and about half are below $30. So the mean will probably be about $30.

b. You can use a scientific calculator to perform the arithmetic necessary to calculate the mean and the standard deviation. However, your calculator may also have a statistics mode that simplifies this calculation. Press [MODE] and [STAT] to put your calculator in statistics mode. *The statistics functions are often second key functions.*

Enter the data by entering each number and then pressing the [Σ+] key. This provides a cumulative sum.

Enter: 27.69 [Σ+] 27 [Σ+] 33.73 [Σ+] ... 35.70 [Σ+]

Wrong entries can be deleted by using [Σ−] .

The display will show how many numbers were entered.

To find the mean, press [2nd] [x̄] .   *30.87964286*

The mean is $30.88.

c. Find the standard deviation.

Enter: [2nd] [σn] *3.793188898*

The standard deviation is $3.79.

When studying the standard deviation of a set of data, it is very important to keep the mean in mind. For example, suppose a company that sells video equipment found that the standard deviation of monthly prices of their equipment over the last two years was $50. If the mean of the prices over those two years was $200, then the standard deviation indicates a significant variation or change. However, if the mean was $900, then the standard deviation of $50 indicates a much smaller variation.

You can use a graphing calculator to calculate the mean and standard deviation of a set of data.

EXPLORATION

GRAPHING CALCULATORS

First, enter the elements of the data set into the calculator as follows. Press the ⬛ STAT ⬛ key and choose 1: Edit. Now you may enter the data into List 1 (L1). After each entry, press ⬛ENTER⬛. After the data entry is complete, press ⬛ STAT ⬛ and choose ⬛ CALC ⬛. Then choose 1: 1-Var Stats. Enter the name of the list that you are using. Press ⬛2nd⬛ ⬛ L1 ⬛. Then press ⬛ENTER⬛. $\bar{x}$ and σ_x will be displayed on the screen along with other information.

Your Turn

a. Solve Example 1 by using a graphing calculator.

b. Practice the above process until you can explain it to a friend in your own words. Then write out your explanation.

c. Find the mean and standard deviation of the set of data below by using a graphing calculator.

16, 42, 21, 19, 18.6, 41, 37, 24, 29.2, 26, 35

Example **2**

Agriculture

Corn production (in billions of bushels) in the United States from 1984 to 1993 is shown in the graph below. Find the mean corn production and the standard deviation of the data.

U.S. Corn Production in Billions of Bushels

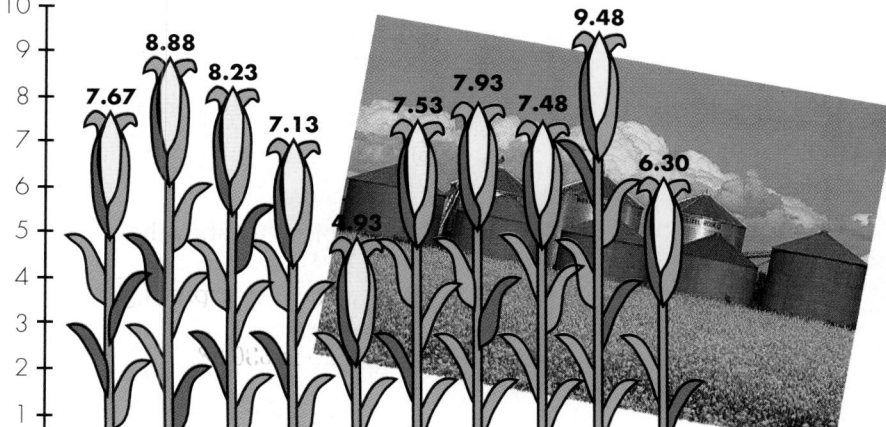

10, 9, 8.88, 8.23, 7.67, 7.13, 7.53, 7.93, 7.48, 9.48, 6.30, 4.93

1984 1985 1986 1987 1988 1989 1990 1991 1992 1993

Source: *Weatherwise* Magazine, 1994

(continued on the next page)

We can use a spreadsheet to do the calculations as follows.

Year	Billions of Bushels	Difference from Mean	Square of Difference
1984	7.67	0.114	0.012996
1985	8.88	1.324	1.752976
1986	8.23	0.674	0.454276
1987	7.13	−0.426	0.181476
1988	4.93	−2.626	6.895876
1989	7.53	−0.026	0.000676
1990	7.93	0.374	0.139876
1991	7.48	−0.076	0.005776
1992	9.48	1.924	3.701776
1993	6.30	−1.256	1.577536
Total	75.56	Total	14.72324
Mean = B12/10		7.556	
SD = $\sqrt{(D12/10)}$		1.2133935	

The mean is 7.556, and the standard deviation is 1.213.

So the average corn production in the 10-year period 1984–1993 was 7.556 billion bushels. In most of the years, the production was between 7.556 − 1.213 or 6.343 billion and 7.556 + 1.213 or 8.769 billion bushels.

CHECK FOR UNDERSTANDING

Communicating Mathematics

Study the lesson. Then complete the following.

1. **Explain** the meaning of standard deviation in your own words.

2. Refer to Example 1. Find the number of ticket prices that were within one standard deviation of the mean. What percent of the total number of ticket prices is this?

3. **Write** a paragraph describing the events that occurred in the summer of 1993 that resulted in lower than usual corn production in the United States.

4. Over the last 12 months, the mean number of full-time employees in a city government was 2783 with a standard deviation of 43.

   a. What does this say about the variation of the number of employees in the government?

   b. If the standard deviation were 430, what would that say about the variation of the number of employees?

Find the mean and standard deviation to the nearest hundredth for each set of data.

5. {110, 70, 20, 40, 10}

6. {48, 36, 40, 29, 45, 51, 38, 47, 39, 37}

7. {43, 56, 78, 81, 47, 42, 34, 22, 78, 98, 38, 46, 54, 67, 58, 92, 55}

8.

Stem	Leaf
4	3 5 6 8
5	2 4 5 6
6	1 2 4 5 5 6 7 7 7 $5|2 = 5.2$

9. Quality Control   A coffee machine is designed to dispense 250 mL for each cup. The actual measures in milliliters for a sample of 10 cups were 251, 246, 252, 249, 250, 248, 246, 253, 250, and 251.

   a. Find the standard deviation of the amounts.

   b. Do you think the variation is large or small?

   c. How do you think the variation would change if someone poured each cup of coffee, rather than used a machine?

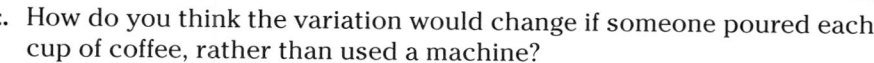

EXERCISES

Find the mean and standard deviation to the nearest hundredth for each set of data.

10. {45, 65, 145, 85, 25, 25}

11. {400, 300, 325, 275, 425, 375, 350}

12. {5, 4, 5, 5, 5, 5, 6, 6, 6, 6, 7, 7, 7, 7, 8, 9}

13. {234, 345, 123, 368, 279, 876, 456, 235, 333, 444}

14. {13, 14, 15, 16, 17, 18, 19, 20, 21, 23, 67, 56, 34, 99, 44, 55}

15.

Stem	Leaf
4	4 5 6 7 7
5	3 5 6 7 8 9
6	7 7 8 9 9 9 $4|5 = 45$

16.

Stem	Leaf
5	7 7 7 8 9
6	3 4 5 5 6 7
7	2 3 4 5 6 $6|3 = 6.3$

17.

Stem	Leaf
3	0 0 1 2 4
•	5 6 6 6 8 9
4	1 1 3 4 4
•	5 5 6 7 $3|4 = 3.4$

18.

Stem	Leaf
4	1 3 9
5	2 3 6 9
6	4 4 5 7 8
7	2 4 7 $5|2 = 52$

19. {76, 78, 89, 90, 34, 56, 50}

20. {321, 322, 323, 324, 325, 326, 327, 328, 329, 330}

21. Suppose you have a set of data in which all elements are the same. What is the standard deviation of this data? Explain your answer.

22. Sports The weights, in pounds, of the starting players for the basketball teams of three high schools are given below.

Jonesboro: 150, 145, 120, 168, 175
Hillview: 124, 157, 195, 205, 177
Murrayville: 146, 155, 176, 186, 199

a. Find the standard deviation for the weights of the players on the Jonesboro basketball team.

b. Find the standard deviation for the weights of the players on the Hillview basketball team.

c. Find the standard deviation for the weights of the players on the Murrayville basketball team.

d. Which team has the most variation in weight? How do you think this variation will impact their play?

23. Airlines The on-time performance of the nation's nine largest airlines declined in 1993, although complaints also dropped.

Airline	Percent of flights arriving on time	Complaints per 100,000 fliers
Southwest	88.7%	0.29
Northwest	84.8	0.68
TWA	83.0	1.09
United	80.5	0.64
American	80.0	0.38
Continental	77.9	1.93
Delta	75.4	0.45
USAir	73.5	0.83
America West	71.9	0.74

a. What are the mean and standard deviation of percent of flights arriving on time?

b. What are the mean and standard deviation of the complaints?

c. Write a paragraph that explains the relationship between on-time flights and passenger complaints.

24. Sales In 1991, *Zillions* magazine tested jeans for durability and quality in their laboratory. The following table gives the top five lab test winners for men's and women's jeans along with the price of each pair.

Men's	Price	Women's	Price
Wrangler ProRodeo Cowboy Cut	$20	Sears Jeans That Fit	$19
Wrangler American Hero	$17	Wrangler ProRodeo Cowboy Cut	$26
Levi's 509	$31	Chic Heavenly Blues	$48
Wrangler Rustler	$15	PS Gitano	$21
J. C. Penney Long Haul	$23	Gap Straight Leg	$30

a. Find the mean and standard deviation of the prices of the men's jeans.

b. Find the mean and standard deviation of the prices of the women's jeans.

c. Which had the greatest variation in price?

d. Survey 10 students in your school lunchroom and determine the brand of jeans that they wear. Compare your results with the tests above.

e. Suppose during end-of-year clearance sales, the stores lower the prices of all women's jeans by $5. How will the new standard deviation compare with the original standard deviation? Why do you think this occurs?

25. Food Professionally-trained food testers working for *Zillions* magazine dipped into several brands of potato chips three times a day, four days a week, for five weeks in 1994. The result of their work is listed in the table below.

Very Good	Price per ounce (¢)	Good	Price per ounce (¢)
Lady Lee	14	Ruffles Light	25
Cape Cod	20	Kroger	12
Jays	18	Wise	17
Cape Cod Unsalted	20	Charles Chips	21
Eagle Idaho Russet	23	Pringles Original	21
Albertsons	12	Lay's	21
Ruffles	21	Pringles Light Original	24
Lay's Crunch Taters	23	Golden Flake	17
Vons	13	Pathmark	10
		Michael Season's	28
		Wise Cottage Fries	21
		Pringles Idaho Rippled	22
		Barrell O'Fun	20
		New York Deli	24
		Keebler Ripplin's	19
		O'Boisie's	25
		Lay's Unsalted	21

a. Find the mean and standard deviation of the prices of the chips rated "Very Good."

b. Find the mean and standard deviation of the prices of the chips rated "Good."

Mixed Review

26. Solve $(3x - 9)(x + 12) < 0$. (Lesson 6–7)

27. Architecture Look at the diagram below of a suspension bridge.

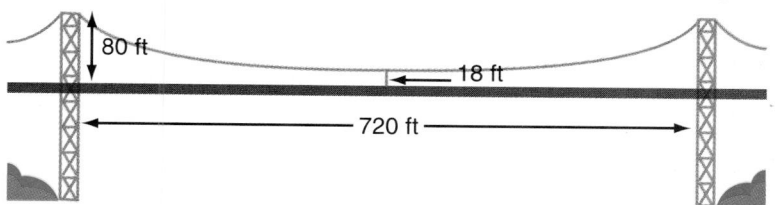

The equation $y = 0.00048x^2 + 18$ can be used to describe the cable hanging between the two upright supports, where x represents the horizontal distance along the roadbed from the lowest point of the cable and y represents the distance up from the roadbed. If a vertical cable is to be installed from the roadbed to the hanging cable and is 48 feet long, how far from the lowest point of the hanging cable will it be? (Lesson 6–2)

28. Simplify $(2r^2 + r - 3) \div (2r + 3)$. (Lesson 5-3)

29. Write an augmented matrix for the system of equations. Then solve.
(Lesson 4–7)

$3x + 4y = 22$
$7x - y = 10$

30. Given the function $f(x, y) = 6x - 3y$, find $f(-5, 2)$. (Lesson 3–5)

Integration: Statistics
The Normal Distribution

APPLICATION
Basketball

In professional basketball, a power forward is responsible for rebounding and scoring close to the basket. So, it makes sense that most power forwards are very tall. Forty NBA power forwards and their heights are listed in the table below.

Name	Height	Name	Height	Name	Height
Barkley	78″	Hill	81″	Perkins	81″
Brickowski	82″	Johnson	79″	Pinckney	81″
Bryant	81″	Jones	80″	Reid	81″
Cage	81″	Laettner	83″	Rodman	80″
Campbell	83″	Long	80″	Smith	82″
Coleman	82″	Lynch	80″	Thorpe	82″
Cummings	81″	Malone	81″	Tisdale	81″
A. Davis	81″	Mason	79″	Vaught	81″
D. Davis	83″	McDaniel	79″	Weatherspoon	78″
Ellis	80″	Mills	82″	Br. Williams	83″
Gilliam	81″	Nance	82″	Bu. Williams	80″
Grant	82″	Oakley	81″	J. Williams	82″
Green	81″	Owens	81″	Willis	84″
Gugliotta	82″				

One way of analyzing data is to consider how frequently each value occurs. The table below shows the frequencies of the heights of the forty basketball players. The graph below visually displays the frequencies of the heights in the table.

Height	Frequency
78″	2
79″	3
80″	6
81″	15
82″	9
83″	4
84″	1

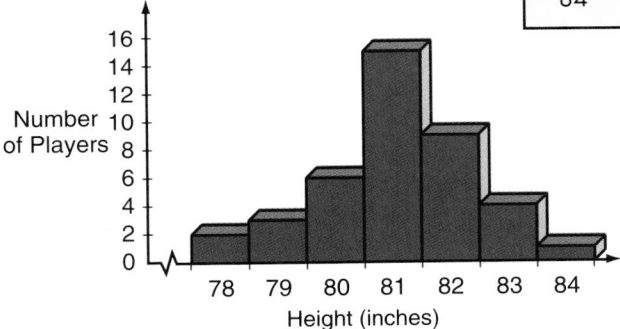

The bar graph shows a **frequency distribution** of the heights. That is, it shows how the heights are spread out over the range from 78 inches to 84 inches. A graph like this one that shows a frequency distribution is called a **histogram.**

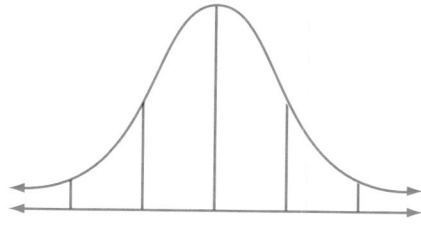

Curves are very useful in displaying information. You used parabolas to represent related data. Curves are also used to show frequency distributions, especially when the distribution contains a large number of values. While the curves may be of any shape, many distributions have graphs shaped like the one at the left. Many distributions with this type of graph are **normal distributions.**

The curve of the graph of a normal distribution is symmetric and is often called a **bell curve.** The shape of the curve indicates that the frequencies in a normal distribution are concentrated around the center portion of the distribution. What does this tell you about the mean?

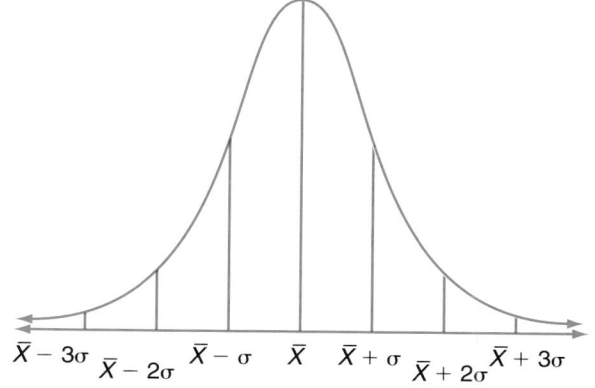

Normal distributions have these properties.

1. The graph is maximized at the mean.
2. The mean, median, and mode are about equal.
3. The data are symmetrical about the mean.
4. About 68% of the values are within one standard deviation from the mean.
5. About 95% of the values are within two standard deviations from the mean.
6. About 99% of the values are within three standard deviations from the mean.

Suppose the scores on the mathematics component of the Scholastic Assessment Test (SAT) of 1,000,000 students are recorded and the frequency of those scores is normally distributed. If the mean score is 500 and the standard deviation is 90, then the graph at the right approximates the curve for the frequency distribution of the scores.

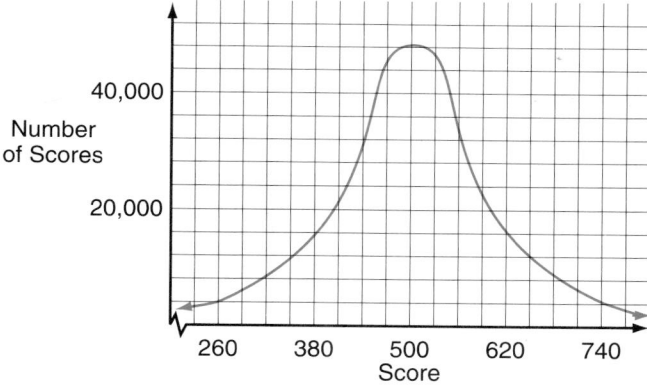

As shown by the graph, the mean is the most frequent score. Of the 1,000,000 students, the following is true.

- About 680,000 scored between 410 and 590 points.
- About 950,000 students scored between 320 and 680 points.
- About 990,000 students scored between 230 and 770 points.

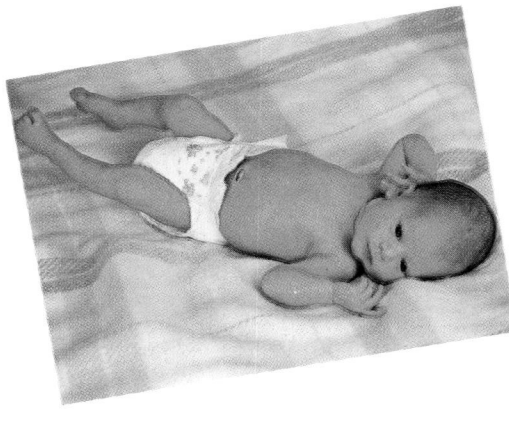

Normal distributions occur quite frequently in real life. In addition to test scores, the lengths of newborn babies, cholesterol levels, the useful life and size of manufactured items, and production levels can be represented by normal distributions. In all of these cases, the number of data items must be large for the distribution to be normal.

Example **1** The useful lives of 10,000 batteries are normally distributed. The mean useful life is 20 hours, and the standard deviation is 4 hours.
  a. Sketch a normal curve showing the useful life at one, two, and three standard deviations from the mean.
  b. How many batteries will last between 16 and 24 hours?
  c. How many batteries will last less than 12 hours?

  a. Draw a normal curve with 20 hours as the mean.
  one standard deviation from the mean: $\overline{x} - 4$ and $\overline{x} + 4$ or 16–24 hours

  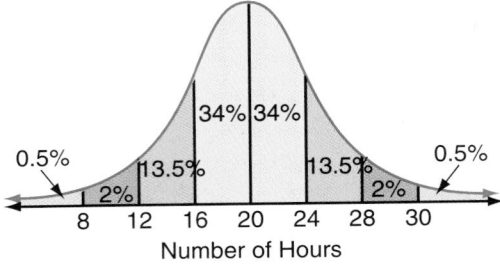

  two standard deviations from the mean: $\overline{x} - 2(4)$ and $\overline{x} + 2(4)$ or 12–28 hours

  three standard deviations from the mean: $\overline{x} - 3(4)$ and $\overline{x} + 3(4)$ or 8–30 hours

  b. The percentage of batteries lasting between 16 and 24 hours is 34% + 34% or 68% of the batteries.
  10,000 × 68% = 6800 batteries

  c. The percentage of batteries lasting less than 12 hours is 0.5% + 2% or 2.5%.
  10,000 × 2.5% = 250 batteries

You can use normal distributions to estimate the probabilities of certain events occurring.

Example **2**

Medicine

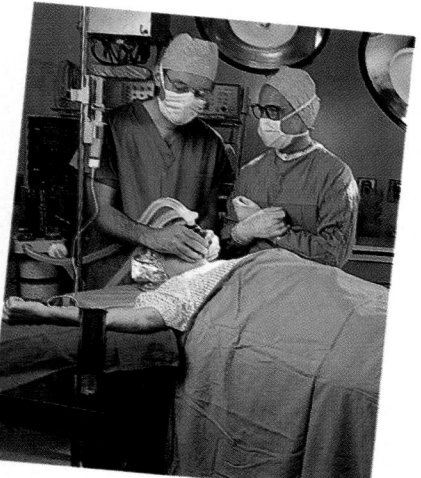

The correct number of milligrams of anesthetic an anesthesiologist must administer to a patient is normally distributed. The mean is 100 milligrams, and the standard deviation is 20 milligrams.
  a. Of a sample of 200 patients, about how many people require more than 120 milligrams of anesthetic for a response?
  b. What is the probability that a patient chosen at random will require between 80 and 120 milligrams?

  a. This frequency distribution is shown by the curve below. The percentages represent the percentages of patients requiring the dosage within the given interval.

  The percentage of people requiring more than 120 milligrams of anesthetic is 13.5% + 2% + 0.5% or 16%.
  $$200 \times 16\% = 32$$
  So about 32 of the 200 patients require more than 120 milligrams of anesthetic.

  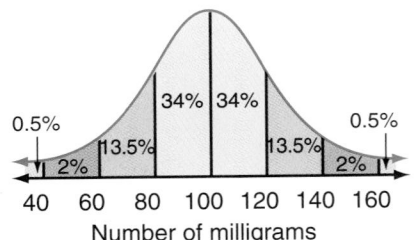

  b. The percentage of people requiring between 80 and 120 milligrams of anesthetic is 68%. So the probability that a patient chosen at random will require such amounts of anesthetic is 68%.

The histogram at the right shows the number of urban cities in 1990 in the United States for varying population sizes. You can see that the resulting curve displays a **skewed** distribution. A skewed curve that is high at the left and has a tail at the right like this one is *positively skewed*. A skewed curve that is high at the right and has a tail to the left is *negatively skewed*.

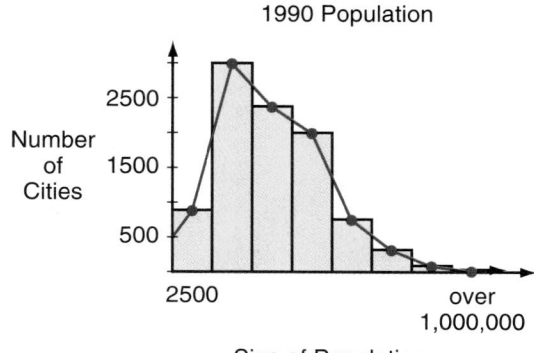

1990 Population

Number of Cities

2500

1500

500

2500   over 1,000,000

Size of Population

CHECK FOR UNDERSTANDING

Communicating Mathematics

Study the lesson. Then complete the following.

1. **Refer** to the application at the beginning of the lesson. Explain why the heights of the power forwards approximate a normal distribution. Explain why the histogram approximates a normal curve.

2. **Sketch** a negatively skewed graph. Describe a situation in which you would expect data to be distributed this way.

3. **Compare and contrast** the means and standard deviations of the graphs.

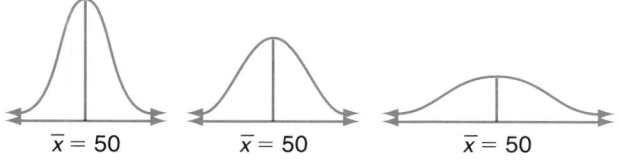

$\bar{x} = 50$ $\bar{x} = 50$ $\bar{x} = 50$

MATH JOURNAL

4. Explain why you think that the SAT scores of the students in your class would be skewed or why they would be normal.

Guided Practice

5. The table at the right shows female mathematics SAT scores in 1990.

   a. State whether the tables of data would result in a graph that is positively skewed, is negatively skewed, or appears to be normally distributed.

   b. Sketch a histogram of the data.

Scores	Percent of Females
200–290	5
300–390	18
400–490	26
500–590	26
600–690	18
700–800	7

6. Mrs. Sung gave a test in her trigonometry class. The scores were normally distributed with a mean of 85% and a standard deviation of 3%.

   a. What percentage would you expect to score between 82% and 88%?

   b. What percentage would you expect to score between 88% and 91%?

   c. What is the probability that a student chosen at random scored between 79% and 91%?

7. **Quality Control** The useful life of a radial tire is normally distributed with a mean of 30,000 miles and a standard deviation of 5000 miles. The company makes 10,000 tires a month.

   a. About how many tires will last between 25,000 and 35,000 miles?
   b. About how many tires will last more than 40,000 miles?
   c. About how many tires will last less than 25,000 miles?
   d. What is the probability that if you buy a radial tire at random, it will last between 20,000 and 35,000 miles?

Practice **State whether the tables of data would result in a graph that is positively skewed, negatively skewed, or appears to be normally distributed. Then sketch a histogram of the data.**

8. the number of hours of TV students watch in a week

Hours	Percentage
0–5	16
6–10	24
11–15	5
16–20	8
21–25	4
26+	7

Source: *The Elementary Mathematician*

9. the record low temperatures in the fifty states

Temperature (°F)	Number of States
−80 to −66	2
−65 to −51	10
−50 to −36	17
−35 to −19	16
−18 to −2	4
−1 to 15	1

Source: *The World Almanac, 1995*

10. The shelf life of a particular dairy product is normally distributed with a mean of 12 days and a standard deviation of 3.0 days.

   a. About what percentage of the products last between 9 and 15 days?

   b. About what percentage of the products last between 12 and 15 days?

   c. About what percentage of the products last 3 days or less?

   d. About what percentage of the products last 15 or more days?

11. The vending machine in the basement of McMicken Hall usually dispenses about 6 oz of soft drink. Lately, it is not working properly and the variability in how much of the soft drink it dispenses has been getting greater. The amounts are normally distributed with a standard deviation of 0.2 oz.

   a. What percent of the time will you have more than 6 oz of soft drink?

   b. What percent of the time will you have less than 6 oz of soft drink?

   c. What percent of the time will you have between 5.6 and 6.4 oz of soft drink?

12. The Floppy Disk Company makes 3.5" floppy disks for disk drives that are 3.7" wide. The size of a manufactured disk is normally distributed with a standard deviation of 0.1". The company manufactures 1000 disks every hour.

   a. What percentage of the disks would you expect to be greater than 3.7"?

   b. In one hour, how many disks would you expect to be between 3.4" and 3.7"?

   c. About how many disks will be unable to fit in the disk drive?

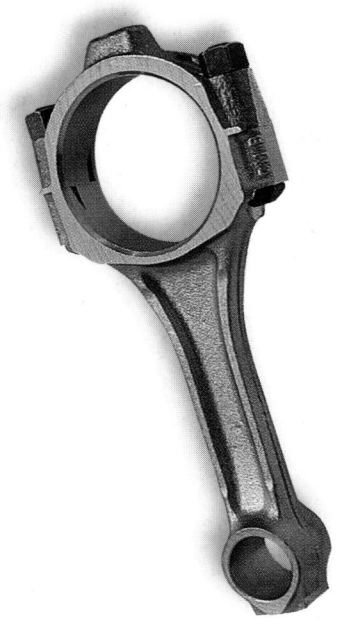

13. The diameter of the connecting rod in a certain imported sports car must be between 1.480 and 1.500 cm, inclusive, to be usable. The diameters of the connecting rods are normally distributed with a mean of 1.495 cm and a standard deviation of 0.005 cm.

   a. Of 1000 connecting rods, how many measure at least 1.495 cm?

   b. Of 1000 connecting rods, how many measure between 1.485 cm and 1.500 cm?

   c. What percentage of rods will have to be discarded?

   d. What is the probability that a rod chosen at random will be usable?

14. Suppose the weights of the books used in your school were normally distributed.

   a. Where on the normal curve would you find the weight of your Algebra 2 textbook?

   b. Which textbook would be three standard deviations to the left of the mean (extremely light)?

   c. Which textbook would be three standard deviations to the right of the mean (extremely heavy)?

15. Humor   The following quote appeared in *Chance* magazine in Winter, 1995.

> Someone asked an accountant, a mathematician, an engineer, a statistician, and an actuary how much 2 + 2 was. The accountant said "4." The mathematician said "it depends on your number base." The engineer took out his slide-rule and said "approximately 3.99." The statistician consulted his tables and said "I am 95% confident that it lies between 3.95 and 4.05." The actuary said "What do you want it to add up to?"

How did each person's perspective affect his or her answer? Describe the information the statistician used to give his answer.

16. Health   A recent study showed that the systolic blood pressure of high school students ages 14–17 is normally distributed with a mean of 120 mm Hg and a standard deviation of 12 mm Hg. Suppose a high school has 800 students.

   a. About what percentage of the students have blood pressure below 108 mm Hg?

   b. About how many students have blood pressure between 108 and 144 mm Hg?

17. Noise Level   Airplane pilots often suffer from hearing loss as a result of being exposed to high noise levels. A team of researchers measured the cockpit noise levels of 16 commercial aircraft. The results are listed in the table below.

Plane	Decibels of Noise Level	Plane	Decibels of Noise Level
1	80	9	85
2	83	10	80
3	83	11	75
4	86	12	75
5	72	13	74
6	90	14	77
7	87	15	80
8	83	16	82

Source: *Archives of Environmental Health*

   a. Calculate the mean and standard deviation of the data.

   b. Construct a histogram for the noise levels of the planes. Use the intervals 70-73, 74-77, 78-81, 82-85, 86-89, and 90-93.

   c. Do you think the data appear to be normally distributed? Explain.

18. **Physiology** Nineteenth-century Belgian scholar Lambert Adolphe Jacques Quetelet discovered that if you take a large group of people and measure a physical characteristic such as height, weight, or arm length, the data will be nearly normally distributed. Below is a set of data Quetelet collected in 1846. They are the chest measurements of 5738 Scottish soldiers.

Chest Measures (inches)	Number of men	Chest Measures (inches)	Number of men
33	3	41	934
34	18	42	658
35	81	43	370
36	185	44	92
37	420	45	50
38	749	46	21
39	1073	47	4
40	1079	48	1

Source: *Lettres sur la Theorie des Probabilites, appliquee aux Sciences Morales et Politiques*

The mean measurement to the nearest inch is 40 inches and the standard deviation is 2 inches.
a. Construct a histogram for the chest measurements of Scottish soldiers.
b. If the data represented a perfectly normal distribution, what percentage of Scottish soldiers would have chest measurements that were within two standard deviations of the mean?
c. What percentage of soldiers actually had chest measurements that were within two standard deviations of the mean?

19. **Entertainment** The 5th Anniversary Issue of *Entertainment Weekly* gave a summary of grades their critics have awarded the movies reviewed since the magazine's premiere issue. The grades were 96 As, 285 Bs, 267 Cs, 136 Ds, and 36 Fs.
a. Construct a histogram for these data and sketch a curve showing the distribution.
b. Do the data appear to be normally distributed? Explain.

Mixed Review

20. Find the mean and the standard deviation for the following set of data.
7, 16, 9, 4, 12, 3, 9, 4 (Lesson 6–8)

21. **Football** The price of a Super Bowl ticket from Super Bowl I in 1967 to Super Bowl XXIX in 1995 can be described by the function $y = \frac{2}{5}x^2 - 6x + 32$, where y represents the price and x represents the year, with $x = 1$ representing 1967. (Lesson 6–6)
a. Write the equation in the form $y = a(x - h)^2 + k$ and graph the function.
b. What was the approximate price of tickets for Super Bowl XXIX in 1995?
c. What will be the approximate price of tickets for Super Bowl XXXV in 2001?

22. Write a quadratic equation with the roots $2 \pm \sqrt{3}$. (Lesson 6–5)

23. Simplify $(-x - 8)(3x + 4)$. (Lesson 5–2)

24. Solve $6(3x - 5y) + (2 + 8x) = -11x$ for x. (Lesson 2–2)

VOCABULARY

After completing this chapter, you should be able to define each term, property, or phrase and give an example or two of each.

Algebra

axis of symmetry   (p. 335)
boundary   (p. 378)
completing the square
   (p. 347)
constant term   (p. 334)
discriminant   (p. 356)
factoring   (p. 341)
linear term   (p. 334)
parabola   (p. 335)

quadratic equation   (p. 336)
quadratic formula   (p. 354)
quadratic function   (p. 334)
quadratic inequality   (p. 378)
quadratic term   (p. 334)
roots   (p. 336)
vertex   (p. 335)
zero product property
   (p. 341)
zeros   (p. 335)

Statistics

bell curve   (p. 393)
frequency distribution   (p. 392)
histogram   (p. 392)
normal distribution   (p. 393)
skewed   (p. 395)
standard deviation   (p. 384)

Problem Solving

guess and check   (p. 342)

UNDERSTANDING AND USING THE VOCABULARY

Choose the letter of the term that best matches each statement or phrase.

1. the graph of any quadratic function
2. a process whereby the middle term of a quadratic equation of the form $ax^2 + bx + c = 0$ is altered to help solve the quadratic equation
3. the vertical line passing through the vertex of a parabola and dividing the parabola into two mirror images
4. a function described by an equation of the form $f(x) = ax^2 + bx + c$, where $a \neq o$
5. the solutions of an equation
6. a bell-shaped symmetric graph with about 68% of the items within one standard deviation from the mean, about 95% of the items within two standard deviations from the mean, and about 99% of the items within three standard deviations of the mean

a. axis of symmetry
b. completing the square
c. discriminant
d. factoring
e. normal distribution
f. parabola
g. quadratic formula
h. quadratic function
i. roots
j. standard deviation

7. the process whereby a polynomial of degree 2 or more is simplified into the product of monomials, binomials, or a combination thereof
8. in the quadratic formula, the expression under the radical sign, $b^2 - 4ac$
9. The solution(s) of a quadratic equation of the form $ax^2 + bx + c = 0$ with $a \neq 0$ are given by $x = \dfrac{-b \pm \sqrt{b^2 - 4ac}}{2a}$.
10. For a set of data with n values, if x_i represents a value such that $1 \leq i \leq n$, and $\bar{x}$ represents the mean, then $\sqrt{\dfrac{(x_1 - \bar{x})^2 + (x_2 - \bar{x})^2 + \cdots + (x_n - \bar{x})^2}{n}}$ represents this value.

SKILLS AND CONCEPTS

OBJECTIVES AND EXAMPLES

Upon completing this chapter, you should be able to:

- solve quadratic equations by graphing (Lesson 6–1)

Solve $x^2 + 4x - 12 = 0$ by graphing.

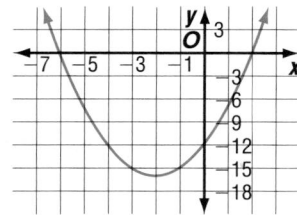

The graph crosses the x-axis at -6 and 2. Thus, the solutions of the equation are -6 and 2.

- solve quadratic equations by factoring (Lesson 6–2)

Solve $x^2 + 9x + 20 = 0$.

$(x + 4)(x + 5) = 0$

$x + 4 = 0 \quad$ or $\quad x + 5 = 0$

$\quad x = -4 \qquad\qquad x = -5$

The solutions are -4 and -5.

- solve quadratic equations by completing the square (Lesson 6–3)

Solve $x^2 + 10x - 39 = 0$ by completing the square.

$x^2 + 10x + \square = 39 + \square$

$x^2 + 10x + 25 = 39 + 25$

$\qquad (x + 5)^2 = 64$

$\qquad\quad x + 5 = \pm 8$

$\quad x + 5 = 8 \quad$ or $\quad x + 5 = -8$

$\qquad\quad x = 3 \qquad\qquad x = -13$

The solutions are -13 and 3.

REVIEW EXERCISES

Use these exercises to review and prepare for the chapter test.

Solve each equation by graphing.

11. $x^2 + 6x - 40 = 0$

12. $x^2 - 2x - 15 = 0$

13. $a^2 - 8a - 20 = 0$

14. $3m^2 + 9m + 6 = 0$

15. $x^2 + 12x + 35 = 0$

16. $0.5x^2 + 0.5x - 15 = 0$

Solve each equation by factoring.

17. $x^2 - 4x - 32 = 0$

18. $3x^2 + 6x + 3 = 0$

19. $5y^2 = 80$

20. $2c^2 + 18c - 44 = 0$

21. $d^2 + 29d + 100 = 0$

22. $v^3 = 49v$

23. $25x^3 - 25x^2 = 36x$

24. $r^2 - 3r - 70 = 0$

Solve each equation by completing the square.

25. $-5x^2 - 5x + 9 = 0$

26. $k^2 + 6k - 4 = 0$

27. $b^2 + 4 = 6b$

28. $n^2 - 10n = 23$

29. $h^2 - 4h - 7 = 0$

30. $5x^2 - 15x - 9 = 2$

OBJECTIVES AND EXAMPLES

• solve quadratic equations by using the quadratic formula   (Lesson 6–4)

Solve $x^2 - 5x - 66 = 0$.

$$x = \frac{-b \pm \sqrt{b^2 - 4ac}}{2a}$$

$$= \frac{-(-5) \pm \sqrt{(-5)^2 - 4(1)(-66)}}{2(1)}$$

$$= \frac{5 \pm 17}{2}$$

$$x = \frac{5 + 17}{2} \text{ or } 11 \quad \text{and} \quad x = \frac{5 - 17}{2} \text{ or } -6$$

• find a quadratic equation to fit a given condition (Lesson 6–5)

Write an equation that has roots of $-\frac{5}{2}$ and 3.

$$s_1 = -\frac{5}{2} \qquad s_2 = 3$$

$$s_1 + s_2 = -\frac{5}{2} + 3 = \frac{1}{2} = -\frac{b}{a}$$

$$s_1 s_2 = -\frac{5}{2} \cdot 3 = -\frac{15}{2} = \frac{c}{a}$$

Therefore $a = 2$, $b = -1$, and $c = -15$.
The equation is $2x^2 - x - 15 = 0$.

• graph quadratic functions of the form $y = a(x - h)^2 + k$ (Lesson 6–6)

Name the vertex, axis of symmetry, and direction of opening for the graph of $f(x) = 3x^2 + 42x + 142$.

$$\begin{aligned} f(x) &= 3x^2 + 42x + 142 \\ &= 3(x^2 + 14x) + 142 \\ &= 3(x^2 + 14x + 49) + 142 - 3(49) \\ &= 3(x + 7)^2 - 5 \end{aligned}$$

So $a = 3$, $h = -7$, and $k = -5$. The vertex is at $(-7, -5)$, and the axis of symmetry is $x = -7$. Since $a = 3$, the graph opens upward.

• graph quadratic inequalities   (Lesson 6–7)

Graph $y \leq -2x^2 - 28x - 89$.

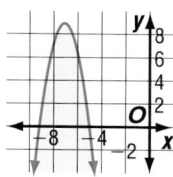

REVIEW EXERCISES

Solve each equation by using the quadratic formula.

31. $x^2 + 2x + 7 = 0$

32. $x + 2x^2 + 1 = -1 - x$

33. $-x^2 + 5x - 9 = 0$

34. $-2x^2 + 12x - 5 = 0$

35. $3c^2 + 7c - 2 = 0$

36. $8b^2 - b - 15 = 0$

Find a quadratic equation that has the given roots.

37. $7, -6$

38. $11, 14$

39. $-\frac{13}{2}, -4$

40. $\frac{3}{4}, \frac{9}{2}$

41. $-2.5, 5.25$

42. $-0.25, 0.25$

Write each equation in the form $f(x) = a(x - h)^2 + k$ if not already in that form. Name the vertex, axis of symmetry, and direction of opening for the graph of each quadratic function. Then graph it.

43. $f(x) = -6(x + 2)^2 + 3$

44. $f(x) = 4(x - 5)^2 - 7$

45. $f(x) = 5x^2 - 35x + 58$

46. $f(x) = -9x^2 + 54x - 8$

47. $f(x) = -\frac{1}{3}x^2 + 8x$

48. $f(x) = 0.25x^2 - 6x - 16$

Graph each inequality.

49. $y > x^2 - 5x + 15$

50. $y < -3x^2 + 48$

51. $y \leq 4x^2 - 36x + 17$

52. $y \geq -x^2 + 7x - 11$

53. $y < x^2 + 5x + 6$

54. $y \geq 3x^2 - 15x + 22$

OBJECTIVES AND EXAMPLES

- find the standard deviation for a set of data
  (Lesson 6–8)

To find the standard deviation of this set of data, follow the steps below.

1. Find the mean of the data.
2. Find the difference between each value in the set of data and the mean.
3. Square each difference.
4. Find the mean of the squares.
5. Take the principal root of this mean.

- solve problems involving normally-distributed data   (Lesson 6–9)

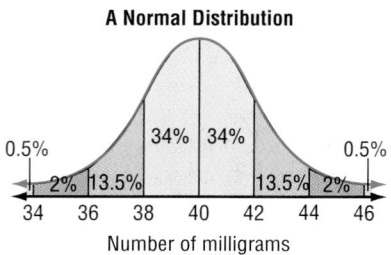

A Normal Distribution

Number of milligrams

REVIEW EXERCISES

Find the mean and the standard deviation to the nearest hundredth for each set of data.

55. {100, 156, 158, 159, 162, 165, 170, 190}

56. {56, 56, 57, 58, 58, 58, 59, 61}

57. {302, 310, 331, 298, 348, 305, 314, 284, 321, 337}

58. {3.4, 4.2, 8.6, 5.1, 3.6, 2.8, 7.1, 4.4, 5.2, 5.6}

59. Mr. Byrum gave an exam to his 30 Algebra 2 students at the end of the first semester. The scores were normally distributed with a mean of 78% and a standard deviation of 6%.
   a. What percentage of the class would you expect to have scored between 72% and 84% on the test?
   b. What percentage of the class would you expect to have scored between 90% and 96% on the test?
   c. Approximately how many students scored between 84% and 90%?
   d. What percentage of the class would you expect to score less than 60%?
   e. Approximately how many students scored between 72% and 84%?

APPLICATIONS AND PROBLEM SOLVING

60. **Space Exploration**   The Apollo 11 spacecraft propelled the first men to the moon and contained three stages of rockets. The first stage dropped off 2 min 40 s after takeoff and the second stage ignited. The initial velocity of the second stage was 2760 m/s with a constant acceleration of 200 m/s². How long did it take the second stage to travel 7040 m?   (Lesson 6–3)

61. **Physics**   The Empire State Building is 1250 feet tall. If an object is thrown upward from the top of the building at an initial velocity of 35 feet per second, its height t seconds after it is thrown is given by the function $h(t) = -16t^2 + 35t + 1250$. How long will it be before the object hits the ground?   (Lesson 6–2)

62. **Work**   The monthly incomes of 10,000 workers at the ProComm plant are distributed normally. Suppose the mean monthly income is $1250 and the standard deviation is $250.   (Lesson 6–9)
   a. How many workers earn more than $1500 per month?
   b. How many workers earn less than $750 per month?
   c. What percentage of the workers earn between $500 and $1750 per month?
   d. What percentage of the workers earn less than $1750 per month?

A practice test for Chapter 6 is provided on page 917.

ALTERNATIVE ASSESSMENT

COOPERATIVE LEARNING PROJECT

Statistics   So far in this text, you have learned several methods of statistical analysis.

- *Standard deviation* is the average measure of how much each value in a set of data differs from the mean.
- A *bar graph* that shows the frequency distribution of data is called a histogram.
- In a *stem-and-leaf plot*, each piece of data is separated into two number columns that are used to form the stem and leaf.
- The *normal distribution* is a symmetric curve of a graph that is often bell-shaped.

In this project, you will have the opportunity to gather data and complete your own statistical analysis using one of the four methods listed above. Your teacher will separate you into groups of four and assign each of you a method of statistical analysis. Once you are assigned to a group and a method, your group should decide which teacher in the school you would like to approach to provide you with a list of his or her latest test scores. Tell the teacher that this is a math project and that only a list of the scores is necessary to complete the project. Be sure to tell the teacher that no names are to be given with any of the test scores. Also, have the teacher provide you with the method by which the test scores were achieved: for example, multiple-choice, true-false, matching, essay, and so on.

Once you have obtained the test scores, begin your analysis of the scores using the method assigned to you by your teacher. Remember, statistics can be used to prove any point an individual wants to make.

When your statistical analysis is complete, share your results with the rest of your group. Are there any similarities and/or differences between the various methods of analysis?

Now share your results with someone in another group who used the same method of analysis as you did. Are there any similarities and/or differences between methods of the same type of analysis? Can any conclusions be drawn about the statistical analysis and the method of testing used by the teacher? Does one method of testing yield better statistical results than another?

THINKING CRITICALLY

Find values of k that meet the following conditions.

a. $x^2 - 7x + k = 0$ has two real roots.

b. $kx^2 - 5x + 7 = 0$ has two imaginary roots.

c. $3x^2 - kx + 15 = 0$ has one root.

PORTFOLIO

Select an item from this chapter that shows your best work and place it in your portfolio. Explain why you believe it to be your best work and how you came to choose this particular piece.

SELF EVALUATION

Every day you are required to make decisions. Some may be as simple as what you should wear to school or what kind of cereal you want to have for breakfast. Others such as what classes to take in school next year or where to apply for a part-time job are a little tougher.

Assess yourself. How do you handle decision-making? It may be helpful to make a list of all your choices and write the pros and cons of each one. Also remember to seek input from friends and family members who have experience in the area. Once you make your decision, accept responsibility for it and move forward.

SECTION ONE: MULTIPLE CHOICE

There are eight multiple-choice questions in this section. After working each problem, write the letter of the correct answer on your paper.

1. Evaluate $(4.5 \times 10^4)(3.33 \times 10^2)$. Express the answer in scientific notation and in standard notation.

   A. 1.4985×10^6; 1,498,500

   B. 14.985×10^6; 14,985,000,000

   C. 1.4985×10^7; 149,850,000,000

   D. 1.4985×10^7; 14,985,000

2. Which matrix below represents translated $\Delta D'E'F'$ if ΔDEF with $D(7, -2)$, $E(4, 5)$, and $F(-3, 4)$ is moved 5 units left and 1 unit up?

   A. $\begin{bmatrix} 8 & 5 & -2 \\ -7 & 0 & -1 \end{bmatrix}$

   B. $\begin{bmatrix} 2 & -1 & -8 \\ -1 & 6 & 5 \end{bmatrix}$

   C. $\begin{bmatrix} -1 & 6 & 5 \\ 8 & 5 & 2 \end{bmatrix}$

   D. $\begin{bmatrix} 2 & -1 & -8 \\ -7 & 0 & -1 \end{bmatrix}$

3. Choose the equation that represents a parabola that is 1 unit to the right and 8 units below the parabola with equation $f(x) = 5x^2$.

   A. $f(x) = 5(x - 8)^2 + 1$

   B. $f(x) = 5x^2 - 8$

   C. $f(x) = 5(x - 1)^2 - 8$

   D. $f(x) = 5(x - 1)^2 + 8$

4. Find the area of a rectangle with length $3\sqrt{3} - \sqrt{2}$ units and width $\sqrt{3} + 3\sqrt{2}$ units.

   A. $3 + 8\sqrt{6}$

   B. 27

   C. $11\sqrt{6}$

   D. $3 + 8\sqrt{5}$

5. Simplify $4(5x + 2y) + 9(x - 2y)$.

   A. $10xy$

   B. $29x - 10y$

   C. $54x + 13y$

   D. $29x$

6. Choose the system of inequalities whose solution is represented by the graph below.

   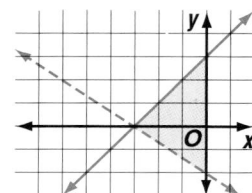

   A. $y > -\dfrac{3}{2}x - 2$
   $x < 0$
   $y \le \dfrac{1}{2}x - 6$

   B. $-2 < y \le 3$
   $x > -3$

   C. $y \le x + 3$
   $y > -\dfrac{2}{3}x - 2$
   $x \le 0$

   D. $y \ge -2x$
   $y < -4x + 3$

7. Choose the false statement regarding the value of the discriminant of a quadratic equation.

   A. If it is a perfect square, two real rational roots exist.

   B. If it is zero, infinitely many roots exist.

   C. If it is negative, two complex roots exist.

   D. If it is not a perfect square, two real irrational roots exist.

8. Find the next number in the pattern 2, 8, 18, 32, 50, __?__.

   A. 54

   B. 68

   C. 98

   D. 72

SECTION TWO: SHORT ANSWER

This section contains seven questions for which you will provide short answers. Write your answer on your paper.

9. Find the mean and the standard deviation for {13.7, 15.0, 13.7, 16.9, 13.6, 14.3, 14.8, 14.8, 15.1, 15.4, 14.9}.

10. Joseph works at Angelino's Pizza. His last three orders were 5 slices of pizza, 2 salads, and 2 sodas for $9.75; 3 slices of pizza, 2 salads, and 1 soda for $7.15; and 2 slices of pizza, 1 salad, and 1 soda, for a total of $4.35. What are the individual prices for pizza, salad, and soda at Angelino's?

11. Simplify $\dfrac{(4 + 3i)^2}{(3 - i)^2}$.

12. Determine the slope of the line that passes through $(2, 0)$ and $(-3, 5)$.

13. An object is fired upwards from the top of a 200-foot tower at a velocity of 80 feet per second. The height of the object t seconds after firing is given by the formula $h(t) = -16t^2 + 80t + 200$. Find the maximum height reached by the object and the time that height is reached.

14. Two and one-half years ago, Miko deposited the $1500 she earned at a summer job into her bank account. Her account earns 7.5% interest annually. Now she is withdrawing the money and the interest to buy a car. Use the formula $A = P(1 + r)^t$, where A is the amount of money in the account after t years if the interest rate is r and the beginning balance is P, to find how much money Miko has to buy the car.

15. Determine the range, quartiles, and interquartile range of the data. Then make a box-and-whisker plot.

Stem	Leaf
4	1 3 9
5	2 3 6 9
6	4 4 5
7	2 4 7 7	2 = 72

SECTION THREE: COMPARISON

This section contains five comparison problems that involve comparing two quantities, one in column A and one in column B. In certain questions, information related to one or both quantities is centered above them. All variables used represent real numbers.

Compare quantities A and B below.

• Write A if quantity A is greater.

• Write B if quantity B is greater.

• Write C if the two quantities are equal.

• Write D if there is not enough information to determine the relationship.

Column A	Column B
16. $(a - 10)(a - 3) \le 0$	$a^2 + 4a - 21 < 0$

$$i^x = -i$$

Column A	Column B
17. 8	x

Column A	Column B
18. $\dfrac{2(3 + 7)^2}{2^3 + 3 \cdot 4}$	$(6 + 7 \cdot 2) - 10 - 8 \div 2 + 4$

Column A	Column B
19. $\dfrac{m^3 + 7m^2 + 17m + 35}{m^2 + 2m + 7}$	$\dfrac{3m^3 - m^2 - 14m + 8}{3m^2 + 5m - 4}$

Column A	Column B
20. $\sqrt[3]{2x + 1} = 3$	$\sqrt[3]{2}\left(3\sqrt[3]{4} + 2\sqrt[3]{32}\right) = x$

Analyzing Conic Sections

Objectives

In this chapter, you will:

- find the distance between two points in the coordinate plane,
- find the midpoint of a line segment in the coordinate plane,
- write equations of conic sections having certain properties,
- graph conic sections,
- use simulations to solve problems, and
- solve systems of quadratic equations and inequalities.

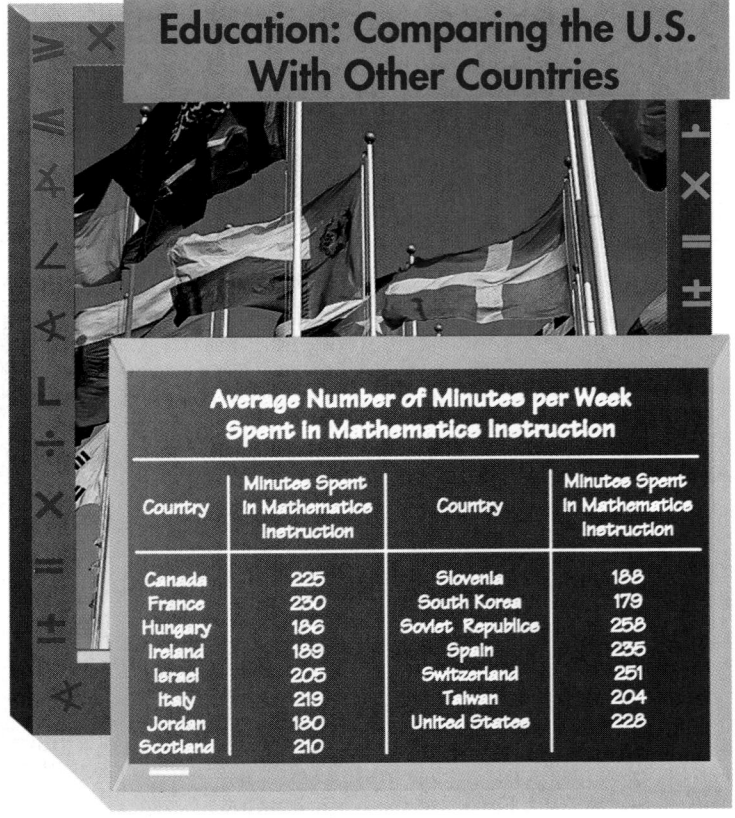

Education: Comparing the U.S. With Other Countries

	Average Number of Minutes per Week Spent in Mathematics Instruction		
Country	Minutes Spent in Mathematics Instruction	Country	Minutes Spent in Mathematics Instruction
Canada	225	Slovenia	188
France	230	South Korea	179
Hungary	186	Soviet Republics	258
Ireland	189	Spain	235
Israel	205	Switzerland	251
Italy	219	Taiwan	204
Jordan	180	United States	228
Scotland	210		

Source: Bureau of the Census, *Statistical Abstract of the United States*

The best and the brightest of U.S. high school students can match up with their international competition in mathematics. How about in geographic knowledge? Can America's educational system still produce the best? Do you know where the Windward Passage is or the location of the Amur River?

TIME *Line*

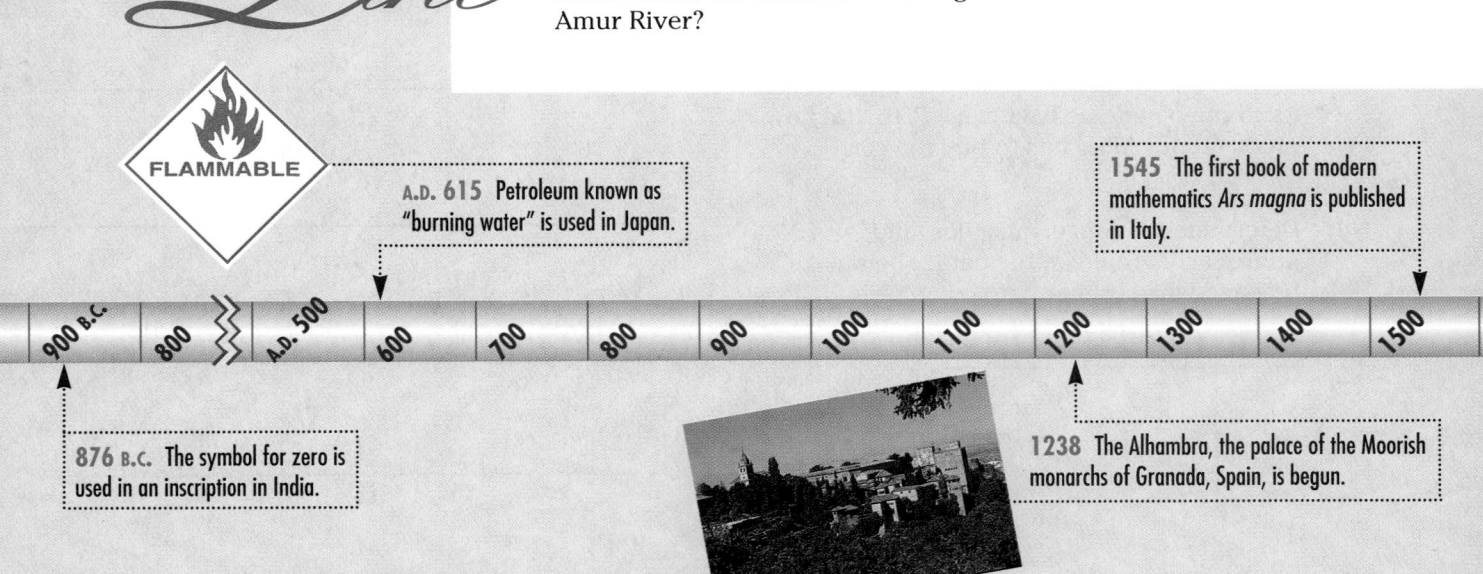

FLAMMABLE

A.D. 615 Petroleum known as "burning water" is used in Japan.

1545 The first book of modern mathematics *Ars magna* is published in Italy.

| 900 B.C. | 800 | A.D. 500 | 600 | 700 | 800 | 900 | 1000 | 1100 | 1200 | 1300 | 1400 | 1500 |

876 B.C. The symbol for zero is used in an inscription in India.

1238 The Alhambra, the palace of the Moorish monarchs of Granada, Spain, is begun.

CHRIS GAL
Mic

Chris Galeczka knew that the official languages of Afghanistan are Pashtu and Dari. This knowledge earned him a $25,000 scholarship and a first-place victory in the 1995 National Geography Bee held in Washington, D.C. The 13-year-old from Sterling Heights, Michigan, was chosen from some six million students competing in every state and six territories. These were narrowed to 57 finalists for the contest at the National Geographic Society headquarters, hosted by *Jeopardy!*'s Alex Trebek.

Knowing every country, their capitals, languages, religions, and governments paid off for Chris. His next competition was an international geography contest, also sponsored by the National Geographic Society, at Epcot at Walt Disney World in Orlando, Florida.

Chapter Project

Organize a match of Conic Jeopardy to test your classmates' tournament skills.

- Use four categories for your Jeopardy game: parabolas, circles, ellipses, and hyperbolas.

- Write five problems for each category in varying degrees of difficulty.

- Construct a Jeopardy grid on the chalkboard and assign points to the various levels.

- Challenge your classmates to a game of Conic Jeopardy.

- Use different sets of questions for each round of the game.

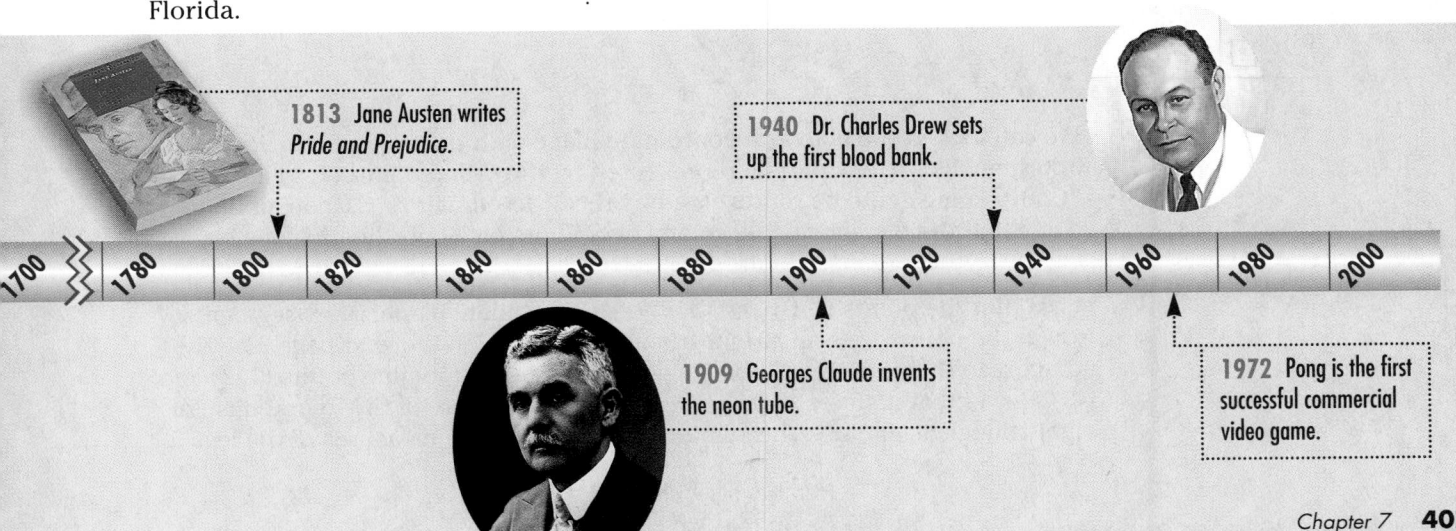

1813 Jane Austen writes *Pride and Prejudice.*

1940 Dr. Charles Drew sets up the first blood bank.

1700 1780 1800 1820 1840 1860 1880 1900 1920 1940 1960 1980 2000

1909 Georges Claude invents the neon tube.

1972 Pong is the first successful commercial video game.

Integration: Geometry
The Distance and Midpoint Formulas

APPLICATION
Aviation

Each July, Browning, Montana, hosts the North American Indian Days. Browning is located just east of Glacier National Park and is on one of the four reservations for the Blackfeet Indians. The other three reservations are located in Canada.

Yoomee wants to fly her small plane from Miles City, Montana, to Browning to participate in the celebration. First, she needs to estimate the distance from Miles City to Browning. A square grid is superimposed on the map of Montana. Each side of a square on the grid represents 70 miles. Miles City appears at 7C on the map, and Browning appears at 3A on the map. How can we use this grid to help the pilot estimate the distance from Miles City to Browning?

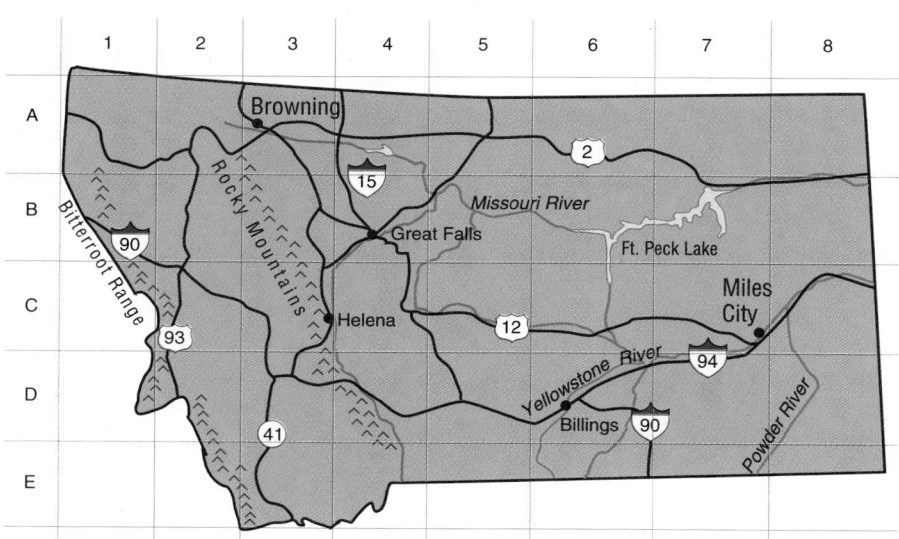

We can look at the map as a coordinate plane with the origin in the lower left corner. Miles City could be represented by the ordered pair (7, 3) instead of 7C. Browning could be represented by the ordered pair (3, 5) instead of 3A. You can find the distance between any two points on a coordinate plane by using the Pythagorean theorem.

First plot the points (7, 3) and (3, 5) on a coordinate plane. Draw segments through the points to form a right triangle as shown on the next page. Use the Pythagorean theorem to find the length of the segment joining points (7, 3) and (3, 5). Recall that it states that the square of the measure of the hypotenuse of a right triangle is equal to the sum of the squares of the measures of the legs $(a^2 + b^2 = c^2)$.

You can find the horizontal or vertical distance between two points by using absolute value. The distance between two points on a number line whose coordinates are a and b is $|a - b|$ or $|b - a|$. The length of the horizontal leg is $|7 - 3|$ or 4 units, and the length of the vertical leg is $|3 - 5|$ or 2 units.

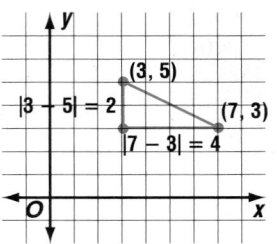

$$c^2 = a^2 + b^2 \quad \textit{Pythagorean theorem}$$

$$d^2 = 4^2 + 2^2 \quad \textit{Replace c with d, a with 4, and b with 2.}$$

$$d^2 = 16 + 4$$

$$d^2 = 20$$

$$d = \sqrt{20}$$

$$d \approx 4.47 \quad \textit{Distance is nonnegative.}$$

The map distance is about 4.47 units. Each unit equals 70 miles. So, the actual distance is about 4.47(70) or 312.9 miles.

The solution for this application suggests a method for finding the distance between any two points on a coordinate plane. Suppose (x_1, y_1) and (x_2, y_2) name two points in the coordinate plane. We can form a right triangle by drawing a vertical line through point (x_1, y_1) and a horizontal line through point (x_2, y_2). The lines will intersect at the third vertex of the triangle, (x_1, y_2). *Why?*

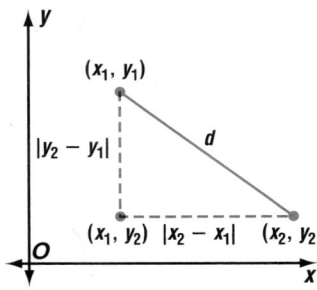

Use the Pythagorean theorem to find the distance, d, between the two points.

$$c^2 = a^2 + b^2 \quad \textit{Pythagorean theorem}$$

$$d^2 = |x_2 - x_1|^2 + |y_2 - y_1|^2 \quad \textit{Replace c with d, a with } |x_2 - x_1|, \textit{ and b with } |y_2 - y_1|.$$

$$d^2 = (x_2 - x_1)^2 + (y_2 - y_1)^2 \quad \textit{Why can } (x_2 - x_1)^2 \textit{ be substituted for } |x_2 - x_1|^2?$$

$$d = \sqrt{(x_2 - x_1)^2 + (y_2 - y_1)^2}$$

Distance Formula for Two Points in a Plane	**The distance between two points with coordinates (x_1, y_1) and (x_2, y_2) is given by $d = \sqrt{(x_2 - x_1)^2 + (y_2 - y_1)^2}$.**

Example ❶ Find the distance between points at $(-5, 7)$ and $(9, -11)$.

$$d = \sqrt{(x_2 - x_1)^2 + (y_2 - y_1)^2} \quad \textit{Distance formula}$$

$$= \sqrt{[9 - (-5)]^2 + (-11 - 7)^2} \quad \textit{Replace } x_2 \textit{ with 9, } x_1 \textit{ with } -5, y_2 \textit{ with } -11, \textit{ and } y_1 \textit{ with 7.}$$

$$= \sqrt{(14)^2 + (-18)^2}$$

$$= \sqrt{196 + 324}$$

$$= \sqrt{520} \text{ or } 2\sqrt{130}$$

The distance is $2\sqrt{130}$ or about 22.8 units.

Example **2** Show that $P(5, 3)$ is the midpoint of the segment joining $M(-1, 6)$ and $N(11, 0)$.

First, we must show that P is on $\overline{MN}$. It is sufficient to show that $\overline{MP}$ and $\overline{MN}$ have the same slope.

slope of $\overline{MP} = \dfrac{6 - 3}{-1 - 5}$

$= \dfrac{3}{-6}$ or $-\dfrac{1}{2}$

slope of $\overline{MN} = \dfrac{6 - 0}{-1 - 11}$

$= \dfrac{6}{-12}$ or $-\dfrac{1}{2}$

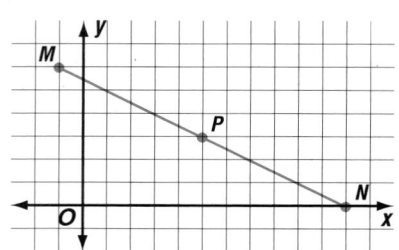

Next, we must show that the distance from M to P is the same as the distance from N to P. The length of $\overline{MP}$ is represented by MP, and the length of $\overline{MN}$ by MN.

distance from M to P

$MP = \sqrt{(-1 - 5)^2 + (6 - 3)^2}$

$= \sqrt{(-6)^2 + 3^2}$

$= \sqrt{36 + 9}$

$= \sqrt{45}$ or $3\sqrt{5}$

distance from N to P

$NP = \sqrt{(11 - 5)^2 + (0 - 3)^2}$

$= \sqrt{6^2 + (-3)^2}$

$= \sqrt{36 + 9}$

$= \sqrt{45}$ or $3\sqrt{5}$

Since the distances are the same, P is the midpoint of $\overline{MN}$.

In Example 2, the coordinates of the endpoints of $\overline{MN}$ are $(-1, 6)$ and $(11, 0)$. Find the mean of the x-coordinates and the mean of the y-coordinates. What do you observe? This suggests the following.

Midpoint of a Line Segment	If a line segment has endpoints at (x_1, y_1) and (x_2, y_2), then the midpoint of the line segment has coordinates $\left(\dfrac{x_1 + x_2}{2}, \dfrac{y_1 + y_2}{2}\right)$.

Example **3** Circle P has a diameter $\overline{AB}$. If A is at $(-3, -5)$ and the center is at $(2, 3)$, find the coordinates of B.

Geometry

Explore Read the problem. You know the coordinates for one endpoint of a diameter and the coordinates for the center of the circle. You want to know the coordinates for the other endpoint of the diameter.

Plan The center of the circle is the midpoint of the diameter. Use the formula for the midpoint of a line segment to find the coordinates of the other endpoint. Let (x, y) be the coordinates of B.

Solve $(x, y) = \left(\dfrac{x_1 + x_2}{2}, \dfrac{y_1 + y_2}{2}\right)$

$(2, 3) = \left(\dfrac{-3 + x}{2}, \dfrac{-5 + y}{2}\right)$

$2 = \dfrac{-3 + x}{2}$ and $3 = \dfrac{-5 + y}{2}$

$4 = -3 + x \qquad\qquad 6 = -5 + y$

$7 = x \qquad\qquad\quad 11 = y$

The coordinates of B are $(7, 11)$.

Examine If (2, 3) is the center of the circle, then it must be equidistant from (−3, −5) and (7, 11). Use the distance formula to check your answer.

$$d = \sqrt{(-3-2)^2 + (-5-3)^2} = \sqrt{89} \text{ or about } 9.43$$
$$d = \sqrt{(7-2)^2 + (11-3)^2} = \sqrt{89} \text{ or about } 9.43$$

The distances are the same, so (2, 3) must be the midpoint of the diameter or the center of the circle.

CHECK FOR UNDERSTANDING

Communicating Mathematics

Study the lesson. Then complete the following.

1. a. **Describe** how you would find the coordinates of the endpoint of a segment given the midpoint and the other endpoint.

   b. **Demonstrate** your method by finding the endpoint of a segment with midpoint (−4, 13) and endpoint (20, 31).

2. **Describe** a situation in which a point is equidistant from the endpoints of a segment but is not the midpoint.

3. **Explain** why it is not important which endpoint is chosen as (x_1, y_1) and which is chosen as (x_2, y_2) when using the distance formula.

Guided Practice

Find the distance between each pair of points with the given coordinates.

4. (7, 8), (−4, 9)   5. (0.5, 1.4), (1.1, 2.9)   6. $(2\sqrt{3}, -5), (-3\sqrt{3}, 9)$

Find the midpoint of each line segment if the coordinates of the endpoints are given.

7. (8, 9), (−3, −4.5)   8. $(-3\sqrt{2}, -4\sqrt{5}), (8\sqrt{2}, 9\sqrt{5})$

9. **Geometry** Show that the triangle with vertices A(−3, 0), B(−1, 4), and C(1, −2) is isosceles.

10. **Geometry** Triangle MNO has vertices M(3, 5), N(−2, 8), and O(7, −4). Find the coordinates of the midpoints of each side.

EXERCISES

Practice

Find the distance between each pair of points with the given coordinates.

11. (−4, 9), (1, −3)   12. (−4, −10), (−3, −11)

13. (9, −2), (12, −14)   14. (0.23, 0.4), (0.68, −0.2)

15. $(-2\sqrt{7}, 10), (4\sqrt{7}, 8)$ 16. $(2\sqrt{3}, 4\sqrt{3}), (2\sqrt{3}, -\sqrt{3})$

17. $\left(-3, \frac{-2}{11}\right), \left(5, \frac{9}{11}\right)$ 18. $\left(\frac{2\sqrt{3}}{3}, \frac{\sqrt{5}}{4}\right), \left(\frac{-2\sqrt{3}}{3}, \frac{\sqrt{5}}{2}\right)$

Find the midpoint of each line segment if the coordinates of the endpoints are given.

19. $(8, 3), (16, 7)$

20. $(5, 9), (12, 18)$

21. $(-5, 3), (-3, -7)$

22. $(6, -5), (-2, -7)$

23. $(0.45, 7), (-0.3, -9)$

24. $(-3, -12), (-8, 0.34)$

Find the value of *a* so that the distance between points with the given coordinates is 10 units.

25. $(-7, 3), (a, 11)$

26. $(7, 2), (-1, a)$

27. $(6, 3), (8, a)$

28. $(-8, 8), (a, 11)$

Given the coordinates of one endpoint of $\overline{AB}$ and its midpoint, *M*, find the coordinates of the other endpoint.

29. $A(9, 3), M(4, 2)$

30. $B(2, 5), M(-1, 7)$

31. $M(-0.8, 3.85), B(2.7, 4.9)$

32. $M\left(\frac{9}{16}, \frac{5}{4}\right), A\left(\frac{1}{4}, 3\right)$

33. A graph of line AB is shown at the right.
    a. Write an equation for this line.
    b. Find the length of $\overline{AB}$.
    c. Find the coordinates of the midpoint of $\overline{AB}$.

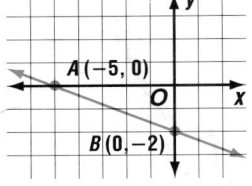

34. Triangle CAT has vertices $C(4, 9), A(8, -9),$ and $T(-6, 5)$.
    a. Find the coordinates of the midpoints of each side of the triangle.
    b. The *median* of a triangle is a segment that joins a vertex of the triangle and the midpoint of the side opposite that vertex. Find the length of the median from point C to $\overline{TA}$.
    c. Find the perimeter of $\triangle CAT$.
    d. Find the perimeter of the triangle formed in part a above.
    e. How do the perimeters in parts c and d compare?

35. Find the perimeter of the quadrilateral shown at the right.

36. Find the center of the circle whose diameter has endpoints at $(9, 0)$ and $(11, -14)$.

37. The vertices of a parallelogram are $M(2, 2)$ $A(12, 2), T(16, 8),$ and $H(6, 8)$.
    a. Find the length of each diagonal.
    b. Find the coordinates of the point of intersection of the diagonals.

38. Triangle *MAY* is a right triangle.

    a. Find the midpoint of the hypotenuse. Call it point *Q*.

    b. Classify $\triangle AQY$ according to the length of its sides. Include sufficient evidence to support your conclusion.

    c. Classify $\triangle AQM$ according to its angles.

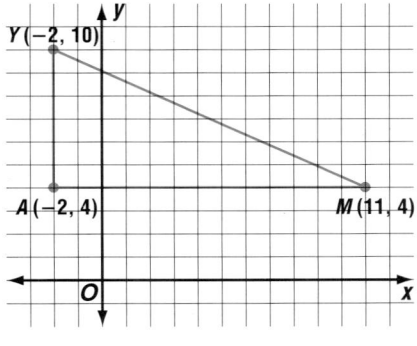

Programming

39. The graphing calculator program at the right can help you find the distance between points with coordinates (x_1, y_1) and (x_2, y_2).

    a. Use the program to find the distance between the points at (1, 5) and (5, 2).

    b. Use the program to find the distance between the points at (−3, 6) and (−9, −12).

    c. Modify the program to find the midpoint of a segment.

    d. Test your program by finding the midpoint of the segment with endpoints at (12, 8) and (−5, −11).

```
PROGRAM:DISTFORM
: ClrHome
: Disp "X1 ="
: Input A
: Disp "Y1 ="
: Input B
: Disp "X2 ="
: Input C
: Disp "Y2 ="
: Input E
: C−A→X
: E−B→Y
: √(X²+Y²)→D
: Disp "DISTANCE"
: Disp "BETWEEN"
: Disp "(X1, Y1) AND"
: Disp "(X2, Y2) IS ="
: Disp D
```

Critical Thinking

40. Find the coordinates of the point that is three-fourths of the way from $P(-1, 12)$ to $Q(5, -10)$.

Applications and Problem Solving

41. Transportation A grid is superimposed on a map of a portion of the state of Florida.

    a. About how far is it from Orlando, Florida, to Tallahassee, Florida, if each unit on the grid represents 70 miles?

    b. How long will it take a plane to fly from Tallahassee, Florida, to Daytona Beach, Florida, if its speed averages 180 miles per hour?

Lake Eola, Orlando, Florida

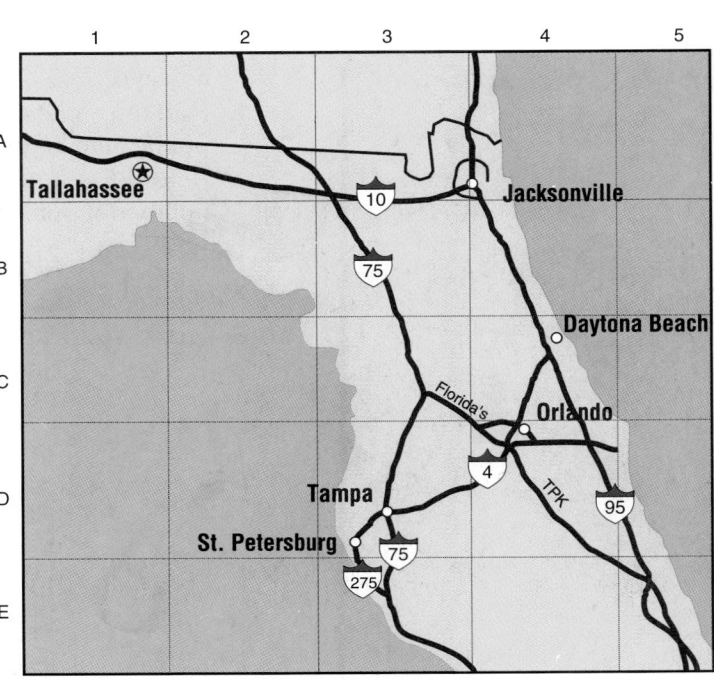

42. Sports In a field goal attempt, a football was kicked from a hash mark on the twenty-yard line. If the ball passed midway between the goal posts, what was the ground distance traveled by the ball from the point that it was kicked to the point on the ground below where it crossed the goal post?

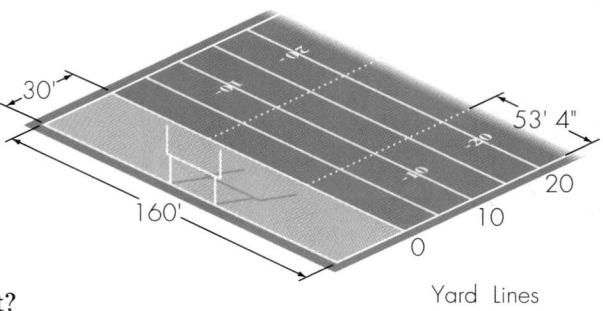

43. Geography The U.S. Geological Survey (USGS) has determined the official center of the United States.

Belle Fourche, South Dakota   The Center of the Nation

   a. Describe a method that might be used to determine the geographical center of the continental United States.

   b. Use a map of the continental United States to determine the geographical center.

   c. USGS has located the geographical center of the continental United States in Kansas. How does the result of your method compare to result of the USGS method?

   d. How could you determine the geographical center of the United States if the land masses of Alaska and Hawaii are included?

   e. USGS has declared the geographical center of the *entire* United States to be near Belle Fourche, South Dakota. Would you agree with this location? Explain.

   f. How do you think the geographical center compares to the population center?

Mixed Review

44. Conservation On September 3, 1964, President Lyndon B. Johnson signed the Wilderness Act, thereby establishing a national wilderness preservation system of the public lands of the United States. The public lands in Illinois are listed in the table at the right. (Lesson 6–8)

Park	Number of Acres
Bald Knob	5863
Bay Creek	2866
Burden Falls	3671
Clear Springs	4730
Crab Orchard	4050
Garden of the Gods	3268
Lusk Creek	4466
Panther Den	685

   a. Find the mean of the public land acreage.

   b. Find the standard deviation of the public land acreage.

   c. Does the standard deviation indicate a large or small variation in the number of public land acres? Explain.

45. Find $(t^2 - 3t + 2) \div (t + 2)$ by using synthetic division. (Lesson 5–3)

46. Simplify $(s + 3)^2$. (Lesson 5–2)

47. Solve the system of equations by using augmented matrices. (Lesson 4–7)

$2a - b + c = 44$

$-a + 3b - 2c = -53$

$5a - 6b - c = 19$

48. Find $3\begin{bmatrix} 4 & -2 \\ 5 & 7 \end{bmatrix} + 2\begin{bmatrix} -3 & 5 \\ -4 & 3 \end{bmatrix}$. (Lesson 4–2)

49. Photography The perimeter of a rectangular picture is 86 inches. Twice the width exceeds the length by 2 inches. What are the dimensions of the picture? (Lesson 3–2)

50. Find a value of a for which the graph of $y = ax + 9$ is perpendicular to the graph of $x + 3y = 14$. (Lesson 2–3)

51. Evaluate $2|-3x| - 9$ if $x = 5$. (Lesson 1–5)

Parabolas

Law Enforcement

Have you ever driven around a corner and observed a police officer pointing a radar gun at your car? The radar gun has a reflector that sends out and receives rays. This reflector is in the shape of a *parabola*.

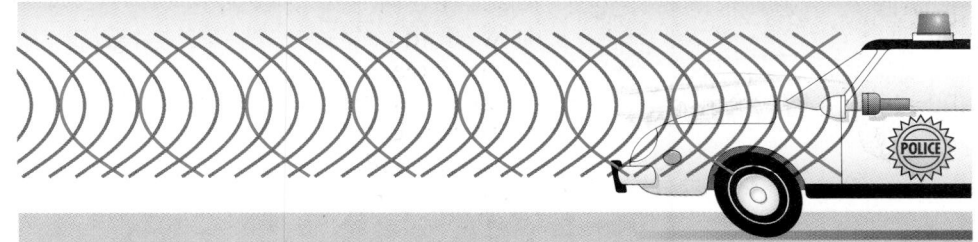

A parabola is a shape formed by slicing a double cone on a slant as shown at the right. Any figure that can be formed by slicing a double cone is called a **conic section**. Other conic sections are shown below.

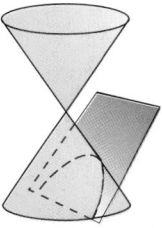

parabola

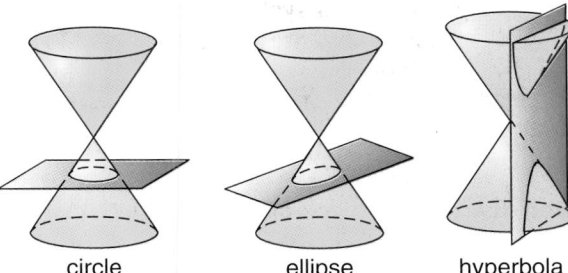

circle   ellipse   hyperbola

A parabola can be defined in terms of the location of a point called the **focus** and a line called the **directrix**.

Definition of a Parabola	**A parabola is the set of all points in a plane that are the same distance from a given point called the *focus* and a given line called the *directrix*.**

The directrix is named by the equation of the line.

The parabola at the right has a focus at (3, 4) and a directrix with equation $y = -2$. We can use the distance formula and the definition of a parabola to find the equation of this parabola.

Let (x, y) be any point on the parabola. The distance from this point to the focus must be the same as its distance from the directrix. The distance from a point to a line is measured along the perpendicular from the point to the line.

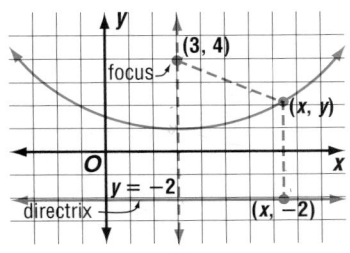

distance between (x, y) and $(3, 4)$ = distance between (x, y) and $(x, -2)$

$$\sqrt{(x-3)^2 + (y-4)^2} = \sqrt{(x-x)^2 + [y-(-2)]^2}$$

$$(x-3)^2 + (y-4)^2 = (0)^2 + (y+2)^2 \quad \textit{Square each side.}$$

$$(x-3)^2 + y^2 - 8y + 16 = y^2 + 4y + 4$$

$$(x-3)^2 + 12 = 12y$$

$$\frac{1}{12}(x-3)^2 + 1 = y$$

The equation of a parabola with focus at $(3, 4)$ and directrix with equation $y = -2$ is $y = \frac{1}{12}(x-3)^2 + 1$. The equation of the *axis of symmetry* for this parabola is $x = 3$. Notice that the axis of symmetry and the directrix are perpendicular to each other. The axis of symmetry intersects the parabola at a point called the *vertex*. The vertex of this parabola is at $(3, 1)$. The parabola opens upward, since $\frac{1}{12}$ is positive.

In general, given the equation of a parabola written in the form $y = a(x-h)^2 + k$, the vertex is at (h, k), and the equation of the axis of symmetry is $x = h$. The parabola opens upward if $a > 0$ and opens downward if $a < 0$.

Example ❶ **Write $y = 2x^2 + 12x + 14$ in the form $y = a(x - h)^2 + k$. Name the vertex, the axis of symmetry, and the direction of opening of the parabola.**

$$y = 2x^2 + 12x + 14$$

$$y = 2(x^2 + 6x) + 14 \qquad \textit{Factor 2 from the x terms.}$$

$$y = 2(x^2 + 6x + \square) + 14 - 2\square \qquad \textit{Complete the square on the right side.}$$

$$y = 2(x^2 + 6x + 9) + 14 - 2(9) \qquad \textit{The 9 added when you complete the square}$$

$$y = 2(x + 3)^2 - 4 \qquad \textit{is multiplied by 2.}$$

$$y = 2[x - (-3)]^2 - 4 \qquad (h, k) = (-3, -4)$$

The vertex of this parabola is located at $(-3, -4)$, and the equation of the axis of symmetry is $x = -3$. The parabola opens upward.

MODELING MATHEMATICS

Parabolas

Materials: wax paper

You can make parabolas by folding wax paper.

Your Turn

a. Start with a sheet of wax paper that is about 15 inches long and 12 inches wide. Make a line that is perpendicular to the sides of the sheet by folding the sheet near one end. Open up the paper again. This line is the directrix. Mark a point about midway between the sides of the sheet so that the distance from the directrix is about 1 inch. This point is the focus.

Put the focus on top of any point on the directrix and crease the paper. Make about 20 more creases by placing the focus on top of other points on the directrix. The lines form the outline of a parabola.

b. Start with a new sheet of wax paper. Form another outline of a parabola with a focus that is about 3 inches from the directrix.

c. On a new sheet of paper, form a third outline of a parabola with a focus that is about 5 inches from the directrix.

d. Compare the shapes of the three parabolas. How does the distance between the focus and the directrix affect the parabola?

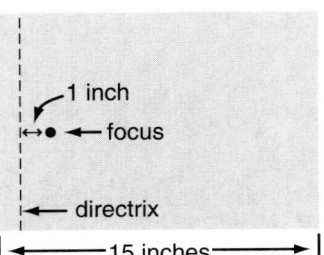

The line segment through the focus of a parabola and perpendicular to the axis of symmetry is called the **latus rectum**. The endpoints of the latus rectum lie on the parabola. In the figure at the right, the latus rectum is $\overline{AB}$. The length of the latus rectum of the parabola with equation $y = a(x - h)^2 + k$ is $\left|\dfrac{1}{a}\right|$ units. The endpoints of the latus rectum are $\left|\dfrac{1}{2a}\right|$ units from the focus.

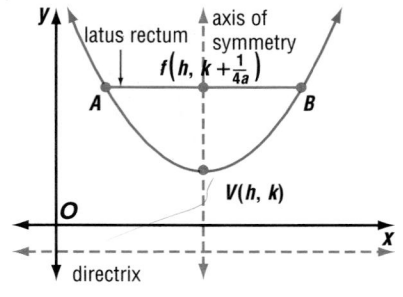

Equations of parabolas with a vertical axis of symmetry are in the form $y = a(x - h)^2 + k$ and are functions. Equations of parabolas with a horizontal axis of symmetry are in the form $x = a(y - k)^2 + h$ and are not functions.   *Why?*

Information about Parabolas

form of equation	$y = a(x - h)^2 + k$	$x = a(y - k)^2 + h$
axis of symmetry	$x = h$	$y = k$
vertex	(h, k)	(h, k)
focus	$\left(h, k + \dfrac{1}{4a}\right)$	$\left(h + \dfrac{1}{4a}, k\right)$
directrix	$y = k - \dfrac{1}{4a}$	$x = h - \dfrac{1}{4a}$
direction of opening	upward if $a > 0$ downward if $a < 0$	right if $a > 0$ left if $a < 0$
length of latus rectum	$\left|\dfrac{1}{a}\right|$ units	$\left|\dfrac{1}{a}\right|$ units

Example 2 **Graph $3x - y^2 = 8y + 31$.**

First write the equation in the form $x = a(y - k)^2 + h$.

$3x - y^2 = 8y + 31$ *There is a y^2 term.*

$\quad 3x = y^2 + 8y + 31$ *Isolate the y terms.*

$\quad 3x = (y^2 + 8y + \square) + 31 - \square$ *Complete the square.*

$\quad 3x = (y^2 + 8y + 16) + 31 - 16$

$\quad 3x = (y + 4)^2 + 15$

$\quad\quad x = \dfrac{1}{3}(y + 4)^2 + 5$ $(h, k) = (5, -4)$

Then use the following information to draw the graph.

vertex: $(5, -4)$

axis of symmetry: $y = -4$

focus: $\left(5 + \dfrac{1}{4\left(\frac{1}{3}\right)}, -4\right)$ or $\left(5\frac{3}{4}, -4\right)$

directrix: $x = 5 - \dfrac{1}{4\left(\frac{1}{3}\right)}$ or $4\frac{1}{4}$

direction of opening: right, since $a > 0$

length of latus rectum: $\left|\dfrac{1}{\frac{1}{3}}\right|$ or 3 units

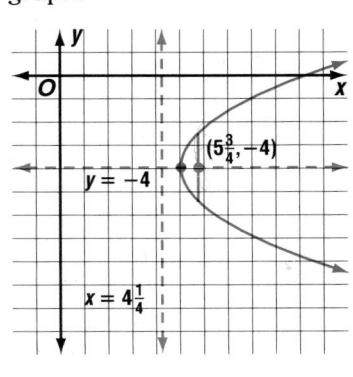

APPLICATION

Solar Energy

③ Solar energy may be harnessed by using parabolic mirrors. The mirrors reflect the rays from the sun to the focus of the parabola. In the Mojave Desert in California, such mirrors are used to heat oil that flows through tubes placed at the focus. The focus of each parabolic mirror at this facility is 6.25 feet above the vertex. The latus rectum is 25 feet long.

a. **Assume that the focus is at the origin and write an equation for the parabola formed by each mirror.**

b. **Then graph the equation.**

a. In order for the mirrors to collect the sun's energy, the parabola must open upward. Therefore, the vertex must be below the focus.

focus: $(0, 0)$ vertex: $(0, -6.25)$

The measure of the latus rectum is 25. So $25 = \left|\frac{1}{a}\right|$, and $a = \frac{1}{25}$.

Write the equation using the form $y = a(x - h)^2 + k$.

An equation for the parabola formed by each mirror is $y = \frac{1}{25}x^2 - 6.25$.

b. Now use all of the information to draw a graph.

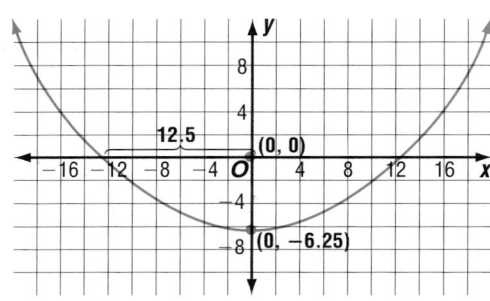

Example **④** **Write an equation for the parabola shown below.**

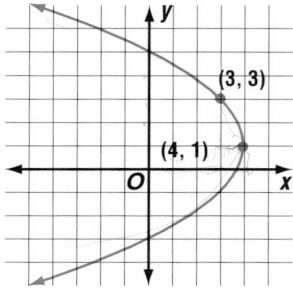

The vertex is at $(4, 1)$. Use the equation $x = a(y - k)^2 + h$.

$$x = a(y - 1)^2 + 4$$

The parabola passes through the point at $(3, 3)$. Use this information to solve for a.

$$3 = a(3 - 1)^2 + 4 \quad x = 3, y = 3$$
$$3 = 4a + 4$$
$$-1 = 4a$$
$$-\frac{1}{4} = a$$

An equation for the parabola is $x = -\frac{1}{4}(y - 1)^2 + 4$.

Communicating Mathematics

Study the lesson. Then complete the following.

1. **Describe** the relationships among the directrix, the focus, the vertex, the axis of symmetry, and the latus rectum.

2. **You Decide**   Ralph says that the graphs of the equations $y = 4(x - 3)^2 - 7$ and $y = 4(3 - x)^2 - 7$ are different. Shanice says they are the same. Who is right? Explain your answer.

3. **Compare and contrast** the equations of parabolas that open right or left and those that open up or down. Tell which equations, if any, are functions.

MODELING MATHEMATICS

4. Suppose you used wax paper folding to create outlines of two parabolas. If one has a focus 10 centimeters from the directrix and the other has a focus 5 centimeters from the directrix, how will the parabolas compare?

Guided Practice

Express each equation in the form $y = a(x - h)^2 + k$ or $x = a(y - k)^2 + h$.

5. $y = 2x^2 - 12x + 6$

6. $y = \frac{1}{2}x^2 + 12x - 8$

7. $x = 3y^2 + 5y - 9$

8. $x = y^2 + 14y + 20$

Name the coordinates of the vertex and focus, the equations of the axis of symmetry and directrix, and the direction of opening of the parabola with the given equation. Then find the length of the latus rectum and graph the parabola.

9. $y = (x - 3)^2 - 4$

10. $y = 2(x + 7)^2 + 3$

11. $y = 3x^2 - 8x + 6$

12. $y = \frac{2}{3}x^2 - 6x + 12$

13. Write an equation for the graph shown at the right.

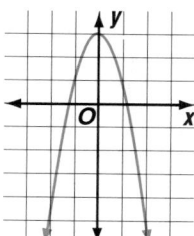

14. Write an equation of the parabola with its focus at $(3, 8)$ and $y = 4$ as the equation for its directrix. Then draw the graph.

15. Write an equation of the parabola with its vertex at $(5, -1)$ and its focus at $(3, -1)$. Then draw the graph.

Practice Name the coordinates of the vertex and focus, the equations of the axis of symmetry and directrix, and the direction of opening of the parabola with the given equation. Then find the length of the latus rectum and graph the parabola.

16. $-6y = x^2$

17. $3(y - 3) = (x + 6)^2$

18. $-2(x - 4) = (y - 1)^2$

19. $4(x - 2) = (y + 3)^2$

20. $(y - 8)^2 = -4(x - 4)$

21. $y = x^2 - 12x + 20$

22. $x = y^2 - 14y + 25$

23. $y = -2x^2 + 5x - 10$

24. $x = 5y^2 - 25y + 60$

25. $y = 3x^2 - 24x + 50$

26. $\frac{1}{2}(y + 1) = (x - 8)^2$

27. $x = -\frac{1}{3}y^2 - 12y + 15$

28. $y = \frac{1}{2}x^2 - 3x + \frac{19}{2}$

29. $x = \frac{1}{4}y^2 - \frac{1}{2}y - 3$

Write an equation for each graph.

30.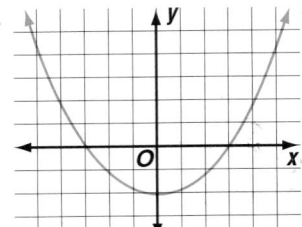

31.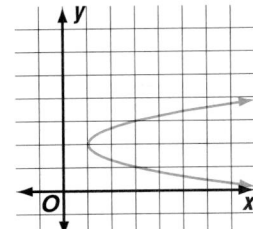

The coordinates of the focus and equation of the directrix of a parabola are given. Write an equation for each parabola. Then draw the graph.

32. $(4, -3); y = 4$

33. $(-3, -2); y = -6$

34. $(3, 0); x = -2$

35. $(4, -3); y = 6$

36. $(10, -4); x = 5$

37. $(4, 0); x = -2$

Write an equation of each parabola described below. Then draw the graph.

38. vertex, $(0, 1)$; focus, $(0, 5)$

39. vertex, $(8, 6)$; focus, $(2, 6)$

40. focus, $(-4, -2)$; directrix, $x = -8$

41. vertex, $(1, 7)$; directrix, $y = 3$

42. vertex, $(-7, 4)$; axis of symmetry, $x = -7$; measure of latus rectum, 6; $a < 0$

43. vertex, $(4, 3)$; axis of symmetry, $y = 3$; measure of latus rectum, 4; $a > 0$

44. a. Draw the graph of $x = 3y^2 + 4y + 1$.
 b. Find the x-intercept(s).
 c. Find the y-intercept(s).
 d. What is the equation of the axis of symmetry?
 e. What are the coordinates of the vertex?

45. The graph of the equation
$y = -\frac{1}{8}(x - 8)^2 + 2$ is shown
at the right.

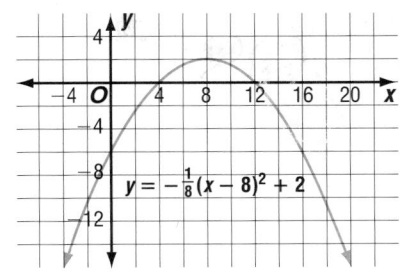

What values of x make each
statement true?

a. $-\frac{1}{8}(x - 8)^2 + 2 = 0$

b. $-\frac{1}{8}(x - 8)^2 + 2 > 0$

c. $-\frac{1}{8}(x - 8)^2 + 2 < 0$

46. Suppose two different parabolas have their vertex at $(-3, 1)$ and contain the point with coordinates $(-1, 0)$. Find their equations.

Applications and Problem Solving

47. Manufacturing   An automobile headlight contains a parabolic reflector. A special bulb with two filaments is used to produce the high and low beams. The filament placed at the focus produces the high beam and the filament placed off the focus produces the low beam. The equation of the cross section of the reflector is $y = \frac{1}{12}x^2$. How far from the vertex should the filament for the high beam be placed?

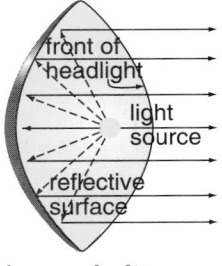

48. Communications   A microphone is placed at the focus of a parabolic reflector to collect sounds for the television broadcast of the World Cup Soccer final game. The focus of the parabola that is a cross section of the reflector is 6 inches from the vertex. The latus rectum is 24 inches in length. Assume the focus is at the origin and the parabola opens to the right. Write an equation for the cross section.

49. Space Science   A spacecraft is in a circular orbit 150 kilometers above Earth. Once it obtains the velocity needed to escape Earth's gravity, the spacecraft will follow a parabolic path with the focus at the center of Earth, as shown below. Suppose it obtains its escape velocity above the North Pole. Assume the center of Earth is at the origin and the radius of Earth is 6400 kilometers. Write an equation for the parabolic path of the spacecraft.

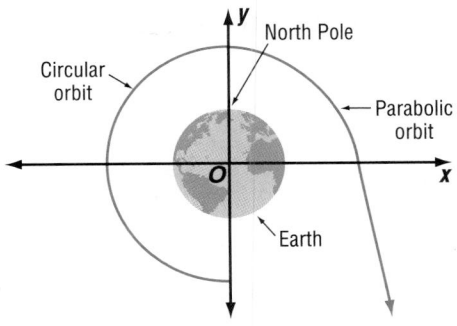

50. Baseball   When a ball is thrown, the path it travels is a parabola. Suppose a baseball is thrown from ground level, reaches a maximum height of 50 feet, and hits the ground 200 feet from where it was thrown. Assuming this situation could be modeled on a coordinate plane with the focus of the parabola at the origin, find the equation of the (parabolic) path of the ball. (*Hint:* The focus is on ground level.)

51. **Draw a Diagram**   A hole in a compact disc needs to be at the center of the disc. If the hole is to be $\frac{7}{16}$ inch in diameter and the diameter of the disc is $4\frac{3}{4}$ inches, how far from the edge of the disc will the edge of the hole be placed?   (Lesson 7–1)

52. Write the equation of a parabola with position 3 units to the right of the parabola with equation $f(x) = x^2$.   (Lesson 6–6)

53. Solve $x^2 + 14x - 12 = 0$ by completing the square.   (Lesson 6–3)

54. Simplify $(3 + 2i)(4 + 5i)$.   (Lesson 5–9)

55. **Health**   Ty's heart rate is usually 120 beats per minute when he runs. If he runs for 2 hours every day, about how many beats will his heart make during the equivalent of two weeks of exercise sessions? Express the answer in both decimal and scientific notation.   (Lesson 5–1)

56. Find $-3\begin{bmatrix} \frac{5}{6} & 3 \\ -2 & \frac{2}{9} \end{bmatrix}$.   (Lesson 4–1)

57. Given $f(x, y) = -3y + 4x$, find $f(-7, -4)$.   (Lesson 3–5)

58. **Business**   The Friendly Fix-It Company charges \$35 for any in-home repair. In addition, the technician charges \$10 per hour after the first half-hour. What would be the cost C of an in-home repair of h hours?   (Lesson 2–4)

59. If $a = 2$, $b = -6$, $c = 3$, and $a^3b^2 + 4ac + 2d \geq 6c^2 - 4ab$, solve for d.   (Lesson 1–6)

WORKING ON THE In·ves·ti·ga·tion

Refer to the Investigation on pages 328–329.

the River Canyon Bridge

After much discussion, your group of designers decides to proceed with the design proposed by the company's consultant. The consultant's suggestion is a general one, and your group must now design a bridge with more exact specifications.

1 Verify that the arch in each of your designs is parabolic. Use the definition of a parabola as your guide.

2 Suppose the directrix of your parabola is the roadway above it. The focus would be inside the parabola. Experiment with different foci to see how the degree of steepness of the parabola changes.

3 Show four different designs in which the steepness of the parabola is different. Indicate the distance between the focus and the directrix in each design.

4 Write an equation for each of the parabolic designs you test.

5 Make a blueprint of each of the four designs. Include the lengths of the struts and how far apart they are located.

Add the results of your work to your Investigation Folder.

Circles

What YOU'LL LEARN

- To write equations of circles, and
- to graph circles having certain properties.

Why IT'S IMPORTANT

You can graph circles to solve problems involving seismology and aviation.

F Y I

The strongest earthquake in American history measured 8.4 on the Richter scale. It occurred on March 27, 1964 near Prince William Sound, Alaska. It killed 131 people and caused an estimated $750 million in property damage.

APPLICATION
Seismology

When an earthquake occurs, the most serious property damage usually occurs at or near the center of the quake, called the *epicenter*. The damage is usually less severe as the distance from the epicenter increases. Shock waves radiate from the epicenter in a circular pattern. The earthquake in California on January 17, 1994, resulted in severe damage to buildings and roads near the epicenter, which was near the city of Northridge.

A **circle** is the set of all points in a plane that are equidistant from a given point in the plane, called the **center**. Any segment whose endpoints are the center and a point on the circle is a **radius** of the circle. *The measure of a radius is also called a radius.*

Assume that (x, y) names any point on the circle at the right. The center is at (h, k), and the radius is represented by r. We can find an equation for the circle by using the distance formula.

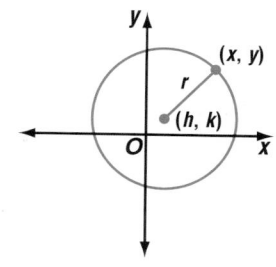

$$\sqrt{(x_2 - x_1)^2 + (y_2 - y_1)^2} = d$$

Substitute the coordinates of (x, y) and (h, k) into the formula. The distance from the center at (h, k) to a point at (x, y) on the circle is the radius r.

$$\sqrt{(x - h)^2 + (y - k)^2} = r$$

$$(x - h)^2 + (y - k)^2 = r^2 \quad \textit{Square each side.}$$

Equation of a Circle	The equation of a circle with center (h, k) and radius r units is $$(x - h)^2 + (y - k)^2 = r^2.$$

Example ❶ The University of Southern California (USC) is located about 4 kilometers west and about 4.5 kilometers south of downtown Los Angeles. A seismograph on the campus indicated that an earthquake occurred, and it is estimated that the epicenter of the quake was about 60 kilometers from the university. Assume that the origin of a coordinate plane is located at the center of Los Angeles. Write an equation of the set of points that could be the epicenter of the quake and draw the graph.

Any point that is 60 kilometers from the university could be the epicenter. So the circle whose center is at the university and whose radius is 60 kilometers is the solution set. The center of the circle is at $(-4, -4.5)$.

$$[x - (-4)]^2 + [y - (-4.5)]^2 = (60)^2$$
$$(x + 4)^2 + (y + 4.5)^2 = 3600$$

The equation is
$(x + 4)^2 + (y + 4.5)^2 = 3600$.

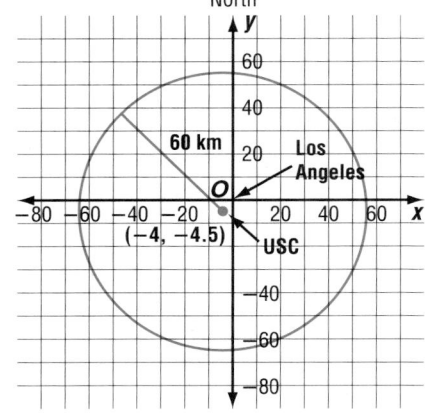

The equation $x^2 + y^2 + 2x - 12y = 35$ also describes a circle. If we were to complete the square for each variable and write the equation in the form $(x - h)^2 + (y - k)^2 = r^2$, the result would be $(x + 1)^2 + (y - 6)^2 = 72$. This circle has its center at $(-1, 6)$, and its radius is $\sqrt{72}$ or $6\sqrt{2}$ units long.

Example ❷ Find the center and radius of a circle with equation $x^2 + y^2 + 16x - 22y - 20 = 0$. Then graph the circle.

Solve by completing the square.

$$x^2 + y^2 + 16x - 22y - 20 = 0$$
$$x^2 + 16x + \square + y^2 - 22y + \square = 20 + \square + \square$$
$$x^2 + 16x + 64 + y^2 - 22y + 121 = 20 + 64 + 121$$
$$(x + 8)^2 + (y - 11)^2 = 205$$

The center of the circle is at $(-8, 11)$.

The radius is $\sqrt{205}$ or about 14.32 units.

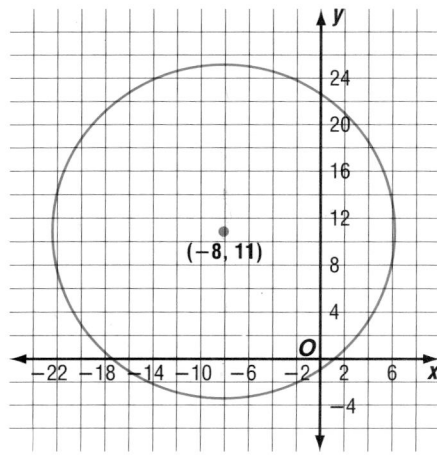

A line in the plane of a circle can intersect the circle in zero, one, or two points. A line that intersects the circle in exactly one point is said to be **tangent** to the circle. The line and the circle are tangent to each other at this point.

Example ❸ **Write an equation of the circle with its center at $(-6, 11)$ and that is tangent to the y-axis.**

To visualize the circle, draw a sketch. Since the circle is tangent to the y-axis, its radius is 6 units.

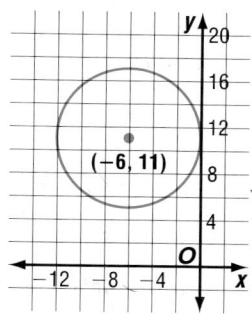

The equation is $(x + 6)^2 + (y - 11)^2 = 36$.

Example ❹ **Write an equation of a circle if the endpoints of a diameter are at $(-7, 11)$ and $(5, -10)$.**

The center of the circle is the midpoint of the diameter.

$$(h, k) = \left(\frac{x_1 + x_2}{2}, \frac{y_1 + y_2}{2}\right).$$

$$= \left(\frac{-7 + 5}{2}, \frac{11 + (-10)}{2}\right)$$

$$= \left(\frac{-2}{2}, \frac{1}{2}\right) \text{ or } \left(-1, \frac{1}{2}\right)$$

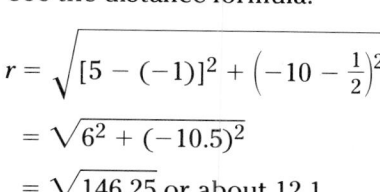

The radius is the distance from the center to one endpoint of the diameter. Use the distance formula.

$$r = \sqrt{[5 - (-1)]^2 + \left(-10 - \frac{1}{2}\right)^2}$$

$$= \sqrt{6^2 + (-10.5)^2}$$

$$= \sqrt{146.25} \text{ or about } 12.1$$

The radius of the circle is about 12.1 units, and r^2 is 146.25.

An equation of the circle is $(x + 1)^2 + \left(y - \frac{1}{2}\right)^2 = 146.25$.

EXPLORATION
GRAPHING CALCULATORS

Graph the two equations below on the same screen.

$$y = \sqrt{9 - x^2} \qquad\qquad y = -\sqrt{9 - x^2}$$

Your Turn

a. Describe the graph formed by the union of these two graphs.

b. Write an equation for the union of the two graphs.

c. Most graphing calculators cannot graph the equation $x^2 + y^2 = 49$ directly. Describe a way to use a graphing calculator to graph the equation. Then graph the equation.

d. Use a graphing calculator to graph the equation $(x - 2)^2 + (y + 1)^2 = 4$.

e. Do you think it is easier to graph the equation in Step d using graph paper and a pencil or using a graphing calculator? Explain.

CHECK FOR UNDERSTANDING

Communicating Mathematics

Study the lesson. Then complete the following.

1. **Describe** the similarities and differences between the graphs of $(x + 3)^2 + (y - 4)^2 = 16$ and $(x - 3)^2 + (y - 2)^2 = 16$.

2. **Concentric circles** are defined as circles with the same center, but *not* necessarily the same radius. Write equations of two concentric circles. Then graph the circles.

3. **Explain** why the phrase "in a plane" is included in the definition of a circle. What would be defined if the phrase were *not* included?

4. **You Decide**   Jean says that you can take the square root of each side of an equation. Therefore, she decides that $(x - 2)^2 + (y + 3)^2 = 36$ and $(x - 2) + (y + 3) = 6$ are equivalent equations. Marco says that the equations are *not* equivalent. Who is right? Explain your answer.

5. How many axes of symmetry does a circle have? Explain your answer.

MATH JOURNAL

6. The circle with equation $(x - a)^2 + (y - b)^2 = r^2$ lies in the first quadrant and is tangent to both the *x*-axis and the *y*-axis. Make a sketch of the circle. Describe the possible values of *a*, *b*, and *r*. Do the same for a circle in Quadrants II, III, and IV.

Guided Practice

Write an equation for each circle if the coordinates of the center and the length of the radius are given.

7. center $(-12, 0)$, $r = \sqrt{23}$ units

8. center $(8, -9.5)$, $r = \frac{1}{2}$ unit

Find the coordinates of the center and the radius of each circle whose equation is given. Then draw the graph.

9. $(x - 4)^2 + (y - 1)^2 = 9$

10. $x^2 + (y - 14)^2 = 34$

11. $(x - 4)^2 + y^2 = \frac{16}{25}$

12. $\left(x + \frac{2}{3}\right)^2 + \left(y - \frac{1}{2}\right)^2 = \frac{8}{9}$

13. $x^2 + y^2 + 8x - 6y = 0$

14. $x^2 + y^2 + 4x - 8 = 0$

15. Write an equation for the graph below.

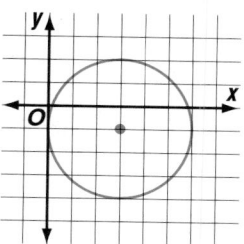

16. Write an equation for a circle that has its center at $(4, -2)$ and passes through $(5, 3)$.

EXERCISES

Practice

Write an equation for each circle if the coordinates of the center and length of the radius are given.

$x^2 + (y-3)^2 = 49$

17. center, $(-1, -5)$; $r = 2$ units

18. center, $(0, 3)$; $r = 7$ units

19. center, $(-8, 7)$; $r = \frac{1}{2}$ unit

20. center, $(-3, -9)$; $r = \frac{5}{6}$ unit

21. center, $(0.5, 0.7)$; $r = 13.5$ units

22. center, $(\sqrt{2}, 3\sqrt{7})$; $r = 0.25$ unit

Find the coordinates of the center and the radius of each circle whose equation is given. Then draw the graph.

23. $x^2 + (y + 2)^2 = 4$

24. $x^2 + y^2 = 144$

25. $(x - 3)^2 + (y - 1)^2 = 25$

26. $(x + 3)^2 + (y + 7)^2 = 81$

27. $(x - 3)^2 + y^2 = 16$

28. $(x - 3)^2 + (y + 7)^2 = 50$

29. $(x + \sqrt{5})^2 + y^2 - 8y = 9$

30. $x^2 + y^2 + 4x = 9$

31. $x^2 + y^2 + 6y = -50 - 14x$

32. $x^2 + y^2 - 6y - 16 = 0$

33. $x^2 + y^2 + 2x - 10 = 0$

34. $x^2 + y^2 - 18x - 18y + 53 = 0$

35. $4x^2 + 4y^2 + 36y + 5 = 0$

36. $x^2 + y^2 + 9x - 8y + 4 = 0$

37. $x^2 + y^2 - 3x + 8y = 20$

38. $x^2 - 12x + 84 = -y^2 + 16y$

39. $x^2 + y^2 + 2x + 4y = 9$

40. $x^2 + 2\sqrt{7}x + 7 + (y - \sqrt{11})^2 = 11$

Write an equation for each graph.

41.

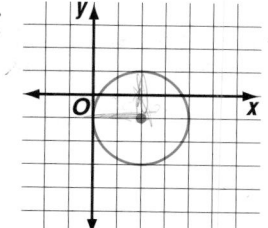

42.

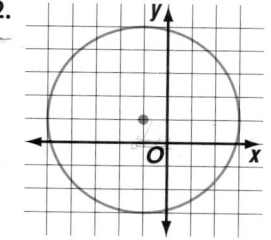

43.

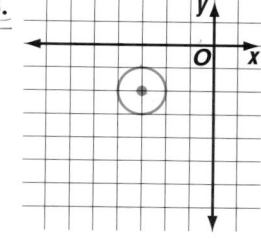

Write an equation for each circle described below.

44. The circle has its center at $(8, -9)$ and passes through the point at $(21, 22)$.

45. The circle passes through the origin and has its center at $\left(-\sqrt{13}, 42\right)$.

46. The endpoints of a diameter are at $(11, 18)$ and $(-13.5, -19)$.

47. The circle is tangent to the y-axis and has its center at $(-8, -7)$.

48. The circle is tangent to $x = -3$, $x = 5$, and the x-axis. The center of the circle is in the first quadrant.

49. The circle is tangent to the y-axis and has a radius of 3 units. The center of the circle is in the third quadrant and lies on the graph of $y = 2x$.

50. A *unit circle* is a circle with a radius of 1 unit.
   a. Write an equation for a unit circle with its center at the origin.
   b. Graph the equation.
   c. Find the area of the circle.
   d. Find the circumference of the circle.

51. Draw the circle whose equation is $x^2 + 6x + y^2 + 4y + 9 = 0$. What is the equation of the line that is tangent to the circle at $(-3, 0)$?

52. Geometry The equation for the circle at the right is $x^2 + y^2 = 25$. The triangle is formed by the y-axis and the lines whose equations are
$y = -\frac{1}{2}x + 5$ and $y = 2x - 5$.

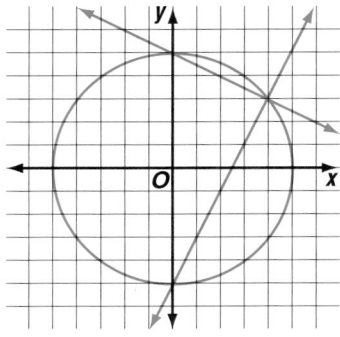

   a. Find the coordinates of the vertices of the triangle.
   b. Recall from geometry that if a triangle is inscribed in a circle and one of its sides is a diameter of the circle, the triangle is a right triangle. Using this information, do you think the triangle in the diagram is a right triangle? Explain.

Graphing Calculator

Write the equations needed to graph each equation on a graphing calculator. Then graph the equations on a graphing calculator.

53. $x^2 + y^2 = 4$

54. $(x + 3)^2 + (y - 1)^2 = 8$

Critical Thinking

55. Consider the graphs whose equations are of the form $(x - 3)^2 + (y - a)^2 = 64$. Assign three different values for a and graph each equation. Describe all graphs whose equations have this form.

Applications and Problem Solving

56. Air Traffic Control The radar for a county airport control tower is located at $(5, 10)$ on the map. It can detect a plane up to 20 miles away. Write an equation for outside limits that a plane can be detected.

57. Satellites A satellite is in a circular orbit 25,000 miles above Earth.
   a. Write an equation for the orbit of this satellite if the origin is at the center of Earth. Use 8000 miles as the diameter of Earth.
   b. Draw a sketch of Earth and the orbit to scale. Label your sketch.

58. Sports The maintenance personnel for the Onalaska, Wisconsin, School District needed to mark off an arc, $\overarc{AB}$, as a boundary on an empty lot to be converted to a baseball field. The figure at the right shows the details. The center C of the circle was located within a building on an adjacent lot. The supervisor needed to know what set of points the crew might connect to form the arc.

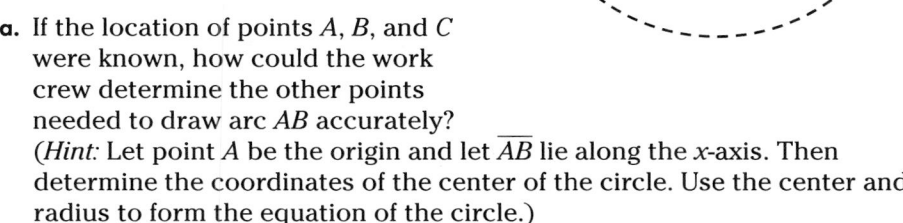

a. If the location of points A, B, and C were known, how could the work crew determine the other points needed to draw arc AB accurately? (*Hint:* Let point A be the origin and let $\overline{AB}$ lie along the x-axis. Then determine the coordinates of the center of the circle. Use the center and radius to form the equation of the circle.)

b. Explain how the equation can be used to draw arc AB.

Mixed Review

59. Write $y^2 = 6x$ in the form $x = a(y - k)^2 + h$. (Lesson 7–2)

60. Solve $x^2 - 3x + 1 = 0$. Then find the sum and product of the roots to check your solutions. (Lesson 6–5)

61. Simplify $\dfrac{3}{4 - i}$. (Lesson 5–10)

62. Business Dawn is writing a computer program to find the salaries of her employees after their annual raise. The percentage of increase is represented by p. Marty's salary is $23,450 now. Write a polynomial to represent Marty's salary in one year and another to represent Marty's salary after three years. Assume that the rate of increase will be the same for the three years. (Lesson 5–2)

63. Solve $\begin{vmatrix} x^2 & x \\ 3 & 1 \end{vmatrix} = 4$. (Lesson 4–4)

64. Solve the system of equations. (Lesson 3–7)

$2a + b = 2$
$5a = 15$
$a + b + c = -1$

65. The sum of Kari's age and her mother's age is 52. Kari's mother is 20 years older than Kari. How old is each? (Lesson 3–2)

66. What is the slope of the line perpendicular to the line that passes through $(5, 1)$ and $(8, 2)$? (Lesson 2–3)

67. Health The optimum heart rate is the rate that a person should achieve during exercise to be most beneficial. The prediction equation $r = 0.6(220 - a)$, where a represents age, can be used to find a person's optimum heart rate r. If Josh is 20 years old, find his optimum heart rate. (Lesson 2–2)

68. Simplify $\sqrt{9} \div \sqrt{4}$. (Lesson 1–1)

MODELING MATHEMATICS

7-4A Drawing Ellipses

A Preview of Lesson 7-4

Materials: thumbtacks string cardboard

grid paper

To draw an ellipse, tie a knot in a piece of string and loop it around the thumbtacks as shown.

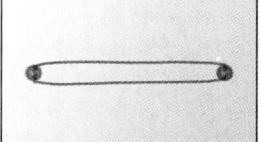

Place your pencil in the string as shown in the diagram. Begin at some point. Keep the string tight and draw a curve. Continue drawing until you return to your starting point. Pick a point on the circle and measure the distance between the point and each thumbtack. Add these numbers together. Pick two more points and repeat this process. What is true about the sum of the distances from the two thumbtacks for all points on the curve?

The two thumbtacks are located at the foci of the ellipse. *Foci is the plural of focus.*

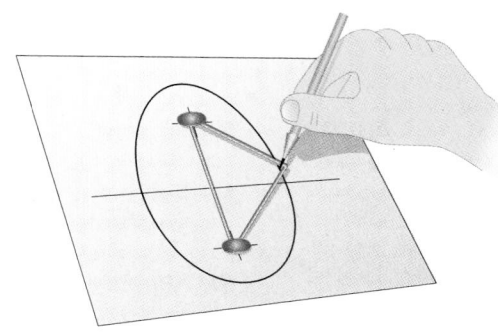

Activity Place a large piece of grid paper on the cardboard.

a. Place the thumbtacks at (8, 0) and (−8, 0). Choose a string of appropriate length. Draw the ellipse.

b. Repeat part a, but place the thumbtacks at (5, 0) and (−5, 0). Use the same piece of string and draw an ellipse. How does this ellipse compare to the one drawn in part a?

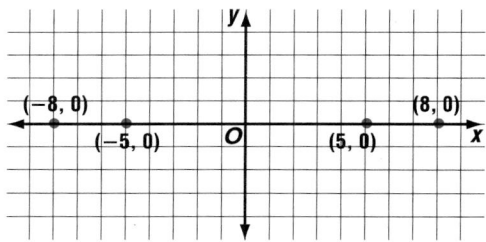

Model Place the thumbtacks at each set of points and draw an ellipse.

1. A(12, 0) and B(−12, 0)   **2.** C(2, 0) and D(−2, 0)   **3.** E(14, 4) and F(−10, 4)

Write

4. Write a paragraph describing what happens to the shape of an ellipse when each of the following changes are made.

a. The thumbtacks are moved closer together or farther apart.

b. Both thumbtacks are placed at the same point.

c. The length of the piece of string is changed.

Ellipses

Museums

A cross section of the whispering chamber at the Museum of Science and Industry in Chicago is shaped like an **ellipse.** In this chamber, a person standing at a focus point can hear a person standing at the other focus point whispering, even though they are 43.42 feet apart.

Museum of Science and Industry, Chicago, Illinois

What YOU'LL LEARN

• To write equations of ellipses, and

• to graph ellipses having certain properties.

Why IT'S IMPORTANT

You can graph ellipses to solve problems involving medicine and astronomy.

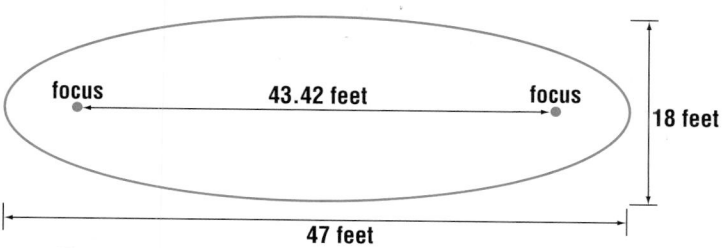

focus   43.42 feet   focus   18 feet

47 feet

The architect for the whispering chamber used an important property of ellipses. Rays emanating from one of the two foci of an ellipse are reflected to the other focus. As you discovered in Lesson 7–4A, the sum of the distances from the two foci of an ellipse is always the same.

Definition of Ellipse	**An ellipse is the set of all points in a plane such that the sum of the distances from the foci is constant.**

The ellipse at the right has foci at $(8, 0)$ and $(-8, 0)$. The sum of the distances from any point with coordinates (x, y) on the ellipse to the foci is 24 units.

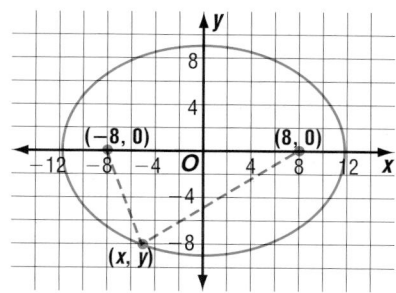

Use the distance formula and the definition of an ellipse to find the equation of the ellipse. Let (x, y) be the coordinates of any point on the ellipse. The sum of the distance between the points at (x, y) and $(8, 0)$ and the distance between the points at (x, y) and $(-8, 0)$ is 24 units.

$$\sqrt{(x + 8)^2 + y^2} + \sqrt{(x - 8)^2 + y^2} = 24$$

$$\sqrt{(x + 8)^2 + y^2} = 24 - \sqrt{(x - 8)^2 + y^2}$$

$$(x + 8)^2 + y^2 = 576 - 48\sqrt{(x - 8)^2 + y^2} + (x - 8)^2 + y^2 \quad \textit{Square each side.}$$

$$x^2 + 16x + 64 + y^2 = 576 - 48\sqrt{(x - 8)^2 + y^2} + x^2 - 16x + 64 + y^2$$

$$32x - 576 = -48\sqrt{(x - 8)^2 + y^2} \quad\quad\quad\quad\quad \textit{Simplify.}$$

$$2x - 36 = -3\sqrt{(x - 8)^2 + y^2} \quad\quad\quad\quad\quad \textit{Divide each side by 16, the GCF.}$$

$$4x^2 - 144x + 1296 = 9[(x - 8)^2 + y^2] \quad\quad\quad\quad \textit{Square each side.}$$

$$4x^2 - 144x + 1296 = 9x^2 - 144x + 576 + 9y^2 \quad\quad \textit{Distributive property}$$

$$5x^2 + 9y^2 = 720 \quad\quad\quad\quad\quad\quad\quad\quad\quad \textit{Simplify.}$$

$$\frac{x^2}{144} + \frac{y^2}{80} = 1 \quad\quad\quad\quad\quad\quad\quad\quad \textit{Divide each side by 720.}$$

The equation of this ellipse is $\frac{x^2}{144} + \frac{y^2}{80} = 1$.

Every ellipse has two axes of symmetry. The points at which the ellipse intersect the axes define two segments with endpoints on the ellipse. The longer segment is called the **major axis**, and the shorter segment is called the **minor axis**. The foci always lie on the major axis. The intersection of the two axes is the **center** of the ellipse.

Study the ellipse at the right. The sum of the distances from the foci to any point on the ellipse is 2a units. The distance from the center to either focus is c units. We can calculate the value of b by using the Pythagorean theorem, $b^2 = a^2 - c^2$.

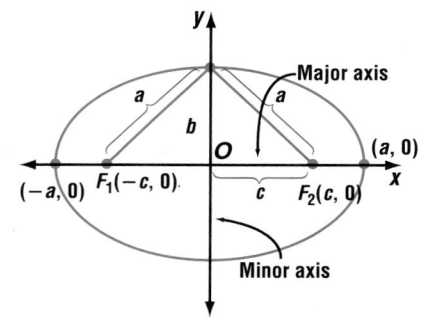

If we found the equation of this general ellipse, we would have the standard equation of an ellipse.

The length of the major axis is 2a.
The length of the minor axis is 2b.
Notice that a > b.

Standard Equations of Ellipses with Center at the Origin	• If an ellipse has foci at (−c, 0) and (c, 0) and the sum of the distances from the foci to any point on the ellipse is 2a units, then the standard equation of an ellipse is $\frac{x^2}{a^2} + \frac{y^2}{b^2} = 1$, where $b^2 = a^2 - c^2$. *In this case, the major axis is horizontal.* • If an ellipse has foci at (0, −c) and (0, c) and the sum of the distances from the foci to any point on the ellipse is 2a units, then the standard equation of an ellipse is $\frac{x^2}{b^2} + \frac{y^2}{a^2} = 1$, where $b^2 = a^2 - c^2$. *In this case, the major axis is vertical.*

For the equation of an ellipse, $a^2 > b^2$. You can decide if the foci are on the x-axis or the y-axis by observing the equation. If a^2 is the denominator of the x^2 term, the foci are on the x-axis. If a^2 is the denominator of the y^2 term, the foci are on the y-axis.

Example ➊ **Find the coordinates of the foci and the lengths of the major and minor axes of an ellipse whose equation is $16x^2 + 4y^2 = 144$. Then draw the graph.**

Write the equation in standard form.

$16x^2 + 4y^2 = 144$

$\dfrac{x^2}{9} + \dfrac{y^2}{36} = 1$ *Divide each side by 144.*

Since $36 > 9$, the foci are on the y-axis, with $a = 6$ and $b = 3$.

$b^2 = a^2 - c^2$

$9 = 36 - c^2$

$c^2 = 27$

$c = 3\sqrt{3}$ or about 5.2

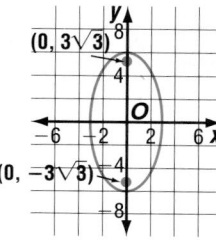

The foci are at $(0, 3\sqrt{3})$ and $(0, -3\sqrt{3})$.

The length of the major axis, $2a$, is 12 units.

The length of the minor axis, $2b$, is 6 units.

You can use this information to draw the ellipse.

Example ➋ **Write the equation of the ellipse shown below.**

The length of the major axis is the distance between the points at $(7, 0)$ and $(-7, 0)$. This distance is 14 units.

$2a = 14$

$a = 7$

The foci are located at $(4, 0)$ and $(-4, 0)$. The value of c is 4.

$b^2 = a^2 - c^2$

$b^2 = 7^2 - 4^2$ or 33

Now, write the equation in the form   *The major axis is horizontal.*

$\dfrac{x^2}{a^2} + \dfrac{y^2}{b^2} = 1$, where $a^2 = 49$ and $b^2 = 33$.

The equation of the ellipse is $\dfrac{x^2}{49} + \dfrac{y^2}{33} = 1$

Planets, satellites, moons, and comets all have orbits that have the shape of a conic section. The most common orbit is in the form of an ellipse.

Example **3**

APPLICATION

Astronomy

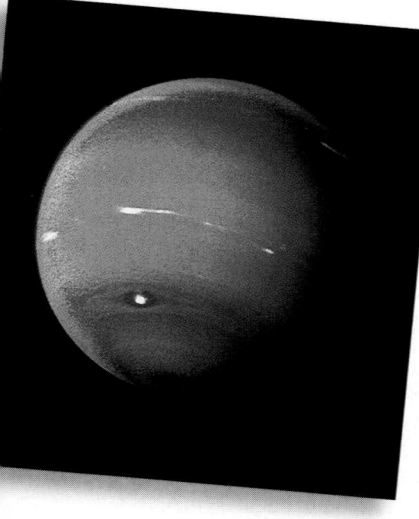

Mercury follows an ellipse-shaped, or *elliptical*, orbit around the sun. At its closest point, Mercury is 29.0 million miles from the center of the sun. At its farthest point, Mercury is 43.8 million miles from the center of the sun. Assume that the center of this orbit lies at the origin, the center of the sun is a focus of this ellipse, and the sun lies on the *x*-axis.

a. Draw a sketch of Mercury's orbit.
b. Write an equation for the orbit.

a. A sketch of Mercury's orbit is shown at the right.

b. The length of the major axis is 29.0 + 43.8 or 72.8 million miles. Use this information to find the value of *a*.

$$2a = 72,800,000$$
$$a = 36,400,000$$

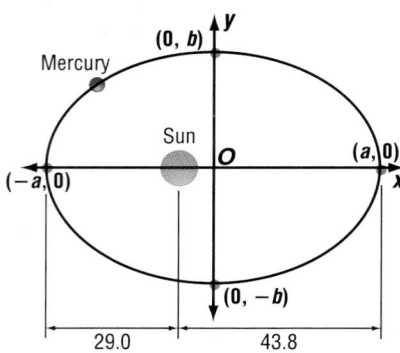

The distance from the center of the orbit to the center of the sun is 36.4 − 29.0 or 7.4 million miles. This is the value of *c*, the distance from the center to the focus.

$$b^2 = a^2 - c^2$$
$$= (36,400,000)^2 - (7,400,000)^2$$
$$= 1,324,960,000,000,000 - 54,760,000,000,000$$
$$= 1,270,200,000,000,000 \text{ or } 1.2702 \times 10^{15}$$

Now we can write the equation.

$$\frac{x^2}{a^2} + \frac{y^2}{b^2} = 1$$

$$\frac{x^2}{(36,400,000)^2} + \frac{y^2}{1,270,200,000,000,000} = 1$$

$$\frac{x^2}{1.32496 \times 10^{15}} + \frac{y^2}{1.2702 \times 10^{15}} = 1 \quad (36,400,000)^2 = 1.32496 \times 10^{15}$$

An equation for the orbit is $\dfrac{x^2}{1.32496 \times 10^{15}} + \dfrac{y^2}{1.2702 \times 10^{15}} = 1$.

A circle can be used to find the foci of an ellipse.

MODELING MATHEMATICS

Locating Foci

Materials: grid paper compass

Your Turn

a. On a coordinate plane, draw an ellipse with its center at the origin. Let the endpoints of the major axis be at (−9, 0) and (9, 0), and let the endpoints of the minor axis be at (0, −5) and (0, 5).

b. Estimate the locations of the foci and mark these points.

c. Use a compass to draw a circle with center at (0, 0) and radius of 9 units.

d. Draw the line *y* = 5 and mark the points at which the line intersects the circle.

e. Draw perpendicular lines from the points of intersection to the *x*-axis. The foci are located at the points where the perpendicular lines intersect the *x*-axis.

f. Draw another ellipse and locate its foci using this method. Why does this method work?

LOOK BACK

You can refer to Lesson 4-2 for more information on translations.

An ellipse with its center at the origin is represented by an equation in the form $\frac{x^2}{a^2} + \frac{y^2}{b^2} = 1$ or $\frac{x^2}{b^2} + \frac{y^2}{a^2} = 1$. The ellipse could be translated h units to the left or right and k units up or down. This would move the center to the point (h, k). Such a move would be equivalent to replacing x with $(x - h)$ and replacing y with $(y - k)$.

Standard Equations of Ellipses with Center at (h, k)

- The standard equation of an ellipse with its center at (h, k) and with a horizontal major axis is $\frac{(x - h)^2}{a^2} + \frac{(y - k)^2}{b^2} = 1$.

- The standard equation of an ellipse with center at (h, k) and with a vertical major axis is $\frac{(x - h)^2}{b^2} + \frac{(y - k)^2}{a^2} = 1$.

Example ④ Graph $\frac{(x + 4)^2}{25} + \frac{(y - 3)^2}{4} = 1$.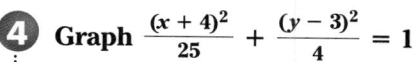

The graph has the same shape as the graph of $\frac{x^2}{25} + \frac{y^2}{4} = 1$. The center, however, is at $(-4, 3)$.

Draw the graph of $\frac{x^2}{25} + \frac{y^2}{4} = 1$.
Then translate the graph 4 units to the left and 3 units up.

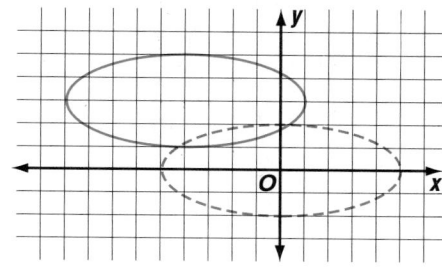

Sometimes equations of ellipses are written in expanded form as they were for circles. You need to complete the square for each variable to write the equation in standard form.

Example ⑤ An equation of an ellipse is $x^2 + 9y^2 - 4x + 54y + 49 = 0$. Find the coordinates of the center and foci and the lengths of the major and minor axes. Then draw the graph.

$$x^2 + 9y^2 - 4x + 54y + 49 = 0$$
$$(x^2 - 4x + \square) + 9(y^2 + 6y + \square) = -49 + \square + 9(\square) \quad \textit{Complete the squares.}$$
$$(x^2 - 4x + 4) + 9(y^2 + 6y + 9) = -49 + 4 + 9(9)$$
$$(x - 2)^2 + 9(y + 3)^2 = 36 \qquad \textit{Standard form}$$
$$\frac{(x - 2)^2}{36} + \frac{(y + 3)^2}{4} = 1$$

The coordinates of the center are $(2, -3)$.

Since $36 > 4$, the foci are on the horizontal axis.

$b^2 = a^2 - c^2$
$4 = 36 - c^2 \quad$ *$a = 6$ and $b = 2$*
$c^2 = 32$
$c = 4\sqrt{2}$ or about 5.66

(continued on the next page)

The foci are at $(2 + 4\sqrt{2}, -3)$ and $(2 - 4\sqrt{2}, -3)$ or about $(7.66, -3)$ and $(-3.66, -3)$.

Since $a = 6$, the length of the major axis is $2a$ or 12 units. Since $b = 2$, the length of the minor axis is $2b$ or 4 units.

Use this information to draw the graph.

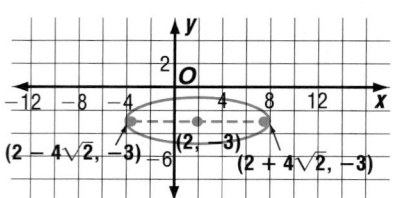

Communicating Mathematics

Study the lesson. Then complete the following.

1. **Describe** how you can tell from the equation of an ellipse which is the major and minor axis.

2. **Describe** how circles and ellipses are related.

3. **Explain** why one side of an equation is always 1 when the equation of an ellipse is written in standard form.

4. **Explain** how the length of the major axis was determined in Example 3.

MODELING MATHEMATICS

5. Draw an ellipse. Then use a circle and lines to find the foci of the ellipse.

Guided Practice

The equation of an ellipse is given. Find the coordinates of the center and state whether the major axis is vertical or horizontal.

6. $\dfrac{x^2}{5} + \dfrac{y^2}{20} = 1$

7. $\dfrac{(x-4)^2}{42} + \dfrac{(y+6)^2}{23} = 1$

Write an equation for each ellipse in standard form.

8. $10x^2 + 2y^2 = 40$

9. $x^2 + 6y^2 - 2x + 12y - 23 = 0$

Find the coordinates of the center and foci and the lengths of the major and minor axes for each ellipse whose equation is given. Then draw the graph.

10. $\dfrac{x^2}{9} + \dfrac{y^2}{18} = 1$

11. $\dfrac{(x-1)^2}{20} + \dfrac{(y+2)^2}{4} = 1$

12. $4x^2 + 8y^2 = 32$

13. $x^2 + 25y^2 - 8x + 100y + 91 = 0$

14. Write an equation for the graph at the right.

15. Write an equation of an ellipse whose endpoints of the major axis are at $(0, 10)$ and $(0, -10)$ and whose foci are at $(0, 8)$ and $(0, -8)$.

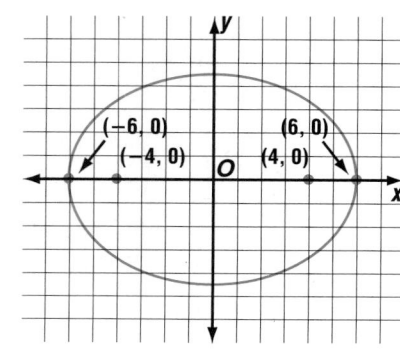

Practice **Write an equation for each ellipse.**

16.

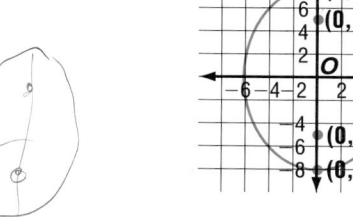

17.

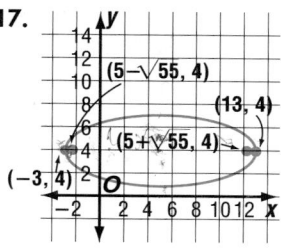

18.

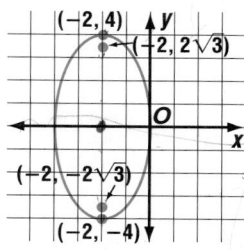

Find the coordinates of the center and foci, and the lengths of the major and minor axes for each ellipse whose equation is given. Then draw the graph.

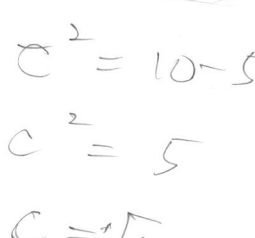

19. $\dfrac{x^2}{5} + \dfrac{y^2}{10} = 1$

20. $\dfrac{x^2}{25} + \dfrac{y^2}{9} = 1$

21. $\dfrac{(x-5)^2}{121} + \dfrac{(y+11)^2}{144} = 1$

22. $\dfrac{(x+8)^2}{144} + \dfrac{(y-2)^2}{81} = 1$

23. $36x^2 + 81y^2 = 2916$

24. $27x^2 + 9y^2 = 81$

25. $16x^2 + 9y^2 = 144$

26. $3x^2 + 9y^2 = 27$

27. $3x^2 + y^2 + 18x - 2y + 4 = 0$

28. $x^2 + 5y^2 + 4x - 70y + 209 = 0$

29. $7x^2 + 3y^2 - 28x - 12y = -19$

30. $16x^2 + 25y^2 + 32x - 150y = 159$

31. $9x^2 + 16y^2 - 18x + 64y = 71$

32. $4x^2 + 9y^2 + 16x - 18y - 11 = 0$

Write an equation for each ellipse described below.

33. The endpoints of the major axis are at $(10, 2)$ and $(-8, 2)$. The foci are at $(6, 2)$ and $(-4, 2)$.

34. The major axis is 20 units in length and parallel to the y-axis. The minor axis is 6 units in length. The center is located at $(4, 2)$.

35. The foci are at $(12, 0)$ and $(-12, 0)$. The endpoints of the minor axis are at $(0, 5)$ and $(0, -5)$.

36. The endpoints of the major axis are at $(-11, 5)$ and $(7, 5)$. The endpoints of the minor axis are at $(-2, 9)$ and $(-2, 1)$.

37. The endpoints of the major axis are at $(2, 12)$ and $(2, -4)$. The endpoints of the minor axis are at $(4, 4)$ and $(0, 4)$.

38. The major axis is 16 units long and parallel to the x-axis. The center is at $(5, 4)$ and minor axis is 9 units long.

Critical Thinking **For each equation, replace k with four different values greater than one. Graph the four resulting equations on a graphing calculator. Describe what happens to the graph as the value of k increases.**

39. $\dfrac{x^2}{k^2} + \dfrac{y^2}{1} = 1$

40. $\dfrac{x^2}{25} + \dfrac{y^2}{4} = k$

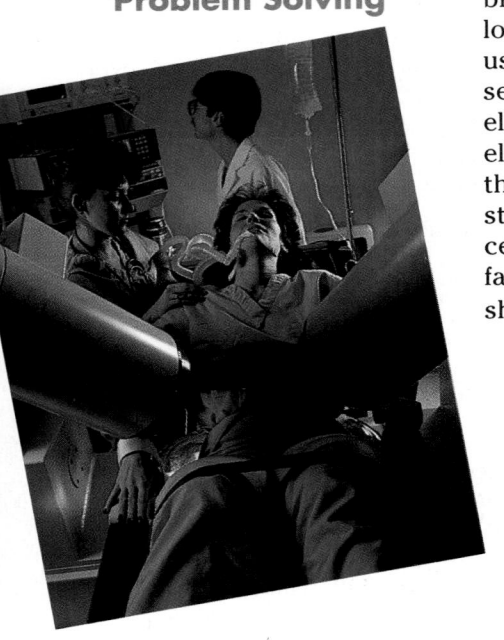

41. Medicine   A lithotriper can be used to break up kidney stones so that they no longer pose a health threat. This instrument uses the properties of ellipses. An electrode sends shock waves out from one focus of an ellipse. The waves are then reflected off an elliptical surface to the other focus where the kidney stone is located, shattering the stone. Suppose that the length of the major axis of the ellipse is 40 centimeters and the length of the minor axis is 20 centimeters. How far from the kidney stone should the electrode be placed in order to shatter it?

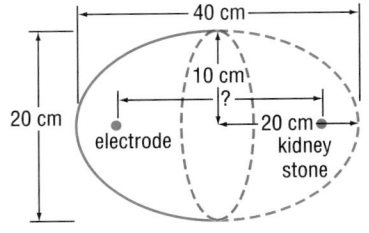

42. National Landmarks   The United States Capitol building contains an elliptical room. It is 96 feet in length and 46 feet in width.

   a. Write an equation to describe the shape of the room. Assume that it is centered at the origin and the major axis is horizontal.

   b. John Quincy Adams discovered that he could overhear the conversations being held at the opposing party leader's desk if he stood in a certain spot in the elliptical chamber. Describe the position of the desk and how far away Adams had to stand to overhear.

43. Astronomy   Find the equation of the elliptical orbit of Mars as it travels around the sun. Assume the radius of the sun is 400,000 miles.

Planet	Shortest Distance from Sun (miles)	Greatest Distance from Sun (miles)
Mercury	28,600,000	43,400,000
Venus	66,800,000	67,700,000
Earth	91,400,000	94,500,000
Mars	128,500,000	155,000,000

44. Museums   Find the equation of the cross section of the whispering chamber described in the application at the beginning of this lesson. The length of the major axis is 47 feet, and the length of the minor axis is 18 feet. Assume that the center of the ellipse is at the origin and the major axis is horizontal.

45. Space Science   The space shuttle travels in an elliptical orbit around Earth. The center of Earth is one focus of the ellipse, and the high and low points of the orbit are both on the major axis. Suppose a shuttle is orbiting Earth so that its high point is 200 miles above Earth's surface and its low point is 100 miles above the surface. Let the x-axis be the major axis.

   a. Find the equation of the path of the shuttle, using the center of the ellipse, not the center of Earth, as the origin. Note that Earth's diameter is about 8000 miles. Draw a diagram and label it.

   b. Find the equation of the path of the shuttle, using the center of Earth as the origin. (*Hint:* Where would the center of the ellipse be then?)

46. Write an equation for the circle whose center is at (6, 2) and whose radius is 5 units.   (Lesson 7–3)

47. Graph $y > x^2 - 7x + 10$.   (Lesson 6–7)

48. Simplify $\left(\sqrt[6]{5a^{\frac{7}{4}}b^{-\frac{2}{3}}}\right)^{12}$. (Lesson 5–7)

49. Technology Computers today are built to perform millions of operations per second. It takes an electric impulse one billionth of a second to travel 8 inches. One billionth of a second, also known as a *nanosecond*, is the measurement used to measure time on computers. One nanosecond equals 10^{-9} seconds. How much time does it take an electric impulse to travel half an inch? Write your answer in scientific notation. (Lesson 5–1)

50. Solve the system of equations by using matrices. (Lesson 4–6)

$2x - y - 5z = 3$

$x + 4y - 2z = 3$

$5x + 3y + 2z = 1$

51. Solve the system of equations by using Cramer's rule. (Lesson 3–3)

$6a + 7b = -10.15$

$9.2a - 6b = 69.944$

52. Find the slope of the line perpendicular to the line that passes through $(0, 0)$ and $(4, -2)$. (Lesson 2–4)

53. Statistics During a cold spell lasting 43 days, the following high temperatures (in °F) were recorded in Chicago. Find the median, mode, and mean of the temperatures. (Lesson 1–3)

26	17	12	5	4	25	17	23	13	6	25
19	27	22	26	20	31	24	12	27	16	27
16	30	7	31	16	5	29	18	16	22	29
8	31	13	24	5	-7	20	29	18	12	

SELF TEST

Find the distance between each pair of points with the given coordinates. (Lesson 7–1)

1. $(9, 5), (4, -7)$ **2.** $(0, -5), (10, -3)$

Find the midpoint of each line segment if the coordinates of the endpoints are given. (Lesson 7–1)

3. $(8, 0), (-5, 12)$ **4.** $(5, -7), (3, -1)$

Name the coordinates of the vertex and focus, the equations of the axis of symmetry and directrix, and the direction of opening of the parabola with the given equation. Then find the length of the latus rectum and graph the parabola. (Lesson 7–2)

5. $y^2 = 6x$ **6.** $y = x^2 + 8x + 20$

Find the coordinates of the center and the radius of each circle whose equation is given. Then draw the graph. (Lesson 7–3)

7. $x^2 + (y - 4)^2 = 49$ **8.** $3x^2 + 3y^2 + 6y + 9x = 2$

9. An equation of an ellipse is $8x^2 + 4y^2 - 16x - 20y = 7$. Write this equation in standard form. (Lesson 7–4)

10. Write an equation of an ellipse whose foci are at $(3, 8)$ and $(3, -6)$ and whose major axis is 18 units long. (Lesson 7–4)

Hyperbolas

What YOU'LL LEARN

- To write equations of hyperbolas, and
- to graph hyperbolas having certain properties.

Why IT'S IMPORTANT

You can graph hyperbolas to solve problems involving forestry and navigation.

APPLICATION
Navigation

During World War I and II, the Long Range Navigational system (LORAN) was developed and used. This system is based on the shape of a **hyperbola**.

Two stations send out different signals at the same time. A ship receives these signals and notes the difference between the time it received one of the signals and the time it received the other. This information is used to locate the ship on a hyperbola with foci located at the two stations. Another set of signals can locate the ship on another hyperbola, and the ship's location is the intersection of the two hyperbolas.

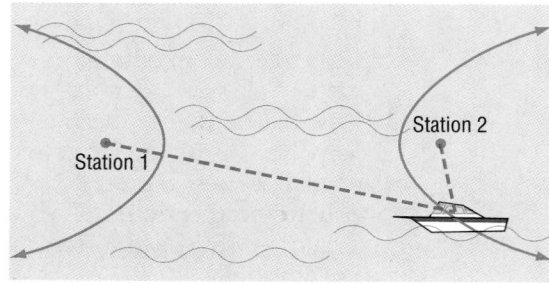

Since this system does *not* rely on land sightings, it can be used for successful navigation at night or for long missions over the ocean.

GLOBAL CONNECTIONS

In 1044, Tseng Kong-liang described the use of magnetized iron "fish" that float in water and can be used to find south. The Chinese began to use the south-pointing compass for navigation about this time. Later, its use spread to the Arab world and then to Europe.

Definition of Hyperbola	A hyperbola is the set of all points in a plane such that the absolute value of the difference of the distances from any point on the hyperbola to two given points, called the *foci*, is constant.

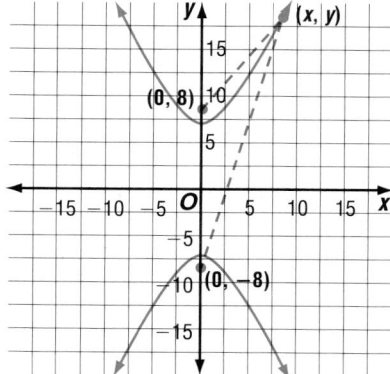

The hyperbola at the left has foci at $(0, 8)$ and $(0, -8)$. The absolute value of the differences of the distances from the foci to any point on the hyperbola is 14. We can use the distance formula and the definition of a hyperbola to find the equation of this hyperbola.

Let (x, y) be the coordinates of any point on the hyperbola. The distance between the points at (x, y) and $(0, 8)$ minus the distance between the points at (x, y) and $(0, -8)$ is ± 14 units.

$$\underset{\substack{\text{distance between} \\ (x,\,y) \text{ and } (0,\,8)}}{} - \underset{\substack{\text{distance between} \\ (x,\,y) \text{ and } (0,\,-8)}}{} = \pm 14$$

$$\sqrt{x^2 + (y-8)^2} - \sqrt{x^2 + (y+8)^2} = \pm 14$$

$$\sqrt{x^2 + (y-8)^2} = \pm 14 + \sqrt{x^2 + (y+8)^2}$$

$$x^2 + (y-8)^2 = 196 \pm 28\sqrt{x^2 + (y+8)^2} + x^2 + (y+8)^2 \qquad \textit{Square each side.}$$

$$x^2 + y^2 - 16y + 64 = 196 \pm 28\sqrt{x^2 + (y+8)^2} + x^2 + y^2 + 16y + 64$$

$$-32y - 196 = \pm 28\sqrt{x^2 + (y+8)^2} \qquad \textit{Simplify.}$$

$$8y + 49 = \pm 7\sqrt{x^2 + (y+8)^2} \qquad \textit{Divide each side by } -4.$$

$$64y^2 + 784y + 2401 = 49[x^2 + (y+8)^2] \qquad \textit{Square each side.}$$

$$64y^2 + 784y + 2401 = 49x^2 + 49y^2 + 784y + 3136 \qquad \textit{Distributive property}$$

$$15y^2 - 49x^2 = 735 \qquad \textit{Simplify.}$$

$$\frac{y^2}{49} - \frac{x^2}{15} = 1 \qquad \textit{Divide each side by 735.}$$

The equation of the hyperbola is $\dfrac{y^2}{49} - \dfrac{x^2}{15} = 1$.

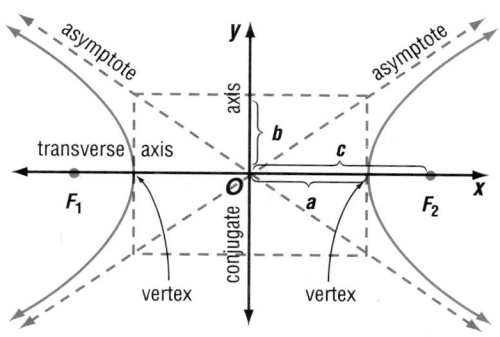

Let's take a closer look at the parts of a hyperbola. The midpoint of the segment connecting the foci of a hyperbola is the **center** of the hyperbola. The point on each branch of the hyperbola that is nearest the center is a **vertex**. As a hyperbola recedes from the center, the branches approach lines called the **asymptotes**.

A hyperbola has many similarities to an ellipse. The distance to the center from a vertex is a units. The distance to the center from a focus is c units. There are two axes of symmetry. The **transverse axis** is a segment of length $2a$ whose endpoints are the vertices of the hyperbola. The **conjugate axis** is a segment of length $2b$ units that is perpendicular to the transverse axis at the center. The lengths of a, b, and c are related differently for a hyperbola than for an ellipse. For a hyperbola, $a^2 + b^2 = c^2$.

Standard Equations of Hyperbolas with Center at the Origin	• **If a hyperbola has foci at $(-c, 0)$ and $(c, 0)$ and if the absolute value of the difference of the distances from any point on the hyperbola to the two foci is $2a$ units, then the standard equation of the hyperbola is $\dfrac{x^2}{a^2} - \dfrac{y^2}{b^2} = 1$, where $c^2 = a^2 + b^2$.** *In this case, the transverse axis is horizontal.* • **If a hyperbola has foci at $(0, -c)$ and $(0, c)$ and if the absolute value of the difference of the distances from any point on the hyperbola to the two foci is $2a$ units, then the standard equation of the hyperbola is $\dfrac{y^2}{a^2} - \dfrac{x^2}{b^2} = 1$, where $c^2 = a^2 + b^2$.** *In this case, the transverse axis is vertical.*

It's easier to graph a hyperbola if the asymptotes are drawn first. To draw the asymptotes, use the values of a and b and draw a rectangle with dimensions $2a$ and $2b$. The point of intersection of the diagonals is the center of the hyperbola. The diagonals are a subset of the asymptotes. The chart below shows the equations of the asymptotes for different hyperbolas.

Equation of Hyperbola	$\dfrac{x^2}{a^2} - \dfrac{y^2}{b^2} = 1$	$\dfrac{y^2}{a^2} - \dfrac{x^2}{b^2} = 1$
Equation of Asymptote	$y = \pm\dfrac{b}{a}x$	$y = \pm\dfrac{a}{b}x$
Transverse Axis	horizontal	vertical

Example ① **Write an equation of a hyperbola with foci at (6, 0) and (−6, 0) if the length of the transverse axis is 8 units. Then draw the graph.**

Half the length of the transverse axis is the distance, a, from the center to a vertex of the hyperbola. The distance from the center to a focus point is represented by c. In this hyperbola, $c = 6$, and $a = 4$.

$c^2 = a^2 + b^2$

$6^2 = 4^2 + b^2$

$36 = 16 + b^2$

$b^2 = 20$

$b = \pm 2\sqrt{5}$ or about ± 4.5

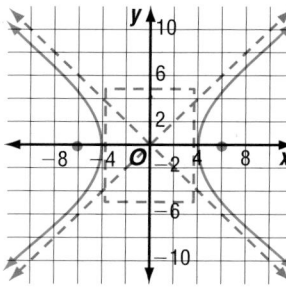

An equation of the hyperbola is $\dfrac{x^2}{16} - \dfrac{y^2}{20} = 1$.

Example ② **A comet travels along a path that is one branch of a hyperbola. The equation of the hyperbola is $\dfrac{y^2}{225} - \dfrac{x^2}{400} = 1$. Find the coordinates of the vertices and foci and the equations of the asymptotes. Then draw the graph.**

APPLICATION

Astronomy

The center of the hyperbola is at (0, 0). It has a vertical transverse axis, and $a = \sqrt{225}$ or 15 and $b = \sqrt{400}$ or 20.

Since $a = 15$, the distance from the center to each vertex is 15 units. Thus, the vertices are at (0, 15) and (0, −15).

To find the foci, first find the value of c.

$c^2 = a^2 + b^2$

$c^2 = 225 + 400$

$c^2 = 625$

$c = 25$

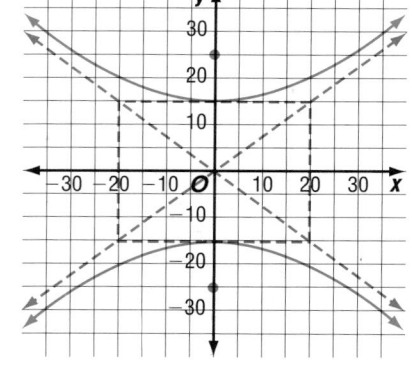

The foci are at (0, 25) and (0, −25).

The equations of the asymptotes are $y = \pm\dfrac{3}{4}x$. Use the information to draw the graph.

So far, we have studied hyperbolas that are centered at the origin. The center of a hyperbola may be located somewhere other than at the origin. If the center is at (h, k), then the standard equation is found by replacing x with $(x - h)$ and y with $(y - k)$. The slopes of the asymptotes are calculated in the same way as if the hyperbola was centered at the origin. Keep in mind that the asymptotes always pass through the center of the hyperbola.

Standard Equations of Hyperbolas with Center at (h, k)	• The equation of a hyperbola with center at (h, k) and with a horizontal transverse axis is $\dfrac{(x - h)^2}{a^2} - \dfrac{(y - k)^2}{b^2} = 1$. • The equation of a hyperbola with center at (h, k) and with a vertical transverse axis is $\dfrac{(y - k)^2}{a^2} - \dfrac{(x - h)^2}{b^2} = 1$.

Example ③ Draw the graph of $\dfrac{(x + 2)^2}{16} - \dfrac{(y - 5)^2}{25} = 1$.

The graph is congruent to the graph of $\dfrac{x^2}{16} - \dfrac{y^2}{25} = 1$, but its center is at $(-2, 5)$.

Draw the graph of $\dfrac{x^2}{16} - \dfrac{y^2}{25} = 1$. Then translate the graph 2 units left and 5 units up.

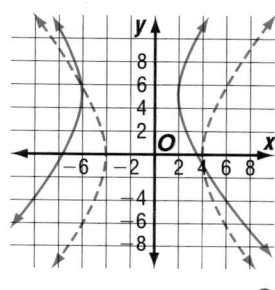

Example ④ Write the equation of the hyperbola shown at the right.

The center of the hyperbola is located at $(-4, 1)$, and the transverse axis is vertical.

The vertices are located 3 units above and below the center, so $a = 3$. The foci are located 5 units above and below the center, so $c = 5$.

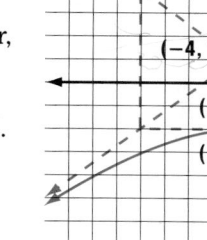

Use this information to find the value of b^2.
$c^2 = a^2 + b^2$
$5^2 = 3^2 + b^2$
$b^2 = 16$

Now write the equation.
$\dfrac{(y - k)^2}{a^2} - \dfrac{(x - h)^2}{b^2} = 1$ *Standard equation of hyperbola with vertical transverse axis*

$\dfrac{(y - 1)^2}{3^2} - \dfrac{[x - (-4)]^2}{16} = 1$ *Substitute values for a^2, b^2, h, and k.*

The equation of the hyperbola is $\dfrac{(y - 1)^2}{9} - \dfrac{(x + 4)^2}{16} = 1$.

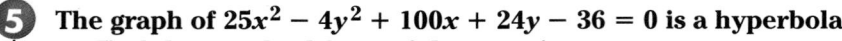

Just as equations of circles and ellipses are sometimes written in expanded form, so are the equations of hyperbolas. You need to complete the square for each variable to write the equation in standard form.

Example ⑤ **The graph of $25x^2 - 4y^2 + 100x + 24y - 36 = 0$ is a hyperbola.**
  a. Find the standard form of the equation.
  b. Find the coordinates of the vertices and foci.
  c. Find the equations of the asymptotes.
  d. Draw the graph.

a. $$25x^2 - 4y^2 + 100x + 24y - 36 = 0$$

$$25(x^2 + 4x + \square) - 4(y^2 - 6y + \square) = 36 + 25(\square) - 4(\square)$$ *Complete the squares.*

$$25(x^2 + 4x + 4) - 4(y^2 - 6y + 9) = 36 + 25(4) - 4(9)$$

$$25(x + 2)^2 - 4(y - 3)^2 = 100$$

$$\frac{(x + 2)^2}{4} - \frac{(y - 3)^2}{25} = 1$$

The standard form of the equation is $\dfrac{(x + 2)^2}{4} - \dfrac{(y - 3)^2}{25} = 1$.

b. The center of this hyperbola is at $(-2, 3)$, and the transverse axis is horizontal. Since the value of a is 2, the vertices are 2 units to the right and left of the center along the transverse axis. The vertices are at $(0, 3)$ and $(-4, 3)$.

Use $a = 2$ and $b = 5$ to find c.
$c^2 = a^2 + b^2$
$c^2 = 2^2 + 5^2$
$c^2 = 29$
$c = \sqrt{29}$

The foci are at $(-2 + \sqrt{29}, 3)$ and $(-2 - \sqrt{29}, 3)$ or about $(3.39, 3)$ and $(-7.39, 3)$.

c. The asymptotes pass through the center at $(-2, 3)$. The slopes of the asymptotes are $\pm\dfrac{b}{a}$ or $\pm\dfrac{5}{2}$.

$y - y_1 = m(x - x_1)$ $y - y_1 = m(x - x_1)$

$y - 3 = \dfrac{5}{2}[x - (-2)]$ *Let $(x, y) = (-2, 3)$.* $y - 3 = -\dfrac{5}{2}[x - (-2)]$

$y - 3 = \dfrac{5}{2}x + 5$ $y - 3 = -\dfrac{5}{2}x - 5$

$y = \dfrac{5}{2}x + 8$ $y = -\dfrac{5}{2}x - 2$

The equations of the asymptotes are $y = \dfrac{5}{2}x + 8$ and $y = -\dfrac{5}{2}x - 2$.

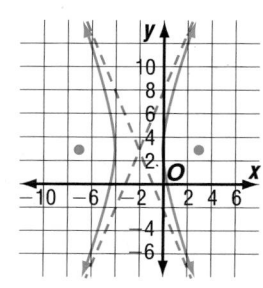

d. Use this information to draw the graph.

Communicating Mathematics

Study the lesson. Then complete the following.

1. **Compare and contrast** hyperbolas and parabolas.

2. **Compare and contrast** hyperbolas and ellipses.

3. How would you know if the graph of an equation is an ellipse or a hyperbola?

4. **Describe** the steps you would follow to draw the graph of the equation $\dfrac{(y-5)^2}{36} - \dfrac{(x+2)^2}{9} = 1$.

Guided Practice

State whether the graph of each equation is an ellipse or a hyperbola.

5. $\dfrac{x^2}{24} - \dfrac{y^2}{36} = 1$

6. $\dfrac{x^2}{100} + \dfrac{y^2}{25} = 1$

7. $\dfrac{y^2}{20} - \dfrac{x^2}{32} = 1$

Write an equation for each hyperbola in standard form.

8. $6x^2 - 12y^2 = 108$

9. $y^2 - 3x^2 + 6x + 6y = 18$

Find the coordinates of the vertices and the foci and the slopes of the asymptotes for each hyperbola whose equation is given. Then draw the graph.

10. $\dfrac{y^2}{18} - \dfrac{x^2}{20} = 1$

11. $x^2 - 36y^2 = 36$

12. $\dfrac{(y+6)^2}{20} - \dfrac{(x-1)^2}{25} = 1$

13. $5x^2 - 4y^2 - 40x - 16y = 36$

14. Write an equation for the graph at the right.

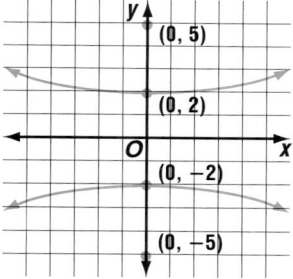

15. A hyperbola is centered at the origin with a horizontal transverse axis. The value of a is 1, and the value of b is 4. Write an equation for the hyperbola.

Practice

Write an equation for each hyperbola.

16.

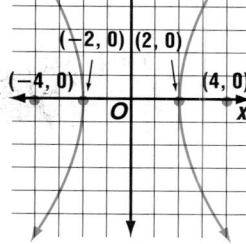

17.

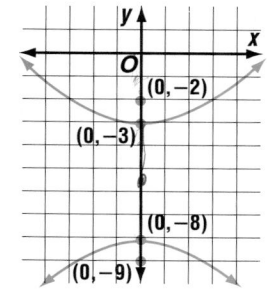

18.

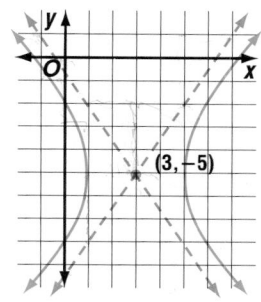

Find the coordinates of the vertices and foci and the slopes of the asymptotes for each hyperbola whose equation is given. Then draw the graph.

19. $\dfrac{x^2}{81} - \dfrac{y^2}{49} = 1$

20. $\dfrac{y^2}{36} - \dfrac{x^2}{4} = 1$

21. $\dfrac{x^2}{9} - \dfrac{y^2}{25} = 1$

22. $\dfrac{y^2}{16} - \dfrac{x^2}{25} = 1$

23. $x^2 - 2y^2 = 2$

24. $y^2 = 36 + 4x^2$

25. $x^2 - y^2 = 4$

26. $2x^2 - 6y^2 = 12$

27. $\dfrac{(y-4)^2}{16} - \dfrac{(x+2)^2}{9} = 1$

28. $\dfrac{(y-3)^2}{25} - \dfrac{(x-2)^2}{16} = 1$

29. $\dfrac{(x+1)^2}{4} - \dfrac{(y+3)^2}{9} = 1$

30. $\dfrac{(x+6)^2}{36} - \dfrac{(y+3)^2}{9} = 1$

31. $(x+3)^2 - 4(y-2)^2 = 4$

32. $9y^2 - 120x^2 - 54y = 999$

33. $y^2 - 3x^2 + 6y + 6x = 18$

34. $4x^2 - 25y^2 - 8x - 96 = 0$

Write an equation for each hyperbola described below.

35. The hyperbola is centered at $(2, -3)$ and has a horizontal transverse axis. The value of a is 7 and the value of b is 2.

36. The hyperbola is centered at $(-4, 5)$ and has a vertical transverse axis. The value of a is 4 and the value of b is 9.

37. The vertices of the hyperbola are at $(-5, 0)$ and $(5, 0)$. The conjugate axis has a length of 12 units.

38. The vertices of the hyperbola are at $(0, -4)$ and $(0, 4)$. The conjugate axis has a length of 14 units.

An equation of the form $xy = c$ is a hyperbola with the x- and y-axes as asymptotes. Sketch the graph of each hyperbola.

39. $xy = 3$ **40.** $xy = 8$ **41.** $xy = -5$ **42.** $xy = -12$

Graphing Calculator

For the equation $xy = c$, replace c with different values. Graph the resulting equations on a graphing calculator. Answer each question.

43. Suppose $c > 0$. How are the graphs similar and how are they different?

44. Suppose $c < 0$. How are the graphs similar and how are they different?

45. What happens to the graph if $c = 0$?

Critical Thinking

46. A hyperbola with a horizontal transverse axis contains the point at $(4, 3)$. The equations of the asymptotes are $y - x = 1$ and $y + x = 5$. Write the equation of the hyperbola.

Applications and Problem Solving

47. Chemistry Boyle's Law states that if the temperature of a gas is constant, then the volume of the gas is inversely proportional to the pressure exerted by the gas. This law is represented by the equation $PV = k$. In this equation, P represents the pressure, V represents the volume, and k is a constant. The constant for a certain gas is 22,500. Graph $PV = k$ for the gas.

48. **Forestry** A forest ranger at an outpost and another ranger at the primary station both heard an explosion. The outpost and the primary station are 6 kilometers apart.

   a. If one ranger heard the explosion 6 seconds before the other, write an equation that describes all the possible locations of the explosion. Place the two ranger stations on the *x*-axis with the midpoint between the stations at the origin. The transverse axis is horizontal. (*Hint:* The speed of sound is about 0.35 km per second.)

   b. Draw a sketch of the possible locations of the explosion. Include the ranger stations in the drawing.

Mixed Review

49. Write the equation of the ellipse whose foci are at $(5, 4)$ and $(-3, 4)$. The major axis is 10 units long. (Lesson 7–4)

50. Name the vertex and the axis of symmetry for the graph of $f(x) = (x + 2)^2$. (Lesson 6–6)

51. Solve $2q^2 + 11q = 21$ by factoring. (Lesson 6–2)

52. **Astronomy** Venus has an average distance of 1.08×10^8 kilometers from the sun. Saturn has an average distance of 1.428×10^9 kilometers from the sun. About how much closer to the sun is Venus? (Lesson 5–1)

53. **Statistics** The number of years of life expected at birth for women in certain countries is given below. Make a box-and-whisker plot of the data, labeling any outliers. (Lesson 4–8)

77.2	76.8	76.0	74.3	77.5	78.8	78.4	75.4	77.5
76.0	73.7	75.6	77.2	79.5	79.5	75.0	72.9	76.2
79.9	79.6	74.0	77.6	75.6	75.9	73.2		

54. Solve the system of equations. (Lesson 3–7)

   $r + s + t = 15$

   $r + t = 12$

   $s + t = 10$

55. If $h(x) = [5x - 4]$, find $h(-1.5)$. (Lesson 2–6)

56. **Business** Derringer Cleaners charges \$52 to clean a wedding dress. If the equation relating time spent in hours to the cost in dollars is $C = 12 + 20t$, find the time spent cleaning the dress. (Lesson 2–1)

57. Simplify $7x + 8y + 9y - 5x$. (Lesson 1–2)

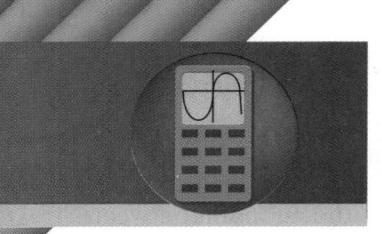

7-6A Graphing Technology
Conic Sections

A Preview of Lesson 7-6

Most conic sections are relations, not functions. Since most graphing calculators plot functions, we must manipulate the equations before entering them into the calculator. For example, the equation for a circle, $x^2 + y^2 = 16$, cannot be entered directly into the calculator since it requires that the equation be entered in a Y= format.

Example **Graph $x^2 + y^2 = 16$ in the standard viewing window.**

First solve the equation for y.

$$x^2 + y^2 = 16$$
$$y^2 = 16 - x^2$$
$$y = \pm \sqrt{16 - x^2}$$

Now enter the equations $y = \sqrt{16 - x^2}$ and $y = -\sqrt{16 - x^2}$ separately, since there is no $\pm$ key.

Enter: | Y= | | 2nd | | √ | | (| 16

| − | | X,T,θ,n | | x² | |) | | ENTER |

| (−) | | 2nd | | √ | | (| 16

| − | | X,T,θ,n | | x² | |) | | ZOOM | 6

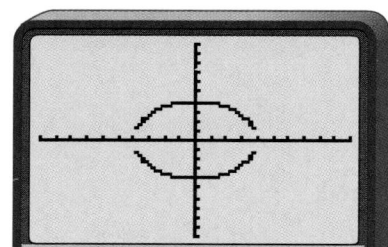

The graph appears to be an ellipse. The calculator screen can be set so that units on the x- and y-axes are equal length. To do this quickly, enter | ZOOM | 5. The graph no longer appears distorted and shows the circle.

To save time entering equations, you can define Y1 as $\sqrt{16 - x^2}$ and Y2 as −Y1. To do this, enter Y1 as shown above. Then move the cursor to Y2 and enter the following.

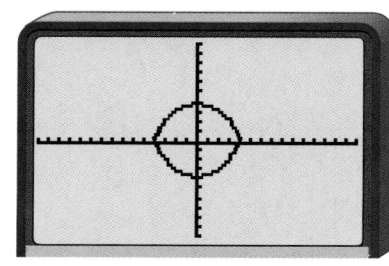

Enter: | (−) | | 2nd | | Y-VARS | 1 1

Example **2** Graph the hyperbola $9y^2 + 36y - x^2 + 6x - 54 = 0$ in the $[-10, 15]$ by $[-0, 10]$ viewing window.

Rewrite the equation by completing the square and solving for y.

$$y = -2 \pm \sqrt{\frac{81 + (x-3)^2}{9}}$$

To enter the equations into the calculator efficiently, let Y1 = $\sqrt{\frac{81 + (x-3)^2}{9}}$, Y2 = -2 + Y1, and Y3 = -2 − Y1.

Enter: Y= 2nd √ ((81

+ (X,T,θ,n − 3)

x²) ÷ 9) ENTER

(−) 2 + 2nd Y-VARS

1 1 ENTER

(−) 2 − 2nd Y-VARS 1 1

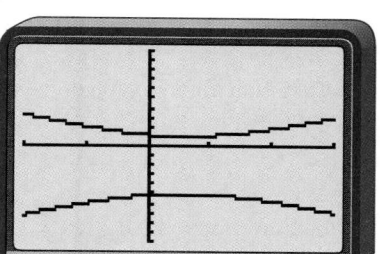

Then deselect Y1 and enter GRAPH. *To deselect Y1, press* Y= *and highlight the equals sign next to Y1. Then press* ENTER.

EXERCISES

Use a graphing calculator to graph each conic section. Name the conic section and sketch the graph that appears.

1. $y = x^2 + 9x - 12$
2. $(x - 3)^2 + y^2 = 25$
3. $16x^2 + 4y^2 = 48$
4. $10x^2 - 7y^2 - 70 = 0$
5. $25x^2 + 4y^2 - 24y = 64$
6. $x^2 - 8x + y^2 = 25$
7. $4x^2 + y^2 - 100 = 0$
8. $y^2 + 5 = x^2 + 2y + 1$
9. $y^2 - 12y - x + 25 = 0$
10. $(y + 1)^2 - x^2 - 4 = 0$

Conic Sections

APPLICATION
Aeronautics

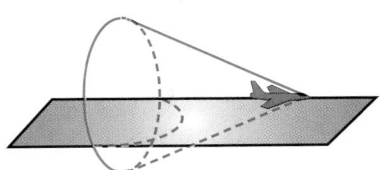

The F-15 Eagle can reach speeds in excess of Mach 2.5. Mach 1 is the speed of sound, and Mach 2.5 is 2.5 times the speed of sound.

A plane like the F-15 Eagle, flying faster than the speed of sound, produces a shock wave in the shape of a cone. When the shock wave hits the ground, a sonic boom is heard. If the plane is flying parallel to the ground, the sonic boom is heard at points that form one branch of a hyperbola. What shape would the points form if the plane is climbing and *not* flying parallel to the ground? What shape would be formed if the plane could fly vertically?

Recall that parabolas, circles, ellipses, and hyperbolas are called conic sections, because they are the cross sections formed when a double cone is sliced by a plane.

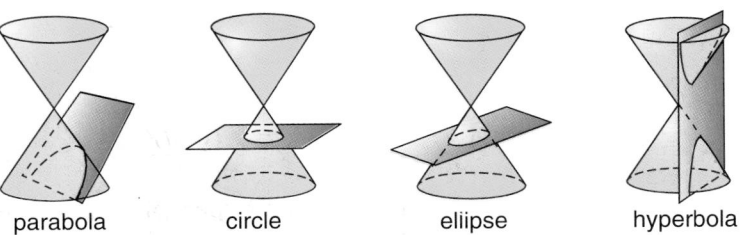

parabola   circle   eliipse   hyperbola

The conic sections can all be described by a general quadratic equation.

Equation of a Conic Section	The equation of a conic section can be written in the form $Ax^2 + Bxy + Cy^2 + Dx + Ey + F = 0$, where A, B, and C are *not* all zero.

You can identify the conic section that is represented by a given equation by writing the equation in one of the standard forms you have learned.

Conic Section	Standard Form of Equation
parabola	$y = a(x - h)^2 + k$ or $x = a(y - k)^2 + h$
circle	$(x - h)^2 + (y - k)^2 = r^2$
ellipse	$\dfrac{(x - h)^2}{a^2} + \dfrac{(y - k)^2}{b^2} = 1$ or $\dfrac{(x - h)^2}{b^2} + \dfrac{(y - k)^2}{a^2} = 1$ $a \neq b$
hyperbola	$\dfrac{(x - h)^2}{a^2} - \dfrac{(y - k)^2}{b^2} = 1$ or $\dfrac{(y - k)^2}{a^2} - \dfrac{(x - h)^2}{b^2} = 1$ or $xy = c$, when $c \neq 0$

What YOU'LL LEARN

- To write equations of conic sections in standard form,
- to identify conic sections from their equations, and
- to use simulation to solve problems.

Why IT'S IMPORTANT

You can graph conic sections to solve problems involving aeronautics and space science.

Example ① Identify the graph of $y^2 - 3x + 6y + 12 = 0$ as a parabola, a circle, an ellipse, or a hyperbola. Then graph the equation.

Complete the square to write the equation in standard form.

$y^2 - 3x + 6y + 12 = 0$

$$3x = y^2 + 6y + 12$$
$$3x = (y^2 + 6y + \square) + 12 - \square$$
$$3x = (y^2 + 6y + 9) + 12 - 9$$
$$3x = (y + 3)^2 + 3 \quad \textit{Factor.}$$
$$x = \frac{1}{3}(y + 3)^2 + 1 \quad \textit{Divide by 3.}$$

This equation has the standard form of $x = a(y - k)^2 + h$. So the graph of the equation $y^2 - 3x + 6y + 12 = 0$ is a parabola with vertex at $(1, -3)$ and opening to the right.

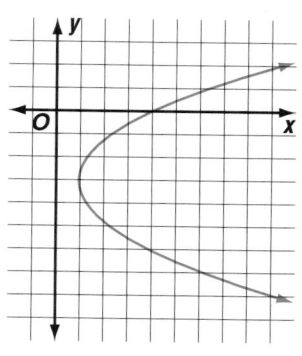

Example ②

The shock wave generated by a jet plane intersects the ground in a curve with equation $x^2 - 14x + 4 = 9y^2 - 36y$. What is the shape of the curve? Sketch the curve.

To determine the shape of the curve, complete the squares and write the equation in standard form.

$$x^2 - 14x + 4 = 9y^2 - 36y$$
$$x^2 - 14x - 9y^2 + 36y = -4$$
$$(x^2 - 14x + \square) - 9(y^2 - 4y + \square) = -4 + \square - 9(\square) \quad \textit{Complete the squares.}$$
$$(x^2 - 14x + 49) - 9(y^2 - 4y + 4) = -4 + 49 - 9(4)$$
$$(x - 7)^2 - 9(y - 2)^2 = 9$$
$$\frac{(x - 7)^2}{9} - \frac{(y - 2)^2}{1} = 1 \quad \textit{Divide each side by 9.}$$

The curve is a hyperbola. The center of the hyperbola is at $(7, 2)$. Since the value of a is 3, the vertices are 3 units to the right and left of the center along the transverse axis. The vertices are located at $(4, 2)$ and $(10, 2)$. The value of b is 1.

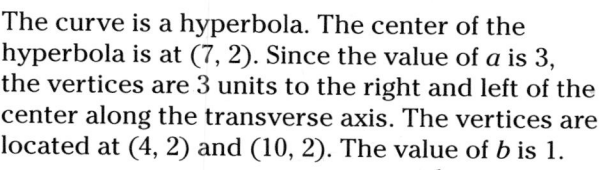

The slopes of the asymptotes are $\pm\frac{1}{3}$. Use all of this information to sketch the curve.

Of course, the shock wave will only generate one half of the hyperbola.

You can easily determine the type of conic section represented by an equation of the form $Ax^2 + Bxy + Cy^2 + Dx + Ey + F = 0$ when $B = 0$ by looking at A and C.

Conic Section	Relationship of A and C
parabola	$A = 0$ or $C = 0$, but not both.
circle	$A = C$
ellipse	A and C have the same sign and $A \neq C$.
hyperbola	A and C have opposite signs.

Conic sections can be modeled by slicing a double cone. Sometimes it helps to **use a simulation** to model other mathematical situations that are difficult to solve directly.

Example **3**

PROBLEM SOLVING

Use a Simulation

DeLIGHTful Cereal has placed contest tickets that are printed with two consecutive letters from the name of the cereal in their cereal boxes, that is, DE, LI, GH, TF, and UL. To win an official Olympic poster, you must collect all five different tickets whose letters spell the name of the cereal. The total number of each kind of ticket is the same. How many boxes would you expect to have to buy to win the contest?

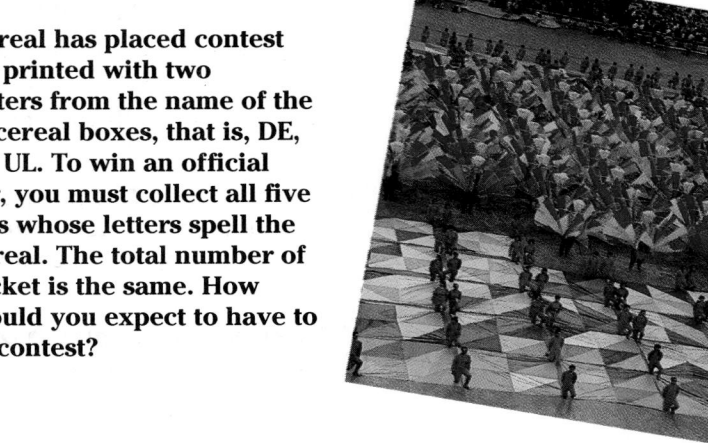

Explore   Read the problem. In order to win the poster, you must collect five different tickets. If you are very lucky, you could win after buying only five boxes of cereal. However, most people will need to buy more than five boxes. How many boxes will you have to buy to win the poster?

The roll of a die simulates buying one box of cereal.

Plan   It would be expensive and inconvenient, but you could buy lots of boxes of cereal to see how many it takes to win. Instead, you could simulate these purchases by assigning each of the tickets a number from 1 to 5. Then roll a die to simulate buying one box and to determine the prize in the box. If you roll a 6, ignore it. Keep rolling until all 5 numbers appear. Several simulations should be tried.

Solve   Five simulations are recorded below.

Rolls	Number of Rolls (Boxes Bought) to Win
5 1 4 3 4 3 5 3 3 1 4 2	12
4 3 2 5 5 4 4 4 3 2 4 4 1	13
1 2 1 5 5 5 4 3	8
3 5 5 4 1 3 1 1 3 1 2	11
4 3 5 1 3 4 3 4 3 4 1 1 3 2	14

Find the mean to estimate of the number of boxes you would need to buy.

$$\frac{12 + 13 + 8 + 11 + 14}{5} = 11.6 \text{ or about } 12$$

Our estimate is that you would have to buy about 12 boxes to win the poster.

Examine   When you simulate a situation in this way, the more trials you conduct, the better your estimate. You may wish to conduct more trials to see if the answer is reasonable.

Communicating Mathematics

Study the lesson. Then complete the following.

1. **Describe** how a parabola is formed by slicing a cone.

2. **Explain** how to slice a double cone so that the cross section will be two straight intersecting lines.

3. **Describe** a problem that might be solved by using a model or simulation.

 MATH JOURNAL

4. **Assess Yourself** Draw an example of each of the four conic sections you have studied. Write a sentence or two describing the properties of each of these curves. Which conic section is your "favorite?" Explain your answer.

Guided Practice

State whether the graph of each equation is a parabola, a circle, an ellipse, or a hyperbola.

5. $x^2 + y^2 = 20$

6. $x = (y - 5)^2 + 9$

7. $\dfrac{x^2}{23} - \dfrac{(y - 8)^2}{34} = 1$

8. $\dfrac{(y - 7)^2}{3} + \dfrac{(x + 2)^2}{2} = 1$

Write each equation in standard form. State whether the graph of the equation is a parabola, a circle, an ellipse, or a hyperbola. Then graph the equation.

9. $y = x^2 + 3x + 1$

10. $y^2 - 2x^2 - 16 = 0$

11. $x^2 + y^2 = x + 2$

12. $x^2 + 4y^2 + 2x - 24y + 33 = 0$

13. Write $x + 2 = x^2 + y$ in standard form. Then graph the equation.

Practice

Write each equation in standard form. State whether the graph of the equation is a parabola, a circle, an ellipse, or a hyperbola. Then graph the equation.

14. $6x^2 + 6y^2 = 162$

15. $x^2 = 8y$

16. $4x^2 + 2y^2 = 8$

17. $4y^2 - x^2 + 4 = 0$

18. $(x - 1)^2 + 9(y - 4)^2 = 36$

19. $y + 4 = (x - 2)^2$

20. $x^2 + y^2 + 6y + 13 = 40$

21. $x^2 - y^2 + 8x = 16$

22. $x^2 + y^2 + 4x - 6y = -4$

23. $y + x^2 = -(8x + 23)$

24. $3x^2 + 4y^2 + 8y = 8$

25. $(y - 4)^2 = 9(x - 4)$

26. $x^2 - 8y + y^2 + 11 = 0$

27. $25y^2 + 9x^2 - 50y - 54x = 119$

28. $x^2 + 4y^2 - 11 = 2(4y - x)$

29. $9y^2 + 18y = 25x^2 + 216$

30. $x^2 + y^2 = 2x + 8$

31. $6x^2 - 24x - 5y^2 - 10y - 11 = 0$

The graph of an equation of the form $Ax^2 + Bxy + Cy^2 + Dx + Ey + F = 0$ is either a conic section or a *degenerate case*. The degenerate cases for the conic sections are stated below. Graph each equation and identify the result.

32. $4x^2 - y^2 = 0$

33. $4y^2 + 3x^2 + 32y - 6x = -67$

34. $x^2 - x = 0$

Conic	Degenerate Case
ellipse or circle	isolated point
hyperbola	two intersecting lines
parabola	two parallel lines or one line

35. The program below determines the type of graph represented by an equation of the form $Ax^2 + Bxy + Cy^2 + Dx + Ey + F = 0$, where $B = 0$.

```
PROGRAM: CONICS
: Prompt A,C,D,E,F          : End
: If A≠0                    : Lbl 2
: Then                      : If A=C
: Goto 1                    : Then
: End                       : Disp "CIRCLE"
: If D=0                    : Stop
: Then                      : End
: Disp"DEGENERATE","CASE"   : If AC>0
: Goto 2                    : Then
: End                       : Disp "ELLIPSE"
: Goto 2                    : Stop
: Lbl 1                     : End
: If C≠0                    : If AC<0
: Then                      : Then
: D²/(4A)+E²/(4C)→G         : Disp "HYPERBOLA"
: End                       : Stop
: If G≠F                    : End
: Then                      : Disp "PARABOLA"
: Goto 2
```

Use the program to determine the type of conic section each equation represents.

a. $12x^2 + 36x + 16y^2 + 32y - 5 = 0$ **b.** $25x^2 - 4y^2 = 100$

c. $x^2 + 12x + y^2 - 8y = -44$ **d.** $(y + 3)^2 = -12(x - 2)$

36. Graph $\dfrac{x^2}{16} - \dfrac{y^2}{9} = 1$ on a graphing calculator. Then graph $\dfrac{x^2}{16} - \dfrac{y^2}{9} = k$ for $k = \dfrac{1}{2}, \dfrac{1}{4}, \dfrac{1}{8}, \dfrac{1}{16}, \dfrac{1}{32}$, and so on. Answer each question.

a. What happens as the value of k approaches 0?

b. Identify the graph when $k = 0$.

37. The equation of an ellipse is $\dfrac{x^2}{16} + \dfrac{y^2}{4} = 1$. The directrix of a parabola is tangent to the ellipse at one endpoint of the minor axis. If the focus of the parabola is located at the other endpoint of the minor axis, write all possible equations of the parabola.

38. Space Science The orbits of comets follow the paths of the conic sections. For example, Halley's Comet follows an elliptical orbit with the sun located at one of its foci. What type of orbits pass by Earth only once?

39. Use a Simulation A baseball card manufacturer is packaging a puzzle piece in each package of cards. The total number of each piece is the same.

a. If there are 6 different puzzle pieces, design a simulation to determine how many packages of baseball cards you would need to buy to collect all the pieces.

b. Use your simulation to estimate the number of packages you would need to buy to collect all the pieces.

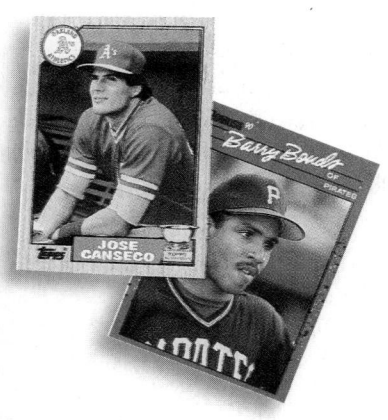

40. Write an equation of the hyperbola with center at (5, 4) if its transverse axis is vertical, $a = 6$, and $b = 4$. (Lesson 7–5)

41. Identify the quadratic term, the linear term, and the constant term of $f(x) = 4x^2 - 8x - 2$. (Lesson 6–1)

42. Simplify $(m^5n^{-3})^2m^2n^7$. (Lesson 5–1)

43. Evaluate the determinant of $\begin{bmatrix} 4 & -7 & 2 \\ 3 & -3 & 3 \\ 2 & 7 & 5 \end{bmatrix}$. (Lesson 4–4)

44. Manufacturing The Oklahoma City division of SuperSports, Inc. produces footballs and basketballs. It takes 4 hours on machine A and 2 hours on machine B to make a football. Producing a basketball requires 6 hours on machine A, 6 hours on machine B, and 1 hour on machine C. Machine A is available 120 hours per week, machine B is available 72 hours per week, and machine C is available 10 hours per week. If the company makes $3 profit on each football and $2 profit on each basketball, how many of each should they make to maximize their profit? (Lesson 3–6)

45. Graph $4y - x \le 6$. (Lesson 2–7)

46. Evaluate $7 - 4^2 + 27 - 1$. (Lesson 1–1)

WORKING ON THE In·ves·ti·ga·tion

Refer to the Investigation on pages 328–329.

the **River Canyon Bridge**

Once you have determined the parabolic equation for an arch, you can find other equations that might be helpful in determining the dimensions of the bridge.

1 Using the equations for the parabolas, determine a general formula for finding the length of a strut from any point under the roadway. Explain how you found the formula. Then find specific formulas for all four of your designs.

2 Use your formula to find the lengths of the struts on each bridge. How do these measurements compare with your initial blueprints?

3 For each design, determine the locations on the canyon walls at which the ends of the arch will be anchored. State each location in terms of the distance from the top of the canyon (where the roadway is located). Explain how you determined each location.

4 Revise your blueprints as necessary with your new data.

Add the results of your work to your Investigation Folder.

7-6B Conic Sections

An Extension of Lesson 7-6

Materials: conic graph paper

Recall that a parabola is the set of all points that are equally distant from the focus and the directrix.

You can draw a parabola based on this definition by using special conic graph paper. This graph paper contains a series of concentric circles equally spaced from each other and a series of parallel lines tangent to each circle.

Number the circles consecutively beginning with the smallest circle. Number the lines with consecutive integers as shown in the sample at the right. Be sure that line 1 is tangent to circle 1.

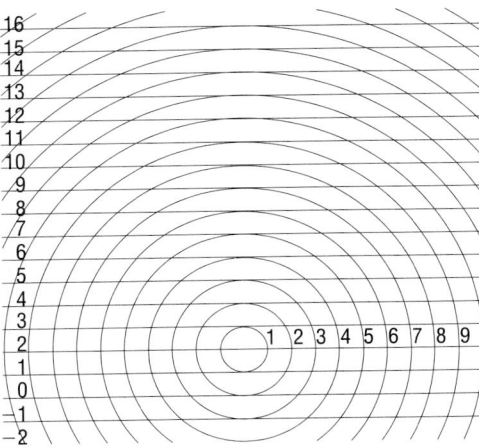

Activity 1

Mark the point at the intersection of circle 1 and line 1. Mark both points that are on line 2 and circle 2. Continue this process marking both points on line 3 and circle 3, and so on. Now connect the points with a smooth curve.

Look at the diagram at the right. What is the graph? Note that every point on the graph is equally distant from the center of the small circle and from the line labeled 0. The center of the small circle is the focus of the parabola, and line 0 is the directrix.

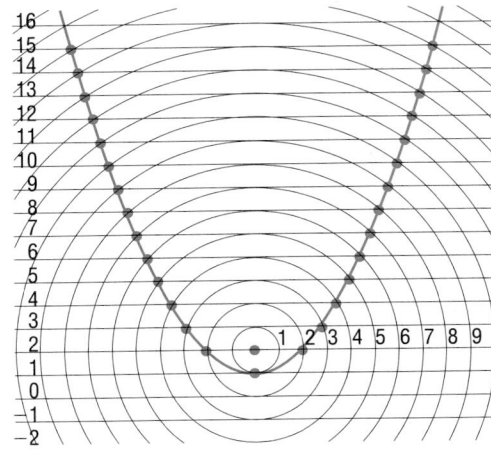

Activity 2

An ellipse is the set of points such that the sum of their distances from two fixed points is constant. The two fixed points are called the foci.

Use graph paper like that shown below. It contains two small circles and a series of concentric circles from each. The concentric circles are tangent to each other as shown.

Choose the constant 13. Mark the points at the intersections of circle 9 and circle 4, because 9 + 4 = 13. Continue this process until you have marked the intersection of all circles whose sum is 13.

Connect the points to form an ellipse. The foci are the centers of the two small circles on the graph paper.

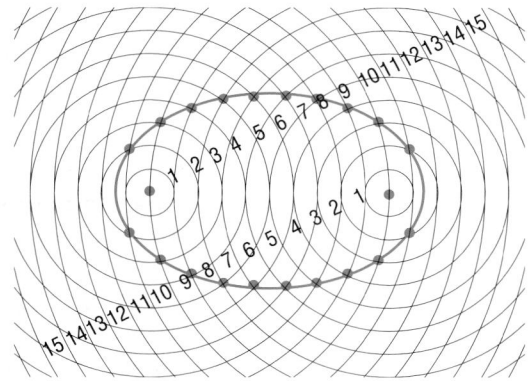

Activity 3

A hyperbola is the set of points such that the difference of their distances from two fixed points is constant. The two fixed points are called the foci.

Choose the same graph paper that you used for the ellipse. Choose the constant 7. Mark the points at the intersections of circle 9 and circle 2, because 9 − 2 = 7. Continue this process until you have marked the intersection of all circles whose difference is 7.

Connect the points to form a hyperbola.

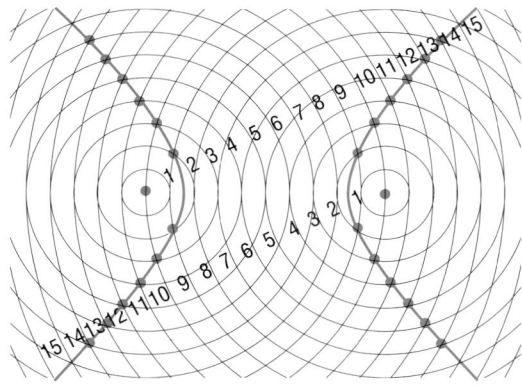

Model

1. Use the paper from Activity 1. Mark the intersection of line 0 and circle 2. Then mark the two points on line 1 and circle 3, the two points on line 2 and circle 4, and so on. Draw the new parabola. Continue this process and make as many parabolas as you can on one sheet of conic graph paper. The focus is always the center of the small circle. Why are the resulting graphs parabolas?

2. In Activity 2, we drew an ellipse such that the sum of the distances from two fixed points was 13. Choose 9, 10, 11, 12, 14, and so on for that sum, and draw as many ellipses as you can on one piece of conic graph paper. What happens as the sum increases? decreases?

3. In Activity 3, we drew a hyperbola such that the difference from two fixed points was 7. Choose other numbers and draw as many hyperbolas as you can on one piece of graph paper. What happens as the difference increases? decreases?

7-7A Graphing Technology
Solving Quadratic Systems

A Preview of Lesson 7–7

As you know, the graphing calculator is capable of graphing several equations on the screen at one time. You can use this capability with the TRACE or "intersect" features to determine approximate solutions of a system of quadratic equations, if a solution exists.

Example **Solve the system of equations with a graphing calculator. Round the answer to the nearest hundredth.**

$$2y = 12 - x^2$$
$$x^2 - y^2 = 25$$

First, solve each equation for y.

$$2y = 12 - x^2 \qquad\qquad\qquad x^2 - y^2 = 25$$
$$y = \frac{1}{2}(12 - x^2) \qquad\qquad x^2 - 25 = y^2$$
$$\qquad\qquad\qquad\qquad\qquad \pm\sqrt{x^2 - 25} = y$$

Now, enter the equations into the calculator and graph in the standard viewing window. Any points where the graphs intersect represent solutions to the system of equations.

Enter: [Y=] .5 [(] 12 [−] [X,T,θ,n]

[x²] [)] [ENTER]

[2nd] [√] [(] [X,T,θ,n] [x²] [−]

25 [)] [ENTER]

[(−)] [2nd] [Y-VARS] 1 2

[ZOOM] 6

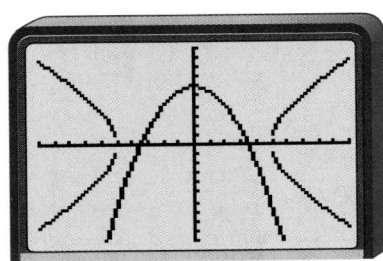

We see that the graphs do not intersect. Therefore, there is no solution to this system of equations.

Example **2** Solve the system of equations with a graphing calculator. Round the coordinates to the nearest hundredth.

$$x^2 + y^2 = 25$$
$$x^2 - y^2 = 1$$

Solve each equation for y.

$$x^2 + y^2 = 25 \qquad\qquad x^2 - y^2 = 1$$
$$y^2 = 25 - x^2 \qquad\qquad x^2 - 1 = y^2$$
$$y = \pm\sqrt{25 - x^2} \qquad\qquad \pm\sqrt{x^2 - 1} = y$$

Now, enter the equations into the calculator and graph in the "square" viewing window. Any points where the graphs intersect represent solutions to the system of equations.

Enter:  25

ENTER

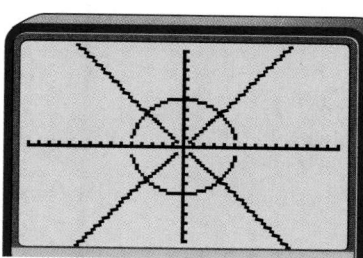

We see from the graph that there are four points of intersection. Use ZOOM and TRACE to determine the points of intersection. The solutions are approximately $(-3.61, 3.46)$, $(3.61, 3.46)$, $(3.61, -3.46)$, and $(-3.61, -3.46)$.

EXERCISES

Use a graphing calculator to solve each system of equations. Round the coordinates to the nearest hundredth.

1. $x^2 + y^2 = 16$
   $y = 2x^2 - 2$

2. $9x^2 - 4y^2 = 36$
   $x^2 + 4y^2 = 36$

3. $x^2 + 9y^2 = 9$
   $y = x^2 - 1$

4. $x^2 + y^2 = 64$
   $9y^2 - 4x^2 = 1$

5. $(x - 1)^2 + y^2 = 9$
   $x^2 + 64y^2 = 64$

6. $y = -x^2 + 7$
   $y = x^2 - 7$

7. $x = y^2 - 10y + 25$
   $x^2 + y^2 = 25$

8. $x^2 + y^2 = 1$
   $x^2 + y^2 = 45$

Solving Quadratic Systems

What YOU'LL LEARN

- To solve systems of equations involving quadratics graphically and algebraically, and
- to solve systems of inequalities involving quadratics graphically.

Why IT'S IMPORTANT

You can use quadratic systems of equations to solve problems involving chemistry and advertising.

CONNECTION
Chemistry

An important concept in chemistry, Boyle's Law, states that if the temperature of a gas is constant, the pressure exerted by the gas varies inversely as the volume. So, $PV = k$, where P represents pressure in kilopascals, V represents volume in cubic decimeters, and k is a constant. A medical technologist provides oxygen for patients with respiratory problems through a tank containing compressed oxygen. The constant for oxygen at 25°C is 504. The volume of the tank is 12 cubic decimeters. What is the pressure of the oxygen in the tank?

We can solve the system of equations described above by graphing to find the pressure in the tank. First write the system of equations.

$PV = k$ *Boyle's Law*
$PV = 504$ *The constant for oxygen at 25° C is 504.*
$V = 12$ *The volume of the tank is 12 dm³.*

Now graph $PV = 504$ and $V = 12$.

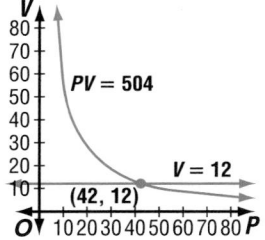

The point at which the line intersects the hyperbola is the solution.

The pressure is 42 kilopascals when the volume is 12 cubic decimeters.

Since a negative pressure or volume is impossible, we will only consider positive values of P and V.

If the graphs of a system of equations are a conic section and a straight line, the system will have zero, one, or two solutions.

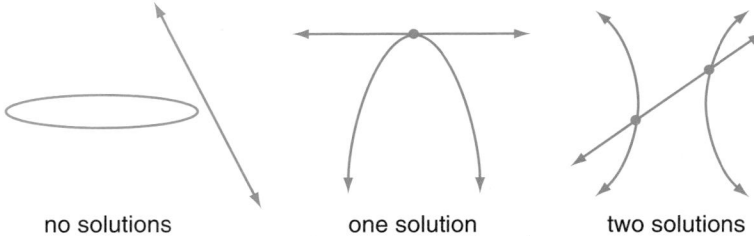

no solutions   one solution   two solutions

We have solved systems of linear equations graphically and algebraically. We can use similar methods to solve systems involving quadratic equations.

Example **Solve the system of equations.**

$x^2 + 4y^2 = 25$

$2y + x = 1$

Method 1: Graphing

The graph of the equation $x^2 + 4y^2 = 25$ is an ellipse with its center at the origin, a major axis of 10 units, and a minor axis of 5 units. The graph of the equation $2y + x = 1$ is a line with slope $-\frac{1}{2}$ and y-intercept $\frac{1}{2}$. The graph shows that there are two solutions for this system of equations. You can use the graph to estimate solutions of $(-3, 2)$ and $\left(4, -1\frac{1}{2}\right)$.

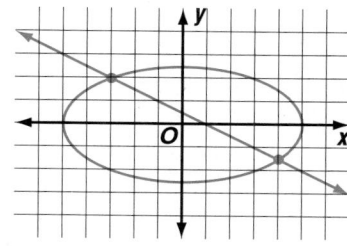

Method 2: Algebra

Use substitution to solve the system algebraically. First rewrite $2y + x = 1$ as $x = 1 - 2y$.

$$x^2 + 4y^2 = 25$$

$(1 - 2y)^2 + 4y^2 = 25$ *Substitute $1 - 2y$ for x.*

$8y^2 - 4y - 24 = 0$ *Simplify.*

$2y^2 - y - 6 = 0$ *Divide each side by 4.*

$(2y + 3)(y - 2) = 0$ *Factor.*

$2y + 3 = 0$ or $y - 2 = 0$ *Zero product property*

$y = -\frac{3}{2}$ $y = 2$

Now solve for x.

$x = 1 - 2y$ $x = 1 - 2y$

$x = 1 - 2\left(-\frac{3}{2}\right)$ $x = 1 - 2(2)$

$x = 4$ $x = -3$

The solutions of the system of equations are $\left(4, -1\frac{1}{2}\right)$ and $(-3, 2)$.

Compare these to the graphical estimate.

If the graphs of a system of equations are two conic sections, the system will have zero, one, two, three, or four solutions.

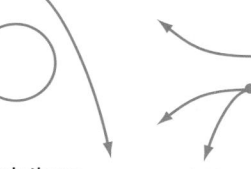

no solutions

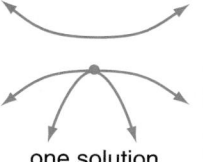

one solution

two solutions

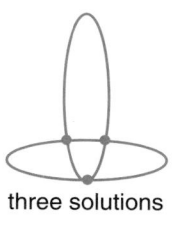

three solutions

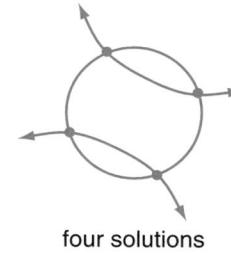

four solutions

Example 2 Solve the system of equations.

$5x^2 + y^2 = 30$

$y^2 - 16 = 9x^2$

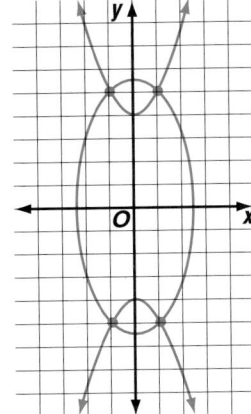

Method 1: Graphing

The graph of the equation $5x^2 + y^2 = 30$ is an ellipse with its center at the origin, a major axis of $2\sqrt{30}$ units, and a minor axis of $2\sqrt{6}$ units. The graph of the equation $y^2 - 16 = 9x^2$ is a hyperbola with its center at the origin, $a = 4$ and $b = \frac{4}{3}$.

The graph shows that there are four solutions for this system of equation, which appear to be $(1, 5)$, $(-1, 5)$, $(1, -5)$, and $(-1, -5)$.

Method 2: Algebra

Use the elimination method to solve the system algebraically.

$$
\begin{array}{ll}
5x^2 + y^2 = 30 \\
y^2 - 16 = 9x^2
\end{array}
\qquad \rightarrow \qquad
\begin{array}{rl}
5x^2 + y^2 = & 30 \\
\underline{(+)\; 9x^2 - y^2 = -16} \\
14x^2 \quad\;\; = & 14 \\
x^2 = & 1 \\
x = & \pm 1
\end{array}
$$

Substitute 1 and -1 for x and solve for y.

$$
\begin{array}{ll}
5x^2 + y^2 = 30 & \qquad 5x^2 + y^2 = 30 \\
5(1)^2 + y^2 = 30 & \qquad 5(-1)^2 + y^2 = 30 \\
\quad\;\; y^2 = 25 & \qquad\qquad\;\; y^2 = 25 \\
\quad\;\;\; y = \pm 5 & \qquad\qquad\;\;\; y = \pm 5
\end{array}
$$

The solutions are $(1, 5)$, $(-1, 5)$, $(1, -5)$, and $(-1, -5)$, which agree with the graph.

Example 3 Solve the system of equations by any method. Use a graphing calculator to determine if the results are reasonable.

$x + 4 = (y - 2)^2$

$2y + x = 0$

Rewrite $2y + x = 0$ as $x = -2y$.

$$
\begin{array}{ll}
x + 4 = (y - 2)^2 \\
-2y + 4 = (y - 2)^2 & \textit{Substitute } -2y \textit{ for x.} \\
-2y + 4 = y^2 - 4y + 4 \\
y^2 - 2y = 0 \\
y(y - 2) = 0 & \textit{Factor.}
\end{array}
$$

$y = 0$ or $y - 2 = 0$ *Zero product property*

$\qquad\qquad\quad y = 2$

If $y = 0$, then $x = -2(0)$ or 0. If $y = 2$, then $x = -2(2)$ or -4. Thus, the solutions are $(0, 0)$ and $(-4, 2)$.

Check by graphing in the standard viewing window. First solve each equation for y.

$$x + 4 = (y - 2)^2 \qquad 2y + x = 0$$
$$\pm\sqrt{x + 4} = y - 2 \qquad 2y = -x$$
$$\pm\sqrt{x + 4} + 2 = y \qquad y = -\frac{x}{2}$$

Enter: [Y=] [(−)] [X,T,θ,n] [÷] 2 [ENTER]

[2nd] [√] [(] [X,T,θ,n] [+] 4

[)] [+] 2 [ENTER] [(−)] [2nd] [√]

[(] [X,T,θ,n] [+] 4 [)] [+]

2 [ZOOM] 6

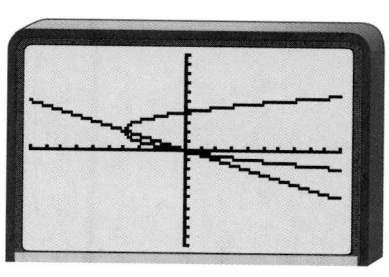

Example ④

Mirrors

A hyperbolic mirror is a mirror in the shape of one branch of a hyperbola. Such a mirror reflects light rays directed at one focus toward the other focus. Suppose a hyperbolic mirror is modeled by the upper branch of the hyperbola whose equation is $\frac{y^2}{9} - \frac{x^2}{16} = 1$. A light source is located at $(-8, 0)$. Where should the light from this source hit the mirror so that the light will be reflected to $(0, -5)$?

The foci of the hyperbola are at $(0, 5)$ and $(0, -5)$. In order for the light to be reflected to the focus at $(0, -5)$, it must be directed at the other focus at $(0, 5)$. The equation of the line that passes through $(-8, 0)$ and $(0, 5)$ is $y = \frac{5}{8}x + 5$. Solve the system of equations $\frac{y^2}{9} - \frac{x^2}{16} = 1$ and $y = \frac{5}{8}x + 5$ by using substitution.

$$\frac{y^2}{9} - \frac{x^2}{16} = 1$$
$$16y^2 - 9x^2 = 144 \quad \textit{Multiply each side by 144.}$$
$$16\left(\frac{5}{8}x + 5\right)^2 - 9x^2 = 144 \quad \textit{Substitute } \left(\frac{5}{8}x + 5\right) \textit{ for } y.$$
$$\frac{25}{4}x^2 + 100x + 400 - 9x^2 = 144 \quad \textit{Simplify.}$$
$$25x^2 + 400x + 1600 - 36x^2 = 576 \quad \textit{Multiply each side by 4.}$$
$$11x^2 - 400x - 1024 = 0$$

Use the quadratic formula.

$$x = \frac{-(-400) \pm \sqrt{(-400)^2 - 4(11)(-1024)}}{2(11)} \quad a = 11, b = -400, \text{ and } c = -1024.$$

$$= \frac{200 \pm 24\sqrt{89}}{11}$$

The value of x is about 38.8 or about -2.4. The closest point for the light to hit the mirror is where x equals about -2.4 and y equals about $\frac{5}{8}(-2.4) + 5$ or 3.5.

You have learned how to solve systems of linear inequalities by graphing. You can use a similar method when the inequalities involve conic sections.

Example **Solve the system of inequalities by graphing.**

$$9x^2 + y^2 < 81$$
$$x^2 + y^2 \geq 16$$

The graph of $9x^2 + y^2 < 81$ is the interior of the ellipse $9x^2 + y^2 = 81$. This region is shaded in blue.

The graph of $x^2 + y^2 \geq 16$ is the circle $x^2 + y^2 = 16$ and its exterior. This region is shaded in yellow.

The intersection of these two graphs represents the solutions for the system of inequalities.

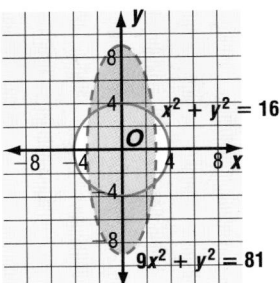

CHECK FOR UNDERSTANDING

Communicating Mathematics

Study the lesson. Then complete the following.

1. **Describe** a situation in which you would choose to solve a system of quadratic equations algebraically rather than graphically. How would a graph of the equations help you to find the algebraic solutions?

2. **Describe** the algebraic solution if the graphs of a system of quadratic equations do *not* intersect.

3. **Write** equations for two different conic sections that intersect at the point (2, 4).

4. The blue region of the graph at the right represents the solution of $x^2 + y^2 \leq 25$ and the yellow region represents the solution of $4y + x^2 \leq 25$. What does the green region represent?

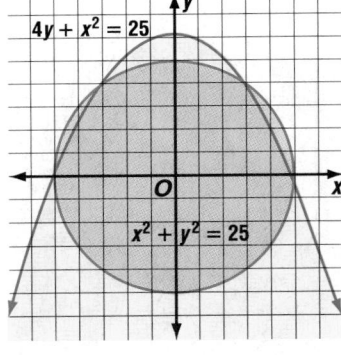

5. List the possible number of intersections for each pair of graphs and then draw an example for each possibility.

   a. a circle and a line   **b.** a parabola and a circle

   c. an ellipse and a hyperbola   **d.** a parabola and a hyperbola

Guided Practice

Name the type of conic section represented by each equation. Then solve the system algebraically.

6. $x^2 + y^2 = 25$
   $y - x = 1$

7. $x^2 + 3y^2 = 12$
   $x^2 - y^2 = 9$

8. $2x^2 - 2y^2 = 72$
   $4y^2 + x^2 = 25$

9. $3x = 8y^2$
   $8y^2 - 2x^2 = 16$

Solve each system of equations algebraically. Check your solutions with a graphing calculator.

10. $y = x + 2$
$y = x^2$

11. $5x^2 + y^2 = 30$
$9x^2 - y^2 = -16$

Solve each system of inequalities by graphing.

12. $x^2 + y^2 < 25$
$4x^2 - 9y^2 < 36$

13. $y^2 < x$
$x^2 - 4y^2 < 16$

14. Write the system of equations represented by the graph at the right.

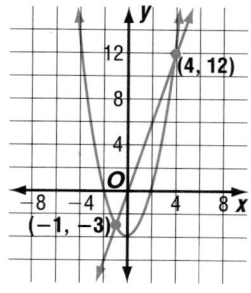
(4, 12)
12
8
4
O
−8 −4   4   8 x
(−1, −3)

EXERCISES

Practice

Solve each system of equations algebraically. Check your solutions with a graphing calculator.

15. $y = 6$
$y^2 = x^2 + 9$

16. $y = 2x^2$
$y = x + 3$

17. $y^2 = x^2 - 25$
$x^2 - y^2 = 7$

18. $4x^2 + y^2 = 100$
$4x + y^2 = 20$

19. $x^2 + y^2 = 64$
$x^2 + 64y^2 = 64$

20. $x + 4 = (y - 1)^2$
$x + y + 1 = 0$

21. $y^2 = x^2 - 7$
$x^2 + y^2 = 25$

22. $y = 7 - x$
$y^2 + x^2 = 9$

23. $x^2 + 2y^2 = 33$
$x^2 + y^2 - 19 = 2x$

24. $x = (y - 3)^2 + 2$
$y + x = 5$

25. $\frac{x^2}{30} + \frac{y^2}{6} = 1$
$x = y$

26. $\frac{x^2}{36} - \frac{y^2}{4} = 1$
$x = y$

27. $x^2 + y^2 = 36$
$y = x + 2$

28. $3x^2 - y^2 = 9$
$x^2 + 2y^2 = 10$

29. $x^2 + y^2 = 36$
$8y = x^2 - 79$

30. $y = -x^2 + 3$
$x^2 + 4y^2 = 36$

31. $x^2 + 2y^2 = 16$
$y^2 + 2x^2 = 17$

32. $3x^2 - 20y^2 - 12x + 80y - 96 = 0$
$3x^2 + 20y^2 = 80y + 48$

Solve each system of inequalities by graphing.

33. $x^2 + y^2 \geq 4$
$x^2 + y^2 \leq 36$

34. $x^2 + y^2 \leq 25$
$x + 2y \geq 1$

35. $x + y = 4$
$9x^2 - 4y^2 \geq 36$

36. $4x^2 + 9y^2 \geq 36$
$4y^2 + 9x^2 \leq 36$

37. $(y - 3)^2 \geq x + 2$
$x^2 \leq y + 4$

38. $(x + 2)^2 + 16(y + 3)^2 \geq 16$
$x + y = 0$

Write the system of equations or inequalities represented by each graph.

39.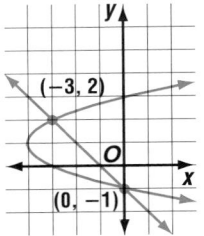

40.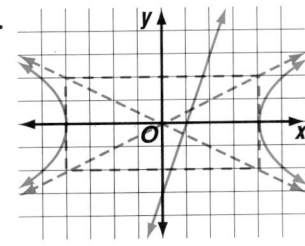

41.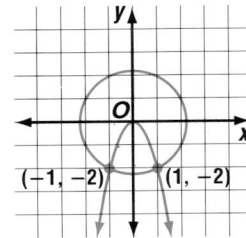

42.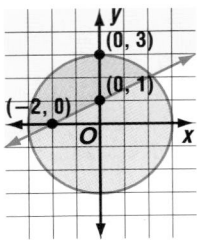

43.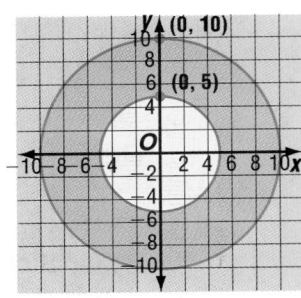

44.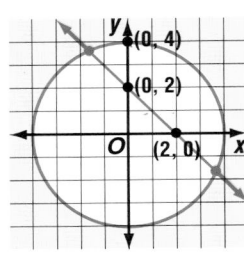

Graphing Calculator

Write a system of equations that will satisfy each condition stated. Use a graphing calculator to verify that you are correct.

45. two circles that intersect in three points

46. two parabolas that intersect in two points

47. a hyperbola and a circle that intersect in three points

48. a circle and an ellipse that do not intersect

49. a circle and an ellipse that intersect in four points

50. a hyperbola and an ellipse that intersect in two points

Critical Thinking

51. Solve each system of equations.

   a. $x + y^2 = 2$
   $2y - 2\sqrt{2} = x(\sqrt{2} + 2)$

   b. $x^2 + y^2 = 1$
   $y = 3x + 1$
   $x^2 + (y + 1)^2 = 4$

Applications and Problem Solving

52. **Advertising** The corporate logo for CBS Inc. is shown at the right. It is similar to a small circle and its interior, an ellipse and its exterior, and a large circle and its interior. Write three inequalities that, when graphed, will model the CBS logo on a coordinate plane.

53. **Seismology** Three tracking stations have detected an earthquake in an area. The first station is located at the origin on the map. Each grid in the map represents one square mile. The second tracking station is located at (0, 30) while the third station is located at (35, 18). The epicenter was 50 miles from the first station, 40 miles from the second station, and 13 miles from the third station. Where was the epicenter of the earthquake located?

54. **Astronomy** In the early 1600s, Johann Kepler studied the orbits of the planets and determined that they were elliptical. It is now known that orbits can take the shape of any of the conic sections. The equations of different orbits are listed below. State the shape of each orbit. (Lesson 7–6)

    a. $x^2 + y^2 = 75,000$

    b. $y - x^2 = 3x + 5$

55. Find the standard deviation for {5.2, 5.7, 6.0, 5.6, 2.4}. (Lesson 6–8)

56. Graph $2x^2 - x - 6 < y$. (Lesson 6–7)

57. Simplify $\dfrac{5ab^2 - 4ab + 7a^2b}{ab}$. (Lesson 5–3)

58. Find $[2\ \ -5\ \ 7] - [-3\ \ 8\ \ -1]$. (Lesson 4–2)

59. Solve the system of equations algebraically. (Lesson 3–2)

    $\dfrac{1}{2}x + \dfrac{1}{3}y = 2$

    $x - y = -1$

60. Find a and b so there is no solution to the system $5x - 4y = 12$ and $bx + ay = 3$. (Lesson 3–1)

61. **Statistics** The chart at the right shows the average amount of time each day that people watch television for the given years. (Lesson 2–5)

    a. Draw a scatter plot and find a prediction equation for the data.

    b. Predict the viewing time for the year 2000.

62. Solve $3m + 2 = 7n - 4$ for m. (Lesson 2–2)

63. Solve $-\dfrac{16}{19}k = 8$. (Lesson 1–4)

Year	Average Viewing Time
1950	4 h 35 min
1955	4 h 51 min
1960	5 h 6 min
1965	5 h 29 min
1970	5 h 56 min
1975	6 h 7 min
1980	6 h 36 min
1985	7 h 10 min

Mathematics and SOCIETY

Asteroids and Comets

The article below appeared in *Science News* on February 5, 1994.

AN ESTIMATED 2,000 ASTEROIDS LARGER than 1 kilometer follow orbits that cross that of Earth. Are any of these rocky missiles headed on a collision course with our planet? The impact, equivalent to the destructive power of 10,000 megatons of TNT, would dramatically disrupt life on our planet. . . . Clark R. Chapman of the Planetary Science Institute in Tucson, Ariz., and David Morrison of NASA's Ames Research Center in Mountain View, Calif., calculate that there is a 1-in-10,000 chance that a 2-km-wide asteroid or comet will collide with Earth during the next century. . . . A rocky body measuring 100 meters across could hit home about once every 100 to 200 years. ∎

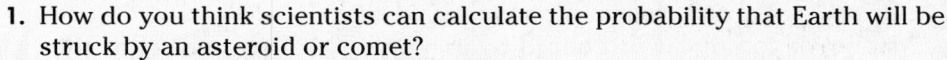

1. How do you think scientists can calculate the probability that Earth will be struck by an asteroid or comet?

2. What types of damage could result if a large asteroid or comet collided with Earth? What would affect the amount of damage?

3. If we discovered that a large asteroid or comet were on a collision course with Earth, what, if anything, could be done about it?

the *River Canyon Bridge*

Refer to the Investigation on pages 328–329.

When a bridge project is proposed to a department of highways, the department employs engineers, geologists, surveyors, engineers, and electronic draftsmen to work together to develop the plans for the bridge. Cost factors are examined as well as the feasibility of the building site. Traffic patterns are studied and weight limits are imposed. Many of the skills you have used in the exploration are used by the professionals to determine the correct bridge for the canyon highway.

Analyze

You have conducted experiments and organized your data in various ways. It is now time to analyze your findings and state your conclusions.

> **PORTFOLIO ASSESSMENT**
>
> You may want to keep your work on this Investigation in your portfolio.

1 Select one of your designs or create a new design that represents the best bridge for this location.

2 Build a two-dimensional model of the bridge using the building materials described in the Investigation. Cut a large rectangle from the cardboard box. Draw in a picture of the canyon. Build the model on top of the cardboard by using glue to anchor the pipe cleaners and string. Measure and calculate accurately, following your blueprint as closely as possible.

3 How does this model compare with the one you originally built in the Investigation?

Write

Once your model is completed, finish the project by completing the following.

4 Write a narrative describing the features and dimensions of your bridge. Justify why this design was chosen and why others were discarded.

5 Explain how you designed your model and how you used mathematics to plan its construction. Discuss the design of your bridge and why you believe it is sturdy and structurally sound.

6 Explain the process you used to construct your model.

A completed project should include:

- an accurately drawn blueprint, complete with labeled dimensions of the bridge and the parabolic equations that represent the arch of the bridge,

- a model of the bridge over the canyon, constructed to scale and following the designs and dimensions of your blueprint, and a well-written narrative describing the features, dimensions, and process of construction as outlined above.

VOCABULARY

After completing this chapter, you should be able to define each term, property, or phrase and give an example or two of each.

Algebra

asymptotes   (p. 441)
center of a circle   (p. 423)
center of an ellipse   (p. 432)
center of an hyperbola   (p. 441)
circle   (p. 423)
concentric circles   (p. 426)
conic section   (p. 415)
conjugate axis   (p. 441)

directrix   (p. 415)
distance formula   (p. 409)
ellipse   (p. 431)
foci of an ellipse   (p. 431)
foci of a hyperbola   (p. 440)
focus of a parabola   (p. 415)
hyperbola   (p. 440)
latus rectum   (p. 417)
major axis   (p. 432)
midpoint formula   (p. 410)

minor axis   (p. 432)
parabola   (p. 415)
radius   (p. 423)
tangent   (p. 425)
transverse axis   (p. 441)
vertex   (p. 441)

Problem Solving

use a simulation
   (p. 452)

UNDERSTANDING AND USING THE VOCABULARY

Tell whether each statement is true or false. If the statement is false, correct it to make it true.

1. An ellipse is the set of all points in a plane such that the sum of the distances from two given points in the plane, called the foci, is constant.

2. The equation of a circle is $(x - k)^2 - (y - h)^2 = r^2$.

3. The transverse axis is the line segment of a hyperbola of length $2a$ that has its endpoints at the vertices of the hyperbola.

4. The major axis is the longer of the two axes of symmetry of an ellipse.

5. The formula used to find the distance between two points is given by $d = \sqrt{(x_2 - x_1)^2 + (y_2 - y_1)^2}$.

6. A tangent line is a line that intersects a circle in exactly two points.

7. A parabola is the set of all points that are the same distance from a given point called the directrix and a given line called the focus.

8. The radius is the distance from the center of a circle to any point on the circle.

9. The conjugate axis is the line segment parallel to the transverse axis.

10. A conic section is formed by slicing a hollow double cone with a plane.

11. A hyperbola is the set of all points in a plane such that the absolute values of the sum of the distances from any point on the hyperbola to two given points is twice the distance of the conjugate axis.

12. An asymptote is a line that a curve approaches before changing directions.

13. The midpoint formula is given by the following: $\left(\dfrac{x_1 - x_2}{2}, \dfrac{y_1 - y_2}{2} \right)$.

14. The set of all points in a plane that are equidistant from a given point in a plane, called the center, form a circle.

15. The equation of a conic section is given by $Ax^2 + Bxy + Cy^2 + Dx + Ey + F = 0$.

SKILLS AND CONCEPTS

OBJECTIVES AND EXAMPLES

Upon completing this chapter, you should be able to:

- find the distance between two points in the coordinate plane (Lesson 7–1)

Find the distance between the points at $(6, -4)$ and $(-3, 8)$.

$$d = \sqrt{(x_2 - x_1)^2 + (y_2 - y_1)^2}$$
$$= \sqrt{(-3 - 6)^2 + [8 - (-4)]^2}$$
$$= \sqrt{81 + 144}$$
$$= \sqrt{225} \text{ or } 15$$

- find the midpoint of a line segment in the coordinate plane (Lesson 7–1)

Find the midpoint of the segment whose endpoints are at $(-5, 9)$ and $(11, -1)$.

$$\left(\frac{x_1 + x_2}{2}, \frac{y_2 + y_1}{2}\right) = \left(\frac{-5 + 11}{2}, \frac{9 + (-1)}{2}\right)$$
$$= \left(\frac{6}{2}, \frac{8}{2}\right) \text{ or } (3, 4)$$

- graph parabolas having certain properties (Lesson 7–2)

Graph $4y - x^2 = 14x - 27$.

First write the equation in the form $y = a(x - h)^2 + k$.

$$4y - x^2 = 14x - 27 \quad \rightarrow \quad y = \frac{1}{4}(x + 7)^2 - 19$$

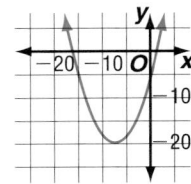

REVIEW EXERCISES

Use these exercises to review and prepare for the chapter test.

Find the distance between each pair of points with the given coordinates.

16. $(-2, 10)$ and $(-2, 13)$
17. $(8, 5)$ and $(-9, 4)$
18. $\left(3\sqrt{3}, 5\sqrt{3}\right)$ and $\left(-4\sqrt{3}, -7\sqrt{3}\right)$
19. $(7, -3)$ and $(1, 2)$
20. $(-13, 16)$ and $(5, -8)$

Find the midpoint of each line segment if the coordinates of the endpoints are given.

21. $(1, 2)$ and $(4, 6)$
22. $(-8, 0)$ and $(-2, 3)$
23. $(7a, -5b)$ and $(a, -3b)$
24. $\left(\frac{3}{5}, -\frac{7}{4}\right)$ and $\left(\frac{1}{4}, -\frac{2}{5}\right)$

Name the coordinates of the vertex and focus, the equations of the axis of symmetry and directrix, and the direction of opening of the parabola with the given equation. Then find the length of the latus rectum and graph the parabola.

25. $(x - 1)^2 = 12(y - 1)$
26. $(y + 6) = 16(x - 3)^2$
27. $x^2 - 8x + 8y + 32 = 0$
28. $x = 16y^2$

OBJECTIVES AND EXAMPLES

- graph circles having certain properties
  (Lesson 7–3)

  Graph $x^2 + y^2 + 8x - 24y + 16 = 0$.

  First write the equation in the form $(x - h)^2 + (y - k)^2 = r^2$: $(x + 4)^2 + (y - 12)^2 = 144$.

  Then draw the graph.

  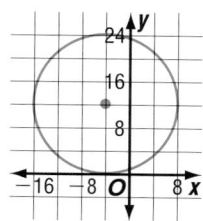

REVIEW EXERCISES

Find the coordinates of the center and the radius of each circle whose equation is given. Then draw the graph.

29. $x^2 + y^2 = 169$

30. $(x + 5)^2 + (y - 11)^2 = 49$

31. $x^2 + y^2 - 6x + 16y - 152 = 0$

32. $x^2 + y^2 + 6x - 2y - 15 = 0$

- graph ellipses having certain properties
  (Lesson 7–4)

  Graph $x^2 + 3y^2 - 16x + 24y + 31 = 0$.

  First write the equation in standard form.
  $$\frac{(x - 8)^2}{81} + \frac{(y + 8)^2}{27} = 1$$
  Then draw the graph.

  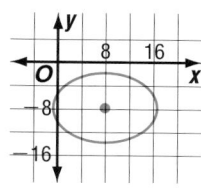

Find the coordinates of the center and foci, and the lengths of the major and minor axes for each ellipse whose equation is given. Then draw the graph.

33. $49x^2 + 16y^2 = 784$

34. $9x^2 + 4y^2 = 36$

35. $25x^2 + 64y^2 = 1600$

36. $\frac{x^2}{16} + \frac{y^2}{25} = 1$

- graph hyperbolas having certain properties
  (Lesson 7–5)

  Graph $9x^2 - 4y^2 + 18x + 32y - 91 = 0$.

  First write the equation in standard form.
  $$\frac{(x + 1)^2}{4} - \frac{(y - 4)^2}{9} = 1$$
  Then draw the graph.

  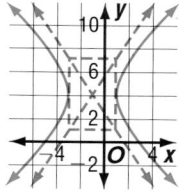

Find the coordinates of the vertices and foci and the slopes of the asymptotes for each hyperbola whose equation is given. Then draw the graph.

37. $9y^2 - 4x^2 = 36$

38. $25x^2 - 4y^2 = 100$

39. $9y^2 - 16x^2 = 144$

40. $x^2 - 25y^2 = 25$

OBJECTIVES AND EXAMPLES

• identify conic sections from their equations
(Lesson 7–6)

$$Ax^2 + Bxy + Cy^2 + Dx + Ey + F = 0 \quad (B = 0)$$

Conic Section	Relationship of A and C
parabola	A = 0 or C = 0, but not both.
circle	A = C
ellipse	A and C have the same sign and A ≠ C.
hyperbola	A and C have opposite signs.

• solve systems of equations involving quadratics
graphically and algebraically (Lesson 7–7)

A system of equations whose graphs are a line
and a conic section can have zero, one, or two
solutions. A system of equations whose graphs
are two conic sections can have zero, one, two,
three, or four solutions.

REVIEW EXERCISES

**State whether the graph of each equation is
a parabola, a circle, an ellipse, or a
hyperbola.**

41. $7x^2 + 9y^2 = 63$

42. $11x^2 + 11y^2 = 55$

43. $(x - 4)^2 = 6y$

44. $5y^2 - 13x^2 = 81$

45. $x^2 + y^2 = 289$

**Solve each system of equations
algebraically. Check your solution with a
graphing calculator.**

46. $x^2 + y^2 - 18x + 24y + 200 = 0$
    $4x + 3y = 0$

47. $4x^2 + y^2 - 48x - 2y + 129 = 0$
    $x^2 + y^2 - 2x - 2y - 7 = 0$

48. $x^2 + y^2 + 2x - 12y + 12 = 0$
    $y + x = 0$

APPLICATIONS AND PROBLEM SOLVING

49. Transportation A piano delivery truck
traveled 45 miles north on Interstate 71 before
turning east on Interstate 70. After traveling
60 miles on I-70, about how far is the truck from
its original starting point? (Lesson 7–1)

50. Telecommunications The cross-section of a
satellite dish is in the shape of a parabola. For
this dish, the receiver is located at the focus
and is 3 feet above the vertex. Find an equation
for the cross-section. Assume that the vertex is
at the origin. (Lesson 7–2)

51. Aviation The control
tower for the Metro
Blairsville Airport is
located at (9, 23) on the
county map. The radar
used at the airport can
detect airplanes up to
45 miles away. Write an
equation for the position of the most distant
plane that the people in the control tower can
detect in terms of the county map. (Lesson 7–3)

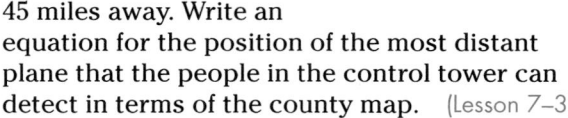

**A practice test for Chapter 7 is provided on
page 918.**

ALTERNATIVE ASSESSMENT

PERFORMANCE ASSESSMENT TASK

Architectural Geometry   You have been commissioned to design a circular theater with circular rows surrounding a circular central stage. The floor of the building is a square 1500 feet on each side. You are interested in maximizing the seating capacity of this theater, without sacrificing the comfort of the patrons who will purchase theater tickets.

Consider three different options.
- a stage with a 75-foot diameter
- a stage with a 100-foot diameter
- a stage with a 150-foot diameter

The marketing division of your company believes that an average price of $25 per ticket is the best price for this theater. Write a proposal for the width of the rows, taking into careful consideration the advantages and disadvantages of the different widths. Compare your row width with each size of stage. Which size of stage along with which row width will allow for the most patrons and the greatest comfort?

The people who will see your proposals are impressed with facts and figures, so be sure to include tables and graphs to illustrate which proposal you believe would be best for the theater and its patrons.

THINKING CRITICALLY

- Write equations for two parabolas that do not intersect.
- Write equations for a line and a parabola that intersect only at the point (2, 5).
- Write equations for a parabola and a circle that intersect the line $y = 2x + 3$ in the same point.
- Write equations for a line, a parabola, and a circle that each contain the points (3, 0) and (0, −3).

PORTFOLIO

Describe what you have learned about conic sections, including how they are related. List the four different types of conic sections, comparing each standard equation to the others.

Write an example of an equation for each type of conic section. Then draw each on a separate coordinate grid, labeling important characteristics such as center, foci, vertex, axes of symmetry, and so on. Place these in your portfolio.

SELF EVALUATION

"I dare you." Did you ever hear or say these words when you were younger? How daring are you? A good problem solver is willing to try new ways of approaching problems.

Assess yourself. Do you try new ways to solve problems? Taking risks and not using familiar ways of solving problems can be scary. However, if you learn to take risks, you will be rewarded with a thorough understanding of the concepts you are taught.

In·ves·ti·ga·tion

MATERIALS NEEDED

2 sizes of coffee cans

wooden chopsticks

duct tape

large nail

hammer

metal snips

metric ruler

funnel

metric measuring cup

Imagine you work for a major petroleum company as a technical support representative for all of the gas stations in a certain region. Gasoline is stored in underground cylindrical tanks just below the surface of the station. The tanks are positioned with their circular bases perpendicular to the surface. An opening on the top center of the tank is used to fill and measure the gasoline in the tank. A pipe at the bottom of the tank transfers gasoline to the pumps.

A dipstick measures the amount of gasoline left in the storage tank. It is lowered through the opening in the storage tank until it touches the bottom of the tank. Then it is removed. The wet part of the stick indicates how many gallons of gasoline are left in the storage tank according to the markings on the stick. It is similar to the oil dipstick used in a car.

The owner of one of the gas stations that you service has misplaced his dipstick for one of his tanks. He could order a replacement stick, but since the tank is an odd size, the dipstick is very expensive. He asks you to design a stick with the appropriate scale marked on it.

In this Investigation, you will begin to research the problem by experimenting with coffee cans and wooden chopsticks. The coffee cans will represent the gas tanks, and the chopsticks will represent dipsticks.

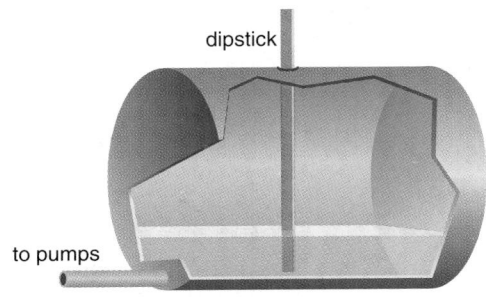

dipstick

to pumps

SET UP

1 Make a model of a gas tank by using a coffee can with the plastic lid taped securely onto the can with duct tape.

2 Punch a hole in the side of the coffee can using a nail and hammer. Using a pair of metal snips, make the hole big enough to insert a funnel.

COFFEE CAN EXPERIMENT

radius of can:	length of can:		capacity of can:	
water in can (mL)	measure on dipstick (cm)	% of can filled	% of dipstick covered	% change in capacity
50 mL				
100 mL				
⋮				

3 Copy the chart above onto a sheet of paper.

COLLECT DATA

4 Measure the radius of the lid and the length of the coffee can. Record these data in your chart.

5 Use a funnel and a metric measuring cup to pour 50 mL of water into the coffee can through the hole on its side. Lower the chopstick into the hole until it touches the bottom and pull it out. With a metric ruler, measure the distance on the chopstick that is wet and record the measurement in your chart.

6 Repeat this process, adding 50 mL at a time, until the coffee can is full. Did you use all of the last 50 mL? What is the capacity of the coffee can in mL? Record this value in your chart.

7 Estimate how much water is in the can when it is one-half full, one-fourth full, and three-fourths full. Make a note of these numbers.

8 For each 50-mL increment, calculate the ratio of the amount of water in the coffee can to the capacity of the coffee can. Convert these ratios to percents of a full coffee can. Record these data in your chart. Were your estimates correct for the amount of water when the can is one-half full, one-fourth full, and three-fourths full?

9 For each increment, calculate the ratio of the measure on the dipstick to the total length of the dipstick. Convert these ratios to percents of a fully-covered dipstick. Record these data in your chart.

10 Make a drawing to illustrate the measurement scale for a dipstick for this coffee can. Include markings for one-fourth, one-half, and three-fourths full.

11 How does the height of the water change as the amount of water in the can changes? Determine the change in the percent of water in the tank at each 50-mL increment. Record these data in your chart.

Make an Investigation Folder in which you can store all of your work on this Investigation for future use. Be sure to keep your chart and materials in your Investigation Folder.

You will continue working on this Investigation throughout Chapters 8 and 9.

Fill It Up! Investigation

Working on the Investigation
Lesson 8–3, p. 499

Working on the Investigation
Lesson 8–8, p. 534

Working on the Investigation
Lesson 9–1, p. 555

Working on the Investigation
Lesson 9–4, p. 575

Closing the Investigation
End of Chapter 9, p. 584

Exploring Polynomial Functions

Objectives

In this chapter, you will:

- find factors and zeros of polynomial functions,
- approximate real zeros of polynomial functions,
- graph polynomial functions,
- find the composition of functions,
- determine the inverses of functions or relations, and
- work backward to solve problems.

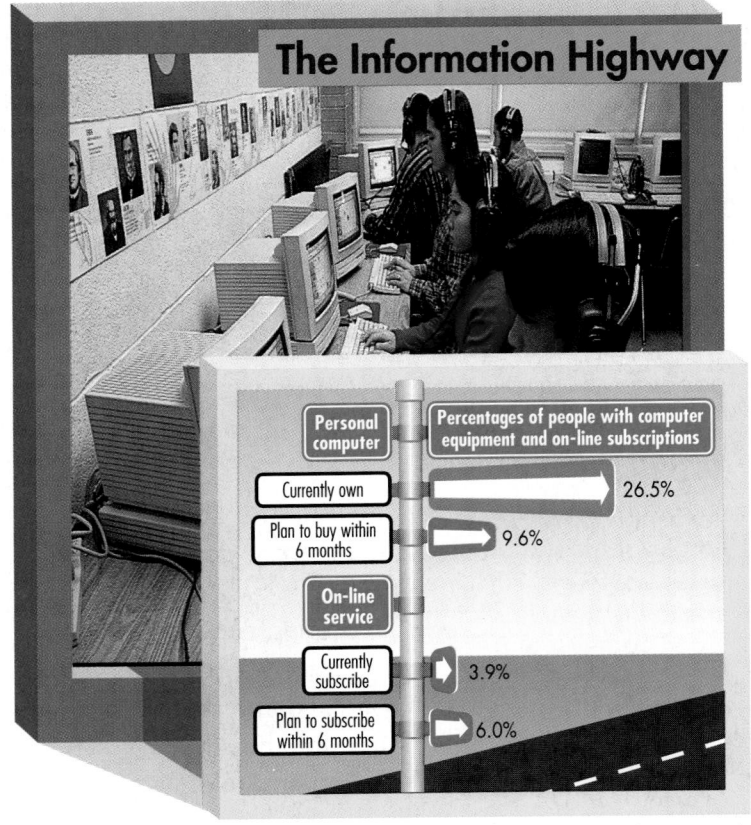

The Information Highway

Personal computer	Percentages of people with computer equipment and on-line subscriptions	
Currently own		26.5%
Plan to buy within 6 months		9.6%
On-line service		
Currently subscribe		3.9%
Plan to subscribe within 6 months		6.0%

Source: Bruskin/Goldring research poll of 1015 people

The Internet began in 1969 as a network for military sites, universities, and research institutions. Since the beginning, the Internet has been free and open, having no restrictions on who has access or what information is shared. Now Congress may decide to regulate the Net with the Communications Decency Act. Should the government protect minors from harm or should cyberspace be free of government regulation?

TIME*Line*

470 B.C. Hippasus of Greece discovers the dodecahedron, a regular solid with 12 regular pentagons as its faces.

1360 Francs are first coined in France.

600 B.C.   500   400   300   A.D. 900   950   1000   1050   1100   1150   1200   1250   1300

A.D. 1174 Earliest horse races begin in England.

At age 21, while at the University of Illinois' National Center for Supercomputer Applications, **Marc Andreessen** helped create the Internet browsing program Mosaic, which helps people point and click their way around the World Wide Web. In 1994, he joined Jim Clark, the founder of Silicon Graphics, to form a new company, Netscape, in California. He improved the Mosaic program, now called Navigator, and captured 70% of the Internet market, approaching 10 million customers. In the summer of 1995, Netscape went public on the NASDAQ market, and Marc became the owner of $565 million worth of stocks in the company.

Chapter Project

In small groups, plan an Internet research project. Have one person do research on the history of the Internet, its origins, early participants, and social implications for society. Another person should research the current Internet and its uses, business, educational, and personal. Investigate procedures for going online, who provides access, costs, and benefits of linking in cyberspace. The question of government regulation could be researched listing arguments for and against censorship and the political history involved. Group members should present and discuss their findings.

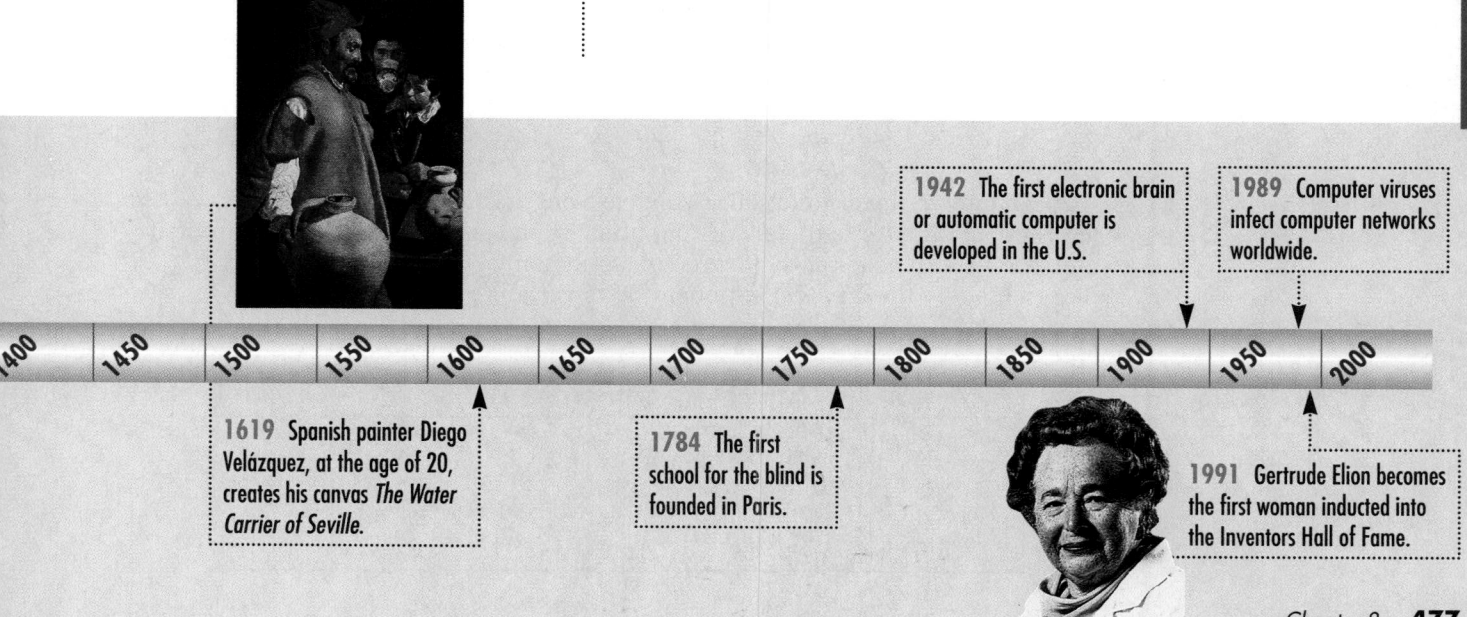

1942 The first electronic brain or automatic computer is developed in the U.S.

1989 Computer viruses infect computer networks worldwide.

1619 Spanish painter Diego Velázquez, at the age of 20, creates his canvas *The Water Carrier of Seville*.

1784 The first school for the blind is founded in Paris.

1991 Gertrude Elion becomes the first woman inducted into the Inventors Hall of Fame.

Polynomial Functions

What YOU'LL LEARN

- To evaluate polynomial functions, and
- to identify general shapes of the graphs of polynomial functions.

Why IT'S IMPORTANT

You can use polynomial functions to solve problems involving biology and energy.

CONNECTION
Biology

Calvin and Hobbes   by Bill Watterson

Calvin has to give up his ambition of migrating with the wildebeests when he learns that they live on another continent. The Serengeti Plain is a wild game reserve in Africa where herds of wildebeests, or gnus, roam. The population of wildebeests on the Serengeti Plain can be described by the function $f(x) = -0.125x^5 + 3.125x^4 + 58,000$, where x represents the number of years since 1990. The graph at the right shows this function. The expression $-0.125x^5 + 3.125x^4 + 58,000$ is a **polynomial in one variable.** *You will find the population of wildebeests in Example 2.*

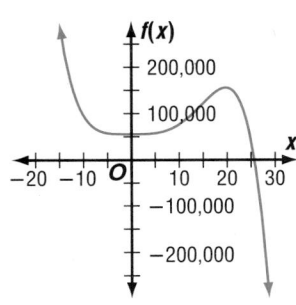

Definition of a Polynomial in One Variable	A polynomial of degree n in one variable x is an expression of the form $a_0x^n + a_1x^{n-1} + \ldots + a_{n-2}x^2 + a_{n-1}x + a_n$, where the coefficients $a_0, a_1, a_2, \ldots, a_n$ represent real numbers, a_0 is not zero, and n represents a nonnegative integer.

In Chapter 2, linear functions, which are degree 1, were identified and graphed. In Chapter 6, quadratic functions, which are degree 2, were identified and graphed. In general, the degree of a polynomial in one variable is determined by the greatest exponent of its variable.

Remember that $4 = 4x^0$ and $x + 8 = x^1 + 8x^0$.

Polynomial	Expression	Degree
Constant	4	0
Linear	$x + 8$	1
Quadratic	$3x^2 + 4x - 3$	2
Cubic	$4x^3 - 5$	3
General	$a_0x^n + a_1x^{n-1} + \ldots + a_{n-2}x^2 + a_{n-1}x + a_n$	n

Example **1** Determine if each expression is a polynomial in one variable. If so, determine its degree.

a. $6x^4 + 3x^2 + 4x - 8$

This is a polynomial in one variable, x. The degree is 4.

b. $9x^3y^5 + 2x^2y^6 - 4$

This is not a polynomial in one variable. It contains two variables, x and y.

c. $t^{-3} + 4t^2 - 1$

This is not a polynomial, because the variable has a negative exponent.

d. $5x^7 + 3x^2 + \dfrac{2}{x}$

This is not a polynomial, because the term $\dfrac{2}{x}$ cannot be written in the form x^n, where n is a nonnegative integer.

When a polynomial equation is used to represent a function, the function is a **polynomial function.** For example, the equation $f(x) = 4x^2 - 5x + 2$ describes a quadratic polynomial function, and the equation $p(x) = 2x^3 + 4x^2 - 5x + 7$ describes a cubic polynomial function. These and other polynomial functions can be defined by the following general rule.

Definition of a Polynomial Function	A polynomial function of degree n can be described by an equation of the form $P(x) = a_0x^n + a_1x^{n-1} + \ldots + a_{n-2}x^2 + a_{n-1}x + a_n$, where the coefficients $a_0, a_1, a_2, \ldots, a_{n-1}$, and a_n represent real numbers, a_0 is not zero, and n represents a nonnegative integer.

If you know an element in the domain of any polynomial function, you can find the corresponding value in the range. Remember that if $f(x)$ is the function and 4 is an element in the domain, the corresponding element in the range is $f(4)$. To find $f(4)$, evaluate the function for $x = 4$.

Example **2** Refer to the application at the beginning of the lesson. Use the polynomial function to estimate the population of wildebeests in 1995.

CONNECTION
Biology

x represents the number of years since 1990, $1995 - 1990$ or 5.

$$f(x) = -0.125x^5 + 3.125x^4 + 58,000$$
$$f(5) = -0.125(5)^5 + 3.125(5)^4 + 58,000 \quad \textit{Replace x with 5.}$$
$$= -390.625 + 1953.125 + 58,000 \text{ or } 59,562.5 \quad \textit{Evaluate.}$$

Therefore, in 1995, there were approximately 59,562 wildebeests.

Example **3** a. Find $p(a + 2)$ if $p(x) = x^3 - 2x + 1$.

$$p(a + 2) = (a + 2)^3 - 2(a + 2) + 1 \quad \textit{Substitute a + 2 for x.}$$
$$= a^3 + 6a^2 + 12a + 8 - 2a - 4 + 1$$
$$= a^3 + 6a^2 + 10a + 5 \quad \textit{(continued on the next page)}$$

b. Find $-2p(a) + p(a + 1)$ **if** $p(x) = x^3 + 3x^2 - 5.$

$$-2p(a) + p(a + 1) = [-2(a^3 + 3a^2 - 5)] + [(a + 1)^3 + 3(a + 1)^2 - 5]$$
$$= -2a^3 - 6a^2 + 10 + a^3 + 3a^2 + 3a + 1 + 3(a^2 + 2a + 1) - 5$$
$$= -2a^3 - 6a^2 + 10 + a^3 + 3a^2 + 3a + 1 + 3a^2 + 6a + 3 - 5$$
$$= -a^3 + 9a + 9$$

Remember that the x-coordinate of a point at which the graph crosses the x-axis is called a <u>zero</u> of the function. On the coordinate plane, these zeros are real numbers.

The graphs of several polynomial functions are shown below. Notice how many times the graph of each function intersects the *x*-axis. In each case, this is the maximum number of real zeros the function may have. How does the degree compare to the maximum number of real zeros?

Constant function

$f(x) = 2$

Degree 0

Linear function

$f(x) = \frac{3}{2}x - 3$

Degree 1

Quadratic function

$f(x) = x^2 + 2x - 3$

Degree 2

Cubic function

$f(x) = x^3 - 5x + 2$

Degree 3

Quartic function

$f(x) = x^4 - 3x^3 - 2x^2 + 7x + 1$

Degree 4

Quintic function

$f(x) = x^5 - 5x^3 + 4x$

Degree 5

These are the general shapes of the graphs for polynomial functions with degree greater than 0 and positive **leading coefficients**. The leading coefficient is the coefficient of the term with the highest degree. So, for the function $f(x) = 5x^3 + 2x^2 + 8x - 1$, the leading coefficient is 5.

Notice that the simplest polynomial graphs have equations in the form $f(x) = x^n$, where *n* is a positive number. Note the general shapes of the graphs for even-degree polynomial functions and odd-degree polynomial functions.

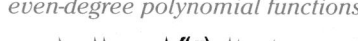

even-degree polynomial functions   *odd-degree polynomial functions*

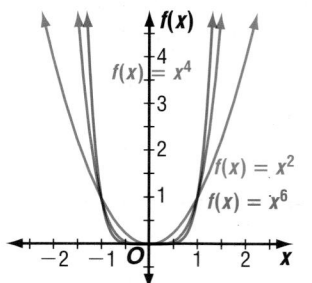
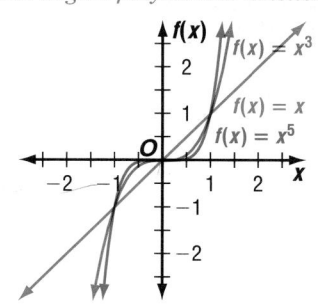

LOOK BACK

Refer to Lesson 6-1 for a review of roots of quadratic equations.

Note that the even-degree functions are tangent to the x-axis at the origin. When this happens, the function has two zeros or roots that are the same number. For example, $x^2 - 6x + 9 = 0$ can be factored as $(x - 3)(x - 3) = 0$. So 3 is the root of the equation.

An even-degree function may or may not intersect the x-axis, depending on its location in the coordinate plane. If it does not intersect the x-axis, its roots are all imaginary. An odd-degree function always crosses the x-axis at least once. *Why?*

Example Determine if each graph represents an odd-degree function or an even-degree function. Then state how many real zeros each function has.

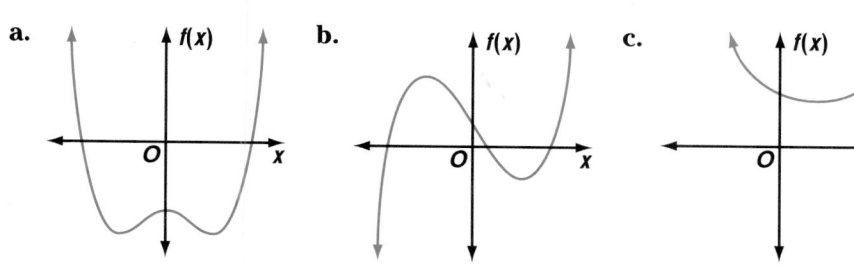

Graph	left-most y values	right-most y values	degree of function	times graph crosses x-axis	number of real zeros
a.	positive	positive	even	2	2
b.	negative	positive	odd	3	3
c.	positive	positive	even	0	0

In Chapter 2, you studied families of graphs of linear equations. In Chapter 6, you studied families of parabolas. Some families of graphs of polynomial equations are shown below. The equation below each graph is the equation of the parent graph for that family.

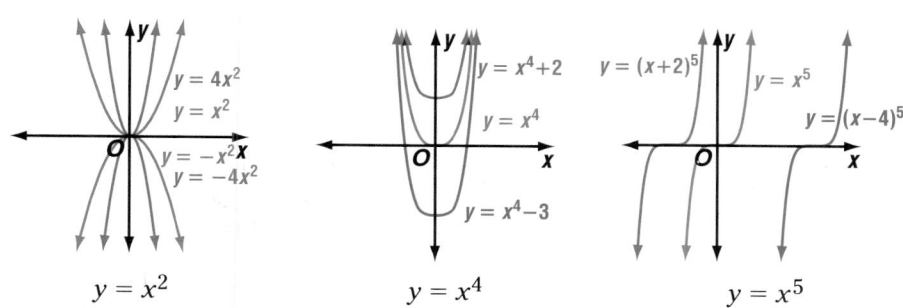

$$y = x^2$$ $$y = x^4$$ $$y = x^5$$

CHECK FOR UNDERSTANDING

Communicating Mathematics

Study the lesson. Then complete the following.

1. Refer to Example 4. What appears to be the degree of each function?

2. **Describe** the characteristics of the graphs of odd-degree and even-degree polynomial functions whose leading coefficients are positive.

3. **State** how many real zeros are possible for each polynomial function.

   a. quartic   **b.** linear   **c.** quadratic   **d.** quintic   **e.** cubic

4. **Sketch** the graph of an odd-degree function with a positive leading coefficient and three real roots.

5. **You Decide** Carlos explains to his friend Zach, "The graphs of odd-degree polynomial functions always intersect the *x*-axis an odd number of times, and the graphs of even-degree functions always intersect the *x*-axis an even number of times." Zach doesn't believe this is always the case. Who is correct? Give an example to support your answer.

 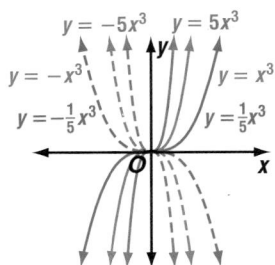

6. Look at the family of graphs at the right for the function $f(x) = x^3$. Investigate the relationship between the similar functions and their graphs.

Guided Practice

Find the degree of each polynomial in one variable. If it is not a polynomial in one variable, explain why.

7. $4t^3 + 8t^2 + 2t - 1$ 8. $9ab^2 + 4ab + 3$ 9. $7x^3 - 8x^5 + 8x - 7$

Identify each polynomial function as *linear*, *quadratic*, *cubic*, *quartic*, or *quintic*. State the degree and how many real zeros are possible.

10. $f(x) = 6x^3 + 8x + 7$ 11. $f(x) = 3x^5 + 7x^4 - 5x^3 + 6x + 9$

12. $f(x) = 2x + 5$ 13. $f(x) = 5x^3 + 6x^2 - 8x^4 - 10x + 17$

Match the polynomial and its functional value.

14. $p(x) = 3x^2 + 4x + 5$ a. $p(4) = -166$
15. $p(x) = x^4 - 7x^3 + 8x - 6$ b. $p(5) = 100$
16. $p(x) = 7x^2 - 9x + 10$ c. $p(-4) = -259$
17. $p(x) = 4x^3 - 2x^2 - 6x + 5$ d. $p(-2) = 56$

18. Refer to the graph at the right.
   a. Determine whether the degree of the function is even or odd.
   b. How many real zeros does the polynomial function have?

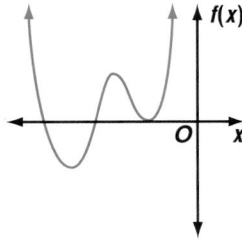

Find $p(2)$ and $p(-1)$ for each function.

19. $p(x) = 2x^2 + 6x - 8$ 20. $p(x) = -3x^4 + 1$

Find $f(x + h)$ for each function.

21. $f(x) = 2x - 3$ 22. $f(x) = 4x^2$

23. **Energy** The power generated by a windmill is a function of the speed of the wind. The approximate power is given by the function $P(s) = \dfrac{s^3}{1000}$, where *s* represents the speed of the wind in kilometers per hour. Find the units of power generated by a windmill when the wind speed is 25 kilometers per hour.

Practice

Determine whether the degree of the function represented by each graph is even or odd. How many real zeros does each polynomial function have?

24.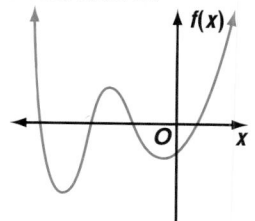

25.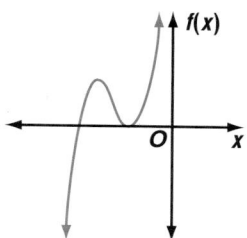

26.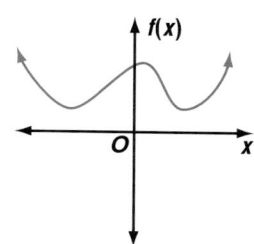

Find $p(3)$ and $p(-2)$ for each function.

27. $p(x) = 5x + 6$

28. $p(x) = x^2 - 2x + 1$

29. $p(x) = 2x^3 - x^2 - 3x + 1$

30. $p(x) = x^5 - x^2$

31. $p(x) = -x^4 + 53$

32. $p(x) = x^5 + 5x^4 - 15x^2 - 8$

Find $f(x + h)$ for each function.

33. $f(x) = x + 2$

34. $f(x) = x - 4$

35. $f(x) = 5x^2$

36. $f(x) = x^2 - 2x + 5$

37. $f(x) = 3x^2 + 7$

38. $f(x) = x^3 + x$

Find $4[p(x)]$ for each function.

39. $p(x) = x^2 + 5$

40. $p(x) = 6x^3 - 4x^2 + 2$

41. $p(x) = \dfrac{x^3}{4} + \dfrac{x^2}{16} - 2$

Find an equation for each graph.

42.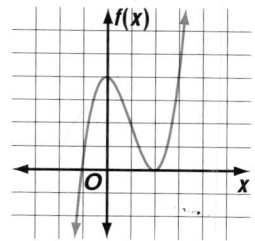

43.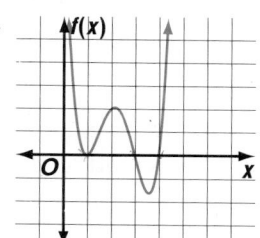

44.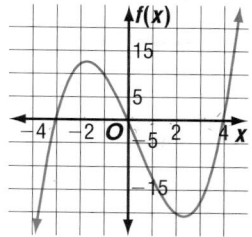

45. Sketch a graph of a polynomial function $f(x)$ that has the indicated number and type of zeros.

   a. 5 real
   b. 3 real, 2 imaginary
   c. 4 imaginary

Find $2p(a) + p(a - 1)$ for each function.

46. $p(x) = 4x + 1$

47. $p(x) = x^2 + 3$

48. $p(x) = x^2 - 5x + 8$

Find $2[f(x + 3)]$ for each function.

49. $f(x) = 2x + 9$

50. $f(x) = x^2 - 6$

51. $f(x) = x^2 + 3x + 12$

Graphing Calculator

52. Sketch a graph that matches each description. Write an equation for the graph of each function.

   a. a quadratic function with zeros at -1 and 2.
   b. a cubic function with zeros at -2, -1, and 2.
   c. a quartic function with zeros at -2, -1, 1, and 2.
   d. a quintic function with zeros at -2, -1, 0, 1, and 2.

53. Although a fourth-degree function can have as many as four real zeros, $P(x) = x^4 + x^2 + 1$ has no real zeros. Can you explain why?

54. The graph of the polynomial function $f(x) = ax(x - 4)(x + 1)$ goes through the point at $(5, 15)$.

   a. Find the value of a.

   b. Sketch the graph of the function.

**Applications and
Problem Solving**

55. Biology   The intensity of light emitted by a firefly can be determined by the polynomial function $L(t) = 10 + 0.3t + 0.4t^2 - 0.01t^3$, where t is the temperature in Celsius and $L(t)$ is the light intensity in lumens. If the temperature is $30°$ C, find the light intensity.

56. Patterns   If you look at a cross section of a honeycomb, you see a pattern of hexagons. This pattern has one hexagon surrounded by six more hexagons. Surrounding these is a third "ring" of 12 hexagons, and so on. Assume that the pattern continues.

   a. Find the number of hexagons in the 4th ring.

   b. Make a table that shows the total number of hexagons in the first ring, the first two rings, the first three rings, and the first four rings.

   c. Show that the polynomial function $h = 3r^2 - 3r + 1$ gives the total number of hexagons when $r = 1, 2, 3$. Identify the domain and range of this function.

   d. Use the equation to find the total number of hexagons in a honeycomb with 12 rings.

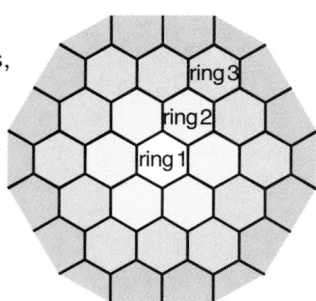

**Mixed
Review**

57. Seismology   Two tracking stations have detected an earthquake. The first station determined that the epicenter was 25 miles away. The second station determined that the epicenter was 42 miles away. If the first station is located at the origin and the second station is 50 miles due east of the first station, where could the epicenter have been?   (Lesson 7–7)

58. Write the standard form of the equation $x^2 + 4y^2 = 4$. Graph the equation and state whether the graph is a parabola, a circle, an ellipse, or a hyperbola.   (Lesson 7–6)

59. Find two numbers whose difference is -40 and whose product is a minimum.   (Lesson 6–4)

60. Solve $\sqrt{n + 12} - \sqrt{n} = 2$.   (Lesson 5–8)

61. Biology   The average human has between 1,600,000 and 1,700,000 sweat glands, mostly located in the palms of the hand and soles of the feet. Express both of these figures in scientific notation.   (Lesson 5–1)

62. Determine the dimensions of the product $A_{4 \times 2} \cdot B_{2 \times 3}$.   (Lesson 4–3)

63. Find $\begin{bmatrix} -9 & 6 \\ 5 & 19 \end{bmatrix} - \begin{bmatrix} -3 & 18 \\ -4 & 12 \end{bmatrix}$.   (Lesson 4–2)

64. In which octant does the point at $(7, -2, 9)$ lie?   (Lesson 3–7)

65. Solve the system of equations by using either the substitution or elimination method.   (Lesson 3–2)

$$2x - y = 36$$
$$3x - \frac{1}{2}y = 26$$

66. Evaluate $2\left| -3x \right| - 9$ if $x = 5$.   (Lesson 1–5)

The Remainder and Factor Theorems

What YOU'LL LEARN

- To find factors of polynomials by using the factor theorem and synthetic division.

Why IT'S IMPORTANT

You can use the remainder and factor theorems to find factors of polynomials that model situations in engineering and architecture.

APPLICATION
Sports

A javelin is usually thrown from about shoulder height, around 5 feet off the ground. It is not unusual for the javelin to start off with an upward velocity of about 74 feet per second. Based on this information, one can determine that the height of a javelin t seconds after it is thrown can be described by the function $h(t) = -16t^2 + 74t + 5$, if the effect of air resistance is ignored. The 16 in the function is associated with the strength of the Earth's gravity; if the javelin were thrown on another planet, there would be a different number in the equation.

The graph of $h(t)$ is shown at the right. Notice that when $t = 0$, the value of $h(t)$ is 5. Suppose we find the height of the javelin after 3 seconds.

$$h(t) = -16t^2 + 74t + 5$$

$$h(3) = -16(3)^2 + 74(3) + 5 \quad \textit{Replace t with 3.}$$

$$= -144 + 222 + 5$$

$$= 83$$

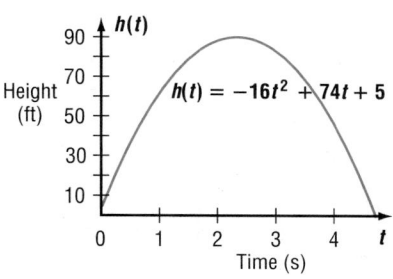

After 3 seconds, the height of the javelin is 83 feet.

Divide the polynomial in the function by $t - 3$, and compare the remainder to $h(3)$.

<table>
<tr><td>

Method 1: Long Division

</td><td>

Method 2: Synthetic Division

</td></tr>
</table>

Method 1: Long Division

$$
\begin{array}{r}
-16t + 26 \\
t - 3\overline{)-16t^2 + 74t + 5} \\
\underline{-16t^2 + 48t} \\
26t + 5 \\
\underline{26t - 78} \\
83
\end{array}
$$

Method 2: Synthetic Division

$$
\begin{array}{r}
\underline{3}\begin{array}{|rrr}
-16 & 74 & 5 \\
 & -48 & 78 \\
\hline
-16 & 26 & \begin{array}{|r}83\end{array}
\end{array}
\end{array}
$$

Notice that the value of $h(3)$ is the same as the remainder when the polynomial is divided by $t - 3$. This illustrates the **remainder theorem.**

FYI

In the 1980s, the aerodynamic design of javelins was so advanced that athletes began throwing them into stands of spectators and out of stadiums by accident. The men's javelins were re-engineered for the 1996 Olympics, so as not to endanger spectators or to make stadiums obsolete in size.

LOOK BACK

Refer to Lesson 5-3 for a review of synthetic division.

The Remainder Theorem	If a polynomial $f(x)$ is divided by $x - a$, the remainder is the constant $f(a)$, and $$\text{dividend} = \text{quotient} \cdot \text{divisor} + \text{remainder}$$ $$f(x) = q(x) \cdot (x - a) + f(a),$$ where $q(x)$ is a polynomial with degree one less than the degree of $f(x)$.

Example **1** Let $f(x) = 3x^4 - x^3 + 2x - 6$. Show that $f(2)$ is the remainder when $f(x)$ is divided by $x - 2$.

Use synthetic division to divide by $x - 2$.

$$
\begin{array}{r|rrrrr}
2 & 3 & -1 & 0 & 2 & -6 \\
  &   & 6 & 10 & 20 & 44 \\
\hline
  & 3 & 5 & 10 & 22 & \,|\,38
\end{array}
$$ *Long division could also be used.*

The quotient is $3x^3 + 5x^2 + 10x + 22$ with a remainder of 38.

Now find $f(2)$. $f(2) = 3(2)^4 - (2)^3 + 2(2) - 6$
$$= 48 - 8 + 4 - 6 \text{ or } 38$$

Thus, $f(2) = 38$, the same number as the remainder after division by $x - 2$.

As illustrated in Example 1, synthetic division can be used to find the value of a function. When synthetic division is used to find the value of a function, it is called **synthetic substitution**. This is a convenient way of finding the value of a function, especially when the degree of the polynomial is greater than 2.

Example **2** If $f(x) = x^4 + 2x^3 - 10x^2 + 5x - 7$, find $f(8)$.

Method 1: Synthetic Substitution

When $f(x)$ is divided by $x - 8$, the remainder is $f(8)$.

$$
\begin{array}{r|rrrrr}
8 & 1 & 2 & -10 & 5 & -7 \\
  &   & 8 & 80 & 560 & 4520 \\
\hline
  & 1 & 10 & 70 & 565 & \,|\,4513
\end{array}
$$

The remainder is 4513. Thus, by synthetic substitution, $f(8) = 4513$.

Method 2: Direct Substitution

$f(8) = (8)^4 + 2(8)^3 - 10(8)^2 + 5(8) - 7$
$$= 4096 + 1024 - 640 + 40 - 7 \text{ or } 4513$$

By substitution, $f(8) = 4513$, the same result as by synthetic substitution.

Consider $f(x) = x^4 + x^3 - 13x^2 - 25x - 12$. If $f(x)$ is divided by $x - 4$, then the remainder is 0. Therefore, 4 is a zero of $f(x)$.

$$
\begin{array}{r|rrrrr}
4 & 1 & 1 & -13 & -25 & -12 \\
  &   & 4 & 20 & 28 & 12 \\
\hline
  & 1 & 5 & 7 & 3 & \,|\,0
\end{array}
$$

The quotient of $f(x)$ and $x - 4$ is $x^3 + 5x^2 + 7x + 3$.

Check: $f(x) = x^4 + x^3 - 13x^2 - 25x - 12$
$$f(4) \stackrel{?}{=} (4)^4 + (4)^3 - 13(4)^2 - 25(4) - 12$$
$$0 \stackrel{?}{=} 256 + 64 - 208 - 100 - 12$$
$$0 = 0 \ \checkmark$$

From the results of the division and by using the remainder theorem, we can make the following statement.

$$\underset{\text{dividend}}{x^4 + x^3 - 13x^2 - 25x - 12} \;=\; \underset{\text{quotient}}{(x^3 + 5x^2 + 7x + 3)} \;\cdot\; \underset{\text{divisor}}{(x - 4)} \;+\; \underset{\text{remainder}}{0}$$

Since the remainder is 0, $x - 4$ is a factor of $x^4 + x^3 - 13x^2 - 25x - 12$. This illustrates the **factor theorem**, which is a special case of the remainder theorem.

The Factor Theorem	The binomial $x - a$ is a factor of the polynomial $f(x)$ if and only if $f(a) = 0$.

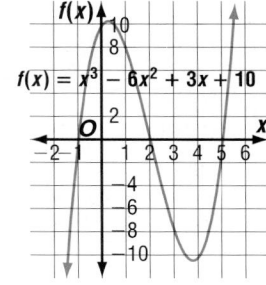

$f(x) = x^3 - 6x^2 + 3x + 10$

Suppose you wanted to find the zeros of $f(x) = x^3 - 6x^2 + 3x + 10$. From the graph at the right, you can see that the graph crosses the x-axis at -1, 2, and 5. These are the zeros of the function. Using these zeros and the zero product property, we can express the polynomial in factored form.

$$f(x) = (x + 1)(x - 2)(x - 5)$$

Many polynomial functions are not easily graphed and once graphed, the exact zeros are often difficult to determine. The factor theorem can help you find all the factors of a polynomial. Suppose we wanted to determine if $x - 3$ is a factor of $x^3 + 5x^2 - 12x - 36$ and, if it is, what the other factors are.

Let $f(x) = x^3 + 5x^2 - 12x - 36$. The binomial $x - 3$ is a factor of the polynomial if 3 is a zero. Use the factor theorem.

$$\begin{array}{r|rrrr} 3 & 1 & 5 & -12 & -36 \\ & & 3 & 24 & 36 \\ \hline & 1 & 8 & 12 & 0 \end{array}$$

Since the remainder is 0, $x - 3$ is a factor of the polynomial. Further, since $x - 3$ is a factor of the polynomial, it follows that the remainder is 0.

When you divide a polynomial by one of its binomial factors, the quotient is called a **depressed polynomial.** The polynomial $x^3 + 5x^2 - 12x - 36$ can be factored as $(x - 3)(x^2 + 8x + 12)$. The polynomial $x^2 + 8x + 12$ is the depressed polynomial, which also may be factorable.

$$x^2 + 8x + 12 = (x + 2)(x + 6)$$

So, $x^3 + 5x^2 - 12x - 36 = (x - 3)(x + 2)(x + 6)$.

Example 3

APPLICATION

Engineering

When a certain type of plastic is cut into sections, the length of each section determines the strength of the plastic. The function $f(x) = x^4 - 14x^3 + 69x^2 - 140x + 100$ can describe the relative strength of a section of length x feet. Sections of plastic x feet long, where $f(x) = 0$, are extremely weak. After testing the plastic, engineers discovered that sections 2 feet long and 5 feet long were extremely weak.
a. Show that $x - 2$ and $x - 5$ are factors of the polynomial function.
b. Find other lengths of plastic that are extremely weak, if they exist.

(continued on the next page)

a.
$$\begin{array}{r|rrrrr} 2 & 1 & -14 & 69 & -140 & 100 \\ & & 2 & -24 & 90 & -100 \\ \hline & 1 & -12 & 45 & -50 & 0 \end{array}$$

The remainder is 0, so $x - 2$ is a factor of $x^4 - 14x^3 + 69x^2 - 140x + 100$.

So, $x^4 - 14x^3 + 69x^2 - 140x + 100 = (x^3 - 12x^2 + 45x - 50)(x - 2)$.

$$\begin{array}{r|rrrr} 5 & 1 & -12 & 45 & -50 \\ & & 5 & -35 & 50 \\ \hline & 1 & -7 & 10 & 0 \end{array}$$

The remainder is 0, so $(x - 5)$ is a factor of $x^4 - 14x^3 + 69x^2 - 140x + 100$.

So, $x^4 - 14x^3 + 69x^2 - 140x + 100 = (x - 2)(x - 5)(x^2 - 7x + 10)$.

b. Are there other values of x at which $f(x) = 0$? If so, sections of plastic with these lengths will be extremely weak.

Factor the depressed polynomial, if possible.
$x^2 - 7x + 10 = (x - 2)(x - 5)$

So, $x^4 - 14x^3 + 69x^2 - 140x + 100 = (x - 2)(x - 5)(x - 2)(x - 5)$ or $(x - 2)^2 (x - 5)^2$.

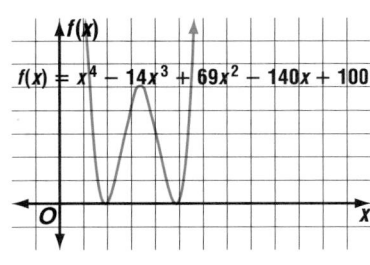

$f(x) = x^4 - 14x^3 + 69x^2 - 140x + 100$

The graph of the polynomial function touches the x-axis at 2 and 5. Thus, the only lengths of plastic that are extremely weak are 2 feet and 5 feet long.

CHECK FOR UNDERSTANDING

Communicating Mathematics

Study the lesson. Then complete the following.

1. If the divisor is a factor of a polynomial, then what is the remainder after division?

2. **You Decide**   Jack tells Mayuko that if $x - 6$ is a factor of a polynomial $f(x)$, then $f(6) = 0$. Mayuko argues that if $x - 6$ is a factor of $f(x)$, then $f(-6) = 0$. Who is correct? Explain.

3. **a. State** the zeros of the polynomial $P(x)$ whose graph is shown at the right.

   b. Describe what $P(x) + 100$ would look like. How many real zeros would it have?

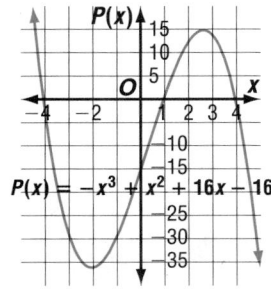

$P(x) = -x^3 + x^2 + 16x - 16$

4. **State** the degree of each polynomial. Then state the degree of the depressed polynomial that would result from dividing the polynomial by one of its binomial factors.

   a. $4x^4 + 3x^3 - 5x^2 + 8$ **b.** $5x^2 + 7x^5 - 8x - 2$

5. **Assess Yourself**   A depressed polynomial of degree 2 is the result of synthetic substitution. What options do you have to find the remaining factors? Which would you choose to use first? Explain.

Divide using synthetic division and write your answer in the form dividend = quotient · divisor + remainder. Is the binomial a factor of the polynomial?

6. $(x^3 - 4x^2 + 2x - 6) \div (x - 4)$ 7. $(x^4 - 16) \div (x - 2)$

Use synthetic substitution to find g(2) and g(−1) for each function.

8. $g(x) = x^3 - 5x + 2$ 9. $g(x) = x^4 - 6x - 8$

Given a polynomial and one of its factors, find the remaining factors of the polynomial. Some factors may not be binomials.

10. $x^3 + 2x^2 - x - 2; x - 1$ 11. $x^3 - 6x^2 + 11x - 6; x - 2$

12. $2x^3 + 7x^2 - 53x - 28; (x + 7)$ 13. $x^4 + 2x^3 + 2x^2 - 2x - 3; x + 1$

14. Use the graph of the polynomial function at the right to determine at least one binomial factor of the polynomial. Then find all factors of the polynomial.

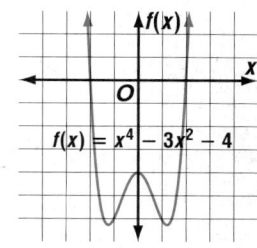

$f(x) = x^4 - 3x^2 - 4$

15. Use synthetic substitution to show that $x - 8$ is a factor of $x^3 - 4x^2 - 29x - 24$. Then find any remaining factors.

EXERCISES

Divide using synthetic division and write your answer in the form dividend = quotient · divisor + remainder. Is the binomial a factor of the polynomial?

16. $(x^3 - 6x^2 + 2x - 4) \div (x - 2)$ 17. $(x^3 - 2x^2 - 5x + 6) \div (x - 3)$

18. $(2x^3 + 8x^2 - 3x - 1) \div (x - 2)$ 19. $(x^3 + 27) \div (x + 3)$

20. $(2x^3 + x^2 - 8x + 16) \div (x + 4)$ 21. $(6x^3 + 9x^2 - 6x + 2) \div (x + 2)$

22. $(x^3 - 64) \div (x - 4)$ 23. $(4x^4 - 2x^2 + x + 1) \div (x - 1)$

Use synthetic substitution to find f(2) and f(−1) for each function.

24. $f(x) = x^3 - 2x^2 - x + 1$ 25. $f(x) = 2x^2 - 8x + 6$

26. $f(x) = x^3 + 2x^2 - 3x + 1$ 27. $f(x) = x^3 - 8x^2 - 2x + 5$

28. $f(x) = 5x^4 - 6x^2 + 2$ 29. $f(x) = 3x^4 + x^3 - 2x^2 + x + 12$

Given a polynomial and one of its factors, find the remaining factors of the polynomial. Some factors may not be binomials.

30. $x^3 - x^2 - 5x - 3; x + 1$ 31. $x^3 - 3x + 2; x - 1$

32. $x^3 + x^2 - 16x - 16; x - 4$ 33. $6x^3 - 25x^2 + 2x + 8; 3x - 2$

34. $2x^3 + 17x^2 + 23x - 42; 2x + 7$ 35. $x^4 + 2x^3 - 8x - 16; x + 2$

36. $8x^4 + 32x^3 + x + 4; 2x + 1$ 37. $16x^5 - 32x^4 - 81x + 162; x - 2$

Use the graph of each polynomial function to determine at least one binomial factor of the polynomial. Then find all of the factors.

38. $f(x) = x^5 + x^4 - 3x^3 - 3x^2 - 4x - 4$

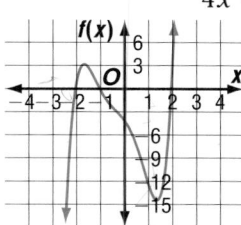

39. $f(x) = x^4 + 7x^3 + 15x^2 + 13x + 4$

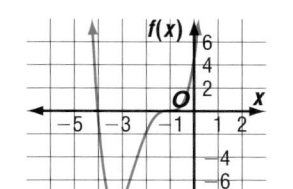

Find values for _k_ so that each remainder is 3.

40. $(x^2 - x + k) \div (x - 1)$

41. $(x^2 + kx - 17) \div (x - 2)$

42. $(x^3 + 4x^2 + x + k) \div (x + 2)$

43. $(x^2 + 5x + 7) \div (x + k)$

Critical Thinking

44. Consider the function $f(x) = x^3 + 2x^2 - 5x - 6$.

  a. Use synthetic substitution to find $f(-4)$, $f(-2)$, $f(0)$, $f(2)$, and $f(4)$.

  b. On a coordinate plane, graph the ordered pairs of the form $(x, f(x))$ you found and connect them to make a smooth curve.

  c. How many times does the graph cross the _x_-axis? Does this agree with what you learned in Lesson 8-1 about graphs of polynomial functions?

Applications and Problem Solving

45. **Architecture** According to the _Guinness Book of Records,_ the 1454-foot Sears Tower in Chicago, Illinois, is the tallest office building in the world. It has 110 floors and 18 elevators. The elevators traveling from one floor to the next do not travel at a constant speed. Suppose the speed of an elevator in feet per second is given by the function $F(t) = -0.5t^4 + 4t^3 - 12t^2 + 16t$, where _t_ is the time in seconds.

  a. Find the speed of the elevator at 1, 2, and 3 seconds.

  b. It takes 4 seconds for the elevator to go from one floor to the next. Use synthetic substitution to find $f(4)$. Explain what this means.

46. **Technology** The graph at the right shows how the cost of computer viruses in the U.S. has drastically increased since 1990. The function $C(x) = 0.03x^3 - 0.02x^2 + 0.2x + 0.1$ estimates this cost over the interval shown, with $x = 0$ representing the year 1990. Estimate the cost of computer viruses in the year 1999.

Computer Bugs (in billions)

$0.1 $0.3 $0.7 $1.4 $2.7

'90   '91   '92   '93   '94

Source: National Computer Security Association

Mixed Review

47. **Art** Joyce Jackson purchases works of art for an art gallery. Two years ago, she bought a painting for $20,000, and last year, she bought one for $35,000. If these paintings appreciate at 14% per year, how much are the two paintings worth now?   (Lesson 8–1)

48. Find the coordinates of the vertices and foci and the slopes of the asymptotes for the hyperbola given by the equation $\dfrac{y^2}{81} - \dfrac{x^2}{25} = 1$.   (Lesson 7–5)

49. Is $(4, -4)$ a solution to the quadratic inequality $-y \le -x^2 + 5x$?   (Lesson 6–7)

50. Solve $2x^2 - 5x + 4 = 0$ by using the quadratic formula.   (Lesson 6–4)

51. Find $(6x^3 - 5x^2 - 12x - 4) \div (3x + 2)$.   (Lesson 5–3)

52. Find $7\begin{bmatrix} -1 & 4 \\ 8 & -6 \end{bmatrix} + 2\begin{bmatrix} 6 & -5 \\ 1 & 8 \end{bmatrix}$.   (Lesson 4–2)

53. Solve $\begin{bmatrix} 2x \\ y + 1 \end{bmatrix} = \begin{bmatrix} y \\ 3 \end{bmatrix}$ for _x_ and _y_.   (Lesson 4–1)

54. If $a = -1$, $b = 7$, $c = 4$, and $-a^3b^2 + 2ab + 3d \ge \frac{1}{2}c^3$, solve for _d_.   (Lesson 1–6)

8–3A Graphing Technology
Polynomial Functions

A Preview of Lesson 8–3

You can use a graphing calculator to graph polynomial functions and approximate the real zeros of the function. When using a calculator to approximate zeros, it is important to view a complete graph of the function before zooming in on a certain point. Otherwise, zeros may be overlooked because they were not in the viewing window. Remember that a complete graph of a function shows all the characteristics of the graph such as all x- and y-intercepts, relative maximum and minimum points, and the end behavior of the graph.

Example **1** **Use a graphing calculator to obtain a complete graph of $f(x) = 2x^3 + 6x^2 - 14x + 12$. Then approximate each real zero to the nearest hundredth.**

Let's try graphing in the standard viewing window.

Enter: [Y=] 2 [X,T,θ,n] [∧] 3 [+]

6 [X,T,θ,n] [x²] [−] 14

[X,T,θ,n] [+] 12 [ZOOM] 6

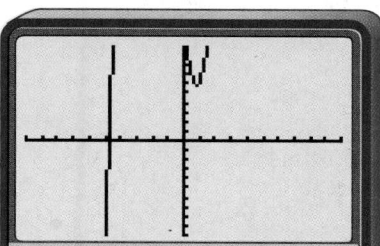

We see that this viewing window does not contain a complete graph. Change the viewing window to $[-10, 10]$ by $[-10, 60]$ with a scale factor of 1 for the x-axis and 5 for the y-axis.

This window can accommodate the complete graph.

LOOK BACK

Refer to Lesson 6-1A for information on using the automatic ROOT feature.

According to the graph, there is one x-intercept (real zero) for this function. There are three zeros for any third-degree polynomial, so two of the zeros for this function must be imaginary. Use ZOOM, TRACE, or ROOT to approximate the real zero.

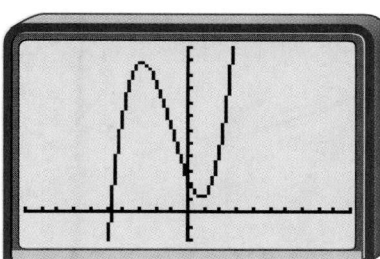

The only real zero is approximately -4.74.

When using a graphing calculator to approximate real zeros, it is helpful to know that a function with degree n has at most n real zeros. Thus, a function with degree 5 has at most five real zeros. If you can see five x-intercepts in the viewing window, you know you have found all of the zeros and that they are all real. However, if there are fewer than five x-intercepts, there are duplicate real zeros or the zeros are not all real. Complex or imaginary zeros occur in conjugate pairs, so a fifth-degree function may have one, three, or five real zeros.

Example **2** Use a graphing calculator to obtain a complete graph of $f(x) = 3x^5 - 5x^4 - 2x^3 + x^2 - 6x + 8$. Then approximate each real zero to the nearest hundredth.

First, try graphing in the standard viewing window.

Enter: 3 [X,T,θ,n] [∧] 5 [−] 5 [X,T,θ,n]

[∧] 4 [−] 2 [X,T,θ,n] [∧] 3 [+]

[X,T,θ,n] [x²] [−] 6 [X,T,θ,n] [+]

8 [ZOOM] 6

The standard viewing window does not accommodate the complete graph. The view shown at the right uses the window $[-5, 5]$ by $[-10, 15]$.

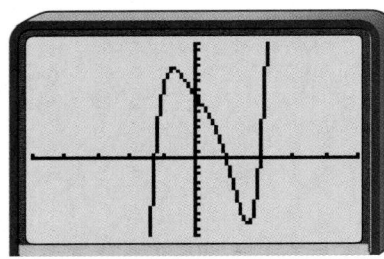

According to the graph, there are three real zeros for this function. Use ZOOM, TRACE, or ROOT to approximate the real zeros. They are approximately -1.24, 0.93, and 2.

EXERCISES

Use a graphing calculator to obtain a complete graph of each polynomial function. Describe your viewing window and state the number of real zeros.

1. $f(x) = 2x^3 - 3x^2 - 12x + 17$
2. $f(x) = 3x^4 - 8x^3 - 35x^2 + 72x + 47$
3. $j(x) = 0.1x^4 + x^3 - x^2 + 3x + 18$
4. $g(x) = x^5 - 4x^4 + 2x^3 - 7x + 15$

Graph each function so that a complete graph is shown. Then approximate each of the real zeros to the nearest hundredth.

5. $f(x) = x^4 - 3x^2 - 6x - 2$
6. $h(x) = 2x^5 + 3x - 2$
7. $c(x) = 3x^{13} + 4x^3 + 2$
8. $m(x) = 2x^8 + 4x^2 + 1$
9. $p(x) = 8x^5 - 20x^3 + 73x^2 + 28x - 4$
10. $f(x) = x^5 + x^4 - 8x^3 - 10x^2 + 7x - 4$

Graphing Polynomial Functions and Approximating Zeros

INTEGRATION
Geometry

In her geometry class, Hillary Borchers has a project in which she must create cracker shapes that form a tessellation in the same manner as some Keebler crackers. For her art class, she must design a special packaging box for the crackers. In order to best display the tessellation, the bottom of the box will be a square, and it will be open at the top so that the crackers can be seen through cellophane.

Hillary is to use a 108 square-inch sheet of a special metallic paper to cover the sides and bottom of the box. What will be the dimensions of the box if it is to hold a maximum volume of crackers? You can use a graph to solve this problem.

First, write polynomial equations to describe the surface area and volume of the box. Let x represent the length of the side of the square on the bottom, and let h represent the height of the box.

$$\underbrace{\textit{Surface Area}} = \underbrace{\textit{area of the base}} + \underbrace{\textit{area of the four sides}}$$

$$SA = x^2 + 4xh$$
$$108 = x^2 + 4xh \quad \textit{Replace SA with 108.}$$

$$\underbrace{\textit{Volume}} = \underbrace{\textit{area of the base}} \cdot \underbrace{\textit{height}}$$
$$V = x^2 \cdot h$$

In order to find the volume by graphing, we need to express the volume in terms of one variable. First, solve the surface area formula for h in terms of x.

$$x^2 + 4xh = 108$$
$$4xh = 108 - x^2$$
$$h = \frac{108 - x^2}{4x}$$

Then, substitute the value of h in the formula for volume.

$$V = x^2 h$$

$$= x^2 \left(\frac{108 - x^2}{4x} \right) \quad \textit{Replace h with } \frac{108 - x^2}{4x}.$$

$$= \frac{108x^2 - x^4}{4x} \quad \textit{Simplify.}$$

$$= 27x - \frac{x^3}{4}$$

Let $V(x) = 27x - \frac{x^3}{4}$.

Make a table of values.

x	V(x)
-11	**35.75**
-10	**-20**
-9	-60.75
-8	-88
-7	-103.25
-6	-108
-5	-103.75
-4	-92

A zero is between x = -11 and x = -10.

x	V(x)
-3	-74.25
-2	-52
-1	-26.75
0	**0**
1	26.75
2	52
3	74.25
4	92

A zero is at x = 0.

x	V(x)
5	103.75
6	108
7	103.25
8	88
9	60.75
10	**20**
11	**-35.75**
12	-108

Greatest value for V(x)

A zero is between x = 10 and x = 11.

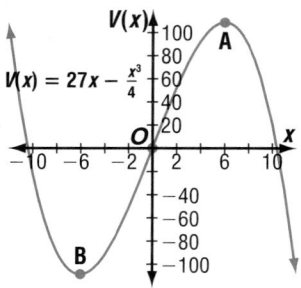

$$V(x) = 27x - \frac{x^3}{4}$$

Sketch the graph of the function $V(x)$ by connecting those points with a smooth curve. The graph will cross the x-axis somewhere between the pairs of x values where the corresponding $V(x)$ values change sign. Since the x-intercepts are zeros of the function, there is a zero between each pair of these x values. This strategy is called the **location principle.**

The Location Principle	**Suppose $y = f(x)$ represents a polynomial function and a and b are two numbers such that $f(a) < 0$ and $f(b) > 0$. Then the function has at least one real zero between a and b.**

The plurals of maximum and minimum are maxima and minima.

The graph above shows the shape of the graph of a general third-degree polynomial function. Point A on the graph is a **relative maximum** of the cubic function, since no other nearby points have a greater y-coordinate. Likewise, point B is a **relative minimum,** since no other nearby points have a lesser y-coordinate. You can also see from the tables of values where there is a relative maximum and a relative minimum. You can use this information to help graph functions that have imaginary zeros.

You can use the coordinates of the relative maximum to determine the point at which the box has the maximum volume. In the tables of values, the point at $(6, 108)$ appears to have the greatest y-coordinate. To check whether it is truly the relative maximum, compute the y values for an x value on either side of this point.

Find $V(5.9)$ and $V(6.1)$.

$$V(x) = 27x - \frac{x^3}{4} \qquad\qquad V(x) = 27x - \frac{x^3}{4}$$

$$V(5.9) = 27(5.9) - \frac{5.9^3}{4} \qquad V(6.1) = 27(6.1) - \frac{6.1^3}{4}$$

$$\approx 107.96 \qquad\qquad\qquad \approx 107.95$$

Since both y values are less than the y value for the maximum, the point at $(6, 108)$ is a relative maximum. So, in order for the box to have a maximum volume, the side of the box has to be 6 inches long. The box would have a maximum volume of 108 cubic inches.

To determine the height of the box, substitute 6 for x in the surface area equation.

$$h = \frac{108 - x^2}{4x}$$

$$= \frac{108 - 6^2}{4(6)} \quad \textit{Replace x with 6.}$$

$$= 3$$

The box must be 3 inches tall to have maximum volume. Thus, the dimensions of Hillary's cracker box are 6 inches by 6 inches by 3 inches.

Example **1**

CONNECTION

Physics

Under certain conditions, the velocity of an object as a function of time is described by the function $V(t) = 9t^3 - 93t^2 + 238t - 120$. Approximate the zeros of $V(t)$ to the nearest tenth and draw the graph.

Evaluate the function for several successive values of t to locate the zeros. Then plot the points and connect them to form a smooth graph.

t	$V(t)$
0	**−120**
1	**34**
2	56
3	**0**
4	−80
5	−130
6	**−96**
7	**76**

zero between t = 0 and t = 1

← *zero at t = 3*

zero between t = 6 and t = 7

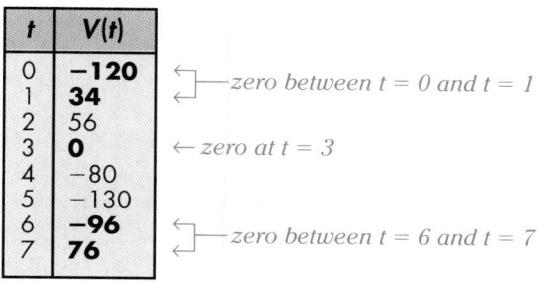

$V(t) = 9t^3 - 93t^2 + 238t - 120$

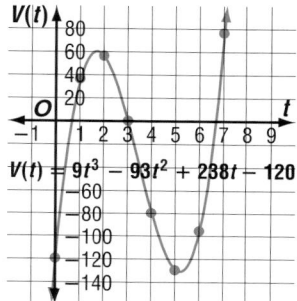

One zero lies between 0 and 1. Another zero is 3. A third zero lies between 6 and 7.

To approximate the zeros to the nearest tenth, you have to repeat the process of evaluating $V(t) = 9t^3 - 93t^2 + 238t - 120$ for successive values of t expressed in tenths, as we did in the application at the beginning of the lesson. Using a scientific calculator will help find these values more easily.

To evaluate $V(0.5)$, do the following.

Enter: 9 $\boxed{\times}$ .5 $\boxed{y^x}$ 3 $\boxed{-}$ 93 $\boxed{\times}$ .5 $\boxed{x^2}$ $\boxed{+}$ 238 $\boxed{\times}$ .5 $\boxed{-}$ 120 $\boxed{=}$ *−23.125*

Following this procedure for the rest of the values in the chart, you will find that the zeros approximated to the nearest tenth are 0.7 and 6.7.

t	$V(t)$
0.5	−23.125
0.6	−8.736
0.7	4.117
6.5	−30.625
6.6	−12.816
6.7	6.697

zero

zero

Example ❷ Graph $f(x) = x^3 - 5x^2 + 3x + 12$.

In order to graph the function, you need to find several points and then connect them to make a smooth curve. Since $f(x)$ is a third-degree polynomial function, it will have 3 or 1 real zeros. Also, its left-most points will have negative values for y, and its right-most points will have positive values for y.

Make a table and evaluate several successive values of x to locate the zeros and to find the relative maximum and relative minimum.

x	$f(x)$
-2	-22
-1	3
0	**12**
0.5	**12.375**
1	**11**
2	6
2.9	**3.039**
3	**3**
4	**8**

— *zero between $x = -2$ and $x = -1$*

} *indicates a relative maximum*

} *indicates a relative minimum*

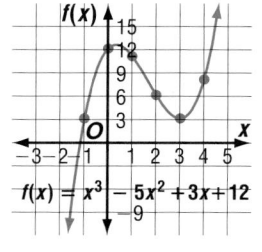

$f(x) = x^3 - 5x^2 + 3x + 12$

The function has one relative maximum and one relative minimum. The values of $f(0.5)$ and $f(2.9)$ were calculated to approximate the maximum and minimum more closely. There is a zero between -2 and -1. *Use a graphing calculator to check the graph.*

A graphing calculator can be helpful in finding the relative maximum and relative minimum of a function.

To find the relative maximum and relative minimum of $f(x) = x^3 - 6x^2 + 6x + 5$, press [Y=] and enter the equation. Then press [ZOOM] 6. The graph appears to have a relative minimum between 3 and 4 and a relative maximum between 0 and 1. To find the actual relative minimum, follow these steps.

Enter: [MATH] 6 [2nd] [Y-VARS] 1 [ENTER] [,] [X,T,θ,n] [,] 3 [,] 4 [)] [ENTER] *3.414214414*

Thus, there is a relative minimum at $x \approx 3.41$.

Your Turn

a. Find the y-coordinate of the relative minimum to the nearest hundredth.

b. Find the coordinates of the relative maximum of the function to the nearest hundredth. (*Hint*: Use the fMax feature by pressing [MATH] 7.)

c. Graph the function $f(x) = x^3 + x^2 - 7x - 3$, and find the relative maximum and relative minimum to the nearest hundredth.

Communicating Mathematics

Study the lesson. Then complete the following.

1. **State** the greatest number of relative minima that are possible for each condition.
   a. a third-degree polynomial with a positive leading coefficient
   b. a third-degree polynomial with a negative leading coefficient
   c. a fourth-degree polynomial with a positive leading coefficient
   d. a fourth-degree polynomial with a negative leading coefficient

2. Refer to the application at the beginning of the lesson. Why did we not choose one of the negative zeros for the volume of the box?

3. **Sketch** a graph of each polynomial.
   a. even-degree polynomial function with one relative maximum and two relative minima
   b. odd-degree polynomial function with one relative maximum and one relative minimum; the leading coefficient is negative
   c. even-degree polynomial function with four relative maxima and three relative minima
   d. odd-degree polynomial function with three relative maxima and three relative minima; the left-most points are negative

4. Consider the function $f(x) = x^4 - 8x^2 + 10$.
   a. Evaluate $f(x)$ for successive integers between -4 and 4 inclusive.
   b. Between what successive integers do the zeros appear? Approximate those zeros to the nearest tenth.
   c. State the ranges of x values where the values of $f(x)$ are negative and ranges where the values of $f(x)$ are positive.
   d. State the relative maximum(s) and relative minimum(s).
   e. Graph the function.

Guided Practice

Approximate the real zeros of each function to the nearest tenth.

5. $f(x) = x^3 - x^2 + 1$

6. $g(x) = x^4 + 3x^3 - 5$

Graph each function.

7. $f(x) = x^3$

8. $f(x) = x^3 - x^2 - 4x + 4$

9. $f(x) = -3x^3 + 20x^2 - 36x + 16$

10. $f(x) = x^4 - 7x^2 + x + 5$

State whether each graph is of odd degree or even degree. State the number of relative minima and relative maxima.

11.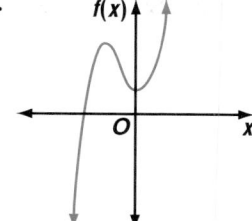

12.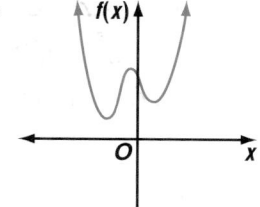

13.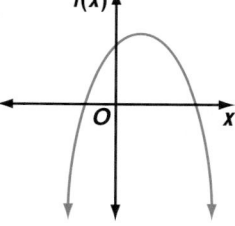

14. **Pharmacy** A syringe is to deliver an injection of 2 cubic centimeters of medication. If the plunger is pulled out two centimeters to have the proper dosage, approximate the radius of the inside of the syringe to the nearest hundredth of a centimeter. Use the formula for the volume of a cylinder, $V = \pi r^2 h$.

Practice

Approximate the real zeros of each function to the nearest tenth.

15. $f(x) = x^3 - 2x^2 + 6$

16. $h(x) = 2x^5 + 3x - 2$

17. $r(x) = x^5 - 6$

18. $g(x) = x^3 + 1$

19. $f(x) = x^4 + 2x^3 - x^2 - 3$

20. $p(x) = x^3 + 2x^2 - 3x - 5$

21. $n(x) = 3x^3 - 16x^2 + 12x + 6$

22. $h(x) = x^4 - 4x^2 + 2$

Graph each function.

23. $f(x) = 4x^6$

24. $f(x) = 3x^5$

25. $f(x) = x^3 - x$

26. $f(x) = -x^3 - 4x^2$

27. $f(x) = x^3 + 5$

28. $f(x) = x^4 - 81$

29. $f(x) = 15x^3 - 16x^2 - x + 2$

30. $f(x) = x^4 - 10x^2 + 9$

31. $f(x) = -x^4 + x^3 + 8x^2 - 3$

32. $f(x) = x^3 - x^2 - 8x + 12$

Approximate the real zeros of each function to the nearest tenth. Then use the functional values to graph the function.

33. $r(x) = x^5 + 4x^4 - x^3 - 9x^2 + 3$

34. $g(x) = x^4 - 9x^3 + 25x^2 - 24x + 6$

35. $h(x) = x^3 - 3x^2 + 2$

36. $f(x) = x^3 + 5x^2 - 9$

37. $f(x) = x^4 + 7x + 1$

38. $p(x) = x^5 + x^4 - 2x^3 + 1$

Graphing Calculator

39. a. Graph $y = x^2(x - 2)(x + 3)$ and $y = 4x^2(x - 2)(x + 3)$.

   b. Compare and contrast the graphs.

40. Find the relative maxima and relative minima of each function.

   a. $f(x) = x^3 - 4x^2 + 8$

   b. $f(x) = x^3 + 3x^2 - 12x$

Critical Thinking

41. Study the graphs for Exercises 23–32. Write a statement comparing the graphs of functions of even degree with those of functions of odd degree.

Applications and Problem Solving

42. Geometry A function that represents the volume of a pyramid with a height of the same measure as the side of its square base is $V(s) = \frac{1}{3}s^3$.

   a. Graph the function.

   b. Find the zeros of the function.

   c. Find the maximum and minimum of the function.

   d. Make a conjecture about how all of this data relates.

43. Aerospace Engineering The space shuttle has an external tank for the fuel that the main engines need for the launch. This tank is shaped like a capsule, a cylinder with a hemispherical dome at either end. The cylindrical part of the tank has a volume of 1170 cubic meters and a height of 17 meters more than the radius of the tank. What are the dimensions of the tank to the nearest tenth of a meter? (*Hint:* Use the formula for the volume of a cylinder.)

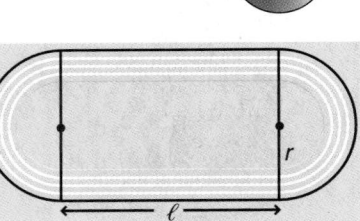

44. Physical Fitness An indoor running track is being built at a physical fitness center. It will consist of a rectangular region with a semicircle on each end. If the perimeter of the room is to be a 200-meter running track, find the dimensions that will make the area of the rectangular region as large as possible.

45. **Business** The Energy Booster Company keeps their stock of Health Aid in a rectangular tank with sides that measure $(x - 1)$ cm, $(x + 3)$ cm, and $(x - 2)$ cm. Suppose they would like to bottle their Health Aid in $x - 3$ containers of the same size. How many cubic centimeters of liquid will remain unbottled? (Lesson 8–2)

46. Find the coordinates of the center and foci, and the lengths of the major and minor axes of the ellipse whose equation is $\frac{x^2}{4} + \frac{y^2}{25} = 1$. (Lesson 7–4)

47. Name the coordinates of the vertex and the equation of the axis of symmetry for the graph of $y = (x - 3)^2 - 11$. (Lesson 6–6)

48. Solve $x^2 - 20x = -75$ by factoring. (Lesson 6–2)

49. Solve $-3y^2 = 18$. (Lesson 5–9)

50. Solve the system of equations by using augmented matrices. (Lesson 4–7)

$$a + \frac{1}{2}b - 3c = 19$$
$$\frac{1}{2}a - b + 2c = -16$$
$$5a + 2b - 2c = 50$$

51. Graph the line that passes through $(-2, 1)$ and is perpendicular to $5 + 3x = -2y$. (Lesson 2–3)

52. Simplify $\sqrt{64} \div \sqrt{4}$. (Lesson 1–1)

WORKING ON THE Investigation

Refer to the Investigation on pages 474–475.

Select a coffee can that is a different size from the one you have been using. Repeat the experiment with the new coffee can using the same 50-mL increments. Record your measurements and data in a new chart like the one on page 475.

1 What is the capacity of the new coffee can?

2 How much water is in the can when the can is one-half full, one-fourth full, and three-fourths full? Make a note of these numbers.

3 Complete the entire chart. Were your calculations for the capacity of the can when it is one-half full, one-fourth full, and three-fourths full correct?

4 Make a drawing to illustrate the measurement scale of a dipstick for this coffee can. Describe the similarities and differences between the measurements of the two coffee can experiments. How are the scales alike and how are they different?

5 Use the information in your chart to make a table of ordered pairs (x, y), where x represents the amount of water in the coffee can and y represents the percent of a full can. Graph the ordered pairs. Describe the graph.

6 Use the data from the first experiment to make another graph like the one described above. Describe the graph. What are the similarities or differences between the two experiments in terms of the graphs you have drawn?

Add the results of your work to your Investigation Folder.

8-3B Graphing Technology
Modeling Real-World Data

An Extension of Lesson 8–3

As we saw in Lesson 2-5B, you can use a graphing calculator to generate a scatter plot of data points and then determine the linear equation for the graph that best fits the plotted points. You can also use a graphing calculator to model data whose curve of best fit is a polynomial function.

Example

The table at the right shows how much time it takes each eight-hour work day to pay one day's worth of taxes. Draw a scatter plot and curve of best fit that shows how the year is related to hours worked.

First, convert hours in the table to minutes.

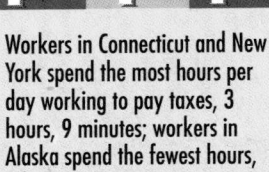
Year	Hours: Minutes
1930	0:52
1940	1:29
1950	2:02
1960	2:20
1970	2:32
1980	2:40
1990	2:45

Year	Number of Minutes
1930	52
1940	89
1950	122
1960	140
1970	152
1980	160
1990	165

Next, set the window parameters. The values of the data suggest that you should use a viewing window [1920, 2000] by [0, 200] with a scale factor of 5 for the *x*-axis and 10 for the *y*-axis.

Then enter the data. Press STAT 1 to display lists for storing data. (If old data has been previously stored, clear the lists.) The years will be entered into list L1.

To clear a list, highlight the list heading and press CLEAR ENTER .

Enter: 1930 ENTER 1940 ENTER 1950 ENTER ... 1990 ENTER

Use the ▶ key to move the cursor to column L2 to enter the number of minutes worked to pay taxes.

Enter: 52 ENTER 89 ENTER 122 ENTER ... 165 ENTER

Now draw the scatter plot.

The Plot 1 setting should be a scatter plot using lists L1 and L2.

Enter: 2nd [STAT PLOT] 1 [ENTER] [GRAPH]

Next, compute and graph the equation of the curve of best fit. Try a cubic curve for this equation.

Enter: [STAT] [▶] 7 [2nd] [L1] [,]

[2nd] [L2] [ENTER] [Y=]

[VARS] 5 [▶] [▶] 7 [GRAPH]

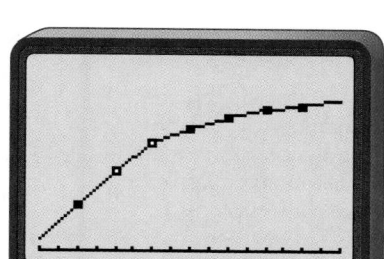

The ZOOM IN feature allows you to move a cursor along the graph or scatter plot and read the coordinates of the points. Press ZOOM 2 and any of the arrow keys to observe what happens.

Approximately how many minutes should you expect to work each day to pay taxes in the year 2000?

In the year 2000, you should expect to work approximately 168.29 minutes each day to pay taxes.

EXERCISES

When an earthquake occurs, seismic waves are detected thousands of kilometers away from the epicenter within a matter of minutes. The table at the right gives the travel time of a primary seismic wave and the corresponding distance from the epicenter for several minutes.

Travel Time (min)	Distance (km)
1	400
2	800
5	2500
7	3900
10	6250
12	8400
13	10,000

1. Use a graphing calculator to draw a scatter plot for the data. State the viewing windows and scale factors that you used.

2. Calculate and graph curves of best fit that show how travel time is related to the distance. Try LinReg, QuadReg, CubicReg, and QuartReg.

3. Write the equation for the curve you think best fits the data. Describe the fit of the graph to the data.

4. Based on the graph of a QuartReg curve, how far away from the epicenter will the wave be felt $8\frac{1}{2}$ minutes after the quake occurs? About how far are you from the epicenter if you feel the wave 15 minutes after the quake?

Roots and Zeros

What YOU'LL LEARN

- To find the number and type of zeros of a polynomial function.

Why IT'S IMPORTANT

You can qualify the number of roots for polynomials that model situations in marketing and physiology.

Tourist Attractions in France

1. Euro Disney
2. Pompidou Centre (art and culture center)
3. Eiffel Tower
4. Parc de La Villette (City of Science)
5. Musée du Louvre

APPLICATION
Marketing

Mrs. Botti's French class is selling long-stemmed roses to raise money for their trip to France. The students are making boxes in which they will deliver the flowers. They want the boxes to have square ends (width and height the same), but the length should be 12 inches longer than the width so the very long flowers will fit. They want the volume of each box to be 256 cubic inches, so that each box can hold enough moistened packing material to keep the roses fresh. Find the dimensions for a box that satisfy all these requirements.

We can find the dimensions of the box by writing a polynomial equation. Then we can use the factor theorem and synthetic substitution. First define each dimension of the box in terms of the width w.

w = width = height $w + 12$ = length

Then let $V(w)$ be the function defining the volume.

$V(w) = w \cdot w \cdot (w + 12)$ *volume = ℓwh*

$256 = w^3 + 12w^2$ *Substitute 256 for V(w) and multiply.*

$0 = w^3 + 12w^2 - 256$ *Subtract 256 from each side.*

Study the chart below. We will use a shortened form of synthetic substitution for several values of w to search for the solutions to $0 = w^3 + 12w^2 - 256$. The values for w are in the first column of the chart. Beside each value is the last line of the synthetic substitution. Recall that the first three numbers are the coefficients of the depressed polynomial. The last number in each row is the remainder.

w	1	12	0	-256
1	1	13	13	-243
2	1	14	28	-200
3	1	15	45	-121
4	1	16	64	0
5	1	17	85	169

A remainder of 0 occurs when $w = 4$. This means that $w - 4$ is a factor of the polynomial. The depressed polynomial is $w^2 + 16w + 64$.

The polynomial $w^3 + 12w^2 - 256$ can be factored as $(w - 4)(w^2 + 16w + 64)$. The trinomial $w^2 + 16w + 64$ can be further factored as $(w + 8)(w + 8)$, or $(w + 8)^2$. Thus, the solutions of the equation $0 = (w - 4)(w + 8)^2$ are $w = 4$ and $w = -8$. Since negative widths are not possible when designing a box, the width and height of the box should be 4 inches, and the length should be $4 + 12$ or 16 inches.   *Do these dimensions produce the correct volume?*

In Chapter 6, you learned that a *zero* of a function $f(x)$ is any value a such that $f(a) = 0$. This zero is also a *root*, or *solution*, of the equation formed when $f(x) = 0$. When the function is graphed, the real zeros of the function will be the *x*-intercepts of the graph.

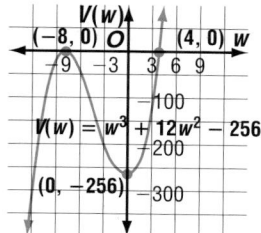

In the equation $w^3 + 12w^2 - 256 = 0$, the roots are 4 and -8. The graph of $V(w) = w^3 + 12w^2 - 256$ touches or crosses the horizontal axis at those two points.

When you solve a polynomial equation with degree greater than zero, it may have one or more real roots, or no real roots (the roots are imaginary). Since real numbers and imaginary numbers both belong to the set of complex numbers, all polynomial equations with degree greater than zero have at least one root in the set of complex numbers. This is the **fundamental theorem of algebra.**

Fundamental Theorem of Algebra	**Every polynomial equation with degree greater than zero has at least one root in the set of complex numbers.**

The following corollary of the fundamental theorem of algebra is an even more powerful tool for problem solving.

Corollary	**A polynomial equation of the form $P(x) = 0$ of degree n with complex coefficients has exactly n roots in the set of complex numbers.**

Notice that $w^3 + 12w^2 - 256 = 0$ appears to have only two roots, even though it is a third-degree equation. However, remember that the factored version of the equation was $(w - 4)(w + 8)^2 = 0$. The fact that $w + 8$ appears twice among the factors of $w^3 + 12w^2 - 256$ means -8 is a "double root." Since it is understood that -8 is counted twice among the roots of the equation, we know that $w^3 + 12w^2 - 256 = 0$ really has three roots: -8, -8, and 4. Thus, this third-degree polynomial equation has three roots, which verifies the corollary above. The roots of a polynomial equation may be different real numbers or they may be complex. For example, $x^3 + x = 0$ has three roots: 0, i, and $-i$. *You should verify this by factoring.*

Many problems can be solved by using any one of a number of different strategies. Sometimes it takes more than one strategy to solve a problem.

Example **1** **Find all roots of $0 = x^3 + 3x^2 - 10x - 24$.**

We can find the roots of the equation by combining some of the strategies that you have previously learned. Let's *list some possibilities* for roots and then eliminate those that are not roots. Suppose we begin with integral values from -4 to 4 and use the shortened form of synthetic substitution. Because this polynomial function has degree 3, the equation has three roots. However, some of them may be imaginary.

The related function of the equation is $f(x) = x^3 + 3x^2 - 10x - 24$. You can use either synthetic substitution or a scientific calculator to find $f(a)$ quickly. Find $f(-4)$.

Method 1: Synthetic Substitution

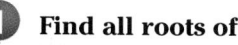

$$
\begin{array}{r|rrrr}
-4 & 1 & 3 & -10 & -24 \\
   &   & -4 & 4 & 24 \\
\hline
   & 1 & -1 & -6 & 0 \\
\end{array}
$$

(continued on the next page)

Method 2: Scientific Calculator

Enter: 4 $\boxed{+/-}$ $\boxed{STO\blacktriangleright}$ $\boxed{X}$ 1 $\boxed{+}$ 3 $\boxed{=}$ -1

$\boxed{X}$ $\boxed{RCL}$ $\boxed{+}$ 10 $\boxed{+/-}$ $\boxed{=}$ -6

$\boxed{X}$ $\boxed{RCL}$ $\boxed{+}$ 24 $\boxed{+/-}$ $\boxed{=}$ 0

The display shown after each $\boxed{=}$ gives the second, third, and fourth coefficients of the depressed polynomial. To evaluate $f(x)$ for other values for x, simply change the first number entered in the series of keystrokes shown above.

x	1	3	-10	-24
-4	1	-1	-6	**0**
-3	1	0	-10	6
-2	1	1	-12	**0**
-1	1	2	-12	-12
0	1	3	-10	-24
1	1	4	-6	-30
2	1	5	0	-24
3	1	6	8	**0**

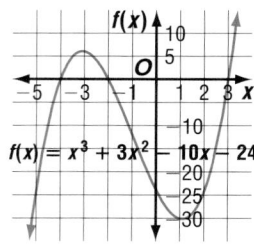

$f(x) = x^3 + 3x^2 - 10x - 24$

The zeros occur at $x = -4$, $x = -2$, and $x = 3$. The graph of the function verifies that there are three real roots.

LOOK BACK

Refer to Lesson 6-4 for information on imaginary roots.

Remember when you solved a quadratic equation like $x^2 + 9 = 0$, there were always two imaginary roots. In this case, $3i$ and $-3i$ are the roots. These numbers are a *conjugate pair*. In any polynomial function, if an imaginary number is a zero of that function, its conjugate is also a zero. This is called the **complex conjugates theorem.**

Complex Conjugates Theorem	Suppose a and b are real numbers with $b \neq 0$. If $a + bi$ is a zero of a polynomial function, then $a - bi$ is also a zero of the function.

Example 2 Find all zeros of $f(x) = x^3 - 5x^2 - 7x + 51$ if $4 - i$ is one zero of $f(x)$.

Since $4 - i$ is a zero, $4 + i$ is also a zero, according to the complex conjugates theorem. So, both $x - (4 - i)$ and $x - (4 + i)$ are factors of the polynomial $x^3 - 5x^2 - 7x - 51$.

$f(x) = [x - (4 - i)][x - (4 + i)](\underline{\ ?\ })$

$= [x^2 - (4 + i)x - (4 - i)x + (4 - i)(4 + i)](\underline{\ ?\ })$ *Multiply.*

$= (x^2 - 4x - ix - 4x + ix + 16 - i^2)(\underline{\ ?\ })$ *Simplify.*

$= (x^2 - 8x + 17)(\underline{\ ?\ })$ *Remember that $-i^2 = 1$.*

Since $f(x)$ has degree 3, there are three factors. Use division to find the third factor.

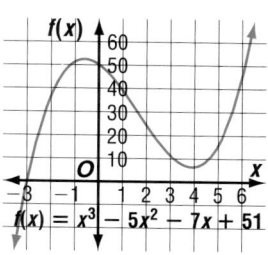

$$\begin{array}{r} x + 3 \\ x^2 - 8x + 17{\overline{\smash{\big)}\,x^3 - 5x^2 - 7x + 51}} \end{array}$$

$$\underline{x^3 - 8x^2 + 17x} \qquad \textit{Subtract.}$$
$$3x^2 - 24x + 51$$
$$\underline{3x^2 - 24x + 51} \qquad \textit{Subtract.}$$
$$0$$

Therefore, $f(x) = (x^2 - 8x + 17)(x + 3)$. Since $x + 3$ is also a factor, -3 is also a zero. The three zeros are $4 - i$, $4 + i$, and -3. The graph verifies the nature of the zeros.

French mathematician René Descartes made more discoveries about zeros of polynomial functions. His rule of signs is given below.

Descartes' Rule of Signs

If $P(x)$ is a polynomial function whose terms are arranged in descending powers of the variable,
- **the number of positive real zeros of $P(x)$ is the same as the number of changes in sign of the coefficients of the terms, or is less than this by an even number, and**
- **the number of negative real zeros of $P(x)$ is the same as the number of changes in sign of the coefficients of the terms of $P(-x)$, or is less than this by an even number.**

Example ③ State the number of positive and negative real zeros for $p(x) = 4x^5 + 3x^4 - 2x^3 + 5x^2 - 6x + 1$.

Use Descartes' rule of signs. Count the number of changes in sign for the coefficients of $p(x)$.

$$p(x) = 4x^5 \quad + \quad 3x^4 \quad - \quad 2x^3 \quad + \quad 5x^2 \quad - \quad 6x \quad + \quad 1$$
$$ 4 \qquad\quad 3 \qquad\quad -2 \qquad\quad 5 \qquad\quad -6 \qquad\quad 1$$
$$\qquad\quad \textit{no} \qquad \textit{yes} \qquad \textit{yes} \qquad \textit{yes} \qquad \textit{yes}$$

Since there are four sign changes, there are either 4, 2, or 0 positive real zeros.

Find $p(-x)$ and count the number of changes in signs for its coefficients.

$$p(-x) = 4(-x)^5 + 3(-x)^4 - 2(-x)^3 + 5(-x)^2 - 6(-x) + 1$$
$$= -4x^5 + 3x^4 + 2x^3 + 5x^2 + 6x + 1$$
$$\textit{yes} \qquad \textit{no} \qquad \textit{no} \qquad \textit{no} \qquad \textit{no}$$

Since there is one sign change, there is exactly 1 negative real zero.

Thus, the function $p(x)$ has either 4, 2, or 0 positive real zeros and exactly 1 negative real zero.

Using a graphing calculator or sketching the graph may help in determining the nature of the zeros of a function.

In Example 3, since $p(x)$ has degree 5, it has five zeros. Using the information in the example, you can make a chart of the possible combinations of real and imaginary zeros.

Number of Positive Real Zeros	Number of Negative Real Zeros	Number of Imaginary Zeros	
4	1	0	$4 + 1 + 0 = 5$
2	1	2	$2 + 1 + 2 = 5$
0	1	4	$0 + 1 + 4 = 5$

Example **Write the polynomial function of least degree with integral coefficients whose zeros include 7 and $3 + 2i$.**

If $3 + 2i$ is a zero, then $3 - 2i$ is also a zero. *Why?*

Use the zero product property to write a polynomial equation that has these zeros, $7, 3 + 2i$, and $3 - 2i$, as roots.

$0 = (x - 7)[x - (3 + 2i)][x - (3 - 2i)]$ *Why is $x - 7$ a factor?*

So, $f(x) = (x - 7)[x - (3 + 2i)][x - (3 - 2i)]$

$\quad = (x - 7)[x^2 - (3 - 2i)x - (3 + 2i)x + (3 + 2i)(3 - 2i)]$

$\quad = (x - 7)(x^2 - 3x + 2ix - 3x - 2ix + 9 - 4i^2)$

$\quad = (x - 7)(x^2 - 6x + 13)$

$\quad = x^3 - 13x^2 + 55x - 91$

CHECK FOR UNDERSTANDING

Communicating Mathematics

Study the lesson. Then complete the following.

1. **Write** an example of each function. List the possibilities for its zeros.

   a. quadratic   **b.** cubic   **c.** quartic

2. **Describe** the complex conjugate theorem. Use $6 + 7i$ as an example.

3. **State** the zeros of a polynomial function if $(x + 6)$ and $[x - (5 + i)]$ are factors of the polynomial.

4. **a. Write** a polynomial function $p(x)$ whose coefficients have four sign changes.

   b. Find the number of sign changes that $p(-x)$ has.

   c. Describe the nature of the zeros.

5. The graph of $f(x) = x^3 - 8$ is shown at the right.

   a. Describe the nature of the zeros.

   b. Find the zeros of the function.

6. A classmate has been out ill and missed learning Descartes' rule of signs. Write an example and explain Descartes' rule of signs to your fellow classmate.

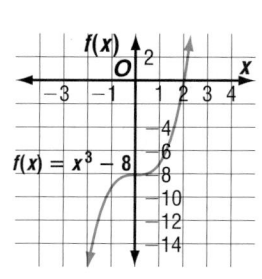

Guided Practice

Find $f(-x)$ for each function given.

7. $f(x) = 6x^4 - 3x^3 + 5x^2 - x + 2$ 8. $f(x) = x^7 - x^3 + 2x - 1$

State the number of positive real zeros, negative real zeros, and imaginary zeros for each function.

9. $f(x) = x^3 - 6x^2 + 1$

10. $f(x) = x^4 + 5x^3 + 2x^2 - 7x - 9$

Given a function and one of its zeros, find all of the zeros of the function.

11. $h(x) = x^3 - 6x^2 + 10x - 8; 4$

12. $g(x) = x^3 + 6x^2 + 21x + 26; -2$

13. $f(x) = x^3 + 7x^2 + 25x + 175; 5i$

14. $p(x) = x^4 - 9x^3 + 24x^2 - 6x - 40; 3 - i$

Write the polynomial function of least degree with integral coefficients that has the given zeros.

15. $-4, 1, 5$

16. $9, 1 + 2i$

17. **Manufacturing** The volume of a candy carton is 120 in³. To hold the correct number of candy bars, the carton must be 3 inches longer than it is wide. The height is 2 inches less than the width. Find the dimensions of the carton.

EXERCISES

Practice

State the number of positive real zeros, negative real zeros, and imaginary zeros for each function.

18. $f(x) = 5x^3 + 8x^2 + 4x + 3$

19. $g(x) = x^4 + x^3 + 2x^2 + 3x + 1$

20. $h(x) = 4x^3 - 6x^2 + 8x - 5$

21. $f(x) = x^4 - 9$

22. $r(x) = x^5 - x^3 - x + 1$

23. $g(x) = x^{14} + x^{10} + x^9 + x + 1$

24. $p(x) = x^5 - 6x^4 - 3x^3 + 7x^2 - 8x + 1$

25. $f(x) = x^{10} - x^8 + x^6 - x^4 + x^2 - 1$

Given a function and one of its zeros, find all of the zeros of the function.

26. $p(x) = x^3 + 2x^2 - 3x + 20; -4$

27. $f(x) = x^3 - 4x^2 + 6x - 4; 2$

28. $v(x) = x^3 - 3x^2 + 4x - 12; 2i$

29. $h(x) = 4x^4 + 17x^2 + 4; 2i$

30. $g(x) = 2x^3 - x^2 + 28x + 51; -\frac{3}{2}$

31. $q(x) = 2x^3 - 17x^2 + 90x - 41; \frac{1}{2}$

32. $f(x) = x^3 - 3x^2 + 9x + 13; 2 + 3i$

33. $r(x) = x^4 - 6x^3 + 12x^2 + 6x - 13; 3 + 2i$

34. $h(x) = x^4 - 15x^3 + 70x^2 - 70x - 156; 5 - i$

Write the polynomial function of least degree with integral coefficients that has the given zeros.

35. $-2, 1, 3$

36. $2, 4i$

37. $4i, 3, -3$

38. $3, 1 + i$

39. $2i, 3i, 1$

40. $6, 2 + 2i$

Critical Thinking

41. If $f(x) = x^3 + kx^2 - 7x - 15$, find the value of k so that $-2 - i$ is a zero of $f(x)$.

42. Suppose a fifth-degree polynomial has exactly two x-intercepts. Describe the nature of the roots of the function. Sketch some examples to support your reasoning.

43. Medicine Doctors can measure cardiac output in patients at high risk for a heart attack by monitoring the concentration of dye injected into a vein near the heart. A normal heart's dye concentration is approximated by $d(x) = -0.006x^4 - 0.140x^3 - 0.053x^2 + 1.79x$, where x is the time in seconds.

  a. Find all real zeros by graphing. Then verify them by using synthetic division.

  b. Which root makes sense for an answer to this problem? Why?

44. Physiology During a five-second respiratory cycle, the volume of air in liters in the human lungs can be described by the function $A(t) = 0.1729t + 0.1522t^2 - 0.0374t^3$, where t is the time in seconds. Find the volume of air held by the lungs at 3 seconds.

45. Graph $f(x) = x^3 - 5x + 7$. (Lesson 8–3)

46. Find the center and radius of a circle whose equation is $x^2 + (y - 3)^2 - 4x - 77 = 0$. (Lesson 7–3)

47. Write a quadratic equation that has roots 3 and -5. (Lesson 6–5)

48. Find $\begin{bmatrix} -2 & \frac{2}{3} \\ -\frac{1}{4} & 3 \end{bmatrix} + \begin{bmatrix} 5 & \frac{4}{9} \\ \frac{1}{2} & -9 \end{bmatrix}$. (Lesson 4–2)

49. Design Marco is designing a new dartboard. The center of the board is defined by the inequality $|x| + |y| \leq 2$. Draw the graph of this inequality to see what Marco's new dartboard will look like. (Lesson 3–4)

50. Name which ordered pairs, $(7, -3)$, $(-4, -1)$, or $(12, -6)$, satisfy $-2|x| - 5y < 3$. (Lesson 2–7)

51. Find the value of $f(12)$ when $f(x) = \dfrac{19}{23 - x}$. (Lesson 2–1)

52. Evaluate $-4|-5x| + 17$ if $x = 2$. (Lesson 1–5)

SELF TEST

Find each value if $p(x) = 4x^3 - 3x^2 + 2x - 5$. (Lesson 8–1)

1. $p(a^2)$　　　　　　　　　　　　**2.** $p(x + 1)$

Given a polynomial and one of its factors, find the remaining factors of the polynomial. (Lesson 8–2)

3. $x^3 + x^2 - 24x + 36; x - 3$　　　　**4.** $2x^3 + 13x^2 + x - 70; x - 2$

Graph each function. (Lesson 8–3)

5. $g(x) = x^5 - 5$　　　　　　　　**6.** $h(x) = x^3 - x^2 + 4$

State the number of positive real zeros, negative real zeros, and imaginary zeros for each function. (Lesson 8–4)

7. $f(x) = x^3 + 8x^2 - 7x + 10$　　　　**8.** $f(x) = 6x^4 + 18x^3 + 4x - 9$

9. Determine whether the degree of the function represented by the graph at the right is even or odd. How many real zeros does the polynomial function have? (Lesson 8–1)

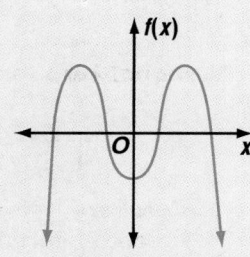

10. Manufacturing The height of a certain juice can is 4 times the radius of the top of the can. Determine the dimensions of the can if the volume is approximately 17.89 cubic inches. (*Hint*: The formula for the volume of a right circular cylinder is $V = \pi r^2 h$.) (Lesson 8–4)

Rational Zero Theorem

What YOU'LL LEARN

- To identify all possible rational zeros of a polynomial function by using the rational zero theorem, and
- to find zeros of polynomial functions.

Why IT'S IMPORTANT

You can use the rational zero theorem to find zeros of polynomials that model situations in finance and food production.

APPLICATION
Architecture

The largest pyramid in the United States is the Luxor Hotel and Casino in Las Vegas, Nevada. The volume of this unique hotel and casino is 28,933,800, or about 29 million cubic feet. The height of the pyramid is 148 feet less than the length of the building. The base of the building is square. What are the dimensions of this building?

The formula for the volume of a pyramid is $V = \frac{1}{3}Bh$, where B represents the area of the base and h represents the height. Let's set up an equation to find the dimensions of the pyramid. Let s represent the length of one side of the base of the pyramid. Then the height is $s - 148$.

$$V = \frac{1}{3}Bh$$

$$28{,}933{,}800 = \frac{1}{3}(s^2)(s - 148)$$

$$86{,}801{,}400 = s^2\,(s - 148) \quad \text{\textit{Multiply each side by 3.}}$$

$$86{,}801{,}400 = s^3 - 148s^2 \quad \text{\textit{Distributive property}}$$

$$0 = s^3 - 148s^2 - 86{,}801{,}400$$

We could use synthetic substitution to test possible zeros. But the numbers are so large that we might have to test hundreds of possible zeros before we find one. In situations like this, the **rational zero theorem** can give us some direction in testing possible zeros. This theorem and a corollary are stated below.

Rational Zero Theorem	Let $f(x) = a_0x^n + a_1x^{n-1} + \ldots + a_{n-1}x + a_n$ represent a polynomial function with integral coefficients. If $\frac{p}{q}$ is a rational number in simplest form and is a zero of $y = f(x)$, then p is a factor of a_n and q is a factor of a_0.
Corollary (Integral Zero Theorem)	If the coefficients of a polynomial function are integers such that $a_0 = 1$ and $a_n \neq 0$, any rational zeros of the function must be factors of a_n.

Let $V(s) = s^3 - 148s^2 - 86{,}801{,}400$ be the related function for $0 = s^3 - 148s^2 - 86{,}801{,}400$. All coefficients are integers, $a_0 = 1$, and $a_n = 86{,}801{,}400$. The graph of $V(x)$ is shown at the right.

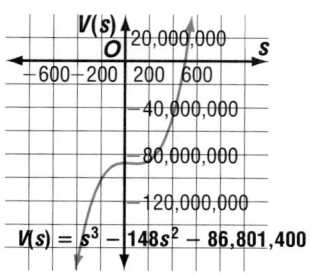

According to the integral zero theorem, any rational zeros must be factors of 86,801,400.

$$86{,}801{,}400 = 2^3 \times 3^2 \times 5^2 \times 7 \times 83^2$$

So the possible zeros in this case are ±1 through ±10, ±12, ±14, ±15, $\pm18, \dots, \pm175$, ±300, ±450, ±489, ±498, and so on, to $\pm86{,}801{,}400$.

According to Descartes' rule of signs, there will be only one positive real zero and no negative real zeros. The graph of this function crosses the x-axis one time. We can use synthetic substitution to test for possible zeros, and we can stop testing when we find the first zero. Let's make a chart. Since $s - 148 = h$ and h must be positive, we need to consider only values for s that are greater than 148.

s	1	-148	0	$-86{,}801{,}400$
175	1	27	4725	$-85{,}974{,}525$
300	1	152	45,600	$-73{,}121{,}400$
450	1	302	135,900	$-25{,}646{,}400$
498	1	350	174,300	**0**

One zero is 498. Thus, $s - 498$ is a factor of the polynomial, and 498 is a root of the equation. The dimensions are 498 feet by 498 feet by $498 - 148$ or 350 feet. *Verify the dimensions by substituting them into the formula for the volume of a pyramid.*

Example ❶ **List all possible rational zeros of $f(x) = 3x^3 + 9x^2 + x - 10$, and state whether they are positive or negative.**

Since $a_0 \neq 1$, we cannot use the integral zero theorem. If $\dfrac{p}{q}$ is a rational root, then p is a factor of -10 and q is a factor of 3. The possible values of p are ±1, ±2, ±5, and ±10. The possible values of q are ±1 and ±3. So all the possible rational zeros are as follows.

±1, ±2, ±5, ±10, $\pm\dfrac{1}{3}$, $\pm\dfrac{5}{3}$, and $\pm\dfrac{10}{3}$

Now use Descartes' rule of signs.

$f(x) = 3x^3 + 9x^2 + x - 10$

Since there is one sign change, there is one positive real zero.

$f(x) = -3x^3 + 9x^2 - x - 10$

Since there are two sign changes, there are two or no negative real zeros.

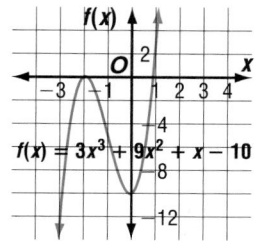

The graph of the function shown above verifies that there is one positive real zero and two negative real zeros. *Note that the two negative real zeros are the same number.*

Example 2

INTEGRATION
Geometry

The volume of a rectangular solid is 1430 cubic centimeters. The width is 1 centimeter less than the length, and the height is 2 centimeters greater than the length. Find the dimensions of the solid.

Explore Read the problem and define the variable.

Let ℓ represent the length of the solid.

Plan Write an equation.

Volume = length × width × height

$V = \ell(\ell - 1)(\ell + 2)$

Solve $V = \ell(\ell - 1)(\ell + 2)$

$1430 = \ell^3 + \ell^2 - 2\ell$ *Replace V with 1430.*

$0 = \ell^3 + \ell^2 - 2\ell - 1430$ *Subtract 1430 from each side.*

Possible rational zeros are $\pm1, \pm2, \pm5, \pm10, \pm11$, and ±13. Since measures must be positive and according to Descartes' rule of signs there is one positive real zero, we can stop testing possible zeros when we find the first one. Let's make a table and test each possible rational zero.

$\frac{p}{q}$	1	1	-2	-1430
1	1	2	0	-1430
2	1	3	4	-1422
5	1	6	28	-1290
10	1	11	108	-350
11	1	12	130	**0**

One zero is 11. The other dimensions are $11 - 1$ or 10 cm and $11 + 2$ or 13 cm.

Examine Check to see if the dimensions are correct.

$10 \times 11 \times 13 = 1430$ ✓

You have learned many rules to help you determine the number and characteristics of the zeros of a function. Example 3 shows how many of them can be used.

Example 3

Find all zeros of $f(x) = 4x^4 - 13x^3 - 13x^2 + 28x - 6$.

• From the corollary to the fundamental theorem of algebra, we know there are exactly 4 complex roots.

• According to Descartes' rule of signs, there are either 3 or 1 positive real zeros and exactly 1 negative real zero.

• According to the rational zero theorem, the possible rational zeros are $\pm\frac{1}{4}, \pm\frac{1}{2}, \pm\frac{3}{4}, \pm1, \pm\frac{3}{2}, \pm2, \pm3$, and ±6.

• Use synthetic substitution and a chart to find at least one zero.

$\frac{p}{q}$	4	-13	-13	28	-6
$\frac{1}{4}$	4	-12	-16	24	0

One zero is $\frac{1}{4}$.

(continued on the next page)

The depressed polynomial after division by $x - \frac{1}{4}$ is

$4x^3 - 12x^2 - 16x + 24$. Now use a synthetic division chart with this polynomial.

x	4	-12	-16	24
$\frac{1}{2}$	4	-10	-21	$\frac{27}{2}$
$\frac{3}{4}$	4	-9	$-\frac{91}{4}$	$\frac{111}{16}$
1	4	-8	-24	**0**

Another zero is 1.

The new depressed polynomial is $4x^2 - 8x - 24$. Use the quadratic formula to find other possible zeros.

$$x = \frac{-(-8) \pm \sqrt{(-8)^2 - 4(4)(-24)}}{2(4)} \qquad a = 4,\ b = 8,\ c = -24$$

$$= \frac{8 \pm \sqrt{448}}{8} \text{ or } 1 \pm \sqrt{7}$$

The zeros are $\frac{1}{4}$, 1, $1 + \sqrt{7}$, and $1 - \sqrt{7}$.

The approximate values of the irrational zeros are 3.65 and -1.65. So, there are 3 positive real zeros and 1 negative zero.

The graph of the function shown at the right crosses the x-axis 4 times, confirming that there are 4 real roots.

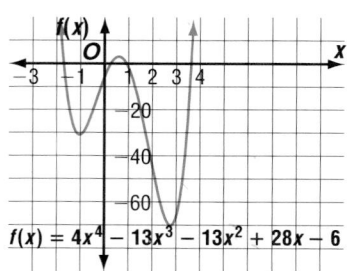

$f(x) = 4x^4 - 13x^3 - 13x^2 + 28x - 6$

CHECK FOR UNDERSTANDING

Communicating Mathematics

Study the lesson. Then complete the following.

1. a. **Explain** when you can use the integral zero theorem to determine possible rational zeros for a polynomial function.

   b. Why is it helpful to use the rational zero theorem while finding the zeros of a polynomial function?

2. Refer to Example 2. When testing possible zeros, would starting with a number greater than 1 have made more sense? Explain.

3. **Write** a polynomial function with four possible rational zeros.

4. a. **Explain** why there cannot be three positive zeros for
   $p(x) = x^3 + 4x^2 - 3x + 2$.

   b. **Explain** why there cannot be four positive zeros for
   $p(x) = x^3 + 2x^2 + 3x + 1$.

5. Refer to the application at the beginning of the lesson. List all of the possible roots between 0 and 100.

6. Write a polynomial function that has possible rational zeros of ± 1, ± 3, $\pm \frac{1}{2}$, and $\pm \frac{3}{2}$.

Guided Practice

List all of the possible rational zeros for each function.

7. $h(x) = x^3 + 8x + 6$

8. $d(x) = 6x^3 + 6x^2 - 15x - 2$

Find all of the rational zeros for each function.

9. $f(x) = x^3 - x^2 - 34x - 56$

10. $p(x) = x^3 - 3x - 2$

11. $g(x) = x^4 - 3x^3 + x^2 - 3x$

12. $h(x) = 6x^3 + 11x^2 - 3x - 2$

13. Find all of the zeros of $h(x) = 9x^5 - 94x^3 + 27x^2 + 40x - 12$.

14. Write a polynomial function of least degree that has zeros -3, 2, and 5.

15. **Geometry**   The volume of the figure at the right is 384 cm³. Find the dimensions.

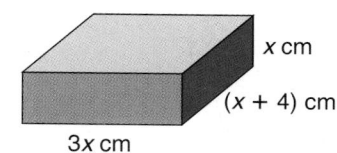

x cm

$(x + 4)$ cm

$3x$ cm

EXERCISES

Practice

List all of the possible rational zeros for each function.

16. $f(x) = x^3 + 6x + 2$

17. $p(x) = x^4 - 10$

18. $n(x) = x^5 + 6x^3 - 12x + 18$

19. $p(x) = 3x^3 - 5x^2 - 11x + 3$

20. $f(x) = 3x^4 + 15$

21. $h(x) = 9x^6 - 5x^3 + 27$

Find all of the rational zeros for each function.

22. $p(x) = x^3 - 5x^2 - 22x + 56$

23. $f(x) = x^3 + x^2 - 80x - 300$

24. $g(x) = x^4 - 3x^3 - 53x^2 - 9x$

25. $h(x) = 2x^3 - 11x^2 + 12x + 9$

26. $f(x) = 2x^5 - x^4 - 2x + 1$

27. $p(x) = x^4 + 10x^3 + 33x^2 + 38x + 8$

28. $n(x) = x^4 + x^2 - 2$

29. $t(x) = x^4 - 13x^2 + 36$

30. $h(x) = x^4 - 3x^3 - 5x^2 + 3x + 4$

31. $p(x) = x^3 + 3x^2 - 25x + 21$

32. $f(x) = x^5 - 6x^3 + 8x$

33. $g(x) = 48x^4 - 52x^3 + 13x - 3$

Find all of the zeros of each function.

34. $f(x) = 6x^3 + 5x^2 - 9x + 2$

35. $p(x) = 6x^4 + 22x^3 + 11x^2 - 38x - 40$

36. $g(x) = 5x^4 - 29x^3 + 55x^2 - 28x$

37. $p(x) = x^5 - 2x^4 - 12x^3 - 12x^2 - 13x - 10$

Critical Thinking

38. Suppose k and $2k$ are zeros of $f(x) = x^3 + 4x^2 + 9kx - 90$. Find k and all three zeros of $f(x)$.

Applications and Problem Solving

39. **Stock Market**   In 1994, IBM's research lab discovered a flaw in Intel's Pentium™ chip that could have caused an error as often as once every 24 days. Intel's stock was affected on the day the flaw was discovered, as shown in the graph at the right. The function $f(x) = -0.002x^4 + 0.05x^3 - 0.3x^2 - 0.4x + 63$ can be used to model Intel's stock prices at time x, where $x = 0$ represents 9:30 A.M., $x = 1$ represents 10:00 A.M., and so on.

a. Use $f(x)$ to estimate the price of Intel's stock at 2:30 P.M.

b. Compare this value to an estimate of the value from the graph.

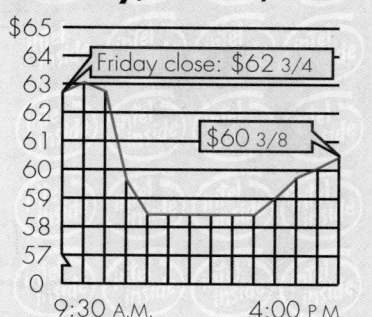

Price of Intel Stock
Monday, Dec.12,1994

Friday close: $62 3/4

$60 3/8

9:30 A.M.   4:00 P.M.

Source: *Bloomberg Business News*

40. Food Production I.C. Dreams makes ice cream cones. The volume of each cone is about 5.24 cubic inches, and the height is 4 inches more than the radius of the opening of the cone. Find the dimensions of the cone. Use the formula for the volume of a cone, $V = \frac{1}{3}\pi r^2 h$.

41. Patterns The diagrams below show the number of regions formed by connecting the points on a circle.

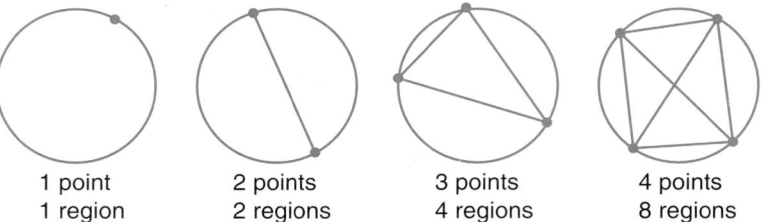

| 1 point | 2 points | 3 points | 4 points |
| 1 region | 2 regions | 4 regions | 8 regions |

The number of regions formed by connecting n points of a circle can be described by the function $f(n) = \frac{1}{24}(n^4 - 6n^3 + 23n^2 - 18n + 24)$.

a. Find the number of regions formed by connecting 5 points of a circle. Draw a diagram to verify your solution.

b. How many points would you have to connect to form 99 regions?

Mixed Review

42. Write a polynomial function of least degree with integral coefficients that has -2 and $2 + 3i$ as zeros.   (Lesson 8–4)

43. Write $2y^2 = 14x$ in the form $x = a(y - k)^2 + h$.   (Lesson 7–2)

44. Solve $c^2 - 9c - 58 = -7c + 5$ by factoring.   (Lesson 6–2)

45. Physics A model airplane is fixed on a string so that it flies around in a circle. The designers of the plane want to find the time it takes for the airplane to make a complete circle. They know that the formula $F_c = m\left(\frac{4\pi^2 r}{T^2}\right)$ describes the force required to keep the airplane going in a circle, m represents the mass of the plane, r represents the radius of the circle, and T represents the time for a revolution. Solve the formula for T. Write the answer in simplest radical form.   (Lesson 5–8)

46. Use augmented matrices to solve the system of equations.   (Lesson 4–7)
$$5x - 7y + z = 29$$
$$-2x - 3y + 5z = 20$$
$$x - 9y + 3z = 13$$

47. Solve $\begin{bmatrix} -5x \\ 9x + 4 \end{bmatrix} = \begin{bmatrix} 15y \\ -31y \end{bmatrix}$ for x and y.   (Lesson 4–1)

48. Given the function $f(x, y) = 9x - 3y$, find $f(-3, 7)$.   (Lesson 3–6)

49. Marcia and Roberto want to build a ramp that they can use while rollerblading. If they want the ramp to have a base of 8 feet and slope of $\frac{1}{2}$, how tall will their ramp be?   (Lesson 2–3)

50. Margie is 6 years older than Max. Moira is 19 years younger than Margie. If Max is 17, how old is Moira?   (Lesson 1–4)

Using Quadratic Techniques to Solve Polynomial Equations

APPLICATION
Finance

On his seventeenth birthday, Montel received $100. On his eighteenth birthday, he received $150. One year ago, on his nineteenth birthday, he received $200. Montel put his birthday money into an account paying 6% interest, compounded annually, and did not withdraw or add any additional money. Determine the amount of money currently in his account.

We can use the formula for compound interest, $A = P(1 + r)^t$, where P is the original amount of money deposited, r is the interest rate (written as a decimal), and t is the number of years invested. The amount of money currently in his account is the sum of the amounts he received on his last three birthdays, plus interest.

The interest rate is 6%, so $r = 0.06$. Let $x = 1 + r$ or 1.06, and let $T(x)$ represent the total amount of money currently in the account. Find $T(1.06)$.

Total	=	money from 17th birthday	+	money from 18th birthday	+	money from 19th birthday

$$T(x) = 100x^3 + 150x^2 + 200x$$
$$T(1.06) = 100(1.06)^3 + 150(1.06)^2 + 200(1.06) \quad \textit{Replace x with 1.06.}$$
$$= \$499.64 \quad \text{The amount of money in Montel's account is \$499.64.}$$

Note that the polynomial function contains a factor that is a quadratic since $T(x) = 100x^3 + 150x^2 + 200x$ or $50x(2x^2 + 3x + 4)$.

In some cases, we can rewrite polynomial equations and use quadratic techniques to solve them. For example, $x^4 - 38x^2 + 72 = 0$ can be written as $(x^2)^2 - 38(x^2) + 72 = 0$. Equations that can be written in the form $a[f(x)]^2 + b[f(x)] + c = 0$ are said to be in **quadratic form.**

Definition of Quadratic Form	For any numbers a, b, and c, except $a = 0$, an equation that can be written as $a[f(x)]^2 + b[f(x)] + c = 0$, where $f(x)$ is some expression in x, is in **quadratic form.**

Example ① Solve $x^4 - 17x^2 + 16 = 0$.

The graph of $y = x^4 - 17x^2 + 16$ crosses the x-axis 4 times, so there are 4 real zeros.

$$x^4 - 17x^2 + 16 = 0$$
$$(x^2)^2 - 17(x^2) + 16 = 0 \quad \textit{Quadratic form}$$
$$(x^2 - 16)(x^2 - 1) = 0$$
$$(x - 4)(x + 4)(x - 1)(x + 1) = 0$$

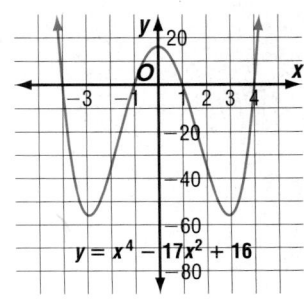

(continued on the next page)

Use the zero product property.

$x - 4 = 0$ or $x + 4 = 0$ or $x - 1 = 0$ or $x + 1 = 0$

$x = 4$ $x = -4$ $x = 1$ $x = -1$

The roots are -4, 4, -1, and 1, which are verified on the graph.

You can solve cubic equations with the quadratic formula if a quadratic factor can be found.

Example ② Solve $x^3 + 64 = 0$.

$$x^3 + 64 = 0$$

$$(x + 4)(x^2 - 4x + 16) = 0 \quad \textit{Factor.}$$

Use the zero product property.

$x + 4 = 0$ or $x^2 - 4x + 16 = 0$

$x = -4$ $x = \dfrac{4 \pm \sqrt{(-4)^2 - 4(1)(16)}}{2(1)}$

$$= \dfrac{4 \pm \sqrt{-48}}{2} \text{ or } 2 \pm 2i\sqrt{3}$$

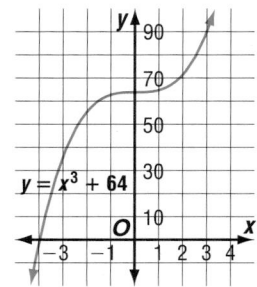

The roots are -4 and $2 \pm 2i\sqrt{3}$.

The only real root is -4.

The graph of the related function crosses the x-axis only once at -4.

p.274 purple box

You have studied the rule $(a^m)^n = a^{mn}$ in Chapter 5. This property of exponents is often used to solve equations that have terms with rational exponents.

Example ③

Finance

Isabel earned \$1000 from her summer job, and on August 1 she decided to put it in the bank to save it for a cruise she wants to take during the next spring break, which starts on April 1. The cruise costs \$1046, but Isabel figures that if her money earns some interest she may have enough money by April. As she shops around for interest rates at various banks, what interest rate should she be looking for so that her \$1000 on August 1 will grow to \$1046 by April 1? Use the interest formula $A = P(1 + r)^t$.

Let x represent $1 + r$ and substitute the known values into the formula:

$A = \$1046$, $P = \$1000$, and $t = 8$ months or $\frac{2}{3}$ year.

$$A = P(1 + r)^t$$

$$1046 = 1000x^{\frac{2}{3}} \quad \textit{Substitute.}$$

$$1.046 = x^{\frac{2}{3}} \quad \textit{Divide each side by 1000.}$$

$$(1.046)^3 = \left(x^{\frac{2}{3}}\right)^3 \quad \textit{Cube each side.}$$

$$1.14 = x^2$$

$$\pm\sqrt{1.14} \text{ or } \pm 1.07 = x \quad \textit{Take the square root of each side. Why } \pm?$$

Since $x = 1 + r$, then $r = 0.07$ or $r = -2.07$. Since interest rates cannot be negative, the interest rate is 0.07 or 7%.

Some equations involving rational exponents can be written in quadratic form.

Example **4** Solve $x^{\frac{1}{2}} - 8x^{\frac{1}{4}} + 15 = 0$.

$$x^{\frac{1}{2}} - 8x^{\frac{1}{4}} + 15 = 0$$

$$\left(x^{\frac{1}{4}}\right)^2 - 8\left(x^{\frac{1}{4}}\right) + 15 = 0 \quad \textit{Quadratic form}$$

$$\left(x^{\frac{1}{4}} - 5\right)\left(x^{\frac{1}{4}} - 3\right) = 0 \quad \textit{Factor.}$$

Use the zero product property.

$$x^{\frac{1}{4}} - 5 = 0 \quad \text{or} \quad x^{\frac{1}{4}} - 3 = 0$$

$$x^{\frac{1}{4}} = 5 \qquad\qquad x^{\frac{1}{4}} = 3$$

$$\left(x^{\frac{1}{4}}\right)^4 = 5^4 \qquad \left(x^{\frac{1}{4}}\right)^4 = 3^4$$

$$x = 625 \qquad\qquad x = 81$$

Check:
$$x^{\frac{1}{2}} - 8x^{\frac{1}{4}} + 15 = 0$$

$$625^{\frac{1}{2}} - 8(625)^{\frac{1}{4}} + 15 \stackrel{?}{=} 0$$

$$25 - 40 + 15 \stackrel{?}{=} 0$$

$$0 = 0 \quad\checkmark$$

$$81^{\frac{1}{2}} - 8(81)^{\frac{1}{4}} + 15 = 0$$

$$9 - 24 + 15 \stackrel{?}{=} 0$$

$$0 = 0 \quad\checkmark$$

The real roots are 81 and 625.

Example **5** Solve $x - 2\sqrt{x} - 3 = 0$.

$$x - 2\sqrt{x} - 3 = 0$$

$$\left(\sqrt{x}\right)^2 - 2(\sqrt{x}) - 3 = 0 \quad \textit{Quadratic form}$$

$$\sqrt{x} = \frac{-b \pm \sqrt{b^2 - 4ac}}{2a} \quad \textit{Use the quadratic formula.}$$

$$\sqrt{x} = \frac{2 \pm \sqrt{(-2)^2 - 4(1)(-3)}}{2(1)} \quad \textit{a = 1, b = -2, and c = -3}$$

$$\sqrt{x} = \frac{2 \pm \sqrt{16}}{2}$$

$$\sqrt{x} = 3 \qquad \text{or} \qquad \sqrt{x} = -1$$

$$x = 9$$

There is no real number x such that $\sqrt{x} = -1$. The only real solution is 9.

CHECK FOR UNDERSTANDING

Communicating Mathematics

Study the lesson. Then complete the following.

1. **Explain** why the graph in Example 2 crosses the x-axis only once when three roots are given.

2. **Explain** the steps you would take to solve $\sqrt{x^4 + 48} = 4x$.

3. **Write** three examples of equations that are not quadratic but can be written in quadratic form. Then write them in quadratic form.

Guided Practice

Factor each polynomial. Identify the quadratic factor if one exists.

4. $x^4 - 3x^3 + 6x^2$

5. $2x^3 + 7x^2 - 8x$

6. $4m^3 + 9m - 16m^2$

7. $y^3 - y^5 - 100y$

8. $x^7 + x^{\frac{7}{2}} + x^5$

9. $x^3 - 729$

Write each equation in quadratic form if possible. If not, explain why not.

10. $3r + 7\sqrt{r} = 11$

11. $a^8 + 10a^4 - 16 = 0$

Solve each equation.

12. $x - 16x^{\frac{1}{2}} = -64$

13. $m^4 + 7m^3 + 12m^2 = 0$

14. $3m^{\frac{3}{2}} - 81 = 0$

15. $y^3 = 26.6y - 3.2y^2$

16. Geometry The width of a rectangular prism is w centimeters. The height is 2 centimeters less than the width. The length is 4 centimeters more than the width. If the volume of the prism is 8 times the measure of the length, find the dimensions of the prism.

EXERCISES

Practice

Write each equation in quadratic form if possible. If not, explain why not.

17. $x^8 + 10x^4 = -13.2$

18. $11x^4 + 3x = -8$

19. $84n^4 - 62n^2 = 0$

20. $7q + 8\sqrt{q} = 13$

21. $5y^4 + 7y = 8$

22. $11n^4 = -44n^2$

Solve each equation.

23. $x^3 - 3x^2 - 10x = 0$

24. $n^3 + 12n^2 + 32n = 0$

25. $b^3 = 1331$

26. $m^4 - 7m^2 + 12 = 0$

27. $z - 8\sqrt{z} - 240 = 0$

28. $y^3 - 729 = 0$

29. $y^{\frac{2}{3}} - 9y^{\frac{1}{3}} + 20 = 0$

30. $r - 19r^{\frac{1}{2}} + 60 = 0$

31. $6.25m^3 - 12.25m = 0$

32. $y^{\frac{1}{3}} = 7.5$

33. $m^5 + 1.5m^4 = 15.04m^3$

34. $p^{\frac{2}{3}} + 11p^{\frac{1}{3}} + 28 = 0$

35. Write an equation for a polynomial that has roots -3, 0, and 2.

Critical Thinking

36. Write an explanation about how you would solve the equation $|a - 3|^2 - 9|a - 3| = -8$. Then solve the equation.

Applications and Problem Solving

37. Geometry The formula for the area of an ellipse is $A = \pi ab$. Find the measure of a and b to the nearest hundredth of an inch if an ellipse has an area of 8.85 square inches and the measure of b is 2.3 inches greater than a.

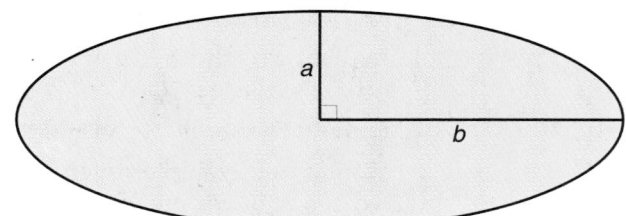

38. Geometry A piece of wire is cut into two pieces. One piece is bent into the shape of a square, and the other into the shape of an equilateral triangle. The side of the square and a side of the equilateral triangle have the same length, and the measure of that length, in inches, is an integer. The original piece of wire was less than 50 inches long before it was cut.

a. Find all possible integral measurements for the length of the side of the square and the triangle.

b. What is the shortest possible length for the original piece of wire?

c. What is the longest possible length for the original piece of wire?

39. Aerospace The force of gravity decreases with the square of the distance from the center of Earth. So, as an object moves further from Earth, its weight decreases. The radius of Earth is approximately 3960 miles. The formula relating weight and distance is

$(3960 + r)^2 = \dfrac{3960^2 \cdot W_E}{W_S}$, where W_E represents

the weight of a body on Earth, W_S represents the weight of a body a certain distance from the center of Earth, and r represents the distance of an object above Earth's surface.

a. An astronaut weighs 140 pounds on Earth and 120 pounds in space. How far is he above Earth's surface?

b. An astronaut weighs 125 pounds on Earth. What is her weight in space if she is 99 miles above the surface of Earth?

Mixed Review

40. Manufacturing The volume of a milk carton is 200 cubic inches. The base of the carton is square, and the height is 3 inches more than the length of the base. What are the dimensions of the carton? (Lesson 8–5)

41. Find the value of c such that the points at $(7, 2)$ and $(3, c)$ are 5 units apart. (Lesson 7–1)

42. Agriculture The function $f(x) = -x^2 + 8x$, where x is the number of apple trees planted in a given area and $f(x)$ is the number of pounds of apples produced per day, can be used to determine how many apples are to be planted in a certain area. (Lesson 6–7)

a. Graph $f(x) = -x^2 + 8x$.

b. If Wessel Farm wants to produce at least 12 pounds of apples per day, how many trees should they plant in the area? Write as an inequality.

c. According to the graph of $f(x)$, production is low if few or many trees are planted and production is high if a medium number of trees is planted. Give some possible reasons why this might be true in real life.

43. Physics The formula for finding the time t that it takes an object dropped

from a height of h feet to reach the ground is $t = \sqrt{\dfrac{2h}{g}}$, where g represents

the acceleration due to gravity. All objects in free fall near Earth's surface have an acceleration due to gravity of 32 feet per second squared. If a plant falls off a windowsill 64 feet from the ground, how long will it take the plant to reach the ground? (Lesson 5–5)

44. Find M if $\begin{bmatrix} -9 & 12 \\ 4 & -7 \end{bmatrix} \cdot M = \begin{bmatrix} -9 & 12 \\ 4 & -7 \end{bmatrix}$. (Lesson 4–5)

45. Use Cramer's rule to solve the system of equations. (Lesson 3–3)
$\dfrac{x}{2} - \dfrac{2y}{3} = 2\dfrac{1}{3}$
$3x + 4y = -50$

46. Geography The following numbers are the percent of people in the South American countries who live in urban areas. (Lesson 1–3)
84, 87, 51, 76, 65, 54, 46, 70, 86, 83, 35

a. Make a stem-and-leaf plot of this data.

b. How many countries are less than 60% urban?

c. Argentina is the most urbanized country in South America. What percent of it is urban?

d. Guyana is the least urbanized country in South America. What percent of it is urban?

Composition of Functions

What YOU'LL LEARN

- To find the composition of functions.

Why IT'S IMPORTANT

You can use composition of functions to solve problems involving biology and foreign currency.

CONNECTION
Biology

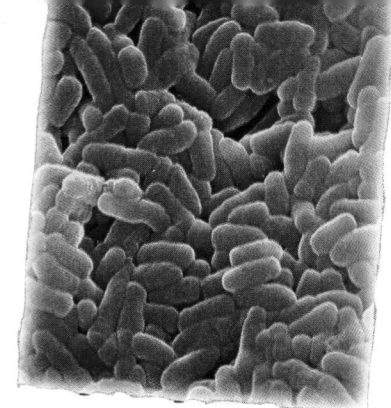

Temperature is measured in different units in different countries. An American scientist and a German scientist are working on incubating a bacterium in their respective countries. They are sharing their findings with each other through the Internet. The last message from the German scientist says that her bacterium died at a temperature of 312° K, which she discovered was not warm enough. The American scientist's temperature for incubation is 98.2° F. Should the American scientist be worried?   *This problem will be solved in Example 5.*

Let $K(x)$ be the function for converting Celsius temperatures to Kelvin, and let $C(x)$ be the function for converting Fahrenheit temperatures to Celsius.

$K(x) = x + 273$ *Converting Celsius to Kelvin*

$C(x) = \dfrac{5}{9}(x - 32)$ *Converting Fahrenheit to Celsius*

Using **composition of functions** is one way to solve the problem.

Composition of Functions	**Suppose f and g are functions such that the range of g is a subset of the domain of f. Then the composite function $f \circ g$ can be described by the equation $[f \circ g](x) = f[g(x)]$.**

$[f \circ g](x)$ and $f[g(x)]$ are both read "f of g of x."

Given two functions f and g, you can find $f \circ g$ and $g \circ f$ if the range of each function is a subset of the domain of the other function.

Example **1** If $f = \{(1, 2), (3, 4), (5, 4)\}$ and $g = \{(2, 5), (4, 3)\}$, find $f \circ g$ and $g \circ f$.

$f[g(2)] = f(5)$ or 4 $g(2) = 5$	$g[f(1)] = g(2)$ or 5 $f(1) = 2$
$f[g(4)] = f(3)$ or 4 $g(4) = 3$	$g[f(3)] = g(4)$ or 3 $f(3) = 4$
	$g[f(5)] = g(4)$ or 3 $f(5) = 4$
$f \circ g = \{(2, 4), (4, 4)\}$	$g \circ f = \{(1, 5), (3, 3), (5, 3)\}$

The composition of functions can be shown by mappings. Suppose $f = \{(6, 2), (4, 5), (0, -2)\}$ and $g = \{(5, 4), (-2, 6), (2, 0)\}$. The composition of these functions is shown below.

$f \circ g$

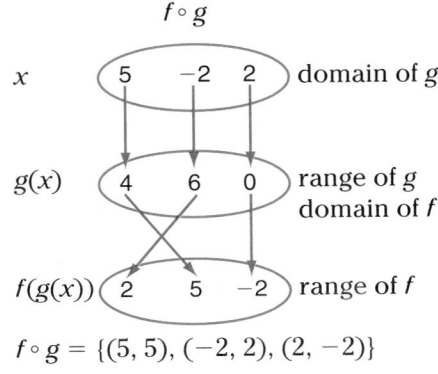

$f \circ g = \{(5, 5), (-2, 2), (2, -2)\}$

$g \circ f$

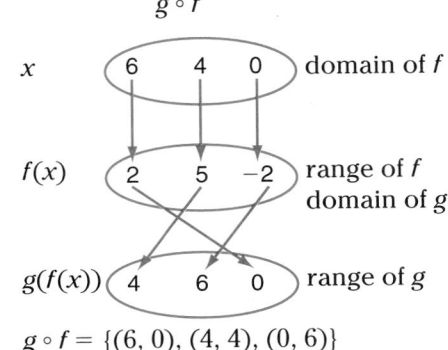

$g \circ f = \{(6, 0), (4, 4), (0, 6)\}$

The composition of two functions may not exist. Look back at the definition of the composition of functions. The composition of functions f and g, $f \circ g$, is defined when the range of g is a subset of the domain of f. If this condition is not met, the composition is not defined.

Example 2

If $h = \{(4, 6), (2, 4), (6, 8), (8, 10)\}$ and $k = \{(4, 5), (6, 5), (8, 12), (10, 12)\}$, find $h \circ k$ and $k \circ h$, if they exist.

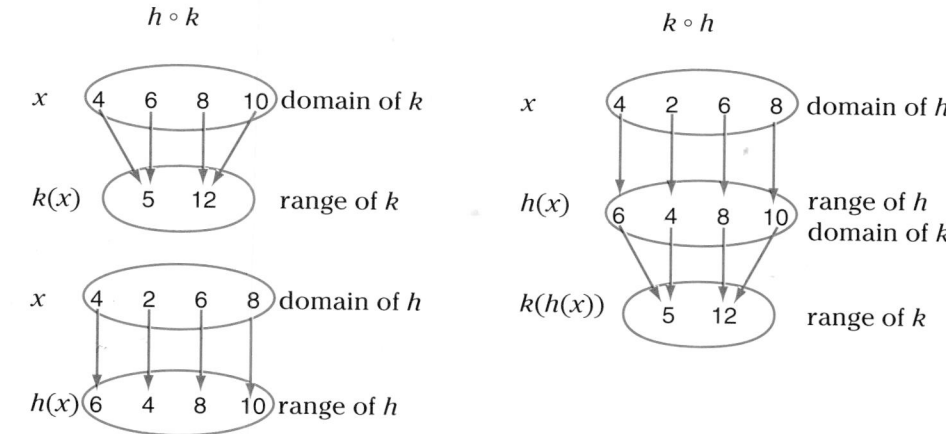

$h \circ k$ does not exist.

The range of k is not a subset of the domain of h. *Why not?*

$k \circ h = \{(4, 5), (2, 5), (6, 12), (8, 12)\}$

Sometimes, when two functions are composed, the graph of the composition resembles the graph of one of the original functions.

Example 3

If $f(x) = x^2 - 4$ and $g(x) = 4x - 1$, find $[f \circ g](x)$.

$[f \circ g](x) = f[g(x)]$

$\qquad = f(4x - 1)$ *Substitute $4x - 1$ for $g(x)$.*

$\qquad = (4x - 1)^2 - 4$ *Evaluate f when x is $(4x - 1)$.*

$\qquad = 16x^2 - 8x + 1 - 4$

$\qquad = 16x^2 - 8x - 3$ *Simplify.*

The graphs of each function are shown below.

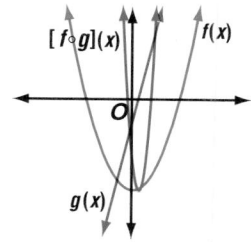

$f(x)$ is quadratic.

$g(x)$ is linear.

$[f \circ g](x)$ is quadratic.

Example **4** If $f(x) = x + 5$ and $g(x) = x^2 - 2$, find $[f \circ g](3)$ and $[g \circ f](3)$.

$$
\begin{aligned}
[f \circ g](3) &= f[g(3)] \\
&= f(3^2 - 2) \qquad \textit{Substitute } 3^2 - 2 \textit{ for } g(3). \\
&= f(7) \qquad\qquad \textit{Simplify.} \\
&= 7 + 5 \text{ or } 12 \quad \textit{Evaluate f when x is 7.}
\end{aligned}
$$

$$
\begin{aligned}
[g \circ f](3) &= g[f(3)] \\
&= g(3 + 5) \qquad \textit{Substitute } 3 + 5 \textit{ for } f(3). \\
&= g(8) \qquad\qquad \textit{Simplify.} \\
&= 8^2 - 2 \text{ or } 62 \quad \textit{Evaluate g when x is 8.}
\end{aligned}
$$

Example **5**

CONNECTION
Biology

Refer to the application at the beginning of the lesson. Should the American scientist be worried that her incubation temperature is not warm enough?

In order to compare the two temperatures, we need to convert the American scientist's temperature from Fahrenheit to Kelvin. To do this, convert the temperature from Fahrenheit to Celsius and then from Celsius to Kelvin. This can be written as $[K \circ C](x)$ or $K[C(x)]$.

$$
\begin{aligned}
[K \circ C](x) &= K[C(x)] \\
&= K\left[\frac{5}{9}(x - 32)\right] \quad \textit{Substitute } \frac{5}{9}(x - 32) \textit{ for } C(x). \\
&= \left[\frac{5}{9}(x - 32)\right] + 273
\end{aligned}
$$

$$
\begin{aligned}
[K \circ C](98.2) &= \left[\frac{5}{9}(98.2 - 32)\right] + 273 \quad \textit{Replace x with 98.2.} \\
&\approx 309.8 \qquad\qquad\qquad\qquad \textit{Simplify}
\end{aligned}
$$

The American scientist is incubating her bacterium at $309.8°$ K. This is $2.2°$ K cooler than the German scientist's incubation temperature in which the bacterium died, so she should be worried.

Iteration is a special type of composition, the composition of a function with itself.

Example **6** Find $f(4)$ if $f(0) = 2$ and $f(n) = f(n - 1) + n$.

$$
\begin{aligned}
f(0) &= 2 \\
f(1) &= f(1 - 1) + 1 = f(0) + 1 = 3 \quad \textit{Use f(0) to find f(1).} \\
f(2) &= f(2 - 1) + 2 = f(1) + 2 = 5 \quad \textit{Use f(1) to find f(2).} \\
f(3) &= f(3 - 1) + 3 = f(2) + 3 = 8 \quad \textit{Use f(2) to find f(3).} \\
f(4) &= f(4 - 1) + 4 = f(3) + 4 = 12 \quad \textit{Use f(3) to find f(4).}
\end{aligned}
$$

Communicating Mathematics

Study the lesson. Then complete the following.

1. **Write** out how you would read $[g \circ h](x)$.

2. In Example 2, what values would the domain of h need for $h \circ k$ to exist?

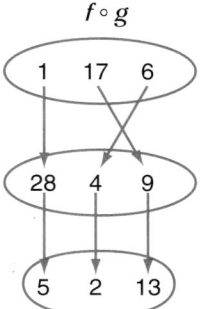

$f \circ g$

3. Look at the mapping of $[f \circ g](x)$ at the right.

   a. State the domain and the range of f and g.

   b. Write the functions f and g as a set of ordered pairs.

   c. Does $[g \circ f](x)$ exist? Explain.

4. Refer to Example 3. What type of function is $[g \circ f](x)$?

5. **Show** that if $f(x) = x^2$ and $g(x) = x - 4$, then $[f \circ g](x) \neq [g \circ f](x)$.

6. **Write** two functions $f(x)$ and $g(x)$ such that $f[g(x)] = x^2 - 6$.

7. **Explain** how you could find $f(2)$ and $f(3)$ if $f(x)$ is defined by $f(x + 1) = \frac{1}{4}f(x)$ and $f(1) = 24$.

8. Draw your family tree. Explain how your family tree could be a composition of functions.

Guided Practice

Find $[f \circ g](2)$ and $[g \circ f](2)$.

9. $f(x) = x + 6$
   $g(x) = x - 3$

10. $f(x) = x^2 + 3$
    $g(x) = x + 1$

Find $g[h(x)]$ and $h[g(x)]$.

11. $g(x) = 4x$
    $h(x) = 2x - 1$

12. $g(x) = x + 2$
    $h(x) = x^2$

If $f(x) = x^2$, $g(x) = 3x$, and $h(x) = x + 2$, find each value.

13. $f[g(1)]$

14. $[f \circ h](4)$

15. $h[f(x)]$

Find the first four iterations of each function, given the initial value.

16. $f(0) = 1, f(n) = f(n - 1) + 3$

17. $f(1) = 3, f(n) = 2f(n - 1)$

18. **Bonus** A sales representative for a furniture manufacturer is paid an annual salary plus a bonus of 3% of her sales *over* $275,000. Let $f(x) = x - 275,000$ and let $h(x) = 0.03x$.

    a. If x is greater than 275,000, is her bonus represented by $f[h(x)]$ or $h[f(x)]$? Explain.

    b. Find her bonus if her sales for the year are $400,000.

Practice

Find $[f \circ g](3)$ and $[g \circ f](3)$.

19. $f(x) = x$
    $g(x) = -x$

20. $f(x) = x^2$
    $g(x) = x^3$

21. $f(x) = x + 1$
    $g(x) = x^2 + 6$

22. $f = \{(1, -7), (2, 3), (3, 0)\}$
    $g = \{(0, 11), (3, 1)\}$

23. $f = \{(-1, 9), (3, 6)\}$
    $g = \{(-5, 3), (6, 12), (3, -1)\}$

24. $f(x) = 7x - 5$
    $g(x) = x^2 - 3x + 7$

Find $g[h(x)]$ and $h[g(x)]$.

25. $g(x) = x + 7$
    $h(x) = x + 4$

26. $g(x) = 5x$
    $h(x) = 2x$

27. $g(x) = x - 2$
    $h(x) = x^2$

28. $g(x) = -2x$
    $h(x) = -3x + 1$

29. $g(x) = x + 1$
    $h(x) = x^3$

30. $g(x) = |x|$
    $h(x) = x - 3$

If $f(x) = x^2$, $g(x) = 4x$, and $h(x) = x - 1$, find each value.

31. $h[g(2)]$

32. $[f \circ g](4)$

33. $[h \circ f](3)$

34. $[f \circ h](-3)$

35. $h[g(-2)]$

36. $h[f(-4)]$

37. $g[f(x)]$

38. $[f \circ g](x)$

39. $[f \circ (g \circ h)](x)$

Find the first four iterations of each function, given the initial value.

40. $f(0) = 4, f(n) = f(n - 1) - 5$

41. $f(1) = 2, f(n) = 3f(n - 1)$

42. $f(1) = 3, f(n) = f(n - 1) + 4n$

43. $f(0) = 0.2, f(n) = n(f(n - 1))$

44. $f(0) = -2, f(n) = 3f(n - 1) - 2n$

45. $f(0) = 6, f(n) = 4f(n - 1) + n^2$

Express $f \circ g$ and $g \circ f$, if they exist, as sets of ordered pairs.

46. $f = \{(1, 1), (0, -3)\}$
    $g = \{(1, 0), (-3, 1), (2, 1)\}$

47. $f = \{(3, 8), (4, 0), (6, 3), (7, -1)\}$
    $g = \{(0, 4), (8, 6), (3, 6), (-1, 8)\}$

Critical Thinking

48. Name two functions f and g such that $f[g(x)] = g[f(x)]$.

49. If $f(0) = 4$ and $f(x + 1) = 3f(x) - 2$, find $f(4)$.

Applications and Problem Solving

50. **Foreign Currency** The British Isles, located northwest of the European mainland, consist of two large islands, Great Britain and Ireland, and many smaller islands. Carolyn Martinez took a trip to the British Isles during the summer of 1995 and needed to exchange American dollars for British pounds. After she arrived in Ireland, she needed to exchange her British pounds for Irish punts. Look at the functions below.

$B(x) = 0.9733x$ *Converting British pounds to Irish punts*

$A(x) = 0.6252x$ *Converting American dollars to British pounds*

a. Find the equation of the composition function $B[A(x)]$. Explain what this composition represents.

b. How many Irish punts will she get for $500?

51. Discounts Jeanette bought a new electric wok that was originally priced at $38. The department store advertised a rebate of $5 as well as a discount of 25% off all small appliances.

a. Express the price of the wok after the rebate and the price after the discount using function notation. Let x represent the price of the wok, $r(x)$ represent the price after the rebate, and $p(x)$ represent the price after the discount.

b. Find $r[p(x)]$ and explain what this value represents.

c. Find $p[r(x)]$ and explain what this value represents.

52. Finance Nashota pays $30 each month on a credit card that charges 1.4% interest monthly. She has a balance of $450. The balance at the beginning of the nth month is given by the following function that is defined recursively.

$f(1) = 450$

$f(n) = f(n - 1) + 0.014f(n - 1) - 30$

Find the balances at the beginning of the first five months.

Mixed Review

53. Write $x^6 + 3x^3 - 10 = 0$ in quadratic form. (Lesson 8–6)

54. Write $y^2 = 6x$ in the form $x = a(y - k)^2 + h$. (Lesson 7–2)

55. Statistics An astronomer made ten measurements in minutes of degrees (') of the angular distance between two stars. The measurements were 11.20', 11.17', 10.92', 11.06', 11.19', 10.97', 11.09', 11.05', 11.22', and 11.03'. Find the mean and the standard deviation of the measurements. (Lesson 6–8)

56. Find the product $(m + 7)^2$. (Lesson 5–2)

57. Find $\begin{bmatrix} 4 & 5 \\ -2 & 9 \\ -1 & 4 \end{bmatrix} \cdot \begin{bmatrix} 5 & 3 & -6 & 0 \\ -2 & 1 & 4 & -1 \end{bmatrix}$. (Lesson 4–3)

Molecular cloud from which new stars emerge

58. Art Cecilia would like to place a picture of her triangle-shaped painting in the center of a page in her portfolio. She would like the longest side on the bottom. The lengths of the sides of the picture are 4 inches, 3 inches, and 3 inches. Suppose the center of her page is represented by the origin. Find a system of inequalities that describes the points her picture would occupy on the page, so that the top and bottom of the picture are at an equal distance above and below the origin, and the left and right corners are at an equal distance from the origin. (Lesson 3–5)

59. What type of special function is $f(x) = x$? (Lesson 2–6)

a. constant   b. identity   c. absolute value   d. step

60. Find a value of b for which the graph of $y = bx - 7$ is perpendicular to the graph of $x - 2y = 18$. (Lesson 2–3)

61. Simplify $-4(3a + 2b) - 3(-7a - 6b)$. (Lesson 1–2)

MODELING MATHEMATICS

8–7B Exploring Iteration

An Extension of Lesson 8–7

Materials: grid paper

Each result of the iteration process is called an **iterate.** To iterate a function $f(x)$, begin with a starting value x_0, find $f(x_0)$, and call the result x_1. Then find $f(x_1)$, and call the result x_2. Find $f(x_2)$ and call the result x_3, and so on.

Activity 1 Find the first three iterates, x_1, x_2, and x_3, of the function $f(x) = \frac{1}{2}x + 5$ for an initial value of $x_0 = 2$.

Step 1 To obtain the first iterate, find the value of the function for $x_0 = 2$.

$f(x_0) = f(2)$

$\quad = \frac{1}{2}(2) + 5$ or 6   So, $x_1 = 6$.

Step 2 To obtain the second iterate x_2, substitute the function value for the first iterate, x_1, for x.

$f(x_1) = f(6)$

$\quad = \frac{1}{2}(6) + 5$ or 8   So, $x_2 = 8$.

Step 3 Now find the third iterate, x_3, by substituting x_2 for x.

$f(x_2) = f(8)$

$\quad = \frac{1}{2}(8) + 5$ or 9   So, $x_3 = 9$.

Therefore, the first three iterates for the function $f(x) = \frac{1}{2}x + 5$ for an initial value of $x_0 = 2$ are 6, 8, 9.

Graphing the iterations of a function can help us understand the process of iteration better. Follow these steps.

- Graph a function $g(x)$ and the line $f(x) = x$ on the coordinate plane.

- Choose an initial value, x_0, and locate the point $(x_0, 0)$.

- Draw a vertical line from $(x_0, 0)$ to the graph of $g(x)$. This will be the segment from the point $(x_0, 0)$ to $(x_0, g(x_0))$.

- Now draw a horizontal segment from this point to the graph of the line $f(x) = x$. This will be the segment from $(x_0, g(x_0))$ to $(g(x_0), g(x_0))$.

- Repeat the process for many iterations.

This process is called **graphical iteration.**

You can think of the line $f(x) = x$ as a mirror that reflects each function value to become the input for the next iteration of the function. The points at which the graph of the function $g(x)$ intersects the graph of the line $f(x) = x$ are called **fixed points.** If you try to iterate the initial value that corresponds to the x-coordinate of a fixed point, the iterates will all be the same.

Four basic paths are possible when a linear function is iterated.

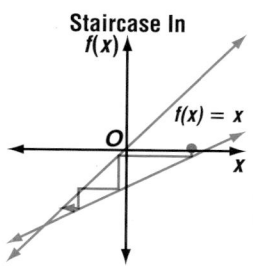

Staircase Out

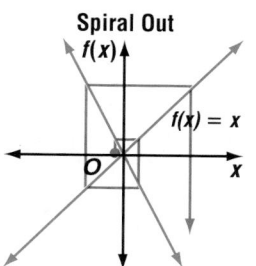

Staircase In

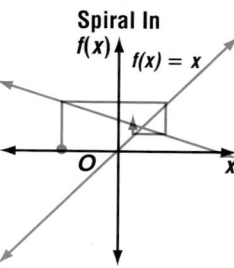

Spiral Out

Spiral In

Activity 2 Perform graphical iteration on the function $g(x) = 4x$ for the first three iterates if the initial value is $x_0 = 0.25$. Which of the four types of paths does the iteration take?

Step 1 To do the graphical iteration, first graph the functions $f(x) = x$ and $g(x) = 4x$.

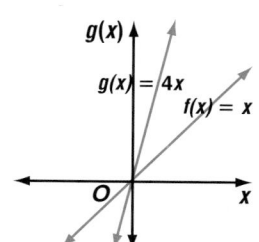

Step 2 Start at the point $(0.25, 0)$ and draw a vertical line to the graph of $g(x) = 4x$. From that point, draw a horizontal line to the graph of $f(x) = x$.

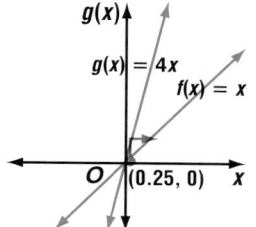

Step 3 Repeat the process from the point on $f(x) = x$. Then repeat again.

The path of the iterations staircases out.

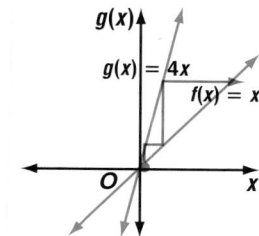

Model Find the first three iterates of each function using the given initial value. If necessary, round your answers to the nearest hundredth.

1. $g(x) = 5x$; $x_0 = 0.2$
2. $g(x) = -2x + 1$; $x_0 = -0.5$
3. $g(x) = 3 - 0.4x$; $x_0 = -4$
4. $g(x) = 3x - 0.5x^2$; $x_0 = 1$

Draw Graph each function and the function $f(x) = x$ on the same set of axes. Then draw the graphical iteration for $x_0 = 1$. State the slope of the linear function and tell what type of path the graphical iteration forms.

5. $g(x) = 4x + 12$
6. $g(x) = \frac{3}{5}x + 2$
7. $g(x) = -2x - 3$
8. $g(x) = 5x - 7$
9. $g(x) = \frac{1}{4}x + 1$
10. $g(x) = -\frac{1}{3}x + 4$

Write 11. Write a paragraph explaining the relationship between the slope of a linear function and the type of path that the graphical iteration forms.

12. What type of path do you think is formed when you perform the graphical iteration on the function $f(x) = 5x - x^2$? How does it compare to the iteration of linear functions?

8-8

Inverse Functions and Relations

The Tāj Mahal, Āgra, India

What YOU'LL LEARN

- To determine the inverse of a function or relation,
- to graph functions and their inverses, and
- to work backward to solve problems.

Why IT'S IMPORTANT

You can use inverses to solve problems involving shopping and world cultures.

APPLICATION
World Cultures

The ancient Hindus loved to do number puzzles. Aryabhata, a mathematician who lived in India during the sixth century A.D., was especially drawn to these puzzles. Look at the number puzzle below.

> Choose a number between 1 and 10.
> Multiply that number by 4.
> Add 6 to the resulting number.
> Divide by 2.
> Then subtract 5.

Aryabhata could have correctly told you your original number. How? Suppose your original number was 8. The chart below shows the steps of the puzzle.

Verbal Instructions	Number	Functional Representation
Choose a number between 1 and 10.	8	$f(x) = x$
Multiply that number by 4.	8 × 4 or 32	$g(x) = 4[f(x)] = 4x$
Add 6.	32 + 6 or 38	$h(x) = g(x) + 6 = 4x + 6$
Divide by 2.	38 ÷ 2 or 19	$j(x) = h(x) \div 2 = \frac{4x + 6}{2}$
Then subtract 5.	19 − 5 or 14	$k(x) = j(x) - 5 = \frac{4x + 6}{2} - 5$

GL●BAL CONNECTIONS

Aryabhata I (476–550), a Hindu mathematician and astronomer, was the most important early scholar in Indian mathematics. In his work *Aryabhatiya*, he gave a value of 3.1416 for π, used the decimal and place-value system, and supplied a variety of rules for algebra.

To tell you the original number, Aryabhata would **work backward** and do the inverse, or opposite, of the steps as he went along. Subtraction is the inverse operation of addition, and division is the inverse operation of multiplication.

The chart below shows the inverse of the puzzle shown above.

Verbal Instructions	Number	Functional Representation
Tell me the number that you ended with.	14	$p(x) = x$
Add 5 to the number.	14 + 5 or 19	$r(x) = p(x) + 5 = x + 5$
Multiply by 2.	19 × 2 or 38	$t(x) = r(x) \times 2 = 2x + 10$
Subtract 6.	38 − 6 or 32	$v(x) = t(x) - 6 = 2x + 4$
Divide by 4.	32 ÷ 4 or 8	$w(x) = \frac{v(x)}{4} = \frac{1}{2}x + 1$

The functions $k(x) = \frac{4x + 6}{2} - 5$ and $w(x) = \frac{1}{2}x + 1$ are **inverse functions.**

Definition of Inverse Functions	Two functions f and g are inverse functions if and only if both of their compositions are the identity function. That is, $$[f \circ g](x) = x \text{ and } [g \circ f](x) = x.$$

You can determine if two functions are inverse functions by finding both of their compositions. If both compositions equal the identity function $h(x) = x$, then the functions are inverse functions.

You can also determine if two functions are inverse functions by graphing. The graphs of a function and its inverse are mirror images, or reflections, of each other with respect to the graph of the identity function $h(x) = x$. Its graph is the line of symmetry.

Example **1** **Determine whether $f(x) = 3x - 4$ and $g(x) = \dfrac{x + 4}{3}$ are inverse functions.**

Method 1: Finding Compositions

Find $[f \circ g](x)$ and $[g \circ f](x)$ to determine whether these functions are inverse functions.

$$[f \circ g](x) = f[g(x)] \qquad\qquad [g \circ f](x) = g[f(x)]$$

$$= f\!\left(\frac{x + 4}{3}\right) \qquad\qquad = g(3x - 4)$$

$$= 3\!\left(\frac{x + 4}{3}\right) - 4 \qquad\qquad = \frac{(3x - 4) + 4}{3}$$

$$= x \qquad\qquad\qquad = x$$

Since both $[f \circ g](x)$ and $[g \circ f](x)$ equal x, then $f(x)$ and $g(x)$ are inverse functions. That is, f is the inverse of g and g is the inverse of f.

Method 2: Graphing

Graph both functions.

Suppose the plane containing the graphs could be folded along the line $h(x) = x$. Then the graphs would coincide. This verifies that the functions are inverses.

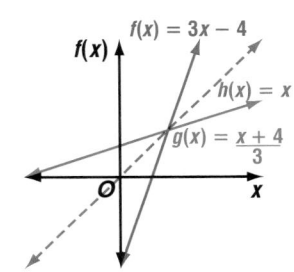

We can write f is the inverse of g and g is the inverse of f using the notation $f = g^{-1}$ and $g = f^{-1}$. The symbol g^{-1} is read "g inverse" or "the inverse of g." By the definition of inverse functions, we can write the following.

The -1 is not an exponent.

$$[f \circ f^{-1}](x) = x \text{ and } [f^{-1} \circ f](x) = x$$

The ordered pairs of inverse functions are related. Use the functions in Example 1 and evaluate $f(5)$. Then find $f^{-1}[f(5)]$.

$f(5) = 3(5) - 4$ or 11 $\qquad\qquad$ $f^{-1}[f(5)] = f^{-1}(11)$

The ordered pair $(5, 11)$ $\qquad\qquad\qquad$ $= \dfrac{(11) + 4}{3}$ or 5
belongs to f.

$\qquad\qquad\qquad\qquad\qquad$ The ordered pair $(11, 5)$
$\qquad\qquad\qquad\qquad\qquad$ belongs to f^{-1}.

So, the inverse of a function can be found by exchanging the domain and range of the function.

Property of Inverse Functions	Suppose f and f^{-1} are inverse functions. Then $f(a) = b$ if and only if $f^{-1}(b) = a$.

To find the inverse of a function f, you can interchange the variables in the equation $y = f(x)$.

Example ② **Find the inverse of $f(x) = 3x + 6$. Then verify that f and f^{-1} are inverse functions.**

Method 1: Algebra

$y = 3x + 6$ *Rewrite $f(x) = 3x + 6$ as $y = 3x + 6$.*

$x = 3y + 6$ *Interchange x and y.*

$y = \dfrac{x - 6}{3}$ *The inverse is also a function.*

The inverse of $f(x) = 3x + 6$ is $f^{-1}(x) = \dfrac{x - 6}{3}$.

Check: To verify that f and f^{-1} are inverses, show that the compositions of f and f^{-1} are identity functions.

$[f \circ f^{-1}](x) = f[f^{-1}(x)]$ $[f^{-1} \circ f](x) = f^{-1}[f(x)]$

$\qquad\qquad = f\left(\dfrac{x - 6}{3}\right)$ $\qquad\qquad = f^{-1}(3x + 6)$

$\qquad\qquad = 3\left(\dfrac{x - 6}{3}\right) + 6$ $\qquad\qquad = \dfrac{(3x + 6) - 6}{3}$

$\qquad\qquad = x$ $\qquad\qquad = x$

The functions are inverses, since both $[f \circ f^{-1}](x)$ and $[f^{-1} \circ f](x)$ equal x.

Method 2: Graphing

Now graph both functions.

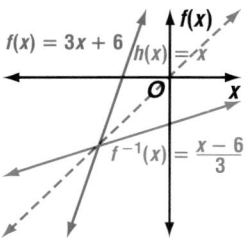

The graphs are reflections of each other over the line $h(x) = x$. This verifies that the functions are inverses.

In Example 2, f and f^{-1} were both functions. However, the inverse of a function is not always a function.

Example ③ **Find the inverse of $f(x) = x^2 + 4$. Determine whether the inverse is a function.**

Rewrite $f(x)$ as $y = x^2 + 4$.

To make the inverse of a quadratic function a function, only nonnegative values of the range are considered. Using only these values, the inverse is called a <u>square root function</u>. In Example 3, the square root function is $y = \sqrt{x - 4}$.

$x = y^2 + 4$ *Interchange x and y.*

$\pm\sqrt{x - 4} = y$ *Solve for y.*

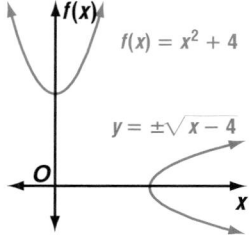

The inverse of $f(x) = x^2 + 4$ is $y = \pm\sqrt{x - 4}$. This inverse is not a function, since the graph does not pass the vertical line test for functions.
Check this result.

MODELING MATHEMATICS

Inverses of Functions

Materials:  geomirror · grid paper · straightedge

Use a full sheet of grid paper. Draw axes in the center of the page, and label each mark on each axis as one unit.

Your Turn

a. Use a straightedge to graph $y = 2x - 8$ on the grid paper. Label the graph with its equation.

b. On the same set of axes, use a straightedge to graph $y = x$ as a dashed line.

c. Place the reflective mirror so that the drawing edge is on the line $y = x$ and carefully plot points that are part of the reflection of the original line with respect to the line of symmetry.

d. Draw a line through the points. This is the inverse of the original function.

e. What is the equation of the inverse?

f. Try this activity with the function $y = x^3 - 5$. Is the inverse also a function? Explain.

You may recall that a relation is a set of ordered pairs. The **inverse relation** is the set of ordered pairs obtained by reversing the coordinates of each original ordered pair.

Definition of Inverse Relations	Two relations are inverse relations if and only if whenever one relation contains the element (a, b), the other relation contains the element (b, a).

Example ④

Geometry

The vertices of square *ABCD* form the relation $\{(2, 4), (8, 4), (2, -2), (8, -2)\}$. Find the inverse of this relation and determine if the resulting ordered pairs are also the vertices of a square.

To find the inverse of this relation, reverse the coordinates of the ordered pairs.

The inverse of the relation is $\{(4, 2), (4, 8), (-2, 2), (-2, 8)\}$.

Plotting the points shows that the ordered pairs are also vertices of a square.

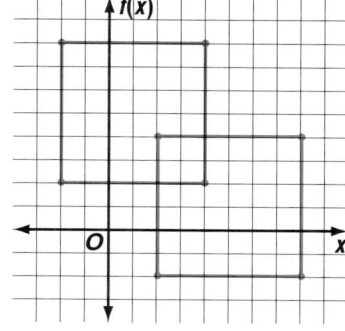

CHECK FOR UNDERSTANDING

Communicating Mathematics

Study the lesson. Then complete the following.

1. **Explain** the difference between inverse functions and inverse relations. Are inverse functions also inverse relations? Explain.

2. **Describe** in your own words how to find the inverse of a function.

3. **Explain** why the inverse relation in Example 4 is not a function.

4. **Explain** how the graph of a function is related to the graph of its inverse.

5. **Sketch** the graph of the inverse of $t(x)$, shown at the right. Is the inverse a function? Explain.

6. Use a reflective mirror and grid paper to graph the inverse of the function $f(x) = \frac{1}{2}x^2 - 3$.

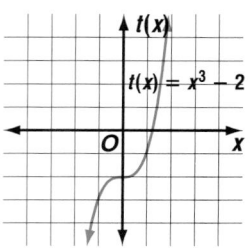

Guided Practice

Find the inverse of each relation and determine whether the inverse is a function.

7. $\{(3, 2), (4, 2)\}$

8. $\{(3, 8), (4, -2), (5, -3)\}$

Find the inverse of each function. Then graph the function and its inverse.

9. $y = 7x$

10. $y = x$

11. $f(x) = x - 6$

12. $y = -2x - 1$

13. Determine whether $f(x) = 6x + 2$ and $g(x) = \frac{x + 2}{6}$ are inverse functions.

14. **Temperature** Refer to the application at the beginning of Lesson 8-7. The formula for converting Fahrenheit to Celsius is $C(x) = \frac{5}{9}(x - 32)$. Find $C^{-1}(x)$, the inverse of $C(x)$, and explain what practical purpose the inverse serves.

EXERCISES

Practice

Find the inverse of each relation and determine whether the inverse is a function.

15. $\{(2, 4), (-3, 1), (2, 8)\}$

16. $\{(-1, -2), (-3, -2), (-1, -4), (0, 6)\}$

17. $\{(1, 3), (1, -1), (1, -3), (1, 1)\}$

18. $\{(6, 11), (-2, 7), (0, 3), (-5, 3)\}$

Find the inverse of each function. Then graph the function and its inverse.

19. $y = 6$

20. $y = 4x$

21. $f(x) = 4x + 4$

22. $g(x) = \frac{1}{2}x + 2$

23. $f(x) = -x$

24. $g(x) = x - 2$

25. $y = x^2 - 9$

26. $y = (x - 9)^2$

27. $f(x) = \frac{2x - 1}{3}$

28. $y = x^2 + 1$

29. $y = (x - 4)^2$

30. $y = (x + 2)^2 - 3$

Determine whether each pair of functions are inverse functions.

31. $f(x) = x + 7$

$g(x) = x - 7$

32. $g(x) = 2x - 3$

$h(x) = -2x + 3$

33. $f(x) = \frac{x - 2}{3}$

$g(x) = 3x - 2$

34. $f(x) = \frac{x - 1}{2}$

$g(x) = 2x + 1$

Sketch the graph of the inverse of each relation. Then determine if the inverse is a function.

35.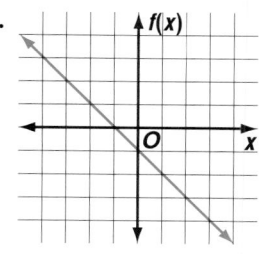

36.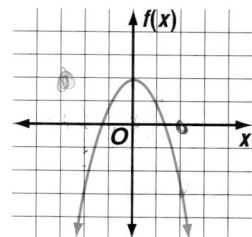

37.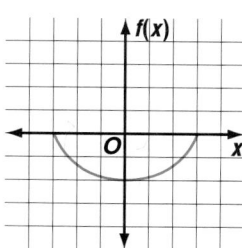

Programming

38. The graphing calculator program at the right evaluates $f[g(x)]$ and $g[f(x)]$ if $f(x) = 2x - 1$ and $g(x) = x^2 + 1$. The calculator prompts you to enter a value for x and finds the values of $f[g(x)]$ and $g[f(x)]$. It also tells you if the functions are inverses. To use this program for other pairs of functions, change the second and third lines to reflect the expressions for $f[g(x)]$ and $g[f(x)]$.

```
Program: INVERSES
: Prompt X
: 2(X² + 1) − 1→A
: (2X − 1)² + 1→B
: Disp "F(G(X)) = ",A
: Disp "G(F(X)) = ",B
: If A = B
: Then
: Disp "INVERSES"
: Else
: Disp "NOT INVERSES"
```

Use this program and a chart to determine if each pair of functions are inverses of each other.

a. $f(x) = x + 1$
 $g(x) = x - 1$

b. $f(x) = \frac{1}{2}x^2 + 4$
 $g(x) = 2x + 8$

c. $g(x) = \frac{1}{5}(x + 7)$
 $h(x) = 5x - 7$

Critical Thinking

39. Find a function that is its own inverse. Can you find more than one?

Applications and Problem Solving

40. Consumerism LaKisha bought a stereo on sale from Electronics Unlimited. She used a $40 gift certificate to help pay for the stereo. If the stereo was on sale at 25% off and the final bill was $522.50 plus tax, what is its regular price?

41. Work Backward Jake asked Cynthia to choose a number between 1 and 20. He told her to add 7 to that number, then multiply by 4, then subtract 6, then divide by 2. Cynthia told Jake that her final number was 35. What was her original number?

42. Sales Sales associates at Electronics Unlimited earn $8 an hour plus a 4% commission on the merchandise they sell. Write a function to describe their weekly income and find how much merchandise they must sell in order to earn $500 in a 40-hour week.

Mixed Review

43. Chemistry While performing an experiment, Joy Chen found the temperature of a solution at different times. She needs to record the temperature in degrees Kelvin, but only has a thermometer with a Fahrenheit scale. Joy knows that a Kelvin temperature is 273 degrees greater than an equivalent Celsius temperature and that the formula $C = \frac{5}{9}(F - 32)$ converts a Fahrenheit temperature to Celsius.

What will she record when the thermometer reads 59° F?
(Lesson 8–7)

44. Write an equation for the parabola with focus at (2, 4) and directrix $y = 6$. Then draw the graph. (Lesson 7–2)

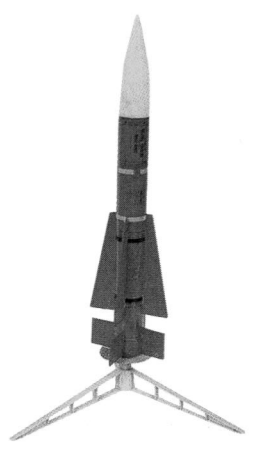

45. Physics A toy rocket is fired upwards from the top of a 200-foot tower at a velocity of 80 feet per second. The height of the rocket t seconds after firing is given by the formula $h(t) = -16t^2 + 80t + 200$. Find the maximum height reached by the rocket and the time at which that height is reached. (Lesson 6–4)

46. Divide $(n^3 - n^2 + 4n + 6)$ by $(n + 1)$ using synthetic division. (Lesson 5–3)

47. Find $-2\begin{bmatrix} 0 & 3 \\ -5 & 3 \end{bmatrix} + \begin{bmatrix} 5 & 3 \\ -3 & 9 \end{bmatrix}$. (Lesson 4–2)

48. Graph $|x| + y \geq 3$. (Lesson 2–7)

49. State the domain and range of the relation $\{(9, 0), (3, 1), (12, 7), (1, -4), (12, 8), (-11, -3), (0, -6)\}$. Is this relation a function? (Lesson 2–1)

50. Manufacturing A company manufactures auto parts. The diameter of a piston cannot vary more than 0.001 cm. Write an inequality to represent the diameter of a piston if its diameter is supposed to be 10 cm. (Lesson 1–7)

WORKING ON THE Investigation

Refer to the Investigation on pages 474–475.

Now that you have performed your own experiments on coffee cans of different sizes, you decide to get some data on the markings on a real gasoline tank from another colleague. The length of the tank is 20 feet and its radius is 10 feet. The table shows the distance in feet measured on the dipstick, the volume in cubic feet of gasoline in the tank, the percent of the tank that is full, and the percent of the dipstick that is wet.

meas. on stick (ft)	vol. of gas in tank (ft³)	% of vol. in tank	% of stick covered	change in vol. %
0	0.0	0.0	0	
1	327.0	5.2	10	
2	894.6	14.2	20	
3	1585.3	25.2	30	
4	2347.0	37.4	40	
5	3141.6	50.0	50	
6	3936.2	62.6	60	
7	4697.8	74.8	70	
8	5388.6	85.8	80	
9	5956.2	94.8	90	
10	6283.2	100.0	100	

1 What are the differences and similarities in your charts and this chart? Are there any differences in the dipstick measurement increments and the volume measurement increments in your data and these data? Explain.

2 What is the volume of this tank in cubic feet?

3 Where on the dipstick would the markings be to show that the tank is one-half full, one-fourth full, and three-fourths full? Does this coincide with the locations at which you found the markings should be for your data?

4 Copy and complete the table.

5 Make a graph of the change in percentage of capacity for each increment. Describe the graph. How does this graph differ from those you made in your experiments? How is this graph similar to the graphs you made in your experiments?

6 Graph the inverse of this relation. Does the original graph represent a function? Does the graph of the inverse represent a function?

Add the results of your work to your Investigation Folder.

8-8B Square Root Functions and Relations

What YOU'LL LEARN

- To graph and analyze square root functions, and
- to graph square root inequalities.

Why IT'S IMPORTANT

You can use square root functions to solve problems involving weather and oceanography.

Weather

The function $d = \sqrt{\dfrac{3h}{2}}$ represents the greatest distance d in miles that a person h feet high can see on a clear day. Suppose Kenyatta is standing on the 102nd floor observation deck of the Empire State Building, 1250 feet high, on a clear day. What is the greatest distance that he can see? *You will solve this problem in Example 2.*

Because the function described above involves a square root, it is called a **square root function.** In order for a square root to be a real number, the radicand cannot be negative. When graphing a square root function, determine when the radicand would be negative and remember to exclude those values from the domain.

In Lesson 8-8, you learned that the inverse of a quadratic function is a square root function if only the nonnegative range is considered. Look at the graph at the right. Notice that for a square root function, negative values are excluded from the range.

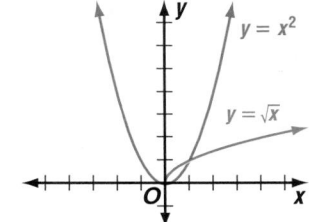

For many families of square root functions, the parent function is $y = \sqrt{x}$.

EXPLORATION

GRAPHING CALCULATORS

To graph square root function like $y = \sqrt{x - 1}$, use the $\boxed{\sqrt{}}$ key.

Enter: $\boxed{Y=}$ $\boxed{2nd}$ $\boxed{\sqrt{}}$ $\boxed{(}$ $\boxed{X,T,\theta,n}$

$\boxed{-}$ 1 $\boxed{)}$ $\boxed{ZOOM}$ 6

Your Turn

a. Graph $y = \sqrt{x}$, $y = \sqrt{x} + 1$, and $y = \sqrt{x} - 2$ in the viewing window $[-2, 8]$ by $[-4, 6]$. State the domain and range of each function and describe the similarities and differences among the graphs.

b. Graph $y = \sqrt{x}$, $y = \sqrt{2x}$, and $y = \sqrt{8x}$ in the viewing window $[0, 10]$ by $[0, 10]$. State the domain and range of each function and describe the similarities and differences among the graphs.

Example **1** Graph $y = \sqrt{6x + 5}$. State the domain, range, and x- and y-intercepts.

Since the radicand cannot be negative, identify the domain.

$$6x + 5 \geq 0$$
$$6x \geq -5$$
$$x \geq -\frac{5}{6}$$

Thus, the x-intercept is $-\frac{5}{6}$. Generate a table of values and graph the function.

x	y
$-\frac{5}{6}$	0
0	2.2
2	4.1
4	5.4
6	6.4
8	7.3
10	8.1

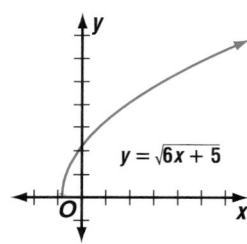

$y = \sqrt{6x + 5}$

The domain is $x \geq -\frac{5}{6}$, and the range is $y \geq 0$. The y-intercept is $\sqrt{5}$ or about 2.24.

Example **2**

Weather

Refer to the application at the beginning of the lesson.

a. Graph the function $d = \sqrt{\frac{3h}{2}}$. State the domain and range.

b. Find the greatest distance that Kenyatta can see.

a. Make a table of values and graph the function.

h	d
0	0
2	$\sqrt{3}$ or 1.73
4	$\sqrt{6}$ or 2.45
6	$\sqrt{9}$ or 3.00
8	$\sqrt{12}$ or 3.46
10	$\sqrt{15}$ or 3.87

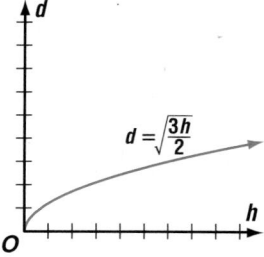

$d = \sqrt{\frac{3h}{2}}$

The domain and range of the function are both all nonnegative real numbers.

b. If the observation deck is 1250 feet high, then the greatest distance Kenyatta can see is $\sqrt{\frac{3(1250)}{2}}$ or about 43.3 miles. *Check this result against the graph.*

You can use what you know about square root functions to graph *square root inequalities*.

Example ③ **Graph** $y < \sqrt{2x + 4}$.

Generate a table of values and graph the relation. Since the boundary should not be included, the graph should be dashed.

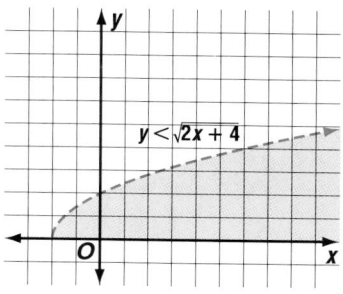

Because the range includes only nonnegative real numbers, the graph is only in the first and second quadrants. Select a point on one of the half-planes and test its ordered pair. For example, use (1, 1). Since $1 < \sqrt{2(1) + 4}$, the half-plane containing (1, 1) should be shaded.

CHECK FOR UNDERSTANDING

Communicating Mathematics

Study the lesson. Then complete the following.

1. **Describe** the differences between the graphs of $y = \sqrt{x} - 6$ and $y = \sqrt{x - 6}$.

2. **Find** the minimum value in the domain for $y = \sqrt{2x + 3}$.

3. Refer to the application at the beginning of the lesson.

   a. Write a mathematical sentence that describes *all* of the distances Kenyatta would be able to see.

   b. Graph the relation that you came up with in part a.

Guided Practice

Graph each function. State the domain and range of each function.

4. $y = -\sqrt{x}$

5. $y = \sqrt{7x}$

6. $y = \sqrt{x - 1} + 5$

7. $y = \sqrt{5x + 1}$

Graph each inequality.

8. $y > \sqrt{x + 9} - 8$

9. $y \le \sqrt{3x + 4} - 4$

10. **Oceanography** A *tsunami* is a large ocean wave generated by an undersea earthquake. The formula for a tsunami's speed s in meters per second is $s = 3.1\sqrt{d}$, where d is the depth of the ocean in meters.

    a. Describe the domain and range of the function.

    b. Determine the speed of a tsunami if the earthquake occurs in a part of the ocean that is 10,000 meters deep.

    c. Graph the function and use the graph to verify your answer to part b.

Practice **Graph each function. State the domain and range of each function.**

11. $y = \sqrt{3x}$

12. $y = -\sqrt{4x}$

13. $y = -2\sqrt{x}$

14. $y = 5\sqrt{x}$

15. $y = \sqrt{x + 2}$

16. $y = \sqrt{x - 7}$

17. $y = -\sqrt{2x + 1}$

18. $y = \sqrt{3x - 2}$

19. $y = \sqrt{x + 6} - 5$

20. $y = \sqrt{x - 5} + 3$

21. $y = \sqrt{7x - 1} + 2$

22. $y = 2\sqrt{3 - 4x} + 6$

Graph each inequality.

23. $y \le -6\sqrt{x}$

24. $y < \sqrt{x + 8}$

25. $y > \sqrt{5x + 7}$

26. $y \ge \sqrt{2x - 7}$

27. $y \ge \sqrt{x - 3} + 4$

28. $y < \sqrt{6x - 2} + 3$

Critical Thinking

29. In Lesson 6-6A, you investigated the role that a, h, and k played in the graph of a quadratic function of the form $y = a(x - h)^2 + k$.

   a. Describe the roles a, h, and k play for the family of square root functions of the form $y = a\sqrt{x - h} + k$.

   b. Use what you found in part a to describe the graphs of the functions below without graphing them. Relate your answers to the graph of the parent function $y = \sqrt{x}$.

   $$y = -2\sqrt{x} \qquad y = \sqrt{x - 4} \qquad y = \sqrt{x} + 3 \qquad y = 3\sqrt{x - 1} + 5$$

Applications and Problem Solving

30. **Geometry**   There are several sizes of ice cream cones available at Johnson's Real Ice Cream Shoppe, but all of them are 5 inches long.

   a. Write a square root function that expresses the radius r of the cones as a function of volume V. Use the formula $V = \frac{1}{3}\pi r^2 h$, where h is the height.

   b. Describe the domain and range of the function.

   c. Determine the volume of a cone that has a radius of 2 inches.

   d. Graph the function and use the graph to verify your answer to part c.

31. **Geometry**   The volume V and the surface area A of a soap bubble are related by the formula $V = 0.094\sqrt{A^3}$.

   a. Describe the domain and range of the function.

   b. Determine the volume of a soap bubble that has a surface area of 12 cm^2.

   c. Graph the function and use the graph to verify your answer to part b.

   d. Does this graph look like the graphs of the other square root functions you have graphed in this lesson? Why or why not?

After completing this chapter, you should be able to define each term, property, or phrase and give an example or two of each.

Algebra
complex conjugates theorem   (p. 504)
composition of functions   (p. 520)
depressed polynomial   (p. 487)
Descartes' rule of signs   (p. 505)
factor theorem   (p. 487)
fundamental theorem of algebra   (p. 503)
integral zero theorem   (p. 509)
inverse function   (p. 528)
inverse relation   (p. 531)
leading coefficient   (p. 480)
location principle   (p. 494)
polynomial function   (p. 479)
polynomial in one variable   (p. 478)
quadratic form   (p. 515)

rational zero theorem   (p. 509)
relative maximum   (p. 494)
relative minimum   (p. 494)
remainder theorem   (p. 485)
square root function   (p. 531)
synthetic substitution   (p. 486)

Discrete Mathematics
fixed points   (p. 526)
graphical iteration   (p. 526)
iterate   (p. 526)
iteration   (p. 522)

Problem Solving
work backward   (p. 528)

UNDERSTANDING AND USING THE VOCABULARY

Choose the letter of the term that best matches each statement or phrase.

1. A point on the graph of a polynomial function that has no other nearby points with lesser y-coordinates is the ___?___.

2. The ___?___ is the factor in the term in a polynomial function with the highest degree.

3. The ___?___ says that in any polynomial function, if an imaginary number is a zero of that function, then its conjugate is also a zero.

4. When a polynomial is divided by one of its binomial factors, the quotient is called a ___?___.

5. A point on the graph of a polynomial function that has no other nearby points with a greater y-coordinate is the ___?___.

6. $f \circ g\,(x) = f[g(x)]$ represents a ___?___.

7. $(x^2)^2 - 17(x^2) + 16 = 0$ is written in ___?___.

8. $f(x) = 6x - 2$ and $g(x) = \frac{x + 2}{6}$ are ___?___ since $[f \circ g](x) = x$ and $[g \circ f](x) = x$.

a. complex conjugates theorem
b. composition of functions
c. depressed polynomial
d. inverse functions
e. leading coefficient
f. quadratic form
g. relative maximum
h. relative minimum

STUDY GUIDE AND ASSESSMENT

SKILLS AND CONCEPTS

OBJECTIVES AND EXAMPLES

Upon completing this chapter, you should be able to:

- evaluate polynomial functions (Lesson 8–1)

Find $p(a + 1)$ if $p(x) = 5x - x^2 + 3x^3$.

$p(a + 1) = 5(a + 1) - (a + 1)^2 + 3(a + 1)^3$

$= 5a + 5 - (a^2 + 2a + 1) + 3(a + 1)(a^2 + 2a + 1)$

$= 5a + 5 - a^2 - 2a - 1 + 3(a^3 + 3a^2 + 3a + 1)$

$= 5a + 5 - a^2 - 2a - 1 + 3a^3 + 9a^2 + 9a + 3$

$= 3a^3 + 8a^2 + 12a + 7$

- find factors of polynomials by using the factor theorem and synthetic division (Lesson 8–2)

Show that $x + 2$ is a factor of $x^3 - 2x^2 - 5x + 6$. Then find any remaining factors.

$$
\begin{array}{r|rrrr}
-2 & 1 & -2 & -5 & 6 \\
   &   & -2 & 8 & -6 \\
\hline
   & 1 & -4 & 3 & 0
\end{array}
$$

The remainder is 0, so $x + 2$ is a factor of $x^3 - 2x^2 - 5x + 6$, and $x^3 - 2x^2 - 5x + 6 = (x + 2)(x^2 - 4x + 3)$.

Since $x^2 - 4x + 3 = (x - 3)(x - 1)$,
$x^3 - 2x^2 - 5x + 6 = (x + 2)(x - 3)(x - 1)$.

- approximate the real zeros of polynomial functions (Lesson 8–3)

The location principle states that if $y = f(x)$ represents a polynomial function, and a and b are two numbers such that $f(a) < 0$ and $f(b) > 0$, then the function has at least one zero between a and b.

REVIEW EXERCISES

Use these exercises to review and prepare for the chapter test.

Find $p(-4)$ and $p(x + h)$ for each function.

9. $p(x) = x - 2$

10. $p(x) = -x + 4$

11. $p(x) = 6x + 3$

12. $p(x) = x^2 + 5$

13. $p(x) = x^2 - x$

14. $p(x) = 2x^3 - 1$

Use synthetic substitution to find $f(3)$ and $f(-2)$ for each function.

15. $f(x) = x^2 - 5$

16. $f(x) = x^2 - 4x + 4$

17. $f(x) = x^3 - 3x^2 + 4x + 8$

18. $f(x) = x^4 - 5x + 2$

Given a polynomial and one of its factors, find the remaining factors of the polynomial. Some factors may not be binomials.

19. $x^3 + 5x^2 + 8x + 4; x + 1$

20. $x^3 + 4x^2 + 7x + 6; x + 2$

21. $x^3 - x^2 - 4x + 4; x + 2$

22. $x^4 - 6x^3 + 22x + 15; x + 1$

Approximate the real zeros of each function to the nearest tenth. Then use the functional values to graph the function.

23. $h(x) = x^3 - 6x - 9$

24. $f(x) = x^4 + 7x + 1$

25. $p(x) = x^5 + x^4 - 2x^3 + 1$

26. $g(x) = x^3 - x^2 + 1$

27. $r(x) = 4x^3 + x^2 - 11x + 3$

28. $f(x) = x^3 + 4x^2 + x - 2$

OBJECTIVES AND EXAMPLES

- find the number and type of zeros of a polynomial function   (Lesson 8–4)

State the number of positive real zeros and negative real zeros for $f(x) = 5x^4 + 6x^3 - 8x + 12$.

Since $f(x)$ has two sign changes, there are 2 or 0 real positive zeros.

$f(-x) = 5x^4 - 6x^3 + 8x + 12$

Since $f(-x)$ has two sign changes, there are 0 or 2 negative real zeros.

REVIEW EXERCISES

State the number of positive real zeros, negative real zeros, and imaginary zeros for each function.

29. $f(x) = 2x^4 - x^3 + 5x^2 + 3x - 9$
30. $f(x) = 7x^3 + 5x - 1$
31. $f(x) = -4x^4 - x^2 - x - 1$
32. $f(x) = 3x^4 - x^3 + 8x^2 + x - 7$
33. $f(x) = x^4 + x^3 - 7x + 1$

- find zeros of polynomial functions   (Lesson 8–5)

Find all of the zeros of $f(x) = x^3 + 7x^2 - 36$.

There are exactly 3 complex zeros.

There are either 1 or 3 positive real zeros and 2 or 0 negative real zeros.

The possible rational zeros are ± 1, ± 2, ± 3, ± 4, ± 6, ± 9, ± 12, ± 18, ± 36.

$$
\begin{array}{r|rrrr}
2 & 1 & 7 & 0 & -36 \\
  &   & 2 & 18 & 36 \\
\hline
  & 1 & 9 & 18 & 0
\end{array}
$$

$x^3 + 7x^2 - 36 = (x - 2)(x^2 + 9x + 18) = (x - 2)(x + 3)(x + 6)$

Therefore, the zeros are 2, -3, and -6.

Find all of the rational zeros for each function.

34. $f(x) = 2x^3 - 13x^2 + 17x + 12$
35. $f(x) = x^4 + 5x^3 + 15x^2 + 19x + 8$
36. $f(x) = x^3 - 3x^2 - 10x + 24$
37. $f(x) = 2x^3 - 5x^2 - 28x + 15$
38. $f(x) = 2x^4 - 9x^3 + 2x^2 + 21x - 10$

- solve nonquadratic equations by using quadratic techniques   (Lesson 8–6)

$x^3 - 3x^2 - 54x = 0$

$x(x^2 - 3x - 54) = 0$

$x(x - 9)(x + 6) = 0$

$x = 0, x = 9, x = -6$

$y - 4\sqrt{y} - 45 = 0$

$(\sqrt{y})^2 - 4(\sqrt{y}) - 45 = 0$

$(\sqrt{y} - 9)(\sqrt{y} + 5) = 0$

$\sqrt{y} - 9 = 0 \quad$ or $\quad \sqrt{y} + 5 = 0$

$\qquad y = 81 \qquad\quad$ no real solution

Solve each equation.

39. $3x^3 + 4x^2 - 15x = 0$
40. $m^4 + 3m^3 = 40m^2$
41. $a^3 - 64 = 0$
42. $r + 9\sqrt{r} = -8$
43. $x^4 - 8x^2 + 16 = 0$
44. $x^{\frac{2}{3}} - 9x^{\frac{1}{3}} + 20 = 0$

OBJECTIVES AND EXAMPLES	REVIEW EXERCISES

find the composition of functions   (Lesson 8–7)

If $f(x) = x^2 - 2$ and $g(x) = 8x - 1$, find $g[f(x)]$ and $f[g(x)]$.

$$g[f(x)] = 8(x^2 - 2) - 1$$
$$= 8x^2 - 16 - 1$$
$$= 8x^2 - 17$$
$$f[g(x)] = (8x - 1)^2 - 2$$
$$= (64x^2 - 16x + 1) - 2$$
$$= 64x^2 - 16x - 1$$

Find $g[h(x)]$ and $h[g(x)]$.

45. $h(x) = 2x - 1$
$g(x) = 3x + 4$

46. $h(x) = x^2 + 2$
$g(x) = x - 3$

47. $h(x) = x^2 + 1$
$g(x) = -2x + 1$

48. $h(x) = -5x$
$g(x) = 3x - 5$

49. $h(x) = x^3$
$g(x) = x - 2$

50. $h(x) = x + 4$
$g(x) = |x|$

determine the inverse of a function or relation
(Lesson 8–8)

Find the inverse of $f(x) = -3x + 1$.

Rewrite $f(x)$ as $y = -3x + 1$. Then interchange the variables and solve for y.

$$x = -3y + 1$$
$$3y = -x + 1$$
$$y = \frac{-x + 1}{3}$$
$$f^{-1}(x) = \frac{-x + 1}{3}$$

Find the inverse of each function. Then graph the function and its inverse.

51. $f(x) = 3x - 4$

52. $f(x) = -2x - 3$

53. $g(x) = \frac{1}{3}x + 2$

54. $f(x) = \frac{-3x + 1}{2}$

55. $y = x^2$

56. $y = (2x + 3)^2$

APPLICATIONS AND PROBLEM SOLVING

57. Financial Planning   Jo Phillips, a financial advisor, is helping Toshi's parents develop a plan to save money for his college education. Toshi will start college in six years. According to Ms. Phillip's plan, Toshi's parents will save $1000 each year for the next three years. During the fourth and fifth years, they will save $1200 each year. During the last year before he starts college, they will save $2000.
(Lesson 8–2)

  a. In the formula $A = P(1 + r)^t$, $A =$ the balance, $P =$ the amount invested, $r =$ the interest rate, and $t =$ the number of years the money has been invested. Use this formula to write a polynomial equation to describe the balance of the account when Toshi starts college.

  b. Find the balance of the account if the interest rate is 6%.

58. Manufacturing   The DrinKone Company makes paper cups that are cone-shaped. The volume of a cone is about 7.07 cubic inches. The diameter of the top of the cone is equal to the height of the cone. Determine the dimensions of the cone. (*Hint:* The formula for the volume of a cone is $V = \frac{1}{3}\pi r^2 h$.)
(Lesson 8–5)

59. Business   The CD Menagerie adds a 100% markup to the wholesale price of compact discs before placing them in the store for sale. If a CD is on sale for 20% off and the customer pays $12 for it, what is its wholesale price?
(Lesson 8–7)

A practice test for Chapter 8 is provided on page 919.

ALTERNATIVE ASSESSMENT

COOPERATIVE LEARNING PROJECT

Designing a Package   In this project, imagine that you are a packaging engineer. Packaging engineers determine the ideal shape and substance for a container based on what it will hold and how it will be used. You are the lead designer for a series of new containers to hold note paper. Each container should have twice the volume of the next smaller size. The container should use as little material as possible because of the need to find new ways to save material, reduce waste, encourage recycling, and limit packaging that could be dangerous for the environment. Prepare your new containers, including models and/or diagrams.

Each container will be made from material that starts out as a flat square or rectangle shape. Each container will then be formed by cutting out a square of the same size from each corner and then the edges will be folded up to form the container.

Follow these steps to accomplish your task.

- Make a model and/or draw diagrams of the containers.

- Write an algebraic equation to describe a relation between the length, width, and height of each of your containers and their volume.

- Write an algebraic equation to describe the amount of material that is left over after the container is constructed.

- Determine the amount of wasted material for each of your containers.

- Prepare a written presentation.

You may also want to investigate why soft drink companies have changed the size of their 24-pack packages to cubes. Focus on the following factors.

- marketing advantages

- customer convenience

- cost savings for materials

THINKING CRITICALLY

Do all polynomial functions with real coefficients have at least one real root if they have odd degree? Explain. Support your answers with examples.

PORTFOLIO

If you are given a problem to solve and there are two or more processes that can be used, how do you determine which one is most efficient for that situation? Long division and synthetic division are both methods used to divide polynomials by a first-degree binomial.

Write a comparative analysis for the two methods.

- Describe the advantages and disadvantages to both methods.

- Describe the set-up that must be used for each method

- List common errors that occur when using each of the methods

- Describe the relationship between the two methods.

Place this in your portfolio.

SELF EVALUATION

What type of learner are you? There are a variety of learning styles that researchers have identified. A few of these models are: auditory (by hearing), visual (by seeing), tactile (by touching), kinesthetic (by moving), and linguistic (by speaking).

Assess yourself. Under which style or styles do you learn best?

- Are you the type of person that is very hands-on?

- Do you like to do modeling or manipulating?

- Do you prefer to see a graph or diagram?

- Do you enjoy listening to an explanation?

- Or are you a good memory person?

Describe how you best learn under a particular style and relate it to both mathematics and your daily life.

COLLEGE ENTRANCE EXAM PRACTICE

CHAPTERS 1–8

There are eight multiple-choice questions in this section. After working each problem, write the letter of the correct answer on your paper.

1. Evaluate $\dfrac{2[15 - 9 \div 3 + 2]}{(7 + 5) \div 4}$.

A. $\dfrac{28}{3}$ **B.** $\dfrac{8}{3}$

C. $\dfrac{4}{5}$ **D.** 2

2. The epicenter of an earthquake is located at (3, 11) on a state map. Severe property damage occurred as far as 7 miles away from the epicenter. Choose the equation for the most distant point at which severe property damage occurred.

A. $(x - 7)^2 + (y - 3)^2 = 49$

B. $(x - 3)^2 + (y - 11)^2 = 49$

C. $\sqrt{(x - 3)^2 + (y - 11)^2} = 49$

D. $(x - 3) + (y - 11) = 7$

3. Which augmented matrix is not equivalent to the augmented matrix below?

$$\begin{bmatrix} 1 & -2 & 1 & \vdots & 6 \\ 3 & 2 & -1 & \vdots & 0 \\ 2 & 1 & -6 & \vdots & -2 \end{bmatrix}$$

A. $\begin{bmatrix} 1 & -2 & 1 & \vdots & 6 \\ 0 & 8 & -4 & \vdots & -18 \\ 4 & 2 & -12 & \vdots & -4 \end{bmatrix}$

B. $\begin{bmatrix} 5 & -5 & -3 & \vdots & 16 \\ 3 & 2 & -1 & \vdots & 0 \\ 1 & -2 & 1 & \vdots & 6 \end{bmatrix}$

C. $\begin{bmatrix} -9 & -6 & 3 & \vdots & 0 \\ 2 & 1 & -6 & \vdots & -2 \\ 1 & -2 & 1 & \vdots & -2 \end{bmatrix}$

D. $\begin{bmatrix} 3 & -6 & 3 & \vdots & 18 \\ 15 & 12 & -7 & \vdots & 0 \\ 8 & 4 & -24 & \vdots & -8 \end{bmatrix}$

4. Casey hit a foul ball straight up over the plate that reached a height of 112 feet. How long will it be before the ball reaches the ground? The formula for the total time is $t = \sqrt{\dfrac{2h}{g}}$, where h is the height of the ball in feet and g is the acceleration due to gravity, or 32 ft/s².

A. $2\sqrt{7}$ seconds **B.** $2\sqrt{\dfrac{4}{7}}$ seconds

C. 14 seconds **D.** 7 seconds

5. Which equation is true for the polynomial function $g(x) = x^2 - x + 5$?

A. $g(-2) = 7$

B. $2[g(a - 1)] = 2a^2 - 6a + 6$

C. $g(h + 3) = h^2 + 5h + 11$

D. $g(b) - g(2b) = -b^2 + b$

6. Which graph represents a quadratic function?

A. 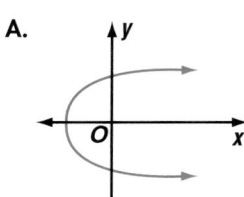 **B.**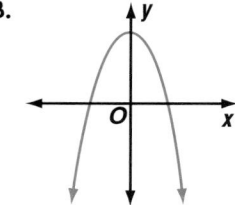

C. 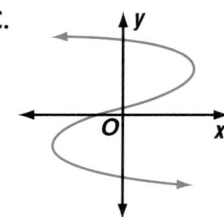 **D.**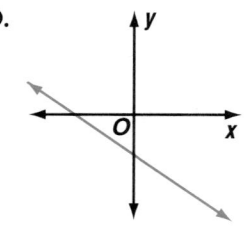

7. Which formula could be used to find the distance between the points at (3, −5) and (−4, −2)?

A. $\sqrt{(-4 - 3)^2 + (-5 - 2)^2}$

B. $\left(\dfrac{3 - 4}{2}, \dfrac{-5 - 2}{2} \right)$

C. $\left(\dfrac{-4 - 3}{2}, \dfrac{-2 + 5}{2} \right)$

D. $\sqrt{(-4 - 3)^2 + [-2 - (-5)]^2}$

8. Choose the system of equations that Lorenzo solved if he used Cramer's rule to solve for y as follows.

$$y = \frac{\begin{vmatrix} 2 & 2 \\ 3 & -5 \end{vmatrix}}{\begin{vmatrix} 2 & -6 \\ 3 & 4 \end{vmatrix}}$$

A. $2x - 6y = 2$
$-5x + 4y = 3$

B. $2x + 2y = -6$
$3x - 5y = 4$

C. $2x - 6y = 2$
$3x + 5y = -5$

D. $-6x + 2y = 2$
$4x - 5y = 3$

SECTION TWO: SHORT ANSWER

This section contains seven questions for which you will provide short answers. Write your answer on your paper.

9. A walkway of uniform width will be constructed along the inside edges of a rectangular lawn that measures 24 meters by 32 meters. The remaining lawn will have an area of 425 square meters. How wide is the walkway?

10. Solve the system of equations by graphing.

$$4x^2 + 9y^2 = 36$$
$$4x^2 - 9y^2 = 36$$

11. Find the values for x and y that make the equation $(x + 2y) + (2x - y)i = 5 + 5i$ true.

12. Write the standard form of the equation $6x^2 - 24x - 5y^2 - 10y = 11$. Then state whether the graph of the equation is a parabola, a circle, an ellipse, or a hyperbola.

13. Find the value of k so that the remainder of $(x^2 + 5x + 7) \div (x + k)$ is 3.

14. Solve the system of equations by using a matrix equation.

$$3x + 2y = 9$$
$$x - 2y = 11$$

15. The camera store at City Mall adds a 40% markup to the wholesale price of its cameras. If Mario chooses a camera that is $325 wholesale and uses a 15% off coupon, how much will he pay for the camera?

SECTION THREE: COMPARISON

This section contains five comparison problems which involve comparing two quantities, one in column A and one in column B. In certain questions, information related to one or both quantities is centered above or between them. All variables used represent real numbers.

Compare quantities A and B below.

- Write A if quantity A is greater.
- Write B if quantity B is greater.
- Write C if the two quantities are equal.
- Write D if there is not enough information to determine the relationship.

Column A	Column B
16. the standard deviation for $\{5, 7, 3, 4, 2, 4, 4, 4,$ $4, 5, 4, 3, 4, 3, 4\}$	the value of $g(-3)$ if $g(x) = \frac{x^2}{2} - 3$
17. the x-intercept of the line that passes through $(6, 1)$ and $(6, 7)$	$[9 \; 4 \; 2] \cdot \begin{bmatrix} 2 \\ 6 \\ 11 \end{bmatrix}$
18. $\qquad\qquad x < 0$ x^2	x^3
19. $\qquad A(3, -2), B(-2, 1), C(5, 5)$ the distance from A to C	the distance from B to C
20. $\sqrt[3]{-8000}$	$-\sqrt[4]{256}$

Exploring Rational Expressions

Objectives

In this chapter, you will:

*   graph rational functions,
*   solve problems involving direct, inverse, and joint variation,
*   simplify rational expressions,
*   solve rational equations, and
*   solve problems by organizing data.

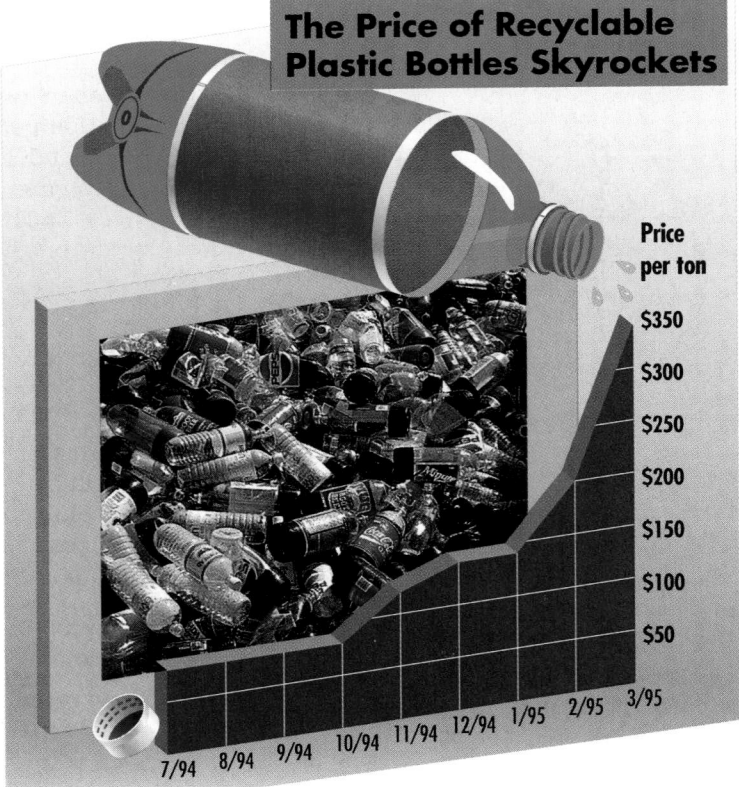

The Price of Recyclable Plastic Bottles Skyrockets

Price per ton

$350
$300
$250
$200
$150
$100
$50

7/94 8/94 9/94 10/94 11/94 12/94 1/95 2/95 3/95

Source: *Oakland Press,* 1995

TIME *Line*

What can one teenager do about the environment? You can clean your personal environment, your room. You can participate in a local effort to clean a neighborhood. Or you could go national and get everyone's help to better our environment.

1878 Cuban poet, essayist, and patriot José Martí writes his collection of poems *Versos Libres (Free Verses).*

190 B.C. 185 180 175 170 A.D. 1300 1400 1500 1600 1700 1800 1900 1930

180 B.C. Hypicle's *De ascensionibus,* a work on astronomy, introduces the 360-degree circle to Greek mathematics.

PEOPLE IN THE NEWS

Chapter Project

Form small groups and have some members research different environment issues such as landfills, the rain forest, pollution, or endangered species. Other members could make a list of ways that people can conserve energy and make a difference in helping the environment in everyday life, such as obeying the speed limit and turning off the water while brushing their teeth. Research recycling programs that are available in your community and find out how to participate in these programs. Create charts, graphs, or posters to visually display the data you collect and present your findings to the class.

When **Melissa Poe** was 9, she became concerned about pollution and the environment and wrote a letter to the President. She then started a club at her school in Nashville, Tennessee, to help protect the environment. At first her club had only six members. Now Poe is 16 and Kids For A Clean Environment (Kids F.A.C.E.®) has over 200,000 members worldwide. They publish a newsletter and created the Kid's Earth Flag, which was unveiled on Earth Day in 1995. It is a banner made of 20,000 cloth squares designed by kids from all over the world.

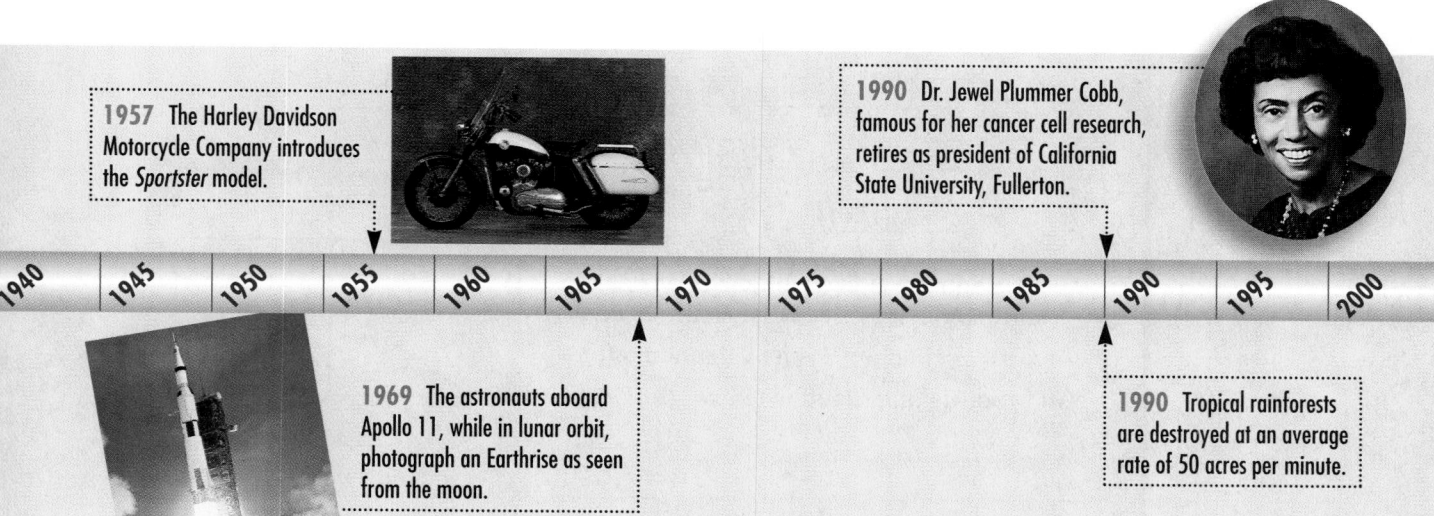

1957 The Harley Davidson Motorcycle Company introduces the *Sportster* model.

1990 Dr. Jewel Plummer Cobb, famous for her cancer cell research, retires as president of California State University, Fullerton.

1969 The astronauts aboard Apollo 11, while in lunar orbit, photograph an Earthrise as seen from the moon.

1990 Tropical rainforests are destroyed at an average rate of 50 acres per minute.

1940　1945　1950　1955　1960　1965　1970　1975　1980　1985　1990　1995　2000

9-1A Graphing Technology
Rational Functions

A Preview of Lesson 9–1

A rational function is of the form $f(x) = \dfrac{p(x)}{q(x)}$, where $p(x)$ and $q(x)$ are polynomial functions and $q(x) \neq 0$. A graphing calculator is a good tool for exploring graphs of rational functions.

Graphs of rational functions may have breaks in **continuity**. This means that, unlike polynomial functions, which can be traced with a pencil that never leaves the paper, a rational function may not be traceable. Breaks in continuity can occur where there is a vertical asymptote or **point discontinuity**. Point discontinuity is like a hole in the graph. Vertical asymptotes and points of discontinuity occur for values of x that make the denominator of a rational function zero.

Example **Graph $y = \dfrac{1}{x}$ in the standard viewing window.** *This is sometimes called the inverse function.*

The inverse function is often the parent function when examining families of rational functions.

Enter: [Y=] 1 [÷] [X,T,θ,n] [ZOOM] 6

There is a break in continuity at $x = 0$. Looking at the equation, we can see that when $x = 0$, the function is undefined. In this case, the graph has a vertical asymptote at $x = 0$. *Sometimes graphing calculators "graph" this line. This is because the calculator connects all points plotted. To eliminate the graph of the asymptote, change the MODE setting from "Connected" to "Dot."*

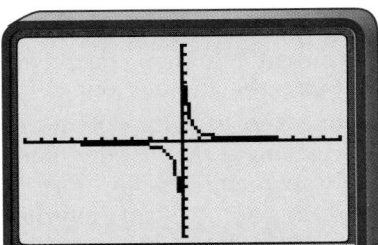

Example **Graph $y = \dfrac{x^2 - 9}{x + 3}$ in the window $[-5, 4.4]$ by $[-7, 2]$ with scale factors of 1. Use the "Dot" mode.**

Enter: [Y=] [(] [X,T,θ,n] [x²] [−] 9

[)] [÷] [(] [X,T,θ,n] [+] 3

[)] [GRAPH]

The graph looks like a line with a break in continuity at $x = -3$. This occurs because if $x = -3$, then $\dfrac{x^2 - 9}{x + 3}$ has a denominator of 0. So y is undefined when $x = -3$. *If you TRACE along the graph to $x = -3$, you will see that there is no corresponding y value.*

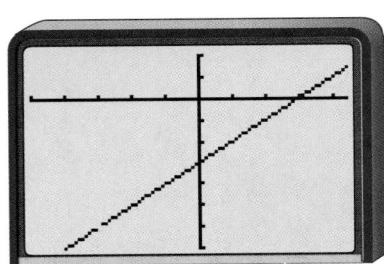

Recall that a complete graph of a function shows all of the important features of the graph, including the end behavior. The end behavior of some rational functions shows that graphs can have horizontal asymptotes.

Example **3** Graph $y = \frac{4x + 2}{x - 1}$ in the standard viewing window. Then find the equation for the horizontal asymptote.

Enter: Y= ((4 X,T,θ,n + 2) ÷
(X,T,θ,n − 1) ZOOM 6

Trace along the graph and observe the y values as x grows larger and as x grows smaller. The y values approach 4. Thus, the equation for the horizontal asymptote is $y = 4$.

What is an equation of the horizontal asymptote for the graph in Example 1?

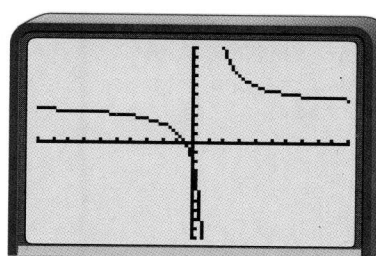

EXERCISES

Use a graphing calculator to graph each function so that a complete graph is shown. Sketch the graph on a sheet of paper. Write the equations of any vertical asymptotes or the x-coordinates of any points of discontinuity.

1. $f(x) = \frac{2}{x - 4}$

2. $g(x) = \frac{x}{x + 2}$

3. $h(x) = \frac{8}{(x + 3)(x - 5)}$

4. $p(x) = \frac{2x}{3x - 6}$

5. $q(x) = \frac{x^2 - 16}{x - 4}$

6. $m(x) = \frac{-(2x + 7)}{(x + 7)(x + 2)}$

Use a graphing calculator to graph each function so that a complete graph is shown. Sketch the graph on a sheet of paper. Trace the graph to determine the equations for any horizontal asymptotes.

7. $f(x) = \frac{7}{x}$

8. $j(x) = \frac{x^2 - 3x + 2}{x^2 + x - 6}$

9. $c(x) = \frac{8x - 5}{2x}$

10. $d(x) = \frac{x - 9}{9 + 2x}$

Graphing Rational Functions

What YOU'LL LEARN

- To graph rational functions.

Why IT'S IMPORTANT

You can graph rational functions to solve problems involving medicine and auto safety.

CONNECTION

Mathematics History

Mathematician Maria Gaetana Agnesi was one of the greatest women scholars of all time. In the analytic geometry section of her book *Analytical Institutions*, Agnesi discussed the characteristics of the equation $x^2y = a^2(a - y)$, called the "curve of Agnesi." The equation can be expressed as $y = \dfrac{a^3}{x^2 + a^2}$.

Because the function described above is the *ratio* of two polynomial expressions, a^3 and $x^2 + a^2$, it is called a **rational function.** A rational function is a function of the form $f(x) = \dfrac{p(x)}{q(x)}$, where $p(x)$ and $q(x)$ are polynomial functions and $q(x) \neq 0$. Here are other examples of rational functions.

$$f(x) = \frac{x}{x-1} \qquad g(x) = \frac{3}{x-3} \qquad h(x) = \frac{x+1}{(x+2)(x-5)}$$

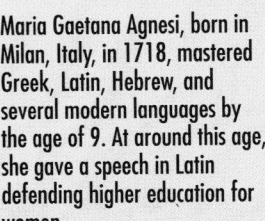

Because rational functions are expressed in the form of a fraction, the denominator of a rational function cannot be zero, since division by zero is not defined. *In the examples above, the functions are not defined at $x = 1$, $x = 3$, and $x = -2$ and $x = 5$, respectively.*

When graphing a rational function, determine when the denominator is 0, and remember to exclude those values from the domain. Sometimes a dashed line is drawn at those locations to show that a function of the form $f(x) = \dfrac{p(x)}{q(x)}$ approaches the values at which $q(x) = 0$, but neither touches nor intersects it.

Look at the graph of $f(x) = \dfrac{2}{x-4}$ shown below. Notice that the graph of $f(x)$ approaches the graph of $x = 4$, but never touches or intersects it.

x	f(x)
-100	-0.0192
-10	-0.1429
0	-0.50
3	-2
3.5	-4
3.9	-20
4	**undefined**
4.1	20
4.5	4
5	2
10	0.3333
100	0.0208

The lines that the graph of a rational function approaches are called **asymptotes.** If a function is not defined when $x = a$, then either there is a line with the equation $x = a$ that is a vertical asymptote or there is a "hole" in the graph at $x = a$. As shown at the right, the graph of $x = 4$ is a vertical asymptote. If the value of a function approaches a number b as the value of $|x|$ increases, the line with equation $f(x) = b$ is a horizontal asymptote. In the graph at the right, as $|x|$ increases, the function approaches 0. So, the graph of $f(x) = 0$ is a horizontal asymptote.

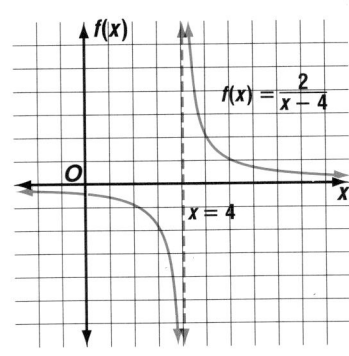

Example ❶ **Graph $f(x) = \dfrac{x}{x + 1}$. State the domain and range.**

If $x = -1$, then $f(x)$ is undefined. Thus, the domain for this function includes all real numbers except -1. By plotting points, we see that this graph has a vertical asymptote with the equation $x = -1$. Study the pattern of values of $f(x)$ as the value of $|x|$ increases to determine the equation of the horizontal asymptote.

LOOK BACK

You can refer to Lesson 7-5 to review asymptotes of hyperbolas.

x	f(x)
0	0
5	0.833
50	0.98
100	0.99
1000	0.999
−0.5	−1
−1.5	3
−5	1.25
−50	1.02
−100	1.0101
−1000	1.001

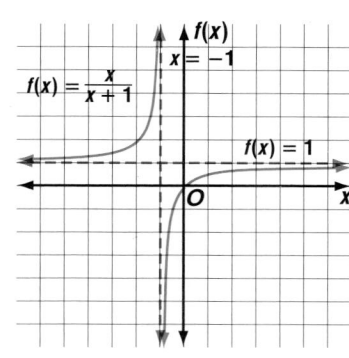

Finding the x- and y-intercepts is often useful when graphing rational functions.

As the value of $|x|$ increases, it appears that the value of the function gets closer and closer to 1. The line with the equation $f(x) = 1$ is a horizontal asymptote of the function, and the range for this function includes all real numbers except 1.

To graph the function, plot points on either side of each asymptote until a pattern is visible. Then sketch the graph.

Using a calculator can help you generate a table of values that pinpoint horizontal asymptotes more easily.

Example ❷ **Graph $f(x) = \dfrac{6}{(x - 2)(x + 3)}$.**

The vertical asymptotes are the lines whose equations are $x = 2$ and $x = -3$. Use a calculator to generate a table of values for $f(x)$. Be sure to generate values near both sides of the asymptotes. Let's look at $x = 10$.

Enter: 6 ÷ ((10 − 2) × (10

+ 3)) = *0.057692307*

The same pattern of keystrokes can be used to find other values of $f(x)$ that approach those of the horizontal asymptote.

$f(10) = 0.057692307$

$f(100) = 0.000594412$

$f(1000) = 0.000005994$

$f(-10) = 0.071428571$

$f(-100) = 0.000606428$

$f(-1000) = 0.000006006$

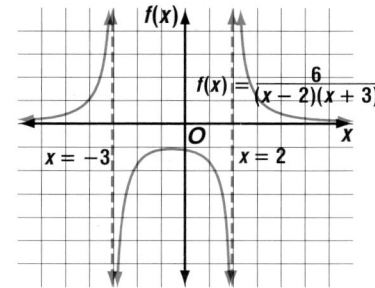

As $|x|$ increases, it appears that the value of the function approaches, but never quite reaches, 0. An equation of the horizontal asymptote is $f(x) = 0$. Plot points on either side of each asymptote. Then sketch the graph.

MODELING MATHEMATICS

Rational Functions

Materials: balances metric measuring cups

different liquids such as water, cooking oil, alcohol, sugar water, and salt water

The density of a material can be expressed as $D = \frac{m}{V}$, where m is the mass of the material in grams and V is the volume in cubic centimeters. By finding the volume and density of 20 grams of each liquid, you can sketch a graph of the function $D = \frac{20}{V}$.

Your Turn

a. Use a balance and metric measuring cups to find the volume of 20 grams of different liquids, such as water, cooking oil, alcohol, sugar water, and salt water.

b. Use the rational function $D = \frac{m}{V}$ to find the density of each liquid.

c. Graph the data by plotting the points (volume, density) on a graph. Then connect the points.

d. From the graph, find the asymptotes.

As you have learned, graphs of rational functions may have point discontinuity rather than vertical asymptotes. The graphs of these functions appear to have "holes." These holes are usually shown as circles on the graph.

Example **Graph $g(x) = \frac{x^2 - 25}{x - 5}$.**

x	g(x)
−1	4
0	5
1	6
2	7
3	8
4	9
5	undefined
6	11
7	12

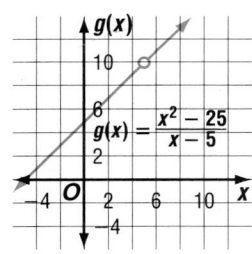

The graph looks like a line with a hole at (5, 10). Remember that the denominator cannot be zero, so $x - 5 \neq 0$, or $x \neq 5$. The rest of the graph looks like the line whose equation is $g(x) = x + 5$ because $\frac{x^2 - 25}{x - 5}$ simplifies to $x + 5$ when $x \neq 5$.

As you saw in Example 3, not all rational functions have graphs that are similar in appearance. Some contain no negative values for $f(x)$.

CONNECTION

Mathematics History

4 Refer to the "curve of Agnesi" in the application at the beginning of the lesson. This bell-shaped curve was discovered by Fermat nearly a hundred years before Agnesi. Graph $f(x) = \dfrac{a^3}{x^2 + a^2}$ if $a = 4$.

$$f(x) = \frac{a^3}{x^2 + a^2}$$

$$= \frac{(4)^3}{x^2 + (4)^2} \text{ or } \frac{64}{x^2 + 16} \quad \textit{Substitute 4 for x.}$$

Since $x^2 + 16$ can never equal 0, the graph of the function has no vertical asymptotes. Use a calculator to estimate the location of the horizontal asymptote.

x	f(x)
−500	0.0003
−50	0.0254
−5	1.5610
0	4
5	1.5610
50	0.0254
500	0.0003

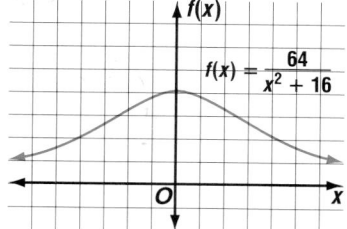

The pattern of values suggests that the equation of the horizontal asymptote will be $f(x) = 0$.

CHECK FOR UNDERSTANDING

Communicating Mathematics

Study the lesson. Then complete the following.

1. Consider $f(x) = \dfrac{1}{x - 3}$.

   a. **State** the domain and range of the function.

   b. **Write** an equation of the vertical asymptote for $f(x)$.

   c. **Write** an equation for the horizontal asymptote of the graph of $f(x)$. Generate a table of values to support your answer.

2. **Describe** the difference between the graphs of $y = \dfrac{x - 4}{x + 2}$ and $y = \dfrac{x^2 - 4}{x + 2}$.

3. **Write** equations for the vertical and horizontal asymptotes of the graph at the right.

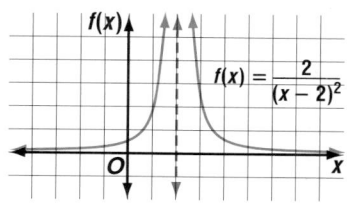

Guided Practice

State the equations of the vertical and horizontal asymptotes for each rational function.

4. $f(x) = \dfrac{1}{x - 2}$

5. $f(x) = \dfrac{2}{(x - 1)(x + 3)}$

Graph each rational function.

6. $f(x) = \dfrac{1}{x}$

7. $g(x) = \dfrac{x}{x - 5}$

8. $f(x) = \dfrac{x - 5}{x + 1}$

9. $g(x) = \dfrac{x - 1}{x - 4}$

10. $f(x) = \dfrac{1}{(x + 2)^2}$

11. $h(x) = \dfrac{8}{(x - 1)(x + 3)}$

12. **Transportation**   A train travels at one velocity V_1 for a given amount of time t_1 and then another velocity V_2 for a different amount of time t_2. The average velocity is given by the formula $V = \dfrac{V_1 t_1 + V_2 t_2}{t_1 + t_2}$.

    a. Let t_1 be the independent variable and let V be the dependent variable. Draw the graph if $V_1 = 60$ mph, $V_2 = 40$ mph, and $t_2 = 8$ hours.

    b. Find V when $t_1 = 9$ hours.

EXERCISES

Practice

State the equations of the vertical and horizontal asymptotes for each rational function.

13. $f(x) = \dfrac{x}{x - 4}$ **14.** $g(x) = \dfrac{6}{(x - 6)^2}$ **15.** $f(x) = \dfrac{4}{(x - 1)(x + 5)}$

16. $f(x) = \dfrac{1}{3x}$ **17.** $h(x) = \dfrac{x - 1}{x - 3}$ **18.** $t(x) = \dfrac{x}{x - 7}$

Graph each rational function.

19. $f(x) = \dfrac{3}{x}$ **20.** $f(x) = \dfrac{1}{x + 2}$ **21.** $f(x) = \dfrac{x}{x - 3}$

22. $f(x) = \dfrac{x + 4}{x - 1}$ **23.** $f(x) = \dfrac{x - 1}{x - 3}$ **24.** $f(x) = \dfrac{-5}{x + 1}$

25. $f(x) = \dfrac{x^2 - 36}{x + 6}$ **26.** $f(x) = \dfrac{-3}{(x - 2)^2}$ **27.** $f(x) = \dfrac{5x}{x + 1}$

28. $f(x) = \dfrac{x^2 - 1}{x - 1}$ **29.** $f(x) = \dfrac{1}{(x + 3)^2}$ **30.** $f(x) = \dfrac{3}{(x - 1)(x + 5)}$

31. $f(x) = \dfrac{-1}{(x + 2)(x - 3)}$ **32.** $f(x) = \dfrac{x}{x^2 - 1}$ **33.** $f(x) = \dfrac{x - 1}{x^2 - 4}$

Critical Thinking

34. Compare and contrast each family of graphs.

    a. $y = \dfrac{1}{x}$ and $y - 7 = \dfrac{1}{x}$ **b.** $y = \dfrac{1}{x}$ and $y = 4\left(\dfrac{1}{x}\right)$ **c.** $y = \dfrac{1}{x}$ and $y = \dfrac{1}{x + 5}$

    d. Without making a table of values, use what you observed in parts a–c to sketch a graph of $y - 7 = 4\left(\dfrac{1}{x + 5}\right)$.

Applications and Problem Solving

35. Medications   For certain medicines, health care professionals may use Young's Rule, $C = \dfrac{y}{y + 12} \cdot D$, to estimate the proper dosage for a child when the adult dosage is known.

    a. If C represents the child's dose, D represents the adult dose, and y represents the child's age in years, use Young's Rule to estimate the dosage of amoxicillin for an eight-year-old child if the adult dosage is 250 mg.

    b. Draw and describe the graph of $C = \dfrac{y}{y + 12}$. Note its shape, asymptotes, domain, and range.

36. Auto Safety   When a car has a front-end collision, the objects in the car (including passengers) keep moving forward until the impact occurs. After impact, objects are repelled. Seat belts and airbags limit how far you are jolted forward. The formula for the velocity you are thrown backward is $V_f = \dfrac{m_1 - m_2}{m_1 + m_2} v_i$, where m_1 and m_2 are the masses of the two objects meeting and v_i is the initial velocity.

    a. Let m_1 be the independent variable and let V_f be the dependent variable. Graph the function if $m_2 = 7$ kg and $v_i = 5$ m/s.

    b. Find the value of V_f when the value of m_1 is 5 kg.

37. Determine whether $f(x) = x$ and $g(x) = -x$ are inverse functions. (Lesson 8–8)

38. Approximate the real zeros of $g(x) = x^4 - 9x^3 + 25x^2 - 24x + 6$ to the nearest tenth. (Lesson 8–3)

39. Write an equation for $6y^2 - 34x^2 = 204$ in standard form. (Lesson 7–5)

40. Statistics The lifetimes of 10,000 light bulbs are normally distributed. The mean lifetime is 300 days, and the standard deviation is 40 days. (Lesson 6–9)

    a. How many light bulbs will last between 260 and 340 days?

    b. How many light bulbs will last between 220 and 380 days?

    c. How many light bulbs will last less than 300 days?

    d. How many light bulbs will last more than 300 days?

    e. How many light bulbs will last more than 380 days?

    f. How many light bulbs will last less than 180 days?

41. Find the value of the discriminant and describe the nature of the roots of $x^2 - 10x + 25 = 0$. Then solve the equation. (Lesson 6–4)

42. Solve the system of equations using augmented matrices. (Lesson 4–7)

$$8x - 3y - 4z = 6$$
$$4x + 9y - 2z = -4$$
$$6x + 12y + 5z = -1$$

43. Find the x- and y-intercepts of $3x - 2y = 12$. (Lesson 2–3)

WORKING ON THE
In·ves·ti·ga·tion

Refer to the Investigation on pages 474–475.

Suppose the gas station owner also told you that the tank is buried $2\frac{1}{2}$ feet below the surface. It has a diameter of $5\frac{3}{4}$ feet and a length of $9\frac{1}{8}$ feet.

1 What is the volume of the tank in cubic feet?

2 Research to find how many gallons are in a cubic foot. What is the volume of the tank in gallons?

3 Where on the dipstick would the marking be showing that the tank was one-half full? How did you arrive at this marking?

4 Using the data from your two experiments and the table of data from the Investigation in Lesson 8–8, describe the scale of the dipstick. How do you use the stick to determine the gallons of gasoline left in the tank?

Add the results of your work to your Investigation Folder.

Direct, Inverse, and Joint Variation

APPLICATION
Meteorology

Many areas of Northern California depend on the snowpack of the Sierra Nevada mountain range for their water supply. The volume of water produced from melting snow varies directly with the volume of snow. Meteorologists have determined that 250 cm³ of snow will melt to 28 cm³ of water. How much water does 900 cm³ of melting snow produce?

The relationship between the volume of the snowpack and the volume of water it produces can be expressed with the equation $y = kx$, where y represents the water produced from x, the amount of the snowpack. The k in this equation is called the **constant of variation.**

To find the value of k, substitute corresponding volumes of snowpack and water in the equation and solve for k.

$$y = kx$$
$$28 = k(250) \quad \text{\textit{Replace y with 28 and x with 250.}}$$
$$\frac{28}{250} = k \qquad \text{\textit{Divide each side by 250. }} \textbf{\textit{Estimate:}} \; \frac{25}{250} = 0.1$$
$$0.112 = k$$

So the relationship between the amount of snowpack and the amount of water it melts to is $y = 0.112x$. Use this equation to find the amount of water produced from 900 cm³ of snow.

$$y = kx$$
$$= 0.112(900) \quad \text{\textit{Replace k with 0.112 and x with 900.}}$$
$$= 100.8 \qquad \textbf{\textit{Estimate:}} \; \textit{(0.1)900 = 90}$$

Snowpack measuring 900 cm³ will produce 100.8 cm³ of water.

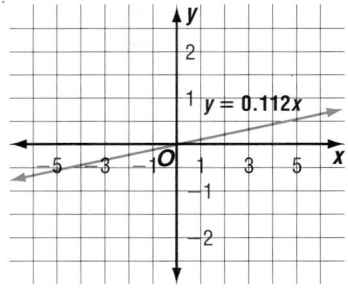

The relationship described above is an example of a **direct variation.** This means that y is a multiple of x. Note that the graph of a direct variation is a line through the origin. An equation of a direct variation is a special case of an equation written in slope-intercept form, $y = mx + b$. When $m = k$ and $b = 0$, $y = mx + b$ becomes $y = kx$. So the slope of a direct variation equation is its constant.

To express a direct variation, we say that *y varies directly as x*. In other words, as *x* increases, *y* increases or decreases at a constant rate.

Direct Variation	**y varies directly as x if there is some nonzero constant k such that y = kx. k is called the *constant of variation.***

If you know that *y* varies directly as *x* and one set of values, you can use a proportion to find the other set of corresponding values.

$$y_1 = kx_1 \text{ and } y_2 = kx_2$$

$$\frac{y_1}{x_1} = k \qquad \frac{y_2}{x_2} = k$$

Therefore, $\frac{y_1}{x_1} = \frac{y_2}{x_2}$.

Using the properties of equality, you can find many other proportions that relate these same *x* and *y* values.

Example If *y* varies directly as *x* and *y* = 9 when *x* is −15, find *y* when *x* = 21.

Use a proportion that relates the values.

$$\frac{y_1}{x_1} = \frac{y_2}{x_2}$$

$$\frac{9}{-15} = \frac{y_2}{21} \qquad \textit{Substitute the known values.}$$

$$-15y_2 = (9)(21) \qquad \textit{Cross multiply.}$$

$$y_2 = -12.6 \qquad \textit{Divide each side by 15.}$$

When *x* = 21, the value of *y* is −12.6.

Many quantities are **inversely proportional** or are said to *vary inversely* with each other. For example, speed and time vary inversely with each other. When you travel to a particular location, as your speed increases, the time it takes to arrive at that location decreases.

Inverse Variation	**y varies inversely as x if there is some nonzero constant k such that xy = k or y = $\frac{k}{x}$.**

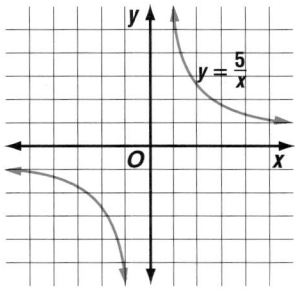

Suppose *y* varies inversely as *x* such that $xy = 5$ or $y = \frac{5}{x}$. The graph of this equation is shown at the left. Notice that in this case, *k* is a positive value, 5, so as the values of *x* increase, the values of *y* decrease.

Just as with direct variation, a proportion can be used with indirect variation to solve problems where some quantities are known. The following proportion is only one of several that can be formed.

$$x_1y_1 = k \text{ and } x_2y_2 = k$$

$$x_1y_1 = x_2y_2 \qquad \textit{Substitution property of equality}$$

$$\frac{x_1}{y_2} = \frac{x_2}{y_1} \qquad \textit{Divide each side by } y_1y_2.$$

Example **2** If y varies inversely as x and $y = 4$ when $x = 12$, find y when $x = 5$.

$$\frac{x_1}{y_2} = \frac{x_2}{y_1}$$

$$\frac{12}{y_2} = \frac{5}{4}$$ *Substitute the known values.*

$$5y_2 = 48$$ *Cross multiply.*

$$y_2 = \frac{48}{5} \text{ or } 9.6$$ *Divide each side by 5.*

When $x = 5$, the value of y is $\frac{48}{5}$ or 9.6.

Another type of variation is **joint variation.** This type of variation occurs when one quantity varies directly as the product of two or more other quantities.

Joint Variation	y varies jointly as x and z if there is some number k such that $y = kxz$, where $x \neq 0$ and $z \neq 0$.

Example **3**

Geometry

The area A of a trapezoid varies jointly as the height h and the sum of its bases b_1 and b_2. Find the equation of joint variation if $A = 48$ in^3, $h = 8$ in., $b_1 = 5$ in., and $b_2 = 7$ in.

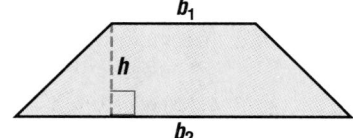

Explore Read the problem and use the known values of A, h, b_1, and b_2 to find the equation of joint variation.

Plan *The area varies jointly as the height and the sum of its bases.*

$$A = k \cdot h \cdot (b_1 + b_2)$$

Solve

$$A = kh(b_1 + b_2)$$

$$48 = k(8)(5 + 7)$$ *Substitute the known values.*

$$48 = k(8)(12)$$

$$48 = 96k$$

$$0.5 = k$$ *Solve for k.*

The equation for the area of a trapezoid is $A = 0.5h(b_1 + b_2)$.

Examine

$$A = 0.5h(b_1 + b_2)$$

$$48 \stackrel{?}{=} 0.5(8)(5 + 7)$$ *A = 48, h = 8, b_1 = 5, and b_2 = 7*

$$48 \stackrel{?}{=} 0.5(8)(12)$$

$$48 = 48 \checkmark$$

Thus, the equation is correct.

Communicating Mathematics

Study the lesson. Then complete the following.

1. **Describe** what happens when the value of y increases if y varies directly as x.

2. **Explain** how k is different from y and x in the equation $y = kx$.

3. **State** whether the graph at the right represents a direct or inverse variation. Explain.

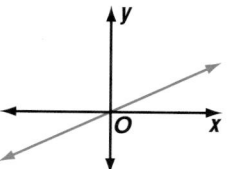

4. **Describe** two quantities in real life that vary directly with each other and two quantities that vary inversely with each other.

Guided Practice

State whether each equation represents a *direct, inverse,* or *joint* variation. Then name the constant of variation.

5. $xy = -5$

6. $y = 2xz$

7. $\frac{x}{y} = 3$

Write an equation for each statement. Then solve the equation.

8. If y varies inversely as x and $y = 5$ when $x = 10$, find y when $x = 2$.

9. If y varies directly as x and $y = 15$ when $x = 3$, find y when $x = 12$.

10. If y varies jointly as x and z and $y = 80$ when $x = 5$ and $z = 8$, find y when $x = 16$ and $z = 2$.

11. **Architecture** Architects have to consider how sound travels when designing large buildings such as theaters, auditoriums, or museums. Sound intensity I is inversely proportional to the square of the distance from the sound source d.

   a. Write an equation that represents this situation.

   b. If d is the independent variable and I is the dependent variable, graph the equation from part a when $k = 16$.

   c. If a person in a theater moves to a seat twice as far from the speakers, compare the new sound intensity to that of the original.

Practice

State whether each equation represents a *direct, inverse,* or *joint* variation. Then name the constant of variation.

12. $\frac{x}{5} = y$

13. $\frac{x}{y} = -7$

14. $x = 4y$

15. $x = \frac{1}{y}$

16. $A = \frac{1}{2}bh$

17. $\frac{2}{3}a = -\frac{1}{2}b$

Write an equation for each statement. Then solve the equation.

18. If y varies directly as x and $y = 12$ when $x = 3$, find y when $x = 16$.

19. If r varies inversely as t and $r = 18$ when $t = -3$, find r when $t = -11$.

20. If y varies directly as x and $x = 6$ when $y = 0.5$, find y when $x = 10$.

21. If y varies inversely as x and $y = 2$ when $x = 25$, find x when $y = 40$.

22. Suppose y varies jointly as x and z. Find y when $x = 8$ and $z = 3$, if $y = 16$ when $z = 2$ and $x = 5$.

23. **Geometry**   The area of a parallelogram varies jointly as its base and height. Parallelogram *DUCK* has a base of 15 meters, a height of 12 meters, and an area of 180 square meters. Find the height of parallelogram *DOVE* if its area is 1615 square meters and its base is 42.5 meters.

24. **Geometry**   The area of a triangle varies jointly as the base and height. Find the equation of joint variation if $A = 100$, $b = 25$, and $h = 8$.

25. If y varies directly as x and $y = 1$ when $x = 5$, find y when $x = 22$.

26. If y varies inversely as x and $x = 14$ when $y = 7$, find x when $y = 2$.

27. If y varies directly as x and $y = \frac{2}{5}$ when $x = \frac{1}{20}$, find y when $x = \frac{1}{2}$.

28. If y varies inversely as x and $y = \frac{1}{8}$ when $x = 16$, find y when $x = \frac{2}{3}$.

Critical Thinking

29. **a.** How does the circumference of a circle vary with respect to its radius?
    b. What is the constant of variation?

30. **a.** How does the volume of a sphere vary with respect to the cube of its radius?
    b. What is the constant of variation?

Applications and Problem Solving

31. **Swimming**   When a person swims underwater, the pressure in his or her ears varies directly with the depth at which he or she is swimming. At 10 feet, the pressure is about 4.3 pounds per square inch (psi).
    a. Find an equation of direct variation that represents this situation.
    b. Find the pressure if the depth is 60 feet.
    c. It is unsafe for amateur divers to swim where the water pressure is more than 65 psi. How deep can an amateur diver safely swim?
    d. Make a table showing the number of pounds of pressure at various depths of water. Use the data to sketch a graph of pressure versus depth.

32. **Aeronautics**   The LEM (Lunar Exploration Module) used by astronauts to explore the moon's surface during the Apollo space missions weighs about 30,000 pounds on Earth. On the moon, there is less gravity so it weighs less, meaning that less fuel is needed to lift off from the moon's surface. The force of gravity on Earth is about 6 times as much as that on the moon. How much does the LEM weigh on the moon?

33. **Relativity**   Have you ever wondered why time seems to pass more quickly as you get older? Suppose the length that a particular period of time *seems* to be is inversely proportional to your age. At your present age, time seems to pass normally. For example, a day seems like a day and a week seems like a week.
    a. How long will a month seem to be when you are three times as old as you are now?
    b. How long did a week seem to be when you were a fifth as old as you are now?
    c. Winona, who is 30 years old, says to her 5-year-old, "You have to wait 10 more minutes before going swimming." In terms of the mother's time scale, how long does 10 minutes seem to her child?

34. Telecommunications It has been found that the average number of daily phone calls C between two cities is directly proportional to the product of the populations P_1 and P_2 of two cities and inversely proportional to the square of the distance d between the cities. That is, $C = \dfrac{kP_1P_2}{d^2}$.

City	Population (1992)
Chicago	2,768,000
Memphis	610,000
New Orleans	490,000
San Antonio	966,000

   a. The distance between San Antonio and New Orleans is about 575 miles. If the average number of daily phone calls between the cities is 29,000, find the value of k and write the equation of variation.

   b. Memphis is about 399 miles from New Orleans. Find the average number of daily phone calls between them.

   c. The average number of daily phone calls between Memphis and Chicago is 112,451. Find the distance between Memphis and Chicago.

   d. Could you use this formula to find the populations or the average number of phone calls between two adjoining cities? Explain.

35. Energy Conservation Many homes lose a significant amount of heat through their windows. The heat loss of a glass window varies jointly as the area of the window and the difference between the outside and inside temperatures. A window 3 feet wide by 5 feet long loses 500 BTU per hour when the temperature outside is 10° cooler than the temperature inside. Find the heat loss through the same window if the difference between the outside and inside temperatures is 30°.

Mixed Review

36. Graph $f(x) = \dfrac{x}{x+1}$. (Lesson 9–1)

37. Find $g[h(x)]$ and $h[g(x)]$ for $g(x) = -x$ and $h(x) = -x$. (Lesson 8–7)

38. Use synthetic substitution to find $f(2)$ and $f(-1)$ for $f(x) = 3x^4 + 8x^2 - 1$. (Lesson 8–2)

39. Write an equation for the ellipse described below. (Lesson 7–4)
The major axis is 12 units long and parallel to the y-axis. The minor axis is 8 units long and the center is at $(-2, 3)$.

40. Solve $d^2 \geq 3d + 28$. (Lesson 6–7)

41. Travel The cruise ship *The Silver Dollar* has been rented to take 100 passengers to the Green Mountain Resort. The fare is $5 per person. The owner of the cruise ship has agreed to reduce the fare by 2¢ for each person taking the cruise for every passenger over 100 passengers in the group. How many passengers will produce a maximum profit for the owner? (Lesson 6–3)

42. Manufacturing A manufacturer of boat motors has specifications for parts with given tolerance limits. If a part is to be 3.2 inches wide with a tolerance of 0.01 inches, this means that it must be at least 3.19 inches wide or at most 3.21 inches wide. This tolerance limit can be expressed by the absolute value inequality $|w - 3.2| \leq 0.01$, where w represents the width of the part. (Lesson 1–7)

   a. Find the maximum and minimum acceptable dimensions of a part that is supposed to be 7.32 centimeters long with a tolerance of 0.002 centimeter.

   b. Find the tolerance if a part must satisfy the inequality $5.18 \leq w \leq 5.24$.

Multiplying and Dividing Rational Expressions

What YOU'LL LEARN

- To simplify rational expressions, and
- to simplify complex fractions.

Why IT'S IMPORTANT

You can use rational expressions to solve problems involving geometry.

INTEGRATION
Geometry

The bases of two parallelograms are also the adjacent sides of rectangle N. Parallelogram L has an area of $2x^2 - 13x + 20$ square meters and a height of $3x + 1$ meters. Parallelogram M has an area of $3x^2 + 10x + 3$ square meters and a height of $x - 4$ meters. Find the area of rectangle N. *This problem will be solved in Example 5.*

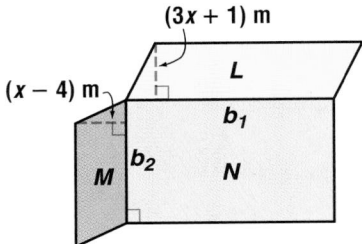

Because variables in algebra represent real numbers, operations with rational numbers and **rational algebraic expressions** are very similar. A rational number can be expressed as the quotient of two integers. A rational algebraic expression can be expressed as the quotient of two polynomials. In either case, the denominator can never be 0.

LOOK BACK

You can refer to Lesson 5-4 to review finding the GCF.

rational numbers		*rational algebraic expressions*		
$\dfrac{3}{7}$,	$\dfrac{175}{100}$, $\dfrac{-7}{11}$	$\dfrac{7}{2x}$,	$\dfrac{x+3}{x-5}$,	$\dfrac{x+5}{x^2+4x-5}$

To write a fraction in simplest form, you divide both the numerator and denominator by their greatest common factor (GCF). To simplify a rational algebraic expression, you use similar properties.

Example

a. Simplify $\dfrac{2x(x+1)}{(x+1)(x^2-4)}$.

b. Under what conditions is this expression undefined?

a. Look for common factors.

$$\frac{2x(x+1)}{(x+1)(x^2-4)} = \frac{2x}{x^2-4} \cdot \frac{\overset{1}{\cancel{x+1}}}{\underset{1}{\cancel{x+1}}} \qquad \textit{How is this similar to simplifying } \tfrac{4}{6}?$$

$$= \frac{2x}{x^2-4} \qquad \frac{x+1}{x+1}=1$$

b. To find when the expression is undefined, completely factor the original denominator.

$$\frac{2x(x+1)}{(x+1)(x^2-4)} = \frac{2x(x+1)}{(x+1)(x-2)(x+2)}$$

The values that would make the denominator equal 0 are -1, 2, and -2. So, the expression is undefined when $x = -1$, $x = 2$, or $x = -2$.

TECHNOLOGY Tips

Verify this result by using a graphing calculator to graph the related equation. Press
ZOOM 4 to see the point of discontinuity.

Sometimes you can factor out -1 in the numerator or denominator to help simplify rational expressions.

Example **2** Simplify $\dfrac{3a^3 - a^4}{2a^3 - 6a^2}$.

$\dfrac{3a^3 - a^4}{2a^3 - 6a^2} = \dfrac{a^3(3 - a)}{2a^2(a - 3)}$ *Factor the numerator and the denominator.*

$= \dfrac{\overset{a}{\cancel{a^3}}(-1)(\cancel{a - 3})}{2\underset{1}{\cancel{a^2}}(\underset{1}{\cancel{a - 3}})}$ $3 - a = -1(-3 + a)$ or $-1(a - 3)$

$= \dfrac{(-1)a}{2}$

$= -\dfrac{a}{2}$ *For what values is the original expression undefined?*

Remember that to multiply two fractions, you first multiply the numerators and then multiply the denominators. To divide two fractions, you multiply by the multiplicative inverse, or reciprocal, of the divisor.

Multiplication	**Division**
$\dfrac{3}{4} \cdot \dfrac{2}{15} = \dfrac{\overset{1}{\cancel{3}} \cdot \overset{1}{\cancel{2}}}{\underset{1}{\cancel{2}} \cdot 2 \cdot \underset{1}{\cancel{3}} \cdot 5}$	$\dfrac{3}{5} \div \dfrac{9}{10} = \dfrac{3}{5} \cdot \dfrac{10}{9}$
$= \dfrac{1}{2 \cdot 5}$ or $\dfrac{1}{10}$	$= \dfrac{\overset{1}{\cancel{3}} \cdot 2 \cdot \overset{1}{\cancel{5}}}{\underset{1}{\cancel{5}} \cdot 3 \cdot \underset{1}{\cancel{3}}}$ or $\dfrac{2}{3}$

The same procedures are used for multiplying and dividing rational expressions. These can be generalized by the following rules.

Multiplying and Dividing Rational Expressions	For all rational expressions $\dfrac{a}{b}$ and $\dfrac{c}{d}$, $\dfrac{a}{b} \cdot \dfrac{c}{d} = \dfrac{ac}{bd}$, if $b \neq 0$ and $d \neq 0$, and $\dfrac{a}{b} \div \dfrac{c}{d} = \dfrac{a}{b} \cdot \dfrac{d}{c} = \dfrac{ad}{bc}$, if $b \neq 0$, $c \neq 0$, and $d \neq 0$.

The following examples show how these rules are used with rational expressions.

Example **3** Simplify each expression.

a. $\dfrac{2a^2}{5b^2c} \cdot \dfrac{3bc^2}{8a^3}$

$\dfrac{2a^2}{5b^2c} \cdot \dfrac{3bc^2}{8a^3} = \dfrac{\overset{1}{\cancel{2}} \cdot \overset{1}{\cancel{a}} \cdot \overset{1}{\cancel{a}} \cdot 3 \cdot \overset{1}{\cancel{b}} \cdot \overset{1}{\cancel{c}} \cdot c}{5 \cdot b \cdot \underset{1}{\cancel{b}} \cdot \underset{1}{\cancel{c}} \cdot \underset{4}{\cancel{8}} \cdot \underset{1}{\cancel{a}} \cdot \underset{1}{\cancel{a}} \cdot a}$

$= \dfrac{3 \cdot c}{5 \cdot b \cdot 4 \cdot a}$ or $\dfrac{3c}{20ab}$

b. $\dfrac{8x^2y}{15a^2b} \div \dfrac{2xy^2}{5ab^4}$

$$\dfrac{8x^2y}{15a^2b} \div \dfrac{2xy^2}{5ab^4} = \dfrac{8x^2y}{15a^2b} \cdot \dfrac{5ab^4}{2xy^2}$$ *Multiply by the reciprocal of the divisor.*

$$= \dfrac{\overset{4}{\cancel{8}} \cdot \overset{1}{\cancel{5}} \cdot \overset{x}{\cancel{x^2}} \cdot \overset{1}{\cancel{y}} \cdot \overset{1}{\cancel{a}} \cdot \overset{b^3}{\cancel{b^4}}}{\underset{3}{\cancel{15}} \cdot \underset{1}{\cancel{2}} \cdot \underset{1}{\cancel{x}} \cdot \underset{y}{\cancel{y^2}} \cdot \underset{a}{\cancel{a^2}} \cdot \underset{1}{\cancel{b}}}$$ *Factor and divide.*

$$= \dfrac{4b^3x}{3ay}$$

You follow these same steps when the rational expressions contain numerators and denominators that are polynomials with two or more terms.

Example **4** Find $\dfrac{x^2 + 2x - 8}{x^2 + 4x + 3} \div \dfrac{x - 2}{3x + 3}$. **Write the answer in simplest form.**

$$\dfrac{x^2 + 2x - 8}{x^2 + 4x + 3} \div \dfrac{x - 2}{3x + 3} = \dfrac{x^2 + 2x - 8}{x^2 + 4x + 3} \cdot \dfrac{3x + 3}{x - 2}$$

$$= \dfrac{(x + 4)\overset{1}{\cancel{(x - 2)}}}{(x + 3)\underset{1}{\cancel{(x + 1)}}} \cdot \dfrac{3\overset{1}{\cancel{(x + 1)}}}{\underset{1}{\cancel{(x - 2)}}}$$

$$= \dfrac{3(x + 4)}{(x + 3)} \text{ or } \dfrac{3x + 12}{x + 3}$$

Example **5** **Refer to the application at the beginning of the lesson. Find the area of rectangle N.**

INTEGRATION
Geometry

The area of a parallelogram is found by using the formula $A = bh$. Since you know the area and height of each parallelogram, you can find the base measures, b_1 and b_2, by dividing the area by the height.

$$b_1 = \dfrac{2x^2 - 13x + 20}{3x + 1} \qquad\qquad b_2 = \dfrac{3x^2 + 10x + 3}{x - 4}$$

$$= \dfrac{(x - 4)(2x - 5)}{(3x + 1)} \qquad\qquad = \dfrac{(3x + 1)(x + 3)}{(x - 4)}$$ *For what values are these expressions undefined?*

The area of the rectangle is found by $A = bh$. In this case, b and h are the measures of the bases of the parallelograms. The area of rectangle N can be found by multiplying the bases b_1 and b_2.

$$A = b_1 \cdot b_2$$

$$= \dfrac{(x - 4)(2x - 5)}{(3x + 1)} \cdot \dfrac{(3x + 1)(x + 3)}{(x - 4)}$$

$$= (2x - 5)(x + 3)$$

$$= 2x^2 + x - 15$$

The area of rectangle N is $2x^2 + x - 15$ square meters.

A **complex fraction** is a rational expression whose numerator and/or denominator contains a rational expression. The expressions below are complex fractions.

$$\frac{\dfrac{z + 4t}{w}}{6z} \qquad \frac{\dfrac{8}{x}}{\dfrac{x}{3 - y}} \qquad \frac{\dfrac{1}{x} + 3}{\dfrac{2}{x} + 5} \qquad \frac{\dfrac{x^2 - 4}{2}}{\dfrac{2 - x}{5}}$$

Remember that a fraction is nothing more than a way to express a division problem. That is, $1 \div 3$ can be expressed as $\frac{1}{3}$. So to simplify any complex fraction, rewrite it as a division expression and use the rules for division.

Example **6** **Simplify** $\dfrac{\dfrac{5a^2 - 20}{2a + 2}}{\dfrac{10a - 20}{4a}}$.

Rewrite the complex fraction as a division expression.

$$\frac{\dfrac{5a^2 - 20}{2a + 2}}{\dfrac{10a - 20}{4a}} = \frac{5a^2 - 20}{2a + 2} \div \frac{10a - 20}{4a}$$

$$= \frac{5a^2 - 20}{2a + 2} \cdot \frac{4a}{10a - 20}$$

$$= \frac{\overset{1}{\cancel{5}}(a + 2)\overset{1}{\cancel{(a - 2)}}}{\underset{1}{\cancel{2}}(a + 1)} \cdot \frac{\overset{\overset{1}{\cancel{2}}}{\cancel{4}}a}{\underset{\underset{1}{\cancel{2}}}{\cancel{10}}\underset{1}{\cancel{(a - 2)}}}$$

$$= \frac{a(a + 2)}{a + 1}$$

CHECK FOR UNDERSTANDING

Communicating Mathematics

Study the lesson. Then complete the following.

1. Suppose the numerator of a rational expression is a polynomial and the denominator of a rational expression is a different polynomial. Will factoring the polynomials necessarily provide a way to simplify the expression? Explain your answer.

2. **Explain** under what conditions a rational polynomial expression is not defined.

3. **State** the multiplicative inverse of $\frac{9a}{13b}$.

4. **State** the greatest common factor of $ab - bc$ and $3xy + 4tr$.

Guided Practice

Find the GCF of the numerator and denominator for each expression. Then simplify the expression.

5. $\dfrac{30xy}{12x^2}$

6. $\dfrac{-3xy^4}{21x^2y^2}$

7. $\dfrac{c + 5}{2c + 10}$

Simplify each expression.

8. $\dfrac{m^3}{3n} \div \left(-\dfrac{m^4}{9n^2}\right)$

9. $\dfrac{3ab}{4ac} \cdot \dfrac{6a^2}{3b^2}$

10. $-\dfrac{-3}{5a} \div \left(-\dfrac{9}{15ab}\right)$

11. $\left(\dfrac{3a^2}{a+2}\right)\left(\dfrac{a+2}{a^2}\right)$

12. $\dfrac{5}{m-3} \div \dfrac{10}{m-3}$

13. $\left(\dfrac{4a+4}{3}\right)\left(\dfrac{1}{a+1}\right)$

14. $\dfrac{w^2-11w+24}{w^2-18w+80} \cdot \dfrac{w^2-15w+50}{w^2-9w+20}$

15. $\dfrac{\dfrac{2y}{y^2-4}}{\dfrac{3}{y^2-4y+4}}$

16. **Geometry** The area of a triangle can be expressed as $4x^2 - 2x - 6$ square meters. The height of the triangle is $x + 1$ meters. Find the length of the base of the triangle.

EXERCISES

Practice

Simplify each expression.

17. $\dfrac{45xy^3}{20y^7}$

18. $\dfrac{(-3x^2y)^3}{9x^2y^2}$

19. $\dfrac{5x-5}{x^2-1}$

20. $\dfrac{p^3}{2q} \div \dfrac{-p^2}{4q}$

21. $\dfrac{y^2}{x+2} \div \dfrac{y}{x+2}$

22. $\dfrac{3h}{h+1}\left(\dfrac{1}{(h-2)}\right)$

23. $\dfrac{2a^2}{5b^2c} \cdot \dfrac{3bc^2}{8a^2}$

24. $\dfrac{35}{16x^2} \div \dfrac{21}{4x}$

25. $\dfrac{2x^3y}{z^5} \div \left(-\dfrac{4xy}{z^3}\right)^2$

26. $\dfrac{(ab)^2}{c} \cdot \dfrac{cx^2}{xa^3b}$

27. $\left(\dfrac{2x}{y}\right)^2 \cdot \dfrac{5}{6x}$

28. $\dfrac{t+3}{t-1} \cdot \dfrac{t-1}{t}$

29. $\dfrac{(xy)}{a^3} \div \dfrac{x^2y^3}{(ab)^3}$

30. $\dfrac{4a^3b}{7c^2d^3} \cdot \dfrac{21c^3d}{16abc^2}$

31. $\dfrac{9x^2y^3}{(5xyz)^2} \div \dfrac{(3xy)^3}{20x^2y}$

32. $\dfrac{3x+6}{7x-7} \cdot \dfrac{14x-14}{5x+10}$

33. $\dfrac{3x^2-3}{2x^2+8x+6} \div \dfrac{5x^2-10x+5}{4x+12}$

34. $\dfrac{4x^2-4}{9(x+1)^2} \cdot \dfrac{3x+3}{2x-2}$

35. $\dfrac{12x+6}{21x^2-21} \div \dfrac{6x^2+9x+3}{7x^3-7x^2}$

36. $\dfrac{12x^2+6x-6}{4(x+1)^2} \div \dfrac{6x-3}{2x+10}$

37. $\dfrac{5x^2+10x-75}{4x^2-24x-28} \cdot \dfrac{2x^2-10x-28}{x^2+7x+10}$

38. $\dfrac{\dfrac{x+y}{2x-y}}{\dfrac{x+y}{2x+y}}$

39. $\dfrac{\dfrac{m+n}{5}}{\dfrac{m^2+n^2}{15}}$

40. $\dfrac{\dfrac{6y^2-6}{8y^2+8y}}{\dfrac{3y-3}{4y^2+4y}}$

41. $\dfrac{\dfrac{5x^2-5x-30}{45-15x}}{\dfrac{6+x-x^2}{4x-12}}$

Critical Thinking

42. Simplify $\dfrac{x^{-1}+y^{-1}}{x^{-1}-y^{-1}}$.

43. **Geometry** The bases of two parallelograms are also the adjacent sides of a rectangle C. Parallelogram A has an area of $12x^2 + 2x - 2$ square feet and height of $2x - 5$ feet. Parallelogram B has an area of $2x^2 - 3x - 5$ square feet and height of $3x - 1$ feet. Find the area of rectangle C.

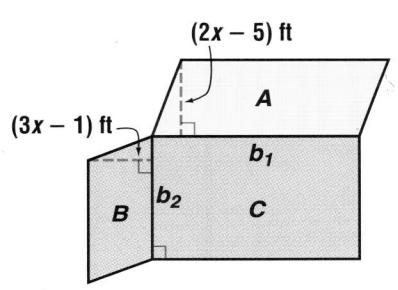

44. **Geometry** The lengths of the sides of a right triangle can be expressed as $x + 2$ in., $x + 9$ in., and $x + 10$ in. Find the lengths of the sides.

45. **Chemistry** Boyle's Law states that the volume of a gas V varies inversely with applied pressure P. This is shown by the formula $P_1 V_1 = P_2 V_2$. Suppose a helium-filled balloon has a volume of 16 m^3 at sea level. The pressure at sea level is 1 atmosphere. The balloon rises to a point in the air where the pressure is 0.75 atmosphere. What is its volume? (Lesson 9–2)

46. Solve $x^4 + 5x^3 + 6x^2 = 0$. (Lesson 8–6)

47. Find $f(x + h)$ for $f(x) = x^2 - \frac{1}{2}x$. (Lesson 8–1)

48. Find the coordinates of the center and the radius of the circle whose equation is $(x + 2)^2 + (y - 1)^2 = 81$. Then draw the graph. (Lesson 7–3)

49. **Statistics** Corporate Car Leasing leases cars to companies for use by their employees. The miles per gallon ratios for the cars leased to three of their clients are listed below. (Lesson 6–8)

Boxes to Go: 22, 14, 33, 11, 25, 11, 22, 14, 36, 35, 28, 20, 36, 15, 21, 12, 22, 10

Fitright Shoes: 32, 16, 22, 24, 23, 13, 23, 31, 15, 21, 24, 27, 30, 21, 12, 24

TLC, Ltd.:   23, 28, 16, 30, 12, 22, 11, 33, 25, 28, 21, 25, 16, 30, 12, 29, 18, 24, 13, 25

a. Find the standard deviation for the Boxes to Go data.
b. Find the standard deviation for the Fitright Shoes data.
c. Find the standard deviation for the TLC, Ltd. data.
d. Which of the companies had the least variation?

50. **City Planning** In the Winston Woods Park, a rectangular playground was planned that was to be 30 meters long by 20 meters wide. When the neighborhood association received a government grant for the playground, they decided to double the area of the playground by adding strips of the same width to one side and one end of the playground. (Lesson 6–2)

a. How wide will the strips have to be?
b. What are the new dimensions of the playground?

51. Evaluate the determinant of $\begin{bmatrix} 2 & -1 & -6 \\ 5 & 0 & 3 \\ -3 & 2 & 11 \end{bmatrix}$. (Lesson 4–4)

52. Banking Donna Bowers has a total of $4000 in her savings account and in a certificate of deposit (CD). Her savings account earns 6.5% interest annually. The CD pays 8% if the money is invested for five years. How much does she have in each investment if her interest earnings for the year will be $297.50? (Lesson 3–3)

SELF TEST

State the equations of the vertical and horizontal asymptotes for the rational functions whose graphs are shown below. (Lesson 9–1)

1.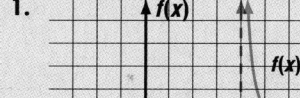

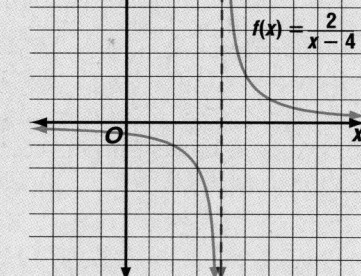

$f(x) = \dfrac{2}{x-4}$

2.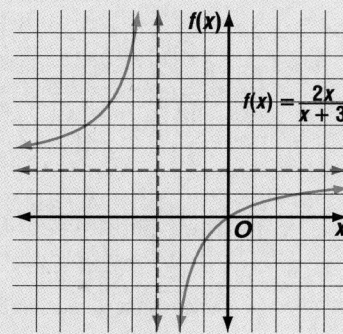

$f(x) = \dfrac{2x}{x+3}$

Graph each rational function. (Lesson 9–1)

3. $f(x) = \dfrac{3}{x+2}$

4. $f(x) = \dfrac{-2}{x^2 - 6x + 9}$

Solve. (Lesson 9–2)

5. If x varies directly as y and $y = \dfrac{1}{5}$ when $x = 11$, find x when $y = \dfrac{2}{5}$.

6. If n varies inversely as m and $m = -8$ when $n = -2$, find m when $n = \dfrac{2}{3}$.

7. Diamonds A jewelry store is having a sale on diamonds. A $\frac{1}{2}$-carat diamond is $640, and a $1\frac{1}{2}$-carat diamond is $1920. How much do you think a 2-carat diamond will cost? (Lesson 9–2)

Simplify each expression. (Lesson 9–3)

8. $\dfrac{4xy}{2yz} \cdot \dfrac{11x^2 y}{5y^2}$

9. $\dfrac{48}{6a + 42} \cdot \dfrac{7a + 49}{16}$

10. $\dfrac{w^2 + 5w + 4}{6} \div \dfrac{w + 1}{18w + 24}$

Adding and Subtracting Rational Expressions

What YOU'LL LEARN

- To find the least common denominator of two or more algebraic expressions, and
- to add and subtract rational expressions.

Why IT'S IMPORTANT

You can use rational expressions to solve problems involving electricity and phgotography.

APPLICATION

Electricity

An *electrical conductor* is any piece of material that allows electricity to flow through it. Resistance is measured in units called *ohms*. A resistor is an electrical conductor that is manufactured to have a certain resistance. If two resistors are connected in *parallel*, then the resistance of the combination R is given by the formula $\frac{1}{R} = \frac{1}{R_1} + \frac{1}{R_2}$, where R_1 and R_2 are the resistances of the resistors. What is the effective resistance of a 30-ohm resistor and a 20-ohm resistor that are connected in parallel? *You will solve this problem in Exercise 42.*

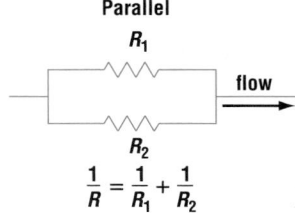

Parallel
R_1
flow
R_2
$\frac{1}{R} = \frac{1}{R_1} + \frac{1}{R_2}$

Electricity-Producing Countries (billion kw/hr)
1. U.S., 3079.09
2. Russia, 1068.00
3. Japan, 888.09
4. China, 677.55
5. Germany, 573.75

This formula involves the addition of two rational expressions. Remember from arithmetic that to add (or subtract) fractions, they must first be written as equivalent fractions with a common denominator. The least common denominator (LCD) is usually used. The LCD is the least common multiple (LCM) of the denominators.

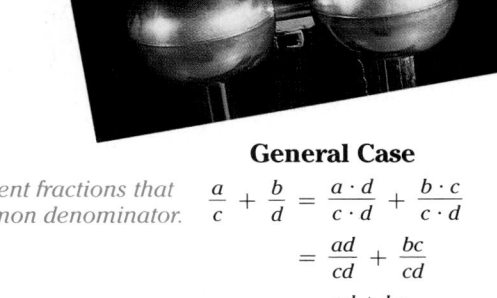

Specific Case

$$\frac{4}{5} + \frac{3}{7} = \frac{4 \cdot 7}{5 \cdot 7} + \frac{3 \cdot 5}{7 \cdot 5}$$ *Find equivalent fractions that have a common denominator.*

$$= \frac{28}{35} + \frac{15}{35}$$

$$= \frac{43}{35}$$ *Add the numerators.*

General Case

$$\frac{a}{c} + \frac{b}{d} = \frac{a \cdot d}{c \cdot d} + \frac{b \cdot c}{c \cdot d}$$

$$= \frac{ad}{cd} + \frac{bc}{cd}$$

$$= \frac{ad + bc}{cd}$$

You can use this method to add and subtract rational expressions as well.

Example ❶ **Simplify** $\dfrac{2x}{5ab^3} + \dfrac{4y}{3a^2b^2}$.

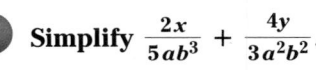

$$\frac{2x}{5ab^3} + \frac{4y}{3a^2b^2} = \frac{2x(3a)}{5ab^3(3a)} + \frac{4y(5b)}{3a^2b^2(5b)}$$ *The LCD is $15a^2b^3$. Find equivalent fractions that have this denominator.*

$$= \frac{6ax}{15a^2b^3} + \frac{20by}{15a^2b^3}$$ *Simplify each numerator and denominator.*

$$= \frac{6ax + 20by}{15a^2b^3}$$ *Add the numerators.*

Sometimes the common denominator is not easily recognized, especially when working with algebraic rational expressions. Just as in arithmetic, the LCD for two algebraic expressions must contain each factor of each denominator raised to the highest power that occurs in either denominator. You can find the LCD of more complicated fractions by factoring first, as in Example 2.

Example **2** Simplify $\dfrac{x}{x^2 + 5x + 6} - \dfrac{2}{x^2 + 4x + 4}$.

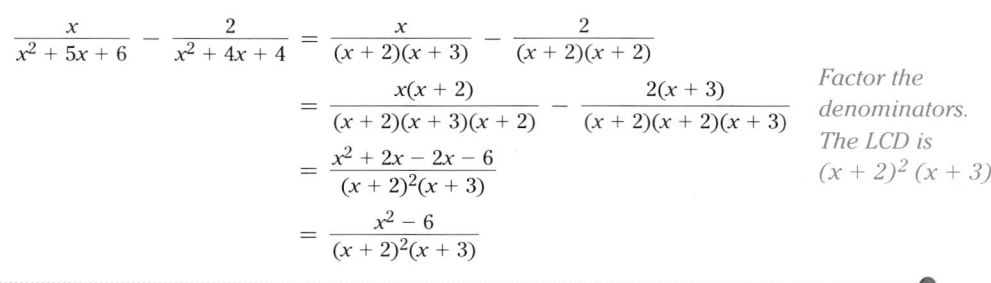

$$\frac{x}{x^2 + 5x + 6} - \frac{2}{x^2 + 4x + 4} = \frac{x}{(x + 2)(x + 3)} - \frac{2}{(x + 2)(x + 2)}$$

Factor the denominators. The LCD is $(x + 2)^2 (x + 3)$.

$$= \frac{x(x + 2)}{(x + 2)(x + 3)(x + 2)} - \frac{2(x + 3)}{(x + 2)(x + 2)(x + 3)}$$

$$= \frac{x^2 + 2x - 2x - 6}{(x + 2)^2(x + 3)}$$

$$= \frac{x^2 - 6}{(x + 2)^2(x + 3)}$$

In Example 2, you saw that the numerator was expressed as a binomial but the denominator was left as a product of factors. When you simplify the numerator, you sometimes discover that the polynomial contains a factor common to the denominator. Thus the rational expression can be further simplified.

Example **3** Simplify $\dfrac{x - 5}{2x - 6} - \dfrac{x - 7}{4x - 12}$.

$$\frac{x - 5}{2x - 6} - \frac{x - 7}{4x - 12} = \frac{x - 5}{2(x - 3)} - \frac{x - 7}{4(x - 3)}$$

Factor the denominators.

$$= \frac{(2)(x - 5) - (x - 7)}{4(x - 3)}$$

Since 2 is a factor of 4, it is not necessary to include an extra factor of 2 in the LCD. The LCD is $4(x - 3)$.

$$= \frac{2x - 10 - x + 7}{4(x - 3)}$$

Combine like terms in the numerator.

$$= \frac{x - 3}{4(x - 3)} \text{ or } \frac{1}{4}$$

Simplify, since $x - 3$ is a factor of the numerator and denominator.

The *Mathematics Exploration Toolkit (MET)* can be used to simplify rational expressions.

The CALC commands listed below are helpful when simplifying rational expressions.

SIMPLIFY (simp)   FACTOR (fac)   REDUCE (red)

Use the SIMPLIFY command to combine rational expressions. Once combined, use the FACTOR command to factor the numerator and denominator. In this form, the REDUCE command can be used to divide out any factors common to the numerator and denominator.

Simplify $\dfrac{y^2 - 4}{y^2} \div \dfrac{y + 2}{y}$.

Enter: $((y^\wedge 2 - 4)/y^\wedge 2)/((y + 2)/y)$ $\dfrac{y^2 - 4}{y^2} \div \dfrac{y + 2}{y}$

simp $\dfrac{y^2 - 4}{y^2} \cdot \dfrac{y}{y + 2}$

fac $\dfrac{y^3 - 4y}{y^3 + 2y^2}$

fac $\dfrac{y(y - 2)(y + 2)}{yy(y + 2)}$

red $\dfrac{y - 2}{y}$

Your Turn

Use CALC commands to simplify each rational expression.

a. $\dfrac{x^3 - 2x^2}{x^4 - x^2}$ b. $\dfrac{8}{2y - 16} - \dfrac{y}{8 - y}$ c. $\dfrac{2y^2 - y - 1}{y^2 - 1} \cdot \dfrac{y + 1}{2y^2 + y}$

d. $\dfrac{2a}{3a - 15} + \dfrac{-16a + 20}{3a^2 - 12a - 15}$ e. $\dfrac{x}{x - 1} - \dfrac{x - 1}{x} - \dfrac{1}{x^2 - x}$

Sometimes simplifying complex fractions involves adding or subtracting rational expressions. One way to simplify a complex fraction is to simplify the numerator and the denominator separately and then simplify the resulting expressions.

Example **4** **Simplify** $\dfrac{x + \frac{x}{3}}{x - \frac{x}{6}}$.

$\dfrac{x + \frac{x}{3}}{x - \frac{x}{6}} = \dfrac{\frac{3x}{3} + \frac{x}{3}}{\frac{6x}{6} - \frac{x}{6}}$ *The LCD of the numerator is 3.*

The LCD of the denominator is 6.

$= \dfrac{\frac{4x}{3}}{\frac{5x}{6}}$ *Simplify the numerator and denominator.*

$= \dfrac{4x}{3} \div \dfrac{5x}{6}$ *Write the complex fraction as a division problem.*

$= \dfrac{4x}{3} \cdot \dfrac{6}{5x}$

$= \dfrac{4}{1} \cdot \dfrac{2}{5}$ or $\dfrac{8}{5}$

Example **5**

APPLICATION

Photography

To take sharp, clear pictures, a photographer must focus the camera precisely. The distance from the object to the lens *p* and the distance from the lens to the film *q* must be accurately calculated to ensure a sharp image. The focal length of the lens is *f*. These measurements are shown in the diagram below.

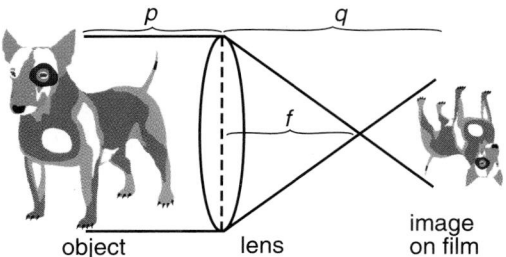

object   lens   image on film

The formula that relates these measures is $\frac{1}{p} + \frac{1}{q} = \frac{1}{f}$. Katanya has a camera with a focal length of 10 cm. If the lens is 12 cm from the film, how far should the dog be from the lens so that the picture will be in focus?

$$\frac{1}{p} + \frac{1}{q} = \frac{1}{f}$$

$$\frac{1}{p} = \frac{1}{f} - \frac{1}{q} \qquad \textit{Subtract } \tfrac{1}{q} \textit{ from each side.}$$

$$\frac{1}{p} = \frac{1}{10} - \frac{1}{12} \qquad \textit{f = 10 and q = 12}$$

$$= \frac{1 \cdot 6}{10 \cdot 6} - \frac{1 \cdot 5}{12 \cdot 5} \qquad \textit{Find equivalent fractions that have a common denominator.}$$

$$= \frac{6}{60} - \frac{5}{60}$$

$$= \frac{1}{60} \qquad \textit{Subtract.}$$

In this case, because $\frac{1}{p} = \frac{1}{60}$, $p = 60$. Therefore, Katanya's dog should be 60 cm from the lens.

fabulous

FIRSTS

James Conway Farley (1854–1910)

James Conway Farley of Richmond, Virginia, was the first African American to gain recognition as a photographer. Of the many photographs he made, only one remains that is attributed to him.

CHECK FOR UNDERSTANDING

Practice **Study the lesson. Then complete the following.**

1. **Find** each of the following for $x^2 + 5x + 6$ and $x^2 + x - 2$.
   a. the LCM   **b.** the GCF

2. **Write** $\frac{3y - 4}{5}$ as the difference of two fractions.

3. **You Decide** Christine says that the sum of $\frac{a}{x}$ and $\frac{a}{b}$ is $\frac{a}{x + b}$ because the numerators are the same. André says "No way—it doesn't work that way." Christine asks "Why not?" What should André's answer be?

M*ATH* J*OURNAL*

4. **Assess Yourself** Explain how to find the LCD of two rational algebraic expressions and explain when it is necessary to find the LCD.

Find the LCD for each pair of denominators.

5. $10x^2, 35xy^2$

6. $x(x-2), x^2-4$

Simplify each expression.

7. $\dfrac{6}{ab} + \dfrac{8}{a}$

8. $\dfrac{2}{x^2y} - \dfrac{1}{xy}$

9. $\dfrac{1}{(x+1)} + 2$

10. $\dfrac{7}{y-8} - \dfrac{6}{8-y}$

11. $\dfrac{6}{x^2+4x+4} + \dfrac{5}{x+2}$

12. $\dfrac{x}{x-y} + \dfrac{y}{y^2-x^2} + \dfrac{2x}{x+y}$

13. Photography Refer to the lens formula in Example 5. Write the expression $\dfrac{1}{p} + \dfrac{1}{q}$ as a single rational expression.

EXERCISES

Find the LCD for each pair of denominators.

14. $12y^2, 6x^2$

15. $4w-12, 2w-6$ $4w \cdot 12$

16. $36x^2y, 20xyz$

17. $x^2-y^2, x^2(x+y)$

18. $(x+2)(x+1), x^2-1$

19. $2x-10, 2x^2-4x-30$

Simplify each expression.

20. $\dfrac{3m+2}{m+n} + \dfrac{4}{2m+2n}$

21. $5 + \dfrac{x-3}{x+2}$

22. $\dfrac{5}{3a} - \dfrac{2}{7a} - \dfrac{1}{2a}$

23. $\dfrac{y}{y-4} - \dfrac{3}{4-y}$

24. $\dfrac{m}{m^2-4} + \dfrac{2}{3m+6}$

25. $y-3 + \dfrac{1}{y-3}$

26. $x+1 + \dfrac{1}{x+1}$

27. $\dfrac{x}{x+3} - \dfrac{6x}{x^2-9}$

28. $\dfrac{5}{x+3} - \dfrac{2}{x-2}$

29. $\dfrac{m}{m^2-m-20} + \dfrac{2}{m+4}$

30. $\dfrac{5}{x^2-3x-28} + \dfrac{7}{2x-14}$

31. $\dfrac{x}{x^2+2x+1} - \dfrac{x+2}{x+1} - \dfrac{3x}{x+1}$

32. $\dfrac{1}{x^2-9x+20} - \dfrac{5}{x^2-10x+25}$

33. $-\dfrac{18}{9xy} + \dfrac{7}{2x} - \dfrac{2}{3x^2}$

34. $\dfrac{m^2+n^2}{m^2-n^2} + \dfrac{m}{n-m} + \dfrac{n}{m+n}$

35. $3 + \dfrac{x}{x+2} - \dfrac{2}{x^2-4}$

36. $\dfrac{x-4}{x^2+2x-8} - \dfrac{x+2}{x^2-16}$

37. $\dfrac{x+1}{x-1} + \dfrac{x+2}{x-2} + \dfrac{x}{x^2-3x+2}$

38. $\dfrac{(x+y)\left(\dfrac{1}{x} - \dfrac{1}{y}\right)}{(x-y)\left(\dfrac{1}{x} + \dfrac{1}{y}\right)}$

39. $\dfrac{\dfrac{1}{x+2} + \dfrac{1}{x-5}}{\dfrac{2x^2-x-3}{x^2-3x-10}}$

Programming

40. The graphing calculator program at the right computes the GCF and LCM for a pair of positive integers.

Use the program to find the GCF and LCM for each pair of integers.

a. 41, 3
b. 1078, 1547
c. 199, 24
d. 187, 221
e. 182, 1690
f. 766, 424

```
PROGRAM: GCFLCM
:Input "INTEGER",A
:Input "INTEGER",B
:A→E
:B→F
:Lbl R
:If A=B
:Goto 5
:If A<B
:Goto 4
:A−B→A
:Goto R
:Lbl 4
:B−A→B
:Goto R
:Lbl 5
:Disp "GCF IS",A
:Disp "LCM IS",E*F/A
```

Critical Thinking

41. Show that the product of two numbers equals the product of their LCM and GCF.

Applications and Problem Solving

42. Electricity   Refer to the application at the beginning of the lesson. Use the formula $\frac{1}{R} = \frac{1}{R_1} + \frac{1}{R_2}$ to find the effective resistance of a 30-ohm resistor and a 20-ohm resistor that are connected in parallel.

43. Biology   After a person eats something that contains sugar, the pH or acid level A of their mouth can be determined by the formula $A = -\frac{20.4t}{t^2 + 36} + 6.5$, where t is the number of minutes that have elapsed since the food was eaten.

a. Sketch a graph of the equation.

b. Estimate the acid level after 30 minutes.

c. If normal pH is 6.5, use the graph to describe what happens to the pH level after the first hour.

d. As the pH level decreases, the acidity level increases. After eating sugar, when is the acidity level the highest?

Mixed Review

44. Simplify $-\frac{x^2 - y^2}{x + y} \cdot \frac{1}{x - y}$.   (Lesson 9–3)

45. Find all the zeros of $f(x) = 8x^3 - 36x^2 + 22x + 21$.   (Lesson 8–5)

46. Astronomy In the early 1600s, Johann Kepler studied the orbits of the planets and determined that they are elliptical. It is now known that orbits can take the shape of any of the conic sections. The equations of different orbits are given below. State the shape of each orbit. (Lesson 7–6)

a. $x^2 + y^2 = 75{,}000$ b. $x^2 + 5x = y^2 - 6y - 1$ c. $x^2 + y^2 - 4x = 9$

47. Write an equation of a parabola with focus at $(11, -1)$ and whose directrix is $y = 2$. (Lesson 7–2)

48. Write an equation of the parabola shown at the right. (Lesson 6–6)

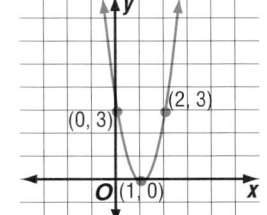

49. Solve $(x - 1)^2 - 4 = 0$ by graphing. (Lesson 6–1)

50. Fire Fighting The Woodsville City Fire Department is planning to buy some new hoses. These hoses must be powerful enough to propel water at least 75 feet into the air. The ad for the hose they are considering says that the water flows from the hose at a velocity as high as 72 feet per second. Use the formula $v = \sqrt{2gh}$, where v is the velocity of the water, g is the acceleration due to gravity (32 ft/s^2), and h is the maximum height of the water flow, to determine whether this hose will be suitable. (Lesson 5–8)

51. Solve $\dfrac{5}{6}x - 15 = 20y$ for x. (Lesson 2–2)

WORKING ON THE

In·ves·ti·ga·tion

Refer to the Investigation on pages 474–475.

When you found the percent of the tank that is full and the percent of the dipstick that is wet, you used ratios. Ratios in a general algebraic form are rational expressions.

1 Use your data to write a rational expression for finding the ratio for the percent of the tank that is full. Verify your expression.

2 Use your data to write a rational expression for finding the ratio for the percent of the dipstick that is wet. Verify your expression.

3 Determine how you calculated the percent of change in volume. Express this calculation as a rational expression.

4 Use the data from the table in Lesson 8–8 to verify your rational expression. Can this expression be used for any set of data regardless of the dimensions of the tank? Explain.

Add the results of your work to your Investigation Folder.

Solving Rational Equations and Inequalities

What YOU'LL LEARN

• To solve rational equations and inequalities.

Why IT'S IMPORTANT

You can use rational equations and inequalities to solve problems involving art, cycling, and engineering.

APPLICATION

Cycling

Ken Hiroshi has started bicycling because his doctor told him it was great exercise. One thing his doctor suggested was that Ken find his normal bicycling speed, because for cycling to be an ideal exercise, you can't go too fast or too slow. Ken is having difficulty measuring his speed, because he always has to take wind into consideration. On a particular day, the wind added 3 km/hr to his rate when he was cycling with the wind, and subtracted 3 km/hr from his rate on his return trip. Ken discovered that in the same amount of time he could cycle 36 km with the wind, he could go only 24 km against the wind. What is his normal bicycling speed with no wind at all?

Recall that the formula that relates rate, time, and distance is $r \cdot t = d$. If r is Ken's rate with no wind, then his rate with the wind is $r + 3$ and his rate against the wind is $r - 3$. If t is his time for the trip each way, then using rate × time = distance, $(r - 3)t = 24$ and $(r + 3)t = 36$.

Solve each equation for t.

$$(r - 3)t = 24 \qquad\qquad (r + 3)t = 36$$
$$t = \frac{24}{r - 3} \qquad\qquad\quad t = \frac{36}{r + 3}$$

The two expressions equal to t are equal to each other.

$$\frac{24}{r - 3} = \frac{36}{r + 3}$$

An equation that contains one or more rational expressions is called a **rational equation**. It is easiest to solve a rational equation if the fractions are eliminated. This can be done by multiplying each side of the equation by the least common denominator (LCD). Remember that when you multiply each side by the LCD, each term on each side must be multiplied by the LCD.

To solve $\frac{24}{r - 3} = \frac{36}{r + 3}$, multiply each side by the least common denominator $(r - 3)(r + 3)$.

$$\frac{24}{\cancel{(r-3)}^{1}}\,\cancel{(r-3)}\,(r+3) = \frac{36}{\cancel{(r+3)}^{1}}\,(r-3)\cancel{(r+3)}$$
$$24(r + 3) = 36(r - 3)$$
$$24r + 72 = 36r - 108$$
$$-12r = -180$$
$$r = 15$$

Without the wind, Ken would travel at a rate of 15 kilometers per hour.

Check by graphing $t = \dfrac{24}{r - 3}$ and $t = \dfrac{36}{r + 3}$ on the same coordinate axes. The intersection point is at (15, 2).

Remember, when you solve an equation by multiplying each side by a polynomial, you must make sure that you are not multiplying by 0. Even a false equation such as $3 = 2$ becomes true if you multiply each side by 0; the equation $0 \cdot 3 = 0 \cdot 2$ is true because both sides equal 0. This is the reason why checking your solutions in the *original* equation is so important.

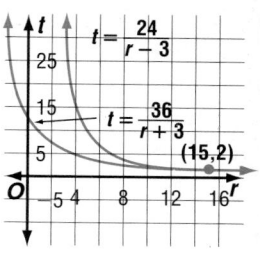

Example 1 Solve each equation. Check the solution.

a. $\dfrac{x + 1}{3(x - 2)} = \dfrac{5x}{6} + \dfrac{1}{x - 2}$

$$\dfrac{x + 1}{3(x - 2)} = \dfrac{5x}{6} + \dfrac{1}{x - 2} \qquad \text{\textit{The LCD is } } 6(x - 2).$$

$$6(x - 2)\,\dfrac{x + 1}{3(x - 2)} = 6(x - 2)\left(\dfrac{5x}{6} + \dfrac{1}{x - 2}\right)$$

$$\overset{2}{\underset{1}{6}}(x - 2)\,\dfrac{x + 1}{\underset{1}{3}(x - 2)} = \overset{1}{6}(x - 2)\left(\dfrac{5x}{\underset{1}{6}}\right) + 6(x - 2)\,\dfrac{1}{x - 2}$$

$$2(x + 1) = (x - 2)(5x) + 6$$

$$2x + 2 = 5x^2 - 10x + 6$$

$$5x^2 - 12x + 4 = 0$$

$$(5x - 2)(x - 2) = 0$$

$$x = \dfrac{2}{5} \quad \text{or} \quad x = 2$$

Check: Let $x = \dfrac{2}{5}$.
Let $x = 2$.

$$\dfrac{\frac{2}{5} + 1}{3\left(\frac{2}{5} - 2\right)} \overset{?}{=} \dfrac{5\left(\frac{2}{5}\right)}{6} + \dfrac{1}{\frac{2}{5} - 2}$$

$$\dfrac{\frac{7}{5}}{3\left(-\frac{8}{5}\right)} \overset{?}{=} \dfrac{2}{6} + \dfrac{1}{\frac{-8}{5}}$$

$$\dfrac{\frac{7}{5}}{-\frac{24}{5}} \overset{?}{=} \dfrac{2}{6} + \left(-\dfrac{5}{8}\right)$$

$$\left(\dfrac{7}{5}\right)\left(-\dfrac{5}{24}\right) \overset{?}{=} \dfrac{8}{24} + \left(-\dfrac{15}{24}\right)$$

$$-\dfrac{7}{24} = -\dfrac{7}{24} \quad \checkmark$$

$$\dfrac{2 + 1}{3(2 - 2)} \overset{?}{=} \dfrac{5(2)}{6} + \dfrac{1}{2 - 2}$$

$$\dfrac{3}{3(0)} \overset{?}{=} \dfrac{10}{6} + \dfrac{1}{0}$$

When you check the value 2, you get a zero in the denominator. So, 2 must be eliminated as a solution.

The solution is $\dfrac{2}{5}$.

b. $\dfrac{r + 2}{2r + 1} = \dfrac{r}{3} + \dfrac{3}{4r + 2}$

$$\dfrac{r + 2}{2r + 1} = \dfrac{r}{3} + \dfrac{3}{4r + 2} \qquad \text{\textit{The LCD is } } 6(2r + 1).$$

$$6(2r + 1)\left(\dfrac{r + 2}{2r + 1}\right) = 6(2r + 1)\left(\dfrac{r}{3} + \dfrac{3}{2(2r + 1)}\right)$$

$$\overset{}{\underset{1}{6}}(2r + 1)\left(\dfrac{r + 2}{2r + 1}\right) = \overset{2}{6}(2r + 1)\left(\dfrac{r}{3}\right) + \overset{3}{6}(2r + 1)\left(\dfrac{3}{2(2r + 1)}\right)$$

(continued on the next page)

$$6(r + 2) = 2(2r + 1)(r) + (3)(3)$$
$$6r + 12 = 4r^2 + 2r + 9$$
$$4r^2 - 4r - 3 = 0$$
$$(2r - 3)(2r + 1) = 0$$
$$r = \frac{3}{2} \text{ or } r = -\frac{1}{2}$$

Check:
$$\frac{r + 2}{2r + 1} = \frac{r}{3} + \frac{3}{4r + 2}$$

$$\frac{\frac{3}{2} + 2}{2\left(\frac{3}{2}\right) + 1} = \frac{\frac{3}{2}}{3} + \frac{3}{4\left(\frac{3}{2}\right) + 2}$$

$$\frac{\frac{7}{2}}{4} \overset{?}{=} \frac{1}{2} + \frac{3}{8}$$

$$\frac{7}{8} \overset{?}{=} \frac{4}{8} + \frac{3}{8}$$

$$\frac{7}{8} = \frac{7}{8} \quad \checkmark$$

When you check the value $-\frac{1}{2}$, you get a zero in the denominator. So, $-\frac{1}{2}$ must be eliminated as a solution.

The solution is $\frac{3}{2}$.

You can use what you now know about solving rational equations to solve *rational inequalities*.

Example **2** **Solve** $\dfrac{x-2}{x} < \dfrac{x-4}{x-6}$.

$$\frac{x-2}{x} < \frac{x-4}{x-6}$$

$$\frac{x-2}{x} - \frac{x-4}{x-6} < 0 \qquad \textit{Subtract } \frac{x-4}{x-6} \textit{ from each side.}$$

$$\frac{(x-6)(x-2) - x(x-4)}{x(x-6)} < 0 \ [x(x-6)] \qquad \textit{The LCD is } x(x-6).$$

$$\frac{x^2 - 2x - 6x + 12 - x^2 + 4x}{x(x-6)} < 0 \qquad \textit{Apply the distributive property.}$$

$$\frac{-4x + 12}{x(x-6)} < 0 \qquad \textit{Simplify.}$$

$$\left(-\frac{1}{4}\right) \frac{-4(x-3)}{x(x-6)} > 0 \left(-\frac{1}{4}\right)$$

$$\frac{x-3}{x(x-6)} > 0 \qquad x \neq 0, 6, \text{ or } 3$$

On a number line, identify excluded values and values that make the inequality untrue. Test values on each part of the number line to check the result.

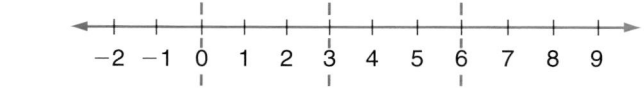

-2 -1 0 1 2 3 4 5 6 7 8 9

Try $x = -1$.	Try $x = 1$.	Try $x = 5$.	Try $x = 8$.
$\dfrac{x-2}{x} < \dfrac{x-4}{x-6}$	$\dfrac{x-2}{x} < \dfrac{x-4}{x-6}$	$\dfrac{x-2}{x} < \dfrac{x-4}{x-6}$	$\dfrac{x-2}{x} < \dfrac{x-4}{x-6}$
$\dfrac{-1-2}{-1} < \dfrac{-1-4}{-1-6}$	$\dfrac{1-2}{1} < \dfrac{1-4}{1-6}$	$\dfrac{5-2}{5} < \dfrac{5-4}{5-6}$	$\dfrac{8-2}{8} < \dfrac{8-4}{8-6}$
$3 \not< \dfrac{5}{7}$	$-1 < \dfrac{3}{5} \ \checkmark$	$\dfrac{3}{5} \not< -1$	$\dfrac{3}{4} < 2 \ \checkmark$
false	true	false	true

So, $0 < x < 3$ or $x > 6$.

To solve rational inequalities, complete the following steps.

1. State the excluded values.
2. Solve the inequality.
3. Test values to find the solution.

Some real-world problems can be solved with rational equations.

Example ③

APPLICATION

Engineering

Construction of the English Channel Tunnel

Most tunnels are drilled using tunnel-boring machines that begin at both ends of the tunnel. Suppose a new underwater tunnel is being built and one tunnel-boring machine alone can finish the tunnel in 4 years. A different type of machine can tunnel to the other side in 3 years. If both machines start at opposite ends and work at the same time, when will the tunnel be finished?

Explore In one year, machine A can complete $\frac{1}{4}$ of the tunnel. In 2 years, it can complete $\frac{1}{4} \cdot 2$ or $\frac{2}{4}$ of the tunnel. In t years, it can complete $\frac{1}{4} \cdot t$ or $\frac{t}{4}$ of the tunnel.

In one year, machine B can complete $\frac{1}{3}$ of the tunnel. Using the same pattern as above, in t years it can complete $\frac{1}{3} \cdot t$ or $\frac{t}{3}$ of the tunnel.

Plan In t years, machine A can complete $\frac{t}{4}$ of the tunnel and, in that same time, machine B can complete $\frac{t}{3}$ of the tunnel. Together, they can complete the entire tunnel.

$$\underbrace{\frac{t}{4}}_{machine\ A} + \underbrace{\frac{t}{3}}_{machine\ B} = \underbrace{1}_{entire\ tunnel}$$

Solve
$$\frac{t}{4} + \frac{t}{3} = 1$$
$$12\left(\frac{t}{4} + \frac{t}{3}\right) = 12(1) \qquad \textit{Multiply by the LCD.}$$
$$3t + 4t = 12 \qquad \textit{Distributive property}$$
$$7t = 12 \qquad \textit{Combine like terms.}$$
$$t = \frac{12}{7} \text{ or } 1\frac{5}{7} \quad \textit{Divide each side by 7.}$$

The tunnel can be completed in $1\frac{5}{7}$ years or about 1 year 9 months.

Examine Machine A digs through $\frac{1}{4}t$ or $\frac{1}{4}\left(\frac{12}{7}\right)$ of the tunnel: $\frac{1}{4}\left(\frac{12}{7}\right) = \frac{3}{7}$.

Machine B digs through $\frac{1}{3}t$ or $\frac{1}{3}\left(\frac{12}{7}\right)$ of the tunnel: $\frac{1}{3}\left(\frac{12}{7}\right) = \frac{4}{7}$.

Since $\frac{3}{7} + \frac{4}{7} = \frac{7}{7}$ or 1, the solution checks.

Sometimes it is helpful to **organize the data** that is given in a problem. One way of doing this is by making a drawing. In this way you can evaluate how to write an equation that helps you solve the problem.

Example ④

A car travels 300 km in the same time that a freight train travels 200 km. The speed of the car is 20 km/h more than the speed of the train. Find the speed of the car and the speed of the train.

Draw a diagram to display the information that you know.

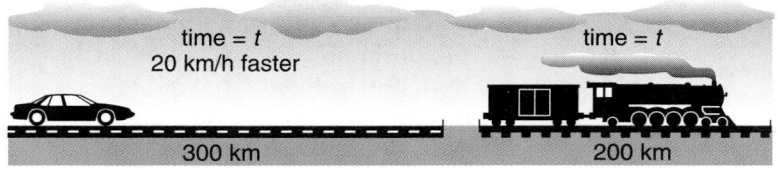

time = t
20 km/h faster

time = t

300 km

200 km

Remember that $d = rt$. Since both vehicles travel the same amount of time, rewrite the formula in terms of t. That is, $\frac{d}{r} = t$. Then you can equate the formulas for the two vehicles in terms of t. Let r represent the speed of the train. So $r + 20$ represents the speed of the car.

$$\underbrace{car's\ time} = \underbrace{train's\ time}$$

$$\frac{distance}{rate} = \frac{300}{r + 20} \qquad \frac{distance}{rate} = \frac{200}{r}$$ *Both vehicles travel the same time.*

$$\frac{300}{r + 20} = \frac{200}{r}$$

$$r(r + 20)\left(\frac{300}{r + 20}\right) = r(r + 20)\left(\frac{200}{r}\right)$$ *Multiply each side by the LCD, $r(r + 20)$.*

$$300r = 200r + 4000$$

$$100r = 4000$$ *Subtract $200r$ from each side.*

$$r = 40$$ *Divide by 100.*

So, the speed of the freight train is 40 km/h, and the speed of the car is 40 + 20 or 60 km/h.

Communicating Mathematics

Study the lesson. Then complete the following.

1. **Explain** why the equation $x + \dfrac{1}{x-1} = 1 + \dfrac{1}{x-1}$ has no solution.

2. **Explain** why it is necessary to test values when solving rational inequalities.

3. **a. State** what you would multiply each side of $\dfrac{x}{x-3} + \dfrac{1}{3} = 1$ by in order to solve the equation.
   b. What value(s) of x cannot be a solution?

4. **You Decide**   Rick solved the rational equation $\dfrac{(x+3)^2}{5} = x + 3$ and told his friend Janine that the solution was 2. Janine said "You're right, Rick, but there's another solution to the equation." "That's impossible," said Rick, showing her his solution. Look at Rick's solution at the right. Who is correct? Explain.

$$\frac{(x+3)^2}{5} = x + 3$$

$$\frac{(x+3)^2}{5(x+3)} = 1$$

$$\frac{(x+3)}{5} = 1$$

$$(x+3) = 5$$

$$x = 2$$

MATH JOURNAL

5. Describe other methods you could use to organize data to make it easier to solve problems.

Guided Practice

Find the LCD for each equation. State what values should be excluded as possible solutions. Then solve the equation and check your solution.

6. $\dfrac{2}{y+4} + y = \dfrac{1}{5}$

7. $\dfrac{1}{m-4} = \dfrac{2}{m-2}$

Solve each equation or inequality. Check your solutions.

8. $\dfrac{y}{y+1} = \dfrac{2}{3}$

9. $\dfrac{x}{3} - \dfrac{2}{5} = 1$

10. $b^2 + \dfrac{17b}{6} = \dfrac{1}{2}$

11. $\dfrac{2x}{3} - \dfrac{x+3}{6} > 2$

12. **Construction**   The Delaware Demolition Company wants to build a brick wall to hide the area where they store wrecked cars from public view. One bricklayer can build this wall in 5 days. Another bricklayer can do the job in 4 days. If the company hires both of them to work together, how long will it take them to finish the wall?

Practice

Find the LCD for each equation. State what values should be excluded as possible solutions. Then solve the equation and check your solution.

13. $\dfrac{x+2}{2} - \dfrac{3}{4} = x$

14. $\dfrac{1}{a} + \dfrac{1}{2} = \dfrac{2}{a}$

15. $\dfrac{6}{m} = \dfrac{9}{m^2}$

16. $\dfrac{x}{x-3} + \dfrac{1}{3} = 1$

17. $\dfrac{3}{a-6} - \dfrac{1}{a-2} = 3$

18. $\dfrac{3y}{2+y} - \dfrac{5}{7} = 4$

Solve each equation or inequality. Check your solutions.

19. $\dfrac{x+1}{3} + \dfrac{x-1}{3} = \dfrac{4}{3}$

20. $\dfrac{5+7z}{8} - \dfrac{15+3z}{10} < 2$

21. $y + 5 \le \dfrac{6}{y}$

22. $\dfrac{1}{t-1} + \dfrac{1}{t+2} = \dfrac{1}{2}$

23. $\dfrac{1}{m+2} - \dfrac{1}{3-m} = -\dfrac{1}{6}$

24. $\dfrac{1}{2y+1} + \dfrac{1}{y+1} \ge \dfrac{8}{15}$

25. $\dfrac{1}{9} + \dfrac{1}{2a} = \dfrac{1}{a^2}$

26. $\dfrac{1}{1-x} = 1 - \dfrac{x}{x-1}$

27. $\dfrac{3}{x^2+3x} + \dfrac{x+2}{x+3} = \dfrac{1}{x}$

28. $\dfrac{6}{y^2+2y} - \dfrac{y+1}{y+2} = \dfrac{2}{y}$

29. $\dfrac{1}{x+4} = \dfrac{2}{x^2+3x-4} - \dfrac{1}{1-x}$

30. $\dfrac{3}{b^2+5b+6} + \dfrac{b-1}{b+2} = \dfrac{7}{b+3}$

Critical Thinking

31. Solve for a if $\dfrac{1}{a} - \dfrac{1}{b} = c$.

32. Find the values of A and B, if $\dfrac{A}{z+2} + \dfrac{B}{2z-3} = \dfrac{5z-11}{2z^2+z-6}$.

Applications and Problem Solving

33. **Basketball**   Selena has scored 11 free throws in 20 attempts. Her coach has told her that she needs to bring her free-throw average (number of free throws made ÷ number of attempts) up to at least 70% or the coach will assign someone else to play her position. Selena wants to increase her average as quickly as possible.
   a. Write an inequality to determine the number of consecutive free throws Selena must score to bring her average up to 0.70.
   b. Solve the inequality.
   c. Interpret your answer in terms of Selena's problem.

34. **Statistics**   A number x is said to be the *harmonic mean* of y and z if $\dfrac{1}{x}$ is the average of $\dfrac{1}{y}$ and $\dfrac{1}{z}$.
   a. Find y if $x = 8$ and $z = 20$.
   b. Find x if $y = 5$ and $z = 8$.

35. **Organize Data**   During one fourth of the time it took her to travel from Denver to Cheyenne, Alicia drove in a snowstorm at an average speed of 40 mph. She drove the rest of the time at an average speed of 65 mph. What was her average speed for the entire trip?

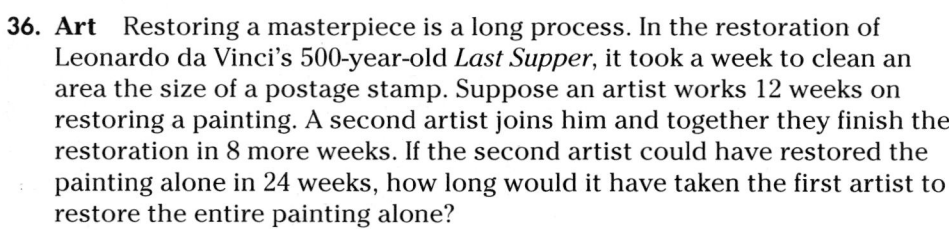

36. **Art**   Restoring a masterpiece is a long process. In the restoration of Leonardo da Vinci's 500-year-old *Last Supper*, it took a week to clean an area the size of a postage stamp. Suppose an artist works 12 weeks on restoring a painting. A second artist joins him and together they finish the restoration in 8 more weeks. If the second artist could have restored the painting alone in 24 weeks, how long would it have taken the first artist to restore the entire painting alone?

37. **Number Theory**   The ratio of 8 less than a number to 28 more than that number is 2 to 5. What is the number?

38. Simplify $\dfrac{3}{a-2} + \dfrac{2}{a-3}$. (Lesson 9–4)

39. State the number of positive real zeros, negative real zeros, and imaginary zeros for $f(x) = x^3 + 1$. (Lesson 8–4)

40. Solve the system of equations. (Lesson 7–7)

$x + y + 7 = 0$

$x^2 + y^2 = 25$

41. Geometry Find the distance between the points at $(-0.5, 1)$ and $(-2.2, -0.3)$. (Lesson 7–1)

42. Write a quadratic equation that has 3 and $\dfrac{1}{2}$ as roots. (Lesson 6–5)

43. Environment The Natural Ideas store needs to order cube-shaped boxes to package commemorative globes designed for the 30th anniversary of Earth Day on April 22, 2000. The globes were made by school children in the community. The formula for the radius r of a sphere is $r = \sqrt[3]{\dfrac{3V}{4\pi}}$, where V is the volume. If the volume of each globe is 175 in^3, what is the minimum size (to the nearest inch) that the boxes can be inside to fit around the globe? (Lesson 5–6)

44. Solve $4x + 3 < -9$ or $7 < 2x - 11$. (Lesson 1–7)

Mathematics and SOCIETY

Prime Factoring

The excerpt below appeared in an article in *Science News* on May 7, 1994.

IT'S EASY TO MULTIPLY TWO LARGE PRIME numbers to obtain a larger number as the answer. But the reverse process—factoring a large number to determine its components—presents a formidable challenge. The problem appears so hard that the difficulty of factoring underlies the so-called RSA method of encrypting digital information. Last week, an international team of computer scientists, mathematicians, and other experts succeeded in finding the factors of a 129-digit number suggested 17 years ago as a test of the security of the RSA cryptographic scheme. This feat and other work now complicate encoding schemes used for national and commercial security. The effort required the use of more than 600 computers scattered throughout the world. ∎

1. If you were in charge of security for a company and you found out about the work described above, how do you think you would feel? Why?

2. Since 1977, what developments have occurred to enable the huge task of factoring RSA-129 to be performed?

3. What do you think about using a direct challenge in order to get a large problem solved? Can you think of an example from your own knowledge or experience where a challenge was used to inspire action?

In·ves·ti·ga·tion

Refer to the Investigation on pages 474–475.

Gasoline stations usually have several tanks buried beneath the concrete surface. These tanks contain different octanes of gasoline. The tanks may differ in size depending on the demand for each octane of gasoline. In many areas, these tanks are refilled on a weekly basis.

The national average consumption of gasoline is 687 gallons of fuel per year per vehicle. The price of gasoline rises and falls with the global demand for barrels of crude oil. With the exception of the gasoline shortages in the 1970s, gasoline has been in abundance since 1950. However, the price of gasoline has risen with the price of crude oil.

PORTFOLIO ASSESSMENT

You may want to keep your work on this Investigation in your portfolio.

Analyze

You have conducted experiments and organized your data in various ways. It is now time to analyze your findings and state your conclusions.

1 Look over the data and graphs that you made from the two experiments. What generalizations can you make about these experiments that can be applied to any cylindrical tank?

2 Use your graphs and data to determine a general formula for finding the scale markings on a dipstick for a tank one-half full, one-fourth full, one-eighth full, five-eighths full, and three-fourths full given any cylindrical tank with radius r and height h.

3 Explain and justify your calculations and formula. Use a specific case as an example that your general process and formula work.

4 Draw scatter plots of the three sets of data. Let the vertical axis represent the dipstick reading in feet and let the horizontal axis represent the volume of the tank as percent full. Describe patterns in the data.

Write

Your superior has received several calls similar to the one you received from the franchise owner. He has asked you for a summary of your findings and your report to the franchise owner.

5 Summarize the procedure you used to investigate the problem of designing the scale of a dipstick in a cylindrical tank. Explain your findings from the experiments you conducted.

VOCABULARY

After completing this chapter, you should be able to define each term, property, or phrase and give an example or two of each.

Algebra

asymptotes   (p. 550)

complex fraction   (p. 565)

constant of variation   (p. 556)

continuity   (p. 548)

direct variation   (p. 556)

dividing rational expressions   (p. 563)

inversely proportional   (p. 557)

inverse variation   (p. 557)

joint variation   (p. 558)

multiplying rational expressions   (p. 563)

point discontinuity   (p. 548)

rational algebraic expressions   (p. 562)

rational equation   (p. 576)

rational function   (p. 550)

rational inequalities   (p. 578)

Problem Solving

organize data   (p. 580)

UNDERSTANDING AND USING THE VOCABULARY

State whether each sentence is *true* or *false*. If false, replace the underlined word or number to make a true sentence.

1. The equation $y = \dfrac{x^2 - 1}{x + 1}$ has a(n) <u>asymptote</u> at $x = -1$.

2. The complex fraction $\dfrac{\frac{4}{5}}{\frac{2}{3}}$ can be reduced to $\dfrac{6}{5}$.

3. The equation $y = 3x$ is an example of <u>direct</u> variation.

4. The equation $y = \dfrac{x^2}{x + 1}$ is a(n) <u>polynomial</u> equation.

5. The graph of $y = \dfrac{4}{x - 4}$ has a(n) <u>variation</u> at $x = 4$.

6. The equation $b = \dfrac{2}{a}$ is a(n) <u>inverse</u> variation.

7. On the graph of $y = \dfrac{x - 5}{x + 2}$, there is a break in continuity at <u>$x = 2$</u>.

8. The formula for the area of a triangle, $A = \dfrac{1}{2} bh$, is a(n) <u>inverse</u> variation.

SKILLS AND CONCEPTS

OBJECTIVES AND EXAMPLES

Upon completing this chapter, you should be able to:

- graph rational functions (Lesson 9–1)

  Graph $f(x) = \dfrac{5}{x(x+4)}$.

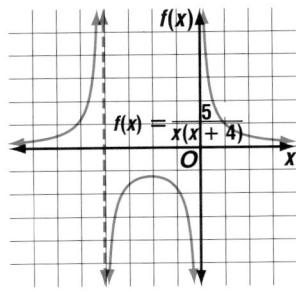

REVIEW EXERCISES

Use these exercises to review and prepare for the chapter test.

State the equations of the vertical and horizontal asymptotes for each rational function. Then graph each rational function.

9. $f(x) = \dfrac{4}{x-2}$

10. $f(x) = \dfrac{x}{x+3}$

11. $f(x) = \dfrac{2}{x}$

12. $f(x) = \dfrac{x-4}{x+3}$

13. $f(x) = \dfrac{5}{(x+1)(x-3)}$

- solve problems involving direct, inverse, and joint variation (Lesson 9–2)

  direct: $y = kx$
  inverse: $xy = k$
  joint: $y = kxz$

  If y varies inversely as x and $x = 14$ when $y = -6$, find x when $y = -11$.
  $$\frac{x_1}{y_2} = \frac{x_2}{y_1}$$
  $$\frac{14}{-11} = \frac{x_2}{-6}$$
  $$-11x_2 = -84$$
  $$x_2 = \frac{84}{11} \text{ or } 7.\overline{63}$$

  The value of x when $y = -11$ is $\dfrac{84}{11}$ or $7.\overline{63}$.

Write an equation for each statement. Then solve the equation.

14. If y varies directly as x and $y = 21$ when $x = 7$, find x when $y = -5$.

15. If y varies inversely as x and $y = 9$ when $x = 2.5$, find y when $x = -0.6$.

16. If y varies inversely as x and $x = 28$ when $y = 18$, find x when $y = 63$.

17. If y varies directly as x and $x = 28$ when $y = 18$, find x when $y = 63$.

18. If y varies jointly as x and z and $x = 2$ and $z = 4$ when $y = 16$, find y when $x = 5$ and $z = 8$.

19. If y varies jointly as x and z and $x = 4$ and $z = 2$ when $y = 25$, find x when $y = 12$ and $z = 20$.

OBJECTIVES AND EXAMPLES

● simplify rational expressions   (Lesson 9–3)

$$\frac{3x}{2y} \cdot \frac{8y^3}{6x^2} = \frac{\overset{1}{\cancel{3}} \cdot \overset{1}{\cancel{x}} \cdot \overset{4}{\cancel{8}} \cdot \overset{y^2}{\cancel{y^3}}}{\underset{1}{\cancel{2}} \cdot \underset{1}{\cancel{y}} \cdot \underset{2}{\cancel{6}} \cdot \underset{x}{\cancel{x^2}}}$$

$$= \frac{2y^2}{x}$$

$$\frac{x^2 - 4}{x^2 - 9} \div \frac{x + 2}{x - 3} = \frac{x^2 - 4}{x^2 - 9} \cdot \frac{x - 3}{x + 2}$$

$$= \frac{\overset{1}{\cancel{(x+2)}}(x-2)\overset{1}{\cancel{(x-3)}}}{(x+3)\underset{1}{\cancel{(x-3)}}\underset{1}{\cancel{(x+2)}}}$$

$$= \frac{x - 2}{x + 3}$$

REVIEW EXERCISES

Simplify each expression.

20. $\dfrac{-4ab}{21c} \cdot \dfrac{14c^2}{22a^2}$

21. $\dfrac{y - 2}{a - 3} \cdot (a - 3)$

22. $\dfrac{a^2 - b^2}{6b} \div \dfrac{a + b}{36b^2}$

23. $\dfrac{5x(x + y)}{a} \div \dfrac{25x^3(x + y)}{a^2}$

24. $\dfrac{y^2 - y - 12}{y + 2} \div \dfrac{y - 4}{y^2 - 4y - 12}$

25. $\dfrac{x^2 + 3x - 10}{x^2 + 8x + 15} \cdot \dfrac{x^2 + 5x + 6}{x^2 + 4x + 4}$

● Simplify complex fractions   (Lesson 9–3)

$$\frac{\dfrac{p^2 + 7p}{3p}}{\dfrac{49 - p^2}{3p - 21}} = \frac{p^2 + 7p}{3p} \div \frac{49 - p^2}{3p - 21}$$

$$= \frac{p^2 + 7p}{3p} \cdot \frac{3p - 21}{49 - p^2}$$

$$= \frac{\overset{1}{\cancel{p}}\overset{1}{\cancel{(p+7)}}}{\underset{11}{\cancel{3p}}} \cdot \frac{\overset{1}{\cancel{3}}\overset{1}{\cancel{(p-7)}}}{-\underset{1}{\cancel{(p+7)}}\underset{1}{\cancel{(p-7)}}}$$

$$= -1$$

Simplify each expression.

26. $\dfrac{\dfrac{1}{x}}{\dfrac{2x}{17}}$

27. $\dfrac{\dfrac{1}{n^2 - 6n + 9}}{\dfrac{n + 3}{2n^2 - 18}}$

28. $\dfrac{\dfrac{x^2 + 7x + 10}{x + 2}}{\dfrac{x^2 + 2x - 15}{x + 2}}$

● add and subtract rational expressions
(Lesson 9–4)

$$\frac{14}{x + y} - \frac{9x}{x^2 - y^2} = \frac{14}{x + y} - \frac{9x}{(x + y)(x - y)}$$

$$= \frac{14(x - y) - 9x}{(x + y)(x - y)}$$

$$= \frac{14x - 14y - 9x}{(x + y)(x - y)}$$

$$= \frac{5x - 14y}{(x + y)(x - y)}$$

Simplify each expression.

29. $\dfrac{-9}{4a} + \dfrac{7}{3b}$

30. $\dfrac{x + 2}{x - 5} + 6$

31. $\dfrac{x - 1}{x^2 - 1} + \dfrac{2}{5x + 5}$

32. $\dfrac{7}{y} - \dfrac{2}{3y}$

33. $\dfrac{7}{y - 2} - \dfrac{11}{2 - y}$

34. $\dfrac{3}{4b} - \dfrac{2}{5b} - \dfrac{1}{2b}$

35. $\dfrac{m + 3}{m^2 - 6m + 9} - \dfrac{8m - 24}{9 - m^2}$

OBJECTIVES AND EXAMPLES

- solve rational equations   (Lesson 9–5)

Solve $\dfrac{1}{x-1} + \dfrac{2}{x} = 0$.

The LCD for the two denominators is $x(x-1)$.
Multiply each side of the equation by the LCD.

$$\frac{1}{x-1} + \frac{2}{x} = 0$$

$$x(x-1)\left(\frac{1}{x-1} + \frac{2}{x}\right) = x(x-1)(0)$$

$$x(x-1)\left(\frac{1}{x-1}\right) + x(x-1)\left(\frac{2}{x}\right) = x(x-1)(0)$$

$$1(x) + 2(x-1) = 0$$

$$x + 2x - 2 = 0$$

$$3x - 2 = 0$$

$$3x = 2$$

$$x = \frac{2}{3}$$

REVIEW EXERCISES

Solve each equation. Check your solutions.

36. $\dfrac{3}{y} + \dfrac{7}{y} = 9$

37. $1 + \dfrac{5}{y-1} = \dfrac{7}{6}$

38. $\dfrac{3x+2}{4} = \dfrac{9}{4} - \dfrac{3-2x}{6}$

39. $\dfrac{1}{r^2-1} = \dfrac{2}{r^2+r-2}$

40. $\dfrac{x}{x^2-1} + \dfrac{2}{x+1} = 1 + \dfrac{1}{2x-2}$

APPLICATIONS AND PROBLEM SOLVING

41. Physics The current I in an electrical circuit varies inversely with the resistance R in the circuit.   (Lesson 9–2)

  a. Use the table below to write an equation relating the current and the resistance.

I (amperes)	0.5	1.0	1.5	2.0	2.5	3.0	5.0
R (ohms)	12	6.0	4.0	3.0	2.4	2.0	1.2

  b. What is the constant of variation?

42. Number Theory The denominator of a fraction is 1 less than twice the numerator. If 7 is added to both the numerator and denominator, the resulting fraction has a value of $\dfrac{7}{10}$. Find the original fraction.

(Lesson 9–5)

43. Auto Mechanics When air is pumped into a tire, the pressure required varies inversely as the volume of the air. If the pressure is 30 lb/in^2 when the volume is 140 in^3, find the pressure when the volume is 100 in^3.   (Lesson 9–2)

A practice test for Chapter 9 is provided on page 920.

ALTERNATIVE ASSESSMENT

COOPERATIVE LEARNING PROJECT

Saving for College In this chapter, you explored rational functions. You graphed rational functions, simplified rational expressions, and solved rational equations. These equations were helpful in solving application problems.

In this project, you will organize a painting job. College Craft is the name of a painting company that uses college students to do its painting in the summer. The following list shows the students that Frank Anstine, the manager, knows will be returning this summer to paint, their experience, and their hourly rate of pay.

> José - 3 years experience - $12.00
> Brittany - 1 year experience - $8.00
> John - 1 year experience - $8.00

He also has two new students joining the company. They don't have any experience, and he will pay each of them $6.00/hour.

Mr. Anstine has received a job that will involve painting the outside of a house, including the trim. This job will take approximately 45 hours of time for one person to do it. Of this time, 15 hours would be in trim work. Brittany and José are the two best trim painters of the experienced students. Who should Mr. Anstine put on the paint and trim jobs in order to get the best work but at a low cost to him? He uses several students for each job in order to get the job done in less days.

Follow these steps to accomplish your task.

- Organize the data that you have.

- Determine a rational model to describe the amount of time it will take for various combinations of students to work.

- Determine what each of these combinations will cost Frank to pay the workers.

- Determine what needs to be changed in your model when changing the worker combinations.

- Use graphs to help show your conclusion.

- Write a paragraph describing the problem and your solution.

THINKING CRITICALLY

- Explain why, in the definition of joint variation, there is the restriction that $x \neq 0$ and $z \neq 0$.

- Are $\frac{1}{4x} = \frac{1}{7x}$ and $4x = 7x$ equivalent equations? Explain your answer. What happens when you multiply each side of the first equation by the LCD?

PORTFOLIO

Write an application problem for each of the three types of variations: direct, inverse, and joint. Solve each problem and then describe why your problems apply to each of the variations and what your answers mean in reference to the problems. Place this in your portfolio.

SELF EVALUATION

Comparing various methods and outcomes can help when making a decision about a problem. To compare is to examine the qualities of two or more things in order to discover similarities or differences.

Assess yourself. Do you use comparison when determining a strategy to use? Do you compare time, efficiency, and completeness of a method when determining how to solve a problem? Give an example of two different methods that can be used on a certain math and/or daily life problem. Write an explanation of the comparison of the two methods and which one would be best to use for that specific problem.

In·ves·ti·ga·tion

MATERIALS NEEDED

- centimeter ruler
- goggles
- masking tape
- meterstick or measuring tape
- washers
- packaging tape
- paper clip
- rubber bands (3 different sizes)
- small paper cup
- wooden dowels

Have you ever considered bungee jumping? Modern bungee jumping dates back to April Fool's Day 1979 when members of the Oxford Dangerous Sports Club of Britain jumped off the 245-foot Clifton Bridge in Bristol, England, with bungee cords attached to their ankles. They were wearing tuxedos and top hats in honor of the occasion.

Suppose you work for a sporting goods manufacturer. Your company has decided to produce bungee jumping equipment. They believe the sport will "take off." You will lead a team of engineers who will investigate the specifications and equipment needed to produce bungee jumping equipment. You have been asked to create a detailed report of your findings. You will need to conduct research experiments before beginning the design process.

You decide that you need to conduct a number of experiments on the elasticity of the bungee. In the first experiment you will shoot a rubber band bungee and measure how far it will fly when stretched various lengths. You will also explore the maximum spring potential of the bungee. Realizing that the weight of a person is also a factor, you will experiment with weight and the springiness of the materials used.

In this Investigation, you will conduct research experiments and then analyze the data to formulate a detailed report. Your design team consists of three people. For the protection of everyone in the classroom, each person must wear protective goggles or eyeglasses during the experiments.

Make an Investigation Folder in which you can store all of your work on this Investigation for future use.

DISTANCE STRETCHED (NEAREST CENTIMETER)	DISTANCE SHOT (NEAREST TENTH OF A METER)	DIFFERENCE FROM PREVIOUS SHOT
(natural state)	0	0.0

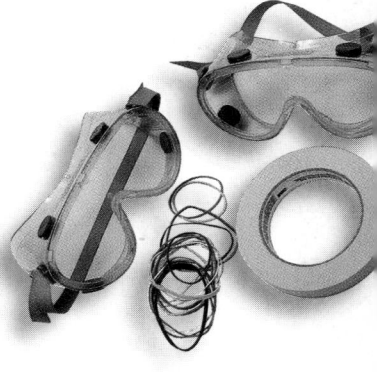

SETUP

1 Make three copies of the table. Select three rubber bands of different sizes.

2 Use masking tape to mark a starting line for your group on the floor. Mark a parallel line five meters away.

3 Select one rubber band for the first set of experiments. Measure the length of the rubber band in centimeters in its natural state (not stretched). Record this length in the first row of column 1.

4 Place one end of the rubber band over the end of a centimeter ruler. Hold the ruler chest-high, parallel to the floor, with the shooting edge directly above the starting line marked on the floor. Pull back the rubber band to a little past its natural state. For example, if a rubber band measures 5 centimeters in its natural state, then pull the rubber band so it stretches to 6 centimeters.

DATA COLLECTION

5 Use the table to record the length to which the rubber band was stretched. Measure how far the rubber band flew by measuring the distance from the starting line to the spot where the rubber band landed. Use a meterstick or measuring tape to measure the distance shot to the nearest tenth of a meter. Record this distance in column 2. Note that the first measurement will be in centimeters and the second in meters.

6 Shoot the rubber band again, increasing the distance the rubber band is stretched by 1 centimeter and measuring the distance the rubber band travels. Record the measurements in columns 1 and 2 of your table.

7 Continue to shoot the rubber band, increasing the distance it is stretched by 1 centimeter for each trial, until the rubber band will no longer stretch or it breaks. Record the distances in column 2. *Column 3 will be used later in the Investigation.*

8 Repeat this experiment using the other two rubber bands. Record the data in your other two tables.

You will continue working on this Investigation throughout Chapters 10 and 11.

Be sure to keep your tables, graphs, and other materials in your Investigation Folder.

Boing!! Investigation

Working on the Investigation Lesson 10–3, p. 616

Working on the Investigation Lesson 10–6, p. 630

Working on the Investigation Lesson 11–1, p. 655

Working on the Investigation Lesson 11–5, p. 682

Closing the Investigation End of Chapter 11, p. 702

Exploring Exponential and Logarithmic Functions

Objectives

In this chapter, you will:

- simplify expressions and solve equations involving real exponents,
- write exponential equations in logarithmic form and vice versa,
- evaluate expressions and solve equations involving logarithms,
- find common and natural logarithms and antilogarithms, and
- solve equations with variable exponents by using logarithms.

AIDS Cases Worldwide

Numbers in Millions

Source: *Weekly Epidemiological Record*

Many teenagers think they are immortal. When the news mentions tragedy, whether violence involving guns, traffic fatalities, or death from AIDS, many teens believe it will never happen to them. The best remedy for this "live forever" attitude is knowledge. Information about firearm safety, teen driving statistics, and the AIDS epidemic is vital for your own personal safety.

TIME*Line*

580 B.C. Greek mathematician and philosopher Pythagoras is born.

1094 Gondolas are introduced in Venice.

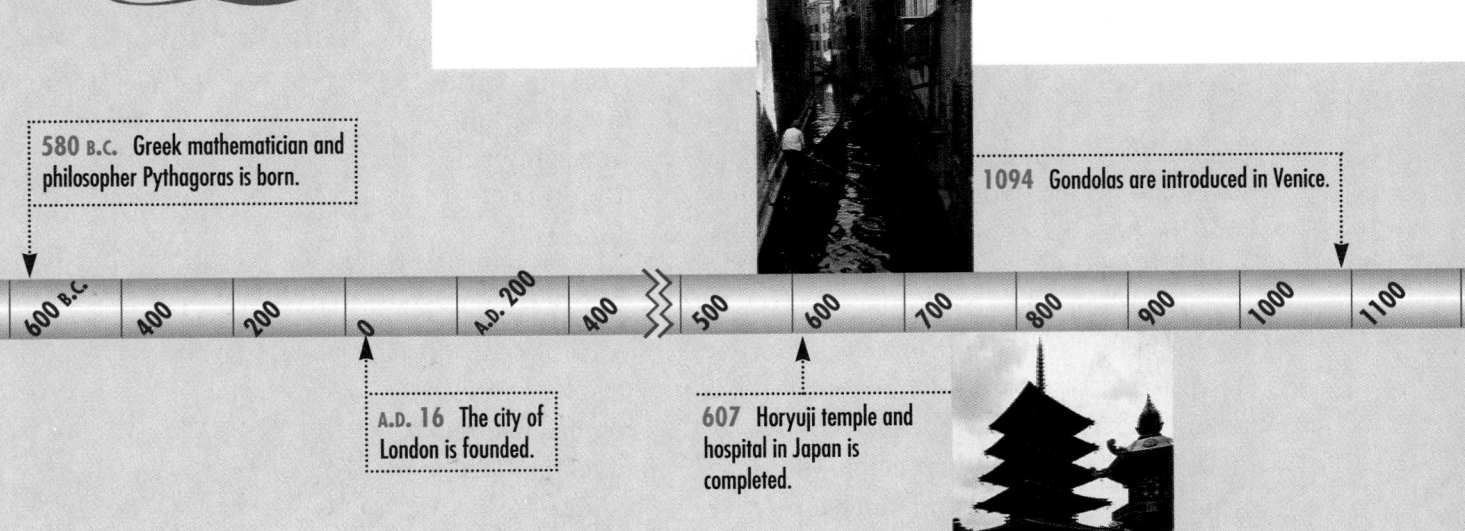

A.D. 16 The city of London is founded.

607 Horyuji temple and hospital in Japan is completed.

Chapter Project

Research the number of AIDS cases in the United States for each year since 1980.

- Draw a graph for these data.
- Use the coordinates of two points on your graph to write an exponential equation to model the growth in the number of AIDS cases.

- Use your equation to predict the number of AIDS cases in the United States in the year 2010.
- Write a paper about the research being done concerning AIDS. Are there any prospects for a cure for this disease? What can be done to slow the growth in the number of cases each year?

Neil Willenson is a 25-year-old political activist from Wisconsin who wanted to do something to help children with AIDS. He read Ryan White's book about his struggle with AIDS, and this encouraged him to start a summer camp for young children with the disease. He began raising funds. In 1993, Camp Heartland opened and 74 children had a free summer camp experience.

Since then, over a million dollars in private donations have helped hundreds of children with HIV or AIDS have fun at summer camp. Neil was awarded the $25,000 Arthur Ashe Award which he donated to Camp Heartland. He believes "kids need a safe haven where they can go and feel 100% accepted."

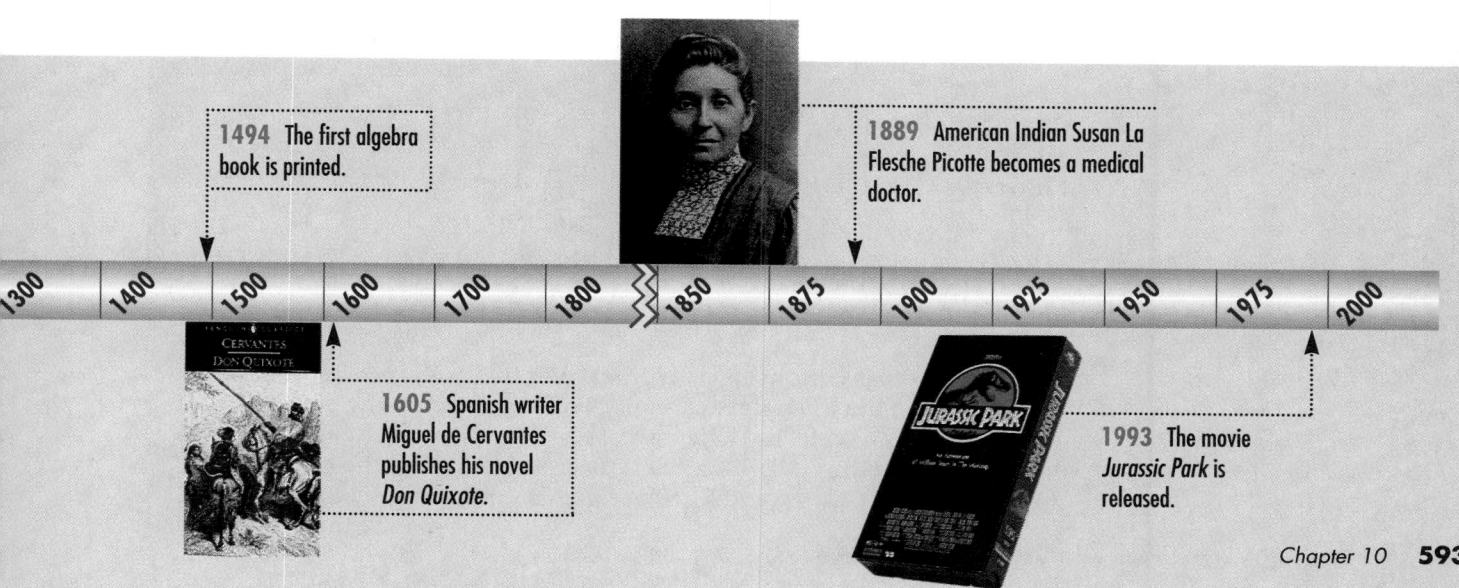

1494 The first algebra book is printed.

1889 American Indian Susan La Flesche Picotte becomes a medical doctor.

1300   1400   1500   1600   1700   1800   1850   1875   1900   1925   1950   1975   2000

1605 Spanish writer Miguel de Cervantes publishes his novel *Don Quixote*.

1993 The movie *Jurassic Park* is released.

10–1A Graphing Technology
Exponential and Logarithmic Functions

A Preview of Lesson 10–1

You can draw graphs of exponential and logarithmic functions with a graphing calculator. An **exponential function** is a function of the form $y = a^x$, where $a > 0$ and $a \neq 1$. A **logarithmic function** is a function of the form $x = a^y$, where $a > 0$ and $a \neq 1$. It is denoted $y = \log_a x$ and is the inverse of the exponential function $y = a^x$.

Example Graph $y = 2^x$ and $y = \left(\frac{1}{2}\right)^x$ in the standard viewing window. Then describe any similarities and differences.

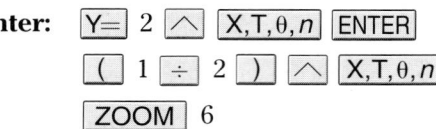

Enter: [Y=] 2 [∧] [X,T,θ,n] [ENTER]
[(] 1 [÷] 2 [)] [∧] [X,T,θ,n]
[ZOOM] 6

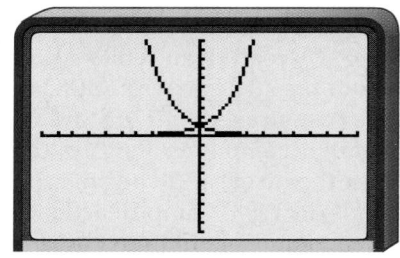

The two graphs are reflections of each other with the *y*-axis serving as the axis of symmetry. Both graphs pass through the point (0, 1), and both have the *x*-axis as a horizontal asymptote.

TECHNOLOGY *Tips*

Remember to clear the Y = list before entering a new equation.

The TI-83 has $y = \log_{10} x$ as a built-in function. Enter [Y=] [LOG] [X,T,θ,n] [GRAPH] to view this graph.

To graph logarithmic functions with bases other than 10, you must use the **change of base formula,** $\log_a x = \dfrac{\log_{10} x}{\log_{10} a}$. *You will use the change of base formula in Lesson 10–6.*

Example Graph $y = 2^x$ and $y = \log_2 x$ in the standard viewing window. Then describe any similarities and differences.

Use the change of base formula to graph $y = \log_2 x$.

$$\log_2 x = \frac{\log_{10} x}{\log_{10} 2}$$

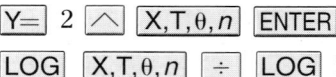

Enter: [Y=] 2 [∧] [X,T,θ,n] [ENTER]
[LOG] [X,T,θ,n] [÷] [LOG] 2
[ZOOM] 6

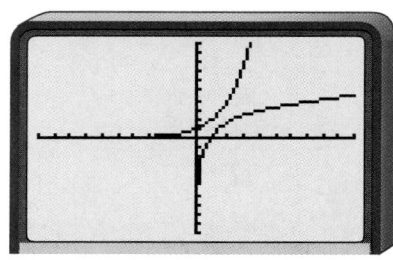

Since these two functions are inverses, the domain of $y = 2^x$, real numbers, is the range of $y = \log_2 x$, and the range of $y = 2^x$, positive real numbers, is the domain of $y = \log_2 x$.

The graphs are reflections of each other over the line $y = x$. TRACE each function and examine the points carefully. For any point on the graph of $y = \log_2 x$, there is a point on the graph of $y = 2^x$ whose coordinates are reversed. For example, (1, 0), (4, 2), and (8, 3) are points on the graphs of $y = \log_2 x$, and (0, 1), (2, 4), and (3, 8) are points on the graph of $y = 2^x$.

There are two methods that can be used with a graphing calculator to find a solution to exponential equations.

Method 1: Graph each side of the equation as a separate function and estimate the point at which they intersect.

Method 2: Rewrite the equation so that one side equals zero. Graph the related function and find the x-intercept.

Example ❸ Find the solution of $3^x = 2^{5x-1}$ to the nearest hundredth. Use the viewing window [−3, 3] by [−3, 3].

Use the second method to solve the equation.
Rewrite the equation as $3^x - 2^{5x-1} = 0$. Graph the related function $y = 3^x - 2^{5x-1}$.

Enter: $\boxed{\text{Y=}}$ 3 $\boxed{\wedge}$ $\boxed{\text{X,T,}\theta\text{,}n}$

$\boxed{-}$ 2 $\boxed{\wedge}$ $\boxed{(}$ 5 $\boxed{\text{X,T,}\theta\text{,}n}$

$\boxed{-}$ 1 $\boxed{)}$ $\boxed{\text{GRAPH}}$

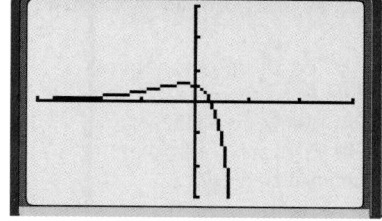

Now use ZOOM-IN or ROOT to approximate the x-intercept.

The solution is about 0.29.

Check this solution using the first method.

Enter: $\boxed{\text{Y=}}$ 3 $\boxed{\wedge}$ $\boxed{\text{X,T,}\theta\text{,}n}$ $\boxed{\text{ENTER}}$

$\boxed{\text{Y=}}$ 2 $\boxed{\wedge}$ $\boxed{(}$ 5 $\boxed{\text{X,T,}\theta\text{,}n}$ $\boxed{-}$

1 $\boxed{)}$ $\boxed{\text{GRAPH}}$

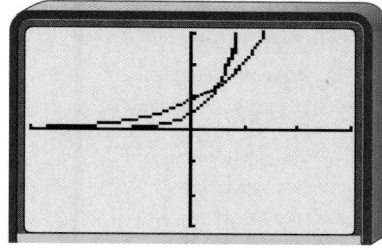

Use ZOOM-IN or INTERSECT to find the approximate value of x at the point of intersection.

The solution of 0.29 is correct.

EXERCISES

Use a graphing calculator to obtain a complete graph of each function. Then sketch the graph on a sheet of paper.

1. $y = 10^x$　　　　　　　　　　　2. $y = 3.5^x$

3. $y = 0.1^x$　　　　　　　　　　　4. $y = 0.05^x$

5. $y = \log_4 x$　　　　　　　　　　6. $y = \log_{0.3} x$

Solve each equation graphically. Round solutions to the nearest hundredth.

7. $4^x = 8$　　　　　　　　　　　8. $3.2^x = 52.5$

9. $2.1^{x-5} = 9.7$　　　　　　　　10. $0.65^{x+3} = 3^{2x-1}$

11. $2^x = x^2$　　　　　　　　　　12. $1.5^x = 2500$

13. $\log_{10} x = 0.23$　　　　　　　14. $\log_2 (x + 2) = \log_{0.5} 2$

15. $\log_9 (x + 4) = \log_2 x$　　　　16. $3^{4x-7} = 4^{2x+3}$

Real Exponents and Exponential Functions

What YOU'LL LEARN

• To simplify expressions and solve equations and inequalities involving real exponents.

Why IT'S IMPORTANT

You can use exponential functions to solve problems involving disease control and animal behavior.

APPLICATION
Disease Control

In the movie *Outbreak*, a dreadful disease is predicted to spread across the United States in just a matter of days. How could a disease spread so quickly?

Suppose that every 8 hours, a sick person infects 2 people before the disease is diagnosed and he or she is quarantined. Consider each 8 hours as one time period. At the beginning, one individual is infected. During the first time period, this person infects 2 people. During the second time period, the first person is quarantined, but the 2 people he or she infected each infect 2 more people. During the third time period, the 2 people are quarantined, and the 4 people they infected are each responsible for infecting 2 more people. During the third time period, 4×2 or 8 people are infected.

This pattern can be summarized in the table at the right. The number of people infected during a time period y can be expressed as a function of time where x is the number of 8-hour periods. This function, $y = 2^x$, is an **exponential function**.

During the end of the 21st time period (the end of the first week), 2^{21} or 2,097,152 new people will be infected!

Time Period	Number Infected	Pattern
0	1	2^0
1	$1 \times 2 = 2$	2^1
2	$2 \times 2 = 4$	2^2
3	$4 \times 2 = 8$	2^3
4	$8 \times 2 = 16$	2^4
.	.	.
.	.	.
.	.	.
x	y	2^x

Let's take a close look at the graph of $y = 2^x$. Make a table of values to help draw the curve. *Note that negative values of x have no meaning in the example above.*

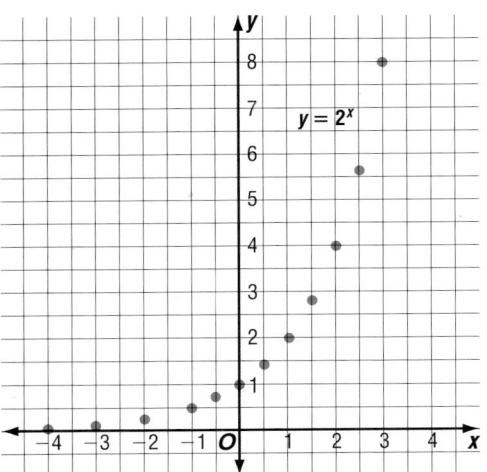

x	2^x or y	y
-4	$2^{-4} = \frac{1}{16}$	0.06
-3	$2^{-3} = \frac{1}{8}$	0.13
-2	$2^{-2} = \frac{1}{4}$	0.25
-1	$2^{-1} = \frac{1}{2}$	0.50
$-\frac{1}{2}$	$2^{-\frac{1}{2}} = \frac{1}{2}\sqrt{2}$	0.71
0	$2^0 = 1$	1

x	2^x or y	y
$\frac{1}{2}$	$2^{\frac{1}{2}} = \sqrt{2}$	1.41
1	$2^1 = 2$	2
$\frac{3}{2}$	$2^{\frac{3}{2}} = 2\sqrt{2}$	2.83
2	$2^2 = 4$	4
$\frac{5}{2}$	$2^{\frac{5}{2}} = 4\sqrt{2}$	5.66
3	$2^3 = 8$	8

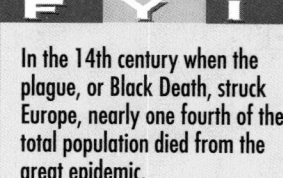
Since 2^x has not been defined when x is irrational, there are "holes" in the graph of $y = 2^x$. We can expand the domain of $y = 2^x$ to include irrational numbers. The domain will then include all real numbers. Draw the graph of $y = 2^x$ with no "holes" in the graph. The set of points on the graph is complete, or *continuous*, and the graph is a smooth curve. You can use the graph of $y = 2^x$ to estimate the value of 2^x when x is any real number.

Use the graph to estimate the value of $2^{\sqrt{3}}$, which is the value of y when $x = \sqrt{3}$.

$1.7 < \sqrt{3} < 1.8$ since $\sqrt{3} \approx 1.732$.

From the graph, the value of y is approximately 3.3.

Use a calculator to check this answer.

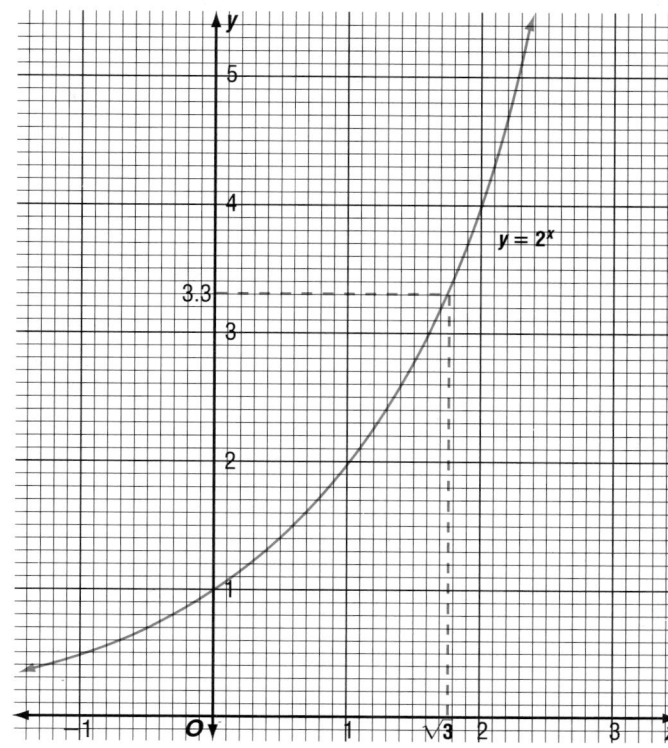

Enter: $\quad 2 \;\boxed{y^x}\; 3 \;\boxed{\sqrt{x}}$

$\boxed{=} \quad 3.32199709$

The calculator verifies the estimation from the graph.

$2^{\sqrt{3}} \approx 3.3$

All properties of rational exponents apply to real exponents.

Example **1** Simplify each expression.

a. $7^{\sqrt{2}} \cdot 7^{\sqrt{3}}$

$7^{\sqrt{2}} \cdot 7^{\sqrt{3}} = 7^{\sqrt{2} + \sqrt{3}}$ *Product of powers property*
You can use your calculator to verify the result.

b. $\left(8^{\sqrt{3}}\right)^{\sqrt{5}}$

$\left(8^{\sqrt{3}}\right)^{\sqrt{5}} = 8^{\sqrt{3} \cdot \sqrt{5}}$ *Power of a power property*

$= 8^{\sqrt{15}}$

In general, an exponential function can be written in the form $y = ab^x$.

| **Definition of Exponential Function** | An equation of the form $y = a \cdot b^x$, where $a \neq 0$, $b > 0$, and $b \neq 1$, is called an **exponential function with base** b. |

A family of exponential functions is graphed at the right. Notice that $a = 1$ in each case.

$y = 7^x$ *When $b > 1$, the value of y increases as the value of x increases.*

$y = 2^x$

$y = \left(\dfrac{3}{4}\right)^x$ *When $0 < b < 1$, the value of y increases as the value of x decreases.*

$y = \left(\dfrac{1}{2}\right)^x$

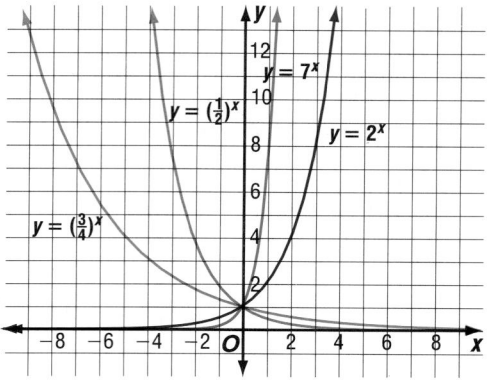

Another family of exponential functions is graphed at the left. Study this family of graphs.

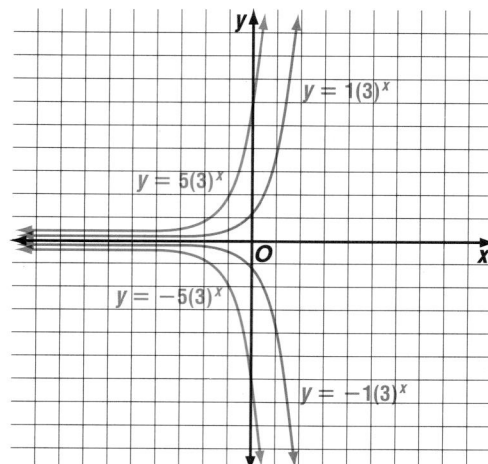

$y = 1(3)^x$
$y = 5(3)^x$ *Notice that the y-intercept*
$y = -1(3)^x$ *of each graph equals the*
$y = -5(3)^x$ *value of a.*

The graph of $y = a \cdot b^x$ is the reflection of $y = -a \cdot b^x$ across the x-axis.

Exponential functions are frequently used to model population growth.

Example ❷   During the nineteenth century, rabbits were brought to Australia. Since the rabbits had no natural enemies on that continent, their population increased rapidly. Suppose there were 65,000 rabbits in Australia in 1865 and 2,500,000 rabbits in 1867.

APPLICATION
Animal Control

a. **Write an exponential equation that could be used to model the rabbit population in Australia. Write the equation in terms of the number of years elapsed since 1865.**
b. **Estimate the Australian rabbit population in 1872.**

a. For the year 1865, the time t equals 0, and the initial population p equals 65,000. Substitute these values in the standard exponential equation to find the value of a.

$$p = ab^t$$
$$65,000 = ab^0 \quad \text{\textit{Replace p with 65,000 and t with 0.}}$$
$$65,000 = a(1) \quad b^0 = 1$$
$$65,000 = a$$

For the year 1867, the time t equals 2, and the population p equals 2,500,000. Substitute these values in the standard exponential equation to find the value of b.

$$p = ab^t$$
$$2,500,000 = 65,000b^2 \quad \text{\textit{Replace p with 2,500,000, a with 65,000, and t with 2.}}$$
$$38.46 \approx b^2 \quad \text{\textit{Division property of equality}}$$
$$6.20 \approx b$$

The equation that models the rabbit population is $p = 65,000(6.20)^t$.

b. For the year 1872, the time t equals 7.

$$p = 65{,}000(6.20)^t$$
$$= 65{,}000(6.20)^7$$
$$\approx 22{,}890{,}495{,}000$$

According to the equation, the rabbit population was about 22,890,495,000 in 1872.

The following property is very useful when solving equations involving exponential functions.

Property of Equality for Exponential Functions	Suppose b is a positive number other than 1. Then $b^{x_1} = b^{x_2}$ if and only if $x_1 = x_2$.

This property also holds for inequalities.

Example 3   **a.** Solve $64 = 2^{3n+1}$.

$$64 = 2^{3n+1}$$
$$2^6 = 2^{3n+1}$$
$$6 = 3n + 1$$
$$\frac{5}{3} = n$$

The solution is $\frac{5}{3}$.

Check: $64 = 2^{3n+1}$
$$64 \stackrel{?}{=} 2^{3\left(\frac{5}{3}\right)+1}$$
$$64 \stackrel{?}{=} 2^{5+1}$$
$$64 \stackrel{?}{=} 2^6$$
$$64 = 64 \checkmark$$

b. Solve $6^{2n-1} > \frac{1}{216}$.

$$6^{2n-1} > \frac{1}{216}$$
$$6^{2n-1} > 6^{-3}$$
$$2n-1 > -3$$
$$n > -1$$

The solution is $n > -1$.

Check: Try $n = 0$.
$$6^{2n-1} > \frac{1}{216}$$
$$6^{2(0)-1} \stackrel{?}{>} \frac{1}{216}$$
$$\frac{1}{6} > \frac{1}{216} \checkmark$$

CHECK FOR UNDERSTANDING

Communicating Mathematics

Study the lesson. Then complete the following.

1. **You Decide**   Todd says that $y = x^2$ is an exponential function. Juan disagrees. Who is correct? Explain.

2. **Describe** the domains and ranges of functions of the form $y = a \cdot b^x$ when:
   a. $a = 1, b > 0$.   **b.** $a > 1, b > 0$.   **c.** $a < 1, b > 0$.

3. **Compare** the graphs of $y = 2^{-x}$, $y = \left(\frac{1}{2}\right)^x$, and $y = (0.5)^x$.

4. **Name** the y-intercepts for the graph of $y = b^x$, where b is any real number.

5. **Explain** why 1 is excluded as a value for a in the property of equality for exponential functions.

6. **Explain** how you could use a calculator to determine which of the following values is the best approximation for the value of x in the equation $3 = 2.3^x$. Give the best approximation.
   a. 1.2   **b.** 1.3   **c.** 1.4   **d.** 1.5

Guided Practice

Use the graph of $y = 2^x$ on page 597 or a calculator to approximate each expression to the nearest tenth.

7. $2^{\sqrt{5}}$

8. $4^{0.6}$

Use the rule of exponents to simplify each expression.

9. $5^{\sqrt{2}} \cdot 5^{3\sqrt{2}}$

10. $\left(3^{\sqrt{5}}\right)^{\sqrt{5}}$

11. $27^{\sqrt{5}} \div 3^{\sqrt{5}}$

Find the value of a if the graph of an exponential function of the form $y = a \cdot 2^x$ passes through the given point.

12. $A(2, 12)$

13. $B(3, -16)$

Solve each equation or inequality.

14. $3^n = 81$

15. $2^{2n} \le \dfrac{1}{16}$

16. $\left(\dfrac{1}{7}\right)^{b-3} = 343$

17. **Biology** Mitosis is a process of cell duplication in which one cell divides into two. The *Escherichia coli* is one of the fastest growing bacteria. It can reproduce itself in 15 minutes. If you begin with one *Escherichia coli* cell, how many cells will there be in one hour?

Escherichia coli

EXERCISES

Practice

Use the graph of $y = 2^x$ on page 597 or a calculator to evaluate each expression to the nearest tenth.

18. $2^{1.7}$

19. $2^{-0.5}$

20. $2^{\sqrt{2}}$

21. $2^{-1.1}$

22. $16^{0.4}$

23. $8^{-0.3}$

Simplify each expression.

24. $\left(2^{\sqrt{2}}\right)^{\sqrt{8}}$

25. $4^{\sqrt{2}} \cdot 4^{2\sqrt{2}}$

26. $7^{3\sqrt{2}} \div 7^{\sqrt{2}}$

27. $\left(y^{\sqrt{3}}\right)^{\sqrt{12}}$

28. $5^{\sqrt{3}} \cdot 5^{\sqrt{27}}$

29. $64^{\sqrt{7}} \div 2^{\sqrt{7}}$

30. $2\left(3^{\sqrt{2}}\right)\left(3^{-\sqrt{2}}\right)$

31. $\left(a^{\sqrt{5}}\right)^{\sqrt{20}}$

32. $\left(m^{\sqrt{3}} + n^{\sqrt{2}}\right)^2$

Find the value of a if the graph of an exponential function of the form $y = a \cdot 3^x$ passes through the given point.

33. $C(2, 36)$

34. $D(-1, 15)$

35. $E(4, -81)$

36. $F(-2, -2)$

37. $G(5, 27)$

38. $H\left(-3, \dfrac{1}{9}\right)$

Solve each equation or inequality.

39. $3^{4x} = 3^{3-x}$

40. $5^{n-3} \ge \dfrac{1}{25}$

41. $\dfrac{1}{32} = 2^{1-m}$

42. $9^{2p} = 27^{p-1}$

43. $16^n > 8^{n+1}$

44. $\left(\dfrac{1}{9}\right)^m = 81^{m+4}$

45. $2^x \cdot 4^{x+5} = 4^{2x-1}$

46. $2^{5x} \cdot 16^{1-x} = 4^{x-3}$

47. $25^x = 5^{x^2-15}$

48. On the same coordinate plane, graph $y = 4^x$, $y = -(4)^x$, and $y = \left(\dfrac{1}{4}\right)^x$. Compare the graphs.

Use a graphing calculator to graph each of the following.

49. $y = 2.1^x$ **50.** $y = 0.5(2.1)^x$ **51.** $y = -0.2(2.1)^x$

Critical Thinking

52. Using a calculator, approximate to the nearest tenth the value of x when $y = 2$ in the expression $y = 2.3^x$.

Applications and Problem Solving

53. Look for a Pattern A large piece of paper is cut in half, and one of the resulting pieces is placed on top of the other. Then the pieces in the stack are cut in half and placed on top of each other. Suppose this procedure is repeated several times.

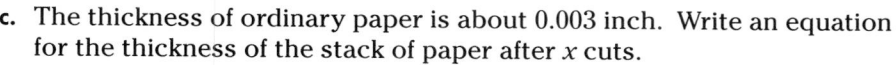

   a. How many pieces will be in the stack after the first cut? after the second cut? after the third cut? after the fourth cut?

   b. Use the pattern in step a to write an equation for the number of pieces in the stack after x cuts.

   c. The thickness of ordinary paper is about 0.003 inch. Write an equation for the thickness of the stack of paper after x cuts.

   d. How thick will the stack of paper be after 30 cuts?

54. Animal Behavior Studies show an animal will defend an area in square yards that is directly proportional to the 1.31 power of the animal's weight in pounds.

   a. If a 45-pound beaver will defend 170 square yards, write an equation for the area a defended by a beaver weighing w pounds.

   b. Thousands of years ago, some beavers grew to be 11 feet long and weighed 430 pounds. Use your equation in step a to determine the area defended by these animals.

55. Atmospheric Pressure Atmospheric pressure decreases at higher altitudes. In the equation $P = 14.7(10)^{-0.02h}$, P represents the atmospheric pressure in pounds per square inch, and h represents the altitude above sea level in miles.

   a. The elevation of Boston, Massachusetts, is about sea level. Find the atmospheric pressure in Boston.

   b. The elevation of Denver, Colorado, is about 1 mile. Find the atmospheric pressure in Denver.

   c. The highest mountain in the world is Mt. Everest in Nepal, which is about 5.5 miles above sea level. Find the atmospheric pressure at the top of this mountain.

   d. About 6 million years ago, the Mediterranean Sea dried up. The bottom of the valley formed by this dry sea was about 1.9 miles below sea level. Find the atmospheric pressure at the bottom of this valley.

   e. Graph the equation for the atmospheric pressure. Explain the meaning of the points on the graph that have negative values for h.

56. **City Planning** The school board in Orlando, Florida, needs to project the population growth for the remainder of the century to plan for the construction of new schools. One of the formulas the board can use is $P = 164{,}693(2.7)^{0.007t}$, where t represents the number of years since 1990 and 164,693 was the population of Orlando according to the 1990 census. Use a calculator to determine how large the population will be in the year 2000.

57. **Forestry** The diameter of the base of a tree trunk in centimeters varies directly with the $\frac{3}{2}$ power of its height in meters.

   a. A young sequoia tree is 6 meters tall, and the diameter of its base is 19.1 centimeters. Use this information to write an equation for the diameter d of the base of a sequoia tree if its height is h meters high.

   b. One of the oldest living things on Earth is the General Sherman Tree in Sequoia National Park in California. This sequoia is between 2200 and 2500 years old. If it is about 83.8 meters high, find the diameter at its base.

Mixed Review

58. Solve $\dfrac{6}{a-7} = \dfrac{a-49}{a^2-7a} + \dfrac{1}{a}$. (Lesson 9–5)

59. State the number of positive real zeros, negative real zeros, and imaginary zeros for $f(x) = -x^4 - x^2 - x - 1$. (Lesson 8–4)

60. **Communication** A microphone is placed at the focus of a parabolic reflector to collect sounds for the television broadcast of a football game. The focus of the parabola that is the cross section of the reflector is 5 inches from the vertex. The latus rectum is 20 inches long. Assuming that the focus is at the origin and the parabola opens to the right, write the equation of the cross section. (Lesson 7–2)

61. **Cartography** Edison is located at (9, 3) on the road map. Kettering is located at (12, 5) on the same map. Each side of a grid on the map represents 10 miles. Use the distance formula to approximate the distance between Edison and Kettering. (Lesson 7–1)

62. Solve $x^2 + 14x - 12 = 0$ by completing the square. (Lesson 6–3)

63. Find the multiplicative inverse of $3 - 4i$. (Lesson 5–10)

64. Simplify $\dfrac{(3+\sqrt{5})}{(1+\sqrt{2})}$. (Lesson 5–6)

65. If $A = \begin{bmatrix} -2 & 3 \\ 1 & 10 \\ 0 & -6 \end{bmatrix}$ and $B = \begin{bmatrix} 9 & 3 \\ 1 & 4 \end{bmatrix}$, find AB. (Lesson 4–3)

66. Solve the system of equations. (Lesson 3–7)

   $r + s + t = 15$
   $r + t = 12$
   $s + t = 10$

67. Use the vertical line test to determine whether the relation graphed at the right is a function. (Lesson 2–1)

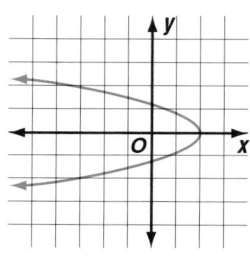

68. State the property illustrated by the following equation. $4 + (a + r) = (4 + a) + r$ (Lesson 1–2)

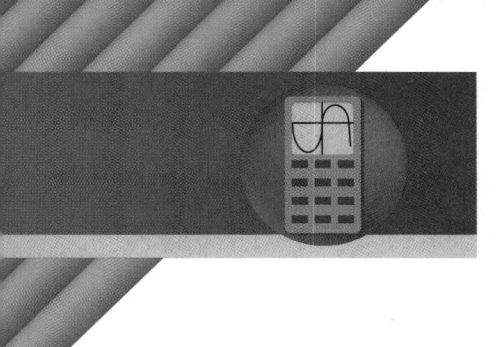

10–1B Graphing Technology
Curve Fitting with Real-World Data

An Extension of Lesson 10–1

As we have seen in earlier chapters, we are often confronted with data for which we need to find an equation that best describes the information.

Example

In 1985, Kayla received $30.00 from her grandparents for her fifth birthday. Her mother deposited it into a bank account for her. Both Kayla and her mother forgot about the money and made no further deposits or withdrawals. The table to the right shows the account balance for several years.

Elapsed Time (years)	Balance
0	$30.00
5	$41.10
10	$56.31
15	$77.16
20	$105.71
25	$144.83
30	$198.43

a. Use a graphing calculator to enter the data and draw a scatter plot that shows how the account balance is related to time.
b. If Kayla discovers the account with the birthday money on her 50th birthday, how much will she have in the account?

a. Enter the elapsed time data into the L1 list on the STAT menu and enter the balance data into the L2 list. Be sure to clear the Y= list. Press 2nd [STAT PLOT] and select plot 1. Make sure that plot 1 is on, scatter plot is chosen, the *x* list is L1, and the *y* list is L2. Use the viewing window [0, 35] with a scale factor of 5 by [0, 300] with a scale factor of 20. Press GRAPH .

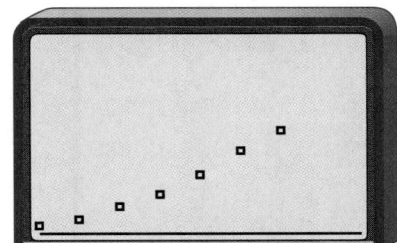

LOOK BACK

You can refer to Lessons 2-5B and 8-3B for information on entering data and graphing scatter plots on the graphing calculator.

We see from the data that the equation that best fits the data must be a curve. This means the equation is probably polynomial or exponential. Let's try an exponential model. To determine the exponential equation that best fits the data, use the exponential regression feature of the calculator.

Enter: STAT ▶ ALPHA A 2nd L1 , 2nd L2 ENTER

The equation is $y = 29.99908551(1.065001351)^x$.

The calculator also reports an *r*-value of 0.999999998. Recall that this number is a correlation coefficient that indicates how well the equation fits the data. A perfect fit would be $r = 1$. Therefore, we can conclude that this equation is indeed a good fit for the data.

(continued on the next page)

To check this equation visually, overlap the graph of the equation with the scatter plot.

Enter:  5

 7

GRAPH

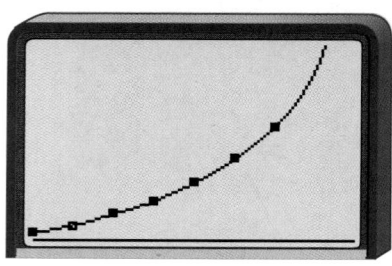

b. On Kayla's 50th birthday, the money will have been in the account for 50 − 5 or 45 years. From the graphics screen, enter 2nd CALC 1 45 ENTER. (Be sure your viewing window is large enough to include $x = 45$.) The calculator returns a y-value of 510.34651. Kayla will have $510.35 in the account when she is 50 years old.

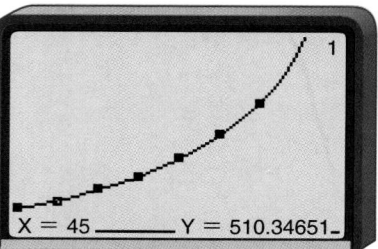
X = 45 _____ Y = 510.34651

EXERCISES

According to the World Almanac, the population per square mile in the United States has changed dramatically over a period of years.

Year	People per Square Mile	Year	People per Square Mile
1790	4.5	1890	17.8
1800	6.1	1900	21.5
1810	4.3	1910	26.0
1820	5.5	1920	29.9
1830	7.4	1930	34.7
1840	9.8	1940	37.2
1850	7.9	1950	42.6
1860	10.6	1960	50.6
1870	10.9	1970	57.5
1880	14.2	1980	64.0

1. Use a graphing calculator to draw a scatter plot of the data. Then calculate and graph the curve of best fit that shows how the year is related to the number of people per square mile. Use ExpReg for this example.

2. Write the equation of best fit. Write a sentence that describes the fit of the graph to the data.

3. Based on the graph, estimate the population density for 2000. Check this using the CALC value.

4. Do you think there are any other types of equations that would be good models for this data? Why or why not?

5. **History** What event occurred between 1800 and 1810 that would account for the sudden big decrease in population per square mile?

Logarithms and Logarithmic Functions

What YOU'LL LEARN

- To write exponential equations in logarithmic form and vice versa,
- to evaluate logarithmic expressions, and
- to solve equations and inequalities involving logarithmic functions.

Why IT'S IMPORTANT

You can use logarithmic functions to solve problems involving chemistry and geology.

APPLICATION
Geology

Logarithms are exponents. They were once used to simplify calculations, but the advent of calculators and computers caused calculation with logarithms to be used less and less.

An example of logarithms at work is the Richter scale. The Richter scale is used to measure the strength of an earthquake. It is a logarithmic scale based on the powers of ten. The table below gives the effects of earthquakes of various intensities.

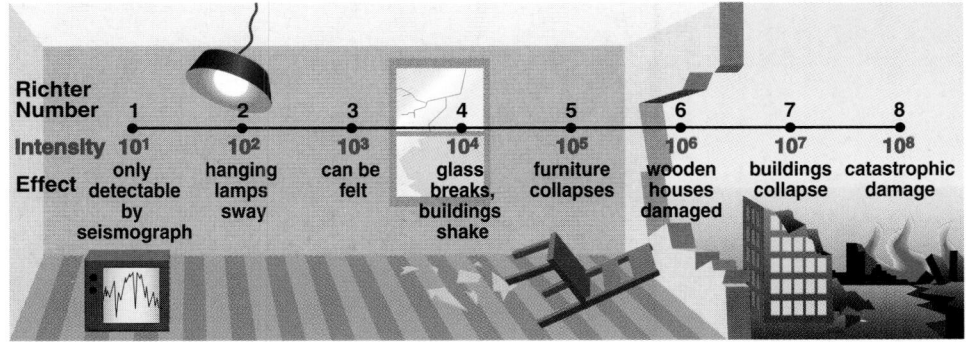

Richter Number	1	2	3	4	5	6	7	8
Intensity	10^1	10^2	10^3	10^4	10^5	10^6	10^7	10^8
Effect	only detectable by seismograph	hanging lamps sway	can be felt	glass breaks, buildings shake	furniture collapses	wooden houses damaged	buildings collapse	catastrophic damage

The 1906 San Francisco earthquake measured 8.3 on the Richter scale. The Loma Prieta earthquake that interrupted the 1989 World Series in San Francisco measured 7.1. *We will compare the magnitudes of these two famous earthquakes in Example 2.*

The tables below show two related exponential equations. You will recognize the equation in the table on the left as an exponential function.

Given the exponent, x, compute the power of 2 as y.

x	$2^x = y$	y
-1	$2^{-1} = y$	?
2	$2^2 = y$	?
3	$2^3 = y$	?
6	$2^6 = y$	?

Given x as the power of 2, compute the exponent, y.

y	$2^y = x$	x
?	$2^y = \frac{1}{2}$	$\frac{1}{2}$
?	$2^y = 4$	4
?	$2^y = 8$	8
?	$2^y = 64$	64

In the relation shown in the table on the right, $2^y = x$, the exponent y is called the **logarithm**, base 2, of x. This relation is written $\log_2 x = y$ and is read "the log base 2 of x is equal to y." The logarithm corresponds to the exponent. Study the diagram below.

Exponential Equation   **Logarithmic Equation**

$$n = b^p \qquad\qquad p = \log_b n$$

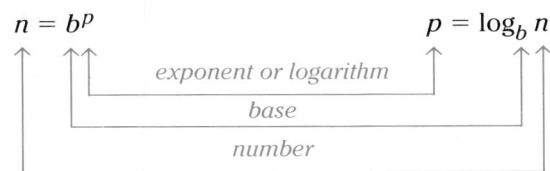

exponent or logarithm

base

number

Definition of Logarithm	Suppose $b > 0$ and $b \neq 1$. For $n > 0$, there is a number p such that $\log_b n = p$ if and only if $b^p = n$.

The chart below shows some equivalent exponential and logarithmic equations.

Exponential Equation	Logarithmic Equation
$5^2 = 25$	$\log_5 25 = 2$
$10^5 = 100{,}000$	$\log_{10} 100{,}000 = 5$
$8^0 = 1$	$\log_8 1 = 0$
$2^{-4} = \frac{1}{16}$	$\log_2 \frac{1}{16} = -4$
$9^{\frac{1}{2}} = 3$	$\log_9 3 = \frac{1}{2}$

You can find the value of a variable in a logarithmic equation $\log_b x = y$ when values for two of the variables are known.

Example **Solve each equation.**

a. $\log_9 x = \frac{3}{2}$

$\log_9 x = \frac{3}{2}$

$9^{\frac{3}{2}} = x$ *Definition of logarithm*

$(3^2)^{\frac{3}{2}} = x$

$3^3 = x$ *Power of a power*

$27 = x$

b. $\log_4 256 = y$

$\log_4 256 = y$

$4^y = 256$ *Definition of logarithm*

$(2^2)^y = 2^8$

$2^{2y} = 2^8$ *Power of a power*

$2y = 8$ *Property of equality for exponential functions*

$y = 4$

Example **2**

Geology

Refer to the application at the beginning of the lesson. Compare the magnitude of the 1906 quake to the 1989 quake.

Let x represent the measure of the 1906 quake and let y represent the measure of the 1989 quake.

Quake	Richter Value as Exponents	Richter Value as Logarithms
1906	$10^{8.3} = x$	$8.3 = \log_{10} x$
1989	$10^{7.1} = y$	$7.1 = \log_{10} y$

Use the Richter values as exponents to find the ratio of the magnitude of the 1906 quake to the 1989 quake.

$\frac{x}{y} = \frac{10^{8.3}}{10^{7.1}}$

$\quad = 10^{1.2}$ *Division of powers*

$\quad \approx 15.8$

Therefore, the intensity of the 1906 earthquake was approximately 16 times greater than that of the 1989 quake.

Let's look at the graphs of an exponential function and its corresponding logarithmic function. In fact, you can use a table of values for $y = 2^x$ to make a table of values for $x = 2^y$.

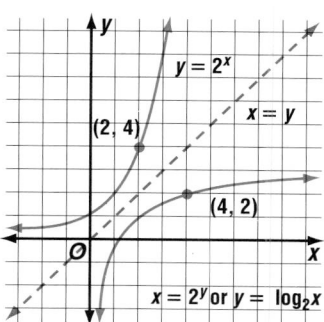

$y = 2^x$

x	y
-3	$\frac{1}{8}$
-2	$\frac{1}{4}$
-1	$\frac{1}{2}$
0	1
1	2
2	4
3	8

$x = 2^y$ or $y = \log_2 x$

x	y
$\frac{1}{8}$	-3
$\frac{1}{4}$	-2
$\frac{1}{2}$	-1
1	0
2	1
4	2
8	3

The x and y values are reversed.

For every point (a, b) on the graph of $y = 2^x$, there is a point on the graph of $y = \log_2 x$ with coordinates (b, a).

Notice that the graphs are reflections of each other over the line $y = x$. The fact that x and y switch places is apparent in the domains and ranges.

LOOK BACK

You can refer to Lesson 8-8 for information on inverse functions.

	For $y = 2^x$	For $x = 2^y$ or $y = \log_2 x$
Domain	all real numbers	positive real numbers
Range	positive real numbers	all real numbers

The relations are inverses of each other. Using the vertical line test, you can see that no vertical line can intersect the graph of $y = \log_2 x$ in more than one place, so $y = \log_2 x$ is a function, called a **logarithmic function**.

Definition of Logarithmic Function	An equation of the form $y = \log_b x$, where $b > 0$ and $b \neq 1$, is called a **logarithmic function**.

LOOK BACK

You can refer to Lesson 8-7 for information on composition of functions.

Since the exponential function $y = b^x$ and the logarithmic function $y = \log_b x$ are inverses of each other, their composites are the identity function. Let $f(x) = \log_b x$ and $g(x) = b^x$. For $f(x)$ and $g(x)$ to be inverses, it must be true that $f(g(x)) = x$ and $g(f(x)) = x$.

$$f(g(x)) = x \qquad\qquad g(f(x)) = x$$
$$f(b^x) = x \qquad\qquad g(\log_b x) = x$$
$$\log_b b^x = x \qquad\qquad b^{\log_b x} = x$$

Example ③ **Evaluate each expression.**

a. $\log_5 5^3$

b. $6^{\log_6(2x+5)}$

$\log_5 5^3 = 3$ $\log_b b^x = x$

$6^{\log_6(2x+5)} = 2x + 5$ $b^{\log_b x} = x$

A property similar to the property for exponential functions applies to the logarithmic functions.

Property of Equality for Logarithmic Functions	Suppose $b > 0$ and $b \neq 1$. Then $\log_b x_1 = \log_b x_2$ if and only if $x_1 = x_2$.

This property also holds for inequalities.

Example **4** **Solve each equation or inequality.**

a. $\log_{10} (t^2 - 6) = \log_{10} t$

$$\log_{10} (t^2 - 6) = \log_{10} t$$
$$t^2 - 6 = t \qquad \text{\textit{Property of equality for logarithmic functions}}$$
$$t^2 - t - 6 = 0$$
$$(t - 3)(t + 2) = 0$$
$$t = 3 \quad \text{or} \quad t = -2 \qquad \text{\textit{Zero product property}}$$

Eliminate -2, because $\log_b x$ is defined only if $x > 0$. Thus, the solution is 3.

b. $\log_3 (3x - 5) \geq \log_3 (x + 7)$

$$\log_3 (3x - 5) \geq \log_3 (x + 7)$$
$$3x - 5 \geq x + 7$$
$$2x \geq 12$$
$$x \geq 6$$

CHECK FOR UNDERSTANDING

Communicating Mathematics

Study the lesson. Then complete the following.

1. **Give** another name for the exponent y in the equation $2^y = x$.

2. **Write** an example of a logarithmic function.

3. **Describe** the domain of $y = 2^x$.

4. **Describe** the range of $y = \log_2 x$.

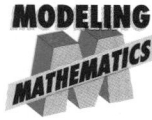

5. **Draw** a graph of $y = 3^x$. Place a geomirror along the line represented by the equation $y = x$. Use the reflection seen in the geomirror to draw the graph of $y = \log_3 x$.

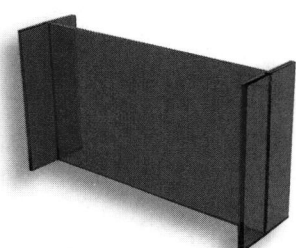

Guided Practice

Write each equation in logarithmic form.

6. $3^3 = 27$ 7. $2^5 = 32$ 8. $4^{-2} = \frac{1}{16}$

Write each equation in exponential form.

9. $\log_5 125 = 3$ 10. $\log_8 4 = \frac{2}{3}$ 11. $\log_{10} 0.001 = -3$

Evaluate each expression.

12. $\log_3 \frac{1}{27}$ 13. $\log_{16} 4$ 14. $5^{\log_5 25}$

Solve each equation or inequality.

15. $\log_7 y = -2$ 16. $\log_b 64 = 3$

17. $\log_{\frac{1}{3}} 27 = x$ 18. $\log_5 (2x - 3) > \log_5 (x + 2)$

19. $\log_{10} (x^2 + 36) = \log_{10} 100$ 20. $\log_3 3^{(2x - 1)} = 7$

21. **Geology** How much stronger is an earthquake with a Richter scale rating of 7 than an aftershock with a rating of 4?

Practice

Evaluate each expression.

22. $\log_{10} 1000$

23. $\log_5 25$

24. $\log_{14} 196$

25. $\log_3 \frac{1}{81}$

26. $\log_2 \frac{1}{128}$

27. $\log_{36} 6$

28. $\log_8 8^4$

29. $3^{\log_3 243}$

30. $7^{\log_7 (x+3)}$

Solve each equation or inequality.

31. $\log_2 x = 5$

32. $\log_3 27 = y$

33. $\log_b 9 = 2$

34. $\log_5 \sqrt{5} = y$

35. $\log_{25} x = \frac{3}{2}$

36. $\log_b 0.01 = -2$

37. $\log_{\frac{1}{10}} x = -3$

38. $\log_{3x} 125 = 3$

39. $\log_{x+2} 16 = 2$

40. $\log_8 (3x - 1) = \log_8 (2x^2)$

41. $\log_2 (4x + 10) - \log_2 (x + 1) < 3$

42. $\log_{10} (x^2 + 16) = \log_{10} 80$

43. $4^{\log_4 (x-1)} = -0.5$

44. $\log_2 2^{(3x+2)} = 14$

45. $3^{\log_3 10} = x$

46. $\log_{10} (\log_8 8) = x$

47. $\log_4 (\log_2 16) = y$

Graph each pair of equations on the same axes.

48. $y = \log_5 x$ and $y = 5^x$

49. $y = \log_{\frac{1}{3}} x$ and $y = \left(\frac{1}{3}\right)^x$

50. Study the graphs in Exercises 48 and 49. What is the relationship of each pair of graphs?

51. Graph $y = \log_{10} x$, $y = \log_5 x$, $y = \log_{\frac{1}{3}} x$ and $y = \log_{\frac{1}{2}} x$ on the same set of axes. Assume that the parent graph is the graph of $y = \log_{10} x$. Describe this family of graphs in terms of the parent graph.

Show that each statement is true.

52. $\log_4 4 + \log_4 16 = \log_4 64$

53. $\log_4 16 = 2 \log_4 4$

54. $\log_2 8 \cdot \log_8 2 = 1$

55. $\log_{10} [\log_3 (\log_4 64)] = 0$

Critical Thinking

56. If $x = \log_{10} 2460$, the value of x is between two consecutive integers. Name these integers and explain how you determined the values.

Applications and Problem Solving

57. Geology The Seattle quake of April 29, 1965, measured 7.0 on the Richter scale. The San Francisco quake of 1989 measured 7.1. How many times more severe was the San Francisco quake than the Seattle quake?

58. Chemistry The pH of a solution is a measure of its acidity and is written as a logarithm to the base 10. A low pH indicates an acidic solution, and a high pH indicates a basic solution. Neutral water has a pH of 7. Acid rain has a pH of 4.2. How many more times acidic is the acid rain than neutral water?

Mixed Review

59. Simplify $11^{\sqrt{5}} \cdot 11^{\sqrt{45}}$. (Lesson 10–1)

60. Physics The volume of any gas varies inversely with its pressure as long as the temperature remains constant. If a helium-filled balloon has a volume of 3.4 cubic decimeters at a pressure of 120 kilopascals, what is its volume at 101.3 kilopascals? (Lesson 9–2)

61. If $f = \{(2, 1), (-1, 6), (3, 2)\}$ and $g = \{(2, 2), (6, -1), (1, 5)\}$, express $f \circ g$ and $g \circ f$, if they exist, as sets of ordered pairs. *(Lesson 8–7)*

62. Write the equation of the hyperbola in the graph at the right. *(Lesson 7–5)*

63. Write a quadratic equation that has roots 6 and -6. *(Lesson 6–5)*

64. Solve $x^2 + 6x = -9$ by factoring. *(Lesson 6–2)*

65. **Physics** Find the time t (in seconds) that it takes for a free-falling object to fall a distance s of 200 feet. Use the formula $t = \frac{1}{4}\sqrt{s}$. *(Lesson 5–6)*

66. Evaluate the determinant of $\begin{bmatrix} 6 & 5 & -2 \\ -3 & 0 & 6 \\ 1 & 4 & 2 \end{bmatrix}$ *(Lesson 4–4)*

67. Solve the system of inequalities by graphing. *(Lesson 3–4)*

$y - x \leq 3$

$y \geq x - 2$

68. Find the y-intercept and the x-intercept of the graph of $3x + 5y = 30$. *(Lesson 2–3)*

69. Solve $8x + 5 < 7x - 3$. *(Lesson 1–6)*

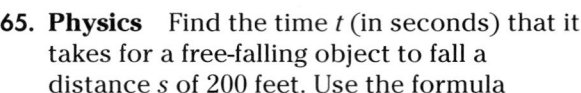

Mathematics and SOCIETY

Earthquake!

The excerpt below appeared in an article in *Science News* on October 15, 1994.

WHEN CHARLES RICHTER INVENTED the concept of seismic magnitude, he made it easy to compare earthquakes. Anyone who can count to 10 will recognize that a magnitude 7.0 shock packs a bigger punch than a 6.0 quake. But the question "How much bigger?" is not so easily answered. In the original definition of magnitude, a 1-point increase meant that peak waves recorded by a Wood-Anderson seismometer jumped by a factor of 10. . . . Seismologists themselves compare earthquakes using seismic moments. . . . But moments are expressed in unwieldy numbers, such as 2×10^{27} newton meters—clearly not an appealing figure for the public. Pat Jorgenson, a USGS spokeswoman in Menlo Park, Calif., says she would prefer to discuss quakes in terms of something people can comprehend. . . . In that vein, a magnitude 1.0 earthquake would equal roughly 6 ounces of TNT. For a magnitude 5.0, think of 1000 tons of TNT. . . . The largest recorded earthquake, of moment magnitude 9.5, in Chile in 1960, equaled about 3 billion tons of TNT. ■

1. Between one-unit intervals on the Richter scale, the magnitude of earthquake strength increases by a factor of 10. What is the mathematical name for this type of scale?

2. The largest earthquakes can be several billion times stronger than the smallest. What problems might arise if you tried to design a simpler, more understandable measuring scale?

3. Choose or invent your own unit of measurement to measure quakes.

Properties of Logarithms

10-3

CONNECTION
Chemistry

What YOU'LL LEARN

- To simplify and evaluate expressions using properties of logarithms, and
- to solve equations involving logarithms.

Why IT'S IMPORTANT

You can use logarithms to solve problems involving biology and medicine.

The levels of the acidity of the foods we eat is a concern to some health-conscious consumers. Most of the foods that we consume tend to be more acidic than basic. The pH scale measures acidity; a low pH indicates an acidic solution, and a high pH indicates a basic solution. It is another example of a logarithmic scale based on powers of ten.

pH Levels of Product	Common Products pH Level
Lemon juice	2.1
Sauerkraut	3.5
Tomatoes	4.2
Black coffee	5.0
Milk	6.4
Pure water	7.0
Eggs	7.8
Milk of magnesia	10.0

Black coffee has a pH of 5, while neutral water has a pH of 7. Black coffee is one hundred times more acidic than neutral water, since $10^{7-5} = 10^2$ or 100.

Since logarithms are exponents, the properties of logarithms can be derived from the properties of exponents that you already know. Recall that the product of powers is found by adding exponents.

$$\log_2 (8 \cdot 32) = \log_2 (2^3 \cdot 2^5) \qquad \log_2 8 + \log_2 32 = \log_2 2^3 + \log_2 2^5$$
$$= \log_2 (2^{3+5}) \qquad\qquad\qquad = 3 + 5$$
$$= 3 + 5$$

So, $\log_2 (8 \cdot 32) = \log_2 8 + \log_2 32$. This example indicates that the logarithm of a product is the sum of the logarithms of the factors.

Product Property of Logarithms	**For all positive numbers m, n, and b, where $b \neq 1$, $\log_b mn = \log_b m + \log_b n$.**

To prove this property, let $b^x = m$ and $b^y = n$.

Then $\log_b m = x$ and $\log_b n = y$.

$$b^x b^y = mn$$
$$b^{x+y} = mn \qquad \textit{Multiplying powers}$$
$$\log_b b^{x+y} = \log_b mn \qquad \textit{Property of equality for logarithmic functions}$$
$$x + y = \log_b mn \qquad \textit{Definition of inverse functions}$$
$$\log_b m + \log_b n = \log_b mn \qquad \textit{Replace x with } \log_b m \textit{ and y with } \log_b n.$$

Example Given $\log_3 5 \approx 1.4650$, find each logarithm.
a. $\log_3 45$ **b. $\log_3 25$**

$\log_3 45 = \log_3 (3^2 \cdot 5)$ $\log_3 25 = \log_3 (5 \cdot 5)$
$= \log_3 3^2 + \log_3 5$ $= \log_3 5 + \log_3 5$
$\approx 2 + 1.4650$ or 3.4650 $\approx 1.4650 + 1.4650$ or 2.9300

To find the quotient of powers, you subtract the exponents.

$$\log_2 \frac{32}{8} = \log_2 \frac{2^5}{2^3} \qquad\qquad \log_2 32 - \log_2 8 = \log_2 2^5 - \log_2 2^3$$
$$= \log_2 (2^{5-3}) \qquad\qquad\qquad\qquad\qquad = 5 - 3$$
$$= 5 - 3 \qquad\qquad\qquad\qquad\qquad\qquad\qquad = 2$$
$$= 2$$

So, $\log_2 \frac{32}{8} = \log_2 32 - \log_2 8$. This example indicates that the logarithm of a quotient can be found by subtracting the logarithm of the denominator from the logarithm of the numerator.

Quotient Property of Logarithms	**For all positive numbers m, n, and b, where $b \neq 1$, $\log_b \frac{m}{n} = \log_b m - \log_b n$.**

To prove this property, let $b^x = m$ and $b^y = n$.

Then $\log_b m = x$ and $\log_b n = y$.

$$\frac{b^x}{b^y} = \frac{m}{n}$$
$$b^{x-y} = \frac{m}{n} \qquad \textit{Division of powers}$$
$$\log_b b^{x-y} = \log_b \frac{m}{n} \qquad \textit{Property of equality for logarithmic functions}$$
$$x - y = \log_b \frac{m}{n} \qquad \textit{Definition of inverse functions}$$
$$\log_b m - \log_b n = \log_b \frac{m}{n} \qquad \textit{Replace x with } \log_b m \textit{ and y with } \log_b n.$$

Example ❷ **Given $\log_4 5 \approx 1.1610$ and $\log_4 15 \approx 1.9534$, find each logarithm.**

a. $\log_4 \frac{5}{16}$ **b. $\log_4 3$**

$$\log_4 \frac{5}{16} = \log_4 \frac{5}{4^2} \qquad\qquad \log_4 3 = \log_4 \frac{15}{5}$$
$$= \log_4 5 - \log_4 4^2 \qquad\qquad = \log_4 15 - \log_4 5$$
$$\approx 1.1610 - 2 \text{ or } -0.8390 \qquad \approx 1.9534 - 1.1610 \text{ or } 0.7924$$

Example ❸

Chemistry

The pH of a substance is the concentration of hydrogen ions, [H^+], measured in moles of hydrogen per liter of substance. It is given by the formula pH $= \log_{10} \frac{1}{[H^+]}$. Find the amount of hydrogen in a liter of acid rain that has a pH of 4.2.

Explore   Read the problem. You know the formula for finding pH and the pH of the rain. You want to find the amount of hydrogen in a liter of this rain.

Plan   Write the equation. Then, solve for [$H+$].

Solve

$$\text{pH} = \log_{10} \frac{1}{[H+]}$$ *Replace pH with 4.2.*

$$4.2 = \log_{10} \frac{1}{[H+]}$$

$$4.2 = \log_{10} 1 - \log_{10} [H+]$$ *Quotient property of logarithms*

$$4.2 = 0 - \log_{10} [H+]$$ *$\log_{10} 1 = 0$*

$$4.2 = -\log_{10} [H+]$$

$$-4.2 = \log_{10} [H+]$$

$$10^{-4.2} = [H+]$$ *Definition of logarithm*

There are $10^{-4.2}$, or about 0.000063, moles of hydrogen in a liter of this rain.

Examine   Check the solution.

$$4.2 \overset{?}{=} \log_{10} \frac{1}{10^{-4.2}}$$

$$4.2 \overset{?}{=} \log_{10} 1 - \log_{10} 10^{-4.2}$$

$$4.2 \overset{?}{=} 0 - (-4.2)$$

$$4.2 = 4.2 \quad \checkmark$$

The power of a power is found by multiplying the two exponents.

$$\log_3 9^4 = \log_3 (3^2)^4 \qquad\qquad 4 \log_3 9 = (\log_3 9) \cdot 4$$
$$= \log_3 3^{2 \cdot 4} \qquad\qquad\qquad\quad = (\log_3 3^2) \cdot 4$$
$$= 2 \cdot 4 \qquad\qquad\qquad\qquad\quad = 2 \cdot 4$$

So, $\log_3 9^4 = 4 \log_3 9$. This example indicates that the logarithm of a power is the product of the logarithm and the exponent.

Power Property of Logarithms	**For any real number p and positive numbers m and b, where $b \neq 1$,** $\log_b m^p = p \cdot \log_b m.$

You will prove this property in Exercise 3.

Example 4 illustrates how to use the properties of logarithms to solve equations involving logarithms.

Example **4** **Solve each equation.**

a. $2 \log_3 6 - \frac{1}{4} \log_3 16 = \log_3 x$

$$2 \log_3 6 - \frac{1}{4} \log_3 16 = \log_3 x$$

$$\log_3 6^2 - \log_3 16^{\frac{1}{4}} = \log_3 x \quad \text{\textit{Power property of logarithms}}$$

$$\log_3 36 - \log_3 2 = \log_3 x$$

$$\log_3 \frac{36}{2} = \log_3 x \quad \text{\textit{Quotient property of logarithms}}$$

$$\log_3 18 = \log_3 x$$

$$18 = x \quad \text{\textit{Property of equality for logarithmic functions}}$$

The solution is 18.

b. $\log_{10} z + \log_{10} (z + 3) = 1$

$$\log_{10} z + \log_{10} (z + 3) = 1$$
$$\log_{10} z(z + 3) = 1 \qquad \text{\textit{Product property of logarithms}}$$
$$z(z + 3) = 10^1 \qquad \text{\textit{Definition of logarithm}}$$
$$z^2 + 3z - 10 = 0$$
$$(z + 5)(z - 2) = 0$$
$$z + 5 = 0 \quad \text{or} \quad z - 2 = 0 \qquad \text{\textit{Zero product property}}$$
$$z = -5 \qquad\qquad z = 2$$

Check: $\log_{10} z + \log_{10} (z + 3) = 1$

$\log_{10} (-5) + \log_{10} (-5 + 3) \overset{?}{=} 1$

$\log_{10} (-5) + \log_{10}(-2) \overset{?}{=} 1$

Both $\log_{10} (-5)$ and $\log_{10} (-2)$ are

undefined, so -5 is not a solution.

$\log_{10} z + \log_{10} (z + 3) = 1$

$\log_{10} 2 + \log_{10} (2 + 3) \overset{?}{=} 1$

$\log_{10} 2 + \log_{10} 5 \overset{?}{=} 1$

$\log_{10} (2 \cdot 5) \overset{?}{=} 1$

$\log_{10} 10 \overset{?}{=} 1$

$1 = 1 \quad \checkmark$

The only solution is 2.

CHECK FOR UNDERSTANDING

Communicating Mathematics

Study the lesson. Then complete the following.

1. **Describe** what the pH of a solution indicates and how it is related to logarithms.

2. **Name** the properties that serve as guidelines to derive the properties of logarithms.

3. **Prove** the power property of logarithms.

MATH JOURNAL

4. **Assess Yourself** List three properties of exponents and their related properties of logarithms. Do you find the properties of exponents or the properties of logarithms easier to understand? Do you have difficulty understanding or applying any of these properties?

Guided Practice

Express each logarithm as the sum or difference of simpler logarithmic expressions.

5. $\log_4 x^2 y$

6. $\log_3 (xy)^3$

7. $\log_5 \dfrac{ac}{b}$

8. $\log_2 r^{\frac{1}{3}} t$

Use $\log_2 3 \approx 1.585$ and $\log_2 7 \approx 2.807$ to evaluate each expression.

9. $\log_2 \dfrac{7}{3}$

10. $\log_2 36$

11. $\log_2 0.75$

Solve each equation.

12. $\log_2 5 + \log_2 x = \log_2 15$

13. $\log_5 16 - \log_5 2t = \log_5 2$

14. $\log_{10} 7 + \log_{10} (n - 2) = \log_{10} 6n$

15. $\log_2 (y + 2) - 1 = \log_2 (y - 2)$

16. **Chemistry** If the pH level of tomato juice is 4.1 and pH level of baking soda is 8.5, how much more acidic is tomato juice than baking soda?

Practice

Use $\log_3 2 \approx 0.6310$ and $\log_3 7 \approx 1.7712$ to evaluate each expression.

17. $\log_3 4$

18. $\log_3 49$

19. $\log_3 \frac{7}{2}$

20. $\log_3 18$

21. $\log_3 \frac{2}{3}$

22. $\log_3 54$

23. $\log_3 108$

24. $\log_3 \frac{18}{49}$

25. $\log_3 \frac{7}{9}$

Solve each equation.

26. $\log_3 2 + \log_3 7 = \log_3 x$

27. $\log_5 42 - \log_5 6 = \log_5 k$

28. $\log_5 m = \frac{1}{3} \log_5 125$

29. $\log_{10} y = \frac{1}{4} \log_{10} 16 + \frac{1}{2} \log_{10} 49$

30. $\log_9 5 + \log_9 (n + 1) = \log_9 6n$

31. $3 \log_5 x - \log_5 4 = \log_5 16$

32. $2 \log_3 y + \log_3 0.1 = \log_3 5 + \log_3 2$

33. $\log_{10} a + \log_{10} (a + 21) = 2$

34. $\log_6 48 - \log_6 \frac{16}{5} + \log_6 5 = \log_6 5x$

35. $\log_3 64 - \log_3 \frac{8}{3} + \log_3 2 = \log_3 4r$

36. $\log_6 (b^2 + 2) + \log_6 2 = 2$

37. $\log_3 (5z + 5) - \log_3 (z^2 - 1) = 0$

Solve for a.

38. $\log_n a = \log_n (y + 3) - \log_n 3$

39. $\log_b 2a - \log_b x^3 = \log_b x$

40. $\log_x a^2 + 5 \log_x y = \log_x a$

41. $\log_b 4 + 2 \log_b a = 2 \log_b (n + 1)$

Write each expression as one logarithm.

42. $2 \log_b x + \frac{1}{3} \log_b (x + 2) - 4 \log_b (x - 3)$

43. $\log_b (xy^2) + 2 \log_b \frac{x}{y} - 3 \log_b \left(yx^{\frac{2}{3}} \right)$

Critical Thinking

44. If $\log_m y = \log_m a - \log_m b - \log_m c$, express y in terms of a, b, and c.

Applications and Problem Solving

45. Medicine The pH of a person's blood can be found by using the Henderson-Hasselbach formula. The formula is $\text{pH} = 6.1 + \log_{10} \frac{B}{C}$, where B represents the concentration of bicarbonate, which is a base, and C represents the concentration of carbonic acid, which is an acid. Most people have a blood pH of about 7.4.

a. Use a property of logarithms to write the equation without a fraction.

b. A pH of 7 is neutral, and pH numbers less than 7 represent acidic solutions. pH levels greater than 7 represent basic solutions. Is blood normally an acid, a base, or a neutral?

c. Use a scientific calculator to find the pH of a person's blood if the concentration of bicarbonate is 25 and concentration of carbonic acid is 2.

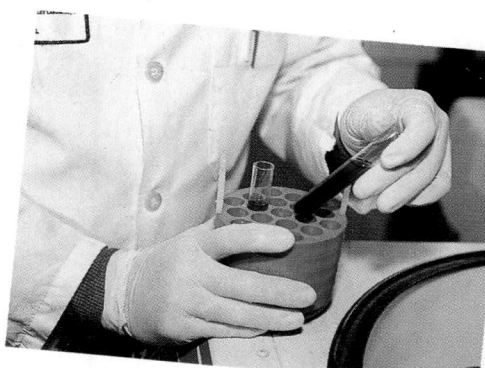

46. Biology The formula for the energy needed to transport a substance from the outside of a living cell to the inside of that cell is $E = 1.4(\log_{10} C_2 - \log_{10} C_1)$, where E represents the energy in kilocalories per gram molecule, C_1 represents the concentration outside the cell, and C_2 represents the concentration inside the cell.

   a. Use the properties of logarithms to write the value of E as one logarithm.

   b. If the concentration inside the cell is three times the concentration outside the cell, find the energy for a substance to travel from the outside to the inside.

Mixed Review

47. Solve $\log_3 243 = y$. (Lesson 10–2)

48. Find $\dfrac{x+2}{x+3} \div \dfrac{x^2+x-12}{x^2-9}$. Write the answer in simplest form. (Lesson 9–3)

49. Find $p(a + 1)$ if $p(x) = 4x - x^2 - 4$. (Lesson 8–1)

50. Geometry Find the coordinates of the midpoint of $\overline{AB}$ with $A(5, 5)$ and $B(6, -7)$. (Lesson 7–1)

51. State whether $x^2 + 8x + 64$ is a perfect square. (Lesson 6–3)

52. Energy A circular cell must deliver 18 watts of energy. If each square centimeter of the cell that is in sunlight produces 0.01 watt of energy, how long must the radius of the cell be? (Lesson 5–8)

53. Factor $x^3 + 2x^2 - 35x$. (Lesson 5–4)

54. Solve the system of equations. (Lesson 3–2)

$$2x + 3y = 2$$
$$3x - 4y = -14$$

WORKING ON THE
In·ves·ti·ga·tion

Refer to the Investigation on pages 590–591.

Refer to the data generated from your shooting experiments with the three rubber bands. Examine the tables you created. Explain any patterns you notice from the tables.

1 Create a coordinate plane with the horizontal axis representing the length the rubber band was stretched and the vertical axis representing the distance the rubber band flew for each rubber band. Graph the data for each rubber band. Draw a best-fit line or curve.

2 Describe the shape of each of the graphs. Are any of the graphs functions? If they are functions, describe the type of function you think each one is and justify your answer.

3 Refer to the third column of each table. Compute the difference between the distance each shot flew and the distance the previous shot flew. What pattern, if any, do you notice? What is the relationship between the data in the second column and the data in the third column?

4 Compare the graphs of the three rubber bands. Are the graphs the same? Explain. If there are differences, describe them.

Add the results of your work to your Investigation Folder.

Common Logarithms

What YOU'LL LEARN

- To identify the characteristic and the mantissa of a logarithm, and
- to find common logarithms and antilogarithms.

Why IT'S IMPORTANT

You can use common logarithms to solve problems involving astronomy and acoustics.

APPLICATION
Acoustics

One of the more useful logarithms is base 10, because our number system is base 10. Base 10 logarithms are called **common logarithms.** These are usually written without the subscript 10, so $\log_{10} x$ is written as $\log x$.

Common logarithms are used in the measure of sound. The loudness L, in decibels, of a particular sound is defined as $L = 10 \log \frac{I}{I_0}$, where I is the intensity of the sound and I_0 is the minimum intensity of sound detectable by the human ear. Soft recorded music is about 4000 times the minimum intensity of sound detectable by the human ear. Use the definition of logarithms to find the loudness in decibels.

$$L = 10 \log \frac{I}{I_0}$$
$$= 10 \log \frac{4000\,I_0}{I_0} \quad \textit{Substitution}$$
$$= 10 \log 4000$$
$$\approx 10(3.602)$$
$$\approx 36$$

Soft recorded music is about 36 decibels. Other common sounds and their approximate decibels levels are listed in the chart below.

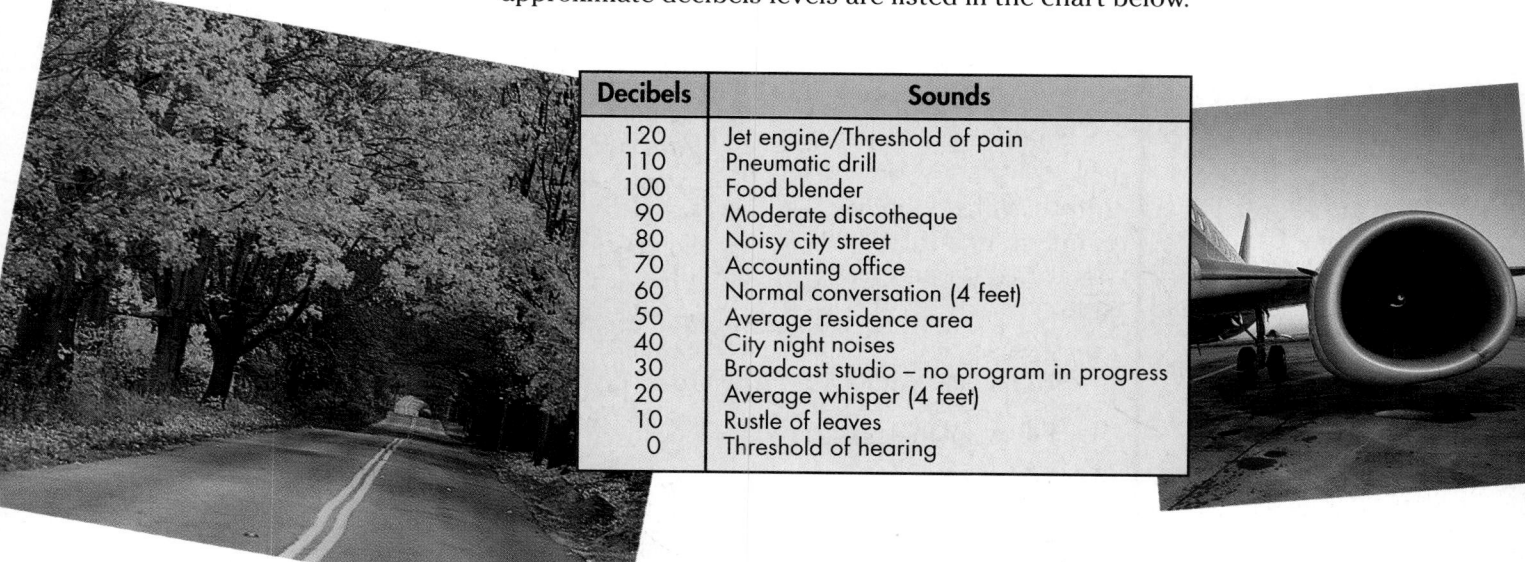

Decibels	Sounds
120	Jet engine/Threshold of pain
110	Pneumatic drill
100	Food blender
90	Moderate discotheque
80	Noisy city street
70	Accounting office
60	Normal conversation (4 feet)
50	Average residence area
40	City night noises
30	Broadcast studio – no program in progress
20	Average whisper (4 feet)
10	Rustle of leaves
0	Threshold of hearing

Expressing a number in scientific notation is helpful when working with common logarithms.

Example **If log 1.2 ≈ 0.0792, find each of the following.**
a. **log 120**   b. **log 0.12**

$$\begin{aligned} \log 120 &= \log (1.2 \times 10^2) \\ &= \log 1.2 + \log 10^2 \\ &\approx 0.0792 + 2 \text{ or } 2.0792 \end{aligned}$$

$$\begin{aligned} \log 0.12 &= \log (1.2 \times 10^{-1}) \\ &= \log 1.2 + \log 10^{-1} \\ &\approx 0.0972 + (-1) \end{aligned}$$

F Y I

A Scottish lord by the name of John Napier is credited with the invention of logarithms. He started his work with logarithms in 1594, and first published a table of logarithms in 1614.

Look closely at the results of Example 1. Notice that log 120 and log 0.12 have the same decimal part, or **mantissa,** 0.0792, but different integer parts, 2 and −1. The integer part of the common logarithm of a number is called its **characteristic** and indicates the magnitude of the number. The characteristic is the exponent of 10 when the original number is expressed in scientific notation.

$$\log (1.2 \times 10^2) = \log 1.2 + \log 10^2$$
$$= 0.0792 + 2$$

mantissa   characteristic

The mantissa is usually expressed as a positive number. To avoid negative mantissas, we rewrite the negative mantissa as the difference of a positive number and an integer, usually 10.

Example **2** **Use a scientific calculator to find log 0.0038. Write the result with a positive mantissa.**

The ⎡LOG⎤ key is used to find common logarithms.

Enter: .0038 ⎡LOG⎤   *−2.420216403*

The value of log 0.0038 is approximately −2.4202.

To write the logarithm with a positive mantissa, add and subtract 10.

$(-2.4202 + 10) - 10 = 7.5798 - 10$ *10 − 10 = 0 and a + 0 = a.*

The characteristic of the logarithm is 7 − 10 or −3. Thus, the mantissa is 0.5798.

Sometimes an application of logarithms requires that you use the inverse of logarithms, exponentiation. When you are given the logarithm of a number and asked to find the number, you are finding the **antilogarithm.** That is, if log $x = a$, then $x = $ antilog a.

Example **Use a scientific calculator to find the antilogarithm of 3.073.**

To find the antilogarithm on your calculator, use the ⎡10ˣ⎤ key.

Enter: 3.073 ⎡2nd⎤ ⎡10ˣ⎤ *1183.041556*

The antilogarithm of 3.073 is approximately 1183.
This means that $10^{3.073} \approx 1183$.

Example

④ On June 15, 1995, Ted Nugent with Bad Company played at the Polaris Amphitheater in Columbus, Ohio. Several miles away, the intensity of the music at the concert registered 66.6 decibels. How many times the minimum intensity of sound detectable by the human ear was this sound, if I_0 is defined to be 1?

APPLICATION

Acoustics

Use the formula $L = 10 \log \dfrac{I}{I_0}$ given at the beginning of the lesson.

$$L = 10 \log \dfrac{I}{I_0}$$
$$66.6 = 10 \log \dfrac{I}{I_0}$$
$$6.66 = \log I$$
$$\text{antilog } 6.66 = \text{antilog } (\log I)$$
$$10^{6.66} = I$$
$$4{,}570{,}882 \approx I$$

The sound several miles away from the rock concert was approximately 4,570,000 times the minimum intensity of sound detectable by the human ear.

CHECK FOR UNDERSTANDING

Communicating Mathematics

Study the lesson. Then complete the following.

1. **Describe** the function you are performing when you take the antilogarithm of a value, such as x.

2. **Name** the base used by the calculator ⎡LOG⎤ key. What are these logarithms called?

3. When a number is expressed in scientific notation, its exponent of 10 corresponds to what part of its common logarithm?

4. The word logarithm is actually a contraction of "<u>log</u>ical <u>arithm</u>etic."

   a. **Explain** why using logarithms was once considered a logical way to multiply or divide some numbers.

   b. **Explain** why logarithms are no longer used to multiply or divide numbers.

   c. **Name** some uses for common logarithms today.

Guided Practice

If log 875 = 2.9420, find each number.

5. characteristic of log 875

6. log 8.75

Use a scientific calculator to find the logarithm for each number rounded to four decimal places. Then state the mantissa and characteristic.

7. 13.7

8. 0.056

Use a scientific calculator to find the antilogarithm of each logarithm rounded to four decimal places.

9. 0.4573

10. -2.1477

11. **Astronomy** The *parallax* of a star is the difference in direction of the star as seen from two widely separate points. The brightness of a star as observed from Earth is its apparent magnitude. Interstellar space is measured in parsecs. One parsec is about 19.2 trillion miles. The absolute magnitude of a star is the magnitude that a star would have if it were 10 parsecs from Earth. For stars more than 30 parsecs from Earth, the formula relating the parallax p, the absolute magnitude M, and the apparent magnitude m is $M = m + 5 + 5 \log p$. The star M35 in the constellation Gemini has an apparent magnitude of 5.3 and a parallax of about 0.018. Find the absolute magnitude of star M35.

EXERCISES

Practice

If log 6500 = 3.8129, find each number.

12. mantissa of log 6500
13. characteristic of log 6500
14. antilog 3.8129
15. log 6.5
16. $10^{3.8129}$
17. mantissa of log 0.065

Use a scientific calculator to find the logarithm for each number rounded to four decimal places. Then state the mantissa and characteristic.

18. 64.7
19. 900.4
20. 0.047
21. 6.377
22. 0.0035
23. 0.0007

Use a scientific calculator to find the antilogarithm for each logarithm rounded to four decimal places.

24. 0.3142
25. 2.1495
26. -0.2615
27. -1.8143
28. $0.5734 - 3$
29. $7.1394 - 10$

Critical Thinking

30. Try to find $(-3)^3$ on your calculator. Some calculators will say "*ERROR*," even though $(-3)^3 = -27$. Can you think of a reason why they might do this? Explain.

Application and Problem Solving

31. **Acoustics** Mary Esperanza had a new muffler installed on her car. As a result, the noise level of the engine of her car dropped from 85 decibels to 73 decibels.

    a. How many times the minimum intensity of sound detectable by the human ear was the car with the old muffler if I_0 is defined to be 1?
    b. How many times the minimum intensity of sound detectable by the human ear is the car with the new muffler?
    c. Find the percent of decrease of the intensity of the sound with the new muffler.

32. **Geology** As you know, the Richter scale is a logarithmic scale. An earthquake that measures 6 on the Richter scale is 10^6 times as intense as the weakest earthquake perceptible by a seismograph.

    a. How much more intense is an earthquake that measures 5.3 on the Richter scale than the weakest perceptible earthquake?
    b. How much more intense was the San Francisco earthquake of 1989, a 7.1 on the Richter scale, than its strongest aftershock, a 4.3 on the Richter scale?

33. Solve $2 \log_6 3 + 3 \log_6 2 = \log_6 x$. (Lesson 10–3)

34. Determine whether $f(x) = \dfrac{x-1}{2}$ and $g(x) = 2x + 1$ are inverse functions. (Lesson 8–8)

35. Write the equation of the parabola shown at the right. (Lesson 6–6)

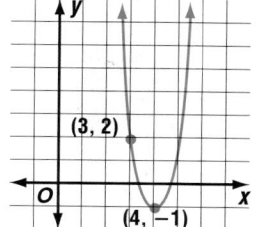

(3, 2)

(4, −1)

36. Astronomy When a solar flare occurs on the sun, it sends out light waves that travel through space at a speed of 1.08×10^9 kilometers per hour. If a satellite in space detects the flare 2 hours after its occurrence, how far is the satellite from the sun? (Lesson 5–1)

37. Find $3 \begin{bmatrix} 4 \\ 1 \\ 7 \end{bmatrix} + 2 \begin{bmatrix} 3 \\ -2 \\ 6 \end{bmatrix} - 5 \begin{bmatrix} -2 \\ 3 \\ 6 \end{bmatrix}$. (Lesson 4–2)

38. Manufacturing Denim Duds makes denim jackets and jeans. Each garment must be cut from a pattern and sewn. There are 40 worker-hours per day available for cutting and 52 worker-hours per day for sewing. The chart below shows the number of hours for each operation needed to make both garments, as well as the profit on the garment.

Garment	Cutting Hours	Sewing Hours	Profit
jacket	1	4	$14
jeans	2	2	$8

How many of each garment should the company make to maximize profit? (Lesson 3–6)

SELF TEST

Solve each equation. (Lesson 10–1)

1. $5^{3y+4} = 5^y$

2. $2^{x+3} = \dfrac{1}{16}$

Evaluate each expression. (Lesson 10–2)

3. $\log_{10} 10{,}000$

4. $\log_3 \dfrac{1}{243}$

5. $\log_{25} 5$

6. Seismology The earthquake that occurred in southern Peru in 1991 registered a 5.6 on the Richter scale. In 1990, an earthquake occurred in northern Peru that registered a 6.4 on the Richter scale. How much more intense was the 1990 earthquake in northern Peru that the 1991 earthquake in southern Peru? (Lesson 10–2)

Solve each equation. (Lesson 10–3)

7. $2 \log_6 4 - \dfrac{1}{3} \log_6 8 = \log_6 y$

8. $\log_2 (9t + 5) - \log_2 (t^2 - 1) = 2$

Use a scientific calculator to find the logarithm for each number rounded to four decimal places. Then state the mantissa and the characteristic. (Lesson 10–4)

9. 600.6

10. 0.189

Natural Logarithms

First Countries/Cities to Issue Stamps

1. United Kingdom, May, 1840
2. New York City, Feb., 1842
3. Zurich, Switzerland, March, 1843
4. Brazil, Aug., 1843
5. Geneva, Switzerland, Oct., 1843

APPLICATION
Postal Service

In 1989, the *Scott Postage Stamp Catalog* listed more than 2400 different postage stamps. As indicated in the graph at the right, the United States Post Office took 77 years to issue its first 600 stamps, 37 years to issue the next 600, approximately half as long to issue the third 600, and so on.

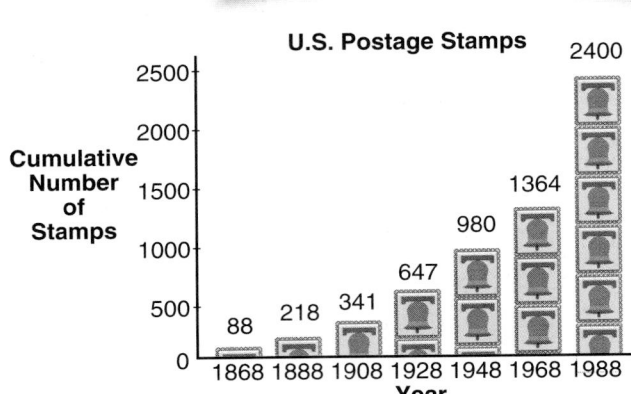

U.S. Postage Stamps

Cumulative Number of Stamps

These figures suggest that the time needed to issue a fixed number of stamps has been decreasing exponentially. In fact, the number of stamps issued is growing exponentially and is approximately modeled by the formula $S = 83e^{0.024t}$, where S represents the cumulative number of stamps issued, t represents the number of years since 1848, and e is a special irrational number.

In 1998, 150 years after the first stamp was issued, about how many different U.S. postage stamps will have been issued? *This problem will be solved in Example 3.*

The number e used in this **exponential growth** problem is used extensively in science and mathematics. It is an irrational number whose value is approximately 2.718. e is the base for the **natural logarithms,** which are abbreviated ln. The natural logarithm of e is 1. All properties of logarithms that you have learned apply to the natural logarithms as well. The key marked ⃞LN on your calculator is the natural logarithm key.

Example **1** **Use a scientific calculator to find ln 3.925.**

    Enter: 3.925 ⃞LN *1.367366351*

    The natural logarithm of 3.925 is approximately 1.3674.

You can take antilogarithms of natural logarithms as well. The symbol for the antilogarithm of x is antiln x.

Example **2** **a. Find x if ln x ≈ 3.4825**

    ln x ≈ 3.4825
       x ≈ antiln 3.4825

 Enter: 3.4825 ⃞2nd ⃞e^x *32.5409729*
So x is approximately 32.5410.

b. Find e if ln e = 1.

    ln e = 1
      e = antiln 1

 Enter: 1 ⃞2nd ⃞e^x *2.718281828*
So e is approximately 2.7183.

Equations involving e are easier to solve using natural logarithms, rather than using common logarithms, since ln e = 1.

Example

Postal Service

If your scientific calculator has no e^x key, use the INV key and then the LN key.

3 Use the formula $S = 83e^{0.024t}$ and natural logarithms to determine about how many different U.S. postage stamps will have been issued from 1848 to 1998.

t = 1998 − 1848 or 150 years

$S = 83e^{0.024t}$	
$S = 83e^{0.024(150)}$	*t = 150*
$S = 83e^{3.6}$	*Simplify.*
$\ln S = \ln(83e^{3.6})$	*Take the natural logarithm of each side.*
$\ln S = \ln 83 + 3.6 \ln e$	*Power and product properties of logarithms*
$\ln S = \ln 83 + 3.6$	*Since e is the base for natural logarithms,*
$\ln S \approx 8.018841$	*ln e = 1.*
$\text{antiln}(\ln S) \approx \text{antiln } 8.018841$	*Take the antilogarithm of each side.*
$S \approx 3038$	

So, approximately 3038 stamps will have been issued by 1998.

When interest is compounded *continuously*, the amount of money A in an account after t years is found using the formula $A = Pe^{rt}$, where P represents the amount of the principal and r represents the annual interest rate.

Example

Finance

4 Mr. and Mrs. Franco are planning to take a cruise for their twenty-fifth wedding anniversary. They have six years to save $3500 for the cruise. If the six-year certificate of deposit they buy now pays 8% interest compounded continuously, how much should they invest now in order to have $3500 for the cruise?

$A = Pe^{rt}$	
$3500 = Pe^{(0.08)(6)}$	*Replace A with 3500, r with 0.08, and t with 6.*
$3500 = Pe^{0.48}$	*Simplify.*
$\ln 3500 = \ln(Pe^{0.48})$	*Take the natural logarithm of each side.*
$\ln 3500 = \ln P + (0.48)\ln e$	*Power and product properties of logarithms*
$\ln 3500 = \ln P + 0.48$	*Since e is the base for natural logarithms,*
$7.680518 \approx \ln P$	*ln e = 1.*
$\text{antiln } 7.680518 \approx \text{antiln}(\ln P)$	*Take the antilogarithm of each side.*
$2165.74 \approx P$	

Mr. and Mrs. Franco should invest $2165.74 to earn enough for the trip.

CHECK FOR UNDERSTANDING

Communicating Mathematics

Study the lesson. Then complete the following.

1. **Name** the base of natural logarithms.

2. **Describe** a situation for which you should choose to use natural logarithms instead of common logarithms to solve a problem.

3. **Describe** a situation that could be represented by the equation $A = 2500e^{0.08t}$.

Use a scientific calculator to find each value rounded to four decimal places.

4. ln 3.12

5. ln 0.045

6. ln 0.772

7. antiln 0.2594

8. antiln 2.0175

9. antiln −1.5771

10. Finance Atepa's grandparents opened a savings account for her when she was born. They placed $1000 in an account that paid $6\frac{1}{2}\%$ interest compounded continuously. Atepa is now 16 years old and would like to buy a used car that costs $2500. Does she have enough money in her account to buy the car? Explain.

EXERCISES

Practice

Use a scientific calculator to find each value rounded to four decimal places.

11. ln 7.95

12. ln 1.34

13. ln 57.3

14. ln 0.958

15. ln 2.7183

16. ln 10,000

17. ln 0.005

18. ln 1.002

19. ln 0.01

20. antiln 0.782

21. antiln 0

22. antiln 2.6049

23. antiln −0.112

24. antiln −4.567

25. antiln 1.005

26. antiln −2.003

27. antiln 1.55

28. antiln −1.679

Graph each pair of equations on the same axis.

29. $y = \ln x$ and $y = e^x$

30. $y = \log x$ and $y = \ln x$

31. Compare and contrast the graphs in Exercise 29.

32. Compare and contrast the graphs in Exercise 30.

Critical Thinking

33. The great Swiss mathematician Leonhard Euler, for whom the number e is named, defined e as the sum of the series $1 + \frac{1}{1} + \frac{1}{1 \cdot 2} + \frac{1}{1 \cdot 2 \cdot 3} + \frac{1}{1 \cdot 2 \cdot 3 \cdot 4} + \ldots$.

   a. Calculate the value of e using six terms in the series.
   b. Calculate the value of e using eight terms in the series.
   c. Which value is more accurate?
   d. Find the percent of the change in the values.

Applications and Problem Solving

34. Finance Kang is saving money to go on a trip to Europe after his college graduation. He will finish college five years from now. If the five-year certificate of deposit he buys pays 7.25% interest compounded continuously, how much should he invest now in order to have $3000 for the trip?

35. Sales The sales of CD-ROM drives have been increasing since 1984. Suppose the number of CD-ROM drives sold S in a given year is approximated by the formula $S = 75,000e^{0.83t}$, where t is the number of years since 1988.

   a. Estimate the number of these drives that will be sold in the year 2000.
   b. Draw a graph of $S = 75,000e^{0.83t}$. How does your graph compare to the graph at the right?

CD-ROM Drive Sales

4.8 million

1.5 million

75,000   240,000

1988   1990   1992   1993

36. Physics The intensity of light decreases as it passes through sea water. The equation $\ln \dfrac{I_0}{I} = 0.014d$ relates the intensity of light I at the depth of d centimeters with the intensity of light I_0 in the atmosphere. Find the depth of the water where the intensity of the light is half the intensity of the light in the atmosphere.

Mixed Review

37. Use a calculator to find each value, rounded to four decimal places. (Lesson 10–4)

    a. log 77.3   **b.** log 0.0056   **c.** antilog 3.5567   **d.** antilog (6.7891 − 10)

38. Simplify $\dfrac{\dfrac{x^2}{x^2 - 25y^2}}{\dfrac{x}{5y - x}}$. (Lesson 9–3)

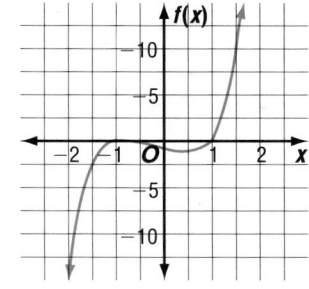

39. Use the graph of the polynomial function $f(x) = x^5 + x^4 - x - 1$ at the right to determine at least one of the binomial factors of the polynomial. Then find all factors of the polynomial. (Lesson 8–2)

40. Manufacturing The weights of boxes of cereal filled by a machine are normally distributed. The mean weight is 510 grams with a standard deviation of 4 grams. (Lesson 6–9)

    a. Of 1000 boxes, how many weigh at least 510 grams?

    b. Of 1000 boxes, how many weigh between 502 and 514 grams?

    c. A machine at the end of the production line checks the weight of the boxes before they are shipped to the stores. If a box weighs more than 522 grams, it is sent back to the beginning of the line to be emptied and reused. Of 1000 boxes, how many will be sent back for this reason?

    d. The machine that checks the weight of the boxes sends back boxes that weigh less than 502 grams. Of 1000 boxes, how many will be sent back for this reason?

41. Photography Shina Murakami is a professional photographer. She has a photograph that is 4 inches wide and 6 inches long. She wishes to make a print of the photograph for a competition. The area of the new print is to be five times the area of the original. If Ms. Murakami is going to add the same amount to the length and the width of the photograph, what will the dimensions of the new print be? (Lesson 6–4)

42. Solve $\begin{vmatrix} x & 5 & 2 \\ -6 & 4 & 1 \\ 3 & 1 & x \end{vmatrix} = x^2 + 22x - 1$. (Lesson 4–4)

43. Graph the system of inequalities. Name the coordinates of the vertices of the polygon formed. Find the maximum and minimum values of the function. (Lesson 3–5)

$x \geq 0$

$y \geq 0$

$y \leq 3 - x$

$3x + y \leq 6$

$f(x, y) = 2x + 4y$

10-6

Solving Exponential Equations

What YOU'LL LEARN

- To solve equations with variable exponents by using logarithms,
- to evaluate expressions involving logarithms with different bases, and
- to solve problems by using estimation.

Why IT'S IMPORTANT

You can use exponential equations to solve problems involving sports and demographics.

APPLICATION
Demographics

The population of the United States is continually growing. In 1992, the population in the United States was 249 million, and the exponential growth rate was 0.9% per year. An equation of the form $P(t) = P_0 e^{kt}$ can be used to model the growth pattern of many things, ranging from growing investments to growing populations. In this case, P_0 is the population at time 0, $P(t)$ represents the population at time t, and k is a positive constant that depends on the situation. The constant k is often called the **exponential growth rate.**

Equations of the type described above are called **exponential equations.** Exponential equations are equations in which the variables appear as exponents. These equations can be solved using the property of equality for logarithmic functions. **Use estimation** to make sure you're on the right track.

Example ➊ Suppose $6^x = 42$.
  a. **Estimate the value of x.**
  b. **Solve the equation.**

PROBLEM SOLVING
Use Estimation

  a. Since $6^2 = 36$ and $6^3 = 216$, the value of x is between 2 and 3. The value of x should be much closer to 2 than 3.

  b.
$$6^x = 42$$

$$\log 6^x = \log 42 \quad \textit{Property of equality for logarithmic functions}$$

$$x \log 6 = \log 42 \quad \textit{Power property of logarithms}$$

$$x = \frac{\log 42}{\log 6} \quad \textit{Divide each side by log 6.}$$

$$x \approx \frac{1.6232}{0.7782}$$

$$x \approx 2.086$$

The solution is approximately 2.086, which is consistent with the estimate.

Check: Use a scientific calculator to find the value of $6^{2.086}$.

Enter: 6 $\boxed{x^y}$ 2.086 $\boxed{=}$ 41.99750672

The answer is reasonable.

APPLICATION
Demographics

Example 2 Refer to the application at the beginning of the lesson. How many years after 1992 will it take for the U.S. population to reach 300 million if the exponential growth rate remains at 0.9%?

$$P(t) = P_0 e^{kt}$$

$$300 = 249 e^{0.009t} \qquad P_0 = 249,\ k = 0.009,\ and\ P(t) = 300$$

$$\frac{300}{249} = e^{0.009t} \qquad \text{Divide each side by 249.}$$

$$\ln \frac{300}{249} = \ln e^{0.009t} \qquad \text{Take the natural logarithm of each side.}$$

$$\ln \frac{300}{249} = 0.009t \qquad \text{Definition of natural logarithm}$$

$$\ln 300 - \ln 249 = 0.009t \qquad \text{Quotient property of logarithms}$$

$$5.7038 - 5.5175 \approx 0.009t$$

$$0.1863 \approx 0.009t$$

$$20.7 \approx t$$

The population in the United States will reach 300,000,000 about 21 years after 1992, or in 2013.

Check: Use a graphing calculator to graph the equation $y = 249 e^{0.009x}$. Use TRACE to find the value of x when $y = 300$.

The value of x is about 20.7. The answer is correct.

`X=20.744681 Y=300.11179`

In some equations, variables are found in more than one exponent.

Example 3 Solve $8^{2x-5} = 5^{x+1}$. $6^{x+2} = 17.2$

$$8^{2x-5} = 5^{x+1}$$

$$\log 8^{2x-5} = \log 5^{x+1} \qquad \text{Property of equality for logarithmic functions}$$

$$(2x - 5) \log 8 = (x + 1) \log 5 \qquad \text{Power property of logarithms}$$

$$2x \log 8 - 5 \log 8 = x \log 5 + \log 5 \qquad \text{Distributive property}$$

$$2x \log 8 - x \log 5 = \log 5 + 5 \log 8$$

$$x(2 \log 8 - \log 5) = \log 5 + 5 \log 8 \qquad \text{Distributive property}$$

$$x = \frac{\log 5 + 5 \log 8}{2 \log 8 - \log 5}$$

$$x \approx \frac{0.6990 + 5(0.9031)}{2(0.9031) - 0.6990}$$

$$x \approx 4.7095$$

The solution is approximately 4.7095. *Check this result.*

It is possible to evaluate expressions involving logarithms with different bases. Since calculators are usually not programmed with all possible bases for logarithms, the **change of base formula** can be very helpful.

Change of Base Formula	**For all positive numbers a, b, and n, where $a \neq 1$ and $b \neq 1$,** $$\log_a n = \frac{\log_b n}{\log_b a}.$$

Example ④ Express each logarithm in terms of common logarithms. Then approximate its value to three decimal places.

a. $\log_8 77$

$\log_a n = \dfrac{\log_b n}{\log_b a}$ *Change of base formula*

$\log_8 77 = \dfrac{\log 77}{\log 8}$ *$a = 8$, $n = 77$, $b = 10$*

≈ 2.0889

The value of $\log_8 77$ is approximately 2.089.

b. $\log_{16} 64$

$\log_a n = \dfrac{\log_b n}{\log_b a}$ *Change of base formula*

$\log_{16} 64 = \dfrac{\log 64}{\log 16}$ *$a = 16$, $n = 64$, $b = 10$*

$= 1.5$

The value of $\log_{16} 64$ is 1.5. *Why is this an exact value?*

CHECK FOR UNDERSTANDING

Communicating Mathematics

Study the lesson. Then complete the following.

1. **You Decide** Karen sees the exponent in the equation $36 = x^5$ and decides to use logarithms to solve the equation. Tisha tells her that this is not an exponential equation and she does not need logarithms to solve the equation. Who is correct? Explain.

2. **Describe** a situation when you might use the change of base formula.

3. Could you use the change of base formula to express a logarithm in terms of natural logarithms? Explain.

Guided Practice

Find the value of each logarithm to three decimal places.

4. $\log_4 22$

5. $\log_{12} 95$

Use logarithms to solve each equation. Round to three decimal places.

6. $5^x = 52$

7. $8^{2a} = 124$

8. $2.1^{t-5} = 9.32$

9. $y = \log_4 125$

10. $2^{2x+3} = 3^{3x}$

11. $2^n = \sqrt{3^{n-2}}$

12. **Finance** If \$1500 is placed in an account that pays 6.5% interest compounded continuously, how long will it take for the money in the account to double? Use the formula $A = Pe^{rt}$.

Practice

Find the value of each logarithm to three decimal places.

13. $\log_5 16$

14. $\log_6 82$

15. $\log_3 125$

16. $\log_2 100$

17. $\log_{12} 25$

18. $\log_4 48$

Use logarithms to solve each equation. Round to three decimal places.

19. $9^b = 45$

20. $2^x = 30$

21. $5^p = 34$

22. $3.1^{a-3} = 9.42$

23. $6^{x+2} = 17.2$

24. $8.2^{n-3} = 42.5$

25. $x = \log_5 61.4$

26. $8^{y-2} = 7.28$

27. $t = \log_8 200$

28. $5^{s+2} = 15.3$

29. $9^{z-4} = 6.28$

30. $7.6^{a-2} = 41.7$

31. $3.5^{3x+1} = 65.4$

32. $20^{x^2} = 70$

33. $8^{x^2-2} = 32$

34. $5.8^{x^2-3} = 82.9$

35. $9^a = 2^a$

36. $5^{x-1} = 3^x$

37. $7^{t-2} = 5^t$

38. $16^{d-4} = 3^{3-d}$

39. $8^{x-2} = 5^x$

40. $5^{3y} = 8^{y-1}$

41. $5^{5a-2} = 2^{2a+1}$

42. $8^{2y} = 52^{4y+3}$

43. $40^{3x} = 5^{2x+1}$

44. $4^n = \sqrt{5^{n-2}}$

45. $\sqrt[3]{2^{x-1}} = 8^{x-2}$

Critical Thinking

46. Let x be any real number and let a, b, and n be positive real numbers with $a \neq 1$ and $b \neq 1$. Show that if $x = \log_a n$, then $x = \dfrac{\log_b n}{\log_b a}$.

Applications and Problem Solving

47. **Demographics**   The population of Antlers, Oklahoma, is about 2500. Suppose it is growing at an exponential rate of 3%. Use the formula $P(t) = P_0 e^{kt}$ to determine approximately how long it will take the town's population to double.

48. **Olympics**   Since the modern Olympics began in 1896 with 42 events, the number of events has continued to grow. Suppose the exponential growth rate of the number of events is 1.9%.

Summer Olympics Events

Year	City	Number
1968	Mexico City	172
1972	Munich	196
1976	Montreal	199
1980	Moscow	200
1984	Los Angeles	223
1988	Seoul	237
1992	Barcelona	257
1996	Atlanta	271

Source: Atlanta Committee for the Olympic Games

   a. Use the formula $E = 42e^{kt}$, where E is the number of events, k is the exponential growth rate, and t is the time in years since 1896, to determine how long after the first modern Olympics we can expect there to be 350 Summer Olympic events.

   b. Draw the graph of $E = 42e^{0.019t}$. How does your graph compare to the chart above?

Mixed Review

49. Use a calculator to find antiln -6.083, rounded to four decimal places.   (Lesson 10–5)

50. **Construction**   A painter works on a job for 10 days and is then joined by an associate. Together they finish the job in 6 more days. The associate could have done the job in 30 days. How long would it have taken the painter to do the job alone?   (Lesson 9–5)

51. Solve $b^4 - 5b^2 + 4 = 0$.   (Lesson 8–6)

52. State whether the graph of $4y^2 - x^2 - 24y + 6x = 11$ is a parabola, a circle, an ellipse, or a hyperbola.   (Lesson 7–6)

53. Solve $(x + 2)(x + 9) > 0$. (Lesson 6–7)

54. Art Morgan needs to paint a landscape for art class. She has 6 feet of framing material to frame the finished painting. What should the dimensions of her canvas be for the painting to have the maximum area? (Lesson 6–1)

55. Find the values of x and y for which the sentence $2x + 5yi = 4 + 15i$ is true. (Lesson 5–9)

56. Write the system of equations as a matrix equation. Then solve the system. $6x + 5y = 8$ (Lesson 4–6)
$3x - y = 7$

57. Write an equation in standard form for the line that passes through the point at (4, 6) and is perpendicular to the line whose equation is $y = \frac{2}{3}x + 5$. (Lesson 2–4)

WORKING ON THE In·ves·ti·ga·tion

Refer to the Investigation on pages 590–591.

Boing!!

Your manager has asked you to explore the maximum spring potential of the bungee. She is interested in finding the longest distance a jumper would bounce back for any length. She believes that the new bungee product would dominate the market if the spring was greater than that of other bungee equipment. The jumper would therefore experience a greater bounce at the end of his or her drop.

You need to analyze the spring of the rubber bands you tested. Define the *elastic potential* as the ratio of the distance bounced to the length of the stretch. In this manner, you can find the point at which the rubber band has the greatest spring in relationship to the amount it is stretched.

1 Examine one of your graphs. How can the elastic potential be illustrated on that graph using your definition? Is there a numerical value that describes the elastic potential? Explain.

2 Draw new graphs for all three rubber bands using the ratio on the vertical axis instead of the distance bounced. Connect the points in order with line segments. Determine the slope of each segment. Record the slope. Describe the slope in terms of a ratio. What does the ratio indicate?

3 What is the maximum elastic potential for each of the rubber bands? At the maximum elastic potential, what is the distance stretched and the resulting shot distance for each rubber band?

4 Now make three graphs with the horizontal axis representing the distance stretched and the vertical axis representing the elastic potential. Are the graphs logarithmic or exponential in shape? Do each of the graphs verify your predictions for the maximum elastic potentials?

5 Make a recommendation regarding the spring of the rubber bands. Indicate the optimum stretch length for each rubber band you tested. Justify your recommendation.

Add the results of your work to your Investigation Folder.

Growth and Decay

What YOU'LL LEARN

- To use logarithms to solve problems involving growth and decay.

Why IT'S IMPORTANT

You can solve growth and decay problems to learn more about paleontology and communication.

fabulous FIRSTS

Willard F. Libby (1908–1980)

W.F. Libby of the University of Chicago won the Nobel Prize for Chemistry in 1960 for his method to use carbon-14 for age determination in archaeology. In 1947, he was the first scientist to explain the formation of carbon-14 in the atmosphere and develop a radioactive test to date organic deposits.

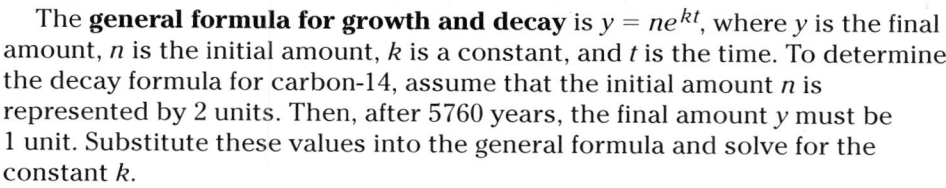

APPLICATION
Paleontology

Paleontologists study life of past geological periods by studying fossil remains. They use carbon-14 (C^{14}) to estimate the age of fossils. Carbon-14 decays with time. In 5760 years, just half of the mass of this substance will remain. This period is called its *half-life*. To find the age of a fossil with just $\frac{1}{5}$ of the carbon-14 remaining, paleontologists must use the decay formula for this substance.

The **general formula for growth and decay** is $y = ne^{kt}$, where y is the final amount, n is the initial amount, k is a constant, and t is the time. To determine the decay formula for carbon-14, assume that the initial amount n is represented by 2 units. Then, after 5760 years, the final amount y must be 1 unit. Substitute these values into the general formula and solve for the constant k.

$$y = ne^{kt} \qquad \text{General formula for growth and decay}$$
$$1 = 2e^{k(5760)} \qquad y = 1, n = 2, t = 5760$$
$$0.5 = e^{5760k}$$
$$\ln 0.5 = \ln e^{5760k} \qquad \text{Take the natural logarithm of each side.}$$
$$\ln 0.5 = 5760k \ln e \qquad \text{Power property of logarithms}$$
$$\ln 0.5 = 5760k \qquad \ln e = 1$$
$$\frac{\ln 0.5}{5760} = k$$
$$-0.00012 \approx k$$

The equation for the decay of carbon-14 is $y = ne^{-0.00012t}$, where t is given in years.

Now use this formula to determine the age of a fossil that has $\frac{1}{5}$ of its carbon-14 remaining. Assume that the initial amount is 5 units and the final amount is 1 unit. Then, solve the equation for t.

$$y = ne^{-0.00012t} \qquad \text{Formula for the decay of carbon-14}$$
$$1 = 5e^{-0.00012t} \qquad y = 1, n = 5$$
$$0.2 = e^{-0.00012t}$$
$$\ln 0.2 = \ln e^{-0.00012t} \qquad \text{Take the natural logarithm of each side.}$$
$$\ln 0.2 = -0.00012t \ln e \qquad \text{Power property of logarithm}$$
$$\ln 0.2 = -0.00012t \qquad \ln e = 1$$
$$\frac{\ln 0.2}{-0.00012} = t$$
$$13,412 \approx t$$

The fossil is about 13,400 years old.

The general formula for growth and decay also describes materials that grow exponentially. In this case, the value of k will be positive.

Example 1

CONNECTION
Biology

Bacteria usually reproduce by a process known as *binary fission*. In this type of reproduction, one bacterium divides, forming two bacteria. Under ideal conditions, some bacteria can reproduce every 20 minutes. Find the constant *k* for the growth of these types of bacteria under ideal conditions and write the growth equation.

$$y = ne^{kt}$$
$$2 = 1e^{k(20)} \quad \textit{One bacterium can produce two in 20 minutes.}$$
$$2 = e^{20k}$$
$$\ln 2 = \ln e^{20k} \quad \textit{Take the natural logarithm of each side.}$$
$$\ln 2 = 20k \ln e \quad \textit{Power property of logarithms}$$
$$\ln 2 = 20k \quad \textit{ln e = 1}$$
$$\frac{\ln 2}{20} = k$$
$$0.0347 \approx k$$

The value of the constant for the bacteria is approximately 0.0347. The growth equation is $y = ne^{0.0347t}$, where *t* is given in minutes.

Certain assets, such as cars, houses, and business equipment, appreciate or depreciate, that is, increase or decrease in value, with time. The formula $V_n = P(1 + r)^n$, where V_n is the new value, *P* is the initial value, *r* is the fixed rate of appreciation or depreciation, and *n* is the number of years, can be used to compute the value of an asset. The value of *r* for a depreciating asset will be negative, and the value of *r* for an appreciating asset will be positive.

Example 2

APPLICATION
Agriculture

The Thomas family includes several generations of farmers. They have an opportunity to buy 50 acres adjacent to their farm for $800 per acre. In the past, the price of farmland has gone up 3% a year. If this continues, how long will it be before the land is worth $1000 per acre?

Explore Read the problem. The problem gives the value of the land now and the annual percent of increase in the price of land. It asks you to find when the land will be worth $1000 an acre.

Plan Substitute 1000 for V_n, 800 for *P*, and 0.03 for *r* in the formula $V_n = P(1 + r)^n$. Then, use logarithms to solve for *n*.

Solve
$$V_n = P(1 + r)^n$$
$$1000 = 800(1 + 0.03)^n \quad \textit{V}_n = 1000, P = 800, r = 0.03$$
$$1.25 = 1.03^n$$
$$\log 1.25 = \log 1.03^n \quad \textit{Take the common logarithm of each side.}$$
$$\log 1.25 = n \log 1.03 \quad \textit{Product property of logarithms}$$
$$\frac{\log 1.25}{\log 1.03} = n$$
$$7.549 \approx n$$

The value of the land will be $1000 per acre in about $7\frac{1}{2}$ years.

Examine Use the formula to find the value of the land in 7.5 years.
$$V_n = P(1 + r)^n$$
$$= 800(1 + 0.03)^{7.5} \quad \textit{P = 800, r = 0.03, n = 7.5}$$
$$\approx 999 \quad \textit{Use a calculator to evaluate the expression.}$$

The value of the V_n is about 1000, so the answer seems reasonable.

Logarithms can be used to solve problems involving an exponential function $y = ab^x$.

Example 3

APPLICATION
Broadcasting

The FM frequencies of radio transmissions range from 88 to 108 megahertz. However, these frequencies are not marked uniformly along the display of a radio receiver. Assume that the frequencies are an exponential function of the distance from the left end of the display. Write an equation for the FM frequencies along a display that is 15 centimeters long.

Let x represent the distance from the left end, and let y represent the frequency. Since 88 megahertz is 0 centimeters from the left end and 108 megahertz is 15 centimeters from the left end, the ordered pairs (0, 88) and (15, 108) are solutions to the function $y = ab^x$. Use (0, 88) to solve for a.

$$y = ab^x$$
$$88 = ab^0 \qquad x = 0, y = 88$$
$$88 = a(1) \qquad b^0 = 1$$
$$88 = a$$

Then use (15, 108) to solve for b.

$$y = ab^x$$
$$108 = 88b^{15} \qquad \textit{a = 88, x = 15, and y = 108}$$
$$\log 108 = \log (88b^{15}) \qquad \textit{Take the common logarithm of each side.}$$
$$\log 108 = \log 88 + \log b^{15} \qquad \textit{Product property of logarithms}$$
$$\log 108 = \log 88 + 15 \log b \qquad \textit{Power properties of logarithms}$$
$$\log 108 - \log 88 = 15 \log b$$
$$\frac{\log 108 - \log 88}{15} = \log b$$
$$0.0059 \approx \log b$$
$$\text{antilog } 0.0059 \approx \text{antilog } (\log b) \qquad \textit{Take the antilogarithm of each side.}$$
$$1.0137 \approx b$$

The relationship of the frequencies and the distance from the left side of the display is approximated by the equation $y = 88(1.0137)^x$, where y is the frequency in megahertz and x is the distance in centimeters.

CHECK FOR UNDERSTANDING

Communicating Mathematics

Study the lesson. Then complete the following.

1. **Describe** a situation in which the constant k in the formula for growth and decay is positive and one in which k is negative. Describe the situation if the value of k is zero.

2. **Explain** why the natural logarithm was used in Example 1 and the common logarithm was used in Examples 2 and 3. Could you solve Example 1 using common logarithms? Could you solve Examples 2 and 3 by using natural logarithms? Explain.

Guided Practice

3. **Broadcasting** Refer to Example 3.
   a. Find the FM frequency that corresponds to the point that is 3.5 centimeters from the left side of the display.
   b. Find the location on the radio display that corresponds to the FM frequency of 100 megahertz.

4. **Business** Zeller Industries bought a computer for $4600. It is expected to depreciate at a steady rate of 20% a year. When will the value have depreciated to $2000?

5. **Chemistry** The half life of radium (Ra^{226}) is 1620 years.
   a. Find the constant k in the formula $y = ne^{kt}$ for radium (Ra^{226}) when t is given in years.
   b. Write the equation for the decay of radium (Ra^{226})
   c. Suppose a 20-gram sample of radium (Ra^{226}) is sealed in a box. Find the mass of the radium after 5000 years.
   d. When will a sample of radium (Ra^{226}) be one fourth of its original mass?
   e. When will a 20-gram sample of radium (Ra^{226}) be completely gone? Explain.

EXERCISES

Applications and Problem Solving

6. **Real Estate** Mr. and Mrs. Sawyer bought a condominium for $75,000. Assuming that its value will appreciate 6% a year, how much will the condo be worth in five years when the Sawyers are ready to move?

7. **Medicine** Radioactive iodine is used to determine the health of the thyroid gland. It decays according to the equation $y = ne^{-0.0856t}$, where t is in days. Find the half-life of this substance.

8. **Broadcasting** The AM frequencies of radio transmissions range from 535 to 1705 kilohertz. Assume that the frequencies are an exponential function of the distance from the left of the display on a radio.
   a. Write an equation for the AM frequencies along a display that is 15 centimeters long.
   b. Find the AM frequency that corresponds to the point that is 4.5 centimeters from the left side of the display.
   c. Find the location on the radio display that corresponds to the AM frequency of 1000 kilohertz.

9. **Population Growth** The city of Knoxville, Tennessee, grew from a population of 546,488 in 1980 to a population of 585,960 in 1990.
   a. Use this information to write a growth equation for Knoxville, where t is the number of years after 1980.
   b. Use your equation to predict the population of Knoxville in 2010.
   c. What factors might affect the population growth of a city such as Knoxville?

10. **Space Travel** A radioisotope is used as a power source for a satellite. The power output is given by the equation $P = 50e^{-\frac{t}{250}}$, where P is the power in watts and t is the time in days.
    a. Find the power available after 100 days.
    b. Ten watts of power are required to operate the equipment in the satellite. How long can the satellite continue to operate?

11. **Paleontology** A paleontologist finds a bone that might be a dinosaur bone. In the laboratory, she finds that the radiocarbon found in this bone is $\frac{1}{10}$ of that found in living bone tissue. Could this bone have belonged to a dinosaur? Explain. (*Hint:* The dinosaurs lived from 220 million years ago to 63 million years ago.)

12. **Real Estate**   The Diaz family bought a new house 10 years ago for $80,000. The house is now worth $140,000. Assuming a steady rate of growth, what was the yearly rate of appreciation?

13. **Cooking**   The amount of time needed to cook scrambled eggs in the microwave depends on the number of eggs being cooked. The chart at the right shows the suggested times for cooking eggs in a certain microwave.

Number of Eggs	Cooking Time (min)
2	$1\frac{3}{4}$
4	3

   a. Assume that the number of minutes is a function of some power of the number of eggs. Write a general equation of the form $t = an^b$, where t is the time in minutes, n is the number of eggs, and a and b are constants. (*Hint:* Use a system of equations to solve for the constants.)

   b. Find the amount of time needed to cook 3 eggs and to cook 5 eggs.

14. **Communication**   A rumor can spread very quickly. The number of people H who have heard a rumor can be approximated by the equation

$$H = \frac{P}{1 + (P - S)e^{-0.4t}}$$ where P is the total population, S is the number of people who start the rumor, and t is the time in minutes. During lunch period, two students decide to start a rumor that the principal will let the students out of school one hour early that day. If there are 1600 students in the school, how much time will pass before half of the students have heard the rumor?

Programming

15. The graphing calculator program at the right uses the formula

$S = R\left[\dfrac{(1 + I)^N - 1}{I}\right]$ to find how many

payments are needed to accumulate a given amount of money when payments are made at regular intervals and interest is compounded at the end of each payment period. In the formula, S represents the money accumulated after the last payment, R represents the amount of each payment, I represents the interest rate per payment period (represented as a decimal), and N represents the number of payments per year. In the program, Y represents the number of payments per year, and A represents the annual interest rate.

```
PROGRAM: PAYMENTS
: Prompt S,R,A,Y
: A/Y→I
: log(SI/R+1)/(log(1+I))→N
: If N=int (N+1)
: Then
: Goto 1
: End
: int(N+1)→N
: Lbl 1
: Disp "PAYMENTS NEEDED:",N
```

Use the program to determine the number of payments needed to accumulate the indicated amount of money, given the amount of each payment, the annual interest rate, and the number of payments each year.

   a. $4000, $300, 9%, 2

   b. $7500, $400, 8.5%, 4

   c. $8995, $156, 9.25%, 12

   d. $14,600, $195, 8.75%, 12

   e. $96,000, $850, 9.65%, 12

   f. $90,000, $425, 9.65%, 26

Critical Thinking

16. Compare the formulas for exponential growth, $y = ne^{kt}$, and for continuously compounded interest, $A = Pe^{rt}$. Explain how the formulas are related.

17. Solve $7^{x-2} = 5^{3-x}$. (Lesson 10–6)

18. Find the equation for the ellipse shown at the right. Then give the coordinates of the foci. (Lesson 7–4)

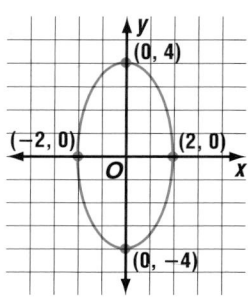

19. Physics The distance s an object travels can be computed when the initial speed v_i, time elapsed t, and the rate of constant acceleration a is known. The formula that relates these factors is $s = v_i t + \frac{1}{2} at^2$. Michael drives a red sports car on a race track at an initial velocity of 24 ft/s and begins to accelerate at a constant rate of 8 ft/s². (Lesson 6–3)
  a. How long will it take him to travel a distance of 100 feet?
  b. How long will it take him to travel a distance of 200 feet?
  c. How long will it take him to travel a distance of 300 feet?
  d. Study your answers to parts a–c. As the distance doubles, does the amount of time double? Explain your answer.

F Y I

The baseball stadium with largest seating capacity is Mile High Stadium, home of the Colorado Rockies. Wrigley Field, home of the Chicago Cubs, has the smallest seating capacity of the major league fields.

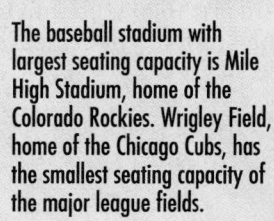

20. Baseball The seating capacities of the major league baseball stadiums are listed below. (Lesson 4–8)

34,142	38,710	40,625	42,400	43,739	44,702	47,313	48,000
48,041	49,292	50,516	52,003	52,416	52,952	53,192	54,816
55,601	55,883	56,000	56,227	57,545	58,727	59,002	59,702
62,000	62,382	64,593	76,100				

  a. Find the range of the data.
  b. Find the quartiles of the data.
  c. Find the interquartile range of the data.
  d. Find any outliers in the data.
  e. Draw a box-and-whisker plot for the data.

21. Geometry The formula for the area of a trapezoid is $A = \frac{h}{2}(b_1 + b_2)$, where A represents the measure of the area, h represents the measure of the altitude, and b_1 and b_2 represent the measures of the bases. Find the measure of the area of the trapezoid whose height is 8 centimeters and whose bases measure 12 centimeters and 20 centimeters. (Lesson 1–2)

VOCABULARY

After completing this chapter, you should be able to define each term, property, or phrase and give an example or two of each.

Algebra
antilogarithm   (p. 618)
change of base formula   (pp. 594, 628)
characteristic   (p. 618)
common logarithms   (p. 617)
exponential equations   (p. 626)
exponential function   (pp. 594, 596)
exponential growth   (p. 622)
exponential growth rate   (p. 626)
general formula for growth and decay
    (p. 631)
logarithm   (p. 605)
logarithmic function   (pp. 594, 607)

mantissa   (p. 618)
natural logarithms   (p. 622)
power property of logarithms   (p. 613)
product property of logarithms   (p. 611)
property of equality for exponential
    functions   (p. 599)
property of equality for logarithmic
    functions   (p. 608)
quotient property of logarithms   (p. 612)

Problem Solving
use estimation   (p. 626)

UNDERSTANDING AND USING THE VOCABULARY

Choose the letter that best answers each question.

1. To solve $\log_8 (2y + 3) = \log_8 (y - 4)$, what property would you use?

2. What do the following have in common?
   log 6500 and log 6.5

3. What kind of function is $y = \frac{1}{2}(5)^x$?

4. Name the property shown in each example.
   a. $3 \log_2 5 = \log_2 5^3$
   b. $\log_4 2x = \log_4 2 + \log_4 x$
   c. $\log_3 \frac{4}{5} = \log_3 4 - \log_3 5$

5. What do the following have in common?
   log 0.28 and log $\frac{3}{4}$

6. What kind of function is $y = \log_2 x$?

7. To solve $9^{2p} = 27^{p-1}$, what property would you use?

a. characteristic
b. exponential function
c. logarithmic function
d. mantissa
e. power property of logarithms
f. product property of logarithms
g. property of equality for exponential functions
h. property of equality for logarithmic functions
i. quotient property of logarithms

OBJECTIVES AND EXAMPLES

Upon completing this chapter, you should be able to:

- simplify expressions and solve equations involving real exponents (Lesson 10–1)

  Simplify $16^{\sqrt{12}} \div 8^{\sqrt{3}}$.

  $$16^{\sqrt{12}} \div 8^{\sqrt{3}} = (2^4)^{\sqrt{12}} \div (2^3)^{\sqrt{3}}$$
  $$= 2^{8\sqrt{3}} \div 2^{3\sqrt{3}}$$
  $$= 2^{8\sqrt{3} - 3\sqrt{3}}$$
  $$= 2^{5\sqrt{3}}$$

- write exponential equations in logarithmic form and vice versa (Lesson 10–2)

  Write $3^3 = 27$ in logarithmic form.

  $$3^3 = 27$$
  $$3 = \log_3 27$$

  Write $\log_4 64 = 3$ in exponential form.

  $$\log_4 64 = 3$$
  $$64 = 4^3$$

- evaluate logarithmic expressions (Lesson 10–2)

  Evaluate $\log_3 3^5$.

  $$\log_3 3^5 = x$$
  $$3^x = 3^5$$
  $$x = 5$$

- solve equations involving logarithmic functions (Lesson 10–2)

  Solve $\log_b 16 = 4$.   Solve $\log_3 10 = \log_3 (2x)$.

  $\log_b 16 = 4$ $\log_3 10 = \log_3 (2x)$

  $\quad b^4 = 16$ $10 = 2x$

  $\quad b^4 = 2^4$ $5 = x$

  $\qquad b = 2$

REVIEW EXERCISES

Use these exercises to review and prepare for the chapter test.

Simplify each expression.

8. $3^{\sqrt{2}} \cdot 3^{\sqrt{2}}$ 22.3el   9. $\left(x^{\sqrt{5}}\right)^{\sqrt{20}}$

10. $\dfrac{49^{\sqrt{2}}}{7^{\sqrt{12}}}$ 0.29   11. $\left(8^{\sqrt{3}}\right)\left(8^{-2\sqrt{3}}\right)\left(8^{4\sqrt{3}}\right)$

Solve each equation.

12. $2^{6x} = 4^{5x+2}$ 13. $49^{3p+1} = 7^{2p-5}$

14. $9^{x^2} = 27^{x^2-2}$ 15. $9^x = \dfrac{1}{81}$

Write each equation in logarithmic form.

16. $7^3 = 343$ 17. $5^{-2} = \dfrac{1}{25}$

18. $4^0 = 1$ 19. $4^{\frac{3}{2}} = 8$

Write each equation in exponential form.

20. $\log_4 64 = 3$ $64 = 4^3$ 21. $\log_8 2 = \dfrac{1}{3}$

22. $\log_6 \dfrac{1}{36} = -2$ $\frac{1}{36} = 6^{-2}$ 23. $\log_6 1 = 0$

Evaluate each expression.

24. $6^{\log_6 7}$

25. $\log_{10} 10^{-3}$

26. $\log_{64} 8$

27. $\log_{12} 144$

Solve each equation.

28. $\log_b 9 = 2$ 29. $\log_4 x = \dfrac{1}{2}$

30. $\log_3 x = -3$ 31. $\log_7 2401 = x$

32. $\log_7 (x^2 + x) = \log_7 12$

33. $\log_6 12 = \log_6 (5x - 3)$

34. $\log_8 (3y - 1) = \log_8 (y + 4)$

35. $\log_2 (x^2 + 6x) = \log_2 (x - 4)$

OBJECTIVES AND EXAMPLES

• simplify and evaluate expressions using properties of logarithms   (Lesson 10–3)

Use $\log_{12} 9 \approx 0.884$ and $\log_{12} 18 \approx 1.163$ to evaluate $\log_{12} 2$.

$$\log_{12} 2 = \log_{12}\left(\frac{18}{9}\right)$$
$$= \log_{12} 18 - \log_{12} 9$$
$$\approx 1.163 - 0.884$$
$$\approx 0.279$$

• solve equations involving logarithms (Lesson 10–3)

Solve $2\log_5 6 - \frac{1}{3}\log_5 27 = \log_5 x$.

$$2\log_5 6 - \frac{1}{3}\log_5 27 = \log_5 x$$
$$\log_5 6^2 - \log_5 27^{\frac{1}{3}} = \log_5 x$$
$$\log_5 36 - \log_5 3 = \log_5 x$$
$$\log_5 \frac{36}{3} = \log_5 x$$
$$\log_5 12 = \log_5 x$$
$$12 = x$$

• find common logarithms and antilogarithms (Lesson 10–4)

Use a scientific calculator to find the logarithm of 56.4.

Use the ⎡LOG⎤ key.

Enter: 56.4 ⎡LOG⎤   *1.751279104*

Use a scientific calculator to find the antilogarithm of 2.738.

Use the ⎡10ˣ⎤ key.

Enter: 2.738 ⎡2nd⎤ ⎡10ˣ⎤   *547.0159629*

• find natural logarithms of numbers   (Lesson 10–5)

Use a scientific calculator to find the natural logarithm of 56.4.

Use the ⎡LN⎤ key.

Enter: 56.4 ⎡LN⎤   *4.032469158*

Use a scientific calculator to find the natural antilogarithm of 2.738.

Use the ⎡eˣ⎤ key.

Enter: 2.738 ⎡2nd⎤ ⎡eˣ⎤   *15.45604208*

REVIEW EXERCISES

Use $\log_9 7 \approx 0.8856$ and $\log_9 4 \approx 0.6309$ to evaluate each expression.

36. $\log_9 28$

37. $\log_9 49$

38. $\log_9 144$

39. $\log_9 15.75$

Solve each equation.

40. $\log_3 x - \log_3 4 = \log_3 12$

41. $\log_2 y = \frac{1}{3}\log_2 27$

42. $\log_5 7 + \frac{1}{2}\log_5 4 = \log_5 x$

43. $2\log_2 x - \log_2 (x + 3) = 2$

44. $\log_7 m = \frac{1}{3}\log_7 64 + \frac{1}{2}\log_7 121$

Use a scientific calculator to find the logarithm for each number rounded to four decimal places.

 $\log(x)$

45. 46.56 *1.6680*

46. 678.1 *2.8313*

47. 0.00468 *-2.3298*

48. 0.183 *-0.7375*

Use a scientific calculator to find the antilogarithm for each number rounded to four decimal places.

10^x

49. 2.75 *562.3413*

50. 1.999 *99.7700*

51. −0.567 *0.2710*

52. −3.47 *3.3884*

Use a scientific calculator to find each value rounded to four decimal places.

$\ln$

53. $\ln 2.3$

54. $\ln 9.25$

55. $\ln 50$

56. $\ln 0.05$

e^x

57. antiln 1.9755

58. antiln 2.246

59. antiln −0.489

60. antiln −4.5

OBJECTIVES AND EXAMPLES

• evaluate expressions involving logarithms with different bases (Lesson 10–6)

Approximate the value of $\log_8 72$ to three decimal places.

$$\log_a n = \frac{\log_b n}{\log_b a}$$

$$\log_8 72 = \frac{\log 72}{\log 8}$$

$$= \frac{1.8573}{0.9031}$$

$$\approx 2.057$$

REVIEW EXERCISES

Approximate the value of each logarithm to three decimal places.

61. $\log_5 15$

62. $\log_4 100$

63. $\log_{12} 15$

64. $\log_2 36$

65. $\log_9 108$

66. $\log_{11} 104$

• solve equations with variable exponents by using logarithms (Lesson 10–6)

Solve $3^{x-4} = 5^{x-1}$.

$$3^{x-4} = 5^{x-1}$$

$$\log 3^{x-4} = \log 5^{x-1}$$

$$(x - 4)\log 3 = (x - 1)\log 5$$

$$x\log 3 - 4\log 3 = x\log 5 - \log 5$$

$$x\log 3 - x\log 5 = 4\log 3 - \log 5$$

$$x(\log 3 - \log 5) = 4\log 3 - \log 5$$

$$x = \frac{4\log 3 - \log 5}{\log 3 - \log 5}$$

$$x \approx -5.4520$$

Use logarithms to solve each equation.

67. $2^x = 53$

68. $\log_4 11.2 = x$

69. $2.3^{x^2} = 66.6$

70. $3^{4x-7} = 4^{2x+3}$

71. $6^{3y} = 8^{y-3}$

72. $x = \log_{20} 1000$

73. $12^{x-4} = 4^{2-x}$

74. $2.1^{x-5} = 9.32$

APPLICATIONS AND PROBLEM SOLVING

75. **Atmospheric Pressure**   Atmosphere pressure can be determined by using the equation $P = 14.7(10)^{-0.02h}$, where P is the atmospheric pressure in pounds per square inch and h is the altitude above sea level in miles. Find the atmospheric pressure at an altitude of 3 miles above sea level. (Lesson 10–1)

76. **Finance**   Mr. and Mrs. Grauser invested $500 at 6.5% compounded continuously. (Lesson 10–5)
    a. Find the value of the investment after 7 years.
    b. When will the Grausers' investment triple?

77. **Biology**   For a certain strain of bacteria, k is 0.872 when t is measured in days. How long will it take 9 bacteria to increase to 738 bacteria? (Lesson 10–7)

78. **Chemistry**   Radium-226 decomposes radioactively. Its half-life, the time it takes for half of the sample to decompose, is 1800 years. Find the constant k in the decay formula for this compound. (Lesson 10–7)

A practice test for Chapter 10 is provided on page 921.

ALTERNATIVE ASSESSMENT

COOPERATIVE LEARNING PROJECT

Buying a House Exponential equations in the form of formulas enable people to make major financial decisions.

In this project, you will develop a report for a married couple that is buying their first house. Jenna and Jarod had a meeting with a mortgage banker. He told them that the rule of thumb for borrowing money for a mortgage is that their monthly payment should be no more than 28% of their gross monthly income. Jenna makes $35,000 a year as a nurse, and Jarod makes $25,000 a year as a teacher. They are thinking about waiting at least three more years to start a family. At that time, Jenna hopes to take a year off work and then resume her work part-time.

Use the monthly payment formula

$$MP = P \left[\frac{\frac{r}{12}\left(1 + \frac{r}{12}\right)^{12T}}{\left(1 + \frac{r}{12}\right)^{12T-1}} \right],$$ where P = amount of

mortgage, r = interest rate, and T = term of mortgage in years, to write a report for Jenna and Jarod about whether to use a 15-year or 30-year mortgage and how much they should plan on borrowing, after putting 20% down, based on their salaries, jobs, future plans, and term of the mortgage.

Consider these ideas while accomplishing your task.

- Calculate how much their monthly payment should be based on the rule of thumb and their circumstances.
- Analyze what calculations will be needed.

- Prepare a spreadsheet or chart to organize the various calculations.
- Determine the price range of a house for them considering a 15-year mortgage.
- Determine the price range of a house for them considering a 30-year mortgage.
- Write a report explaining all of their options.
- Give a recommendation and explain why.

THINKING CRITICALLY

- Show algebraically and graphically that $f(x) = \log_b x$ and $g(x) = b^x$ are inverse functions.
- On some calculators, $\log x$ and 10^x are on the same key position. Why is this? Give examples of other functions that share this relationship.

PORTFOLIO

Evaluating what you don't understand about a concept is important in achieving a comprehensive understanding of that subject. Think about the concept you found most difficult to understand in this chapter. Write about why you found that concept difficult to understand and explain how you managed to understand it. What methods did you use in order to gain understanding and did they work? Place this in your portfolio.

SELF EVALUATION

When searching for an error, you can use three categories to organize your search. These categories are misunderstanding, misapplied strategy, and miscalculation. When misunderstanding is the cause for the error, it is due to a failure to comprehend the overall concept of the problem. When a misapplied strategy is the cause for the error, it comes from using the wrong format. When miscalculation is the cause for the error, it is due to errors in calculating or checking.

Assess yourself. Do you categorize your errors? When looking for the reason for an error, do you go through a checklist to determine the error or do you randomly search for the problem? Give three error categories that could be used in your daily life and explain what each one means.

There are eight multiple-choice questions in this section. After working each problem, write the letter of the correct answer on your paper.

1. Which is a quadratic equation?

   A. $x^3 + 2x + 3 = 0$

   B. $3x - 7 = 0$

   C. $2x^2 - 9x + 7 = 0$

   D. $5x^2 + 3xy - 1 = 0$

2. **Geometry** A mathematics professor called the hardware store to order fencing to outline his pentagon-shaped garden. He gave the store manager the measures of each side as $\frac{1}{x}$, $\frac{2}{x-2}$, $\frac{3}{x}$, $\frac{4}{x-2}$, and $\frac{x}{x-2}$. The manager panicked until he found an algebra student to help him out. What was the perimeter of the professor's garden?

   A. $\frac{9+x}{x(x-2)}$

   B. $\frac{x^2 + 10x - 8}{x(x-2)}$

   C. $\frac{x^2 + 9x - 6}{x}$

   D. $\frac{x+2}{x-2}$

3. Find the second row of
   $$\begin{bmatrix} 0 & 9 \\ 4 & 3 \\ -2 & 7 \end{bmatrix} \cdot \begin{bmatrix} 2 & -1 & -2 \\ 0 & 8 & -5 \end{bmatrix}.$$

   A. $[0 \ \ 72 \ \ -45]$

   B. $[-4 \ \ 58 \ \ -31]$

   C. $[8 \ \ 20 \ \ -23]$

   D. $[72 \ \ 20 \ \ 58]$

4. **Geometry** Find the area of the figure below.

   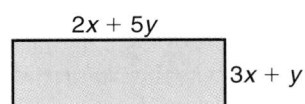

   A. $(5x^2 + 10xy + 5y^2)$ square units

   B. $(6x^2 + 17xy + 5y^2)$ square units

   C. $(21x^2y^2)$ square units

   D. $(11x^2 + 17xy + y^2)$ square units

5. Sally's office has a system to let people know when the department will have a meeting. Sally calls three people, then those three people each call three other people, and so on, until the whole department is notified. If it takes 10 minutes for a person to call three people and the whole department is notified within 30 minutes, how many people will be notified in the last round?

   A. 3 people

   B. 30 people

   C. 9 people

   D. 27 people

6. Choose the graph of $\frac{x^2}{25} + \frac{y^2}{16} = 1$.

   A.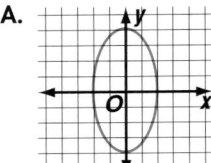

   B.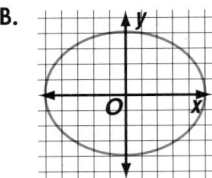

   C.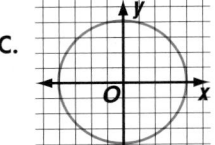

   D.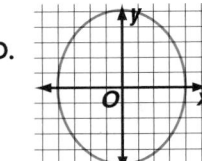

7. In planning a trip, the distance you travel varies jointly as the time and rate of speed. LaDonna Metcalf must travel 396 miles in 8 hours to meet a client. She travels 6 hours at 55 mph. She stops for a half an hour to rest and eat lunch. What is the minimum speed at which she must travel to meet her appointment?

   A. 66 mph

   B. 99 mph

   C. 44 mph

   D. 50 mph

8. **Geometry** The perimeter of Mr. Baxter's back yard is 152 feet. He plans to use wood fencing along one length of the yard, and wire fencing along the other three boundaries. If the length exceeds twice the width by 7 feet, how much wood fencing will Mr. Baxter require?

   A. 53 feet

   B. 23 feet

   C. 99 feet

   D. 138 feet

This section contains ten questions for which you will provide short answers. Write your answer on your paper.

9. Show that $\log_3 27 + \log_3 3 = \log_3 81$.

10. The graphs of $2y + x = 6$ and $y = 2x + 3$ contain two sides of a rectangle. If one vertex of the rectangle has coordinates $(8, 4)$, draw the rectangle.

11. Solve $6t^2 + 28t - 10 = 0$. Then find the sum and product of the roots to check the solution.

12. After conducting a survey on raising taxes for schools in Northridge, a statistician said the number of women in favor of the tax levy could be expressed by $\dfrac{3 + 10t^2 - 17t}{5t^2 + 4t - 1}$. The number of men in favor of the levy can be expressed by $\dfrac{4t^2 - 9}{3 + 5t + 2t^2}$. Find the ratio of women to men in simplest form.

13. Graph the system of equations below and state its solution and the type of system it represents (consistent and independent, consistent and dependent, or inconsistent).

$2x - 7 = 3y$
$x - y = 5$

14. Physics The formula for finding centripetal force F_c, the inward force that must be applied to keep an object moving in a circle, is $F_c = \dfrac{mv^2}{r}$. In this equation, m represents the mass of the object, v represents the velocity, and r represents the radius of the circular path. Solve the formula for the velocity and write the result in simplified form.

15. Find the value of the discriminant of the equation $2x^2 + x - 3 = 0$ and then describe the nature of its roots.

16. Solve the system of equations.
$3x + 4y = -7$
$2x + y = -3$

17. Luis' grandfather invested $150 at 5% interest compounded quarterly. When the account was recently given to Luis, it contained $6230. How long ago did Luis' grandfather invest the $150?

18. Solve $\log_4 (x + 3) + \log_4 (x - 3) = 2$.

This section contains five comparison problems which involve comparing two quantities, one in column A and one in column B. In certain questions, information related to one or both quantities is centered above them. All variables used represent real numbers.

Compare quantities A and B below.

- Write A if quantity A is greater.
- Write B if quantity B is greater.
- Write C if the two quantities are equal.
- Write D if there is not enough information to determine the relationship.

Column A	Column B
$x^2 = 100$ $y^2 = 36$	
19. x	y
20. antiln 0.288	x, if $2000 = 5^{0.045x}$
$f(x) = x + 4$ $g(x) = x + 9$	
21. $f[g(x)]$	$g[f(x)]$
22. $\dfrac{1}{3}$	$\dfrac{1}{3\sqrt{3}}$
23. y, if	x, if
$\dfrac{2y - 5}{6} - \dfrac{y - 5}{4} = \dfrac{3}{4}$	$\dfrac{x - 4}{x - 2} = \dfrac{x - 2}{x + 2} + \dfrac{1}{x - 2}$

Investigating Sequences and Series

Objectives

In this chapter, you will:

- find the next number in a sequence by looking for a pattern,
- find terms in arithmetic and geometric sequences,
- find sums of arithmetic and geometric series, and
- use the binomial theorem to find terms of a binomial expansion.

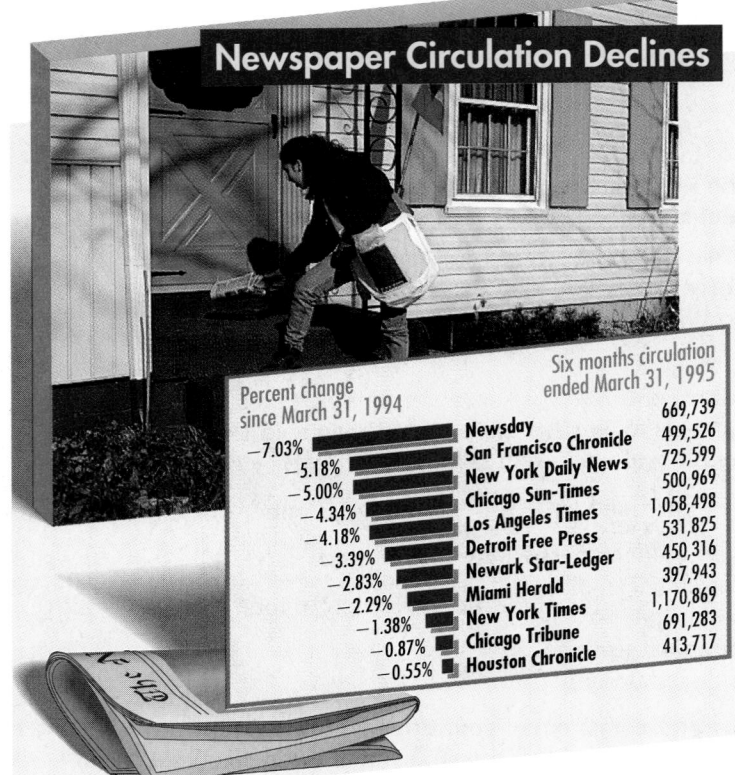

Newspaper Circulation Declines

Percent change since March 31, 1994		Six months circulation ended March 31, 1995
−7.03%	Newsday	669,739
−5.18%	San Francisco Chronicle	499,526
−5.00%	New York Daily News	725,599
−4.34%	Chicago Sun-Times	500,969
−4.18%	Los Angeles Times	1,058,498
−3.39%	Detroit Free Press	531,825
−2.83%	Newark Star-Ledger	450,316
−2.29%	Miami Herald	397,943
−1.38%	New York Times	1,170,869
−0.87%	Chicago Tribune	691,283
−0.55%	Houston Chronicle	413,717

Source: Audit Bureau of Circulations

In 1993, the U.S. Department of Education organized the most comprehensive study of literacy ever done of Americans. The results were shocking, indicating that half of all adults have serious deficiencies in their ability to function and make practical use of the basic skills of reading, writing, and arithmetic.

TIME *Line*

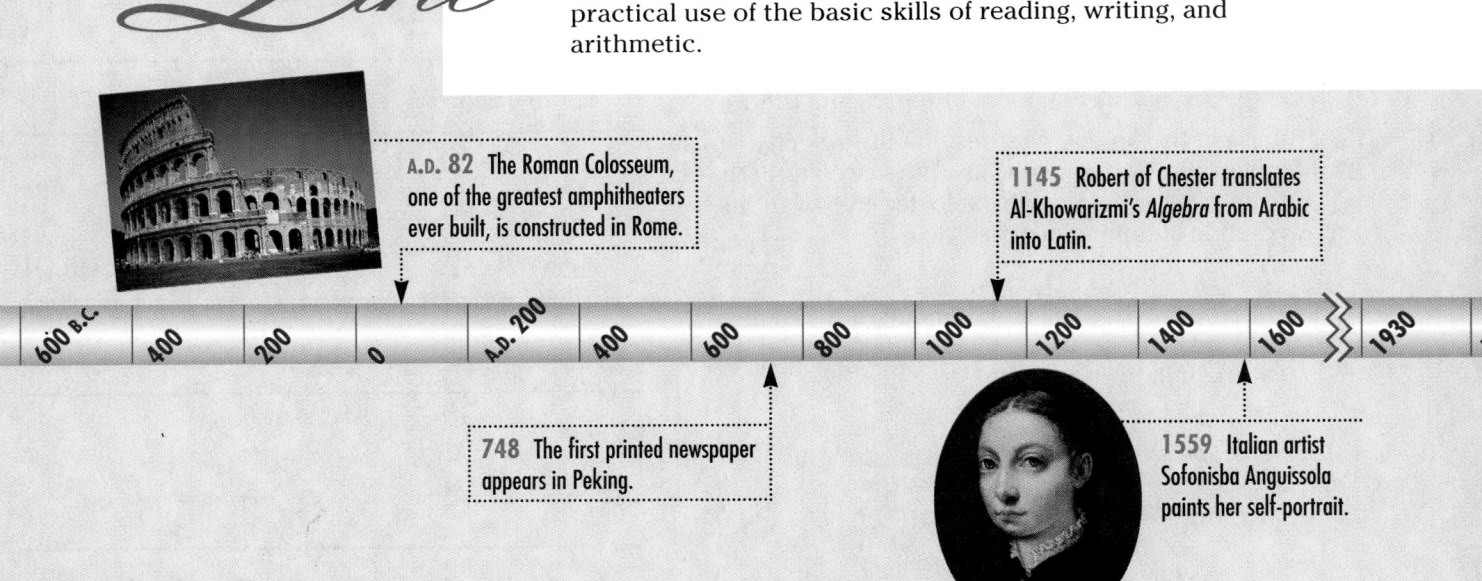

A.D. 82 The Roman Colosseum, one of the greatest amphitheaters ever built, is constructed in Rome.

1145 Robert of Chester translates Al-Khowarizmi's *Algebra* from Arabic into Latin.

600 B.C.   400   200   0   A.D. 200   400   600   800   1000   1200   1400   1600   1930

748 The first printed newspaper appears in Peking.

1559 Italian artist Sofonisba Anguissola paints her self-portrait.

Chapter Project

In small groups, research different volunteer programs in your community that help promote literacy and an appreciation for reading. There may be programs at the library, at colleges or universities in your area, or sponsored by youth groups. Develop a volunteer project or club at your school that would encourage learning in various age groups such as senior citizens or children in elementary school. Make an outline listing the types of services that you could provide to encourage literacy and write a report that includes the time, materials, such as books and magazines, and volunteers needed for the project.

One method of combating America's literacy problem is to demonstrate a passion for reading. **Lynnea Fajardo,** a student at Red Mountain High School in Mesa, Arizona, is a literacy volunteer. She is president of her school's RIF (Reading Is Fundamental) Club, and volunteers her free time promoting literacy by distributing free books to students, libraries, hospitals, and other locations in her community. The RIF Club participates in reading projects, creates open reading corners, and with a program called Reading Buddies, is spreading the love of books to elementary students.

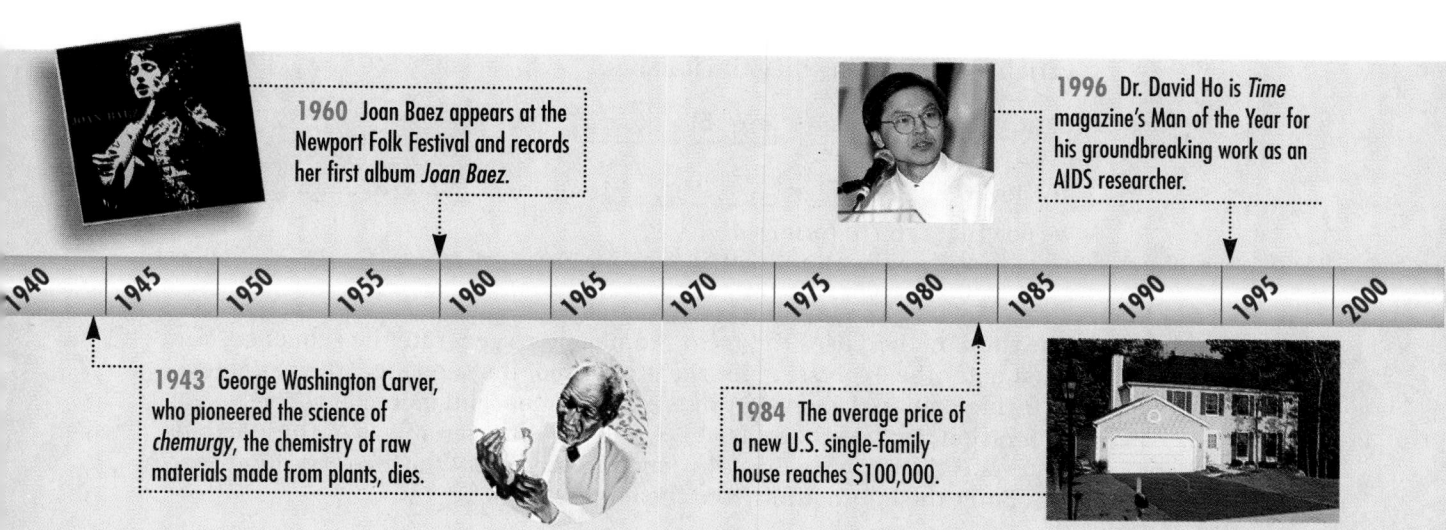

1960 Joan Baez appears at the Newport Folk Festival and records her first album *Joan Baez*.

1996 Dr. David Ho is *Time* magazine's Man of the Year for his groundbreaking work as an AIDS researcher.

1940 1945 1950 1955 1960 1965 1970 1975 1980 1985 1990 1995 2000

1943 George Washington Carver, who pioneered the science of *chemurgy*, the chemistry of raw materials made from plants, dies.

1984 The average price of a new U.S. single-family house reaches $100,000.

11–1A Graphing Technology
Arithmetic Sequences

A Preview of Lesson 11–1

Graphing calculators have several features that allow us to investigate sequences. Both numerical and graphical techniques can be used. In Example 1, numerical techniques are used to find the desired term.

Example **Find the 5th term of the arithmetic sequence 17, 21, 25, ...**

Method 1

Begin by storing the first term of the sequence as x.

Enter: 17 [STO▶] [X,T,θ,n] [ENTER]

Now, to generate the second term, store $17 + 4$ as x.

Enter: [X,T,θ,n] [+] 4 [STO▶] [X,T,θ]
[ENTER]

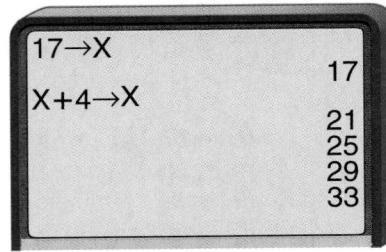

Continue to press [ENTER] to generate successive terms of the sequence.

Remember, whatever the current value for x, [ENTER] computes $x + 4$. The 5th term is 33.

Method 2

First determine the formula for the nth term of the sequence. Then use the table feature of the calculator to generate values. The first term of the sequence is 17, and the common difference is 4. So the formula for the nth term is $a_n = 17 + (n - 1)4$.

Enter this formula into the calculator as $Y1 = 17 + (x - 1)4$.

Enter: [Y=] 17 [+] [(] [X,T,θ,n] [−] 1 [)] 4

Now set up the table with the independent variable representing the term number.

Enter: [2nd] [TblSet] 1 [ENTER] 1 [ENTER]

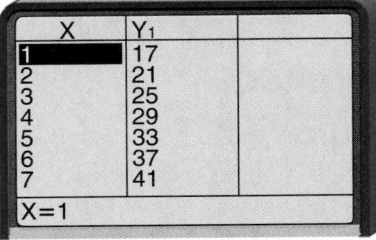

To view the table, enter [2nd] [TABLE]. Again, we see that the 7th term is 41.

Method 3

The third method uses the "seq(" command to generate the sequence. You must enter the expression for the nth term of the sequence, the variable to be incremented, starting value, ending value, and increment. The expression for the nth term is $17 + (N - 1)4$, the variable is N, the starting value is 1, the ending value is 10, and the increment for the term number is 1. Enter the information from the home screen.

You will learn how to find the formula for sequences in Lessons 11–1 and 11–3.

Enter: [2nd] [LIST] 5 17 [+] [(]

[ALPHA] [N] [-] 1 [)] 4 [,]

[ALPHA] [N] [,] 1 [,] 10 [,]

1 [)] [ENTER]

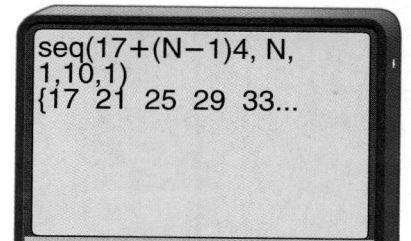

seq(17+(N−1)4, N, 1,10,1)
{17 21 25 29 33...

The calculator generates a list of values for the first 10 terms of the sequence.

Use the right arrow key to see the first seven terms. The seventh term is 41.

You can use a graphing calculator to generate the graph of a sequence. The graph will be a series of plotted points. The x-coordinates represent the term numbers, and the y-coordinates represent the values of the terms. You can then find the terms of a sequence by using the graph.

Example **2** Use graphing to find the 5th and 25th terms of the sequence −112, −101, −90, −79, ...

Use the Edit option on the STAT menu feature to enter the term numbers 1, 2, 3, 4 into L1 and the value of the terms −112, −101, −90, −79 into L2.

We want to plot only those points that represent terms of the sequence. So, from the MODE menu, select Dot. Now set a viewing window for [0, 94] with a scale factor of 10 and [−120, 200] with a scale factor of 25. This viewing window is used so that only integer values appear.

TECHNOLOGY *Tips*

You can also graph the line containing the terms of an arithmetic sequence by entering the formula for the nth term as Y1 and then pressing [GRAPH].

Enter: [STAT] [▶] 5 [2nd] [L1]

[,] [2nd] [L2] [ENTER] [Y=]

[VARS] 5 [▶] [▶] 7 [GRAPH]

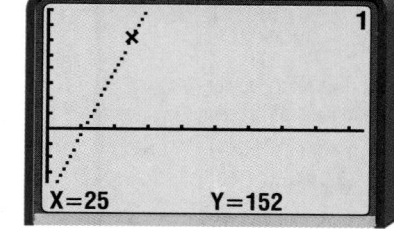

X=25   Y=152

Use the TRACE feature to find the terms. The 5th term is −68, and the 25th term is 152.

EXERCISES

Find the 11th term of each sequence by using a numerical technique.

1. 47, 54, 61, 68, 75, 82, ...
2. 4.5, 3.75, 3, 2.25, 1.5, ...
3. 2.132, 3.13, 4.128, 5.126, ...
4. −57, −59.5, −62, −64.5, −67, ...

Find the 18th term of each sequence by graphing.

5. −20, −16, −12, −8, −4, ...
6. −0.8, 0, 0.8, 1.6, 2.4, ...
7. 27, 18, 9, 0, −9, ...
8. 3, $\frac{13}{5}$, $\frac{11}{5}$, $\frac{9}{5}$, ...
9. Use a graphing calculator to find the missing terms in the sequence 148, 146.83, 145.66, __?__, __?__, __?__, 140.98.
10. 115 is the __?__ th term of the sequence 17, 20.5, 24, 27.5, ...

Arithmetic Sequences

What **YOU'LL LEARN**

- To find the next term in a sequence by looking for a pattern,
- to find the nth term of an arithmetic sequence,
- to find the position of a given term in an arithmetic sequence, and
- to find arithmetic means.

Why **IT'S IMPORTANT**

You can use arithmetic sequences to solve problems involving antiques and broadcasting.

APPLICATION
Broadcasting

In radio broadcasting, autumn is one of the most critical periods for ratings. Radio stations try to pull in as many listeners as they can by using a variety of gimmicks. In the fall of 1995, radio station WBNS 97.1 had a contest in which listeners had a chance to win $1000 every hour. In order to win, listeners needed to call in and correctly answer a contest question. The contest started with $1000, and $97 was added for the next caller each time the previous caller answered the question incorrectly. Suppose you were the 18th caller and the first to answer the question correctly. How much money would you win? *This problem will be solved in Example 3.*

We can use a table to show the amount of prize money available after each caller answers incorrectly.

Number of Callers	1	2	3	4	5	6	7	8
Prize Money	$1000	$1097	$1194	$1291	$1388	$1485	$1582	$1679

The graph shows the information from the table. This graph represents a *discrete* function. That is, the domain is made up of distinct values and there is no continuity between those values. The range is the set of numbers representing the amount of prize money. This set of numbers is an example of a **sequence.** Each number in a sequence is called a **term.** The first term is symbolized by a_1, the second term by a_2, and so on to a_n, the nth term. *What kind of figure would you have if the points on the graph were connected?*

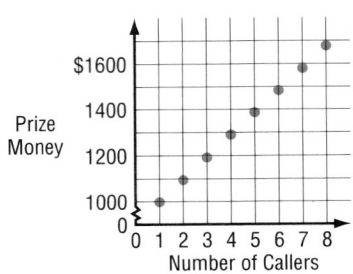

You can find the next term in a sequence by looking for a pattern. One way to do this is to find the difference of consecutive terms.

Example ①

PROBLEM SOLVING

Look for a Pattern

The table at the right shows the cost of mailing letters first class in the U.S. in 1995. If the cost continues to increase at the same rate for each ounce, find how much it costs to mail letters that weigh 6, 7, and 8 ounces.

Weight Not Exceeding (ounces)	Cost
1	$0.32
2	0.55
3	0.78
4	1.01
5	1.24

Find the difference of consecutive terms.

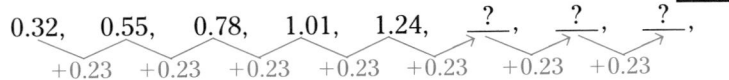

0.32,   0.55,   0.78,   1.01,   1.24,   ?,   ?,   ?,   ...
+0.23   +0.23   +0.23   +0.23   +0.23   +0.23   +0.23

The difference between each term is 0.23. The next three terms are $1.24 + 0.23$ or 1.47, $1.47 + 0.23$ or 1.70, and $1.70 + 0.23$ or 1.93. So, it costs $1.47, $1.70, and $1.93 to mail letters that weigh 6, 7, and 8 ounces, respectively.

The sequence shown in the table in Example 1 contains five terms. Therefore, $a_1 = 0.32$ and $a_5 = 1.24$. Each term of the sequence can be found by adding 0.23 to the previous term. A sequence of this type is called an **arithmetic sequence.** The number added to find the next term of an arithmetic sequence is called the **common difference** and is symbolized by the variable d.

Definition of Arithmetic Sequence	An arithmetic sequence is a sequence in which each term after the first is found by adding a constant, called the common difference d, to the previous term.

To find the next terms in a sequence that you know is arithmetic, you do not need to look for a pattern. Instead, find the common difference by subtracting any term from its succeeding term. Then add the common difference to the last term you are given to find successive terms.

Example ❷ **Find the next four terms of the arithmetic sequence 91, 83, 75, ...**

Find the common difference d by subtracting two consecutive terms.

$83 - 91 = -8$ and $75 - 83 = -8$ So, $d = -8$.

Now add -8 to the last term of the sequence, and then continue adding until the next four terms are found.

$$75 + (-8) = 67$$
$$67 + (-8) = 59$$
$$59 + (-8) = 51$$
$$51 + (-8) = 43$$

Therefore, the next four terms of the sequence are 67, 59, 51, and 43.

There is a pattern in the way terms of an arithmetic sequence are formed. It is possible to develop a formula that expresses each term of an arithmetic sequence in terms of the first term a_1 and the common difference d. Let's use the terms of the sequence in Example 2.

Sequence	numerical	91	83	75	67	...	
	symbols	a_1	a_2	a_3	a_4	...	a_n
Expressed in Terms of d and the First Term	numerical	$91 + 0(-8)$	$91 + 1(-8)$	$91 + 2(-8)$	$91 + 3(-8)$	...	$91 + (n-1)(-8)$
	symbols	$a_1 + 0 \cdot d$	$a_1 + 1 \cdot d$	$a_1 + 2 \cdot d$	$a_1 + 3 \cdot d$	...	$a_1 + (n-1)d$

The following formula generalizes this pattern for any sequence.

Formula for the nth Term of an Arithmetic Sequence	The nth term a_n of an arithmetic sequence with first term a_1 and common difference d is given by $$a_n = a_1 + (n-1)d,$$ where n is a positive integer.

Example **3**

APPLICATION

Broadcasting

Refer to the application at the beginning of the lesson. How much money would you win if you were the 18th caller and you answered the contest question correctly?

If $a_1 = 1000$ and $d = 97$, a_{18} represents the cash the 18th caller will win. So, $n = 18$.

Find a_{18} using $a_n = a_1 + (n - 1)d$.

$a_n = a_1 + (n - 1)d$

$a_{18} = 1000 + (18 - 1)97$ *Substitute the known values.*

$a_{18} = 1000 + 1649$ or 2649 *Simplify.*

If you were the 18th caller and you answered the question correctly, you would win $2649.

Jack L. Cooper (1889–)

African-American radio pioneer Jack Cooper was the first to originate a community news broadcast about African Americans in Chicago during the 1920s. He was also the first to play popular African-American music.

Sometimes you may know two terms of a sequence, but they are not consecutive terms of that sequence. The terms between any two nonconsecutive terms of an arithmetic sequence are called **arithmetic means.** In the sequence below, 32, 41, and 50 are the three arithmetic means between 23 and 59.

$$14, \mathbf{23}, 32, 41, 50, \mathbf{59}, 68, 77$$

Example **4**

a. **Find the four arithmetic means between 18 and 78.**

b. **Graph the sequence using the *x*-axis for the number of the term and the *y*-axis for the term itself.**

a. You can use the *n*th term formula to find the common difference. In the sequence 18, _?_ , _?_ , _?_ , _?_ , 78, 18 is a_1 and 78 is a_6.

$a_n = a_1 + (n - 1)d$

$a_6 = 18 + (6 - 1)d$ $a_1 = 18$ *and* $n = 6$

$78 = 18 + 5d$ $a_6 = 78$

$60 = 5d$ *Subtract 18 from each side.*

$12 = d$ *Divide each side by 5.*

Now use the value of *d* to find the four arithmetic means.

$18 + 12 = 30$ $30 + 12 = 42$ $42 + 12 = 54$ $54 + 12 = 66$

The arithmetic means are 30, 42, 54, and 66.

b.

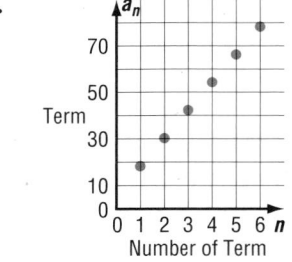

You can write a linear equation for the nth term of an arithmetic sequence.

Example **5** **Write an equation for the nth term of the arithmetic sequence 6, 13, 20, 27, ...**

In this sequence, $a_1 = 6$ and $d = 7$. Use the nth term formula to write the equation.

$a_n = a_1 + (n - 1)d$

$a_n = 6 + (n - 1)7$ *Substitute the known values.*

$a_n = 6 + 7n - 7$

$a_n = 7n - 1$ *Simplify.*

The equation is $a_n = 7n - 1$. The graph of the line given by this equation contains the terms of the arithmetic sequence. *Compare the slope of the line described by this equation and the value of the common difference.*

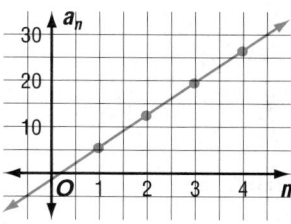

MODELING MATHEMATICS

Arithmetic Sequences

Materials: isometric dot paper

Look at the figures below. The length of the side of each cube is 1 centimeter. Copy the figures on isometric dot paper. Be sure to draw correctly.

Figure 1   Figure 2   Figure 3

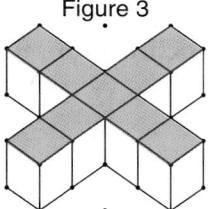

Your Turn

a. Based on the pattern, draw Figure 4 on the dot paper.

b. Find the volumes of the four figures.

c. Suppose the number of cubes in the pattern continues. Write an equation to represent the volume of Figure n.

d. What would the volume of Figure 12 be?

CHECK FOR UNDERSTANDING

Communicating Mathematics

Study the lesson. Then complete the following.

1. **Explain** how to determine whether a list of numbers is an arithmetic sequence.

2. **Explain** whether 24 is a term in the sequence represented by $a_n = 5 + (n - 1)2$.

3. Refer to the graph of the arithmetic sequence at the right.

   a. Write the first five terms of the sequence.

   b. Explain why the points are not connected.

   c. What is the equation of the line that passes through these points?

   d. State the slope of the line in part c and explain how it relates to the nature of the sequence.

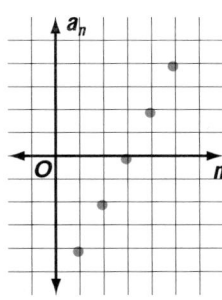

4. **You Decide** Andrea says that 220 is a term of the sequence $a_n = 16 - (n - 1)12$. Janice says that this is impossible. Who is correct? Explain.

Guided Practice

Find the next four terms of each arithmetic sequence.

5. 12, 16, 20, ...

6. 3, 1, −1, ...

Find the first five terms of each arithmetic sequence described.

7. $a_1 = 5, d = 3$

8. $a_1 = 14, d = -2$

Find the nth term of each arithmetic sequence.

9. $a_1 = 3, d = -5, n = 24$

10. $a_1 = -5, d = 7, n = 13$

11. Complete: 68 is the __?__th term of −2, 3, 8, ...

12. Find a_{13} for the arithmetic sequence −17, −12, −7, ...

13. **a.** Find the three arithmetic means between 44 and 92.
    b. Graph the sequence using the x-axis for the number of the term and the y-axis for the term itself.

14. Write an equation for the nth term of the arithmetic sequence −26, −15, −4, 7, ...

15. **Consumerism** Jamila bought a snowboard priced at $545. She put $145 down and made equal monthly payments. At the end of every month, she was given a statement of the balance owed. For the first four months, the balances were $361, $322, $283, and $244. If she paid the same amount each month, what was her balance at the end of 8 months?

EXERCISES

Practice

Find the next four terms of each arithmetic sequence.

16. 9, 16, 23, ...

17. 31, 24, 17, ...

18. $\frac{1}{4}, \frac{3}{4}, \frac{5}{4}, ...$

19. −7.8, −3.8, 0.2, ...

Find the first five terms of each arithmetic sequence described.

20. $a_1 = 12, d = -3$

21. $a_1 = 41, d = 5$

22. $a_1 = \frac{4}{3}, d = -\frac{1}{3}$

23. $a_1 = \frac{5}{8}, d = \frac{3}{8}$

Find the nth term of each arithmetic sequence.

24. $a_1 = 3, d = 7, n = 14$

25. $a_1 = -4, d = -9, n = 20$

26. $a_1 = 5, d = \frac{1}{3}, n = 12$

27. $a_1 = \frac{5}{2}, d = -\frac{3}{2}, n = 11$

28. $a_1 = 35, d = 3, n = 101$

29. $a_1 = 20, d = 4, n = 81$

Complete each statement.

30. 170 is the __?__th term of −4, 2, 8, ...

31. 124 is the __?__ th term of −2, 5, 12, ...

32. −14 is the __?__ nd term of $2\frac{1}{5}, 2, 1\frac{4}{5}, ...$

Find the indicated term in each arithmetic sequence.

33. a_{12} for −17, −13, −9, ...

34. a_{21} for 121, 118, 115, ...

35. a_{43} for 5, 9, 13, 17, ...

36. a_{12} for 8, 3, −2, ...

Find the arithmetic means in each sequence. Then graph each sequence using the x-axis for the number of the term and the y-axis for the term itself.

37. 55, _?_, _?_, _?_, 115

38. 10, _?_, _?_, −8

39. _?_, −5, _?_, _?_, 4, _?_

40. 3, _?_, _?_, _?_, _?_, _?_, 27

Find the values of y that make each sequence arithmetic.

41. −4, 2, 8, $3y + 5$, ...

42. 5, 9, $2y − 1$, ...

43. $y + 2$, 6, y, ...

44. $y + 8$, $4y + 6$, $3y$, ...

Write an equation for the nth term of each arithmetic sequence.

45. 7, 16, 25, 34, ...

46. 18, 11, 4, −3, ...

47. a. The first term of an arithmetic sequence is 7, and each term is 4 more than the previous term. Find the value of the eighth term.

   b. Write a formula for the nth term of this sequence.

48. The first three pentagonal numbers are shown below.

   a. Make drawings to find the next three pentagonal numbers.

   b. Do the pentagonal numbers form an arithmetic sequence? Write an equation representing the nth pentagonal number.

   c. What is the common difference?

   d. Justify whether 397 is a pentagonal number.

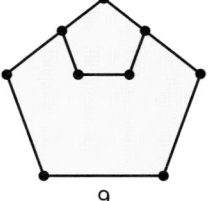

1   5   9

49. The fifth term of an arithmetic sequence is 19, and the 11th term is 43. Find the first term and the 87th term

50. Find three numbers that have a sum of 36, a product of 276, and form an arithmetic sequence.

Graphing Calculator

Use a graphing calculator to find the 12th term of each sequence. Then graph the sequence. Sketch the graph on grid paper.

51. −31, −24, −17, −10, ...

52. 317, 313, 309, 305, ...

53. −16, −13, −10, −7, ...

54. 23, 29, 35, 41, 47, ...

Critical Thinking

55. Use an arithmetic sequence to find how many multiples of 7 are between 36 and 391.

56. The first three terms of an arithmetic sequence are u, v, and w. Express w in terms of u and v. Justify your answer.

Applications and Problem Solving

57. Geology Geologists have calculated that the continents of Europe and America are drifting apart at an average of 12 miles every 1 million years. That's an average of 0.75 inch a year. If the continents continue to drift apart at the same rate, how many inches will they drift in 50 years? (*Hint:* $a_1 = 0.75$.)

58. Physics People used to believe that the heavier an object was, the faster it would fall. Galileo proved that this was incorrect. He dropped two different weights simultaneously from the Leaning Tower of Pisa, and they both hit the ground at the same instant. When an object is dropped from a tall building, no matter how much it weighs, it falls 16 feet in the first second, 48 feet in the second second, and 80 feet in the third second. How many feet would a falling object fall in the tenth second?

59. Look for a Pattern Look at the figures below.

Figure 1   Figure 2   Figure 3

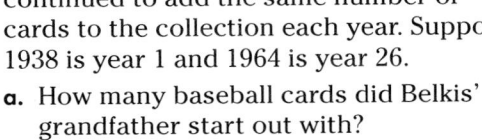

a. Describe the pattern and draw what you think Figure 4 should look like.
b. How many rectangles are in Figure 1? (*Hint:* There are more than 5.)
c. How many rectangles are there in Figures 2 and 3?
d. How many rectangles would there be in Figure 50?
e. How many rectangles would there be in Figure *n*?

60. Antiques In 1996, Belkis received a collection of 574 baseball cards that was started by her grandfather in 1938. Each year, her grandfather added the same number of cards to his collection. When Belkis' father received the collection in 1964, there were 254 cards. Belkis' father continued to add the same number of cards to the collection each year. Suppose 1938 is year 1 and 1964 is year 26.

a. How many baseball cards did Belkis' grandfather start out with?
b. How many cards were added to the collection each year?
c. If Belkis continues to add the same number of cards to the collection each year, how many cards will she have in 2010?

Mixed Review

61. Biology The general formula for growth is $y = ne^{kt}$, where y is the final amount, n is the initial amount, k is a constant, and t is the time. A culture of a certain bacteria will grow from 500 to 4000 bacteria in 90 minutes. Find the constant k for this bacteria if t is in hours. (Lesson 10–7)

62. Solve $\log_3 729 = x$. (Lesson 10–2)

63. Simplify $\dfrac{y^2 - y}{w^2 - y^2} + \dfrac{y^2 - 2y + 1}{1 - y}$. (Lesson 9–4)

64. If $f(x) = 3x$ and $g(x) = x - 1$, find $f[g(x)]$. (Lesson 8–7)

65. State whether the graph of $x = (y + 4)^2 - 6$ is a parabola, a circle, an ellipse, or a hyperbola. (Lesson 7–6)

66. Geometry Find the perimeter of a quadrilateral with vertices at (4, 5), (−4, 6), (−5, −8), and (6, 3). (Lesson 7–1)

67. Name the vertex, axis of symmetry, and direction of opening for the graph of $f(x) = -4(x - 9)^2$. (Lesson 6–6)

68. Solve $\sqrt[3]{x + 5} + 6 = 4$ (Lesson 5–8)

69. If $A = \begin{bmatrix} -3 & 5 \\ 1 & -4 \end{bmatrix}$, find A^{-1}. (Lesson 4–5)

70. Determine the slope of the line that passes through $(1, -3)$ and $(0, -5)$. (Lesson 2–3)

71. Solve $4 + |2x| > 0$. (Lesson 1–7)

WORKING ON THE In·ves·ti·ga·tion

Refer to the Investigation on pages 590–591.

The weight of a jumper has an effect on the stretching of the bungee. You decide to create an experiment to measure the degree of stretch in relation to weight. You will need three rubber bands of different sizes, a small paper cup, a paper clip, 50 or more washers, a centimeter ruler, a wooden dowel rod, and tape.

1 Straighten the paper clip and use it to punch two small holes on opposite sides of the cup about a centimeter below the rim of the cup. Loop one of the rubber bands inside the cup and slide the paper clip through the two holes and through the loop of the rubber band, so that the cup could be held by the rubber band like a bucket on a rope.

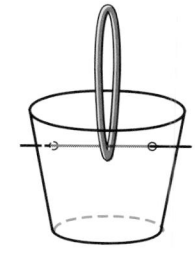

2 Tape a wooden dowel to the top of the table so that about 5 inches of the dowel hangs over the edge of the table. Slide the free end of the rubber band over the dowel so that the cup hangs from the dowel. Make a table with columns labeled *length of rubber band* and *number of washers*. Measure the distance from the top of the cup to the base of the dowel. Record this as your initial length with zero washers.

3 Put five washers in the bottom of the cup. Measure the length of the rubber band again from the top of the cup to the base of the dowel. Record the length and number of washers.

4 Continue adding washers, five at a time, and measuring the length of the rubber band. Record the data as you add each set of washers.

5 Study the numbers in your table. What patterns do you observe? Are you able to continue adding washers until you reach the length that matches the maximum elastic potential that you found earlier?

Add the results of your work to your Investigation Folder.

Arithmetic Series

APPLICATION
World Cultures

Kwanzaa is an African-American harvest festival celebrating the new year. During the festival, a special ritual is performed involving the lighting of seven candles which are called the *mushumaa saba*. Each of these candles symbolizes a different human quality. On the first night, one of the candles is lit and then blown out. On the second night, a new candle and the candle from the previous night are lit and then blown out. For seven nights, this pattern of lighting a new candle and relighting all the candles from the previous nights continues. What is the total number of lightings during this festival?

One candle is lit on the first night, two candles are lit on the second night, and so forth. These lightings can be represented by the sequence 1, 2, 3, 4, 5, 6, 7.

The total would be the sum of the terms in the sequence.

first night	second night	third night	fourth night	fifth night	sixth night	seventh night	
1 +	2 +	3 +	4 +	5 +	6 +	7 =	28

The indicated sum of the terms of a sequence is called a **series.** The series shown above is an **arithmetic series.**

The lists below show examples of arithmetic sequences and their corresponding arithmetic series.

Arithmetic Sequences	**Arithmetic Series**
4, 7, 10, 13, 16	$4 + 7 + 10 + 13 + 16$
$-10, -4, 2$	$-10 + (-4) + 2$
$\frac{2}{7}, \frac{6}{7}, \frac{10}{7}, \frac{14}{7}$	$\frac{2}{7} + \frac{6}{7} + \frac{10}{7} + \frac{14}{7}$

The symbol S_n is used to represent the sum of the first n terms of a series. For example, S_3 means the sum of the first three terms of a series. In the series $4 + 7 + 10 + 13 + 16$, S_3 would be $4 + 7 + 10$ or 21.

If a series has a large number of terms, it is not convenient to list all the terms and then find their sum. To develop a general formula for the sum of any arithmetic series, let's consider the series of candle lightings.

$$S_7 = 1 + 2 + 3 + 4 + 5 + 6 + 7$$

Suppose we write S_7 in two different orders and find the sum.

$$
\begin{array}{rl}
S_7 = & 1 + 2 + 3 + 4 + 5 + 6 + 7 \\
+\;\; S_7 = & 7 + 6 + 5 + 4 + 3 + 2 + 1 \\
\hline
2 \cdot S_7 = & 8 + 8 + 8 + 8 + 8 + 8 + 8
\end{array}
$$

7 sums of 8

$$2 \cdot S_7 = 7(8)$$
$$S_7 = \frac{7}{2}(8) \quad \textit{Divide each side by 2.}$$

Now let's analyze what these numbers represent in terms of S_n. In the equation $S_7 = \frac{7}{2}(8)$, 7 represents n and 8 represents the sum of the first and last terms, $a_1 + a_n$. Thus, we can replace the equation with the formula $S_n = \frac{n}{2}(a_1 + a_n)$. This formula can be used to find the sum of any arithmetic series.

Sum of an Arithmetic Series	The sum S_n of the first n terms of an arithmetic series is given by $S_n = \frac{n}{2}(a_1 + a_n)$, where n is a positive integer.

Example **1** **Find the sum of the first 50 positive even integers.**

$$S_n = \frac{n}{2}(a_1 + a_n)$$
$$S_{50} = \frac{50}{2}(2 + 100) \quad \textit{$a_1 = 2$, $n = 50$, and $a_n = a_{50} = 100$}$$
$$= 25(102) \text{ or } 2550$$

The sum of the first 50 positive even integers is 2550. *Check this result.*

In Lesson 11–1, you learned that in an arithmetic sequence, $a_n = a_1 + (n - 1)d$. Using this formula and substitution gives us another version of the formula for the sum of an arithmetic sequence.

$$S_n = \frac{n}{2}(a_1 + a_n)$$
$$= \frac{n}{2}\{a_1 + [a_1 + (n - 1)d]\} \quad \textit{Substitute $a_1 + (n - 1)d$ for a_n.}$$
$$= \frac{n}{2}[2a_1 + (n - 1)d] \quad \textit{Combine like terms.}$$

You can use this formula when you do not know the value of the last term.

Example **2** **Find the sum of the first 40 terms of an arithmetic series in which $a_1 = 70$ and $d = -21$.**

The series is $70 + 49 + 28 + 7 + \ldots$.

$$S_n = \frac{n}{2}[2a_1 + (n - 1)d] \quad \textit{Use the formula for S_n.}$$
$$= \frac{40}{2}[2(70) + (40 - 1)(-21)] \quad \textit{Substitute the known values.}$$
$$= -13{,}580$$

3 Refer to Exercise 58 in Lesson 11–1. A free-falling object falls 16 feet in the first second, 48 feet in the second second, 80 feet in the third second, and so on. How many feet would a free-falling object fall in 20 seconds if air resistance is ignored?

To find the total distance fallen by an object, add the first 20 terms of the sequence 16, 48, 80, ...

$$S_n = \frac{n}{2}[2a_1 + (n-1)d]$$
$$= \frac{20}{2}[2(16) + (20-1)32] \quad \text{\textit{n = 20, a}}_1 \text{\textit{ = 16, d = 32}}$$
$$= 10(32 + 608)$$
$$= 6400$$

A free-falling object would fall 6400 feet in 20 seconds.

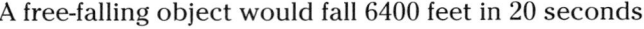

It is sometimes necessary to use both the sum formula and the *n*th term formula to solve a problem. You must analyze the information you are given and then decide which formula to use first.

Example **4** Find the first three terms of an arithmetic series in which $a_1 = 13$, $a_n = 157$, and $S_n = 1445$.

Step 1 Use $S_n = \frac{n}{2}(a_1 + a_n)$ first to find *n* since a_1, a_n, and S_n are known.

$$S_n = \frac{n}{2}(a_1 + a_n)$$
$$1445 = \frac{n}{2}(13 + 157)$$
$$1445 = 85n$$
$$n = 17$$

Step 2 Find *d*.

$$a_n = a_1 + (n-1)d$$
$$157 = 13 + (17-1)d$$
$$144 = 16d$$
$$d = 9$$

Step 3 Determine a_2 and a_3.

$a_2 = 13 + 9$ or 22 $a_3 = 22 + 9$ or 31

The first three terms are 13, 22, and 31.

Writing out a series is often time-consuming and lengthy. To simplify this, mathematicians use a more concise notation called **sigma** or **summation notation.** $2 + 4 + 6 + 8 + ... + 20$ can be expressed as $\sum\limits_{n=1}^{10} 2n$. This expression is read *the sum of 2n as n increases from 1 to 10.*

last value of n
$$\sum_{n=1}^{10} 2n$$
first value of n   *formula for the related sequence*

When using sigma notation, the variable defined below the Σ (sigma) is called the **index of summation.** The upper number is the *upper limit* of the index. To generate the terms of the series, successively replace the index of summation with consecutive integers as values of *n*. In this series, the values of *n* are 1, 2, 3, and so on, through 10.

Example **5** Write the terms of $\sum\limits_{k=3}^{7} (2k + 5)$ and find the sum.

Method 1	**Method 2**
Replace k with 3 and then with 4, 5, 6, and 7 to find the terms.	Use $S_n = \frac{n}{2}(a_1 + a_n)$.
$\sum\limits_{k=3}^{7} (2k + 5) = [2(3) + 5] + [2(4) + 5] + [2(5) + 5] + [2(6) + 5] + [2(7) + 5]$	$n = 5, a_1 = 11$ $a_n = a_5 = 2(7) + 5$ or 19 $S_n = \frac{5}{2}(11 + 19)$ $= 75$
$= 11 + 13 + 15 + 17 + 19$	
$= 75$	

The sum of the series is 75.

You can use the sum and sequence functions on a graphing calculator to find the sum of an arithmetic series.

EXPLORATION

GRAPHING CALCULATORS

To find the sum of $\sum\limits_{n=3}^{12} (4n + 1)$, enter the following.

Enter: [2nd] [LIST] [▶] 5   *Retrieves the sum function.*

[2nd] [LIST] 5   *Retrieves the sequence function.*

After the "sum seq(" prompt, enter the expression, variable, first value of n, last value of n, and step size, or $(4n + 1, n, 3, 12, 1)$.   *Step size will always be 1.*

Enter: 4 [ALPHA] [N] [+] 1 [,] [ALPHA] [N] [,] 3 [,] 12 [,] 1

[)] [ENTER]   *310*

The sum of the series is 310.

Your Turn

Use a graphing calculator to find the sum of each series.

a. $\sum\limits_{n=1}^{20} (4 + 3n)$ **b.** $\sum\limits_{k=3}^{9} (5k + 4)$ **c.** $\sum\limits_{t=2}^{35} (4t - 17)$

Just as a polynomial can be expressed in more than one form, the summation of a series can be expressed in different ways. The summation of the series in Example 5 is expressed as $\sum\limits_{k=3}^{7} (2k + 5)$. It can also be expressed as $\sum\limits_{k=3}^{7} [7 + 2(k - 1)]$ or $\sum\limits_{k=6}^{10} (2k - 1)$.   *Why?*

CHECK FOR UNDERSTANDING

Communicating Mathematics

Study the lesson. Then complete the following.

1. **Define** the term *indicated sum*.

2. **Explain** the purpose of the index of summation.

3. **Explain** when it is necessary to use both versions of the sum formula to solve a problem.

State the first term, the common difference, the last term, and the number of terms for each arithmetic series.

4. $7 + 1 + (-5) + (-11)$

5. $6 + 7.4 + 8.8 + 10.2 + 11.6$

6. $a_1 + 35 + 38 + 41 + 44$

7. $a_1 + 12 - 6 - 24 - 42 - 60 + a_7$

Find S_n for each arithmetic series described.

8. $a_1 = 4, a_n = 100, n = 25$

9. $a_1 = 40, n = 20, d = -3$

10. $a_1 = 132, d = -4, a_n = 52$

11. $d = 5, n = 16, a_n = 72$

Find the sum of each arithmetic series.

12. $5 + 11 + 17 + \dots + 95$

13. $38 + 35 + 32 + \dots + 2$

14. Write the terms of $\sum\limits_{n=1}^{7} 2n$ and find the sum.

15. a. Compute the sum of the first 1000 positive even integers.
   b. Compute the sum of the multiples of 3 from 3 to 999.

EXERCISES

Find S_n for each arithmetic series described.

16. $a_1 = 7, a_n = 79, n = 8$

17. $a_1 = 58, a_n = -7, n = 26$

18. $a_1 = 43, n = 19, a_n = 115$

19. $a_1 = 76, n = 21, a_n = 176$

20. $a_1 = 7, d = -2, n = 9$

21. $a_1 = 3, a_n = -38, n = 8$

22. $a_1 = 5, d = \frac{1}{2}, n = 13$

23. $a_1 = 12, d = \frac{1}{3}, n = 13$

24. $a_1 = 91, d = -4, a_n = 15$

25. $d = 7, n = 18, a_n = 72$

26. $d = -3, n = 21, a_n = -64$

27. $a_1 = -2, d = \frac{1}{3}, a_n = 9$

Find the sum of each arithmetic series.

28. $6 + 13 + 20 + 27 + \dots + 97$

29. $7 + 14 + 21 + 28 + \dots + 98$

30. $34 + 30 + 26 + \dots + 2$

31. $16 + 10 + 4 + \dots + (-50)$

Write the terms of each arithmetic series and find the sum.

32. $\sum\limits_{n=1}^{6} (2n + 11)$

33. $\sum\limits_{t=19}^{23} (5t - 3)$

34. $\sum\limits_{k=7}^{11} (42 - 9k)$

Find the first three terms of each arithmetic series.

35. $a_1 = 17, a_n = 197, S_n = 2247$

36. $n = 19, a_n = 103, S_n = 1102$

37. $n = 31, a_n = 78, S_n = 1023$

38. $a_1 = -13, a_n = 427, S_n = 18{,}423$

39. Write the series $\frac{1}{5} + \frac{2}{5} + \frac{3}{5} + \dots + \frac{12}{5}$ in summation notation and find the sum.

Graphing Calculator

Use a graphing calculator to find the sum of each arithmetic series.

40. $\sum\limits_{n=21}^{75} (2n + 5)$

41. $\sum\limits_{n=10}^{50} (3n - 1)$

42. a. Evaluate $\displaystyle\sum_{a=3}^{6} (a-2)^2$ and $\displaystyle\sum_{a=1}^{4} a^2$.

   b. What do you notice?

   c. Why does this work?

43. Construction   A company responsible for laying the foundation of a new high-rise office building must pay $5000 a day for the first five days that the completion of the foundation is late. On the sixth day and each day thereafter, the penalty is increased by $200 a day. If the company was penalized $65,600, how many days late were they in laying the foundation?

44. Entertainment   As of March, 1994, Andrew Lloyd Webber's musical *Phantom of the Opera* had been performed on Broadway 2576 times. Some theaters that present the musical have seats with limited viewing. For example, not everyone who attended the show at the Ohio Theatre in 1995 could see the famous scene where the chandelier comes crashing to the stage. Suppose only the people in the first 24 rows had 100% visibility. In this section there are 20 seats in the first row, and each subsequent row has one more seat than the row in front of it. How many seats are there in the Ohio Theater where people can see the whole show?

45. Aeronautics   A rocket rises 20 feet in the first second, 60 feet in the second second, and 100 feet in the third second. If it continues at this rate, how many feet will it rise in the 20th second?   (Lesson 11–1)

46. Solve $4.3^{3x+1} = 78.5$.   (Lesson 10–6)

47. Use a calculator to find the natural logarithm of 0.056, rounded to four decimal places.   (Lesson 10–5)

48. Construction   Mike Welch can paint his house in 15 hours. His friend Joe can paint the house in 20 hours. If they work together, how long will it take them to paint the house?   (Lesson 9–5)

49. Determine whether the graph at the right represents an odd-degree function or an even-degree function. Then state how many real zeros the function has.   (Lesson 8–1)

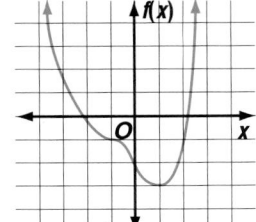

50. Solve the system of equations.   (Lesson 7–7)

$$x - y = -2$$
$$\frac{(x-2)^2}{16} + \frac{y^2}{16} = 1$$

51. Write $f(x) = (x+2)^2 - 6$ in quadratic form. Identify the quadratic term, the linear term, and the constant term.   (Lesson 6–1)

52. Statistics   The U.S. Department of Agriculture recommends two to four servings of fruit daily. Fruit is a good source of fiber as well as vitamins and minerals. The chart at the right displays the fiber content for one serving size of each fruit listed.   (Lesson 4–8)

   a. Find the range of the data.

   b. Find the quartiles of the data.

   c. Find the interquartile range of the data.

   d. Name any outliers in the data.

   e. Make a box-and-whisker plot of the data.

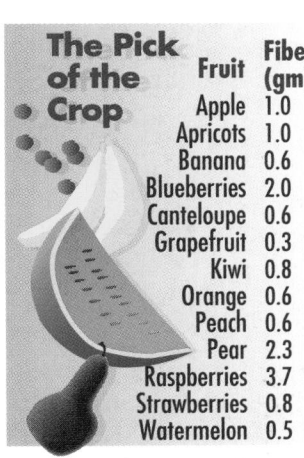

The Pick of the Crop	Fruit	Fiber (gm)
	Apple	1.0
	Apricots	1.0
	Banana	0.6
	Blueberries	2.0
	Canteloupe	0.6
	Grapefruit	0.3
	Kiwi	0.8
	Orange	0.6
	Peach	0.6
	Pear	2.3
	Raspberries	3.7
	Strawberries	0.8
	Watermelon	0.5

Source: *Vitality*, Sept. 1995

Geometric Sequences

APPLICATION
Animation

Animation is the process of using drawings in a sequential manner to simulate movement. A full-feature film requires hundreds of thousands of hand-painted scenes called *cels*. In 1994, Disney's *The Lion King* used a combination of painted cels and computer graphics to minimize the number of cels needed for the picture. For example, one set of cels was produced to picture a wildebeest running over a hillside. The computer technicians took those cels, duplicated them electronically, and produced the effect of a herd of hundreds of wildebeests running over the same hillside.

To produce a vanishing effect in an animation film, the character's image is reduced 50% with each consecutive cel. Suppose the original area of a character's image is defined as 1 (representing 100%) and the vanishing effect is applied over 8 cels of film. What happens to the area of the character with each successive reduction?

Number of Reductions	0	1	2	3	4	5	6	7	8
Area of Image	1	$\frac{1}{2}$	$\frac{1}{4}$	$\frac{1}{8}$	$\frac{1}{16}$	$\frac{1}{32}$	$\frac{1}{64}$	$\frac{1}{128}$	$\frac{1}{256}$
Percent of Original	100	50	25	12.5	6.25	3.125	1.5625	0.78125	0.390625

Pattern $\times\frac{1}{2}$ $\times\frac{1}{2}$ $\times\frac{1}{2}$ $\times\frac{1}{2}$ $\times\frac{1}{2}$ $\times\frac{1}{2}$ $\times\frac{1}{2}$ $\times\frac{1}{2}$

Let's graph the function represented by the number of reductions and the percent of the original. Notice that these points follow the same pattern as the exponential function $y = \left(\frac{1}{2}\right)^x$, except that this graph is discrete.

Notice that with each successive reduction the image is one-half the size of the previous image. Each term of the sequence is found by multiplying the previous term by $\frac{1}{2}$. This is an example of a **geometric sequence.** The number by which each term is multiplied is called the **common ratio** and is symbolized by the variable *r*.

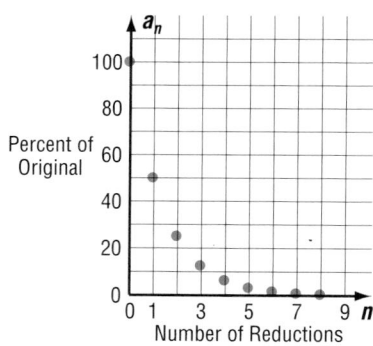

Definition of Geometric Sequence	A geometric sequence is one in which each term after the first is found by multiplying the previous term by a constant called the common ratio, *r*.

As with arithmetic sequences, you can name the terms of a geometric sequence using a_1, a_2, a_3, and so on. Suppose the nth term is defined as a_n. Then its previous term is a_{n-1}. So, $a_n = r(a_{n-1})$. Thus, $r = \dfrac{a_n}{a_{n-1}}$. That is, the common ratio is found by dividing any term by its previous term. This can be used to find any term in a geometric sequence.

Example **Find the next two terms of the geometric sequence 3, 12, 48, ...**

First find the common ratio. Let 3 be a_{n-1} and let 12 be a_n.

$$r = \frac{a_n}{a_{n-1}}$$

$$= \frac{12}{3} \text{ or } 4 \qquad \text{The common ratio is 4.}$$

The fourth term is $a_4 = r \cdot a_3$, so $a_4 = 4(48)$ or 192.

The fifth term is $a_5 = r \cdot a_4$, so $a_5 = 4(192)$ or 768.

The next two terms of the sequence are 192 and 768.

CAREER CHOICES

A **computer scientist** develops and designs the hardware, the processor chips, and video cards that display the computer graphics for computer animation.

Computer scientists have a background in mathematics, science and engineering, and usually a degree in computer science.

For more information, contact:

Association for Computer Machinery
1515 Broadway
New York, NY 10036

We have seen that each term of a geometric sequence can be expressed in terms of r and its previous term. However, it is also possible to develop a formula that expresses each term of a geometric sequence in terms of r and the first term a_1. Study the patterns shown in the table below for the sequence 3, 12, 48, 192, ...

Sequence							
	numerical	3	12	48	192	. . .	
	symbols	a_1	a_2	a_3	a_4	. . .	a_n
Expressed in Terms of r and the Previous Term	numerical	3	3(4)	12(4)	48(4)	. . .	
	symbols	a_1	$a_1 \cdot r$	$a_2 \cdot r$	$a_3 \cdot r$	. . .	$a_{n-1} \cdot r$
Expressed in Terms of r and the First Term	numerical	3 $3(4^0)$	3(4) $3(4^1)$	3(16) $3(4^2)$	3(64) $3(4^3)$	. . .	
	symbols	$a_1 \cdot r^0$	$a_1 \cdot r^1$	$a_1 \cdot r^2$	$a_1 \cdot r^3$	. . .	$a_1 \cdot r^{n-1}$

The three values in the last column of the table all describe the nth term of a geometric sequence. This leads us to the following formula for finding any term of a geometric sequence.

Formula for the nth term of a Geometric Sequence	The nth term a_n of a geometric sequence with first term a_1 and common ratio r is given by either formula. $$a_n = a_{n-1} \cdot r \qquad \text{or} \qquad a_n = a_1 \cdot r^{n-1}$$

Since a_n can represent any term of the sequence, you can use these formulas to find the value of any term in a geometric sequence.

Example **2** Write the first six terms of a geometric sequence in which $a_1 = 3$ and $r = 2$.

Method 1: Use $a_n = a_{n-1} \cdot r$.

$a_1 = 3$

$a_2 = 3 \times 2$ or 6

$a_3 = 6 \times 2$ or 12

$a_4 = 12 \times 2$ or 24

$a_5 = 24 \times 2$ or 48

$a_6 = 48 \times 2$ or 96

Method 2: Use $a_n = a_1 \cdot r^{n-1}$.

$a_1 = 3 \cdot 2^{1-1}$ or 3

$a_2 = 3 \cdot 2^{2-1}$ or 6

$a_3 = 3 \cdot 2^{3-1}$ or 12

$a_4 = 3 \cdot 2^{4-1}$ or 24

$a_5 = 3 \cdot 2^{5-1}$ or 48

$a_6 = 3 \cdot 2^{6-1}$ or 96

You can also use the formula for the nth term to find a given term when you know the common ratio and one term of the geometric sequence, but not the first term of the sequence.

Example **3** Find the ninth term of a geometric sequence in which $a_3 = 63$ and $r = -3$.

Method 1

Start with the third term and use the common ratio to find the ninth term.

$a_3 = 63$

$a_4 = 63 \cdot (-3)$ or -189

$a_5 = -189 \cdot (-3)$ or 567

$a_6 = 567 \cdot (-3)$ or -1701

$a_7 = -1701 \cdot (-3)$ or 5103

$a_8 = 5103 \cdot (-3)$ or $-15{,}309$

$a_9 = -15{,}309 \cdot (-3)$ or $45{,}927$

Method 2

First find the value of a_1.

$a_3 = a_1 \cdot r^{3-1}$

$63 = a_1 \cdot (-3)^2$

$\frac{63}{9} = a_1$

$7 = a_1$

Use the formula $a_n = a_1 \cdot r^{n-1}$ to find a_9.

$a_9 = a_1 \cdot r^{9-1}$

$ = 7 \cdot (-3)^8$ or $45{,}927$

The ninth term of the sequence is 45,927.

Geometric sequences can be modeled in ways other than lists of numbers.

Example **4**

The illustration below shows a sequence of 30°-60°-90° triangles. Draw the next triangle in the sequence.

Notice that the measure of each side of the triangle is twice the measure of the corresponding side in the previous triangle. Thus, the measures form a geometric sequence and $r = 2$.

Multiply each side of the last triangle by 2 to find the measures of the sides of the fourth triangle and draw the triangle.

$8 \cdot 2 = 16$

$4 \cdot 2 = 8$

$4\sqrt{3} \cdot 2 = 8\sqrt{3}$

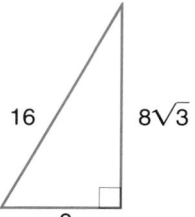

In Lesson 11–1A, you learned to use a graphing calculator to find and graph the terms of an arithmetic sequence. You can use the same skills with geometric sequences.

EXPLORATION — GRAPHING CALCULATORS

Use a graphing calculator to find the first 12 terms of a geometric sequence in which $a_1 = 4$ and $r = 0.5$. Then graph the sequence.

First clear the L1, L2, and Y= lists. Use the seq(command to enter the values of n into L1. Be sure to enter the information from the home screen.

Enter: [2nd] [LIST] 5 [ALPHA] N [,] [ALPHA] N [,] 1 [,] 12 [,] 1 [)] [STO▸] [2nd] [L1] [ENTER]   {1 2 3 4 5 6 7 …

Use the general formula with the seq(command to calculate the terms of the sequence and enter them into L2.

Enter: [2nd] [LIST] 5 4 [×] .5 [∧] [(] [ALPHA] N [−] 1 [)] [,] [ALPHA] N [,] 1 [,] 12 [,] 1 [)] [STO▸] [2nd] [L2] [ENTER]   {4 2 1 .5 .25 …

Press [STAT] 1 to view the lists containing the sequence terms. Take note of the range of values in list L2 so that you can set the proper window for graphing the sequence.

After entering the WINDOW settings, press [2nd] [STAT PLOT] 1. Make sure the plot is turned on, the scatter plot is selected, L1 is the Xlist, and L2 is the Ylist. Then press [GRAPH] .

Your Turn

Find the first ten terms of a geometric sequence in which $a_1 = 1$ and $r = -2$. Then graph the sequence.

In Lesson 11–1, you learned that the missing terms between two nonconsecutive terms in an arithmetic sequence were called *arithmetic means*. Likewise, the missing term(s) between two nonconsecutive terms in a geometric sequence are called **geometric means**. In the sequence 3, 12, 48, 192, 768, …, the three geometric means between 3 and 768 are 12, 48, and 192. You can use the common ratio to find the geometric means in a given sequence.

Example **5** **a. Find the three geometric means between 3.4 and 2125.**
b. Graph the sequence using the *x*-axis for the number of the term and the *y*-axis for the term itself.

a. You can use the *n*th term formula to find the value of *r*. In the sequence 3.4, ___?___ , ___?___ , ___?___ , 2125, 3.4 is a_1 and 2125 is a_5.

$$a_n = a_1 \cdot r^{n-1}$$
$$a_5 = a_1 \cdot r^{5-1} \quad \text{\textit{n = 5}}$$
$$2125 = 3.4 \cdot r^4 \quad \text{\textit{a_5 = 2125 and a_1 = 3.4}}$$
$$625 = r^4 \quad \text{\textit{Divide each side by 3.4.}}$$
$$\pm 5 = r \quad \text{\textit{Take the fourth root of each side.}}$$

There are two possible common ratios, so there are two possible sets of geometric means. Use each value of *r* to find the missing terms.

$r = 5$	$r = -5$
$a_1 = 3.4\,(5)$ or 17	$a_1 = 3.4\,(-5)$ or -17
$a_2 = 17(5)$ or 85	$a_2 = -17(-5)$ or 85
$a_3 = 85(5)$ or 425	$a_3 = 85(-5)$ or -425

Check: Will each value of *r* give you the correct 5th term?

$$425(5) = 2125 \quad \checkmark \qquad\qquad -425(-5) = 2125 \quad \checkmark$$

The geometric means are 17, 85, and 425 or -17, 85, and -425.

b.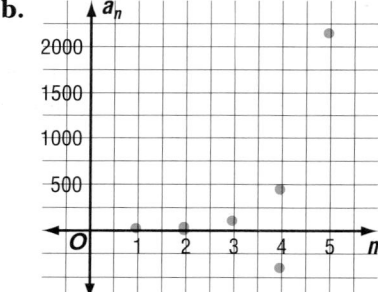

CHECK FOR UNDERSTANDING

Communicating Mathematics

Study the lesson. Then complete the following.

1. **Explain** how a geometric sequence differs from an arithmetic sequence.

2. **Define** common ratio.

3. **Explain** how you know if a list of numbers forms a geometric sequence.

4. **Explain** how you would find the two geometric means between 324 and 12.

MATH JOURNAL

5. **Assess Yourself** The points of the graph of the geometric sequence in the Exploration are part of the graph of an exponential function. Do you think this statement holds true for the graphs of all geometric sequences? Explain. Give an example to support your answer.

Determine whether each sequence is geometric. If so, find the common ratio.

6. 5, 20, 80, 320

7. 3, −15, 75, −375

8. 5, 20, 35, 50

9. $\frac{2}{3}, \frac{4}{9}, \frac{8}{27}, \frac{16}{81}$

Find the next two terms of each geometric sequence.

10. 20, 30, 45, ...

11. $-\frac{1}{4}, \frac{1}{2}, -1$

12. Find the first five terms of the geometric sequence in which $a_1 = -2$ and $r = 3$.

Find the nth term of each geometric sequence.

13. $a_1 = 7, n = 4, r = 2$

14. $a_3 = 32, n = 6, r = -0.5$

15. Find a_9 for the geometric sequence 60, 30, 15, ...

16. **a.** Find the two geometric means between 1 and 8.

b. Graph the sequence using the x-axis for the number of the term and the y-axis for the term itself.

17. **Literature** In one of Grimm's Fairy Tales, Rumpelstiltskin has the ability to spin straw into gold. Suppose on the first day, he spun 5 pieces of straw into gold, and each day thereafter he spun twice as much. How many pieces of straw would he have spun into gold by the end of a week?

EXERCISES

Practice

Find the next two terms of each geometric sequence.

18. 405, 135, 45, ...

19. 81, 108, 144, ...

20. 16, 24, 36, ...

21. 162, 108, 72, ...

22. $\frac{4}{27}, -\frac{4}{9}, \frac{4}{3}, ...$

23. 64, −16, 4, ...

18-40 even

Find the first five terms of each geometric sequence described.

24. $a_1 = 2, \ r = -3$

25. $a_1 = 243, r = \frac{1}{3}$

26. $a_1 = 576, r = -0.5$

Find the nth term of each geometric sequence.

27. $a_1 = \frac{1}{3}, n = 8, r = 3$

28. $a_1 = \frac{1}{64}, n = 9, r = 4$

29. $a_1 = 16,807, n = 6, r = \frac{3}{7}$

30. $a_1 = 4096, n = 8, r = \frac{1}{4}$

31. $a_4 = 16, n = 8, r = 0.5$

32. $a_6 = 3, n = 12, r = 2$

Find the indicated term in each geometric sequence.

33. a_9 for $\frac{1}{5}, 1, 5, ...$

34. a_7 for $\frac{1}{32}, \frac{1}{16}, \frac{1}{8}, ...$

35. a_8 for 4, −12, 36, ...

36. a_6 for 540, 90, 15, ...

Find the geometric means in each sequence. Then graph each sequence using the x-axis for the number of the term and the y-axis for the term itself.

37. 9, _?_, _?_, _?_, 144

38. 4, _?_, _?_, _?_, 324

39. 32, _?_, _?_, _?_, _?_, 1

40. _?_, _?_, 12, _?_, _?_, 96

Find the value(s) of y that makes each sequence geometric.

41. $2, 8, 32, 5y + 3, \ldots$

42. $3, 6, 2y + 18, \ldots$

43. $y + 1, y, y - 4, \ldots$

44. $y + 1, 2y - 1, 4y - 3, \ldots$

Write the formula for the nth term for each graphed sequence.

45.

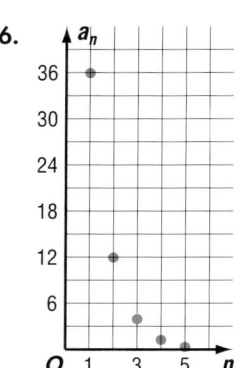

46.

47.

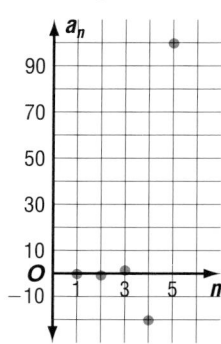

48. The first term of a geometric sequence is 0.6, and each term is 4 times the previous term.

　　a. Find the value of the seventh term.

　　b. Write a formula for the nth term of this sequence.

Graphing Calculator

Use a graphing calculator to list the first n terms of each geometric sequence described. Then graph the sequence. Sketch the graph on grid paper.

49. $a_1 = 2, n = 5, r = 2$

50. $a_1 = 12, n = 10, r = 0.5$

51. $a_1 = 243, n = 8, r = \frac{1}{3}$

Critical Thinking

52. The first three terms of a geometric sequence are a, b, and c. Express c in terms of a and b. Justify your answer.

53. Write an argument to show that $a_n = a_{n-1}r$ and $a_n = a_1 r^{n-1}$ are equivalent.

54. If each term of a geometric sequence is multiplied by the same real number, is the resulting sequence geometric? Explain.

Applications and Problem Solving

55. Ice Sculpture In many states where snow and ice are abundant in the winter months, art students at various universities have ice sculpting contests. They use chain saws, axes, and chisels to create their masterpieces. The events are held outdoors when the temperature is cold enough to prevent rapid thawing. Suppose a warm front passes through during the contest and a 1000-pound sculpture begins to melt. It loses one-fifth of its weight per hour. How much of the sculpture will be left after 6 hours?

56. Medicine Iodine-131 is used medically to study the activity of the thyroid gland. Iodine-131 has a *half-life* of about 8 days. This means that approximately every 8 days, half of the mass of the iodine decays into another element. If a container held a mass of 64 milligrams of iodine-131, how much is left after 40 days?

57. Aquatic Farming In Spain, blue mussels are grown in farms that have produced up to 250,000 pounds of pure seafood per acre per year. This compares with using land to grow cattle at a rate of 100 pounds of beef per acre per year. The farms consist of parallel ropes, called *long lines*, that float on the surface of the ocean. From these long lines, ropes are suspended vertically upon which the mussels grow.

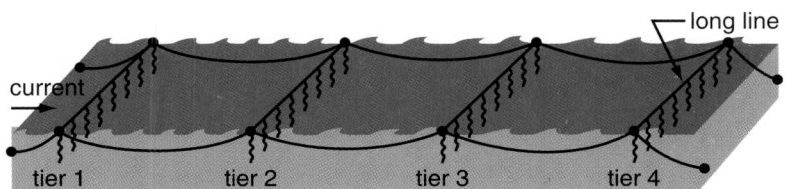

As the water flows through the mussels, they filter out available food particles. If there are equal numbers of mussels in each tier, the concentration of food particles in the water should decrease by the same percent as the water moves from tier to tier.

a. Suppose that the food concentration entering the first tier is 320 mg/L and 99% of the food particles that enter each tier remains as the water leaves the tier. What is the concentration of food entering the 5th tier?

b. After how many tiers will the concentration of the food particles be less than 75% of the original concentration?

c. Assuming there must be at least 200 mg/L of food for a mussel to survive, what is the maximum number of tiers possible in this farm?

Mixed Review

58. Find S_n for an arithmetic series in which $a_1 = 11$, $a_n = 44$, and $n = 23$. (Lesson 11–2)

59. Solve $\log_5 4 + \log_5 x = \log_5 36$. (Lesson 10–3)

60. Real Estate Budget Realty charges a commission of $4800 on the sale of a $90,000 home. At that rate, how much commission would be charged on the sale of a $219,000 home? (Lesson 9–2)

61. Solve $x - 7\sqrt{x} - 8 = 0$. (Lesson 8–6)

62. Graph $f(x) = x^4 - 8x^2 + 10$. (Lesson 8–3)

63. Write an equation of the parabola with a focus at $(7, -7)$ and $x = -2$ as the equation of its directrix. (Lesson 7–2)

64. Solve $(x - 7)(x + 2) > 0$. (Lesson 6–7)

65. Simplify $(5 + i)(2 - 3i)$. (Lesson 5–9)

66. Factor $3y^2 + 5y + 2$. (Lesson 5–4)

67. Airports According to the Airports Council International, the busiest airport in the world is Chicago's O'Hare Airport, and the second busiest is London's Heathrow Airport. Together they handled 55.2 million passengers in the first six months of 1994. If O'Hare handled 7.2 million more than Heathrow, how many were handled by each airport? (Lesson 3–2)

Geometric Series

What YOU'LL LEARN

- To find sums of geometric series,
- to find specific terms in a geometric series, and
- to use sigma notation to express sums.

Why IT'S IMPORTANT

You can use geometric series to solve problems involving aviation and genealogy.

APPLICATION

Genealogy

Shantal is researching her ancestry. She has decided to create a family tree. To help her organize her research, Shantal wanted to figure out how many people she would be researching as she went back from generation to generation. She started her research by asking family members about her grandparents, great-grandparents, and great-great-grandparents. Let's use a diagram to help Shantal calculate her parental lineage for four generations. Let F represent father and M represent mother.

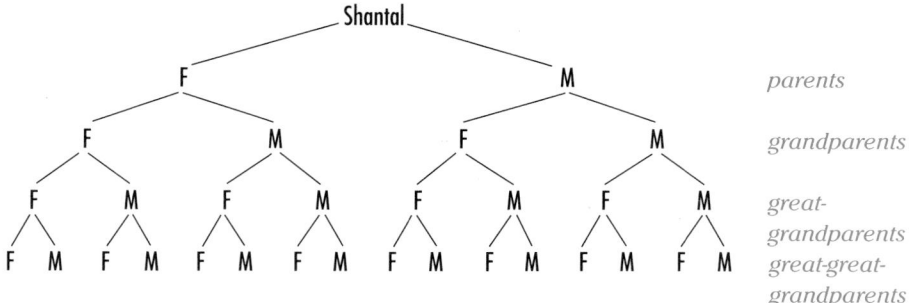

The total is given by the equation below.

$$\underbrace{2}_{parents} + \underbrace{4}_{grandparents} + \underbrace{8}_{great\text{-}grandparents} + \underbrace{16}_{great\text{-}great\text{-}grandparents} = 30$$

What would be the number of people in her parental family tree if she went back 8 previous generations?

Notice that 2, 4, 8, and 16 form a geometric sequence in which $a_1 = 2$ and $r = 2$. The indicated sum of the terms of a geometric sequence is called a **geometric series.** The lists below show examples of geometric sequences and their corresponding geometric series.

Geometric Sequences	Geometric Series
1, 3, 9, 27, 81	$1 + 3 + 9 + 27 + 81$
5, −10, 20	$5 + (-10) + 20$
$4, 1, \frac{1}{4}, \frac{1}{16}$	$4 + 1 + \frac{1}{4} + \frac{1}{16}$

You can develop a formula for finding the sum of a geometric series. Let's explore this formula using the sequence above for 8 terms. As with arithmetic series, S_n represents the sum of n terms.

$$S_8 = 2 + 4 + 8 + 16 + 32 + 64 + 128 + 256 \quad \textit{Now multiply each side by 2.}$$
$$(-)\ 2S_8 = \quad\ \ 4 + 8 + 16 + 32 + 64 + 128 + 256 + 512 \quad \textit{Align the terms.}$$
$$(1-2)S_8 = 2 + 0 + 0 + \ \ 0 + \ \ 0 + \ \ 0 + \ \ \ 0 + \ \ \ 0 - 512 \quad \textit{Subtract the two equations.}$$
$$S_8 = \frac{2 - 512}{1 - 2} \text{ or } 510$$

There are 510 people in Shantal's parental family tree for 8 previous generations.

Now let's analyze what these numbers represent in terms of S_n. In the equation $S_8 = \dfrac{2 - 512}{1 - 2}$, the 2 in the numerator represents a_1, 512 represents a_9, and the 2 in the denominator represents r. Thus, we can replace the equation with the formula $S_8 = \dfrac{a_1 - a_9}{1 - r}$. Since a_9 can also be written as $a_1 r^8$, this expression could also be written as $\dfrac{a_1 - a_1 r^8}{1 - r}$. These rational expressions can be used to find the sum of any geometric series.

Sum of a Geometric Series	**The sum S_n of the first n terms of a geometric series is given by** $$S_n = \frac{a_1 - a_1 r^n}{1 - r} \text{ or } S_n = \frac{a_1(1 - r^n)}{1 - r}, \text{ where } r \neq 1.$$

The formula for the sum of a geometric series restricts the value of r so that division by 0 does not occur. An example of a geometric series in which $r = 1$ is 3, 3, 3, ... To find the sum of n 3s, you would find the product of 3 and n or $3n$. In general terms, the sum of a geometric series in which $r = 1$ would be $n \cdot a_1$.

Example ① Find the sum of the first six terms of the geometric series for which $a_1 = 5$ and $r = -2$.

$$S_n = \frac{a_1 - a_1 r^n}{1 - r}$$

$$= \frac{5 - 5(-2)^6}{1 - (-2)} \qquad a_1 = 5,\ r = -2,\ and\ n = 6$$

$$= \frac{-315}{3} \text{ or } -105$$

The sum of the first six terms of the series is -105. *Check this result.*

Geometric series can be used to explain natural phenomena.

Example ②

Aviation

As a hot-air balloon rises from Earth's surface, the air in the balloon cools. If the air is not reheated, the balloon rises more slowly with each minute of flight. Suppose that after 1 minute, a hot-air balloon rises 120 feet. In each succeeding minute, the balloon rises only 60% as far as it rose in the previous minute. How far will the balloon rise in 8 minutes?

The change in height after each minute is a geometric series in which $a_1 = 120$, $n = 8$, and $r = 0.6$. Use the formula for the sum of a geometric series.

$$S_n = \frac{a_1(1 - r^n)}{1 - r}$$

$$= \frac{120(1 - 0.6^8)}{1 - 0.6}$$

$$\approx 294.961152 \quad Use\ a\ calculator.$$

The balloon will rise about 295 feet in 8 minutes.

How can you find the sum of a geometric series if you know the last term of the series but not how many terms there are in the series? Remember the general formula of the nth term of a geometric sequence, $a_n = a_1 r^{n-1}$. We need to find an equivalent expression that involves r^n.

$$a_n = a_1 r^{n-1}$$
$$a_n \cdot r = a_1 r^{n-1} \cdot r \quad \textit{Multiply each side by r.}$$
$$a_n \cdot r = a_1 r^n \quad\quad r^{n-1}r^1 = r^{n-1+1} \text{ or } r^n$$

Now we can substitute $a_n \cdot r$ for $a_1 r^n$ in the formula for the sum of a geometric series. The formula becomes $S_n = \dfrac{a_1 - a_n r}{1 - r}$.

Example **3** Find the sum of a geometric series for which $a_1 = 729$, $a_n = -3$, and $r = -\dfrac{1}{3}$.

Since we do not know the value of n, use the alternate formula.

$$S_n = \frac{a_1 - a_n r}{1 - r}$$
$$= \frac{729 - (-3)\left(-\frac{1}{3}\right)}{1 - \left(-\frac{1}{3}\right)} \quad a_1 = 729, a_n = -3, \text{ and } r = -\frac{1}{3}$$
$$= \frac{728}{\frac{4}{3}} \text{ or } 546$$

TECHNOLOGY *Tips*

You can use a graphing calculator to generate the terms of a geometric series. Store the first term as *x*. Then generate the second term by storing *x* times the common ratio.

Continue to press ENTER to generate successive terms of the series.

You can also use the formula for the sum of a geometric series to find a given term of the series.

Example **4** Find a_1 in a geometric series for which $S_7 = 70{,}933$ and $r = 4$.

$$S_n = \frac{a_1(1 - r^n)}{1 - r}$$
$$70{,}993 = \frac{a_1(1 - 4^7)}{1 - 4} \quad S_n = 70{,}993 \text{ and } r = 4$$
$$70{,}993 = \frac{-16{,}383 a_1}{-3}$$
$$70{,}993 = 5461 a_1$$
$$13 = a_1$$

MODELING MATHEMATICS

Geometric Series

Materials: 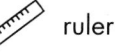 ruler colored pencil

You can use area to model a geometric series.

a. Draw a large equilateral triangle on your paper.

b. Find the midpoint of each side of the triangle and connect the midpoints to form four congruent triangles.

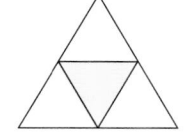

c. Color the center triangle. If the area of the original triangle is 1, what is the area of the colored triangle?

d. Find the midpoints of each side of each uncolored triangle. Connect those midpoints to form congruent triangles and color in each center triangle.

e. Find the total area of the colored triangles.

Your Turn

a. Repeat the process again for each uncolored triangle. What is the resulting shaded area?

b. Write a series for six repetitions of this process.

In Lesson 11–2, you learned that you can write the sum of an arithmetic series by using sigma notation and the general form for an arithmetic sequence. You can also use sigma notation to write the sum of a geometric series.

Example **Write the terms of $\sum_{n=1}^{5} 4(3)^{n-1}$ and find the sum.**

There are two ways in which you can find the sum.

Method 1	Method 2
$\sum_{n=1}^{5} 4(3)^{n-1} = 4(3^{1-1}) + 4(3^{2-1}) + 4(3^{3-1}) +$	Use $S_n = \dfrac{a_1(1-r^n)}{1-r}$
$\qquad 4(3^{4-1}) + 4(3^{5-1})$	$a_1 = 4, r = 3, n = 5$
$\qquad = 4(1) + 4(3) + 4(9) + 4(27) + 4(81)$	$S_n = \dfrac{4(1-3^5)}{1-3}$
$\qquad = 4 + 12 + 36 + 108 + 324$	$\qquad = 484$
$\qquad = 484$	

The sum of the series is 484.

CHECK FOR UNDERSTANDING

Communicating Mathematics

Study the lesson. Then complete the following.

1. **Explain** how a geometric series is similar to an arithmetic series.

2. **Determine** which form of the formula for the sum of a geometric series is most appropriate to use for each series described.
   a. $r = -0.5, n = 5, a_1 = 48$ **b.** $a_n = 3, r = -0.5, a_1 = 48$ **c.** $5, 5, 5, \ldots$

3. **Explain** how you would write $1 - 3 + 9 - 27 + 81 - 243$ in sigma notation.

4. Refer to the Modeling Mathematics activity. Suppose the triangle you started with was an isosceles triangle. In what ways, if any, would the series for the area of colored portion change?

Guided Practice

State the first term, the common ratio, the last term, and the number of terms for each geometric series.

5. $6 + (-18) + 54 + (-162)$

6. $7 + 3.5 + 1.75 + 0.875$

7. $a_1 + 12 + 36 + 108 + 324$

8. $a_1 + 18 - 36 + 72 + a_5$

Find S_n for each geometric series described.

9. $a_1 = 12, r = -3, a_5 = 972$

10. $a_1 = 3, r = -5, a_n = 46{,}875$

11. $a_1 = 5, r = 2, n = 14$

12. $a_1 = 243, n = 5, r = -\dfrac{2}{3}$

Find the sum of each geometric series.

13. $54 + 36 + 24 + 16 + \ldots$ to 6 terms

14. $3 - 6 + 12 + \ldots$ to 7 terms

15. Write the terms of $\sum_{n=1}^{7} 81\left(\dfrac{1}{3}\right)^{n-1}$ and find the sum.

16. **Meteorology** The Brazos River begins at the junction of the Salt and Double Mountain Forks Rivers in Stonewall County, Texas, and empties into the Gulf of Mexico. A 5-day rainstorm caused the river to rise. After the first day, the river rose one inch. Each day, the rise in the river tripled. How much has the river risen after 5 days?

Practice

Find S_n for each geometric series described.

17. $a_1 = 625, r = \frac{3}{5}, n = 5$

18. $a_1 = 4, r = 0.5, n = 8$

19. $a_1 = 5, r = 3, n = 12$

20. $a_1 = 2401, n = 5, r = -\frac{1}{7}$

21. $a_1 = 4, n = 5, r = -3$

22. $a_1 = 625, n = 8, r = 0.4$

23. $a_1 = 1296, a_5 = 1, r = -\frac{1}{6}$

24. $a_1 = 3, a_8 = 384, r = 2$

25. $a_1 = 343, r = -\frac{1}{7}, a_4 = -1$

26. $a_1 = 64, a_{10} = -\frac{1}{8}, r = -\frac{1}{6}$

27. $a_1 = 4, a_6 = 0.125, r = 0.5$

28. $a_2 = -36, a_5 = 972, n = 7$

29. $a_2 = 250, a_3 = 100, n = 8$

30. $a_3 = -36, a_6 = -972, n = 10$

Find the sum of each geometric series.

31. $12 + 12 + 12 + \ldots$ to 12 terms

32. $7 + 21 + 63 + \ldots$ to 10 terms

33. $2401 - 343 + 49 - \ldots$ to 5 terms

34. $\frac{1}{9} - \frac{1}{3} + 1 - \ldots$ to 6 terms

Write the terms of each geometric series and find the sum.

35. $\sum_{n=1}^{9} 5(2)^{n-1}$

36. $\sum_{1}^{8} 64\left(\frac{3}{4}\right)^{n-1}$

37. $\sum_{n=1}^{6} 2(-3)^{n-1}$

Find a_1 for each geometric series described.

38. $S_n = -364, r = -3, n = 6$

39. $S_n = \frac{215}{64}, r = -\frac{1}{2}, n = 7$

40. $S_n = 315, a_n = 5, r = 0.5$

41. $S_n = 165, a_n = 48, r = -\frac{2}{3}$

Express each series in sigma notation and find the sum.

42. $75 + 15 + 3 + \ldots$ to 10 terms

43. $243 + 162 + 108 + 72 + \ldots$ to 12 terms

Graphing Calculator

Use a graphing calculator to find the sum of each geometric series.

44. $\sum_{n=1}^{10} \frac{1}{3}\left(\frac{4}{3}\right)^{n-1}$

45. $\sum_{n=1}^{20} 3(-2)^{n-1}$

Critical Thinking

46. Copy the table below and use a calculator to complete.

Sequence	Sum of First 5 Terms	Sum of First 10 Terms	Sum of First 15 Terms	Sum of First 20 Terms
A: $75 + 15 + 3 + \ldots$				
B: $1 + 2 + 4 + 8 + \ldots$				
C: $a_1 = 25, r = 2$				
D: $a_1 = 4, r = 0.5$				

a. What pattern(s) do you notice as the number of terms included in the sum get larger?

b. Predict what type of number you would get in each case if the table were extended to the first 50 and first 100 terms.

c. Write a general statement about the sum of a geometric series as the number of terms included increases.

47. Savings Yolanda read an article about a surefire plan to have enough money saved for a comfortable early retirement. According to the plan, on the first day she puts aside a penny. Each day thereafter, she contributes an amount that is double the previous day's amount.

   a. If Yolanda stays true to this plan, how much would she have set aside at the end of 10 days? 20 days?

   b. Why is the surefire plan not a feasible one?

48. Landscaping Raheem is helping his father install a fence across their backyard. He uses a sledgehammer to drive pointed fence posts into the ground. On his first stroke, he drives the post 4 inches into the ground. The soil is denser the deeper he drives, so each stroke after the first, he can only drive the post 40% of the distance he did on the previous swing of the hammer. After 10 strokes, how far has he driven the post into the ground?

49. Find the missing terms of the geometric sequence $\underline{\ ?\ }$, $\underline{\ ?\ }$, $\frac{3}{2}$, 9, 54. (Lesson 11–3)

50. Find the nth term of an arithmetic sequence in which $a_1 = -3$, $d = -9$, and $n = 11$. (Lesson 11–1)

51. Solve $6^x = 216$. (Lesson 10–1)

52. Find all zeros of $g(x) = 12x^4 + 4x^3 - 3x^2 - x$. (Lesson 8–5)

53. Find the center and radius of the circle whose equation is $x^2 + y^2 - 3x + 8y = 20$. (Lesson 7–3)

54. Statistics Ramón is studying the effects of different fertilizers on tree growth. The heights in centimeters of the trees he planted last year are 49, 54, 61, 49, 54, 51, 56, and 58. Find the standard deviation of the heights. (Lesson 6–8)

55. Given $f(x, y) = 3x + 2y$, find $f(5, -2)$. (Lesson 3–5)

56. Solve $|x - 15| < 45$. (Lesson 1–7)

SELF TEST

Find the nth term of each arithmetic sequence. (Lesson 11–1)

1. $a_1 = 7$, $d = 3$, $n = 14$

2. $a_1 = 2$, $d = \frac{1}{2}$, $n = 8$

Find the sum of each arithmetic series described. (Lesson 11–2)

3. $6 + 12 + 18 + \ldots + 96$

4. $\sum_{n=1}^{30} (2n - 1)$

5. Find the missing arithmetic means in 24, $\underline{\ ?\ }$, $\underline{\ ?\ }$, $\underline{\ ?\ }$, 48. Then graph the sequence. (Lesson 11–1)

6. Consumerism A display of soup cans at Grocery Mart is stacked in the shape of a pyramid. There are 6 cans in the top row, 8 cans in the next row, 10 cans in the next row, and so on. The display contains 12 rows of cans. How many cans of soup are in the display? (Lesson 11–2)

Find the nth term of each geometric sequence. (Lesson 11–3)

7. $a_1 = 4$, $n = 3$, $r = 5$

8. $a_1 = 243$, $n = 5$, $r = -\frac{1}{3}$

9. Find the missing geometric means in 3, $\underline{\ ?\ }$, $\underline{\ ?\ }$, $\underline{\ ?\ }$, 48. Then graph the sequence. (Lesson 11–3)

Find the sum of each geometric series described. (Lesson 11–4)

10. $a_1 = 5$, $r = 3$, $n = 12$

11. $\sum_{n=1}^{6} 2(-3)^{n-1}$

What YOU'LL LEARN

- To find sums of infinite geometric series.

Why IT'S IMPORTANT

You can use infinite geometric series to solve problems involving physics and ballooning.

Physics

Have you ever had hot cocoa get cold before you finished drinking it? It was a similar problem that led British chemist Sir James Dewar to develop the vacuum flask, commonly called the thermos bottle. A thermos bottle is a double-walled glass vessel with silver coating to reflect the radiation of heat into the bottle to keep the liquid hot or away from the bottle to preserve the cold. The space between the glass walls is a partial vacuum. Since it contains so little air, a good vacuum does not conduct heat or cold. In this manner, the flask can keep liquids at the proper temperature for several hours.

Pumps can be used to create a vacuum. Suppose a pump removes 25% of the air from a sealed 1-liter container on each stroke of its piston. A geometric series could be used to calculate the total amount of air being removed from the container. Since the goal is to remove all of the air and $r = 0.75$, this is an example of an **infinite geometric series.**

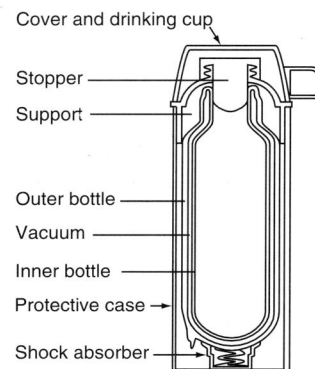

Cover and drinking cup

Stopper

Support

Outer bottle

Vacuum

Inner bottle

Protective case

Shock absorber

Stroke	Air Removed By Pump	Amount of Air Remaining
1	$\frac{1}{4}$	$\frac{3}{4}$
2	$\frac{1}{4}$ of $\frac{3}{4}$, or $\frac{1}{4}\left(\frac{3}{4}\right)$	$\frac{3}{4}$ of $\frac{3}{4}$, or $\left(\frac{3}{4}\right)^2$
3	$\frac{1}{4}$ of $\left(\frac{3}{4}\right)^2$, or $\frac{1}{4}\left(\frac{3}{4}\right)^2$	$\frac{3}{4}$ of $\left(\frac{3}{4}\right)^2$ or $\left(\frac{3}{4}\right)^3$
n	$\frac{1}{4}$ of $\left(\frac{3}{4}\right)^{n-1}$, or $\frac{1}{4}\left(\frac{3}{4}\right)^{n-1}$	$\frac{3}{4}$ of $\left(\frac{3}{4}\right)^{n-1}$, or $\left(\frac{3}{4}\right)^n$

The series representing the total amount of air removed can be written as $\sum_{n=1}^{\infty} \frac{1}{4}\left(\frac{3}{4}\right)^{n-1}$. The amount of air removed after n strokes, S_n, is called a **partial sum** of an infinite series, because it is the sum of a certain number of terms and not the entire series. A graph can help us see a pattern in the partial sums.

n	S_n
1	0.25
2	0.4375
3	0.578125
4	0.6835938
5	0.7626953
10	0.9436865
50	0.9999994
100	1

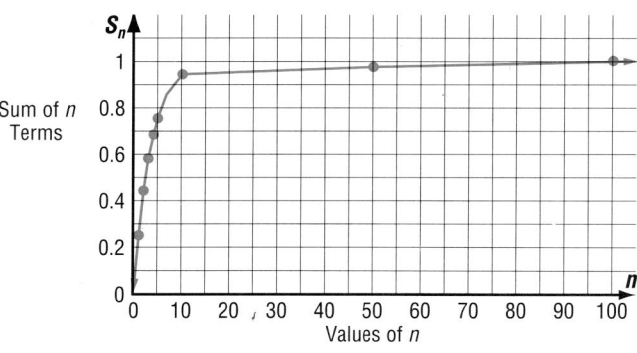

Notice that as n increases, the sum levels off and approaches a limit. This pattern is characteristic of infinite geometric series in which $|r| < 1$.

Let's look at the formula for the sum of a geometric series and determine how it can be used to find a formula for the sum of an infinite geometric series.

$$S_n = \frac{a_1 - a_1 r^n}{1 - r}$$

$$= \frac{a_1}{1 - r} - \frac{a_1 r^n}{1 - r} \qquad \textit{Rewrite as the difference of two rational expressions.}$$

$$= \frac{\frac{1}{4}}{1 - \frac{3}{4}} - \frac{\frac{1}{4}\left(\frac{3}{4}\right)^n}{1 - \frac{3}{4}} \qquad a_1 = \frac{1}{4} \textit{ and } r = \frac{3}{4}$$

$$= \frac{\frac{1}{4}}{\frac{1}{4}} - \frac{\frac{1}{4}\left(\frac{3}{4}\right)^n}{\frac{1}{4}} \textit{ or } 1 - \left(\frac{3}{4}\right)^n$$

As the value of n increases, the value of the second term decreases and approaches 0. Thus, the sum actually has the same value as the first term, 1. That is, gradually the pump is removing all of the air from the container, although at no point does it finish doing so. If the pump operated forever, it would remove all of the air. In fact, after as few as 50 strokes, it has removed nearly all of the air. The table on the previous page supports this conclusion. This suggests the following definition.

Sum of an Infinite Geometric Series	**The sum S of an infinite geometric series where $-1 < r < 1$ is given by** $$S = \frac{a_1}{1 - r}.$$

An infinite geometric series in which $|r| \geq 1$ does not have a sum. Consider the series $1 + 2 + 4 + 8 + 16 + \ldots$. In this series, $a_1 = 1$ and $r = 2$. The table at the right shows some of the partial sums of this series. As the value of n increases, the sum becomes increasingly greater and has no limit. With each additional term of the series, the sum grows without bound. That is, the sum does not approach a particular value. *If $r < -1$, what happens to the sum?*

n	S_n
5	31
10	1023
15	32,767
20	1,048,575

Example ❶ **Find the sum of each infinite geometric series, if it exists.**

a. $\frac{2}{3} + \frac{4}{9} + \frac{8}{27} + \ldots$

First find the value of r to determine if a sum exists.

$a_1 = \frac{2}{3}$ and $a_2 = \frac{4}{9}$, so $r = \frac{\frac{4}{9}}{\frac{2}{3}}$ or $\frac{2}{3}$. Since $\left|\frac{2}{3}\right| < 1$, a sum exists. Now use

the formula for the sum of an infinite geometric series.

$$S = \frac{a_1}{1 - r}$$

$$= \frac{\frac{2}{3}}{1 - \frac{2}{3}} \qquad a_1 = \frac{2}{3} \textit{ and } r = \frac{2}{3}$$

$$= \frac{\frac{2}{3}}{\frac{1}{3}}$$

$$= \frac{2}{3} \cdot \frac{3}{1} \textit{ or } 2$$

b. $1 - 3 + 9 - 27$

$a_1 = 1$ and $a_2 = -3$, so $r = \frac{-3}{1}$. Since $|-3| > 1$, no sum exists.

If no further force is exerted on a pendulum when it is set in motion, the distance it swings back and forth becomes less and less. The total distance it swings is an example of an infinite geometric series.

Example **2**

CONNECTION

Physics

F Y I

If a pendulum can be modeled by an infinite geometric series, then theoretically, it should never stop swinging. In reality, the swing of the pendulum is affected by air friction, so that eventually it does stop moving.

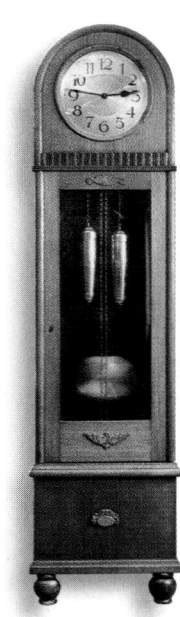

The spring in Juanita's old grandfather clock is broken. When you try to set the pendulum in motion by holding it against the wall of the clock and letting go, it follows a swing pattern of 25 cm, 20 cm, 16 cm, and so on until it comes to rest. What is the total distance the pendulum swings before coming to rest?

The swing pattern of the clock's pendulum forms the infinite series $25 + 20 + 16 + \ldots$. First determine if a sum exists.

$a_1 = 25$ and $a_2 = 20$, so $r = \frac{20}{25}$ or $\frac{4}{5}$. Since $\left|\frac{4}{5}\right| < 1$, a sum exists.

$$S = \frac{a_1}{1-r}$$

$$= \frac{25}{1 - \frac{4}{5}}$$

$$= \frac{25}{\frac{1}{5}}$$

$$= 25 \cdot 5 \text{ or } 125$$

The distance traveled by the pendulum before it stops is 125 cm.

The sum of an infinite geometric series can be used to express a repeating decimal as a rational number in the form $\frac{a}{b}$. Remember that repeating decimals such as $0.\overline{2}$ and $0.\overline{47}$ represent $0.22222\ldots$ and $0.4747474747\ldots$, respectively. Each of the expressions can be written as an infinite geometric series.

Example **3**

Express $0.\overline{12}$ as a rational number of the form $\frac{a}{b}$.

Rewrite the repeating decimal as a sum.

$0.\overline{12} = 0.121212\ldots$

$$= 0.12 + 0.0012 + 0.000012 + \ldots$$

$$= \frac{12}{100} + \frac{12}{10,000} + \frac{12}{1,000,000} + \ldots$$

In this series, $a_1 = \frac{12}{100}$ or 0.12, and each term is $\frac{1}{100}$ of the preceding term,

so $r = \frac{1}{100}$ or 0.01. You can determine the sum of this series by using either the decimal or fractional values for a_1 and r.

Method 1

$$S = \frac{a_1}{1-r}$$

$$= \frac{\frac{12}{100}}{1 - \frac{1}{100}}$$

$$= \frac{\frac{12}{100}}{\frac{99}{100}}$$

$$= \frac{12}{100} \cdot \frac{100}{99}$$

$$= \frac{12}{99} \text{ or } \frac{4}{33}$$

Method 2

Let $S = 0.121212\ldots$

$100S = 12.121212\ldots$ *Multiply each side by 100.*

$99S = 12$ *Subtract S from each side.*

$S = \frac{12}{99} \text{ or } \frac{4}{33}$ *Divide each side by 99.*

In Lesson 11–4, we used sigma notation to write an expression for the partial sum S_n. You can also use sigma notation to write an expression for the sum of an infinite series. The mathematical symbol ∞, or *infinity*, means endless. Since there is no last term in an infinite series, we use the symbol ∞ in place of n as the upper limit of the sum to indicate that the sum goes on forever.

Example **④** **Evaluate** $\displaystyle\sum_{n=1}^{\infty} 35\left(-\frac{1}{4}\right)^{n-1}$.

Remember that sigma notation uses the general form of the nth term of the geometric series, or $a_1 r^{n-1}$. Thus, in this series, $a_1 = 35$ and $r = -\frac{1}{4}$. Use the formula for the sum of an infinite geometric series.

$$S = \frac{a_1}{1 - r}$$

$$= \frac{35}{1 - \left(-\frac{1}{4}\right)}$$

$$= \frac{35}{\frac{5}{4}}$$

$$= \frac{35}{1} \cdot \frac{4}{5} \text{ or } 28$$

Thus, $\displaystyle\sum_{n=1}^{\infty} 35\left(-\frac{1}{4}\right)^{n-1} = 28$.

CHECK FOR UNDERSTANDING

Communicating Mathematics

Study the lesson. Then complete the following.

1. **Write** the formula for the sum of an infinite geometric series. Then explain how this formula is derived from $S_n = \dfrac{a_1 - a_1 r^n}{1 - r}$.

2. **Determine** if the sum of an infinite series in which $r = \dfrac{11}{10}$ can be computed. Explain your answer.

3. **a.** **Substitute** the values $a_1 = 3$ and $r = 0.8$ into the formula for a partial sum of n terms.

   b. Describe the value of $(0.8)^n$ as n increases without bound.

   c. As n increases in $(0.8)^n$, what happens to the value of the numerator in S_n?

   d. As n increases without bound, what happens to the value of S_n?

4. **Estimate** the sum S_n of the geometric series whose partial sums are graphed at the right.

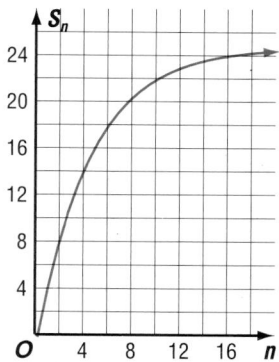

5. **You Decide** Lynn and Trent disagreed on whether the geometric series $\dfrac{1}{2} + \dfrac{3}{8} + \dfrac{9}{24} + \dots$ had a sum. Lynn said the series had no sum since $r = \dfrac{1}{2} \div \dfrac{3}{8}$ or $\dfrac{8}{6}$ and $\left|\dfrac{8}{6}\right| > 1$. Trent still disagreed. Who is correct? Explain.

6. Describe how you would write the sum of a given infinite geometric series in sigma notation. Include examples.

Guided Practice

Find a_1 and r for each series. Then find the sum, if it exists.

7. $36 + 24 + 16 + \dots$

8. $16 - 24 + 36 - \dots$

9. $6 - 4 + \frac{8}{3} - \dots$

10. $\frac{1}{4} + \frac{1}{6} + \frac{2}{9} + \dots$

11. $16 + 24 + 36 + \dots$

12. $\sum_{n=1}^{\infty} 40\left(\frac{3}{5}\right)^{n-1}$

Express each decimal as a rational number of the form $\frac{a}{b}$.

13. $0.\overline{5}$

14. $0.\overline{37}$

15. $0.\overline{175}$

16. Hot-Air Ballooning A hot-air balloon rises 80 feet in its first minute of flight. If in each succeeding minute the balloon rises only 90% as far as in the previous minute, what will be the balloon's maximum altitude?

EXERCISES

Practice

Find the sum of each infinite geometric series, if it exists.

17. $a_1 = 4, r = \frac{5}{7}$

18. $a_1 = 12, r = -\frac{3}{5}$

19. $a_1 = 18, r = 0.6$

20. $a_1 = 14, r = \frac{7}{3}$

21. $15 + 10 + \frac{20}{3} + \dots$

22. $\frac{5}{3} + \frac{25}{3} + \frac{125}{3} + \dots$

23. $1 + \frac{2}{3} + \frac{4}{9} + \dots$

24. $3 + 1.8 + 1.08 + \dots$

25. $18 - 12 + 8 - \dots$

26. $12 - 18 + 25 - \dots$

27. $\frac{5}{3} + \frac{10}{9} + \frac{20}{27} + \dots$

28. $\frac{3}{2} - \frac{3}{4} + \frac{3}{8} - \dots$

29. $\sum_{n=1}^{\infty} 48\left(\frac{2}{3}\right)^{n-1}$

30. $\sum_{n=1}^{\infty} \left(\frac{3}{8}\right)\left(\frac{3}{4}\right)^{n-1}$

31. $\sum_{n=1}^{\infty} -24\left(-\frac{3}{5}\right)^{n-1}$

Express each decimal as a rational number of the form $\frac{a}{b}$.

32. $0.\overline{7}$

33. $0.\overline{1}$

34. $0.\overline{36}$

35. $0.\overline{82}$

36. $4.\overline{6}$

37. $0.4\overline{5}$

38. $0.2\overline{31}$

39. $0.\overline{99}$

Write each infinite geometric series in sigma notation. Then find the sum, if it exists.

40. $10 - 1 + 0.1 + \dots$

41. $3 + 27 + 243 + \dots$

42. $1 - 0.5 + 0.25 - \dots$

43. The sum of an infinite geometric series is 81, and its common ratio is $\frac{2}{3}$. Find the first three terms of the series.

44. The sum of an infinite geometric series is 125, and the value of r is 0.4. Find the first three terms of the series.

45. The common ratio of an infinite geometric series is $\frac{11}{16}$, and its sum is $76\frac{4}{5}$. Find the first four terms of the series.

46. The first term of an infinite geometric series is -8, and its sum is $-13\frac{1}{3}$. Find the first four terms of the series.

47. The infinite series discussed in this lesson are infinite geometric series. Is it possible to have infinite arithmetic series? If so, do these series have sums? Explain and give examples.

48. Draw a Diagram Some physics students were experimenting with the rebound effect of a rubber ball. They used a motion detector and a 10-foot clear glass tube to determine that when a ball was dropped from this height, it rebounds or bounces 6 feet. The motion detector also recorded that on each consecutive bounce of the ball, it rebounded in the same proportion as that of the first bounce.

  a. Draw a diagram to illustrate the pattern of the vertical distances traveled by the ball until it stops bouncing.
  b. Write an infinite geometric series to represent the total of the distances the ball travels in a downward motion. Then find the sum.
  c. Write an infinite geometric series to represent the total of the distances the ball travels upward. Then find the sum.
  d. How do the two series in parts c and d compare?
  e. Find the total vertical distance traveled by the bouncing ball before it stops bouncing.

49. Child's Play Rebeca's little sister likes for her to push her in the swing at the park in their neighborhood. The other day, Rebeca pulled the swing back and let it go. She would have kept pushing, but she suddenly saw a friend at the other end of the park. The swing traveled a total distance of 10 feet before heading back the other way. Each swing afterwards was only 80% as long as the previous one. Find the total distance the swing traveled before it stopped.

50. Geometry The perimeter of square *ABCD* at the right is 40 cm. If the midpoints of each side were connected, a smaller square would result. Suppose the process of connecting midpoints of sides of squares and drawing new squares were continued indefinitely.

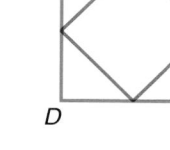

  a. Write an infinite geometric series to represent the sum of the perimeters of the squares formed by this process.
  b. Find the sum of all of the perimeters of all the squares formed.

51. Use sigma notation to express $1 - 3 + 9 - 27 + 81 - 243$. (Lesson 11–4)

52. Physics A vacuum pump removes $\frac{1}{5}$ of the air from a sealed container on each stroke of its piston. What percent of the air remains after five strokes of the piston? (Lesson 11–3)

53. Simplify $\dfrac{3y + 1}{2y - 10} + \dfrac{1}{y^2 - 2y - 15}$. (Lesson 9–4)

54. Astronomy A satellite is in an elliptical orbit with the center of Earth at one focus. The major axis of the orbit is 28,900 miles long, and the center of Earth is 8000 miles from the center of the ellipse. Assuming that the center of the ellipse is the origin and the foci lie on the *x*-axis, write the equation of the path of the satellite. (Lesson 7–4)

55. **Statistics** Suppose 500 items are normally distributed. (Lesson 6–9)
    a. How many items are within one standard deviation from the mean?
    b. How many items are within two standard deviations from the mean?
    c. How many items are within three standard deviations from the mean?
    d. How many items are within one standard deviation less than the mean?
    e. How many items are within two standard deviations greater than the mean?

56. **Geometry** The area of rectangle $ABCD$ is represented by $6x^2 + 38x + 56$. Its width is represented by $2x + 8$. Find the length of rectangle $ABCD$.
    (Lesson 5–3)

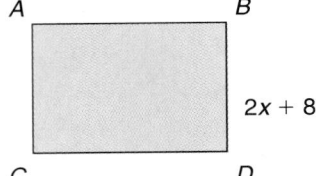

57. Solve the system of equations using augmented matrices. (Lesson 4–7)

    $a + \frac{1}{2}b - 3c = 19$

    $\frac{1}{2}a - b + 2c = -16$

    $5a + 2b - 2c = 50$

58. **Retailing** A salesman receives \$25 for every vacuum cleaner he sells. If he sells more than 10 vacuum cleaners a week, he will receive an additional \$1.75 for each successive sale until he is paid a maximum of \$46 per vacuum cleaner. How many must he sell to reach this maximum?
    (Lesson 3–5)

59. Find the slope-intercept form of the equation of the line that has a slope of $\frac{3}{4}$ and passes through the point at $(8, 2)$. (Lesson 2–4)

WORKING ON THE In·ves·ti·ga·tion

Refer to the Investigation on pages 590–591.

A scatter plot can help you analyze data and search for any patterns that may exist.

1 Draw a scatter plot for the data from the table in the Working on the Investigation feature in Lesson 11–1 on page 655. Then draw a best-fit line or curve. What do you notice about the graph? What relationship exists between the weight (number of washers) in the cup and the length of the rubber band? What type of a function does it portray?

2 Write an equation to relate the weight in the cup and the length the rubber band stretches.

3 What would you predict would happen in all aspects of the experiment if you continued to add washers?

4 Repeat the experiment with the other two rubber bands.

5 Create two separate scatter plots and best-fit lines or curves of the weights and the lengths of the other two rubber bands. Describe the graphs. Are the data points linear? How does the length compare as the weight in the cup gets heavier? Do each of the graphs illustrate a similar effect with more weight?

Add the results of your work to your Investigation Folder.

Recursion and Special Sequences

APPLICATION
Nature

What **YOU'LL LEARN**

- To recognize and use special sequences, and
- to iterate functions.

Why **IT'S IMPORTANT**

You can use recursion and special sequences to solve problems involving inflation and nature.

The seeds in the center of a sunflower show an example of a pattern often found in nature. The number of counterclockwise spirals is 8, and the number of clockwise spirals is 13. These are two numbers in a famous pattern of numbers called the **Fibonacci sequence,** named after its discoverer, Leonardo Fibonacci, who presented it in 1201. The sequence is shown below.

$$1, 1, 2, 3, 5, 8, 13, 21, 34, 55, 89, 144, \ldots$$

Notice the pattern in this sequence. After the second number, each number in the sequence is the sum of the two numbers that precede it. That is, $2 = 1 + 1$, $3 = 2 + 1, 5 = 3 + 2, 8 = 5 + 3, 13 = 8 + 5$, and so on.

first term	a_1		1
second term	a_2		1
third term	a_3	$a_1 + a_2$	$1 + 1 = 2$
fourth term	a_4	$a_2 + a_3$	$1 + 2 = 3$
fifth term	a_5	$a_3 + a_4$	$2 + 3 = 5$
$\vdots$	$\vdots$	$\vdots$	$\vdots$
nth term	a_n	$a_{n-2} + a_{n-1}$	

The formula $a_n = a_{n-2} + a_{n-1}$ is an example of a **recursive formula.** This means that each succeeding term is formulated from one or more previous terms.

Definition of Recursive Formula	**A recursive formula has two parts:** • **the value(s) of the first term(s), and** • **a recursion equation that shows how to find each term from the term(s) before it.**

$a_n = a_1 + (n - 1)d$ and $a_n = a_1 r^{n-1}$ are not recursive. These sequences are determined by the number of the term n rather than by the preceding term.

A recursive formula for a sequence describes how to find the nth term from the term(s) before it.

Sequence	Sequence Type	Recursive Formula
9, 13, 17, ...	arithmetic	$a_{n+1} = a_n + 4, a_1 = 9, n \geq 1$
7, 21, 63, ...	geometric	$a_{n+1} = a_n \cdot 3, a_1 = 7, n \geq 1$
1, 1, 2, 3, 5, 8, ...	Fibonacci	$a_n = a_{n-1} + a_{n-2}, a_1 = 1, a_2 = 1, n \geq 3$

Example **1** **Find the first five terms of the sequence in which $a_1 = 5$ and $a_{n+1} = 2a_n + 3, n \geq 1$.**

$a_1 = 5$ and $a_{n+1} = 2a_n + 3$

$a_{1+1} = 2a_1 + 3$ $n = 1$ | $a_{3+1} = 2a_3 + 3$ $n = 3$
     $a_2 = 2(5) + 3$ or 13 | $a_4 = 2(29) + 3$ or 61

$a_{2+1} = 2a_2 + 3$ $n = 2$ | $a_{4+1} = 2a_4 + 3$ $n = 4$
     $a_3 = 2(13) + 3$ or 29 | $a_4 = 2(61) + 3$ or 125

The first five terms of the sequence are 5, 13, 29, 61, and 125.

MODELING MATHEMATICS

Special Sequences

Materials: penny, nickel, and dime

The object of the game *Tower of Hanoi* is to move a stack of n coins from one position to another with the fewest moves, a_n. There are three positions and the following rules must be followed.

- You can only move one coin at a time.
- A coin must be placed on top of another coin, not underneath.
- A smaller coin may be placed on top of a larger coin, but not vice versa. For example, a nickel may not be placed on top of a dime.

Draw 3 circles on a sheet of paper, as shown below.

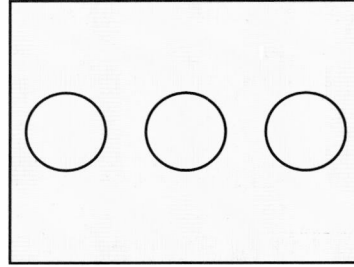

Your Turn

a. Place a dime on the first circle. What is the least number of moves that you can take to get it to the second circle?

b. Place a dime and a penny on the first circle, with the dime on top. What is the least number of moves that you can take to get the stack to another circle? (Remember, a penny cannot be placed on top of a dime.)

c. Place a nickel, penny, and dime on the first circle. What is the least number of moves that you can take to get the stack to another circle?

d. The least number of moves a_n required to move a stack of n coins can be represented by the function $a_n = 2^n - 1$. Find the number of moves required to move a stack of 4 coins and a stack of 5 coins.

LOOK BACK

Refer to Lesson 8–7B for an introduction to iteration.

Recall that iteration is a special type of recursion. Iteration is the process of composing a function with itself repeatedly. To understand the concept of iteration more clearly, use a calculator to find $\sqrt{2}$.

$$\sqrt{2} \approx 1.414213562$$

Then take the square root of that square root.

$$\sqrt{1.414213562} = 1.189207115.$$

Repeat the process three more times.

$$\sqrt{\sqrt{\sqrt{\sqrt{\sqrt{2}}}}} = 1.021897149$$

GLOBAL CONNECTIONS

The most popular board game on the African Continent is *mancala* and its hundreds of variations. This game of movement and capture using stones and cups is very mathematical in nature and has been played for centuries.

Each successive use of the square root function depends on the output of the previous root. In symbols, this can be represented by $f(f(f(f(f(x)))))$, where $f(f(x)) = f \circ f(x)$, the composition of functions. This example illustrates a recursion that composes a function with itself repeatedly. Note that all iterations are recursions, but not vice versa. For example, the Fibonacci sequence, which is a recursion, requires the input of two preceding terms. It is not the composition of a function to itself.

Iteration needs an initial value, x_0. To iterate a function $f(x)$, find the function value $f(x_0)$ of the initial value x_0. The value of $f(x_0)$ is x_1. The second iterate is the value of the function performed on the output, that is, $f(f(x_0))$ or $f(x_1)$.

Example **2** **Find the first three iterates x_1, x_2, x_3 of the function $f(x) = 3x + 1$ for an initial value of $x_0 = 1$.**

To find the first iterate x_1, find the value of the function for $x_0 = 1$.

$f(x_0) = f(1)$
$ = 3(1) + 1$ or 4 $\qquad$ So, $x_1 = 4$.

To find the second iterate x_2, substitute x_1 for x.

$f(x_1) = f(4)$
$ = 3(4) + 1$ or 13 $\qquad$ So, $x_2 = 13$.

Substitute x_2 for x to find the third iterate.

$f(x_2) = f(13)$
$ = 3(13) + 1$ or 40 $\qquad$ So, $x_3 = 40$.

The first three iterates of the function $f(x) = 3x + 1$ for an initial value of $x_0 = 1$ are 4, 13, and 40.

Iteration sequences may describe a wide variety of real-world situations. One use of iteration involves compound interest.

Example **3**

APPLICATION
Savings

Haloke's parents started a savings account for her when she was born. They invested $500 in an account that pays 6% interest compounded annually. Find the balance of the account after each of the first three years.

Explore $\quad$ We need to find an equation that describes the balance of the account after 3 years.

Plan $\quad$ Let b_n represent the balance after each year. So $b_0 = 500$. The interest is found by multiplying the balance of the account after the $(n - 1)$st year by the annual interest rate, which is 6%.

$$\underbrace{new\ balance}_{b_n} = \underbrace{previous\ balance}_{b_{n-1}} + \underbrace{accumulated\ interest}_{0.06 \cdot b_{n-1}}$$

Solve $\quad b_1 = b_0 + 0.06b_0$
$ = 500 + (0.06)(500)$
$ = 530 \qquad\qquad$ *balance after the first year*

$ b_2 = b_1 + 0.06b_1$
$ = 530 + 0.06(530)$
$ = 561.80 \qquad\quad$ *balance after the second year*

$ b_3 = b_2 + 0.06b_2$
$ = 561.80 + 0.06(561.80)$
$ = 595.51 \qquad\quad$ *balance after the third year*

Examine $\quad$ In the formula $A = P(1 + r)^t$, A is the balance, P is the amount invested, r is the interest rate, and t is the time in years that the money was invested. Substitute the known values into the formula.

$A = P(1 + r)^t$
$ = 500(1 + 0.06)^3 \quad$ *P = 500, r = 0.06, and t = 3*
$ = 595.51 \quad \checkmark$

Communicating Mathematics

Study the lesson. Then complete the following.

1. Which formula for the nth term of an arithmetic sequence is recursive: $a_n = a_1 + (n - 1)d$ or $a_n = a_{n-1} + d$? Explain how you know.
2. **Explain** how the composition of functions and iteration are related. Use the function $f(x) = 5x - 1$ as an illustration.
3. Refer to the Modeling Mathematics activity. Write the function $a_n = 2^n - 1$ as a recursive function in terms of a_{n-1}.

Guided Practice

Find the first six terms of each sequence.

4. $a_1 = 12, a_{n+1} = a_n - 3$

5. $a_1 = 1, a_2 = 2, a_{n+2} = 4a_{n+1} - 3a_n$

Find the first three iterates of each function, using the given initial values.

6. $f(x) = 3x - 4, x_0 = 3$

7. $f(x) = x^2 + 2, x_0 = -1$

Find the first five iterates of $f(x) = 2x - 8$ for each initial value.

8. $x_0 = -7$

9. $x_0 = 0.4$

10. **Inflation** If inflation is 3%, the cost $c(x)$ next year for a loaf of Italian bread can be found by the function $c(x) = x(1.03)^n$, where x is the cost this year and n is the number of years from now. If the cost this year is $1.80, what will the cost be ten years from now, if inflation continues at the same rate?

Practice

Find the first six terms of each sequence.

11. $a_1 = 9, a_{n+1} = 2a_n - 4$

12. $a_1 = -6, a_{n+1} = a_n + 3$

13. $a_1 = 13, a_{n+1} = a_n + 5$

14. $a_1 = 4, a_2 = -3, a_{n+2} = a_{n+1} + 2a_n$

15. $a_1 = 1, a_{n+1} = \dfrac{n}{n+1} \cdot a_n$

16. $a_1 = 6, a_{n+1} = a_n + (n + 3)$

Find the first three iterates of each function, using the given initial values.

17. $f(x) = 9x - 2, x_0 = 2$

18. $f(x) = 4x - 3, x_0 = 2$

19. $f(x) = 3x + 5, x_0 = -4$

20. $f(x) = 2x^2 - 5, x_0 = -1$

21. $f(x) = 3x^2 - 4, x_0 = 1$

22. $f(x) = x^2 + 3x + 1, x_0 = 1$

Find the first five iterates of $f(x) = 4x - 5$ for each initial value.

23. $x_0 = 6$

24. $x_0 = -2$

25. $x_0 = \dfrac{1}{2}$

26. $x_0 = 0.85$

27. If $a_0 = 7$ and $a_{n+1} = a_n + 12$, find the value of a_5.

28. If $a_0 = 1$ and $a_{n+1} = -2.1$, then what is the value of a_4?

Critical Thinking

29. Iterate the function $f(x) = 0.1x + 0.3$ several times with the initial value of $x_0 = 0.3$. Describe the pattern that emerges.

Applications and Problem Solving

30. Finance Aletha Johnson is taking out a mortgage loan for $120,000 to buy a new house. Her monthly payments are $942.55, and the function $B_n = B_{n-1}(1.0075) - 942.55$ describes the balance of the loan at the end of each month. Find the balance of the loan for the first eight months.

31. Look for a Pattern Look at the figures representing triangular numbers shown below.

Figure 1   Figure 2   Figure 3   Figure 4   Figure 5

a. Write a sequence of the first five triangular numbers.

b. Write a recursive formula for the nth triangular number, a_n.

c. What is the 80th triangular number?

32. Special Sequences The Lucas sequence was developed from the Fibonacci sequence. In the Lucas sequence, $L_1 = F_1$ and $L_n = F_{n+1} + F_{n-1}$ for $n \geq 2$. Find the first eight terms of the Lucas sequence.

Mixed Review

33. Find the sum of the infinite series $9 + 6 + 4 + \ldots$. (Lesson 11–5)

34. Physics The intensity of illumination on a surface varies inversely as the square of the distance from the light source. A surface is 12 meters from a light source. How far must the surface be from the source to receive twice as much illumination? (Lesson 9–2)

35. Write the equation of the hyperbola graphed at the right. (Lesson 7–5)

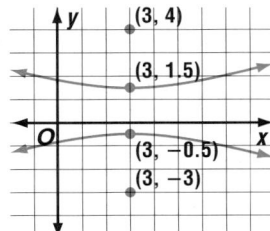

36. Find $3 \begin{bmatrix} 4 \\ 1 \\ 7 \end{bmatrix} + 2 \begin{bmatrix} 3 \\ -2 \\ 6 \end{bmatrix} - 5 \begin{bmatrix} -2 \\ 3 \\ 6 \end{bmatrix}$. (Lesson 4–2)

37. Statistics The numbers of job-related injuries at a construction site for each month of 1995 are listed below. (Lesson 1–3)

| 10 | 13 | 15 | 39 | 21 | 24 |
| 19 | 16 | 39 | 17 | 23 | 25 |

a. Make a line plot of the numbers of injuries.

b. Make a stem-and-leaf plot of the number of injuries.

c. Find the median, mode, and mean of the numbers of injuries.

Fractals

11-7

What YOU'LL LEARN

- To define and draw fractals, and
- to write a recursive formula for the perimeter or area of a fractal.

Why IT'S IMPORTANT

You can use fractals to solve problems involving geology and literature.

APPLICATION
Nature

Nature is filled with many intricate shapes. Euclidean geometry with its points, planes, and polygons cannot begin to describe the shapes we see from day to day. Clouds, leaves, wood, coastlines, rust, and broccoli are far too complicated for Euclidean geometry. Yet, patterns are imbedded in these shapes. In this century, a new branch of mathematics called **fractal geometry** is providing models of nature's designs. Some fractals that model natural patterns are shown below.

One of the most obvious fractal patterns occurs in a tree. Consider how the main trunk of the tree splits into large branches. The large branches of a tree split into smaller branches. Similarly, the smaller branches split into twigs, and so on.

MODELING MATHEMATICS

Fractal Trees

Materials: isometric dot paper

In this activity, you will make a fractal tree and investigate some of its patterns.

Your Turn

a. At the bottom center of your dot paper, draw a vertical line segment to represent the tree trunk. From the endpoint, draw two branches half as long. The angle between the branches should be 120°.

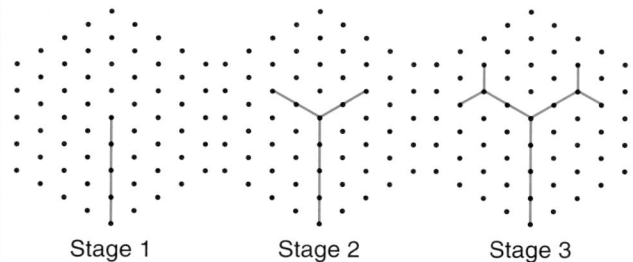

Stage 1 Stage 2 Stage 3

b. Continue drawing branches until they become too small to draw. The first three stages are shown.
c. Can you find small parts of the tree that look like the entire tree? If so, circle them on your drawing.
d. Suppose the length of the tree trunk is 1 unit. How many branches are $\frac{1}{2}$ unit? What is the total length of the branches that are $\frac{1}{2}$ unit long? Find the total lengths for branches that are $\frac{1}{4}$, $\frac{1}{8}$, and $\frac{1}{16}$ unit.
e. If you could continue to draw branches, what is the total length of the branches of the tree?

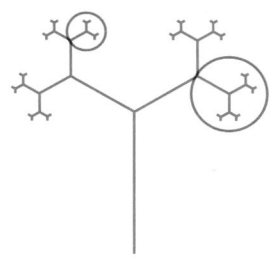

When small parts of a fractal tree are magnified, the detail is not lost. In fact, the magnified part looks the same as the entire structure. This characteristic is called **self-similarity**. Self-similar objects are those in which we can find replicas of the entire shape or object embedded over and over again inside the object in different sizes.

One of the characteristics of a fractal is that it exhibits self-similarity. Other characteristics can be described as follows.

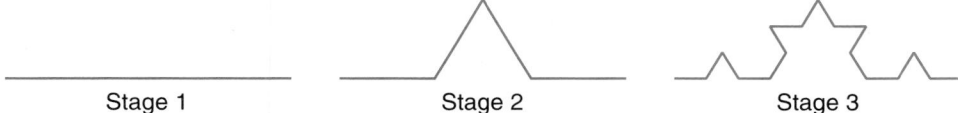

Definition of Fractal	A fractal is a geometric figure that has self-similarity, is created using a recursive process, and is infinite in structure.

The key to constructing a fractal lies in the replication of a pattern. Suppose you start with a rule that describes a pattern. At each stage in the process, you apply this rule to smaller and smaller parts of the figure. To construct a fractal, this iteration would be applied over and over, without end. In 1904, Swedish mathematician Helge von Koch used such a method to produce a classic fractal, now called the Koch curve.

To construct a Koch curve, start with a line segment. The iterative rule is to construct an equilateral triangle on the middle third of each segment, removing the base of the triangle.

Stage 1   Stage 2   Stage 3

When a Koch curve is applied to an equilateral triangle, a Koch snowflake is produced.

Example **Draw the first four stages of the Koch snowflake.**

Stage 1   Stage 2   Stage 3   Stage 4

In addition to producing fascinating visual patterns, number patterns are also evident in fractals. These patterns can be described by recursive formulas. Consider the number of segments at each stage of the Koch snowflake. Let's begin by finding a pattern in the Koch curve shown at the right.

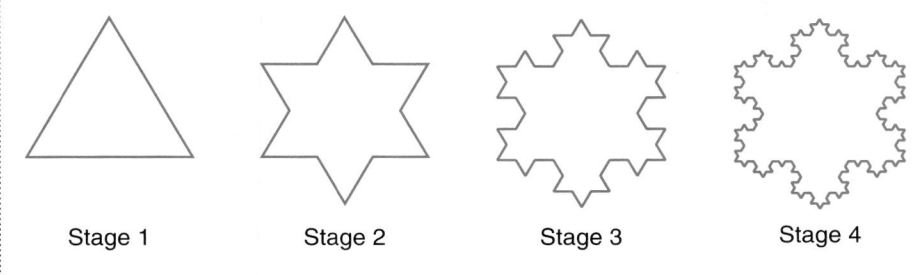

1 segment

4 segments

16 segments

Stage	1	2	3	...
Segments	1	4	16	...

Since the Koch snowflake is formed by applying the Koch curve to the three segments of an equilateral triangle, the number of segments in a Koch snowflake is as follows.

Stage	1	2	3	...
Segments	3 × 1 or 3	3 × 4 or 12	3 × 16 or 48	...

At each successive stage, one line segment is replaced by four smaller segments. Therefore, the pattern can be described by the recursive formula $a_1 = 3$ and $a_{n+1} = 4a_n$. It appears that the number of segments becomes infinite as the number of stages increases.

Example ❷

INTEGRATION

Geometry

Refer to the Koch snowflake in Example 1.

a. **If the length of each segment of the equilateral triangle in Stage 1 is 9 units, find a recursive formula that describes how the perimeter of the snowflake changes as the number of stages increases.**

b. **Find the perimeter of the Koch snowflake at Stage 5.**

c. **Describe in words how the perimeter of the Koch snowflake changes as the number of stages increases.**

a. Remember that a Koch snowflake is formed by constructing an equilateral triangle on the middle third of each segment. First, determine the perimeter through Stage 3.

Stage (n)	1	2	3	...
Number of segments	3	12	48	...
Length of each segment	9	3	1	...
Perimeter (a_n)	27	36	48	...

As the number of stages increases, the perimeter increases by a factor of $\frac{4}{3}$. Therefore, a recursive formula for the perimeter is $a_1 = 27$, $a_{n+1} = \frac{4}{3}a_n$.

b. From the chart above, you know that $a_3 = 48$. Find a_4 and a_5.

$$a_4 = \frac{4}{3}a_3 \qquad a_5 = \frac{4}{3}a_4$$
$$= \left(\frac{4}{3}\right)48 \qquad = \left(\frac{4}{3}\right)64$$
$$= 64 \qquad = 85.3$$

The perimeter of the Koch snowflake at Stage 5 is about 85.3 units.

c. It appears that the perimeter of the Koch snowflake increases without bound. The common ratio or growth factor is $\frac{4}{3}$ and $\left|\frac{4}{3}\right| > 1$.

Fractal geometry is a branch of *chaos theory*, which is being used to study and make sense of natural phenomena traditionally thought to neither have patterns nor be capable of being described through mathematical modeling. Such phenomena include the study of weather patterns, earthquake occurrences, and fluctuations in the stock market. Let's see how chaos theory works by playing a chaos game.

First, draw an equilateral triangle on a piece of paper. Label the vertices of the triangle T, R, and L to represent top, right, and left. Choose any point x_0 in the interior of the triangle as the starting point. Roll a die to choose T, R, or L at random. Then plot the point that is halfway to the vertex that has been selected. Continue to roll a die to choose vertices and plot points.

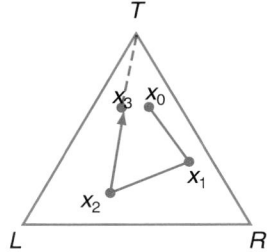

Do you see a pattern? Probably not! It seems unlikely that a random succession of midpoints would produce anything but a lot of points. Even if you could take the time to plot 500 such points, you may not see a pattern. However, here is where a computer or graphing calculator is an invaluable tool. In the following Exploration, you will program a graphing calculator to play a chaos game.

EXPLORATION

GRAPHING CALCULATORS

The following program will generate 3000 points in a chaos game.

```
PROGRAM: SIERPINS
:FnOff
:ClrDraw
:PlotsOff
:AxesOff
:0→Xmin: 1→Xmax
:0→Ymin: 1→Ymax
:rand→X: rand→Y
:For (K, 1, 3000)
:rand→N
:If N≤1/3
:Then
:.5X→X
:.5Y→Y
:End
```

```
:If 1/3<N and N≤2/3
:Then
:.5(.5+X)→X
:.5(1+Y)→Y
:End
:If 2/3<N
:Then
:.5(1+X)→X
:.5Y→Y
:End
:Pt-On (X, Y)
:End
:StorePic 6
```

Your Turn

a. Describe the pattern produced by this chaos game.

b. Do you think this pattern might be a fractal? Explain your reasoning.

Sierpinski's triangle is also called Sierpinski's gasket.

The picture generated in the Exploration above is a fractal called *Sierpinski's triangle,* named after the Polish mathematician Waclaw Sierpinski, who introduced it in 1916. Sierpinski's triangle can be constructed by using the following recursive process.

1. Start with an equilateral triangle.

2. Connect the midpoints of the sides with line segments.

3. Remove the middle triangle.

4. Apply steps 1–3 to each of the remaining triangles repeatedly.

Example **3** **a.** Draw the first four stages of Sierpinski's Triangle.
b. If the area of the triangle at Stage 1 is 64 square units, find a recursive formula that describes how the area of the triangle changes as the number of stages increases.
c. Describe in words how the area of the triangle changes as the number of stages increases.

a.

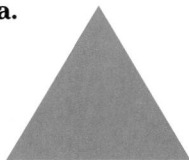  

Stage 1   Stage 2   Stage 3   Stage 4

b. Make a table showing the area of each stage.

Stage	1	2	3	...
Area	64	48	36	...

As the number of stages increases, the area decreases by a factor of $\frac{3}{4}$. Therefore, a recursive formula is $a_1 = 64$, $a_{n+1} = \frac{3}{4}a_n$.

c. It appears that the area remaining in the triangle approaches zero. The common ratio or growth factor is $\frac{3}{4}$ and $\left|\frac{3}{4}\right| < 1$.

CHECK FOR UNDERSTANDING

Communicating Mathematics

Study the lesson. Then complete the following.

1. **Explain** how a cloud or a tree can be considered a fractal.

2. **Explain** how recursion is used in creating fractals.

3. **Explain** how self-similarity is evident in a fern.

Guided Practice

4. The figure at the right shows the first two stages of a fractal formed by constructing a square on the middle third of the segment and removing the base of the square. Draw the next two stages.

         Stage 1   Stage 2

5. Refer to Example 3.
   a. Write a recursive formula that describes how the number of shaded triangles changes as the number of stages increases.
   b. Find the number of shaded triangles at Stage 5.

6. **Botany** Botanists have found that the angle between the main branches of a tree and its trunk remain constant in each species. How is this finding related to fractal geometry?

Practice

Draw the next stage of a fractal formed by replacing each segment with the pattern shown.

7.

8.

9. A fractal is formed by trisecting the sides of a square, which forms 9 smaller squares, and then removing the middle square.

Stage 1   Stage 2

   a. Draw Stages 3 and 4.

   b. Write a recursive formula that describes how the number of squares changes as the number of stages increases.

   c. Find the number of squares at Stage 6.

   d. Suppose the area at Stage 1 is 81 square units. Write a recursive formula that describes how the area of the square carpet changes as the number of stages increases.

   e. Describe how the area of one of the squares changes as the number of stages increases.

Critical Thinking

10. The perimeter of a Koch snowflake becomes infinite as the number of stages increases. Is the area of a Koch snowflake infinite or finite? Explain your reasoning.

Applications and Problem Solving

11. **Literature** In *Jurassic Park*, mathematician Ian Malcolm uses chaos theory. Find a copy of the book or a book review and describe the role that chaos theory plays in the book.

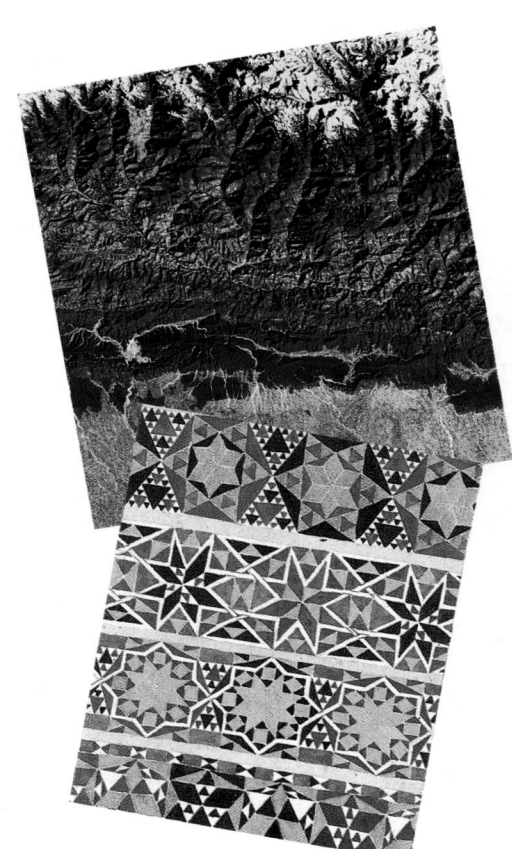

12. **Geology** The photograph at the right shows the foothills of the Himalayas as seen from the Landsat-1 satellite. Describe the fractal self-similarity of the foothills.

13. **Art** The photograph at the right shows Escher's studies of Sierpinski's triangle patterns on the twelfth century pulpit of the Ravello cathedral, designed by Nicola di Bartolomeo of Foggia. Create your own design using Sierpinski's triangle.

Mixed Review

14. Find the first five terms of the sequence in which $a_1 = 3$, $a_{n+1} = 2a_n + 5$. (Lesson 11–6)

15. Use a calculator to find the natural logarithm of 0.056, rounded to four decimal places. (Lesson 10–5)

16. **Electronics** Sharon Weisman is an electrical engineer designing the electrical circuits for a new office building. There are three basic things to be considered in an electrical circuit: the flow of the electrical current I, the resistance to the flow Z called impedance, and electromotive force E called voltage. These quantities are related in the formula $E = I \cdot Z$. The current of the circuit Ms. Weisman is designing is to be $(35 - j40)$ amperes. Electrical engineers use the letter j to represent the imaginary unit. Find the impedance of the circuit if the voltage is to be $(430 - j330)$ volts. (Lesson 5–10)

17. Write $\begin{bmatrix} 5 & 1 \\ 2 & -3 \end{bmatrix} \cdot \begin{bmatrix} x \\ y \end{bmatrix} = \begin{bmatrix} 26 \\ 41 \end{bmatrix}$ as a system of linear equations. (Lesson 4–6)

18. Solve $|x - 7| = 12$. (Lesson 1–5)

Mathematics and SOCIETY

Chaos on the Tilt-a-Whirl

The excerpt below appeared in an article in *Science News* on February 26, 1994.

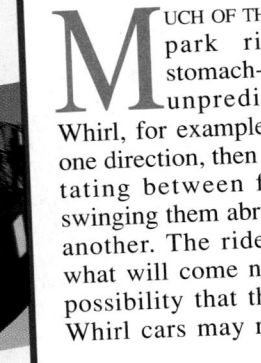

MUCH OF THE FUN OF AN AMUSEMENT-park ride arises from its stomach-churning, mind-tingling unpredictability. The Tilt-a-Whirl, for example, spins its passengers in one direction, then another, sometimes hesitating between forays and sometimes swinging them abruptly from one motion to another. The rider never knows exactly what will come next. . . . Intrigued by the possibility that the motion of the Tilt-a-Whirl cars may represent an example of chaotic behavior, Richard L. Kautz of the National Institute of Standards and Technology . . . and Brett M. Huggard of Northern Arizona University . . . worked out a mathematical equation to describe the forces acting on each car. . . . Chaotic motion occurs at intermediate speeds, close to the 6.5 revolutions per minute at which the ride actually operates. . . . At intermediate speeds, the jumbled mixture of car rotations never repeats itself exactly, which gives the Tilt-a-Whirl its lively and unpredictable behavior. ∎

1. Tilt-a-Whirl riders can often be seen throwing their weight from side to side at certain places in the ride cycle. Why might they be doing this?

2. Contrast the type of ride you experience on a Tilt-a-Whirl with that which you experience on a roller coaster or merry-go-round. What accounts for these differences?

3. In designing an amusement-park ride, what kinds of necessary information could a mathematical model provide?

The Binomial Theorem

11-8

What YOU'LL LEARN

- To expand powers of binomials by using Pascal's triangle and the binomial theorem, and
- to find specific terms of binomial expansions.

Why IT'S IMPORTANT

You can use the binomial theorem to solve problems involving probability and games.

INTEGRATION
Probability

Carol Perez plays in a bowling league on Saturday mornings. Her scores are improving. At the present time, she averages 2 strikes per game (10 frames). Today, her team is playing the number one team, and Carol feels "up" for the competition. She wants to score 4 strikes in the first game. What is the probability that she will make 4 strikes in the first game? To answer this question, you need to understand the binomial theorem.   *The approach and solution to this problem will be considered in Example 6.*

You have observed patterns in both geometric and arithmetic sequences. Observe some of the patterns that appear when $(a + b)^n$ is expanded for $n = 0, 1, 2, 3,$ and 4.

$$(a + b)^0 = 1a^0b^0$$
$$(a + b)^1 = 1a^1b^0 + 1a^0b^1$$
$$(a + b)^2 = 1a^2b^0 + 2a^1b^1 + 1a^0b^2$$
$$(a + b)^3 = 1a^3b^0 + 3a^2b^1 + 3a^1b^2 + 1a^0b^3$$
$$(a + b)^4 = 1a^4b^0 + 4a^3b^1 + 6a^2b^2 + 4a^1b^3 + 1a^0b^4$$

What is the pattern of the exponents of a and b in each row?

Here is a list of some of the patterns seen in the expansion of $(a + b)^n$.

1. The exponent of $(a + b)^n$, n, is the exponent of a in the first term and the exponent of b in the last term.

2. In successive terms, the exponent of a decreases by one. It is n in the first term and zero in the last term.

3. In successive terms, the exponent of b increases by one. It is zero in the first term and n in the last term.

4. The sum of the exponents in each term is n.

5. The coefficients are symmetric. They increase at the beginning and decrease at the end to the expansion.

The coefficients form a pattern that is often displayed in a triangular formation. This is known as **Pascal's triangle.** Notice that each row is formed by starting and ending with 1. Then each coefficient is the sum of the pair of coefficients above it in the previous row.

$$(a + b)^0 \qquad\qquad\qquad 1$$
$$(a + b)^1 \qquad\qquad\qquad 1 \qquad 1$$
$$(a + b)^2 \qquad\qquad\quad 1 \qquad 2 \qquad 1$$
$$(a + b)^3 \qquad\qquad 1 \qquad 3 \qquad 3 \qquad 1$$
$$(a + b)^4 \qquad\quad 1 \qquad 4 \qquad 6 \qquad 4 \qquad 1$$
$$(a + b)^5 \qquad 1 \qquad 5 \qquad 10 \qquad 10 \qquad 5 \qquad 1$$

F Y I

Pascal's triangle is named for a French mathematician Blaise Pascal (1623–1662). However, this triangle appeared in a Chinese publication in A.D. 1303. A Persian mathematician, Omar Khayyam, appeared to have knowledge of the triangle about A.D. 1100. Also, an Italian mathematician, Niccolo Tartaglia, claimed he invented the triangle in 1556.

Example **1** **Use the pattern in Pascal's triangle to write $(x + y)^6$ in expanded form.**

The next line of Pascal's triangle is listed below.

$$1 \quad 6 \quad 15 \quad 20 \quad 15 \quad 6 \quad 1 \quad \textit{(a + b)}^6 \textit{ has 7 terms.}$$

$$(a + b)^6 = 1x^6y^0 + 6x^5y^1 + 15x^4y^2 + 20x^3y^3 + 15x^2y^4 + 6x^1y^5 + 1x^0y^6$$

$$= x^6 + 6x^5y + 15x^4y^2 + 20x^3y^3 + 15x^2y^4 + 6xy^5 + y^6$$

Another way to show the coefficients is by writing them in terms of the previous coefficient, as you would in a sequence.

$(a + b)^0$				1	
$(a + b)^1$			1	$\frac{1}{1}$	
$(a + b)^2$		1	$\frac{2}{1}$	$\frac{2 \cdot 1}{1 \cdot 2}$	
$(a + b)^3$	1	$\frac{3}{1}$	$\frac{3 \cdot 2}{1 \cdot 2}$	$\frac{3 \cdot 2 \cdot 1}{1 \cdot 2 \cdot 3}$	
$(a + b)^4$	1 $\frac{4}{1}$	$\frac{4 \cdot 3}{1 \cdot 2}$	$\frac{4 \cdot 3 \cdot 2}{1 \cdot 2 \cdot 3}$	$\frac{4 \cdot 3 \cdot 2 \cdot 1}{1 \cdot 2 \cdot 3 \cdot 4}$	

Eliminate common factors that are shown in color. The coefficients are symmetrical.

This pattern provides the coefficients of $(a + b)^n$ for any value of n where n is a nonnegative integer. This pattern is summarized in the **binomial theorem.**

The Binomial Theorem	**If n is a positive integer, then** $(a + b)^n = 1a^nb^0 + \frac{n}{1}a^{n-1}b^1 + \frac{n(n-1)}{1 \cdot 2}a^{n-2}b^2 + \ldots + 1a^0b^n.$

Example **2** **Use the binomial theorem to write $(a - b)^7$ in expanded form.**

The expansion will have eight terms. Find the first four terms using the sequence $1, \frac{7}{1}, \frac{7 \cdot 6}{1 \cdot 2}, \frac{7 \cdot 6 \cdot 5}{1 \cdot 2 \cdot 3}$. Then use symmetry to find the remaining terms.

$$(a - b)^7 = a^7(-b)^0 + \frac{7}{1}a^6(-b)^1 + \frac{7 \cdot 6}{1 \cdot 2}a^5(-b)^2 + \frac{7 \cdot 6 \cdot 5}{1 \cdot 2 \cdot 3}a^4(-b)^3 + \ldots$$

$$= a^7 - 7a^6b^1 + 21a^5b^2 - 35a^4b^3 + \ldots$$

$$= a^7 - 7a^6b + 21a^5b^2 - 35a^4b^3 + 35a^3b^4 - 21a^2b^5 + 7ab^6 - b^7$$

Note that in terms having the same coefficients, the exponents are reversed, as in $21a^5b^2$ and $21a^2b^5$.

By definition, 0! = 1.

The pattern in the factors of the coefficients in Example 2 are parts of special products called **factorials.** The product of $4 \cdot 3 \cdot 2 \cdot 1$ can be expressed as 4! and is read *4 factorial.* For all positive integer values of *n*, $n! = n(n-1)(n-2)(n-3) \ldots 2 \cdot 1$.

Example **Evaluate $\frac{9!}{4!5!}$.** *how to enter into calculator* $\frac{9!}{5! \times 4!}$

You can evaluate factorials by using the $\boxed{x!}$ key on a scientific calculator.

$$\frac{9!}{4!5!} = \frac{9 \cdot 8 \cdot 7 \cdot 6 \cdot \overset{1}{\cancel{5 \cdot 4 \cdot 3 \cdot 2 \cdot 1}}}{4 \cdot 3 \cdot 2 \cdot 1 \cdot \underset{1}{\cancel{5 \cdot 4 \cdot 3 \cdot 2 \cdot 1}}}$$

Note that $9! = 9 \cdot 8 \cdot 7 \cdot 6 \cdot 5!$.
So $\frac{9!}{4!5!} = \frac{9 \cdot 8 \cdot 7 \cdot 6 \cdot 5!}{4!5!}$ *or* $\frac{9 \cdot 8 \cdot 7 \cdot 6}{4 \cdot 3 \cdot 2 \cdot 1}$.

$$= \frac{9 \cdot 8 \cdot 7 \cdot 6}{4 \cdot 3 \cdot 2 \cdot 1} \text{ or } 126$$

EXPLORATION

GRAPHING CALCULATORS

You can use a graphing calculator to find the value of 6!. To do this, first press 6. Then press $\boxed{\text{MATH}}$ and highlight PRB. Four items will show on your screen. You want the factorial function, so use the direction keys to highlight 4. Press $\boxed{\text{ENTER}}$ and then $\boxed{\text{ENTER}}$ again. Your screen should show that 6! is equal to 720.

Your Turn

Use a graphing calculator to evaluate each expression.

a. 12!   **b.** 7!   **c.** $\frac{10!}{5!}$ **d.** $\frac{15!}{8!7!}$

In Example 2, notice that products like $\frac{7 \cdot 6 \cdot 5}{1 \cdot 2 \cdot 3}$ can be written as a quotient of factorials. In this case, $\frac{7 \cdot 6 \cdot 5}{1 \cdot 2 \cdot 3} = \frac{7!}{3!4!}$. Using this same method, we can rewrite the series that equals $(a + b)^7$ using factorials.

$$(a + b)^n = \frac{7!}{0!7!}a^7 + \frac{7!}{1!6!}a^6b^1 + \frac{7!}{2!5!}a^5b^2 + \frac{7!}{3!4!}a^4b^3 + \frac{7!}{4!3!}a^3b^4 +$$
$$\frac{7!}{5!2!}a^2b^5 + \frac{7!}{6!1!}a^1b^6 + \frac{7!}{7!0!}b^7$$

This same pattern can be used to write the series using sigma notation.

$$(a + b)^7 = \sum_{k=0}^{7} \frac{7!}{k!(7 - k)!}a^{7-k}b^k$$

The binomial theorem can also be written both in factorial notation and in sigma notation.

$$(a + b)^n = \frac{n!}{0!(n - 0)!}a^n + \frac{n!}{1!(n - 1)!}a^{n-1}b^1 + \frac{n!}{2!(n - 2)!}a^{n-2}b^2 + \ldots$$
$$= \sum_{k=0}^{n} \frac{n!}{k!(n - k)!}a^{n-k}b^k$$

Example ④ **Express $(3m + d)^5$ using sigma notation. Then expand and simplify the expression.**

$$(3m + d)^5 = \sum_{k=0}^{5} \frac{5!}{k!(5-k)!}(3m)^{5-k}d^k$$

$$= \frac{5!}{0!5!}(3m)^5 d^0 + \frac{5!}{1!4!}(3m)^4 d^1 + \frac{5!}{2!3!}(3m)^3 d^2 + \frac{5!}{3!2!}(3m)^2 d^3 +$$

$$\frac{5!}{4!1!}(3m)^1 d^4 + \frac{5!}{5!0!}(3m)^0 d^5$$

$$= \frac{5 \cdot 4 \cdot 3 \cdot 2 \cdot 1}{1 \cdot 5 \cdot 4 \cdot 3 \cdot 2 \cdot 1}(3m)^5 + \frac{5 \cdot 4 \cdot 3 \cdot 2 \cdot 1}{1 \cdot 4 \cdot 3 \cdot 2 \cdot 1}(3m)^4 d +$$

$$\frac{5 \cdot 4 \cdot 3 \cdot 2 \cdot 1}{2 \cdot 1 \cdot 3 \cdot 2 \cdot 1}(3m)^3 d^2 + \frac{5 \cdot 4 \cdot 3 \cdot 2 \cdot 1}{3 \cdot 2 \cdot 1 \cdot 2 \cdot 1}(3m)^2 d^3 +$$

$$\frac{5 \cdot 4 \cdot 3 \cdot 2 \cdot 1}{4 \cdot 3 \cdot 2 \cdot 1 \cdot 1}(3m)d^4 + \frac{5 \cdot 4 \cdot 3 \cdot 2 \cdot 1}{5 \cdot 4 \cdot 3 \cdot 2 \cdot 1 \cdot 1}d^5$$

$$= 243\,m^5 + 405m^4 d + 270m^3 d^2 + 90m^2 d^3 + 15md^4 + d^5$$

Sometimes a particular term in the expansion of a binomial is needed. Note that in the sigma notation form of the binomial theorem, $k = 0$ for the first term, $k = 1$ for the second term, and so on. In general, the value of k is always one less that the number of the term you are seeking.

Example ⑤ **Find the seventh term of $(p + q)^{11}$.**

First use the binomial theorem to write the general form of the expansion.

$$(p + q)^{11} = \sum_{k=0}^{11} \frac{11!}{k!(11-k)!}p^{11-k}q^k$$

In the seventh term, $k = 6$ since k starts at 0.

The seventh term, $\frac{11!}{6!(11-6)!}p^{11-6}q^6$, is $\frac{11 \cdot 10 \cdot 9 \cdot 8 \cdot 7}{5 \cdot 4 \cdot 3 \cdot 2 \cdot 1}p^5 q^6$ or $462p^5 q^6$.

The binomial theorem can also be used to compute probability.

Example ⑥

INTEGRATION
Probability

Refer to the probability question at the beginning of the lesson. What is the probability that Carol will get 4 strikes in the game?

Let s represent the probability of getting a strike in a given frame, and let n represent the probability of not getting a strike in the frame. Since there are 10 frames in a game, we can use the binomial theorem to find any term in the expansion of $(s + n)^{10}$.

$$(s + n)^{10} = \sum_{k=0}^{10} \frac{10!}{k!(10-k)!}s^{10-k}n^k$$

To find the probability of 4 strikes, compute the term in which the exponent of s is 4. Since $10 - 6 = 4$, find the term in which $k = 6$, the seventh term.

$$\frac{10!}{6!4!}s^4 n^6 \text{ or } 210s^4 n^6$$

With an average of two strikes per game, the probability that Carol will get a strike in a given frame is 2 out of 10 or 0.2. The probability that Carol will not get a strike is 8 out of 10 or 0.8. Evaluate the expression for the seventh term if $s = 0.2$ and $n = 0.8$.

$$210s^4 n^6 \quad \rightarrow \quad 210(0.2)^4(0.8)^6 \approx 0.088$$

The probability that Carol will make 4 strikes in the game is about 0.088 or 8.8%.

Communicating Mathematics

Study the lesson. Then complete the following.

1. **Explain** how to form additional rows of Pascal's triangle.

2. Is the sequence pattern for the coefficients of $(a + b)^n$ an arithmetic sequence, a geometric sequence, or neither? Explain.

3. The sixth term of $(a + b)^8$ is $\frac{8!}{5!3!}a^3b^5$.

   a. **Explain** the relationship between the number of the term and the exponents.

   b. **Explain** the relationship among the exponents, the degree of the expansion, and the factorials.

   c. **Explain** how this expression relates to the general form for a term $\frac{n!}{k!(n-k)!}a^{n-k}b^k$.

4. **Name** three methods for computing the coefficients of a binomial expansion.

Guided Practice

Use a calculator to evaluate each expression.

5. $8!$
6. $\frac{13!}{9!}$
7. $\frac{12!}{2!10!}$

Expand each binomial.

8. $(p + q)^5$
9. $(t + 2)^6$
10. $(x - y)^4$

Find the indicated term of each expansion.

11. fourth term of $(a + b)^8$
12. fifth term of $(2a + 3b)^{10}$

13. **Geometry** Use the binomial theorem to write an expression for the volume of the cube at the right.

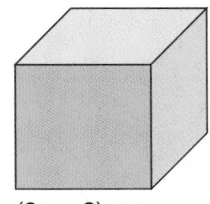

$(2x - 3)$ cm

Practice

15-37 odd

Use a calculator to evaluate each expression.

14. $9!$
15. $5!$
16. $13!$
17. $\frac{9!}{7!}$
18. $\frac{7!}{4!}$
19. $\frac{15!}{11!}$
20. $\frac{12!}{8!4!}$
21. $\frac{10!}{4!6!}$
22. $\frac{14!}{5!9!}$

Expand each binomial.

23. $(r + s)^7$
24. $(a - b)^3$
25. $(m - a)^5$
26. $(2a + b)^6$
27. $(2b - x)^4$
28. $(3x - 2y)^5$
29. $(3x + 2y)^4$
30. $\left(\frac{a}{2} + 2\right)^5$
31. $\left(3 + \frac{m}{3}\right)^5$

Find the indicated term of each expression.

32. fourth term of $(x + 2)^7$
33. seventh term of $(x + y)^{12}$
34. sixth term of $(x - y)^9$
35. fourth term of $(2x + 3y)^9$
36. fifth term of $(2a + 3b)^{10}$
37. fifth term of $\left(\frac{2}{5} + \frac{3}{5}\right)^{10}$

Simplify.

38. $\dfrac{(k + 1)!}{k!}$

39. $(k + 2)!(k + 3)$

40. $\dfrac{(n + 3)!}{(n + 1)!}$

41. $\dfrac{8!(n - 4)!}{8 \cdot 6!(n - 3)!}$

Programming

42. The program at the right generates the coefficients in Pascal's triangle from row 0 through row n for $(a + b)^n$. You must input the values of n and press enter to generate each new row.

Find the coefficients of the terms of each expansion.

a. $(a + b)^7$
b. $(x + y)^8$
c. $(r + s)^{10}$

```
PROGRAM: PASCAL
: Disp "ENTER THE
  VALUE OF N"
: Input Y
: For(N,0,Y)
: ClrHome
: Disp N
: Disp " "
: For (R,0,N)
: 1→C
: If N<N−R+1
: Goto 1
: For(X,N,N−R+1,-1)
: CX/(N−X+1)→C
: End
: Lbl 1
: Disp C
: End
: Pause
: End
```

Critical Thinking

43. Suppose $(a + b)$ is raised to some positive integer power and one term in the binomial series is $170{,}544a^{15}b^7$.

a. Which term of the series is $170{,}544a^{15}b^7$?
b. To what power was $(a + b)$ raised?
c. What is the next term in the series?

Applications and Problem Solving

44. Games At the Centerville Centennial Celebration, the young children play various games. A diagram of one game is shown at the right. In this game, children drop ball bearings down a chute. A pattern of nails causes the ball bearings to take various paths down to the five sections at the bottom. At each nail, there is an equal chance for the ball bearing to go either way. Since the bearings have only two choices at each level, this is called a binomial distribution.

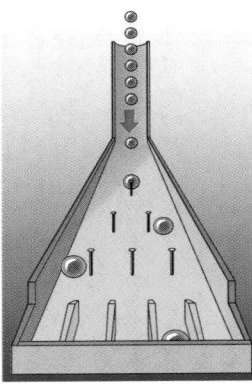

a. If 16 ball bearings are dropped through the chute, how many would you expect to end in each section?
b. If 64 ball bearings are dropped through the chute, how many would you expect to end in each section?
c. If another level of a nails are added to the game, describe the expected distribution of 64 ball bearings.
d. Describe how this game is related with the Pascal's triangle.

45. Look for a Pattern Find the sum of each of the first five rows of Pascal's triangle. Predict the sum of the tenth row of Pascal's triangle.

46. Football Steve Young is a quarterback with the San Francisco 49ers. In 1993, he had a pass completion rate of 0.68. What is the probability that he would complete exactly 3 out of his next 4 pass attempts?

47. Pets Study the graph at the right. Eight Austrian teenagers have gathered to play a game called curling. If the teenagers are from different households, what is the probability that exactly half of the teenagers come from households that own one or more cats?

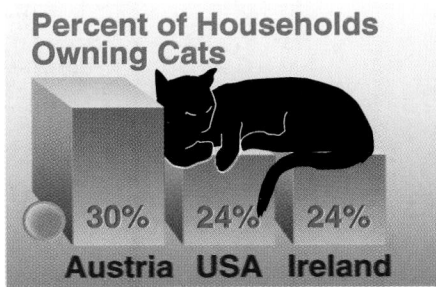

Percent of Households Owning Cats

30% 24% 24%
Austria USA Ireland

Source: The Gallup Organization

48. Employment A company has a large number of qualified job applicants for a specific job. Suppose 40% of the applicants are women and 60% are men. The company decides to hire 7 of the applicants and 5 of the applicants chosen are men.

  a. If the applicants were picked at random, what is the probability that the 5 out of the 7 would be men?

  b. Do you think that the company is biased in its hiring practices? Explain.

  c. Researchers who study possible bias in hiring practices frequently look for probabilities that are 5% or less. Would this situation cause a researcher to look into the hiring practices of the company? Explain.

Curling

Mixed Review

49. Find the sum of the infinite series $\frac{1}{2} + \frac{1}{3} + \frac{2}{9} + \frac{4}{27} + \ldots$ (Lesson 11–5)

50. The third term of an arithmetic sequence is 14, and the ninth term is -1. Find the first four terms of the sequence.   (Lesson 11–1)

51. Biology The number of a certain type of bacteria can increase from 80 to 164 in 3 hours. Find the approximate value of k in the growth formula $y = ne^{kt}$, where t is given in hours.   (Lesson 10–7)

52. Solve $\log_3 243 = x$.   (Lesson 10–2)

53. Auto Mechanics When air is pumped into a tire, the pressure required varies inversely as the volume of the air. If the pressure is 30 lb/in^2 when the volume is 140 in^3, find the pressure when the volume is 100 in^3.   (Lesson 9–2)

54. If $f(x) = x^2 + 6$ and $g(x) = 3x - 4$, find $[f \circ g](x)$.   (Lesson 8–7)

55. Is the graph of $x^2 + y^2 - 8x + 6y + 24 = 0$ a parabola, a circle, an ellipse, or a hyperbola?   (Lesson 7–6)

56. Solve $5 - \sqrt{b + 2} = 0$.   (Lesson 5–8)

57. Write an equation of the line that passes through the point at (4, 6) and is parallel to the line whose equation is $y = \frac{2}{3}x + 5$.   (Lesson 2–4)

In·ves·ti·ga·tion

Boing!!

Refer to the Investigation on pages 590–591.

Review the conclusions you have drawn from your experiments with the rubber bands and review the instructions given to you by your manager as you begin to close this Investigation.

Analyze

You have conducted experiments and organized your data in various ways. It is now time to analyze your findings and state your conclusions.

> **PORTFOLIO ASSESSMENT**
>
> You may want to keep your work on this Investigation in your portfolio.

1 Review your data and create a summary chart of the different experiments and how they relate to one another.

2 What information does this chart reflect? Does it give information about factors that would contribute to the performance of the bungee equipment? Explain.

3 What can you conclude about the weight of the falling object and the effect it will have on the length that the bungee is stretched? Do you think gravity will have any effect on this? Explain.

Write

The summary report to your manager should explain your process for investigating bungee jumping equipment and what you found from your investigations.

4 Explain the experiments you used to analyze bungee jumping. Use your graphs and charts to summarize your findings.

5 For each size rubber band, describe the optimum length that each rubber band should be stretched to get the most spring for the effort. Explain how you found the elastic potential for the rubber bands and which length rubber band has the greatest elastic potential.

6 Determine the weight required to stretch each rubber band to the length at which it reaches its greatest elastic potential. Explain your findings.

7 Using all of these factors and data, write a report to your manager.
- Explain the factors that contribute to a bungee providing an optimum spring.
- List specific size bungees, the weight that would allow the bungee to stretch to a desired length for the best spring and the elastic potential that would be achieved.
- Describe a general relationship between the size of the bungee and the weight of a person using the bungee.
- Describe how these factors relate to the elastic potential.
- Determine what effect repeated use of the bungee has on the bungee's elasticity and safety.
- Summarize the report with an overall recommendation to your manager.

VOCABULARY

After completing this chapter, you should be able to define each term, property, or phrase and give an example or two of each.

Discrete Mathematics

arithmetic means   (p. 650)

arithmetic sequence
(p. 649)

arithmetic series   (p. 656)

binomial theorem   (p. 696)

common difference   (p. 649)

common ratio   (p. 662)

factorials   (p. 697)

Fibonacci sequence   (p. 683)

formula for *n*th term of an
arithmetic sequence
(p. 649)

formula for *n*th term of a
geometric sequence
(p. 663)

geometric means
(p. 665)

geometric sequence
(p. 662)

geometric series
(p. 670)

index of summation
(p. 658)

infinite geometric series
(p. 676)

partial sum   (p. 676)

Pascal's triangle   (p. 695)

recursive formula   (p. 683)

sequence   (p. 648)

series   (p. 656)

sigma notation   (p. 658)

sum of an arithmetic series
(p. 657)

sum of an infinite geometric
series   (p. 677)

sum of a geometric series
(p. 671)

summation notation
(p. 658)

term   (p. 648)

Geometry

fractal geometry   (p. 688)

self-similarity   (p. 689)

UNDERSTANDING AND USING THE VOCABULARY

Choose the correct letter that best completes each statement.

1. A(n) _____ of an infinite series is the sum of a certain number of terms, but not the entire series.

2. If a sequence has a common ratio, then it is a(n) _____.

3. Using _____, the series $2 + 5 + 8 + 11 + 14$ can be written as $\sum\limits_{n=1}^{5} (3n - 1)$.

4. Eleven and 17 are the two _____ between 5 and 23 in the sequence 5, 11, 17, 23.

5. Using the _____, $(a - 2)^4$ can be expanded to $a^4 - 8a^3 + 24a^2 - 32a + 16$.

6. The _____ of the sequence $3, 2, \frac{4}{3}, \frac{8}{9}, \frac{16}{27}$ is $\frac{2}{3}$.

7. The sequence $\frac{1}{2}, 3, 5\frac{1}{2}, 8, 10\frac{1}{2},$ is a(n) _____, and $2\frac{1}{2}$ is the _____.

8. In the 7th row of Pascal's triangle, the third term is _____.

9. The _____ $11 + 16.5 + 22 + 27.5 + 33$ has a sum of 110.

10. A(n) _____ is expressed as $n! = n(n - 1)(n - 2) \cdots 2 \cdot 1$.

a. arithmetic means

b. arithmetic sequence

c. arithmetic series

d. binomial theorem

e. common difference

f. common ratio

g. factorial

h. geometric means

i. geometric sequence

j. geometric series

k. partial sum

l. sigma notation

m. 15

n. 20

OBJECTIVES AND EXAMPLES	REVIEW EXERCISES

OBJECTIVES AND EXAMPLES

Upon completing this chapter, you should be able to:

• find the *n*th term of an arithmetic sequence (Lesson 11–1)

If $a_1 = -17$, $d = 4$, and $n = 12$, find a_{12}.

$a_n = a_1 + (n - 1)d$

$a_{12} = -17 + (12 - 1)4$

$a_{12} = -17 + 44$

$a_{12} = 27$

REVIEW EXERCISES

Use these exercises to review and prepare for the chapter test.

Find the *n*th term of each arithmetic sequence.

11. $a_1 = 6$, $d = 8$, $n = 5$

12. $a_1 = -5$, $d = 7$, $n = 22$

13. $a_1 = 5$, $d = -2$, $n = 9$

14. $a_1 = -2$, $d = -3$, $n = 15$

15. $a_1 = 4$, $d = 3$, $n = 32$

16. $a_1 = 8$, $d = -5$, $n = 10$

• find the position of a given term in an arithmetic sequence (Lesson 11–1)

-3 is what term of the sequence 7, 5, 3, ... ?

$-3 = 7 + (n - 1)(-2)$

$-3 = 7 - 2n + 2$

$n = 6$

Complete each statement.

17. 72 is the ___?___ th term of -5, 2, 9, ...

18. -37 is the ___?___ th term of 1, -1, -3, -5, ...

19. 49 is the ___?___ th term of 4, 9, 14, ...

20. $-\frac{17}{4}$ is the ___?___ th term of $2\frac{1}{4}$, 2, $1\frac{3}{4}$, ...

• find arithmetic means (Lesson 11–1)

Find the two arithmetic means between 4 and 25.

$25 = 4 + (4 - 1)d$

$25 = 4 + 3d$

$7 = d$

Therefore, $4 + 7 = 11$ and $11 + 7 = 18$.

The arithmetic means are 11 and 18.

Find the arithmetic means in each sequence.

21. -7, ___?___, ___?___, ___?___, 9

22. 12, ___?___, ___?___, 4

23. ___?___, 6, ___?___, ___?___, -3, ___?___

24. ___?___, 49, ___?___, ___?___, 28

• find sums of arithmetic series (Lesson 11–2)

Find S_n for the arithmetic series described by $a_1 = 34$, $a_n = 2$, and $n = 9$.

$S_n = \frac{n}{2}(a_1 + a_n)$

$S_9 = \frac{9}{2}(34 + 2)$

$S_9 = \frac{9}{2}(36)$

$S_9 = 162$

Find S_n for each arithmetic series described.

25. $a_1 = 12$, $a_n = 117$, $n = 36$

26. $4 + 10 + 16 + ... + 106$

27. $a_1 = 85$, $a_n = 25$, $n = 21$

28. $10 + 4 + (-2) + ... + (-50)$

29. Write the terms of the arithmetic series $\sum_{n=2}^{13} (3n + 1)$ and find the sum.

30. Find the first three terms of an arithmetic series if $a_1 = 3$, $a_n = 24$, and $S_n = 108$.

OBJECTIVES AND EXAMPLES

• find the nth term of a geometric sequence
(Lesson 11–3)

If $a_1 = 7$, $r = 3$, and $n = 5$, find a_5.

$a_n = a_1 r^{n-1}$

$a_5 = 7(3)^{5-1}$

$a_{12} = 7(81)$

$a_{12} = 567$

REVIEW EXERCISES

Find the indicated term in each geometric sequence.

31. a_6 for $\frac{2}{3}, \frac{4}{3}, \frac{8}{3}, \ldots$
32. a_n for $a_1 = 2$, $r = 2$, and $n = 5$
33. a_n for $a_1 = 7$, $r = 2$, and $n = 4$
34. a_n for $a_1 = 243$, $r = -\frac{1}{3}$, and $n = 5$

• find geometric means   (Lesson 11–3)

Find the two geometric means between 1 and 8.

$8 = 1r^{4-1}$

$8 = r^3$

$2 = r$

Therefore, $1(2) = 2$ and $2(2) = 4$.

The geometric means are 2 and 4.

Find the geometric means in each sequence.

35. $3, \underline{\ ?\ }, \underline{\ ?\ }, \underline{\ ?\ }, 48$
36. $7.5, \underline{\ ?\ }, \underline{\ ?\ }, \underline{\ ?\ }, 120$
37. $8, \underline{\ ?\ }, \underline{\ ?\ }, \underline{\ ?\ }, \underline{\ ?\ }, \frac{1}{4}$
38. $5, \underline{\ ?\ }, \underline{\ ?\ }, \underline{\ ?\ }, 80$
39. $-2, \underline{\ ?\ }, -98, \underline{\ ?\ }, \underline{\ ?\ }$

• find sums of geometric series   (Lesson 11–4)

Find S_n for the geometric series described by $a_1 = 7$, $r = 3$, and $n = 14$.

$S_n = \dfrac{a_1 - a_1 r^n}{1 - r}$

$S_{14} = \dfrac{7 - 7(3^{14})}{1 - 3}$

$\quad\ = 16{,}740{,}388$

Find S_n for each geometric series described.

40. $a_1 = 12$, $r = 3$, $n = 5$
41. $a_1 = 4$, $r = -\frac{1}{2}$, $n = 6$
42. $a_1 = 256$, $r = 0.75$, $n = 9$
43. $a_1 = 1$, $a_5 = \frac{1}{16}$, $r = -\frac{1}{2}$
44. $a_1 = 625$, $a_5 = 81$, $r = \frac{3}{5}$

• find specific terms in a series   (Lesson 11–4)

Find a_1 if $S_5 = 2.75$ and $r = -2$.

$2.75 = \dfrac{a_1(1 - (-2)^5)}{1 - (-2)}$

$\dfrac{11}{4} = \dfrac{a_1(33)}{3}$

$33 = 132a_1$

$\dfrac{1}{4} = a_1$

Find a_1 for each geometric series described.

45. $S_n = 1031$, $r = \frac{2}{5}$, $n = 5$
46. $S_n = 30$, $r = -2$, $n = 4$
47. $S_n = -61$, $r = -1$, $n = 5$
48. $S_n = 244$, $r = -3$, $n = 5$

OBJECTIVES AND EXAMPLES

• find sums of infinite geometric series   (Lesson 11–5)

Find the sum of the infinite geometric series described by $a_1 = 18$ and $r = -\frac{2}{7}$.

$$S = \frac{a_1}{1 - r}$$

$$S = \frac{18}{1 - \left(-\frac{2}{7}\right)} = \frac{18}{\frac{9}{7}} \text{ or } 14$$

REVIEW EXERCISES

Find the sum of each infinite geometric series, it if exists.

49. $a_1 = 6$ and $r = \frac{11}{12}$

50. $\frac{1}{8} - \frac{3}{16} + \frac{9}{32} - \frac{27}{64} + \ldots$

51. $a_1 = -2$ and $r = -\frac{5}{8}$

52. $10 - \frac{5}{2} + \frac{5}{8} + \ldots$

• iterate functions   (Lesson 11–6)

Find the first three iterates of the function $f(x) = -5x - 1$ for an initial value of $x_0 = -1$.

$$f(x_0) = f(-1)$$
$$= -5(-1) - 1 \text{ or } 4 \qquad \text{So, } x_1 = 4.$$
$$f(x_1) = f(4)$$
$$= -5(4) - 1 \text{ or } -21 \qquad \text{So, } x_2 = -21.$$
$$f(x_2) = f(-21)$$
$$= -5(-21) - 1 \text{ or } 104 \qquad \text{So, } x_3 = 104.$$

The first three iterates are 4, −21, and 104.

Find the first three iterates of each function, using the given initial values.

53. $f(x) = -2x + 3, x_0 = 1$

54. $f(x) = 7x - 4, x_0 = 2$

55. $f(x) = x^2 - 6, x_0 = -1$

56. $f(x) = -2x^2 - x + 5, x_0 = -2$

• expand powers of binomials by using Pascal's triangle and the binomial theorem   (Lesson 11–8)

Use the binomial theorem to express $(a - 2b)^4$ in expanded form.

$$(a - 2b)^4 = a^4 + 4a^3(-2b) + \frac{4(3)}{1 \cdot 2} a^2(-2b)^2$$
$$+ \frac{4(3)(2)}{1 \cdot 2 \cdot 3} a(-2b)^3 + (-2b)^4$$
$$= a^4 - 8a^3b + 24a^2b^2 - 32ab^3 + 16b^4$$

Expand each binomial.

57. $(x + y)^3$

58. $(x - 2)^4$

59. $(3r + s)^5$

Find the indicated term of each expression.

60. fourth term of $(x + 2y)^6$

61. second term of $(4x - 5)^{10}$

62. seventh term of $(x - y)^{15}$

APPLICATIONS AND PROBLEM SOLVING

63. **Fine Arts**   A layered sculpture is arranged so that there are 5 diamonds on the first layer, 7 diamonds on the second layer, 9 diamonds on the third layer, and so on. How many diamonds are there on the twentieth layer?   (Lesson 11–2)

A practice test for Chapter 11 is provided on page 922.

64. **Fractals**   A side of an equilateral triangle is 20 inches long. The midpoints of its sides are joined to form a smaller equilateral triangle. If this process is continued infinitely, find the sum of the perimeters of the triangles. (Lesson 11–7)

ALTERNATIVE ASSESSMENT

COOPERATIVE LEARNING PROJECT

Peach Farming In this project, you will analyze a planting situation for a tree farmer. A peach farmer in Georgia calculates that he can produce 54 boxes of peaches per tree if he plants 40 trees per acre. If he plants one more tree per acre, the yield drops one box per tree due to congestion. His farm is 160 acres.

Find the number of trees per acre he should plant in order to maximize the total production. What would his total production of boxes be if he would plant that number of trees per acre? Would it be more productive and cost efficient for him to plant 43 trees or 51 trees? Explain.

What type of sequence is the number of boxes of peaches per acre for successive numbers of trees? Write a recursive formula to describe it.

Incorporate these ideas in your project.

- Formulate a chart for your data.
- Look for patterns.
- Determine a maximization for total production.
- Look at all the avenues in tree farming and write a recommendation for the farmer.

THINKING CRITICALLY

In the Fibonacci sequence, each term after the second term is found by adding the two previous terms. Is this sequence an arithmetic sequence? Explain your answer.

Use 1, 2, 3, ... as the basis to generate at least two different sequences. For each sequence generated, give a rule and the first six terms in the sequence.

PORTFOLIO

Throughout this course, you have been working in groups to solve problems. What roles do you play in the group?

- Are you a leader who takes charge?
- Do you help to keep your group on task?
- Do you ask questions?
- Do you just listen and copy down answers?
- In what ways are you a good group member?
- How could you do better?
- Did your group work well together? Why?

Place this in your portfolio.

SELF EVALUATION

The problem-solving strategy of looking for a pattern is an important skill that we all need to develop. However, expecting a pattern in a problem to automatically jump out does not often happen. Changing certain conditions in a problem situation and looking for a pattern to emerge is more likely to happen.

Assess yourself. Do you look for patterns? Do you change certain conditions in order to help you look for patterns? Think of a situation in mathematics and also in your daily life where looking for a pattern would be an appropriate strategy to use in order to help solve the situation. Explain how it was helpful.

COLLEGE ENTRANCE EXAM PRACTICE

CHAPTERS 1–11

SECTION ONE: MULTIPLE CHOICE

There are ten multiple-choice questions in this section. After working each problem, write the letter of the correct answer on your paper.

1. Find the center and radius of the circle whose equation is $(x + 8)^2 + (y - 3)^2 = 25$.

A. $(8, -3)$; 5

B. $(-8, 3)$; 5

C. $(3, 8)$; 5

D. $(-3, 8)$; 5

2. Solve for x in the equation $\begin{vmatrix} x & 7 & 5 \\ 0 & 3 & 4 \\ 3 & 2 & -2 \end{vmatrix} = 11$.

A. -6

B. 84

C. -28

D. 2

3. Find the first three terms of the arithmetic series for which $a_1 = 6$, $a_n = 306$, and $S_n = 1716$.

A. 6, 36, 66

B. 6, 17, 28

C. 6, 25, 44

D. 6, 66, 726

4. Simplify $\left(4^{\sqrt{3}}\right)^{\sqrt{2}}$.

A. $4^{\sqrt{5}}$

B. 4096

C. $4^{\sqrt{6}}$

D. $4^{\sqrt{\frac{3}{2}}}$

5. Solve the system of equations.

$3x - 2y + 2z = -2$

$x - 3y + z = -2$

$2x - y + 4z = 7$

A. $(-2, 1, 3)$

B. $(2, 7, -4)$

C. $(1, 4, 3)$

D. $\left(-2, \frac{1}{3}, -\frac{5}{3}\right)$

6. If $A_{4 \times 3}$ is multiplied by $B_{3 \times 5}$, what are the dimensions of the product?

A. 3×3

B. 4×5

C. 5×4

D. 4×3

7. The last term of an arithmetic sequence is 207, the common difference is 3, and the number of terms is 14. Choose the equation that would find the first term of the sequence.

A. $x = 207 + (3 - 1)14$

B. $14 = 207 + (x - 1)3$

C. $207 = x + (14 - 1)3$

D. $3 = 207 + (14 - 1)x$

8. Simplify $\dfrac{\frac{3x}{4x - 1}}{1 + \frac{3x}{x - 1}}$.

A. 1

B. $\dfrac{3x(x - 1)}{(4x - 1)^2}$

C. $\dfrac{3x}{x - 1}$

D. $\dfrac{1}{4x - 1}$

9. Which two radicals are like radicals?

A. $\sqrt{5}$ and $\sqrt[3]{5}$

B. $7\sqrt{x}$ and $\sqrt{7x}$

C. $\sqrt{4x}$ and $\sqrt{4y}$

D. $2\sqrt[3]{9}$ and $3\sqrt[3]{9}$

10. Simplify $(5ab^2)(a^3b)(-3c^2) + (7bc)(3ac)(a^3b)^2$.

A. $6a^{11}b^6c^4$

B. $2a^4b^3c^2 + 21a^7b^3c^2$

C. $-15a^4b^3c^2 + 21a^7b^3c^2$

D. $-15a^3b^2c^2 + 10a^6b^3c$

SECTION TWO: SHORT ANSWER

This section contains ten questions for which you will provide short answers. Write your answer on your paper.

11. Jackie is in charge of building a set for the school play. She wants each rectangular window to have an area of 315 square inches. She also wants each window to be 6 inches taller than it is wide. What are the dimensions of the window?

12. A piece of machinery valued at $75,000 depreciates at a steady rate of 8% yearly. When will the value be $15,000? Use $V_n = P(1 + r)^n$.

13. The volume of a rectangular box is 2475 cubic units. The length of the box is three units more than twice the width of the box. The height is two units less than the width. Find the dimensions of the box.

14. The teaching staff of Fairmeadow High School informs its members of school cancellation by telephone. The principal calls 2 teachers, each one of whom calls 2 other teachers, and so on. In order to inform the entire staff, 6 rounds of calls are made. Counting the principal, find how many people are on staff at Fairmeadow High School.

15. Find $(a^4 - 5a^3 - 13a^2 + 53a + 60) \div (a + 1)$ by using synthetic division.

16. A ball dropped 120 feet bounces $\frac{2}{3}$ of the height from which it fell on each bounce. How far will it travel before coming to rest?

17. The hourly wages of eight employees of the Sequoia Insurance Company are $4.45, $5.50, $6.30, $11.00, $5.50, $17.20, $12.20, and $7.80. Find the median, mode, mean, and standard deviation of the wages of these employees.

18. Find the radius r of a sphere whose surface area S is 616 square inches. Use the formula $r = \frac{1}{2}\sqrt{\dfrac{S}{\pi}}$.

19. The sum of Josh and Yuji's ages is 32. Twice Josh's age is 4 more than double Yuji's age. Find their ages.

20. Television signals travel at the speed of light. If the speed of light is 3.00×10^5 kilometers per second, how long would it take for signals broadcasting from a television station to reach a house 36 kilometers away?

SECTION THREE: COMPARISON

This section contains five comparison problems that involve comparing two quantities, one in column A and one in column B. In certain questions, information related to one or both quantities is centered above them. All variables used represent real numbers.

Compare quantities A and B below.

- Write A if quantity A is greater.
- Write B if quantity B is greater.
- Write C if the two quantities are equal.
- Write D if there is not enough information to determine the relationship.

Column A	Column B
$\dfrac{a}{b} = \dfrac{7}{3}$	
21. $12a$	$28b$
22. $5^{x+2} = 15.3$	$9^{x-4} = 6.28$
$x > 1$	
23. $\dfrac{2x^3 + 15x^2 + 22x - 15}{x + 3}$	$\dfrac{x^3 + 3x^2 - 7x + 1}{x - 1}$
$\dfrac{a + b}{a^{\frac{1}{2}} + b}$	
24. a	b
25. the inverse of $f(x) = \dfrac{5x + 2}{2}$	$\dfrac{x - 1}{2\sqrt{5}} \cdot \dfrac{4}{\sqrt{5}}$

Investigating Discrete Mathematics and Probability

Objectives

In this chapter, you will:

- solve problems by solving simpler, related problems,
- solve problems involving permutations and combinations,
- find probabilities of events,
- determine the odds of success or failure of events, and
- use experiments and simulation to solve problems.

Most teenagers begin to earn money at minimum wage. Yet some enterprising teens have the courage and determination to start their own businesses. Have you ever thought about starting your own business? What are your interests? Is there a need for a certain service or business in your community?

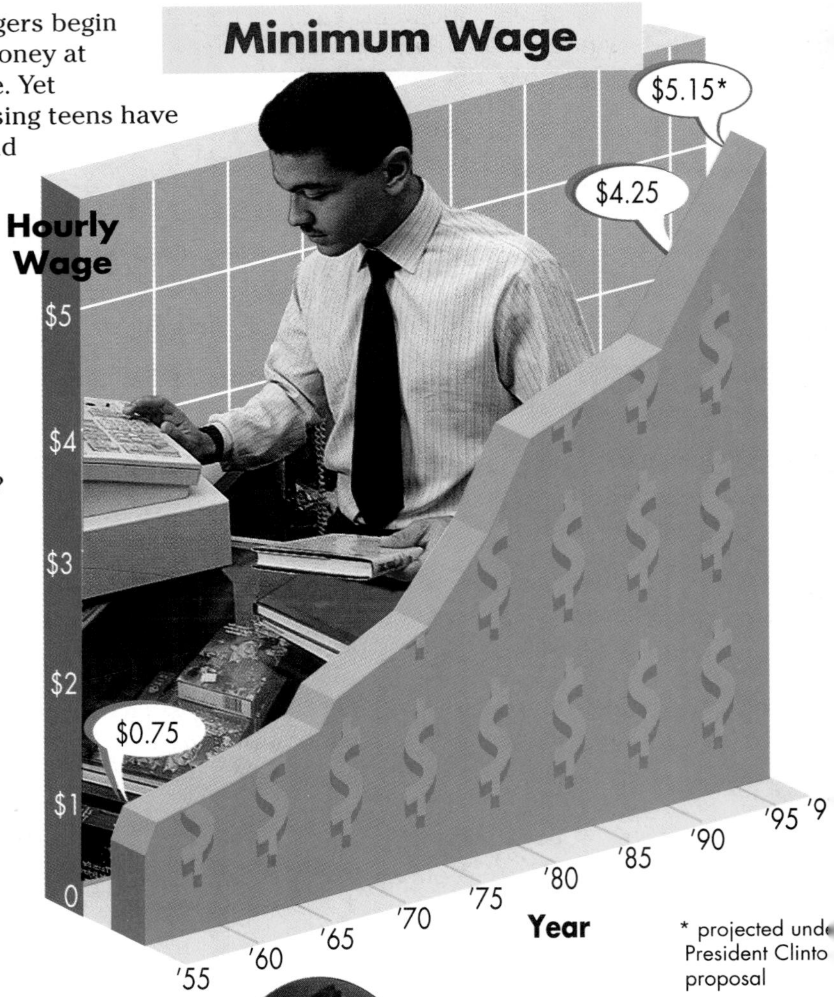

Minimum Wage

$5.15*

$4.25

Hourly Wage

$5

$4

$3

$2

$1

$0.75

0

'55 '60 '65 '70 '75 '80 '85 '90 '95 '9

Year

* projected unde President Clinto proposal

TIME Line

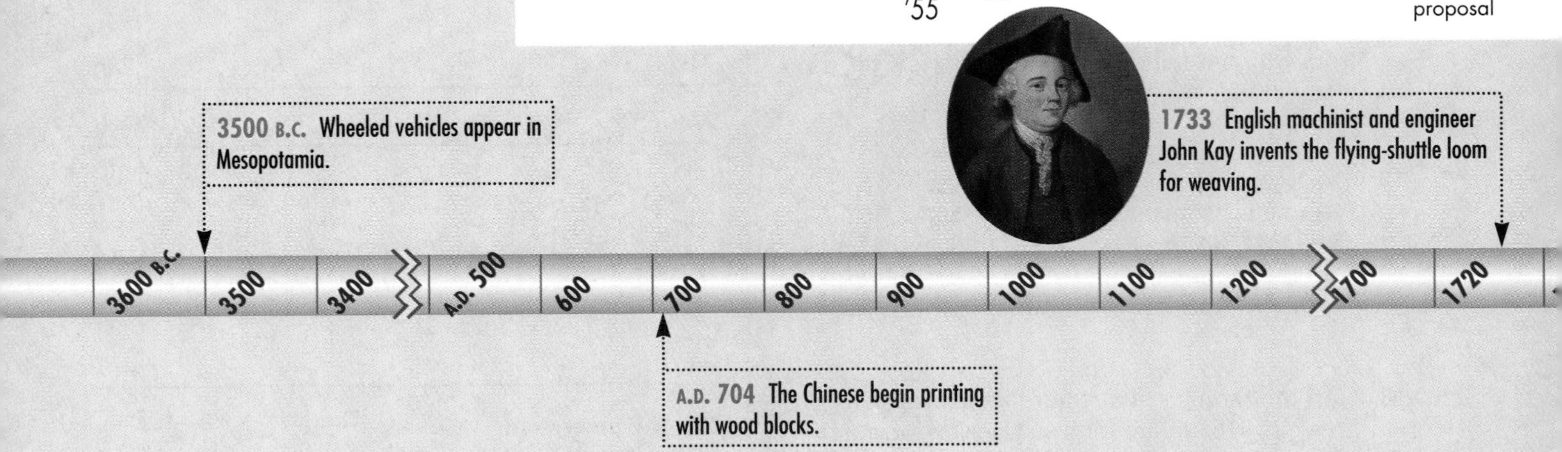

3500 B.C. Wheeled vehicles appear in Mesopotamia.

1733 English machinist and engineer John Kay invents the flying-shuttle loom for weaving.

3600 B.C. 3500 3400 A.D. 500 600 700 800 900 1000 1100 1200 1700 1720

A.D. 704 The Chinese begin printing with wood blocks.

PEOPLE IN THE NEWS

I n response to the riots in L.A. in 1992, a group of forty students in Tammy Bird's biology class at Crenshaw High School in Los Angeles decided to start an organic garden. The garden flourished, and the group donated some of the produce to needy families and sold the rest at a farmers' market.

The students formed a company called **"Food from the 'Hood"** and started a college scholarship fund. Local business leaders helped them develop and sell a salad dressing called "Straight out 'the Garden," now sold in over 2000 stores in 23 states. Expected profits this year could be $50,000, all going to the scholarship fund. As Terie Smith, 15, says, "We showed that a group of inner-city kids can and did make a difference."

Chapter Project

O rganize into cooperative groups of entrepreneurs. Your group should investigate various ways high school students could use their talents and expertise to start their own business. Use the research facilities of your library, information gathered from local business and community groups or government agencies, and decide what services or products are needed in your community. Decide what steps are needed to get started, if entrepreneurial training programs are available in your area, and the benefits of using them. List the advantages and disadvantages of running your own business and report on your findings.

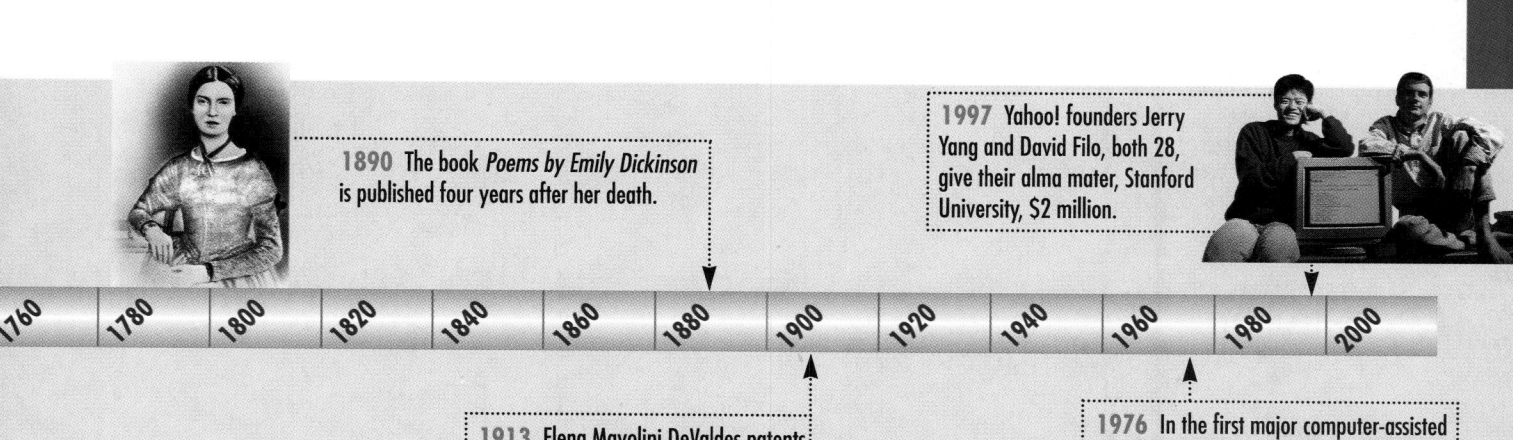

1890 The book *Poems by Emily Dickinson* is published four years after her death.

1997 Yahoo! founders Jerry Yang and David Filo, both 28, give their alma mater, Stanford University, $2 million.

1760　1780　1800　1820　1840　1860　1880　1900　1920　1940　1960　1980　2000

1913 Elena Mayolini DeValdes patents a bottle stopper that is a forerunner of today's tamperproof packaging.

1976 In the first major computer-assisted proof, it is shown that any map can be colored with four colors so that no two regions with the same color are adjacent.

The Counting Principle

What YOU'LL LEARN

- To solve problems by using the fundamental counting principle, and
- to solve problems by using the strategy of solving a simpler problem.

Why IT'S IMPORTANT

You can use the counting principle to solve problems involving communication and linguistics.

APPLICATION
Telecommunications

Area codes were invented in the 1940s when the Bell Telephone Company realized they could not hire enough operators to handle all of the long-distance calls that were being made. At that time, Bell researchers developed a system of 3-digit codes that would automatically route calls to long distance. These area codes were used for North America, including Canada, the Caribbean, and Bermuda, but not Mexico.

With this system, the first digit could be any number from 2 through 9. Only 1 or 0 could be used as the middle digit. The last digit could be any number between 0 and 9. How many area code combinations were possible?
This problem will be solved in Example 2.

Sometimes a problem is easier to solve if you first **solve a simpler problem.** After solving the simpler problem, you can use the concepts that you used to solve the original problem.

Example ① Suppose your state is adding a new area code. The first digit must be a 6 or 7, the second digit must be a 0 or 1, and the third digit can be a 3, 4 or 5. How many area codes are possible?

PROBLEM SOLVING
Solve a Simpler Problem

F Y I

The Bell system predicted in the 1940s that there would be enough combinations to last until the year 2000. They were about 5 years off. In addition to now dialing 1 before the area code, starting in 1995, the middle digits of area codes can now be digits other than 1 or 0.

First Digit	Second Digit	Third Digit	Possible Area Codes
	0	3	603
		4	604
		5	605
6	1	3	613
		4	614
		5	615
	0	3	703
		4	704
		5	705
7	1	3	713
		4	714
		5	715

Thus, your state has a total of 12 choices for a new area code.

Choosing the three numbers for an area code are called **independent events** because one choice does not affect the others. The *tree diagram* shown in Example 1 illustrates all of the different choices your state has in choosing the final area code.

You can find this same total number of choices without drawing a diagram.

	First Digit	Second Digit	Third Digit
Choices	6 or 7	0 or 1	3, 4, or 5
Number of Choices	2	2	3

The total number of choices can be found by multiplying the number of choices for each decision. The total number of choices is $2 \cdot 2 \cdot 3$ or 12. This example illustrates the **fundamental counting principle**.

Fundamental Counting Principle	If event *M* can occur in *m* ways and is followed by an independent event *N* that can occur in *n* ways, then the event *M* followed by the event *N* can occur in *m · n* ways.

This principle can be extended to any number of events. You can use this principle and what you learned in Example 1 to find the total number of area code combinations referred to in the application at the beginning of the lesson.

Example 2

Using the Bell researchers' 3-digit system from the 1940s, how many total area code combinations were possible?

Since each choice of digits is not affected by the previous choice, these are *independent events*.

	First Digit	Second Digit	Third Digit
Choices	2, 3, 4, 5, 6, 7, 8, 9	0, 1	0, 1, 2, 3, 4, 5, 6, 7, 8, 9
Number of Choices	8	2	10

The total number of combinations can be found by multiplying the number of choices for each digit. The total number of area code combinations is $8 \cdot 2 \cdot 10$ or 160.

Example 3

The first group of three digits in a telephone number is called a prefix.

Justin works part-time delivering pizzas for a local restaurant. His manager provides him with a pager so she can contact him at any time, even while he is making deliveries. To activate the pager, a 7-digit pager number is assigned to it. If the prefix is 337, how many 7-digit numbers are available?

Since each of the last four digits can be used any number of times, there are 10 choices for each. These are *independent events*.

Digit in Pager Number	4th	5th	6th	7th
Number of Choices	10	10	10	10

There are $10 \cdot 10 \cdot 10 \cdot 10$ or 10,000 possible pager numbers available.

Some applications involve **dependent events**. That is, the outcome of one event *does affect* the outcome of another event.

Example **4**

Tara opened a savings account so she could deposit the money she earns from her paper route. She received an automatic teller machine card with her account. Tara needs to choose a 4-digit PIN (personal identification number) for her new card, but she may not use any digit more than once. From how many different 4-digit PINs can she choose?

After the first digit is chosen, it cannot be chosen again. So there are only nine choices for the second digit. After the second digit is chosen, there are only eight choices for the third digit and seven choices for the fourth digit. These are *dependent events*.

Digit in PIN	1st	2nd	3rd	4th
Number of Choices	10	9	8	7

Tara can choose $10 \cdot 9 \cdot 8 \cdot 7$ or 5040 different PIN numbers.

Example **5**

Solve the problem in the comic below.

PEANUTS®

PEANUTS reprinted by permission of United Feature Syndicate, Inc.

LOOK BACK

You can refer to Lesson 11-8 for information on factorials.

You can help Peppermint Patty overcome her math anxiety. Finding out how many ways nine different books can be arranged on a shelf is an example of a *dependent event*.

Books	1st	2nd	3rd	4th	5th	6th	7th	8th	9th
Number of Choices	9	8	7	6	5	4	3	2	1

There are $9 \cdot 8 \cdot 7 \cdot 6 \cdot 5 \cdot 4 \cdot 3 \cdot 2 \cdot 1$ or 362,880 ways the books can be arranged on a shelf. *Note that $9 \cdot 8 \cdot 7 \cdot 6 \cdot 5 \cdot 4 \cdot 3 \cdot 2 \cdot 1 = 9!$.*

Communicating Mathematics

Study the lesson. Then complete the following.

1. **Explain** the fundamental counting principle in your own words.

2. **Describe** the difference between independent and dependent events. Give an example of each.

3. **Draw** a tree diagram to illustrate all the different choices when spinning both spinners at the same time.

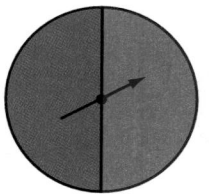 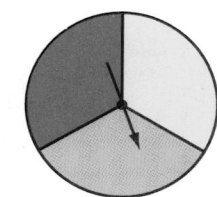

4. **Explain** the advantages of using a simpler problem to solve problems.

MATH JOURNAL

5. Now that the middle digit in an area code can be any number between 0 or 9, show how many area code combinations are possible.

Guided Practice

Draw a tree diagram to illustrate all of the possibilities.

6. the possibilities for boys and girls in a family with two children (*Hint:* Having a boy, then a girl is different from having a girl, then a boy.)

7. sweatshirts that come in sizes small, medium, and large and in the colors red, navy, and white

State whether the following events are *independent* or *dependent*.

8. choosing the color and size of a pair of shoes

9. choosing the winner and loser of a chess game

Solve each problem.

10. **School** At Dublin Coffman High School, Cecelia is taking six different classes. Assuming that each of these classes is offered each period, how many different schedules might she have?

11. **Tests** Alberto's math quiz has eight true-false questions. How many different choices for giving answers to the eight questions are possible?

Practice

Draw a tree diagram to illustrate all of the possibilities.

12. two pennies are tossed and a number cube is rolled

13. boys and girls in a family with three children

14. buying tennis, basketball, aerobic, running, or cross-country shoes in black, white, blue, or red

15. ordering a hamburger rare, medium, or well done with either ketchup, mayonnaise, cheese, onion, or tomato as your choice of topping

16. choosing a phone that comes in a wall or desk model in black, almond, or transparent that has a redial or hold button

State whether the events are *independent* or *dependent*.

17. choosing a president, vice president, secretary, and treasurer for Student Council

18. selecting a fiction book and a non-fiction book at the library

19. The letters A through Z are written on pieces of paper and placed in a jar. Four of them are selected one after the other without replacing any of the pieces of paper.

20. Each of six people guess the total number of points in a basketball game. They write down the guess without telling what it is.

Solve each problem.

21. Suppose five points in a plane represent towns that are connected by roads. Starting at any one town, how many different routes are there so that you visit each town exactly once?

22. How many different batting orders does a baseball team of nine players have if the pitcher bats last?

23. The letters r, s, t, v, and w are to be used to form 5-letter passwords for an office security system. How many passwords can be formed if the letters can be used more than once in any password?

24. In Ohio, a standard license plate has three letters followed by three digits. The first letter cannot be I or O, and the last digit cannot be zero. How many possible plates are there?

25. How many ways can six books be arranged on a shelf if one of the books is a dictionary and it must be on an end?

26. For a particular model of truck, a truck dealer offers 5 versions of that model, 16 body colors, and 8 truck cab colors. How many different possibilities are available for that model?

In this text, when referring to a deck of cards, we mean a standard deck of 52 cards.

27. Suppose five cards are drawn from a standard deck of cards. Three are red and two are black.
    a. How many possibilities are there for this hand?
    b. Suppose exactly one of the black cards is a face card. Now how many possibilities are there? *A face card is a jack, queen, or king.*

Critical Thinking

28. Write a problem that uses the fundamental counting principle and has an outcome of 2340. Explain in your own words how you went about finding a problem to fit the criteria.

Applications and Problem Solving

29. **Golf** A golf club manufacturer makes irons with 7 different shaft lengths, 3 different grips, 5 different lies, and 2 different club head materials. How many different combinations are offered?

30. Linguistics An employee of the U.S. Public Health Service has organized a list of 30 words from which impressive-sounding 3-word phrases can be formed. If you choose one word at random from each of these three columns, you can make a distinctive phrase.

Column 1	Column 2	Column 3
balanced	digital	capability
compatible	incremental	concept
functional	logistical	contingency
integrated	management	flexibility
optional	monitored	hardware
parallel	organizational	mobility
responsive	policy	options
synchronized	reciprocal	programming
systematized	third-generation	projection
total	transitional	time-phase

a. How many different 3-word phrases can be made using this list?

b. How many different 3-word phrases could be made if each column had only five words in it?

c. How many different 3-word phrases could be made if each column had sixteen words in it?

d. Write a sentence using the 3-word phrase that you think sounds most impressive.

31. Dining Antonio's Italian Cuisine offers an Early Bird Special for customers who dine before 6:30 P.M. This offer includes an appetizer, a soup, and an entree, all for $6.95. There are 4 choices of appetizers, 3 soups, and 8 entrees. How many different meals are available under this offer?

32. Solve a Simpler Problem How many 5-digit numbers exist between 65,000 and 69,999 if each number has no repeated digits?

33. Communication When determining call letters for a radio station, the first letter must be either W or K. If no letters repeat, how many ways can the 4 call letters of a station be arranged?

Mixed Review

34. Expand $(2m - 3)^6$. (Lesson 11–8)

35. Solve $\ln 9.5 = \ln (e^{0.2x})$. (Lesson 10–5)

36. Money The average American household income is $36,000. Of this average, $11,160 is spent on housing, $6480 on transportation, $5400 on food, $3240 on personal insurance and pensions, $2160 on clothing and services, $1800 on health care, and $5760 on miscellaneous expenses. Make a circle graph to display this information. (Lesson 9–5)

37. Graph $f(x) = x^6$. (Lesson 8–3)

38. Electronics The headlights on a car contain parabolic reflectors. A special lightbulb with two filaments is used to produce the high and low beams. The filament placed at the focus produces the high beam, and the low beam is produced by the filament placed off the focus. If the equation of the parabola that is the cross section of the reflector is $y = \frac{1}{10}x^2$, where should the filament for the high beam be placed? (Lesson 7–2)

39. Simplify $\dfrac{\sqrt{3} + n\sqrt{6}}{4 - \sqrt{n}}$. (Lesson 5–6)

What YOU'LL LEARN

- To solve problems involving linear and circular permutations.

Why IT'S IMPORTANT

You can use permutations to solve problems involving retailing and cheerleading.

APPLICATION
Entertainment

Have you ever wondered why certain movies are shown at one theater but not at another? The selection of movies depends on what movies the film buyer purchases from a film distributor. The film buyer views all of the movies and then looks at the theaters in his or her area. The movies are selected based upon such things as the size and location of the theater, the clientele that frequents the theater, the type of movie, and the amount of money the movie will generate.

Suppose a film buyer has 11 films for an 8-screen theater. Since the screens are in rooms with different seating capacities, the film buyer must decide on which screens to show the movies, based on expected ticket sales. How many different arrangements are there for the 8 screens to show the 11 movies?

One way to organize this problem is to make a *decision chart*. Ask yourself, "How many decisions do I need to make in this problem?" In this case, the theater has 8 movie screens, so there are eight decisions to be made, a decision for each screen.

$\underline{11}$	$\underline{10}$	$\underline{9}$	$\underline{8}$	$\underline{7}$	$\underline{6}$	$\underline{5}$	$\underline{4}$
1st Screen	2nd Screen	3rd Screen	4th Screen	5th Screen	6th Screen	7th Screen	8th Screen

After one of the 11 movies is chosen for the first screen, there are only 10 movie choices left for the second screen, and so on, until the eighth screen is filled. The number of choices for each movie screen is affected by the choice of the previous screen, so these are dependent events. There are $11 \cdot 10 \cdot 9 \cdot 8 \cdot 7 \cdot 6 \cdot 5 \cdot 4$ or 6,652,800 movie arrangements possible.

The eight movie screens in the example above are in a certain order. When a group of objects or people are arranged in a certain order, the arrangement is called a **permutation.** In a permutation, the *order* of the objects is very important. The arrangement of objects in a line is called a **linear permutation**.

Notice that $11 \cdot 10 \cdot 9 \cdot 8 \cdot 7 \cdot 6 \cdot 5 \cdot 4$ is the product of the first 8 factors of 11!. We can write an equivalent expression in terms of 11!.

$$11 \cdot 10 \cdot 9 \cdot 8 \cdot 7 \cdot 6 \cdot 5 \cdot 4 = 11 \cdot 10 \cdot 9 \cdot 8 \cdot 7 \cdot 6 \cdot 5 \cdot 4 \cdot \frac{3 \cdot 2 \cdot 1}{3 \cdot 2 \cdot 1}$$

$$= \frac{11 \cdot 10 \cdot 9 \cdot 8 \cdot 7 \cdot 6 \cdot 5 \cdot 4 \cdot 3 \cdot 2 \cdot 1}{3 \cdot 2 \cdot 1} \text{ or } \frac{11!}{3!}$$

Notice that the denominator of $\frac{11!}{3!}$ is the same as $(11 - 8)!$.

Sleepless in Seattle

Aladdin

The number of ways to arrange 11 things taken 8 at a time is written as $P(11, 8)$. Thus, $P(n, r)$ is read, "the permutation of n objects taken r at a time" and is defined in the following manner.

Definition of $P(n, r)$	The number of permutations of n objects taken r at a time is defined as follows. $$P(n, r) = \frac{n!}{(n - r)!}$$

Example ① A group of 5 teens went to the movie theater. They found a row with 7 empty seats. How many different ways can the teens be seated in the row?

You must find the number of permutations of 7 seats, taken 5 at a time.

$$P(n, r) = \frac{n!}{(n - r)!}$$

$$P(7, 5) = \frac{7!}{(7 - 5)!} \qquad n = 7, r = 5$$

$$= \frac{7 \cdot 6 \cdot 5 \cdot 4 \cdot 3 \cdot \overset{1}{2} \cdot \overset{1}{1}}{\underset{1}{2} \cdot \underset{1}{1}} \text{ or } 2520$$

There are 2520 ways five teens can be seated in a row of 7 seats.

TECHNOLOGY Tips

You can use a graphing calculator to find permutations. The nPr function can be found by pressing MATH ◄ 2.

In Example 1, you may have noticed that the factors of $(n - r)!$ are contained in $n!$. Instead of writing all the factors of each term, you could also have evaluated the expression in the following way.

$$\frac{7!}{(7 - 5)!} = \frac{7 \cdot 6 \cdot 5 \cdot 4 \cdot 3 \cdot \overset{1}{2!}}{\underset{1}{2!}} \qquad \frac{2!}{2!} = 1$$

$$= 7 \cdot 6 \cdot 5 \cdot 4 \cdot 3 \text{ or } 2520$$

Example ②

APPLICATION

Retailing

A manager of Camelot Music is reducing some of the prices of CDs for a special promotion. She has 5 pop/rock CDs, 4 rap CDs, and 4 jazz CDs that she wants to arrange on a shelf for this sale. How many ways can these CDs be arranged on a shelf if they are ordered according to type?

First consider how many different ways each type of CD can be arranged.

Remember, by definition, 0! = 1.

The pop/rock CDs can be arranged in $P(5, 5)$ or 5! different ways.

The rap CDs can be arranged in $P(4, 4)$ or 4! different ways.

The jazz CDs can be arranged in $P(4, 4)$ or 4! different ways.

(continued on the next page)

Now consider how many ways the 3 types can be arranged. There are 3 types of CDs, so $P(3, 3) = 3!$ ways.

The total number of ways the CDs can be arranged is the product of these four permutations.

$$5! \cdot 4! \cdot 4! \cdot 3! = 414,720$$

There are 414,720 ways to arrange the CDs according to type.

The Hawaiian language is made up of only 12 letters, the vowels a, e, i, o, u and the consonants h, k, l, m, n, p, and w. The five letters in the word *aloha* can be arranged in $P(5, 5)$ or 5! ways. However, some of these 5! or 120 arrangements look the same because there are two *a*'s. If you label the *a*'s as a_1 and a_2, then $a_1 loha_2$ is different from $a_2 loha_1$. However, without subscripts, the two arrangements look the same. To account for this in the final count of possible permutations, divide $P(5, 5)$ or 120 by the number of arrangements of *a*.

The two *a*'s can be arranged in $P(2, 2)$ or 2! ways.

$$\frac{P(5, 5)}{P(2,2)} = \frac{5!}{2!}$$
$$= \frac{5 \cdot 4 \cdot 3 \cdot 2!}{2!} \text{ or } 60$$

Thus, there are 60 ways to arrange the letters in *aloha*.

When some objects are alike, use the rule below to find the number of permutations of those objects.

Permutations with Repetitions	The number of permutations of *n* objects of which *p* are alike and *q* are alike is $\frac{n!}{p!q!}$.

This rule can be extended for any number of objects that are repeated.

Example **How many different ways can the letters of the word *PERPENDICULAR* be arranged?**

The first and fourth letter, P, are the same.

The second and fifth letter, E, are the same.

The third and last letter, R, are the same.

So, you need to find the permutations of 13 letters, of which 3 sets of letters are the same. You must divide 13! by 2! 3 times.

$$\frac{13!}{2!2!2!} = \frac{13 \cdot 12 \cdot 11 \cdot 10 \cdot 9 \cdot 8 \cdot 7 \cdot 6 \cdot 5 \cdot 4 \cdot 3 \cdot 2!}{2!2!2!} \text{ or } 778,377,600$$

There are 778,377,600 ways to arrange the letters.

Sometimes in a permutation the arrangement of objects is in a circle. This is called a **circular permutation**.

Example The disk jockey at station WXYZ is setting up some of the music she will be playing during her shift. She is loading a CD tray with 6 different compact discs. How many different ways can these discs be arranged on the tray?

Each tray is a circle. Let *a* represent the first disc, let *b* represent the second disc, and so on. Three possible arrangements of the compact discs are shown below.

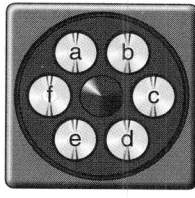 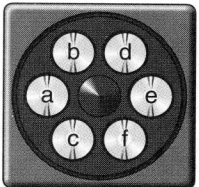 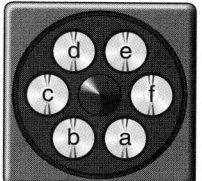

How does the arrangement change as the first tray is turned? Which arrangement is *really* different from the other two?

When 6 different objects are placed in a line, there are 6! or 720 arrangements of the 6 objects taken 6 at a time. However, when the 6 objects are arranged in a circle, some of the arrangements are alike. These arrangements fall into groups of 6. Once an arrangement is determined, the other five members of the group are formed by rotating the circle. Then you can rearrange the items and another group of 6 is formed. Thus, the total number of *really* different arrangements of 6 objects around a circle is $\frac{1}{6}$ of the total number of arrangements in a line.

$$\frac{1}{6} \cdot 6! = \frac{6 \cdot 5 \cdot 4 \cdot 3 \cdot 2 \cdot 1}{6}$$
$$= 5 \cdot 4 \cdot 3 \cdot 2 \cdot 1$$
$$= 5! \text{ or } 120 \qquad \textit{Note that 5! = (6 − 1)!.}$$

There are (6 − 1)! or 120 possible arrangements of the discs.

Singles of All Time in the U.S.

1. *White Christmas,* Bing Crosby, 1942
2. *I Want to Hold Your Hand,* Beatles, 1964
3. *Hound Dog/Don't Be Cruel,* Elvis Presley, 1956
4. *It's Now or Never,* Elvis Presley, 1960
5. *I Will Always Love You,* Whitney Houston, 1992

When *n* objects are arranged in a circle, use the following rule to find the number of permutations of those objects.

Circular Permutations	If *n* distinct objects are arranged in a circle, then there are $\frac{n!}{n}$ or (n − 1)! permutations of the objects around the circle.

Suppose *n* objects are in a circular arrangement, but the position of the objects is related to a fixed point. Rotating the circle will relate a different object to the fixed point and will make a new arrangement of the objects. Because of this fixed point, the permutations are now considered linear. The number of permutations for a circular permutation with a fixed point is *n*!.

5 Let each circle represent an empty cul-de-sac ready to be developed into 5 different lots with each lot having a different home design. Let the labeled points represent the 5 home designs. Let the arrow represent the home nearest the opening to the cul-de-sac. How many arrangements are possible?

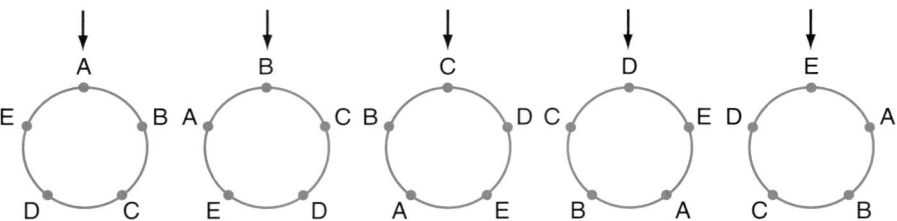

These arrangements can be considered different because in each one, a different house sits next to the cul-de-sac opening. Thus, there are $P(5, 5)$ or 5! arrangements relative to a fixed point.

$$5! = 5 \cdot 4 \cdot 3 \cdot 2 \cdot 1 \text{ or } 120$$

There are 120 possible arrangements.

A cul-de-sac is a street closed at one end.

Suppose three keys are placed on a key ring. How many different arrangements are possible? Using the formula for circular permutations, it appears that there are at most $(3 - 1)!$ or 2 different arrangements of keys on the ring. But what happens if the first key ring arrangement is turned over?

When the key ring is turned over, the first arrangement becomes the second arrangement. Then there is really only one arrangement of the three keys. These two arrangements are **reflections** of each other. As a result, there are only half as many arrangements when reflections are possible.

$$\frac{(3 - 1)!}{2} = \frac{2 \cdot 1}{2} \text{ or } 1$$

Reflections also occur in linear arrangements. Suppose five juniors are being inducted into the National Honor Society. Sitting on one side of the gym are Marisa, Brandon, Tyree, Cheryl, and Haley, in order from left to right. However, an observer on the opposite side of the gym sees the order from left to right as Haley, Cheryl, Tyree, Brandon, and Marisa. If you were asked how many possible arrangements of the students there were for these students, your answer would be half of 5! or 60 arrangements.

CHECK FOR UNDERSTANDING

Communicating Mathematics

Study the lesson. Then complete the following.

1. **Explain** what 6! means.

2. **Discuss** the important characteristic that a counting problem has to have in order to classify it as a permutations problem.

3. **Explain** why 5! gives the correct solution to the possible number of ways to arrange the letters in the word *VIDEO*.

4. **Explain** the difference between a linear permutation and a circular permutation.

5. **You Decide** Brad says there are *twice* as many ways for 8 people to stand in a line than there are for them to stand around in a circle. Denise says there are *eight times* as many ways. Who is correct? Explain your answer.

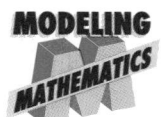

6. Model the meaning of $P(4, 4)$ using colored counters or candies. What is its value?

Guided Practice

How many different ways can the letters of each word be arranged?

7. MATH   8. FUN

9. SEE   10. PENCIL

Determine whether each arrangement is *linear* or *circular*. Then determine if it is also a *reflection* and find the number of arrangements.

11. batting order of a baseball team with 9 players

12. a group of 6 children playing Ring Around The Rosy

13. **Pageants** The Miss Teen USA pageant has fifty-two contestants. The judges choose the next Miss Teen USA and her three runners-up.

    a. Is order important? Why?

    b. Make a decision chart for this problem.

    c. In how many different ways can the next Miss Teen USA and her three runners-up be chosen?

EXERCISES

Practice

How many different ways can the letters of each word be arranged?

14. LEVEL   15. FLORIDA   16. POP

17. PARALLEL   18. ALASKA   19. FREE

20. PEGGY   21. STUDY   22. MISSISSIPPI

23. ALGEBRA   24. ESSENTIAL   25. REPETITION

Determine whether each arrangement is linear or circular. Then determine if it is also a reflection and find the number of arrangements.

26. ten beads on a necklace with a clasp

27. a basketball huddle of 5 players

28. six nickels in a circle on a table

29. eight pizza toppings placed on a revolving tray

30. seven shoppers in line at a checkout counter

31. six charms on a bracelet with no clasp

Evaluate each expression.

32. $\dfrac{P(6, 4)}{P(5, 3)}$ 33. $\dfrac{P(6, 3)P(4, 2)}{P(5, 2)}$ 34. $\dfrac{P(12, 6)}{P(12, 3)P(8, 2)}$

Solve for *n*.

35. $P(n, 4) = 40[P(n - 1, 2)]$ 36. $P(n, 4) = 3[P(n, 3)]$

37. $n[P(5, 3)] = P(7, 5)$ 38. $208P(n, 2) = P(16, 4)$

39. Geometry How many ways can 6 points be labeled *A* through *F* on the circle at the right, relative to its *x*-intercept?

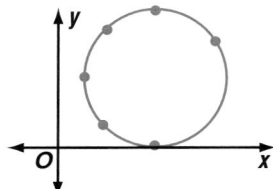

Critical Thinking

40. According to the writers of the CD-ROM game *Trivial Pursuit*, you can play chess 25×10^{120} times without repeating the same total moves. Explain how you could get that result. List all the factors you need to consider. You may want to draw a diagram to help explain your thinking.

Applications and Problem Solving

41. Cheerleading Seven football cheerleaders printed large letters on cards that spell out their school's mascot, COUGARS. Each card has one letter on it and each cheerleader is supposed to hold up one card. At half-time they realize that someone has mixed up the cards. How many ways are there to arrange the cards?

42. Emergency Service The table at the right shows common medical emergency telephone numbers for six countries. How many other 3-digit emergency telephone numbers are possible if any digit can be repeated?

43. Photography A photographer is taking a picture of a bride and a groom together with 6 attendants. How many ways can he arrange the 8 people in a line if the bride and groom stand in the middle?

Country	Telephone Number
Australia	001
Germany	110
Great Britain	999
Italy	113
Japan	119
United States	911

Source: *Health,* 1994

Where U.S. Consumers Buy PCs

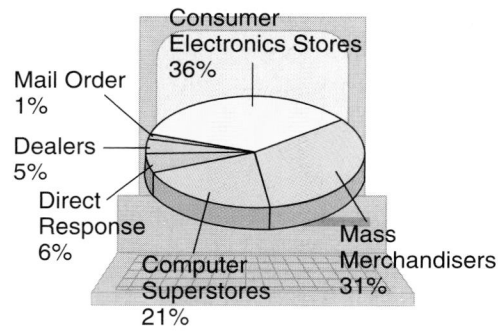

Consumer Electronics Stores 36%
Mail Order 1%
Dealers 5%
Direct Response 6%
Computer Superstores 21%
Mass Merchandisers 31%

44. Statistics Look at the circle graph at the left.
   a. How many different arrangements of the wedges are possible?
   b. How many different arrangements are possible if the largest section of the graph stays in the same place?

45. Electricity Eight switches are connected on a circuit so that if any one or more of the switches are closed, the light will go on. How many combinations of open and closed switches exist that will permit the light to go on?

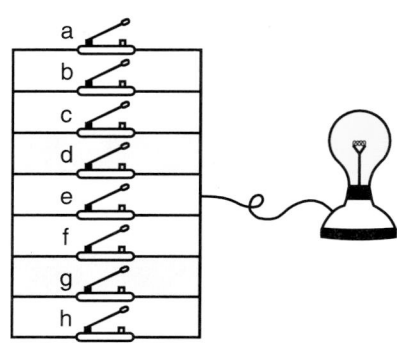

46. Puzzles Refer to the puzzle below.

a. How many ways are there to arrange the letters in each of the first two scrambled words?

b. How many ways are there to arrange the letters in each of the last two scrambled words?

c. Unscramble each of the words to find the letters in the circles. How many ways are there to arrange these letters?

d. Unscramble the circled letters to find the answer to the riddle.

e. How many different seating arrangements are possible for the jurors?

Mixed Review

47. How many ways can you have 50¢ if you have at least one quarter? (Lesson 12–1)

48. Find the sum of $\frac{2}{3} + \frac{1}{3} + \frac{1}{6} + \cdots$. (Lesson 11–5)

49. **Biology** An amoeba divides into two amoebas once every hour. How long would it take for a single amoeba to become a colony of 4096 amoebas? (Lesson 10–2)

50. If y varies inversely as x, and $y = -8$ when $x = 1.5$, find x when $y = -3$. (Lesson 9–2)

51. State the number of possible positive real zeros, negative real zeros, and imaginary zeros of $f(x) = 3x^5 - 8x^2 + 1$. (Lesson 8–4)

52. **Animals** The fastest-recorded physical action of any living thing is the wing beat of the common midge. This tiny insect normally beats its wings at a rate of 133,000 times per minute. How many times would the midge beat its wings in an hour at this rate? Write your answer in scientific notation. (Lesson 5–1)

53. Find $\begin{bmatrix} 3 & -1 \\ 2 & 5 \end{bmatrix} \cdot \begin{bmatrix} 4 & -1 & -2 \\ -3 & 5 & 4 \end{bmatrix}$. (Lesson 4–3)

54. Graph the system of inequalities and name the vertices of the polygon formed. Then find the maximum and minimum values of $f(x, y) = 4x - 3y$ for the region. (Lesson 3–5)

$x \le 5$

$y \ge -3x$

$2y \le x + 7$

$y \ge x - 4$

55. Write a mathematical expression for the verbal expression *the theater can hold no more than 400 people.* (Lesson 1–6)

Combinations

12-3

What YOU'LL LEARN

- To solve problems involving combinations.

Why IT'S IMPORTANT

You can use combinations to solve problems involving social studies and the lottery.

APPLICATION

Lottery

How would you like to win $27 million? That's how much the Virginia state lottery was worth in 1992 when a group of Australian investors tried to buy every combination of numbers— all 7.1 million of them! The investors won by buying 5.6 million sets of numbers, one of which was the winner. For each ticket, they chose 6 numbers out of 44. The order of the numbers selected was not important, only the set of the numbers themselves. An arrangement, or listing, in which order is not important, is called a **combination.** *You will verify the number of combinations in Exercise 44.*

Let's look at a simpler combination. Suppose a lottery was held in which each person chose three numbers from the numbers 0–9 and the order did not matter. Since order is not important, we are counting combinations, *not* permutations, of 10 objects taken 3 at a time. The combination of 10 objects taken 3 at a time is written as $C(10, 3)$. You already know how to count $P(10, 3)$.

$$P(10, 3) = \frac{10!}{(10 - 3)!}$$
$$= 10 \cdot 9 \cdot 8 \text{ or } 720$$

But these permutations are not all different combinations. This group includes $(1, 4, 8)$, $(1, 8, 4)$, $(4, 1, 8)$, $(4, 8, 1)$, $(8, 1, 4)$, and $(8, 4, 1)$. Since these are all the same combination, we are counting this combination, and all others as well, 3! or 6 times when we calculate $P(10, 3)$. Thus, combinations should be counted as follows.

$$C(10, 3) = \frac{P(10, 3)}{3!}$$
$$= \frac{10!}{(10 - 3)! \cdot 3!} \qquad P(10, 3) = \frac{10!}{(10 - 3)!}$$
$$= \frac{10!}{7! \cdot 3!} \text{ or } 120$$

The three numbers can be chosen from the 10 numbers in 120 ways.

Notice that 7! and $(10 - 3)!$ are equivalent. This suggests the following definition.

Definition of $C(n, r)$	**The number of combinations of *n* distinct objects taken *r* at a time is defined as follows.** $$C(n, r) = \frac{n!}{(n - r)!r!}$$

The basic difference between a permutation and a combination is that order is considered in a permutation and order is *not* considered in a combination.

Example **1** **The principal at Cobb County High School wants to start a peer mediation group to work with discipline problems. He needs to narrow down his choice to six students from a group of nine students. How many ways can a group of six be selected?**

In this case, order does not matter. The combination of nine objects or people taken six at a time is written as $C(9, 6)$.

$$C(9, 6) = \frac{9!}{(9 - 6)! \cdot 6!} \qquad C(n, r) = \frac{n!}{(n - r)!r!}$$

$$= \frac{9!}{3! \cdot 6!} \text{ or } 84$$

The six peer mediation members can be chosen from the nine students in 84 ways.

TECHNOLOGY *Tips*

You can use a graphing calculator to find combinations. The nCr function can be found by pressing `MATH` `◄` `3`.

Example **2** **A basket contains 4 acorn squash, 5 gourds, and 8 pumpkins. How many ways can 2 acorn squash, 1 gourd, and 2 pumpkins be chosen?**

This involves the product of three combinations, one for each type of item.

$C(4, 2)$ Two of 4 acorn squash will be chosen.
$C(5, 1)$ One of 5 gourds will be chosen.
$C(8, 2)$ Two of 8 pumpkins will be chosen.

We can multiply the combinations because of the fundamental counting principle.

$$C(4, 2) \cdot C(5, 1) \cdot C(8, 2) = \frac{4!}{(4 - 2)!2!} \cdot \frac{5!}{(5 - 1)!1!} \cdot \frac{8!}{(8 - 2)!2!}$$

$$= \frac{4!}{2!2!} \cdot \frac{5!}{4!1!} \cdot \frac{8!}{6!2!}$$

$$= \frac{\overset{1}{\cancel{2}}}{\underset{1}{\cancel{4}} \cdot 3 \cdot \cancel{2!}} \cdot \frac{5 \cdot \cancel{4!}}{\cancel{4!}} \cdot \frac{8 \cdot 7 \cdot \cancel{6!}}{\cancel{6!} \cdot \cancel{2}}$$

$$= 3 \cdot 5 \cdot 8 \cdot 7 \text{ or } 840$$

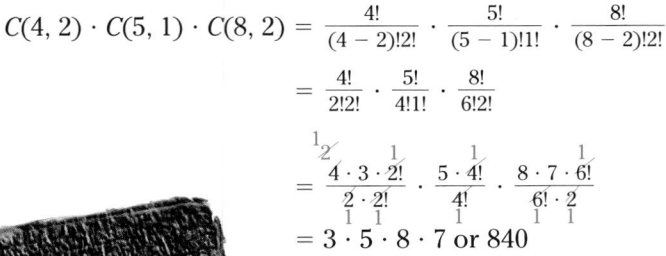

You can also use a scientific calculator to evaluate the expression.

Enter: 4 `x!` `×` 5 `x!` `×` 8 `x!` `÷` `(`

2 `x!` `×` 2 `x!` `×` 4 `x!` `×`

6 `x!` `×` 2 `x!` `)` `=` *840*

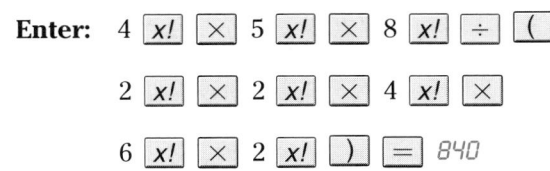

There are 840 different ways to choose the items.

Example **3**

Geometry

Find the total number of diagonals that can be drawn in an octagon.

Each diagonal has two endpoints. Suppose one has endpoints *B* and *G*. Since segments *BG* and *GB* are the same, order is not important. The combination of 8 points taken 2 at a time gives the total number of segments connecting any two points.

$C(8, 2) = \frac{8!}{6!2!}$ or 28

However, 28 is not our answer. Since eight of the segments connecting the points are sides of the octagon, you must subtract 8 from 28, the number of combinations.

Thus, the total number of diagonals in an octagon is 28 − 8 or 20.

Some applications can involve *both* permutations and combinations. The example below involves the use of a standard deck of 52 playing cards. Remember, there are four suits, each suit containing 13 cards.

Example **4**

CONNECTIONS

Cards and card games probably originated in China between the 7th and 10th centuries. They were brought to Europe by Venetian explorers in the 13th century. The 52-card modern deck derives from the French in the 15th century.

Eight cards are drawn from a standard deck of 52 cards. How many 8-card hands having 5 cards of one suit and 3 cards of another suit can be formed?

First consider how many ways 2 suits can be chosen from 4 suits. Since a different number of cards is being selected from each suit, order is important. Then consider the combinations possible with each suit.

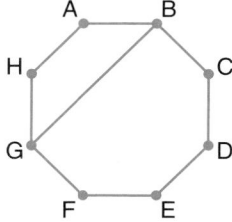

$P(4, 2)$ Select 2 suits from 4 suits.
$C(13, 5)$ Select 5 cards from 13 cards in one suit.
$C(13, 3)$ Select 3 cards from 13 cards in the other suit.

$P(4, 2) \cdot C(13, 5) \cdot C(13, 3) = \frac{4!}{2!} \cdot \frac{13!}{8!5!} \cdot \frac{13!}{10!3!}$
$= 12 \cdot 1287 \cdot 286$ or 4,416,984

There are 4,416,984 eight-card hands having 5 cards of one suit and 3 cards of another.

CHECK FOR UNDERSTANDING

Communicating Mathematics

Study the lesson. Then complete the following.

1. **Describe** the difference between a permutation and a combination.

2. **Write** an expression to represent the possible number of starting teams of five basketball players that can be formed from 11 players.

3. **Draw** a diagram to verify that the answer for Example 3 is 20 diagonals.

4. **Explain** how you would find out how many committees of 3 men and 2 women can be formed from a group of 4 men and 5 women.

5. Your principal announces that everyone should buy a permutation lock for their lockers. Your friend leans over to you and says, "Doesn't she mean a combination lock?" How would you answer your friend?

Guided Practice

Determine whether each situation involves a permutation or a combination.

6. choosing a class president, vice president, and secretary

7. four tennis players from a group of nine

8. eight toppings for ice cream

Evaluate each expression.

9. $C(4, 2)$

10. $C(7, 2)$

11. $C(3, 2) \cdot C(8, 3)$

12. $C(8, 5) \cdot C(7, 3)$

Solve each problem.

13. **Floral Design** A bucket at DeSantis Florists contains 8 red tulips, 5 white daisies, and 4 yellow tulips. How many bouquets could be created so that each bouquet has 2 red tulips, 1 white daisy, and 2 yellow tulips?

14. **Volleyball** How many starting volleyball teams of 6 members can be formed from a bench of 12 talented players?

EXERCISES

Practice

Determine whether each situation involves a permutation or a combination.

15. a classroom seating chart

16. finding the diagonals of a polygon

17. the batting order of the Pittsburgh Pirates

18. 10 books on a library shelf

19. a hand of five cards from a deck of 52 cards

20. a seven-person committee from your class

21. first, second, and third chairs for six clarinets in a band

22. six outfits chosen from fourteen outfits to be modeled

Evaluate each expression.

23. $C(5, 2)$

24. $C(10, 5)$

25. $C(8, 4)$

26. $C(24, 21)$

27. $C(12, 7)$

28. $C(10, 4)$

29. $C(12, 4) \cdot C(8, 3)$

30. $C(9,3) \cdot C(6, 2)$

31. $C(10, 4) \cdot C(5, 3)$

32. $C(8,2) \cdot C(5,1) \cdot C(4, 2)$

INTEGRATION
Geometry

33. Suppose there are 8 points in a plane such that no three points are collinear. How many distinct triangles could be formed with 3 of these points as vertices?

34. A circle has nine randomly-placed points. In how many different ways can you form each polygon listed below?

    a. triangle   **b.** octagon   **c.** pentagon

    d. quadrilateral   **e.** decagon   **f.** hexagon

Solve for *n*.

35. $C(n, 5) = C(n, 7)$

36. $C(11, 8) = C(11, n)$

37. $C(14, 3) = C(n, 11)$

38. $C(n, 7) = C(n, 2)$

Critical Thinking

39. Prove $C(n, r) = \dfrac{P(n, r)}{r!}$.

Applications and Problem Solving

40. School The biology class is preparing for its final exam. From a class of 22 girls and 16 boys, how many study groups of 2 girls and 3 boys can be formed?

41. Committees A five-member recycling committee is being formed from a group of 8 freshmen and 10 sophomores. How many committees can be formed given each condition?

a. all freshmen

b. all sophomores

c. 1 freshman, 4 sophomores

d. 3 freshmen, 2 sophomores

42. Solve a Simpler Problem Ten points are marked on a circle. How many line segments can be drawn between any two of the points?

43. Business A box of Anthony Thomas candy contains 9 dark chocolate creams, 6 milk chocolate creams, and 4 milk chocolates with nuts. How many ways can 5 candies be selected to meet each condition?

a. all milk chocolate creams

b. all nuts

c. all dark chocolate creams

d. 2 of one kind, 3 of another

e. 2 nuts, 2 milk chocolate creams, 1 dark chocolate cream

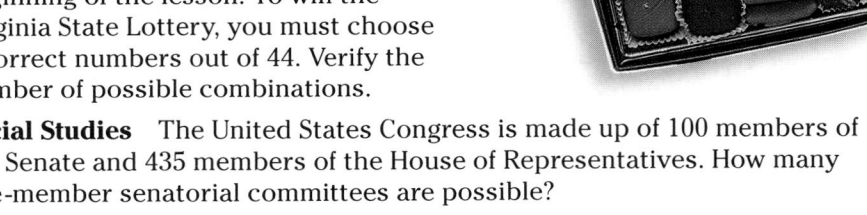

44. Lottery Refer to the application at the beginning of the lesson. To win the Virginia State Lottery, you must choose 6 correct numbers out of 44. Verify the number of possible combinations.

45. Social Studies The United States Congress is made up of 100 members of the Senate and 435 members of the House of Representatives. How many five-member senatorial committees are possible?

46. World Cultures *Hanafuda* is a popular card game invented in Japan in the eighteenth century. The deck is made up of 12 suits, with each suit having four cards depicting a plant related to one of the 12 months of the year. How many 7-card hands can be formed so that 3 are from one suit and 4 are from another?

Month	Plant
January	Pine tree
February	Plum tree
March	Cherry tree
April	Wisteria
May	Iris
June	Peony
July	Bush clover
August	Pampas grass
September	Chrysanthemum
October	Maple
November	Willow
December	Paulownia tree

Mixed Review

47. **Community Service** As a community project, the members of the Spanish Club at Lakota High School are preparing circular vegetable trays to be delivered to homebound senior citizens in their neighborhood. Each tray is made up of 5 items: broccoli, carrots, celery, green peppers, and cucumbers. How many different ways can these items be arranged on a tray if each item must fit in its own section? (Lesson 12–2)

48. Find $\sum_{s=1}^{4} 2s\left(-\frac{1}{2}\right)^2$. (Lesson 11–4)

49. **Finance** Graham's grandparents started a savings account for him when he was born. They invested $100 in an account with 8% annual interest compounded annually. (Lesson 10–1)
   a. Write an exponential equation to express the amount of money in the account on Graham's nth birthday.
   b. How much is in the account on his 16th birthday?

50. Write a quadratic equation that has roots $-2, -7$. (Lesson 6–5)

51. **Chemistry** The mole is a standard unit of measure in chemistry. One mole of any compound contains 6.02×10^{23} molecules. How many molecules are in 19.9 moles of ammonia? (Lesson 5–1)

52. Graph the system of equations and state its solution. (Lesson 3–1)
   $x + y = 1$
   $3x - 2y = -7$

53. Find the slope and y-intercept of the graph of $3x - 4y = -10$. (Lesson 2–4)

54. Solve $|x + 4| = 5$. (Lesson 1–5)

"Fowl" Advertising

The article below appeared in *Teacher Magazine* in March, 1995.

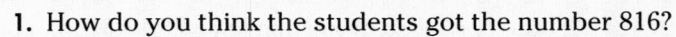

A NATIONAL RESTAURANT CHAIN THAT boasts about its tasty chicken is eating crow after a high school mathematics class cried foul over a television ad. The ad shows football star Joe Montana standing at the counter at a Boston Chicken restaurant puzzling over side-dish choices; an announcer says that more than 3,000 combinations can be created by choosing three of the . . . 16 side-dish offerings. But Bob Swaim, a math teacher at Souderton Area High School near Philadelphia, and his class did the math and found that the correct number was 816. "We goofed," says . . . a spokesman for Boston Chicken. "Apparently we didn't listen to our high school math teachers." The company has, however, listened to Swaim and corrected its ads. For their part, the students were awarded free meals and $500 to expand the math menu at Souderton. ■

1. How do you think the students got the number 816?
2. When you hear celebrities, media persons, or people in authority quote numbers or statistics, do you assume they must be correct? What factors might be preventing you from listening more critically and raising more questions?
3. Why do you think advertisers often use famous people and celebrities in ads for their products?
4. Have you ever bought a product because of a celebrity who helped to advertise it? Why or why not?

12-4

Probability

What YOU'LL LEARN

- To find the probability of an event, and
- to determine the odds of success and failure of an event.

Why IT'S IMPORTANT

You can use probability to solve problems involving darts, genetics, and music.

APPLICATION
Earthquakes

On January 17, 1994, an earthquake registering 6.7 on the Richter scale struck Northridge in the densely-populated San Fernando Valley of Los Angeles. The earthquake, which lasted only 40 seconds, left 20,000 people homeless and caused approximately $20 billion in damages. Geologists say there is an 86% chance that a quake of even greater magnitude will strike the southern part of California sometime in the next three decades. What are the chances that such an earthquake will hit within 15 years? within 5 years? We can use **probability** to measure the chances of an event occurring.

Mathematicians often use coin tossing and dice rolling to illustrate probability. When you toss a coin, there are only two possible outcomes—heads or tails. A desired outcome is called a **success.** Any other outcome is called a **failure**.

Probability of Success and of Failure	If an event can succeed in s ways and fail in f ways, then the probabilities of success, $P(s)$, and of failure, $P(f)$, are as follows. $$P(s) = \frac{s}{s+f} \qquad P(f) = \frac{f}{f+s}$$

What does the sum of s and f represent?

If an event cannot fail, it has a probability of 1. If an event cannot succeed, it has a probability of 0. The probability of an event occurring is always between 0 and 1, inclusive. In fact, the sum of $P(s)$ and $P(f)$ is always equal to 1. Thus, they are called **complements.** So if $P(s) = \frac{1}{5}$, then $P(f) = 1 - \frac{1}{5}$ or $\frac{4}{5}$. This property is often used in finding the probability of events.

Example **1**

The term __at random__ means that an outcome is chosen without any preference.

Like any ratio, probability can be expressed as a fraction, decimal, or percent.

A bag of M&M's® contains 12 red, 11 yellow, 5 green, 6 orange, 5 blue, and 16 brown candies. What is the probability that if you choose an M&M from the bag without looking, or *at random*, you will choose a yellow M&M?

The probability of choosing a yellow M&M is written $P(yellow)$.

There are 11 ways to choose a yellow M&M, and there are $12 + 5 + 6 + 5 + 16$ or 44 ways *not* to choose a yellow M&M. So, $s = 11$ and $f = 44$.

$$P(yellow) = \frac{s}{s+f} \qquad \text{\textit{Replace s with 11 and f with 44.}}$$
$$= \frac{11}{11+44}$$
$$= \frac{11}{55} \text{ or } \frac{1}{5}$$

The probability of selecting a yellow M&M is $\frac{1}{5}$ or 20%.

Many people enjoy playing games where players of equal skill have the same chance of winning.

A *fair game* is one in which each player has an equal chance of winning. In an *unfair game*, players do *not* have an equal chance of winning. In the game *Scissors, Paper, Stone*, the winner is decided by the following rules.

• scissors cut paper
• paper wraps stone
• stone breaks scissors

If both players pick the same object, the round is a draw.

Your Turn

a. Play 27 rounds of *Scissors, Paper, Stone* with a partner. Record the number of times each player wins.

b. How many different outcomes are possible?

c. How many ways can you win?

d. How many ways can your partner win?

e. Is *Scissors, Paper, Stone* a fair game? Explain.

In a different version of *Scissors, Paper, Stone*, there are 3 players. If all 3 players match, player A gets a point; if 2 players match, player B gets a point; and if none of the players match, player C gets a point.

f. Play 27 rounds of this version of *Scissors, Paper, Stone* with two other people. Record the number of times each player wins.

g. What is the probability that A will win? B? C?

h. Is this a fair game? Explain.

The counting methods you used in finding permutations and combinations are often used in determining probability.

Example **Suppose Miguel draws 5 cards from a standard deck of 52 cards. What is the probability that his hand contains 2 cards of one suit and 3 cards of another suit?**

First, find how many 5-card hands meet these conditions.

$P(4, 2)$ Select 2 suits among 4. Order is important since different numbers of cards are to come from each suit.

$C(13, 2)$ Select 2 cards from a suit containing 13 cards.

$C(13, 3)$ Select 3 cards from the other suit.

Now use the counting principle.

$$P(4, 2) \cdot C(13, 2) \cdot C(13, 3) = \frac{4!}{2!} \cdot \frac{13!}{11!2!} \cdot \frac{13!}{10!3!} \text{ or } 267,696$$

So, the number of successes is 267,696.

Now find the total number of possible 5-card hands.

$$C(52, 5) = \frac{52!}{47!5!} \text{ or } 2,598,960$$

There are 2,598,960 ways to choose a 5-card hand, so $s + f = 2,598,960$.

$$\frac{s}{s + f} = \frac{267,696}{2,598,960} \text{ or about } 0.103 \quad \textit{Use a calculator.}$$

The probability of drawing a 5-card hand with 2 cards of one suit and 3 cards of another suit is about 10%.

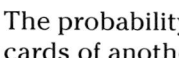

Another way to measure the chance of an event occurring is with **odds**. The odds of an event can be expressed as the ratio of successes to failures.

<table>
<tr>
<td>Definition
of Odds</td>
<td>The odds of the successful outcome of an event can be expressed as the ratio of the number of ways it can succeed to the number of ways it can fail.

    Odds in favor = number of successes : number of failures
    Odds against = number of failures : number of successes</td>
</tr>
</table>

Example ❸

APPLICATION

Genetics

Color blindness is the genetic inability to distinguish between certain colors. The most common form is a red-green deficiency, in which a person sees all colors in tones of yellow and blue. Eight out of 100 males and 1 out of 1000 females have some form of color blindness.

a. What are the odds of a male being color-blind?

b. What are the odds of a female being color-blind?

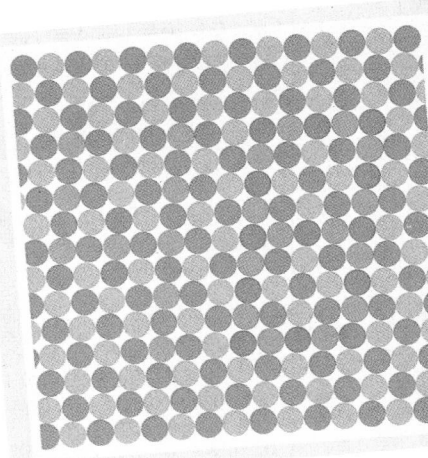

Color blindness test

a. Eight out of 100 males are color-blind. So the number of successes (being color-blind) is 8. The number of males who are not color-blind is $100 - 8$ or 92. So the number of failures is 92.

    odds of a male being color-blind = $s:f$
                                   = 8:92 or 2:23

The odds of a male being color-blind are 2 to 23.

b. One out of 1000 females is color-blind. So the number of successes is 1. The number of females who are not color-blind is $1000 - 1$ or 999. So the number of failures is 999.

    odds of a female being color-blind = $s:f$
                                          = 1:999

The odds of a female being color-blind are 1 to 999.

Sometimes you may have to use permutations and combinations when finding odds.

Example ❹

A committee to organize the school prom has 6 seniors and 5 juniors. If a subcommittee of 4 students is selected at random to choose the music for the prom, what are the odds that it will contain 2 seniors and 2 juniors?

Since a committee is being formed, order is not important. Use combinations to find the odds.

There are $C(6, 2)$ ways to choose 2 seniors from 6 seniors. There are $C(5, 2)$ ways to choose 2 juniors from 5 juniors. Use the counting principle to find the number of successes (a senior or a junior).

$$C(6, 2) \cdot C(5, 2) = \frac{6!}{4!2!} \cdot \frac{5!}{3!2!} \text{ or } 150$$

So, the number of successes is 150.

Now find the number of possible subcommittees of 4 people out of a group of 11 people.

$$C(11, 4) = \frac{11!}{7!4!} \text{ or } 330$$

The number of 4-person subcommittees that do not meet the conditions is $330 - 150$ or 180. Thus, the odds of selecting a subcommittee of 2 juniors and 2 seniors are 150:180 or 5:6.

CHECK FOR UNDERSTANDING

Communicating Mathematics

Study the lesson. Then complete the following.

1. **Define** the term *probability*.

2. **Explain** whether or not $P(s)$ can equal $\frac{5}{4}$.

3. **State** the odds of getting a 4 when rolling a die.

4. **Determine** the odds that an event will *not* occur if the odds that the event *will* occur are 2:5.

5. **You Decide** LaToya says that if she tosses three coins, she can get exactly two heads. Donovan bets he could do something even harder—get exactly two heads if he tosses 4 coins. LaToya argues that it is *easier* to get two heads out of four. Who is right? Explain how you reached your conclusion.

MATH JOURNAL

6. **Assess Yourself** Describe an event in your life that has a probability of 1 and an event that has a probability of 0.

Guided Practice

State the odds of an event occurring, given the probability of the event.

7. $\frac{3}{4}$

8. $\frac{2}{9}$

State the probability of an event occurring, given the odds of the event.

9. 6:5

10. 1:1

11. The probability of Cathy earning a college scholarship is $\frac{4}{5}$. What are the odds that she will *not* earn a scholarship?

12. **Genealogy** The odds that an American is of English ancestry are 1:9. What is the probability that an American is of English ancestry?

Suppose you select 2 letters at random from the word *Pacific*. Find each probability.

13. P(2 vowels)

14. P(2 consonants)

15. P(1 vowel, 1 consonant)

16. **Genetics** Find the probability of a couple having a left-handed child, given the following odds.
    a. If both parents are left-handed, the odds are 2 to 1.
    b. If only one parent is left-handed, the odds are 1 to 6.
    c. If neither parent is left-handed, the odds are 1 to 16.

Practice

State the odds of an event occurring, given the probability of the event.

17. $\frac{1}{2}$ **18.** $\frac{3}{8}$ **19.** $\frac{11}{12}$

20. $\frac{4}{7}$ **21.** $\frac{1}{5}$ **22.** $\frac{4}{11}$

State the probability of an event occurring, given the odds of the event.

23. 6:1   **24.** 3:7   **25.** 5:6

26. 9:8   **27.** 1:8   **28.** 7:9

29. A bag of Jelly-Belly® jelly beans contains 40 Polynesian Punch and 120 Blueberry jelly beans.

  a. If you draw one jelly bean out of the bag, find the probability that it is Polynesian Punch.

  b. If you add 40 Polynesian Punch jelly beans to the original bag and draw out a bean, what is the probability that it is Polynesian Punch?

  c. How many Polynesian Punch jelly beans do you need to add to the original bag to double the original probability of drawing a Polynesian Punch jelly bean? Explain.

Alma has 4 gray kittens and 7 white kittens. She randomly picks up 2 to give to her nieces. Find the probability of each selection. Then find the odds of that selection.

30. P(2 gray kittens)   **31.** P(2 white kittens)   **32.** P(1 of each color)

Sonia is moving and all of her CDs are mixed up in a box. Twelve CDs are rock, eight are jazz, and five are classical. If she reaches in the box and selects three at random, find each probability.

33. P(all jazz)   **34.** P(all rock)

35. P(1 classical, 2 jazz)   **36.** P(2 classical, 1 rock)

37. P(1 classical, 1 jazz, 1 rock)   **38.** P(2 jazz, 1 reggae)

INTEGRATION
Geometry

39. A square target 15 cm on a side, like the one shown at the right, contains 40 non-overlapping circles each 2 cm in diameter. Find the probability that a dart thrown at random hits one of the circles.

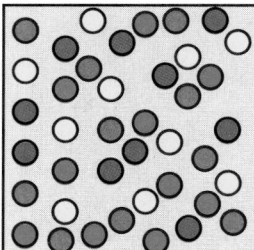

40. A red, blue, and green die are rolled. The number on each die represents a side length of a triangle. So, 4, 2, 4 would represent an isosceles triangle.

  a. How many combinations can be rolled?

  b. What is the probability that you could build an equilateral triangle?

  c. What is the probability that you could build a nonequilateral isosceles triangle?

41. Find the probability that a point, chosen at random, belongs to the shaded region of the figure at the right. Write your answer in terms of π.

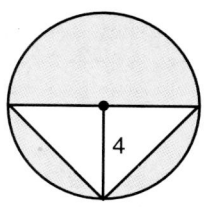

Critical Thinking

42. What is the probability that $x^2 + kx + 16$ will factor if $0 \leq k \leq 10$ and k is an integer chosen at random?

43. Two octahedral, or eight-sided dice are rolled. What is the probability that the sum of the numbers on the dice equals 9?

Applications and Problem Solving

44. Music In the summer of 1994, 250,000 people attended the Woodstock 25th-Anniversary Concert in Saugerties, New York. Approximately 5000 were treated for injuries, which was less than the insurance companies covering the concert had anticipated. Out of 50 randomly selected concert-goers, how many would you expect were treated for injuries?

45. Geography A state is chosen at random from the 50 states. Find each probability.

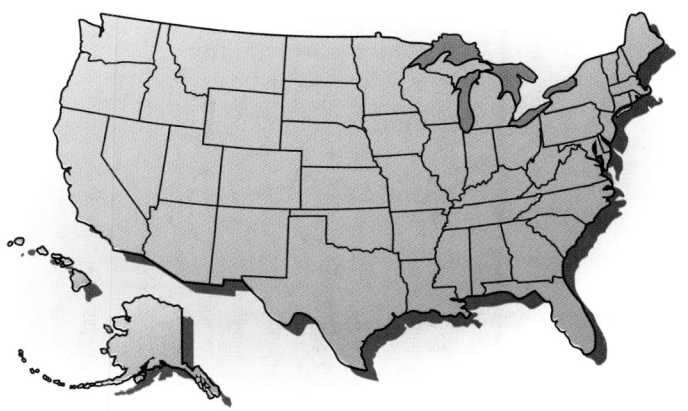

  a. P(next to the Pacific Ocean)
  b. P(has at least one representative in the House of Representatives)
  c. P(has at least five neighboring states)
  d. P(has three U.S. senators)

46. Finance The state of Florida has a Lotto drawing in which 6 numbers out of 49 are drawn at random. The proceeds from the lottery help to finance education in the state. What is the probability that you will win one of the weekly Lotto drawings if you buy one ticket?

47. Genetics Guinea pigs have genes that can produce four types of fur: black short (BBSS, BBSs, BbSS, or BbSs), black long (BBss or Bbss), white short (bbSS or bbSs), or white long (bbss). Use the Punnett square below to find the probability that two parents with black short fur (both with BbSs fur genes) will produce an offspring that has white short fur.

	BS	**Bs**	**bS**	**bs**
BS	BBSS	BBSs	BbSS	BbSs
Bs	BBSs	BBss	BbSs	Bbss
bS	BbSS	BbSs	bbSS	bbSs
bs	BbSs	Bbss	bbSs	bbss

48. Darts According to the *Guinness Book of World Records*, John Lowe won a record prize of £102,000, or approximately $178,500, for his performance in a darts event in England on October 13, 1984. A dart, thrown at random, hits the dart board shown at the right. Find each probability. (*Hint:* What is the area of each ring?)

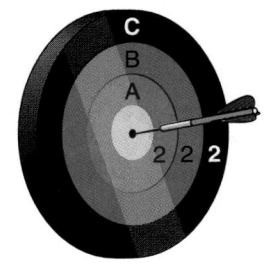

a. $P(A)$ **b.** $P(B)$ **c.** $P(C)$

Mixed Review

49. Music The Groveport Community Children's Chorus now has six altos and eight sopranos. For the songs they are performing at the spring concert, they need two alto soloists and two soprano soloists. How many ways can these four soloists be selected at random? (Lesson 12–3)

50. How many ways can 5 books be placed on a shelf? (Lesson 12–2)

51. Find the missing terms of ___?___, ___?___, ___?___, 3, 9, 27. (Lesson 11–3)

52. Approximate the real zeros of $g(x) = x^4 - 4x^2 + 3$ to the nearest tenth. (Lesson 8–3)

53. Write $f(m) = (2m - 5)^2$ in quadratic form. (Lesson 6–1)

54. Simplify $\sqrt[4]{5} + 6\sqrt[4]{5} - 2\sqrt[4]{5}$. (Lesson 5–6)

55. Which of the points, $(0, 0)$, $(1, 2)$, or $(-3, 1)$, satisfy the inequality $x + 2y \le 7$? (Lesson 2–7)

56. Find the slope of a line that is perpendicular to the line whose equation is $x = 4y + 7$. (Lesson 2–3)

57. Solve $2 \le \frac{x}{3} + 5 \le 13$. (Lesson 1–7)

SELF TEST

1. At the Burger Bungalow, you can order your hamburger with or without cheese, with or without onions or pickles, and either rare, medium, or well-done. (Lesson 12–1)
   a. What type of pictorial representation could you use to calculate the number of choices?
   b. How many different hamburgers are possible?

2. For a particular model of car, a dealer offers 6 versions of that model, 18 body colors, and 7 upholstery colors. How many different possibilities are available for that model? (Lesson 12–1)

3. Five algebra and four geometry books are to be arranged on a shelf. How many ways can they be arranged if all the algebra books must be together? (Lesson 12–2)

4. Government How many ways can the 100 United States senators seat themselves in a 100-seat auditorium if there are no restrictions? Write your answer in factorial form. (Lesson 12–2)

Determine whether each situation involves a permutation or combination. (Lesson 12–3)

5. 8 guests seated around a table for dinner

6. a hand of 5 cards from a standard deck of cards

Evaluate each expression. (Lessons 12–2 and 12–3)

7. $P(12, 3)$ **8.** $C(8, 3)$

9. How many 6-person volleyball teams can be chosen from a group of 13 athletes? (Lesson 12–3)

10. Games In bridge, all 52 cards in a deck are dealt among four players. How many different hands are possible? Write your answer in factorial form. (Lesson 12–3)

Multiplying Probabilities

What YOU'LL LEARN

- To find the probability of two or more independent or dependent events.

Why IT'S IMPORTANT

You can multiply probabilities to solve problems involving spelling and transportation.

APPLICATION

Transportation

One study found that when an airplane is more than 80% full, there is a greater chance the flight will not depart on schedule because of the time it takes to stow all the carry-on bags. As a result, the flight may not arrive on schedule. A travel agent informs a customer making flight reservations that the probability that her flight to Nashville will arrive on schedule is 85%. Her flight to Denver is on a different airline and has an 80% chance of arriving on schedule. What is the probability that both her flights will arrive on schedule?   *This problem will be solved in Example 1.*

If there are two events, as in the application above, you can find the probability of *both* events occurring if you know the probability of *each* event occurring. You can use an *area diagram* to model the probability of two events occurring at the same time.

MODELING MATHEMATICS

Area Diagrams

Suppose there are 1 red and 3 blue pens and 1 black and 2 yellow pencils in a drawer. The area diagram below represents the probabilities of choosing pairs of pens and pencils if one of each is chosen at random. For example, rectangle A represents drawing 1 blue pen and 1 yellow pencil.

Pens

	Blue $\frac{3}{4}$	Red $\frac{1}{4}$
Pencils — Yellow $\frac{2}{3}$	A	B
Black $\frac{1}{3}$	C	D

P(blue pen and yellow pencil)

$= P$(blue pen) $\cdot$ P(yellow pencil)

$= \frac{3}{4} \cdot \frac{2}{3}$ or $\frac{1}{2}$

Your Turn

a. Find the probabilities of rectangles B, C, and D and explain what each area represents.

b. What is the length and width of the whole square? What is the area? Why does the area necessarily have to have this value?

c. Suppose you have a bouquet of roses and daisies: 1 lavender, 2 red, and 3 pink roses and 1 yellow and 3 white daisies. Make an area diagram that represents the probabilities of randomly selecting 1 rose and 1 daisy from the bouquet. Label the diagram and describe what each rectangle represents.

Since your first choice (a pen) does not affect your second choice (a pencil), these outcomes are *independent*.

Probability of Two Independent Events	If two events *A* and *B* are independent, then the probability of both events occurring is found as follows. $P(A \text{ and } B) = P(A) \cdot P(B)$

Example 1

Refer to the application at the beginning of the lesson. The probability that the first flight arrives on schedule is 85% or $\frac{17}{20}$. The probability that the second flight arrives on schedule is 80% or $\frac{4}{5}$. What is the probability that both flights arrive on schedule?

The two events are independent since the outcome of one flight does not affect the outcome of the other flight. Let A be the event that the first flight arrives on schedule. Let B be the event that the second flight arrives on schedule.

$$P(A \text{ and } B) = P(A) \cdot P(B)$$

$$= \frac{17}{20} \cdot \frac{4}{5} \text{ or } \frac{17}{25}$$

The probability of both flights arriving on schedule is $\frac{17}{25}$ or 68%.

Example 2

Every Friday, the physical education classes go bowling at a local bowling alley. On one shelf at the bowling alley, there are 5 black and 3 green bowling balls. One student selects a bowling ball at random from the shelf and then puts it back because it is too heavy. A second student then selects a ball from the same shelf. What is the probability that each student picked a green bowling ball?

The events are independent since the first bowling ball is placed back on the shelf. The outcome of the second selection is not affected by the results of the first selection.

$$P(\text{both green}) = P(\text{green}) \cdot P(\text{green})$$

$$= \frac{3}{8} \cdot \frac{3}{8} \text{ or } \frac{9}{64}$$

The probability that both students selected a green bowling ball is $\frac{9}{64}$ or about 0.14. You can verify the result with a tree diagram.

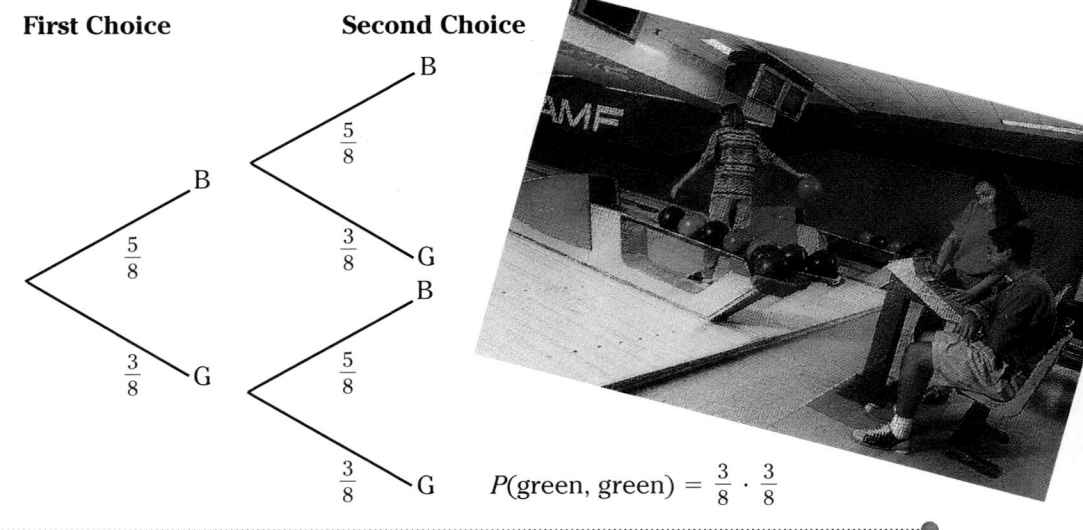

First Choice   **Second Choice**

$P(\text{green, green}) = \frac{3}{8} \cdot \frac{3}{8}$

In Example 2, what is the probability that both students selected a green bowling ball if the first selection is not put back on the shelf? These events are dependent because the outcome of the first event affects the second selection.

Suppose the first selection is green.

first selection   *second selection*

$P(\text{green}) = \dfrac{3}{8}$ $P(\text{green}) = \dfrac{2}{7}$ *Notice that when the green bowling ball is removed, there is not only one less green ball but also one less ball on the shelf.*

$P(\text{both green}) = P(\text{green}) \cdot P(\text{green following green})$

$$= \dfrac{3}{8} \cdot \dfrac{2}{7} \text{ or } \dfrac{3}{28}$$

The probability that both students selected a green ball is $\dfrac{3}{28}$ or about 0.11.

Probability of Two Dependent Events	**If two events *A* and *B* are dependent, then the probability of both events occurring is found as follows.** **$P(A \text{ and } B) = P(A) \cdot P(B \text{ following } A)$**

Example 3

Mark has 2 quarters, and he needs 50¢ more to pay the turnpike toll. There are 4 dimes, 8 quarters, and 9 nickels in his glove compartment. If he reaches in and selects two coins at random without replacing the first one, find the probability of each event described below.

Because the first coin is not replaced, these events are dependent. Therefore, $P(A \text{ and } B) = P(A) \cdot P(B \text{ following } A)$. Let *d* represent the dime, *q* the quarter, and *n* the nickel.

a. a quarter, then a nickel

$P(q, \text{ then } n) = P(q) \cdot P(n \text{ following } q)$

$P(q, \text{ then } n) = \dfrac{8}{21} \cdot \dfrac{9}{20} \text{ or } \dfrac{6}{35}$

The probability is $\dfrac{6}{35}$ or about 0.17.

b. two quarters

$P(q, \text{ then } q) = P(q) \cdot P(q \text{ following } q)$

$P(q, \text{ then } q) = \dfrac{8}{21} \cdot \dfrac{7}{20} \text{ or } \dfrac{2}{15}$

The probability is $\dfrac{2}{15}$ or about 0.13.

Example 4

From a deck of 52 cards, 3 cards are randomly chosen. They are a 10, a jack, and another 10, in that order.

a. Find the probability of this event occurring if the cards are replaced after each selection.

When the cards are replaced, the events are independent.

$P(10, \text{jack}, 10) = P(10) \cdot P(\text{jack}) \cdot P(10)$

$P(10, \text{jack}, 10) = \dfrac{4}{52} \cdot \dfrac{4}{52} \cdot \dfrac{4}{52} \text{ or } \dfrac{1}{2197}$

The probability is $\dfrac{1}{2197}$ or about 0.0005.

b. Find the probability of the event occurring if the cards are not replaced.   *Will this probability be greater or less than when the cards are replaced? Why?*

When the cards are not replaced, the events are dependent.

$P(10, \text{jack}, 10) = P(10) \cdot P(\text{jack following } 10) \cdot P(10 \text{ following jack following } 10)$

$P(10, \text{jack}, 10) = \dfrac{4}{52} \cdot \dfrac{4}{51} \cdot \dfrac{3}{50} \text{ or } \dfrac{2}{5525}$

The probability is $\dfrac{2}{5525}$ or about 0.0004.

Communicating Mathematics

Study the lesson. Then complete the following.

1. **Explain** how to find the probability of two independent events.

2. **Describe** what is meant by *dependent events*.

3. **Determine** whether events A and B are dependent or independent if a blue die and a red die are rolled and A is the event that the blue die shows 6, and B is the event that the red die shows an even number. Explain your answer.

4. **Write** an example of two real-life events that are dependent.

5. **Examine** the problem below and explain your conclusions. You are given a bag containing 10 marbles.

   a. Ten times you draw a marble, record its color, and put it back. If you don't record any black marbles, can you conclude that there aren't any black marbles in the bag?

   b. If you do this 50 times and you don't record any black marbles, can you conclude that there aren't any black marbles in the bag?

   c. How many times do you have to repeat the drawing and replacing of marbles to be absolutely certain that there aren't any black marbles in the bag? Explain.

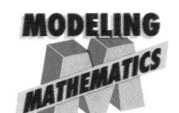

6. **Model** the problem in Example 2 using an area diagram.

Guided Practice

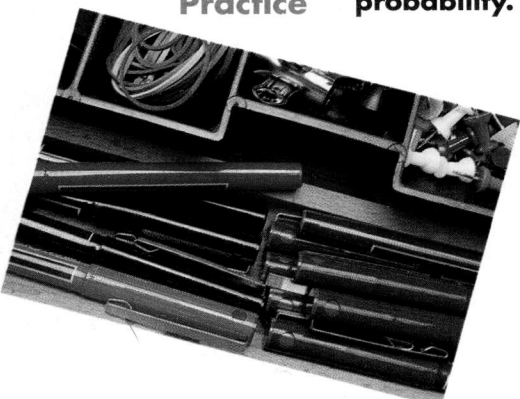

Determine if the events are *independent* or *dependent*. Then find the probability.

7. Minal has 7 blue pens, 3 black pens, and 2 red pens in her desk drawer. If she selects three pens at random with no replacement, what is the probability that she will first select a blue pen, then a black pen, and then another blue pen?

8. A green die and a red die are tossed. What is the probability that a 2 shows on the green die and a 6 shows on the red die?

9. José's wallet contains three $1 bills, four $5 bills, and two $10 bills. If three bills are selected in succession, find the probability of selecting one of each if:
   a. each bill is replaced.   b. no bills are replaced.

There are 8 movie videos, 3 exercise videos, and 5 cartoon videos on the shelf. Suppose two videotapes are to be selected at random from the shelf. Find each probability.

10. P(selecting 2 movie videos), if no replacement occurs

11. P(selecting 2 movie videos), if replacement occurs

12. P(selecting an exercise video, then a cartoon video), if no replacement occurs

Practice

Determine if the events are *independent* or *dependent*. Then find the probability.

13. There are 3 glasses of diet cola and 5 glasses of regular cola on the counter. Susan drinks 2 of them at random. What is the probability that she drank 2 glasses of diet cola?

14. A bowl contains 4 peaches and 5 apricots. Monica randomly selects one, puts it back, and then randomly selects another. What is the probability that both selections were apricots?

15. When Tricia plays her video game, the odds are 3 to 4 that she will reach the highest level of the game. What is the probability that she will reach the highest level the next four games?

The Scrabble® tiles *A*, *B*, *G*, *I*, *M*, *R*, and *S* are placed face down in the lid of the game. If two tiles are chosen at random, find each probability.

16. P(selecting 2 consonants), if replacement occurs

17. P(selecting 2 consonants), if no replacement occurs

18. P(selecting the same letter twice), if no replacement occurs

19. Suppose you roll one red die and one green die and get a sum of 8.
    a. List the different ways in which this can occur.
    b. Suppose you know the sum is 8 but not the number on each die. Explain why the probability that you rolled two 4s would be $\frac{1}{5}$.

Corinne takes her 3-year-old son into an antique shop. There are 4 statues, 3 picture frames, and 3 vases on a shelf. The 3-year-old accidentally knocks 2 items off the shelf and breaks them. Find each probability.

20. P(breaking 2 vases)

21. P(breaking 2 statues)

22. P(breaking a picture frame, then a vase)

23. P(breaking a picture frame and a vase)

24. Suppose you spin the spinner at the right two times.
    a. Sketch a tree diagram showing all of the possibilities and use it to find the probability of spinning a red and then a blue.
    b. Sketch an area diagram of the outcomes.
    c. Shade the region on your area diagram corresponding to getting the same color twice.
    d. What is the probability that you get the same color on both spins?
    e. If you know that you got the same color twice, what is the probability that the color was red?

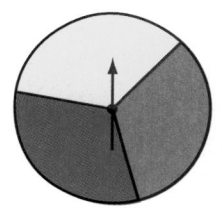

Two dice are rolled. Find each probability.

25. $P(2 \text{ and } 3)$

26. $P(\text{no } 6\text{s})$

27. $P(\text{two } 4\text{s})$

28. $P(1 \text{ and any number})$

29. $P(\text{two numbers alike})$

30. $P(\text{two different numbers})$

31. A bag contains 7 red, 4 blue, and 6 yellow marbles. If 3 marbles are selected in succession, find the probability that all three are different colors if:

  a. no replacement occurs.

  b. replacement occurs each time.

32. A box of chocolates contains 8 cherry cremes, 3 lemon cremes, and 5 marshmallow cremes. Arlene randomly chooses three chocolates. Find the probability that she will select one of each if:

  a. no chocolates are replaced.

  b. each chocolate is replaced.

A standard deck of 52 cards contains 4 suits of 13 cards each. Find each probability if 13 cards are drawn and no replacement occurs.

33. $P(\text{all clubs})$

34. $P(\text{all black cards})$

35. $P(\text{all one suit})$

36. $P(\text{all face cards})$

37. In a game using the spinners below, you are allowed to move around the game board if you get your color on *both* spinners.

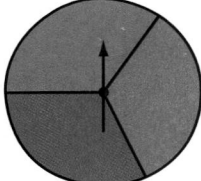 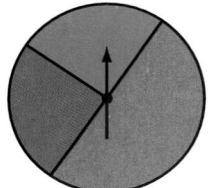

  a. Chet always chooses the color blue because that is his favorite color. Use an area model to find the probability that Chet can move his marker.

  b. Hoshi always chooses red. What is the probability that she can move?

  c. How many outcomes are there in which no one gets to move? Name them.

Critical Thinking

38. On a recent *Today* show, two women were profiled who had known each other since childhood in England. After one of the women married, she moved to California with her husband, and the women lost contact for 50 years. Then one day, while standing in line at a restaurant in California, the women struck up a conversation and each of them realized that she was speaking to her long-lost friend. After listening to their story, Katie Couric, the host of *Today,* pointed out that the chances of this happening must be "one in a million." One of the women, however, said that she believed the chances were probably closer to "one in a billion." Describe the steps you would take to compute the odds of this occurring. Be sure to list all of the factors that would have to be considered.

39. Literature   The following quote is from *The Mirror Crack'd*, which was written by Agatha Christie in 1962.

> "I think you're begging the question," said Haydock, "and I can see looming ahead one of those terrible exercises in probability where six men have white hats and six men have black hats and you have to work it out by mathematics how likely it is that the hats will get mixed up and in what proportion. If you start thinking about things like that, you would go round the bend. Let me assure you of that!"

a. If the twelve white and black hats are all mixed up and each man randomly chooses a hat, what is the probability that the first three men get their own hats?

b. Find the probability that the first five men get their own hats, if the hats are replaced after each selection.

40. Spelling   In 1994, Ned Andrews won the 67th National Spelling Bee by spelling the word "antediluvian." Suppose Ned has a 93% chance of spelling any given word in a contest correctly.

a. What is the probability that he spells the first five words in a contest correctly?

b. What is the probability that he spells the first three words correctly and then misspells the fourth and fifth words?

c. If each contestant is given 30 words to spell, what is the probability that Ned will spell all of his words right?

41. Probability   A gumball machine contains 7 red gumballs, 8 orange gumballs, 9 purple gumballs, 7 white gumballs, and 5 yellow gumballs. Tyson had three quarters with which to buy three gumballs. Find each probability.   (Lesson 12–4)

a. P(3 red gumballs)

b. P(2 white gumballs, 1 purple gumball)

c. P(1 purple gumball, 1 orange gumball, 1 yellow gumball)

42. Find the sum of $50 + 33 + 16 + \ldots + (-52)$.   (Lesson 11–2)

43. Solve $\dfrac{1}{y+1} - \dfrac{3}{y-3} = 2$.   (Lesson 9–5)

44. Solve $x^2 - 22x = -117$ by factoring.   (Lesson 6–2)

45. Archaeology   Since carbon-14 is present in all living organisms and decays at a predictable rate after death, archaeologists use the amount of carbon-14 left in a fossil to estimate the age of the fossil. This is commonly called *carbon dating*. The approximate number of milligrams A of carbon-14 left in a fossil after 5000 years can be found using the formula

$A = A_0 \, (2.7)^{-\frac{3}{5}}$, where A_0 is the initial amount of carbon-14 in the organism. Find the amount of carbon-14 left in an organism that contained 500 milligrams of carbon-14.   (Lesson 5–7)

46. Graph the system of equations and state its solution. Then state whether the system is *consistent and independent, consistent and dependent,* or *inconsistent.*   (Lesson 3–1)

$3x - y = 4$

$9x - 6 = 3y$

47. What is the slope of a line perpendicular to the line that passes through $(7, 4)$ and $(0, -4)$?   (Lesson 2–3)

48. Find the value of $12a^2 + bc$ if $a = 3$, $b = 7$, and $c = -2$.   (Lesson 1–1)

Adding Probabilities

What YOU'LL LEARN

- To find the probability of mutually exclusive or inclusive events.

Why IT'S IMPORTANT

You can add probabilities to solve problems involving meteorology and recycling.

APPLICATION
Meteorology

The Born Loser®

THE BORN LOSER reprinted by permission of Newspaper Enterprise Association, Inc.

In the cartoon above, the forecaster predicted a 100% chance of rain for the weekend. Do you think this forecast is correct? Why or why not? *You will solve this problem in Exercise 4.*

When there are two events, it is important to understand how they are related before finding the probability of one or the other event occurring. Suppose you draw a card from a standard deck of 52 cards. What is the probability of drawing a jack or a queen? Since a card cannot be both a jack *and* a queen, the events are **mutually exclusive**. That is, the two events cannot occur at the same time. The probability of drawing a jack or a queen is found by adding their individual probabilities.

$$P(\text{drawing a jack or queen}) = P(\text{drawing a jack}) + P(\text{drawing a queen})$$
$$= \frac{4}{52} + \frac{4}{52}$$
$$= \frac{8}{52} \text{ or } \frac{2}{13}$$

The probability of drawing a jack or a queen is $\frac{2}{13}$.

Probability of Mutually Exclusive Events	If two events, *A* and *B*, are mutually exclusive, then the probability that either *A* or *B* occurs is the sum of their probabilities. $$P(A \text{ or } B) = P(A) + P(B)$$

What is the probability of drawing a queen or a diamond from a deck of cards? Since it is possible to draw a card that is both a queen and a diamond, these events are *not* mutually exclusive. They are called **inclusive** events. In this case, you must adjust the formula for mutually exclusive events much like you did to account for duplication in some permutations.

$P(\text{queen})$	$P(\text{diamond})$	$P(\text{diamond queen})$
$\frac{4}{52}$	$\frac{13}{52}$	$\frac{1}{52}$
1 queen in each suit	*diamonds*	*queen of diamonds*

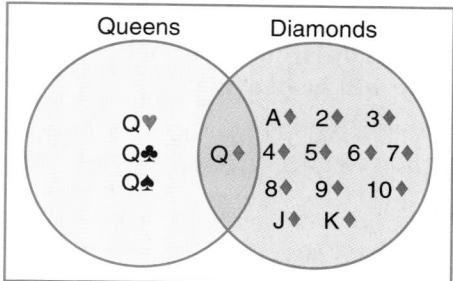

Queens   Diamonds

Q♥
Q♣   Q♦ A♦ 2♦ 3♦
Q♠   4♦ 5♦ 6♦ 7♦
                8♦ 9♦ 10♦
                J♦ K♦

The probability of drawing a queen is counted twice, once for a queen and once for a diamond. To find the correct probability, you must subtract P(queen of diamonds) from the sum of their individual probabilities.

P(queen or diamond) = P(queen) + P(diamond) − P(queen of diamonds)

$$= \frac{4}{52} + \frac{13}{52} - \frac{1}{52} \text{ or } \frac{4}{13}$$

The probability of drawing a queen or a diamond is $\frac{4}{13}$.

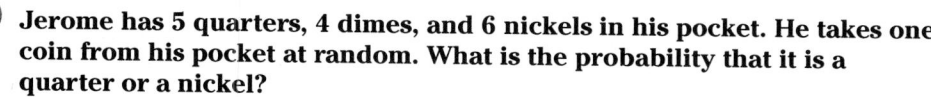

Probability of Inclusive Events	**If two events, A and B, are inclusive, then the probability that either A or B occurs is the sum of their probabilities decreased by the probability of both occurring.**
	$$P(A \text{ or } B) = P(A) + P(B) - P(A \text{ and } B)$$

Example ❶

Jerome has 5 quarters, 4 dimes, and 6 nickels in his pocket. He takes one coin from his pocket at random. What is the probability that it is a quarter or a nickel?

These are mutually exclusive events since a coin cannot be a quarter *and* a nickel. Since P(quarter and nickel) = 0, find the sum of the individual probabilities.

P(quarter or nickel) = P(quarter) + P(nickel)

$$= \frac{5}{15} + \frac{6}{15} \text{ or } \frac{11}{15}$$

The probability of selecting a quarter or a nickel is $\frac{11}{15}$.

Example ❷

APPLICATION

Recycling

F Y I

Every two weeks, Americans throw away enough glass bottles and jars to fill the twin towers of the World Trade Center in New York.

In 1973, Americans recycled only 15% of all aluminum cans. Today, 64% or 59.5 billion cans are recycled each year, many through curbside recycling. In one community, a survey of 300 people was conducted to determine how many would participate in a curbside recycling program. Of the people surveyed, 134 said they would recycle aluminum cans, and 108 said they would recycle glass. Of those people, 62 said they would recycle both. If a member of the community were selected at random, what is the probability that he or she would participate in a community recycling program by recycling aluminum *or* glass?

Since it is possible to recycle both aluminum and glass, these events are inclusive.

P(recycle aluminum) = $\frac{134}{300}$ P(recycle glass) = $\frac{108}{300}$

P(recycle aluminum and glass) = $\frac{62}{300}$

P(recycle aluminum or glass) = $\frac{134}{300} + \frac{108}{300} - \frac{62}{300}$ or $\frac{180}{300}$

The probability that a person chosen would recycle aluminum *or* glass is $\frac{180}{300}$ or $\frac{3}{5}$.

Example ③ There are 5 boys and 6 girls on the yearbook staff. A committee of 5 people is being selected at random to design the front cover of the book. What is the probability that the committee will have at least 3 boys?

At least 3 boys means that the committee may have 3, 4, or 5 boys. It is not possible to select a group of 3 boys, a group of 4 boys, and a group of 5 boys all in the same 5-member committee.

$P(\text{at least 3 boys}) = P(\text{3 boys}) + P(\text{4 boys}) + P(\text{5 boys})$

$$= \underset{\text{3 boys, 2 girls}}{\frac{C(5, 3) \cdot C(6, 2)}{C(11, 5)}} + \underset{\text{4 boys, 1 girl}}{\frac{C(5, 4) \cdot C(6,1)}{C(11, 5)}} + \underset{\text{5 boys, 0 girls}}{\frac{C(5, 5) \cdot C(6, 0)}{C(11, 5)}}$$

$$= \frac{150}{462} + \frac{30}{462} + \frac{1}{462} \text{ or } \frac{181}{462}$$

The probability of at least 3 boys on the committee is $\frac{181}{462}$ or about 0.392.

CHECK FOR UNDERSTANDING

Communicating Mathematics

Study the lesson. Then complete the following.

1. **Describe** the difference between *mutually exclusive* and *inclusive* events.

2. **Write** an example of three inclusive events that could occur in everyday life.

3. **Draw** a Venn diagram to illustrate the events in Example 2.

 MATH JOURNAL

4. Refer to the comic at the beginning of the lesson.
   a. Why is the forecaster's prediction incorrect?
   b. What do you need to know to find the correct probability of rain for the weekend?

Guided Practice

5. Determine if each event of drawing a card from a standard deck of cards is *mutually exclusive* or *inclusive*. Then find the probability.
   a. $P(5 \text{ or ace})$
   b. $P(\text{jack or diamond})$

6. Six coins are dropped onto the floor. Find each probability.
   a. $P(\text{at least 4 heads})$
   b. $P(\text{3 tails or 2 heads})$
   c. $P(\text{4 tails or 1 head})$
   d. $P(\text{all heads or all tails})$

7. The letters of the alphabet are placed in a bag. What is the probability of selecting a vowel or a letter from the word *equation*?

8. **School** The enrollment at Southburg High School is 1400. Suppose 550 students take French, 700 take algebra, and 400 take both French and algebra. What is the probability that a student selected at random takes French or algebra?

Practice

Determine if the events in Exercises 9–12 are *mutually exclusive* or *inclusive*. Then find each probability.

9. There are 3 physics books, 4 math books, and 2 history books on a shelf. If a book is randomly selected, what is the probability of selecting a physics book or a history book?

10. A card is drawn from a deck of cards. What is the probability that it is a black card or a face card?

11. One die is tossed. What is the probability of tossing a 5 or a number greater than 3?

12. In the drama club, 7 of the 20 girls are seniors, and 4 of the 14 boys are seniors. What is the probability of randomly selecting a boy or a senior to represent the drama club at a national performing arts symposium?

Juanita has 9 rings in her jewelry box. Five are gold and 4 are silver. If she randomly selects 3 rings to wear to a party, find each probability.

13. P(exactly 2 silver)

14. P(all 3 gold or all 3 silver)

15. P(at least 2 gold)

16. P(at least 1 silver)

Two cards are drawn from a standard deck of 52 cards. Find each probability.

17. P(both kings or both black)

18. P(both kings or both face cards)

19. P(both face cards or both red)

20. P(both either red or a king)

Seven women and six men walk into a computer store at the same time. There are five salespeople available to help them. Find the probability that a salesperson will first help:

21. P(4 women, 1 man or 4 men, 1 woman)

22. P(3 women, 2 men or 3 men, 2 women)

23. P(all women or all men)

24. P(at least 3 women)

The numbers 1 through 30 are written on Ping-Pong™ balls and placed in one wire cage. The numbers 20 through 45 are also written on Ping-Pong™ balls and placed in a different wire cage. One ball is chosen at random from each spinning cage. Find each probability.

25. P(each is a 25)

26. P(neither is a 20)

27. P(at least one is a 30)

28. P(each is greater than 15)

Critical Thinking

29. Suppose there are three inclusive events, A, B, and C. List all the events you would need to consider in order to calculate $P(A \text{ or } B \text{ or } C)$ and describe how you would calculate the probability.

30. The problem below appeared in Lewis Carroll's book, *Pillow-Problems Thought Out During Sleepless Nights*.

> A bag contains one counter, known to be either white or black. A white counter is put in, the bag shaken, and a counter drawn out, which proves to be white. What is now the chance of drawing a white counter?

Analyze the problem and explain how you arrived at your answer.

Applications and Problem Solving

31. International Volunteers During the 1990s, approximately 6000 people volunteered to serve in the Peace Corps. If 1800 of the volunteers were women in their 20s, use the graph at the right to find the probability that a Peace Corps volunteer in the 1990s was either a woman or a person in his or her 20s.

Peace Corps Volunteers

Women	52%
In their 20s	74%
Single	93%
Holders of a bachelor's degree	96%

Source: Peace Corps

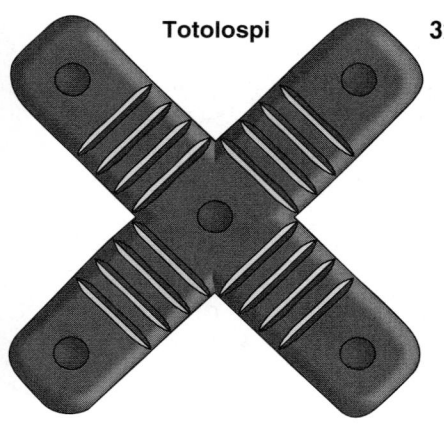

Totolospi

32. World Cultures *Totolospi* is a Hopi game of chance that is played by adults as well as children. The players use cane dice, which have both a flat side and a round side, and a counting board inscribed in stone, like the one shown at the left. When tossing 3 cane dice, if three round sides land up, the player advances 2 lines. If three flat sides land up, the player advances 1 line. If a combination is thrown, the player loses a turn. The winner is the first to reach the opposite end of the arm on which he or she is playing.

a. Draw a tree diagram showing all of the possibilities when throwing 3 cane dice.

b. Find each probability.
   P(advancing 2 lines)
   P(advancing 1 line)
   P(advancing at least 1 line)
   P(losing a turn)

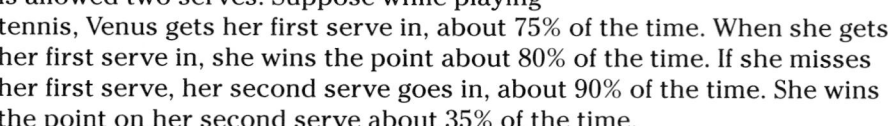

33. Tennis Fourteen-year-old Venus Williams learned to play tennis at a public park in Compton, California. She now competes in junior tennis tournaments all around the country. On each point in tennis, a player is allowed two serves. Suppose while playing tennis, Venus gets her first serve in, about 75% of the time. When she gets her first serve in, she wins the point about 80% of the time. If she misses her first serve, her second serve goes in, about 90% of the time. She wins the point on her second serve about 35% of the time.

a. Draw a tree diagram of the situation.

b. Find the probability that Venus Williams wins a point when she is serving.

c. If you know she won a point while serving, what is the probability that she made her first serve?

34. Law Enforcement A law enforcement agency hired a technical researcher to find the following probabilities concerning drivers.

Probability of:	
A driver being intoxicated	0.02
An intoxicated driver:	
having an unimpeded trip	0.99911
having an accident	0.00045
being arrested	0.00044
having his case dismissed	0.30
being convicted after arrest	0.70
A driver not being intoxicated	0.98
An unintoxicated driver:	
having an unimpeded trip	0.99984
having an accident	0.00016

Find the following probabilities. Round answers to seven decimal places. (Lesson 12–5)

a. P(being intoxicated and having an unimpeded trip)

b. P(being unintoxicated and having an unimpeded trip)

c. P(being intoxicated, arrested, and convicted)

d. P(being intoxicated, arrested, and dismissed)

35. Probability A gumball machine contains 7 red gumballs, 8 orange gumballs, 9 purple gumballs, 7 white gumballs, and 5 yellow gumballs. Tyson had three quarters with which to buy three gumballs. Find each probability. (Lesson 12–4)

a. P(1 red gumball, then another red gumball, then another red gumball)

b. P(1 purple gumball, then 1 orange gumball, then 1 yellow gumball)

c. P(1 white gumball, then another white gumball, then 1 purple gumball)

36. Find the 57th term of the arithmetic sequence 6, 15, 24, 33, ... (Lesson 11–1)

37. Simplify $\dfrac{\frac{3x+5}{3x+1}-2}{3+\frac{3x}{1-2x}}$. (Lesson 9–3)

38. Find $f(2)$ for $f(x) = -3x^3 + 2$. (Lesson 8–1)

39. History The elliptical chamber in the United States Capitol building is 46 feet wide and 96 feet long. (Lesson 7–4)

a. Write an equation to describe the shape of the room. Assume that it is centered at the origin and the major axis is horizontal.

b. John Quincy Adams discovered that he could overhear the conversations being held at the opposing party leader's desk if he stood in a certain spot in the elliptical chamber. Describe the position of the desk and how far Adams had to stand to overhear.

40. Solve $\begin{bmatrix} 3 & -5 \\ 5 & -7 \end{bmatrix} \cdot \begin{bmatrix} x \\ y \end{bmatrix} = \begin{bmatrix} 4 \\ 8 \end{bmatrix}$ by using inverse matrices. (Lesson 4–6)

Binomial Experiments and Simulations

What YOU'LL LEARN

- To use binomial experiments to find probabilities, and
- to use simulation to solve various probability problems.

Why IT'S IMPORTANT

You can use binomial experiments to solve problems involving basketball and biology.

APPLICATION
Basketball

In 1995, Arkansas played UCLA for the NCAA championship. During the regular season, Clint McDaniel, one of Arkansas' top scorers, made about 7 out of every 9 free throws he attempted. Thus, the probability that McDaniel makes a free throw is $\frac{7}{9}$. In the championship game, McDaniel attempted 4 free throws. He made 3 and missed 1. Based on his regular season record, what was the probability of making 3 of 4?

Let S stand for scoring when he attempts a free throw. Let M stand for missing when he attempts a free throw.

The possible ways of scoring on 3 free throws and missing 1 free throw are shown at the right. This shows the combination of 4 things (free throws) taken three at a time (scores), or $C(4, 3)$.

M,	S,	S,	S
S,	M,	S,	S
S,	S,	M,	S
S,	S,	S,	M

LOOK BACK

You can refer to Lesson 11-8 for more information on binomial expansion.

The terms of the binomial expansion of $(S + M)^4$ can be used to find the probabilities of each combination of scores and misses.

$$(S + M)^4 = S^4 + 4S^3M + 6S^2 M^2 + 4SM^3 + M^4$$

Coefficient	Term	Meaning
$C(4, 4) = 1$	S^4	1 way to score all 4 times
$C(4, 3) = 4$	$4S^3M$	4 ways to score 3 times and miss 1 time
$C(4, 2) = 6$	$6S^2 M^2$	6 ways to score 2 times and miss 2 times
$C(4, 1) = 4$	$4SM^3$	4 ways to score 1 time and miss 3 times
$C(4, 0) = 1$	M^4	1 way to miss all 4 times

The probability that McDaniel scores on a free throw is $\frac{7}{9}$. So, the probability that he misses is $\frac{2}{9}$. To find the probability of scoring 3 out of 4 free throws, substitute $\frac{7}{9}$ for S and $\frac{2}{9}$ for M in the term $4S^3M$.

$$4S^3M = 4\left(\frac{7}{9}\right)^3\left(\frac{2}{9}\right)$$

$$= \frac{2744}{6561} \text{ or about } 0.418$$

So the probability of Clint McDaniel making 3 out of 4 free throws during the championship game was about 42%, given his regular season record.

Problems that can be solved using binomial expansion are called **binomial experiments**.

Conditions of a Binomial Experiment	A binomial experiment exists *if and only if* these conditions occur. • **There are exactly two possible outcomes for any trial.** • **There is a fixed number of trials.** • **The trials are independent.** • **The probability of each trial is the same.**

Example ❶ **Dwight forgot to read the newspaper to prepare for the social studies quiz on current events, so he had to guess on all five true/false questions on the quiz. What is the probability that he will get 3 answers right and 2 answers wrong?**

There are only two possible outcomes for each question, right or wrong. There are five questions on the quiz, and they are independent. The probability of guessing an answer correctly is $\frac{1}{2}$. This problem meets all the conditions outlined above. Thus, it is a binomial experiment.

Let R represent guessing the answer correctly, and let W represent guessing the answer incorrectly. When $(R + W)^5$ is expanded, the term R^3W^2 represents 3 correct answers and 2 incorrect answers. The coefficient of R^3W^2 is $C(5, 3)$.

P(3 correct, 2 incorrect)

$\qquad = C(5, 3)R^3W^2$ *Replace R with P(R) and W with P(W).*

$\qquad = \frac{5 \cdot 4}{2 \cdot 1}\left(\frac{1}{2}\right)^3\left(\frac{1}{2}\right)^2$ or $\frac{5}{16}$ *Both P(R) and P(W) equal $\frac{1}{2}$.*

Thus, the probability that Dwight will get 3 answers right and 2 answers wrong is $\frac{5}{16}$ or about 0.313.

Example ❷ **An article in *USA Today* reported that approximately 1 out of 6 cars sold in 1994 was green. Suppose a salesperson sells 7 cars per week. What is the probability that he or she sells at least 3 green cars in a week?**

Consumerism

Explore   There are two possible outcomes for car color: green or not green. Since 1 out of 6 cars sold was green, the probability of selling a car that is green is $\frac{1}{6}$. The probability of selling a car that is *not* green is $\frac{5}{6}$.

Let G represent the probability that a car sold is green.

Let N represent the probability that a car sold is *not* green.

Plan   Look at the binomial expansion of $(G + N)^7$.

$(G + N)^7 = G^7 + 7G^6N + 21G^5N^2 + 35G^4N^3 + 35G^3N^4 +$
$\qquad\qquad 21G^2N^5 + 7GN^6 + N^7$

The probability of selling at least 3 green cars equals the sum of the probabilities of selling 3, 4, 5, 6, or all 7 green cars.

(continued on the next page)

Solve $P(\text{at least 3 green cars})$

$$= G^7 + 7G^6N + 21G^5N^2 + 35G^4N^3 + 35G^3N^4$$

$$= \left(\frac{1}{6}\right)^7 + 7\left(\frac{1}{6}\right)^6\left(\frac{5}{6}\right) + 21\left(\frac{1}{6}\right)^5\left(\frac{5}{6}\right)^2 + 35\left(\frac{1}{6}\right)^4\left(\frac{5}{6}\right)^3 + 35\left(\frac{1}{6}\right)^3\left(\frac{5}{6}\right)^4$$

$$= \frac{1}{279,936} + \frac{35}{279,936} + \frac{525}{279,936} + \frac{4375}{279,936} + \frac{21,875}{279,936} \text{ or } \frac{8937}{93,312}$$

The probability that the salesperson sells at least 3 cars is $\frac{8937}{93,312}$, or about 0.096.

Examine   Check to see if the answer makes sense. The probability of selling 0, 1, or 2 green cars is 0.904 and $1 - 0.904$ is 0.096. Thus, the answer is correct.

Up to this point in this text, all of the probabilities we have found have been **theoretical probabilities.** They are determined using mathematical methods and provide an idea of what might happen in a given situation.

Experimental probability is determined by performing tests or experiments and observing the outcomes. One method for finding experimental probability is **simulation.** In a simulation, a device is used to model the event, and you observe how the model responds to the conditions listed in a given problem. This process saves long and difficult samplings.

Example 3

Biology

While studying genetics, students in a biology class perform a lab experiment in which they each plant 4 corn seeds to determine the plants' gene combinations. The Punnett square at the right shows that green parent plants (Gg) produce green plants (GG and Gg) and albino, or white, plants (gg). What is the probability that at least one of a student's plants will be white?

	G	**g**
G	GG	Gg
g	Gg	gg

LOOK BACK

You can refer to Lesson 7-6 for information on simulation.

From the Punnett square, we know $P(\text{white}) = \frac{1}{4}$ and $P(\text{green}) = \frac{3}{4}$. We could spin a spinner like the one below four times to simulate the colors of four plants. The green section (G) represents the probability that a plant will be green. The white section (W) represents the probability that a plant will be white.

Now spin the spinner 4 times and record your results. Below are the results of 20 trials.

GGWG	GGGW	GWGW	GWWG	**GGGG**
GGGG	WGWW	GGGW	WGGG	GWGG
WWGG	**GGGG**	WGGG	GGWG	**GGGG**
GWGG	**GGGG**	**GGGG**	WGGG	WGGW

The trials in blue represent at least one white plant.

In our simulation, 14 of 20 trials yielded at least one W, or at least one white plant. Therefore, based on this simulation, the probability that at least one plant out of four will be white is $\frac{14}{20}$ or about 0.7.

Other devices can be used to create simulations. For example, to find the probability that two out of three children are boys, you could toss three coins and let heads be one gender and tails be the other or roll three dice and let odd numbers be one gender and even numbers be the other. Another way to use the dice is to let the numbers 1, 2, and 3 represent boys and 4, 5, and 6 represent girls.

MODELING MATHEMATICS

Experimental vs. Theoretical Probability

Materials: individual-size packages of candy (M&M's®, Skittles®, or Reese's Pieces®)

Work in groups of four.

- Each person should open a bag of candy, but *not* look into the bag. Each should then remove one piece of candy from the bag. Record the color of the candy on a group chart and then put the candy back in the bag. (Don't eat the candy until this experiment is finished!)
- Repeat this task 20 times and record your findings.
- Determine the experimental probability of each color for your group. Then determine the experimental probability of each color for the entire class.
- Empty your bags and count the actual number of items of each color. Determine the theoretical probability of each color for your group. Then

determine the theoretical probability of each color for the entire class.

Your Turn

a. Compare the experimental to the theoretical probabilities. Which pair of probabilities were closer to each other: your individual probabilities, your group's probabilities, or your class's probabilities? Why do you think this is the case?

b. How different do you think the results would be if you dumped a large bag of candy into a bowl on each table and conducted the experiment again?

CHECK FOR UNDERSTANDING

Communicating Mathematics

Study the lesson. Then complete the following.

1. **List** the conditions that must be satisfied for a problem to be classified as a binomial experiment.

2. **Draw** a tree diagram to answer the question in the application at the beginning of the lesson.

3. **Name** some objects that could be used to simulate a given situation.

4. **Describe** a situation in which a simulation would not be useful.

5. **Explain** how increasing the number of trials in a simulation affects the results.

6. Refer to Example 3.
   a. Use binomial expansion to find the probability that at least one of a student's plants will be white.
   b. Compare your results with those obtained by simulation.

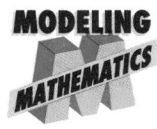

7. **Model** the following problem using any type of device you choose and use simulation to find the probability. If a family has 4 children, what is the probability that at least 2 are boys?

Determine whether each situation represents a binomial experiment. Solve those that represent a binomial experiment.

8. What is the probability of 1 head and 2 tails if Jordan tosses a coin 3 times?

9. What is the probability of Vanessa drawing 4 jacks from a deck of cards for each condition?
   a. She replaces the card each time.   b. She does not replace the card.

10. Four cans of root beer, 8 cans of diet cola, and 6 cans of orange soda are placed in a cooler. Two cans are randomly selected with replacement after the first selection. Find each probability.
    a. both diet   b. both root beer   c. both orange
    d. 1 diet, 1 root beer   e. 1 diet, 1 orange   f. 1 root beer, 1 orange

11. **Traffic Control**   The probability that a traffic light at Morse Road is green is $\frac{3}{5}$. What is the probability that exactly 4 of the next 7 cars will have to stop?

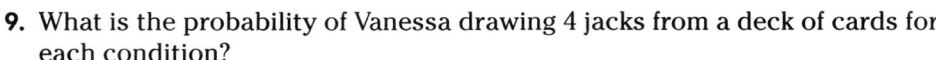

EXERCISES

Practice

Find each probability if a coin is tossed four times.

12. P(all tails)   13. P(2 heads, 2 tails)   14. P(at least 3 tails)

Find each probability if a die is rolled five times.

15. P(only one 5)   16. P(at least three 5s)   17. P(no more than two 5s)

As an apartment manager, Anne Lewis is responsible for showing prospective renters the different model apartments. When showing a model, the probability of pulling out the correct key from her set of apartment keys is $\frac{1}{4}$. If she shows 5 models in a day, find each probability.

18. P(never the correct key)   19. P(correct at least 4 times)

20. P(no more than 4 times correct)   21. P(correct exactly 3 times)

Prisana guesses at all 10 true/false questions on her economics test. Find each probability.

22. P(6 correct)   23. P(at least half correct)   24. P(all wrong)

Luis Gonzalez of the Houston Astros has a batting average of 0.300 (meaning 300 hits in 1000 times at bat). Find each probability for the next 5 times at bat.

25. P(exactly 2 hits)   26. P(at least 2 hits)   27. P(at least 4 hits)

If a thumbtack is dropped, the probability of its landing point up is 0.4. If 12 tacks are dropped, find each probability.

28. P(exactly 4 points up)   29. P(all points up)   30. P(at least 5 points up)

31. Football Luke Monroe is a quarterback on the football team at Jacksonville High School. In his freshman season, he has completed $\frac{2}{3}$ of his passes. Assume he will do the same in his sophomore year. Use a simulation to find the probability of completing at least 6 of 10 passes for an entire game if he has already completed 4 of 5 passes in the first half.

32. World Cultures The Cayuga Indians played a game of chance called *Dish*, in which they used 6 smoothed and flattened peach stones blackened on one side by burning. They placed the peach stones in a wooden bowl and tossed them. The winner was the first person to get a prearranged number of points. The table below shows the points that were given for each toss. Assume that each face (black or neutral) has an equal chance of showing up.

 a. Copy and complete the table by finding the probability of each outcome.

 b. Find the probability that a player gets at least 1 point for a toss. (*Hint:* Find $P(\text{at least 5 black}) + P(\text{at least 5 neutral})$.)

Outcomes	Points	Probability
6 black	5	
5 black, 1 neutral	1	
4 black, 2 neutral	0	
3 black, 3 neutral	0	
2 black, 4 neutral	0	
1 black, 5 neutral	1	
6 neutral	5	

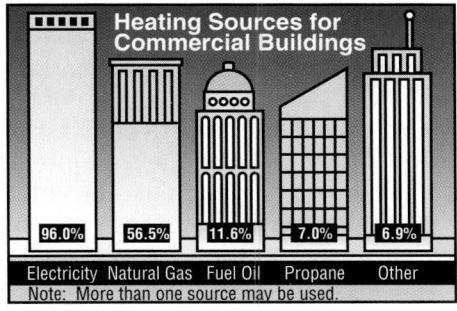

Heating Sources for Commercial Buildings

Electricity	Natural Gas	Fuel Oil	Propane	Other
96.0%	56.5%	11.6%	7.0%	6.9%

Note: More than one source may be used.

Source: Energy Information Administration, 1992

33. Energy The graph at the left displays the various heating sources for commercial buildings. Find the probability that on a street with 6 commercial buildings, at least 1 is heated with propane.

34. Probability There are 8 girls and 8 boys on the faculty advisory committee. Three are juniors. Find the probability of selecting a boy or a girl from the committee who is not a junior. (Lesson 12–6)

35. Use a calculator to find the common logarithm of 349.948, rounded to four decimal places. Then state the characteristic and the mantissa. (Lesson 10–4)

36. Graph $y = \dfrac{-3x}{x-1}$. (Lesson 9–1)

37. Use the distance formula to find the distance between $(9, 6)$ and $(8, 0)$. (Lesson 7–1)

38. Business Images Camera Shop can sell 21 KS-2 cameras a month at $120 each. The owner estimates that for each $5 decrease in price, they could sell three more of these cameras a month. The cameras cost the store $75 each. (Lesson 6–3)

 a. If the store sells all of the cameras bought each month, what should they charge for a camera to maximize profit?

 b. What is the maximum profit?

39. Find $\begin{bmatrix} -3 & 14 & 12 \\ -2 & -1 & 7 \end{bmatrix} + \begin{bmatrix} 1 & -5 & 10 \\ 22 & 13 & -8 \end{bmatrix}$. (Lesson 4–2)

Sampling and Testing Hypotheses

What YOU'LL LEARN

- To determine an unbiased sample,
- to find margins of sampling error, and
- to test hypotheses by designing and conducting experiments.

Why IT'S IMPORTANT

Sampling and hypothesis testing are used to help analyze and describe data.

APPLICATION

Sports

Do you think that professional athletes, celebrities, and entertainers are overpaid, underpaid, or are paid just about the right amount? A survey taken by the Roper Starch marketing firm in May, 1994, of a group of 1996 men and women ages 18 and over found that 87% of the respondents said that they thought pro athletes were overpaid.

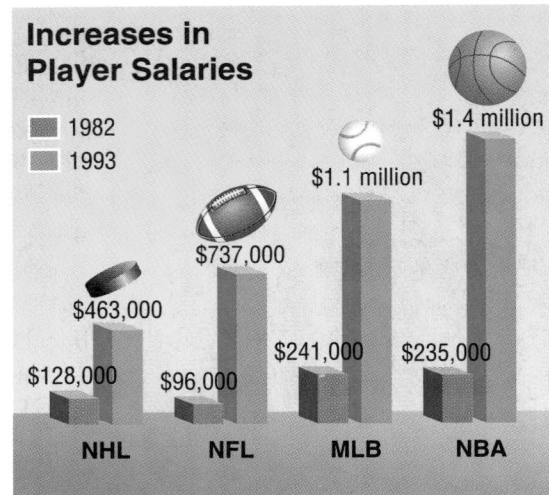

Increases in Player Salaries

1982
1993

$1.4 million
$1.1 million
$737,000
$463,000
$241,000 $235,000
$128,000 $96,000

NHL   NFL   MLB   NBA

Source: *USA TODAY* research

When opinion polling organizations or marketing firms want to find out how the public feels about some issue, they do not have the time or money to ask everyone. Instead, they obtain their results by asking a small portion of the people in the population in which they are interested. To be sure that the results are representative of the population and are unbiased, they need to make sure that this portion is a **random** or **unbiased sample** of the population. A sample of size n is random when every possible sample of size n has an equal chance to be selected.

Example **Do you think that the following methods will produce a random sample?**

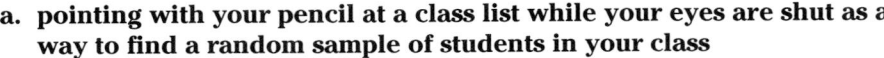

a. pointing with your pencil at a class list while your eyes are shut as a way to find a random sample of students in your class

This would probably not result in a random sample because you will tend to point toward the middle of the page. So, those at the beginning or end of the page will have less of a chance to be selected.

b. putting the names of all seniors in a hat, then drawing names from the hat to determine a random sample of seniors

This would result in a random sample because each senior will have an equal chance to be selected.

c. selecting one person whose last name begins with each letter of the alphabet to find a random sample of students in your grade

This would not result in a random sample because there are many more people with last names beginning with some letters than with others. For example, there are many more last names beginning with s or t than with q or z. Those people whose last names begin with s would have a smaller chance of being selected than those whose last names begin with q.

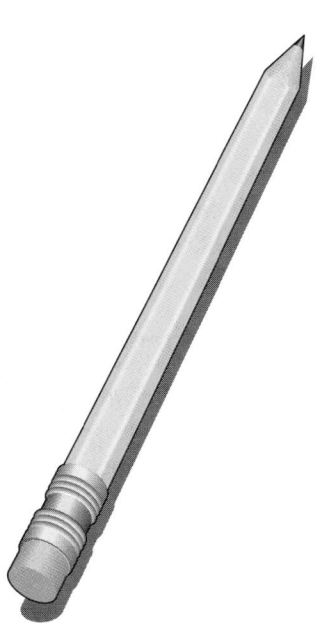

Suppose you take repeated random samples of a population. How close will the sample results be to the population results? It is reasonable to think that the sample data will not be exactly like the population data, but they should be similar. As the size of the sample increases, it more accurately reflects the population. If you sampled only three people and two prefer Brand A, you could say, "Two out of three people choose Brand A over any other brand," but you would not be giving a true picture of how the total population would respond. The **sampling error** is the difference between the sample results and the true population results.

Margin of Sampling Error	**If the percentage of people in a sample responding in a certain way is p and the size of the sample is n, then 95% of the time, the percentage of the population responding in that same way will be within $p \pm ME$, where** $$ME = 2\sqrt{\frac{p(1 - p)}{n}}.$$ **ME is called the _margin of sampling error_.**

That is, the probability is 0.95 that $p \pm ME$ will contain the true population results.

Example ② In a _Washington Post_ survey of 1003 randomly-selected adults published on May 26, 1995, 61% of those surveyed said they regret things they didn't do in their lives. What is the margin of error?

$$ME = 2\sqrt{\frac{p(1 - p)}{n}}$$

$$= 2\sqrt{\frac{0.61(1 - 0.61)}{1003}} \qquad p = 61\% \text{ or } 0.61; n = 1003$$

$$\approx 0.030802 \qquad \text{Use a calculator.}$$

0.030802 would be reported as a 3% margin of error. This means that there is a 95% probability that the value of p in the population is between $61 - 3$ or 58% and $61 + 3$ or 64% and that 58% to 64% of adults regret things they didn't do in their lives.

Example ③

Sports

Refer to the application at the beginning of the lesson. In a survey taken in May, 1994, 87% of the people surveyed stated that they thought pro athletes were overpaid. This survey had a margin of error of 4%.

a. What does the 4% indicate about the results?

b. How many people were surveyed?

a. The 4% means that there is a 95% probability that the value of p in the population is between $87 - 4$ or 83% and $87 + 4$ or 91% and that 83% to 91% of the population believes that pro athletes are overpaid.

b. $$ME = 2\sqrt{\frac{p(1 - p)}{n}}$$

$$0.04 = 2\sqrt{\frac{0.87(1 - 0.87)}{n}} \qquad ME = 0.04 \text{ and } p = 0.87$$

$$0.02 = \sqrt{\frac{0.87(0.13)}{n}} \qquad \text{Divide each side by 2.}$$

$$0.0004 = \frac{0.87(0.13)}{n} \qquad \text{Square each side.}$$

$$n = 282.75 \qquad \text{Use a calculator.}$$

Since it is not possible for 0.75 of a person to exist, round down. Thus, there were about 282 people in the survey.

A **hypothesis** is a statement to be tested. You can design an experiment to help you determine whether a hypothesis is true or false.

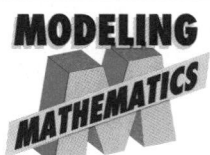

MODELING MATHEMATICS

Design an Experiment

Design an experiment to test the following hypothesis.
People react to sound and touch at the same rate.

Materials: meterstick stopwatch grid paper

Separate the class into two groups. You can measure reaction time by having someone drop a meterstick and then having someone else catch it between their fingers. The distance the stick falls will be directly proportional to their reaction time. Half of the class will investigate the time it takes to react when someone is told the stick has dropped. The other half will measure the time it takes to react when the catcher is alerted by a touch.

Step 1   Describe the variables that can be controlled.

Factors such as the length of the meterstick, the height from which it is dropped, the position of the person catching the stick, the number of practice runs, and whether to use one try or to take the average of several tries can be controlled.

Step 2   Describe the variables that should be randomized.

Factors such as whether boys or girls have a different reaction time should be randomized. Randomly assign boys and girls to each treatment.

Step 3   Conduct the experiment.

Conduct the experiment in each group and record the results. Organize the results so that they can be compared.

Step 4   Organize and summarize your results.

Based on the results of your experiment, do you think your hypothesis is true?

CHECK FOR UNDERSTANDING

Communicating Mathematics

Study the lesson. Then complete the following.

1. **Give** an example of a biased sample and a random sample.

2. **Explain** what happens to the margin of sampling error when the size of the sample *n* increases. Why does this happen?

Guided Practice

Determine whether each situation represents a random sampling. Write *yes* or *no* and explain.

3. collecting 1-digit numbers from license plates on cars on the interstate

4. surveying students in the advanced chemistry classes to determine the average time students in your school study each week

Find the margin of sampling error in Exercises 5 and 6. Explain what it indicates about the results.

5. In a survey of 520 randomly-selected high school students, 68% of those surveyed stated that they were involved in extracurricular activities at their school.

6. In a survey of 1730 randomly-selected adults, 45% agreed that the results of call-in polls are believable.

7. **Media**   According to a recent survey in *American Demographics*, 77% of Americans age 12 or older said they listen to the radio every day. Suppose the survey had a margin of error of 5%.

   a. What does the 5% indicate about the results?

   b. How many people were surveyed?

8. Design an experiment to test the following hypothesis.
   Students who eat breakfast regularly score higher on math tests than students who do not eat breakfast regularly.

EXERCISES

Practice

Determine whether each situation represents a random sampling. Write *yes* or *no* and explain.

9. obtaining a list of teenage girls' first names from your high school yearbook

10. calling every twentieth person listed in the telephone book to determine which political candidate is favored

11. asking every tenth person coming out of a health spa how many times a week they exercise to determine how often people in the city exercise

12. finding the heights of all the boys in a freshman gym class to determine the average height of all the boys in your school

13. taking a poll during lunch to find how many students in your school would participate in a car wash fund-raiser

Find the margin of sampling error in Exercises 14 and 15. Explain what it indicates about the results.

14. A poll conducted for the Robert Wood Johnson Foundation asked people to name the most serious problem facing the country. Forty-six percent of the randomly-selected people said crime. Find the margin of error if 800 people were randomly selected.

15. Although skim milk has as much calcium as whole milk, only 33% of 2406 adults surveyed in *Shape* magazine said skim milk is a good calcium source.

16. According to *Vitality* magazine, in a recent survey of randomly-selected adults who smoked, 34% of those surveyed said they have tried to quit smoking. If this survey had a margin of error of 4%, how many people were surveyed?

Programming

17. The graphing calculator program at the right creates a population of numbers with a mean of *M* and a standard deviation of *S*. It then draws a random sample of *N* data from the population and calculates the mean and standard deviation of that sample. You must enter *M*, *S*, and *N* when the program prompts you to do so.   *It may take a little longer for the program to calculate a large sample.*

   a. Run the program for a population with a mean of 50 and a standard deviation of 5 for sample sizes of 5, 10, 50, 100, and 200. Record the data in a table.

```
PROGRAM: SAMPLE
: Prompt M,S,N
: 0→A:0→B
: For (K,1,N)
: 0→R
: For(L,1,12)
: rand+R→R
: End
: iPart (S*(R−6))+M→X
: A+X→A:B+X*X→B
: End
: B−A*A/N→B
: Disp "SAMPLE MEAN=",A/N
: Disp "SAMPLE S.D.=",
  √ (B/N)
: Stop
```

(continued on the next page)

b. Find the average of the sample means and the average of the sample standard deviations. How closely do these averages match the population mean and standard deviation?

c. Look at the data for each of your samples. What conclusion can you draw about the size of the sample and how representative the sample is of the population?

Critical Thinking

18. The following excerpt appeared in *Chance* magazine in 1993.

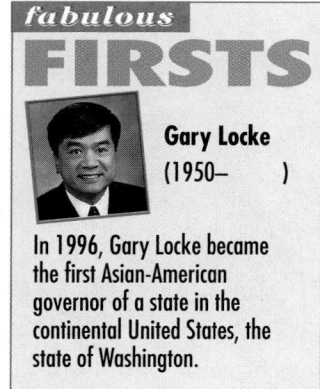

fabulous
FIRSTS

Gary Locke
(1950–   )

In 1996, Gary Locke became the first Asian-American governor of a state in the continental United States, the state of Washington.

> How big was Bill Clinton's lead on October 20, the day after the third presidential debate of the 1992 campaign? Nineteen points, reported NBC and *Wall Street Journal.* Fourteen points, said *U.S. News and World Report.* . . .And CNN gave us two answers—12 points in their poll done by the Gallup organization and 7 points in their poll conducted by Yankelovich-Clancy-Schulman. . . .News organizations report the results of their election surveys with a familiar caveat: The poll has a "margin of error," usually plus or minus three percentage points. If the survey were conducted repeatedly on the same population and if the only source of variability in the polls were random sampling, then in only 1 out of 20 tries would the results differ from those reported by more than 3%. . . .But they produced estimates of Clinton's lead that differed by as much as *12* percentage points.

List some possible explanations for the differences in these poll results.

Applications and Problem Solving

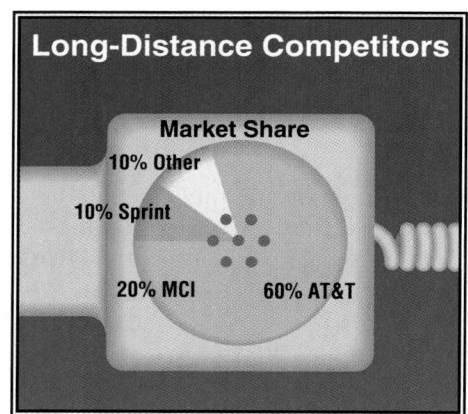

Long-Distance Competitors

Market Share
10% Other
10% Sprint
20% MCI
60% AT&T

Source: *USA TODAY* research

19. Politics   To vote in a presidential election in most states, a person must be registered at least 30 days before election day. According to the postelection validation of registration and voting, about 90% of those people who were registered actually voted on election day. Suppose this postelection survey, which has been part of the National Election Study since 1964, had a margin of error of 2%.

a. What does the 2% indicate the results?

b. How many people were surveyed?

20. Consumerism   A recent article in *USA Today* reported the percentage of telephone customers that use various long-distance companies. The data are shown at the left. Suppose 1500 telephone customers were randomly selected for the survey. Find the margin of error to the nearest tenth of a percent for each of the survey results.

a. AT&T   **b.** MCI   **c.** Sprint

21. Design an experiment to test the following hypothesis.

People who exercise regularly fall asleep quicker at night than those who don't get regular exercise.

Mixed Review

22. Probability   Chris guessed at all ten questions on a true-false test. What is the probability that all of the guesses were correct?   (Lesson 12–7)

23. Solve $2 \log_6 3 + 3 \log_6 2 = \log_6 x$.   (Lesson 10–3)

24. Use synthetic substitution to find $f(3)$ and $f(-2)$ for all $f(x) = 2x^2 - 8x + 6$.   (Lesson 8–2)

25. Solve $x^2 \le 6$.   (Lesson 6–7)

26. Solve $|m - 4| + 2 \ge 0$.   (Lesson 1–7)

VOCABULARY

After completing this chapter, you should be able to define each
term, property, or phrase and give an example or two of each.

Probability

binomial experiments (p. 753)

complements (p. 732)

experimental probability
  (p. 754)

failure (p. 732)

inclusive events (p. 747)

mutually exclusive events
  (p. 746)

odds (p. 734)

probability (p. 732)

simulation (p. 754)

success (p. 732)

theoretical probability (p. 754)

Geometry

area diagram (p. 739)

Problem Solving

solve a simpler problem
  (p. 712)

Statistics

hypothesis (p. 760)

random sample (p. 758)

sampling error (p. 759)

unbiased sample (p. 758)

Discrete Mathematics

circular permutation
  (p. 721)

combination (p. 726)

dependent events (p. 714)

fundamental counting
  principle (p. 713)

independent events (p. 713)

linear permutation (p. 718)

permutation (p. 718)

reflections (p. 722)

UNDERSTANDING AND USING THE VOCABULARY

Choose the letter of the term that best matches each statement or phrase.

1. two events whose outcomes may be the same

2. an illustration used to show the total number of possible
   outcomes

3. the number of possibilities of n objects, taken r at a time and
   defined as $C(n, r) = \dfrac{n!}{(n - r)!r!}$

4. the number of ways that n objects can be arranged in a circle and
   defined by $(n - 1)!$

5. a statement that a given population characteristic is true

6. the desired outcome of an event

7. a sample in which every member of the population has an equal
   chance to be selected

8. If one event can occur in m ways and another in n ways, then
   the number of ways that both can occur is $m \cdot n$.

9. the number of possibilities of n objects arranged in a line and
   defined by $P(n, r) = \dfrac{n!}{(n - r)!}$

10. two events in which the outcome can never be the same

11. the ratio of the number of ways an event can succeed to the number of ways it can fail

12. the difference between the sample result and the true population results

a. circular permutation

b. combination

c. fundamental counting principle

d. hypothesis

e. inclusive events

f. linear permutation

g. mutually exclusive events

h. odds

i. random sample

j. sampling error

k. success

l. tree diagram

OBJECTIVES AND EXAMPLES	REVIEW EXERCISES

Upon completing this chapter, you should be able to:

Use these exercises to review and prepare for the chapter test.

- solve problems by using the fundamental counting principle (Lesson 12–1)

  How many different license plates are possible with two letters followed by three digits?

  The license plate consists of five separate symbols, chosen one at a time. There are 26 different possibilities for each letter. There are 10 different possibilities for each number. Thus, the number of license plates possible for this country is as follows.

  $26 \cdot 26 \cdot 10 \cdot 10 \cdot 10 = 26^2 \cdot 10^3$ or 676,000

13. How many different batting orders does a girls' fast-pitch softball team of ten players have if it does not matter who bats last?

14. The letters a, c, e, g, i, and k are used to form 6-letter passwords for a movie theater security system. How many passwords can be formed if the letters can be used more than once in any given password?

15. Using all ten digits 0-9, how many 4-digit patterns can be formed if each number can only be used once?

- solve problems involving linear and circular permutations (Lesson 12–2)

  Eleven keys are to be placed on a key ring. How many different ways can the keys be placed on the key ring if the key ring has a clasp?

  Since there is a clasp, this is treated as a linear permutation. However, since the key ring can be turned over, it is still reflective. So, the number of arrangements will be half that of a linear permutation.

  $\frac{11!}{2} = 19,958,400$

How many different ways can the letters of each word be arranged?

16. LINEAR   **17.** PERMUTATION

18. CIRCULAR   **19.** REFLECTIVE

20. Four people are taking a road trip to Montreal, Canada. Two sit in the front seat and two sit in the back seat. Three of the people agree to share the driving. In how many different arrangements can the four people sit?

21. How many ways can 8 people be seated at a round table?

- solve problems involving combinations (Lesson 12–3)

  A basket contains 3 apples, 6 oranges, 7 pears, and 9 peaches. How many ways can 1 apple, 2 oranges, 6 pears, and 2 peaches be selected?

  This involves the product of four combinations, one for each type of fruit.

  $C(3, 1) \cdot C(6, 2) \cdot C(7, 6) \cdot C(9, 2)$

  $= \frac{3!}{(3-1)!1!} \cdot \frac{6!}{(6-2)!2!} \cdot \frac{7!}{(7-6)!6!} \cdot \frac{9!}{(9-2)!2!}$

  $= 3 \cdot 15 \cdot 7 \cdot 36$ or 11,340

  There are 11,340 different ways to choose the fruit from the basket.

22. A college basketball team has 13 players. In how many different ways can the coach choose the five starting players, assuming that each player can play any position?

23. A pizza shop offers 8 different toppings How many different 2-topping pizzas can be made?

24. Geometry   Find the number of diagonals in a polygon that has 16 sides.

OBJECTIVES AND EXAMPLES

• find the probability of an event (Lesson 12–4)

A bag of golf tees contains 23 red, 19 blue, 16 yellow, 21 green, 11 orange, 19 white, and 17 black tees. What is the probability that if you choose a tee from the bag at random, you will choose a green tee?

There are 21 ways to choose a green tee and $23 + 19 + 16 + 11 + 19 + 17$ or 105 ways not to choose a green tee. So, s is 21 and f is 105.

$$P(\text{green tee}) = \frac{s}{s + f}$$

$$= \frac{21}{21 + 105} \text{ or } \frac{1}{6}$$

The probability is 1 out of 6 or about 16.7%.

• find the probability of two or more independent or dependent events (Lesson 12–5)

There are 3 dimes, 2 quarters, and 5 nickels in Robert's pocket. If he reaches in and selects three coins at random without replacing any of the coins, what is the probability that he will choose a dime, then a quarter, then a nickel?

Let d = dime, q = quarter, and n = nickel.

$$P(d, \text{then } q, \text{then } n) = \frac{3}{10} \cdot \frac{2}{9} \cdot \frac{5}{8} \text{ or } \frac{1}{24}$$

The probability is $\frac{1}{24}$ or about 4.2%.

• find the probability of mutually exclusive or inclusive events (Lesson 12–6)

Trish has four $1 bills and six $5 bills. She takes three bills from her wallet at random. What is the probability that Trish will pull at least two $1 bills from her wallet?

$P(\text{at least two \$1 bills})$
$= P(\text{two \$1, one \$5}) + P(\text{three \$1, no \$5})$

$$= \frac{C(4, 2) \cdot C(6, 1)}{C(10, 3)} + \frac{C(4, 3) \cdot C(6, 0)}{C(10, 3)}$$

$$= \frac{\frac{4! \cdot 6!}{(4 - 2)!2!(6 - 1)!1!}}{\frac{10!}{(10 - 3)!3!}} + \frac{\frac{4! \cdot 6!}{(4 - 3)!3!(6 - 0)!0!}}{\frac{10!}{(10 - 3)!3!}}$$

$$= \frac{36}{120} + \frac{4}{120} \text{ or } \frac{1}{3}$$

The probability is $\frac{1}{3}$ or about 0.333.

REVIEW EXERCISES

25. There are 5 girls and 4 boys on the Bruins student council. A committee of three is to be selected at random to study the council's plans for a vending machine. What is the probability that the three selected are all girls?

26. A sports cooler for the high school football team contains several types of colas: 14 regular, 6 cherry, 10 diet, 7 diet cherry, 8 caffeine-free, and some caffeine-free diet. You pick one of the colas without looking at the type of drink. The probability that it is a diet drink is $\frac{3}{7}$. How many caffeine-free diet colas are in the sports cooler?

27. There are 28 pieces of paper, numbered 1 through 28, in a box. You pick a piece of paper, replace it, and then pick another piece of paper. What is the probability that the first number was greater than 14 and the second number was greater than 25 or less than 6?

28. For a state lottery game, plastic balls numbered 1 through 46 are placed in a drum. Six balls are drawn at random. What is the probability that all six are odd numbers, if no replacement occurs?

Determine whether each event is *mutually exclusive* or *inclusive*. Then find the probability.

29. There are 10 algebra 1 books, 15 geometry books, 20 algebra 2 books, and 15 pre-calculus books on a shelf. If a book is randomly selected, what is the probability of selecting a geometry book or a pre-calculus book?

30. A card is drawn from a standard deck of cards. What is the probability the card will be a king or less than a six?

31. A die is rolled. What is the probability of rolling a 6 or a number less than 3?

32. Two cards are drawn from a standard deck of 52 cards. What is the probability that both are jacks or both are red?

OBJECTIVES AND EXAMPLES

• use binomial experiments to find probabilities
(Lesson 12–7)

To practice for a jigsaw puzzle competition, Laura and Julian put together 4 jigsaw puzzles. The probability that Laura puts in the last piece is $\frac{3}{5}$, and the probability that Julian puts in the last piece is $\frac{2}{5}$. What is the probability that Laura will put in the last piece of at least two puzzles?

The probability equals the sum of the probabilities of putting in the last piece in 2, 3, or all 4 puzzles.

$$P = L^4 + 4L^3W + 6L^2W^2$$
$$= \left(\tfrac{3}{5}\right)^4 + 4\left(\tfrac{3}{5}\right)^3\left(\tfrac{2}{5}\right) + 6\left(\tfrac{3}{5}\right)^2\left(\tfrac{2}{5}\right)^2$$
$$= \tfrac{81}{625} + \tfrac{216}{625} + \tfrac{216}{625} \text{ or } 0.8208$$

The probability is 82.08%.

REVIEW EXERCISES

33. Find the probability of getting 7 heads in 8 tosses of a coin. The probability of success (getting a head) in a single toss is $\frac{1}{2}$. Thus, the probability of failure (getting a tail) is $\frac{1}{2}$.

34. Find the probability that a family with seven children will have exactly five boys. Assume that the probability of having a boy is $\frac{1}{2}$.

35. A die is rolled twelve times. Find the probability of each of the following.

    a. a 3 twelve times

    b. exactly one 3 in twelve tries

    c. six 3s out of twelve tries

• find margins of sampling error   (Lesson 12–8)

In a survey taken at a local high school, 75% of the student body stated that they thought school lunches should be free. This survey had a margin of error of 3%. How many people were surveyed?

$$ME = 2\sqrt{\frac{p(1 - p)}{n}}$$
$$0.03 = 2\sqrt{\frac{0.75(1 - 0.75)}{n}} \quad \textit{ME = 0.03 and p = 0.75}$$
$$n = 833\tfrac{1}{3} \quad \textit{Use a calculator.}$$

Since it is not possible for $\frac{1}{3}$ of a person to exist, there were about 833 people in the survey.

36. In a poll asking people to name their most valued freedom, 51% of the randomly selected people said it was the freedom of speech. Find the margin of sampling error if 625 people were randomly selected.

37. According to a recent survey of mothers with children who play sports, 63% of them would prefer that their children not play football. Suppose the margin of error is 4.5%. How many mothers were surveyed?

APPLICATIONS AND PROBLEM SOLVING

38. **Combination Lock**   You have forgotten the combination lock to your locker in the health club. There are 36 numbers on the lock, and the correct combination is R ☐ – L ☐ – R ☐. How many combinations may be necessary to try in order to open the lock? (*Hint:* Numbers such as 3-3-3 may be repeated.)   (Lesson 12–1)

39. **Auto Manufacturing**   An auto manufacturer produces 9 models, each available in 7 different colors, with 5 different upholstery fabrics, and 4 different interior colors. How many varieties of automobiles are available?
(Lesson 12–1)

A practice test for Chapter 12 is provided on page 923.

ALTERNATIVE ASSESSMENT

COOPERATIVE LEARNING PROJECT

Probability In this chapter, you learned ways of grouping objects by combinations and permutations. You learned how to find the probability of an event. And you learned how to predict events based on probability. Many examples were given of the ways probability is involved in our everyday lives. A lot of these examples come from the games we play.

YAHTZEE® is a game that involves the rolling of 5 six-sided dice and keeping score by how many dice show the same number. SCRABBLE® involves the random selection of wooden tiles. Each wooden tile has an inscribed letter that is assigned a point value. The object of the game is to make words and obtain high point values based on where the word is placed on the board. TRIVIAL PURSUIT® is another game played with dice. The object of this game is to roll two dice, move a game piece around the board and answer a trivia question on a given category based on where the game piece lands. As you might suspect, almost every game involves probability.

In this project, you will design and construct a game based on probability. All games have certain features in common. The object of the game is to win, so your game should have a starting point and an ending point. But not all games involve a playing board. Some games involve playing pieces, others do not. Some games are played with dice, others with spinners, and still others with marbles.

Follow these steps to design and construct your game.

- Brainstorm with your group to choose the features your game will have.
- Outline a plan you can follow.
- Use materials you can easily find. It's okay to borrow dice, spinners, marbles, and so on from other games. Be original; don't design your game so it looks like another game that already exists.
- Carry out your plan.
- Determine the probabilities of the events which will take place in your game.
- Write several paragraphs describing how you designed and play your game. Include how you constructed the game, the rules of the game, and how someone can win the game.

THINKING CRITICALLY

- Write a probability problem involving binomial expansion.
- Have your fellow students see if they can solve the problem. Make sure you can solve the problem yourself. Explain how and why you chose this problem.

PORTFOLIO

Select your favorite word problem from this chapter and place it in your portfolio with a note explaining why it is your favorite word problem.

SELF EVALUATION

We all have to cope with the unpredictability of life. When you plan a camping trip, you consider the possibility of bad weather. When you ride your bike, you consider the possibility of a flat tire. When you participate in sports, you consider the possibility of winning or losing. Understanding probability can help you make decisions. But, it does not guarantee you will always like the outcome of certain events. Adaptability to whatever events might come is just as important as understanding the probability of those same events.

Assess yourself. How adaptable are you? How do you react when the unpredictability of life "rains on your parade?" List two or three ways you can make a conscious effort to be more adaptable when you do not like the outcome of an event. Understand that knowing probability is important, but being adaptable to the unpredictability of life is equally important.

In·ves·ti·ga·tion

Scream Machines!

225-foot drop, Kennywood's Steel Phantom

MATERIALS NEEDED

- drawing paper
- file folder
- heavy cardboard
- masking tape
- protractor
- ruler
- small toy car
- straws

Roller coasters and other amusement park rides that make you scream with fear and excitement are designed to provide safe, quick rides for crowds of people, while maximizing the thrill for riders. The thrill comes from tall heights, sudden drops, high speeds, tight curves, and sensory effects (such as dark tunnels, special lighting, wind, and sounds).

As the coaster cars are pulled up the first hill by motors, potential energy *PE* (stored energy due to location) is gained by the cars. This motorized pull on the first hill is the only external energy applied to the cars so that they can complete the remainder of the ride. *PE* becomes kinetic energy *KE* (the energy of motion) as the cars "fall" down the hills, accelerated by gravity. Gravity is a vertical force, so a steep hill helps the cars gain speed.

To design a good roller coaster, you would want to maximize falling heights and include steep angles of fall. For the cars to be able to negotiate curves, certain factors must be considered, such as the speed of the cars as they approach a curve, the angle of the curve, and the banking needed to hold the cars on the track. You would also want to consider that friction during the ride constantly diminishes the potential and kinetic energy.

In this Investigation, your team of four engineers will be given the task of designing a roller coaster that will thrill the rider, fit into a designated land space, have a ride time of at least 2 minutes, and be safe.

Make an Investigation Folder in which you can store all of your work on this Investigation for future use.

Six Flags Over Texas

1 Think about the shapes of roller coasters that you have ridden or have seen in magazines, books, or amusement park advertisements. Find pictures of the different types of coasters you wish to consider. Sketch how the supporting beams and trusses of each roller coaster appear.

2 Study the designs with your team. Based on the information you have gathered, decide upon a roller coaster design that your team could build using straws, tape, and strips of lightweight cardboard. Sketch a scale model of the coaster, labeling the height of each hill, the angle of each vertical drop, and the angle of curvature for each curve. Also label the angles for positioning the trusses to support the structure.

BUILD THE ROLLER COASTER

3 Build a scale model of your coaster design using straws and masking tape as building materials to construct the supporting structure. Cut a file folder into strips for the lightweight cardboard needed for the track. Make sure the strips are wide enough to support a toy car. Use a piece of heavy cardboard measuring 2 feet by 2 feet to represent the land on which the coaster is anchored. Include a legend on the base of your model that states the scale factor.

Mean Streak, Cedar Point

4 Test your completed model by releasing a small toy car from the top of the first hill. Can the car run the entire track? If not, how should your design be altered so the car can complete the full ride?

5 Revise your model and retest the model with the car. Record your observations and corrective measures after each trial of the car.

You will continue working on this Investigation throughout Chapters 13 and 14.

Be sure to keep your tables, graphs, and other materials in your Investigation Folder.

Scream Machines! Investigation

Working on the Investigation
Lesson 13–2, p. 785

Working on the Investigation
Lesson 13–6, p. 810

Working on the Investigation
Lesson 14–1, p. 834

Working on the Investigation
Lesson 14–5, p. 859

Closing the Investigation
End of Chapter 14, p. 868

Exploring Trigonometric Functions

Objectives

In this chapter, you will:

- find values of trigonometric functions,
- solve problems by using right triangle trigonometry,
- examine solutions to problems, and
- solve triangles by using the law of sines and law of cosines.

The Sounds of Music

Type	Year Introduced
Record Player	1877
33-RPM Records	1948
45-RPM Records	1949
Transistor Radio	1957
Audio Cassette	1965
Walkman	1979
CDs	1982
CD Walkman	1986

Source: *Encyclopedia Brittanica, 1994*

Music has been an artistic expression of humans since the dawn of humanity. Every society, from the earliest bands of people spreading across the globe, to modern post-industrial countries, has created music to express its identity. Music is universal, important, and a very pleasurable activity, both for listening and for creating. How important is music in your life?

TIME *Line*

3000 B.C. Stonehenge, a circle of giant megalithic stones, is built in southern England.

A.D. 1636 Pierre de Fermat discovers the second known pair of amicable numbers: 17,296 and 18,416.

| 3500 B.C. | 3000 | 2500 | 2000 | A.D. 1620 | 1640 | 1660 | 1680 | 1700 | 1720 | 1740 | 1760 | 1780 |

2500 The oldest surviving work of literature, the *Epic of Gilgamesh*, relates the story of the most popular Mesopotamian hero.

1789 First U.S. Congress meets.

PEOPLE IN THE NEWS

Chapter Project

You probably think of music as being sounds or notes. But did you know that music can also be represented mathematically? In fact, trigonometry can be used to connect music and physics, seemingly unrelated subjects.

Do research on the ways music, trigonometry, and physics are related. Devise a creative presentation to share your research with your classmates. Some possibilities include a poster, a video, or an experiment.

One student who understands passion for music is **AnnaMaría Padilla**. She is only 16 years old, yet she is a student at the University of New Mexico earning a degree in business administration while also enrolled at Santa Fe Community College, studying for an associate of arts degree. She is a flamenco and classical guitarist who believes that teenagers should pursue their dreams with energy and enthusiasm. So strong is her conviction that she has recorded a CD and has written and published a book titled *Why Wait? Graduate!* that encourages students to pursue their educational goals. She hopes that, in sharing the secrets of her success, young people will see alternatives that they could also pursue.

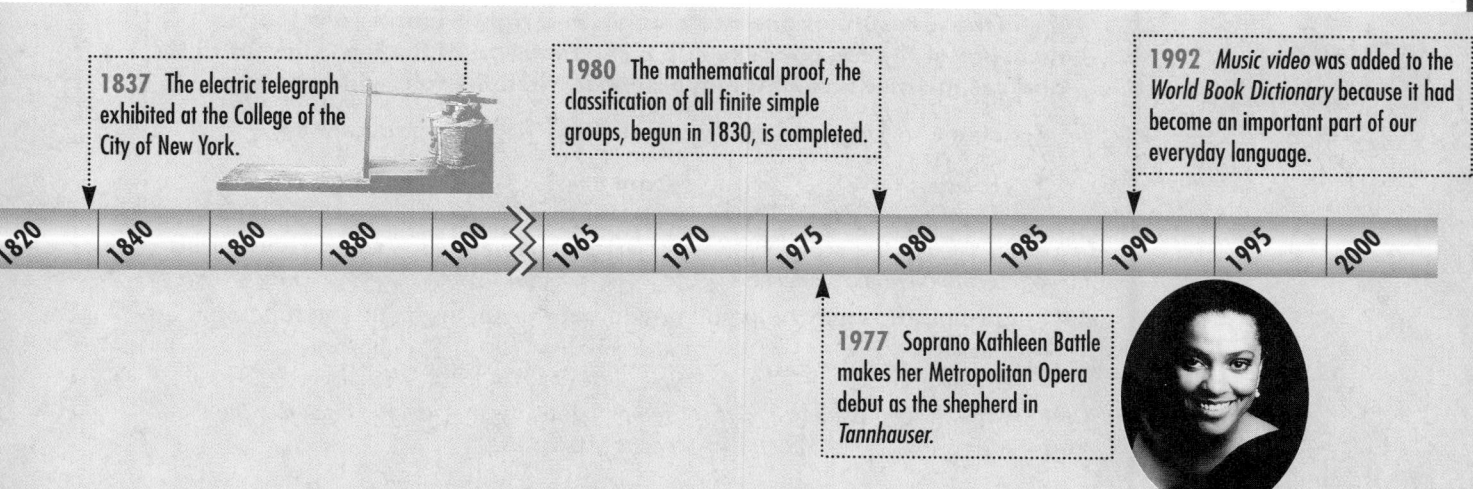

1837 The electric telegraph exhibited at the College of the City of New York.

1980 The mathematical proof, the classification of all finite simple groups, begun in 1830, is completed.

1992 *Music video* was added to the *World Book Dictionary* because it had become an important part of our everyday language.

1977 Soprano Kathleen Battle makes her Metropolitan Opera debut as the shepherd in *Tannhauser.*

13-1

An Introduction to Trigonometry

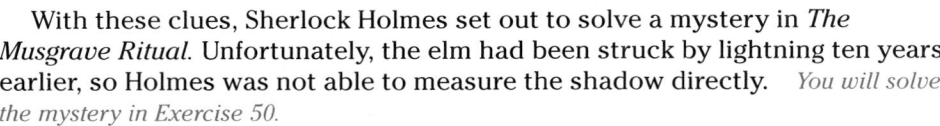

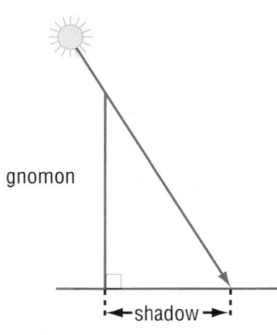

What YOU'LL LEARN

- To find values of trigonometric functions for acute angles, and
- to solve problems involving right triangles.

Why IT'S IMPORTANT

You can use trigonometry to solve problems involving surveying and literature.

"Where was the sun? Over the oak.
Where was the shadow? Under the elm."

With these clues, Sherlock Holmes set out to solve a mystery in *The Musgrave Ritual*. Unfortunately, the elm had been struck by lightning ten years earlier, so Holmes was not able to measure the shadow directly.   *You will solve the mystery in Exercise 50.*

Sherlock Holmes was not the first to use "shadow reckoning." The study of **trigonometry** probably began when early astronomers used the length of a shadow cast by a stick, called a *gnomon,* to determine the time of day. Egyptians relied on sundials as early as 1500 B.C., and there is evidence that astronomers in China, Mesopotamia, and India also understood that as the sun rises in the sky, it casts a unique shadow. In other words, the shadow is a function of the time of day.

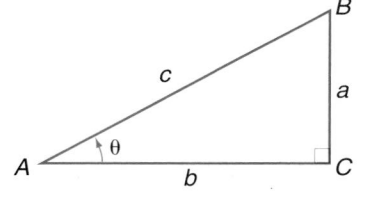

The word *trigonometry* was derived from two Greek words—*trigon* meaning triangle and *metra* meaning measurement. So, trigonometry began as the study of the relationships between the angles and sides of a right triangle.

Consider right triangle *ABC* shown below in which the measure of the acute angle *A* is identified with the Greek letter *theta,* θ.

The hypotenuse of the triangle is side $\overline{AB}$.
Its length is *c* units.
The leg opposite ∠*A* is $\overline{BC}$.
Its length is *a* units.
The leg adjacent to ∠*A* is $\overline{AC}$.
Its length is *b* units.

Using these sides, you can define six **trigonometric functions—sine, cosine, tangent, secant, cosecant,** and **cotangent,** which are abbreviated as sin, cos, tan, sec, csc, and cot, respectively.

Trigonometric Functions	**If θ is the measure of one acute angle in a right triangle, *a* is the measure of the leg opposite θ, *b* is the measure of the leg adjacent to θ, and *c* is the measure of the hypotenuse, then the following are true.**

$$\text{sine } \theta = \frac{a}{c} \qquad \text{cosine } \theta = \frac{b}{c} \qquad \text{tangent } \theta = \frac{a}{b}$$

$$\text{cosecant } \theta = \frac{c}{a} \qquad \text{secant } \theta = \frac{c}{b} \qquad \text{cotangent } \theta = \frac{b}{a}$$

Notice that the sine, cosine, and tangent functions are reciprocals of the cosecant, secant, and cotangent functions, respectively.

The following ratios may help you remember the sin, cos, and tan functions.

$$\sin \theta = \frac{\text{opposite}}{\text{hypotenuse}} \qquad \cos \theta = \frac{\text{adjacent}}{\text{hypotenuse}} \qquad \tan \theta = \frac{\text{opposite}}{\text{adjacent}}$$

SOH-CAH-TOA is a mnemonic device for remembering the first letter of each word in the ratios. For example, SOH refers to sin-opposite-hypotenuse.

The domain of each of these trigonometric functions is the set of all acute angles θ of a right triangle. The values of the functions depend only on the measure of θ and not on the size of the right triangle. For example, consider sin θ in the figure at the right.

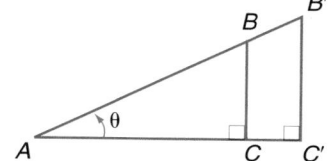

Using △ABC | Using △AB'C'

$$\sin \theta = \frac{BC}{AB} \qquad \sin \theta = \frac{B'C'}{AB'}$$

Notice that the triangles are similar because they are two right triangles that share a common angle, θ. Since they are similar, the ratios of corresponding sides are equal. That is, $\frac{BC}{AB} = \frac{B'C'}{AB'}$. Therefore, you will find the same value for sin θ regardless of which triangle you use.

Example **Find the values of the six trigonometric functions for angle θ. Round each value to four decimal places.**

For angle θ, the hypotenuse is $\overline{AB}$, the opposite leg is $\overline{BC}$, and the adjacent leg is $\overline{AC}$.

$\sin \theta = \frac{5}{13}$ or 0.3846 $\qquad$ $\csc \theta = \frac{13}{5}$ or 2.6000

$\cos \theta = \frac{12}{13}$ or 0.9231 $\qquad$ $\sec \theta = \frac{13}{12}$ or 1.0833

$\tan \theta = \frac{5}{12}$ or 0.4167 $\qquad$ $\cot \theta = \frac{12}{5} = 2.4000$

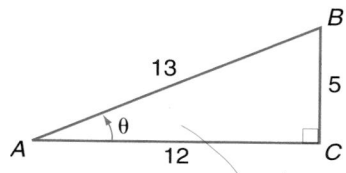

Example **Find tan A when cos $A = \frac{2}{3}$. Round to four decimal places.**

For convenience of notation, we refer to the angle with vertex A as angle A (∠A) and use A to stand for its measurement.

Since $\cos A = \frac{2}{3}$, the measure of the leg adjacent to ∠A is 2, and the measure of the hypotenuse is 3. Draw and label a right triangle. Then use the Pythagorean theorem to find the measure of the leg opposite ∠A.

$\cos A = \frac{adjacent}{hypotenuse}$

$a^2 + 2^2 = 3^2$

$a^2 = 5$

$a = \sqrt{5}$

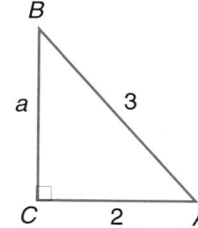

Now, find tan A.

$\tan A = \frac{\sqrt{5}}{2} \qquad \tan A = \frac{opposite}{adjacent}$

≈ 1.1180

Angles that measure 30°, 45°, and 60° occur frequently in trigonometry. You can find the values of the trigonometric functions for these angles by using the special characteristics of a 45°–45° right triangle and a 30°–60° right triangle.

To find the trigonometric values for a 45° angle, use an isosceles right triangle, $\triangle ABC$. Let the length of each leg be 1 unit. Use the Pythagorean theorem to find the length of the hypotenuse.

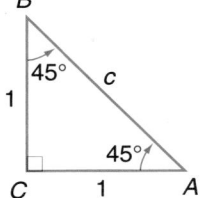

$$1^2 + 1^2 = c^2$$
$$2 = c^2$$
$$\sqrt{2} = c$$

The length of the hypotenuse is $\sqrt{2}$ units.

Therefore, the sine, cosine, and tangent values for 45° are as follows.

$\sin 45° = \dfrac{1}{\sqrt{2}}$ or $\dfrac{\sqrt{2}}{2}$ $\qquad$ $\cos 45° = \dfrac{1}{\sqrt{2}}$ or $\dfrac{\sqrt{2}}{2}$ $\qquad$ $\tan 45° = \dfrac{1}{1}$ or 1

To find the trigonometric values for a 30° angle, use an equilateral triangle, $\triangle XYZ$. Let the length of each side be 2 units. The altitude $\overline{ZW}$ separates $\triangle XYZ$ into two 30°–60° right triangles. Since $\overline{ZW}$ is the perpendicular bisector of $\overline{XY}$, the length of $\overline{XW}$ is 1 unit. Find the length of altitude $\overline{ZW}$.

$x^2 + 1^2 = 2^2$ $\qquad$ *Pythagorean theorem*
$$x^2 = 3$$
$$x = \sqrt{3}$$

The length of altitude $\overline{ZW}$ is $\sqrt{3}$ units.

Therefore, the sine, cosine, and tangent values for 30° are as follows.

$\sin 30° = \dfrac{1}{2}$ $\qquad\qquad$ $\cos 30° = \dfrac{\sqrt{3}}{2}$ $\qquad\qquad$ $\tan 30° = \dfrac{1}{\sqrt{3}}$ or $\dfrac{\sqrt{3}}{3}$

To find the sine, cosine, and tangent for 60°, use $\triangle XWZ$ shown above.

$\sin 60° = \dfrac{\sqrt{3}}{2}$ $\qquad$ $\cos 60° = \dfrac{1}{2}$ $\qquad$ $\tan 60° = \dfrac{\sqrt{3}}{1}$ or $\sqrt{3}$

Before hand-held calculators became accessible, students had to rely on "trig tables" to find the values of trigonometric functions for angles other than 30°, 45°, and 60°. Today, you can use the $\boxed{\text{SIN}}$, $\boxed{\text{COS}}$, and $\boxed{\text{TAN}}$ keys on your calculator to find these values.

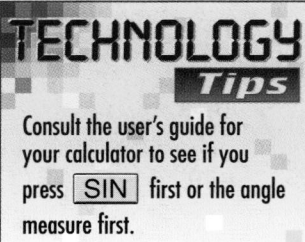

TECHNOLOGY *Tips*

Consult the user's guide for your calculator to see if you press $\boxed{\text{SIN}}$ first or the angle measure first.

EXPLORATION

CALCULATORS

In this Exploration, you will use a calculator to investigate the behavior of the sine and cosine functions for different values of θ. *Be sure your calculator is in degree mode.*

Your Turn

a. Choose five values of θ between 0° and 90° and evaluate sin θ for each of them. Write your answers as ordered pairs (θ, sin θ).

b. Graph the ordered pairs on a coordinate plane.

c. Use your graph to estimate the value of sin 0° and sin 90°. Check your answer with a calculator.

d. Explain how the value of sin θ changes as θ increases from 0° to 90°.

e. Repeat steps a–d using the cosine function.

You can also use the inverse capabilities of a calculator to find the measure of an angle when you know one of its trigonometric ratios. *You will learn more about inverses in Lesson 13–7.*

Example Find x if $\sin x = 0.7590$. Round to the nearest degree.

ENTER: 0.7590 $\boxed{\text{SIN}^{-1}}$ *49.37611923*   *The* $\boxed{\text{SIN}^{-1}}$ *key may be a second function on your calculator.*

Therefore, x is approximately $49°$.

If you know the measure of any two sides of a right triangle or the measure of one side and one acute angle, you can determine the measures of all the sides and angles of the triangle. This process of finding the missing measures is known as **solving a triangle.**

Example **4** **Solve each triangle. Round measures of sides to the nearest tenth and measures of angles to the nearest degree.**
a. $\triangle ABC$

You know the measures of the sides. You need to find A and B.

Find A. $\sin A = \dfrac{12}{15}$ $\sin A = \dfrac{opposite}{hypotenuse}$

$\sin A = 0.8$

Use a calculator and the $\sin^{-1}$ function to find the angle whose sine is 0.8.

0.8 $\boxed{\text{SIN}^{-1}}$ *53.130102*

To the nearest degree, $A \approx 53°$.

Find B. $53° + B \approx 90°$ *Angles A and B are complementary.*

$B \approx 37°$

Therefore, $A \approx 53°$ and $B \approx 37°$.

b. $\triangle XYZ$

You know the measure of the hypotenuse and one acute angle. You need to find x, y, and Y.

Find x and y. $\sin 63° = \dfrac{x}{23}$ $\cos 63° = \dfrac{y}{23}$

$0.8910 \approx \dfrac{x}{23}$ $0.4540 \approx \dfrac{y}{23}$

$x \approx 20.5$ $y \approx 10.4$

Find Y. $63° + Y = 90°$

$Y = 27°$

Therefore, $Y = 27°$, $x \approx 20.5$, and $y \approx 10.4$.

The gnomon dates from 3500 B.C. and the first Egyptian sundials from 1500 B.C. Next came the hemispherical sundial from Babylon about 300 B.C. More complex sundials like the hemicyclium, which uses conic sections, were created later by the Greeks.

Trigonometry has many practical applications in the real world. Among the most important is the ability to find distances or lengths that cannot be measured directly.

Example **5**

APPLICATION

Surveying

Utah's Bryce Canyon National Park contains some of the most colorful rock formations on Earth. For over 60 million years, water and ice have worn the canyon rocks into unusual shapes in shades of red, copper, pink, and cream. Some parts of the canyon are 1000 feet deep.

To find the distance across Bryce Canyon at a particular point, a surveyor sets up a transit at point C and sights a rock formation across the canyon at point B. Then the surveyor turns the transit 90° and sights point A that is 100 feet away. Using the transit at point A, the surveyor determines that the measure of $\angle A$ is 84°. Find the distance across the canyon from B to C.

Let a represent the distance, in feet, from B to C.

$\tan 84° = \dfrac{a}{100}$ $\tan = \dfrac{opposite}{adjacent}$

$9.5144 \approx \dfrac{a}{100}$

$a \approx 951.44$

The distance across Bryce Canyon at this point is about 951 feet.

Some applications of trigonometry use an **angle of elevation** or **angle of depression**. In the figure at the right, the angle formed by the line of sight from the boat and a horizontal line is called an angle of elevation. The angle formed by the line of sight from the parasail and a horizontal line is called an angle of depression.

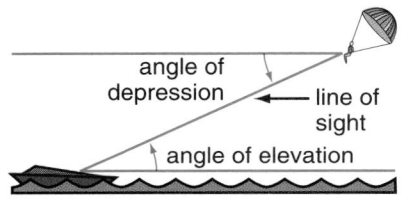

angle of depression — line of sight

angle of elevation

The line of sight is a transversal intersecting the two horizontal lines. The angle of elevation and the angle of depression are alternate interior angles. Since the horizontal lines are parallel, the angle of elevation and the angle of depression are congruent.

Example **6**

APPLICATION

Geology

A geologist measured a 43° angle of elevation to the top of a volcano crater. After moving 0.25 kilometers farther away, the angle of elevation was 38°. How high is the top of the volcano crater?

The figure at the right shows two right triangles that share a common height. Let h represent the height of the crater in kilometers. Let x represent the side adjacent to the angle whose measure is 43°. Write a system of equations in two variables.

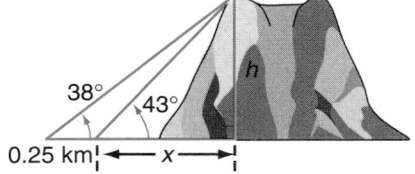

$38°$ $43°$ h

$0.25 \text{ km} \longleftarrow x \longrightarrow$

$\tan 43° = \dfrac{h}{x}$ $\tan 38° = \dfrac{h}{0.25 + x}$

$h = x \tan 43°$ $h = (0.25 + x) \tan 38°$

First, solve for x. $x \tan 43° = (0.25 + x) \tan 38°$ *Substitution*

$x \tan 43° = 0.25 \tan 38° + x \tan 38°$ *Distributive property*

$x \tan 43° - x \tan 38° = 0.25 \tan 38°$ *Subtract $x \tan 38°$ from each side.*

$x(\tan 43° - \tan 38°) = 0.25 \tan 38°$ *Distributive property*

$$x = \frac{0.25 \tan 38°}{\tan 43° - \tan 38°}$$

$$x \approx 1.29$$

Now, find h.　　$h = x \tan 43°$

$$h \approx 1.29(0.9325)$$

$$h \approx 1.20$$

Therefore, the volcano crater is about 1.20 kilometers high.

CHECK FOR UNDERSTANDING

Communicating Mathematics

Study the lesson. Then complete the following.

1. **Identify** the hypotenuse, the leg adjacent to θ, and the leg opposite θ in the triangle at the right.

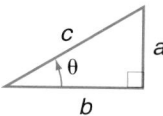

2. **Define** the word *trigonometry.*

3. **Evaluate** the six trigonometric functions of θ in the triangle at the right. Round your answers to four decimal places.

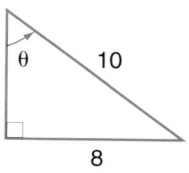

4. **Draw and label** a figure that shows an angle of depression from a person on a Ferris wheel who is looking at a friend waiting in line.

*M*ATH *J*OURNAL

5. Write a paragraph explaining how you can use trigonometry to find the height of a flagpole. Include a drawing with your explanation.

Guided Practice

Suppose θ is an acute angle of a right triangle. For each function, find the values of the remaining five trigonometric functions of θ. Round to four decimal places.

6. $\sin \theta = \dfrac{\sqrt{3}}{2}$

7. $\tan \theta = 2$

Write an equation involving sin, cos, or tan that can be used to find *x*. Then solve the equation. Round measures of sides to the nearest tenth and measures of angles to the nearest degree.

8.
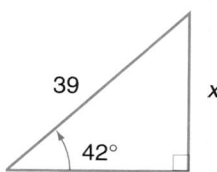

9.
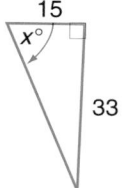

Find the value of *x*. Round to the nearest degree.

10. $\sin x = 0.7364$

11. $\cos x = 0.9912$

Solve each right triangle. Round measures of sides to the nearest tenth and measures of angles to the nearest degree.

12.
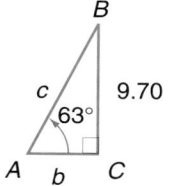

13.
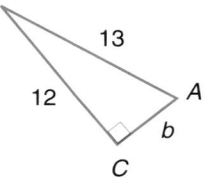

14.
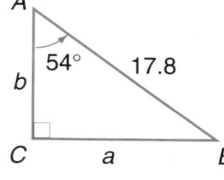

15. **Traveling** In a sightseeing boat near the base of the Horseshoe Falls at Niagara Falls, a passenger estimates the angle of elevation to the top of the falls to be 30°. If the Horseshoe Falls are 173 feet high, what is the distance from the boat to the base of the falls?

Practice

Suppose θ is an acute angle of a right triangle. For each function, find the values of the remaining five trigonometric functions of θ. Round to four decimal places.

16. $\tan \theta = \frac{12}{5}$

17. $\cos \theta = \frac{1}{4}$

18. $\cot \theta = 2$

19. $\csc \theta = \frac{5}{2}$

20. $\sec \theta = 3$

21. $\sin \theta = 0.5$

Write an equation involving sin, cos, or tan that can be used to find x. Then solve the equation. Round measures of sides to the nearest tenth and measures of angles to the nearest degree.

22.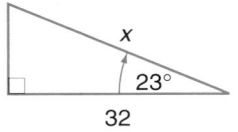

23.

24.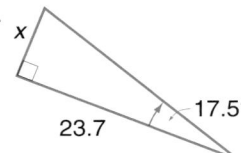

25.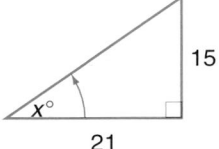

26.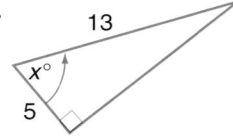

27.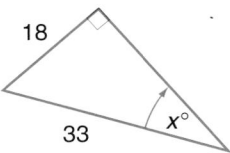

Find the value of x. Round to the nearest degree.

28. $\tan x = 0.5923$

29. $\cos x = 0.5269$

30. $\tan x = 0.2126$

31. $\sin x = 0.9998$

32. $\cos x = 0.9998$

33. $\sin x = 0.5000$

Solve each right triangle. Assume that C represents the right angle and c is the hypotenuse. Round measures of sides to the nearest tenth and measures of angles to the nearest degree.

34.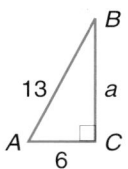

35.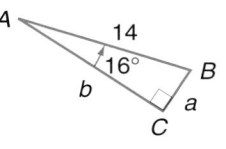

36.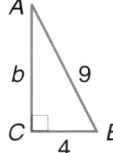

37. $B = 18°, a = \sqrt{15}$

38. $A = 56°, c = 16$

39. $A = 45°, c = 7\sqrt{2}$

40. $c = 25, A = 15°$

41. $B = 30°, b = 11$

42. $\tan A = \frac{7}{8}, a = 7$

43. $a = 7, A = 27°$

44. $\tan B = \frac{8}{6}, b = 8$

45. $\sin A = \frac{1}{3}, a = 5$

Geometry

46. In square *ABCD* at the right, the midpoint of side $\overline{AD}$ is *E*. Find the values of *x*, *y*, and *z* to the nearest tenth of a degree.

47. Isosceles triangle *RST* below at the right has base $\overline{TS}$ measuring 10 centimeters and base angles each measuring 39°. Find the length of the altitude $\overline{QR}$.

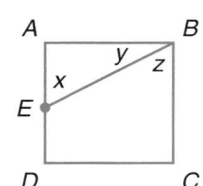

Critical Thinking

48. Describe a set of given conditions for which it would be impossible to solve a right triangle.

49. Explain why the sine and cosine of an acute angle are never greater than 1 but the tangent of an acute angle may be greater than 1.

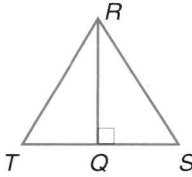

50. Literature   Refer to the application at the beginning of the lesson. Sherlock Holmes needed to find the length of the shadow cast by the elm. He was able to determine that the elm tree was 64 feet tall before it was struck by lightning. At the appropriate time, Holmes used a 6-foot rod to cast a shadow. Its shadow was 9 feet. What was the length of the shadow of the elm?

51. Broadcasting   Dolores and Bill are standing 100 feet apart and in a straight line with the WWV television tower. The angle of elevation from Bill to the tower is 30° and the angle of elevation from Dolores is 20°. Find the height of the television tower to the nearest foot.

52. Skiing   The Aerial run in Snowbird, Utah, is 8395 feet long. Its vertical drop is 2900 feet. If the slope were constant, estimate the angle of elevation that the run makes with the horizontal.

53. History   The Great Pyramid in Egypt has a square base, 230 meters on each side. The triangular faces of the pyramid make an angle of 51.8° with the base. Suppose you want to make a model of the pyramid for your history project. What measure should you use for the base angle of each triangle?

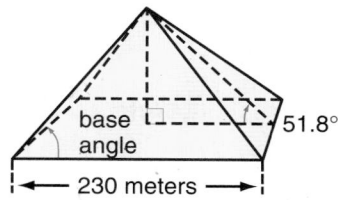

54. Aviation   When an airplane is flying, the air pressure creates a force called the *lift*, which is perpendicular to the wings. If a plane banks for a turn, this lift is separated into a horizontal and vertical force. The horizontal force is what is responsible for the turn, and the measure of the vertical force is the plane's weight.

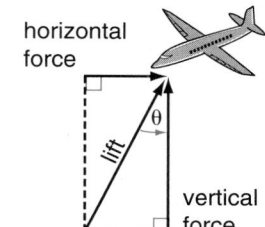

  a. Suppose a plane weighs 500,000 pounds. Find the measure of the lift and horizontal component for a banking angle, θ, of 20°.

  b. If the maximum lift that the wings can sustain is 650,000 pounds, what is the maximum banking angle?

55. Statistics   Determine whether a sample of people attending a concert is a random sample for a survey of people's favorite performer.   (Lesson 12–8)

56. Four of every 7 pitches thrown by Elias Ramos are strikes. What is the probability that 4 of the next 5 pitches will be strikes?   (Lesson 12–7)

57. Probability   State the probability of an event occurring given that the odds of the event are $\frac{5}{1}$.   (Lesson 12–4)

58. Expand the binomial $(r + s)^6$.   (Lesson 11–8)

59. Find the nth term of a geometric sequence in which $a_1 = 4$, $n = 3$, and $r = 5$.   (Lesson 11–3)

60. Solve $\log_4(x + 2) + \log_4(x - 4) = 2$.   (Lesson 10–3)

61. Simplify $\left(x^{\sqrt{2}}\right)^{\sqrt{8}}$.   (Lesson 10–1)

62. Use the distance formula to find the distance between the points at $(3, 3)$ and $\left(\sqrt{3}, \sqrt{3}\right)$.   (Lesson 7–1)

63. Chemistry   A chemist performed an experiment that yields 1.8×10^{24} molecules of ethanol. The mole is the standard unit of measure for the chemical quantity of a substance. There are 6.02×10^{23} molecules in a mole. How many moles of ethanol did the experiment yield?
(Lesson 5–1)

64. What property of real numbers is demonstrated by $x(a + b) = xa + xb$?
(Lesson 1–2)

Angles and Angle Measure

What YOU'LL LEARN

* To change radian measure to degree measure and vice versa, and
* to identify coterminal angles.

Why IT'S IMPORTANT

You can use angles to solve problems involving astronomy and geography.

CONNECTION
Geography

When French novelist Jules Verne wrote *Around the World in Eighty Days* in 1873, traveling that great distance in such a short time was unheard of. Do you think you could travel westward around the world in just one day? How fast would you have to travel to accomplish this? *A problem like this will be solved in Exercise 50.*

The answers to these questions depend on your position on Earth. Cartographers use a grid that contains circles through the poles, called *longitude* lines, and circles parallel to the equator, called *latitude* lines. In the figure at the right, point P is located by traveling north from point Q on the equator through a central angle of $a°$ to a circle of latitude, and then west along that circle through an angle of $b°$.

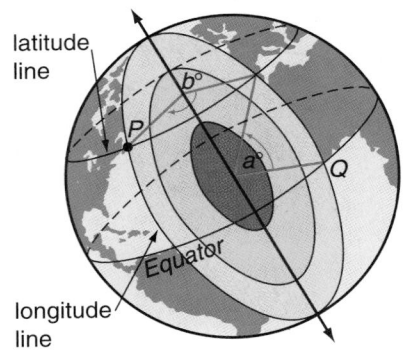

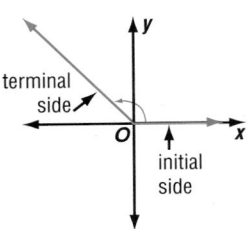

On a coordinate plane, an angle may be generated by the rotation of two rays that share a fixed endpoint at the origin. One ray, called the **initial side** of the angle, is fixed along the positive x-axis. The other ray, called the **terminal side** of the angle, can rotate about the center. An angle positioned so that its vertex is at the origin and its initial side is along the positive x-axis is said to be in **standard position.**

The measure of an angle is determined by the amount of rotation from the initial side to the terminal side. If the rotation is in a counterclockwise direction, the measure of the angle is positive. If the rotation is in a clockwise direction, the measure of the angle is negative.

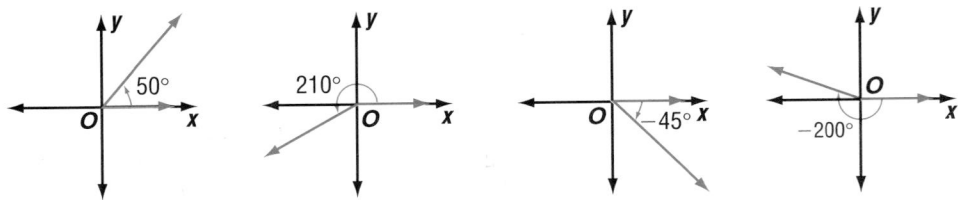

When terminal sides rotate, they may sometimes make one or more revolutions. An angle whose terminal side has made exactly one revolution has a measure of 360°.

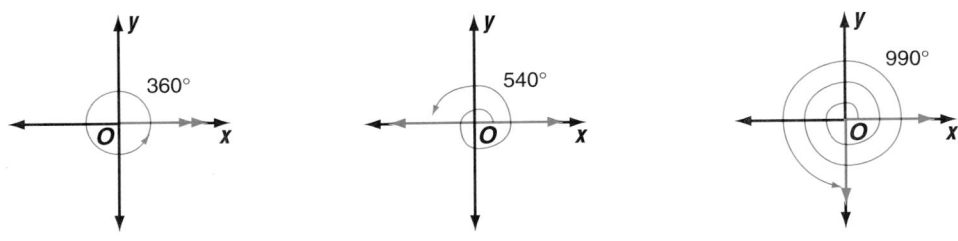

Eratosthenes, an astronomer who lived in Greece in the third century B.C., is credited with providing the first accurate measure of Earth's circumference. He found that at noon on the day of the summer solstice, the sun was directly over the city of Syene. At the same time, in Alexandria, which is north of Syene, the sun was 7°12′ south of being directly overhead. If the distance between the two cities was 5000 stadia, find Eratosthenes' measure of the circumference of Earth.

The stadium (singular form of stadia) was an ancient unit of measurement equal to about 0.098 mile.

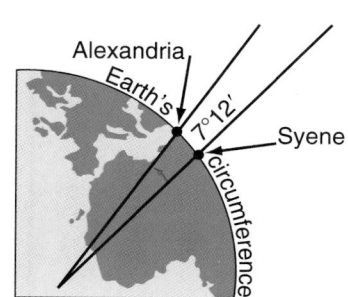

Since there are 360° in a full rotation around Earth, the following proportion can be written.

$$\frac{360°}{7°12′} = \frac{c}{5000}$$

$$(7°12′)c = (360°)5000$$

$$c = \frac{(360°)5000}{7°12′}$$

$$c = \frac{(360°)5000}{7.2°} \qquad 7°12′ = \left(7\frac{12}{60}\right)° \ or \ 7.2°$$

$$c = 250,000$$

Eratosthenes' measure of Earth's circumference was 250,000 stadia or 24,500 miles. This is only 158 miles less than the currently accepted value.

As you have seen, the degree is commonly used in applications involving surveying and navigation. However, in the late 1800s, mathematicians began to see the need for another unit of measure, called a **radian,** that would simplify certain mathematical and physical formulas.

The definition of a radian is based on the concept of a **unit circle,** which is a circle of radius 1 unit with its center at the origin of a coordinate system. The radian measure of an angle is based on the length of an arc on the unit circle. In the figure at the right, θ is in standard position so that the rays of the angle intercept an arc with length 1 unit. The measure of this angle is defined to be 1 radian.

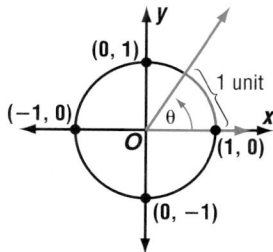

As with degrees, the measure of an angle in radians is positive if its rotation is counterclockwise. The measure is negative if the rotation is clockwise.

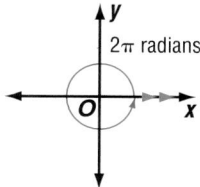

The circumference of any circle is $2\pi r$, where r is the radius measure. So the circumference of a unit circle is $2\pi(1)$ or 2π units. Therefore, an angle representing one complete revolution of the circle measures 2π radians.

Several common angles and their radian measure are shown below.

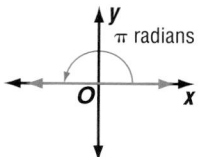 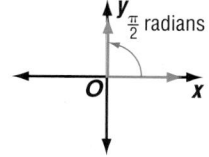 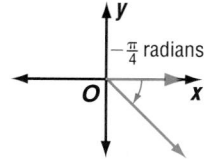 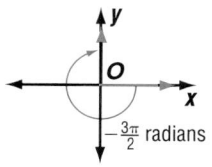

To find a relationship between degree measure and radian measure, consider one complete counterclockwise rotation. The degree measure is 360 while the radian measure of the same angle is 2π. Therefore, the following equation is true.

$$2\pi \text{ radians} = 360°$$

To change angle measures from radians to degrees or vice-versa, solve the equation above in terms of both units.

$$2\pi \text{ radians} = 360°$$
$$\frac{2\pi \text{ radians}}{2\pi} = \frac{360°}{2\pi}$$
$$1 \text{ radian} = \frac{180°}{\pi}$$

$$2\pi \text{ radians} = 360°$$
$$\frac{2\pi \text{ radians}}{360} = \frac{360°}{360}$$
$$\frac{\pi \text{ radians}}{180} = 1°$$

1 radian ≈ 57.3°

If you know the degree measure of an angle and you need to find the radian measure, multiply the number of degrees by $\frac{\pi \text{ radians}}{180°}$.

Example ❷ **Change each degree measure to radian measure.**

a. 45°

$$45° \left(\frac{\pi \text{ radians}}{180°} \right) = \frac{45\pi}{180} \text{ radians or } \frac{\pi}{4} \text{ radians}$$

b. −225°

$$-225° \left(\frac{\pi \text{ radians}}{180°} \right) = \frac{-225\pi}{180} \text{ radians or } -\frac{5\pi}{4} \text{ radians}$$

c. 450°

$$450° \left(\frac{\pi \text{ radians}}{180°} \right) = \frac{450\pi}{180} \text{ radians or } \frac{5\pi}{2} \text{ radians}$$

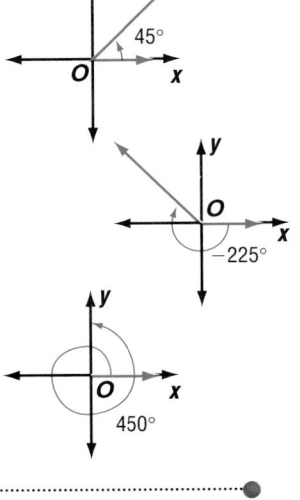

If you know the radian measure of an angle and you need to find the degree measure, multiply the number of radians by $\frac{180°}{\pi \text{ radians}}$.

Example ❸ **Change each radian measure to degree measure.**

The word <u>radian</u> is usually omitted when angles are expressed in radian measure.

a. $\frac{\pi}{3}$ radians

$$\frac{\pi}{3} \text{ radians} \left(\frac{180°}{\pi \text{ radians}} \right) = \left(\frac{180°\pi}{3\pi} \right) \text{ or } 60°$$

b. $-\frac{3\pi}{4}$ radians

$$-\frac{3\pi}{4} \text{ radians} \left(\frac{180°}{\pi \text{ radians}} \right) = \left(\frac{540°\pi}{4\pi} \right) \text{ or } -135°$$

c. 3 radians

$$3 \text{ radians} \left(\frac{180°}{\pi \text{ radians}} \right) = \left(\frac{540°}{\pi} \right) \text{ or about } 172°$$

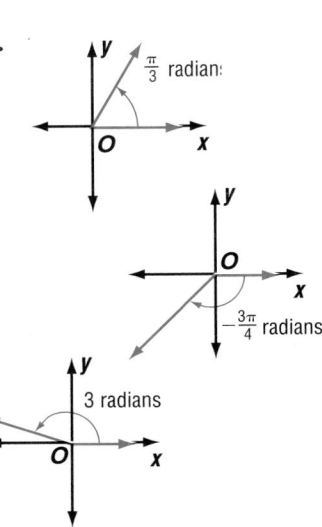

If you graph a 390° angle and a 30° angle in standard position on the same coordinate plane, you will notice that the terminal side of the 390° angle is the same as the terminal side of the 30° angle. When two angles in standard position have the same terminal sides, they are called **coterminal angles.**

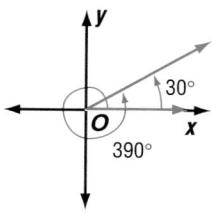

Every angle has infinitely many coterminal angles.

Notice that $390° - 30° = 360°$. In degree measure, coterminal angles differ by an integral multiple of 360°. You can find an angle that is coterminal to a given angle by adding or subtracting a multiple of 360°. In radian measure, a coterminal angle is found by adding or subtracting a multiple of 2π.

Example **4** **Find one positive and one negative angle that are coterminal with each angle.**

a. 135°

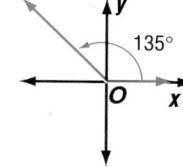

A positive angle is $135° + 360°$ or $495°$.

A negative angle is $135° - 360°$ or $-225°$.

b. $\dfrac{11\pi}{4}$

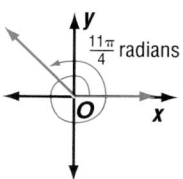

A positive angle is $\dfrac{11\pi}{4} - 2\pi$ or $\dfrac{3\pi}{4}$.

A negative angle is $\dfrac{11\pi}{4} - 4\pi$ or $-\dfrac{5\pi}{4}$.

CHECK FOR UNDERSTANDING

Communicating Mathematics

Study the lesson. Then complete the following.

1. **Describe** when an angle is in standard position.

2. **Draw** a unit circle and label an angle that measures 1 radian.

3. **Choose** the angle that measures $\dfrac{\pi}{2}$ radians.

a. b. c. d.

4. **Explain** why the measures of coterminal angles differ by a multiple of 360°.

5. **You Decide** Eduardo thinks that an angle of 20° is coterminal with an angle of 2000°. Toshi thinks the angles are not coterminal. Who is correct? Explain your reasoning.

Guided Practice

Match each angle with its measure. There may be more than one correct answer.

6.   7.   8.   9.

a. 270°   b. $\dfrac{\pi}{2}$ c. 2π d. 90°

e. 180°   f. $-\dfrac{\pi}{2}$ g. $-\pi$ h. $-90°$

Change each degree measure to radian measure.

10. $45°$ **11.** $-120°$ **12.** $540°$

Change each radian measure to degree measure.

13. $\frac{2\pi}{3}$ **14.** $-\frac{7\pi}{4}$ **15.** 5

Find one positive angle and one negative angle that are coterminal with each angle.

16. $-60°$ **17.** $\frac{\pi}{4}$ **18.** $750°$

19. Time Find both the degree and radian measures of the angle through which the hour hand on a clock rotates from 3:00 P.M. to 5:00 P.M.

EXERCISES

Practice

Change each degree measure to radian measure.

20. $-90°$ **21.** $180°$ **22.** $135°$

23. $1200°$ **24.** $-315°$ **25.** $-800°$

26. $150°$ **27.** $540°$ **28.** $3600°$

Change each radian measure to degree measure.

29. π **30.** $-\frac{\pi}{2}$ **31.** $-\frac{8\pi}{3}$

32. $\frac{3\pi}{4}$ **33.** 5π **34.** $\frac{5\pi}{2}$

35. 7 **36.** 3.5 **37.** -1.5

Find one positive angle and one negative angle that are coterminal with each angle.

38. $-120°$ **39.** $310°$ **40.** -5π

41. $\frac{9\pi}{4}$ **42.** $-450°$ **43.** $720°$

44. $\frac{\pi}{8}$ **45.** $-900°$ **46.** $-\frac{8\pi}{3}$

Critical Thinking

47. If (a, b) is on the unit circle with center at the origin, prove that each point is also on the unit circle.

a. $(a, -b)$ **b.** (b, a) **c.** $(b, -a)$

Applications and Problem Solving

48. Astronomy Earth rotates on its axis once every 24 hours.

a. How long does it take Earth to rotate through an angle of $300°$?

b. How long does it take Earth to rotate through an angle of $\frac{2\pi}{3}$ radians?

49. Physics When an object travels on a circular path like the one shown at the right, its *angular velocity*, ω, is the rate at which θ changes. Angular velocity is defined by the equation $\omega = \frac{\theta}{t}$, where θ is usually expressed in radians and t represents time. Find the angular velocity in radians per second of a point on a bicycle tire if it completes 2 revolutions in 3 seconds.

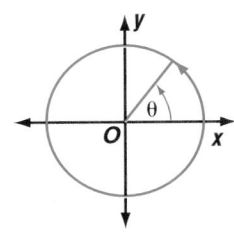

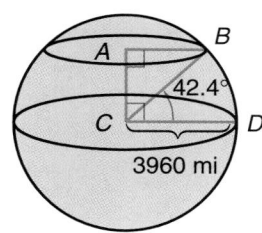

50. Geography Suppose a plane tried to circle Earth at the latitude of Boston, Massachusetts, in one day.

   a. Boston is located at 42.4°N latitude. In the figure at the right, the circle through point B represents the latitude line through Boston. Find the measure of $\angle ACB$.

   b. Use $\triangle ABC$ to find the radius AB. (*Hint:* The radius of Earth is about 3960 miles.)

   c. What is the circumference of the latitude line passing through Boston?

   d. Suppose a plane tried to circle Earth from Boston in one day. At what speed, in miles per hour, would it have to travel?

Mixed Review

51. Navigation The top of a lighthouse is 120 meters above sea level. From the top of the lighthouse, the measurement of the angle of depression to a boat on the ocean is 43°. How far is the boat from the foot of the lighthouse? (Lesson 13–1)

52. Probability Two dice are rolled. Find $P(3 \text{ and } 4)$. (Lesson 12–5)

53. Geometry Find the total number of diagonals that can be drawn in a decagon, shown at the right. (Lesson 12–3)

54. Find the sum of a geometric series for which $a_1 = 48$, $a_n = 3$, and $r = \frac{1}{2}$. (Lesson 11–4)

55. Solve $\dfrac{7}{m-3} = \dfrac{m+4}{m-3}$. (Lesson 9–5)

56. For the functions $g(x) = x - 1$ and $h(x) = x^2$, find $g[h(x)]$. (Lesson 8–7)

57. Write a quadratic equation that has roots 6 and 4. (Lesson 6–5)

58. Use a matrix equation to solve the system of equations. (Lesson 4–6)

$$5x + 3y = -5$$
$$7x + 5y = -11$$

59. What property is illustrated by $(11a + 3b) + 0 = (11a + 3b)$? (Lesson 1–2)

WORKING ON THE

Refer to the Investigation on pages 768–769.

The *Mean Streak* roller coaster at Cedar Point Amusement Park in Sandusky, Ohio, is one of the fastest and tallest wooden roller coasters in the world. Built in 1991, it is 5427 feet long. Its first hill is 161 feet high with a vertical drop of 155 feet and 52° angle of descent. It has banked curves and a course that crisscrosses the structure nine times. The 12 hills and valleys give you a ride that lasts about 2.5 minutes, reaching a top speed of 65 mph.

1 Make a sketch of the first hill of the *Mean Streak*. Label L_v as the vertical height of the hill and θ as the angle of descent. Let L_h represent the horizontal distance from the valley before the hill to the valley after the hill. Use a right triangle to find the value of L_h.

2 Discuss what determines the *Mean Streak's* top speed of 65 mph, and estimate where and how long a roller coaster would go at that speed.

Add the results of your work to your Investigation Folder.

Trigonometric Functions of General Angles

What YOU'LL LEARN

- To find values of trigonometric functions for general angles, and
- to use trigonometric identities to find values of trigonometric functions.

Why IT'S IMPORTANT

You can use trigonometry to solve problems invoving basketball and optics.

APPLICATION
Optics

Rainbows that result from light passing through a prism are caused by the refraction of light. When the light passes from one medium into another, the light ray is bent and the rainbow results. In this case, the light passes from air into glass. According to Snell's law, the angle at which the light ray approaches the prism, called the *angle of incidence,* and the angle at which the light ray is bent, called the *angle of refraction,* are related by the formula $2 \sin I = 3 \sin r$. Find the angle of refraction for this prism if the angle of incidence is 60°. *This problem will be solved in Example 5.*

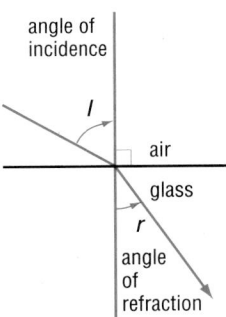

In Lesson 13–1, you found the values of trigonometric functions whose domains were the set of all acute angles of a right triangle. In this lesson, we will extend the domain to include angles of any measure.

Place an angle, θ, in standard position as shown at the right. The trigonometric functions of an angle in standard position may be defined in terms of the ordered pair for *any* point $P(x, y)$ on its terminal side and the distance r between that point and the origin. By the Pythagorean theorem, $r = \sqrt{x^2 + y^2}$.

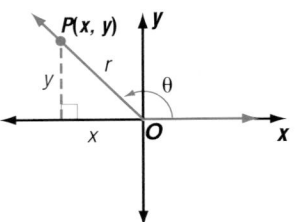

Trigonometric Functions of an Angle in Standard Position	**For any angle in standard position with measure θ, a point $P(x, y)$ on its terminal side, and $r = \sqrt{x^2 + y^2}$, the trigonometric functions of θ are as follows.**

$$\sin \theta = \frac{y}{r} \qquad \cos \theta = \frac{x}{r} \qquad \tan \theta = \frac{y}{x}$$

$$\csc \theta = \frac{r}{y} \qquad \sec \theta = \frac{r}{x} \qquad \cot \theta = \frac{x}{y}$$

Example **1** The terminal side of an angle θ in standard position passes through $P(8, -15)$. Find the exact values of sin θ, cos θ, and tan θ.

You know $x = 8$ and $y = -15$. You need to find r.

$$r = \sqrt{x^2 + y^2}$$
$$= \sqrt{8^2 + (-15)^2}$$
$$= \sqrt{289} \quad \text{or} \quad 17$$

Now, write the ratios.

$$\sin \theta = \frac{y}{r} \qquad \cos \theta = \frac{x}{r} \qquad \tan \theta = \frac{y}{x}$$
$$= \frac{-15}{17} \qquad \quad = \frac{8}{17} \qquad \quad = \frac{-15}{8}$$

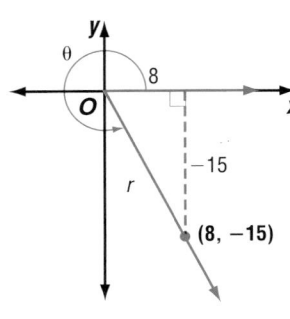

(8, −15)

As illustrated in Example 1, the values of the trigonometric functions may be either positive or negative. Sometimes the value is 0 or it is undefined. Since r is always positive, the signs of the functions are determined by the signs of x and y. These signs are determined by the quadrant in which the terminal side of θ lies.

LOOK BACK

You can refer to Lesson 2-1 for information about quadrants.

The chart at the right summarizes the signs of the trigonometric functions for each quadrant. The domain of the sine and cosine functions is the set of real numbers, because $\sin \theta$ and $\cos \theta$ are defined for any angle θ. However, since division by zero is undefined, there are several angle measures that are excluded

Function	Quadrant			
	I	II	III	IV
$\sin \theta$ or $\csc \theta$	+	+	−	−
$\cos \theta$ or $\sec \theta$	+	−	−	+
$\tan \theta$ or $\cot \theta$	+	−	+	−

from the domain of the tangent, cotangent, secant, and cosecant functions. For example, angles like $90°$, π, or $-270°$ have their terminal sides on an axis where x or y is equal to zero. These angles are called **quadrantal angles.**

Example **Find the values of the six trigonometric functions for an angle in standard position that measures 90°.**

Choose $P(0, 1)$ on the terminal side of the angle.
Therefore, $x = 0$, $y = 1$, and $r = 1$.

$\sin 90° = \dfrac{y}{r}$ or 1 $\qquad$ $\cos 90° = \dfrac{x}{r}$ or 0

$\tan 90° = \dfrac{y}{x}$ $\quad$ Since division by zero is undefined, $\tan 90°$ is undefined.

$\cot 90° = \dfrac{x}{y}$ or 0 $\qquad$ $\csc 90° = \dfrac{r}{y}$ or 1

$\sec 90° = \dfrac{r}{x}$ $\quad$ Since division by zero is undefined, $\sec 90°$ is undefined.

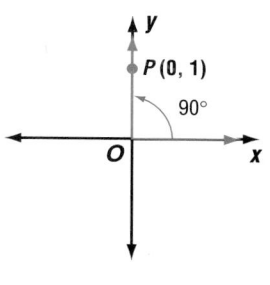

You can find the trigonometric functions of other special angles by using relationships from geometry. In a $30°$–$60°$ right triangle, the lengths of the sides are in the ratio $1:\sqrt{3}:2$. In a $45°$–$45°$ right triangle, the lengths of the sides are in the ratio $1:1:\sqrt{2}$.

Example **Find the exact value of each trigonometric function.**
a. tan 300°

Sketch the angle in standard position. Its terminal side lies in Quadrant IV. Notice the $30°$–$60°$ right triangle.

Choose $P(x, y)$ on the terminal side of the angle so that $r = 2$. It follows that $x = 1$ and $y = -\sqrt{3}$.
$\tan 300° = \dfrac{y}{x}$ or $-\sqrt{3}$. $\quad -\sqrt{3} \approx -1.7321$

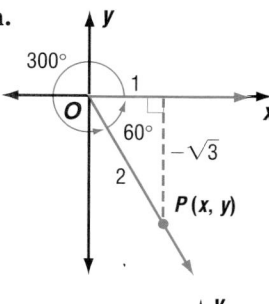

Verify the values with a calculator.

b. $\sin \dfrac{3\pi}{4}$

The terminal side of the angle lies in Quadrant II. Notice the $45°$–$45°$ right triangle. Choose $P(x, y)$ so that $x = -1$ and $y = 1$. Therefore, $r = \sqrt{2}$.

$\sin \dfrac{3\pi}{4} = \dfrac{y}{r}$ or $\dfrac{\sqrt{2}}{2}$ $\quad \dfrac{\sqrt{2}}{2} \approx 0.7071$

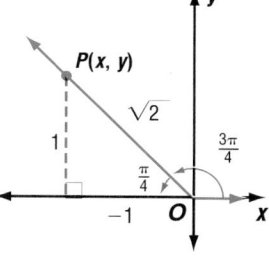

In the Exploration in Lesson 13–1, you used a calculator to investigate the behavior of the sine and cosine functions for different values of θ between 0° and 90°. In the following Exploration, you will investigate the behavior of these functions between 0° and 360°.

EXPLORATION   CALCULATORS

As the value of θ increases from 0° to 90°, sin θ increases from 0 to 1 and cos θ decreases from 1 to 0. What happens for angle measures greater than 90°?

Your Turn

a. Choose twenty values of θ between 0° and 360° and evaluate sin θ for each of them. Write your answers as the ordered pair (θ, sin θ).

b. Graph the ordered pairs on a coordinate plane.

c. Explain how the value of sin θ changes as θ increases from 0° to 360°.

d. Predict the behavior of sin θ from 360° to 720°.

e. Repeat steps a–d using the cosine function.

There are many relationships among the trigonometric functions that can be derived from their definitions. These **trigonometric identities** are true for *all* values of the variable(s) for which the expressions are defined. Consider the ratio $\dfrac{\sin \theta}{\cos \theta}$.

$$\frac{\sin \theta}{\cos \theta} = \frac{\frac{y}{r}}{\frac{x}{r}} \qquad \sin \theta = \frac{y}{r}, \cos \theta = \frac{x}{r}$$

$$= \frac{y}{r} \cdot \frac{r}{x} \text{ or } \frac{y}{x} \qquad \frac{y}{r} \div \frac{x}{r} = \frac{y}{r} \cdot \frac{r}{x}$$

By definition, $\tan \theta = \dfrac{y}{x}$. Therefore, $\dfrac{\sin \theta}{\cos \theta} = \tan \theta$. This and other trigonometric identities are defined as follows.

Trigonometric Identities	**The following trigonometric identities hold for all values of θ except those for which any function is undefined.**

$$\frac{\sin \theta}{\cos \theta} = \tan \theta \qquad \frac{\cos \theta}{\sin \theta} = \cot \theta$$

These are sometimes called the <u>quotient identities</u>.

$$\csc \theta = \frac{1}{\sin \theta} \qquad \sec \theta = \frac{1}{\cos \theta} \qquad \cot \theta = \frac{1}{\tan \theta}$$

These are sometimes called the <u>reciprocal identities</u>.

If you know the values of the sine and cosine functions, you can use the trigonometric identities to find the values of all six trigonometric functions.

Example 4 If $\sin \theta = \dfrac{3}{5}$ and $\cos \theta = -\dfrac{4}{5}$, the terminal side of θ must lie in Quadrant II. Find the exact values of tan θ, csc θ, cot θ, and sec θ.

First, use the quotient identity to find tan θ.

$$\tan \theta = \frac{\sin \theta}{\cos \theta}$$

$$= \frac{\frac{3}{5}}{-\frac{4}{5}} \text{ or } -\frac{3}{4} \qquad \frac{3}{5} \div \left(-\frac{4}{5}\right) = \frac{3}{5} \cdot \left(-\frac{5}{4}\right) \text{ or } -\frac{3}{4}$$

Use the reciprocal identities to find $\cot\theta$, $\sec\theta$, and $\csc\theta$.

$$\cot\theta = \frac{1}{\tan\theta} \quad\text{or}\quad -\frac{4}{3} \qquad \sec\theta = \frac{1}{\cos\theta} \quad\text{or}\quad -\frac{5}{4} \qquad \csc\theta = \frac{1}{\sin\theta} \quad\text{or}\quad \frac{5}{3}$$

In Lesson 13–1, you used trigonometric functions to solve problems dealing with surveying, navigation, and construction. Trigonometric functions also appear in formulas for optics, electronics, physics, and many other applications.

Example

Optics

5 Refer to the application at the beginning of the lesson.

a. **The angle of incidence I and the angle of refraction r are related by the formula $2\sin I = 3\sin r$. Find the angle of refraction if $I = 60°$.**

$2\sin I = 3\sin r$

$2\sin 60° = 3\sin r$ *Replace I with 60°.*

$\dfrac{2\sin 60°}{3} = \sin r$

$0.5774 \approx \sin r$ *sin 60° ≈ 0.8660*

$35.3° \approx r$

The angle of refraction is about 35°.

b. **The general form of Snell's law is $n = \dfrac{\sin I}{\sin r}$, where n is the index of refraction. Suppose a beam of light moves from a vacuum to glass. If the angle of refraction is 30° and the index of refraction is $\sqrt{2}$, what is the angle of incidence?**

$n = \dfrac{\sin I}{\sin r}$

$\sqrt{2} = \dfrac{\sin I}{\sin 30°}$ *Replace r with 30° and n with $\sqrt{2}$.*

$\sqrt{2}\sin 30° = \sin I$ *sin 30° = 0.5*

$0.7071 = \sin I$

$45° = I$

The angle of incidence is 45°.

CHECK FOR UNDERSTANDING

Communicating Mathematics

Study the lesson. Then complete the following.

1. **State** the quadrant or quadrants in which $\sin\theta$ and $\cos\theta$ are both positive.

2. **Define** a trigonometric identity.

3. **Explain** why $\sec\theta = \dfrac{1}{\cos\theta}$.

4. **You Decide** Luisa thinks $\sin 13°$ is greater than $\sin 12°$. Henry thinks you can't tell unless you use a calculator or trig table. Who is correct? Explain your reasoning.

5. **Copy and complete** the chart below that shows the value of $\sin\theta$, $\cos\theta$, and $\tan\theta$ for the quadrantal angles. A dash indicates that the function is undefined.

Function	0°	90°	180°	270°
$\sin\theta$	0			−1
$\cos\theta$		0	−1	
$\tan\theta$		—	0	

State whether the value of each function is *positive, negative, zero,* **or** *undefined.*

6. $\sin 200°$ **7.** $\cos \frac{\pi}{2}$ **8.** $\tan \frac{\pi}{4}$

Find the exact values of sin θ, cos θ, and tan θ if the terminal side of θ in standard position contains the given point.

9. $P(1, -8)$ **10.** $P(-3, -4)$

Find the exact value of each trigonometric function.

11. $\cos 120°$ **12.** $\tan \left(-\frac{\pi}{3}\right)$ **13.** $\sin 225°$

Suppose θ is an angle in standard position whose terminal side lies in the given quadrant. For each function, find the exact values of the remaining five trigonometric functions of θ.

14. $\sin \theta = -\frac{4}{5}$; Quadrant IV   **15.** $\tan \theta = 2$; Quadrant I

16. Navigation Ships and airplanes measure distance in nautical miles. The formula 1 nautical mile = $(6077 - 31 \cos 2\theta)$ feet, where θ is the latitude in degrees, can be used to find the approximate length of a nautical mile at a certain latitude. Find the length of a nautical mile where the latitude is 30°.

EXERCISES

State whether the value of each function is *positive, negative, zero,* **or** *undefined.*

17. $\sin (-135°)$ **18.** $\cos 405°$ **19.** $\tan 315°$

20. $\sin 2\pi$ **21.** $\cos \frac{\pi}{4}$ **22.** $\sin \frac{11\pi}{4}$

23. $\tan 90°$ **24.** $\cos 450°$ **25.** $\sin (-45°)$

Find the exact values of sin θ, cos θ, and tan θ if the terminal side of θ in standard position contains the given point.

26. $P(-15, 8)$ **27.** $P(-3, 0)$ **28.** $P(-\sqrt{2}, \sqrt{2})$

29. $P(5, -3)$    **30.** $P(0, 2)$ **31.** $P(4, 4)$

Find the exact value of each trigonometric function.

32. $\cos 150°$ **33.** $\cos \frac{11\pi}{3}$ **34.** $\tan 135°$

35. $\sin 240°$ **36.** $\sin \frac{3\pi}{2}$ **37.** $\cos (-60°)$

38. $\sin (-180°)$ **39.** $\tan 405°$ **40.** $\tan \left(-\frac{5\pi}{6}\right)$

Suppose θ is an angle in standard position whose terminal side lies in the given quadrant. For each function, find the exact values of the remaining five trigonometric functions of θ.

41. $\cos \theta = -\frac{1}{2}$; Quadrant II   **42.** $\sec \theta = \sqrt{3}$; Quadrant IV

43. $\tan \theta = 3$; Quadrant III   **44.** $\sin \theta = -\frac{1}{5}$; Quadrant IV

45. $\cot \theta = -5$; Quadrant II   **46.** $\csc \theta = -3$; Quadrant IV

47. If $\cos \theta = \frac{2}{3}$, find all possible values of $\sin \theta$.

48. If $\sec \theta = -3$, find all possible values of $\sin \theta$ and $\cos \theta$.

49. If $\cos \theta = 0$, find all possible values of $\sin \theta$ and $\tan \theta$.

Suppose θ is an angle in standard position with the given conditions. State the quadrant or quadrants in which the terminal side of θ lies.

50. $\sin \theta > 0$ **51.** $\sin \theta > 0$, $\cos \theta < 0$ **52.** $\tan \theta > 0$, $\cos \theta < 0$

Critical Thinking

53. If θ is any angle for which the functions are defined, prove $\cot \theta = \frac{\cos \theta}{\sin \theta}$.

Applications and Problem Solving

54. Baseball The formula $R = \frac{V_0^2 \sin 2\theta}{g}$ gives the distance of a baseball that is hit at an initial velocity of V_0 feet per second at an angle of θ with the ground. The variable g represents the acceleration due to gravity, which is 32 feet per second^2.

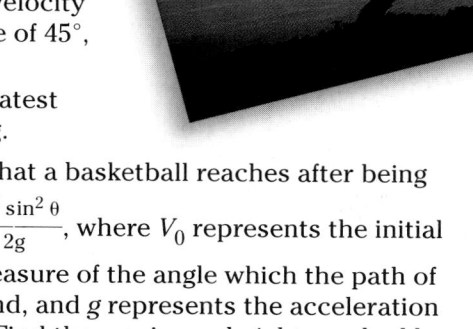

  a. If the ball was hit with an initial velocity of 100 feet per second at an angle of 45°, how far was it hit?

  b. Which angle will result in the greatest distance? Explain your reasoning.

55. Basketball The maximum height that a basketball reaches after being shot is given by the formula $H = \frac{V_0^2 \sin^2 \theta}{2g}$, where V_0 represents the initial velocity, θ represents the degree measure of the angle which the path of the basketball makes with the ground, and g represents the acceleration due to gravity, 32 feet per second^2. Find the maximum height reached by a free-throw if it is shot with an initial velocity of 25 feet per second at an angle of 65°.

Mixed Review

56. Change $\frac{5\pi}{6}$ radians to degrees.   (Lesson 13–2)

57. Solve $\triangle ABC$ shown at the right if $c = 21$ and $b = 18$.
(Lesson 13–1)

58. Probability In homeroom, 3 of the 16 girls have red hair, and 2 of the 15 boys have red hair. What is the probability of selecting a boy or a red-haired person as homeroom representative to student council?
(Lesson 12–6)

59. Find the sum of the infinite geometric series $\frac{4}{3} - \frac{2}{3} + \frac{1}{3} - \frac{1}{6} + \cdots$.
(Lesson 11–5)

60. Name the next four terms of the arithmetic sequence 21, 15, 9, ...
(Lesson 11–1)

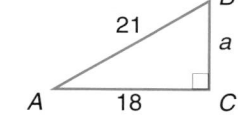

61. Chemistry The pH of a solution is related to the number of gram atoms of hydrogen ions, H^+, by the formula $\text{pH} = \log_{10} \frac{1}{H^+}$. If the pH level of a lake is 5, how much more acidic is it than neutral water that has a pH of 7?
(Lesson 10–3)

62. Simplify $\frac{w+12}{4w-16} - \frac{w+4}{2w-8}$.   (Lesson 9–4)

63. Write the polynomial function of least degree with integral coefficients whose zeros are 6 and $4 - 2i$.   (Lesson 8–4)

64. Find the slope-intercept form of the equation that passes through $(-2, 5)$ and $(3, 1)$ (Lesson 2–4)

65. Solve $-1.6m + 5 = -7.8$.   (Lesson 1–4)

Law of Sines

APPLICATION

Forestry

Yosemite National Park, located in California's Sierra Nevada mountains, is home to beautiful meadows, spectacular waterfalls, and jagged mountains. More than 30 kinds of trees and more than 1300 species of plants can be found in the park.

When drought conditions exist in the park, the Park Service often imposes restrictions on open fires. Suppose two forest rangers, 10 miles apart from each other on a straight service road, both sight an illegal campfire away from the road. Using their radios to communicate with each other, they determine that the fire is between them. The first ranger's line of sight to the fire makes an angle of 34° with the road, and the second ranger's line of sight to the fire makes a 67° angle with the road. How far is the fire from each ranger? What is the shortest distance from the road to the fire? *This problem will be solved in Example 3.*

In Lesson 13–1, you solved problems that involved acute angles of right triangles. It is also possible to use trigonometric functions to solve triangles that do not necessarily contain a right angle. You can even use trigonometric functions to find the area of triangles.

Consider $\triangle ABC$ with height h units and sides with lengths a units, b units, and c units. The area of this triangle is $\frac{1}{2}ch$. Note that $\sin A = \frac{h}{b}$ or $h = b \sin A$. By combining these equations, you can find a new formula for the area of the triangle.

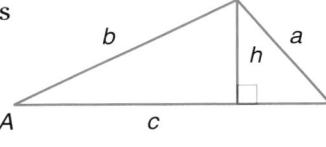

$$\text{Area} = \frac{1}{2}ch$$
$$= \frac{1}{2}c(b \sin A) \quad h = b \sin A$$

You can find two other formulas for the area of the triangle in a similar way.

$$\text{Area} = \frac{1}{2}bc \sin A = \frac{1}{2}ac \sin B = \frac{1}{2}ab \sin C$$

These formulas allow you to find the area of any triangle when you know the measures of two sides and the included angle.

Example **1** **Find the area of $\triangle ABC$ if $b = 35$, $c = 23$, and $A = 36°$.**

$$\text{Area} = \frac{1}{2}bc \sin A$$
$$= \frac{1}{2}(35)(23)\sin 36° \quad \sin 36° \approx 0.5878$$
$$\approx 236.59$$

To the nearest whole unit, the area is 237 square units.

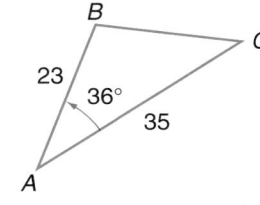

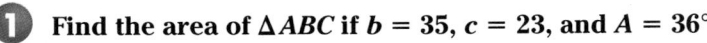

All of the area formulas above represent the area of the same triangle. So, the following must be true.

$$\frac{1}{2}bc \sin A = \frac{1}{2}ac \sin B = \frac{1}{2}ab \sin C$$

The **law of sines** is obtained by dividing each of the expressions above by $\frac{1}{2}abc$.

$$\frac{\sin A}{a} = \frac{\sin B}{b} = \frac{\sin C}{c}$$

Law of Sines	Let $\triangle ABC$ be any triangle with a, b, and c representing the measures of sides opposite angles with measurements A, B, and C, respectively. Then, $$\frac{\sin A}{a} = \frac{\sin B}{b} = \frac{\sin C}{c}.$$

You can apply the law of sines to a triangle if you know
- the measures of two angles and the measure of any side, or
- the measures of two sides and the angle opposite one of the sides.

Example 2   **Use the law of sines to solve the triangle below. Round measures of sides to the nearest tenth.**

You are given the measures of two angles and a side.
First, find the measure of the third angle, $\angle C$.

$$40° + 60° + C = 180°$$
$$C = 80°$$

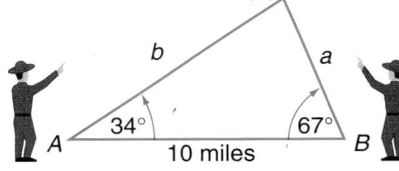

Now, use the law of sines to find b and c.

$$\frac{\sin A}{a} = \frac{\sin B}{b} \qquad\qquad \frac{\sin A}{a} = \frac{\sin C}{c}$$
$$\frac{\sin 40°}{20} = \frac{\sin 60°}{b} \qquad\qquad \frac{\sin 40°}{20} = \frac{\sin 80°}{c}$$
$$b = \frac{20 \sin 60°}{\sin 40°} \qquad\qquad c = \frac{20 \sin 80°}{\sin 40°}$$
$$b \approx 26.9 \qquad\qquad c = 30.6$$

Therefore, $b \approx 26.9$, $c \approx 30.6$, and $C = 80°$.

Example 3   **Refer to the application at the beginning of the lesson.**
a. How far is the fire from each ranger?
b. What is the shortest distance from the road to the fire?

Forestry

First, draw a diagram. You are given the measure of two angles and a side. Find the measure of the third angle, angle C.

$$34° + 67° + C = 180°$$
$$C = 79°$$

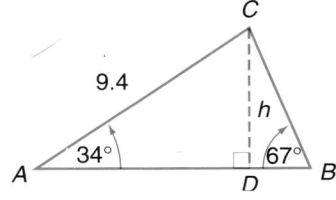

a. Use the law of sines to find a and b, the distance from each ranger to the fire.

$$\frac{\sin 79°}{10} = \frac{\sin 34°}{a} \qquad\qquad \frac{\sin 79°}{10} = \frac{\sin 67°}{b}$$
$$a \approx \frac{10 \sin 34°}{\sin 79°} \qquad\qquad b \approx \frac{10 \sin 67°}{\sin 79°}$$
$$a \approx 5.7 \qquad\qquad b \approx 9.4$$

Therefore, the fire is 9.4 miles from Ranger A and 5.7 miles from Ranger B.

b. The shortest distance from the road to the fire is from point D, which is on the perpendicular from C to the road. Let h represent the measure of segment CD.

$$\sin 34° = \frac{h}{9.4}$$
$$h \approx 5.3$$

The shortest distance from the road to the fire is about 5.3 miles.

Example Use the law of sines to solve the triangle below. Round to the nearest tenth.

You are given the measures of two sides and the angle opposite one of them. First, find the measure of the angle opposite the other given side, $\angle B$.

$$\frac{\sin A}{a} = \frac{\sin B}{b}$$

$$\frac{\sin 42°}{63} = \frac{\sin B}{57}$$

$$\sin B = \frac{57 \sin 42°}{63}$$

$$\sin B \approx 0.6054$$

$$B \approx 37.3°$$

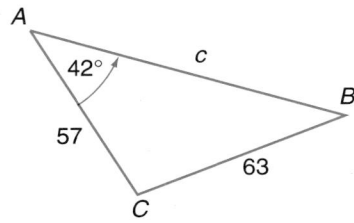

Next, find the measure of the third angle, $\angle C$.

$$37.3° + 42° + C = 180°$$

$$C \approx 100.7°$$

Then, find the measure of c.

$$\frac{\sin A}{a} = \frac{\sin C}{c}$$

$$\frac{\sin 42°}{63} = \frac{\sin 100.7°}{c}$$

$$c = \frac{63 \sin 100.7°}{\sin 42°}$$

$$c \approx 92.5$$

Therefore, $C \approx 100.7°$, $B \approx 37.3°$, and $c \approx 92.5$ units.

When solving a triangle, you must analyze the data to determine whether there is a solution or not. For example, if you are given the measures of two angles and a side, as in Examples 2 and 3, the triangle has a unique solution. However, if you are given the measures of two sides and the angle opposite one of them is given, a single solution may not exist. One of the following will be true.

- No triangle exists, and there is no solution.

- Exactly one triangle exists, and there is one unique solution.

- Two triangles exist, and there are two solutions.

Suppose you are given a, b, and A. First, consider the case where $A < 90°$. If $a < b$, there are three possibilities.

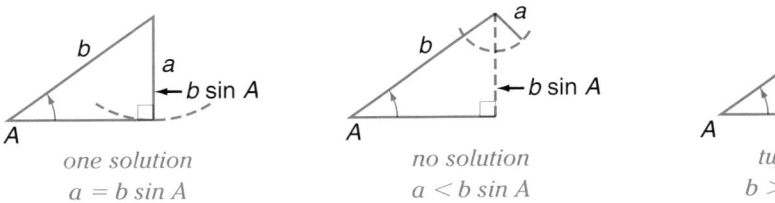

| *one solution* | *no solution* | *two solutions* |
| $a = b \sin A$ | $a < b \sin A$ | $b > a > b \sin A$ |

If $a > b$, there is one unique solution.

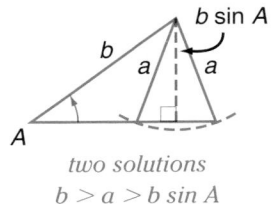

Consider the case where $A \geq 90°$. There are two possibilities.

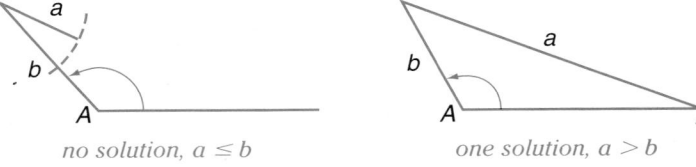

no solution, $a \leq b$ *one solution, $a > b$*

In Example 4, you were given the measures of two sides and the angle opposite one of them. When you solved the triangle, there was one unique solution. The following examples show triangles in which there are no solutions and two solutions.

Example **5** Solve each triangle described below.

a. $A = 35°$, $b = 14$, and $a = 6$

PROBLEM SOLVING

Examine the Solution

Angle A is less than $90°$. Find $b \sin A$ and compare with a.

$b \sin A = 14 \sin 35°$ *$b \sin A$ is the minimum*

       $= 14(0.5736)$ *distance from C to $\overline{AB}$.*

       ≈ 8.03

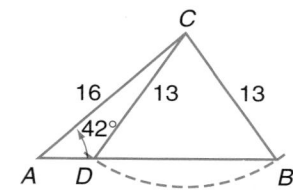

Since $6 < 8.03$, there is no solution.

b. $A = 42°$, $a = 13$, and $b = 16$.

$b \sin A = 16 \sin 42°$

       $\approx 16(0.6691)$

       ≈ 10.71

Since $42° < 90°$ and $10.71 < 13 < 16$, there are two solutions. The two triangles to be solved are $\triangle ABC$ and $\triangle ADC$.

When two solutions exist, it is called the <u>ambiguous case</u>. Why?

Solve $\triangle ABC$.

First, use the law of sines to find B.

$\dfrac{\sin 42°}{13} = \dfrac{\sin B}{16}$ $\dfrac{\sin A}{a} = \dfrac{\sin B}{b}$

$\sin B = \dfrac{16 \sin 42°}{13}$

$B \approx 55.4°$

Find $\angle ACB$. | Find c.

$42° + 55.4° + \angle ACB \approx 180°$ | $\dfrac{\sin 42°}{13} = \dfrac{\sin 82.6°}{c}$

           $\angle ACB \approx 82.6°$ | $c \approx \dfrac{13 \sin 82.6°}{\sin 42°}$

                                     | $c \approx 19.3$

Therefore, $B \approx 55.4°$, $\angle ACB \approx 82.6°$, and $c \approx 19.3$.

Solve $\triangle ADC$.

First, find $\angle ADC$.

$\triangle DBC$ is isosceles, so the measures of the base angles are equal. Therefore, since $\angle B \approx 55.4°$, $\angle CDB \approx 55.4°$.

$\angle ADC$ is supplementary to $\angle CDB$, so $\angle ADC \approx 124.6°$.

Find $\angle ACD$.

$42° + 124.6° + \angle ACD \approx 180°$

               $\angle ACD \approx 13.4°$

Then, use the law of sines to find the measure of segment AD in $\triangle ADC$.

$\dfrac{\sin 42°}{13} = \dfrac{\sin 13.4°}{AD}$

    $AD \approx \dfrac{13 \sin 13.4°}{\sin 42°}$

    $AD \approx 4.5$

Therefore, $AD \approx 4.5$, $\angle ACD \approx 13.4°$, and $\angle ADC \approx 124.6°$.

Communicating Mathematics

Study the lesson. Then complete the following.

1. **State** the law of sines.
2. **Describe** a set of conditions for which the law of sines can be used.
3. **Explain** how you know when a triangle has no solution.
4. **Draw** $\triangle ABC$ if $\angle ABC = 35°$, $BC = 8$, and $\angle BCA = 70°$.
5. **Choose** a value for a so that the triangle at the right has one solution.

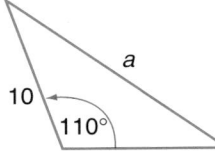

Guided Practice

Write an equation that can be used to find the area of each triangle. Then solve the equation. Round to the nearest tenth.

6.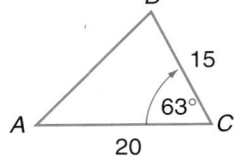

7.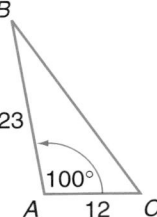

Solve each triangle. Round measures of sides to the nearest tenth and measures of angles to the nearest degree.

8.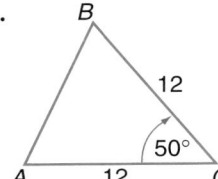

9.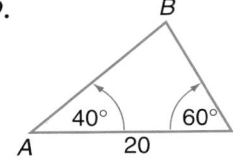

10. $a = 8$, $A = 49°$, $B = 57°$

Determine whether each triangle has no solution, one solution, or two solutions. Then solve each triangle. Note that the triangles may not be drawn to scale.

11.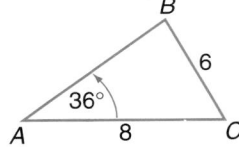

12.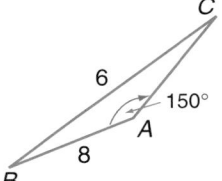

13. $a = 64$, $c = 90$, $C = 98°$

14. The longest side of a triangle is 67 inches. Two angles have measures of 47° and 55°. Solve the triangle.

EXERCISES

Practice

Write an equation that can be used to find the area of each triangle. Then solve the equation. Round to the nearest tenth.

15.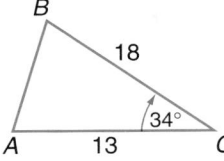

16.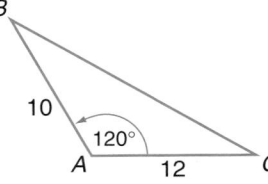

17. $b = 24$, $a = 20$, $C = 73°$
18. $b = 35$, $c = 47$, $A = 67°$
19. $a = 11.5$, $c = 19$, $B = 20°$
20. $a = 9.4$, $c = 13.5$, $B = 95°$

Solve each triangle. Round measures of sides and angles to the nearest tenth.

21.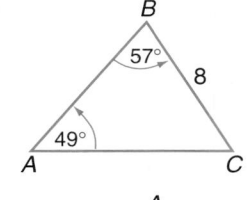

22.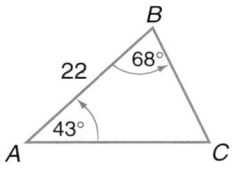

23.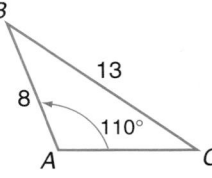

24.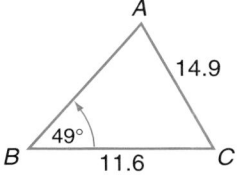

25.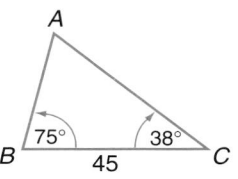

26.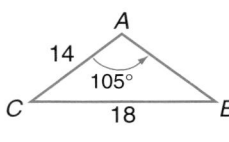

27. $A = 30°, C = 70°, c = 8$

28. $c = 17, b = 15, C = 64°$

29. $a = 14, b = 7.5, A = 103°$

30. $a = 23, A = 73°, C = 24°$

Determine whether each triangle has no solution, one solution, or two solutions. Then solve each triangle. Note that the triangles may not be drawn to scale.

31.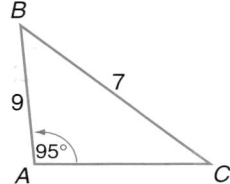

32.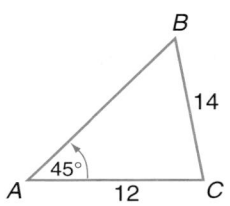

33.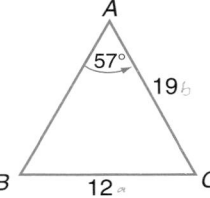

34. $a = 9, b = 20, A = 31°$

35. $a = 12, b = 14, A = 90°$

36. $a = 125, b = 150, A = 25°$

37. $A = 40°, b = 16, a = 10$

38. $a = 18, b = 20, A = 120°$

39. $A = 40°, b = 10, a = 8$

INTEGRATION
Geometry

40. An isosceles triangle has a base of 22 centimeters and exactly one angle measuring 36°. Find its perimeter.

41. The sides of a triangle measure 22, 13, and 8. Find the measure of the smallest angle.

Critical Thinking

42. Prove that the law of sines holds for right triangles.

Applications and Problem Solving

43. Aviation A pilot takes off from Newport News, Virginia, and flies toward the Atlantic Ocean. After reaching point C, the plane develops mechanical difficulties, and the pilot needs to return either to Newport News or Norfolk, Virginia. How far is it to the nearer airport?

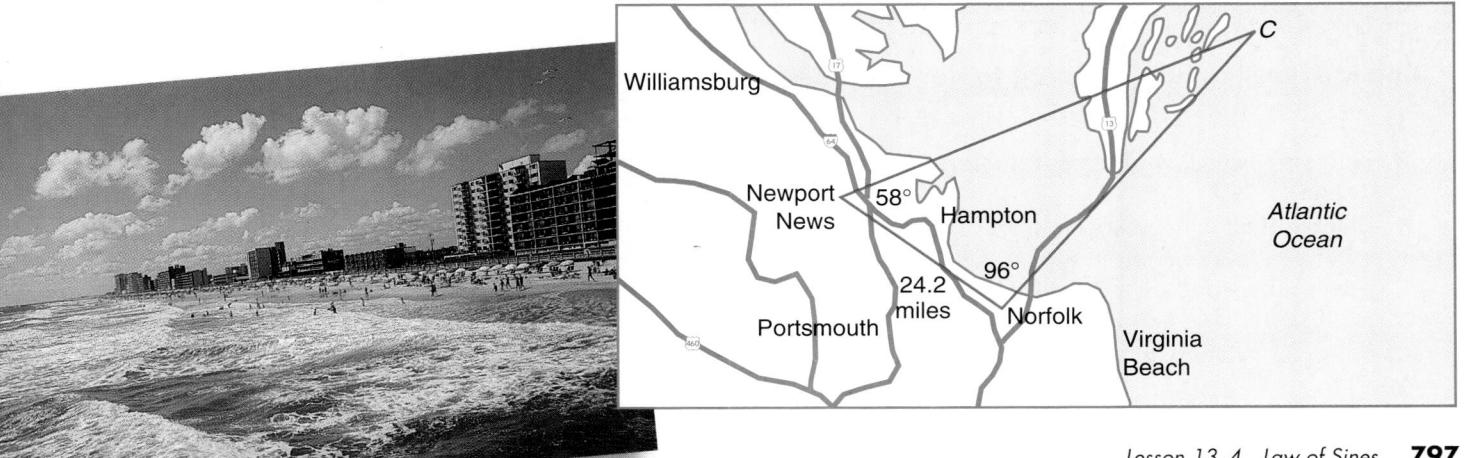

44. Make a Drawing Karen was given an assignment to draw and then construct a triangular model of three steel girders for her engineering class. Two of the girders measured 7 cm and 6 cm, and the angle opposite the 7 cm girder had to be 30°. Can she construct the triangle? If so, how long is the third girder?

45. Communication A low-watt radio station has its transmitter on County Line Road, 10 miles from where it intersects with the interstate highway. If the radio station has a range of 7 miles, between what two distances from the intersection can cars on the interstate hear the radio station?

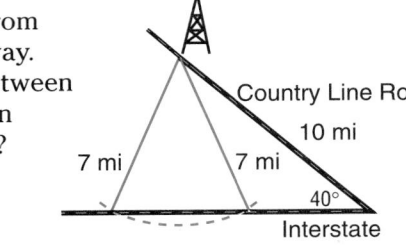

46. Geology A geologist measured a 43° angle of elevation to the top of a volcano crater. After moving 0.25 kilometers farther away, the angle of elevation was 38°.
 a. Use the law of sines to find the height of the top of the volcano crater.
 b. This problem was solved in Example 6 on page 776. Compare and contrast the method used there with the method you used here.

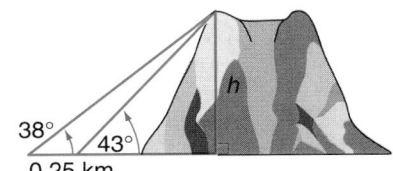

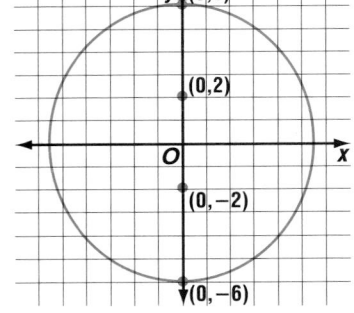

Mixed Review

47. Find the exact values of the six trigonometric functions for an angle in standard position that measures 180°. (Lesson 13–3)

48. Counting How many ways can 6 different books be arranged on a shelf? (Lesson 12–2)

49. Find the sum of the first 50 terms of an arithmetic series where $a_1 = 5$ and $d = 25$. (Lesson 11–2)

50. Finance Use the formula $A = Pe^{rt}$ to determine whether Sean can buy a used car costing $2500 with the $1000 his grandparents invested for him 16 years ago at 7%. (Lesson 10–5)

51. Simplify $\dfrac{3x - 21}{x^2 - 49} \div \dfrac{3x}{x^2 + 7x}$.

52. Write the equation of the ellipse shown at the right. (Lesson 7–4)

53. Factor $2x^2 - 11x - 21$. (Lesson 5–4)

54. Evaluate the determinant of $\begin{bmatrix} 6 & 4 \\ -3 & 2 \end{bmatrix}$. (Lesson 4–4)

SELF TEST

1. Solve the right triangle shown at the right. (Lesson 13–1)

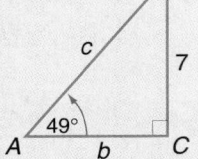

Change each degree measure to radian measure. (Lesson 13–2)

2. 90° **3.** 150° **4.** −135°

Change each radian measure to degree measure. (Lesson 13–2)

5. $\dfrac{3\pi}{2}$ **6.** $-\dfrac{7\pi}{4}$ **7.** 2

8. Find sin θ, cos θ, and tan θ if the terminal side of θ in standard position contains $P(-2, 0)$. (Lesson 13–3)

9. Electronics The power P in watts absorbed by an AC circuit is given by the formula $P = IV \cos \theta$, where I is the current in amps, V is the voltage, and θ is the measure of the phase angle. Find the power absorbed by a circuit if its current is 2 amps, its voltage is 120 volts, and its phase angle is 70°. (Lesson 13–3)

10. Solve the triangle in which $A = 40°$, $b = 12$, and $a = 5$. (Lesson 13–4)

Law of Cosines

APPLICATION
Paleontology

If Jurassic Park were a real place, it would be easy for scientists to study how dinosaurs move from place to place. But, since it's not, scientists are left to study the footprints made by dinosaurs millions of years ago. At dinosaur digs, anthropologists use *locomotor parameters,* which are numbers associated with physical motion.

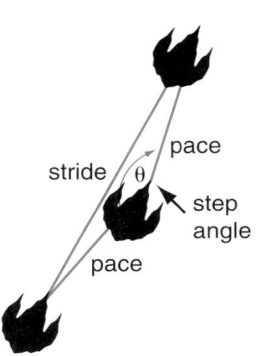

The figure at the right shows footprints of a carnivorous dinosaur taken from the Glen Rose formation in Texas. The *pace* is the distance from the left footprint to the right footprint, and vice versa. The *stride* is the distance from left footprint to the next left footprint or the right footprint to the next right footprint. If an animal walks in such a way that the footprints are directly in line, the stride will be twice the pace. But usually, the footprints show a "zig-zag" pattern that can be described numerically by the *step angle,* θ. An efficient walker has a step angle that approaches 180°, meaning that the animal minimizes zig-zag motion while maximizing forward motion.

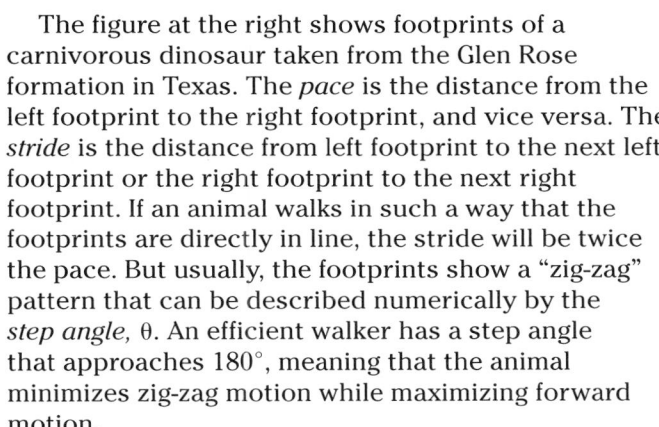

Scene from *Jurassic Park*

Anthropologists use trigonometry to determine the step angle. However, problems such as this, in which you know the measures of the sides of a triangle, cannot be solved using the law of sines. You can solve problems such as this by using the **law of cosines.**

To derive the law of cosines, consider $\triangle ABC$ with height h units and sides with lengths a units, b units, and c units. Suppose segment AD is x units long. Then segment DC is $b - x$ units long. What relationship exists between a, b, c, and A?

$a^2 = (b - x)^2 + h^2$ *Use the Pythagorean theorem for $\triangle DBC$.*

$a^2 = b^2 - 2bx + x^2 + h^2$ *Expand $(b - x)^2$.*

$a^2 = b^2 - 2bx + c^2$ *In $\triangle ADB$, $c^2 = x^2 + h^2$.*

$a^2 = b^2 - 2b(c \cos A) + c^2$ *$\cos A = \frac{x}{c}$, so $x = c \cos A$.*

$a^2 = b^2 + c^2 - 2bc \cos A$

The measure a is now defined in terms of the measures of the other two sides and angle A. You can find two other formulas relating the lengths of sides to the cosine of B and C in a similar way. All three formulas can be summarized as follows.

<table>
<tr><td>Law of Cosines</td><td>Let △ABC be any triangle with a, b, and c representing the measures of sides, and opposite angles with measurement A, B, and C, respectively. Then the following equations are true.
$$a^2 = b^2 + c^2 - 2bc \cos A$$
$$b^2 = a^2 + c^2 - 2ac \cos B$$
$$c^2 = a^2 + b^2 - 2ab \cos C$$</td></tr>
</table>

You can apply the law of cosines to a triangle if you know
- the measures of three sides, or
- the measures of two sides and the included angle.

Example Find c if $a = 15$, $b = 18$, and $C = 34°$.

You are given the measure of two sides and the included angle.
$$c^2 = a^2 + b^2 - 2ab \cos C$$
$$c^2 = 15^2 + 18^2 - 2(15)(18)\cos 34°$$
$$c^2 \approx 101.32$$
$$c \approx 10.07$$

Example Solve each triangle. Round to the nearest tenth.
a. $A = 47°$, $c = 27$, $b = 22$

You are given the measures of two sides and the included angle. First, determine a by using the law of cosines.
$$a^2 = b^2 + c^2 - 2bc \cos A$$
$$a^2 = 22^2 + 27^2 - 2(22)(27)\cos 47°$$
$$a^2 \approx 402.8$$
$$a \approx 20.1$$

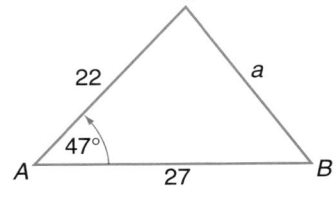

Next, use the law of sines to determine the measure of an angle.
$$\frac{\sin A}{a} = \frac{\sin B}{b}$$
$$\frac{\sin 47°}{20.1} \approx \frac{\sin B}{22} \qquad A = 47°, a \approx 20.1, b = 22$$
$$\sin B \approx \frac{22 \sin 47°}{20.1} \qquad \sin 47° \approx 0.7314$$
$$\sin B \approx 0.8005$$
$$B \approx 53.2°$$

Now, determine the measure of the third angle, $\angle C$.

$$47° + 53.2° + C \approx 180$$
$$C \approx 79.8°$$
Therefore, $a \approx 20.1$, $B \approx 53.2°$, and $C \approx 79.8°$.

b. $p = 29$, $q = 31$, $r = 48$

You are given the measures of three sides.
Use the law of cosines to find the measure of an angle.
$$p^2 = q^2 + r^2 - 2qr \cos P$$
$$29^2 = 31^2 + 48^2 - 2(31)(48)\cos P$$
$$2(31)(48)\cos P = 31^2 + 48^2 - 29^2$$
$$\cos P = \frac{31^2 + 48^2 - 29^2}{2(31)(48)}$$
$$\cos P \approx 0.8145$$
$$P \approx 35.5°$$

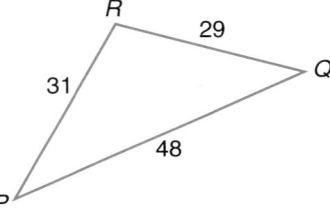

Use the law of sines to determine the measure of another angle.

$$\frac{\sin P}{p} = \frac{\sin Q}{q}$$

$$\frac{\sin 35.5°}{29} = \frac{\sin Q}{31}$$

$$\sin Q \approx \frac{31 \sin 35.5°}{29}$$

$$\sin Q \approx 0.6208$$

$$Q \approx 38.4°$$

Now find the measure of the third angle.

$$35.5° + 38.4° + R \approx 180°$$

$$R \approx 106.1°$$

Therefore, $P \approx 35.5°$, $Q \approx 38.4°$, and $R \approx 106.1°$.

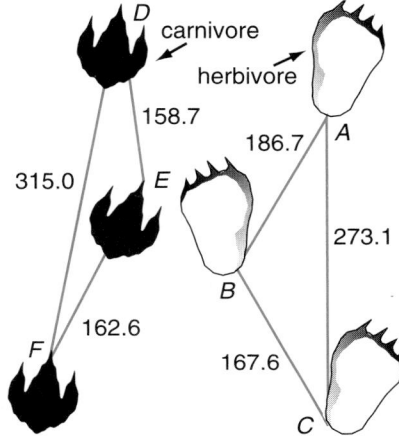

Example ③

At the Glen Rose formation in Texas, an anthropologist measured the pace and stride of footprints made by a bipedal (two-footed), carnivorous (meat-eating) dinosaur and the hindfeet of a herbivorous (plant-eating) dinosaur. The data are shown at the right.

a. Find the step angle for each dinosaur.

b. What can you tell about the motion of each dinosaur from its step angle?

a. Find the step angle for the herbivore, $\angle B$. $b = 273.1$, $a = 167.6$, $c = 186.7$

$$b^2 = a^2 + c^2 - 2ac \cos B \quad \text{Use the law of cosines.}$$

$$(273.1)^2 = (167.6)^2 + (186.7)^2 - 2(167.6)(186.7)\cos B$$

$$2(167.6)(186.7)\cos B = (167.6)^2 + (186.7)^2 - (273.1)^2$$

$$\cos B = \frac{(167.6)^2 + (186.7)^2 - (273.1)^2}{2(167.6)(186.7)}$$

$$\cos B \approx -0.1859$$

$$B \approx 100.7° \quad \textit{The step angle for the herbivore is 100.7°.}$$

Find the step angle for the carnivore, $\angle E$. $e = 315.0$, $d = 162.6$, $f = 158.7$

$$e^2 = d^2 + f^2 - 2df \cos E \quad \text{Use the law of cosines.}$$

$$(315.0)^2 = (162.6)^2 + (158.7)^2 - 2(162.6)(158.7)\cos E$$

$$2(162.6)(158.7)\cos E = (162.6)^2 + (158.7)^2 - (315.0)^2$$

$$\cos E = \frac{(162.6)^2 + (158.7)^2 - (315.0)^2}{2(162.6)(158.7)}$$

$$\cos E \approx -0.9223$$

$$E \approx 157.3° \quad \textit{The step angle for the carnivore is 157.3°.}$$

b. Since the step angle for the carnivore is closer to 180°, it appears as though the carnivore made more forward progress with each step than the sauropod.   *Why do you suppose a step angle close to 180° was important for a carnivore?*

Communicating Mathematics

Study the lesson. Then complete the following.

1. **Describe** a set of conditions for which the law of cosines can be used.

2. **State** which form of the law of cosines you would use to find a in the triangle at the right.

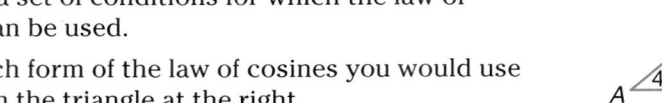

3. **Choose** the triangles that should be solved by beginning with the law of cosines.

a.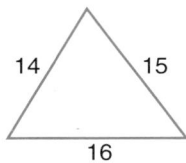

b.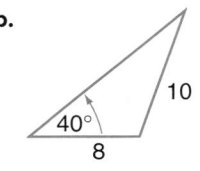

c.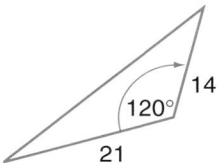

4. **Explain** why you cannot use the law of sines to solve a triangle if you are given $A = 80°$, $b = 20$, and $c = 55$.

5. **Make a chart** that summarizes the conditions necessary to use the law of sines and law of cosines.

MODELING MATHEMATICS

6. Collect data from your classmates or members of the track team and determine their step angles. Compare and contrast the step angles when walking versus running. Collect data from your classmates' pets and compare and contrast the step angles of different kinds of pets.

Determine whether each triangle can be solved by beginning with the law of sines or law of cosines. Then solve each triangle.

7.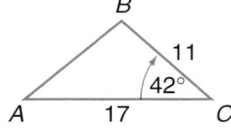

8.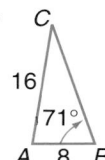

9. $a = 12$, $c = 15$, $A = 34°$

10. $a = 15$, $b = 18$, $c = 19$

11. The sides of a triangle are 6.8 cm, 8.4 cm, and 4.9 cm long. Find the measure of the smallest angle.

Guided Practice

Determine whether each triangle can be solved by beginning with the law of sines or law of cosines. Then solve each triangle.

12.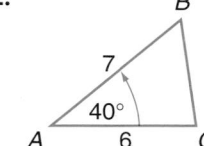

13.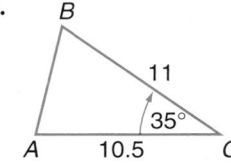

14.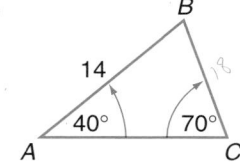

15.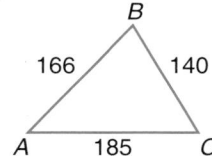

16.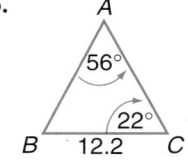

17.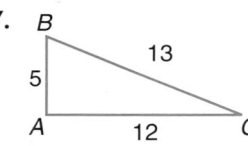

18. $A = 35°$, $b = 16$, $c = 19$

19. $a = 20$, $c = 24$, $B = 47°$

20. $a = 21.5$, $b = 13$, $C = 38.3°$

21. $A = 40°$, $B = 59°$, $c = 14$

22. $a = 51$, $c = 61$, $B = 19°$

23. $a = 13.7$, $A = 25°$, $B = 78°$

24. $a = 15$, $b = 25$, $c = 40$

25. $a = 345$, $b = 648$, $c = 442$

26. $c = 10.3$, $a = 21.5$, $b = 16.7$

27. $A = 28°$, $b = 5$, $c = 4.9$

28. $A = 29°$, $b = 7.6$, $c = 14.1$

29. $a = 8$, $b = 24$, $c = 18$

Geometry

30. The sides of a parallelogram measure 55 cm and 71 cm. Find the length of each diagonal if the larger angle measures 106°.

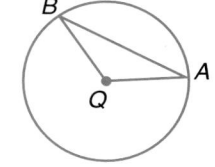

31. Circle Q at the right has a radius of 15 cm. Two radii $\overline{QA}$ and $\overline{QB}$ form an angle of 123°. Find the length of chord AB.

32. The sides of a triangle are 50 meters, 70 meters, and 85 meters. Find the measure of the angle opposite the shortest side.

Programming

33. The graphing calculator program at the right finds the measure of a side of a triangle using the law of cosines. A is the measure of the missing side, B and C are the measures of the second and third sides, and θ is the measure of the angle opposite side A.

   Use the program at the right to find the measure of the missing side.

   a. $B = 2$, $C = 4$, $\theta = 78°$
   b. $B = 9$, $C = 19$, $\theta = 45°$
   c. $B = 5.4$, $C = 6.9$, $\theta = 95°$

```
PROGRAM: Cosine
:Degree
:Disp "INPUT B"
:Input B
:Disp "INPUT C"
:Input C
:Disp "INPUT θ"
:Input θ
:√(B²+C²-2BCcosθ)→A
:Disp "A="
:Disp A
```

Critical Thinking

34. Explain how the Pythagorean theorem is a special case of the law of cosines.

Applications and Problem Solving

35. **Surveying** Two sides of a triangular plot of land have lengths of 400 feet and 600 feet. The measure of the angle between those sides is 46.3°. Find the perimeter and area of the plot.

36. **Emergency Medicine** A medical rescue helicopter has flown 45 miles from its home base to pick up an accident victim and 35 miles from there to the hospital. The angle between the two legs of the trip was 125°. The pilot needs to know how far he is now from his home base so he can decide whether to refuel before returning. How far is the hospital from the helicopter's base?

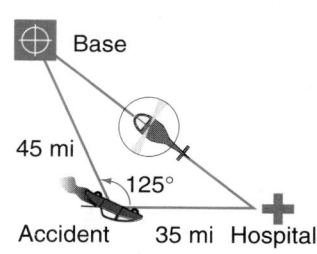

37. **Geography** Use the information in the map below to find the shortest distance from St. Petersburg, Florida, to New Orleans, Louisiana. The angle at Tallahassee, Florida, measures 125°, and the angle at Pensacola, Florida, measures 166°. (*Hint:* First find the distance from St. Petersburg to Pensacola.)

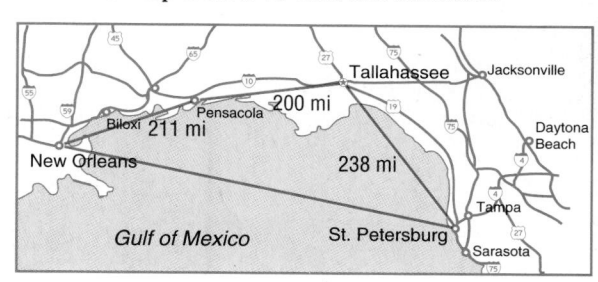

38. Use the law of sines to solve the triangle at the right. (Lesson 13–4)

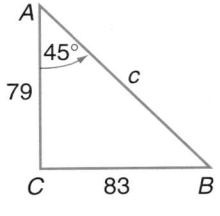

39. Change $-45°$ to radians. (Lesson 13–2)

40. How many 7-letter patterns can be formed from the letters of the word BENZENE? (Lesson 12–2)

41. Graph $f(x) = \dfrac{x}{x - 1}$. (Lesson 9–1)

42. **Geometry** The volume of a rectangular solid is 72 cubic units. The width is twice the height and the length is 7 units more than the height. Find the dimensions of the solid. (Lesson 8–5)

43. Solve $z^2 + 4z = 96$ by completing the square. (Lesson 6–3)

44. Simplify $\left(3 + \sqrt{2}\right)\left(\sqrt{10} + \sqrt{5}\right)$. (Lesson 5–6)

45. **Entertainment** More than 100 years ago, on December 28, 1895, the first motion picture was shown to a public audience. Many of the people who witnessed the film of a train pulling into a station were so afraid that they dove to the floor. By 1907, there were about 5000 nickelodeons in the United States. The admission price in 1907 was 5 cents. (Lesson 2–4)

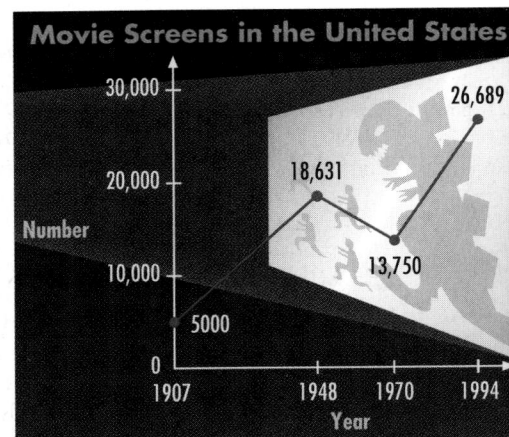

Movie Screens in the United States

a. The graph shows the increase in the number of movie screens in the United States from 1907 until 1994. Write an equation to represent this situation.

b. Predict the number of movie screens there will be in the year 2000.

Mathematics and SOCIETY

The Channel Tunnel

The excerpt below appeared in an article in *Popular Science* in May, 1994.

THIS IS THE CHANNEL TUNNEL. DREAMED of by Napoleon, futilely attempted 100 years ago, and built at a cost greater than most countries' gross national product, the 31-mile underwater rail link between England and continental Europe is finally in place. By 1996, the system will transport some 8 million car passengers and 4.5 million bus passengers each year . . . The owner, Eurotunnel, hopes to entice a large chunk of the tourist traffic that currently uses hovercrafts and ferries to cross the choppy waters of the English Channel . . . Commonly referred to as "The Chunnel," the Channel Tunnel is actually a complex of three parallel passageways, dipping as far as 148 feet below the seabed of the English Channel. ■

1. The two long sections of tunnel were dug toward each other from the British and French sides until they finally met and were linked deep below the English Channel. What types of measurements were needed to ensure that the two sections would meet properly?

2. What types of hazards might Chunnel travelers be exposed to? How might the Chunnel designers have protected travelers against them?

Circular Functions

13-6

What YOU'LL LEARN

- To define and use the trigonometric functions based on the unit circle.

Why IT'S IMPORTANT

You can use circular functions to solve problems involving music and entertainment.

CONNECTION
Music

If you have taken piano lessons, you probably know that the musical scale includes the notes A-B-C-D-E-F-G. If you start at the A below middle C and play the seven white keys that are highlighted in the figure below, there is a one-to-one correspondence between the keys on the piano and the notes of the musical scale. That is, for each highlighted key, there is exactly one musical note or name. However, there are several keys on the piano that are named A, B, C, and so on. If you play all the white keys, there is a many-to-one correspondence between the keys on the piano and the notes of the scale.

Similarly, there is a many-to-one correspondence between angles in standard position and their trigonometric functions. For example, the figure at the right shows that $\sin 30° = \frac{1}{2}$ and $\sin 150° = \frac{1}{2}$. In this lesson, we will now further generalize the trigonometric functions by defining them in terms of the unit circle.

Consider an angle θ in standard position. The terminal side of the angle intersects the unit circle at a unique point, $P(x, y)$. Recall that $\sin \theta = \frac{y}{r}$ and $\cos \theta = \frac{x}{r}$. Since $P(x, y)$ is on the unit circle, $r = 1$. Therefore, $\sin \theta = y$ and $\cos \theta = x$.

Definition of Sine and Cosine	If the terminal side of an angle θ in standard position intersects the unit circle at $P(x, y)$, then $\cos \theta = x$ and $\sin \theta = y$.

Since there is exactly one point $P(x, y)$ for any angle θ, the relations $\cos \theta = x$ and $\sin \theta = y$ are functions of θ. Because they are both defined using a unit circle, they are often called **circular functions.**

Example **1** **Point $P(0.6, 0.8)$ is located on a unit circle. Find $\sin \theta$, $\cos \theta$, and $\tan \theta$.**

$\sin \theta$ is the value of the y-coordinate.

$\sin \theta = 0.8$

$\cos \theta$ is the value of the x-coordinate.

$\cos \theta = 0.6$

$\tan \theta$ can be found by using the identity $\tan \theta = \frac{\sin \theta}{\cos \theta}$.

$\tan \theta = \frac{0.8}{0.6}$

≈ 1.3333

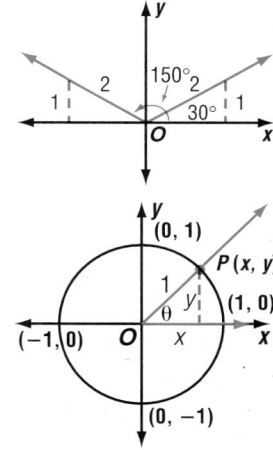

In the Exploration below, you will investigate the behavior of the sine and cosine functions on the unit circle.

Press MODE and highlight Degree and Par. Then use the following range values to set up a viewing window: TMIN = 0, TMAX = 360, TSTEP = 15, XMIN = −2.4, XMAX = 2.35, XSCL = 0.5, YMIN = −1.5, YMAX = 1.55, YSCL = 0.5. Define the unit circle with the definition $X_{1T} = \cos T$ and $Y_{1T} = \sin T$. Press GRAPH .

Your Turn

a. Activate the TRACE function to move around the circle. What does T represent? What does the x-value represent? What does the y-value represent?

b. Determine the sine and cosine of the angles whose terminal sides lie at 0°, 90°, 270°, and 360°.

c. How does the sine function change as you move around the unit circle? How does the cosine function change?

In this chapter, you have found the values of trigonometric functions for acute angles of right triangles, for angles in standard position on a coordinate plane, and now for angles of a unit circle. You have found exact values by using characteristics of special right triangles, and you have found approximate values by using a calculator. And you have done all of this for angles measured in degrees and in radians. This information can be summarized in the chart below. For convenience, the decimal approximations are rounded to the nearest tenth.

degrees	0	30	45	60	90	120	135	150	180	210	225	240	270	300	315	330	360
radians	0	$\frac{\pi}{6}$	$\frac{\pi}{4}$	$\frac{\pi}{3}$	$\frac{\pi}{2}$	$\frac{2\pi}{3}$	$\frac{3\pi}{4}$	$\frac{5\pi}{6}$	π	$\frac{7\pi}{6}$	$\frac{5\pi}{4}$	$\frac{4\pi}{3}$	$\frac{3\pi}{2}$	$\frac{5\pi}{3}$	$\frac{7\pi}{4}$	$\frac{11\pi}{6}$	2π
sin θ	0	$\frac{1}{2}$	$\frac{\sqrt{2}}{2}$	$\frac{\sqrt{3}}{2}$	1	$\frac{\sqrt{3}}{2}$	$\frac{\sqrt{2}}{2}$	$\frac{1}{2}$	0	$-\frac{1}{2}$	$-\frac{\sqrt{2}}{2}$	$-\frac{\sqrt{3}}{2}$	−1	$-\frac{\sqrt{3}}{2}$	$-\frac{\sqrt{2}}{2}$	$-\frac{1}{2}$	0
nearest tenth	0	0.5	0.7	0.9	1	0.9	0.7	0.5	0	−0.5	−0.7	−0.9	−1	−0.9	−0.7	−0.5	0
cos θ	1	$\frac{\sqrt{3}}{2}$	$\frac{\sqrt{2}}{2}$	$\frac{1}{2}$	0	$-\frac{1}{2}$	$-\frac{\sqrt{2}}{2}$	$-\frac{\sqrt{3}}{2}$	−1	$-\frac{\sqrt{3}}{2}$	$-\frac{\sqrt{2}}{2}$	$-\frac{1}{2}$	0	$\frac{1}{2}$	$\frac{\sqrt{2}}{2}$	$\frac{\sqrt{3}}{2}$	1
nearest tenth	1	0.9	0.7	0.5	0	−0.5	−0.7	−0.9	−1	−0.9	−0.7	−0.5	0	0.5	0.7	0.9	1

You will learn more about the graphs of trigonometric functions in Chapter 14.

As is often the case, a graph may be a more effective way of presenting data. In the graphs below, the horizontal axis shows the values of θ, and the vertical axis shows the values of sin θ or cos θ.

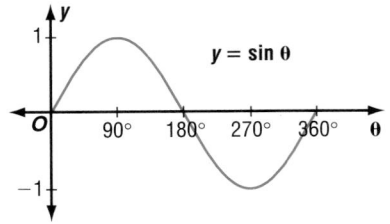

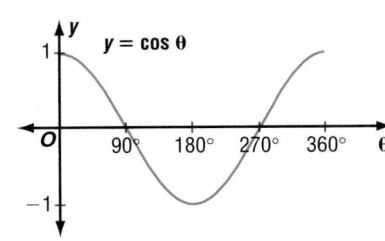

The following chart contains the same information as the earlier chart, but for angles from 360° to 720°. Compare the information in the two charts. As you can see, the values of sin θ and cos θ are the same for two angles that are coterminal.

degrees	360	390	405	420	450	480	495	510	540	570	585	600	630	660	675	690	720
radians	2π	$\frac{13\pi}{6}$	$\frac{9\pi}{4}$	$\frac{7\pi}{3}$	$\frac{5\pi}{2}$	$\frac{8\pi}{3}$	$\frac{11\pi}{4}$	$\frac{17\pi}{6}$	3π	$\frac{19\pi}{6}$	$\frac{13\pi}{4}$	$\frac{10\pi}{3}$	$\frac{7\pi}{2}$	$\frac{11\pi}{3}$	$\frac{15\pi}{4}$	$\frac{23\pi}{6}$	4π
sin θ	0	$\frac{1}{2}$	$\frac{\sqrt{2}}{2}$	$\frac{\sqrt{3}}{2}$	1	$\frac{\sqrt{3}}{2}$	$\frac{\sqrt{2}}{2}$	$\frac{1}{2}$	0	$-\frac{1}{2}$	$-\frac{\sqrt{2}}{2}$	$-\frac{\sqrt{3}}{2}$	-1	$-\frac{\sqrt{3}}{2}$	$-\frac{\sqrt{2}}{2}$	$-\frac{1}{2}$	0
nearest tenth	0	0.5	0.7	0.9	1	0.9	0.7	0.5	0	−0.5	−0.7	−0.9	−1	−0.9	−0.7	−0.5	0
cos θ	1	$\frac{\sqrt{3}}{2}$	$\frac{\sqrt{2}}{2}$	$\frac{1}{2}$	0	$-\frac{1}{2}$	$-\frac{\sqrt{2}}{2}$	$-\frac{\sqrt{3}}{2}$	-1	$-\frac{\sqrt{3}}{2}$	$-\frac{\sqrt{2}}{2}$	$-\frac{1}{2}$	0	$\frac{1}{2}$	$\frac{\sqrt{2}}{2}$	$\frac{\sqrt{3}}{2}$	1
nearest tenth	1	0.9	0.7	0.5	0	−0.5	−0.7	−0.9	−1	−0.9	−0.7	−0.5	0	0.5	0.7	0.9	1

Notice there is a many-to-one correspondence between the angles and their sine and cosine functions.

Every 360°, or 2π radians, represents one complete revolution of the terminal side. As you can see by comparing the two charts, for every 360° or 2π radians, the sine and cosine functions repeat their values. So, we can say that the sine and cosine functions are **periodic.** Each has a **period** of 360° or 2π radians.

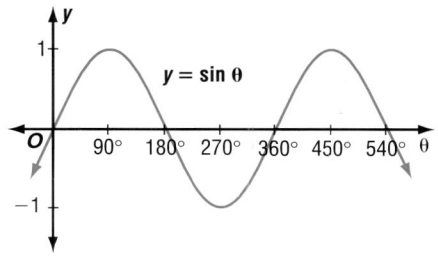

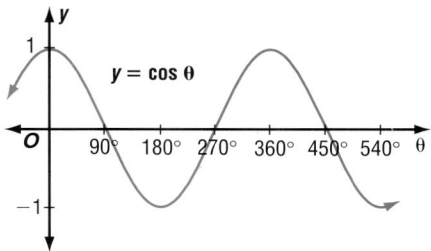

Definition of Periodic Function	**A function is called periodic if there is a number *a* such that *f(x)* = *f(x + a)* for all *x* in the domain of the function. The least positive value of *a* for which *f(x)* = *f(x + a)* is called the period of the function.**

When you look at the graph of a periodic function, you will see a repeating pattern: a shape that repeats over and over as you travel out the *x*-axis. The period is the distance along the *x*-axis from the beginning of the pattern to the point at which it begins again.

For the sine and cosine functions, cos (*x* + 360°) = cos *x*, and sin (*x* + 360°) = sin *x*. In radian measure, cos (*x* + 2π) = cos *x*, and sin (*x* + 2π) = sin *x*. Therefore, the period of the sine and cosine functions is 360° or 2π.

Example Find the exact value of each function.

a. **420°**

$$\sin 420° = \sin(60 + 360)°$$
$$= \sin 60°$$
$$= \frac{\sqrt{3}}{2}$$

b. $\cos\left(-\frac{3\pi}{4}\right)$

$$\cos\left(-\frac{3\pi}{4}\right) = \cos\left(-\frac{3\pi}{4} + 2\pi\right)$$
$$= \cos\left(\frac{5\pi}{4}\right)$$
$$= -\frac{\sqrt{2}}{2}$$

Many real-world situations have characteristics that can be described with periodic functions.

Example ❸ When a string on a guitar is plucked, it is displaced from a fixed point in the middle of the string and vibrates back and forth, producing a musical tone. The exact tone depends on the frequency, or number of cycles per second, that the string vibrates. To produce an A, the frequency is 440 cycles per second, or 440 hertz (Hz).
a. Identify the period of this function.
b. Graph this situation.

a. Since the string vibrates at a frequency of 440 Hz, the period is the time it takes to complete one cycle, or $\frac{1}{440}$ second.

b. Let the horizontal axis represent the time in seconds. Let the vertical axis represent how far the fixed point on the string is displaced from its resting position.

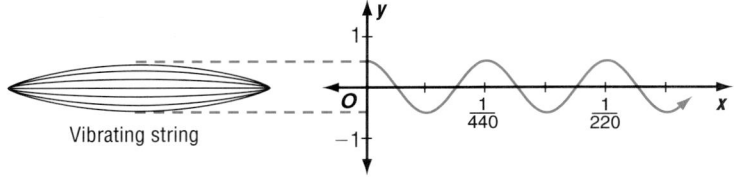

Vibrating string

CHECK FOR UNDERSTANDING

Communicating Mathematics

Study the lesson. Then complete the following.

1. **Define** the sine and cosine functions for the angle at the right.

2. **Look for a pattern** in the chart on page 806.

3. **Compare and contrast** the graphs of the sine and cosine functions on page 807.

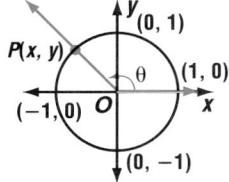

Math Journal

4. **Assess Yourself** Write a paragraph in which you state your understanding of trigonometry as it has developed from the acute angles of a right triangle to angles on a coordinate plane that intersect the unit circle.

Guided Practice

Find sin θ, cos θ, and tan θ for each angle.

5.

$P\left(-\frac{3}{5}, \frac{4}{5}\right)$

6.

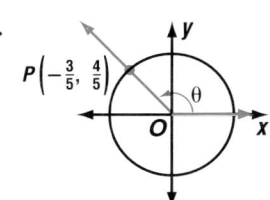

$P\left(-\frac{12}{13}, -\frac{5}{13}\right)$

7. Determine the period of the function that is graphed below.

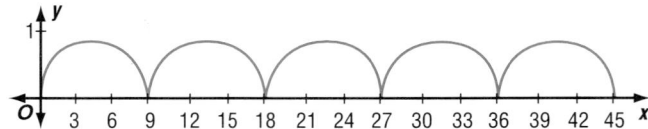

Find the value of each function.

8. $\cos\left(-\frac{3\pi}{4}\right)$

9. $\sin 660°$

10. **Music** Refer to the application at the beginning of the lesson. Determine whether the musical scale on a piano is a periodic function. If so, name the period.

Practice

Find sin θ, cos θ, and tan θ for each angle.

11.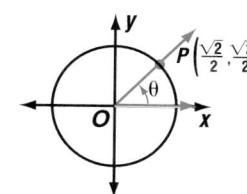

12.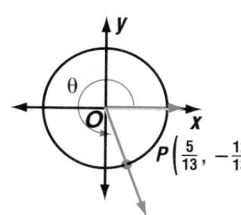

13.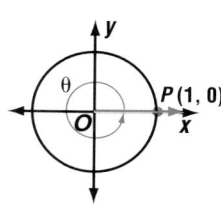

Determine the period of each function.

14.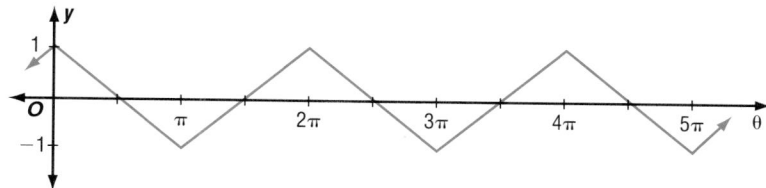

15.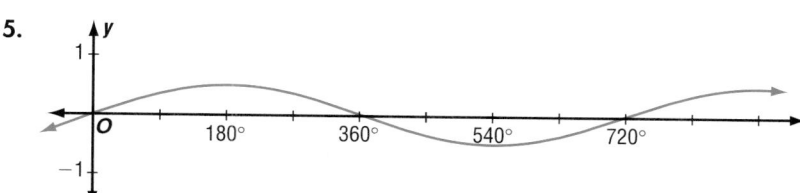

16.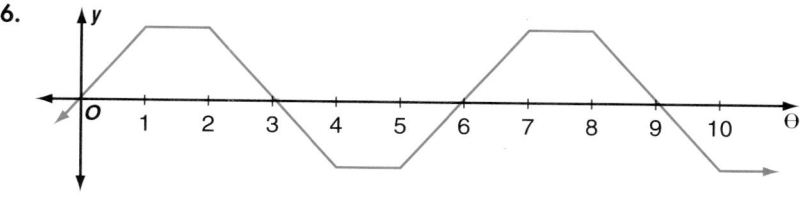

Find the exact value of each function.

17. $\sin 1020°$

18. $\cos (-450°)$

19. $\sin (-180°)$

20. $\sin \left(-\dfrac{13\pi}{6}\right)$

21. $\sin \dfrac{3\pi}{2}$

22. $\cos \dfrac{9\pi}{2}$

23. $4(\sin 30°)(\cos 60°)$

24. $\dfrac{\sin 30° + \cos 60°}{2}$

25. $\dfrac{4 \sin 300° + 2 \cos 30°}{3}$

26. $\sin 30° + \sin 60°$

27. $(\sin 60°)^2 + (\cos 60°)^2$

28. $8(\sin 120°)(\cos 120°)$

Graphing Calculator

29. Use a graphing calculator to graph the functions $y = \sin θ$ and $y = 2 \sin θ$ on the same screen. Predict the shape of the graph of $y = 3 \sin θ$. Check by graphing.

Critical Thinking

30. Determine the domain and range of the functions $y = \sin θ$, $y = \cos θ$, and $y = \tan θ$.

Applications and Problem Solving

31. Entertainment As you ride a Ferris wheel, the height that you are above the ground varies periodically. Consider the height of the center of the wheel to be the starting point. A particular wheel has a diameter of 38 feet and travels at a rate of 4 revolutions per minute. Make a graph in which the horizontal axis represents time and the vertical axis represents height in relation to the starting point.

32. Physics The motion of a weight on a spring varies periodically. Suppose you pull the weight down 3 inches from its equilibrium point and then release it. It bounces above the equilibrium point and then returns below the equilibrium point in 2 seconds. Graph the height of the spring as a function of time.

Mixed Review

33. Aviation A pilot is flying from Chicago to Columbus, a distance of 300 miles. In order to avoid an area of thunderstorms, she alters her course by 15° and flies on this course for 75 miles. How far is she from Columbus? (Lesson 13–5)

75 mi B a

15°

Chicago   300 mi   Columbus

34. Find tan 135°. (Lesson 13–3)

35. Determine whether the figure at the right is a fractal. Explain your reasoning. (Lesson 11–7)

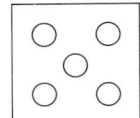

36. Solve $4^x = 24$. (Lesson 10–6)

37. Solve $x^4 - 13x^2 + 36 = 0$. (Lesson 8–6)

38. If $A = \begin{bmatrix} -3 & 5 \\ 1 & -4 \end{bmatrix}$, find A^{-1}. (Lesson 4–5)

39. Use the elimination method to solve the system of equations. (Lesson 3–2)
$$x + 2y = -2$$
$$3x - 2y = 10$$

WORKING ON THE
In·ves·ti·ga·tion

Refer to the Investigaton on pages 768—769.

Scream Machines!

Select a picture of a roller coaster whose support structure is made of triangular shapes. Find a section of the structure that contains at least two different triangular shapes.

1 Make an enlarged scale drawing of the section you chose. Select one triangle and label its angles, *A*, *B*, and *C*, and its sides *a*, *b*, and *c*.

2 Use a ruler to measure the lengths of *b* and *c*. Use a protractor to measure ∠*A*. Include these measurements on your drawing.

3 Use the law of cosines and the law of sines to solve your triangle. Use a ruler and protractor to check your calculations. If calculations do not agree with your measurements, try to explain why.

4 Repeat this activity with a different triangle in the section.

5 Do you think a knowledge of trigonometry would be helpful to a coaster designer? Explain why or why not.

Add the results of your work to your Investigation Folder.

Inverse Trigonometric Functions

APPLICATION
Architecture

Chinese-born architect I. M. Pei is well known for incorporating triangles into the design of his buildings. Among his famous designs are the glass pyramid entrance to the Louvre museum in Paris and the newly opened Rock and Roll Hall of Fame in Cleveland, Ohio.

Architects often use trigonometry when designing buildings. Sometimes the value of some trigonometric function for an angle is known and it is necessary to find the measure of the angle. The concept of inverse functions can be applied to find the inverse of trigonometric functions.

In Lesson 8–8, you learned that the inverse of a function is the relation in which all the values of x and y are reversed. The graphs of $y = \cos x$ and its inverse, $x = \cos y$, are shown below. Notice that the inverse is not a function, since it fails the vertical line test. None of the inverses of the trigonometric functions are functions.

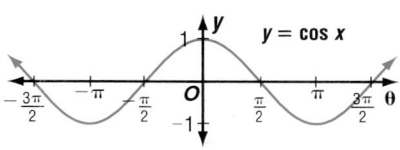

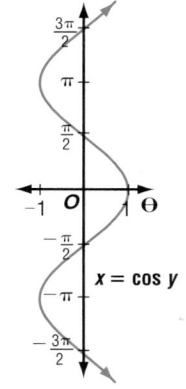

We must restrict the domain of the trigonometric functions so that the inverses are functions. The values in these restricted domains are called **principal values**. Capital letters are used to distinguish trigonometric functions with restricted domains from the usual trigonometric functions.

Definitions of the Cosine, Sine, and Tangent Functions	$y = \text{Cos } x$ if and only if $y = \cos x$ and $0 \le x \le \pi$. $y = \text{Sin } x$ if and only if $y = \sin x$ and $-\frac{\pi}{2} \le x \le \frac{\pi}{2}$. $y = \text{Tan } x$ if and only if $y = \tan x$ and $-\frac{\pi}{2} \le x \le \frac{\pi}{2}$.

The inverse of the Cosine function is called the **Arccosine function** and is symbolized by Cos^{-1} or **Arccos**.

Definition of Inverse Cosine	**Given $y = \text{Cos } x$, the inverse Cosine function is defined by the following equation.** $$y = \text{Cos}^{-1} x \text{ or } y = \text{Arccos } x$$

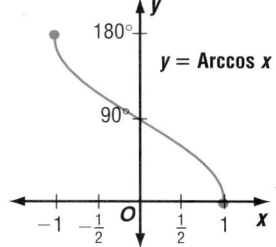

The Arccosine function has the following characteristics.

- Its domain is the set of real numbers from -1 to 1.
- Its range is the set of angle measures from 0 to π, inclusive.
- $\text{Cos } x = y$ if and only if $\text{Cos}^{-1} y = x$.
- $[\text{Cos}^{-1} \circ \text{Cos}](x) = [\text{Cos} \circ \text{Cos}^{-1}](x) = x$. *Recall function composition from Lesson 8–7.*

The definitions of the Arcsine and Arctangent functions are similar to the definition of the Arccosine function.

Definition of Inverse Sine and Inverse Tangent	Given $y = \text{Sin } x$, the inverse Sine function is defined as follows. $y = \text{Sin}^{-1} x$ or $y = \text{Arcsin } x$ Given $y = \text{Tan } x$, the inverse Tangent function is defined as follows. $y = \text{Tan}^{-1} x$ or $y = \text{Arctan } x$

The expressions in each row of the table below are equivalent. You can use these expressions to rewrite and solve trigonometric equations.

$y = \text{Sin } x$	$x = \text{Sin}^{-1} y$ or $x = \text{Arcsin } y$
$y = \text{Cos } x$	$x = \text{Cos}^{-1} y$ or $x = \text{Arccos } y$
$y = \text{Tan } x$	$x = \text{Tan}^{-1} y$ or $x = \text{Arctan } y$

Example ① Solve $\text{Cos } x = \frac{1}{2}$.

If $\text{Cos } x = \frac{1}{2}$, then x is the least value whose cosine is $\frac{1}{2}$.

$x = \text{Arccos } \frac{1}{2}$  $\quad$ 60

Therefore, $x = 60°$. *In radians, $x = \frac{\pi}{3}$.*

Many application problems involve finding the inverse of a trigonometric function.

Example ② Highway curves are usually banked, or tilted inward, so that cars can negotiate the curve more safely. The proper banking angle θ for a car making a turn of radius r feet at a velocity of v feet per second is given by the the equation $\tan \theta = \frac{v^2}{gr}$, where g is the acceleration due to gravity, 32 ft/s². An engineer is designing a curve with a radius of 1000 feet. If the speed limit on the curve will be 55 mph, at what angle should the curve be banked?

First, rewrite 55 miles per hour in feet per seconds.

$$\frac{55 \text{ miles}}{\text{hour}} \cdot \frac{5280 \text{ feet}}{\text{mile}} \cdot \frac{\text{hour}}{3600 \text{ seconds}} \approx 80.7 \text{ feet per second}$$

Now, use the equation $\tan \theta = \frac{v^2}{gr}$.

$$\tan \theta = \frac{80.7^2}{(32)(1000)}$$

$$\theta = \arctan \frac{80.7^2}{(32)(1000)}$$

Use a calculator to find θ.

80.7  $\quad$ 11.503467

Therefore, the curve should be banked at an angle of 11.5°.

You can also use a calculator to find the value of a complicated trigonometric expression.

Example 3

Use a scientific calculator to find $\tan\left(\text{Sin}^{-1}\left(\frac{5}{13}\right)\right)$.

$\boxed{(}$ 5 $\boxed{\div}$ 13 $\boxed{)}$ $\boxed{\text{SIN}^{-1}}$ $\boxed{\text{TAN}}$ 0.4166667

Therefore, $\tan\left(\text{Sin}^{-1}\left(\frac{5}{13}\right)\right) \approx 0.4167$.

CHECK FOR UNDERSTANDING

Communicating Mathematics

Study the lesson. Then complete the following.

1. **Describe** how $y = \text{Sin } x$ and $y = \text{Arcsin } x$ are related.

2. **Explain** why the domains of the trigonometric functions must be restricted before finding the inverse functions.

3. **Explain** how you know when the domain of a trigonometric function is restricted.

Guided Practice

Write each equation in the form of an inverse function.

4. $x = \sin \theta$

5. $\tan y = -3$

Solve each equation.

6. $x = \text{Arcsin } 0$

7. $\text{Arctan } 1 = y$

Find each value.

8. $\sin\left(\text{Cos}^{-1}\left(\frac{2}{3}\right)\right)$

9. $\cos\left(\text{Cos}^{-1}\left(\frac{4}{5}\right)\right)$

10. $\cos\left(\text{Cos}^{-1}\left(\frac{1}{2}\right)\right)$

11. $\text{Tan}^{-1}(-1)$

12. $\text{Sin } \frac{\pi}{6}$

13. $\text{Sin}^{-1}(1)$

14. **Architecture** The support for a roof will be shaped like two right triangles as shown at the right. Each right triangle will have one leg 8 feet long and a hypotenuse of 16 feet. Find θ.

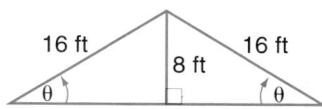
16 ft   16 ft   8 ft θ θ

EXERCISES

Practice

Write each equation in the form of an inverse function.

15. $a = \cos b$

16. $\sin y = x$

17. $\tan \alpha = \beta$

18. $\sin 30° = \frac{1}{2}$

19. $\cos 45° = y$

20. $-\frac{4}{3} = \tan x$

Solve each equation.

21. $x = \text{Cos}^{-1}(0)$

22. $x = \text{Sin}^{-1}\left(\frac{1}{\sqrt{2}}\right)$

23. $y = \text{Arctan } \frac{\sqrt{3}}{3}$

24. $\text{Arctan } 0 = x$

25. $\text{Sin}^{-1}\left(\frac{1}{2}\right) = y$

26. $x = \text{Cos}^{-1}\left(\frac{\sqrt{2}}{2}\right)$

Find each value.

27. $\text{Cos}^{-1}\left(-\frac{1}{2}\right)$

28. $\sin\left(\text{Sin}^{-1}\left(\frac{1}{2}\right)\right)$

29. $\text{Sin}^{-1}\left(\cos \frac{\pi}{2}\right)$

30. $\tan\left(\text{Cos}^{-1}\left(\frac{6}{7}\right)\right)$

31. $\cot\left(\text{Sin}^{-1}\left(\frac{5}{6}\right)\right)$

32. $\cot\left(\text{Sin}^{-1}\left(\frac{7}{9}\right)\right)$

33. $\sin\left(\text{Arctan } \frac{\sqrt{3}}{3}\right)$ **34.** $\cos\left(\text{Arcsin } \frac{3}{5}\right)$ **35.** $\tan(\text{Arctan } 3)$

36. $\text{Sin}^{-1}\left(\tan \frac{\pi}{4}\right)$ **37.** $\text{Arctan } \sqrt{3}$ **38.** $\text{Arccos } \frac{\sqrt{3}}{2}$

39. $\cos\left(\text{Tan}^{-1}\left(\sqrt{3}\right)\right)$ **40.** $\cos\left[\text{Arcsin}\left(-\frac{1}{2}\right)\right]$ **41.** $\cos(\text{Tan}^{-1}(1))$

42. $\cos\left[\text{Cos}^{-1}\left(\frac{\sqrt{2}}{2}\right) - \frac{\pi}{2}\right]$ **43.** $\sin\left(2 \text{ Sin}^{-1}\left(\frac{1}{2}\right)\right)$ **44.** $\sin\left(2 \text{ Cos}^{-1}\left(\frac{3}{5}\right)\right)$

Critical Thinking

45. Prove that $[\text{Cos}^{-1} \circ \text{Cos}](x) = [\text{Cos} \circ \text{Cos}^{-1}](x) = x$.

Applications and Problem Solving

46. Navigation The *Western Princess* sailed due east 24 miles before turning north. When the *Princess* became disabled and radioed for help, the rescue boat needed to know the fastest route to her. The navigator of the *Princess* found that the fastest route for the rescue boat would be 48 miles. The cosine of the angle at which the rescue boat should sail is $\frac{1}{2}$. Find the angle at which the rescue boat should travel to aid the *Western Princess*.

47. Optics When light is polarized, all of the waves are traveling in parallel planes. You may have polarized sunglasses that eliminate glare by polarizing the light. Suppose vertically polarized light with intensity I_o strikes a polarizing filter with its axis at an angle of θ with the vertical. The intensity of the transmitted light I_t and θ are related by the equation $\cos \theta = \sqrt{\dfrac{I_t}{I_o}}$. If one fourth of the polarized light is transmitted through the lens, at what angle is the lens being held?

Mixed Review

48. Determine the period of the function shown at the right. (Lesson 13–6)

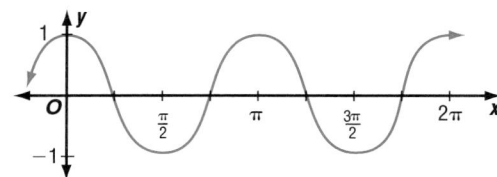

49. Recreation Isabel is flying a kite for which the angle of elevation is $70°$. The string on the kite is 65 meters long. How far is the kite above the ground? (Lesson 13–1)

50. Write the expression $\displaystyle\sum_{k=1}^{10} (2 + k)$ in expanded form and find the sum. (Lesson 11–1)

51. Biology Bacteria from a certain strain grows from 80 to 164 bacteria in 3 hours. Find k for the growth formula for this strain. (*Hint:* The exponential growth formula is $y = ne^{kt}$.) (Lesson 10–7)

52. Find the center and radius of a circle whose equation is $(x - 4)^2 + (y - 9)^2 = 4$. (Lesson 7–3)

53. Solve $x^2 - 7x = 0$ by factoring. (Lesson 6–2)

54. Find AA if $A = \begin{bmatrix} 2 & 7 \\ 0 & -1 \end{bmatrix}$. (Lesson 4–3)

55. Solve the system of inequalities by graphing. (Lesson 3–4)

$x + y > 2$

$y > 3$

VOCABULARY

After completing this chapter, you should be able to define each term, property, or phrase and give an example or two of each.

Geometry
angle of depression (p. 776)
angle of elevation (p. 776)
coterminal angles (p. 783)
initial side (p. 780)
standard position (p. 780)
terminal side (p. 780)

Trigonometry
circular functions (p. 805)
cosecant (p. 772)
cosine (p. 772, 805)
Cosine function (p. 811)
cotangent (p. 772)

inverse Cosine (Cos⁻¹ or Arccos) (p. 811)
inverse Sine (Sin⁻¹ or Arcsin) (p. 812)
inverse Tangent (Tan⁻¹ or Arctan) (p. 812)
law of cosines (p. 799)
law of sines (p. 792)
period (p. 807)
periodic function (p. 807)
principal values (p. 811)
quadrantal angles (p. 787)
quotient identities (p. 788)
radian (p. 781)
reciprocal identities (p. 788)

secant (p. 772)
sine (p. 772, 805)
Sine function (p. 811)
solving a triangle (p. 775)
tangent (p. 772)
Tangent function (p. 811)
trigonometric functions (p. 772)
trigonometric functions of an angle in standard position (p. 786)
trigonometric identities (p. 788)
trigonometry (p. 772)
unit circle (p. 781)

UNDERSTANDING AND USING THE VOCABULARY

State whether each sentence is *true* or *false*. If false, replace the underlined word(s) or number to make a true sentence.

1. When two angles in standard position have the same terminal side, they are called <u>quadrantal</u> angles.

2. The <u>law of sines</u> is used when the measure of two angles and the measure of any side are known.

3. <u>Trigonometric</u> functions can be defined by using a unit circle.

4. $\underline{\csc \theta = \dfrac{1}{\cos \theta}}$ is a reciprocal identity.

5. A <u>radian</u> is the measure of an angle on the unit circle where the rays of the angle intercept an arc with length 1 unit.

6. If the measures of three sides of a triangle are known, then the <u>law of sines</u> can be used to solve the triangle.

7. The period of a function is the distance along the <u>x-axis</u> from the beginning of the pattern to the point at which it begins again.

8. <u>60°</u> is a quadrantal angle.

9. $\underline{\cot \theta = \dfrac{\sin \theta}{\cos \theta}}$

10. In a coordinate plane, the <u>initial</u> side of an angle is the ray that rotates about the center.

SKILLS AND CONCEPTS

OBJECTIVES AND EXAMPLES	REVIEW EXERCISES

Upon completing this chapter, you should be able to:

Use these exercises to review and prepare for the chapter test.

- find values of trigonometric functions for acute angles (Lesson 13–1)

Find the values of the other five trigonometric functions if $\sin \theta = \frac{7}{8}$.

$\cos \theta = \dfrac{\sqrt{15}}{8}$ or 0.4841 $\sec \theta = \dfrac{8\sqrt{15}}{15}$ or 2.0656

$\tan \theta = \dfrac{7\sqrt{15}}{15}$ or 1.8074 $\cot \theta = \dfrac{\sqrt{15}}{7}$ or 0.5533

$\csc \theta = \dfrac{8}{7}$ or 1.1429

Suppose θ is an acute angle of a right triangle. For each function, find the values of the remaining five trigonometric functions of θ. Round to four decimal places.

11. $\cos \theta = \dfrac{2}{15}$ **12.** $\tan \theta = 0.5$

13. $\sin \theta = \dfrac{3}{4}$ **14.** $\sec \theta = 2\dfrac{1}{2}$

Find the value of x. Round to the nearest degree.

15. $\tan x = 0.2679$ **16.** $\cos x = 0.8387$

17. $\sin x = 0.9659$ **18.** $\tan x = 1.0724$

- solve problems involving right triangles (Lesson 13–1)

Solve the right triangle in which $c = 14$ and $A = 42°$.

$\sin 42° = \dfrac{a}{14}$ $\cos 42° = \dfrac{b}{14}$

$14 \sin 42° = a$ $14 \cos 42° = b$

$9.4 \approx a$ $10.4 \approx b$

$B = 90° - 42°$ or $48°$

$a \approx 9.4, b \approx 10.4, B = 48°$

Solve each right triangle. Assume that C represents the right angle and c is the hypotenuse. Round measures of sides to the nearest tenth and measures of angles to the nearest degree.

19. $c = 16, a = 7$

20. $A = 25°, c = 6$

21. $B = 45°, c = 12$

22. $B = 83°, b = \sqrt{31}$

23. $a = 9, B = 49°$

24. $\cos A = \dfrac{1}{4}, a = 4$

- change radian measure to degree measure and vice versa (Lesson 13–2)

Change the degree measure 240° to radians.

$240° \cdot \dfrac{\pi \text{ radians}}{180°} = \dfrac{240\,\pi}{180}$ radians or $\dfrac{4\pi}{3}$ radians

Change the radian measure $\dfrac{\pi}{12}$ to degrees.

$\dfrac{\pi}{12} \text{ radians} \cdot \dfrac{180°}{\pi \text{ radians}} = \dfrac{180°\pi}{12\pi}$ or $15°$

Change each degree measure to radian measure.

25. $255°$ **26.** $-210°$

27. $65°$ **28.** $120°$

Change each radian measure to degree measure.

29. $\dfrac{7\pi}{4}$ **30.** $-\dfrac{5\pi}{12}$

31. -4π **32.** $\dfrac{5\pi}{3}$

OBJECTIVES AND EXAMPLES	REVIEW EXERCISES

find values of trigonometric functions for general angles (Lesson 13–3)

Find the exact value of sin 120°.

Sketch the angle in standard position. Its terminal side lies in Quadrant II. Notice the 30°−60° right triangle. Choose $P(x, y)$ on the terminal side of the angle so that $r = 2$.

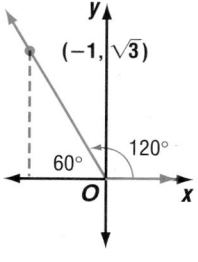

$(-1, \sqrt{3})$

$120°$

$60°$

It follows that $x = -1$ and $y = \sqrt{3}$.

$\sin 120° = \dfrac{y}{r}$ or $\dfrac{\sqrt{3}}{2}$

Find the exact value of each trigonometric function.

33. $\cos 210°$ **34.** $\tan 120°$

35. $\sin \dfrac{5}{4}\pi$ **36.** $\cos 3\pi$

37. $\sec (-30°)$ **38.** $\cot \dfrac{7}{6}\pi$

Find the exact values of sin θ, cos θ, and tan θ if the terminal side of θ in standard position contains the given point.

39. $P(2, 5)$ **40.** $P(15, -8)$

---

solve triangles by using the law of sines (Lesson 13–4)

Solve the triangle in which $A = 54°$, $C = 72°$, and $a = 20$.

$\dfrac{\sin 54°}{20} = \dfrac{\sin 72°}{c}$

$c = \dfrac{20 \sin 72°}{\sin 54°} \approx 23.5$

$B = 180 - (54 + 72)$ or $54°$

$\dfrac{\sin 54°}{b} = \dfrac{\sin 54°}{20}$

$b = 20$

Solve each triangle described. Round measures of sides to the nearest tenth and measures of angles to the nearest degree.

41. $A = 50°$, $b = 12$, $a = 10$

42. $B = 46°$, $C = 83°$, $b = 65$

43. $A = 45°$, $B = 30°$, $b = 20$

44. $A = 105°$, $a = 18$, $b = 14$

---

examine solutions (Lesson 13–4)

How many solutions does the triangle described by $A = 37°$, $a = 6$, and $b = 12$ have?

Angle A is less than 90°.

Find $b \sin A$ and compare with a.

$b \sin A = 12 \sin 37°$

≈ 7.22

$a\ ?\ b\ \sin A$

$6 < 7.22$

Therefore, there is no solution.

Determine whether each triangle described has no solution, one solution, or two solutions. Then solve each triangle.

45. $a = 24$, $b = 36$, $A = 64°$

46. $a = 17$, $b = 21$, $A = 64°$

47. $b = 10$, $c = 15$, $C = 66°$

48. $A = 82°$, $a = 9$, $b = 12$

OBJECTIVES AND EXAMPLES

• solve triangles by using the law of cosines
(Lesson 13–5)

Solve the triangle in which $A = 62°$, $c = 12$, and $b = 15$.

$a^2 = b^2 + c^2 - 2bc \cos A$
$a^2 = (15)^2 + (12)^2 - 2(15)(12) \cos 62°$
$a^2 = 200$
$a \approx 14.1$
$\frac{\sin 62°}{14.1} \approx \frac{\sin C}{12}$
$\sin C \approx \frac{12 \sin 62°}{14.1}$ or about 48.7°
$B \approx 180 - (62 + 48.7)$ or 69.3°

• define and use the trigonometric functions based on the unit circle (Lesson 13–6)

Find the value of $\cos\left(-\frac{7}{4}\pi\right)$.

$\cos\left(-\frac{7}{4}\pi\right) = \cos\left(-\frac{7}{4}\pi + 2\pi\right)$
$\phantom{\cos\left(-\frac{7}{4}\pi\right)} = \cos\left(\frac{1}{4}\pi\right)$
$\phantom{\cos\left(-\frac{7}{4}\pi\right)} = \frac{\sqrt{2}}{2}$

• find values of expressions involving inverse trigonometric functions (Lesson 13–7)

Find $\text{Cos}^{-1}\left[\tan\left(-\frac{\pi}{6}\right)\right]$.

$\text{Cos}^{-1}\left[\tan\left(-\frac{\pi}{6}\right)\right] = \text{Cos}^{-1}\left(-\frac{\sqrt{3}}{3}\right)$
$ A = \text{Cos}^{-1}\left(-\frac{\sqrt{3}}{3}\right)$
$ \cos A = -\frac{\sqrt{3}}{3}$
$ A \approx 125.3°$

REVIEW EXERCISES

Solve each triangle described. Round measures of sides to the nearest tenth and measures of angles to the nearest degree.

49. $C = 65°$, $a = 4$, $b = 7$

50. $b = 2$, $c = 5$, $A = 60°$

51. $a = 6$, $b = 7$, $C = 40°$

52. $B = 24°$, $a = 42$, $c = 6.5$

53. $a = 11$, $b = 13$, $c = 15$

Find the exact value of each function.

54. $\sin(-150°)$

55. $\cos 300°$

56. $(\sin 45°)(\sin 225°)$

57. $\sin \frac{5}{4}\pi$

58. $(\sin 30°)^2 + (\cos 30°)^2$

59. $\frac{4 \cos 150° + 2 \sin 300°}{3}$

Find each value.

60. $\text{Sin}^{-1}(-1)$

61. $\text{Cos}^{-1}\left(\frac{\sqrt{3}}{2}\right)$

62. $\text{Tan}^{-1}\sqrt{3}$

63. $\cos(\text{Sin}^{-1} 1)$

64. $\tan\left(\text{Arcsin}\frac{3}{5}\right)$

65. $\cot\left(\text{Tan}^{-1}\frac{8}{15}\right)$

APPLICATIONS AND PROBLEM SOLVING

66. **Aviation**   A pilot 3000 feet above the ocean notes the measurement of the angle of depression to a ship is 42°. How far is the plane from the ship? (Lesson 13–1)

67. **Geography**   Town and Rich streets meet to form a triangular region with the river bank as shown. Find the measure of the angle that is formed by the two streets. (Lesson 13–5)

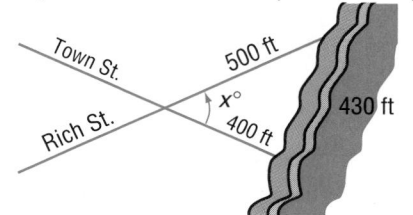

A practice test for Chapter 13 is provided on page 924.

ALTERNATIVE ASSESSMENT

COOPERATIVE LEARNING PROJECT

Starfish Varieties   In this project, you will design your own varieties of starfish. A starfish is a spiny-skinned sea animal that has thick, armlike extensions on its body. Most species have five such "arms" and look somewhat like five-pointed stars. Research starfish to get an idea of the various differences in their size, shape, and color.

Design three different species of starfish of your own. Describe, make a drawing of, give a name to, and color these three species. Include the diameter, angle measure of the end of each extension, radius of each extension, and any other data that would be helpful in visualizing your starfish designs. Your data must be accurate and mathematically sound.

Incorporate these ideas in your project.

- Research starfish.

- Be creative in your designs.

- Write and solve trigonometric relationships involved in the designs.

- Draw a "blueprint" for each design.

- Draw a realistic picture with color for each design.

- Prepare a paragraph describing each design as well as naming it.

THINKING CRITICALLY

- When solving a right triangle, certain information needs to be known. Summarize the combinations of sides and angles of a right triangle that must be known in order to arrive at a solution.

- In which quadrant does $\angle B$ lie, given that $\tan B < 0$ and $\cot B < 0$?

PORTFOLIO

Of the seven lessons in this chapter, pick the one with which you are still having trouble understanding. Describe this lesson and analyze your learning relative to the material. Answer questions like the following.

- What was the lesson about?

- What made sense to you in this lesson?

- What questions do you still have about this lesson?

- How could the methods used in this lesson be better explained?

Place this in your portfolio.

SELF EVALUATION

Comparing strategies and sharing ideas with other students enhances confidence in your abilities and thinking skills. Verbalizing draws out thoughts and questions from both people, which in turn helps to clarify your thinking.

Assess yourself. Do you test your ideas on other people? Do you allow your thoughts and skills to be processed by someone other than yourself before you determine a final solution? Do you present your ideas to another person in a way that makes you defend your thoughts and allows the situation to be fully explored? Describe a time when you used this process and the results you got.

SECTION ONE: MULTIPLE CHOICE

There are eight multiple-choice questions in this section. After working each problem, write the letter of the correct answer on your paper.

1. When Tashia spins the spinner shown below, what is the probability that the result will be a number greater than 3 or an even number?

   A. $\frac{1}{9}$

   B. $\frac{20}{81}$

   C. $\frac{2}{3}$

   D. 1

2. Solve $\sqrt{c + 4} = \sqrt{c + 20} - 2$.

   A. 20

   B. -15

   C. -10

   D. 5

3. Takara insists she needs to go to the mall to buy more clothes. Her mother disagrees. "But I have only 5 shirts, 4 pairs of pants, and 2 pairs of shoes!" Takara's mother still feels she has enough clothes and tells her to list all possible outfits so that she can see for herself. How many different outfits should Takara be able to list?

   A. 40 outfits

   B. 11 outfits

   C. 22 outfits

   D. 60 outfits

4. The Wildcats play 84 games this season. It is now midseason, and they have won 30 games. To win at least 60% of all their games, how many of the remaining games must they win?

   A. 21 games

   B. 50 games

   C. 24 games

   D. 33 games

5. Points X and Y are on opposite sides of a valley. Point C is 60 kilometers from point X. Angle YXC is 108° and angle YCX is 35°. Find the width of the valley.

   A. 63 km

   B. 57.2 km

   C. 37 km

   D. 45.2 km

6. The Worthington Public Library charges a fine of 5¢ per day for overdue books. What will the fine be on a book that is two weeks overdue?

   A. 50¢

   B. 10¢

   C. 60¢

   D. 70¢

7. Michelle and her family took a vacation to Oregon. As Michelle observed Mount Hood from a distance, the sine of the angle at which she looked up was $\frac{40}{41}$. Find the tangent of this angle.

   A. $\frac{40}{9}$

   B. $\frac{41}{9}$

   C. $\frac{9}{41}$

   D. $\frac{9}{40}$

8. Write $\frac{2}{3} + \frac{3}{5} + \frac{4}{7} + \frac{5}{9} + \ldots + \frac{21}{41}$ using sigma notation.

   A. $\displaystyle\sum_{n=2}^{20} \frac{n + 1}{n + 2}$

   B. $\displaystyle\sum_{n=1}^{20} \frac{n + 1}{2n + 1}$

   C. $\displaystyle\sum_{n=1}^{20} \frac{n + 1}{n + 2}$

   D. $\displaystyle\sum_{n=2} \frac{n + 1}{2n + 1}$

SECTION TWO: SHORT ANSWER

This section contains seven questions for which you will provide short answers. Write your answer on your paper.

9. Calvin keeps his CD's in a carousel on top of his stereo. He has 6 jazz, 4 rock, and 2 classical CD's. How many different ways can he arrange those CD's on the carousel?

10. Naren bought 2 slices of pizza, two cartons of milk, and one chocolate chip cookie for lunch. He spent $2.05 on his lunch. Ted brought a sandwich from home and bought a carton of milk and 3 cookies for a total of 75¢. Sarah spent $0.95 on one slice of pizza and a carton of milk. What are the individual prices of a slice of pizza, a carton of milk, and a chocolate chip cookie?

11. Viho is practicing his free throws. He knows the rim of the basket is 10 feet above the floor. From the spot on the floor where he is standing, the angle of elevation to the rim is 33° 33′. Find the distance from Viho's feet to the rim.

12. Find all zeros of the function $f(x) = 4x^4 - 35x^3 + 78x^2 + 28x - 165$.

13. For breakfast, Mrs. Crocker baked a dozen blueberry muffins, six bran muffins, and eight banana-nut muffins and put them on a serving plate on the table. Her daughter, Betty, took two blueberry muffins to school with her. Then Mr. Crocker sat down to breakfast. What is the probability that he will eat a blueberry muffin and a bran muffin?

14. Find the vertices and foci of the hyperbola whose equation is $25x^2 - 4y^2 = 100$.

15. Find the value of each of the trigonometric functions for angle A shown below.

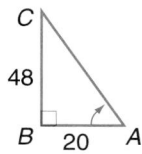

SECTION THREE: COMPARISON

This section contains five comparison problems which involve comparing two quantities, one in column A and one in column B. In certain questions, information related to one or both quantities is centered above them. All variables used represent real numbers.

Compare quantities A and B below.

- Write A if quantity A is greater.
- Write B if quantity B is greater.
- Write C if the two quantities are equal.
- Write D if there is not enough information to determine the relationship.

Column A	Column B
$-2x + 7y = -21$	
16. the y-intercept of the line parallel to the given line and containing $(-3, 4)$	the y-intercept of the line perpendicular to the given line and containing $(-2, 9)$
the value of the missing side	

17.

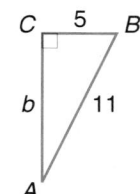

$\log_2 3 = 1.585;\ \log_2 7 = 2.807$	
18. $\log_2 0.75$	$\log_2 \dfrac{36}{49}$
19. the probability that a hockey team will win if the odds of winning are 8:3	the odds of hitting a target if the probability of missing it is $\dfrac{11}{19}$
20. $\dfrac{a+3b}{7} = 3$ $11a - b = -7$	$r + 11 = 8t$ $8(r - t) = 3$

Using Trigonometric Graphs and Identities

Objectives

In this chapter, you will:

- graph trigonometric functions,
- use trigonometric identities,
- solve problems by working backward, and
- solve trigonometric equations.

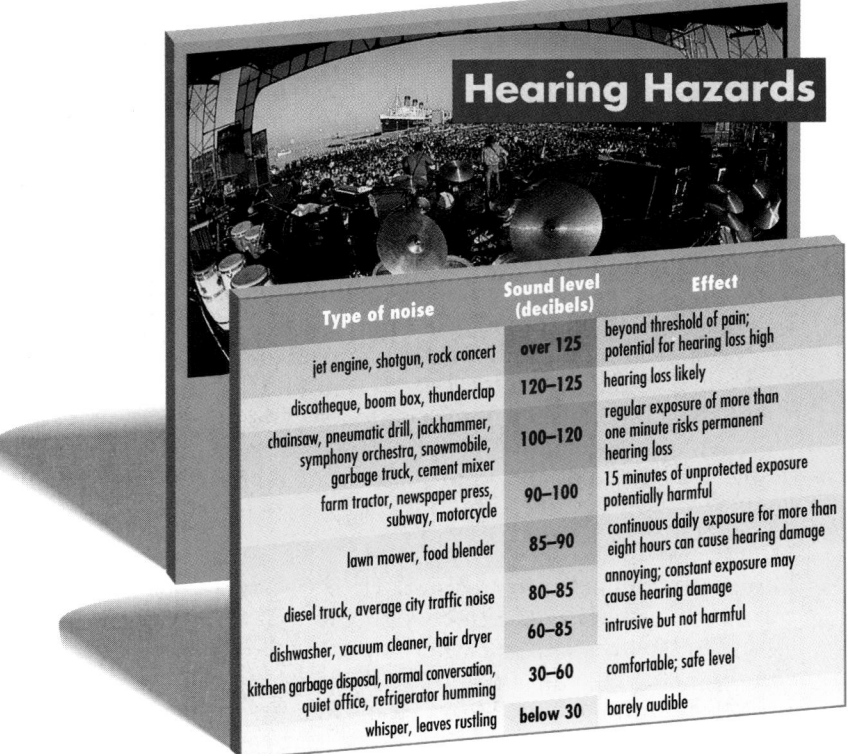

Hearing Hazards

Type of noise	Sound level (decibels)	Effect
jet engine, shotgun, rock concert	over 125	beyond threshold of pain; potential for hearing loss high
discotheque, boom box, thunderclap	120–125	hearing loss likely
chainsaw, pneumatic drill, jackhammer, symphony orchestra, snowmobile, garbage truck, cement mixer	100–120	regular exposure of more than one minute risks permanent hearing loss
farm tractor, newspaper press, subway, motorcycle	90–100	15 minutes of unprotected exposure potentially harmful
lawn mower, food blender	85–90	continuous daily exposure for more than eight hours can cause hearing damage
diesel truck, average city traffic noise	80–85	annoying; constant exposure may cause hearing damage
dishwasher, vacuum cleaner, hair dryer	60–85	intrusive but not harmful
kitchen garbage disposal, normal conversation, quiet office, refrigerator humming	30–60	comfortable; safe level
whisper, leaves rustling	below 30	barely audible

Source: National Institute on Deafness and Other Communication Disorders, National Institutes of Health, January 1990

Warning! Rock concerts may be hazardous to your hearing! There are many causes of deafness and people are affected in varying degrees. More than 24 million people in the United States suffer from some form of deafness, and nearly 2 million of them are profoundly deaf. The gap between hearing and hearing-impaired people has slowly narrowed. In 1988, students at Washington's Gallaudet College led a protest that pressured the institution into hiring its first deaf president.

TIME Line

A.D. 1140 Near Paris, the Abbey of St. Denis is built in the Gothic style.

1631 Thomas Harriot introduces the signs < (less than) and > (greater than).

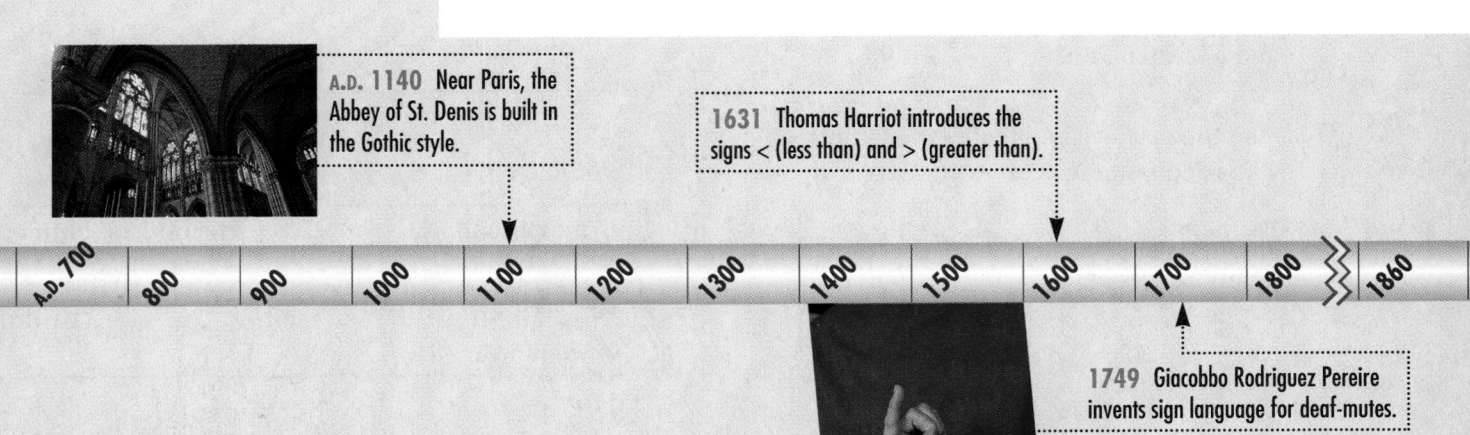

A.D. 700　800　900　1000　1100　1200　1300　1400　1500　1600　1700　1800　1860

1749 Giacobbo Rodriguez Pereire invents sign language for deaf-mutes.

PEOPLE IN THE NEWS

Chapter Project

Write a report that includes a list of hearing-impaired people who have made important contributions in areas such as science, medicine, politics, and the arts. List their contributions and describe how they communicate with hearing people. Include a discussion of the relationship between the vibration made by sound waves and the graphs of trigonometric functions.

At 21, **Heather Whitestone** became the first disabled person to win the Miss America Pageant in 1995. For the talent portion of the contest, Heather danced to music she "heard" by feeling its vibrations. Although she has been almost completely deaf since she was 18 months old, having no hearing in one ear and only 5 percent in the other, she learned to talk. She also learned to read lips and took her own notes while attending Berry High School in Alabama. Heather overcame her obstacles and graduated from high school with a 3.6 GPA.

As Miss America, Heather became an inspiration to many. "When children see me speaking and see me dance," she said, "they will realize they have no excuse for not making their own dreams come true."

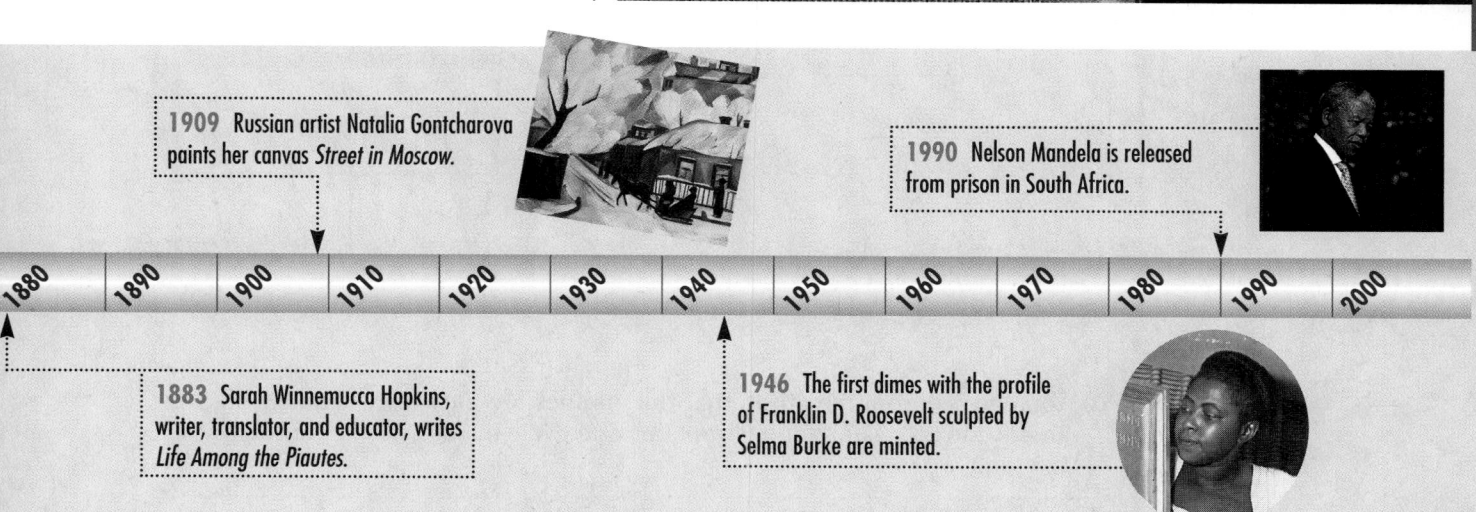

1909 Russian artist Natalia Gontcharova paints her canvas *Street in Moscow*.

1990 Nelson Mandela is released from prison in South Africa.

1883 Sarah Winnemucca Hopkins, writer, translator, and educator, writes *Life Among the Piautes*.

1946 The first dimes with the profile of Franklin D. Roosevelt sculpted by Selma Burke are minted.

14–1A Graphing Technology
Trigonometric Functions

A Preview of Lesson 14–1

Graphing calculators can graph trigonometric functions in both degrees and radians. To select the appropriate angle measure, check the MODE menu of the calculator. In this lesson, only degrees will be used. Press [MODE] and make sure that Degree is selected.

Example **Graph $y = \sin x$ in the trigonometric window. Find the period and the amplitude.**

In the trigonometric window, found by pressing [ZOOM] *7, each tick mark on the x-axis represents 90°. Each tick mark on the y-axis represents one unit.*

Enter: [Y=] [SIN] [X,T,θ,n]

[ZOOM] 7

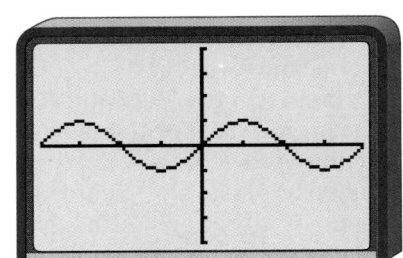

Notice that the y values of the graph are between -1 and 1. The **amplitude** of a periodic function is half the absolute value of the difference between the maximum and minimum values. So, the amplitude of this graph is $\frac{1}{2}|1 - (-1)|$ or 1. The graph completes one cycle in 360°, so the *period* of the function is 360°.

Graph the function using the *radian* setting. Compare and contrast the two graphs.

Example **Graph $y = 3 \cos 2x$ in the trigonometric window. Find the period and amplitude.**

Enter: [Y=] 3 [COS] 2 [X,T,θ,n]

[ZOOM] 7

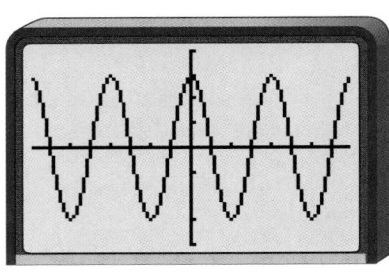

You can see from the graph that the y values are between -3 and 3, so the amplitude is 3. The graph completes one cycle in 180°, so the period of the function is 180°.

Example Graph $y = \cos x$ and $y = \cos (x + 90°)$ in the trigonometric window. Describe any similarities or differences between the two graphs.

Enter: Y= COS X,T,θ,*n* ENTER

COS (X,T,θ,*n* +

90) ZOOM 7

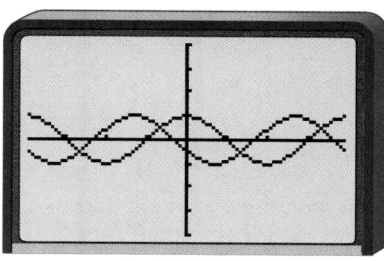

The graph of $y = \cos (x + 90°)$ has the same shape as the graph of $y = \cos x$ except that it is shifted 90° to the left. This is called a **phase shift** of 90°.

You can graph all six trigonometric functions with a graphing calculator. The graphs of $y = \csc x$ and $y = \cot x$ are all defined in terms of $\sin x$, $\cos x$, and $\tan x$. Example 4 illustrates how to obtain these graphs.

Example ❹ Graph $y = \sec x$ in the trigonometric window.

The vertical asymptotes may appear on the viewing screen, as shown at the right.

Recall that $\sec x = \dfrac{1}{\cos x}$

Enter: Y= 1 ÷ COS X,T,θ,*n*

ZOOM 7

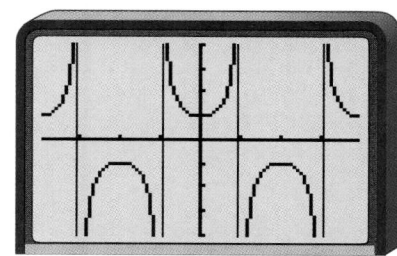

The graph has no amplitude. It takes 360° to complete one cycle. Therefore, the period is 360°. If you trace the graph, you will see that the function is undefined for $x = 90° + k \cdot 180°$, where k is any integer. *Why?*

EXERCISES

Graph each equation with a graphing calculator so that a complete graph is shown. Then sketch each graph on a sheet of paper.

1. $y = \tan x$
2. $y = \cos x$
3. $y = \cos (-x)$
4. $y = \csc x$
5. $y = \cot x$
6. $y = \tan (x - 180°)$
7. $y = \tan 5x$
8. $y = 3 \sec (-x)$
9. $y = 12 \sin (x + 45°)$
10. $y = \frac{1}{2} \csc (2x + 90°)$
11. $y = 3 \tan (90° - x)$
12. $y = 0.1 \sec (360° + 2x)$
13. $y = 0.5 \cot (x + 45°)$
14. $y = 3 \cos \frac{2}{3} x$

Graphing Trigonometric Functions

14-1

What YOU'LL LEARN

- To graph trigonometric functions, and
- to find the amplitude and period for variations of the sine and cosine functions.

Why IT'S IMPORTANT

You can graph trigonometric functions to solve problems involving zoology and music.

APPLICATION
Amusement Parks

The world's first Ferris wheel was built for the World's Columbian Exhibition in Chicago in 1892. The wheel was 250 feet in diameter and had 36 cars that could carry 40 passengers each. Suppose you are seated in a Ferris wheel with a diameter of 50 feet. While the wheel rotates, your vertical position varies between 0 and 25 feet above and below an imaginary horizontal plane that contains a horizontal line through the center of the wheel.

The diagram below illustrates your vertical position as a function of time for a Ferris wheel that makes 1 rotation every 30 seconds.

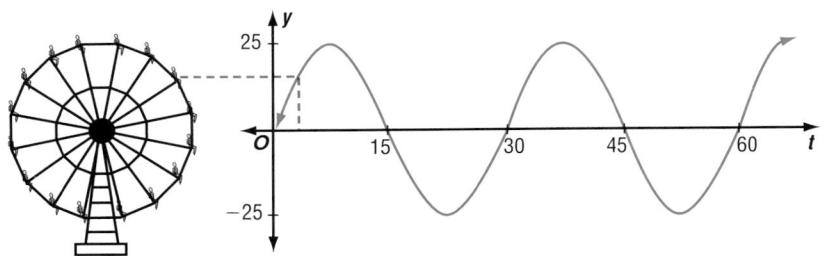

In each rotation, the wheel repeats the pattern that you see in the diagram. Recall that functions which have a graph that repeats a basic pattern are called *periodic functions*.

LOOK BACK

You can refer to Lesson 13-6 for information about periodic functions.

To find the period, start from any point on the graph and proceed to the right until the pattern begins to repeat. The simplest approach is to begin at the origin. Notice that at 30 the graph begins to repeat. Thus, the period of the function is 30 seconds.

To graph $y = \sin \theta$ or $y = \cos \theta$, use values of θ expressed either in degrees or radians. Ordered pairs for points on these graphs are of the form $(\theta, \sin \theta)$ and $(\theta°, \cos \theta)$.

θ	0°	30°	45°	60°	90°	120°	135°	150°	180°	210°	225°	240°	270°	300°	315°	330°	360°
sin θ	0	$\frac{1}{2}$	$\frac{\sqrt{2}}{2}$	$\frac{\sqrt{3}}{2}$	1	$\frac{\sqrt{3}}{2}$	$\frac{\sqrt{2}}{2}$	$\frac{1}{2}$	0	$-\frac{1}{2}$	$-\frac{\sqrt{2}}{2}$	$-\frac{\sqrt{3}}{2}$	-1	$-\frac{\sqrt{3}}{2}$	$-\frac{\sqrt{2}}{2}$	$-\frac{1}{2}$	0
nearest tenth	0	0.5	0.7	0.9	1	0.9	0.7	0.5	0	−0.5	−0.7	−0.9	−1	−0.9	−0.7	−0.5	0
cos θ	1	$\frac{\sqrt{3}}{2}$	$\frac{\sqrt{2}}{2}$	$\frac{1}{2}$	0	$-\frac{1}{2}$	$-\frac{\sqrt{2}}{2}$	$-\frac{\sqrt{3}}{2}$	-1	$-\frac{\sqrt{3}}{2}$	$-\frac{\sqrt{2}}{2}$	$-\frac{1}{2}$	0	$\frac{1}{2}$	$\frac{\sqrt{2}}{2}$	$\frac{\sqrt{3}}{2}$	1
nearest tenth	1	0.9	0.7	0.5	0	−0.5	−0.7	−0.9	−1	−0.9	−0.7	−0.5	0	0.5	0.7	0.9	1

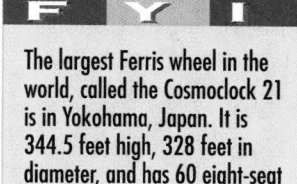
After plotting several points, complete the graphs of $y = \sin \theta$ and $y = \cos \theta$ by connecting the points with a smooth, continuous curve.

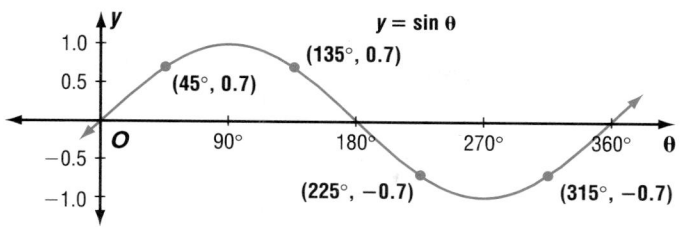

Negative values of θ would be represented to the left of zero.

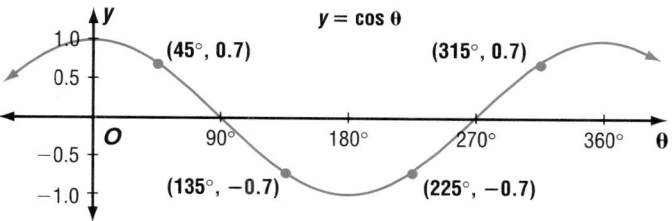

As you recall from your studies of the sine and cosine functions in Chapter 13, each of these functions has a *period* of 360° or 2π radians. That is, the graph of each function repeats itself every 360° or 2π radians. The following example illustrates a sine function that has a period of less than 360°.

Example **Graph $y = \sin 2\theta$. State the period.**

First, complete a table of values.

θ	0°	15°	30°	45°	60°	75°	90°	105°	120°	135°	150°	165°	180°
2θ	0°	30°	60°	90°	120°	150°	180°	210°	240°	270°	300°	330°	360°
sin 2θ	0	$\frac{1}{2}$	$\frac{\sqrt{3}}{2}$	1	$\frac{\sqrt{3}}{2}$	$\frac{1}{2}$	0	$-\frac{1}{2}$	$-\frac{\sqrt{3}}{2}$	-1	$-\frac{\sqrt{3}}{2}$	$-\frac{1}{2}$	0

Then plot the points for ordered pairs of the form (θ, y), and connect to form a curve.

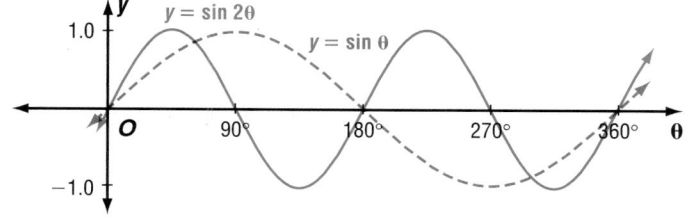

The graph of $y = \sin 2\theta$ repeats every 180° or π radians. Therefore, the period of $y = \sin 2\theta$ is 180° or π radians. *Notice that the period is $\frac{360°}{2}$ or $\frac{2\pi}{2}$.*

The trigonometric function in Example 1 has a maximum value of 1 and a minimum value of -1. The **amplitude** of the graph of a periodic function is the absolute value of half the difference between its maximum value and its minimum value. So, in Example 1, the amplitude of the graph is $\left|\frac{1-(-1)}{2}\right|$ or 1.

You can use a graphing calculator to compare the graphs of $y = \sin b\theta$ for various values of b.

EXPLORATION   GRAPHING CALCULATORS

Set MODE to degrees. Then use the viewing window $[0, 720]$ by $[-1.5, 1.5]$ with a scale factor of 45 for the x-axis and 0.5 for the y-axis.

Your Turn

a. Enter the function $Y1 = \sin X$ into the $Y=$ list. Press $\boxed{\text{GRAPH}}$. How many times does the function reach a maximum point on screen?

b. Enter the function $Y2 = \sin 2X$. Press $\boxed{\text{GRAPH}}$. How many times does the function reach its maximum point?

c. Enter the function $Y3 = \sin (X/2)$. How many times does the function reach its maximum point?

d. Repeat the steps for several other values of b. What conclusion can you draw about the effect of b on the period of the function $y = \sin b\theta$?

Example Graph $y = \frac{1}{2} \cos \theta$. State the amplitude.

θ	0°	30°	60°	90°	120°	150°	180°	210°	240°	270°	300°	330°	360°
$\cos \theta$	1	$\frac{\sqrt{3}}{2}$	$\frac{1}{2}$	0	$-\frac{1}{2}$	$-\frac{\sqrt{3}}{2}$	-1	$-\frac{\sqrt{3}}{2}$	$-\frac{1}{2}$	0	$\frac{1}{2}$	$\frac{\sqrt{3}}{2}$	1
$\frac{1}{2} \cos \theta$	$\frac{1}{2}$	$\frac{\sqrt{3}}{4}$	$\frac{1}{4}$	0	$-\frac{1}{4}$	$\frac{\sqrt{3}}{4}$	$-\frac{1}{2}$	$\frac{\sqrt{3}}{4}$	$-\frac{1}{4}$	0	$\frac{1}{4}$	$\frac{\sqrt{3}}{4}$	$\frac{1}{2}$

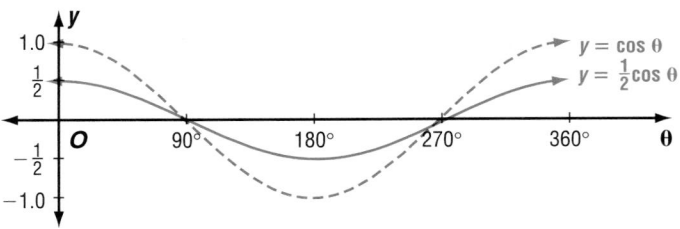

The amplitude of $y = \frac{1}{2} \cos \theta$ is $\frac{1}{2}$. *Notice that the amplitude is $\left|\frac{1}{2}\right|$ or $\frac{1}{2}$.*

There are many applications of trigonometry in everyday life. One of these is the pitch of a pure musical tone that you hear when the tone travels through the air to reach your ear. When viewed on an oscilloscope, the tone produces the graph of a sine or cosine function. The louder the tone, the greater the amplitude of the function.

Example **3**

APPLICATION
Physics

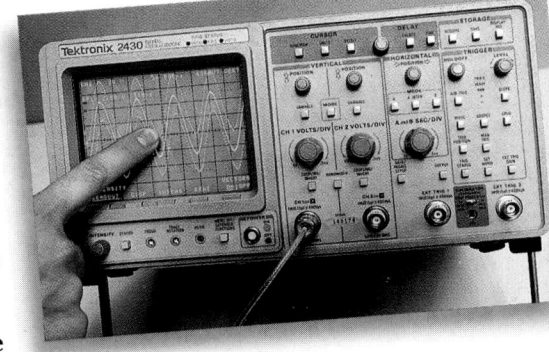

A pure tone is graphically viewed on an oscilloscope screen at various levels of loudness. At one of the levels, the screen displays a graph that can be modeled by the equation $y = -10 \sin 100\pi t$, where y is the pressure variation above or below the atmospheric pressure of air in dynes per square centimeter at time t in seconds.

a. Graph the function and state the amplitude and period.

b. Find the pressure variation when $\frac{1}{16}$ second has elapsed.

a. Since the equation is expressed in terms of radians, use values of θ expressed in radians to complete a table of values and graph the function.

t seconds	0	0.001	0.002	0.003	0.004	0.005	0.006	0.008	0.01	0.012	0.014	0.016	0.018	0.02
$100\pi t$	0	0.1π	0.2π	0.3π	0.4π	0.5π	0.6π	0.8π	π	1.2π	1.4π	1.6π	1.8π	2π
$\sin 100\pi t$	0	0.31	0.59	0.81	0.95	1.0	0.95	0.59	0	-0.59	-0.95	-0.95	-0.59	0
$-10 \sin 100\pi t$	0	-3.1	-5.9	-8.1	-9.5	-10	-9.5	-5.9	0	5.9	9.5	9.5	5.9	0

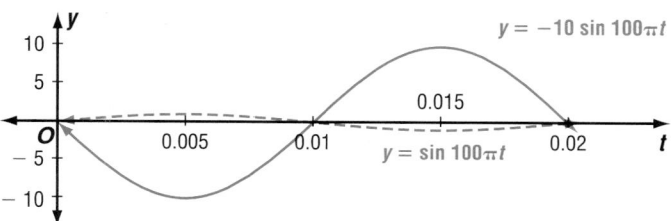

The amplitude of the graph is 10, and the period is 0.02 second.

Notice that the amplitude is $|-10|$ or 10 and the period is $\frac{2\pi}{|100\pi|} = \frac{1}{50}$ or 0.02.

b. Find y when $t = \frac{1}{16}$

$$y = -10 \sin 100\pi t$$
$$= -10 \sin 100\pi \left(\frac{1}{16}\right) \qquad \text{Replace } t \text{ with } \frac{1}{16}.$$
$$= -10 \sin 6.25\pi$$
$$= -10 \sin [3(2\pi) + 0.25\pi] \quad \sin (2k\pi + \theta) = \sin \theta$$
$$= -10 \sin (0.25\pi)$$
$$\approx -10(0.707) \text{ or } -7.1$$

When $\frac{1}{16}$ second has elapsed, the pressure is about 7.1 dynes per square centimeter less than atmospheric pressure (about 1,000,000 dynes per square centimeter).

These examples suggest the following generalizations.

Amplitudes and Periods	For functions of the form $y = a \sin b\theta$ and $y = a \cos b\theta$, the amplitude is $|a|$, and the period is $\dfrac{360°}{|b|}$ or $\dfrac{2\pi}{|b|}$.

The four remaining trigonometric functions—tangent, cotangent, secant, and cosecant—can also be graphed. Make a table of values for $y = \tan \theta$ and graph the function.

θ	0°	30°	45°	60°	90°	120°	135°	150°	180°	210°	225°	240°	270°	300°	315°	330°	360°
$\tan \theta$	0	$\dfrac{\sqrt{3}}{3}$	1	$\sqrt{3}$	nd	$-\sqrt{3}$	-1	$-\dfrac{\sqrt{3}}{3}$	0	$\dfrac{\sqrt{3}}{3}$	1	$\sqrt{3}$	nd	$-\sqrt{3}$	-1	$-\dfrac{\sqrt{3}}{3}$	0

nd = not defined

The tangent function is not defined for 90°, 270°, ... , $90° + k \cdot 180°$, where k is an integer. The graph is separated by vertical asymptotes, indicated by dashed lines. The x-intercepts of the asymptotes are the values for which $y = \tan \theta$ is not defined.

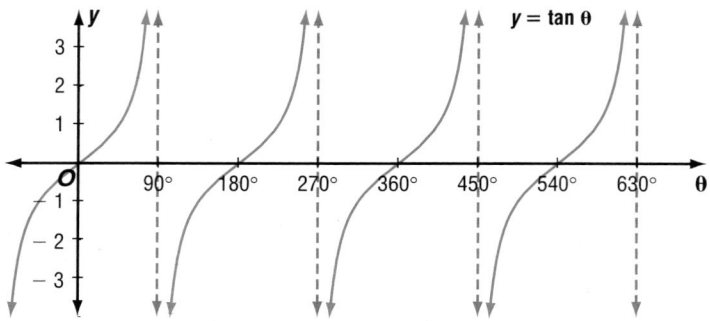

The period of the tangent function is 180° or π radians. Since the tangent function has no maximum or minimum value, it has no amplitude.

The graphs of the secant, cosecant, and cotangent functions are shown below. Compare them to the graphs of the cosine, sine, and tangent functions, which are shown in red.

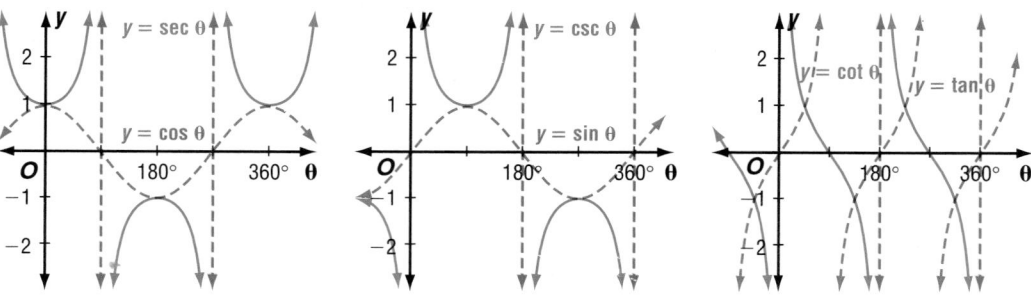

Notice that the period of the secant and cosecant functions is 360° or 2π radians. The period of the cotangent function is 180° or π radians. What are the amplitudes of the secant, cosecant, and cotangent functions?

Example **Graph** $y = -\frac{1}{2} \csc 2\theta$.

θ	0°	15°	30°	45°	60°	75°	90°	105°	120°	135°	150°	165°	180°
2θ	0°	30°	60°	90°	120°	150°	180°	210°	240°	270°	300°	330°	360°
$-\frac{1}{2} \csc 2\theta$	nd	−1	$-\frac{\sqrt{3}}{3}$	$-\frac{1}{2}$	$\frac{\sqrt{3}}{3}$	−1	nd	1	$\frac{\sqrt{3}}{3}$	$\frac{1}{2}$	$\frac{\sqrt{3}}{3}$	1	nd

nd = not defined

The period is $\frac{360°}{|2|}$ or 180°. This function has no maximum or minimum value when θ equals 0°, 90°, 180°, ... , 0° + $k \cdot$ 90°. Thus, it has no amplitude.

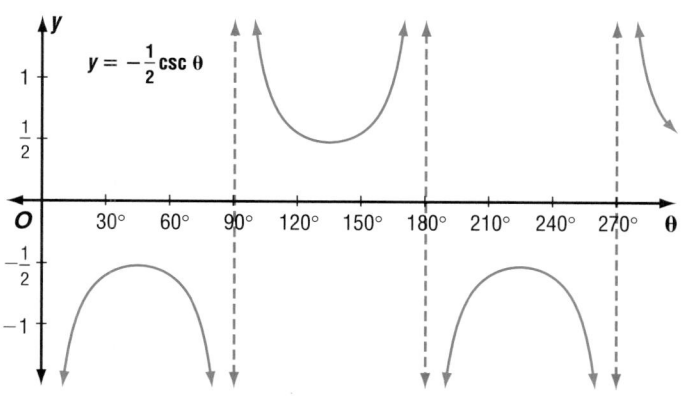

$y = -\frac{1}{2} \csc \theta$

CHECK FOR UNDERSTANDING

Communicating Mathematics

Study the lesson. Then complete the following.

1. **Explain** what it means to say that the period of a function is 180°.

2. **Explain** what the amplitude of a sine or cosine function is.

3. Write your impressions of the function that is graphed at the right.

   a. Does it appear to have an amplitude? If so, what is it?

   b. Does it appear to be periodic? If so, what is the period?

   c. Does it appear to be a sine or cosine function? If so, what is a possible equation?

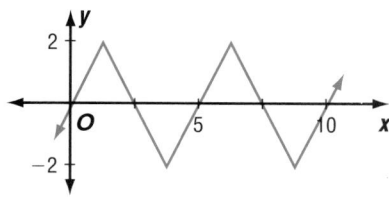

4. **You Decide** Tina and Tanya were observing the graph of the equation $y = 10 \cos 524\pi t$ on the screen of an oscilloscope. They noticed that by rotating one of the dials, they were able to move the graph to the right until the y-intercept changed from 10 to 0. Neither the amplitude nor the period of the graph had changed. Tina concluded that the sine function could be used to describe the graph in its new position. Tanya believed that since the original curve was a cosine graph, only a cosine function could be used. Who was right?

5. Write a paragraph describing ways in which the functions $y = a \sin b\theta$ and $y = a \cos b\theta$ are the same and ways in which they are different.

Guided Practice

Match the graphs on the right with the equations on the left.

6. $y = \sin \theta$

7. $y = 4 \sin \theta$

8. $y = \cos 0.4\theta$

9. $y = \sin 4\theta$

a.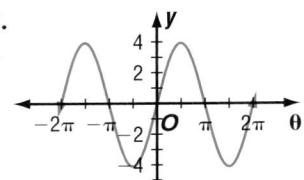

b.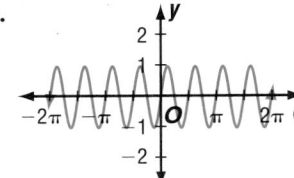

c.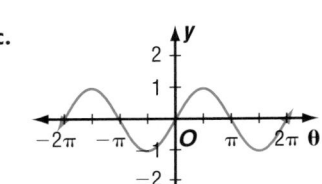

d.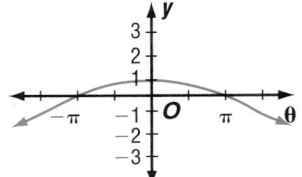

State the amplitude (if it exists) and the period of each function. Then graph each function.

10. $y = 3 \cos \frac{1}{2}\theta$

11. $y = 6 \sin \frac{2}{3}\theta$

12. $y = -2 \sin \theta$

13. $y = \sec 3\theta$

14. $y = 3 \tan \theta$

15. $y = \cot 5\theta$

16. Which of the functions in Exercises 10–15 are defined for all real values of θ?

17. Music When represented on an oscilloscope, the note A above middle C has a period of $\frac{1}{440}$. Which of the following can be an equation for an oscilloscope graph of this note? The amplitude of the graph is K.

a. $y = K \sin 220\pi t$ **b.** $y = K \sin 440\pi t$ **c.** $y = K \sin 880\pi t$

EXERCISES

Practice

Graph each function.

18. $y = \frac{1}{2} \sin \theta$

19. $y = 3 \sin \theta$

20. $y = \cos 3\theta$

21. $y = \sin 4\theta$

22. $y = \cos 2\theta$

23. $y = -3 \sin \theta$

24. $y = \cot \theta$

25. $y = \frac{2}{3} \cos \theta$

26. $y = 3 \sec \theta$

27. $y = \csc 2\theta$

28. $y = \frac{1}{3} \sec \theta$

29. $y = 4 \sin \frac{1}{2}\theta$

30. $y = 4 \cos \frac{3}{4}\theta$

31. $y = 3 \csc \frac{1}{2}\theta$

32. $y = -\frac{1}{2} \cot 2\theta$

33. $2y = \tan \theta$

34. $3y = 2 \sin \frac{1}{2}\theta$

35. $\frac{3}{4}y = \frac{2}{3} \sin \frac{3}{5}\theta$

Write an equation of the given sine or cosine function having the specified characteristics.

	Function	Amplitude	Period
36.	sine	4	360°
37.	cosine	0.6	720°
38.	sine	5	180°
39.	cosine	$\frac{1}{3}$	90°
40.	cosine	4.25	360°
41.	sine	6.7	120°

Write an equation for the graph that is displayed.

42.

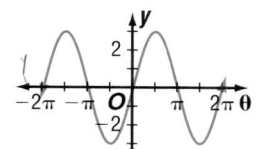

43.

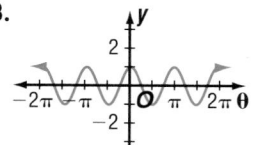

44.

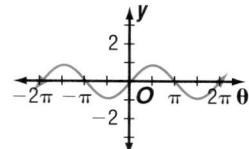

Graphing Calculator

45. Graph Y1 = sin X in the viewing window [0, 720] and [−2, 2]. Use scale factors of 45 for the *x*-axis and 0.5 for the *y*-axis. Then graph the following functions. *Be sure your calculator is in degree mode.*

   a. Y2 = 2 sin X   **b.** Y3 = −2 sin X   **c.** Y4 = $-\frac{1}{2}$ sin X

   d. Examine your results from parts a–c. What conclusions can you draw about the effect of the constant *a* on the graph of the function $y = a \sin \theta$?

Critical Thinking

46. Suppose you were told that for a certain value of *d*, $y = \sin \theta$ and $y = \cos (\theta + d)$ have exactly the same graph. How would you find a value of *d* for which this true?

Applications and Problem Solving

47. Zoology In predator-prey systems, the number of predators and the number of prey tend to vary in a periodic manner. In a certain region with coyotes as predators and rabbits as prey, the rabbit population *R* varied according to the equation $R = 1000 + 250 \sin \frac{1}{2}\pi t$, where *t* is the time in years since January 1, 1990.

   a. Graph the function that describes the rabbit population at time *t*.

   b. What was the rabbit population on January 1, 1990?

   c. What is the maximum rabbit population? On what date was the maximum population first reached?

   d. What is the minimum rabbit population? On what date was the minimum population first reached?

48. Communications The carrier wave for a certain FM station can be modeled by an equation of the form $y = A \sin(10^7 \cdot 2\pi t)$, where t is the time in seconds. Determine the period of the carrier wave.

Mixed Review

49. Trigonometry Find $\sin\left(\text{Cos}^{-1}\frac{15}{17}\right)$. (Lesson 13–7)

50. Cosmotology Sandy Chung carries several colors of lipstick in a cosmetic bag in her purse. The probability of pulling out the color she wants without looking is $\frac{1}{3}$. If she uses lipstick 4 times a day, find the probability that she never pulls out the correct lipstick all day. (Lesson 12–7)

51. Find the missing terms of the geometric sequence $\underline{\ ?\ }$, $\underline{\ ?\ }$, 9.5, 19, 38. (Lesson 11–3)

52. Use a calculator to find ln 56.9 rounded to four decimal places. (Lesson 10–5)

53. Solve $\frac{x-3}{2x} = \frac{x-2}{2x+1} - \frac{1}{2}$. (Lesson 9–5)

54. Graph $\frac{(x-1)^2}{9} - \frac{(y+4)^2}{16} = 1$. (Lesson 7–5)

55. Express $\sqrt[3]{16a^5b^7}$ using rational exponents. (Lesson 5–7)

56. Find the maximum and minimum values of the function $f(x, y) = x - y$ defined for the polygonal region having vertices with coordinates $(0, 0)$, $(0, 5)$, $(3, 4)$, and $(6, 0)$. (Lesson 3–5)

57. Write an expression to demonstrate the distributive property. (Lesson 1–2)

WORKING ON THE Investigation

Refer to the Investigation on pages 768–769.

Scream Machines!

Review the meaning of potential energy *PE* and kinetic energy *KE*. When a car is at its maximum height, it has maximum *PE* and minimum *KE*. When it has just arrived at the bottom of the steepest slope, it has maximum *KE* and little *PE*. The formulas for calculating *PE* and *KE* are

$PE = w \cdot L_v$ and $KE = \frac{1}{2} \cdot \frac{w}{g}s^2$, where w is the

weight of the car, L_v is its vertical height on the track, g is the acceleration of gravity (32 ft/s²), and s is the speed of the cars in ft/s.

1 Sketch what you think the graph of the potential energy of a roller coaster car would be from the beginning to the end of the ride. Let the horizontal axis represent time and the vertical axis represent the amount of energy.

2 Graph the amount of kinetic energy for the cars during the ride on the same coordinate plane with the graph of the potential energy. Discuss any similarities or differences in the shapes of the two graphs. How do these graphs compare to the graph of $y = \sin\theta$?

Add the results of your work to your Investigation Folder.

Trigonometric Identities

APPLICATION
Running

While practicing on a circular track, a runner notices that his body is not perpendicular to the ground. Instead, it leans away from a vertical position. The nonnegative acute angle θ that the runner's body makes with the vertical is called the **angle of incline** and is described by the equation $\tan \theta = \dfrac{v^2}{gR}$, where R is the radius of the track in meters, v is the speed of the runner in meters per second, and g is the acceleration due to gravity, 9.8 meters per second squared.

This is not the only equation that describes the angle of incline in terms of trigonometric functions. Another such equation is $\sin \theta = \cos \theta \cdot \dfrac{v^2}{gR}$, where $0 \le \theta < 90°$.

Are these two equations completely independent of one another or are they merely different versions of the same relationship? To answer this question, recall the following relationship that you learned in Chapter 13.

$$\tan \theta = \frac{\sin \theta}{\cos \theta}, \text{ if } \cos \theta \ne 0$$

This is an example of a **trigonometric identity,** an equation that is true for all values for which every expression in the equation is defined. You know that the above identity is true except for angle measures such as $\theta = 90°, 270°, 450°, ..., 90° + k \cdot 180°$. The cosine of each of these angles is 0, so none of the expressions $\tan 90°$, $\tan 270°$, $\tan 450°$, and so on, are defined.

Now divide each side of the second formula, $\sin \theta = \cos \theta \cdot \dfrac{v^2}{gR}$, by $\cos \theta$, allowing only values of θ for which $\cos \theta \ne 0$. The result is as follows.

$$\frac{\sin \theta}{\cos \theta} = \frac{\cos \theta}{\cos \theta} \cdot \frac{v^2}{gR}$$

$$\tan \theta = 1 \cdot \frac{v^2}{gR} \qquad \textit{Tangent identity}$$

$$\tan \theta = \frac{v^2}{gR}$$

Notice that this is the first angle of inclination formula. The two formulas are not independent. Each is a variation of the other.

In Chapter 13, you also became familiar with other trigonometric identities, such as those below.

$$\cot \theta = \frac{\cos \theta}{\sin \theta} \qquad \sec \theta = \frac{1}{\cos \theta} \qquad \csc \theta = \frac{1}{\sin \theta}$$

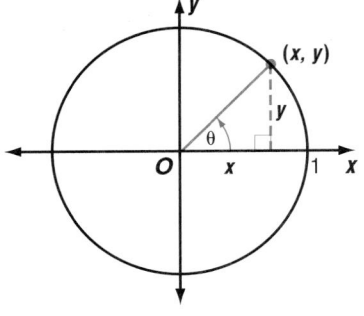

To find another useful trigonometric identity, begin with the equation for the unit circle, $x^2 + y^2 = 1$. By substituting $\cos \theta$ for x and $\sin \theta$ for y, we have the following equation.

$$(\cos \theta)^2 + (\sin \theta)^2 = 1$$

This equation is usually written as follows.

$$\cos^2 \theta + \sin^2 \theta = 1$$

The equation $\cos^2 \theta + \sin^2 \theta = 1$ is an identity because it is true for all values of θ. Some other trigonometric identities are given below.

Basic Trigonometric Identities	**The following trigonometric identities hold true for all values of θ except those for which either side of the equation is undefined.** $$\cos^2 \theta + \sin^2 \theta = 1$$ $$\tan^2 \theta + 1 = \sec^2 \theta$$ $$\cot^2 \theta + 1 = \csc^2 \theta$$

The steps in the following example show the development of a basic trigonometric identity. The expression on one side of the equation is transformed into the exact form of the expression on the other side.

Example ① **Show that $1 + \cot^2 \theta = \csc^2 \theta$.**

$$1 + \cot^2 \theta = \csc^2 \theta$$

$$1 + \left(\frac{\cos \theta}{\sin \theta}\right)^2 = \csc^2 \theta \qquad \textit{Definition of cot } \theta$$

$$1 + \frac{\cos^2 \theta}{\sin^2 \theta} = \csc^2 \theta$$

$$\frac{\sin^2 \theta}{\sin^2 \theta} + \frac{\cos^2 \theta}{\sin^2 \theta} = \csc^2 \theta \qquad 1 = \frac{\sin^2 \theta}{\sin^2 \theta}$$

$$\frac{\sin^2 \theta + \cos^2 \theta}{\sin^2 \theta} = \csc^2 \theta$$

$$\frac{1}{\sin^2 \theta} = \csc^2 \theta \qquad \sin^2 \theta + \cos^2 \theta = 1$$

$$\csc^2 \theta = \csc^2 \theta$$

Therefore, $1 + \cot^2 \theta = \csc^2 \theta$.

You can use trigonometric identities to find values of trigonometric functions. In the next example, β is the Greek letter beta.

Example **2** **Find $\csc \beta$ if $\cot \beta = \frac{4}{5}$ and $180° \le \beta \le 270°$.**

$\csc^2 \beta = 1 + \cot^2 \beta$ *Trigonometric identity*

$\csc^2 \beta = 1 + \left(\frac{4}{5}\right)^2$ *Substitute $\frac{4}{5}$ for $\cot \beta$.*

$\csc^2 \beta = \frac{41}{25}$

$\csc \beta = \pm \sqrt{\frac{41}{25}}$ *Take the square root of each side.*

$\csc \beta = \pm \frac{\sqrt{41}}{\sqrt{25}}$

$\csc \beta = -\frac{\sqrt{41}}{5}$ or about -1.2806 *$\csc \beta$ is negative for values of β between $180°$ and $270°$.*

Example **3** **If $\cot A = 4$, find $\sin A$. Assume that angle A is positive and acute.**

$\csc^2 A = 1 + \cot^2 A$ *Trigonometric identity*

$\csc^2 A = 1 + (4)^2$

$\csc^2 A = 17$

$\frac{1}{\sin^2 A} = 17$ *$\csc A = \frac{1}{\sin A}$*

$\sin^2 A = \frac{1}{17}$

$\sin A = \frac{\sqrt{1}}{\sqrt{17}}$ *$\sin A$ is positive for positive acute angles.*

$\sin A = \frac{\sqrt{17}}{17}$ or about 0.2425

Trigonometric identities can also be used to simplify expressions containing trigonometric functions. Simplifying an expression that contains trigonometric functions means that the expression is written as a numerical value or in terms of a single trigonometric function, if possible.

Example **4** **Simplify $\frac{1}{1 + \cos x} + \frac{1}{1 - \cos x}$.**

$\frac{1}{1 + \cos x} + \frac{1}{1 - \cos x} = \frac{(1 - \cos x) + (1 + \cos x)}{(1 + \cos x)(1 - \cos x)}$

$= \frac{2}{1 - \cos^2 x}$

$= \frac{2}{\sin^2 x}$ *$\cos^2 x + \sin^2 x = 1$, so $1 - \cos^2 x = \sin^2 x$.*

$= 2 \cdot \frac{1}{\sin^2 x}$

$= 2 \csc^2 x$ *$\frac{1}{\sin x} = \csc x$*

A portion of a race track has the shape of a circular arc with a radius of 16.7 meters. As a runner races along the arc, the sine of her angle of incline θ is found to be $\frac{1}{4}$. Find the speed of the runner.

Use the angle-of-incline formula given in the application at the beginning of the lesson, $\tan \theta = \frac{v^2}{gR}$. The value of $\tan \theta$ can be found if $\cos \theta$ is also known. So, find $\cos \theta$ first.

$$\cos^2 \theta + \sin^2 \theta = 1$$
$$\cos^2 \theta = 1 - \sin^2 \theta$$
$$\cos^2 \theta = 1 - \left(\frac{1}{4}\right)^2$$
$$\cos^2 \theta = 1 - \frac{1}{16}$$
$$\cos^2 \theta = \frac{15}{16}$$
$$\cos \theta = \frac{\sqrt{15}}{4} \quad \text{\textit{Take the square root of each side.}}$$
$$\text{\textit{The cosine of an acute angle is positive.}}$$

Use $\cos \theta$ to find $\tan \theta$.

$$\tan \theta = \frac{\sin \theta}{\cos \theta}$$
$$= \frac{\frac{1}{4}}{\frac{\sqrt{15}}{4}} \quad \text{\textit{sin }} \theta = \frac{1}{4} \text{ \textit{and cos} } \theta = \frac{\sqrt{15}}{4}$$
$$= \frac{1}{\sqrt{15}}$$
$$= \frac{\sqrt{15}}{15} \text{ or about } 0.2582$$

Now substitute the value of $\tan \theta$ and the given values to solve the problem.

$$\tan \theta = \frac{v^2}{gR} \qquad \text{\textit{tan }} \theta = 0.2582, g = 9.8, \text{ \textit{and} } R = 16.7$$
$$0.2582 = \frac{v^2}{9.8(16.7)}$$
$$0.2582 = \frac{v^2}{163.66} \qquad \text{\textit{Multiply each side by 163.66.}}$$
$$42.257 \approx v^2$$
$$\pm\sqrt{42.257} \approx v$$
$$\pm 6.5 \approx v$$

Since her speed is positive, the speed of the runner is about 6.5 meters per second.

Communicating Mathematics

Study the lesson. Then complete the following.

1. **Describe** what the coordinates (x, y) represent on the unit circle.

2. **Explain** what it means to simplify an expression containing trigonometric functions.

3. **Determine** in which quadrant(s) the terminal side of angle β is if $\tan \beta = -\frac{4}{3}$. Explain why.

Guided Practice

Solve for values of θ between 0° and 90°.

4. If $\cot \theta = 2$, find $\tan \theta$.

5. If $\sin \theta = \frac{4}{5}$, find $\cos \theta$.

6. If $\cos \theta = \frac{2}{3}$, find $\sin \theta$.

7. If $\cos \theta = \frac{2}{3}$, find $\csc \theta$.

8. Show that $\sin x \sec x = \tan x$ is an identity.

Simplify each expression.

9. $\tan \theta \cos^2 \theta$

10. $\csc^2 \theta - \cot^2 \theta$

11. $\frac{\cos x \csc x}{\tan x}$

12. **History**   Pythagoras is most famous for the theorem that bears his name. The identity $\cos^2 \theta + \sin^2 \theta = 1$ is an example of a *Pythagorean identity*. Why do you think that this identity is classified in this way?

Practice

Solve for values of θ between 90° and 180°.

13. If $\cos \theta = -\frac{3}{5}$, find $\csc \theta$.

14. If $\sin \theta = \frac{1}{2}$, find $\tan \theta$.

15. If $\sin \theta = \frac{3}{5}$, find $\cos \theta$.

16. If $\tan \theta = -2$, find $\sec \theta$.

Solve for values of θ between 180° and 270°.

17. If $\cos \theta = -\frac{3}{5}$, find $\csc \theta$.

18. If $\sec \theta = -3$, find $\tan \theta$.

19. If $\cot \theta = \frac{1}{4}$, find $\csc \theta$.

20. If $\sin \theta = -\frac{1}{2}$, find $\cos \theta$.

Solve for values of θ between 270° and 360°.

21. If $\cos \theta = \frac{5}{13}$, find $\sin \theta$.

22. If $\tan \theta = -1$, find $\sec \theta$.

23. If $\sec \theta = \frac{5}{3}$, find $\cos \theta$.

24. If $\csc \theta = -\frac{5}{3}$, find $\cos \theta$.

Simplify each expression.

25. $\csc \alpha \cos \alpha \tan \alpha$

26. $\cos \alpha \csc \alpha$

27. $\sec^2 \theta - 1$

28. $\sin x + \cos x \tan x$

29. $\dfrac{\tan \beta}{\sin \beta}$

30. $\dfrac{1 - \sin^2 \alpha}{\sin^2 \alpha}$

31. $\tan \beta \cot \beta$

32. $\tan x \csc x$

33. $\sin \beta (1 + \cot^2 \beta)$

34. $\dfrac{1}{\sin^2 \theta} - \dfrac{\cos^2 \theta}{\sin^2 \theta}$

35. $\dfrac{\tan^2 \theta - \sin^2 \theta}{\tan^2 \theta \sin^2 \theta}$

36. $2(\csc^2 \theta - \cot^2 \theta)$

Show that each equation is an identity.

37. $1 + \tan^2 \theta = \sec^2 \theta$

38. $1 + \cot^2 \theta = \csc^2 \theta$

39. $\sec \alpha - \cos \alpha = \sin \alpha \tan \alpha$

40. $\dfrac{\sec \theta}{\csc \theta} = \tan \theta$

Critical Thinking

41. If $\tan \beta = \dfrac{3}{4}$, find $\dfrac{\sin \beta \sec \beta}{\cot \beta}$.

42. Refer to the diagram at the right to help you determine why $\tan^2 \theta + 1 = \sec^2 \theta$ is a Pythagorean identity. It will help you to know that the words tangent and secant are based on the Latin words for "touch" and "cut," respectively.

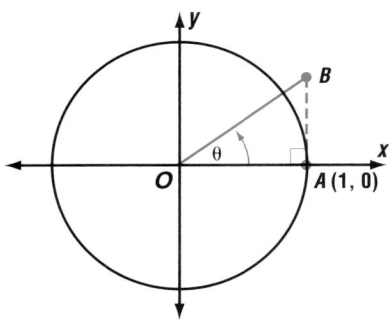

Applications and Problem Solving

43. **Running**   Refer to the application at the beginning of the lesson. Find the measure of the angle of incline in Example 5.

44. **Amusement Parks**   The angle-of-incline formula can be used for a person riding on a merry-go-round. Suppose the sine of the angle of incline of a person riding on an outside horse is $\dfrac{1}{5}$ and the diameter of the merry-go-round is 16 meters. Find the value of the angle of incline and the velocity of the merry-go-round.

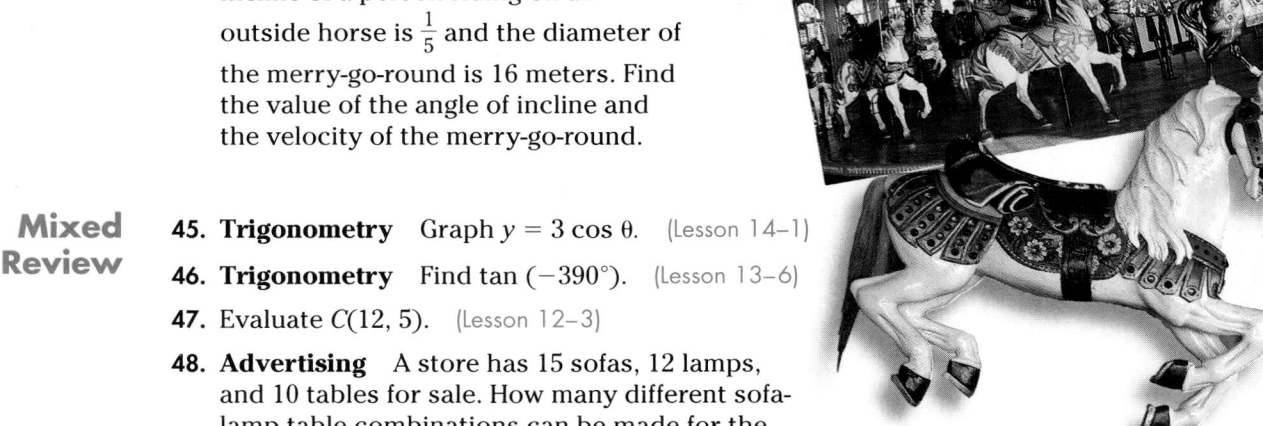

Mixed Review

45. **Trigonometry**   Graph $y = 3 \cos \theta$.   (Lesson 14–1)

46. **Trigonometry**   Find $\tan (-390°)$.   (Lesson 13–6)

47. Evaluate $C(12, 5)$.   (Lesson 12–3)

48. **Advertising**   A store has 15 sofas, 12 lamps, and 10 tables for sale. How many different sofa-lamp-table combinations can be made for the sales brochure?   (Lesson 12–1)

49. Find the next three terms of the sequence 3, 8, 13, ...   (Lesson 11–1)

50. Find all the roots of $x^3 + 3x^2 - 6x - 8 = 0$.   (Lesson 8–4)

51. Solve $y^2 + 4y - 21 = 0$ by factoring.   (Lesson 6–2)

52. Solve for n if $\begin{vmatrix} 4 & 3 & 6 \\ 2 & 2n & 7 \\ -4 & -3n & 3 \end{vmatrix} = -582$.   (Lesson 4–4)

14–3A Graphing Technology
Verifying Trigonometric Identities

A Preview of Lesson 14–3

You can use a graphing calculator to determine whether an equation may be a trigonometric identity. To verify an identity, graph the expressions on each side of the equals sign as two separate functions. If the graphs of the two functions do not match, then the equation is not an identity. If the graphs do coincide, then the equation *may* be an identity. The equation must be verified algebraically to be sure that it is an identity.

Example

Use a graphing calculator to determine whether the equation $\sec^2 x - 1 = \sin^2 x \sec^2 x$ may be an identity. *Be sure your calculator is in degree mode.*

Graph the equations $y = \sec^2 x - 1$ and $y = \sin^2 x \sec^2 x$ in the trigonometric window.

Remember that $\sec x = \dfrac{1}{\cos x}$.

Enter:

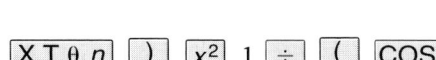

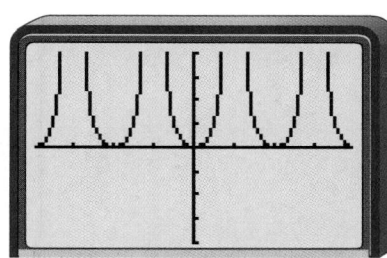

Since the graphs of the two functions coincide, the equation *may* be an identity.

EXERCISES

Use a graphing calculator to determine whether each equation may be an identity.

1. $(1 + \sin x)(1 - \sin x) = \cos^2 x$

2. $\cot x + \tan x = \csc x \cot x$

3. $\dfrac{\sec^2 x}{\tan x} = \sec x \csc x$

4. $\dfrac{1}{\sec x} + \dfrac{1}{\csc x} = 1$

5. $\sin (90° - x) = \cos x$

6. $\dfrac{1}{\sec x \tan x} = \csc x - \sin x$

7. $\dfrac{\tan x}{1 + \tan x} = \dfrac{\sin x}{\sin x + \cos x}$

8. $\cot^2 x (\sec^2 x - 1) = 1$

9. $\cos 2x = 1 - 2 \sin^2 x$

10. $\dfrac{\csc(-x)}{\sec(-x)} = -\cot x$

11. $\cos 3x + 1 = 2 \cos^2 x$

12. $\dfrac{\csc^2 x}{\csc x - 1} = \dfrac{1 + \sin x}{\sin x}$

Verifying Trigonometric Identities

APPLICATION
Physics

The device in the diagram at the right is called a *conical pendulum* because of the conical surface swept out by line segment $\overline{SP}$ as the weight rotates about the line. A formula for the relationship between the length L of the string and the angle θ that the string makes with the vertical line is given by the equation $L = \dfrac{g \sec \theta}{\omega^2}$ where g is the acceleration of gravity and ω (the Greek letter omega) is the angular velocity of the weight about the vertical line in radians per second. Is the equation $L = \dfrac{g \tan \theta}{\omega^2 \sin \theta}$ also an equation for the relationship between L and θ? *This problem will be solved in Example 3.*

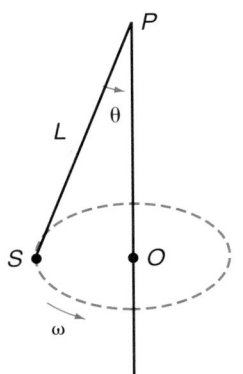

You can use the basic trigonometric identities and the definitions of the trigonometric functions to verify other identities. For example, suppose you wish to know whether $\tan \theta \, (\cot \theta + \tan \theta) = \sec^2 \theta$ is an identity. It is not sufficient to try some value of θ and conclude that the statement is true for all values of θ if it is true for that one. To verify that an equation is an identity, you must consider the general case.

Verifying an identity is like checking the solution of an equation. You do not know whether the expressions on the two sides of the equation are equal. That is what you are trying to verify. So, you must simplify one or both sides *separately* until they are the same. Often, it is easier to work with only one side of the equation. You may choose either side.

$$\tan \theta \, (\cot \theta + \tan \theta) \overset{?}{=} \sec^2 \theta \qquad \textit{Simplify the left side only.}$$

$$\tan \theta \left(\frac{1}{\tan \theta} + \tan \theta \right) \overset{?}{=} \sec^2 \theta \qquad \textit{cot } \theta = \tfrac{1}{\tan \theta}$$

$$1 + \tan^2 \theta \overset{?}{=} \sec^2 \theta \qquad \textit{Distributive property}$$

$$\sec^2 \theta \overset{?}{=} \sec^2 \theta \qquad \textit{1 + tan}^2 \theta = \sec^2 \theta$$

Thus, $\tan \theta \, (\cot \theta + \tan \theta) = \sec^2 \theta$ is an identity.

Example **1** **Verify that** $\frac{\sin^2 x}{1 - \cos x} = 1 + \cos x$ **is an identity.**

Notice that if the denominator, $1 - \cos x$, is multiplied by $1 + \cos x$, the result is $1 - \cos^2 x$, which equals $\sin^2 x$.

$$\frac{\sin^2 x}{1 - \cos x} \overset{?}{=} 1 + \cos x \qquad \textit{Multiply the numerator and} \\ \textit{denominator by } 1 + \cos x.$$

$$\frac{\sin^2 x}{1 - \cos x} \cdot \frac{1 + \cos x}{1 + \cos x} \overset{?}{=} 1 + \cos x \qquad \textit{Simplify.}$$

$$\frac{\sin^2 x (1 + \cos x)}{1 - \cos^2 x} \overset{?}{=} 1 + \cos x \qquad \textit{Simplify.}$$

$$\frac{\sin^2 x (1 + \cos x)}{\sin^2 x} \overset{?}{=} 1 + \cos x$$

$$1 + \cos x = 1 + \cos x$$

The identity could also be verified by working from the right side. The first step would be to multiply $1 + \cos x$ by $\frac{1 - \cos x}{1 - \cos x}$.

When you verify a trigonometric identity, you are really working backward. In Example 1, notice that the last step, $1 + \cos x = 1 + \cos x$, is actually the first step in the reasoning process. Since that step is clearly true, you can conclude that the next-to-last step is also true and write it with the symbol "=" instead of the symbol " $\overset{?}{=}$ ". Reason in the same manner to the step before that, continuing all the way back to the original equation, which now can be written without the " $\overset{?}{=}$ " symbol: $\frac{\sin^2 x}{1 - \cos x} = 1 + \cos x$.

You can use the problem-solving strategy **work backward** to solve other types of problems.

Example **2** **Find the sum of the reciprocals of two numbers whose sum is 9 and whose product is 18.**

PROBLEM SOLVING
Work Backward

Let x and y be the numbers.

We could set up these equations. $\quad x + y = 9$
$$xy = 18$$

Solving this system of equations is complicated. Rather than using this approach, work backward. The desired outcome is $\frac{1}{x} + \frac{1}{y}$.

$$\frac{1}{x} + \frac{1}{y} = \frac{y}{xy} + \frac{x}{xy} \qquad \textit{The LCD is xy.}$$

$$= \frac{x + y}{xy}$$

Looking back to our two original equations, we can see that $x + y = 9$ and $xy = 18$. So, $\frac{x + y}{xy} = \frac{9}{18}$ or $\frac{1}{2}$. The sum of the reciprocals is $\frac{1}{2}$.

Example **3** **Refer to the application at the beginning of the lesson.**

Does $\dfrac{g \sec \theta}{\omega^2} = \dfrac{g \tan \theta}{\omega^2 \sin \theta}$?

$\dfrac{g \sec \theta}{\omega^2} \overset{?}{=} \dfrac{g \tan \theta}{\omega^2 \sin \theta}$ *Simplify the right side.*

$\dfrac{g \sec \theta}{\omega^2} \overset{?}{=} \dfrac{g\left(\dfrac{\sin \theta}{\cos \theta}\right)}{\omega^2 \sin \theta}$ $\tan \theta = \frac{\sin \theta}{\cos \theta}$

$\dfrac{g \sec \theta}{\omega^2} \overset{?}{=} \dfrac{g\left(\dfrac{1}{\cos \theta}\right)}{\omega^2 \cdot 1}$ *Divide the numerator and denominator of the right side by sin θ.*

$\dfrac{g \sec \theta}{\omega^2} = \dfrac{g \sec \theta}{\omega^2}$ $\frac{1}{\cos \theta} = \sec \theta$

So, the formulas are equivalent.

The following suggestions may be helpful as you verify trigonometric identities.

- Start with the more complicated side of the equation. Transform the expression into the form of the simpler side.

or

- Work with each side of the equation at the same time. Transform each expression separately into the same form.

- Substitute one or more basic trigonometric identities to simplify the expression.

- Try factoring or multiplying to simplify the expression.

- Multiply both the numerator and the denominator by the same trigonometric expression.

- If nothing else seems to work, write both sides of the identity in terms of sine and cosine only. Then simplify each side as much as possible.

Example **4** **Verify that $1 - \tan^4 \beta = 2 \sec^2 \beta - \sec^4 \beta$ is an identity.**

$$1 - \tan^4 \beta \overset{?}{=} 2 \sec^2 \beta - \sec^4 \beta$$

$$(1 - \tan^2 \beta)(1 + \tan^2 \beta) \overset{?}{=} \sec^2 \beta \,(2 - \sec^2 \beta) \quad \textit{Factor each side.}$$

$$[1 - (\sec^2 \beta - 1)][\sec^2 \beta] \overset{?}{=} (2 - \sec^2 \beta)(\sec^2 \beta) \quad \textit{1 + tan}^2 \textit{ β = sec}^2 \textit{ β}$$

$$(2 - \sec^2 \beta)(\sec^2 \beta) = (2 - \sec^2 \beta)(\sec^2 \beta) \quad \textit{Simplify.}$$

Thus, the identity is verified.

Communicating Mathematics

Study the lesson. Then complete the following.

1. **Describe** the various methods you can use to identify trigonometric identities.

2. **Verify** the identity in Example 3 by simplifying each side separately into the same trigonometric expression.

Guided Practice

Verify that each of the following is an identity.

3. $\sin \theta \sec \theta \cot \theta = 1$

4. $\tan^2 x \cos^2 x = 1 - \cos^2 x$

5. $\csc y \sec y = \cot y + \tan y$

6. $\tan \alpha \sin \alpha \cos \alpha \csc^2 \alpha = 1$

7. $\dfrac{\sec \beta + \csc \beta}{1 + \tan \beta} = \csc \beta$

8. $\dfrac{1 - 2\cos^2 \beta}{\sin \beta \cos \beta} = \tan \beta - \cot \beta$

9. **Physics** Philip is building a clock that will use a conical pendulum to keep time. If the pendulum rotates at an angular velocity of 8 radians per second, the clock will keep the correct time. Due to the shape of the clock, he wants the pendulum to swing outward at an angle of 40° from the vertical. How long should he make the pendulum, in centimeters? (*Hint:* The acceleration due to gravity is 980 cm/s².)

Practice

Verify that each of the following is an identity.

10. $\sec^2 x - \tan^2 x = \tan x \cot x$

11. $\dfrac{1}{\sec^2 \theta} + \dfrac{1}{\csc^2 \theta} = 1$

12. $\tan^2 \theta - \sin^2 \theta = \tan^2 \theta \sin^2 \theta$

13. $\dfrac{\sec \alpha}{\sin \alpha} - \dfrac{\sin \alpha}{\cos \alpha} = \cot \alpha$

14. $\dfrac{\sin \alpha}{1 - \cos \alpha} + \dfrac{1 - \cos \alpha}{\sin \alpha} = 2\csc \alpha$

15. $\dfrac{\sin \theta}{\sec \theta} = \dfrac{1}{\tan \theta + \cot \theta}$

16. $\dfrac{1 - \cos x}{\sin x} = \dfrac{\sin x}{1 + \cos x}$

17. $\dfrac{\sec \theta + 1}{\tan \theta} = \dfrac{\tan \theta}{\sec \theta - 1}$

18. $\dfrac{1 - \cos x}{1 + \cos x} = (\csc x - \cot x)^2$

19. $\cos^2 x + \tan^2 x \cos^2 x = 1$

20. $\dfrac{\cot \theta + \csc \theta}{\sin \theta + \tan \theta} = \cot \theta \csc \theta$

21. $\dfrac{1 + \tan^2 \theta}{\csc^2 \theta} = \tan^2 \theta$

22. $\dfrac{1 + \sin x}{\sin x} = \dfrac{\cot^2 x}{\csc x - 1}$

23. $\dfrac{\cos y}{1 + \sin y} + \dfrac{\cos y}{1 - \sin y} = 2\sec y$

24. $\cos^4 \theta - \sin^4 \theta = \cos^2 \theta - \sin^2 \theta$

25. $\cot x (\cot x + \tan x) = \csc^2 x$

26. $\dfrac{\tan^2 x}{\sec x - 1} = 1 + \dfrac{1}{\cos x}$

27. $\dfrac{1 + \tan \alpha}{1 + \cot \alpha} = \dfrac{\sin \alpha}{\cos \alpha}$

28. $\sin \theta + \cos \theta = \dfrac{1 + \tan \theta}{\sec \theta}$

29. $1 + \sec^2 x \sin^2 x = \sec^2 x$

Critical Thinking

30. Create a trigonometric identity. Explain the method you used to do this. Then trade with another student and verify each other's identities.

Applications and Problem Solving

31. **Work Backward** After cashing her paycheck, Estrella paid her father back the $15 she had borrowed. She then spent half of the remaining money on clothes, and then spent half of what remained on a concert ticket. She bought a cassette tape for $7.45 and had $10.25 left. What was the amount of Estrella's paycheck?

32. Physics If a ball is hit or kicked from ground level, the maximum height it will reach is given by the formula

$$h = \frac{v_0^2 \sin^2 \theta}{2g},$$ where θ is the measure of the angle between the ground and the initial path of the ball, v_0 is its initial velocity in meters per second, and g is the acceleration due to gravity. The value of g is 9.8 m/s^2.

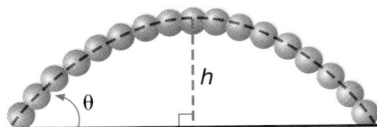

a. Show that the formula $h = \dfrac{v_0^2 \tan^2 \theta}{2g \sec^2 \theta}$ is equivalent to the one given above.

b. If a baseball is hit with an initial velocity of 47 meters per second at an initial angle of 50° from the ground, find the maximum height the ball will reach, to see whether it will stay under the roof of the Astrodome, which is about 60 meters high.

c. For a given initial velocity, a ball's maximum horizontal range is attained when the angle of elevation is 45°. If a ball was thrown to a height of 50 meters and traveled as far horizontally as it possibly could, what was its initial velocity?

Mixed Review

33. Trigonometry Simplify $\dfrac{\sec \alpha}{\sin \alpha} - \dfrac{\sin \alpha}{\cos \alpha}$. (Lesson 14–2)

34. Trigonometry Find the exact value of $\cos \dfrac{17\pi}{3}$. (Lesson 13–3)

35. Probability Two dice are rolled. Find the probability of rolling a 3 and a 4. (Lesson 12–5)

36. Find the sum of the infinite geometric series, $12 + 6 + 3 + \ldots$, if it exists. (Lesson 11–5)

37. Simplify $\left(x^2 y^2\right)^2 x^3 y^3$. (Lesson 5–1)

38. Suppose $h(x) = [3x - 1]$. Find $h(-2.1)$. (Lesson 2–6)

SELF TEST

State the amplitude (if it exists) and period of each function. (Lesson 14–1)

1. $y = \cos 4\theta$

2. $y = 3 \sin \theta$

3. $y = 2 \tan \dfrac{1}{5}\theta$

Graph each function. (Lesson 14–1)

4. $y = \dfrac{1}{2} \cos \theta$

5. $y = 5 \sin \theta$

Solve for values between 90° and 270°. (Lesson 14–2)

6. If $\cot \theta = -\dfrac{2}{5}$, find $\csc \theta$.

7. If $\cos \alpha = -\dfrac{1}{2}$, find $\tan \alpha$.

Verify that each of the following is an identity. (Lesson 14–3)

8. $\sec \theta - \tan \theta \sin \theta = \cos \theta$

9. $(1 - \sin^2 \theta)(1 + \tan^2 \theta) = 1$

10. Work Backward If the sum of two numbers is 4 and the product of the numbers is 7, find the sum of the reciprocals of these numbers. (Lesson 14–3)

Sum and Difference of Angles Formulas

CONNECTION
Physical Science

In Earth's northern hemisphere, the day with the most hours of sunlight occurs around June 22, and the day with the fewest hours of sunlight occurs around December 22. Suppose that E is the amount of light energy reaching a square foot patch of ground when the sun is directly overhead. When the sun is not overhead, the amount of light will depend on the angle that a ray of sunlight makes with the horizon.

Aurora Borealis, or "Northern Lights"

On June 22, the maximum amount of light energy falling on a square foot of ground at a certain location is given by $E \sin (113.5° - \phi)$ where ϕ (the Greek letter phi) is the latitude of the location. How would the amount of light energy that you receive compare with the amount received by other parts of Earth? *This problem will be solved in Example 2.*

It is often helpful to use the formulas for the trigonometric values of the difference or sum of two angles such as $\sin (113.5° - \phi)$. For example, you could find $\sin 15°$ by evaluating $\sin (45° - 30°)$. It is important to realize that $\sin (\alpha - \beta)$ is not the same as $\sin \alpha - \sin \beta$. The following discussion will show how to evaluate expressions like $\sin (\alpha - \beta)$ or $\cos (\alpha + \beta)$.

The figure at the right shows two angles α and β in standard position on the unit circle.

Use the distance formula to find d, where $(x_1, y_1) = (\cos \beta, \sin \beta)$ and $(x_2, y_2) = (\cos \alpha, \sin \alpha)$.

$$d = \sqrt{(\cos \alpha - \cos \beta)^2 + (\sin \alpha - \sin \beta)^2}$$

$$d^2 = (\cos \alpha - \cos \beta)^2 + (\sin \alpha - \sin \beta)^2$$

$$d^2 = (\cos^2 \alpha - 2\cos \alpha \cos \beta + \cos^2 \beta) + (\sin^2 \alpha - 2 \sin \alpha \sin \beta + \sin^2 \beta)$$

$$d^2 = \cos^2 \alpha + \sin^2 \alpha + \cos^2 \beta + \sin^2 \beta - 2 \cos \alpha \cos \beta - 2 \sin \alpha \sin \beta$$

$$d^2 = 1 + 1 - 2 \cos \alpha \cos \beta - 2 \sin \alpha \sin \beta \qquad \text{\small sin}^2 \alpha + \text{\small cos}^2 \alpha = 1 \text{ and}$$
$$\text{\small sin}^2 \beta + \text{\small cos}^2 \beta = 1$$

$$d^2 = 2 - 2 \cos \alpha \cos \beta - 2 \sin \alpha \sin \beta$$

Now find the value of d^2 when the angle having measure $\alpha - \beta$ is in standard position on the unit circle, as shown in the figure below.

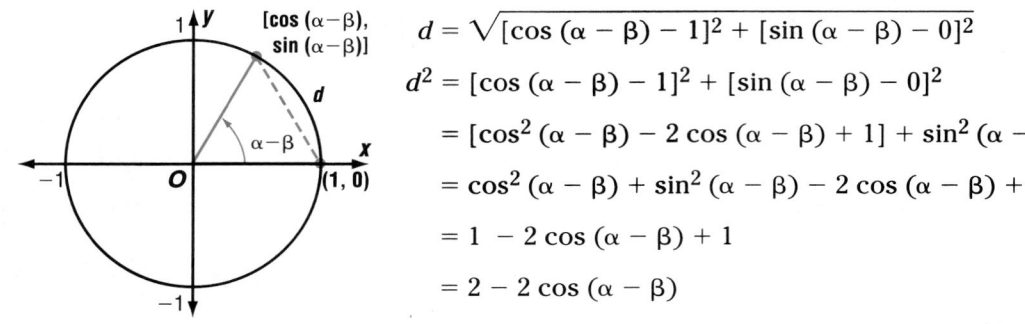

$$d = \sqrt{[\cos (\alpha - \beta) - 1]^2 + [\sin (\alpha - \beta) - 0]^2}$$

$$d^2 = [\cos (\alpha - \beta) - 1]^2 + [\sin (\alpha - \beta) - 0]^2$$

$$= [\cos^2 (\alpha - \beta) - 2 \cos (\alpha - \beta) + 1] + \sin^2 (\alpha - \beta)$$

$$= \cos^2 (\alpha - \beta) + \sin^2 (\alpha - \beta) - 2 \cos (\alpha - \beta) + 1$$

$$= 1 - 2 \cos (\alpha - \beta) + 1$$

$$= 2 - 2 \cos (\alpha - \beta)$$

By equating the two expressions for d^2, you can find a formula for $\cos (\alpha - \beta)$.

$$d^2 = d^2$$

$$2 - 2 \cos (\alpha - \beta) = 2 - 2 \cos \alpha \cos \beta - 2 \sin \alpha \sin \beta$$

$$-1 + \cos (\alpha - \beta) = -1 + \cos \alpha \cos \beta + \sin \alpha \sin \beta \qquad \textit{Divide each side by } -2.$$

$$\cos (\alpha - \beta) = \cos \alpha \cos \beta + \sin \alpha \sin \beta \qquad \textit{Add 1 to each side.}$$

Use the formula for $\cos (\alpha - \beta)$ to find a formula for $\cos (\alpha + \beta)$.

$$\cos (\alpha + \beta) = \cos [\alpha - (-\beta)]$$

$$= \cos \alpha \cos (-\beta) + \sin \alpha \sin (-\beta)$$

$$= \cos \alpha \cos \beta - \sin \alpha \sin \beta \qquad \textit{cos } (-\beta) = \cos \beta; \sin (-\beta) = -\sin \beta$$

You can use a similar method to find formulas for $\sin (\alpha + \beta)$ and $\sin (\alpha - \beta)$.

Sum and Difference of Angles Formulas	**The following identities hold true for all values of α and β.** $\cos (\alpha \pm \beta) = \cos \alpha \cos \beta \mp \sin \alpha \sin \beta$ $\sin (\alpha \pm \beta) = \sin \alpha \cos \beta \pm \cos \alpha \sin \beta$

Notice the symbol $\mp$ in the formula for $\cos (\alpha \pm \beta)$. It means "minus or plus." In the cosine formula, when the sign on the left side of the equation is plus, the sign on the right side is minus; when the sign on the left side is minus, the sign on the right side is plus. The signs match each other in the sine formula.

The following examples show how to find exact values of trigonometric expressions by using the sum and difference formulas.

Example **1** **Find the exact value of each expression.**
    a. sin 105°

        Use the formula $\sin (\alpha + \beta) = \sin \alpha \cos \beta + \cos \alpha \sin \beta$.

$$\sin 105° = \sin (60° + 45°)$$

$$= \sin 60° \cos 45° + \cos 60° \sin 45°$$

$$= \frac{\sqrt{3}}{2} \cdot \frac{\sqrt{2}}{2} + \frac{1}{2} \cdot \frac{\sqrt{2}}{2}$$

$$= \frac{\sqrt{6} + \sqrt{2}}{4} \text{ or about } 0.9659$$

b. cos (−120°)

Use the formula cos (α − β) = cos α cos β + sin α sin β.

$$\cos(-120°) = \cos(60° - 180°)$$
$$= \cos 60° \cos 180° + \sin 60° \sin 180°$$
$$= \frac{1}{2} \cdot (-1) + \frac{\sqrt{3}}{2} \cdot 0$$
$$= -\frac{1}{2}$$

Example ②

Refer to the application at the beginning of the lesson.

a. For what latitude will the light energy per square foot be greatest on June 22?

b. Suppose you live in Portland, Oregon (latitude: 45.5° N; longitude: 122.7° W). What is the maximum amount of light energy that a square foot of ground can receive on June 22 as a percent of that received at the latitude in part a?

a. Since the value of the sine function cannot be greater than 1, the given expression can be no greater than $E \cdot 1$ or E. The sine function equals 1 when $113.5° - \phi = 90°$; that is, when $\phi = 113.5° - 90°$, or $23.5°$. Thus, the greatest intensity of light energy on June 22 occurs at a latitude of 23.5° N.

b. *Explore*   Although information is provided about both the latitude and longitude of Portland, only the latitude is needed to solve the problem.

Plan   Use the difference formula for the sine.

$$\sin(113.5° - \phi)$$
$$= \sin 113.5° \cos \phi - \cos 113.5° \sin \phi$$
$$= 0.9171 \cos \phi - (-0.3987) \sin \phi$$
$$= 0.9171 \cos \phi + 0.3987 \sin \phi$$

Solve   The latitude of Portland is 45.5° N.

$$\sin(113.5° - 45.5°)$$
$$= 0.9171 \cos 45.5° + 0.3987 \sin 45.5°$$
$$= 0.9171 \cdot 0.7010 + 0.3987 \cdot 0.7133$$
$$= 0.6429 + 0.2844$$
$$= 0.9273$$

In Portland, the maximum light energy per square foot is $0.9273E$. Therefore, the light energy is about 93% of that found at a latitude of 23.5°.

Examine   Use a calculator to find sin (113.5° − 45.5°) or sin (68°). The answer is approximately 0.927 or 93%. Thus, the answer is correct.

You can also use the sum and difference formulas to verify identities.

Example **3** **Verify that each of the following is an identity.**
a. $\cos (90° - \theta) = \sin \theta$

$$\cos (90° - \theta) \overset{?}{=} \sin \theta$$
$$\cos 90° \cos \theta + \sin 90° \sin \theta \overset{?}{=} \sin \theta$$
$$0 \cdot \cos \theta + 1 \cdot \sin \theta \overset{?}{=} \sin \theta$$
$$\sin \theta = \sin \theta$$

b. $\cos (180° + \theta) = -\cos \theta$

$$\cos (180° + \theta) \overset{?}{=} -\cos \theta$$
$$\cos 180° \cos \theta - \sin 180° \sin \theta \overset{?}{=} -\cos \theta$$
$$-1 \cdot \cos \theta - 0 \cdot \sin \theta \overset{?}{=} -\cos \theta$$
$$-\cos \theta = -\cos \theta$$

CHECK FOR UNDERSTANDING

Communicating Mathematics

Study the lesson. Then complete the following.

1. **Determine** whether the equation $\cos (x + y) = \cos x + \cos y$ is an identity.

2. **Describe** a method for finding the exact value of sin 15°. Then find the value.

Guided Practice

Find the exact value of each expression.

3. $\cos 75°$

4. $\sin 165°$

5. $\cos 255°$

6. $\cos 80° \cos 20° + \sin 80° \sin 20°$

Verify that each of the following is an identity.

7. $\sin (270° - \theta) = -\cos \theta$

8. $\cos (90° + \theta) = -\sin \theta$

9. $\sin (x + 30°) + \cos (x + 60°) = \cos x$

10. Refer to Example 2. Another way to solve the problem would be to substitute 45.5° into the expression $\sin (113.5° - \theta)$ and find the value of sin 68° by using a calculator. When you do this, you obtain 0.9271838, or about 0.9272. This is slightly less than the result obtained in Example 2. Why is this so?

EXERCISES

Practice

Find the exact value of each expression.

11. $\sin 285°$

12. $\sin 75°$

13. $\cos 195°$

14. $\cos 105°$

15. $\cos 345°$

16. $\cos 165°$

17. $\sin 65° \cos 35° - \cos 65° \sin 35°$

18. $\sin 40° \cos 20° + \cos 40° \sin 20°$

19. $\cos 25° \cos 5° - \sin 25° \sin 5°$

Verify that each of the following is an identity.

20. $\cos(270° - \theta) = -\sin\theta$

21. $\sin(90° + \theta) = \cos\theta$

22. $\sin(180° + \theta) = -\sin\theta$

23. $\sin(90° - \theta) = \cos\theta$

24. $\sin(60° + \theta) + \sin(60° - \theta) = \sqrt{3}\cos\theta$

25. $\sin(x + y)\sin(x - y) = \sin^2 x - \sin^2 y$

26. $\sin\left(\theta + \dfrac{\pi}{3}\right) - \cos\left(\theta + \dfrac{\pi}{6}\right) = \sin\theta$

Use the identity $\tan(\alpha - \beta) = \dfrac{\tan\alpha - \tan\beta}{1 + \tan\alpha\tan\beta}$ to find the exact value of each expression.

27. $\tan(225° - 120°)$

28. $\tan(315° - 120°)$

29. $\tan(30° + 30°)$

30. $\tan 195°$

Graphing
Calculator

Use a graphing calculator to determine the angle x that would make each equation true.

31. $\cos(\alpha + x) = -\sin x$

32. $\sin(\alpha + x) = -\cos x$

33. $\tan(\alpha + x) = -\cot x$

34. $\sin(\alpha + x) = -\sin x$

Critical
Thinking

35. Use the formulas for $\sin(\alpha + \beta)$ and $\cos(\alpha + \beta)$ to derive the formula for $\tan(\alpha + \beta)$. (*Hint:* Divide all terms of the expression by $\cos\alpha\cos\beta$.)

Applications and
Problem Solving

36. Physical Science   Use the formula for $\sin(\alpha - \beta)$ to find the light energy that falls on a square foot of ground in each city on June 22. Express your answer in terms of E, the energy from an overhead sun. Check your answer by substituting directly into the expression $\sin(113.5° - \phi)$.

Anchorage, Alaska

  a. Anchorage, Alaska (Latitude: 61.2° N)

  b. Bangor, ME (Latitude: 44.8° N)

  c. Key West, FL (Latitude: 24.6° N)

  d. your community

37. Physical Science   On December 22, the maximum amount of light energy that falls on a square foot of ground at a certain location is given by $E\sin(113.5° + \phi)$, where ϕ is the latitude of the location. Use the sum formula for sines to find the light energy that falls on a square foot of ground at each location on December 22. Express your answer in terms of E, the energy from an overhead sun. Check your answer by substituting directly into the expression $\sin(113.5° + \phi)$.

  a. Dallas, TX (Latitude: 32.8° N)

  b. Portland, OR (Latitude: 45.5° N)

  c. Equator (Latitude: 0.0° N)

Key West, Florida

38. Geology   Geologist Norma Ayala measures the angle between one side of a rectangular lot and the line from her position to the opposite corner of the lot as 30°. She then measures the angle between that line and the line to the point on the property where a river crosses the property as 45°. Dr. Ayala stands 100 yards from the opposite corner of the property. How far is she from the point at which the river crosses the property line?

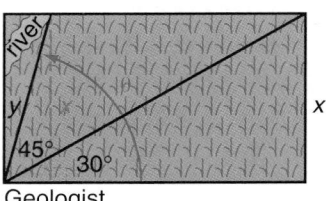

Geologist

Mixed Review

39. Trigonometry   Verify that $\sin \theta \sec \theta \cot \theta = 1$ is an identity. (Lesson 14–3)

40. Golf   Mr. Sanchez estimates that he hooked his last drive 18° to the left of where he intended and that the ball is now 200 yards from the tee. If the hole for which he was aiming is 180 yards from the tee, how far, to the nearest yard, is the ball from the hole?   (Lesson 13–5)

41. Probability   Two cards are drawn from a standard deck of cards. What is the probability of having drawn a black card or an ace?   (Lesson 12–6)

42. Business   At an annual board meeting of a small corporation, it was decided that 8% of the profits for the next year would be divided among the six managers of the corporation. There are two sales managers and four nonsales managers. Fifty percent of the amount would be split equally among all six managers. The other 50% would be split among the four nonsales managers. Let p represent the annual profits of the corporation. (Lesson 9–4)

   a. Write an expression to represent the share of the profits each nonsales manager will receive.

   b. Simplify this expression.

   c. Write an expression in simplest form to represent the share of the profits each sales manager will receive.

43. Graph $f(x) = \dfrac{x}{x - 2}$.   (Lesson 9–1)

44. Use synthetic substitution to find $f(2)$ and $f(-1)$ if $f(x) = x^4 + x^3 + x^2 + x + 1$.   (Lesson 8–2)

45. Construction   A subdivision has 60 lots available. The builder knows from experience that she should plan to build at least three times as many ranch-style houses as colonial-style. If she will make a profit of $5000 on each colonial house and $4500 on each ranch, how many of each kind should she plan to build in this development?   (Lesson 3–6)

46. Statistics   Choose a statistical graph— bar, line, circle, or pictograph— to represent the data shown in the table at the right. Then draw the graph.   (Lesson 1–3)

Technology Used by Teens	
Item	**Percent**
cable TV	66
cellular telephone	35
computer	46
on-line service	17
VCR	96
video game player	77

Source: Chilton Research Services

Double-Angle and Half-Angle Formulas

What YOU'LL LEARN

- To find values of sine and cosine involving double and half angles, and
- to verify identities by using double- and half-angle formulas.

Why IT'S IMPORTANT

You can use double-angle and half-angle formulas to solve problems involving aviation and physics.

APPLICATION
Aviation

A plane that travels at the speed of sound (about 740 miles per hour) is said to be traveling at Mach 1. The Mach number, named after the Austrian physicist Ernst Mach (1838–1916), is defined as the ratio of the speed of the plane to the speed of sound. On October 14, 1947, Charles (Chuck) Yeager became the first person to fly an aircraft faster than Mach 1. In the process, his Bell X-1 rocket airplane created a sonic boom.

Chuck Yeager

When a plane travels at a Mach number greater than 1, a sonic boom is created by sound waves forming a cone that intersects the ground in the outline of a hyperbola. If θ is the measure of the angle at the vertex of the cone, then the Mach number is related to θ by the equation $\sin \frac{\theta}{2} = \frac{1}{M}$, provided that $M > 1$.

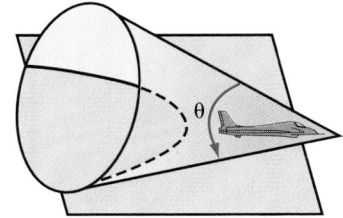

Suppose the measure of the cone's vertex angle is 45°. What is the speed of the plane? *This problem will be solved in Example 3.*

You can use the formula for $\sin (\alpha + \beta)$ to find the sine of twice an angle θ, $\sin 2\theta$, and the formula for $\cos (\alpha + \beta)$ to find the cosine of twice an angle θ, $\cos 2\theta$.

Let θ represent the measure of the angle.

$$\sin 2\theta = \sin (\theta + \theta)$$
$$= \sin \theta \cos \theta + \cos \theta \sin \theta$$
$$= 2 \sin \theta \cos \theta$$
$$\cos 2\theta = \cos (\theta + \theta)$$
$$= \cos \theta \cos \theta - \sin \theta \sin \theta$$
$$= \cos^2 \theta - \sin^2 \theta$$

You can find alternate forms for $\cos 2\theta$ by making substitutions into the expression $\cos^2 \theta - \sin^2 \theta$.

$$\cos^2 \theta - \sin^2 \theta = (1 - \sin^2 \theta) - \sin^2 \theta \quad \textit{Substitute } 1 - \sin^2 \theta \textit{ for } \cos^2 \theta.$$
$$= 1 - 2 \sin^2 \theta$$
$$\cos^2 \theta - \sin^2 \theta = \cos^2 \theta - (1 - \cos^2 \theta) \quad \textit{Substitute } 1 - \cos^2 \theta \textit{ for } \sin^2 \theta.$$
$$= 2 \cos^2 \theta - 1$$

These formulas are called the **double-angle formulas.**

LOOK BACK

Refer to Lesson 7-5 for a review of hyperbolas.

Double-Angle Formulas	The following identities hold true for all values of θ. $\sin 2\theta = 2 \sin \theta \cos \theta \qquad \cos 2\theta = \cos^2 \theta - \sin^2 \theta$ $\cos 2\theta = 1 - 2 \sin^2 \theta$ $\cos 2\theta = 2 \cos^2 \theta - 1$

Example Suppose x is between $90°$ and $180°$ and $\sin x = \frac{1}{2}$. Find the exact value of each expression.

a. $\cos 2x$

Use the identity $\cos 2x = 1 - 2 \sin^2 x$.

$\cos 2x = 1 - 2 \sin^2 x$

$\qquad = 1 - 2\left(\frac{1}{2}\right)^2 \qquad$ *Substitute $\frac{1}{2}$ for sin x.*

$\qquad = \frac{1}{2}$

The value of $\cos 2x$ is $\frac{1}{2}$.

b. $\sin 2x$

Find the value of $\cos x$. Then use the identity $\sin 2x = 2 \sin x \cos x$.

$\cos^2 x = 1 - \sin^2 x \qquad cos^2 x + sin^2 x = 1$

$\qquad = 1 - \left(\frac{1}{2}\right)^2$

$\qquad = \frac{3}{4}$

$\cos x = -\frac{\sqrt{3}}{2} \qquad$ *cosine is negative in the second quadrant.*

$\sin 2x = 2 \sin x \cos x$

$\qquad = 2\left(\frac{1}{2}\right)\left(-\frac{\sqrt{3}}{2}\right) \qquad$ *Substitute $\frac{1}{2}$ for sin x and $-\frac{\sqrt{3}}{2}$ for cos x.*

$\qquad = -\frac{\sqrt{3}}{2}$

The value of $\sin 2x$ is $-\frac{\sqrt{3}}{2}$.

You can also derive formulas to find the cosine and the sine of half a given angle. Let α represent the measure of this angle.

Find $\cos \frac{\alpha}{2}$. $\qquad\qquad$ Find $\sin \frac{\alpha}{2}$.

$2 \cos^2 \theta - 1 = \cos 2\theta \qquad 1 - 2 \sin^2 \theta = \cos 2\theta \qquad$ *Use double-angle formulas.*

$2 \cos^2 \frac{\alpha}{2} - 1 = \cos \alpha \qquad 1 - 2 \sin^2 \frac{\alpha}{2} = \cos \alpha \qquad$ *Substitute $\frac{\alpha}{2}$ for θ and α for 2θ*

$\cos^2 \frac{\alpha}{2} = \frac{1 + \cos \alpha}{2} \qquad \sin^2 \frac{\alpha}{2} = \frac{1 - \cos \alpha}{2} \qquad$ *Solve for the squared term.*

$$\cos \frac{\alpha}{2} = \pm \sqrt{\frac{1 + \cos \alpha}{2}} \qquad \sin \frac{\alpha}{2} = \pm \sqrt{\frac{1 - \cos \alpha}{2}}$$ *Take the square root of each side.*

These are called the **half-angle formulas.** The signs are determined by the function of $\frac{\alpha}{2}$.

Half-Angle Formulas	**The following identities hold true for all values of α.** $$\cos \frac{\alpha}{2} = \pm \sqrt{\frac{1 + \cos \alpha}{2}} \qquad \sin \frac{\alpha}{2} = \pm \sqrt{\frac{1 - \cos \alpha}{2}}$$

Example **2** Suppose $\sin x = -\frac{9}{41}$ and x is between $270°$ and $360°$. Find $\cos \frac{x}{2}$.

Since $\cos \frac{x}{2} = \pm \sqrt{\frac{1 + \cos x}{2}}$, we must find $\cos x$ first.

Use $\cos^2 x + \sin^2 x = 1$.

$$\cos^2 x + \sin^2 x = 1$$

$$\cos^2 x + \left(-\frac{9}{41}\right)^2 = 1 \qquad \textit{Replace } \sin x \textit{ with } -\frac{9}{41}.$$

$$\cos^2 x = 1 - \frac{81}{1681}$$

$$\cos^2 x = \frac{1600}{1681}$$

$$\cos x = \pm\frac{40}{41} \qquad \textit{Take the square root of each side.}$$

Since x is in the fourth quadrant, $\cos x = \frac{40}{41}$.

$$\cos \frac{x}{2} = \pm \sqrt{\frac{1 + \cos x}{2}} \qquad \textit{Half-angle formula}$$

$$= \pm \sqrt{\frac{1 + \frac{40}{41}}{2}} \qquad \textit{Replace } \cos x \textit{ with } \frac{40}{41}.$$

$$= \pm \sqrt{\frac{81}{82}} \textit{ or } \pm \frac{9\sqrt{82}}{82}$$

Since x is between $270°$ and $360°$, $\frac{x}{2}$ is between $135°$ and $180°$. Thus, $\cos \frac{x}{2}$ is negative and equals $-\frac{9\sqrt{82}}{82}$ or about -0.9939.

Example **3** Refer to the application at the beginning of the lesson. A plane traveling at a supersonic speed sends out sound waves that form a cone with a vertex angle of 45°. Find the speed of the plane if Mach 1 is about 740 mph.

APPLICATION
Aviation

Use the equation $\sin \frac{\theta}{2} = \frac{1}{M}$ to find the Mach number. Then use the Mach number to find the speed.

The value of the left side of the Mach equation is found by using a half-angle identity. Since 45° is in the first quadrant, the negative solution is ignored.

$$\sin \frac{\theta}{2} = \sqrt{\frac{1 - \cos \theta}{2}}$$

$$= \sqrt{\frac{1 - \cos 45°}{2}} \quad \textit{Replace } \theta \textit{ with } 45°.$$

$$= \sqrt{\frac{1 - \frac{\sqrt{2}}{2}}{2}}$$

$$= \sqrt{\frac{2 - \sqrt{2}}{4}}$$

$$= \frac{1}{2}\sqrt{2 - \sqrt{2}} \text{ or about } 0.3827$$

Substitute this value into the Mach equation.

$$0.3827 \approx \frac{1}{M}$$

$$0.3827M \approx 1$$

$$M \approx \frac{1}{0.3827} \text{ or about } 2.61$$

Multiply the Mach number by the speed of sound to get the speed of the plane.

$$s \approx 2.61 \cdot 740 \text{ mph or about } 1930 \text{ mph}$$

The speed of the plane is about 1930 miles per hour.

You can also use the double-angle and half-angle formulas to verify identities.

Example **4** Verify that $\cot x = \dfrac{\sin 2x}{1 - \cos 2x}$ is an identity.

$$\cot x = \frac{\sin 2x}{1 - \cos 2x}$$

$$= \frac{2 \sin x \cos x}{1 - (1 - 2 \sin^2 x)} \quad \textit{Substitute } 2 \sin x \cos x \textit{ for } \sin 2x \textit{ and}$$
$$\textit{1 - 2 sin}^2 x \textit{ for cos 2x.}$$

$$= \frac{2 \sin x \cos x}{2 \sin^2 x} \quad \textit{Simplify.}$$

$$= \frac{\cos x}{\sin x}$$

$$= \cot x$$

Communicating Mathematics

Study the lesson. Then complete the following.

1. **Explain** how to find $\sin x$ if $2x$ is in the third quadrant.

2. Refer to Example 2. Find $\sin \frac{x}{2}$ if $\sin x = -\frac{9}{41}$ and x is between $270°$ and $360°$.

3. **You Decide**   Jack notices that $\sin 1° = 0.01745$, $\sin 2° = 0.03490$, and $\sin 4° = 0.0698$ and concludes that if you double an angle measure then you also double its sine. Minya says he is wrong and can give examples to prove it. Who is right? Explain.

Guided Practice

Find the exact values of $\sin 2x$, $\cos 2x$, $\sin \frac{x}{2}$, and $\cos \frac{x}{2}$ for each of the following.

4. $\sin x = \frac{5}{13}$; x is between $90°$ and $180°$.

5. $\cos x = \frac{1}{5}$; x is in Quadrant IV.

Find the exact value of each expression by using the half-angle formulas.

6. $\cos \frac{\pi}{8}$

7. $\sin 22\frac{1}{2}°$

Verify that each of the following is an identity.

8. $(\sin x + \cos x)^2 = 1 + \sin 2x$

9. $\dfrac{1}{\sin x \cos x} - \dfrac{\cos x}{\sin x} = \tan x$

10. **Aviation**   Refer to the application at the beginning of the lesson. The Mach number for a certain plane traveling at supersonic speed is 1.4. Find the measure of the vertex angle of the cone formed by the sound waves that the plane sends out.

Practice

Find the exact values of $\sin 2x$, $\cos 2x$, $\sin \frac{x}{2}$, and $\cos \frac{x}{2}$ for each of the following.

11. $\sin x = \frac{4}{5}$; x is between $90°$ and $180°$.

12. $\cos x = \frac{3}{5}$; x is in Quadrant I.

13. $\cos x = -\frac{1}{3}$; x is between $180°$ and $270°$.

14. $\cos x = -\frac{2}{3}$; x is in Quadrant III.

Find the exact values of sin 2x, cos 2x, sin $\frac{x}{2}$, and cos $\frac{x}{2}$ for each of the following.

15. $\sin x = -\frac{3}{5}$; x is in Quadrant III.

16. $\sin x = -\frac{3}{4}$; x is between 270° and 360°.

17. $\cos x = -\frac{1}{3}$; x is in Quadrant II.

18. $\sin x = -\frac{1}{4}$; x is between 180° and 270°.

Find the exact value of each expression by using the half-angle formulas.

19. $\sin 105°$

20. $\sin 195°$

21. $\sin \frac{7\pi}{8}$

22. $\cos \frac{19\pi}{12}$

Verify that each of the following is an identity.

23. $\cos^2 2x + 4 \sin^2 x \cos^2 x = 1$

24. $\sin^2 \theta = \frac{1}{2}(1 - \cos 2\theta)$

25. $\sin 2x = 2 \cot x \sin^2 x$

26. $\sin^4 x - \cos^4 x = 2 \sin^2 x - 1$

27. $2 \cos^2 \frac{x}{2} = 1 + \cos x$

28. $\tan^2 \frac{x}{2} = \frac{1 - \cos x}{1 + \cos x}$

Critical Thinking

29. Explain the method that you would use to find $\sin x$ if $\sin 4x = \frac{2}{3}$ and the terminal side of $4x$ lies in the second quadrant. Then find $\sin x$. (*Hint:* Notice that the terminal sides of x and $2x$ do not necessarily lie in Quadrant I.)

Applications and Problem Solving

30. **Aviation** A plane traveling at supersonic speed sends out sound waves that form a cone with a vertex angle of 60°. Find the speed of the plane.

31. **Physics** The index of refraction for a medium through which light is passing is the ratio of the velocity of light in free space to the velocity of light in the medium. For light passing through a medium such as glass or diamonds, the index of refraction n is given by the equation $n = \dfrac{\sin\left(\frac{\theta}{2} + \frac{\alpha}{2}\right)}{\sin \frac{\alpha}{2}}$, where α is the deviation angle and θ is the angle of the apex of the medium. If the medium is a zircon with an index of refraction of 1.9 and the angle of the apex of the diamond is 90°, determine the measure of the deviation angle.

32. Geology A geologist stands on a ledge and finds that the angle of depression to a river's surface is 12°. The angle of depression to the riverbed below the surface is 13°. The geologist is 1500 feet from the river's surface. (Lesson 14–4)

a. Write an expression for the sine of the angle between the line from the geologist to the river's surface and the line from the geologist to the riverbed.

b. How far above the riverbed is the surface of the water?

33. Trigonometry Solve △ABC. Round measures of sides and angles to the nearest tenth. (Lesson 13–4)

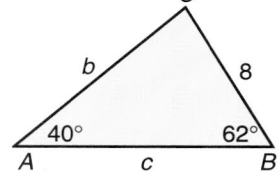

34. Environment Martel Johnson is an environmental research assistant. As part of her job, she makes video tapes of the countries she researches. On a circular carousel, she has six tapes of South America, four tapes of Africa, and two tapes of Australia. How many different ways can she arrange these tapes on the carousel? (Lesson 12–2)

35. Use a calculator to find the antilogarithm of 0.2586, rounded to four decimal places. (Lesson 10–4)

36. Find $\begin{bmatrix} 4 \\ -1 \\ 3 \end{bmatrix} \cdot \begin{bmatrix} 1 & 0 & 0 \\ 0 & 1 & 0 \\ 0 & 0 & 1 \end{bmatrix}$. (Lesson 4–3)

37. Geometry The formula for the volume of a right circular cone is $V = \frac{1}{3}\pi r^2 h$, where r represents the radius of the circular base and h represents the height. Solve the formula for h. (Lesson 1–4)

WORKING ON THE
In·ves·ti·ga·tion

Refer to the Investigation on pages 768–769.

In the same way that a basketball is projected upward by the motion of your hands during a free-throw shot, a roller-coaster car is projected up the next hill by the energy accumulated going down the previous hill. The formula for the height of the car can be determined from the formula $h = (v_i \sin \theta)t - 0.5gt^2$, where h is the height of the object, v_i is the initial velocity, g is the acceleration due to gravity (32 ft/s^2), and t is time (friction is ignored).

1 Suppose the first car of a roller coaster leaves the bottom of the first hill at an initial speed of 60 mph at an angle of 44° from the horizontal. Draw a sketch of this hill. Find the maximum height the coaster reaches on the next hill and at what time. Label your drawing.

2 Find the horizontal distance R the car travels as it follows the contour of the second hill if $R = (v_i \cos \theta)t$.

3 How would these formulas help in altering your roller coaster design? How might friction affect your calculations and plans?

Add the results of your work to your Investigation Folder.

14-6A Graphing Technology
Solving Trigonometric Equations

A Preview of Lesson 14-6.

The coordinates of the points that make up the graph of a trigonometric function represent all of the values that satisfy the function. When you solve a trigonometric equation, you find all values of the variable that satisfy the equation. So, you can solve trigonometric equations by graphing the related trigonometric function and then finding the zeros of that function.

Example **1** Use a graphing calculator to solve $\sin x = 0.8$ if $0° \le x < 360°$.

To get the correct viewing window, be sure the calculator is in Degree mode.

Rewrite the equation as $\sin x - 0.8 = 0$. Then graph the function $f(x) = \sin x - 0.8$ and look for the zeros. Use the viewing window [0, 360] by [−2, 1] with a scale factor of 90 for the *x*-axis and 1 for the *y*-axis.

Enter: Y= SIN X,T,θ,*n*

   − .8 GRAPH

Based on the graph, you can see that there are two roots in the interval

$0° \le x < 360°$. ZOOM in or use 2nd

CALC root to approximate the solutions. The approximate solutions are 53.1° and 126.9°.

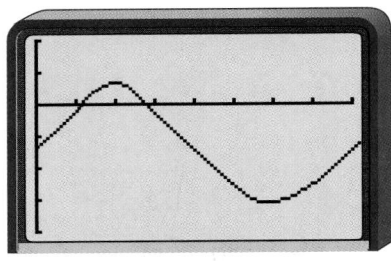

Example **2** Use a graphing calculator to solve $2 \cos x + 3 = 0$ if $0° \le x < 360°$.

The related function is $f(x) = 2 \cos x + 3$. Use the viewing window [0, 360] by [−3, 5] with a scale factor of 90 for the *x*-axis and 1 for the *y*-axis.

Enter: Y= 2 COS X,T,θ,*n*

   + 3 GRAPH

The function has no zeros. Thus, the related equation $2 \cos x + 3 = 0$ has no real solutions. *Note that $\cos x = -\dfrac{3}{2}$ and values of cosine are never less than −1.*

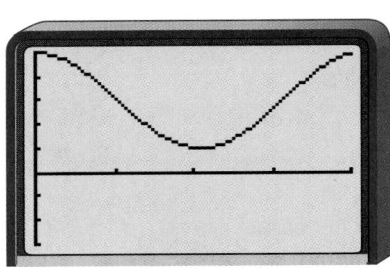

EXERCISES

Use a graphing calculator to solve each equation for the indicated values of x.

1. $\sin x = 0.2$ if $0° \le x < 360°$
2. $0.5 \cos x = 1.4$ if $-720° \le x < 720°$
3. $\sin 2x = \sin x$ if $0° \le x < 360°$
4. $\tan x = \sin x$ if $0° \le x < 360°$
5. $3 \sin 2x - 5 \sin x = 1$ if $-360° \le x < 360°$
6. $\tan^2 x \cos x + 5 \cos x = 0$ if $0° \le x < 360°$

Solving Trigonometric Equations

These are called <u>conditional equations.</u>

CONNECTION

Physical Science

You know that the number of hours of daylight varies with the time of year. This variation can be approximated by a sine function. For example, in parts of northeast United States, the number of hours of daylight d may be represented by $d = 3 \sin \frac{2\pi}{365} t + 12$, where t is the number of days after March 21. On what days would you have $10\frac{1}{2}$ hours of daylight? *This problem will be solved in Example 6.*

Trigonometric identities are true for *all* values of the variable for which the equation is defined. However, most **trigonometric equations**, like most algebraic equations, are true for *some* but not *all* values of the variable.

Example **Solve $\sin^2 x + \cos 2x - \cos x = 0$ if $0° \le x < 360°$.**

$$\sin^2 x + \cos 2x - \cos x = 0$$
$$\sin^2 x + (1 - 2\sin^2 x) - \cos x = 0 \quad \cos 2x = 1 - 2\sin^2 x$$
$$1 - \sin^2 x - \cos x = 0$$
$$\cos^2 x - \cos x = 0 \quad 1 - \sin^2 x = \cos^2 x$$
$$\cos x (\cos x - 1) = 0 \quad \text{Factor out } \cos x.$$

Now use the zero product property.

$\cos x = 0$	or	$\cos x - 1 = 0$
$x = 90°$ or $270°$		$\cos x = 1$
		$x = 0°$

The solutions are $0°$, $90°$, and $270°$.

Trigonometric equations are usually solved for values of the variable between $0°$ and $360°$ or 0 radians and 2π radians. There are solutions outside that interval. These other solutions differ by integral multiples of the period of the function.

Example **Solve $\cos \theta + 1 = 0$ for all values of θ if θ is measured in radians.**

$$\cos \theta + 1 = 0$$
$$\cos \theta = -1$$

Look at the graph of $y = \cos \theta$ to find solutions to $\cos \theta = -1$.

The solutions are $\pi, 3\pi, 5\pi$, and so on, and $-\pi, -3\pi, -5\pi$, and so on. The only solution in the interval 0 radians to 2π radians is π. The period of the cosine function is 2π radians. So the solutions can be written as $\pi + 2k\pi$, where k is any integer.

If an equation cannot be solved easily by factoring, try writing the expression in terms of only one trigonometric function.

Example **3** **Solve $\sin^2 x - 1 = \cos^2 x$.**

$$\sin^2 x - 1 = \cos^2 x$$

$\sin^2 x - 1 = 1 - \sin^2 x$ $\cos^2 x = 1 - \sin^2 x$

$$2 \sin^2 x = 2$$

$\sin^2 x = 1$ *Divide each side by 2.*

$\sin x = \pm 1$ *Take the square root of each side.*

$$x = 90° + k \cdot 360° \text{ or } 270° + k \cdot 360°$$

The solutions are $90° + k \cdot 360°$ and $270° + k \cdot 360°$, where k is any integer.

Some trigonometric equations have *no solution*. In other words, there is no replacement for the variable that will make the sentence true. For example, the equation $\sin x = -3$ has no solution, since all values of $\sin x$ are between -1 and 1, inclusive. Thus, the solution set for $\sin x = -3$ is empty.

Example **4** **Solve $2 \sin^2 \theta - 3 \sin \theta - 2 = 0$ if $0 \le \theta < 2\pi$.**

LOOK BACK

Refer to Lesson 5-4 to review factoring trinomials.

$$2 \sin^2 \theta - 3 \sin \theta - 2 = 0$$

$$(\sin \theta - 2)(2 \sin \theta + 1) = 0$$

$\sin \theta - 2 = 0$ or $2 \sin \theta + 1 = 0$ *Factor*

$\sin \theta = 2$ $2 \sin \theta = -1$ *Zero product property*

There is no solution to $\sin \theta = 2$ since all values of $\sin \theta$ are between -1 and 1, inclusive. $\sin \theta = -\dfrac{1}{2}$

$$\theta = \frac{7\pi}{6} \text{ or } \frac{11\pi}{6}$$

The solutions are $\dfrac{7\pi}{6}$ or $\dfrac{11\pi}{6}$.

Some algebraic operations, such as squaring, may result in answers that are *not* solutions of the original equation. So, it is necessary to check your solutions to trigonometric equations.

Example **5** **Solve $\sin x = 1 + \cos x$ if $0° \le x < 360°$.**

$$\sin x = 1 + \cos x$$

$\sin^2 x = (1 + \cos x)^2$ *Square each side of the equation.*

$1 - \cos^2 x = 1 + 2 \cos x + \cos^2 x$ $\sin^2 x = 1 - \cos^2 x$

$$0 = 2 \cos x + 2 \cos^2 x$$

$0 = 2 \cos x \, (1 + \cos x)$ *Factor.*

$2 \cos x = 0$ or $1 + \cos x = 0$ *Zero product property*

$\cos x = 0$ $\cos x = -1$

$x = 90° \text{ or } 270°$ $x = 180°$

Check:

$$\sin x = 1 + \cos x$$
$$\sin 90° \overset{?}{=} 1 + \cos 90°$$
$$1 \overset{?}{=} 1 + 0$$
$$1 = 1 \quad \checkmark$$

$$\sin x = 1 + \cos x$$
$$\sin 180° \overset{?}{=} 1 + \cos 180°$$
$$0 \overset{?}{=} 1 + (-1)$$
$$0 = 0 \quad \checkmark$$

$$\sin x = 1 + \cos x$$
$$\sin 270° \overset{?}{=} 1 + \cos 270°$$
$$-1 \overset{?}{=} 1 + 0$$
$$-1 \ne 1$$

The solutions are 90° and 180°.

You can use a graphing calculator to visualize the solution of a trigonometric equation as the intersection of two trigonometric graphs.

EXPLORATION

GRAPHING CALCULATORS

Set MODE to degrees. Then use the viewing window [0, 360] by [−2, 2] with a scale factor of 45 for the *x*-axis and 0.5 for the *y*-axis.

Your Turn

a. Graph the two sides of the equation of Example 5. Enter the functions Y1 = cos X and Y2 = 1 + sin X into the Y = list. Then press GRAPH . How many times do the two curves intersect?

b. Use the TRACE function to find the intersection points of the graphs. How are these points related to the results of Example 5?

c. Graph the function $y = \sin^2 x$ as Y1 and the function $y = \sin x - 0.5$ as Y2. Then press GRAPH . What does the graph tell you about the trigonometric equation $\sin^2 x = \sin x - \frac{1}{2}$?

Example ⑥

CONNECTION
Physical Science

Refer to the connection at the beginning of the lesson. In Hartford, Connecticut, the number of hours of daylight can be approximated by the equation $d = 3 \sin \frac{2\pi}{365}t + 12$. The angle measures are in radians, and leap years are not included.

On what days will you have $10\frac{1}{2}$ hours of daylight?

Explore   You know that $d = 10\frac{1}{2}$. You want to find the value of *t*.

Plan   Replace the known values into the formula and solve to find the number of days after March 21. Then use a calendar to find the actual dates.

(continued on the next page)

Solve

$$d = 3 \sin \frac{2\pi}{365}t + 12$$

$$10\frac{1}{2} = 3 \sin \frac{2\pi}{365}t + 12 \quad \text{Replace } d \text{ with } 10\frac{1}{2}.$$

$$-1.5 = 3 \sin \frac{2\pi}{365}t \quad \text{Subtract 12 from each side.}$$

$$-\frac{1}{2} = \sin \frac{2\pi}{365}t \quad \text{Divide each side by 3.}$$

$$\frac{2\pi}{365}t = \frac{7\pi}{6} \quad \text{or} \quad \frac{2\pi}{365}t = \frac{11\pi}{6}$$

$$t = \frac{7\pi}{6} \cdot \frac{365}{2\pi} \qquad\qquad t = \frac{11\pi}{6} \cdot \frac{365}{2\pi}$$

$$t \approx 213 \qquad\qquad\qquad t \approx 335$$

There will be $10\frac{1}{2}$ hours of daylight 213 days after March 21 and 335 days after March 21; that is, on October 20 and February 19.

Examine Check the solutions by substituting $t = 213$ and $t = 335$ into the original equation.

CHECK FOR UNDERSTANDING

Communicating Mathematics

Study the lesson. Then complete the following.

1. **Explain** why the number of solutions to the equation $\sin x = -\frac{1}{2}$ is infinite.

2. **Explain** why the equation $\cos x = -2$ has no solutions.

3. **Assess Yourself** How well do you understand the difference between an identity and a conditional equation? State whether the following sentence is true or false. "An identity is any equation that has an infinite number of solutions." Then explain your answer.

Guided Practice

Find all solutions if $0° \le \theta < 360°$.

4. $\cos^2 \theta = 1$

5. $\sin 2\theta = \frac{1}{2}$

6. $2 \cos^2 \theta + 2 = 5 \cos \theta$

7. $\sin \theta + \sin \theta \cos \theta = 0$

Solve each equation for all values of θ if θ is measured in radians.

8. $\cos 2\theta = \cos \theta$

9. $\cos 2\theta + \cos \theta + 1 = 0$

10. $3 \sin^2 \theta - \cos^2 \theta = 0$

11. $4 \cos^2 \theta - 4 \cos \theta + 1 = 0$

12. Physical Science Using the results of Example 6, tell what days of the year have *at least* $10\frac{1}{2}$ hours of daylight. Explain how you know.

EXERCISES

Practice

Find all solutions if 0° ≤ x < 360°.

13. $4\cos^2 x = 1$ **14.** $2\sin^2 x - 1 = 0$

15. $\sin 2x = 2\cos x$ **16.** $2\cos^2 x = \sin x + 1$

17. $4\sin^2 x - 4\sin x + 1 = 0$ **18.** $\sin 2x = \cos x$

Find all solutions if 0 ≤ θ < 2π.

19. $2\sin\theta = -1$ **20.** $2\cos\theta - 1 = 0$

21. $2\sin\theta = -\sqrt{3}$ **22.** $4\sin^2\theta = 1$

23. $2\sin^2\theta - \sin\theta = 1$ **24.** $2\sin^2\theta = -\sin\theta$

Solve each equation for all values of x if x is measured in degrees.

25. $\sin x = \cos x$ **26.** $\sin^2 x - 2\sin x - 3 = 0$

27. $\tan x = \sin x$ **28.** $\sin x = 1 + \cos x$

29. $3\cos 2x - 5\cos x = 1$ **30.** $\tan^2 x - \sqrt{3}\tan x = 0$

31. $\sin^2 x - \sin x = 0$ **32.** $\cos x \tan x - \sin^2 x = 0$

Solve each equation for all values of θ if θ is measured in radians.

33. $2\sin^2\theta - 3\sin\theta - 2 = 0$ **34.** $\cos 2\theta + 3\cos\theta - 1 = 0$

35. $2\sin^2\theta - \cos\theta - 1 = 0$ **36.** $\cos^2\theta - \frac{5}{2}\cos\theta - \frac{3}{2} = 0$

37. $\cos^2\theta - \frac{7}{2}\cos\theta - 2 = 0$ **38.** $2\cos^2\theta + 3\sin\theta - 3 = 0$

39. $\cos 2\theta = 1 - \sin\theta$ **40.** $\cos\theta = 3\cos\theta - 2$

Graphing Calculator

41. On a graphing calculator, set the mode to degrees. Use a viewing window of [0, 360] by [−2.5, 2.5] with a scale factor of 45 for the *x*-axis and 0.5 for the *y*-axis.

   a. Enter the functions $Y_1 = \sin 2X$ and $Y_2 = 2\cos X$ into the Y= list. Then press $\boxed{\text{GRAPH}}$. How many times do the two curves intersect? For what values of *x* do they intersect?

   b. To which of Exercises 13–18 does the graph correspond? Does the graph agree with your answer to that exercise?

   c. Repeat part a, but replace 2 cos X by 2 cos X sin X. What do you notice about the two curves?

   d. What do parts a–c show you about the difference between an equation that is an identity and one that is not?

42. The graphing calculator program below finds solutions to $2\cos^2 x + 3\cos x - 2 = 0$ within a given interval. The interval is entered in degrees. The program is designed to pause after each solution is found. Press ENTER after each solution to continue. The program will indicate "Done" when finished. You can use this program to solve other trigonometric equations by simply changing the equation within the program.

```
PROGRAM:SOLN
: Disp "FIND THE SOLUTIONS    ; Then
    OF"                      : Disp "ONE SOLUTION IS"
: Disp "2(cos x)² + 3cos     : Disp J
  (x)-2=0"                   : Disp "DEGREES"
: Disp                       : C+1 → C
: Disp "ENTER THE LEAST      : Pause
  VALUE TO BE TESTED"        : End
: Input A                    : End
: Disp "ENTER THE GREATEST   : If C=0
  VALUE TO BE TESTED"        : Then
: Input B                    : Disp "NO SOLUTIONS
: 0 → C                          BETWEEN"
: FOR(J,A,B)                 : Disp A
: J/57.2957795 → Y           : Disp "AND"
: (2(cos Y)²+3cos(Y)         : Disp B
  -2) → F                    : Disp "DEGREES"
: If abs (F)≤0.0000001       : End
```

Make the necessary changes in the program to solve each equation for values of _x_ within the given interval.

a. $2\cos^2 x + 3\cos x - 2 = 0$ between $0°$ and $180°$

b. $2\cos x - \sin 2x = 0$ between $0°$ and $360°$

c. $2\sin^2 x + \sin x - 1 = 0$ between $180°$ and $240°$

d. $\sin x + \cos x \tan^2 x = 0$ between $-360°$ and $0°$

e. $\cos 2x + \sin x = 1$ between $0°$ and $720°$

43. Solve $\dfrac{\tan x - \sin x}{\tan x + \sin x} = \dfrac{\sec x - 1}{\sec x + 1}$ for all values of x if x is measured in radians.

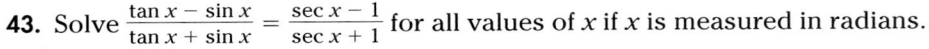

44. Physical Science In New Orleans, Louisiana, the number of hours of daylight is approximately described by the equation $d = 2.3 \sin \dfrac{2\pi}{365}t + 11.7$, where t is the number of days after March 21 and the angle measures are in radians. On what days will you have 12.85 hours of daylight?

45. Physics According to Snell's law, the angle at which light enters water α is related to the angle at which light travels in water β by the equation $\sin \alpha = 1.33 \sin \beta$. Is there any angle at which you could direct a beam of light at the surface of a pool so that the light is not bent? (*Hint:* If the light is not bent, what must be true about α and β?)

46. Trigonometry If x is an angle whose terminal side lies in the second quadrant, in which quadrant does the terminal side for $\frac{x}{2}$ lie? (Lesson 14–5)

47. Trigonometry Find $\cos A$ when $\tan A = \frac{\sqrt{5}}{2}$. (Lesson 13–1)

48. Probability A canister contains 20 pieces of candy: 5 pineapple flavored, 9 cherry flavored, and 6 lemon flavored. Two are selected at random. Find each probability. (Lesson 12–4)

    a. P(2 pineapple)   **b.** P(1 pineapple and 1 lemon)

49. Cuisine A soufflé is a dish made from a sauce, egg yolks, beaten egg whites, and a flavoring of seafood, fruit, or vegetables. When it bakes, the batter rises dramatically and is very light and fluffy when done properly. However, soufflés often fall while baking. Dawn Depew is preparing soufflés for her parents' anniversary party. She averages 4 out of 5 soufflés baking correctly. If she prepares 6 soufflés for the party, what is the likelihood that exactly 4 of them will turn out all right? (Lesson 11–8)

50. Solve $2^{2x} = \frac{1}{8}$. (Lesson 10–1)

51. Solve $b^2 - \frac{3}{4}b + \frac{1}{8} = 0$ by completing the square. (Lesson 6–3)

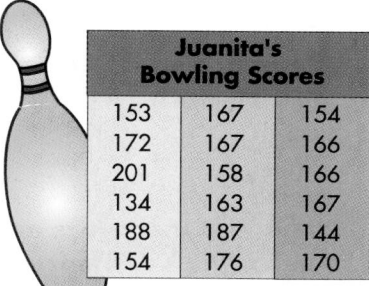

Juanita's Bowling Scores

153	167	154
172	167	166
201	158	166
134	163	167
188	187	144
154	176	170

52. Bowling Juanita's bowling scores for each week of her summer bowling league are shown in the table at the left. (Lesson 4–8)

    a. Make a box-and-whisker plot of the data.

    b. Use the plot to describe the pattern in Juanita's bowling scores for the season.

53. Consumerism Ian can buy one piece of candy for 29¢. He has $6.09 that he can spend. Write an inequality to show how many pieces of candy Ian can buy. Then solve the inequality. (Lesson 2–7)

Mathematics and SOCIETY

Fractal Drums

The excerpt below appeared in an article in *Science News* on September 17, 1994.

IT'S EASY TO DISTINGUISH THE SOUNDS OF different types of drums, even without seeing the instruments. What makes these sounds so readily identifiable is that each drum vibrates at characteristic frequencies. . . . The sounds of drums also suggest important questions in mathematics. . . . It turns out, for example, that different membrane shapes can sometimes generate identical spectra of frequencies.

A number of mathematicians and physicists are studying how the wiggliness of a drum's rim affects its sound, especially in cases where the boundary is so wrinkled . . . that it can be termed a fractal. . . . Such investigations may eventually furnish clues as to why fractals seem to abound in nature, from crazily indented coastlines to the intricate branching of air passages in the human lung. ■

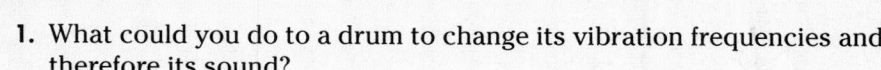

1. What could you do to a drum to change its vibration frequencies and therefore its sound?

2. If you were blindfolded, could you identify some of the physical characteristics (such as softness, size, shape, weight) of the sources of sound that you hear?

3. Given that drums of different shapes can produce identical frequency spectra, what would you say to a geophysicist who is trying to construct a model of Earth's interior from a study of underground seismic vibrations?

In·ves·ti·ga·tion

Refer to the Investigation on pages 768–769.

Amusement parks are a very competitive business. In order to attract large crowds, each park must have at least a ride or two that can be touted as the best, most thrilling, and most unusual. In recent years, some kind of roller coaster has remained at the top of the list of the most popular and exciting rides. As technology, engineering, and material manufacturing improve, progressive amusement parks constantly replace older rides with newer, more exciting ones.

Analyze

You have conducted experiments and organized your data in various ways. It is now time to analyze your findings and state your conclusions.

> **PORTFOLIO ASSESSMENT**
>
> You may want to keep your work on this Investigation in your portfolio.

1 Make a table to summarize your findings when building and revising your model of a roller coaster.

2 Make a list of things you would want to consider if you were to design and build the "best new roller coaster in the world." Include the type of roller coaster it would be.

3 Use your list to make a diagram showing how each item interrelates with others in the list. Include the mathematics you think you would need in considering each item.

4 List the types of specialists you would like to include on your team to build the coaster and how those specialists would contribute to the project.

Write

Amusement park owners and managers are always searching for new ideas to improve their rides. Suppose you have been given the opportunity to submit a proposal for a new roller coaster design that will fit into a 35-acre corner of Mathematica Amusement World. The ride should last at least 2 minutes. An old merry-go-round and a small Ferris wheel are to be removed from the site to make room for your new coaster.

5 Prepare a proposal for the design and building of your roller coaster. Include a scale drawing, the types of building materials, and the style of the ride.

6 Include a report on the rationale for your chosen design in which you address safety features, thrill factors, and unique features that will draw visitors to the amusement park. Also include why this one ride is better than the ones it is replacing.

7 Make sure you name the new roller coaster and tell how this name is reflected in the ride's design. Prepare a visual display to show how the name will appear at the ride's entrance.

VOCABULARY

After completing this chapter, you should be able to define each term, property, or phrase and give an example or two of each.

Trigonometry

amplitude   (p. 824, 828)

angle of incline   (p. 835)

conditional equation   (p. 861)

difference of angles formula   (p. 848)

double-angle formulas   (p. 853)

half-angle formulas   (p. 855)

phase shift   (p. 825)

sum of angles formula   (p. 848)

trigonometric equations   (p. 861)

trigonometric identity   (p. 835)

UNDERSTANDING AND USING THE VOCABULARY

Choose the letter that best matches each phrase.

1. $\cos (x + y) = \cos x \cos y - \sin x \sin y$

2. $\sin 2\theta = 2 \sin \theta \cos \theta$

3. the acute angle described by $\tan \theta = \dfrac{v^2}{gR}$

4. $\sin (x - y) = \sin x \cos y - \cos x \sin y$

5. the formula used to find $\cos 22\frac{1}{2}°$

6. an equation that is not true for all values of the variable

7. an equation that is true for all values for which every expression in the equation is defined

8. from one point on a graph until the pattern begins to repeat; a complete revolution

9. the absolute value of half the difference between the maximum value and the minimum value of a periodic function

a. amplitude
b. angle of incline
c. conditional equation
d. difference of angles formula
e. double-angle formula
f. half-angle formula
g. period
h. sum of angles formula
i. trigonometric identity

SKILLS AND CONCEPTS

OBJECTIVES AND EXAMPLES

Upon completing this chapter, you should be able to:

- graph trigonometric functions (Lesson 14–1)

Graph $y = 2 \cos \theta$.

amplitude $= |2|$ or 2

period $= \dfrac{360°}{|1|}$ or $360°$

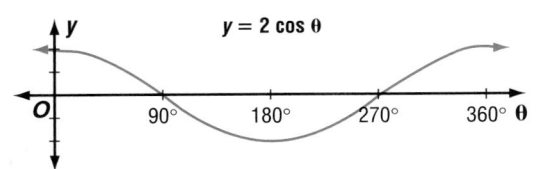

- use trigonometric identities to simplify or evaluate expressions (Lesson 14–2)

Simplify $\sin \theta \cot \theta \cos \theta$.

$$\sin \theta \cot \theta \cos \theta = \frac{\overset{1}{\cancel{\sin \theta}}}{1} \cdot \frac{\cos \theta}{\underset{1}{\cancel{\sin \theta}}} \cdot \frac{\cos \theta}{1}$$

$$= \cos^2 \theta$$

- verify trigonometric identities (Lesson 14–3)

Verify $\tan x + \cot x = \sec x \csc x$.

$$\tan x + \cot x \overset{?}{=} \sec x \csc x$$

$$\frac{\sin x}{\cos x} + \frac{\cos x}{\sin x} \overset{?}{=} \sec x \csc x$$

$$\frac{\sin^2 x + \cos^2 x}{\cos x \sin x} \overset{?}{=} \sec x \csc x$$

$$\frac{1}{\cos x \sin x} \overset{?}{=} \sec x \csc x$$

$$\sec x \csc x = \sec x \csc x$$

REVIEW EXERCISES

Use these exercises to review and prepare for the chapter test.

Graph each function.

10. $y = -\dfrac{1}{2} \cos \theta$

11. $y = 4 \sin 2\theta$

12. $y = \sin \dfrac{1}{2}\theta$

13. $y = 5 \sec \theta$

14. $y = \dfrac{1}{2} \csc \dfrac{2}{3}\theta$

15. $y = \tan 4\theta$

Solve for values of θ between 270° and 360°.

16. If $\csc \theta = \dfrac{-5}{3}$, find $\cot \theta$.

17. If $\sin \theta = -\dfrac{1}{2}$, find $\sec \theta$.

Simplify each expression.

18. $\sin \alpha \csc \alpha - \cos^2 \alpha$

19. $\cos^2 \theta \sec \theta \csc \theta$

20. $\cos \beta + \sin \beta \tan \beta$

21. $\sin \alpha (1 + \cot^2 \alpha)$

Verify that each of the following is an identity.

22. $\dfrac{\sin \theta}{\tan \theta} + \dfrac{\cos \theta}{\cot \theta} = \cos \theta + \sin \theta$

23. $\dfrac{\sin \theta}{1 - \cos \theta} = \csc \theta + \cot \theta$

24. $\cot^2 \theta \sec^2 \theta = 1 + \cot^2 \theta$

25. $\sec x (\sec x - \cos x) = \tan^2 x$

26. $\dfrac{\cos x}{\csc x} - \dfrac{\sin x}{\cos x} = -\sin^2 x \tan x$

27. $\dfrac{\csc \theta + 1}{\cot \theta} = \dfrac{\cot \theta}{\csc \theta - 1}$

OBJECTIVES AND EXAMPLES

• find values of sine and cosine involving sum and difference formulas (Lesson 14–4)

Find the exact value of cos 75°.

$\cos 75° = \cos(30° + 45°)$

$\qquad = \cos 30° \cos 45° - \sin 30° \sin 45°$

$\qquad = \dfrac{\sqrt{3}}{2} \cdot \dfrac{\sqrt{2}}{2} - \left(\dfrac{1}{2}\right) \cdot \dfrac{\sqrt{2}}{2}$

$\qquad = \dfrac{\sqrt{6}}{4} - \dfrac{\sqrt{2}}{4}$

$\qquad = \dfrac{\sqrt{6} - \sqrt{2}}{4}$

REVIEW EXERCISES

Find the exact value of each expression.

28. $\cos 15°$

29. $\cos 285°$

30. $\sin 195°$

31. $\sin 255°$

32. $\sin 165°$

33. $\cos(-210°)$

• verify identities using the sum and difference formulas

Verify that $\cos(\theta - 90°) = \sin \theta$.

$\cos(\theta - 90°) \overset{?}{=} \sin \theta$

$\cos \theta \cos 90° + \sin \theta \sin 90° \overset{?}{=} \sin \theta$

$\cos \theta \cdot 0 + \sin \theta \cdot 1 \overset{?}{=} \sin \theta$

$\sin \theta = \sin \theta$

Verify that each of the following is an identity.

34. $\cos(\theta + 270°) = \sin \theta$

35. $\cos(180° - \alpha) = -\cos \alpha$

• find values of sine and cosine involving double and half angles (Lesson 14–5)

Suppose x is between 90° and 180° and $\cos x = -\dfrac{4}{5}$.

Find the exact value of $\sin 2x$. Find the value of $\sin x$. Then use the identity $\sin 2x = 2 \sin x \cos x$.

$\begin{aligned} \sin^2 x &= 1 - \cos^2 x \\ &= 1 - \left(-\dfrac{4}{5}\right)^2 \\ &= \dfrac{9}{25} \end{aligned}$ $\qquad \begin{aligned} \sin 2x &= 2 \sin x \cos x \\ &= 2 \cdot \dfrac{3}{5} \cdot -\dfrac{4}{5} \\ \sin 2x &= -\dfrac{24}{25} \end{aligned}$

$\sin x = \dfrac{3}{5}$

Find $\sin 2x$, $\cos 2x$, $\sin \dfrac{x}{2}$, and $\cos \dfrac{x}{2}$ for each of the following, given the two angle measures between which the terminal side of x lies.

36. $\sin x = \dfrac{1}{4}$; x is in Quadrant I

37. $\sin x = \dfrac{-5}{13}$, x is in Quadrant III

38. $\cos x = \dfrac{-15}{17}$, x is in Quadrant II

39. $\cos x = \dfrac{12}{13}$, x is in Quadrant IV

OBJECTIVES AND EXAMPLES

- verify identities by using double- and half-angle formulas. (Lesson 14–5)

Verify that $\csc 2x = \dfrac{\sec x}{2 \sin x}$.

$$\csc 2x \overset{?}{=} \dfrac{\sec x}{2 \sin x}$$

$$\dfrac{1}{\sin 2x} \overset{?}{=} \dfrac{\frac{1}{\cos x}}{2 \sin x}$$

$$\dfrac{1}{\sin 2x} \overset{?}{=} \dfrac{1}{2 \sin x \cos x}$$

$$\dfrac{1}{\sin 2x} = \dfrac{1}{\sin 2x}$$

REVIEW EXERCISES

Verify that each of the following is an identity.

40. $\sin^4\theta - \cos^4\theta = -\cos 2\theta$

41. $\dfrac{1}{2} \sin 2x = \dfrac{\tan x}{1 + \tan^2 x}$

42. $\dfrac{\sin 2\theta}{1 - \cos 2\theta} = \cot \theta$

- solve trigonometric equations   (Lesson 14–6)

Solve $\sin 2\theta + \sin \theta = 0$ if $0° \le \theta < 360°$.

$$\sin 2\theta + \sin \theta = 0$$

$$2 \sin \theta \cos \theta + \sin \theta = 0$$

$$\sin \theta (2 \cos \theta + 1) = 0$$

$\sin \theta = 0 \qquad$ or $\quad 2 \cos \theta + 1 = 0$

$\quad \theta = 0° \text{ or } 180° \qquad\qquad \cos \theta = -\dfrac{1}{2}$

$$\theta = 120° \text{ or } 240°$$

Find all solutions if $0° \le \theta < 360°$.

43. $2 \cos^2 \theta + \sin^2 \theta = 2 \cos \theta$

44. $\cos \theta = 1 - \sin \theta$

45. $2 \sin 2\theta = 1$

Solve each equation for all values of x if x is measured in degrees.

46. $6 \sin^2 x - 5 \sin x - 4 = 0$

47. $2 \cos^2 x = 3 \sin x$

48. $2 \sin x \cos x = 1$

APPLICATIONS AND PROBLEM SOLVING

49. Navigation   When a plane is 320 miles from the runway, the angle of depression from the cockpit to the airport is 35°. At the same time, the angle of depression from the cockpit to the center of a lake located between the plane and the airport is 43°. How far is the center of the lake from the airport?   (Lesson 14–4)

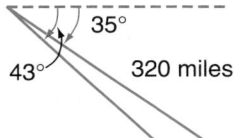

35°

43°   320 miles

50. Music   A tuning fork generates a wave represented by $\sin x + \cos x = y$. If $0 \le x < 2\pi$, what values of x satisfy the equation $\sin x + \cos x = \sqrt{2}$?   (Lesson 14–6)

A practice test for Chapter 14 is provided on page 925.

ALTERNATIVE ASSESSMENT

COOPERATIVE LEARNING PROJECT

Tessellations   In this project, you will develop the idea of tessellations and design your own tessellations. A figure such as a regular hexagon, which can be used repeatedly to cover a surface without gaps or overlaps, is said to tile or *tessellate* the surface. The resulting pattern is called a *tessellation.*

Draw the first six regular polygons. For each one, trace additional copies around one of the vertices to determine if it tessellates. Which regular polygons tessellate and which ones do not tessellate? What conditions must the angles of a regular polygon satisfy in order for the polygon to tessellate?

Below is a convex pentagon. Will this pentagon tessellate? What sides will you try to match up in order to tessellate it? How many angles meet at vertex A for this tessellation? What is the number of degrees in each of these angles? How many angles meet at vertex D? What is the number of degrees in each of these angles?

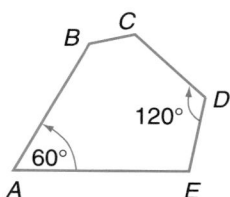

Can nonconvex polygons be tessellated? Can any of the three polygons below form a tessellation? If so, form a portion of that tessellation. Now, create and design a tessellation of your own.

Incorporate these ideas in your activity.

- Research tessellations.
- Experiment, and then verify conditions that need to be satisfied.
- Be creative in your design.
- Prepare a paragraph describing tessellations and their conditions.

THINKING CRITICALLY

- Explain how to derive a formula for cot $(A - B)$.
- Explain how you could use the identity tan $(A + B) = \dfrac{\tan A + \tan B}{1 - \tan A \tan B}$ to derive a formula for tan 2θ. Then derive it.

PORTFOLIO

Graphs can "show" a person information that is not easily understood otherwise. Graph the six trigonometric functions carefully. Compare and contrast these graphs. Classify each of them as discontinuous or continuous. Explain your reasoning for each classification. Place this in your portfolio.

SELF EVALUATION

You may be able to determine how a problem ends and then work in reverse toward the initial state to determine the steps in between.

Assess yourself. Do you work backward when solving problems? What must be known in order to use this strategy? Describe a problem in mathematics and in your daily life where you found this strategy to be helpful.

For Additional Assessment

For Reference

Lesson 1-1   Find the value of each expression.

1. $3(2^2 + 3)$

2. $2(3 + 8) - 3$

3. $5 + 3^2 - 16 + 4$

4. $(5 + 3) - 16 \div 4$

5. $4 + 8(4) \div 2 - 10$

6. $15 \div 3 \times 5 + 1$

7. $[(4 + 8)^2 \div 9] \cdot 5$

8. $5 + 8^2 \div 4 \cdot 3$

9. $5 \cdot 7 - 2(5 + 1) \div 3$

10. $3 + 7^2 - 16 \div 2$

11. $12 + 20 \div 4 - 5$

12. $[7 - (8 - 6)^2] - 1$

13. $\frac{1}{2}(3^2 + 5 \cdot 7) - 8$

14. $\frac{3 \cdot 5 + 3^2}{2^3}$

15. $\frac{6^2 + 4(2^4)}{28 + 9 \times 8}$

Evaluate each expression if $a = -0.5$, $b = 4$, $c = 5$, and $d = -3$.

16. $3b + 4d$

17. $ab^2 + c$

18. $bc + d \div a$

19. $7ab - 3d$

20. $ad + b^2 - c$

21. $d(b + d)^3$

22. $\frac{4a + 3c}{3b}$

23. $\frac{3ab^2 - d^3}{a}$

24. $\frac{5a + ad}{bc}$

Lesson 1-2   Find the value of each expression. Then name the sets of numbers to which each value belongs. (Use N, W, Z, Q, I, and R.)

1. $4.1 + 8.2$

2. $-54 \div 6$

3. $\sqrt{36} - 3$

4. $\sqrt{81} - 4$

Name the property illustrated by each equation.

5. $(4 + 9a)2b = 2b(4 + 9a)$

6. $3\left(\frac{1}{3}\right) = 1$

7. $a(3 - 2) = a \cdot 3 - a \cdot 2$

8. $3 + (-3) = 0$

9. $(j + k) + 0 = (j + k)$

10. $(2a)b = 2(ab)$

Name the additive inverse and the multiplicative inverse of each number.

11. 3

12. $-\frac{1}{8}$

13. $\frac{2}{7}$

14. 0.2

Simplify each expression.

15. $6(2a + 3b) + 5(3a - 4b)$

16. $7s + 9t + 2s - 7t$

17. $4(3x - 5y) - 8(2x + y)$

18. $0.2(5m - 8) + 0.3(6 - 2m)$

19. $\frac{1}{2}(7p + 3q) + \frac{3}{4}(6p - 4q)$

20. $\frac{4}{5}(3v - 2w) - \frac{1}{5}(7v - 2w)$

Lesson 1-3   Find the median, mode, and mean for each set of data.

1. $4, 1, 2, 1, 1$

2. $216, 399, 219, 179, 180, 399$

3. $58, 52, 49, 60, 61, 56, 50, 61$

4. $25.5, 26.7, 20.9, 23.4, 26.8, 24.0, 25.7$

The following high temperatures were recorded during a cold spell in Cleveland lasting thirty-eight days. Use this data for Exercises 5–9.

29°	26°	17°	12°	5°	4°	25°	17°	23°	18°	13°	6°	25°
20°	27°	22°	26°	30°	31°	2°	12°	27°	16°	27°	16°	30°
6°	16°	5°	0°	5°	29°	18°	16°	22°	29°	8°	23°	

5. Make a line plot of the temperatures.

6. Make a stem-and-leaf plot of the temperatures.

7. Find the median of the temperatures.

8. Find the mode of the temperatures.

9. Find the mean of the temperatures.

Lesson 1-4   Write an algebraic expression to represent each verbal expression.

1. twice the sum of a number and seven
2. twelve decreased by the square of a number
3. the product of the square of a number and 6
4. the product of 3 and a number decreased by 1
5. the sum of eight and four times a number
6. the square of the sum of a number and 11

Name the property illustrated by each statement.

7. If $5 \times 6 = 30$, then $30 = 5 \times 6$.
8. If $x + (4 + 5) = 21$, then $x + 9 = 21$.
9. $4g + 99 = 4g + 99$
10. If $a + 1 = 6$, then $3(a + 1) = 3(6)$.
11. If $7x = 42$, then $7x - 5 = 42 - 5$.
12. If $3 + 5 = 8$ and $8 = 2 \times 4$, then $3 + 5 = 2 \times 4$.

Solve each equation.

13. $5t + 8 = 88$
14. $27 - x = -4$
15. $17a = -8 + 9a$
16. $6 = \frac{3x - 6}{3}$
17. $\frac{3a + 3}{4} = \frac{5}{2}$
18. $\frac{3}{4}y = \frac{2}{3}y + 5$
19. $-6 = 3.1s + 6.4$
20. $8s - 3 = 5(2s + 1)$
21. $0.4(p - 9) = 0.3(p + 4)$
22. $2(m - 4) + 5 = 9$
23. $5(a - 1) = 2(a + 5)$
24. $4c - 3 = 7c + 18$
25. $0.5z + 10 = z + 4$
26. $3(k - 2) = k + 4$
27. $4(y + 1) + 7 = y + 17$
28. $\frac{2r - 3}{-7} = 5$
29. $8q - \frac{q}{3} = 46$
30. $\frac{1}{3}(3d - 1) = \frac{1}{2}(d + 2)$

Lesson 1-5   Evaluate each expression if $x = -5$, $y = 3$, and $z = -2.5$.

1. $|2x|$
2. $|-3y|$
3. $|2x + y|$
4. $|y + 5z|$
5. $-|x + z|$
6. $8 - |5y - 3|$
7. $2|x| - 4|2 + y|$
8. $|x + y| - 6|z|$

Solve each equation.

9. $|d + 1| = 7$
10. $|a - 6| = 10$
11. $2|x - 5| = 22$
12. $|t + 9| - 8 = 5$
13. $|p + 1| + 10 = 5$
14. $6|g - 3| = 42$
15. $2|y + 4| = 14$
16. $|3b - 10| = 2b$
17. $|3x + 7| + 4 = 0$
18. $|2c + 3| - 15 = 0$
19. $7 - |m - 1| = 3$
20. $3 + |z + 5| = 10$
21. $4|h + 1| = 32$
22. $2|2x + 3| = 34$
23. $3|a - 5| - 4 = 14$
24. $2|2d - 7| + 1 = 35$
25. $|3t + 6| + 9 = 30$
26. $|d - 3| = 2d + 9$
27. $|4y - 5| + 4 = 7y + 8$
28. $|2b + 4| - 3 = 6b + 1$
29. $|5t| + 2 = 3t + 18$

Lesson 1-6   Solve each inequality. Graph the solution set.

1. $2z + 5 \le 7$
2. $3r - 8 > 7$
3. $-3x > 6$ $x < -2$
4. $0.75\,b < 3$
5. $2(3f + 5) \ge 28$
6. $-33 > 5g + 7$
7. $-3(y - 2) \ge -9$
8. $7a + 5 > 4a - 7$
9. $5(b - 3) \le b - 7$
10. $3(2x - 5) < 5(x - 4)$
11. $2(4m - 1) + 3(m + 4) \ge 6m$
12. $8(2c - 1) > 11c + 22$
13. $5y - 4(2y + 1) \le 2(0.5 - 2y)$
14. $2(d + 4) - 5 \ge 5(d + 3)$
15. $8 - 3t < 4(3 - t)$
16. $-x \ge \frac{x + 4}{7}$
17. $\frac{a + 8}{4} \le \frac{7 + a}{3}$
18. $-y < \frac{y + 8}{3}$
19. $2 + 4(d - 2) \le 3 - (d - 1)$
20. $5(x - 1) - 4x \ge 3(3 - x)$
21. $6s - (4s + 7) > 5 - s$

Define a variable and write an inequality for each problem. Then solve.

22. The product of 7 and a number is greater than 42.
23. Twice a number decreased by 3 is no more than 11.
24. The opposite of four times a number is greater than or equal to 16.
25. Fifty-four is less than the product of 18 and a number.
26. Thirty decreased by a number is less than twice the number plus three.

Lesson 1-7 State an absolute value inequality for each of the following. Then graph each solution set.

1. all numbers between -5.5 and 5.5

2. all number less than -9 and greater than 9

3. all numbers greater than or equal to -6 and less than or equal to 6

Solve each inequality. Graph each solution set.

4. $|a + 3| < 1$

5. $|t - 4| > 1$

6. $|7x| \geq 21$

7. $|8p| \leq 16$

8. $|2y - 5| < 3$

9. $|3d + 6| \geq 3$

10. $2 < n + 4 < 7$

11. $-3 \leq s - 2 \leq 5$

12. $|7d| \geq -42$

13. $|4x - 1| < 5$

14. $|6v + 12| > 18$

15. $7 < 4x + 3 < 19$

16. $3m - 2 < 7$ or $2m + 1 > 13$

17. $|z + 2| \geq z$

18. $5t + 3 \leq -7$ or $4t - 2 \geq 8$

19. $12 + |2q| < 0$

20. $|2r + 4| < 6$

21. $|5w - 3| \geq 9$

22. $|3h| + 15 < 0$

23. $|5n| - 16 \geq 4$

24. $4x + 7 < 5$ or $2x - 4 > 12$

Lesson 2-1 State the domain and range of each relation. Then graph the relation and identify whether it is a function or not. For each function, state whether it is discrete or continuous.

1. $\{(1, 2), (2, 3), (3, 4), (4, 5)\}$

2. $\{(0, 3), (0, 2), (0, 1), (0, 0)\}$

3. $y = x^2$

4. $y = -2x$

5. $x = y + 1$

6. $x = |y|$

A function g includes the ordered pairs $(1, 2)$, $(2, 4)$, $(3, 6)$, and $(4, 8)$. State whether g will still be a function if each ordered pair given below is also included in g. Write *yes* or *no*.

7. $(1, 3)$

8. $(-1, 3)$

9. $(-1, -3)$

10. $(1, -3)$

Find each value if $f(x) = \frac{1}{2}(x + 7)$ and $g(x) = (x + 1)^2 - \frac{2}{x}$.

11. $f(2)$

12. $f(-4)$

13. $f\left(\frac{1}{2}\right)$

14. $f(a + 2)$

15. $g(4)$

16. $g(-2)$

17. $g\left(\frac{1}{2}\right)$

18. $g(b - 1)$

Lesson 2-2 State whether each equation is linear. Write *yes* or *no*.

1. $\frac{x}{2} - y = 7$

2. $x(y + 5) = 0$

3. $g(x) = \frac{2}{x - 3}$

4. $x = 3 + y$

5. $f(x) = 7$

6. $\frac{3}{x} - \frac{1}{4} = \frac{4}{3}$

Write each equation in standard form where A, B, and C are integers whose greatest common factor is 1. Identify A, B, and C.

7. $x + 7 = y$

8. $5x - 7y = \frac{1}{2}$

9. $y = \frac{2}{3}x + 8$

Find the x-intercept and the y-intercept of the graph of each equation.

10. $0.05x + 0.02y = 4$

11. $x = 3y$

12. $x = 7$

Graph each equation.

13. $2y = 3x$

14. $y = x - 4$

15. $2x + y = 6$

16. $3x - 2y = -12$

17. $0.2x - 0.5y = 1$

18. $5x = 20$

19. $3y + 7 = 12$

20. $\frac{x}{2} - \frac{y}{5} = \frac{1}{3}$

21. $\frac{3}{4}y - x = 1$

Lesson 2-3 Find the slope of the line that passes through each pair of points. Then determine whether the line rises to the right, falls to the right, is horizontal or vertical.

1. $(0, 3), (5, 0)$

2. $(2, 3), (5, 7)$

3. $(2, 8), (2, -8)$

Find the slope of the graph of each equation.

4. $2x + y = 8$

5. $x - 5y = -15$

6. $y = 7$

Determine the value of a so that a line through the points with the given coordinates has the given slope.

7. $(5, 0), (a, 9)$; slope $= 3$

8. $(a, 8), (2, -8)$; slope $= \frac{4}{3}$

9. $(a, 3), (7, 7)$; slope undefined

10. $(0, 3), (5, a)$; slope $= -5$

11. $\left(\frac{1}{4}, a\right), \left(\frac{1}{2}, \frac{3}{4}\right)$; slope $= -\frac{1}{8}$

12. $(3.6, 1.2), (4.8, a)$; slope $= -1$

Lesson 2-4 State the slope and y-intercept of the graph of each equation.

1. $y = 3x + 2$

2. $y = 0.12x - 3.75$

3. $-y = 6x + 4$

4. $6y = 3x - 9$

5. $2 - 5x = 5y$

6. $6y + 42 = 5x$

Write an equation in slope-intercept form that satisfies each condition.

7. slope $= \frac{5}{8}$, passes through origin

8. slope $= -0.3$, passes through $(-2, 7)$

9. passes through $(5, 3)$ and $(-5, -3)$

10. x-intercept $= 2$, y-intercept $= 7$

11. passes through $(7, 1)$, parallel to the graph of $y = \frac{5}{3}x - 2$

12. passes through $(7, 1)$, perpendicular to the graph of $y = \frac{5}{3}x - 2$

13. passes through $(0, 5)$, perpendicular to the line that passes through $(7, 8)$ and $(2, 4)$

Lesson 2-5

1. The table below shows the charges for telephone calls of various lengths of time.

Minutes	1	2	3	4	5	6	7	8	9
Cost	$0.20	$0.36	$0.52	$0.68	$0.84	$1.00	$1.16	$1.32	$1.48

a. Draw a scatter plot to show how minutes m and cost c are related.

b. Write a prediction equation that relates the cost of a telephone call and the number of minutes of the call.

c. Using your equation find the approximate cost of a call that lasts 15 minutes.

2. The table below is information from an investor who charted three stocks for several months.

Month	0	1	2	3	4	5	6	7	8
Stock A	$31.72	$32.03	$32.65	$32.55	$33.26	$33.89	$34.39	$34.64	$34.83
Stock B	$13.20	$13.14	$13.42	$13.84	$13.88	$14.21	$14.61	$14.74	$14.91
Stock C	$14.37	$14.43	$14.65	$14.98	$15.49	$15.86	$16.25	$16.96	?

a. Draw three scatter plots to show how the month m and the price p of each stock are related.

b. Write a prediction equation that relates the month with the price of each stock.

c. Use your equation for stock C to find the approximate price of the stock in the eighth month.

Lesson 2-6 If $g(x) = \left[-\frac{x}{3}\right] - 2$, find each value.

1. $g(6)$　　　　　　　**2.** $g(-6)$　　　　　**3.** $g(-3)$　　　　　　**4.** $g(10)$

5. $g\left(\frac{2}{3}\right)$　　　　　　**6.** $g(-25.5)$　　　**7.** $g\left(\frac{9}{2}\right)$　　　　　**8.** $g(23.7)$

Identify each function as C for constant, D for direct variation, A for absolute value, or G for greatest integer function. Then graph each function.

9. $f(x) = -2x + 5$　　　　　　**10.** $f(x) = \frac{17}{4}$　　　　　　**11.** $h(x) = |x| - 4$

12. $g(x) = -2[x + 5]$　　　　　**13.** $g(x) = \frac{x}{2}$　　　　　　**14.** $h(x) = |-2x| - 8$

Graph each pair of equations on the same coordinate plane. Discuss the similarities and differences in the two graphs.

15. $y = |x| - 4$ and $y = |x - 4|$　　　　　　　**16.** $y = [-2x]$ and $y = -[2x]$

17. $y = |x - 5|$ and $y = |5 - x|$　　　　　　　**18.** $y = [x]$ and $y = \left[\frac{x}{2}\right]$

Lesson 2-7 Graph each inequality.

1. $y \geq x - 15$　　　　　　　**2.** $y \leq -3x - 1$　　　　　　**3.** $4y \leq 2x - 3y + 8$

4. $3x > y$　　　　　　　　　**5.** $5x + 3 < 18$　　　　　　　**6.** $x + 2 \geq y - 7$

7. $7 > y - 7$　　　　　　　**8.** $2x < 5 - y$　　　　　　　**9.** $y > \frac{1}{5}x - 8$

10. $2y - 5x \leq 8$　　　　　**11.** $-2x + 5 \leq 3y$　　　　　**12.** $3x + 2y \geq 0$

13. $x - 3 < 5$　　　　　　　**14.** $y > 5x - 3$　　　　　　　**15.** $3x + 4y < 9$

16. $|x| \leq y + 3$　　　　　　**17.** $\frac{y}{2} \leq x - 1$　　　　　　**18.** $2x - 3y \leq 18$

19. $-y < \frac{2x}{3} + 5$　　　**20.** $-y \geq 8 - x$　　　　　　**21.** $|y| < 7$

Lesson 3-1 Graph each system of equations and state its solution. Also, state whether the system is *consistent and independent*, *consistent and dependent*, or *inconsistent*.

1. $x - y = 2$　　　　　　　　**2.** $x + 3y = 18$　　　　　　　**3.** $2x + 6y = 6$

　　$2x - 2y = 10$　　　　　　　$-x + 2y = 7$　　　　　　　$\frac{1}{3}x + y = 1$

4. $x + 3y = 0$　　　　　　　**5.** $2x - y = 7$　　　　　　　**6.** $y = \frac{1}{3}x + 1$

　　$2x + 6y = 5$　　　　　　　$\frac{2}{5}x - \frac{4}{3}y = -2$　　　　　$y = 4x + 1$

7. $\frac{3}{4}x - y = 0$　　　　　　**8.** $2x + 3y = 5$　　　　　　　**9.** $y = \frac{x}{2}$

　　$\frac{y}{3} + \frac{x}{2} = 6$　　　　　　$-6x - 9y = -15$　　　　　$2y = x + 4$

10. $\frac{2}{3}x = \frac{5}{3}y$　　　　　　**11.** $9x - 5 = 7y$　　　　　　**12.** $x - 2y = 4$

　　$2x - 5y = 0$　　　　　　$4\frac{1}{2}x - 3\frac{1}{2}y = 2\frac{1}{2}$　　　　　$y = x - 2$

Lesson 3-2 Solve each system of equations. Use either substitution or elimination.

1. $7x + y = 9$
$5x - y = 15$

2. $2x + 3y = 10$
$x + 6y = 32$

3. $x = 4y - 10$
$5x + 3y = -4$

4. $r + 5s = -17$
$2r - 6s = -2$

5. $2x - 3y = 7$
$3x + 6y = 42$

6. $2a + 5b = -13$
$3a - 4b = 38$

7. $6p + 8q = 20$
$5p - 4q = -26$

8. $\frac{5}{2}x + \frac{1}{3}y = 13$
$\frac{1}{2}x - y = -7$

9. $\frac{2}{7}c - \frac{4}{3}d = 16$
$\frac{4}{7}c + \frac{8}{3}d = -16$

10. $3x - 4y = -27$
$2x + y = -7$

11. $3c + 4d = -1$
$6c - 2d = 3$

12. $5x + 3y = -4$
$7x - y = 36$

13. $x = 2y - 1$
$4x - 3y = 21$

14. $3m + 4n = 28$
$5m - 3n = -21$

15. $7x - y = 35$
$y = 5x - 19$

Lesson 3-3 Find the value of each determinant.

1. $\begin{vmatrix} 7 & 6 \\ 2 & 5 \end{vmatrix}$

2. $\begin{vmatrix} -4 & 5 \\ 6 & 2 \end{vmatrix}$

3. $\begin{vmatrix} 5 & 1 \\ 7 & -2 \end{vmatrix}$

Use Cramer's rule to solve each system of equations.

4. $5x - 3y = 19$
$7x + 2y = 8$

5. $4p - 3q = 22$
$2p + 8q = 30$

6. $-x + y = 5$
$2x + 4y = 38$

7. $2a - 3b = 7$
$5a + 7b = -55$

8. $2m + 6n = -6$
$4m + 3n = -18$

9. $8r + 3s = 5$
$6r - 2s = -9$

10. $\frac{1}{3}x - \frac{1}{2}y = -8$
$\frac{3}{5}x + \frac{5}{6}y = -4$

11. $\frac{1}{4}c + \frac{2}{3}d = 6$
$\frac{3}{4}c - \frac{5}{3}d = -4$

12. $0.3a + 1.6b = 0.44$
$0.4a + 2.5b = 0.66$

13. $\frac{2}{3}m - \frac{5}{3}n = -\frac{1}{3}$
$\frac{5}{9}m + \frac{7}{6}n = 1$

14. $3y = 4x + 28$
$5x + 7y = 8$

15. $4.5x = 3y$
$2(x - 4y) = -20$

Lesson 3-4 Solve each system of inequalities by graphing.

1. $x \le 5$
$y \ge -3$

2. $x + y \le 2$
$y - x \le 4$

3. $x + y < 5$
$x < 2$

4. $y + x < 2$
$y \ge x$

5. $y < 3$
$y - x \ge -1$

6. $y \le x + 4$
$x + y \ge 1$

7. $y < \frac{1}{3}x + 5$
$y < 2x + 1$

8. $y + x \ge 1$
$y - x \ge -1$

9. $|x| > 2$
$|y| \le 5$

10. $|x - 2| \le 3$
$4y - 2x \ge 6$

11. $4x + 3y \ge 12$
$2y - x \ge 1$

12. $y \le -1$
$3x - 2y \ge 6$

13. $y > 1$
$y < -3x + 3$
$y > -3x + 1$

14. $y \ge -\frac{1}{2}x + 1$
$y \le -3x + 5$
$y \le 2x + 2$

15. $2x + 5y < 25$
$y < 3x - 2$
$5x - 7y < 14$

Lesson 3-5 A feasible region has vertices at $(-3, 2)$, $(1, 3)$, $(6, 1)$, and $(2, -2)$. Find the maximum and minimum values of each function.

1. $f(x, y) = 2x - y$

2. $f(x, y) = x + 5y$

3. $f(x, y) = y - 4x$

4. $f(x, y) = -x + 3y$

5. $f(x, y) = 3x - y$

6. $f(x, y) = 2y - 2x$

Graph each system of inequalities. Name the coordinates of the vertices of the feasible region. Find the maximum and minimum values of the given function for this region.

7. $4x - 5y \geq -10$
   $y \leq 6$
   $2x - 5y \leq -10$
   $f(x, y) = x + y$

8. $x \leq 5$
   $y \geq 2$
   $2x - 5y \geq -10$
   $f(x, y) = 3x + y$

9. $x - 2y \geq -7$
   $x + y \geq 8$
   $2x - y \leq 7$
   $f(x, y) = 3x - 4y$

10. $y \leq 4x + 6$
    $x + 4y \leq 7$
    $2x + y \leq 7$
    $x - 6y \leq 10$
    $f(x, y) = 2x - y$

11. $y \geq 0$
    $y \leq 5$
    $y \leq -x + 7$
    $5x + 3y \geq 20$
    $f(x, y) = x + 2y$

12. $y \geq 0$
    $3x - 2y \geq 0$
    $x + 3y \leq 11$
    $2x + 3y \leq 16$
    $f(x, y) = 4x + y$

Lesson 3-6

1. **Manufacturing** A shoe manufacturer makes outdoor and indoor soccer shoes. There is a two-step manufacturing process for both kinds of shoes. Each pair of outdoor shoes requires 2 hours of processing in step one, 1 hour in step two, and produces a profit of $20 for the company. Indoor shoes require 1 hour of processing in step one, 3 hours in step two, and produce a profit of $15 for each pair. The company has 40 hours of labor per day available for step one and 60 hours of labor per day available for step two.

   a. If x represents the number of pairs of outdoor shoes and y the number of indoor shoes produced per day, write a system of inequalities that represents the number of pairs of outdoor and indoor soccer shoes that can be produced in one day.

   b. Draw the graph showing the feasible region.

   c. Write an expression for the profit per day.

   d. How many of each kind of shoe should be produced each day to have the maximum profit? What is the maximum profit?

2. **Manufacturing** A toy manufacturer makes a $3 profit on yo-yos and a $3 profit on tops. Department A requires 3 hours to make parts for 100 yo-yos and 4 hours to make parts for 100 tops. Department B needs 5 hours to make parts for 100 yo-yos and 2 hours to make parts for 100 tops. Department A has 450 hours available and department B has 400 hours available.

   a. How many yo-yos and tops should be made to maximize the profit?

   b. What is the maximum profit the company can make from these two products?

Lesson 3-7 For each system of equations, an ordered triple is given. Determine whether or not it is a solution of the system.

1. $4x + 2y - 6z = -38$
   $5x - 4y + z = -18$
   $x + 3y + 7z = 38; (-3, 2, 5)$

2. $u + 3v + w = 14$
   $2u - v + 3w = -9$
   $4u - 5v - 2w = -2; (1, 5, -2)$

3. $x + y = -6$
   $x + z = -2$
   $y + z = 2; (-4, -2, 2)$

Solve each system of equations.

4. $5a = 5$
   $6b - 3c = 15$
   $2a + 7c = -5$

5. $s + 2t = 5$
   $7r - 3s + t = 20$
   $2t = 8$

6. $2u - 3v = 13$
   $3v + w = -3$
   $4u - w = 2$

7. $4a + 2b - c = 5$
   $2a + b - 5c = -11$
   $a - 2b + 3c = 6$

8. $x + 2y - z = 1$
   $x + 3y + 2z = 7$
   $2x + 6y + z = 8$

9. $2x + y - z = 7$
   $3x - y + 2z = 15$
   $x - 4y + z = 2$

Lesson 4-1 Perform the indicated operations.

1. $\frac{3}{2}[8 \ 4 \ 3]$

2. $-10\begin{bmatrix} 1.26 & 8.95 \\ 2.47 & -3.62 \end{bmatrix}$

3. $-5\begin{bmatrix} 7.5 \\ -3.8 \end{bmatrix}$

Solve for the variables.

4. $[2x \ \ 3y \ \ -z] = [2y \ \ -z \ \ 15]$

5. $\begin{bmatrix} x+y \\ 4x-3y \end{bmatrix} = \begin{bmatrix} 1 \\ 11 \end{bmatrix}$

6. $-2\begin{bmatrix} w+5 & x-z \\ 3y & 8 \end{bmatrix} = \begin{bmatrix} -16 & -4 \\ 6 & 2x+8z \end{bmatrix}$

7. $y\begin{bmatrix} 2 & x \\ 5 & 1 \end{bmatrix} = \begin{bmatrix} 4 & -10 \\ 10 & 2z \end{bmatrix}$

8. $\begin{bmatrix} 2x \\ -y \\ 3z \end{bmatrix} = \begin{bmatrix} 16 \\ 18 \\ -21 \end{bmatrix}$

9. $\begin{bmatrix} x-3y \\ 4y-3x \end{bmatrix} = -5\begin{bmatrix} 2 \\ x \end{bmatrix}$

10. $\begin{bmatrix} x^2+1 & 5-y \\ x+y & y-4 \end{bmatrix} = \begin{bmatrix} 2 & x \\ 5 & 2 \end{bmatrix}$

11. $\begin{bmatrix} x+y & 3 \\ y & 6 \end{bmatrix} = \begin{bmatrix} 0 & 2y-x \\ z^2 & 4-2x \end{bmatrix}$

Lesson 4-2 Perform the indicated operations.

1. $\begin{bmatrix} 3 & 5 \\ -7 & 2 \end{bmatrix} + \begin{bmatrix} -2 & 6 \\ 8 & -1 \end{bmatrix}$

2. $\begin{bmatrix} 45 & 36 \\ 18 & 63 \end{bmatrix} - 9\begin{bmatrix} 5 & 4 \\ 2 & 7 \end{bmatrix}$

3. $4[-8 \ 2 \ 9] - 3[2 \ -7 \ 6]$

4. $\frac{4}{5}\begin{bmatrix} -5 \\ 6 \\ 8 \end{bmatrix} + \frac{1}{4}\begin{bmatrix} 9 \\ -6 \\ 12 \end{bmatrix} - \frac{1}{2}\begin{bmatrix} 5 \\ 4 \\ 7 \end{bmatrix}$

5. $\begin{bmatrix} -3 & 6 & -9 \\ 4 & -3 & 0 \\ 8 & -2 & 3 \end{bmatrix} - \begin{bmatrix} 1 & 5 & 7 \\ 5 & 2 & -6 \\ 3 & 0 & -2 \end{bmatrix}$

6. $5\begin{bmatrix} 3 & 1 & 0 \\ 0 & 0 & 2 \\ 1 & -1 & -1 \end{bmatrix} - 3\begin{bmatrix} 2 & 0 & 3 \\ 1 & 1 & 2 \\ 2 & 1 & -1 \end{bmatrix}$

7. $5\begin{bmatrix} 6 & -2 \\ 5 & 4 \end{bmatrix} - 2\begin{bmatrix} 6 & -2 \\ 5 & 4 \end{bmatrix} + 4\begin{bmatrix} 7 & -6 \\ -4 & 2 \end{bmatrix}$

8. $1.3\begin{bmatrix} 3.7 & 4.8 \\ -5.4 & 9.5 \end{bmatrix} + 4.1\begin{bmatrix} 6.4 & -1.9 \\ -3.7 & -2.8 \end{bmatrix} - 6.2\begin{bmatrix} -0.8 & 5.1 \\ 3.2 & 7.4 \end{bmatrix}$

Lesson 4-3 Find the dimensions of each matrix product.

1. $A_{4\times3} \cdot B_{3\times4}$

2. $R_{2\times5} \cdot S_{2\times5}$

3. $G_{2\times5} \cdot H_{2\times3}$

4. $M_{3\times8} \cdot N_{8\times2}$

5. $X_{4\times3} \cdot Y_{2\times6}$

6. $J_{5\times3} \cdot K_{3\times6}$

7. $C_{m\times n} \cdot D_{n\times p}$

8. $X_{2\times9} \cdot Y_{9\times7}$

Perform the indicated operations, if possible.

9. $[7 \ 2] \cdot \begin{bmatrix} -3 \\ 5 \end{bmatrix}$

10. $\begin{bmatrix} 2 & -4 \\ 0 & 5 \end{bmatrix} \cdot \begin{bmatrix} 1 & 3 \\ -2 & -1 \end{bmatrix}$

11. $\begin{bmatrix} 1 & 3 \\ -2 & -1 \end{bmatrix} \cdot \begin{bmatrix} 2 & -4 \\ 0 & 5 \end{bmatrix}$

12. $\begin{bmatrix} 3 & 2 \\ 5 & 2 \end{bmatrix} \cdot \begin{bmatrix} -8 \\ 15 \end{bmatrix}$

13. $\begin{bmatrix} -1 \\ 2 \\ 1 \end{bmatrix} \cdot \begin{bmatrix} 7 & 6 & 1 \\ 2 & -4 & 0 \end{bmatrix}$

14. $\begin{bmatrix} 0 & 1 & -2 \\ 5 & 3 & -4 \\ -1 & 0 & 0 \end{bmatrix} \cdot \begin{bmatrix} 1 & -3 & 0 \\ 2 & 0 & -1 \\ 0 & 1 & -2 \end{bmatrix}$

15. $\begin{bmatrix} 3 & -2 \\ 4 & 5 \end{bmatrix} \cdot \begin{bmatrix} 1 & 0 \\ 0 & 1 \end{bmatrix}$

16. $\begin{bmatrix} 3 & -2 \\ 4 & -5 \end{bmatrix} \cdot \begin{bmatrix} \frac{5}{7} & -\frac{2}{7} \\ \frac{4}{7} & -\frac{3}{7} \end{bmatrix}$

Lesson 4-4 Determine whether each matrix has a determinant. Write *yes* or *no*. If *yes*, find the value of the determinant.

1. $[3\ 5]$

2. $\begin{bmatrix} 1 & -5 \\ 3 & 4 \end{bmatrix}$

3. $\begin{bmatrix} -1 & 6 & 1 \\ 0 & 5 & 1 \\ -5 & 2 & 3 \end{bmatrix}$

4. $\begin{bmatrix} 2 & 0 \\ 0 & 0 \\ 0 & 2 \end{bmatrix}$

Evaluate each determinant using expansion of minors.

5. $\begin{vmatrix} 2 & -3 & 5 \\ 1 & -2 & -7 \\ -1 & 4 & -3 \end{vmatrix}$

6. $\begin{vmatrix} 0 & -1 & 2 \\ -2 & 1 & 0 \\ 2 & 0 & -1 \end{vmatrix}$

7. $\begin{vmatrix} 4 & 3 & -2 \\ 2 & 5 & -8 \\ 6 & 4 & -1 \end{vmatrix}$

8. $\begin{vmatrix} -3 & 0 & 2 \\ 1 & -2 & -1 \\ 0 & 5 & 0 \end{vmatrix}$

Evaluate each determinant using diagonals.

9. $\begin{vmatrix} 3 & 2 & -1 \\ 2 & 3 & 0 \\ -1 & 0 & 3 \end{vmatrix}$

10. $\begin{vmatrix} 1 & 0 & 0 \\ 0 & 1 & 0 \\ 0 & 0 & 1 \end{vmatrix}$

11. $\begin{vmatrix} 4 & 3 & -2 \\ 2 & 5 & -8 \\ 6 & 4 & -1 \end{vmatrix}$

12. $\begin{vmatrix} 6 & 12 & 15 \\ 9 & 3 & 14 \\ 5 & 6 & 3 \end{vmatrix}$

Use a determinant to find the area of the triangle with the given vertices.

13. $(2, 3), (5, 6), (0, 0)$

14. $(-5, -8), (2, 7), (6, -3)$

15. $(-2, 2), (2, 2), (2, -2)$

Lesson 4-5 Find the inverse of each matrix, if it exists. If it does not exist, explain why.

1. $\begin{bmatrix} 2 & 3 \\ 1 & 1 \end{bmatrix}$

2. $\begin{bmatrix} 3 & 2 \\ 0 & -4 \end{bmatrix}$

3. $[3\ 8]$

4. $\begin{bmatrix} 3 & -6 \\ 2 & -4 \end{bmatrix}$

5. $\begin{bmatrix} 2 & 4 \\ 2 & 3 \end{bmatrix}$

6. $\begin{bmatrix} 8 & 5 \\ 6 & 4 \end{bmatrix}$

7. $\begin{bmatrix} 10 & 3 \\ 5 & 2 \end{bmatrix}$

8. $\begin{bmatrix} -3 & 4 \\ -4 & 8 \end{bmatrix}$

Determine whether each statement is *true* or *false*.

9. $\begin{bmatrix} -7 & -6 \\ 8 & 7 \end{bmatrix} \cdot \begin{bmatrix} -7 & -6 \\ 8 & 7 \end{bmatrix} = I$

10. $\begin{bmatrix} -3 & 4 \\ 2 & -2 \end{bmatrix} \cdot \begin{bmatrix} -2 & -2 \\ -4 & -3 \end{bmatrix} = I$

11. $\begin{bmatrix} 1 & 0 \\ 0 & 1 \end{bmatrix} \cdot \begin{bmatrix} -3 & 7 \\ 1 & 8 \end{bmatrix} = \begin{bmatrix} -3 & 7 \\ 1 & 8 \end{bmatrix}$

Lesson 4-6 Write a matrix equation for each system.

1. $5x + 3y = 6$
   $2x - y = 9$

2. $3x + 4y = -8$
   $2x - 3y = 6$

3. $x + 3y = 1$
   $4x - y = -22$

4. $4x - 3y = -1$
   $5x - 2y = 39$

Matrix M^{-1} is the inverse of the coefficient matrix. Use M^{-1} to solve each matrix equation.

5. $\begin{bmatrix} 3 & 4 \\ 2 & -5 \end{bmatrix} \cdot \begin{bmatrix} x \\ y \end{bmatrix} = \begin{bmatrix} 33 \\ -1 \end{bmatrix}$ $M^{-1} = \frac{1}{23}\begin{bmatrix} 5 & 4 \\ 2 & -3 \end{bmatrix}$

6. $\begin{bmatrix} 1 & 2 & -1 \\ -2 & 3 & 1 \\ 1 & 1 & 3 \end{bmatrix} \cdot \begin{bmatrix} x \\ y \\ z \end{bmatrix} = \begin{bmatrix} 6 \\ 1 \\ 8 \end{bmatrix}$ $M^{-1} = \frac{1}{27}\begin{bmatrix} 8 & -7 & 5 \\ 7 & 4 & 1 \\ -5 & 1 & 7 \end{bmatrix}$

Solve each matrix equation or system of equation using inverse matrices.

7. $\begin{bmatrix} 1 & 1 \\ 0.1 & -0.08 \end{bmatrix} \cdot \begin{bmatrix} x \\ y \end{bmatrix} = \begin{bmatrix} 18,000 \\ 0 \end{bmatrix}$

8. $\begin{bmatrix} -1 & 1 \\ 7 & -6 \end{bmatrix} \cdot \begin{bmatrix} x \\ y \end{bmatrix} = \begin{bmatrix} 0 \\ 3 \end{bmatrix}$

9. $5x - y = 7$
   $8x + 2y = 4$

10. $3x + y = 4$
    $2x + 2y = 3$

11. $6x + 5y = 7$
    $3x - 10y = -4$

12. $3x - 5y = 1$
    $x + 3y = 5$

Lesson 4-7 Write an augmented matrix for each system of equations. Then solve each system.

1. $y = 2x + 3$
$y = 4x - 1$

2. $3x + 4y = -6$
$5x - 3y = 19$

3. $3x + 2y + 4z = 9$
$2x + 5y - 2z = -7$
$4x + y - 3z = -3$

4. $x + 2y - 3z = 12$
$5x - 3y + z = -11$
$2x + y + 4z = -5$

Describe the solution for the system of equations represented by each reduced augmented matrix.

5. $\begin{bmatrix} -2 & 0 & | & -12 \\ 0 & 4 & | & -48 \end{bmatrix}$

6. $\begin{bmatrix} 1 & 2 & 0 & | & 9 \\ 0 & 3 & 1 & | & 6 \\ 0 & 0 & 0 & | & 0 \end{bmatrix}$

7. $\begin{bmatrix} 0 & 0 & 0 & | & -2 \\ 0 & 3 & 0 & | & 6 \\ 0 & 0 & 4 & | & 1 \end{bmatrix}$

Solve each system of equations using augmented matrices.

8. $2x = 3y - 31$
$3x + 5y = 1$

9. $6y = 4x - 9$
$3x = 7y - 2$

10. $7x + 5y = 27$
$5x - 5y = 5$

11. $4x + 2y = 0$
$3x + 5y = 7$

12. $2x - 3y + 4z = -15$
$5x - y - 3z = 8$
$3x - 2y + 5z = -5$

13. $2x - y - 4z = -8$
$4x + y + 3z = 6$
$6x - z = -2$

14. $x - 4y + 2z = 4$
$2x + 3y + 2z = -5$
$2x - 5y + z = 2$

15. $3x - 2y - 4z = 13$
$2x + 3y + 3z = 4$
$2x - 2y - 5z = 5$

Lesson 4-8 Use the box-and-whisker plot to answer each question.

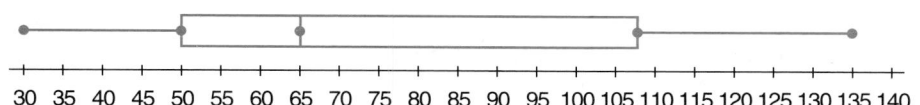

1. What is the range of the data?

2. What percent of the data is greater than 65?

3. What percent of the data is less than 50?

4. What percent of the data is greater than 30 but less than 135?

Find the range, quartiles, interquartile range, and outliers for each set of data. Then make a box-and-whisker plot for each set of data.

5. {24, 19, 24, 22, 13, 20, 17, 24 20, 23, 24, 20, 17, 18, 20, 16, 17, 20, 10, 21, 20, 22, 23, 20, 13, 23}

6. {15, 9, 15, 11, 10, 9, 12, 13, 12, 11, 13, 12, 14, 12, 13, 10, 6, 12, 11, 8, 15, 14, 3, 5, 7, 8, 6, 12, 14}

7. {97, 83, 73, 33, 89, 67, 89, 73, 28, 55, 89, 80, 38, 84, 83, 89, 81, 88, 87, 76, 36, 72, 91, 73, 70, 47}

8. {30.4, 18.9, 11.1, 8.2, 5.5, 5.2, 13.6, 13.2, 8.8, 5.4, 6.2, 8.7}

9. {679, 565, 805, 556, 718, 625, 553, 2064, 496, 1033}

Lesson 5-1 Simplify. Assume that no variable equals 0.

1. $x^7 x^3 x$

2. $(-3)^4 (-3)$

3. $\dfrac{t^{12}}{t}$

4. $\dfrac{6^5}{6^3}$

5. $(m^3)^8$

6. -3^4

7. $\left(\dfrac{x}{5}\right)^2$

8. $(-2y^5)^2$

9. $3x^0$

10. $\dfrac{5^6 a^{x+y}}{5^4 a^{x-y}}$

11. ab^{-1}

12. $\dfrac{b^{-4}}{b^{-5}}$

13. $(y^{-5})^{-7}$

14. $(5x^2 y^{-3})^4$

15. $\dfrac{1}{x^{-3}}$

16. $\dfrac{2^{-1} xy^2}{2^3 y^8}$

Evaluate. Express each answer in both scientific and decimal notation.

17. $(8.95 \times 10^9)(1.82 \times 10^7)$

18. $(-3.1 \times 10^5)(7.9 \times 10^{-8})$

19. $\dfrac{(2.38 \times 10^{13})(7.56 \times 10^{-5})}{(4.2 \times 10^{18})}$

Lesson 5-2 Determine whether each expression is a polynomial. Write *yes* or *no*. Then state the degree of each polynomial.

1. $5r^3 + 7r^2s - 3rs^2$

2. $\frac{xy}{3}$

3. $\frac{a^2}{b}$

4. $2 + \frac{7}{w}$

5. 15

Simplify.

6. $4x^3 + 5x - 7x^2 - 2x^3 + 5x^2 - 7y^2$

7. $(2x^2 - 3x + 11) + (7x^2 + 2x - 8)$

8. $(-3x^2 + 7x + 23) + (-8x^2 - 5x + 13)$

9. $(-3x^2 + 7x + 23) - (-8x^2 - 5x + 13)$

10. $5a^2b(4a - 3b)$

11. $\frac{7u}{w}\left(4u^2w^3 - 5uw + \frac{w}{7u}\right)$

12. $-4x^5(-3x^4 - x^3 + x + 7)$

13. $-5x^4(9 - 2x^2 + 4x^3 - 7x^4)$

14. $(2x - 3)(4x + 7)$

15. $(3x - 5)(-2x - 1)$

16. $(2x - 3)(x^2 + 4x + 7)$

17. $(3x - 5)(4x^2 - 2x - 1)$

18. $(2x + 5)(2x - 5)$

19. $(3x - 7)(3x + 7)$

20. $(x + 4)(x - 4)$

21. $(5 + 2w)(5 - 2w)$

22. $(3a^4 - 5)(3a^4 + 5)$

23. $(-4x - 10)(-4x + 10)$

24. $(3x + 7)^2$

25. $(5x - 2)^2$

26. $(x + 7)^2$

27. $(t - 5)^2$

28. $(7b - 0.8)^2$

29. $(5x - 3x^4)^2$

Lesson 5-3 **Simplify.**

1. $\frac{18r^3s^2 + 36r^2s^3}{9r^2s^2}$

2. $\frac{15v^3w^2 - 5v^4w^3}{-5v^4w^3}$

3. $\frac{x^2 - x + 1}{x}$

4. $(5hb + 5hc) \div (b + c)$

5. $(25c^4d + 10c^3d^2 - cd) \div 5cd$

6. $(16f^{18} + 20f^9 - 8f^6) \div 4f^3$

7. $(33m^5 + 55mn^5 - 11m^3)(11m)^{-1}$

8. $(8g^3 + 19g^2 - 12g + 9) \div (g + 3)$

9. $(p^{21} + 3p^{14} + p^7 - 2)(p^7 + 2)^{-1}$

10. $(15x^3 + 19x^2y - 7xy^2 + 5y^3) \div (3x + 5y)$

11. $(n^3 + 125) \div (n + 5)$

12. $(10z^3 - 90z) \div (2z - 6)$

13. $(8k^2 - 56k + 98) \div (2k - 7)$

14. $(2r^2 + 5r - 3) \div (r + 3)$

15. $(6y^2 + 7y - 3)(2y + 3)^{-1}$

Use synthetic division to find each quotient.

16. $(6a^4 - 22a^3 - 9a^2 + 9a - 17) \div (a - 4)$

17. $(q^4 + 8q^3 + 3q + 17) \div (q + 8)$

18. $(15v^3 + 8v^2 - 21v + 6) \div (5v - 4)$

19. $(10y^4 + 3y^2 - 7)(2y^2 - 1)^{-1}$

20. $(5s^3 + s^2 - 7) \div (s + 1)$

21. $(x^3 + 13x^2 - 12x - 8)(x + 2)^{-1}$

22. $(t^4 - 2t^3 + t^2 - 3t + 2) \div (t - 2)$

23. $(z^4 - 3z^3 - z^2 - 11z - 4) \div (z - 4)$

24. $(3r^4 - 6r^3 - 2r^2 + r - 6) \div (r + 1)$

25. $(2b^3 - 11b^2 + 12b + 9) \div (b - 3)$

Lesson 5-4 Factor completely. If the polynomial is not factorable, write *prime*.

1. $14a^3b^3c - 21a^2b^4c + 7a^2b^3c$

2. $10ax - 2xy - 15ab + 3by$

3. $x^2 + x - 42$

4. $2x^2 + 5x + 3$

5. $6x^2 + 71x - 12$

6. $6x^4 - 12x^3 + 3x^2$

7. $x^2(x + 3) + 2(x + 3)$

8. $x^2 - 6x + 2$

9. $2x^3 + 6x^2 + x + 3$

10. $x^2 - 2x - 15$

11. $6x^2 + 23x + 20$

12. $24x^2 - 76x + 40$

13. $6p^2 - 13pq - 28q^2$

14. $2x^2 - 6x + 3$

15. $x^2 + 49 - 14x$

16. $9x^2 - 64$

17. $9 - t^{10}$

18. $x^2 + 16$

19. $a^4 - 81b^4$

20. $3a^3 + 12a^2 - 63a$

21. $x^3 - 8x^2 + 15x$

22. $x^2 + 6x + 9$

23. $18x^3 - 8x$

24. $3x^2 - 42x + 40$

25. $2x^2 + 4x - 1$

26. $35ac - 3bd - 7ad + 15bc$

27. $5h^2 - 10hj + h - 2j$

28. $16r^2 - 24r + 9$

29. $3a^2 + 6a + 9y$

Lesson 5-5 Use a calculator to approximate each value to three decimal places.

1. $\sqrt{289}$

2. $\sqrt[4]{0.0625}$

3. $\sqrt[3]{-343}$

4. $\sqrt{7832}$

5. $\sqrt[10]{32^4}$

6. $\sqrt[3]{49}$

7. $\sqrt[5]{5}$

8. $-\sqrt[4]{25}$

Simplify.

9. $\sqrt{\left(-\frac{2}{3}\right)^4}$

10. $\sqrt[5]{-32}$

11. $-\sqrt{-144}$

12. $\sqrt[4]{a^{16}b^8}$

13. $\sqrt{9h^{22}}$

14. $\pm\sqrt[4]{81x^4}$

15. $\sqrt[5]{\dfrac{1}{100000}}$

16. $\sqrt[5]{p^{25}q^{15}r^5s^{20}}$

17. $\sqrt[3]{-d^6}$

18. $-\sqrt[4]{(2x^2-y)^8}$

19. $\sqrt[5]{0}$

20. $\pm\sqrt{16m^6\,n^2}$

21. $\sqrt[3]{(2x-y)^3}$

22. $\sqrt[4]{(r+s)^4}$

23. $\sqrt{9a^2+6a+1}$

24. $\sqrt{4y^2+12y+9}$

Lesson 5-6 Simplify.

1. $\sqrt{75}$

2. $7\sqrt{12}$

3. $\sqrt[3]{81}$

4. $\sqrt{5r^5}$

5. $\sqrt[4]{7^8\,x^5y^6}$

6. $3\sqrt{5}+6\sqrt{5}$

7. $\sqrt{18}-\sqrt{50}$

8. $4\sqrt[3]{32}+\sqrt[3]{500}$

9. $\sqrt{12}\sqrt{27}$

10. $\sqrt[3]{6}-\sqrt{6}$

11. $\sqrt{10}(2-\sqrt{5})$

12. $\sqrt{3}(5\sqrt{2}+4\sqrt{7})$

13. $(1-\sqrt{7})(4+\sqrt{7})$

14. $(5+\sqrt{2})(3+\sqrt{3})$

15. $(2+\sqrt{5})(2-\sqrt{5})$

16. $(x-\sqrt{3y})(x+\sqrt{3y})$

17. $(8+\sqrt{11})^2$

18. $(5z-2\sqrt{7})^2$

19. $\sqrt{\dfrac{3m^3}{24n^5}}$

20. $\dfrac{\sqrt{18}}{\sqrt{32}}$

21. $2\sqrt[3]{\dfrac{r^5}{2s^2t}}$

22. $\sqrt{\dfrac{2}{3}}-\sqrt{\dfrac{3}{8}}$

23. $\dfrac{5}{3-\sqrt{10}}$

24. $\dfrac{x+\sqrt{5}}{x-\sqrt{5}}$

Lesson 5-7 Evaluate.

1. $2401^{\frac{1}{4}}$

2. $27^{\frac{4}{3}}$

3. $(-32)^{\frac{2}{5}}$

4. $-81^{\frac{3}{4}}$

5. $(-125)^{-\frac{2}{3}}$

6. $7^{\frac{5}{2}}\cdot 7^{\frac{1}{2}}$

7. $8^{-\frac{2}{3}}\cdot 64^{\frac{1}{6}}$

8. $\left(\dfrac{48}{1875}\right)^{-\frac{5}{4}}$

Express in simplest radical form.

9. $7^{\frac{5}{9}}$

10. $32^{\frac{2}{3}}$

11. $k^{\frac{8}{5}}$

12. $x^{\frac{2}{5}}\cdot x^{\frac{5}{8}}$

13. $m^{\frac{2}{5}}\cdot n^{\frac{4}{5}}$

14. $\left(p^{\frac{5}{4}}\,q^{\frac{7}{2}}\right)^{\frac{8}{3}}$

15. $3^{\frac{9}{2}}\,c^{\frac{3}{2}}$

16. $\dfrac{7^{\frac{3}{4}}}{7^{\frac{5}{3}}}$

Simplify.

17. $\dfrac{1}{t^{\frac{9}{5}}}$

18. $a^{-\frac{8}{7}}$

19. $\dfrac{y^{\frac{1}{2}}}{x^{\frac{1}{2}}+y^{\frac{1}{2}}}$

20. $(a+b)^{-\frac{3}{4}}$

21. $\dfrac{r}{r^{\frac{7}{5}}}$

22. $\dfrac{8}{5^{\frac{1}{2}}+3^{\frac{1}{2}}}$

23. $\dfrac{v^{\frac{11}{7}}-v^{\frac{4}{7}}}{v^{\frac{4}{7}}}$

24. $\left(z^{\frac{5}{3}}\cdot z^{-\frac{9}{2}}\right)^{-\frac{12}{17}}$

Lesson 5-8 Solve each equation. Be sure to check for extraneous solutions.

1. $\sqrt{x} = 16$

2. $\sqrt{z+3} = 7$

3. $\sqrt[4]{a+5} = 1$

4. $5\sqrt{s} - 8 = 3$

5. $\sqrt[4]{m+7} + 11 = 9$

6. $d + \sqrt{d^2 - 8} = 4$

7. $g\sqrt{7} + 8 = g$

8. $\sqrt{x-8} = \sqrt{13+x}$

9. $\sqrt{3x+10} = 1 + \sqrt{2x+5}$

10. $\sqrt{3-x} = \sqrt{11} - \sqrt{x}$

11. $(3x+8)^{\frac{1}{2}} = 2$

12. $\sqrt{3n-1} = \sqrt{4-2n}$

13. $8w + 3 = 4 - w\sqrt{5}$

14. $\sqrt{5y+4} = 8$

15. $2 - 4\sqrt{21-6c} = 6$

16. $\sqrt{3x+25} + \sqrt{10-2x} = 0$

17. $\sqrt{2c+3} - 7 = 0$

18. $\sqrt{3z-5} - 3 = 1$

19. $\sqrt{5y+1} + 6 = 10$

20. $\sqrt{3f+1} - 2 = 6$

21. $\sqrt{y-5} - \sqrt{y} = 1$

Lesson 5-9 Simplify.

1. $\sqrt{-289}$

2. $\sqrt{-\dfrac{25}{121}}$

3. $\sqrt{-625b^8}$

4. $\sqrt{-\dfrac{28f^6}{27g^5}}$

5. $(7i)^2$

6. $(6i)(-2i)(11i)$

7. $(\sqrt{-8})(\sqrt{-12})$

8. $-i^{22}$

9. $i^{17} \cdot i^{11} \cdot i^{26}$

10. $3\sqrt{-24} - 7\sqrt{-96}$

11. $(14 - 5i) + (-8 + 19i)$

12. $(7i) - (2 + 3i)$

13. $(2 + 2i) - (5 + i)$

14. $(7 + 3i)(7 - 3i)$

15. $(8 - 2i)(5 + i)$

16. $(6 + 8i)^2$

17. $(15 - 3i)(15 + 3i)$

18. $(9 + 5i)^2 + (15 - 3i)^2$

Solve each equation.

19. $x^2 + 8 = 3$

20. $\dfrac{4x^2}{49} + 6 = 3$

21. $8x^2 + 5 = 1$

22. $12 - 9x^2 = 38$

Lesson 5-10 Find the conjugate of each number.

1. $3 + i$

2. $\sqrt{2} - 3i$

3. $-8 + 7i$

4. $-i\sqrt{5}$

Find the product of each complex number and its conjugate.

5. $7i$

6. $3 - i$

7. $5 + 3i$

8. $3 - 4i$

Simplify.

9. $\dfrac{2+i}{i}$

10. $\dfrac{8+5i}{4i}$

11. $\dfrac{3-7i}{5+4i}$

12. $\dfrac{2+10i}{6-i}$

13. $\dfrac{9+12i}{8+3i}$

14. $\dfrac{3}{6-2i}$

15. $\dfrac{5i}{3+4i}$

16. $\dfrac{5+3i}{1+i}$

17. $\dfrac{5}{\sqrt{3}+2i}$

18. $\dfrac{3i}{1+i\sqrt{2}}$

19. $\dfrac{2-i\sqrt{3}}{2+i\sqrt{3}}$

20. $\dfrac{1}{\sqrt{5}+i}$

Find the multiplicative inverse of each complex number.

21. $5 + 2i$

22. $\dfrac{7i}{3-4i}$

23. $\dfrac{3-i}{4+i}$

24. $\dfrac{1+i}{1-i}$

Lesson 6-1   Identify the quadratic term, the linear term, and the constant term in each function.

1. $f(x) = 4x^2 - 3x + 8$

2. $g(r) = 2r^2 + 1$

3. $f(z) = (z + 3)^2 - 5$

4. $f(m) = m^2 - 9$

5. $g(a) = (2a - 4)^2 - 16$

6. $f(y) = 5y^2 + 3y$

7. $h(w) = (3w + 5)^2 - 10w$

8. $f(x) = 2 + 7x - 6x^2$

9. $f(x) = 9x^2$

Use the related graph of each equation to determine its solutions.

10. $x^2 - 5x + 6 = 0$

11. $x^2 + 2x + 1 = 0$

12. $x^2 - 2x - 8 = 0$

13. $x^2 + x - 6 = 0$

14. $4x^2 - 9 = 0$

15. $2x^2 + 7x - 4 = 0$

Graph each function. Name the vertex and the axis of symmetry.

16. $f(x) = -x^2$

17. $f(x) = x^2 + 6x - 27$

18. $f(x) = 9 - x^2$

19. $g(x) = x^2 + 5x - 6$

20. $g(x) = x^2 - x + \frac{5}{4}$

21. $f(x) = x^2 + \frac{14}{3}x + \frac{5}{3}$

Lesson 6-2   Solve each equation.

1. $(x + 5)(x - 3) = 0$

2. $(y - 2)(3y + 2) = 0$

3. $s(2s - 1) = 0$

Solve each equation by factoring.

4. $x^2 + 7x + 10 = 0$

5. $x^2 - 4x - 21 = 0$

6. $b^2 = 49$

7. $3z^3 = 75z$

8. $2m^2 + 7m = 9$

9. $7x(x - 1) = 30(x + 1)$

Solve each equation by graphing or by factoring.

10. $8x^2 = 48 - 40x$

11. $5x^2 = 20x$

12. $12d^2 - 71d - 6 = 0$

13. $16x^2 - 64 = 0$

14. $5x^2 - 45x + 90 = 0$

15. $24x^2 - 15 = 2x$

16. $x^2 = 72 - x$

17. $2x^2 + 5x + 3 = 0$

18. $4p^2 + 9 = 12p$

19. $2x^2 - 8x = 0$

20. $8b^2 + 10b = 3$

21. $12p^2 - 5p = 3$

22. $a^2 + 8a + 12 = 0$

23. $x^2 + 9x + 14 = 0$

24. $9g^2 + 1 = 6g$

25. $2x^2 = 6x$

26. $6t^2 + 7t = 3$

27. $g^2 - 4g = 21$

Lesson 6-3   Find the value of c that makes each trinomial a perfect square.

1. $x^2 - 12x + c$

2. $y^2 + 20y + c$

3. $m^2 - 11m + c$

4. $g^2 - \frac{2}{3}g + c$

5. $z^2 + 30z + c$

6. $t^2 - 0.5t + c$

7. $x^2 + \frac{3}{8}x + c$

8. $w^2 + 16w + c$

9. $a^2 - 3a + c$

Find the exact solution for each equation by completing the square.

10. $x^2 + 3x - 4 = 0$

11. $x^2 + 5x = 0$

12. $x^2 + 2x - 63 = 0$

13. $3x^2 - 16x - 35 = 0$

14. $x^2 + 7x + 13 = 0$

15. $5x^2 - 8x + 2 = 0$

16. $x^2 - 6x + 11 = 0$

17. $x^2 - 12x + 36 = 0$

18. $8x^2 + 13x - 4 = 0$

19. $3x^2 + 5x + 6 = 0$

20. $x^2 + 14x - 1 = 0$

21. $4x^2 - 32x + 15 = 0$

22. $3x^2 - 11x - 4 = 0$

23. $x^2 + 8x - 84 = 0$

24. $x^2 - 7x + 5 = 0$

25. $t^2 + 3t - 8 = 0$

26. $a^2 - 5a - 10 = 0$

27. $3z^2 - 12z + 4 = 0$

28. $x^2 + 20x + 75 = 0$

29. $x^2 - 5x - 24 = 0$

30. $2t^2 + t - 21 = 0$

Lesson 6-4 Find the value of the discriminant and describe the nature of the roots (real, complex, rational, irrational) of each quadratic equation. Then solve the equation. Express irrational roots as exact and approximate to the nearest hundredth.

1. $x^2 + 7x + 13 = 0$
2. $6x^2 + 6x - 21 = 0$
3. $5x^2 - 5x + 4 = 0$
4. $16x^2 - 64x + 19 = 0$
5. $9x^2 + 42x + 49 = 0$
6. $4x^2 - 16x + 3 = 0$
7. $2x^2 = 5x + 3$
8. $x^2 + 81 = 18x$
9. $18x^2 = 6x^2 + 5$
10. $4x^2 = 49$
11. $3x^2 - 30x + 75 = 0$
12. $24x^2 + 10x = 43$
13. $9x^2 + 4 = 2x$
14. $7x = 8x^2$
15. $18x^2 = 9x + 45$
16. $y^2 - 4y + 4 = 0$
17. $4x^2 + 8x + 3 = 0$
18. $4y^2 + 16y + 15 = 0$
19. $y^2 - 6y + 13 = 0$
20. $3m^2 = 108m$
21. $x^2 - x + 1 = 0$
22. $n^2 + 4n + 29 = 0$
23. $4a^2 + 3a - 2 = 0$
24. $2x^2 + 5x = 9$
25. $n^2 = 8n - 16$
26. $7b^2 = 4b$
27. $2y^2 + 6y + 5 = 0$
28. $9a^2 - 30a + 25 = 0$
29. $3x^2 - 4x + 2 = 0$
30. $2x^2 + 3x + 3 = 0$

Lesson 6-5 Write a quadratic equation that has the given roots.

1. $9, -7$
2. $\frac{2}{3}, 3$
3. $2 \pm \sqrt{5}$
4. $\frac{-5 \pm 3\sqrt{2}}{3}$
5. $17, 17$
6. $\frac{-6 \pm 2i\sqrt{3}}{5}$
7. $0, 8$
8. $\pm\frac{\sqrt{6}}{4}$

Solve each equation. Check by using the sum and product of the roots.

9. $4x^2 - 3x - 6 = 0$
10. $4x^2 = 625$
11. $81x^2 + 4 = 36x$
12. $x^2 - 7x + 3 = 0$
13. $7x^2 + 3x + 1 = 3$
14. $6x^2 = 13x$
15. $5x^2 + 18x = 0$
16. $x^2 - 8x + 5 = 0$
17. $3x^2 - 16x - 12 = 0$
18. $9n^2 - 1 = 0$
19. $2x^2 - 7x = 15$
20. $15c^2 - 2c - 8 = 0$
21. $7s^2 + 5s - 1 = 0$
22. $12x^2 + 19x + 4 = 0$
23. $x^2 + x - 6 = 0$
24. $m^2 + 5m + 6 = 0$
25. $s^2 + 5s - 24 = 0$
26. $a^2 - 9a + 20 = 0$
27. $x^2 - 12x - 45 = 0$
28. $2m^2 - 10m + 9 = 0$
29. $3s^2 - 11 = 0$

Lesson 6-6 Write each equation in the form $y = a(x - h)^2 + k$ if not already in that form. Then name the vertex, axis of symmetry, and direction of opening for the graph of each quadratic function.

1. $f(x) = -9(x - 7)^2 + 3$
2. $f(x) = (x + 6)^2 - 1$
3. $f(x) = 2(x - 8)^2 - 5$
4. $f(x) = -(x + 1)^2 + 7$
5. $f(x) = -x^2 + 10x - 3$
6. $f(x) = -2x^2 + 16x + 7$
7. $f(x) = 3x^2 + 9x + 8$
8. $f(x) = 8x^2 - 3x + 1$
9. $f(x) = \frac{3}{4}x^2 - 6x - 5$
10. $f(x) = x^2 - 2x + 4$
11. $f(x) = -3x^2 + 18x$
12. $f(x) = -2x^2 - 40x + 10$
13. $f(x) = 2x^2 - 8x + 9$
14. $f(x) = \frac{1}{3}x^2 + 2x + 7$
15. $f(x) = x^2 + 6x + 9$
16. $f(x) = x^2 + 3x + 6$
17. $f(x) = 2x^2 + 8x + 9$
18. $f(x) = x^2 - 8x + 9$
19. $f(x) = -x^2 - 10x + 10$
20. $f(x) = -\frac{2}{3}x^2 + 4x - 3$
21. $f(x) = -2x^2 - 8x - 1$

Write an equation for the parabola that passes through the given points.

22. $(-1, 5), (2, -4), (5, 5)$
23. $(-3, 1), (-2, -1), (-1, -7)$
24. $(-5, -3), (-1, 5), (0, 9.5)$
25. $(0, -8), (2, -2), (4, -8)$
26. $(3, 0), (0, -9), (5, -4)$
27. $(-5, 5), (0, 0), (5, 5)$
28. $(0, -5), (-1, -7), (-4, -1)$
29. $(1, 5), (4, -4), (0, 0)$
30. $(1, 1.2), (0, 0.8), (-2, 3.6)$

Lesson 6-7 Graph each inequality.

1. $y \leq 5x^2 + 3x - 2$

2. $y \geq \frac{1}{2}x^2 - 4x$

3. $y > -3x^2 + 2$

4. $y \geq -x^2 - x + 3$

5. $y \leq \frac{3}{8}x^2 + 2x - 1$

6. $y \leq -5x^2 + 2x - 3$

7. $y > 4x^2 + x$

8. $y \geq -x^2 - 7$

9. $y < -\frac{2}{3}x^2 + x + 6$

Solve each inequality.

10. $(x + 1)(x - 1) < 0$

11. $(2x + 3)(5x - 1) \geq 0$

12. $x^2 - 2x - 8 \leq 0$

13. $-x^2 - 5x - 6 > 0$

14. $-3x^2 \geq 5$

15. $20x^2 + 9x + 1 < 0$

16. $2x^2 \geq 5x + 12$

17. $x^2 + 3x + 4 > 0$

18. $2x - x^2 \leq -15$

19. $13x - 28x^2 + 6 < 0$

20. $3x^2 \leq 36$

21. $x - x^2 \geq 1$

Lesson 6-8 Find the mean and standard deviation to the nearest hundredth for each set of data.

1. {86, 71, 74, 65, 45, 42, 76}

2. {16, 20, 15, 14, 24, 23, 25, 10, 19}

3. {18, 24, 16, 24, 22, 24, 22, 22, 24, 13, 17, 18, 16, 20, 16, 7, 22, 5, 4, 24}

4. {364, 305, 217, 331, 305, 311, 352, 319, 272, 238, 311, 226, 220, 226, 215, 160, 123, 4, 24, 238, 99}

5. {55, 50, 50, 55, 65, 50, 45, 35, 50, 40, 70, 40, 70, 50, 90, 30, 35, 55, 55, 40, 75, 35, 40, 45, 65, 50, 60}

6.
Stem	Leaf
130	1 1 3 9
131	7 7
132	1 2 3 5 6 7 7
133	0 0 1 3
134	1 3 4 4 5 6 8 9
135	0 1 1 1 4 5 8 9
136	1 3 5 $130|3 = 13.03$

7.
Stem	Leaf
0	7 8
1	0 1 1 2 3 3 4 4 5 5 6 6 7 7 9 9
2	0 0 1 1 1 2 2 3 3 3 4 5 $2|1 = 21$

Lesson 6-9

1. The diameters of metal fittings made by a machine are normally distributed. The mean diameter is 7.5 centimeters, and the standard deviation is 0.5 centimeters.
   a. What percentage of the fittings have diameters between 7.0 and 8.0 centimeters?
   b. What percentage of the fittings have diameters between 7.5 and 8.0 centimeters?
   c. What percentage of the fittings have diameters greater than 6.5 centimeters?
   d. Of 100 fittings, how many will have a diameter between 6.0 and 8.5 centimeters?

2. The number of hours of television watched weekly by 3000 families is normally distributed. The mean is 22 hours and the standard deviation is 7.5 hours.
   a. How many families watch at least 22 hours of television per week?
   b. How many families watch television between 7 and 29.5 hours per week?
   c. What percent of the 3000 families watch more than 37 hours of television a week?

3. The scores on a college entrance exam are normally distributed. The mean score is 510, and the standard deviation is 80.
   a. Of the 50,000 people who took the exam, how many scored above 510?
   b. Of the 50,000 people who took the exam, how many scored between 430 and 590?
   c. What percent of the people who took the exam scored below 670?
   d. A student must score above 750 on the college entrance exam to qualify for a full scholarship at Carleton University. What percentage of the people who took the exam will qualify?

Lesson 7-1 Use the distance formula to find the distance between each pair of points.

1. $(5, 7)$, $(3, 19)$

2. $(-2, -1)$, $(5, 3)$

3. $(-3, 15)$, $(7, -8)$

4. $(6, -3)$, $(-4, -9)$

5. $(3.89, -0.38)$, $(4.04, -0.18)$

6. $(5\sqrt{3}, 2\sqrt{2})$, $(-11\sqrt{3}, -4\sqrt{2})$

Find the midpoint of each line segment whose endpoints are given.

7. $(7, -3)$, $(-11, 13)$

8. $(16, 29)$, $(-7, 2)$

9. $(43, -18)$, $(-78, -32)$

10. $(-7.54, 3.42)$, $(4.89, -9.28)$

11. $(-8, 4.19)$, $(0.34, 20)$

12. $(684, -239)$, $(528, -735)$

Find the value of p so that each pair of points is 25 units apart.

13. $(p, 3)$, $(8, 27)$

14. $(5, p)$, $(-2, 6)$

15. $(15, -8)$, $(p, -8)$

16. $(-17, -7)$, $(7, p)$

17. $(8, p)$, $(-3, -5)$

18. $(p, -16)$, $(25, 13)$

19. $(-6.8, p)$, $(3.7, -8.9)$

20. $(p, -28)$, $(-29, -35)$

Given the coordinates of an endpoint of $\overline{AB}$ and its midpoint M, find the coordinates of the other endpoint.

21. $A(5, -16)$, $M(-13, 2)$

22. $M(7.3, 2.8)$, $B(-8.9, -3.4)$

23. $M\left(\frac{5}{8}, \frac{7}{12}\right)$, $A\left(\frac{2}{3}, \frac{5}{4}\right)$

Lesson 7-2 Name the vertex, axis of symmetry, focus, directrix, and direction of opening of the parabola with the given equation. Then find the length of the latus rectum and graph the parabola.

1. $y + \frac{3}{4} = x^2$

2. $\frac{y}{5} = (x + 2)^2$

3. $4(y + 2) = 3(x - 1)^2$

4. $5x + 3y^2 = 15$

5. $y = 2x^2 - 8x + 7$

6. $x = 2y^2 - 8y + 7$

7. $3(x - 8)^2 = 5(y + 3)$

8. $x = 3(y + 4)^2 + 1$

9. $8y + 5x^2 + 15x + 9 = 0$

10. $x = -\frac{1}{5}y^2 + \frac{8}{5}y - 7$

11. $6x = y^2 - 6y + 39$

12. $-8y = x^2$

13. $(x + 3)^2 = \frac{1}{4}(y - 2)$

14. $y = x^2 - 6x + 33$

15. $y = x^2 + 4x + 1$

16. $4(x - 2) = (y + 3)^2$

17. $(y - 8)^2 = -4(x - 4)$

18. $6x = y^2 + 4y + 4$

Lesson 7-3 Write an equation for each circle whose center and radius are given.

1. center $(3, 2)$, $r = 5$ units

2. center $(-5, 8)$, $r = 3$ units

3. center $(1, -6)$, $r = \frac{2}{3}$ units

4. center $(0, 7)$, $r = \sqrt{6}$ units

5. center $(\sqrt{2}, 4)$, $r = 9$ units

6. center $(-8, -\sqrt{10})$, $r = 11$ units

7. center $(0.8, 0.5)$, $r = 0.2$ units

8. center $(-9, 0)$, $r = \frac{5}{7}$ units

9. center $(4, 1)$, $r = 4$ units

Find the center and radius of each circle whose equation is given. Then draw the graph.

10. $x^2 + y^2 = 36$

11. $(x - 5)^2 + (y + 4)^2 = 1$

12. $x^2 + 3x + y^2 - 5y = \frac{1}{2}$

13. $x^2 + y^2 = 14x - 24$

14. $x^2 + y^2 = 2(y - x)$

15. $x^2 + 10x + (y - \sqrt{3})^2 = 11$

16. $x^2 + y^2 = 4x + 9$

17. $x^2 + y^2 + 12x - 10y + 45 = 0$

18. $x^2 + y^2 - 6x + 4y = 156$

19. $x^2 + y^2 - 2(\sqrt{5}x - \sqrt{7}y) = 1$

20. $16(x^2 + y^2) - 8(3x + 5y) + 33 = 0$

Lesson 7-4 Find the coordinates of the center and foci, and the lengths of the major and minor axes for each ellipse whose equation is given. Then draw the graph.

1. $\dfrac{x^2}{36} + \dfrac{y^2}{81} = 1$

2. $\dfrac{x^2}{121} + \dfrac{(y-5)^2}{16} = 1$

3. $\dfrac{(x+2)^2}{12} + \dfrac{(y+1)^2}{16} = 1$

4. $\dfrac{(x-5)^2}{25} + \dfrac{(y+3)^2}{75} = 1$

5. $\dfrac{x^2}{9} + \dfrac{y^2}{36} = 1$

6. $\dfrac{(x-8)^2}{4} + \dfrac{(y+8)^2}{1} = 1$

7. $\dfrac{(x+2)^2}{36} + \dfrac{(y-4)^2}{40} = 1$

8. $\dfrac{(x+8)^2}{121} + \dfrac{(y-7)^2}{64} = 1$

9. $\dfrac{(x-4)^2}{16} + \dfrac{(y+1)^2}{9} = 1$

10. $8x^2 + 2y^2 = 32$

11. $7x^2 + 3y^2 = 84$

12. $9x^2 + 16y^2 = 144$

13. $169x^2 - 338x + 169 + 25y^2 = 4225$

14. $x^2 + 4y^2 + 8x - 64y + 236 = 0$

15. $4x^2 + 5y^2 = 4(6x + 5y + 111)$

16. $169x^2 + y^2 + 2366x = 4y - 8116$

17. $2x^2 + y^2 - 4x + 8y - 6 = 0$

18. $4x^2 + 9y^2 + 24x - 90y = -225$

19. $9x^2 + 10y^2 + 54x + 20y = -1$

20. $9x^2 + 16y^2 - 54x + 64y + 1 = 0$

Lesson 7-5 Find the coordinates of the vertices and foci and the slopes of the asymptotes for each hyperbola whose equation is given. Then draw the graph.

1. $\dfrac{y^2}{25} - \dfrac{x^2}{9} = 1$

2. $\dfrac{x^2}{4} - \dfrac{y^2}{9} = 1$

3. $\dfrac{x^2}{81} - \dfrac{y^2}{36} = 1$

4. $\dfrac{x^2}{9} - \dfrac{y^2}{16} = 1$

5. $\dfrac{y^2}{100} - \dfrac{x^2}{144} = 1$

6. $\dfrac{x^2}{16} - \dfrac{y^2}{4} = 1$

7. $\dfrac{(x-4)^2}{64} - \dfrac{(y+1)^2}{16} = 1$

8. $\dfrac{(y-7)^2}{2.25} - \dfrac{(x-3)^2}{4} = 1$

9. $(x+5)^2 - \dfrac{(y+3)^2}{48} = 1$

10. $x^2 - 9y^2 = 36$

11. $4x^2 - 9y^2 = 36$

12. $49x^2 - 16y^2 = 784$

13. $144x^2 + 1152x - 25y^2 - 100y = 1396$

14. $576y^2 = 49x^2 + 490x + 29449$

15. $23.04y^2 - 46.08y - 1.96x^2 - 3.92x = 24.0784$

16. $25(y+5)^2 - 20(x-1)^2 = 500$

17. $16x^2 - y^2 + 96x + 8y + 112 = 0$

18. $y^2 - 4x^2 - 2y - 16x + 1 = 0$

19. $(y-1)^2 - 4(x-2)^2 = 168$

20. $3x^2 - 12y^2 + 45x + 60y = -60$

Lesson 7-6 Write each equation in standard form. State whether the graph of the equation is a *parabola*, a *circle*, an *ellipse*, or a *hyperbola*. Then graph the equation.

1. $9x^2 - 36x + 36 = 4y^2 + 24y + 72$

2. $x^2 + 4x + 2y^2 + 16y + 32 = 0$

3. $x^2 + 6x + y^2 - 6y + 9 = 0$

4. $9y^2 = 25x^2 + 400x + 1825$

5. $2y^2 + 12y - x + 6 = 0$

6. $x^2 + y^2 = 10x + 2y + 23$

7. $3x^2 + y = 12x - 17$

8. $9x^2 - 18x + 16y^2 + 160y = -265$

9. $x^2 + 10x + 5 = 4y^2 + 16$

10. $\dfrac{(y-5)^2}{4} - (x+1)^2 = 4$

11. $9x^2 + 49y^2 = 441$

12. $4x^2 - y^2 = 4$

13. $x^2 + 4x + y^2 - 8y = 2$

14. $(x+3)^2 = 8(y+2)$

15. $9x^2 + 9y^2 = 9$

16. $y - x^2 = x + 3$

17. $2x^2 - 13y^2 + 5 = 0$

18. $16(x-3)^2 + 81(y+4)^2 = 1296$

19. $4x^2 - y^2 = 16$

20. $x^2 + 5y^2 = 16$

Lesson 7-7 Solve each system of equations algebraically. Check your solutions with a graphing calculator.

1. $4y = x^2 - 4$ $\quad$ _4y·x²-4 -4y_
$x^2 + y^2 = 9$ $\quad$ _0=x²-4x-4_

2. $x = y^2$
$(x + 3)^2 + y^2 = 53$

3. $\frac{x^2}{3} - \frac{(y + 2)^2}{4} = 1$
$x^2 = y^2 + 11$

4. $\frac{(x - 1)^2}{5} + \frac{y^2}{2} = 1$ $\quad$ _(x+2)(x+2)_
$y = x + 1$

5. $x^2 + y^2 = 13$
$x^2 - y^2 = -5$

6. $\frac{x^2}{25} - \frac{y^2}{5} = 1$
$y = x^2 - 4$

7. $x^2 + y = 0$
$x + y = -2$

8. $x^2 - 9y^2 = 36$
$x = y$

9. $5x^2 + y^2 = 30$
$y^2 - 16 = 9x^2$

Solve each system of inequalities by graphing.

10. $\frac{x^2}{16} - \frac{y^2}{1} \geq 1$
$x^2 + y^2 \geq 49$

11. $\frac{x^2}{25} - \frac{y^2}{16} \geq 1$
$y \leq x - 2$

12. $y = x + 3$
$x^2 + y^2 < 25$

13. $4x^2 + (y - 3)^2 \leq 16$
$x - 2y = -1$

Lesson 8-1 Find $p(5)$ and $p(-1)$ for each function.

1. $p(x) = 7x - 3$

2. $p(x) = -3x^2 + 5x - 4$

3. $p(x) = 5x^4 + 2x^2 - 2x$

4. $p(x) = -13x^3 + 5x^2 - 3x + 2$

5. $p(x) = x^6 - 2$

6. $p(x) = \frac{2}{3}x^2 + 5x$

Find $g(a - 2)$ for each function.

7. $g(x) = 7x - 3$

8. $g(x) = 3x + 5$

9. $g(x) = -(x + 2)^2 + 8$

10. $g(x) = -2x^3 + 5x$

Find $-3[f(x)]$ for each function.

11. $f(x) = 4x^2 + 3x - 7$

12. $f(x) = \frac{x}{6} - \frac{2x^3}{9} + 1$

13. $f(x) = 5(x^2 + 2x)$

14. $f(x) = 8 - x$

Lesson 8-2 Divide using synthetic division and write your answer in the form *dividend = quotient · divisor + remainder*. Is the binomial a factor of the polynomial?

1. $(x^3 - x^2 + x + 6) \div (x + 2)$

2. $(5x^3 - 17x^2 + 6x + 2) \div (x - 3)$

3. $(2x^3 - 4x^2 + 3x - 6) \div (x - 4)$

4. $(x^3 - 8) \div (x - 2)$

5. $(x^2 + 6x - 3) \div (x + 1)$

6. $(x^4 + x^3 + x^2 + x + 1) \div (x + 1)$

Use synthetic substitution to find $g(3)$ and $g(-2)$ for each function.

7. $g(x) = 3x^5 - 5x^3 + 2x - 8$

8. $g(x) = -2x^4 + 7x^3 + 8x^2 - 3x + 5$

9. $g(x) = 10x^3 + 2$

10. $g(x) = x^5 + x^4 + x^3 + x^2 + x + 1$

Given a polynomial and one of its factors, find the remaining factors of the polynomial. Some factors may not be binomials.

11. $(x^3 - 8x^2 + x + 42); (x - 7)$

12. $(6x^4 + 13x^3 - 36x^2 - 43x + 30); (x - 2)$

13. $(x^4 + 5x^3 - 27x - 135); (x - 3)$

14. $(2x^3 - 15x^2 - 2x + 120); (2x + 5)$

Lesson 8-3 Approximate the real zeros of each function to the nearest tenth.

1. $f(x) = x^3 - 3x^2 + 8x - 7$

2. $f(x) = 2x^5 + 3x^4 - 8x^2 + x + 4$

3. $f(x) = x^4 - 5x^3 + 6x^2 - x - 2$

4. $f(x) = 2x^6 + 5x^4 - 3x^2 - 5$

5. $f(x) = -x^3 - 8x^2 + 3x - 7$

6. $f(x) = -x^4 - 3x^3 + 5x$

7. $f(x) = x^5 - 7x^4 - 3x^3 + 2x^2 - 4x + 9$

8. $f(x) = x^4 - 5x^3 + x^2 - x - 3$

Graph each function.

9. $f(x) = \dfrac{x^3}{20} + \dfrac{x^2}{40} - 3x + \dfrac{1}{2}$

10. $f(x) = -\dfrac{x^4}{20} - \dfrac{x^3}{2} + \dfrac{3x^2}{4} + 11x + \dfrac{5}{4}$

11. $f(x) = x^4 - 128x^2 + 960$

12. $f(x) = -x^5 + x^4 - 208x^2 + 145x + 9$

13. $f(x) = x^5 - 452x^3 - 183x + 25$

14. $f(x) = 5x^3 - 27x^2 - 37x + 54$

15. $f(x) = 2x^4 - 7x^3 - 19x^2 + 22x + 78$

16. $f(x) = -x^3 + 8x^2 + 12x - 16$

Lesson 8-4 State the number of positive real zeros, negative real zeros, and imaginary zeros for each function.

1. $f(x) = 5x^8 - x^6 + 7x^4 - 8x^2 - 3$

2. $f(x) = 6x^5 - 7x^2 + 5$

3. $f(x) = -2x^6 - 5x^5 + 8x^2 - 3x + 1$

4. $f(x) = 4x^3 + x^2 - 38x + 56$

5. $f(x) = 3x^8 - 15x^5 - 7x^4 - 8x^3 - 3$

6. $f(x) = -x^6 - 8x^5 - 5x^4 - 11x^3 - 2x^2 - 5x - 1$

7. $f(x) = 3x^4 - 5x^3 + 2x^2 - 7x + 5$

8. $f(x) = x^5 - x^4 + 7x^3 - 25x^2 + 8x - 13$

Given a function and one of its zeros, find all of the zeros of the function.

9. $f(x) = x^3 - 7x^2 + 16x - 10;\ 3 - i$

10. $f(x) = x^3 - 4x^2 + 6x - 4;\ 2$

11. $f(x) = x^3 - 16x^2 + 79x - 114;\ 5 - \sqrt{6}$

12. $f(x) = -3x^3 + 6x^2 + 5x - 8;\ 1$

13. $f(x) = 45x^4 - 222x^3 + 209x^2 - 138x - 104;\ \dfrac{2}{3} + i$

14. $f(x) = 4x^4 + 36x^3 + 57x^2 + 225x + 200;\ -\dfrac{5}{2}i$

15. $f(x) = -x^4 + 6x^3 - 19x^2 + 42x - 10;\ 1 + 3i$

Lesson 8-5 List all possible rational zeros for each function.

1. $f(x) = 3x^5 - 7x^3 - 8x + 6$

2. $f(x) = 4x^3 + 2x^2 - 5x + 8$

3. $f(x) = 6x^9 - 7$

4. $f(x) = 12x^4 + 3x^2 - 8x - 100$

Find all the rational zeros for each function.

5. $f(x) = x^4 + 3x^3 - 7x^2 - 27x - 18$

6. $f(x) = 6x^4 - 31x^3 - 119x^2 + 214x + 560$

7. $f(x) = 20x^4 - 16x^3 + 11x^2 - 12x - 3$

8. $f(x) = 2x^4 - 30x^3 + 117x^2 - 75x + 280$

9. $f(x) = 3x^5 - 17x^4 + 33x^3 - 19x^2 - 31x + 21$

10. $f(x) = 2x^6 - 12x^5 + 17x^4 + 6x^3 - 10x^2 + 6x - 9$

11. $f(x) = 48x^5 + 16x^4 - 24x^3 - 8x^2 + 3x + 1$

12. $f(x) = x^5 - x^4 + x^3 + 3x^2 - x$

Find all the zeros of each function.

13. $f(x) = 90x^4 - 99x^3 - 64x^2 + 36x + 16$

14. $f(x) = 2x^4 - 13x^3 - 34x^2 + 65x + 120$

15. $f(x) = 6x^4 + 5x^3 - 8x^2 - 45x - 14$

16. $f(x) = 4x^4 + 19x^2 - 63$

Lesson 8-6 Write each equation in quadratic form if possible. If not, explain why not.

1. $5x^{10} - 6x^5 = 3$

2. $3r + 2\sqrt{r} - 7 = 0$

3. $z^9 = 8z^3$

4. $2y^6 + 3y^4 = 10$

5. $x - 10x^{\frac{1}{2}} + 25 = 0$

6. $x^{\frac{4}{3}} - 7x^{\frac{2}{3}} + 12 = 0$

7. $y^{\frac{1}{2}} - 10y^{\frac{1}{4}} + 16 = 0$

8. $r^{\frac{2}{3}} - 5r^{\frac{1}{3}} + 6 = 0$

9. $x^{\frac{1}{2}} + 7x^{\frac{1}{4}} + 12 = 0$

Solve each equation.

10. $8x^3 + 27 = 0$

11. $5\sqrt[3]{z^2} - \sqrt[3]{z} = 4$

12. $2m^4 = 3m^2 + 5$

13. $3b^5 = 7b^3$

14. $x^3 + 10x^2 + 16x = 0$

15. $y^4 - 3y^2 + 2 = 0$

16. $a^3 = 125$

17. $m - 9\sqrt{m} + 8 = 0$

18. $r^{\frac{2}{3}} - 12r^{\frac{1}{3}} + 20 = 0$

19. $x^{\frac{2}{3}} - 8x^{\frac{1}{3}} + 15 = 0$

20. $m - 11m^{\frac{1}{2}} + 30 = 0$

21. $y^3 - 8y^{\frac{3}{2}} + 16 = 0$

22. $3g^{\frac{2}{3}} - 10g^{\frac{1}{3}} + 8 = 0$

23. $3m + m^{\frac{1}{2}} - 2 = 0$

24. $x^{\frac{1}{2}} - 6x^{\frac{1}{4}} + 8 = 0$

Lesson 8-7 Find $[f \circ g](-4)$ and $[g \circ f](-4)$.

1. $f(x) = 3x + 5$
   $g(x) = x - 3$

2. $f(x) = \sqrt{x}$
   $g(x) = x^2$

3. $f(x) = 2x^2 - 5x + 8$
   $g(x) = \frac{x - 8}{3}$

4. $f(x) = \{(-4, 2), (0, 5), (3, 7)\}$
   $g(x) = \{(-4, 0), (-3, 8), (2, 3)\}$

5. $f(x) = \{(3, 2), (1, -7), (-4, 0)\}$
   $g(x) = \{(-4, 1), (0, 11)\}$

6. $f(x) = x^2 + 1$
   $g(x) = x + 1$

Find $g[h(x)]$ and $h[g(x)]$.

7. $g(x) = 8 - 2x$
   $h(x) = 3x$

8. $g(x) = x^2 - 7$
   $h(x) = 3x + 2$

9. $g(x) = 2x + 7$
   $h(x) = \frac{x - 7}{2}$

10. $g(x) = |2x + 3|$
   $h(x) = 5 - 3x$

If $f(x) = x^2$, $g(x) = 3x$, and $h(x) = x - 1$, find each value.

11. $g[f(1)]$

12. $[f \circ h](3)$

13. $[h \circ f](3)$

14. $[g \circ f](-2)$

15. $g[h(-20]$

16. $f[h(-3)]$

17. $g[f(x)]$

18. $[f \circ (g \circ h)](x)$

Lesson 8-8 Find the inverse of each relation and determine whether the inverse is a function.

1. $\{(-2, 7), (3, 0), (5, -8)\}$

2. $\{(-3, 9), (-2, 4), (3, 9), (-1, 1)\}$ **3.** $\{(1, 5), (2, 3), (4, 3), (-1, 5)\}$

Find the inverse of each function. Then graph the function and its inverse.

4. $f(x) = x - 5$

5. $y = 2x + 8$

6. $g(x) = 2x^2 - 3$

7. $h(x) = \frac{x}{5} + 1$

8. $y = -2$

9. $g(x) = 5 - 2x$

10. $y = -x^2 + 2$

11. $h(x) = -\frac{2}{3}x$

Determine whether each pair of functions are inverse functions. Write *yes* or *no*.

12. $f(x) = \frac{2x - 3}{5}$
   $g(x) = \frac{3x - 5}{3}$

13. $f(x) = 5x - 6$
   $g(x) = \frac{x + 6}{5}$

14. $f(x) = 6 - 3x$
   $g(x) = 2 - \frac{1}{3}x$

15. $f(x) = 3x - 7$
   $g(x) = \frac{1}{3}x + 7$

Lesson 9-1 State the equations of the vertical and horizontal asymptotes for each rational function.

1. $f(x) = \dfrac{1}{x + 4}$

2. $f(x) = \dfrac{x - 2}{x + 3}$

3. $f(x) = \dfrac{5}{(x + 1)(x - 8)}$

4. $f(x) = \dfrac{x}{x + 2}$

5. $f(x) = \dfrac{x^2 - 4}{x + 2}$

6. $f(x) = \dfrac{-4}{x}$

Graph each rational function.

7. $f(x) = \dfrac{1}{x - 5}$

8. $f(x) = \dfrac{3x}{x + 1}$

9. $f(x) = \dfrac{x^2 - 9}{x - 3}$

10. $f(x) = \dfrac{x}{x - 6}$

11. $f(x) = \dfrac{1}{(x - 3)^2}$

12. $f(x) = \dfrac{2}{(x + 3)(x - 4)}$

13. $f(x) = \dfrac{x + 4}{x^2 - 1}$

14. $f(x) = \dfrac{x + 2}{x + 3}$

15. $f(x) = \dfrac{4}{x}$

Lesson 9-2 State whether each equation represents a direct, inverse, or joint variation. Then name the constant of variation.

1. $xy = 10$

2. $x = 6y$

3. $\dfrac{x}{7} = y$

4. $\dfrac{x}{y} = -6$

5. $10x = y$

6. $x = \dfrac{2}{y}$

7. $A = \ell w$

8. $\dfrac{1}{4}b = -\dfrac{3}{5}c$

9. $D = rt$

Write an equation for each statement. Then solve the equation.

10. If y varies directly as x and $y = 16$ when $x = 4$, find y when $x = 12$.

11. If x varies inversely as y and $x = 12$ when $y = -3$, find x when $y = -18$.

12. If m varies directly as w and $m = -15$ when $w = 2.5$, find m when $w = 12.5$.

13. If y varies jointly as x and z and $y = 10$ when $z = 4$ and $x = 5$, find y when $x = 4$ and $z = 2$.

14. If y varies inversely as x and $y = \dfrac{1}{4}$ when $x = 24$, find y when $x = \dfrac{3}{4}$.

15. If y varies jointly as x and z and $y = 45$ when $x = 9$ and $z = 15$, find y when $x = 25$ and $z = 12$.

Lesson 9-3 Simplify each expression.

1. $\dfrac{3x^3}{-2} \cdot \dfrac{-4}{9x}$

2. $\dfrac{21x^2}{-5} \cdot \dfrac{10}{7x^3}$

3. $\dfrac{2u^2}{3} \div \dfrac{6u^3}{5}$

4. $\dfrac{15x^3}{14} \div \dfrac{18x}{7}$

5. $\dfrac{xy^2}{2} \cdot \dfrac{x^2}{2y} \cdot \dfrac{2}{x^2y}$

6. $axy \div \dfrac{ax}{y} \div \dfrac{ay}{x}$

7. $\dfrac{9u^2}{28v} \div \dfrac{27u^2}{8v^2} \div \dfrac{4u^2}{21}$

8. $\dfrac{x^2 - 4}{4x^2 - 1} \cdot \dfrac{2x - 1}{x + 2}$

9. $\dfrac{x^2 - 1}{2x^2 - x - 1} \div \dfrac{x^2 - 4}{2x^2 - 3x - 2}$

10. $\dfrac{2x^2 + x - 1}{2x^2 + 3x - 2} \div \dfrac{x^2 - 2x + 1}{x^2 + x - 2}$

11. $\dfrac{\dfrac{x^4 - y^4}{x^3 + y^3}}{\dfrac{x^3 - y^3}{x + y}}$

Lesson 9-4 Simplify each expression.

1. $\dfrac{12}{7d} - \dfrac{3}{14d}$

2. $\dfrac{x+1}{x} - \dfrac{x-1}{x^2}$

3. $\dfrac{2x+1}{4x^2} - \dfrac{x+3}{6x}$

4. $\dfrac{5}{x} - \dfrac{3}{x+5}$

5. $\dfrac{x}{x-1} + \dfrac{1}{1-x}$

6. $\dfrac{1}{3v^2} + \dfrac{1}{uv} + \dfrac{3}{4u^2}$

7. $\dfrac{1}{x^2-x} + \dfrac{1}{x^2+x}$

8. $\dfrac{1}{x^2-1} - \dfrac{1}{(x-1)^2}$

9. $\dfrac{7x}{13y^2} + \dfrac{4y}{6x^2}$

10. $y - 1 + \dfrac{1}{y-1}$

11. $3m + 1 - \dfrac{2m}{3m+1}$

12. $\dfrac{3x}{x-y} + \dfrac{4x}{y-x}$

13. $\dfrac{6}{4m^2-12mn+9n^2} + \dfrac{2}{2mn-3n^2}$

14. $\dfrac{3}{x^2+5ax+6a^2} + \dfrac{2}{x^2-4a^2}$

15. $\dfrac{4}{a^2-4} - \dfrac{3}{a^2+4a+4}$

16. $\dfrac{4}{3-3z^2} - \dfrac{2}{z^2+5z+4}$

17. $\dfrac{\frac{1}{x+y}}{\frac{1}{x}+\frac{1}{y}}$

18. $\dfrac{1-\frac{1}{x+1}}{1+\frac{1}{x-1}}$

Lesson 9-5 Solve each equation. Check your solutions.

1. $\dfrac{5}{x} + \dfrac{3}{5} = \dfrac{2}{x}$

2. $\dfrac{1}{2+3x} + \dfrac{2}{2-3x} = 0$

3. $\dfrac{x-2}{x} = \dfrac{x-4}{x-6}$

4. $\dfrac{1}{x} + \dfrac{x}{x+2} = 1$

5. $\dfrac{1}{x+1} + \dfrac{1}{x-1} = \dfrac{2}{x^2-1}$

6. $\dfrac{2}{x} + \dfrac{1}{x-2} = 1$

7. $\dfrac{1}{x-3} + \dfrac{1}{x+5} = \dfrac{x+1}{x-3}$

8. $\dfrac{4}{x^2-2x-3} = \dfrac{-x}{3-x} - \dfrac{1}{x+1}$

9. $\dfrac{x}{x+1} + \dfrac{3}{x-3} + 1 = 0$

10. $\dfrac{3x}{x^2+2x-8} = \dfrac{1}{x-2} + \dfrac{x}{x+4}$

11. $\dfrac{5x+2}{x^2-4} = \dfrac{-5x}{2-x} + \dfrac{2}{x+2}$

12. $\dfrac{1}{x-3} + \dfrac{2}{x^2-9} = \dfrac{5}{x+3}$

13. $\dfrac{1}{x^2-1} = \dfrac{2}{x^2+x-2}$

14. $\dfrac{12}{x^2-16} - \dfrac{24}{x-4} = 3$

15. $\dfrac{4}{x-2} - \dfrac{x+6}{x+1} = 1$

Lesson 10-1 Simplify each expression.

1. $\left(4^{\sqrt{2}}\right)\left(4^{\sqrt{8}}\right)$

2. $\left(5^{\sqrt{5}}\right)^{\sqrt{45}}$

3. $\left(x^{\sqrt{3}} + x^{\sqrt{5}}\right)^2$

4. $\left(w^{\sqrt{6}}\right)^{\sqrt{3}}$

5. $27^{\sqrt{5}} \div 3^{\sqrt{5}}$

6. $8^{2\sqrt{3}} \times 4^{\sqrt{3}}$

7. $5\left(2^{\sqrt{7}}\right)\left(4^{-\sqrt{7}}\right)$

8. $\left(y^{\sqrt{x}}\right)^{\sqrt{x}}$

9. $5^{\sqrt{2}} \times 5^{\sqrt{3}}$

10. $\left(6^{\sqrt{5}}\right)^{\sqrt{2}}$

11. $7^{\sqrt{3}} \times 7^{2\sqrt{3}}$

12. $\left(y^{\sqrt{3}}\right)^{\sqrt{27}}$

Solve each equation.

13. $3^x = \dfrac{1}{27}$

14. $5^x = \sqrt{125}$

15. $8^{2+x} = 2$

16. $27^{2x-1} = 3$

17. $4^{2x+5} = 16^{x+1}$

18. $49^{x-2} = 7\sqrt{7}$

19. $6^{x+1} = 36^{x-1}$

20. $10^{x-1} = 100^{4-x}$

21. $\left(\dfrac{1}{5}\right)^{x-3} = 125$

22. $2^{x^2+1} = 32$

23. $36^x = 6^{x^2-3}$

24. $9^{x^2-2x} = 27^{x^2+1}$

Graph each equation.

25. $y = 3.2^x$

26. $y = -(0.5)(3.2)^x$

27. $y = 0.3(3.2)^x$

28. $y = (-10)(3.2)^x$

Lesson 10-2 Evaluate each expression.

1. $\log_4 16$ **2.** $\log_5 125$ **3.** $\log_3 \frac{1}{9}$ **4.** $\log_2 \frac{1}{8}$

5. $\log_6 6\sqrt{6}$ **6.** $\log_8 4$ **7.** $\log_7 \sqrt{49}$ **8.** $\log_{\frac{1}{2}} 8$

Solve each equation for x.

9. $\log_8 x = 2$ **10.** $\log_5 x = 3$ **11.** $\log_{10} x = -\frac{1}{2}$ **12.** $\log_{\frac{1}{9}} x = -\frac{1}{2}$

13. $\log_x 5 = -\frac{1}{2}$ **14.** $\log_x 7 = 1$ **15.** $\log_x 2 = 0$ **16.** $\log_5 (\log_3 x) = 0$

17. $\log_4 (\log_3(\log_2 x)) = 0$ **18.** $\log_2(x^2 - 9) = 4$ **19.** $\log_b (x^2 + 7) = \frac{2}{3} \log_b 64$

20. $\log_{10} (x - 1)^2 = \log_{10} 0.01$ **21.** $\log_{12} (7x - 3) = \log_{12} (5 - x^2)$ **22.** $\log_9 (x^2 + 9x) = \log_9 10$

Lesson 10-3 Use $\log_3 5 = 1.465$ and $\log_3 7 = 1.771$ to evaluate each expression.

1. $\log_3 \frac{7}{5}$ **2.** $\log_3 245$ **3.** $\log_3 35$

Solve each equation.

4. $\log_2 x + \log_2 (x - 2) = \log_2 3$ **5.** $\log_3 x = 2 \log_3 3 + \log_3 5$

6. $\log_5 (x^2 + 7) = \frac{2}{3} \log_5 64$ **7.** $\log_7 (3x + 5) - \log_7 (x - 5) = \log_7 8$

8. $\log_2 (x^2 - 9) = 4$ **9.** $\log_3 (x + 2) + \log_3 6 = 3$

10. $\log_6 x + \log_6 (x - 5) = 2$ **11.** $\log_5 (x + 3) = \log_5 8 - \log_5 2$

12. $2 \log_3 x - \log_3 (x - 2) = 2$ **13.** $\log_6 x = \frac{3}{2} \log_6 9 + \log_6 2$

14. $\log_{10} (x + 6) + \log_{10} (x - 6) = 2$ **15.** $\frac{1}{2} \log_4 (x + 2) + \frac{1}{2} \log_4 (x - 2) = \frac{2}{3} \log_4 27$

16. $\log_3 14 + \log_3 x = \log_3 42$ **17.** $\log_{10} x = \frac{1}{2} \log_{10} 81$

Lesson 10-4 Use a scientific calculator to find the logarithm for each number rounded to four decimal places. Then state the mantissa and characteristic.

1. 55.3 **2.** 0.067 **3.** 334.8 **4.** 0.0008

5. 3356.02 **6.** 0.365 **7.** 99.64 **8.** 5.889

Use a scientific calculator to find the antilogarithm for each number rounded to four decimal places.

9. -2.6659 **10.** 0.0865 **11.** -1.6235 **12.** 0.2554

13. 0.6987 **14.** 0.0045 **15.** -3.0025 **16.** 0.0006

17. 0.2586 **18.** 2.2249 **19.** 1.0024 **20.** -0.2586

Lesson 10-5 Use a scientific calculator to find each value rounded to four decimal places.

1. $\ln 8.25$

2. $\ln 43.5$

3. $\ln 2.6243$

4. $\ln 0.04$

5. antiln -0.1125

6. antiln 1.006

7. antiln -2.445

8. antiln -1

9. $\ln 43.988$

10. antiln -0.0115

11. $\ln 0.0005$

12. antiln 1.25

Lesson 10-6 Approximate the value of each logarithm to three decimal places.

1. $\log_3 21$

2. $\log_4 62$

3. $\log_5 28$

4. $\log_2 25$

5. $\log_{12} 30$

6. $\log_4 63$

7. $\log_7 35$

8. $\log_6 100$

Use logarithms to solve each equation. Approximate the value of each solution to three decimal places.

9. $5^x \cdot 5^{-7x} = 5^{-18}$

10. $3^b = 19$

11. $6^x = 12$

12. $9^{2x+1} = 62.4$

13. $7^{3x-2} = 0.834$

14. $4 = 9^{2x-3}$

15. $6 = 15^{1-x}$

16. $1.76^x = 23.4$

17. $8 = 3 \times 5^x$

18. $3^x = 4 \times 2^x$

19. $2 \times 5^{x+1} = 5 \times 2^{x+2}$

20. $2^{6x-3} = 4 \times 2^{4x}$

Lesson 10-7

1. The number of bacteria B in a culture increases according to the equation $B = B_0 e^{kt}$. The culture starts with 400 bacteria and has 700 bacteria after 3 hours. How many bacteria will there be after 12 hours?

2. Humberto deposited \$785 in an account that pays $8\frac{3}{4}\%$ interest, compounded continuously. How long will it take until Humberto has \$1000?

3. A satellite has a radioactive-isotope power supply. The power output in watts is given by the equation $P = 50e^{(t/250)}$ where P is the power in watts and t is the time in days.

    a. How much power will be available at the end of one year?

    b. What is the half-life of the power supply?

    c. The equipment aboard the satellite requires 10 watts of power to operate properly. What is the operational life of the satellite?

4. A hard-boiled egg has a temperature of 98 degrees Celsius. If it is put into a sink that maintains a temperature of 18 degrees Celsius, its temperature x minutes later is given by $T(x) = 18 + 80 e^{-0.28x}$. Hilda needs her egg to be exactly 30 degrees Celsius for decorating. How long should she leave it in the water?

5. The number of people in the Denver suburban area has grown exponentially since 1950 according to the equation $P = P_0 e^{kt}$. In 1950 there were 350,000 people, and in 1960, there were 705,300 people. How many people lived in the Denver suburban area in 1980?

Lesson 11-1 Find the next four terms of each arithmetic sequence.

1. 9, 7, 5, ...   **2.** 3, 4.5, 6, ...   **3.** 40, 35, 30, ...   **4.** 2, 5, 8, ...

Find the *n*th term of each arithmetic sequence described.

5. $a_1 = 4, d = 5, n = 10$ **6.** $a_1 = -30, d = -6, n = 5$ **7.** $a_1 = \frac{3}{4}, d = -\frac{1}{4}, n = 72$

8. $a_1 = -3, d = 32, n = 8$ **9.** $a_1 = -\frac{1}{5}, d = \frac{3}{5}, n = 17$ **10.** $a_1 = 20, d = -3, n = 16$

Write an equation for the *n*th term of each arithmetic sequence.

11. 3, 5, 7, 9, ...   **12.** 2, −1, −4, −7, ...   **13.** 20, 28, 36, 44, ...

Lesson 11-2 Find S_n for each arithmetic series described.

1. $a_1 = 3, a_n = 20, n = 6$ **2.** $a_1 = 15, a_n = -12, n = 30$ **3.** $a_1 = 90, a_n = -4, n = 10$

4. $a_1 = 16, n = 12, a_n = 14$ **5.** $a_1 = -80, n = 18, a_n = 120$ **6.** $a_1 = -3, n = 14, a_n = -72$

Find the sum of each arithmetic series.

7. 3 + 12 + 21 + 30 + ... + 57   **8.** 1 + 4 + 7 + 10 + ... + 31   **9.** 8 + 16 + 24 + ... + 80

Write the terms of each arithmetic series and find the sum.

10. $\displaystyle\sum_{n=1}^{6} n + 2$ **11.** $\displaystyle\sum_{n=5}^{10} 2n - 5$ **12.** $\displaystyle\sum_{k=1}^{5} 40 - 2k$

13. $\displaystyle\sum_{k=8}^{12} 6 - 3k$ **14.** $\displaystyle\sum_{n=6}^{10} 2 + 3n$ **15.** $\displaystyle\sum_{n=1}^{4} 10n + 2$

Lesson 11-3 Find the next four terms of each geometric sequence.

1. 5, 15, 45, ...   **2.** 2, 10, 50, ...   **3.** 64, 16, 4, ...   **4.** −9, 27, −81, ...

Find the *n*th term of each geometric sequence described.

5. $a_1 = 5, r = 7, n = 6$ **6.** $a_1 = 200, r = -\frac{1}{2}, n = 10$ **7.** $a_1 = 60, r = -2, n = 4$

8. $a_1 = 300, r = \frac{1}{4}, n = 6$ **9.** $a_1 = 8, r = -2, n = 8$ **10.** $a_1 = 1, r = -1, n = 30$

Find the value(s) of *y* that make each sequence geometric.

11. 3, −6, 12, 2y − 12, ...   **12.** 2, −2, 2, 4y + 2, ...   **13.** −10, 50, −250, 20y − 100, ...

Lesson 11-4   Find S_n for each geometric series described.

1. $a_1 = \frac{1}{81}, r = 3, n = 6$

2. $a_1 = 1, r = -2, n = 7$

3. $a_1 = 5, r = 4, n = 5$

4. $a_1 = -27, r = -\frac{1}{3}, n = 6$

5. $a_1 = 1000, r = \frac{1}{2}, n = 7$

6. $a_1 = 125, r = -\frac{2}{5}, n = 5$

Find the sum of each geometric series.

7. $a_1 = 10, r = 3, n = 6$

8. $r = -\frac{1}{5}, n = 5, a_1 = 1250$

9. $r = \frac{1}{3}, a_n = 5, a_1 = 1215$

10. $a_1 = 16, r = \frac{3}{2}, n = 5$

11. $a_1 = 7, r = 2, n = 7$

12. $r = -\frac{1}{2}, a_n = -\frac{3}{2}, n = 6$

13. $a_1 = 16, r = -\frac{1}{2}, n = 10$

14. $a_1 = 243, r = -\frac{2}{3}, n = 5$

15. $a_1 = 5, r = 3, n = 12$

Write the terms of each geometric series and find the sum.

16. $\sum\limits_{k=1}^{5} 2^k$

17. $\sum\limits_{n=0}^{3} 3^{-n}$

18. $\sum\limits_{n=0}^{3} 2(5^n)$

19. $\sum\limits_{k=2}^{5} -(-3)^{k-1}$

20. $\sum\limits_{n=1}^{6} \left(\frac{1}{2}\right)^{n-1}$

21. $\sum\limits_{n=0}^{4} 8\left(\frac{-1}{2}\right)^{n-1}$

Lesson 11-5   Find the sum of each infinite geometric series, if it exists.

1. $54 + 18 + 6 + \ldots$

2. $2 - 2 + 2 - 2 + \ldots$

3. $1000 - 200 + 40 - \ldots$

4. $7 + 3 + \frac{9}{7} + \ldots$

5. $\frac{4}{5} + \frac{2}{25} + \frac{1}{125} + \ldots$

6. $49 + 14 + 4 \ldots$

7. $\frac{3}{4} + \frac{1}{2} + \frac{1}{3} + \ldots$

8. $1 - \frac{1}{4} + \frac{1}{16} - \ldots$

9. $12 - 4 + \frac{4}{3} - \ldots$

10. $3 - 9 + 27 - \ldots$

11. $3 - 2 + \frac{4}{3} - \ldots$

12. $10 - 1 + 0.1 - \ldots$

13. $\sum\limits_{n=1}^{\infty} 3\left(\frac{1}{4}\right)^{n-1}$

14. $\sum\limits_{n=1}^{\infty} 5\left(-\frac{1}{10}\right)^{n-1}$

15. $\sum\limits_{n=1}^{\infty} -\frac{2}{3}\left(-\frac{3}{4}\right)^{n-1}$

Lesson 11-6   Find the first six terms of each sequence.

1. $a_1 = 4, a_{n+1} = 2a_n + 1$

2. $a_1 = 6, a_{n+1} = a_n + 7$

3. $a_1 = 16, a_{n+1} = a_n + (n + 4)$

4. $a_1 = 1, a_{n+1} = \frac{n}{n + 2} \cdot a_n$

Find the first three iterates of each function, using the given initial values.

5. $f(x) = 3x - 1, x_0 = 3$

6. $f(x) = 2x^2 - 8, x_0 = -1$

7. $f(x) = 4x + 5, x_0 = 3$

8. $f(x) = 3x^2 + 1, x_0 = 1$

9. $f(x) = x^2 + 4x + 4, x_0 = 1$

10. $f(x) = x^2 + 9, x_0 = 2$

Lesson 11-7 Draw the next stage of a fractal formed by replacing each segment with the pattern shown.

1.
2.

3. Describe why lightning is an example of fractals found in nature.

Lesson 11-8 Use a calculator to evaluate each expression.

1. $6!$

2. $4!$

3. $\dfrac{13!}{6!}$

4. $\dfrac{12!}{5!}$

5. $\dfrac{10!}{3!\,7!}$

6. $\dfrac{14!}{4!\,10!}$

7. $\dfrac{7!}{2!\,5!}$

8. $\dfrac{9!}{8!}$

Expand each binomial.

9. $(x + y)^3$

10. $(2x - y)^4$

11. $(3r + 4s)^5$

Find the indicated term of each expression.

12. sixth term of $(x + 3)^8$

13. fourth term of $(x - y)^9$

14. fifth term of $(3x + 5y)^{10}$

15. sixth term of $(x + 4y)^7$

Lesson 12-1 Draw a tree diagram to illustrate all of the possibilities.

1. tossing two pennies and rolling two number cubes

2. choosing a denim jacket that comes in dark blue, stone washed, or black that has buttons or snaps

3. ordering a large pizza with thin or thick crust and one topping of either pepperoni, sausage, or vegetables

State whether the events are *independent* or *dependent*.

4. A comedy video and an action video are selected from the video store.

5. The numbers 1–10 are written on pieces of paper and are placed in a hat. Three of them are selected one after the other without replacing any of the pieces of paper.

Solve each problem.

6. On a bookshelf there are 10 different algebra books, 6 different geometry books, and 4 different calculus books. In how many ways can you choose 3 books, one of each kind?

7. In how many different ways can a 10-question true-false test be answered if each question must be answered?

8. How many ways are there of selecting 3 cards, one after the other, from a deck of 26 cards if selected cards are not replaced?

9. A student council has 6 seniors, 5 juniors, and 1 sophomore as members. In how many ways can a 3-member council committee be formed that includes one member of each class?

10. How many license plates of 5 symbols can be made using a letter for the first symbol and digits for the remaining 4 symbols?

Lesson 12-2 How many ways can the letters of each word be arranged?

1. MONDAY
2. EIGHT
3. CINCINNATI
4. INDIANA

Evaluate each expression.

5. $\dfrac{P(5,3)}{P(3,2)}$

6. $\dfrac{P(5,2)\,P(4,3)}{P(6,3)}$

7. $\dfrac{P(10,6)}{P(12,2)\,P(7,2)}$

Lesson 12-3 Determine whether each situation involves a permutation or a combination.

1. choosing a team of 9 players from a group of 20
2. lining up in a cafeteria line
3. choosing 4 books from a list of 12 for a summer reading program
4. arranging the order of songs on a compact disc

Evaluate each expression.

5. $C(8,6)$
6. $C(20,17)$
7. $C(9,4)\cdot C(5,3)$
8. $C(6,1)\cdot C(4,1)\cdot C(9,8)$
9. $C(10,5)\cdot C(8,4)$
10. $C(7,6)\cdot C(3,1)$

Solve each problem.

11. A standard bridge deck of cards consists of 4 suits (diamonds, clubs, hearts, and spades) of 13 cards each. How many 5-card hands can be dealt that includes 4 cards from the same suit and one card from a different suit?

12. At Hamburger Heaven, you can order hamburgers with cheese, onion, pickle, relish, mustard, lettuce, or tomato. How many different combinations of the "extras" can you order, choosing any three of them?

13. Students are required to answer any eight out of ten questions on a certain test. How many different combinations of eight questions can a student choose to answer?

Lesson 12-4 State the probability of an event occurring, given the odds of the event.

1. 5:9
2. 4:8
3. 3:10
4. 2:7
5. 6:13
6. 1:19

A jar contains 3 red, 4 green, and 5 orange marbles. Three marbles are drawn at random. Find the probability of each event.

7. P(all green)
8. P(1 red, 2 not red)
9. P(2 orange, 1 not orange)

A bridge hand consists of 13 cards drawn at random from a 52-card bridge deck. Find the probability of each event.

10. P(4 kings)
11. P(no aces)
12. P(all red cards)

Lesson 12-5   Find each probability.

1. The student council consists of 7 girls and 5 boys, of whom 3 girls and 3 boys are seniors. A 4-person dance committee is chosen.
   a. P(2 girls and 2 boys)
   b. P(3 girls and 1 boy)
   c. P(2 senior boys and 2 senior girls)
   d. P(all seniors of whom at least 2 are girls)

2. The probability that Leon will ask Frank to be his tennis partner is $\frac{1}{4}$, that Paula will ask Frank is $\frac{1}{3}$, and that Ray will ask Frank is $\frac{3}{4}$.
   a. P(Paula will ask and Leon will ask.)
   b. P(Ray and Paula will ask, but Leon will not.)
   c. P(At least two out of the three will ask.)
   d. P(At least one out of the three will ask.)

3. According to the weather reports, the probability of rain on a certain day is 70% in Yellow Falls and 50% in Copper Creek.
   a. P(It will rain in Yellow Falls, but not in Copper Creek.)
   b. P(It will rain in both cities.)
   c. P(It will rain in neither city.)
   d. P(It will rain in at least one of the cities.)

Lesson 12-6   Find each probability.

1. Two letters are chosen at random from the word GEESE and two are chosen at random from the word PLEASE. What is the probability that all four letters are Es or none of the letters is an E?

2. What is the probability of getting 8 or more questions correct on a 10-question true-false test if the questions are answered at random?

3. Three dice are thrown.
   a. P(All three dice show the same number.)
   b. P(Exactly 2 of the dice show the same number.)

4. Two marbles are drawn at random from a bag containing 3 red, 5 blue, and 6 green marbles.
   a. P(at least 1 red marble)
   b. P(at least one green marble)
   c. P(2 marbles of the same color)
   d. P(two marbles of different colors)

Lesson 12-7   Find each probability.

1. A die is rolled and a coin is flipped. What is the probability that the number on the die is even and the coin shows heads?

2. Two marbles are to be drawn at random from a bag containing 8 red marbles and five green marbles. Use a tree diagram to determine the probability that one of the marbles will be red and one will be green.

3. Ten percent of a batch of toothpaste is defective. Five tubes of toothpaste are selected at random from this batch.
   a. P(None are defective.)
   b. P(Exactly one is defective.)
   c. P(At least three are defective.)
   d. P(Less than three are defective.)

4. On a true-false test, Gil guessed on every problem. The test had 80 questions.
   a. P(Each question is right.)
   b. P(Exactly 35 are right.)

Lesson 12-8 Determine whether each situation represents a random sampling. Write *yes* or *no* and explain.

1. finding the most often prescribed pain reliever by asking all of the doctors at your local hospital

2. taking a poll of the most popular baby girl names this year by studying birth announcements in newspapers from different cities across the country

3. polling people who frequent the neighborhood pizza parlor about their favorite restaurant in the city

Find the margin of sampling error in each problem. Explain what it indicates about the results.

4. A poll conducted on the favorite breakfast choice of students in your school showed that 72% of the 2500 students asked indicated oatmeal as their favorite breakfast.

5. Of the 420 people polled at the supermarket, 56% felt that they were easily swayed by the sample items in the aisles and purchased those items, even though they were not intending to when they arrived at the store.

6. Of the 3000 women polled between the ages of 25 and 35, only 45% felt that they consume 1000 to 1500 mg of calcium daily, which is the recommended daily allowance by the National Institute of Health.

Lesson 13-1 Suppose θ is an acute angle of a right triangle. For each function, find the values of the remaining five trigonometric functions of θ. Round your answers to four decimal places.

1. $\cos \theta = \frac{3}{5}$
2. $\tan \theta = \frac{8}{15}$
3. $\sec \theta = \frac{13}{5}$
4. $\csc \theta = \frac{25}{7}$
5. $\cot \theta = 1$

Find the value of x. Round your answers to the nearest degree.

6. $\tan x = 4.1436$
7. $\cos x = 0.3899$
8. $\sin x = 0.8045$
9. $\cos x = 0.0982$
10. $\sin x = 0.5950$
11. $\tan x = 0.1648$
12. $\tan x = 1.2763$
13. $\sin x = 0.2131$
14. $\cos x = 0.7878$
15. $\cos x = 0.4500$
16. $\sin x = 0.3657$
17. $\tan x = 2.8560$

Solve each right triangle. Assume that C represents the right angle and c is the hypotenuse. Round measures of sides and angles to the nearest tenth.

18. $c = 10, a = 8$
19. $a = 2, b = 7$
20. $a = 11, b = 21$
21. $B = 64°, c = 18.2$
22. $a = 33, B = 33°$
23. $c = 6, B = 13°$
24. $b = 42, A = 77°$
25. $A = 57°, c = 18$
26. $A = 35°, a = 7$
27. $B = 36°, c = 18$
28. $\cos A = \frac{3}{5}, c = 10$
29. $\tan B = \frac{1}{2}, b = 7$

Lesson 13-2 Change each degree measure to radian measure.

1. $60°$
2. $270°$
3. $315°$
4. $150°$
5. $-135°$
6. $-315°$
7. $45°$
8. $80°$
9. $24°$
10. $-54°$

Change each radian measure to degree measure.

11. $-\pi$
12. $\frac{9\pi}{4}$
13. $\frac{3\pi}{2}$
14. $-\frac{7\pi}{4}$
15. $\frac{7\pi}{12}$
16. $\frac{9\pi}{10}$
17. $-\frac{17\pi}{30}$
18. 1
19. $-2\frac{1}{3}$
20. $6\frac{1}{2}$

Find one positive angle and one negative angle that are coterminal with each angle.

21. $50°$
22. $-75°$
23. $125°$
24. $-400°$
25. $550°$
26. 3π
27. -2π
28. $\frac{2\pi}{3}$
29. $\frac{12\pi}{5}$
30. $-\frac{9\pi}{5}$

Lesson 13-3 State whether the value of each function is *positive, negative, zero,* or *undefined.*

1. $\sin 145°$
2. $\cos 200°$
3. $\tan 180°$
4. $\tan (-45°)$
5. $\cos 450°$
6. $\tan \left(-\frac{3\pi}{2}\right)$
7. $\sin \frac{5\pi}{3}$
8. $\cos \left(-\frac{3\pi}{4}\right)$
9. $\tan \frac{6\pi}{5}$
10. $\sin \left(-\frac{\pi}{2}\right)$

Find the exact values of $\sin \theta$, $\cos \theta$, and $\tan \theta$ if the terminal side of θ in standard position contains the given point.

11. $P(3, -4)$
12. $P(1, \sqrt{3})$
13. $P(0, -4)$
14. $P(-5, -5)$
15. $P(2.5, 0)$

Find the exact value of each trigonometric function.

16. $\cos 150°$
17. $\sin \left(-\frac{5\pi}{3}\right)$
18. $\tan \frac{7\pi}{6}$
19. $\tan (-300°)$
20. $\cos \frac{7\pi}{4}$

Suppose θ is an angle in standard position whose terminal side lies in the given quadrant. For each function, find the exact values of the remaining five trigonometric functions of θ.

21. $\cos \theta = -\frac{1}{3}$; Quandrant III
22. $\sec \theta = 2$; Quadrant IV
23. $\sin \theta = \frac{2}{3}$; Quardrant II
24. $\tan \theta = -4$; Quandrant IV
25. $\csc \theta = -5$; Quadrant III
26. $\cot \theta = -2$, Quadrant II
27. $\tan \theta = \frac{1}{3}$; Quadrant III
28. $\cos \theta = \frac{1}{4}$; Quadrant I
29. $\csc \theta = -\frac{5}{2}$; Quadrant IV

Lesson 13-4 Write an equation that can be used to find the area of each triangle. Then solve the equation. Round your answer to the nearest tenth.

1. $a = 11, b = 13, C = 31°$
2. $a = 15, b = 22, C = 90°$
3. $a = 12, b = 12, C = 50°$
4. $a = 6, c = 4, B = 52°$
5. $b = 10, c = 17, A = 46°$
6. $b = 4, c = 19, A = 73°$
7. $a = 11, c = 5, B = 55°$
8. $b = 8, c = 12, A = 75°$
9. $a = 12, b = 9, C = 35°$

Determine whether each triangle has no solution, one solution, or two solutions. Then solve each triangle. Round measures of sides and angles to the nearest tenth.

10. $A = 40°, B = 60°, c = 20$
11. $B = 70°, C = 58°, a = 84$
12. $A = 40°, a = 5, b = 12$
13. $A = 58°, a = 26,$ and $b = 29$
14. $A = 38°, B = 63°, c = 15$
15. $a = 6, b = 8, A = 150°$
16. $a = 12, b = 19, A = 57°$
17. $a = 125, b = 150, A = 25°$
18. $a = 64, c = 90, C = 98°$
19. $A = 40°, B = 60°, c = 20$
20. $a = 33, b = 50, A = 132°$
21. $a = 83, b = 79, A = 45°$

Lesson 13-5 Determine whether each triangle can be solved by beginning with the law of sines or law of cosines. Then solve each triangle.

1. $A = 51°, b = 40, c = 45$
2. $a = 10, c = 8, A = 40°$
3. $a = 14, c = 21, B = 60°$
4. $a = 14, b = 15, c = 16$
5. $B = 41°, C = 52°, c = 27$
6. $a = 19, b = 24.3, c = 21.8$
7. $A = 112°, a = 32, c = 20$
8. $b = 8, c = 7, A = 28°$
9. $a = 5, b = 6, c = 7$
10. $C = 35°, a = 11, b = 10.5$
11. $a = 8, A = 49°, B = 58°$
12. $A = 42°, b = 120, c = 160$
13. $c = 14, A = 40°, C = 70°$
14. $a = 10, b = 16, c = 19$
15. $a = 20, c = 24, B = 47°$
16. $a = 10, c = 8, B = 100°$
17. $A = 40°, B = 45°, c = 4$
18. $a = 32, c = 20, A = 112°$
19. $c = 8, b = 16, B = 71°$
20. $b = 100, c = 84, A = 20°$
21. $b = 40, c = 49, B = 53°$

Lesson 13-6 Determine the period of each function.

1.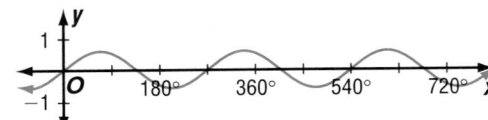

2.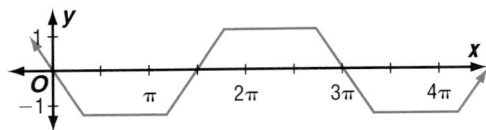

Find the value of each function.

3. $\sin 210°$

4. $\cos 150°$

5. $\cos(-135°)$

6. $\cos \dfrac{3\pi}{4}$

7. $\sin 570°$

8. $\sin 390°$

9. $\sin \dfrac{4\pi}{3}$

10. $\cos \left(-\dfrac{7\pi}{3}\right)$

11. $\cos 30° + \cos 60°$

12. $5(\sin 45°)(\cos 45°)$

13. $\dfrac{\sin 210° + \cos 240°}{2}$

14. $\dfrac{6\cos 120° + 4\sin 150°}{5}$

Lesson 13-7 Write each equation in the form of an inverse function.

1. $\sin m = n$

2. $\tan 45° = 1$

3. $\cos x = \dfrac{3}{2}$

4. $\sin 65° = a$

5. $\tan 60° = \sqrt{3}$

Solve each equation.

6. $y = \operatorname{Sin}^{-1} \dfrac{\sqrt{2}}{2}$

7. $\operatorname{Tan}^{-1} 1 = x$

8. $a = \operatorname{Arccos} \dfrac{\sqrt{3}}{2}$

9. $\operatorname{Arcsin} 0 = x$

10. $y = \operatorname{Cos}^{-1} \dfrac{1}{2}$

Find each value.

11. $\operatorname{Arccos}\left(-\dfrac{\sqrt{2}}{2}\right)$

12. $\operatorname{Sin}^{-1} -1$

13. $\cos\left(\operatorname{Arcsin} \dfrac{\sqrt{2}}{2}\right)$

14. $\tan\left(\operatorname{Sin}^{-1}\left(\dfrac{5}{13}\right)\right)$

15. $\sin 2\left(\operatorname{Arccos} \dfrac{1}{2}\right)$

16. $\sin\left(\operatorname{Arccos} \dfrac{15}{17}\right)$

17. $\sin\left(\operatorname{Tan}^{-1} \dfrac{5}{12}\right)$

18. $\tan\left(\operatorname{Arccos}\left(-\dfrac{\sqrt{3}}{2}\right)\right)$

Lesson 14-1 Graph each function.

1. $y = \csc \dfrac{1}{3}\theta$

2. $y = 2\sec\theta$

3. $y = 2\tan\theta$

4. $y = -3\sin \dfrac{2}{3}\theta$

5. $y = 2\sin \dfrac{1}{5}\theta$

6. $\dfrac{1}{2}y = 3\sin 2\theta$

7. $y = -\dfrac{1}{2}\cos \dfrac{3}{4}\theta$

8. $\dfrac{1}{2}y = 5\csc 3\theta$

9. $y = 2\cot 6\theta$

10. $y = \csc 6\theta$

11. $y = 3\tan \dfrac{1}{3}\theta$

12. $y = \dfrac{4}{3}\cot \dfrac{1}{2}\theta$

Write an equation of the given sine or cosine function having the specific characteristics.

	Function	Amplitude	Period
13.	sine	3	360°
14.	sine	4.8	120°
15.	sine	0.45	60°
16.	sine	5	1080°
17.	sine	$\dfrac{1}{3}$	30°

	Function	Amplitude	Period
18.	cosine	0.8	360°
19.	cosine	5	60°
20.	cosine	$\dfrac{1}{4}$	90°
21.	cosine	3	720°
22.	cosine	6.5	120°

Lesson 14-2 Solve for values of θ between 0° and 90°.

1. If $\cos \theta = \frac{4}{5}$, find $\tan \theta$.

2. If $\sin \theta = \frac{1}{2}$, find $\cos \theta$.

3. If $\sin \theta = \frac{3}{4}$, find $\sec \theta$.

4. If $\tan \theta = 4$, find $\sin \theta$.

5. If $\tan \theta = 4$, find $\cos \theta$.

6. If $\cot \theta = \frac{1}{4}$, find $\csc \theta$.

Solve for values of θ between 180° and 270°.

7. If $\sec \theta = -3$, find $\tan \theta$.

8. If $\cos \theta = -\frac{3}{5}$, find $\csc \theta$.

9. If $\sin \theta = -\frac{1}{2}$, find $\cos \theta$.

10. If $\cot \theta = \frac{1}{4}$, find $\csc \theta$.

11. If $\cot \theta = \frac{3}{5}$, find $\csc \theta$.

12. If $\tan \theta = 4$, find $\cos \theta$.

Simplify each expression.

13. $\dfrac{\sin^2 \theta + \cos^2 \theta}{\sin^2 \theta}$

14. $\dfrac{1 + \tan^2 \theta}{1 + \cot^2 \theta}$

15. $\dfrac{1}{1 + \sin \theta} + \dfrac{1}{1 - \sin \theta}$

16. $\dfrac{1 - \sin^2 \theta}{\sin^2 \theta}$

17. $\csc^2 \theta - \cot^2 \theta$

18. $\cos \theta \csc \theta$

19. $\tan \theta \csc \theta$

20. $\sin \theta \cot \theta$

Lesson 14-3 Verify that each of the following is an identity.

1. $1 + \tan^2 \theta = \sec^2 \theta$

2. $\dfrac{\tan \theta}{\sin \theta} = \sec \theta$

3. $\dfrac{\tan \theta}{\cot \theta} = \tan^2 \theta$

4. $\dfrac{\cos^2 \theta}{1 - \sin \theta} = 1 + \sin \theta$

5. $1 - \cot^4 \theta = 2 \csc^2 \theta - \csc^4 \theta$

6. $\sin^4 \theta - \cos^4 \theta = \sin^2 \theta - \cos^2 \theta$

7. $\cos^2 \theta + \tan^2 \theta \cos^2 \theta = 1$

8. $\dfrac{\sec \theta}{\sin \theta} - \dfrac{\sin \theta}{\cos \theta} = \cot \theta$

9. $\dfrac{\cos \theta}{\sec \theta - 1} + \dfrac{\cos \theta}{\sec \theta + 1} = 2 \cot^2 \theta$

10. $\tan \theta + \cot \theta = \csc \theta \sec \theta$

11. $\dfrac{1 - \cos \theta}{\sin \theta} = \dfrac{\sin \theta}{1 + \cos \theta}$

12. $\cot^2 \theta + \sin^2 \theta = \csc^2 \theta - \cos^2 \theta$

13. $\sec \theta + \tan \theta = \dfrac{\cos \theta}{1 - \sin \theta}$

14. $\dfrac{\cot^2 \theta}{1 + \cot^2 \theta} = 1 - \sin^2 \theta$

15. $\dfrac{\tan \theta - \sin \theta}{\sec \theta} = \dfrac{\sin^3 \theta}{1 + \cos \theta}$

16. $\sin^2 \theta + \sin^2 \theta \tan^2 \theta = \tan^2 \theta$

17. $\dfrac{\sec \theta - 1}{\sec \theta + 1} + \dfrac{\cos \theta - 1}{\cos \theta + 1} = 0$

18. $1 + \sec^2 \theta \sin^2 \theta = \sec^2 \theta$

19. $\tan \theta + \dfrac{\cos \theta}{1 + \sin \theta} = \sec \theta$

20. $\dfrac{\tan \theta}{\sec \theta + 1} = \dfrac{1 - \cos \theta}{\sin \theta}$

21. $\csc \theta - \dfrac{\sin \theta}{1 + \cos \theta} = \cot \theta$

Lesson 14-4 Find the exact value of each expression.

1. $\sin 195°$

2. $\cos 285°$

3. $\sin 255°$

4. $\sin 105°$

5. $\cos 15°$

6. $\sin 15°$

7. $\cos 375°$

8. $\sin 165°$

Verify that each of the following is an identity.

9. $\sin (270° - \theta) = -\cos \theta$

10. $\cos (90° + \theta) = -\sin \theta$

11. $\sin \left(\dfrac{\pi}{2} + x\right) = \cos x$

12. $\sin (x + 30°) + \cos (x + 60°) = \cos x$

13. $\cos (30° - x) + \cos (30° + x) = \sqrt{3} \cos x$

Use the identity $\tan(\alpha + \beta) = \dfrac{\tan \alpha + \tan \beta}{1 - \tan \alpha \tan \beta}$ to find the exact value of each expression.

14. $\tan (60° + 45°)$

15. $\tan (135° + 120°)$

16. $\tan 165°$

17. $\tan 255$

Lesson 14-5 Find the exact values of sin $2x$, cos $2x$, sin $\frac{x}{2}$, and cos $\frac{x}{2}$ for each of the following, given the two measures between which the terminal side of x lies.

1. $\cos x = \frac{7}{25}$; x is between $0°$ and $90°$

2. $\sin x = \frac{2}{5}$; x is between $0°$ and $90°$

3. $\cos x = -\frac{1}{8}$; x is between $180°$ and $270°$

4. $\sin x = -\frac{5}{13}$; x is between $270°$ and $360°$

5. $\sin x = \frac{\sqrt{35}}{6}$; x is between $90°$ and $180°$

6. $\cos x = -\frac{17}{18}$; x is between $90°$ and $180°$

Find the exact value of each expression by using the half-angle formulas.

7. $\sin 15°$

8. $\cos 75°$

9. $\sin \frac{\pi}{8}$

10. $\cos \frac{13\pi}{12}$

Verify that each of the following is an identity.

11. $\dfrac{\sin 2x}{2 \sin^2 x} = \cot x$

12. $1 + \cos 2x = \dfrac{2}{1 + \tan^2 x}$

13. $\csc x \sec x = 2 \csc 2x$

14. $\dfrac{1 - \tan^2 x}{1 + \tan^2 x} = \cos 2x$

15. $\sin 2x \,(\cot x + \tan x) = 2$

16. $\sin^2 x = \frac{1}{2}(1 - \cos 2x)$

17. $\dfrac{\cos x + \sin x}{\cos x - \sin x} = \dfrac{1 + \sin 2x}{\cos 2x}$

18. $\dfrac{\sin \frac{x}{2}}{\cos \frac{x}{2}} = \dfrac{\sin x}{1 + \cos x}$

19. $\cot x = \dfrac{\sin 2x}{1 - \cos 2x}$

Lesson 14-6 Find all solutions if $0° \le x < 360°$.

1. $\cos x = -\dfrac{\sqrt{3}}{2}$

2. $\sin 2x = -\dfrac{\sqrt{3}}{2}$

3. $\cos 2x = 8 - 15 \sin x$

4. $\sin x + \cos x = 1$

5. $2 \sin^2 x + \sin x = 0$

6. $\sin 2x = \cos x$

Find all solutions if $0 \le \theta < 2\pi$.

7. $\tan \theta = 1$

8. $\cos 8\theta = 1$

9. $\sin \theta + 1 = \cos 2\theta$

10. $8 \sin \theta \cos \theta = 2\sqrt{3}$

11. $\cos \theta = 1 + \sin \theta$

12. $2 \cos^2 \theta = \cos \theta$

Solve each equation for all values for x if x is measured in degrees.

13. $2 \sin^2 x - 1 = 0$

14. $\cos x - 2 \cos x \sin x = 0$

15. $\cos 2x \sin x = 1$

16. $(\tan x - 1)(2 \cos x + 1) = 0$

17. $2 \cos^2 x = 0.5$

18. $\sin x \tan x - \tan x = 0$

Solve each equation for all values of θ if θ is measured in radians.

19. $\cos 2\theta \sin \theta = 1$

20. $\sin \frac{\theta}{2} + \cos \frac{\theta}{2} = \sqrt{2}$

21. $\cos 2\theta + 4 \cos \theta = -3$

22. $\sin \frac{\theta}{2} + \cos \theta = 1$

23. $3 \tan^2 \theta - \sqrt{3} \tan \theta = 0$

24. $4 \sin \theta \cos \theta = -\sqrt{3}$

Name the property illustrated by each equation or statement.

1. $(7 \cdot s) \cdot t = 7 \cdot (s \cdot t)$

2. If $(r + s)t = rt + st$, then $rt + st = (r + s)t$.

3. $\left(3 \cdot \frac{1}{3}\right) \cdot 7 = \left(3 \cdot \frac{1}{3}\right) \cdot 7$

4. $(6 - 2)a - 3b = 4a - 3b$

5. $(7 \cdot s) \cdot t = t \cdot (7 \cdot s)$

6. If $5(3) + 7 = 15 + 7$ and $15 + 7 = 22$, then $5(3) + 7 = 22$.

Find the value of each expression.

7. $[2 + 3^3 - 4] \div 2$

8. $(2 + 3)^3 - 4 \div 2$

9. $(4^5 - 4^2) + 4^3$

10. $[5(17 - 2) \div 3] - 2^4$

Evaluate each expression if $a = -9$, $b = \frac{2}{3}$, $c = 8$, and $d = -6$.

11. $\frac{db + 4c}{a}$

12. $\frac{a}{b^2} + c$

13. $2b(4a + a^2)$

14. $\frac{4a + 3c}{3b}$

Name the sets of numbers to which each number belongs. Use N, W, Z, Q, I, and R.

15. $\sqrt{17}$

16. 0.86

17. $\sqrt{64}$

18. $-10 \div 2$

Solve each equation.

19. $5t - 3 = -2t + 10$

20. $2x - 7 - (x - 5) = 0$

21. $5m - (5 + 4m) = (3 + m) - 8$

22. $|8w + 2| + 2 = 0$

23. $12 \left| \frac{1}{2}y + 3 \right| = 6$

24. $2|2y - 6| + 4 = 8$

Solve each inequality. Graph each solution set.

25. $4 > b + 1$

26. $3q + 7 \geq 13$

27. $5(3x - 5) + x < 2(4x - 1) + 1$

28. $|9y - 4| + 8 > 4$

29. $-12 < 7s - 5 \leq 9$

30. $|5 + k| \leq 8$

31. Employment The back-to-back stem-and-leaf plot below shows the median weekly incomes of male and female workers in various occupations. Find the median, mode, and mean of the male workers' incomes and of the female workers' incomes.

Males	Stem	Females
	2	0 1 4 4
5 6	.	6 7 9
0	3	1
7	.	5 6
0 2 2 3	4	1 4
5 8	.	9
1	5	
	.	
	6	
6 7	.	$2 \mid 4 = 240$

Define a variable, write an equation, and solve the problem.

32. Statistics To receive a B in his English class, Dale must earn at least 400 points on five tests. He scored 87, 89, 76, and 77 on his first four tests. What must he score on the last test to receive a B in the class?

33. Car Rental A salesman rented a car that could get 35 miles per gallon. He paid $19.50 a day for the car plus $0.18 per mile. He rented the car for 1 day and paid $33. How many miles did he travel?

Graph each relation. State the domain and the range. Is the relation a function?

1. $\{(-4, -81), (-2, 21), (0, 51), (2, 33), (4, -9)\}$

2. $\{(-5, 0), (-3, 1), (-1, 2), (-3, 3), (-5, 4)\}$

Find values of functions for the given elements of the domain.

3. Find $f(-3)$ if $f(x) = 2x^2 - 3x + 5$.

4. Find $f(5)$ if $f(x) = 11x^3 - x + 1$.

5. Find $f(3.7)$ if $f(x) = 7 - x^2$.

6. Find $f(0)$ if $f(x) = x - 3x^2$.

Graph each equation or inequality.

7. $-2x + 5 \leq 3y$

8. $4x - y + 2 = 0$

9. $x = -4$

10. $y = 2x - 5$

11. $y \leq 10$

12. $x > 6$

13. $-8x + 4y \geq 32$

14. $f(x) = [3x] + 3$

15. $y < 4|x - 1|$

16. $f(x) = 3x - 1$

17. $g(x) = -\frac{1}{2}x$

18. $y = \frac{3}{5}x - 4$

Find the slope of the line that passes through each pair of points.

19. $(8, -4)$ and $(6, 1)$

20. $(-2, 5)$ and $(4, 5)$

21. $(5, 7)$ and $(4, -6)$

22. $(4, 5)$ and $(4, -3)$

23. $(-5, -4)$ and $(5, 2)$

24. $(-3, -5)$ and $(9, -1)$

Write an equation in slope-intercept form that satisfies each condition.

25. slope $= -5$, y-intercept $= 11$

26. slope $= \frac{2}{3}$, passes through $(-6, 15)$

27. passes through $(-3, 7)$ and $(7, -8)$

28. x-intercept $= 9$, y-intercept $= -4$

29. passes through $(4, -2)$ and the origin

30. parallel to $6x - y = 7$, passes through $(-2, 8)$

31. perpendicular to $x + 3y = 7$, passes through $(5, 2)$

The table below shows the expected number of people in the U.S. that are age 100 and over.

Year	1994	1996	1998	2000	2002	2004
Number of People	50,000	56,000	65,000	75,000	94,000	110,000

Source: U.S. Census Bureau

32. Draw a scatter plot of the data to show the relationship. Let x represent the years since 1990.

33. Find a prediction equation to show how time and the number of people aged 100 and over are related.

Graph each system of equations and state its solution.

1. $-4x + y = -5$
$2x + y = 7$

2. $x + y = -8$
$-3x + 2y = 9$

3. $-6x + 3y = 33$
$-4x + y = 16$

Solve each system of equations. Use either the substitution or the elimination method.

4. $3x - 2y = 8$
$y = 6x + 11$

5. $2y = 5x - 1$
$x + y = -1$

6. $-7x + 6y = 42$
$3x + 4y = 28$

Find the value of each second-order determinant.

7. $\begin{vmatrix} -4 & 3 \\ 5 & -2 \end{vmatrix}$

8. $\begin{vmatrix} 1 & -6 \\ 8 & 7 \end{vmatrix}$

9. $\begin{vmatrix} 0 & -4 \\ 2 & 11 \end{vmatrix}$

Use Cramer's rule to solve each system of equations.

10. $4x + 5y = 60$
$5x + 4y = -20$

11. $6x + y = 15$
$x - 4y = -16$

12. $3x - 8y = 23$
$5x + y = 24$

Solve each system of inequalities by graphing.

13. $y \geq x - 3$
$y \geq -x + 1$

14. $x + 2y \geq 7$
$3x - 4y < 12$

15. $|x| > 5$
$x + y < 6$

Graph each system of inequalities. Name the coordinates of the vertices of the feasible region. Find the maximum and the minimum values of the given function for this region.

16. $y \leq 5$
$y \geq -3$
$4x + y \leq 5$
$-2x + y \leq 5$
$f(x, y) = 4x - 3y$

17. $x \geq -10$
$y \geq -6$
$y \leq 1$
$\frac{3}{4}x + y \leq -2$
$y \geq \frac{1}{2}x - 5$
$f(x, y) = 2x + y$

A sporting goods manufacturer makes a $5 profit on soccer balls and a $4 profit on volleyballs. Department S requires 2 hours to make 75 soccer balls and 3 hours to make 60 volleyballs. Department V needs 3 hours to make 75 soccer balls and 2 hours to make 60 volleyballs. Department S has 500 hours available and Department V has 450 hours available.

18. How many soccer balls and volleyballs should be made to maximize the profit?

19. What is the maximum profit the company can make from these two products?

Solve each system of equations.

20. $x + y + z = -1$
$2x + 4y + z = 1$
$x + 2y - 3z = -3$

21. $x + z = 7$
$2y - z = -3$
$-x - 3y + 2z = 11$

22. $x - y + z = 0.5$
$-x - y + z = 0$
$7x - y + 4z = 4.25$

Carla, Beth, and Heidi went on a shopping spree to get ready for college. Carla bought 3 shirts, 4 pairs of pants, and 2 pairs of shoes costing her a total of $149.79. Beth came away with 5 shirts, 3 pairs of pants, and 3 pairs of shoes totaling $183.19. Heidi, not to be outdone, bought 6 shirts, 5 pairs of pants and a pair of shoes. Her total bill came to $181.14. Assume that all the shirts cost the same price, all the pants cost the same price, and all the shoes cost the same price.

23. How much money did each girl pay for a shirt?

24. How much money did each girl pay for a pair of pants?

25. How much money did each girl pay for a pair of shoes?

Perform the indicated operations, if possible.

1. $\begin{bmatrix} 1 & 2 \\ -4 & 3 \\ 5 & 2 \end{bmatrix} \cdot \begin{bmatrix} 5 \\ 4 \end{bmatrix}$

2. $\begin{bmatrix} 2 & -4 & 1 \\ 3 & 8 & -2 \end{bmatrix} - 2\begin{bmatrix} 1 & 2 & -4 \\ -2 & 3 & 7 \end{bmatrix}$

3. $\begin{bmatrix} -0.5 & 0.7 \\ 0.2 & -0.6 \end{bmatrix} + \begin{bmatrix} 3.4 \\ 2.1 \end{bmatrix}$

4. $-4\begin{bmatrix} -5 & 7 \\ 2 & -6 \end{bmatrix} + 0.5\begin{bmatrix} -2 & 8 \\ 2 & -4 \end{bmatrix}$

5. $\begin{bmatrix} -5 & 7 \\ 2 & -6 \end{bmatrix} \cdot \begin{bmatrix} 2 & -4 & 1 \\ 3 & 8 & -2 \end{bmatrix}$

6. $\begin{bmatrix} 6 & 7 \\ -3 & -4 \end{bmatrix} \cdot \begin{bmatrix} -4 & 3 \\ -1 & -2 \\ 2 & 5 \end{bmatrix}$

Determine whether each matrix has a determinant. Write *yes* or *no*. If *yes*, find the value of the determinant. If *no*, explain why not.

7. $\begin{bmatrix} -1 & 4 \\ -6 & 3 \end{bmatrix}$

8. $\begin{bmatrix} -2 & 0 & 5 \\ -3 & 4 & 0 \\ 1 & 3 & -1 \end{bmatrix}$

9. $\begin{bmatrix} 2 & -4 & 1 \\ 3 & 8 & -2 \end{bmatrix}$

10. $\begin{bmatrix} 5 & -3 & 2 \\ -6 & 1 & 3 \\ -1 & 4 & -7 \end{bmatrix}$

Find the inverse of each matrix, if it exists. If it does not exist, explain why not.

11. $\begin{bmatrix} -2 & 5 \\ 3 & 1 \end{bmatrix}$

12. $\begin{bmatrix} -6 & -3 \\ 8 & 4 \end{bmatrix}$

13. $\begin{bmatrix} 5 & -2 \\ 6 & 3 \end{bmatrix}$

14. $\begin{bmatrix} 35 \\ 23 \end{bmatrix}$

Solve each matrix equation or system of equations.

15. $5a + 2b = -49$
    $2a + 9b = 5$

16. $6x - y = -15$
    $5x + 4 = -2y$

17. $2x - y + 3z = 1$
    $x - y + 4z = 0$
    $3x - 2y + z = -5$

18. $\begin{bmatrix} 1 & 8 \\ 2 & -6 \end{bmatrix} \cdot \begin{bmatrix} x \\ y \end{bmatrix} = \begin{bmatrix} -3 \\ -17 \end{bmatrix}$

19. $\begin{bmatrix} 2 & 0 & 1 \\ 4 & 1 & 2 \\ 2 & 0 & 4 \end{bmatrix} \cdot \begin{bmatrix} a \\ b \\ c \end{bmatrix} = \begin{bmatrix} 10 \\ 19 \\ 22 \end{bmatrix}$ and $M^{-1} = \frac{1}{6}\begin{bmatrix} 4 & 0 & -1 \\ -12 & 6 & 0 \\ -2 & 0 & 2 \end{bmatrix}$

Find the range, quartiles, interquartile range, and outliers for each set of data.

20.

Stem	Leaf
0	1 1 8 8 9
1	4 5 5 7 7 7 9
2	1 4 4 5 8 8 8 8 9
3	0 1 3 6 6 7 8
4	4 5 6 9 $2\mid4 = 240$

21. {100, 99, 93, 94, 96, 94, 95, 101, 109, 108, 104, 106, 125, 100, 104, 98, 19}

22. Judi is buying a pair of headphones to use with her stereo receiver. The prices of the 21 different types of stereo headphones sold at the Stereo Studio are $100, $150, $75, $79, $149, $120, $80, $70, $400, $190, $50, $80, $148, $40, $85, $60, $160, $90, $90, $125, and $120. Make a box-and-whisker plot of the data.

For Exercises 23–25, use $\triangle ABC$ whose vertices have coordinates $A(6, 3)$, $B(1, 5)$, and $C(-1, 4)$.

23. Use a determinant to find the area of $\triangle ABC$.

24. Translate $\triangle ABC$ so that the coordinates of B' are (3, 1). What are the coordinates of A' and C'?

25. Find the coordinates of the vertices of a similar triangle whose perimeter is five times that of $\triangle ABC$.

Simplify.

1. $(5b)^4(6c)^2$

2. $(13x - 1)(x + 3)$

3. $(2h - 6)^3$

Evaluate. Express each answer in both scientific and decimal notation.

4. $(3.16 \times 10^3)(24 \times 10^2)$

5. $\dfrac{7,200,000 \cdot 0.0011}{0.018}$

Use synthetic division to find each quotient.

6. $(x^4 - x^3 - 10x^2 + 4x + 24) \div (x - 2)$

7. $(2x^3 + 9x^2 - 2x + 7) \div (x + 2)$

Factor completely. If the polynomial is not factorable, write _prime_.

8. $x^2 - 14x + 45$

9. $2r^2 + 3pr - 2p^2$

10. $x^2 + 2\sqrt{3}x + 3$

Evaluate.

11. $\sqrt{175}$

12. $\left(5 + \sqrt{3}\right)\left(7 - 2\sqrt{3}\right)$

13. $3\sqrt{6} + 5\sqrt{54}$

14. $\dfrac{9}{5 - \sqrt{3}}$

15. $\left(9^{\frac{1}{2}} \cdot 9^{\frac{2}{3}}\right)^{\frac{1}{6}}$

16. $11^{\frac{1}{2}} \cdot 11^{\frac{7}{3}} \cdot 11^{\frac{1}{6}}$

Simplify.

17. $\sqrt[6]{256s^{11}t^{18}}$

18. $v^{-\frac{7}{11}}$

19. $\dfrac{b^{\frac{1}{2}}}{b^{\frac{3}{2}} - b^{\frac{1}{2}}}$

Solve each equation. Be sure to check for extraneous solutions.

20. $\sqrt{b + 15} = \sqrt{3b + 1}$

21. $\sqrt{2x} = x - 4$

22. $\sqrt[4]{y + 2} + 9 = 14$

23. $\sqrt[3]{2w - 1} + 11 = 18$

24. $\sqrt{4x + 28} = \sqrt{6x + 38}$

25. $1 + \sqrt{x + 5} = \sqrt{2x + 5}$

Simplify each complex expression.

26. $(5 - 2i) - (8 - 11i)$

27. $(4 + 3i)(9 - 2i)$

28. $(14 - 5i)^2$

29. Skydiving The approximate time t in seconds that it takes an object to fall a distance d in feet is $t = \sqrt{\dfrac{d}{16}}$. A parachutist falls 11 seconds before the parachute opens. How far does the parachutist fall during this time period?

30. Soccer A soccer field in a city park is rectangular. The length of the field is 120 yards and the width of the field is 60 yards. What is the distance from one corner flag to the opposite corner flag?

31. Electricity A circuit has a current of $(6 + 4j)$ amps and an impedance of $(3 - j)$ ohms. What will be the voltage of the circuit?

32. Geometry Hero of Alexandria, who lived sometime between 150 B.C. and A.D. 250, is credited with a formula for finding the area of any triangle with sides a, b, and c. The formula states that the area of a triangle is equal to $\sqrt{s(s - a)(s - b)(s - c)}$, with $s = \dfrac{1}{2}(a + b + c)$. If the sides of the triangle are 6, 9, and 12 feet, what is the area of the triangle?

33. Write $4^{\frac{1}{4}}f^{\frac{3}{5}}g^{\frac{5}{8}}$ using a single radical.

Solve each equation by graphing.

1. $x^2 + 3x - 40 = 0$

2. $4x^2 - 11x - 3 = 0$

3. $6x^2 - 216 = 0$

Solve each equation by factoring.

4. $-1.6x^2 - 3.2x + 18 = 0$

5. $c^2 + c - 42 = 0$

6. $15x^2 + 16x - 7 = 0$

Solve each equation by completing the square.

7. $b^2 + 8b - 48 = 0$

8. $h^2 + 12h + 11 = 0$

9. $x^2 - 9x - \frac{19}{4} = 0$

Solve each equation by using the quadratic formula.

10. $3c^2 + 7c - 31 = 0$

11. $10v^2 + 3v = 1$

12. $-11x^2 - 174x + 221 = 0$

Find the sum and the product of the roots for each quadratic equation.

13. $2x^2 + 8x - 3 = 0$

14. $5x^2 = 6$

15. $4x^2 + 3x - 12 = 0$

Find a quadratic equation that has the given roots.

16. $-2, 5$

17. $-\frac{8}{3}, \frac{7}{3}$

18. $3 + 4i, 3 - 4i$

Graph each equation. Name the vertex, the axis of symmetry, and the direction of opening for each graph.

19. $y = (x + 2)^2 - 3$

20. $f(x) = x^2 + 10x + 27$

21. $y = (x - 5)^2 - 6$

Graph each inequality.

22. $y \leq x^2 + 6x - 7$

23. $y > -2x^2 + 9$

24. $y \geq 3x^2 - 15x + 22$

Solve each inequality.

25. $(x - 5)(x + 7) > 0$

26. $x^2 - 11x \leq 0$

27. $3d^2 \geq 16$

28. Family A class of 25 students took a survey of the number of brothers and sisters each student had in his or her family. The following is a list in response to this survey: 0, 0, 0, 1, 1, 1, 1, 1, 1, 2, 2, 2, 2, 2, 2, 2, 2, 2, 3, 3, 3, 4, 5, 6.

   a. Find the mean of the number of siblings in this class.

   b. Find the standard deviation for the number of siblings in this class.

   c. Make a histogram representing the number of siblings in this class.

29. Testing The scores on a standardized college entrance exam are normally distributed. The mean score is 525, and the standard deviation is 75.

   a. Of the estimated 65,000 students who took the exam, how many scored above 600?

   b. Of the estimated 65,000 students who took the exam, how many scored below 300?

   c. What percentage of the students who took the exam scored below 675?

Find the distance between each pair of points with the given coordinates.

1. $(-6, 7), (3, 2)$

2. $\left(\frac{1}{2}, \frac{5}{2}\right), \left(-\frac{3}{4}, -\frac{11}{4}\right)$

3. $(8, -1), (8, -9)$

Find the midpoint of each line segment if the coordinates of the endpoints are given.

4. $(7, 1), (-5, 9)$

5. $\left(\frac{3}{8}, -1\right), \left(-\frac{8}{5}, 2\right)$

6. $(-13, 0), (-1, -8)$

State whether the graph of each equation is a *parabola,* a *circle,* an *ellipse,* or a *hyperbola.* Then draw the graph.

7. $x^2 + 4y^2 = 25$

8. $y = 4x^2 + 1$

9. $x^2 = 36 - y^2$

10. $(x + 4)^2 = 7(y + 5)$

11. $4x^2 - 26y^2 + 10 = 0$

12. $25x^2 + 49y^2 = 1225$

13. $-(y^2 - 24) = x^2 + 10x$

14. $5x^2 - y^2 = 49$

15. $25(x - 1)^2 + 121(y + 6)^2 = 3025$

16. $x^2 + 5x + y^2 - 9y = 7$

17. $\frac{1}{3}x^2 - 4 = y$

18. $\frac{y^2}{9} - \frac{x^2}{25} = 1$

Solve each system of equations.

19. $x^2 + y^2 = 100$
    $y = 2 - x$

20. $x^2 - y^2 - 12x + 12y = 36$
    $x^2 + y^2 - 12x - 12y + 36 = 0$

Solve each system of inequalities by graphing.

21. $x^2 + y^2 \le 169$
    $x^2 + y^2 \ge 121$

22. $x^2 + y^2 \le 81$
    $y \ge x - 1$

23. **Seismology**   Three tracking stations have detected an earthquake in the local area. The first station is located at the origin on the county map, where each unit of the grid represents one square mile. The second and third stations are located at $(8, -8)$ and $(11, 10)$, respectively. The epicenter of the earthquake is 5 miles from the first tracking station, 13 miles from the second tracking station, and 10 miles from the third tracking station. What are the coordinates of the epicenter of the earthquake? Explain your answer.

24. **Tunnels**   The opening of a tunnel that goes through the Rocky Mountains is in the shape of a semi-elliptical arch. The arch is 60 feet wide and 40 feet high. Find a graph that models the arch. Then find the height of the arch both 6 feet and 12 feet from the edge of the tunnel.

25. **Architecture**   Two condominium apartment buildings that face each other have a front door awning that is shaped like a hyperbola. The equation of the hyperbola is $25x^2 - 81y^2 = 30{,}625$. Find the distance in meters between the two front door awnings at their closest point.

Is the given binomial a factor of the given polynomial? Show by synthetic division and writing the answer in the form *dividend = quotient · divisor + remainder.*

1. $x^3 - x^2 - 5x - 3; x + 1$

2. $x^3 + 8x + 1; x + 2$

3. $x^4 + 2x^3 - 7x^2 - 8x + 12; x - 2$

For each function, state the number of positive real zeros, negative real zeros, and imaginary zeros.

4. $f(x) = x^3 - x^2 - 14x + 24$

5. $f(x) = 2x^3 - x^2 + 16x - 5$

6. $f(x) = x^4 + x^3 - 9x^2 - 17x - 8$

7. $f(x) = -7x^4 + 2x^3 + x^2 - 16$

Approximate to the nearest tenth the real zeros of each function. Then use the functional values to graph the function.

8. $g(x) = x^3 + 6x^2 + 6x - 4$

9. $h(x) = x^4 + 6x^3 + 8x^2 - x$

10. $f(x) = x^3 + 3x^2 - 2x + 1$

11. $g(x) = x^4 - 2x^3 - 6x^2 + 8x + 5$

Find all the rational zeros for each function.

12. $g(x) = x^3 - 3x^2 + 53x - 9$

13. $h(x) = x^4 + 2x^3 - 23x^2 + 2x - 24$

14. $f(x) = 6x^3 + 4x^2 - 14x + 4$

15. $f(x) = 10x^3 + 43x^2 + 36x - 9$

Find all zeros of each function.

16. $g(x) = x^3 - 3x - 52$

17. $f(x) = x^3 + 4x^2 - 19x - 6$

18. $h(x) = 4x^4 + 11x^3 + 10x^2 - 69x - 54$

19. $g(x) = x^4 - 9x^3 + 11x^2 - 19x - 40$

Solve each equation.

20. $p^3 + 8p^2 = 18p$

21. $16x^4 - x^2 = 0$

22. $r^4 - 9r^2 + 18 = 0$

23. $p^{\frac{3}{2}} - 8 = 0$

24. $2d + 3\sqrt{d} = 9$

25. $z^4 + 6z^3 + 8z^2 = 0$

If $f(x) = 2x - 4$ **and** $g(x) = x^2 + 3$, **find each value.**

26. $g(x + h)$

27. $f[g(x)]$

28. $g[f(x)]$

29. $g[f(-3)]$

Write the polynomial function of least degree with integral coefficients that has the given zeros.

30. $-1, 3, i$

31. $2, -i$

32. $-2, 1 + i$

33. $1, 4, 1 - i$

Simplify each expression.

1. $\dfrac{7ab}{9c} \cdot \dfrac{81c^2}{91a^2b}$

2. $\dfrac{4a}{5b} \cdot \dfrac{15b}{16a}$

3. $\dfrac{6}{x - 5} + 7$

4. $\dfrac{m + 5}{2m + 10}$

5. $\dfrac{7}{5a} - \dfrac{10}{3ab}$

6. $\dfrac{4x}{x^2 - x}$

7. $\dfrac{a^2 - ab}{3a} \div \dfrac{a - b}{15b^2}$

8. $\dfrac{z^2 w - z^2}{z^3 - z^3 w}$

9. $\dfrac{x^2 - y^2}{y^2} \cdot \dfrac{y^3}{y - x}$

10. $\dfrac{4x^2 y}{15a^3 b^3} \div \dfrac{2xy^2}{5ab^3}$

11. $\dfrac{x^2 - 2x + 1}{y - 5} \div \dfrac{x - 1}{y^2 - 25}$

12. $\dfrac{a^2 - b^2}{2a} \div \dfrac{a - b}{6a}$

13. $\dfrac{\dfrac{x^2 - 1}{x^2 - 3x - 10}}{\dfrac{x^2 + 3x + 2}{x^2 - 12x + 35}}$

14. $\dfrac{2x + 2}{x^2 + 5x + 6} \div \dfrac{3x + 3}{x^2 + 2x - 3}$

15. $\dfrac{x - 2}{x - 1} + \dfrac{6}{7x - 7}$

16. $\dfrac{a^3 - b^3}{a + b} \cdot \dfrac{a^2 - b^2}{a^2 + ab + b^2}$

17. $\dfrac{x}{x^2 - 9} + \dfrac{1}{2x + 6}$

18. $\dfrac{\dfrac{x^2 - 5x - 14}{x^2 + 7x + 12}}{\dfrac{3x - 21}{x^2 - 16}}$

State the equations of the vertical and horizontal asymptotes for each rational function. Then graph each rational function.

19. $f(x) = \dfrac{-4}{x - 3}$

20. $f(x) = \dfrac{x}{x + 2}$

21. $f(x) = \dfrac{2}{(x - 2)(x + 1)}$

Solve each equation.

22. $8 - \dfrac{2 - 5x}{4} = \dfrac{4x + 9}{3}$

23. $\dfrac{9}{28} + \dfrac{3}{z + 2} = \dfrac{3}{4}$

24. $\dfrac{3}{x} + \dfrac{x}{x + 2} = \dfrac{-2}{x + 2}$

25. $x + \dfrac{12}{x} - 8 = 0$

26. $\dfrac{5}{6} - \dfrac{2m}{2m + 3} = \dfrac{19}{6}$

27. $\dfrac{x - 3}{2x} = \dfrac{x - 2}{2x + 1} - \dfrac{1}{2}$

28. $\dfrac{2}{x - 2} = 3 - \dfrac{x}{2 - x}$

29. $r + \dfrac{r^2 - 5}{r^2 - 1} = \dfrac{r^2 + r + 2}{r + 1}$

30. Suppose y varies directly as x. If $y = 10$, then $x = -3$. Find y when $x = 20$.

31. Suppose y varies inversely as x. If $y = 9$, then $x = -\dfrac{2}{3}$. Find x when $y = -7$.

32. Suppose g varies directly as w. If $g = 10$, then $w = -3$. Find w when $g = 4$.

33. Suppose y varies jointly as x and z. If $x = 10$ when $y = 250$ and $z = 5$, find x when $y = 2.5$ and $z = 4.5$.

Write each equation in logarithmic form.

1. $6^4 = 1296$

2. $3^7 = 2187$

Write each equation in exponential form.

3. $\log_5 625 = 4$

4. $\log_8 16 = \frac{4}{3}$

Use $\log_4 7 \approx 1.4037$ and $\log_4 3 \approx 0.7925$ to evaluate each expression.

5. $\log_4 21$

6. $\log_4 \frac{7}{12}$

Approximate the value of each logarithm to three decimal places.

7. $\log_{14} 24$

8. $\log_3 50$

Evaluate each expression.

9. $4^{\log_4 3}$

10. $\log_{64} 8$

11. $\log_2 \frac{1}{256}$ $\frac{1}{2^8}$

Simplify each expression.

12. $\left(3^{\sqrt{8}}\right)^{\sqrt{2}}$

13. $81^{\sqrt{5}} \div 3^{\sqrt{5}}$

Solve each equation.

14. $2^{x-3} = \frac{1}{16}$

15. $27^{2p+1} = 3^{4p-1}$

16. $\log_2 128 = y$

17. $\log_m 144 = -2$

18. $\log_3 x - 2\log_3 2 = 3\log_3 3$

19. $\log_9 (x+4) + \log_9 (x-4) = 1$

20. $\log_5 (8r - 7) = \log_5 (r^2 + 5)$

21. $\log_3 3^{(4x-1)} = 15$

22. $\log_2 3 + \log_2 7 = \log_2 x$

23. $\log_5 x = -2$

Find each value rounded to four decimal places.

24. $\ln 9.6$

25. $\log 535$

26. antilog 6.3337

27. antiln 0.4055

Use logarithms to solve each equation.

28. $7.6^{x-1} = 431$

29. $\log_4 37 = x$

30. $3^x = 5^{x-1}$

31. $4^{2x-3} = 9^{x+3}$

32. **Biology**   A certain culture of bacteria will grow from 500 to 4000 bacteria in 1.5 hours. Find the constant k for the growth formula. (Use $y = ne^{kt}$.)

33. **Finance**   Suppose a pilgrim ancestor of Jenny Chambers deposited $10 in a savings account at Provident Savings Bank. The annual interest rate was 4% compounded continuously. The account is now worth $75,000. How long ago was the account started? (Use $A = Pe^{rt}$.)

1. Find the pattern and complete the sequence 3, 3, 6, 18, 72, $\underline{\ ?\ }$, $\underline{\ ?\ }$, $\underline{\ ?\ }$.

2. Find the sixth term of the geometric sequence if $a_1 = 5$ and $r = -2$.

3. How many integers between 26 and 415 are multiples of 9?

4. Find the sum of the arithmetic series if $a_1 = 7$, $n = 31$, and $a_n = 127$.

5. Find the next four terms of the arithmetic sequence 42, 37, 32, ...

6. Find the 27th term of an arithmetic sequence if $a_1 = 2$ and $d = 6$.

7. Find the next two terms of the geometric sequence $\frac{1}{81}, \frac{1}{27}, \frac{1}{9}, ...$

8. Find the sum of the geometric series if $a_1 = 125$, $r = \frac{2}{5}$, and $n = 4$.

9. Find the three arithmetic means between -4 and 16.

10. Find the two geometric means between 7 and 189.

Find each sum.

11. $\displaystyle\sum_{k=3}^{15} (14 - 2k)$

12. $\displaystyle\sum_{n=1}^{\infty} \frac{1}{3}(-2)^{n-1}$

13. Find the sum of the series $91 + 85 + 79 + ... + (-29)$.

14. Find the sum of the series $12 - 6 + 3 - \frac{3}{2} + ...$

15. Expand $(2s - 3t)^5$.

16. Find the third term of $(x + y)^8$.

17. **Design** A landscaper is designing a wall of white brick and red brick. The pattern starts with 20 red bricks on the bottom row. Each row above it contains 3 fewer red bricks than the preceding row. If the top row contains no red bricks, how many rows are there and how many red bricks were used?

18. **Business** Olsen's Electronics invested in computer equipment worth $900,000. The equipment depreciates at the rate of 25% per year of the previous year's value. What will be the value of the equipment at the end of four years?

19. **Physics** In the first three seconds after liftoff, a rocket rises 40 feet, 60 feet, and 80 feet, respectively. If it continues to rise at this rate, how many feet will it rise in the 10th second?

20. **Recreation** One minute after it is released, a gas-filled balloon has risen 100 feet. In each succeeding minute, the balloon rises only 50% as far as it rose in the previous minute. How far will the balloon rise in 5 minutes?

Evaluate each expression.

1. $P(7, 3)$ 2. $C(7, 3)$ 3. $P(13, 5)$ 4. $C(13, 5)$

Solve each problem.

5. How many ways can 9 bowling balls be arranged on the upper rack of a bowling ball rack?

6. From 11 skirts, 9 blouses, 3 belts, and 7 pairs of shoes, how many different outfits can be made?

7. How many ways can the letters of the word *probability* be arranged?

8. How many different soccer teams consisting of 11 players can be formed from 18 players?

9. In a row of 10 parking spaces in a parking lot, how many ways can 4 cars park?

10. How many ways can 6 people be seated at a round table, if it does not matter who sits next to whom?

11. A number is drawn at random from a hat that contains all the numbers from 1 to 100. What is the probability that the number is less than sixteen?

12. Eleven points are equally spaced on a circle. How many pentagons can be formed using these points, five at a time, as vertices?

13. A shipment of ten television sets contains 3 defective sets. How many ways can a hospital purchase 4 of these sets and receive at least 2 of the defective sets?

14. Two cards are drawn in succession from a standard deck of 52 cards without replacement. What is the probability that both cards are greater than 2 and less than 9?

15. In a high school graduating class of 100 students, 52 studied mathematics, 58 studied Spanish, 54 studied computers, 22 studied both mathematics and computers, 25 studied both mathematics and Spanish, 7 studied computers and neither Spanish nor mathematics, 10 studied all three subjects, and 8 did not take any of the three. If a student is selected at random, what is the probability that he took mathematics only?

16. While shooting arrows, William Tell can hit an apple 9 out of 10 times. What is the probability that he will hit it exactly 4 out of 7 times?

17. Ten people are going on a camping trip in 3 cars that hold 5, 2, and 4 passengers, respectively. How many ways is it possible to transport the 10 people to their camp site?

18. From a box containing 5 white golf balls and 3 red golf balls, 3 golf balls are drawn in succession, each being replaced in the box before the next draw is made. What is the probability that all 3 golf balls are the same color?

19. In a ten-question multiple-choice test with four choices for each question, a student who was not prepared guesses on each item. Find the probability that the student gets

   a. six questions right.

   b. at least eight correct.

   c. fewer than eight correct.

20. In a *USA Today*/CNN/Gallup Poll survey of 1208 randomly selected adults published on July 26, 1995, 47% of those surveyed approved of President Clinton's handling of the economy. What is the margin of error? Round your answer to the nearest tenth.

Change each degree measure to radian measure and each radian measure to degree measure.

1. $275°$

2. $-\dfrac{\pi}{6}$

3. $\dfrac{11}{2}\pi$

4. $330°$

5. $-600°$

6. $-\dfrac{7\pi}{4}$

Solve each right triangle. Round measures of sides to the nearest tenth and measures of angles to the nearest degree.

7. $a = 7, A = 49°$

8. $B = 75°, b = 6$

9. $A = 22°, c = 8$

10. $a = 7, c = 16$

Find the exact value of each expression.

11. $\cos(-120°)$

12. $\sin \dfrac{7}{4}\pi$

13. $\tan 135°$

14. $\cot 300°$

15. $\sec\left(-\dfrac{7}{6}\pi\right)$

16. $\csc \dfrac{5\pi}{6}$

17. $\operatorname{Sin}^{-1}\left(-\dfrac{\sqrt{3}}{2}\right)$

18. $\operatorname{Arctan} 1$

19. $\operatorname{Cos}^{-1}(\sin -60°)$

20. Determine the number of possible solutions for a triangle in which $A = 40°$, $b = 10$, and $a = 14$. If a solution exists, solve the triangle.

21. Suppose θ is an angle in standard position whose terminal side lies in Quadrant II. Find the exact values of the remaining five trigonometric functions of θ for $\cos \theta = -\dfrac{\sqrt{3}}{2}$.

22. Geology   From the top of a cliff, a geologist spots a dry riverbed. The measurement of the angle of depression to the riverbed is $70°$. The cliff is 50 meters high. How far is the riverbed from the base of the cliff?

23. Fire Fighting   A firefighter needs to use a 14-meter ladder to enter a window that is 13.5 meters above the ground. At what angle to the ground should the ladder be placed in order to reach the window?

24. Broadcasting   A 40-foot television antenna stands on top of a building. From a point on the ground, the angles of elevation to the top and the bottom of the antenna measure $56°$ and $42°$, respectively. How tall is the building?

25. Geometry   A triangular lot faces two streets that meet at an angle measuring $85°$, as shown at the right. The sides of the lot facing the streets are each 160 feet in length. Find the perimeter of the lot.

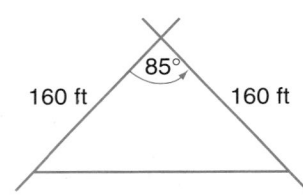

Graph each function.

1. $y = 2 \sin 2x$

2. $y = 2 \cos \frac{1}{5}\theta$

Solve each of the following for values of θ between 180° and 270°.

3. If $\sin \theta = -\frac{1}{2}$, find $\tan \theta$.

4. If $\cot \theta = \frac{3}{4}$, find $\sec \theta$.

Verify that each of the following is an identity.

5. $(\sin x - \cos x)^2 = 1 - \sin 2x$

6. $\dfrac{\cos x}{1 - \sin^2 x} = \sec x$

7. $\dfrac{\sec x}{\sin x} - \dfrac{\sin x}{\cos x} = \cot x$

8. $\dfrac{1 + \tan^2 \theta}{\cos^2 \theta} = \sec^4 \theta$

Evaluate each expression.

9. $\cos 165°$

10. $\sin 255°$

11. $2 \sin 75° \cos 75°$

12. $(1 - \cos^2 \theta) \cot^2 \theta$

13. $\sin \beta (\cos \beta + \sin \beta \tan \beta)$

14. $\sin \alpha \cot \alpha$

15. If x is in Quadrant II and $\cos x = -\frac{1}{6}$, find $\sin 2x$ and $\cos \frac{x}{2}$.

16. If x is in Quadrant I and $\cos x = \frac{3}{4}$, find $\sin \frac{x}{2}$ and $\cos 2x$.

17. If x is in Quadrant III and $\tan x = \frac{5}{12}$, find $\tan 2x$ and $\tan \frac{x}{2}$.

Solve each equation for $0° \leq x < 360°$ if x is measured in degrees.

18. $\sec x = 1 + \tan x$

19. $\cos 2x = \cos x$

20. $\cos 2x + \sin x = 1$

21. $\sin x = \tan x$

22. $2 \sin x \cos x - \sin x = 0$

23. $\sin x + \sin x \cos x = 0$

24. Golf A golf ball leaves the club with an initial velocity of 100 feet per second. The distance the ball travels is found by the formula $d = \dfrac{v_0^2}{g} \sin 2\theta$, where v_0 is the initial velocity, g is the acceleration due to gravity, and θ is the measurement of the angle that the path of the ball makes with the ground. The acceleration due to gravity is 32 feet per second squared.

a. Find the distance the ball travels if the angle between the path of the ball and the ground measures 60°.

b. Find the distance the ball travels if the angle between the path of the ball and the ground measures 45°.

25. Football The approximate distance s in meters that an object will travel if given an initial linear speed v_0 at an angle of elevation θ is given by the formula $s = \dfrac{v_0^2 \sin \theta \cos \theta}{5}$ where v_0 is in meters per second. At what angle must a football be thrown at 20 m/s in order to travel 20 meters? (Disregard the height of the person throwing the ball.)

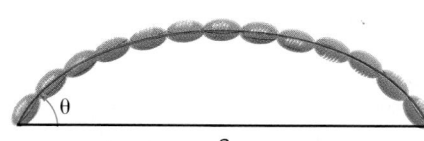

GLOSSARY

A

absolute value (37)   The absolute value of a number is the number of units it is from zero on the number line.

absolute value function (104)   A function described by $y = |x|$ or $f(x) = |x|$.

addition of matrices (195)   If A and B are two $m \times n$ matrices, then $A + B$ is the $m \times n$ matrix in which each element is the sum of the corresponding elements of A and B.

$$\begin{bmatrix} a & b & c \\ d & e & f \\ g & h & i \end{bmatrix} + \begin{bmatrix} j & k & l \\ m & n & o \\ p & q & r \end{bmatrix} = \begin{bmatrix} a+j & b+k & c+l \\ d+m & e+n & f+o \\ g+p & h+q & i+r \end{bmatrix}$$

addition property of equality (28)   For any numbers a, b, and c, if $a = b$, then $a + c = b + c$.

addition property of inequality (43)   For any numbers a, b, and c:

1. if $a > b$, then $a + c > b + c$;

2. if $a < b$, then $a + c < b + c$.

algebraic expressions (8)   An expression that contains at least one variable.

amplitude (824, 828)   The amplitude of the graph of a periodic function is the absolute value of half the difference between its maximum value and its minimum value.

angle of depression (776)   The angle formed by a horizontal line and the line of sight to an object at a lower level.

angle of elevation (776)   The angle formed by a horizontal line and the line of sight to an object at a higher level.

angle of incline (835)   The nonnegative acute angle formed by a vertical line and the line of an object's orientation.

antilogarithm (618)   If $\log x = a$, then $x = \text{antilog } a$.

Arcosine function (811)   The inverse of the Cosine function, symbolized by Cos^{-1} or Arccos.

arithmetic means (650)   The terms between any two nonconsecutive terms of an arithmetic sequence.

arithmetic sequence (648)   A sequence in which each term after the first is found by adding a constant, called the common difference d to the previous term.

arithmetic series (656)   The indicated sum of the terms of an arithmetic sequence.

associative properties (14)   The way three or more numbers are grouped, or associated, does not change their sum or product. That is, for all real numbers a, b, and c, $(a + b) + c = a + (b + c)$ and $(a \cdot b) \cdot c = a \cdot (b \cdot c)$.

asymptotes (441, 550)   Lines that a curve approaches.

augmented matrix (226)   The augmented matrix of a system of equations contains the coefficient matrix of the system with an extra column consisting of the constant terms of the system.

axis of symmetry (335)   The line about which a figure is symmetric.

B

back-to-back stem-and-leaf plot (20)   A back-to-back stem-and-leaf plot is used to compare two sets of data. The same stem is used for the leaves of both plots.

bell curve (393)   A symmetric curve that is the general shape of the graph of a normal distribution, which indicates that the frequencies are concentrated around the center portion of the distribution.

best-fit line (95)   A line drawn on a scatter plot to approximate the linear relationship for a set of data points.

binomial (261)   A polynomial with two unlike terms.

binomial experiments (753)   A binomial experiment exists if and only if there are exactly two possible outcomes for any trial, there is a fixed number of trials, the trials are independent, and the probability of each trial is the same.

binomial theorem (696)   If n is a positive integer, then $(a + b)^n = 1a^nb^0 + \frac{n}{1} a^{n-1}b^1 + \frac{n(n-1)}{1 \cdot 2} a^{n-2}b^2 + \ldots + 1a^0b^n$.

boundary (378)   A line or curve that separates a graph into two parts.

box-and-whisker plot (237)   A pictorial representation of the variability of a set of data that summarizes the data set using its quartiles and extreme values.

C

Cartesian coordinate plane (64)   Composed of the *x*-axis and the *y*-axis which meet at the origin and is divided into four quadrants.

center of a circle (423)   The point from which all points on a circle are equidistant.

center of a hyperbola (441)   The intersection of the conjugate and transverse axes of a hyperbola.

center of an ellipse (432)   The intersection of the major and minor axes of an ellipse.

change of base formula (594, 628)   For all positive numbers *a*, *b*, and *n*, where $a \neq 1$ and $b \neq 1$, $\log_a n = \dfrac{\log_b n}{\log_b a}$.

characteristic (618)   The integer used to express a base 10 logarithm as the sum of an integer and a positive decimal.

circle (423)   The set of all points in a plane that are equidistant from a given point in the plane.

circular functions (805)   The six basic trigonometric functions defined using a unit circle.

circular permutation (721)   If *n* distinct objects are arranged in a circle, then there are $\dfrac{n!}{n}$ or $(n - 1)!$ permutations of the objects around the circle.

coding matrix (212)   A message, where each letter corresponds to a numerical element in a matrix, can be converted to a code by multiplying the matrix by a coding matrix.

coefficient (255)   The numerical factor of a monomial.

column matrix (187)   A matrix that has only one column.

combination (726)   An arrangement, or listing, in which order is not important. The number of combinations of *n* distinct objects taken *r* at a time is $C(n, r) = \dfrac{n!}{(n - r)!r!}$.

common difference (649)   The number *d* added to find the next term of an arithmetic sequence.

common logarithms (617)   Logarithms to the base 10.

common ratio (662)   The number by which each term, after the first, in a geometric sequence is multiplied to obtain the next term.

commutative properties (14)   The order in which two numbers are added or multiplied does not change their sum or product. That is, for all real numbers *a* and *b*, $a + b = b + a$ and $a \cdot b = b \cdot a$.

complements (732)   Two events are complements of each other if the sum of their probabilities is 1.

completing the square (347)   A process used to create a perfect square trinomial.

complex conjugates (317)   Complex numbers of the form $a + bi$ and $a - bi$.

complex conjugate theorem (504)   Suppose *a* and *b* are real numbers with $b \neq 0$. If $a + bi$ is a zero of a polynomial function, then $a - bi$ is also a zero of the function.

complex fraction (565)   A fraction whose numerator and/or denominator contains a fraction.

complex numbers (311)   Any number that can be written in the form $a + bi$, where *a* and *b* are real numbers and *i* is the imaginary unit; *a* is called the real part, and *bi* is called the imaginary part.

composition of functions (520)   Suppose *f* and *g* are functions such that the range of *g* is a subset of the domain of *f*. Then the composite function $f \circ g$ can be described by the equation $[f \circ g](x) = f[g(x)]$.

compound inequality (49)   Two inequalities combined by the words *and* or *or*.

concentric circles (426)   Circles with the same center, but not necessarily the same radius.

conic section (415)   Any figure that can be formed by slicing a double cone with a plane.

conjugate axis (441)   The line segment that is perpendicular to the transverse axis at the center of a hyperbola and is an axis of symmetry of the hyperbola.

conjugate pair (506)   If a polynomial function has two imaginary roots of the form $a + bi$ and $a - bi$, these numbers form a conjugate pair.

conjugates (292)   Binomials in the form $a\sqrt{b} + c\sqrt{d}$ and $a\sqrt{b} - c\sqrt{d}$ where *a*, *b*, *c*, and *d* are rational numbers.

consistent system (127)   A system of equations that has at least one solution.

constant function (74)   A linear function of the form $f(x) = b$, for which the slope is zero.

constant of variation (556)   The constant k in either of the equations $y = kx$ or $y = \frac{k}{x}$.

constants (255)   Monomials that contain no variables.

constant term (334)   In a quadratic function described by $f(x) = ax^2 + bx + c$, c is the constant term.

constraints (153)   The inequalities in a system of inequalities whose graphs form the boundaries of the graph of the system's solution.

continuity (548)   A graph of a function that can be traced with a pencil that never leaves the paper is said to have continuity.

continuous function (67)   A continuous function can be graphed with a line or a smooth curve and has a domain with an infinite number of elements.

coordinate matrix (189)   A matrix containing the coordinates of the vertices of a geometric figure.

coordinate system (64)   Composed of two perpendicular number lines intersecting at their zero points creating a coordinate plane upon which ordered pairs may be graphed.

cosine (805)   If the terminal side of an angle θ in standard position intersects the unit circle at $P(x, y)$, then $\cos \theta = x$.

Cosine (811) $y = \text{Cos } x$ if and only if $y = \cos x$ and $0 \leq < x \leq \pi$.

coterminal angles (783)   Two angles in standard position having the same terminal side are called coterminal angles.

Cramer's rule (141)   A method that uses determinants to solve a system of linear equations.

D

degree **1.** (255)   The degree of a nonzero monomial is the sum of the exponents of its variables.   **2.** (261)   The degree of a polynomial is the degree of the monomial with the greatest degree.

dependent events (714)   The outcome of a dependent event is affected by the outcome of another event.

dependent system (128)   A system of equations that has an infinite number of solutions.

dependent variable (73)   A variable whose value depends upon, or is affected by, the value of another variable.

depressed polynomial (487)   When you divide a polynomial by one of its binomial factors, the quotient is called a depressed polynomial.

Descartes' rule of signs (505)   If $P(x)$ is a polynomial whose terms are arranged in descending powers of the variable,

- the number of positive real zeros of $y = P(x)$ is the same as the number of changes in sign of the coefficients of the terms, or is less than this by an even number, and

- the number of negative real zeros of $y = P(x)$ is the same as the number of changes in sign of the coefficients of the terms of $P(-x)$, or is less than this by an even number.

determinant (141, 205)   A square array of numbers or expressions enclosed between two parallel vertical bars.

dilation (190)   A transformation in which a figure is enlarged or reduced.

dimensions (187)   In a matrix consisting of n rows and m columns, the matrix is said to have dimension $n \times m$ (read "n by m").

direct variation **1.** (103)   A linear function described by $y = mx$ or $f(x) = mx$, where $m \neq 0$.   **2.** (556)   A type of variation where y varies directly as x if there is some constant k such that $y = kx$.

discrete function (65)   A function whose graph consists of points that are not connected.

discrete mathematics (188)   A branch of mathematics that deals with finite or discontinuous quantities.

discriminant (356)   For a quadratic equation $ax^2 + bz + c = 0$, the expression $b^2 - 4ac$ under the radical in the quadratic formula.

distance formula (409)   The distance between two points with coordinates (x_1, y_1) and (x_2, y_2) is given by $d = \sqrt{(x_2 - x_1)^2 + (y_2 - y_1)^2}$.

distributive properties (14)   For all real numbers a, b, and c, $a(b + c) = ab + ac$ and $(b + c)a = ba + ca$.

dividing rational expressions (563)   For all rational expressions $\frac{a}{b}$ and $\frac{c}{d}$, $\frac{a}{b} \div \frac{c}{d} = \frac{a}{b} \cdot \frac{d}{c} = \frac{ad}{bc}$, if $b \neq 0$, $c \neq 0$, and $d \neq 0$.

division property of equality (28)   For any real numbers a, b, and c, if $a = b$ and $c \neq 0$, then $\frac{a}{c} = \frac{b}{c}$.

division property of inequality (44)   For any real numbers a, b, and c:

1. if c is positive and $a < b$, then $\frac{a}{c} < \frac{b}{c}$;

2. if c is positive and $a > b$, then $\frac{a}{c} > \frac{b}{c}$;

3. if c is negative and $a < b$, then $\frac{a}{c} > \frac{b}{c}$;

4. if c is negative and $a > b$, then $\frac{a}{c} < \frac{b}{c}$.

domain (65)   The set of all first coordinates from the ordered pairs of a relation.

double-angle formulas (853)   The following identities hold true for all values of θ.

$\sin 2\theta = 2 \sin \theta \cos \theta$

$\cos 2\theta = \cos^2 \theta - \sin^2 \theta$

$\cos 2\theta = 1 - 2 \sin^2 \theta$

$\cos 2\theta = 2 \cos^2 \theta - 1$

element   1. (141)   A number or variable written within a determinant.   **2.** (186)   Each value in a matrix.

ellipse (431)   The set of all points in a plane such that the sum of the distances from the foci is constant.

empty set (40)   The set having no members, symbolized by { } or $\varnothing$.

equal matrices (188)   Two matrices are considered to be equal if they have the same dimensions and if each element of one matrix is equal to the corresponding element of the other matrix.

equation (27)   A sentence that states that two mathematical expressions are equal.

expansion by minors (205)   A method used to find the value of any third- or higher-order determinant by using determinants of lower order.

experimental probability (754)   A probability determined by performing tests or experiments and observing the outcomes.

exponential equations (626)   An equation in which variables occur in exponents.

exponential function (594, 597)   An equation of the form $y = a \cdot b^x$, where $a \neq 0$, $b > 0$, and $b \neq 1$, is called an exponential function with base b.

exponential growth (622)   Exponential growth occurs when a quantity increases exponentially.

exponential growth rate (626)   The positive constant k in the growth equation $P(t) = P_0 e^{kt}$ is called the exponential growth rate.

extraneous solutions (305)   Solutions that do not satisfy the original equation.

extreme values (304)   Data values that vary greatly from the central group of data values.

factorials (697)   If n is a positive integer, the expression $n!$ is defined as $n! = n(n - 1)(n - 2)(n - 3) \cdots 2 \cdot 1$. By definition, $0! = 1$.

factoring (341)   A method used to solve a quadratic equation that requires using the zero product property.

factors (274)   Numbers, variables, monomials, or polynomials multiplied to obtain a product.

factor theorem (487)   The binomial $x - a$ is a factor of the polynomial $f(x)$ if and only if $f(a) = 0$.

failure (732)   Any outcome other than the desired outcome of an event.

family of graphs (83)   A group of graphs that displays one or more similar characteristics.

feasible region (153)   The area of intersection of the graphs of inequalities in a system of inequalities in which every constraint is met.

Fibonacci sequence (683)   A special sequence often found in nature that is named after its discoverer, Leonardo Fibonacci.

fixed points (526)   The points at which the graph of a function $g(x)$ intersects the graph of the line $f(x) = x$.

FOIL method (263)   An application of the distributive property used to multiply two binomials. The product is the sum of the products of the first, outer, inner and last terms.

formula (9)   A mathematical sentence that expresses a general relationship between certain quantities.

fractal (318, 688)   A geometric figure that has self-similarity, is created using a recursive process, and is infinite in structure.

fractal geometry (688) A new branch of mathematics that provides models of designs in nature.

frequency distribution (392) Shows how data are spread out over the range of values.

function (65) A special type of relation in which each element of the domain is paired with exactly one element of the range.

fundamental counting principle (713) If event M can occur in m ways and is followed by an event N that can occur in n ways, then the event M followed by the event N can occur in $m \cdot n$ ways.

fundamental theorem of algebra (503) Every polynomial equation with degree greater than zero has at least one root in the set of complex numbers.

general formula for growth and decay (631) The formula is $y = ne^{kt}$, where y is the final amount, n is the initial amount, k is a constant, and t is the time.

geometric means (665) The terms between any two nonconsecutive terms in a geometric sequence.

geometric sequence (662) A sequence in which each term after the first is found by multiplying the previous term by a constant called the common ratio, r.

geometric series (670) The indicated sum of the terms of a geometric series.

graphical iteration (526) Graphical iterations produce a visual representation of the process of iteration.

greatest integer function (104) A type of step function described by $f(x) = [x]$ where $[x]$ is the greatest integer *not* greater than x.

guess and check (342) A problem-solving strategy in which several values or combinations of values are tried in order to find a solution to a problem.

half-angle formulas (855) The following identities hold true for all values of α.

$$\cos \frac{\alpha}{2} = \pm \sqrt{\frac{1 + \cos \alpha}{2}}$$

$$\sin \frac{\alpha}{2} = \pm \sqrt{\frac{1 - \cos \alpha}{2}}$$

histogram (392) A bar graph that displays a frequency distribution.

hyperbola (440) The set of all points P in a plane such that the absolute value of the difference of the distances from P to two given points, called the foci, is constant.

hypotenuse (9) The side opposite the right angle in a right triangle.

hypothesis (760) A statement that is to be tested in an experiment.

———— I ————

identify subgoals (296) A problem-solving strategy that uses a series of small steps, or subgoals.

identity function (103) A linear function described by $y = x$ or $f(x) = x$.

identity matrix (213) Any square matrix A that, when multiplied by another matrix B of the same dimensions, equals that same matrix B.

identity properties (14) If a is a real number, then $a + 0 = a = 0 + a$ and $a \cdot 1 = a = 1 \cdot a$.

imaginary unit (310) The imaginary unit i is defined by $i = \sqrt{-1}$.

inclusive events (746) Two events are inclusive if the outcomes of the events may be the same.

inconsistent system (128) A system of equations that has no solution.

independent events (713) The outcome of an independent event is not affected by the outcome of another event.

independent system (127) A system of equations that has exactly one solution.

independent variable (73) The variable whose value does not depend upon, nor is affected by, the value of another variable.

index of summation (658) The variable defined below the Σ in sigma notation.

infinite geometric series (676) The indicated sum of the terms of an infinite geometric sequence.

initial side (780) One of two rays forming an angle on a coordinate plane that is fixed along the positive x-axis.

integral zero theorem (509) If the coefficients of a polynomial function are integers such that $a_0 = 1$ and $a_n \neq 0$, any rational zeros of the function must be factors of a_n.

interquartile range (236)   The difference between the upper and lower quartile of a set of data.

intersection (49)   The graph of a compound inequality containing the word *and* is the intersection of the graphs of the two inequalities.

inverse Cosine (811)   Given $y = \text{Cos } x$, the inverse Cosine function is defined by $y = \text{Cos}^{-1} x$ or $y = \text{Arccos } x$.

inverse functions (528)   Two functions f and g are inverse functions if and only if both their compositions are the identity function. That is, $[f \circ g](x) = x$ and $[g \circ f](x) = x$.

inverse of a matrix (213)   The matrix by which another matrix is multiplied to produce the identity matrix.

inverse properties (14)   If a is a real number, then $a + (-a) = 0 = (-a) + a$ and $a \cdot \frac{1}{a} = 1 = \frac{1}{a} \cdot a$, if $a \neq 0$.

inverse relation (531)   Two relations are inverse relations if and only if whenever one relation contains the element (a, b), the other relation contains the element (b, a).

inverse Sine (812)   Given $y = \text{Sin } x$, the inverse Sine function is defined by $y = \text{Sin}^{-1} x$ or $y = \text{Arcsin } x$.

inverse Tangent (812)   Given $y = \text{Tan } x$, the inverse Tangent function is defined by $y = \text{Tan}^{-1} x$ or $y = \text{Arctan } x$.

inverse variation (557)   A type of variation where y varies inversely as x if there is some constant k such that $xy = k$ or $y = \frac{k}{x}$.

irrational number (13)   Irrational numbers are real numbers that cannot be written as terminating or repeating decimals.

iterate **1.** (318)   To repeat a process or function over and over. **2.** (526)   Each result of the iteration process is called an iterate.

iteration (522)   The process of finding the value of a function for a given number in the domain and generating a sequence of values by using each output as the next input value.

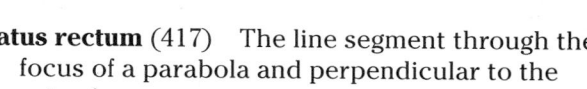

joint variation (558)   A type of variation where y varies jointly as x and z if there is some number k such that $y = kxz$, where $x \neq 0$ and $z \neq 0$.

latus rectum (417)   The line segment through the focus of a parabola and perpendicular to the axis of symmetry.

law of cosines (799)   For any triangle ABC with a, b, and c representing the measures of sides, and opposite angles with measurement A, B, and C, respectively, the following equations are true.

$$a^2 = b^2 + c^2 - 2bc \cos A$$

$$b^2 = a^2 + c^2 - 2ac \cos B$$

$$c^2 = a^2 + b^2 - 2ab \cos C$$

law of sines (793)   Let $\triangle ABC$ be any triangle with a, b, and c representing the measures of sides opposite angles with measurements A, B, and C, respectively. Then,

$$\frac{\sin A}{a} = \frac{\sin B}{b} = \frac{\sin C}{c}.$$

leading coefficient (480)   The coefficient of the term with the highest degree.

legs (9)   The two sides of a right triangle that form the right angle.

like radical expressions (291)   Two or more radical expressions are called like radical expressions if both the indices and the radicands are alike.

like terms (261)   Two or more monomials that are the same or differ only by their numerical coefficients.

linear equation (73)   An equation whose graph is a line.

linear function (74)   A function is linear if it can be defined by $f(x) = mx + b$, where m and b are real numbers.

linear permutation (718)   The arrangement of things in a line.

linear programming (155)   A method for finding the maximum or minimum value of a function in two variables subject to given constraints on the variables.

linear term (334)   In a quadratic function described by $f(x) = ax^2 + bx + c$, bx is the linear term.

line plot (19)   Displays statistical data on a number line so that patterns and variability in data can be determined.

list possibilities (38)   A problem-solving strategy used to find all possible solutions to a problem.

location principle (494)   Suppose $y = f(x)$ represents a polynomial function and a and b are two numbers such that $f(a) < 0$ and $f(b) > 0$. Then the function has at least one real zero between a and b.

logarithm (605)   Suppose $b > 0$ and $b \neq 1$. For $n > 0$, there is a number p such that $\log_b n = p$ if and only if $b^p = n$.

logarithmic function (594, 607)   A function of the form $y = \log_b x$, where $b > 0$ and $b \neq 1$.

look for a pattern (80)   A problem-solving strategy often involving the use of tables to organize information so that a pattern may be determined.

lower extreme (236)   The least value in a set of data.

lower quartile (236)   The median of the lower half of a set of data.

major axis (432)   The longer of the two line segments that form the axes of symmetry for an ellipse.

mantissa (618)   The logarithm of a number between 1 and 10.

mapping (65)   A mapping shows how each member of the domain of a relation is paired with each member of the range.

matrix (186)   A rectangular array of variables or constants in horizontal rows and vertical columns, usually enclosed in brackets.

matrix equation (219)   An equation of the form $AX = C$, where A is the coefficient matrix for a 'system of linear equations, X is the column matrix consisting of the variables of the system, and C is the column matrix consisting of the constant terms of the system.

matrix logic (187)   A problem-solving strategy where information is organized using a matrix and possibilities are eliminated one by one until a solution is eventually found.

mean (21)   The sum of all the values in a set of data divided by the number of values.

measure of central tendency (21)   A number that represents the center or middle of a set of data.

median (21)   The middle value of a set of data. If there are two middle values, the median is the mean of those values.

midpoint formula (410)   If a line segment has endpoints at (x_1, y_1) and (x_2, y_2), then the midpoint of the line segment has coordinates $\left(\dfrac{x_1 + x_2}{2}, \dfrac{y_1 + y_2}{2}\right)$.

minor (205)   The determinant formed when the row and column containing an element of the given determinant are deleted.

minor axis (432)   The shorter of the two line segments that form the axes of symmetry for an ellipse.

mode (21)   The most frequent value in a set of data.

monomials (255)   An expression that is a number, a variable, or the product of a number and one or more variables.

multiplication property of equality (28)   For any real numbers a, b, and c, if $a = b$, then $a \cdot c = b \cdot c$.

multiplication property of inequality (44)   For any real numbers a, b, and c:

1. if c is positive and $a < b$, then $ac < bc$;

2. if c is positive and $a > b$, then $ac > bc$;

3. if c is negative and $a < b$, then $ac > bc$;

4. if c is negative and $a > b$, then $ac < bc$.

multiplying matrices (199)   The product of $A_{m \times n}$ and $B_{n \times r}$ is $(AB)_{m \times r}$. The element in the ith row and the jth column of AB is the sum of the products of the corresponding elements in the ith row of A and the jth column of B.

multiplying rational expressions (563)   For all rational expressions $\dfrac{a}{b}$ and $\dfrac{c}{d}$, $\dfrac{a}{b} \cdot \dfrac{c}{d} = \dfrac{ac}{bd}$, if $b \neq 0$ and $d \neq 0$.

mutually exclusive events (746)   Two events are mutually exclusive if their outcomes can never be the same.

natural logarithms (622)   Natural logarithms are logarithms to the base e.

normal distribution (393)   A data distribution that gives a bell-shaped, symmetric graph. About 68% of the values are within one standard deviation from the mean. About 95% of the values are within two standard deviations from the mean. About 99% of the values are within three standard deviations from the mean.

nth root (282)   For any real numbers a and b, and any positive integer n, if $a^n = b$, then a is an nth root of b.

octants (172)   Three mutually perpendicular planes separate space into eight regions, each called an octant.

odds (734)   The odds of a successful outcome of an event can be expressed as the ratio of the number of ways it can succeed to the number of ways it can fail.

open sentence (27)   A mathematical sentence that contains variables that are to be replaced with numerical values.

ordered pairs (64)   Points in a plane can be located by using ordered pairs of real numbers. An ordered pair consists of the coordinates of the point.

ordered triple (166)   The solution of a system of equations in three variables.

order of operations (7)   To evaluate a numerical expression, the order of operations must be followed. First, simplify the expressions inside grouping symbols. Evaluate all powers. Do all multiplications and divisions from left to right. Finally, do all additions and subtractions from left to right.

organize data (580)   A problem-solving strategy using diagrams, tables, and charts to arrange and evaluate data in order to determine a solution.

origin (64)   The point $(0, 0)$ at which the axes of a coordinate plane intersect.

outlier (237)   Any value in the set of data that is at least 1.5 interquartile ranges beyond the upper and lower quartile.

parabola   1. (332, 335)   The general shape of the graph of a quadratic function.   **2.** (415) The set of all points in a plane that are the same distance from a given point called the focus and a given line called the directrix.

parallel lines (83)   In a plane, lines with the same slope are parallel. All vertical lines are parallel, and all horizontal lines are parallel.

parent graph (83)   The graph of the simplest polynomial of the form $f(x) = x^n$ in a family of graphs.

partial sum (676)   In an infinite series, S_n is called a partial sum because it is the sum of a certain number of terms and not the sum of the entire series.

Pascal's triangle (695)   The pyramid formation of the coefficients of binomial expansion.

periodic function (807)   A function is called periodic if there is a positive number a such that $f(x) = f(x + a)$ for all x in the domain of the function. The least positive value of a for which $f(x) = f(x + a)$ is called the period of the function.

permutation (718)   An arrangement of things in a certain order.

perpendicular lines (83)   In a plane, two oblique lines are perpendicular if and only if the product of their slopes is -1. Any vertical line is perpendicular to any horizontal line.

point discontinuity (548)   Point discontinuity occurs for values of x that make the denominator of a rational function zero and visually appears as breaks in the continuity of the graph of the rational function.

point-slope form (89)   The point-slope form of the equation of a line is $y - y_1 = m(x - x_1)$, where (x_1, y_1) are the coordinates of a point on the line and m is the slope of the line.

polynomial (261)   A monomial or the sum of monomials.

polynomial function (479)   A function that can be described by an equation of the form $P(x) = a_0 x^n + a_1 x^{n-1} + \ldots + a_{n-2} x^2 + a_{n-1} x + a_n$, where the coefficients $a_0, a_1, a_2, \ldots, a_n$ are real numbers, a_0 is not zero, and n is a nonnegative integer.

polynomial in one variable (478)   A polynomial in one variable x is an expression of the form $a_0 x^n + a_1 x^{n-1} + \ldots + a_{n-2} x^2 + a_{n-1} x + a_n$, where the coefficients $a_0, a_1, a_2, \ldots, a_n$ represent real numbers, a_0 is not zero, and n represents a nonnegative integer.

power (255)   An expression in the form x^n.

power property of logarithms (613)   For any real number p and positive numbers m and b, where $b \neq 1$, $\log_b m^p = p \cdot \log_b m$.

prediction equation (95)   The equation of the best-fit line suggested by the data points of a scatter plot. It may be used to estimate, or predict, one of the variables given the other.

prime number (274)   A whole number greater than 1 whose only factors are 1 and itself.

principal root (282)   The principal root is the nonnegative root. If there is a negative root but no nonnegative root, the principal root is the negative root.

principal values (811) The values in the restricted domains of the functions Cosine, Sine, and Tangent are principal values.

probability (732) If an event can succeed in s ways and fail in f ways, then the probabilities of success, $P(s)$, and of failure, $P(f)$, are

$$P(s) = \frac{s}{s+f} \text{ and } P(f) = \frac{f}{f+s}.$$

probability matrix (200) A probability matrix used in situations involving chance.

problem-solving plan (30) A four-step plan used to solve problems that requires one to explore the problem, plan the solution, solve the problem, and examine the solution.

problem-solving strategies (38) Various strategies that are used alone or in combination with each other to solve problems.

product property of logarithms (611) For all positive numbers m, n, and b, where $b \neq 1$, $\log_b mn = \log_b m + \log_b n$.

property of equality for exponential functions (599) Suppose b is a positive number other than 1. Then $b^{x_1} = b^{x_2}$ if and only if $x_1 = x_2$.

property of equality for logarithmic functions (608) Suppose $b > 0$ and $b \neq 1$. Then $\log_b x_1 = \log_b x_2$ if and only if $x_1 = x_2$.

pure imaginary number (310) A complex number of the form bi, where b is real and $b \neq 0$ and i is the imaginary unit.

Pythagorean theorem (9) The Pythagorean theorem states that in a right triangle, if a and b are the measures of the legs and c is the measure of the hypotenuse, then $c^2 = a^2 + b^2$.

quadrantal angles (787) Angles that have their terminal sides on an axis where x or y is equal to zero.

quadrants (64) The four regions into which two perpendicular number lines separate the plane.

quadratic equation (336) An equation that can be written in the form $ax^2 + bx + c = 0$, where $a \neq 0$.

quadratic form (515) For any numbers a, b, and c, except $a = 0$, an equation that can be written as $a[f(x)]^2 + b[f(x)] + c = 0$, where $f(x)$ is some expression in x, is in quadratic form.

quadratic formula (354) The solutions of a quadratic equation of the form $ax^2 + bx + c$, where $a \neq 0$, are given by the quadratic formula, which is $x = \dfrac{-b \pm \sqrt{b^2 - 4ac}}{2a}$.

quadratic function (332, 334) A function described by an equation that can be written in the form $f(x) = ax^2 + bx + c$, where $a \neq 0$.

quadratic inequality (378) An inequality described by $y < ax^2 + bx + c$, where $a \neq 0$.

quadratic term (334) In a quadratic function described by $f(x) = ax^2 + bx + c$, the term ax^2.

quartiles (236) The values in a set of data that separate the data into four sections, each containing 25% of the data.

quotient property of logarithms (612) For all positive numbers m, n, and b, where $b \neq 1$, $\log_b \dfrac{m}{n} = \log_b m - \log_b n$.

radian (781) The measure of an angle that intercepts an arc whose length is one unit.

radical equations (305) Equations that contain radical expressions with variables in the radicand.

radius (423) Any line segment whose endpoints are the center of a circle and a point on the circle.

random sample (758) A sample in which every member of the population has an equal chance to be selected.

range **1.** (65) The set of all second coordinates from the ordered pairs of a relation. **2.** (235) The difference between the greatest and least values in a set of data.

rational algebraic expressions (562) An algebraic expression can be expressed as the quotient of two polynomials where the denominator does not equal zero.

rational equation (576) An equation that contains one or more rational expressions.

rational exponent (298) For any nonzero number b, and any integers m and n, with $n > 1$, $b^{\frac{m}{n}} = \sqrt[n]{b^m} = \left(\sqrt[n]{b}\right)^m$, except when $b > 0$ and n is even.

rational function (550) An equation of the form $f(x) = \dfrac{p(x)}{q(x)}$, where $p(x)$ and $q(x)$ are polynomial functions and $q(x) \neq 0$.

rationalizing the denominator (290)   A process used to eliminate radicals from a denominator or fractions from a radicand.

rational number (13)   A number that can be expressed as a ratio $\frac{m}{n}$, where m and n are integers and n is not zero.

rational zero theorem (509)   Let $f(x) = a_0x^n + a_1x^{n-1} + \ldots + a_{n-1}x + a_n$ represent a polynomial function with integral coefficients $(a \neq 0)$. If $\frac{p}{q}$ is a rational number in simplest form and is a zero of $y = f(x)$, then p is a factor of a_n and q is a factor of a_0.

real numbers (13)   The set of irrational numbers together with the set of rational numbers.

recursive formula (683)   A recursive formula has two parts: the value(s) of the first term(s), and a recursion equation that shows how to find each term from the term(s) before it.

reduced matrix (227)   The solution matrix that results when row operations are performed on an augmented matrix to solve a system of linear equations.

reducing a matrix (227)   The process of performing row operations on an augmented matrix to get the desired solution matrix.

reflections (722)   A reflection occurs when an arrangement of objects can be seen from two different views without changing the arrangement.

reflexive property of equality (28)   For any real number a, $a = a$.

relation (65)   A set of ordered pairs.

relative maximum (494)   A point on the graph of a function is called a relative maximum if no other nearby points have a greater y-coordinate.

relative minimum (494)   A point on the graph of a function is called a relative minimum if no other nearby points have a lesser y-coordinate.

remainder theorem (485)   If a polynomial $f(x)$ is divided by $x - a$, the remainder is the constant $f(a)$, and $f(x) = q(x) \cdot (x - a) + f(a)$, where $q(x)$ is a polynomial with degree one less than the degree of $f(x)$.

root (332, 336)   A solution of an equation.

rotation (201)   A transformation in which a figure is moved around a center point.

row matrix (187)   A matrix that has only one row.

row operations (226)   Operations performed on rows of an augmented matrix to find the solution to a system of linear equations.

sampling error (759)   The difference between the sample results and the true population results.

scalar multiplication (188)   Multiplication of each element of a matrix is multiplied by the same constant.

scatter plot (95)   A scatter plot visually shows the nature of a relationship that is determined both by the shape and closeness of the data.

scientific notation (254)   A number is in scientific notation when it is in the form $a \times 10^n$, where $1 \leq a < 10$ and n is an integer.

second-order determinant (141)   A determinant that has two rows and two columns. Its value can be found by calculating the difference of the products of the two diagonals.

self-similarity (689)   A characteristic of an object in which replicas of an entire shape or object are embedded over and over again inside the object in different sizes.

sequence (649)   A list of numbers in a specific order.

series (656)   The indicated sum of the terms of a sequence.

sigma notation (658)   Notation that uses the Σ symbol to indicate a sum of a series.

simplify (255)   To simplify an expression containing powers means to rewrite the expression without parentheses or negative exponents.

simulation (754)   A method for finding experimental probability in which a device is used to model the event and observations are made of how the model responds to the conditions listed in a given problem.

sine (805)   If the terminal side of an angle θ in standard position intersects the unit circle at $P(x, y)$, then $\sin \theta = y$.

Sine (811) $y = \text{Sin } x$ if and only if $y = \sin x$ and $-\frac{\pi}{2} \leq x \leq \frac{\pi}{2}$.

skewed distribution (395)   A distribution curve that is not symmetric.

slope (82)   The slope of the line that passes through points (x_1, y_1) and (x_2, y_2) is given by $m = \dfrac{y_2 - y_1}{x_2 - x_1}$, where $x_1 \neq x_2$. The slope of a line described by $f(x) = mx + b$ is m.

slope-intercept form (88)   The slope-intercept form of the equation of a line is $y = mx + b$, where m is the slope and b is the y-intercept.

solution (28)   A value that can replace a variable in an equation to make the equation true.

solve a simpler problem (156, 712)   A problem-solving strategy in which a simpler, similar problem is solved to determine the concepts that can be used in solving a more complex problem.

solving a right triangle (775)   The process of finding the measures of all sides and angles of a right triangle.

square matrix (187)   A matrix that has the same number of rows and columns.

square root (281)   For any real numbers a and b, if $a^2 = b$, then a is a square root of b.

square root function (535)   A function that involves a square root.

standard deviation (384)   From a set of data with n values, where x_1 represents the first term and x_n represents the nth term, if $\bar{x}$ represents the mean, then the standard deviation is SD or
$$\sigma_{\bar{x}} = \sqrt{\dfrac{(x_1 - \bar{x})^2 + (x_2 - \bar{x})^2 + \dots + (x_n - \bar{x})^2}{n}}.$$

standard form (73)   The standard form of a linear equation is $Ax + By = C$, where A, B, and C are real numbers and A and B are not both zero.

standard position (780)   An angle positioned so that its vertex is at the origin and its initial side is along the positive x-axis is said to be in standard position.

stem-and-leaf plot (19)   A compact way to display data in which each item is separated into two parts. The stem consists of the digits in the greatest common place value. The leaves contain the other digits of each item of data.

step functions (103)   A function whose graph is a series of disjoint lines or steps.

substitution property of equality (28)   If $a = b$, then a may be replaced by b.

subtraction property of equality (28)   For any numbers a, b, and c, if $a = b$, then $a - c = b - c$.

subtraction property of inequality (43)   For any numbers a, b, and c:
1. if $a > b$, then $a - c > b - c$;
2. if $a < b$, then $a - c < b - c$.

success (732)   A desired outcome of an event.

sum of a geometric series (671)   The sum S_n of the first n terms of a geometric series is given by $S_n = \dfrac{a_1 - a_1 r^n}{1 - r}$ or $S_n = \dfrac{a_1(1 - r^n)}{1 - r}$, where $r \neq 1$.

sum of an arithmetic series (657)   The sum S_n of the first n terms of an arithmetic series is given by $S_n = \dfrac{n}{2}(a_1 + a_n)$, where n is a positive integer.

sum of an infinite geometric series (677)   The sum S of an infinite geometric series where $-1 < r < 1$ is given by $S = \dfrac{a_1}{1 - r}$.

symmetric property of equality (28)   For all real numbers a and b, if $a = b$, then $b = a$.

synthetic division (269)   A simpler method than long division used to divide a polynomial by a binomial.

synthetic substitution (486)   The process of using synthetic division to find the value of a function.

system of equations (126)   A set of equations with the same variables.

system of inequalities (148)   A set of inequalities with the same variables.

T

tangent (425)   A line that intersects a circle in exactly one point is said to be tangent to the circle.

Tangent (811) $y = \text{Tan } x$ *if and only if* $y = \tan x$ and $-\dfrac{\pi}{2} \leq x \leq \dfrac{\pi}{2}$.

term (648)   Each number in a sequence.

terminal side (780)   One of two rays forming an angle on a coordinate plane that rotates about the origin.

terms (261)   The monomials that make up a polynomial.

theoretical probability (754)   Theoretical probability is determined using mathematical methods to provide an idea of what outcomes might occur in a given situation.

transformations (190)   Functions that map points of a graph onto its image.

transition matrix (200)   A matrix that contains information about the transition from one event to another.

transitive property of equality (28)   For all real numbers a, b, and c, if $a = b$ and $b = c$, then $a = c$.

translation (195)   A transformation in which a figure is moved from one location to another on the coordinate plane without changing its size, shape, or orientation.

translation matrix (196)   A matrix that represents the number of units a figure is moved vertically or horizontally from its original position.

transverse axis (441)   One of two line segments that form the axes of symmetry for a hyperbola whose endpoints are the vertices of a hyperbola.

trichotomy property (43)   For any two real numbers, a and b, exactly one of the following statements is true: $a < b$, $a = b$, or $a > b$.

trigonometric equations (861)   An equation involving one or more trigonometric functions which is true for some, but not all, values of the variable.

trigonometric functions (772)   If θ is the measure of one acute angle in a right triangle, a is the measure of the leg opposite θ, b is the measure of the leg adjacent to θ, and c is the measure of the hypotenuse, then

$$\text{sine } \theta = \frac{a}{c} \qquad \text{cosine } \theta = \frac{b}{c}$$
$$\text{tangent } \theta = \frac{a}{b} \qquad \text{cosecant } \theta = \frac{c}{a}$$
$$\text{secant } \theta = \frac{c}{b} \qquad \text{cotangent } \theta = \frac{b}{a}$$

trigonometric identities (788, 835)   A trigonometric equation that is true for all values of the variables.

trigonometry (772)   The study of the relationships between the angles and sides of right triangles.

trinomial (261)   A polynomial with three unlike terms.

unbounded (155)   A function is unbounded if a polygonal region is not formed by the constraints of a system of inequalities.

union (50)   The graph of a compound inequality containing the word *or* is the union of the graphs of the two inequalities.

unit circle (781)   In the coordinate plane, the circle with a radius of one unit and center at the origin.

upper extreme (236)   The greatest value in a set of data.

upper quartile (236)   The median of the upper half of a set of data.

use a simulation (452)   A problem-solving strategy that uses models, or simulations, of mathematical situations that are difficult to solve directly.

variables (27)   Letters used to represent numbers that are not known.

vertex   1. (335)   The point at which a parabola and its axis of symmetry intersect.   **2.** (441) The point on each branch of a hyperbola that is nearest the center of the hyperbola.

vertical line test (65)   If every vertical line drawn on the graph of a relation passes through no more than one point of the graph, then the relation is a function.

work backward (528)   A problem-solving strategy using inverse operations to determine an original value.

x-axis (64)   The horizontal number line that helps to form the coordinate plane.

x-intercept (75)   The x-coordinate of the point at which a graph crosses the x-axis.

y-axis (64)   The vertical number line that helps to form the coordinate plane.

y-intercept (75)   The y-coordinate of the point at which a graph crosses the y-axis.

zero (335)   For any function $f(x)$, if $f(a) = 0$, then a is a zero of the function.

zero product property (341)   For any real numbers a and b, if $ab = 0$, then either $a = 0$, $b = 0$, or both.

SPANISH GLOSSARY

A

absolute value/valor absoluto (37)   El valor absoluto de un número equivale al número de unidades que dicho número dista de cero en la recta numérica.

absolute value function/función valor absoluto (104)   Función descrita por $y = |x|$ o por $f(x) = |x|$.

addition of matrices/suma de matrices (195)   Si A y B son dos matrices de tipo $m \times n$, entonces $A + B$ es la matriz en que cada elemento es la suma de los elementos correspondientes de A y B.

$$\begin{bmatrix} a & b & c \\ d & e & f \\ g & h & i \end{bmatrix} + \begin{bmatrix} j & k & l \\ m & n & o \\ p & q & r \end{bmatrix} = \begin{bmatrix} a+j & b+k & c+l \\ d+m & e+n & f+o \\ g+p & h+q & i+r \end{bmatrix}$$

addition property of equality/propiedad de adición de la igualdad (28)   Para cualquiera de los números a, b y c, si $a = b$, entonces $a + c = b + c$.

addition property of inequality/propiedad de adición de la desigualdad (43)   Para cualquiera de los números a, b y c:

1. si $a > b$, entonces $a + c > b + c$;
2. si $a < b$, entonces $a + c < b + c$.

algebraic expressions/expresiones algebraicas (8)   Una expresión que contiene por lo menos una variable.

amplitude/amplitud (824, 828)   La amplitud de la gráfica de una función periódica es el valor absoluto de la mitad de la diferencia entre su valor máximo y su valor mínimo.

angle of depression/ángulo de depresión (776)   El ángulo formado por una recta horizontal y la línea de visión hasta un objeto a un nivel más bajo.

angle of elevation/ángulo de elevación (776)   El ángulo formado por una recta horizontal y la línea de visión hasta un objeto a un nivel más alto.

angle of incline/ángulo de inclinación (835)   El ángulo agudo no negativo formado por una recta vertical y la línea de orientación de un objeto.

antilogarithm/antilogaritmo (618)   Si $\log x = a$, entonces, $x = $ antilog a.

Arcosine function/función arco coseno (811)   La inversa a la función coseno, simbolizada por Cos^{-1} o Arc cos.

arithmetic mean/media aritmética (650)   Los términos entre dos términos no consecutivos de una secuencia aritmética.

arithmetic sequence/secuencia aritmética (648)   Secuencia en que cada término después del primero se halla sumando una constante, llamada diferencia común d, al término previo.

arithmetic series/serie aritmética (656)   La suma indicada de los términos de una secuencia aritmética.

associative properties/propiedades asociativas (14)   La forma en que tres o más números están agrupados, o asociados, no cambia el resultado de su suma o de su producto. Es decir, para todo número real a, b, y c, $(a + b) + c = a + (b + c)$ y $(a \cdot b) \cdot c = a \cdot (b \cdot c)$.

asymptotes/asíntotas (441, 550)   Curva que se aproxima a unas rectas.

augmented matrix/matriz aumentada (226)   La matriz aumentada de un sistema de ecuaciones contiene la matriz coeficiente del sistema junto con una columna adicional de los términos constantes del sistema.

axis of symmetry/eje de simetría (335)   La línea a partir de la cual una figura es simétrica.

B

back-to-back stem-and-leaf plot/diagrama de tallo y hojas consecutivo (20)   Un diagrama de tallo y hojas consecutivo se usa para comparar dos conjuntos de datos. El mismo tallo se usa para las hojas de ambos diagramas.

bell curve/curva acampanada (393)   Una curva simétrica que tiene la misma forma general que la gráfica de una distribución normal, la cual indica que las frecuencias están concentradas alrededor del centro de la distribución.

best-fit line/línea de mejor ajuste (95)   Una línea que se dibuja en un diagrama de dispersión para aproximar la relación lineal del conjunto de puntos de los datos.

binomial/binomio (261)   Un polinomio con dos términos desemejantes.

binomial experiments/experimentos binómicos (753)   Existe un experimento binómico si y solo si para cualquier prueba hay exactamente dos resultados posibles, hay un número determinado de pruebas, las pruebas son independientes y la probabilidad de cada prueba es la misma.

binomial theorem/teorema binómico (696)   Si n es un entero positivo, entonces, $(a + b)^n = 1a^n b^0 + \frac{n}{1} a^{n-1}b^1 + \frac{n(n - 1)}{1 \cdot 2} a^{n-2}b^2 + ... + 1a^0 b^n$.

boundary/frontera (378)   Una línea o curva que separa una gráfica en dos partes.

box-and-whisker plot/diagrama de caja y patillas (237) Una representación gráfica de la variabilidad de un conjunto de datos que resume el conjunto de datos usando sus valores cuartiles y extremos.

Cartesian coordinate plane/plano de coordenadas cartesianas (64) Plano compuesto del eje x y del eje y, los cuales se cruzan en un punto llamado origen. Está dividido en cuatro cuadrantes.

center of a circle/centro del círculo (423) El punto respecto al cual todos los puntos en un círculo están equidistantes.

center of a hyperbola/centro de la hipérbola (441) La intersección de los ejes conjugado y transversal de una hipérbola.

center of an ellipse/centro de la elipse (432) La intersección de los ejes principales y secundarios de una elipse.

change of base formula/cambio de fórmula de base (594, 628) Para todo número positivo a, b y n, en que $a \neq 1$ y $b \neq 1$, $\log_a n = \dfrac{\log_b n}{\log_b a}$.

characteristic/característica (618) El entero que se usa para expresar un logaritmo de base 10 como la suma de un entero y un decimal positivo.

circle/círculo (423) En un plano, el conjunto de todos los puntos que están equidistantes de un punto dado en el plano.

circular functions/funciones circulares (805) Las seis funciones trigonométricas básicas definidas usando un círculo unidad.

circular permutation/permutación circular (721) Si n número de objetos distintos se arreglan en forma de círculo, entonces hay $\dfrac{n!}{n}$ o $(n - 1)!$ permutaciones de los objetos alrededor del círculo.

coding matrix/matriz de codificación (212) Un mensaje en el cual cada letra corresponde a un elemento numérico en una matriz, y el cual puede ser convertido en código multiplicando la matriz por una matriz de cifrado.

coefficient/coeficiente (255) El factor numérico de un monomio.

column matrix/matriz de columna (187) Una matriz que tiene solo una columna.

combination/combinación (726) Una disposición, o arreglo, en el cual el orden no es importante. El número de combinaciones de n objetos distintos tomados r número de veces a la vez es $C(n, r) = \dfrac{n!}{(n - r)! r!}$.

common difference/diferencia común (649) El número d que se añade para hallar el próximo término de una secuencia aritmética.

common logarithms/logaritmos comunes (617) Logaritmos de base 10.

common ratio/proporción común (662) El número por el cual se multiplica cada término después del primero, para obtener el segundo en una secuencia geométrica.

commutative properties/propiedades conmutativas (14) El orden en que se suman o multiplican dos números no cambia el resultado de su suma o de su producto. Es decir, para todos los números reales a y b, $a + b = b + a$ y $a \cdot b = b \cdot a$.

complements/complementos (732) Dos eventos son complementos uno del otro si la suma de sus probabilidades es 1.

completing the square/completando el cuadrado (347) Un proceso que se usa para crear un cuadrado perfecto trinómico.

complex conjugates/complejos conjugados (317) Números complejos de la forma $a + bi$ y $a - bi$.

complex conjugate theorem/teorema del conjugado complejo (504) Suponiendo que a y b son números reales, en que $b \neq 0$, si $a + bi$ es un cero de una función polinómica, entonces $a - bi$ es también un cero de la función.

complex fraction/fracción compleja (565) Una fracción cuyo numerador y/o denominador contiene una fracción .

complex numbers/números complejos (311) Cualquier número que se puede escribir en la forma $a + bi$, en que a y b son números reales e i es la unidad imaginaria; a se denomina la parte real y bi la parte imaginaria.

composition of functions/composición de funciones (520) Suponiendo que f y g son funciones tales que la amplitud de g es un subconjunto del dominio de f, entonces la función compuesta $[f \circ g](x) = f[g(x)]$.

compound inequality/desigualdad compuesta (49) Dos desigualdades combinadas con las conjunciones y u o.

concentric circles/círculos concéntricos (426) Círculos con el mismo centro, pero no necesariamente el mismo radio.

conic section/sección cónica (415) Cualquier figura formada de la división de un cono doble con un plano.

conjugate axis/eje conjugado (441) El segmento de recta perpendicular al eje transversal en el centro de una hipérbola y que es un eje de simetría de la hipérbola.

conjugate pair/par conjugado (506) Si una función polinómica tiene dos raíces imaginarias de la forma $a + bi$ y $a - bi$, estos números forman un par conjugado.

conjugates/conjugados (292) Binomios de la forma $a\sqrt{b} + c\sqrt{d}$ y $a\sqrt{b} - c\sqrt{d}$ en que a, b, c y d son números racionales.

consistent system/sistema consistente (127)   Un sistema de ecuaciones que tiene por lo menos una solución.

constant function/función constante (74)   Una función lineal de la forma $f(x) = b$, para la cual la pendiente es cero.

constant of variation/constante de variación (556)   La constante k en cualquiera de las ecuaciones $y = kx$ o $y = \frac{k}{x}$.

constants/constantes (255)   Monomios que no contienen variables.

constant term/término constante (334)   En una función cuadrática descrita por $f(x) = ax^2 + bx + c$, el término constante es c.

constraints/restricciones (153)   Las desigualdades en un sistema de desigualdades cuyas gráficas forman las fronteras de la gráfica de la solución del sistema.

continuity/continuidad (548)   Una función que puede ser graficada sin que el lápiz deje el papel se dice que tiene continuidad.

continuous function/función continua (67)   Una función continua se puede trazar con una línea o una curva continua y tiene un dominio con un número infinito de elementos.

coordinate matrix/matriz de coordenadas (189)   Una matriz que contiene las coordenadas de los vértices de una figura geométrica.

coordinate system/sistema de coordenadas (64)   Compuesto de dos rectas numéricas perpendiculares que se intersecan en sus orígenes creando un plano de coordenadas sobre el cual se pueden trazar pares ordenados.

cosine/coseno (805)   Si el lado final de un ángulo θ en posición estándar interseca el círculo unidad en $P(x, y)$, entonces cos θ = x.

Cosine/coseno (811) $y = \cos x$, si y solo si $y = \cos x$ y $0 \leq x \leq \pi$.

coterminal angles/ángulos confinantes (783)   Dos ángulos en posición estándar y que comparten el mismo lado final se llaman ángulos confinantes.

Cramer's rule/regla de Cramer (141)   Un método que usa determinantes para resolver un sistema de dos ecuaciones lineales.

Ⓓ

---

degree/grado   1. (255) El grado de un monomio no nulo es la suma de los exponentes de sus variables. **2.** (261) El grado de un polinomio es el grado del monomio con el grado más alto.

dependent events/eventos dependientes (714)   El resultado de un evento dependiente se ve afectado por el resultado de otro evento.

dependent systems/sistemas dependientes (128)   Un sistema de ecuaciones con un número infinito de soluciones.

dependent variable/variable dependiente (73)   Una variable cuyo valor depende de, o es afectado por el valor de otra variable.

depressed polynomial/polinomio reducido (487)   Cuando uno divide un polinomio entre uno de sus factores binómicos, el cociente se llama polinomio reducido.

Descartes' rule of signs/regla de signos de Descartes (505)   Si $P(x)$ es un polinomio cuyos términos se dan en potencias descendientes de la variable,

- el número de ceros reales positivos de $y = P(x)$ es el mismo que el número de cambios de signo de los coeficientes de los términos, o es menos que esto por un número par, y

- el número de ceros reales negativos de $y = P(x)$, es el mismo que el número de cambios de signo de los coeficientes de los términos de $P(-x)$, o es menos que esto por un número par.

determinant/determinante (141, 205)   Un arreglo cuadrado de números o expresiones encerradas por dos barras verticales paralelas.

dilation/dilatación (190)   Una transformación en la cual se amplía o reduce una figura.

dimensions/dimensiones (187)   En una matriz consistente de n hileras y m columnas, se dice que la matriz tiene dimensión $n \times m$ (lo cual se lee "n por m").

direct variation/variación directa   1. (103) Una función lineal descrita por $y = mx$ o $f(x) = mx$, en que $m \neq 0$. **2.** (556) Un tipo de variación en que y varía directamente con x si existe alguna constante k, tal que $y = kx$.

discrete function/función discreta (65)   Una función cuya gráfica consiste de puntos que no están conectados.

discrete mathematics/matemeaticas discretas (188)   Una rama de las matemáticas que estudia cantidades finitas o interrumpidas.

discriminant/discriminante (356)   Para una expresión cuadrática $ax^2 + bx + c = 0$, la expresión $b^2 - 4ac$ bajo el signo radical en la fórmula cuadrática.

distance formula/fórmula de distancia (409)   La distancia entre dos puntos con coordenadas (x_1, y_1) y (x_2, y_2) es dada por $d = \sqrt{(x_2 - x_1)^2 + (y_2 - y_1)^2}$.

distributive properties/propiedades distributivas (14)   Para todos los números reales a, b y c, $a(b + c) = ab + ac$ y $(b + c)a = ba + ca$.

dividing rational expressions/división de expresiones racionales (563)   Para todas las expresiones racionales $\frac{a}{b}$ y $\frac{c}{d}$, $\frac{a}{b} \div \frac{c}{d} = \frac{a}{b} \cdot \frac{d}{c} = \frac{ad}{bc}$, si $b \neq 0$, $c \neq 0$ y $d \neq 0$.

division property of equality/propiedad de división de la igualdad (28)   Para cualquiera de los números reales *a*, *b* y *c*, si *a* = *b* y *c* ≠ 0, entonces $\frac{a}{c} = \frac{b}{c}$.

division property of inequality/propiedad de división de la desigualdad (44)   Para cualquiera de los números reales *a*, *b* y *c*:

1. si *c* es positivo y *a* < *b*, entonces $\frac{a}{c} < \frac{b}{c}$;

2. si *c* es positivo y *a* > *b*, entonces $\frac{a}{c} > \frac{b}{c}$;

3. si *c* es negativo y *a* < *b*, entonces $\frac{a}{c} > \frac{b}{c}$;

4. si *c* es negativo y *a* > *b*, entonces $\frac{a}{c} < \frac{b}{c}$.

domain/dominio (65)   El conjunto de todas las primeras coordenadas de los pares ordenados de una relación.

double-angle formulas/fórmulas de doble ángulo (853)   Las siguientes identidades son válidas para todos los valores de θ.

$\sin 2\theta = 2 \sin \theta \cos \theta$
$\cos 2\theta = \cos^2 \theta - \sin^2 \theta$
$\cos 2\theta = 1 - 2 \sin^2 \theta$
$\cos 2\theta = 2 \cos^2 \theta - 1$

element/elemento   1. (141) Un número o variable escrito dentro de una determinante.   **2.** (186) Cada valor en una matriz.

ellipse/elipse (431)   El conjunto de todos los puntos en un plano de manera que la suma de las distancias desde los focos es constante.

empty set/conjunto vacío (40)   El conjunto sin ningún elemento, representado con { } o Ø.

equal matrices/matrices idénticas (188)   Se considera que dos matrices son idénticas si poseen las mismas dimensiones y si cada elemento de una matriz es igual al elemento correspondiente de la otra matriz.

equation/ecuación (27)   Un enunciado matemático que dice que dos expresiones matemáticas son iguales.

expansion by minors/expansión de menores (205)   Un método que se usa para hallar el valor de cualquier determinante de tercer orden o más alta mediante el uso de determinantes de orden más bajo.

experimental probability/probabilidad experimental (754)   Una probabilidad que se determina llevando a cabo pruebas o experimentos y observando los resultados.

exponential equation/ecuación exponencial (626)   Una ecuación en que las variables ocurren en forma de exponentes.

exponential function/función exponencial (594, 597)   Una ecuación de la forma $y = a \cdot b^x$, en que $a \neq 0$ y $b \neq 1$, se llama una función exponencial de base *b*.

exponential growth/crecimiento exponencial (622)   El crecimiento exponencial ocurre cuando una cantidad aumenta exponencialmente.

exponential growth rate/tasa de crecimiento exponencial (626)   La constante positiva *k* en la ecuación de crecimiento $P(t) = P_0 e^{kt}$ se llama la tasa de crecimiento exponencial.

extraneous solutions/soluciones extrenuas (305)   Soluciones que no satisfacen la ecuación original.

extreme values/valores extremos (305)   Datos que varían ampliamente de los valores centrales del grupo.

factorial/factorial (697)   Si *n* es un número positivo, la expresión *n*! se define como $n! = n(n-1)(n-2) \cdot (n-3) \cdots 2 \cdot 1$. Por definición, 0! = 1.

factoring/factorización (341)   Un método que se usa para resolver una ecuación cuadrática que requiere el uso de la propiedad del producto de cero.

factors/factores (274)   Números, variables, monomios o polinomios multiplicados para obtener un producto.

factor theorem/teorema del factor (487)   El binomio $x - a$ es un factor del polinomio $f(x)$, si y solo si $f(a) = 0$.

failure/fracaso (732)   Cualquier resultado que no es el resultado deseado de un evento.

family of graphs/familia de gráficas (83)   Un grupo de gráficas que despliegan una o más características semejantes.

feasible region/región factible (153)   El área de intersección de las gráficas de desigualdades en un sistema de desigualdades en que se cumple cada restricción.

Fibonacci sequence/sucesión de Fibonacci (683)   Una sucesión especial que a menudo se encuentra en la naturaleza y la cual fue nombrada en honor a su descubridor, Leonardo Fibonnaci.

fixed points/puntos fijos (526)   Los puntos en donde la gráfica de una función $g(x)$ intersecan la gráfica de la recta $f(x) = x$.

FOIL method/método FOIL (263)   Una aplicación de la propiedad distribuitiva que se usa para multiplicar dos binomios. El producto es la suma de los productos de los términos: primero, exterior, interior y último.

formula/fórmula (9)   Una representación matemática que expresa una relación general entre ciertas cantidades.

fractal/fractal (318, 688)   Una figura geométrica que posee similaridad propia, se crea usando un proceso recursivo y es infinita en estructura.

fractal geometry/geometría fractal (688)   Una nueva rama de las matemáticas que provee modelos de diseños en la naturaleza.

frequency distribution/distribución de frecuencias (392)   Muestra cómo los datos están esparcidos sobre una gama de valores.

function/función (65)   Un tipo especial de relación en que cada elemento del dominio se parea exactamente con un elemento de la amplitud.

fundamental counting principle/principio fundamental de contar (713)   Si el evento M puede ocurrir de m maneras y es seguido por un evento N que ocurre de n maneras, entonces el evento M seguido del evento N puede ocurrir en $m \cdot n$ maneras.

fundamental theorem of algebra/teorema fundamental de álgebra (503)   Cada ecuación polinómica con grado mayor de cero tiene por lo menos una raíz en el conjunto de números complejos.

general formula for growth and decay/formula general para crecimiento y disminución   (631) La fórmula es $y = ne^{kt}$, en que y es la cantidad final, n es la cantidad inicial, k es una constante y t es el tiempo.

geometric means/medias geométricas (665)   Los términos entre cualquier par de términos no consecutivos en una secuencia geométrica.

geometric sequence/sucesión geométrica (662)   Una sucesión en que cada término después del primero se halla multiplicando el término previo por una constante llamada razón común, r.

geometric series/serie geométrica (670)   La suma indicada de los términos en una serie geométrica.

graphical iteration/iteración gráfica (526)   Las iteraciones gráficas producen una representación visual del proceso de iteración.

greatest integer function/función del entero mayor (104)   Un tipo de función escalón descrita por $f(x) = [x]$ en que $[x]$ es el entero mayor, pero *no* mayor que x.

guess and check/conjetura y cotejo (342)   Una estrategia para resolver problemas en la cual se prueban varios valores o combinaciones de valores para hallar una solución al problema.

half-angle formulas/fórmulas de semiángulos (855)   Las siguientes identidades son válidas para todos los valores de α.

$$\cos \frac{\alpha}{2} = \pm \sqrt{\frac{1 + \cos \alpha}{2}}$$

$$\sin \frac{\alpha}{2} = \pm \sqrt{\frac{1 - \cos \alpha}{2}}$$

histogram/histograma (392)   Una gráfica de barras que despliega una distribución de frecuencia.

hyperbola/hipérbola (440)   El conjunto de todos los puntos P en un plano tal que el valor absoluto de la diferencia de las distancias desde P hasta dos puntos dados, llamados focos, es constante.

hypotenuse/hipotenusa (9)   El lado opuesto al ángulo recto en un triángulo rectángulo.

hypothesis/hipótesis (760)   Un enunciado que se prueba en un experimento.

identify subgoals/identificación de submetas (296)   Una estrategia para resolver problemas que utiliza una serie de pasos menores o submetas.

identity function/función identidad (103)   Una función lineal descrita por $y = x$ o $f(x) = x$.

identity matrix/matriz identidad (213)   Cualquier matriz cuadrada A que cuando se multiplica por otra matriz B de la misma dimensión, equivale a esa misma matriz B.

identity properties/propiedades de identidad (14)   Si a es un número real, entonces $a + 0 = a = 0 + a$ y $a \cdot 1 = a = 1 \cdot a$.

imaginary unit/unidad imaginaria (310)   La unidad imaginaria i está definida por $i = \sqrt{-1}$.

inclusive events/eventos inclusivos (746)   Dos eventos son inclusivos si los resultados de los eventos pueden ser los mismos.

inconsistent system/sistema inconsistente (128)   Un sistema de ecuaciones que no tiene solución.

independent events/eventos independientes (713)   El resultado de un evento independiente no se ve afectado por el resultado de otro evento.

independent system/sistema independiente (127)   Un sistema de ecuaciones que tiene exactamente una solución.

independent variable/variable independiente (73)   La variable cuyo valor no depende de, ni es afectado por el valor de otra variable.

index of summation/índice de adición (658)   La variable definida debajo del símbolo Σ en la notación sigma.

infinite geometric series/serie geométrica infinita (676)   La suma indicada de los términos de una sucesión geométrica infinita.

initial side/lado inicial (780)   Uno de los dos rayos que forman un ángulo en un plano de coordenadas fijo a lo largo del eje positivo x.

integral zero theorem/teorema del cero integrado (509)   Si los coeficientes de una función polinómica son números enteros, de modo que $a_0 = 1$ y $a_n \neq 0$, cualquiera de los ceros racionales de la función deben ser factores de a_n.

interquartile range/amplitud intercuartílica (236)   La diferencia entre el cuartil superior y el inferior de un conjunto de datos.

intersection/intersección (49)   La gráfica de una desigualdad compuesta que contiene la palabra *y* es la intersección de las gráficas de las dos desigualdades.

inverse Cosine/coseno inverso (811)   Dado $y = \cos x$, la función del coseno inverso la define $y = \cos^{-1} x$ o $y = \text{arc cos } x$.

inverse functions/funciones inversas (528)   Dos funciones *f* y *g* son funciones inversas si y solo si las composiciones de ambas son la función identidad. Es decir, $[f \circ g](x) = x$ y $[g \circ f](x) = x$.

inverse of a matrix/inverso de una matriz (213)   La matriz por la cual se multiplica otra matriz para producir la matriz identidad.

inverse properties/propiedades del inverso (14)   Si *a* es un número real, entonces $a + (-a) = 0 = (-a) + a$ y $a \cdot \frac{1}{a} = 1 = \frac{1}{a} \cdot a$, si $a \neq 0$.

inverse relation/relación inversa (531)   Dos relaciones son inversas si y solo si siempre que una relación contenga el elmento (a, b), la otra relación contiene el elemento (b, a).

inverse Sine/seno inverso (812)   Dado que $y = \text{seno } x$, la función seno inversa está definida por $y = \text{sen}^{-1} x$ o $y = \text{arc sen } x$.

inverse Tangent/tangente inversa (812)   Dado $y = \text{Tan } x$, la función tangente inversa está definida por $y = \text{Tan}^{-1} x$ o $y = \text{arc tan } x$.

inverse variation/variación inversa (557)   Un tipo de variación en la cual *y* varía inversamente con *x* si existe alguna constante *k*, de modo que $xy = k$ o $y = \frac{k}{x}$.

irrational number/número irracional (13)   Los números irracionales son números reales que no se pueden escribir como decimales periódicos.

iterate/iterar   1. (318) Cuando un proceso o función se repite una y otra vez. **2.** (526) Cada resultado de un proceso de iteración se llama iteración.

iteration/iteración (522)   El proceso mediante el cual se determina el valor de una función para cierto número del dominio y se genera una sucesión de valores usando cada valor de salida como el próximo valor de entrada.

J

joint variation/variación conjunta (558)   Un tipo de variación en el cual *y* varía conjuntamente con *x* y *z*, se existe algún número, *k* de modo que $y = kxz$, en que $x \neq 0$ y $z \neq 0$.

L

latus rectum/lado recto (417)   El segmento de recta a través del foco de una parábola y perpendicular al eje de simetría.

law of cosines/ley de los cosenos (799)   Para cualquier triángulo *ABC* con *a*, *b* y *c* representando las medidas de los lados, y con ángulos opuestos cuyas medidas son *A*, *B* y *C*, respectivamente, las siguientes ecuaciones son ciertas:

$$a^2 = b^2 + c^2 - 2bc \cos A$$
$$b^2 = a^2 + c^2 - 2ac \cos B$$
$$c^2 = a^2 + b^2 - 2ab \cos C.$$

law of sines/ley de los senos (793)   Sea $\triangle ABC$ cualquier triángulo con *a*, *b* y *c* representando las medidas de los lados, y ángulos opuestos con medidas *A*, *B* y *C*, respectivamente. Entonces, $\frac{\sin A}{a} = \frac{\sin B}{b} = \frac{\sin C}{c}$.

leading coefficient/coeficiente guía (480)   El coeficiente del término con el grado más alto.

legs/catetos (9)   Los dos lados de un triángulo rectángulo, los cuales forman el ángulo recto.

like radical expressions/expresiones radicales semejantes (291)   Dos o más expresiones radicales se denominan expresiones radicales semejantes si tanto los índices como los radicandos de ambas son iguales.

like terms/términos semejantes (261)   Dos o más monomios que son iguales o que difieren solo en sus coeficientes numéricos.

linear equation/ecuación lineal (73)   Una ecuación cuya gráfica es una recta.

linear function/función lineal (74)   Se dice que una función es lineal si se puede definir por $f(x) = mx + b$, en la cual *m* y *b* son números reales.

linear permutation/permutación lineal (718)   El despliegue de cosas en una recta.

linear programming/programación lineal (155)   Un método para hallar el valor máximo o mínimo de una función en dos variables sujeto a restricciones dadas en las variables.

linear term/término lineal (334)   En una función cuadrática descrita por $f(x) = ax^2 + bx + c$, *bx* es el término lineal.

line plot/esquema lineal (19)   Despliega datos estadísticos sobre una recta numérica, de modo que se puedan determinar los patrones y las variabilidades en los datos.

list possibilities/listado de posibilidades (38)   Una estrategia para resolver problemas que se usa para hallar todas las soluciones posibles a un problema.

location principle/principio de ubicación (494)   Suponiendo que $y = f(x)$ representa una función polinómica y que *a* y *b* son dos números tales que $f(a) < 0$ y $f(b) > 0$, entonces, la función tiene por lo menos un cero real entre *a* y *b*.

logarithm/logaritmo (605)   Suponiendo que $b > 0$ y $b \neq 1$, para $n > 0$, existe un número p, tal que $\log_b n = p$ si y solo si $b^p = n$.

logarithmic function/función logarítmica (594, 607)   Una función de la forma $y = \log_b x$, en que $b > 0$ y $b \neq 1$.

look for a pattern/busca un patrón (80)   Una estrategia para resolver problemas que a menudo involucra el uso de tablas para organizar la información y determinar un patrón.

lower extreme/extremo bajo (236)   El valor menor en un conjunto de datos.

lower quartile/cuartil inferior (236)   La mediana de la mitad inferior de un conjunto de datos.

major axis/eje mayor (432)   El segmento de recta más largo de los dos segmentos que forman los ejes de simetría de una elipse.

mantissa/mantisa (618)   El logaritmo de un número entre 1 y 10.

mapping/apareo (65)   Un apareo muestra como cada miembro del dominio de una relación se aparea con cada miembro de la amplitud.

matrix/matriz (186)   Una red rectangular de variables o constantes en hileras horizontales y columnas verticales, que por lo general se encierran en corchetes.

matrix equation/ecuación matricial (219)   Una ecuación de la forma $AX = C$, en que A es la matriz coeficiente para un sistema de ecuaciones lineales, X es la matriz de columna que consiste de las variables del sistema y C es la matriz de columna que consiste de los términos constantes del sistema.

matrix logic/lógica matricial (187)   Una estrategia para resolver problemas en la cual la información se organiza en una matriz y se eliminan una por una las posibilidades hasta hallar una solución.

mean/media (21)   La suma de todos los valores en un conjunto de datos dividida entre el número de valores.

measure of central tendency/medida de tendencia central (21)   Un número que representa el centro o la mitad de un conjunto de datos.

median/mediana (21)   El valor del medio de un conjunto de datos. Si existen dos valores de medio, la mediana es la media de los números.

midpoint formula/fórmula de punto medio (410)   Si un segmento de recta tiene extremos en (x_1, y_1) y (x_2, y_2), entonces el punto medio del segmento de recta tiene coordenadas $\left(\dfrac{x_1 + x_2}{2}, \dfrac{y_1 + y_2}{2}\right)$.

minor/menor (205)   La determinante formada cuando se eliminan la hilera y la columna que contienen un elemento de la determinante dada.

minor axis/eje menor (432)   El más corto de los dos segmentos de recta que forman los ejes de simetría de una elipse.

mode/modal (21)   El valor más frecuente en un conjunto de datos.

monomials/monomios (255)   Una expresión que consiste de un número, una variable o el producto de un número y una o más variables.

multiplication property of equality/propiedad de igualdad de la multiplicación (28)   Para cualquiera de los números reales a, b y c, si $a = b$, entonces $a \cdot c = b \cdot c$.

multiplication property of inequality/propiedad de multiplicación de la desigualdad (44)   Para cualquiera de los números reales a, b y c:

1. Si c es positivo y $a < b$, entonces $ac < bc$;
2. Si c es positivo y $a > b$, entonces $ac > bc$;
3. Si c es negativo y $a < b$, entonces $ac > bc$;
4. Si c es negativo y $a > b$, entonces $ac < bc$.

multiplying matrices/multiplicación de matrices (199)   El producto de $A_{m \times n}$ y $B_{n \times r}$ es $(AB)_{m \times r}$. El elemento en la i–esima hilera y la j–esima columna de AB es la suma de los productos de los elementos correspondientes en la i–esima hilera de A y la j–esima columna de B.

multiplying rational expressions/multiplicación de expresiones racionales (563)   Para todas las expresiones racionales $\dfrac{a}{b}$ y $\dfrac{c}{d}$, $\dfrac{a}{b} \cdot \dfrac{c}{d} = \dfrac{ac}{bd}$, si $b \neq 0$ y $d \neq 0$.

mutually exclusive events/eventos mutuamente exclusivos (746)   Dos eventos son mutuamente exclusivos si sus resultados nunca pueden ser iguales.

natural logarithm/logaritmo natural (622)   Los logaritmos naturales son los logaritmos de base e.

normal distribution/distribución normal (393)   Una distribución que resulta en una gráfica simétrica acampanada. Cerca del 68% de los valores se hallan dentro de una desviación estándar de la media. Cerca del 95% de los valores se hallan dentro de dos desviaciones estándares de la media. Cerca del 99% de los valores se hallan dentro de tres desviaciones estándares de la media.

nth root/enésima raíz (282)   Para cualquiera de los números reales a y b y cualquier número entero positivo n, si $a^n = b$, entonces a es una enésima raíz de b.

octants/octantes (172)   Tres planos mutuamente perpendiculares separan el espacio en ocho regiones, cada una de ellas llamada una octante.

odds/posibilidades (734)   Las posibilidades de que un evento tenga un resultado exitoso se pueden expresar como el cociente de la proporción del número de formas en que el evento puede tener éxito dividido entre el número de formas en que puede fracasar.

open sentence/operación abierta (27)   Un enunciado matemático que contiene variables que serán reemplazadas con valores numéricos.

ordered pairs/pares ordenados (64)   Los puntos en un plano se pueden ubicar mediante el uso de pares ordenados de números reales. Un par ordenado consiste de las coordenadas del punto.

ordered triple/triple ordenado (166)   La solución de un sistema de ecuaciones en tres variables.

order of operations/orden de operaciones (7)   Para evaluar una expresión numérica, se debe seguir el orden de operaciones. Primero, simplifica las expresiones dentro de los símbolos de agrupación. Evalúa todas las potencias. Realiza todas las multiplicaciones y las divisiones de izquierda a derecha. Finalmente, realiza todas las sumas y las restas de izquierda a derecha.

organize data/organizar datos (580)   Una estrategia para resolver problemas usando diagramas, tablas y esquemas para ordenar y evaluar datos y así poder hallar una solución.

origin/origen (64)   El punto $(0, 0)$ en donde se intersecan los ejes del plano de coordenadas.

outlier/valor atípico (237)   Cualquier valor en un conjunto de datos que se encuentra por lo menos a 1.5 amplitudes intercuartílicas más allá del cuartillo superior y del inferior.

parabola/parábola 1. (332, 335)   La forma generalizada de la gráfica de una función cuadrática. **2.** (415)   El conjunto de todos los puntos en un plano que se encuentran a la misma distancia de un punto dado llamado el foco y de una recta dada llamada la directriz.

parallel lines/rectas paralelas (83)   En un plano, las rectas que tienen la misma pendiente son paralelas. Todas las rectas verticales son paralelas y todas las rectas horizontales son paralelas.

parent graph/gráfica madre (83)   La gráfica del polinomio más simple de forma $f(x) = x^n$ en una familia de gráficas.

partial sum/suma parcial (676)   En una serie infinita, S_n se llama una suma parcial porque es la suma de cierto número de términos y no la suma de la serie completa.

Pascal's triangle/triángulo de Pascal (695)   La formación en pirámide de los coeficientes de una expansión binómica.

periodic function/función periódica (807)   Una función es periódica si existe un número positivo a tal que $f(x) = f(x + a)$ para toda x en el dominio de la función. El menor valor positivo de a para el cual $f(x) = f(x + a)$ se llama el período de la función.

permutation/permutación (718)   Una disposición de cosas en cierto orden.

perpendicular lines/rectas perpendiculares (83)   En un plano, dos líneas oblicuas son perpendiculares si y solo si el producto de sus pendientes es -1. Cualquier recta vertical es perpendicular a cualquier recta horizontal.

point discontinuity/punto de discontinuidad (548)   El punto de discontinuidad ocurre para los valores de x que hacen que el denominador de una función racional sea cero y aparece a la vista como interrupciones en la continuidad gráfica de la función racional.

point-slope form/forma de punto–pendiente (89)   La forma de punto–pendiente de la ecuación de una recta es $y - y_1 = m(x - x_1)$, en la cual (x_1, y_1) son las coordenadas de un punto sobre la recta y m es la pendiente de la recta.

polynomial/polinomio (261)   Un monomio o la suma de monomios.

polynomial function/función polinómica (479)   Una función que se puede describir con una ecuación de la forma $P(x) = a_0 x^n + a_1 x^{n-1} + ... + a_{n-2} x^2 + a_{n-1} x + a_n$, en la cual los coeficientes $a_0, a_1, a_2, ... a_n$, son números reales, a_0 no es cero y n es un número entero no negativo.

polynomial in one variable/polinomio en una variable (478)   Un polinomio en una variable x es una expresión de la forma $a_0 x^n + a_1 x^{n-1} + ... + a_{n-2} x^2 + a_{n-1} x + a_n$, en la cual los coeficientes $a_0, a_1, a_2, ... a_n$, representan números reales, a_0 no es cero y n representa un número entero no negativo.

power/potencia (255)   Una expresión de la forma x^n.

power property of logarithms/propiedad de potencia de logaritmos (613)   Para cualquier número real p y números positivos m y b, en que $b \neq 1$, $\log_b m^p = p \cdot \log_b m$.

prediction equation/ecuación de predicción (95)   La ecuación de la línea de mejor ajuste sugerida por los puntos de una gráfica de dispersión. Se puede usar para estimar, o predecir, una de las variables a partir de otra dada.

prime number/número primo (274)   Un número entero mayor que 1 cuyos únicos factores son 1 y el número mismo.

principal root/raíz principal (282)   La raíz principal es la raíz no negativa. Si hay una raíz negativa, pero no hay ninguna raíz no negativa, la raíz principal es la raíz negativa.

principal values/valores principales (811)   Los valores de los dominios restringidos de las funciones del coseno, del seno y de la tangente son valores principales.

probability/probabilidad (732)   Si un evento puede suceder de s maneras y puede fracasar de f maneras, entonces las probabilidades de éxito $P(s)$, y de fracaso $P(f)$, son $P(s) = \frac{s}{s + f}$ y $P(f) = \frac{f}{f + s}$.

probability matrix/matriz de probabilidad (200)   Una matriz de probabilidad se usa en situaciones que involucran el azar.

problem solving plan/plan para solucionar problemas (30)   Un plan de cuatro pasos que se usa para solucionar problemas y el cual requiere que uno explore el problema, planifique la solución, resuelva el problema y examine la solución.

problem solving strategies/estrategias para resolver problemas (38)   Varias estrategias que se usan solas o en combinación para resolver problemas.

product property of logarithms/propiedad de producto de logaritmos (611)   Para todos los números positivos m, n y b, en que $b \neq 1$, $\log_b mn = \log_b m + \log_b n$.

property of equality for exponential functions/ propiedad de igualdad para las funciones exponenciales (599)   Suponiendo que b es un número positivo diferente de 1, entonces $b^{x_1} = b^{x_2}$ si y solo si $x_1 = x_2$.

property of equality of logarithmic functions/ propiedad de igualdad de funciones logarítmicas (608)   Suponiendo que $b > 0$ y $b \neq 1$, entonces $\log_b x_1 = \log_b x_2$, si y solo si $x_1 = x_2$.

pure imaginary numbers/números imaginarios puros (310)   Un número complejo de la forma bi, en que b es real y $b \neq 0$, e i es la unidad imaginaria.

Pythagorean theorem/teorema de Pitágoras (9)   El teorema de Pitágoras enuncia que en un triángulo rectángulo, si a y b son las medidas de los catetos y c es la medida de la hipotenusa, entonces, $c^2 = a^2 + b^2$.

quadrantal angles/ángulos de cuadrantes (787)   Ángulos cuyos lados finales están sobre un eje en el cual x o y es igual a cero.

quadrants/cuadrantes (64)   Las cuatro regiones en que dos rectas numéricas perpendiculares separan el plano.

quadratic equation/ecuación cuadrática (336)   Una ecuación que puede ser escrita en la forma $ax^2 + bx + c = 0$, en la cual $a \neq 0$.

quadratic form/forma cuadrática (515)   Para cualquiera de los números a, b y c, excepto $a = 0$, una ecuación que se puede escribir como $a[f(x)]^2 + b[f(x)] + c = 0$, en que $f(x)$ es una expresión en x, está en forma cuadrática.

quadratic formula/fórmula cuadrática (354)   Las soluciones de una ecuación cuadrática de la forma $ax^2 + bx + c = 0$, en la cual $a \neq 0$ son dadas por la fórmula cuadrática que es $x = \dfrac{-b \pm \sqrt{b^2 - 4ac}}{2a}$.

quadratic function/función cuadrática (332, 334)   Una función descrita por una ecuación que se puede escribir en la forma $f(x) = ax^2 + bx + c = 0$, en la cual $a \neq 0$.

quadratic inequality/desigualdad cuadrática (378)   Una desigualdad descrita por $y < ax^2 + bx + c = 0$, en la cual $a \neq 0$.

quadratic term/término cuadrático (334)   En una función cuadrática descrita por $f(x) = ax^2 + bx + c = 0$, el término ax^2.

quartiles/cuartiles (236)   Los valores en un conjunto de datos que separan los datos en cuatro secciones, cada una de las cuales contiene el 25% de los datos.

quotient property of logarithms/propiedad de cociente de logaritmos (612)   Para todos los números positivos m, n y b, en que $b \neq 1$, $\log_b \dfrac{m}{n} = \log_b m - \log_b n$.

R

radian/radián (781)   La medida de un ángulo que interseca un arco cuya longitud es de una unidad.

radical equations/ecuaciones radicales (305)   Ecuaciones que contienen expresiones radicales con variables en el radicando.

radius/radio (423)   Cualquier segmento de recta cuyos extremos son el centro de un círculo y un punto en el círculo.

random sample/muestra al azar (758)   Una muestra en que cada miembro de la población tiene la misma posibilidad de ser seleccionado.

range/amplitud   1. (65) El conjunto de todas las segundas coordenadas de los pares ordenados de una relación.   **2.** (235) La diferencia entre los valores mayor y menor en un conjunto de datos.

rational algebraic expressions/expresiones algebraicas racionales (562)   Una expresión algebraica puede ser expresada como el cociente de dos polinomios en los que el denominador no es igual a cero.

rational equation/ecuación racional (576)   Una ecuación que contiene una o más expresiones racionales.

rational exponent/exponente racional (298)   Para cualquier número no cero b y cualquiera de los números enteros m y n, en que $n > 1$, $b^{\frac{m}{n}} = \sqrt[n]{b^m} = \left(\sqrt[n]{b}\right)^m$, excepto cuando $b > 0$ y n es un número par.

rational function/función racional (550)   Una ecuación en la forma $f(x) = \dfrac{p(x)}{q(x)}$, en que $p(x)$ y $q(x)$ son funciones polinómicas y $q(x) \neq 0$.

rationalizing the denominator/racionalizar el denominador (290)   Un proceso que se usa para eliminar radicales de un denominador o fracciones de un radicando.

rational number/número racional (13)   Un número que se puede expresar como una proporción $\dfrac{m}{n}$, en la cual m y n son enteros y n no es cero.

rational zero theorem/teorema del cero racional (509) Daja que $f(x) = a_0 x^n + a_1 x^{n-1} + ... + a_{n-1} x + a_n$ represente una función polinómica con coeficientes integrales ($a \neq 0$). Si $\frac{p}{q}$ es un número racional en forma reducida y es un cero de $y = f(x)$, entonces p es un factor de a_n y q es un factor de a_0.

real numbers/números reales (13) El conjunto de números irracionales junto con el conjunto de números racionales.

recursive formula/fórmula recursiva (683) Una fórmula recursiva consta de dos partes: el valor o valores del primer término y una ecuación recursiva que muestra cómo hallar cada término a partir del término anterior.

reduced matrix/matriz reducida (227) La matriz que resulta después de efectuar operaciones en las hileras de una matriz aumentada para resolver un sistema de ecuaciones lineales.

reducing a matrix/reducir una matriz (227) El proceso de efectuar operaciones en las hileras de una matriz aumentada para obtener la matriz solución deseada.

reflections/reflexiones (722) Una reflexión ocurre cuando se puede ver una disposición de objetos desde dos vistas diferentes sin cambiar la disposición.

reflexive property of equality/propiedad reflexiva de la igualdad (28) Para cualquier número real a, $a = a$.

relation (65) Un conjunto de pares ordenados.

relative maximum/máximo relativo (494) Un punto en la gráfica de una función se llama máximo relativo si ningún otro punto cercano posee una coordenada y mayor.

relative minimum/mínimo relativo (494) Un punto en la gráfica de una función se llama mínimo relativo si ningún otro punto cercano posee una coordenada y menor.

remainder theorem/teorema del residuo (485) Si un polinomio $f(x)$ se divide entre $x - a$, el residuo es la constante $f(a)$ y $f(x) = q(x) \cdot (x - a) + f(a)$, en que $q(x)$ es un polinomio con un grado menor que el grado de $f(x)$.

root/raíz (332, 336) Una solución de una ecuación.

rotation/rotación (201) Una transformación en la cual se mueve una figura alrededor de un punto central.

row matrix/matriz horizontal (187) Una matriz que solo tiene una hilera.

row operations/operaciones horizontales (226) Operaciones que se efectúan en las hileras de una matriz aumentada para hallar la solución a un sistema de ecuaciones lineales.

sampling error/error del muestreo (759) La diferencia entre los resultados de la muestra y los resultados verdaderos de la población.

scalar multiplication/multiplicación escalar (188) Multiplicación de cada elemento de la matriz por la mizma constante.

scatter plot/diagrama de dispersión (95) Un diagrama de dispersión muestra visualmente la naturaleza de una relación, determinada tanto por la forma como por la proximidad de los datos.

scientific notation/notación científica (254) Un número está en notación científica cuando está en la forma de $a \times 10^n$, en que $1 \leq a < 10$ y n es un número entero.

second order determinant/determinante de segundo orden (141) Una determinante que tiene dos hileras y dos columnas. Su valor se puede hallar calculando la diferencia de los productos de las dos diagonales.

self-similarity/autosemejanza (689) Una característica de un objeto en el cual las replicas de una forma u objeto completo son incrustadas repetidamente dentro del objeto, en diferentes tamaños.

sequence/sucesión (649) Una lista de números en un orden específico.

series/serie (656) La suma indicada de los términos de una secuencia.

sigma notation/notación sigma (658) Notación que usa el símbolo Σ para indicar la suma de una serie.

simplify/simplificar (255) La simplificación de una expresión que contiene potencias significa el escribir de nuevo la expresión sin paréntesis o exponentes negativos.

simulation/simulación (754) Un método para hallar la probabilidad experimental en el cual se utiliza un dispositivo para modelar el evento y se hacen observaciones de la forma en que el modelo responde a las condiciones enumeradas en un problema dado.

sine/seno (805) Si el lado final de un ángulo θ en posición estándar interseca el círculo unidad en $P(x, y)$, entonces sen $\theta = y$.

Sine/seno (811) $y = $ sen x, si y solo si $y = $ sen x y $-\frac{\pi}{2} \geq x \leq \frac{\pi}{2}$.

skewed distribution/distribución asimétrica (395) Una curva de distribución que no es simétrica.

slope/pendiente (82) La pendiente de la recta que pasa por los puntos (x_1, y_1) y (x_2, y_2) está dada por $m = \frac{y_2 - y_1}{x_2 - x_1}$, en que $x_2 \neq x_2$. La pendiente de una recta descrita por $f(x) = mx + b$ es m.

slope-intercept form/forma de pendiente—intersección (88) La forma de pendiente—intersección de la ecuación de una recta es $y = mx + b$, en la cual m es la pendiente y b es la intersección con el eje y.

solution/solución (28)   Un valor que puede reemplazar una variable en una ecuación para satisfacer la ecuación.

solve a simpler problem/resuelve un problema más simple (156, 712)   Una estrategia para resolver problemas en la cual se resuelve un problema similar más simple para determinar los conceptos que se pueden usar en la solución de un problema más complejo.

solving a right triangle/solución de un triángulo rectángulo (775)   El proceso de hallar las medidas de todos los lados y angulos de un triángulo rectángulo.

square matrix/matriz cuadrada (187)   Una matriz con el mismo número de hileras y columnas.

square root/raíz cuadrada (281)   Para cualquiera de los números reales a y b, si $a^2 = b$, entonces a es una raíz cuadrada de b.

square root function/función de raíz cuadrada (535)   Una función que involucra una raíz cuadrada.

standard deviation/desviación estándar (384)   De un conjunto de datos con n valores, en el que x_1 representa el primer término y x_n representa el *ené*simo término, si $\overline{x}$ representa la media, entonces la desviación estándar es SD o

$$\sigma_x = \sqrt{\frac{(x_1 - x)^2 + (x_2 - x)^2 + \ldots + (x_n - x)^2}{n}}.$$

standard form/forma estándar (73)   La forma estándar de una ecuación lineal es $Ax + By = C$, en la cual A, B y C son números reales y A y B no son cero.

standard position/posición estándar (780)   Se dice que un ángulo se encuentra en posición estándar si está colocado en una posición tal que su vértice está en el origen y su lado inicial está a lo largo del eje x positivo.

stem-and-leaf plot/gráfica de tallo y hojas (19)   Una forma compacta de exhibir datos en la cual cada artículo está separado en dos partes. El tallo consiste de los dígitos en el mayor valor de posición común. Las hojas contienen los otros dígitos de cada artículo de los datos.

step fuction/función escalón (103)   Una función cuya gráfica es una serie de rectas o escalones desunidos.

substitution property of equality/propiedad de sustitución de la igualdad (28)   Si $a = b$, entonces, a se puede reemplazar por b.

subtraction property of equality/propiedad de sustracción de la igualdad (28)   Para cualquiera de los números a, b y c, si $a = b$, entonces $a - c = b - c$.

subtraction property of inequality/propiedad de sustracción de la desigualdad (43)   Para cualquiera de los números a, b y c:

1. si $a > b$, entonces $a - c > b - c$;
2. si $a < b$, entonces $a - c < b - c$.

success/éxito (732)   El resultado deseado de un evento.

sum of geometric series/suma de la serie geométrica (671)   La suma S_n de los primeros términos n de una serie geométrica está dada por $S_n = \dfrac{a_1 - a_1 r^n}{1 - r}$ o $S_n = \dfrac{a_1(1 - r^n)}{1 - r}$, en que $r \neq 1$.

sum of arithmetic series/suma de una serie aritmética (657)   La suma S_n de los primeros términos n de una serie aritmética es dada por $S_n = \dfrac{n}{2}(a_1 + a_n)$, en que n es un número entero positivo.

sum of an infinite geometric series/suma de una serie geométrica infinita (677)   La suma S de una serie geométrica infinita en que $-1 < r < 1$ está dada por $S = \dfrac{a_1}{1 - r}$.

symmetric property of equality/propiedad simétrica de la igualdad (28)   Para todos los números reales a y b, si $a = b$, entonces $b = a$.

synthetic division/división sintética (269)   Un método más sencillo que la división larga, el cual se usa para dividir un polinomio entre un binomio.

synthetic substitution/sustitución sintética (486)   El proceso de usar división sintética para hallar el valor de una función.

system of equations/sistema de ecuaciones (126)   Un conjunto de ecuaciones con las mismas variables.

system of inequalities/sistema de desigualdades (148)   Un conjunto de desigualdades con las mismas variables.

T

tangent/tangente (425)   Una recta que interseca a un círculo en exactamente un punto se dice que es tangente al círculo.

Tangent/tangente (811) $y = \tan x$ si y solo si $y = \tan x$ y $-\dfrac{\pi}{2} \leq x \leq \dfrac{\pi}{2}$.

term/término (648)   Cada número en una secuencia.

terminal side/lado final (780)   Uno de los dos rayos que forman un ángulo en un plano de coordenadas que gira alrededor del origen.

terms/términos (261)   Los monomios que componen un polinomio.

theoretical probability/probabilidad teórica (754)   La probabilidad teórica se determina usando métodos matemáticos para proveer una idea de los resultados que pueden ocurrir en una situación dada.

transformations/transformaciones (190)   Funciones que relacionan puntos de una gráfica con su imagen.

transition matrix/matriz de transición (200)   Una matriz que contiene información sobre la transición de un evento a otro.

transitive property of equality/propiedad transitiva de la igualdad (28)   Para todos los números reales a, b y c, si $a = b$ y $b = c$, entonces $a = c$.

translation/traslación (195)   Una transformación en la cual una figura se mueve desde una ubicación a otra en el plano de coordenadas, sin cambiar su tamaño, forma u orientación.

translation matrix/matriz de traslación (196)   Una matriz que representa el número de unidades que una figura se mueve vertical u horizontalmente a partir de su posición original.

transverse axis/eje transversal (441)   Uno de los dos segmentos de recta que forman el eje de simetría de una parábola cuyos extremos son los vértices de una hipérbola.

trichotomy property/propiedad tricotomática (43)   Para cualquiera de los números reales, a y b, exactamente uno de los siguientes enunciados es válido: $a < b$, $a = b$ o $a > b$.

trigonometric equations/ecuaciones trigonométricas (861)   Una ecuación que involucra una o más funciones trigonométricas la cual es válida para algunos, pero no todos, los valores de la variable.

trigonometric functions/funciones trigonométricas (772)   Si θ es la medida de un ángulo agudo en un triángulo rectángulo, a es la medida del cateto opuesto a θ, b es la medida del cateto adyacente a θ y c es la medida de la hipotenusa, entonces

seno $\theta = \dfrac{a}{c}$ coseno $\theta = \dfrac{b}{c}$

tangente $\theta = \dfrac{a}{b}$ cosecante $\theta = \dfrac{c}{a}$

secante $\theta = \dfrac{c}{b}$ cotangente $\theta = \dfrac{b}{a}$

trigonometric identities/identidades trigonométricas (788)   Una ecuación trigonométrica que es válida para todos los valores de las variables.

trigonometry/trigonometría (772)   El estudio de las relaciones entre los ángulos y los lados de los triángulos rectángulos.

trinomial/trinomio (261)   Un polinomio con tres términos no semejantes.

unbounded/no acotado (155)   Una función es no acotada si no se forma una región poligonal por las restricciones de un sistema de desigualdades.

union/unión (50)   La gráfica de una desigualdad compuesta que contiene la palabra o es la unión de las gráficas de las dos desigualdades.

unit circle/círculo unidad (781)   En el plano coordenado, el círculo cuyo radio mide una unidad y cuyo centro se encuentra en el origen.

upper extreme/extremo superior (236)   El mayor valor en un conjunto de datos.

upper quartile/cuartil superior (236)   La mediana de la mitad superior de un conjunto de datos.

use a simulation/usa una simulación (452)   Una estrategia para resolver problemas que usa modelos, o simulaciones, de situaciones matemáticas que son difíciles de resolver directamente.

variables/variables (27)   Letras usadas para representar números desconocidos.

vertex/vértice   1. (335) El punto en que se intersecan una parábola y su eje de simetría.   **2.** (441) El punto de cada rama de una hipérbola que está más cercano al centro de la hipérbola.

vertical line test/prueba de recta vertical (65)   Si cada recta vertical dibujada en la gráfica de una relación pasa por un solo punto de la gráfica, entonces la relación es una función.

work backward/trabaja al revés (528)   Una estrategia para resolver problemas que usa operaciones inversas para determinar un valor original.

x-axis/eje x (64)   La recta numérica horizontal que ayuda a formar el plano de coordenadas.

x-intercept/intersección con el eje x (75)   La coordenada x del punto sobre el cual una gráfica cruza el eje x.

y-axis/eje y (64)   La recta numérica vertical que ayuda a formar el plano de coordenadas.

y-intercept/intersección con el eje y (75)   La coordenada y del punto sobre el cual una gráfica cruza el eje y.

zero/cero (335)   Para cualquier función $f(x)$, si $f(a) = 0$, entonces a es un cero de la función.

zero product property/propiedad del producto cero (341)   Para cualquiera de los números reales a y b, si $ab = 0$, entonces $a = 0$, $b = 0$, o ambos.

SELECTED ANSWERS

CHAPTER 1 ANALYZING EQUATIONS AND INEQUALITIES

Page 6 Lesson 1–1A
1. 9.17 **3.** 8,998,924.67 **5.** 72 **7.** 40.88 **9.** 10, 0, -17.78, -20.56, 26.67

Pages 10–12 Lesson 1–1
7. 54 **9.** 47 **11.** 9 **13.** 28 **15.** 37°C **17.** $A = (a + 6)(a - 6)$ in^2 **19.** 19 **21.** -3 **23.** 6 **25.** 7 **27.** $-\frac{17}{6}, -2.8\overline{3}$
29. -49 **31.** 32 **33.** 9.1 **35.** -9 **37.** -0.75 **39.** 15
41. -1.875 **43.** 19.25 **45.** 117 **47.** 37.5 **49.** \$292.50
51. \$3555.89 **53a.** 528 **53b.** 52.5 **53c.** 28.8 **55a.** $S = 22a^2 + 24a$ **55b.** 448 square units **55c.** 994.48 square units **57a.** 2, 3, 5, 7, 11, 13, 17, 19, 23, 29, 31, 37, 41, 43, 47, 53, 57, 59, 61, 67, 71, 73, 79, 83 **57b.** Answers will vary; sample answers are 3, 7, 31. **57c.** 496, 8128

Pages 16–18 Lesson 1–2
7. -1; R, Q, Z **9.** 7.550; R, I **11.** true
13. additive inverse **15.** $-7, \frac{1}{7}$ **17.** $20c + 4d$ **19.** \$36.00
21. -7; R, Q, Z **23.** -42; R, Q, Z **25.** -1; R, Q, Z
27. $2\frac{1}{4}$; R, Q **29.** $\frac{1}{2}$; R, Q **31.** $\sqrt{67}$; R, I **33.** true
35. true **37.** true **39.** commutative ($+$) **41.** additive inverse **43.** commutative ($\times$) **45.** multiplicative inverse
47. -0.2; 5 **49.** 1; -1 **51.** $3\frac{5}{7}; \frac{7}{26}$ **53.** $32c - 46d$
55. $\frac{11}{6}x - \frac{7}{4}y$ **57.** $4.4m - 2.9n$ **59a.** no **59b.** yes
59c. This method works because the difference between a number and the number whose digits are reversed is 9.
61. 0.09 **63.** 22 **65.** -3.5 **67.** 1440 times **69.** \$737
71. 22

Pages 23–26 Lesson 1–3
7. 8; no mode; 7.8

9a.

Stem	Leaf
2	2 3 4 4 7 8 8 9 9
3	1 1 3 4 4 4 9
4	1 3|1 = 31,000

9b. about \$30,059 **9c.** \$29,000 **9d.** \$34,000 **11.** 34; no mode; 35.4 **13.** 65; 50 and 65; about 63.8 **15.** 12.9; no mode; 12.975 **17.** 43; 43; about 47.29 **19.** 71; 71 and 88; 73 **21a.** Mean; it is higher. **21b.** Mode; it is lower and is what most employees make. It reflects the most representative worker. **23a.** The graphs of the data look different because a different scale is used in each graph. **23b.** Sample answer: Graph A might be used by an employer to show an employee she cannot get a big raise. It appears that sales are steady but not rising drastically enough to warrant a big raise. **23c.** Sample answer: Graph B might be used by a company owner to show a prospective buyer. It looks like there is a dramatic rise in sales.

25a.

Employed	Stem	Unemployed
	0	2 2 2 3 3 3 4 4 6 6
	1	2
	2	
0 4 8	3	
6	4	
3 4 7	5	
4	6	
	7	
0 8	8	
	9	
	10	
	11	
	12	
	13	
4	14	1|2 = 1,200,000

25b. Answers will vary. Sample answers: The number of employed people in these states is much higher than the number of unemployed people. **25c.** Divide the number of unemployed people by the total number of employed and unemployed people.

25d.

California	7.7%
Florida	6.5%
Illinois	6.2%
Massachusetts	6.4%
Michigan	5.1%
New Jersey	6.7%
New York	6.5%
North Carolina	5.0%
Ohio	4.9%
Pennsylvania	6.0%
Texas	5.9%

27. \$20,390; no mode; \$46,867.08; The median is most representative of the data since there is such a wide range of costs.

29a.

```
          × ××      ××        ××     × ×  ×××
        +++++++++++++++++++++++++++++++++++++++
          60   65   70   75   80   85   90   95
```

29b. 79.5; 91; 77.58
29c.

```
              × ××      ××        ××     × × ×××
        +++++++++++++++++++++++++++++++++++++++
          60   65   70   75   80   85   90   95
```

29d. They each increase by 5 since the temperature increased by 5 degrees; different. **29e.** All points on the original line plot would be shifted 5 to the right to obtain the new line plot. **31.** $-10a + 6$ **33.** -7 **35.** 419.5°C

Pages 31–34 Lesson 1–4
7. $5x - 3$ **9.** multiplication ($=$) **11.** -5 **13.** -13
15. -21 **17.** $\frac{3V}{\pi r^2} = h$ **19.** $14 - x^2$ **21.** $4(n + n^2)$
23. $7 + 3n$ **25.** reflexive ($=$) **27.** addition ($=$)
29. subtraction ($=$) **31.** 2 **33.** 2 **35.** $\frac{1}{12}$ **37.** -4
39. $-\frac{409}{13}$ **41.** -16 **43.** 2 **45.** 19 **47.** $\frac{35}{2}$ **49.** $\frac{z}{x} - 2$
51. $\frac{5a - 9}{6}$ **53.** $\frac{Fr^2}{Gm}$ **55.** $30x + 15(5x) = 420$;

four adult tickets; twenty student tickets **57.** $2x + 18 = 32$; \$7 **59.** the product of twice a number and the sum of the number and four added to twice the sum of the number and six **61.** Yes; the last day's trip will take him only 10.2 hours, which will bring the driver's total time to 64.95 hours. **63.** 7920 miles **65.** 217.5; 399; about 265.3

67.

January	Stem	July
1	1	
1 2 6	2	
0 5	3	
1 1 4	4	
1 3 7	5	6
	6	5
	7	0 3 3 5 8 9
$5\|3\| = 35°$	8	1 2 3 6

69. $15a - 19$ **71.** additive identity **73.** \$17,400 **75.** -4

Page 34 Self Test
1. -5 **3.** -1; R, Q, Z **5.** $\sqrt{41}$, 6.403; R, I **7.** -4 **9.** 2

Page 36 Lesson 1–5A
1. 0.75 **3.** -30.83 **5.** 3.67 **7.** $-2, 8$ **9.** $44, -24$

Pages 40–42 Lesson 1–5
7. 5 **9.** $13, -23$ **11.** $18, -6$ **13.** $42, -48$
15a. $|x - 16| = 0.2$ **15b.** $x = 16.2, x = 15.8$ **17.** 5 **19.** 1
21. -1.6 **23.** -6.6 **25.** $20, -14$ **27.** $31, -53$ **29.** $20, -2$
31. $14, -8$ **33.** no solution **35.** $5, -\frac{19}{3}$ **37.** $\frac{11}{8}, -\frac{25}{24}$
39. no solution **41.** 4 **43a.** $-1.2, 3.2$ **43b.** $-1.7, 0, 1.7$
43c. $-0.7, 2$ **43d.** $-2.2, -0.6, 2.2$ **45a.** $|x - 697| = 5$
45b. $x = 702$ or $x = 692$ **47.** TAW, TAX, TAY, TBW, TBX, TBY, TCW, TCX, TCY, UAW, UAX, UAY, UBW, UBX, UBY, UCW, UCX, UCY, VAW, VAX, VAY, VBW, VBX, VBY, VCW, VCX, VCY **49.** $-\frac{15}{2}$ **51.** $p + 5$ **53.** 143; 141; about 143.7
55. -24; Z, Q, R

Pages 47–48 Lesson 1–6
5. $\{a \mid a \geq 1.75\}$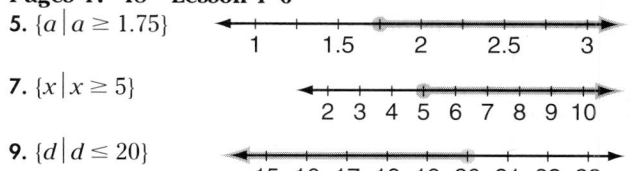

7. $\{x \mid x \geq 5\}$

9. $\{d \mid d \leq 20\}$

11. $\varnothing$ **13.** $n + 15 \geq 27$

15. $\{r \mid r < -5\}$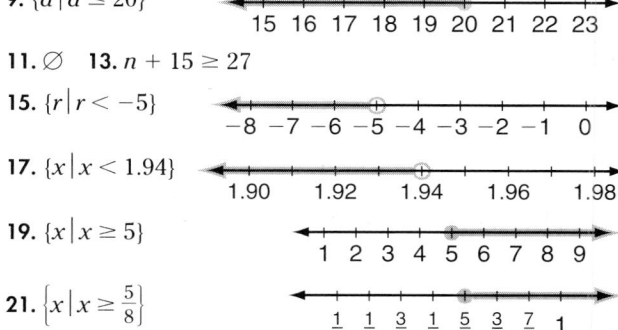

17. $\{x \mid x < 1.94\}$

19. $\{x \mid x \geq 5\}$

21. $\left\{x \mid x \geq \frac{5}{8}\right\}$

23. $\{x \mid x > 2.25\}$

25. $\{z \mid z \leq 5\}$

27. $\left\{m \mid m > \frac{4}{9}\right\}$

29. $\{b \mid b > 2\}$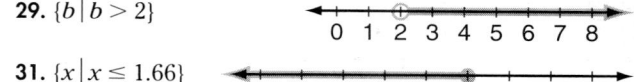

31. $\{x \mid x \leq 1.66\}$

33. $\varnothing$

35. $\left\{x \mid x \leq -\frac{1}{2}\right\}$

37. $\left\{w \mid w \leq \frac{4}{3}\right\}$

39. $\frac{3}{4}x - 25 \geq 8$ **41.** $57 > 0.5x$ **43.** $62 < -6x$

45. $\{r \mid r < 0.4375\}$ **47.** $\{s \mid s > -3\}$ **49a.** $750 < x < 990$
49b. 990 Calories **51.** $15, -7$ **53.** 11.2 **55.** about 6.309
57. distributive **59.** 0.6 amperes

Pages 52–54 Lesson 1–7
5. $|x| < 18$ **7.** $|x| < 4$

9. $\{x < 2 \text{ or } x > 8\}$

11. $\{x \mid x > 1 \text{ or } x < -5\}$

13. $\{x \mid x < -18 \text{ or } x > 10\}$

15. $|x| < 7$

17. $|x| > 11$

19. $|x| < 8$

21. $|x| < 3$ **23.** $|x| \geq 4$ **25.** $|x + 1| > 2$

27. $\left\{x \mid -\frac{5}{4} \leq x \leq \frac{5}{4}\right\}$

29. $\{x \mid x < -5 \text{ or } x > 5\}$

31. $\left\{x \mid x \geq \frac{7}{3} \text{ or } x \leq -\frac{7}{3}\right\}$

33. $\left\{x \mid x > \frac{1}{2} \text{ or } x < -\frac{1}{2}\right\}$

35. $\varnothing$ **37.** $\left\{x \mid x < -4 \text{ or } x > -\frac{10}{3}\right\}$

39. all reals **41.** $\varnothing$

43. $\{x \mid x < -2 \text{ or } x \geq 1\}$

45. $\{x \mid -1 \leq x \leq 1\}$ **47a.** $|x - 58{,}000| \leq 18{,}000$

49. $\{x \mid x < 6\}$

51. $|x| < -8$ **53.** $1, \frac{5}{3}$ **55a.** $(0.50)6 + (0.35)x = 10$
55b. 20   **57.** 20.7; 21.6 and 20.7; 21.03

Page 55   Chapter 1   Highlights
1. j   **3.** f   **5.** b   **7.** l   **9.** c   **11.** i

Pages 56–58   Chapter 1   Study Guide and Assessment
13. 2   **15.** 41   **17.** 4   **19.** -8, Z, Q, R   **21.** $\sqrt{5}$, I, R
23. 54.978, I, R   **25.** multiplicative inverse   **27.** $2a - 4b$
29. 25; no mode; 40.8   **31.** 6.1; 6.1; 6.0   **33.** -21 **35.** 2
37. -20 **39.** 6   **41.** $\frac{A-p}{pr}$ **43.** $\frac{3a^2-1}{2c}$ **45.** $17, -7$
47. 11, 25   **49.** no solution   **51.** $-11, -1$ **53.** $\{z \mid z > 4\}$
55. $\{t \mid t \le 11\}$ **57.** $\{y \mid y > -35\}$
59. $\{z \mid z \ge 6\}$

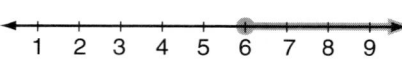

61. $\{x \mid 3 < x < 6\}$

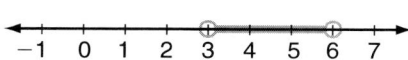

63. $\left\{y \mid \frac{5}{3} < y \le 5\right\}$

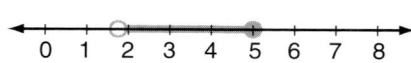

65. $\{x \mid -5 \le x \le -1\}$

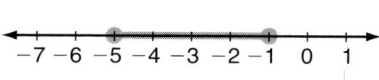

67. $\varnothing$ **69.** $\{x \mid -9 < x < 9\}$

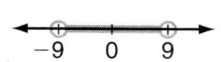

71. 30 cm by 45 cm   **73.** 97¢   **75.** 3777, none, 4330.4

952   *Selected Answers*

CHAPTER 2   GRAPHING LINEAR RELATIONS AND FUNCTIONS

Pages 68–71   Lesson 2–1
7. function   **9.** not a function
11. function; discrete   **13.** not a function

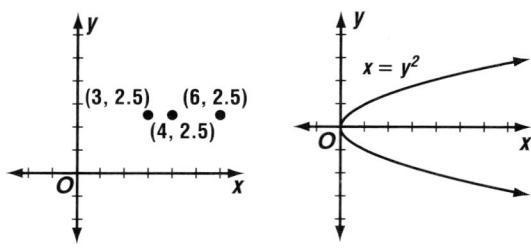

15. -7 **17.** not a function   **19.** not a function
21. not a function   **23.** not a function
25. D = {4, 6, 3},   **27.** D = {3, 4, 5, 6},
R = {5}; function; discrete   R = {3, 4, 5, 6};
                                    function; discrete

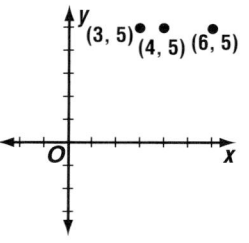   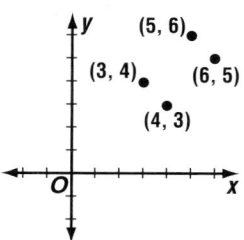

29. D = {all reals},
R = {all reals}; function;
continuous

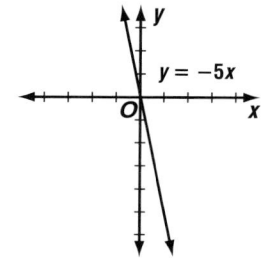

31. no   **33.** yes   **35.** 6   **37.** -3 **39.** $25n^2 - 5n$ **41.** -12
43. Sample answer: $f(x) = 2.5x$

45. 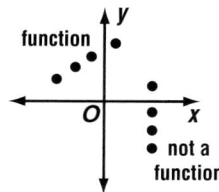   **47.**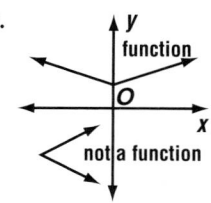

49. all real numbers except 3 and -3 **51a.** D = {year},
R = {stock price}

51.

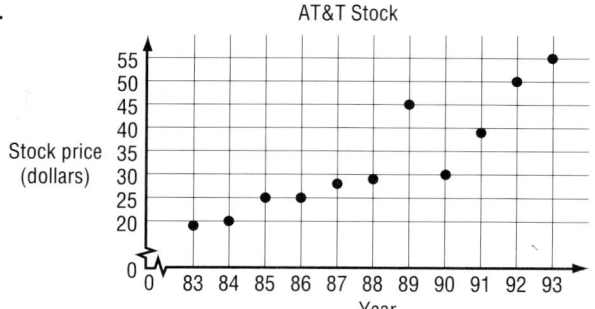

51c. yes   51d. Sample answer: Yes; even when it dropped in value, it rose even higher.   53. D = $\{-3 \le x \le 3\}$, R = $\{-3 \le y \le 3\}$; no   55. $\{y \mid -8 < y < 6\}$ 57. $x < 5.1$ 59a. $2.85   59b. $29.82   61. $x + 2x^2 - 7$ 63. $31a + 10b$

Page 72   Lesson 2–2A

1.
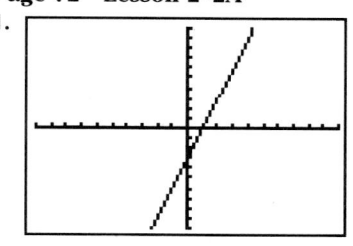

3.
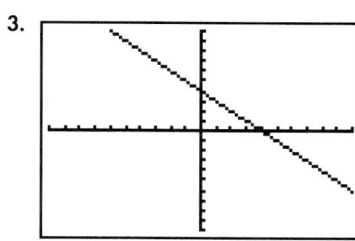

5.
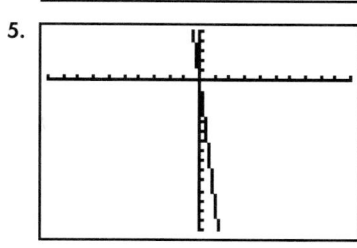

7.
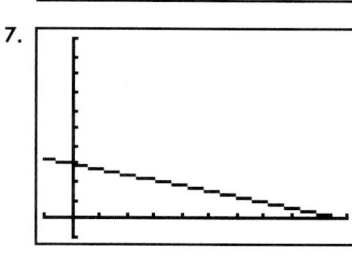

9.
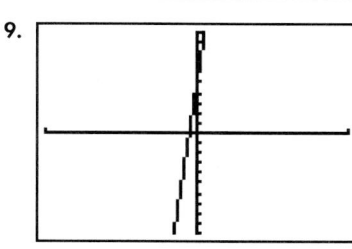

Pages 76–78   Lesson 2–2
7. yes   9. $2x - 5y = 3$; 2, -5, 3   11. y:9, x:$\frac{3}{2}$

13. y:-2, x:2   15.
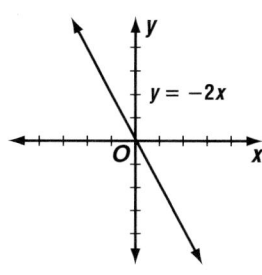

17. $71.80   19. no   21. yes   23. $12x - y = 0$; 12, -1, 0

25. $2x - y = 5$; 2, -1, 5   27. $y = 40$; 0, 1, 40   29. y:0, x:0
31. y:none, x:8   33. y:5, x:3

35.
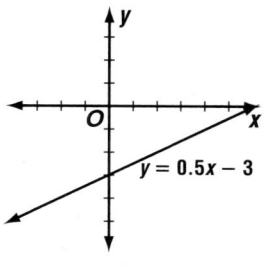

39.
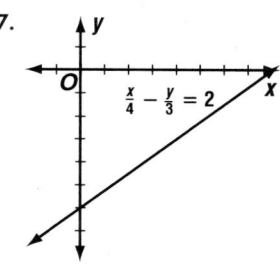

43.
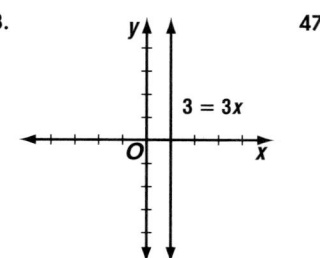

47.
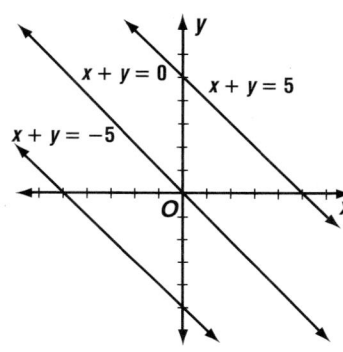

49a. same slope, different y-intercepts, look parallel

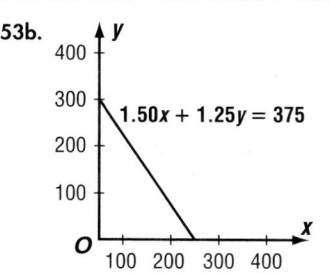

49b. Sample answer: $x + y = 2$ 51a. $d(t) = 2380t$
51b. 119 feet   53a. $1.50x + 1.25y = 375$

53b.
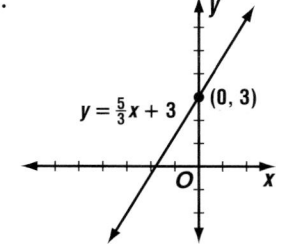

53c. Discrete; it is possible to list the elements of the domain.

53d. yes   55. true for all x 57. 0, $-\frac{10}{3}$ 59. 90 or higher
61. $3s + 14$

Page 79   Lesson 2–2B
1. 1.5   3. 5   5. 0   7. 11.5   9. 0.7

Pages 84–87   Lesson 2–3
7. 1   9. 0; horizontal   11. 1; rises   13. -1 15. $\frac{3}{4}$; $-\frac{4}{3}$

17.

19. 13; rises **21.** 0; horizontal **23.** undefined; vertical
25. undefined **27.** $-\frac{2}{3}$ **29.** 0 **31.** 9 **33.** 1

35.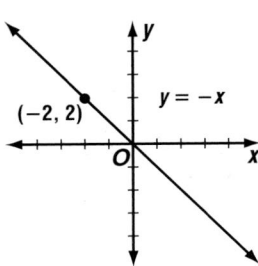

37.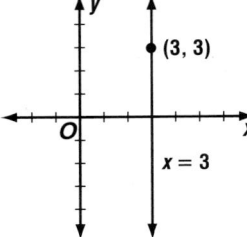

39.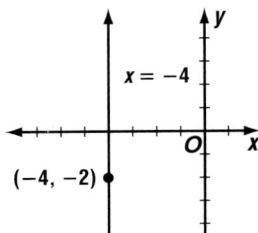

41.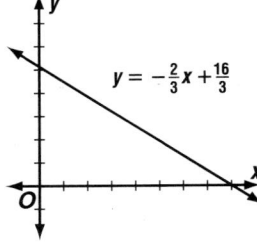

43. No; the product of the slopes is not -1.
45a. Graphs have the same y-intercept and slope is positive; as slope increases, the line gets steeper.
45b. Graphs have the same y-intercept and slope is negative; as the absolute value of the slope increases, the line gets steeper. **47.** -1 **49a.** 36 **49b.** 36
51. $2x + y = 4$ **53.** $-1 < x < 3$ **55.** 1, -9
57. 91; 99; no mode **59.** 9

Page 87 Self Test
1a. D = {0, 5, 10, 15, 20, 25, 30, 35, 40}R = {30, 27, 16, 9, 4, 1, -2, -4, -5}

1b.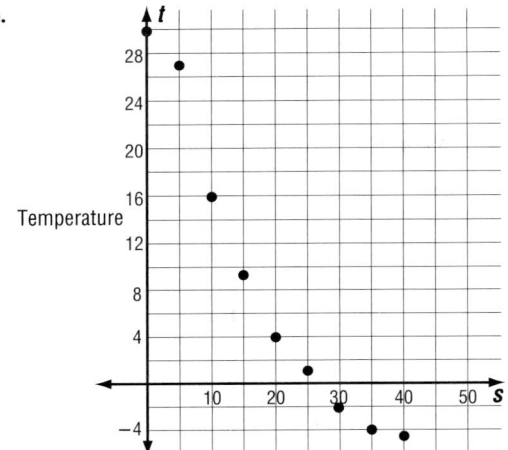
Wind speed

3. $6x + y = 4$ **5.**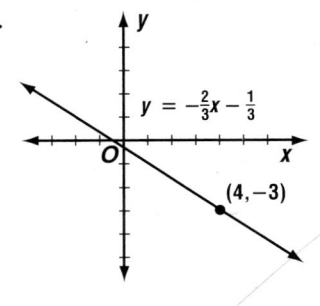

Page 91–94 Lesson 2–5
7. $y = 7x - 3$ **9.** $y = -\frac{1}{3}x + 4$ **11.** $m = 2$; b, 3
13. $y = \frac{5}{4}x + 7$ **15.** $y = \frac{1}{3}x - 2$ **17.** $y = -\frac{5}{2}x + 16$
19. $y = -x - 2$ **21.** $m = -\frac{2}{3}$; b, -4 **23.** $m = -0.3$; b, -6
25. $m = -\frac{3}{5}$; b, 6 **27.** $y = \frac{4}{5}x$ **29.** $y = -\frac{5}{3}x + \frac{29}{3}$
31. $y = -0.5x - 2$ **33.** $y = -\frac{4}{5}x + \frac{17}{5}$ **35.** $y = \frac{3}{2}x$
37. $y = \frac{3}{4}x - \frac{1}{4}$ **39.** $y = \frac{2}{3}x + \frac{10}{3}$ **41.** $y = -x - 4$ **43.** 7

45. 4 **47.** The lines are parallel with different y-intercepts. **49.** The lines are the same. **51.** 8.6 billion hours **53.** $y = 8.33x$ **55a.** $y = 1.45x + 3.2$ **55b.** 16.25 million **55c.** The slope is the average increase in subscribers; the y-intercept is the number of subscribers in 1990. **57.** no **59.** no solution **61.** 1 **63.** $5.42; $5.00; $5.00 **65.** 272.16

Pages 98–100 Lesson 2–5
7. 173 lb **9a.**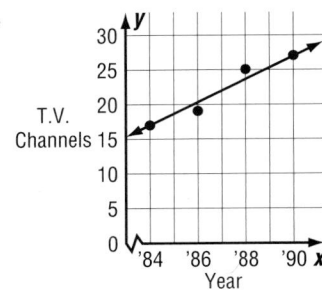

9b. $m = \frac{5}{3}$; y-intercept, 17 **9c.** about 42

11a.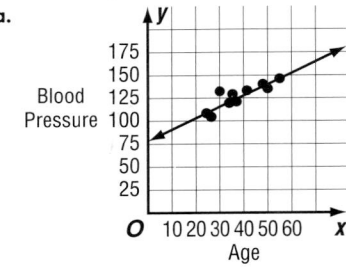

11b. Sample answer: $y = \frac{4}{3}x + 76$ **11c.** Sample answer based on the equation for 10b: 148
13a. Sample answer: $y = -50x + 226$

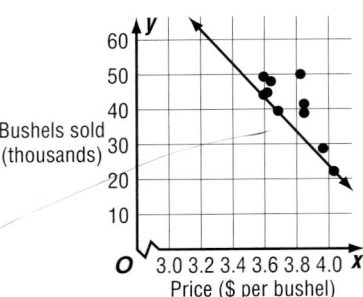

13b. 31,000 bushels **13c.** $4.01 **15.** $y = -\frac{1}{3}x + 4$, $y = 3x - 6$, $y = 3$ **17.** $\{x \mid x < -7 \text{ or } x > -1\}$ **19.** $4x + 8$

1.

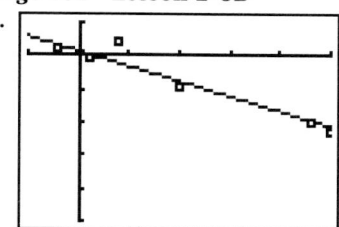

3a. $23,878.56

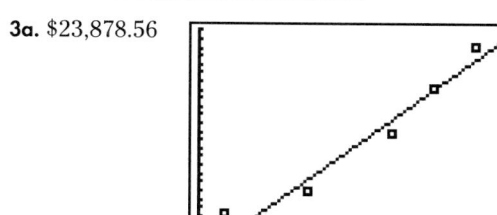

3b. $35,014.52

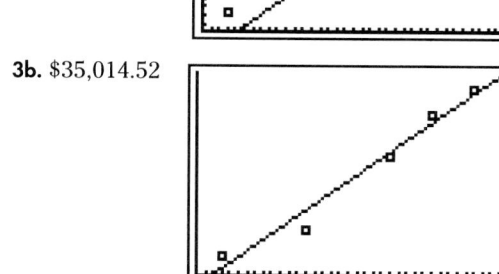

3c. The women's average income has increased at a greater rate than the men's; however, the men's average income was greater than the women's in 1950 and . remains greater in 1990.

Pages 106–108   Lesson 2–6

7. G

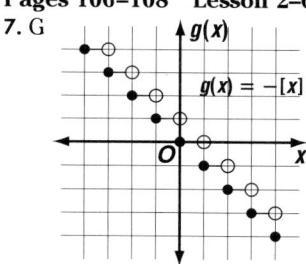

9. C

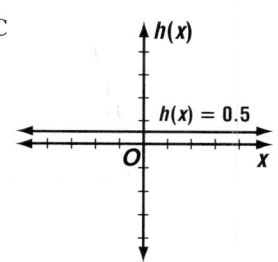

11. 10   **13.** 7   **15.** −2

17. D

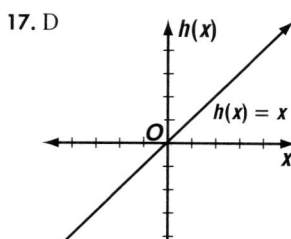

19. C

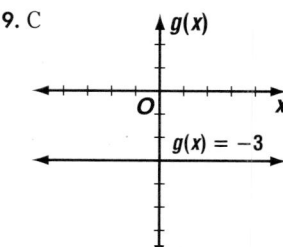

23. D

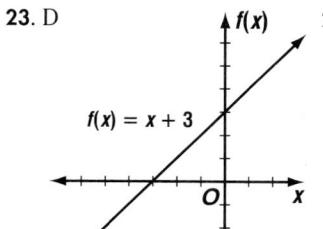

25. G

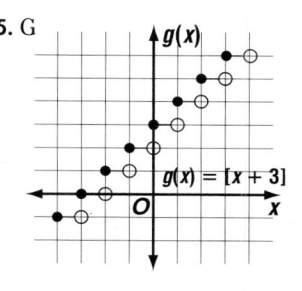

29. same graph translated 4 units right

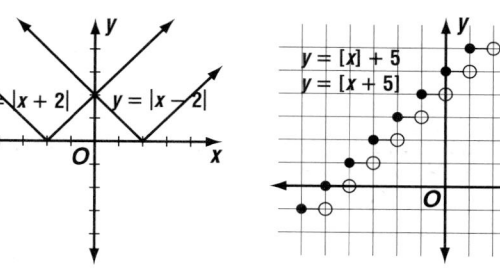

33. Graphs are the same.

37.

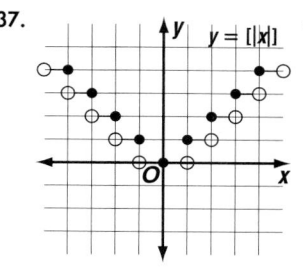

39.

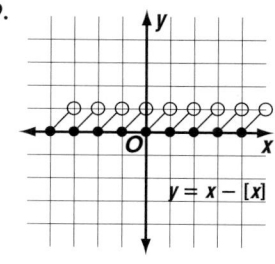

41. $x = \{-1, 5\}$

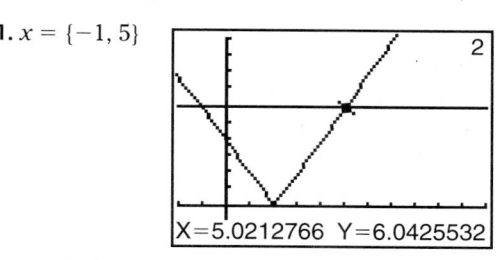

X=5.0212766 Y=6.0425532

43. $y = |x|$: D = {all real numbers}, R = {all positive real numbers and zero}; $x = |y|$: D = all positive real numbers and zero}, R = {all real numbers}

45a.

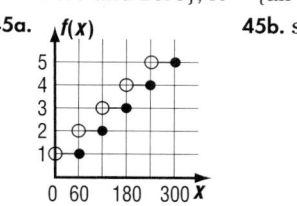

45b. step function

47a. Sample answer: $y = -0.72x + 1460$

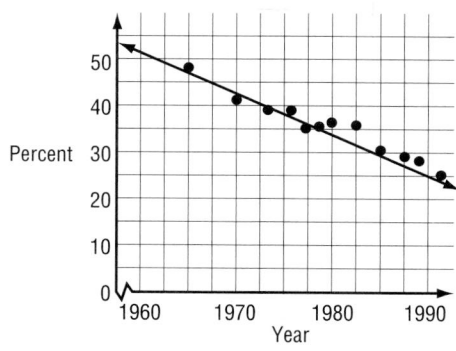

47b. 20%   **49.**

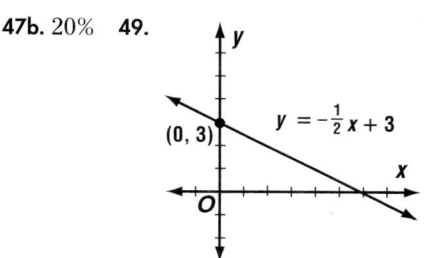

Page 109 Lesson 2–7A

1.
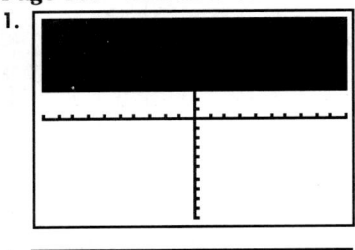

3.
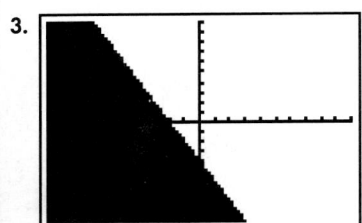

5.
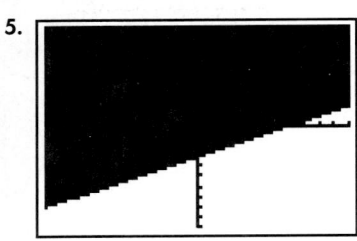

7.
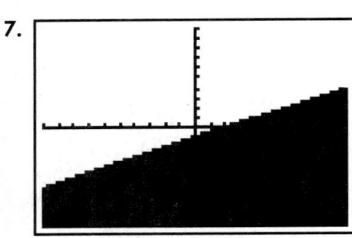

9.
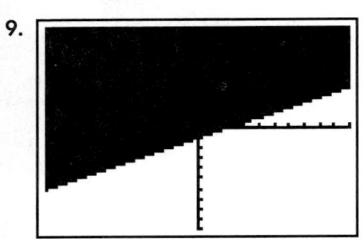

13.
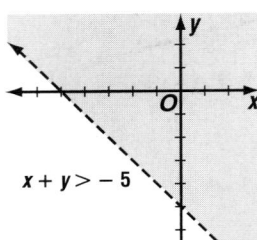

$x + y > -5$

17.
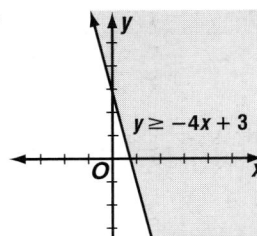

$y \geq -4x + 3$

21.
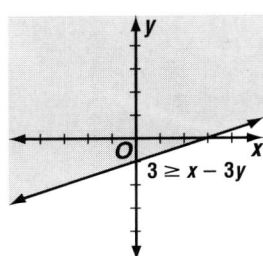

$3 \geq x - 3y$

25. $x < -2$
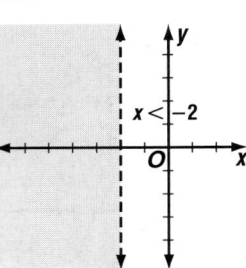

$x < -2$

27.
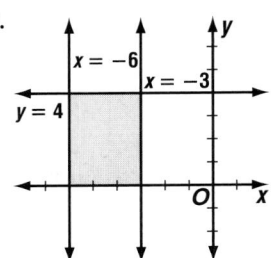

$x = -6$ $x = -3$ $y = 4$

29.
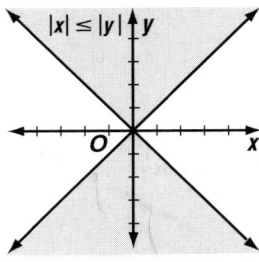

$|x| \leq |y|$

31.
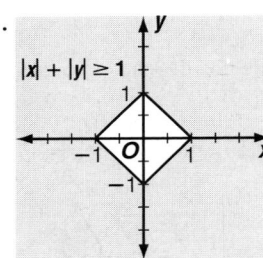

$|x| + |y| \geq 1$

33.
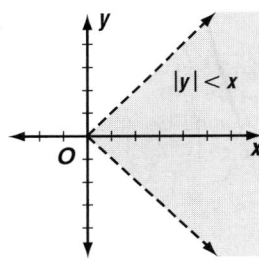

$|y| < x$

35a. $5x + 4y \geq 2500$

35b.
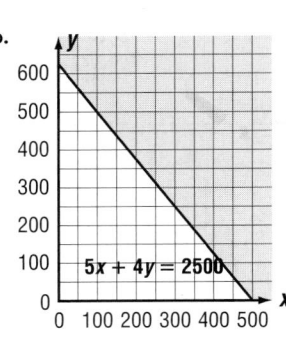

$5x + 4y = 2500$

35c. yes **37a.** $m + v \geq 1200$

37b.
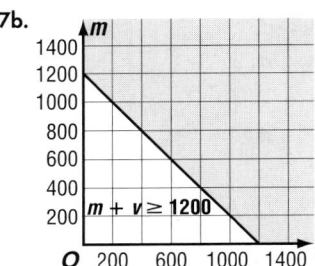

$m + v \geq 1200$

37c. D = {$200 \leq m \leq 800$},
R = {$200 \leq v \leq 800$}

Pages 113–114 Lesson 2–7
5. $(0, 0), (3, -4)$ **7.** $(3, -4)$

9.
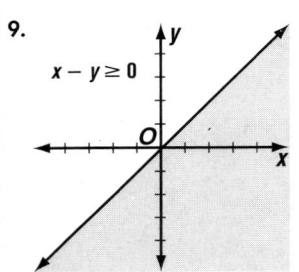

$x - y \geq 0$

11.
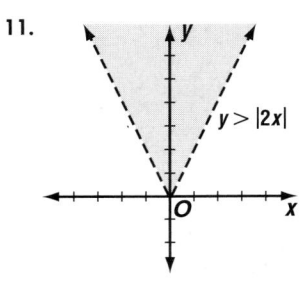

$y > |2x|$

39a.
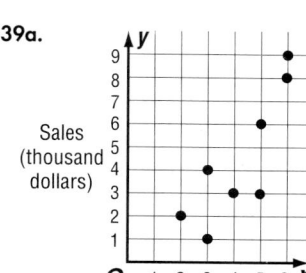

39b. $y = 1333x$
39c. $10,664

41.
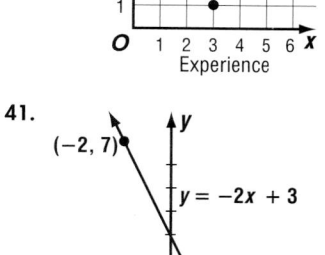

43. $(7, 2)$

Page 115   Chapter 2   Highlights
1. prediction equation   **3.** constant function   **5.** absolute value function   **7.** parallel lines   **9.** perpendicular lines
11. slope

Page 116–118   Chapter 2   Study Guide and Assessment
13. $D = \{1, 3, 5, 7, 9\}$,
$R = \{1, 2, 3, 4, 5\}$; yes

15. $D = \{$all reals$\}$,
$R = \{$all reals$\}$; yes

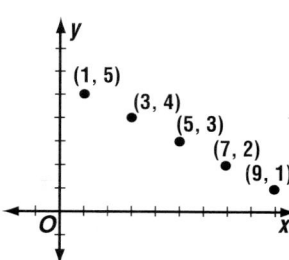

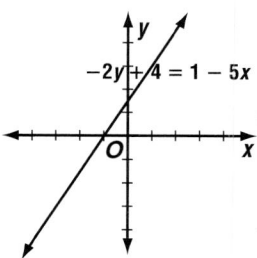

17. -3 **19.** $16v^2 - 10v - 9$

21. yes

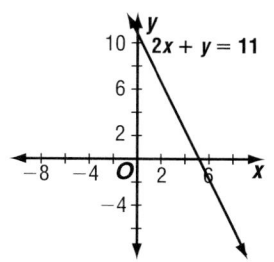

23. no   **25.** $5x + 2y = -4$ **27.** $5x + y = -20$
29. $x - y = 9$ **31.** $-\frac{3}{11}$ **33.** $y = \frac{3}{4}x + \frac{27}{2}$
35. $y = -\frac{5}{3}x - 3$ **37.** $y = -3x + 5$ **39.** $y = -\frac{3}{4}x + \frac{17}{4}$

41a.
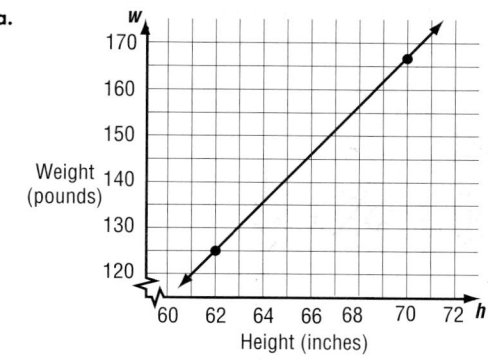

41b. $w = 5.25h - 200.5$ **41c.** 203.75 pounds   **41d.** about 67.75 inches tall

43. G

45. D

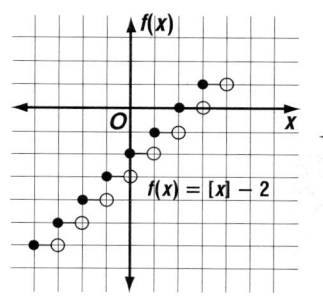

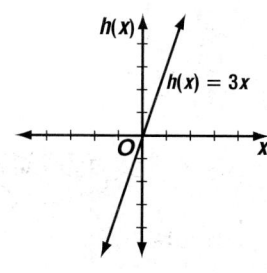

47. A

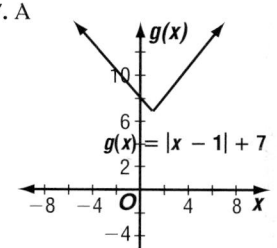

49.
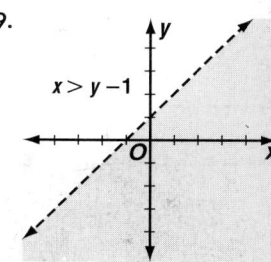

51.
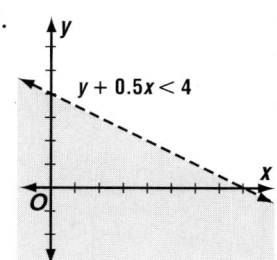

53. $117.20

CHAPTER 3   SOLVING SYSTEMS OF LINEAR EQUATIONS AND INEQUALITIES

Page 125   Lesson 3–1A
1. $(2, 4)$ **3.** $(1, 2)$ **5.** $(3.40, -2.58)$ **7.** $(-0.03, 1.03)$

Pages 129–132   Lesson 3–1
7. 0; inconsistent

z

9. $y = 5x - 3$; consistent, dependent

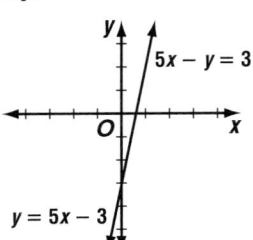

11. no solution; inconsistent

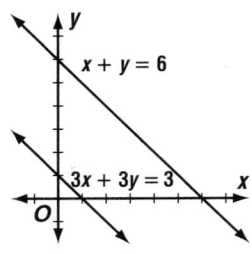

13. a   **15.** 1; consistent, independent; $(-3, 0)$
17. 1; consistent; independent; $(-1, 3)$

19. $(0, -8)$; consistent, independent

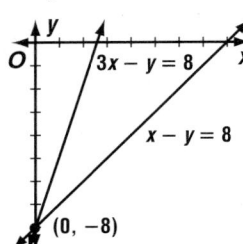

23. $x + 2y = 4$; consistent, dependent

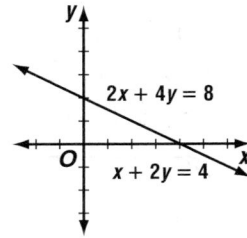

27. $(-9, 3)$; consistent, independent

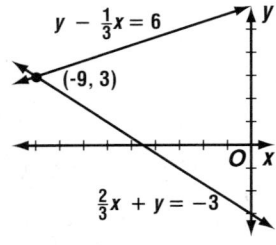

31. $(-5, 4)$; consistent, independent

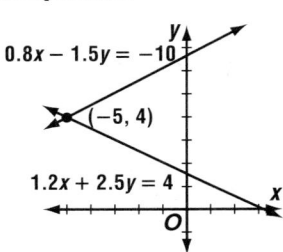

33. $m = \frac{5}{2}, n = -\frac{3}{2}$ **35.** $m = \frac{3}{2}, n \neq 4$ **37.** $(-6, 3)$

39. $(2, -3)$ **41.** $(-5.56, -12.00)$ **43a.** $\frac{a}{d} = \frac{b}{e} = \frac{c}{f}$

43b. $\frac{a}{d} \neq \frac{b}{e}$ **43c.** $\frac{a}{d} = \frac{b}{e} \neq \frac{c}{f}$ **45.** 3 field goals and 4 points after   **47a.** butter   **47b.** about 1957   **47c.** Yes, the consumption of margarine has decreased in recent years.

49.

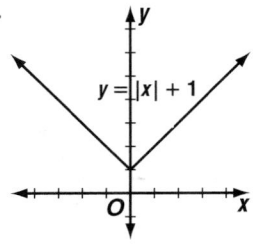

51a. Sample answer: $y = -0.003x + 47$
51b. about 38%

53.

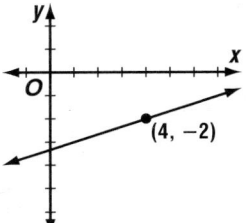

55. $3x - y = 9$

Pages 137–140   Lesson 3–2
7. $(-5, 4)$ **9.** $(-3, 5)$ **11.** $2x + 3y = -15$ **13.** $\left(\frac{10}{3}, \frac{8}{3}\right)$
15. $(3, -2)$ **17.** $(6, 2)$ **19.** $\left(-3, \frac{1}{2}\right)$ **21.** $(-9, 1)$
23. $(8, -5)$ **25.** $(9, 8)$ **27.** $\left(\frac{1}{2}, -\frac{1}{2}\right)$ **29.** no solution
31. $(-7, 2)$ **33.** $(8, -3)$ **35.** $(12, -10)$ **37.** $(-5, 4)$
39. $(1, 4), (-2, -1), (5, 2)$ **41a.** $(9, 1)$ **41b.** $(6, -1)$
41c. $(4, -28)$ **41d.** $(10, 25)$ **43.** 210 cups of hot chocolate, 85 cups of coffee   **45a.** $p = 17.99 + 0.43d$
45b. $p = 3.67 + 0.95d$ **45c.** during the year 2265
47. Sample answer: 20¢ a mile or $15 plus 10¢ a mile
49. step function   **51a.** Sample answer: $y = \frac{7}{8}x + 8$
51b. about 42¢   **53.** 9   **55.** 14.45; 13; no mode

Pages 144–146   Lesson 3–3
5. -3 **7.** $-\frac{6}{5}$ **9.** $(2, -1)$ **11.** $\left(-\frac{17}{23}, -\frac{16}{23}\right)$ **13.** -46
15. -29 **17.** -29 **19.** 0   **21.** 22.79   **23.** $(6, 3)$
25. $\left(\frac{2}{3}, -1\right)$ **27.** $\left(-\frac{3}{4}, 3\right)$ **29.** $(3, 10)$ **31.** $\left(\frac{2}{3}, \frac{5}{6}\right)$
33. $(2, -3)$ **35.** In both cases, the denominator is 0. If the numerator is also 0, there is an infinite number of solutions. If the numerator is not 0, there are no solutions.   **37.** 28.2 min   **39.** $(-1, 2)$

41. $(-1, 3)$; consistent and independent

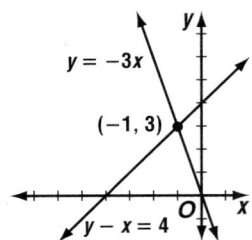

43. Sample answer: $y = -x + 2$ **45.** 120 units^2

Page 147   Lesson 3–4A
1.

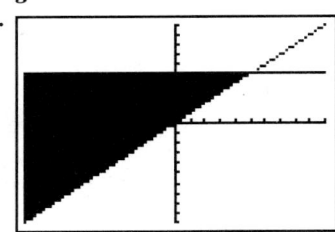

3.

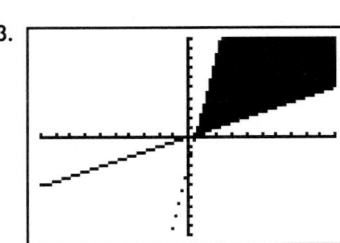

5.

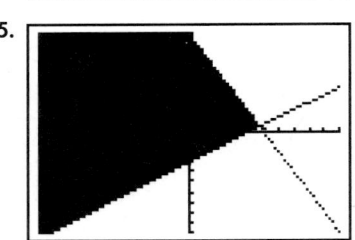

7.

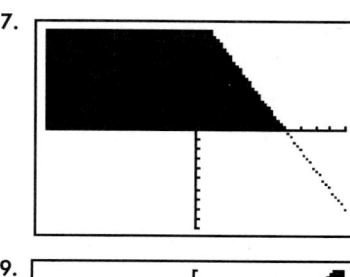

9.

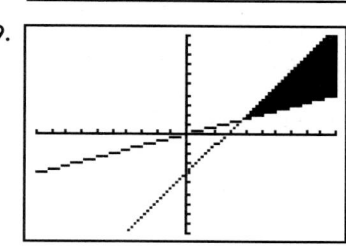

Pages 150–152   Lesson 3–4
7. c

9.

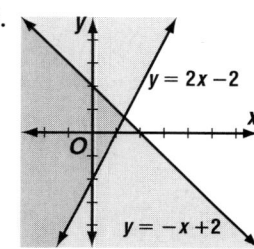

11.

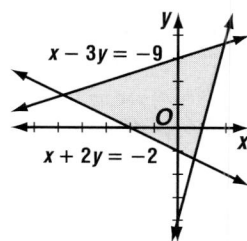

13.

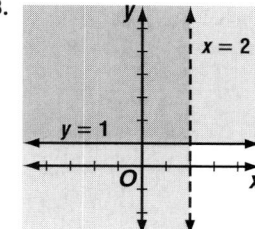

17.

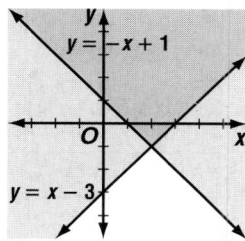

21.

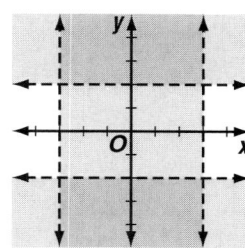

25.

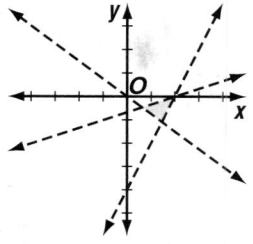

29. Sample answer: $x + y \le 4$, $x - y \le 4$, $y \ge -1$
31. 42 square units
33. $s \ge 7$ and $s + a \le 16$ **35.** $(-4, -4)$

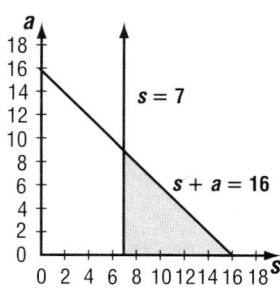

37. no solution   **39.** $3

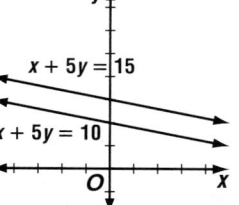

Page 152   Self Test
1. $(-3, -4)$ **3.** $(-3, -9)$ **5.** -61 **7.** $\left(\frac{1}{2}, -2\right)$
9.

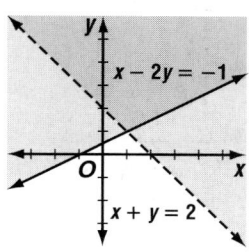

Pages 157–159   Lesson 3–5
5. max: $f(4, 1) = 15$ min: $f(-3, 2) = 14$

7. vertices: $(0, 2)$, $(4, 3)$, $\left(\frac{7}{3}, -\frac{1}{3}\right)$;
max: $f(4, 3) = 25$;
min: $f(0, 2) = 6$

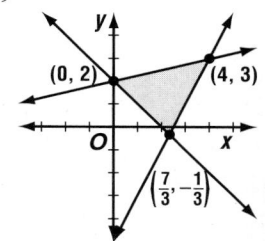

9a. Gauss used the strategy of solving a simpler problem as in Example 3.   **9b.** 2,001,000   **11.** max: $f(3, 5) = 19$ min: $f(-1, -2) = -7$ **13.** max: $f(-1, -2) = 4$ min: $f(3, 5) = -11$

15. vertices: $(1, 0)$, $(3, 0)$, $(1, 4)$;
max: $f(3, 0) = 9$;
min: $f(1, 0) = 3$

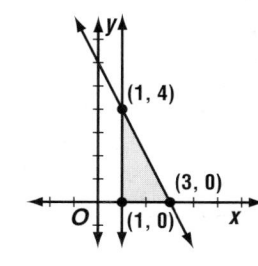

17. vertices: $(0, 1)$,
$(1, 3)$, $(6, 3)$, $(10, 1)$;
max: $f(10, 1) = 31$;
min: $f(0, 1) = 1$

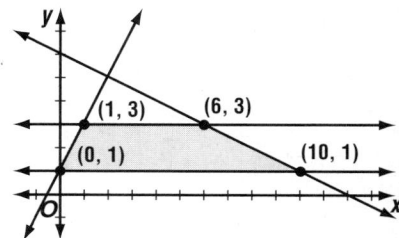

19. vertices: (0, 2), (4, 3), (2, 0);
max: $f(4, 3) = 13$;
min: $f(2, 0) = 2$

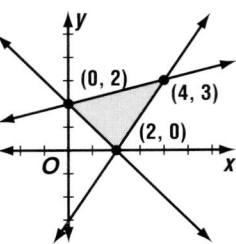

21. vertices: (2, 1), (4, 1),
(4, 4), (2, 3);
max: $f(4, 1) = 1$;
min: $f(4, 4) = -8$

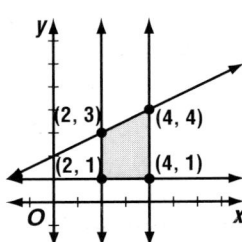

23. vertices: (5, 0), (0, 0), (0, 2), (2, 4), (5, 1);
max: $f(5, 0) = 25$; min: $f(0, 2) = -6$ **25.** vertices:
$(-4.08, -2), (-1.58, 1), (-0.4, 1), (-0.4, -2)$;
max: $f(-0.4, 1) = 0.4$; min: $f(-4.08, -2) = -20.32$
27. $164 **29.** Sample answer: No, he could only increase
his income by $100. **31.** 120 telephone lines **33.** (5, −3)
35. no **37.** $54 - 32b$

Pages 162–164   Lesson 3–6
5a. $w \geq 0, d \geq 0, 8w + 5d \leq 80, 2w + 5d \leq 50$

5b.

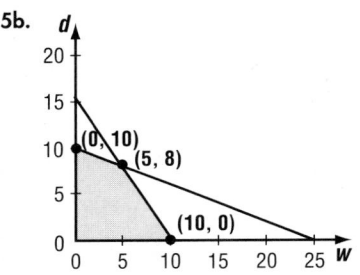

5c. $f(w, d) = 17w + 29d$ **5d.** 5 Wallbangers and
8 Dingbats; $317 **7a.** $20 **7b.** No, the vertex at (10, 30)
still produces the least cost. **9a.** $a \geq 0, b \geq 0, 4a + b \leq$
$32, a + 6b \leq 54$ **9b.** 14 gal **11a.** 72 **11b.** Sample
answer: Take a speed-reading course so that she can read
and answer the multiple-choice questions in less time.

13. vertices: (0, 3), (0, 6),
(2, 5), (1, 3);
max: $f(1, 3) = -3$;
min: $f(0, 6) = -12$

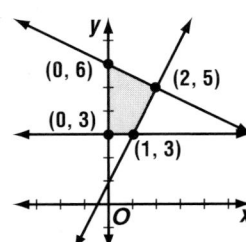

15. 55 mph **17.** $|x| < 4$

Pages 169–171   Lesson 3–7
3. yes **5.** (4, 1, 6) **7.** (1, 4, 9) **9.** Sample answer: $x +$
$y + z = -9, x + y - z = 1, x - y + z = -13$ **11.** no
13. (−2, 5, −4) **15.** (0, 1, 2) **17.** (4, −3, 1) **19.** (−2, 3, 6)
21. $\left(\frac{2}{3}, \frac{1}{2}, -\frac{1}{3}\right)$ **23.** $\left(\frac{1}{2}, \frac{1}{3}, \frac{1}{4}\right)$ **25.** 15, 3, −6 **27.** 16, 7, −5
29. 1-year, $2500; 2-year, $3500; 3-year, $9000

31. 3-pointers, 4; 2-pointers, 9; free throws, 5 **33.** a
35a. $y = 0.25x + 20; y = 0.25x + 35$ **35b.** parallel
35c. $15 **37.** 12, $-\frac{12}{5}$

Page 173   Lesson 3–7B
1.   1

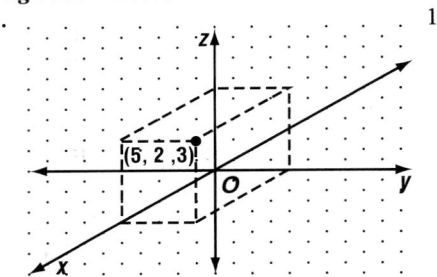

3.   none

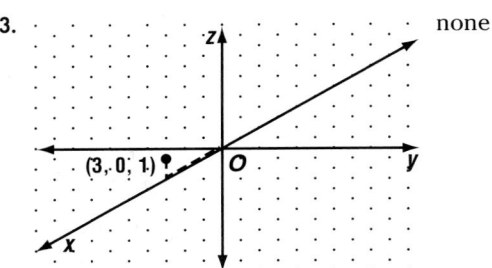

5. (1, 0, 0), (0, 4, 0), (0, 0, 2)

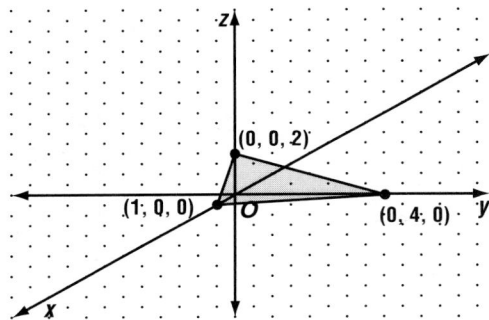

7. (1, 0, 0), (0, −3, 0), $\left(0, 0, \frac{1}{2}\right)$

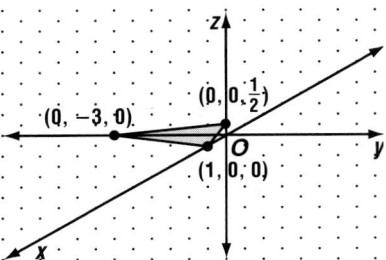

9. (−3, 0, 0), none, (0, 0, 2)

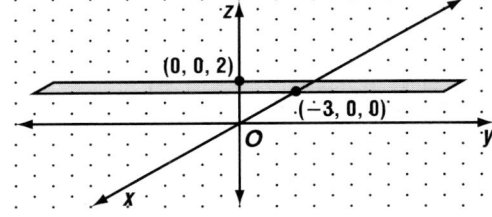

960 *Selected Answers*

11. Sample answer: $5x - 5y + 2z = 10$ **13.** Two-dimensional graphs are divided into quadrants and three-dimensional graphs are divided into octants.

Page 175 Chapter 3 Highlights
1. c **3.** f **5.** a **7.** i **9.** d **11.** k

Pages 176–178 Chapter 3 Study Guide and Assessment
13. no solution; inconsistent

15. $20y - 13x = 10$; consistent, dependent

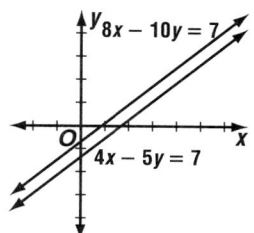

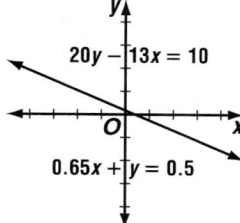

17. $(17.4, -13.6)$ **19.** $(-6, 2)$ **21.** $(3, 2)$ **23.** $\left(\frac{1}{2}, \frac{1}{2}\right)$

25. $\left(2, \frac{13}{8}\right)$ **27.** $\left(-\frac{3}{7}, -\frac{3}{7}\right)$ **29.**

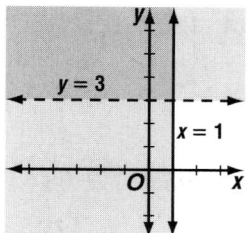

31.

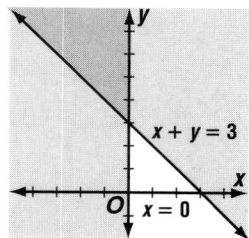

33.

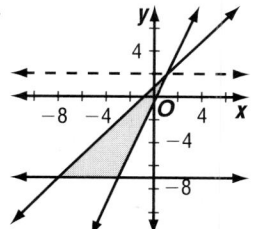

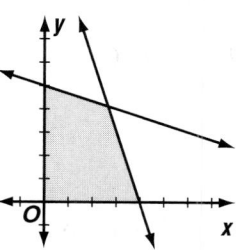

35. vertices: $(0, 5)$, $(3, 4)$, $(4, 0)$, $(0, 0)$; max: $f(3, 4) = 17$; min: $f(0, 0) = 0$

37. $(1, 2, 3)$ **39.** $(3, -1, 5)$ **41.** burger: \$1.89, fries: \$0.99, cola: \$0.79

CHAPTER 4 USING MATRICES

Page 185 Lesson 4–1A
1. $\begin{bmatrix} -1 & -4 \\ 3 & -6 \\ -7 & 2 \end{bmatrix}$ **3.** -10 **5.** $\begin{bmatrix} -.5 & -.5 & .5 \\ .4 & 1 & -.2 \\ .4 & 0 & -.2 \end{bmatrix}$

7. $\begin{bmatrix} 47 & 2 \\ -28 & 44 \end{bmatrix}$ **9.** $\begin{bmatrix} 32 & 14 & 41 \\ -4 & 5 & -10 \end{bmatrix}$ **11.** $\varnothing$

13. $\begin{bmatrix} .020715619 & 9.416195857\text{E-4} \\ .0131826742 & .0221280603 \end{bmatrix}$ **15.** $\begin{bmatrix} 2153 & 182 \\ -2548 & 1880 \end{bmatrix}$

17. $\begin{bmatrix} 277 & 130 \\ -89 & 34 \end{bmatrix}$

Pages 191–193 Lesson 4–1
7. $\begin{bmatrix} -14 & -6 & 2 \end{bmatrix}$ **9.** $x = 2.5, y = 1, z = 3$

11. $S = \begin{array}{c} \\ \text{turkey} \\ \text{ham} \\ \text{roast beef} \end{array}\begin{array}{cc} \text{plain} & \text{cheese} \\ \left[15 \right. & \left. 12 \right] \\ 8 & 10 \\ 8 & 11 \end{array}$ **13.** $\begin{bmatrix} 15 & -6 & 21 \\ -9 & 24 & 12 \end{bmatrix}$

15. $\begin{bmatrix} 2 & -\frac{5}{3} \end{bmatrix}$ **17.** $\begin{bmatrix} -6.5 & 0 & -25.5 \\ -2 & -5 & -12.5 \end{bmatrix}$ **19.** $x = 3, y = -\frac{1}{3}$

21. $x = 3, y = -5, z = 6$ **23.** $x = \pm 5, y = 9, z = 6$
25. $(0, 0), (0, 3), (1.25, 0)$ **27.** The perimeter is one-half the original perimeter; the triangle is rotated $180°$.
29. $x = 3, y = 5, m = 10, r = 2$ **31.** no
33.

	male	female
exercise walking	21%	38%
swimming	27%	28%
bicycle riding	25%	22%
camping	23%	18%
bowling	20%	18%
fishing	25%	11%
exer. with equip.	17%	17%
basketball	18%	6%
aerobic exercising	5%	19%
golf	17%	5%

35. 100 adult, 50 student **37.** 0 **39a.** 30; 26; about 30.4
39b. They are all located in the Southwestern U.S.

Pages 197–198 Lesson 4–2
5. $\begin{bmatrix} 1 & 10 \\ -7 & 5 \end{bmatrix}$ **7.** $\begin{bmatrix} 21 & -2 \end{bmatrix}$ **9a.**

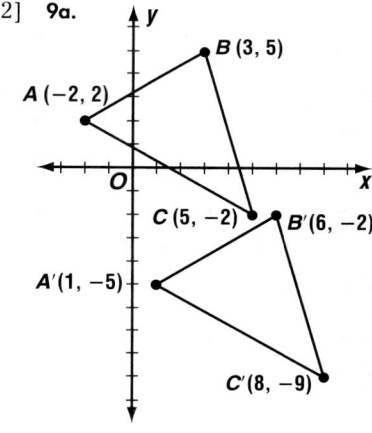

9b. $\begin{bmatrix} 3 & 3 & 3 \\ -7 & -7 & -7 \end{bmatrix}$ **9c.** $\begin{bmatrix} 1 & 6 & 8 \\ -5 & -2 & -9 \end{bmatrix}$ **11.** $\begin{bmatrix} 4 & 20 & 1 \end{bmatrix}$

13. $\begin{bmatrix} -4 & -15 \\ 1.5 & -2 \end{bmatrix}$ **15.** $\begin{bmatrix} -19 & -6 \\ 20 & -11 \\ -20 & 34 \end{bmatrix}$

17a. $\begin{bmatrix} 4 & 4 & 4 & 4 \\ -2 & -2 & -2 & -2 \end{bmatrix}$

17b. $B'(10, -1), T'(1, -7), U'(7, 3)$ **19.** $x = 2, y = 7, z = -2$ **21a.** Sample answer: No; births and deaths are opposite occurrences. **21b.** Yes; $B - D$ represents the population increase for 1992. **21c.** Find $0.99D$.

23. $\begin{bmatrix} -28 & 20 & -44 \\ 8 & -16 & 36 \end{bmatrix}$

25. 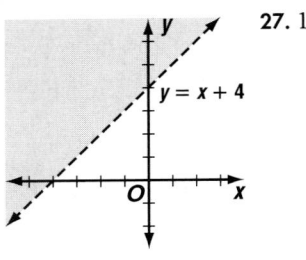 **27.** 12

Pages 202–204   Lesson 4–3

7. 3×2 **9.** $\begin{bmatrix} -39 \\ 18 \end{bmatrix}$ **11.** $A'(-2, -5)$, $B'(-4, 3)$, $C'(4, 1)$;

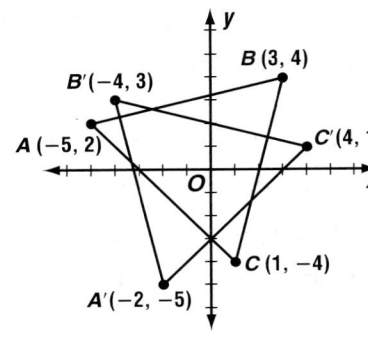

13. not defined   **15.** 3×1 **17.** not defined
19. $\begin{bmatrix} 1 & -25 & 2 \\ 29 & 1 & -30 \end{bmatrix}$ **21.** not possible to evaluate
23. $\begin{bmatrix} 0 & 64 & -40 \\ 9 & 11 & -11 \\ -3 & 39 & -23 \end{bmatrix}$ **25.** No; multiplication is not commutative.   **27.** not defined   **29.** $\begin{bmatrix} -39 & 9 \\ 5 & 16 \end{bmatrix}$
31. rotates $180°$ **33.** reflects over y-axis   **35.** $w = 1$, $x = 0$, $y = 0$, $z = 1$; the original matrix   **37.** about 39%
39a. $22,800   **39b.** sold, $22,735.50; lost, $62.50
41. $\begin{bmatrix} -12 & 6 \\ 30 & -42 \end{bmatrix}$ **43.** $(-9, -7)$ **45.** $(8, 0)$ and $(0, -2)$

Pages 208–211   Lesson 4–4

5. no   **7.** yes, 1   **9.** 45   **11.** yes, 0   **13.** no   **15.** yes, -13
17. 16 units^2 **19.** -28 **21.** 0   **23.** -141 **25.** 11 units^2
27. $x = -6$ **29.** 29 units^2 **31.** $x = 22$
33. Sample answer: $\begin{bmatrix} 1 & 1 & 1 \\ 1 & 1 & 1 \\ 1 & 1 & 1 \end{bmatrix}$ **35.** about 68 mi^2
37. $x = 1$, $y = 3$, $z = 5$ **39.** $(4, 1)$ **41.** about 5.7

Page 211   Self Test

1. Sue - Lou; Elisa - Mateo; Tamara - Bob   **3.** $\begin{bmatrix} 4 & -0.5 \\ 0 & -1.5 \end{bmatrix}$
5. $\begin{bmatrix} -15 & 6 \\ 19 & 43 \\ -6 & -24 \end{bmatrix}$ **7.** $\begin{bmatrix} -15 & 6 \\ 19 & 43 \\ -6 & -24 \end{bmatrix}$ **9.** -119

Pages 216–218   Lesson 4–5

7. not square   **9.** $-\frac{1}{27}\begin{bmatrix} 4 & -1 \\ -7 & -5 \end{bmatrix}$ **11.** $\frac{1}{5}\begin{bmatrix} 1 & 0 \\ 0 & 5 \end{bmatrix}$

13. det = 0   **15.** $\frac{1}{34}\begin{bmatrix} 7 & 3 \\ -2 & 4 \end{bmatrix}$ **17.** $\frac{1}{32}\begin{bmatrix} 1 & 5 \\ -6 & 2 \end{bmatrix}$
19. $-\frac{1}{12}\begin{bmatrix} 6 & 0 \\ -5 & -2 \end{bmatrix}$ **21.** true   **23.** true   **25.** true   **27.** true
29. true   **31.** Let $\begin{bmatrix} a & b \\ c & d \end{bmatrix}$ represent A and let $\begin{bmatrix} 1 & 0 \\ 0 & 1 \end{bmatrix}$ represent I. $\begin{bmatrix} a & b \\ c & d \end{bmatrix} \cdot \begin{bmatrix} 1 & 0 \\ 0 & 1 \end{bmatrix} = \begin{bmatrix} a & b \\ c & d \end{bmatrix}$ $\begin{bmatrix} 1 & 0 \\ 0 & 1 \end{bmatrix} \cdot \begin{bmatrix} a & b \\ c & d \end{bmatrix} = \begin{bmatrix} a & b \\ c & d \end{bmatrix}$
Therefore, $A \cdot I = I \cdot A = A$.   **33a.** ALLATSIX   **33b.** MEET ME AT THE MALL AT SIX   **35.** 6, 8, 10 units   **37.** 7   **39.** no solution

Pages 223–225   Lesson 4–6

5. $\begin{bmatrix} 4 & -7 \\ 3 & 5 \end{bmatrix} \cdot \begin{bmatrix} x \\ y \end{bmatrix} = \begin{bmatrix} 2 \\ 9 \end{bmatrix}$ **7.** $\begin{bmatrix} -3 & 1 \\ 1 & 2 \end{bmatrix} \cdot \begin{bmatrix} x \\ y \end{bmatrix} = \begin{bmatrix} 0 \\ -21 \end{bmatrix}$
9. $\left(-1, \frac{9}{2}\right)$ **11.** infinitely many   **13.** $\begin{bmatrix} 3 & -7 \\ 6 & 5 \end{bmatrix} \cdot \begin{bmatrix} m \\ n \end{bmatrix} =$ $\begin{bmatrix} -43 \\ -10 \end{bmatrix}$ **15.** $\begin{bmatrix} -1 & -1 \\ 2 & -1 \end{bmatrix} \cdot \begin{bmatrix} x \\ y \end{bmatrix} = \begin{bmatrix} 0 \\ 0 \end{bmatrix}$ **17.** $\begin{bmatrix} 2 & -3 \\ 4 & -1 \end{bmatrix} \cdot \begin{bmatrix} x \\ y \end{bmatrix} =$ $\begin{bmatrix} 0 \\ 5 \end{bmatrix}$ **19.** $\left(\frac{2}{9}, -\frac{4}{3}, -\frac{1}{3}\right)$ **21.** $\left(\frac{1}{2}, \frac{1}{3}\right)$ **23.** $(-3, 5)$ **25.** $\left(\frac{5}{3}, \frac{1}{2}\right)$
27. $(0, -4)$

29. no solution

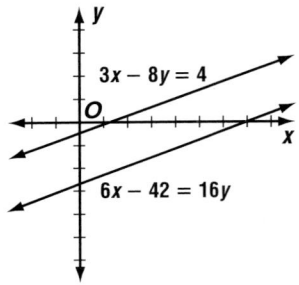

31. $(2.5, 3)$ **33.** 80 mL of 60%; 120 mL of 40%   **35.** true
37. $y < 0$ or $x < 0$ **39.** 7.82 tons per square inch

Pages 229–231   Lesson 4–7

7. $\begin{array}{l} 3x - 5y = 25 \\ 2x + 4y = 24 \end{array}$ **9.** $\begin{bmatrix} 4 & -7 & | & -19 \\ 3 & 2 & | & 22 \end{bmatrix}$ **11.** no solution

13. $\begin{bmatrix} 4 & -3 & | & 5 \\ 2 & 9 & | & 6 \end{bmatrix}$, $\left(\frac{3}{2}, \frac{1}{3}\right)$ **15.** $\begin{bmatrix} 2 & -1 & 4 & | & 6 \\ 1 & 5 & -2 & | & -6 \\ 3 & -2 & 6 & | & 8 \end{bmatrix}$, $(-6, 2, 5)$

17. $\begin{bmatrix} 2 & 1 & 1 & | & 2 \\ -1 & -1 & 2 & | & 7 \\ -3 & 2 & 3 & | & 7 \end{bmatrix}$, $(0, -1, 3)$ **19.** $\left(-\frac{1}{4}z + 1, 2z + 5, z\right)$

21. no solution   **23.** $(2, -3)$ **25.** $\left(-\frac{1}{6}, \frac{1}{2}, -\frac{2}{3}\right)$ **27.** no solution   **29.** a plane   **31.** 17 small, 24 medium, 11 large
33. $\frac{1}{6}\begin{bmatrix} -1 & 5 \\ -2 & 4 \end{bmatrix}$ **35.** a step function   **37.** 449

Page 234   Lesson 4–7B

1. $(-4, 5)$ **3.** $\left(-\frac{1}{7}, -\frac{3}{7}\right)$ **5.** $\left(\frac{5}{4}, -\frac{1}{2}, \frac{1}{4}\right)$ **7.** $(-2, -1, 3)$
9. $(-1, -6, 8)$

Pages 240–244 Lesson 4–8
7a. 20 **7b.** 26 **7c.** 25% **7d.** 21 and 28 **9.** range = 19; $Q_1 = 25$; $Q_2 = 32.5$; $Q_3 = 38$; IR = 13; no outliers

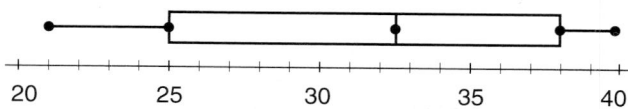

11a. 50% **11b.** 75% **11c.** 50% **11d.** The least value and lower quartile are same number.
13. range = 43; $Q_1 = 31$; $Q_2 = 46$; $Q_3 = 59$; IR = 28; no outliers

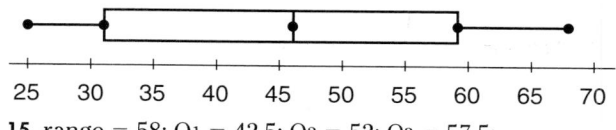

15. range = 58; $Q_1 = 42.5$; $Q_2 = 52$; $Q_3 = 57.5$; IR = 15; outlier = 81

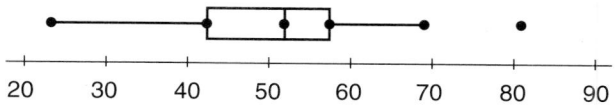

17. range = 2.7; $Q_1 = 13.95$; $Q_2 = 14.9$; $Q_3 = 15.85$; IR = 2; no outliers

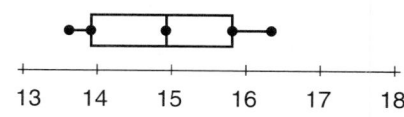

19a. Springfield

S.F.

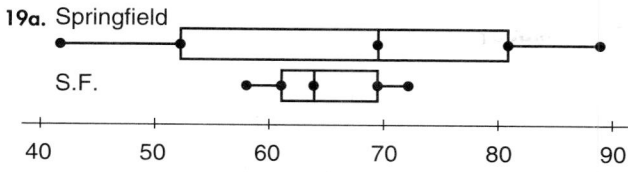

19b. The median average high temperature for San Francisco is 63.5°; the median average high temperature for Springfield is 69.5°. **21.** Sample answer: {1, 4, 4, 4, 5, 5, 5, 6, 6, 15}

23a.

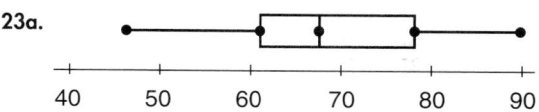

23b. Stem-and-leaf: advantage—all data are shown; disadvantage—difficult to see how data are dispersed. Box-and-whisker: advantage—easy to see range, median, and so on; disadvantage—data are lost.

25a.

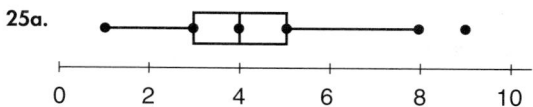

27b. Median is the same, box is longer. **27c.** plain air-popped popcorn, Skittles **29.** [−54, 6] **31.** $\frac{2}{3}$ **33.** $1440

Page 245 Chapter 4 Highlights
1. identity matrix **3.** scalar multiplication
5. determinant **7.** dimensions **9.** equal matrices

Pages 246–248 Chapter 4 Study Guide and Assessment
11. $\begin{bmatrix} 24 & -9 & 6 \\ 12 & 3 & 21 \end{bmatrix}$ **13.** $\begin{bmatrix} 2\frac{1}{2} & -4 \end{bmatrix}$ **15.** $\begin{bmatrix} 5.2 & 20.4 \\ -8 & -14.8 \\ -11.2 & 18 \end{bmatrix}$

17. $x = -5, y = -1$ **19.** $x = -1, y = 0$ **21.** $\begin{bmatrix} -3 & 0 \\ -2 & -6 \end{bmatrix}$

23. $\begin{bmatrix} 1 & -2 \\ -14 & 9 \end{bmatrix}$ **25.** $[-18]$ **27.** not possible to evaluate

29. −52 **31.** −36 **33.** −35 **35.** $\frac{1}{2}\begin{bmatrix} 7 & -6 \\ -9 & 8 \end{bmatrix}$

37. $\frac{1}{12}\begin{bmatrix} -1 & 2 \\ 3 & 6 \end{bmatrix}$ **39.** not possible **41.** (2, 1) **43.** (−1, 2)

45. (4, 2) **47.** $\left(-\frac{1}{2}, 1, 6\right)$ **49.** 90; 25, 50, 75; 50; no outliers

51. 170; 1025, 1075, 1125; 100; no outliers **53.** batteries, $74; spark plugs, $58; wiper blades, $48

CHAPTER 5 EXPLORING POLYNOMIALS AND RADICAL EXPRESSIONS

Pages 259–260 Lesson 5–1
5. $\frac{3w}{z^4}$ **7.** $81a^4$ **9.** $-8x^2$ **11.** $4s^2t^3$ **13.** $\frac{8}{b^3c^3}$
15. 3.86×10^5 **17.** 5×10^0; 5 **19.** b^8 **21.** m^9 **23.** an
25. ab **27.** $18r^5s^5$ **29.** $10m^3n^3$ **31.** $-\frac{2}{3}m^3n^7$ **33.** $-108x^4$
35. $-15m^4n^3p^2$ **37.** $\frac{-1}{4y^4}$ **39.** $-\frac{m^4n^9}{3}$ **41.** $\frac{a^4}{16b^4}$ **43.** $2a^4b^2$
45. 4 **47.** 7.865×10^8 **49.** 1.25×10^{-3} **51.** 3.331×10^{-2}
53. 4.02×10^{-5}; 0.0000402 **55.** 9.025×10^7; 90,250,000
57. 3.1×10^8; 310,000,000 **59.** 6 **61.** 3 **63.** $-\frac{21}{2}$ **65.** x^{10}
67. about 1.53×10^4 seconds or 4.25 hours
69. $\begin{bmatrix} 1 & 0 & 0 \\ 0 & 1 & 0 \\ 0 & 0 & 1 \end{bmatrix}$ **71.** 1, 5, 10 **73.**

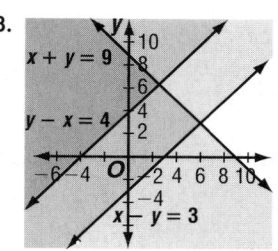

75. $y = 4x - 6$ **77.** **79.** $3\frac{1}{4}$

Pages 264–266 Lesson 5–2
7. yes, 3 **9.** $11x + y$ **11.** $6xy + 18y$ **13.** $x^2 + 9x + 18$
15. $9m^2 - 1$ **17a.** $(4a - 2c)$ yd **17b.** $(a^2 + 3ac + 2c^2)$ yd^2
19. yes, 3 **21.** yes, 7 **23.** yes, 6 **25.** $3z^2 - 2z - 21$
27. $10m^2 + 5m - 15$ **29.** $r^2 - r + 6$ **31.** $4gf^2 - 4fbh$
33. $15m^3n^3 - 30m^4n^3 + 15m^5n^6$ **35.** $-2x^2 - 3xy - 6y^2$
37. $(28m - 12n)$ ft **39.** $q^2 - 2q - 35$ **41.** $25 - r^2$
43. $6y^2 + 5y - 56$ **45.** $g^2 - 2 + \frac{1}{g^4}$ **47.** $y^2 - 6xy + 9x^2$
49. $4p^2 + 4pq^3 + q^6$ **51.** $(10y^2 + 6y)$ ft^2 **53.** $\frac{9xy^2}{2}$ m^2
55. $x^3 - y^3$ **57.** $x^4 + x^2 - 20$ **59.** $y^4 - 2x^2y^2 + x^4$
61a. $(a + b + c)^2$, $a^2 + 2ab + b^2 + 2ac + 2bc + c^2$
61b. $4a + 4b + 4c$ **63a.** $\frac{10x + 14}{9x - 2}$ **63b.** about 18.5°
65. 9 out of 16, RR; 6 out of 16, Rr; 1 out of 16, rr.

67. $(2, 1, -1)$ **69.** $\begin{bmatrix} -14 & -15 \\ 4 & 20 \end{bmatrix}$ **71.** $(12, 2)$ **73.** $(0, 0)$, $(-1, -3)$ **75.** 42.778, 42, 39

Pages 270–273   Lesson 5-3

5. $5y - 4 + 7x$ **7.** $a - 12$ **9.** $a^2 - ab + b^2$ **11.** $3x^3 - 9x^2 + 7x - 6$ **13.** $2x + 5$ **15.** $2xy - 7x^2$ **17.** $4s^2 + 3rs - 5r$ **19.** $-a^2b + a - \frac{2}{b}$ **21.** $x - 15$ **23.** $g + 5$
25. $3t^2 - 2t + 3$ **27.** $5y^2 - 4y + 4 - \frac{11}{y+1}$ **29.** $2y + 7 + \frac{5}{y-3}$ **31.** $x^2 + 3x + 9$ **33.** $3d^2 + 2d + 3 - \frac{2}{3d-2}$
35. $2c^2 + c + 5 + \frac{6}{c-2}$ **37.** $-w^2$ **39.** $2m^3 + m^2 + 3m - 1 + \frac{5}{m-3}$ **41.** $y^4 - 2y^3 + 4y^2 - 8y + 16$ **43.** $y = 2b^2 - b - 1 + \frac{4}{b+1}$ **45.** $r^3 - 9r^2 + 27r - 28$ and $r - 3$ **47.** $\frac{1}{8}$ in.
49. $a^2 - 2ab + b^2$ **51.** $(-1, 2, -3)$ **53.** 18 **55.** 5
57. $-6, 20$

Pages 278–280   Lesson 5-4

7. $-5x(3x + 1)$ **9.** $x(x + y + 3)$ **11.** $(y - 5)(y + 2)$
13. $3(h - 4)(h + 4)$ **15.** $(g + 20)(g^2 - 20g + 400)$
17. yes, $2y^2 + y - 1$ or $(2y - 1)(y + 1)$ **19.** $2a^2b(5a - 6b)$
21. prime **23.** $(y - 10)(y - 2)$ **25.** $(y + 6)(y + 1)$
27. $(x^2 + y)(x^2 - y)$ **29.** $3(n + 8)(n - 1)$ **31.** $3(a - 3b)$ $(a + 3b)$ **33.** $(x + 7)(x - 2)$ **35.** $5(x^2 + 3x - 2)$
37. $3(a + 3)(a + 5)$ **39.** prime **41.** $(2a - 5b)(2x + 7y)$
43. $(9y + 7)(9y - 7)$ **45.** $2x^2(x + 1)^2$ **47.** $(y^2 + 4)$ $(y + 2)(y - 2)$ **49.** $(8a + 3)(a + b + c)$ **51.** $(x + 3y)$ $(3x + 5)(x - 1)$ **53.** $\frac{n-7}{n+3}$

55. correct

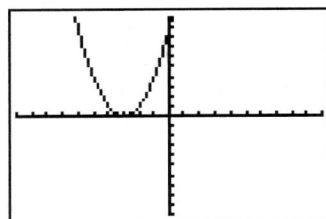

57. incorrect; $(x + 2)(x^2 - 2x + 4)$

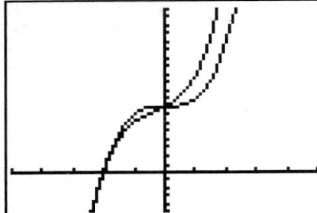

59. incorrect; $(2x + 1)(x - 3)$ **61a.** $(x^2 + 2xy + y^2)$ units^2
61b. 961 units^2

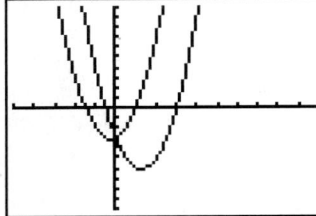

63. $t^2 - 2t + 1$ **65.** 2.592×10^{10} km **67.** $\begin{bmatrix} 0 & -4 \\ -7 & \end{bmatrix}$
69. $(7, -2)$ **71.** $y = \frac{3}{2}x - 1$ **73.** distributive

Pages 285–287   Lesson 5-5

7. -2.844 **9.** 3 **11.** -2 **13.** m **15.** $-5|x^3|$ **17.** $|3m - 2n|$ **19.** -5 **21.** -12 **23.** 0.9 **25.** 5 **27.** 18.574 **29.** 4
31. 14 **33.** 3 **35.** $\frac{1}{2}$ **37.** 10 **39.** -0.5 **41.** y **43.** $-|x|$
45. $6|g^3|$ **47.** $m^2n^3z^4$ **49.** $-3a^3b^4$ **51.** $s + t$
53. $-|x + 1|$ **55.** $\pm|s - t|$ **57.** no real roots
59. $x = 0, y = 0$ **61.** about 127.3 feet **63.** $25b^2$
65. $xy^3 + y + \frac{1}{x}$ **67.** 4 **69.** max: $f(-3, -3) = 3$; min: $f(0, 5) = -10$ **71.**

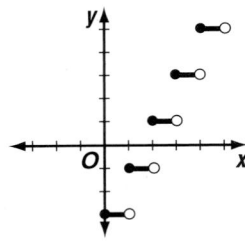

73a. 11.0, 3.3;10.1, 1.9 **73b.** 5.8, 3.0; 8.0, 3.8 **73c.** Sample answer: Teens have more leisure time than adults.

Page 287   Self Test

1. 9.3×10^7 miles **3.** $-108x^8y^3$ **5.** $n^3 - n^2 - 5n + 2$
7. $m^2 - 3 - \frac{19}{m-4}$ **9.** $8(r - 2s^2)(r^2 + 2rs^2 + 4s^4)$
11. $|2n + 3|$

Pages 293–295   Lesson 5-6

9. $4\sqrt{5}$ **11.** $4x^2y$ **13.** $-42\sqrt{15}$ **15.** $\sqrt[3]{9}$
17. $\left|\frac{y}{y-3}\right|\sqrt{y-3}$ **19.** $15\sqrt[3]{5} - 6\sqrt[3]{3}$ **21.** $49 - 11y$
23. $4\sqrt{2}$ **25.** $2\sqrt[3]{2}$ **27.** $2\sqrt[3]{4}$ **29.** $y\sqrt{y}$ **31.** $\sqrt{3}$
33. $3xy^2\sqrt{10x}$ **35.** $-60\sqrt{30}$ **37.** $48\sqrt{7}$ **39.** 0
41. $-42\sqrt{2} - 15\sqrt{14}$ **43.** $3\sqrt[3]{2x}$ **45.** 34 **47.** $8 - 2\sqrt{15}$
49. $5\sqrt{2}$ **51.** $\frac{a^2\sqrt{b}}{b^2}$ **53.** $\frac{16\sqrt{10}}{5}$ **55.** $\frac{5\sqrt{6} - 3\sqrt{2}}{22}$
57. $\frac{\sqrt{x^2 - 1}}{x - 1}$ **59.** $|x| + x^2 + x^4$ **61.** $2y^2\sqrt{70}$ in.
63. when x and y are not negative **65a.** Akikta: 58.09, Francisco: 56.91 **65b.** Akikta **67.** 80 ft/s or about 55 mph **69.** $(a + b)(1 + 3a - 3b)$ **71.** $-6x^3 - 4x^2y + 13xy^2$ **73.** $\begin{bmatrix} 4 & 3 \\ 1 & 3 \end{bmatrix}$ **75.** $(2, 3)$ **77.** 9 **79.** -63

Pages 300–302   Lesson 5-7

7. $27^{\frac{1}{4}}$ **9.** $16^{\frac{1}{3}}a^{\frac{5}{3}}b^{\frac{7}{3}}$ **11.** $\frac{1}{2}$ **13.** 2 **15.** $xy^2z\sqrt[6]{x^3y^2z^3}$
17. $\frac{x^{\frac{2}{3}}}{5x}$ **19.** $\frac{c(a - 4b)^{\frac{1}{2}}}{a - 4b}$ **21.** $y^{\frac{2}{3}}$ **23a.** $\frac{1}{2}$ **23b.** $\frac{3}{5}$ **23c.** $\frac{1}{3}$
25. $\frac{1}{10}$ **27.** $81^{\frac{13}{6}}$ or $9^{\frac{13}{3}}$ or $3^{\frac{26}{3}}$ **29.** $\frac{1}{16}$ **31.** 8 **33.** $4 \cdot 6^{\frac{1}{3}}$
35. $\frac{8 \cdot 3^{\frac{1}{2}}}{3}$ **37.** $m\sqrt{2}$ **39.** $\frac{w^{\frac{1}{5}}}{w}$ **41.** $\frac{x^{\frac{1}{6}}}{x}$ **43.** $\frac{b^{\frac{5}{12}}}{8b}$
45. $\frac{x^2y + 3}{x}$ **47.** $\frac{pqr^{\frac{2}{3}}}{r}$ **49.** $\frac{b^{\frac{2}{3}}}{b} - b^{\frac{1}{3}}$ **51.** $11mn^3$
53. $a - a^{\frac{1}{3}}b^{\frac{2}{3}}$ **55.** $\sqrt[6]{r^3s^2}$ **57.** $13\sqrt[6]{13}$ **59.** $\sqrt{3}$
61. 0.010000001 **63.** 3.51 **65.** 1.12 **67a.** 831 vibrations per second **67b.** 247 vibrations per second **69.** $4x - 20$
71. $(ab - 3)(a^2b^2 + 3ab + 9)$ **73.** $(2, -4)$ **75.** 2, 1, 3
77. 83 **79.** $\{x \mid 3 < x < 6\}$

Pages 307–309   Lesson 5-8

5. -7 **7.** 13 **9.** 7 **11.** $0 \le n < 4$ **13.** $r = \pm\sqrt{y^2 - s^2}$

15. 64 **17.** $-\dfrac{7\sqrt{5}}{30}$ **19.** 13 **21.** 4 **23.** no solution **25.** $\dfrac{54}{4}$

27. $x \le -\dfrac{57}{5}$ **29.** $\dfrac{-15 - 5\sqrt{3}}{6}$ **31.** 9 **33.** $k \ge 0$ **35.** 5.41

37. no solution **39.** 3 **41.** $a \ge 5$ **43.** $\ell = g \cdot \dfrac{T^2}{4\pi^2}$

45. $p = \dfrac{m^6 g^2}{r}$ **47a.** $h = \dfrac{\sqrt{S^2 - r^4\pi^2}}{\pi r}$ **47b.** about 13.4

49. Yes, $\sqrt[k]{\sqrt[m]{b}} = \left(b^{\frac{1}{m}}\right)^{\frac{1}{k}} = b^{\frac{1}{km}} = \sqrt[km]{b}$ **51.** about 282 feet

53. $t = \dfrac{2\pi r\sqrt{GMr}}{GM}$ **55.** $5^{\frac{3}{7}}$ **57.** $|x + 5|$ **59.** y^{10} **61.** 34

63a. $y = 0.05x + 15$ **63b.** $17.10

Pages 314–316 Lesson 5–9
7. $7i\,|mn|\sqrt{2}$ **9.** $-180\sqrt{3}$ **11.** 1 **13.** $5 + 5i$ **15.** 13

17.
$$5x^2 + 40 = 0$$
$$5(-2i\sqrt{2})^2 + 40 = 0$$
$$40i^2 + 40 = 0$$
$$40(-1) + 40 = 0$$
$$0 = 0$$

19. $\pm 4i$ **21.** $\dfrac{2}{3}i$ **23.** $10k^2 i$ **25.** $\dfrac{3i|x|}{5y^4}\sqrt{x}$ **27.** $-60i$

29. $-11\sqrt{2}$ **31.** -12 **33.** $4i\sqrt{3}$ **35.** i **37.** -1

39. $7 + 2i$ **41.** 2 **43.** $15 - i\sqrt{3} - i\sqrt{5}$ **45.** $14 + 5i$

47. 7 **49.** $-2 + 36i$ **51.** $\pm 2i\sqrt{2}$ **53.** $\pm 2i\sqrt{3}$ **55.** $\pm i\dfrac{\sqrt{5}}{2}$

57. $148 - 222i$ **59.** $20 + 15i$ **61.** 2, 3 **63.** $\dfrac{67}{11}, \dfrac{19}{11}$

65a. $y = -6x^2 - 30$, $y = 5x^2 + 40$, $y = 3x^2 + 18$, $y = 7x^2 + 84$
65b.

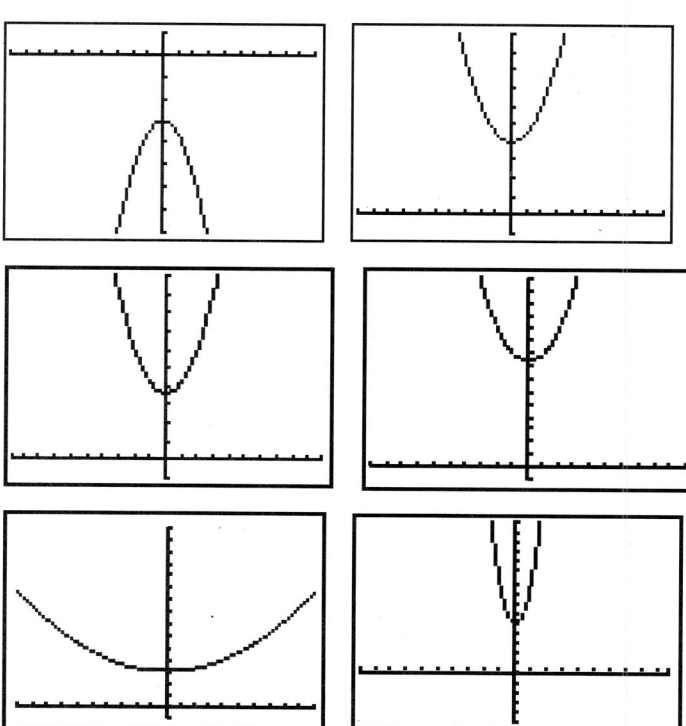

65c. Answers may vary. Sample answers: None of the graphs intercept the x-axis. All of the graphs are parabolas. **65d.** Since none of the graphs intercept the x-axis, none of the functions have real roots. This is confirmed by the solutions that are all imaginary.

67.
$$x^2 - 4x + 13 = 0$$
$$(2 - 3i)^2 - 4(2 - 3i) + 13 \overset{?}{=} 0$$
$$(4 - 12i + 9i^2) - 8 + 12i + 13 \overset{?}{=} 0$$
$$4 - 9 - 8 + 13 \overset{?}{=} 0$$
$$0 = 0$$

Yes, $2 - 3i$, because solutions involving imaginary numbers come in conjugate pairs.

69a.
$$AB = D \cdot (AB)$$
$$\begin{bmatrix} 0 & 1 \\ 1 & 0 \end{bmatrix} \cdot \begin{bmatrix} 0 & -i \\ i & 0 \end{bmatrix} \overset{?}{=} \begin{bmatrix} -1 & 0 \\ 0 & -1 \end{bmatrix} \cdot \left(\begin{bmatrix} 0 & -i \\ i & 0 \end{bmatrix} \cdot \begin{bmatrix} 0 & 1 \\ 1 & 0 \end{bmatrix} \right)$$
$$\begin{bmatrix} i & 0 \\ 0 & -i \end{bmatrix} \overset{?}{=} \begin{bmatrix} -1 & 0 \\ 0 & -1 \end{bmatrix} \cdot \begin{bmatrix} -i & 0 \\ 0 & i \end{bmatrix}$$
$$\begin{bmatrix} i & 0 \\ 0 & -i \end{bmatrix} \overset{?}{=} \begin{bmatrix} i & 0 \\ 0 & -i \end{bmatrix}$$

69b.
$$CB = D \cdot (BC)$$
$$\begin{bmatrix} 1 & 0 \\ 0 & -1 \end{bmatrix} \cdot \begin{bmatrix} 0 & -i \\ i & 0 \end{bmatrix} \overset{?}{=} \begin{bmatrix} -1 & 0 \\ 0 & -1 \end{bmatrix} \cdot \left(\begin{bmatrix} 0 & -i \\ i & 0 \end{bmatrix} \cdot \begin{bmatrix} 1 & 0 \\ 0 & -1 \end{bmatrix} \right)$$
$$\begin{bmatrix} 0 & -i \\ -i & 0 \end{bmatrix} \overset{?}{=} \begin{bmatrix} -1 & 0 \\ 0 & -1 \end{bmatrix} \cdot \begin{bmatrix} 0 & i \\ i & 0 \end{bmatrix}$$
$$\begin{bmatrix} 0 & -i \\ -i & 0 \end{bmatrix} = \begin{bmatrix} 0 & -i \\ -i & 0 \end{bmatrix}$$

71. $\dfrac{6(11 + 2\sqrt{3})}{109}$ **73.** 3.06 seconds **75.** $(2, 5, 4)$ **77.** sofa, $1180; loveseat, $590; table, $330 **79.** $5x - 6y = 30$
81. 19

Pages 319–321 Lesson 5–10
5. $-7i$ **7.** 100 **9.** $\dfrac{7i}{2}$ **11.** $2 - i$ **13.** $\dfrac{5 + i}{2}$

15. $\left(\dfrac{7 + 3i}{1}\right)\left(\dfrac{7 - 3i}{58}\right) = \dfrac{58}{58} = 1$ **17.** $12 - i$ **19.** $10 + 4i$

21. $6 - i\sqrt{7}$ **23.** 29 **25.** 2 **27.** 68 **29.** $\dfrac{7 - 3i}{2}$ **31.** $\dfrac{3 + 6i}{5}$

33. $\dfrac{9 - 6i}{26}$ **35.** $\dfrac{1 + i}{2}$ **37.** i **39.** $\dfrac{-3 - 21i}{10}$ **41.** $\dfrac{3 + i\sqrt{2}}{11}$

43. $\dfrac{6 + 5i}{61}$ **45.** $\dfrac{x - yi}{x^2 + y^2}$ **47.** $2 - 6i$, $-94 - 72i$ **49.** $\dfrac{2 - 3i\sqrt{5}}{7}$

51. $\dfrac{-1 - i}{2}$ **53.** $\dfrac{16 + 63i}{50}$ **55.** $\overline{z \cdot w} = \overline{(a + bi) \cdot (c + di)}$
$$= \overline{ac + adi + bci - bd}$$
$$= \overline{(ac - bd) + (ad + bc)i}$$
$$= (ac - bd) - (ad + bc)i$$
$$= ac - adi - bd - bci$$
$$= ac - adi + bdi^2 - bci$$
$$= a(c - di) - bi(-di + c)$$
$$= (a - bi)(c - di)$$
$$= \bar{z} \cdot \bar{w}$$

57a. $\dfrac{159 + 95j}{17}$ **57b.** $\dfrac{-37 + 134j}{5}$ **57c.** $\dfrac{25 - 25j}{2}$
57d. $\dfrac{1310 - 920j}{41}$ **59.** $(1 + i, -1 + 2i)$, $(-1 + 2i, -4 - 4i)$, $(-4 - 4i, -1 + 32i)$, $(-1 + 32i, -1024 - 64i)$

61. $\dfrac{7\sqrt{2} - 14}{2}$ **63.** 36 inches **65.** $\left(\dfrac{1}{4}, 6, -\dfrac{1}{6}\right)$

67. Yes; each element of the domain is paired with exactly one element of the range. **69.** $-\dfrac{11}{3}$

Pages 323 Chapter 5 Highlights
1. scientific notation **3.** rationalizing the denominator
5. irrational numbers **7.** coefficient **9.** extraneous solution **11.** square root **13.** degree of a polynomial
15. fractals

Pages 324–326   Chapter 5   Study Guide and Assessment

17. $\frac{1}{t^3}$ **19.** $8xy^4$ **21.** $-\frac{3}{2}$ **23.** $170,000,100; 1.7 \times 10^8$
25. $900; 9 \times 10^2$ **27.** $4x^2 + 22x - 34$ **29.** $d^2 - 2d - 15$
31. $4a^4 + 24a^2 + 36$ **33.** $8b^3 - 36b^2c + 54bc^2 - 27c^3$
35. $10x^3 - 5x^2 + 9x - 9$ **37.** $5x^2 + 3x + 1$ **39.** $2(5a^2 - 1)$
$(a - 2)$ **41.** $(s + 8)(s^2 - 8s + 64)$ **43.** ± 16 **45.** no real
roots **47.** $\pm(x^4 - 3)$ **49.** $2m^2$ **51.** $2\sqrt{2}$ **53.** $-5\sqrt{3}$
55. $20 + 8\sqrt{6}$ **57.** $\frac{3 - \sqrt{5}}{4}$ **59.** $\frac{1}{9}$ **61.** $\frac{9}{4}$ **63.** $\frac{xyz^{\frac{2}{3}}}{z}$
65. 343 **67.** $-\sqrt{3}$ **69.** $1, 3$ **71.** 8 **73.** $16i$ **75.** $-65 -$
$10i$ **77.** $12\sqrt{ab}$ **79.** i **81.** 7 **83.** $\frac{7 - 24i}{25}$ **85.** $\frac{-1 + 2i\sqrt{2}}{3}$

87. No, she is not telling the truth. She was going
approximately 49 mph.

CHAPTER 6   EXPLORING QUADRATIC FUNCTIONS AND INEQUALITIES

Page 333   Lesson 6–1A
1. Sample answer: Xmin $= -10$, Xmax $= 10$,   Ymin $=$
-10, Ymax $= 40$ **3.** Sample Answer: Xmin $= -50$,
Xmax $= 10$,   Ymin $= -2000$, Ymax $= 500$ **5.** Sample
Answer: Xmin $= -10$, Xmax $= 10$,   Ymin $= -80$, Ymax $=$
5 **7.** no solution **9.** $-3.14, -38.53$

Pages 338–340   Lesson 6–1
7. $x^2; x; -4$ **9.** $(2, 0), x = 2$

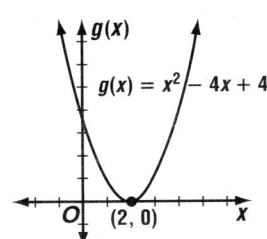

11. $(-3, 0), x = -3$ **13.** 2

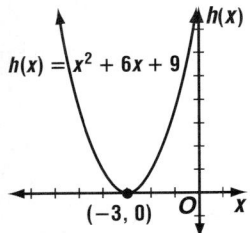
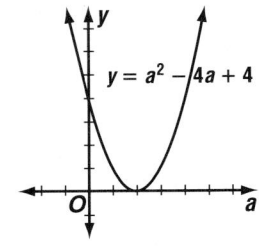

15. Between $x = 6$ and $x = 9$; since the value of y is
negative at $x = 6$ and positive at $x = 9$, the graph must
cross the x-axis between 6 and 9. **17.** $3n^2; 0; -1$
19. $z^2; 3z; 0$ **21.** $9t^2; 6t; -7$ **23.** -4
25. $(0, 0), x = 0$ **27.** $(4\frac{1}{2}, -11\frac{1}{4}), x = 4\frac{1}{2}$

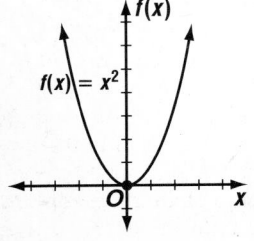
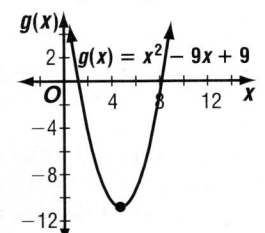

29. $(0, -9), x = 0$ **31.** $(-10, -7), x = -10$

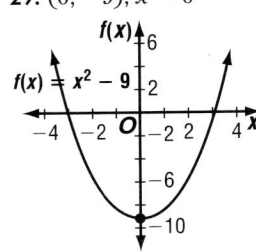
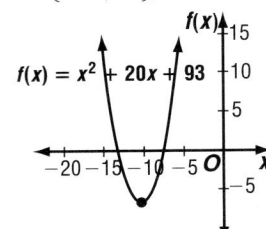

33. $(-1\frac{1}{2}, -3\frac{1}{5}), x = -1\frac{1}{2}$

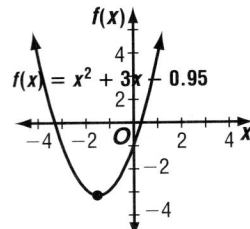

35. $6, -4$ **37.** -2

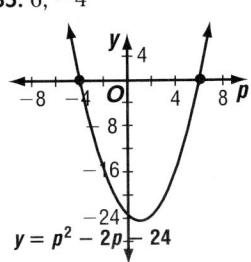
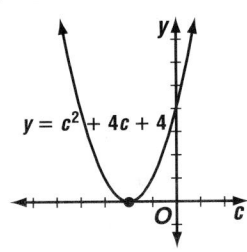

39. $-1.5, 3$ **41.** $-4, \frac{3}{2}$

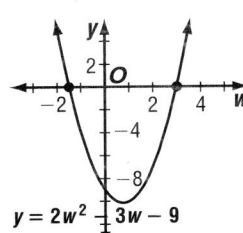
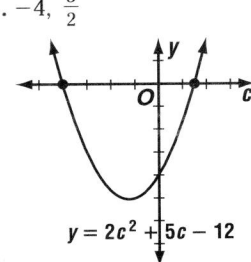

43a. There are no roots because the graph does not
touch the x-axis. **43b.** Sample answer: There cannot be
any negative prime numbers since the graph has no
negative y values.

45a.

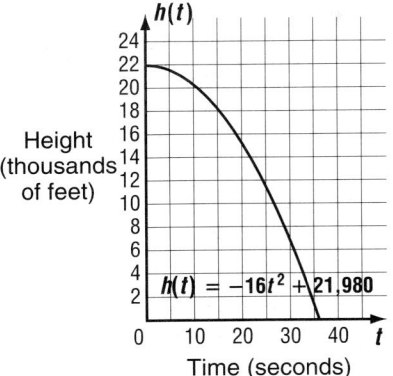

45b. about 37 seconds **47.** 7 **49.** -57 **51.** $y = -3x -$
$16, 3x + y = -16$ **53.** 21
Pages 344–345   Lesson 6–2
5. $-\frac{7}{3}, -5$ **7.** 8 **9.** $2, -\frac{7}{3}$ **11.** $-4, -1$ **13.** $-3, 3$

15. $4, -3$ **17.** 6 **19.** $4, -1$ **21.** $-\frac{1}{6}, \frac{1}{3}$ **23.** $-\frac{4}{3}$ **25.** $5, -8$

27. $-\frac{1}{4}, 3$ **29.** $-\frac{3}{4}, -\frac{4}{3}$ **31.** $0, 3, -3$ **33.** $0, -\frac{6}{7}, \frac{2}{5}$

35. $y = 0.25(x + 8)(x - 4)$ or $y = 0.25x^2 + x - 8$

37a. about 45 minutes   **37b.** 0.76 hour

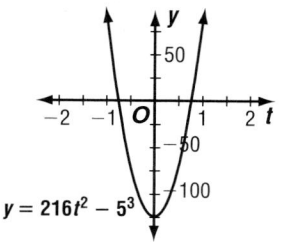

$y = 216t^2 - 5^3$

39. 8 chords **41.** $\pm 4i$ **43.** $\frac{1}{8}\begin{bmatrix} 6 & 2 \\ 3 & 4 \end{bmatrix}$ **45.** $\frac{4 - 27n}{5}$

Pages 350–352 Lesson 6–3

7. $\frac{49}{4}$ **9.** $\frac{5}{3}, -\frac{1}{4}$ **11.** $\frac{7 \pm \sqrt{33}}{2}$ **13.** $-1 \pm i\sqrt{5}$ **15.** 1

17. 400 **19.** 2500 **21.** $-6, 3$ **23.** $4 \pm \sqrt{5}$ **25.** $-\frac{9}{2}$

27. $-2 \pm i\sqrt{7}$ **29.** $\frac{-7 \pm i\sqrt{35}}{6}$ **31.** $\frac{3 \pm \sqrt{89}}{2}$ **33.** $\frac{7}{3}, -\frac{5}{4}$

35. $\frac{-b \pm \sqrt{b^2 - 4ac}}{2a}$ **37a.** ± 16 **37b.** 49 **37c.** 9 **37d.** 25

37e. ± 4 **37f.** 25

39a.

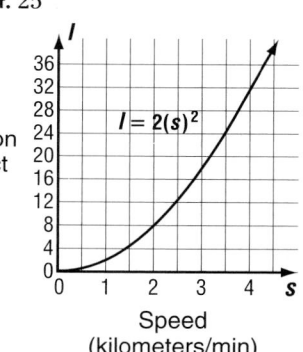

$I = 2(s)^2$

Collision Impact

Speed (kilometers/min)

39b. $2; 8; 32$ **39c.** The impact of the collision quadruples.

41a. 40 ft **41b.** 49 ft **41c.** 3.5 s **43.** $5mn^2 \cdot \sqrt[4]{m}$

45.

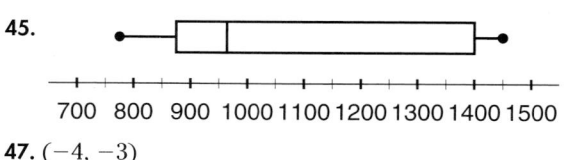

700 800 900 1000 1100 1200 1300 1400 1500

47. $(-4, -3)$

Pages 357–358 Lesson 6–4

7. $3, -5, -2; 49$ **9.** $1, 5, 7; -3$ **11.** $576; 2$ R, Q; $6, -6$

13. $-15; 2$ C; $\frac{-5 \pm i\sqrt{15}}{2}$ **15a.** 2 roots, $-1 \pm \sqrt{6}$ (1.4 and -3.4) **15b.** 1.4 and -3.4 **17.** $0; 1$ R, Q; 3 **19.** $49;$

2 R, Q; $-2, \frac{1}{3}$ **21.** $73; 2$ R, I; $\frac{-11 \pm \sqrt{73}}{6}$; $-0.41, -3.26$

23. $36; 2$ R, Q; $0, 6$ **25.** $-116; 2$ C; $\frac{4 \pm i\sqrt{29}}{5}$ **27.** $225; 2$ R,

Q; $7, -\frac{1}{2}$ **29.** $-31; 2$ C; $\frac{9 \pm i\sqrt{31}}{8}$

31a.

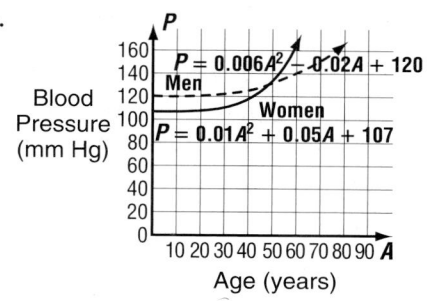

Blood Pressure (mm Hg)

$P = 0.006A^2 - 0.02A + 120$
Men
Women
$P = 0.01A^2 + 0.05A + 107$

10 20 30 40 50 60 70 80 90 A
Age (years)

The graph representing the normal blood pressure of women is more narrow than the graph of men's blood pressure. Initially, women's blood pressure is lower than men's, but it increases at a faster rate.

31b. 121 mm Hg **31c.** 50 years **33.** 1 **35.** $|x + 3|$
37. $(-19, -29)$

Page 358 Self Test

1. $-3, -1$

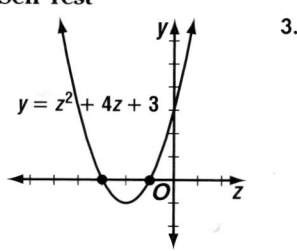

$y = z^2 + 4z + 3$

3. $4, -9$

5. $-11, 5$ **7.** $56, 2$ real, irrational roots **9.** $\frac{9}{2}, 8$

Pages 362–364 Lesson 6–5

5. $12, 22$ **7.** $0, 25$ **9.** $x^2 - 8x + 11 = 0$ **11.** $3x^2 + 19x - 14 = 0$ **13.** $7, -7$ **15.** $\frac{9}{4}, -\frac{9}{4}$ **17.** $x^2 + 3x - 54 = 0$

19. $8x^2 - 21x + 10 = 0$ **21.** $25x^2 - 4 = 0$ **23.** $3x^2 + 14x + 8 = 0$ **25.** $x^2 - 8x + 13 = 0$ **27.** $16x^2 + 16x + 29 = 0$ **29.** $-4, -\frac{3}{2}$ **31.** $0, 8$ **33.** $\frac{1}{2}, -\frac{2}{3}$ **35.** $4 \pm \sqrt{34}$

37. $9, -\frac{5}{4}$ **39a.** $12x^2 - 48x + 13 = 0$ **39b.** $42x^2 - 7x + 10 = 0$ **41.** -6 **43.** 88.2 m/s **45.** $7, -5$

47. $(-4, -21); x = -4$ **49.** $(2x - 3)(2x + 3)$ **51.** $\frac{21}{8}$

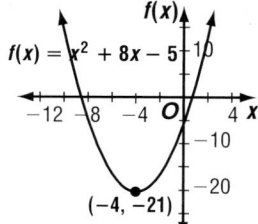

$f(x) = x^2 + 8x - 5$

$(-4, -21)$

Page 366 Lesson 6–6A

1. The value of k determines the vertical position of the graph. As you change the k, the graph will slide up or down the coordinate plane. Examples will vary. **3.** Both graphs have the same shape and vertex, but the first graph opens upward and the second graph opens downward. Examples will vary. **5.** Both graphs have the same shape, but the graph of $y = (x - 8)^2$ is 8 units to the right of the graph of $y = x^2$. **7.** Both graphs have the same shape, but the graph of $y = x^2 - 11$ is 11 units below the graph of $y = x^2$. **9.** The graph of $y = -2x^2$ opens downward and is more narrow than the graph of $y = x^2$. **11.** Both graphs have the same shape, but the graph of $y = -\frac{1}{3}x^2 + 2$ is 2 units above the graph of $y = -\frac{1}{3}x^2$.   **13.** Both graphs have the same shape, but the graph of $y = (x + 2)^2 - 4$ is 5 units below the graph of $y = (x + 2)^2 + 1$.   **15.** The graph of $y = \frac{1}{2}(x - 4)^2 - 5$ is 8 units below and is wider than the graph of $y = 2(x - 4)^2 + 3$.

Pages 372–375   Lesson 6–6
7. $(0, -6)$, $x = 0$, up   **9.** $f(x) = -3(x - 2)^2 + 12$; $(2, 12)$, $x = 2$, down   **11.** $y = -\frac{1}{2}(x + 2)^2 - 3$ **13.** $y = 3x^2 - 4x + 7$ **15.**

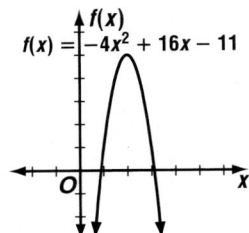

$f(x) = -4x^2 + 16x - 11$

17. $f(x) = 3x^2 - 5$

19. $(-11, -6)$, $x = -11$, down   **21.** $\left(\frac{1}{2}, \frac{1}{4}\right)$, $x = \frac{1}{2}$, up
23. $f(x) = -(x + 2)^2 + 12$; $(-2, 12)$, $x = -2$, down
25. $f(x) = -6(x - 2)^2 + 24$; $(2, 24)$, $x = 2$, down
27. $f(x) = -2(x + 5)^2$; $(-5, 0)$, $x = -5$, down
29. $f(x) = \frac{1}{3}(x - 6)^2 + 3$; $(6, 3)$, $x = 6$, up
31. $y = -\frac{3}{4}(x - 4)^2 + 1$ **33.** $y = \frac{1}{3}x^2 + 5$
35. $y = -3(x - 5)^2 + 4$ **37.** $y = \frac{1}{2}x^2 - 2x - 1$
39. $y = \frac{25}{3}x^2 + \frac{73}{3}x + 6$

41.

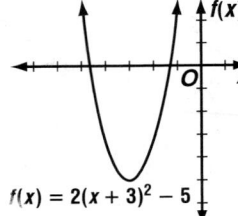

$f(x) = 2(x + 3)^2 - 5$

43.

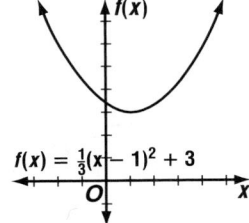

$f(x) = \frac{1}{3}(x - 1)^2 + 3$

45.

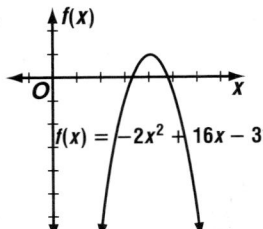

$f(x) = -2x^2 + 16x - 31$

47.

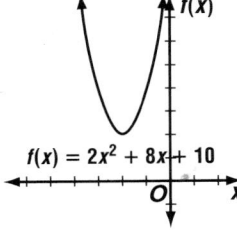

$f(x) = 2x^2 + 8x + 10$

49.

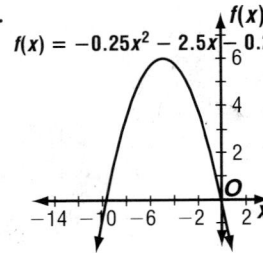

$f(x) = -0.25x^2 - 2.5x + 0.25$

51. $f(x) = -2(x - 2)^2 + 9$
53. $h = -\frac{b}{2a}$, $k = \frac{4ac - b^2}{4a}$ or $c - \frac{b^2}{4a}$

55a. Since the domain is the number of dots on a side, it is the set of integers greater than 0. Likewise, since the range is the total number of dots, it is also the set of integers greater than 0.

55b. $f(x) = \frac{1}{2}x(x + 1)$ **55c.** 0

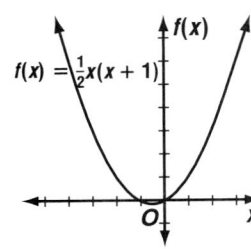

$f(x) = \frac{1}{2}x(x + 1)$

57. 5 ft   **59.** $(1, 4)$, $x = 1$

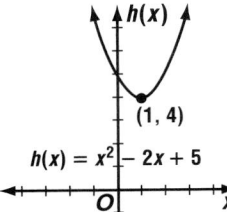

$(1, 4)$

$h(x) = x^2 - 2x + 5$

61. $3x^2 - 10x - 8$ **63.** $\begin{bmatrix} 6 & 0 & -24 \\ 14 & -\frac{2}{3} & -8 \end{bmatrix}$ **65.** $3x + 4y = -8$

Page 377   Lesson 6–7A
1.

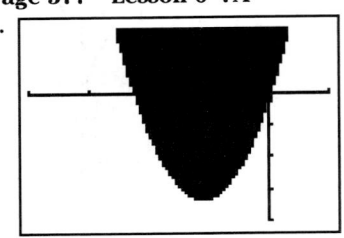

3.

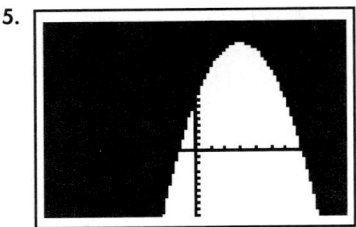

5.

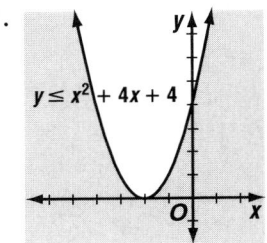

7. $\{x \mid x < -7 \text{ or } x > 3\}$ **9.** $\{x \mid x \le -0.5 \text{ or } x \ge 0\}$
11. $\{x \mid -0.42 < x < 9.42\}$ **13.** $w \ge 3.66$, $\ell \ge 7.66$

Pages 381–383   Lesson 6–7
7. yes   **9.** no
11.

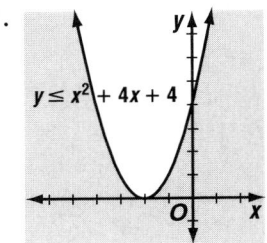

$y \le x^2 + 4x + 4$

13.

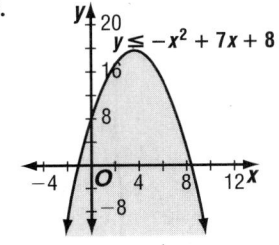

$y \le -x^2 + 7x + 8$

15. $-2 \le x \le 6$ **17.** $x = 5$ **19.** $\{n \mid n \ge 2.5 \text{ or } n \le -3.8\}$
21. {all reals} **23.** $\left\{b \mid -\frac{3}{2} < b < 2\right\}$ **25.** $0 \le w \le 5$
27.

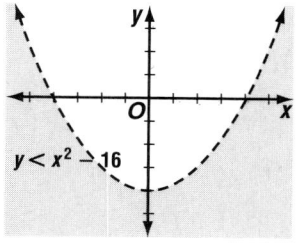

29.

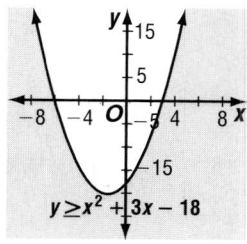

31.

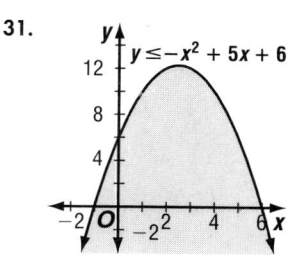

33.

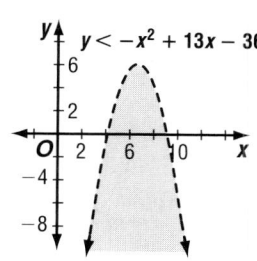

35.

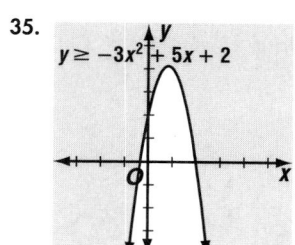

37.

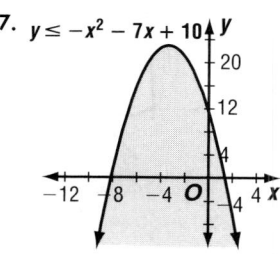

39. $\{x \mid x < -3 \text{ or } x > 6\}$ **41.** $\{q \mid q \le -6 \text{ or } q \ge 4\}$
43. $\left\{x \mid -4 \le x \le \frac{3}{2}\right\}$ **45.** $\{w \mid w \le 0 \text{ or } w \ge 2\}$
47. $\left\{g \mid -\frac{1}{2} < g < 3\right\}$ **49.** $\{n \mid -\sqrt{3} \le n \le \sqrt{3}\}$
51. $\left\{t \mid \frac{-1 - \sqrt{10}}{2} \le t \le \frac{-1 + \sqrt{10}}{2}\right\}$ **53.** $\{x \mid x \le -4 \text{ or } -2 \le x \le 8\}$ **55.** $\{x \mid x \le -3 \text{ or } -2 \le x \le 1 \text{ or } x \ge 2\}$ **57a.** No, the maximum height of the ball is 66 feet. **57b.** 78 feet per second **59.** 1 decrease of \$0.20
61a. $f(x) = -\frac{2}{315}(x - 315)^2 + 630$ **61b.** 630 ft
61c. See students' work for comparisons.

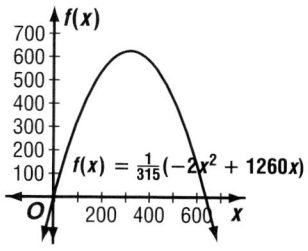

63. $\frac{6 \pm \sqrt{146}}{11}$ **65.** $33 + 20\sqrt{2}$ **67.** $-15a^2 + 14a - 3$
69. $(4, -2)$ **71.** all reals

Pages 389–391 Lesson 6–8
5. 50; 36.3 **7.** 58.18; 20.6 **9a.** 2.24 **9b.** small
9c. Answers will vary. **11.** 350; 50 **13.** 369.30; 194.14
15. 57.4; 9.1 **17.** 3.9; 0.55 **19.** 67.57; 19.68 **21.** 0; no variation from the mean **23a.** 79.52; 5.16 **23b.** 0.78; 0.47
25. 18.22¢; 3.99¢ **25b.** 20.47¢; 4.43¢ **27.** 250 ft

29. $\begin{bmatrix} 3 & 4 & | & 22 \\ 7 & -1 & | & 10 \end{bmatrix}$, $(2, 4)$

Pages 395–398 Lesson 6–9
5a. normally distributed

5b.

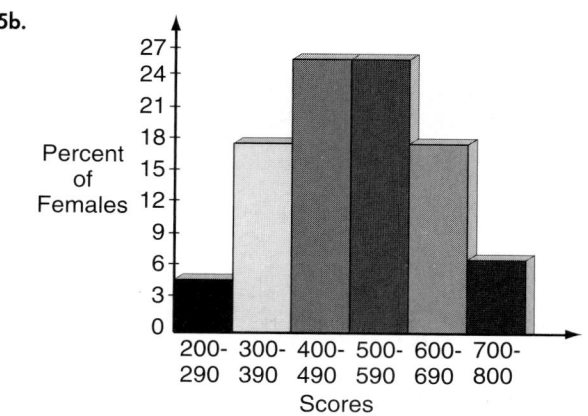

7a. 6800 tires **7b.** 250 tires **7c.** 1600 tires **7d.** 81.5%
9. normally distributed

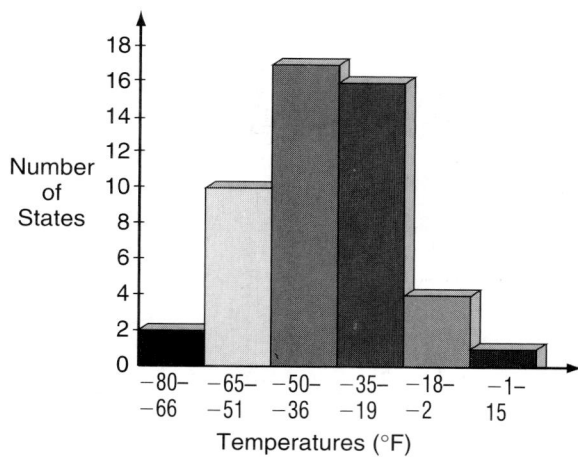

11a. 50% **11b.** 50% **11c.** 95% **13a.** 500 rods
13b. 815 rods **13c.** 16% **13d.** 84% **15.** Sample answer: The statistician was using standard deviations.
17a. 80.75; 4.94

17b.

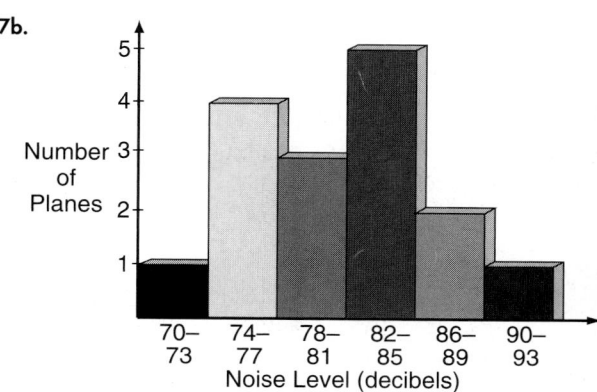

17c. Sample answer: The data do not appear to be normally distributed since they are not symmetrical about the mean.

19a.

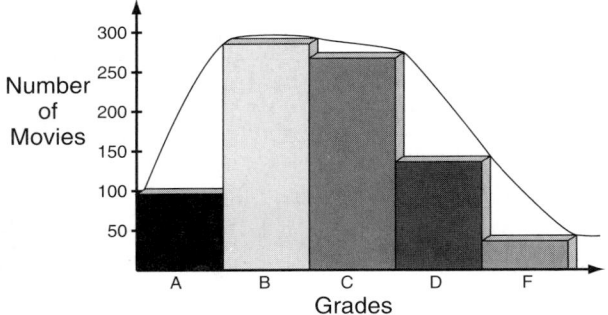

19b. no, positively skewed   **21a.** $y = 0.4(x - 7.5)^2 + 9.5$
21b. $194.40   **21c.** $312   **23.** $-3x^2 - 28x - 32$

Page 399   Chapter 6   Highlights
1. f   **3.** a   **5.** i   **7.** d   **9.** g

Pages 400–402   Chapter 6   Study Guide and Assessment
11. $-10, 4$ **13.** $-2, 10$

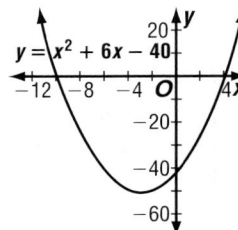
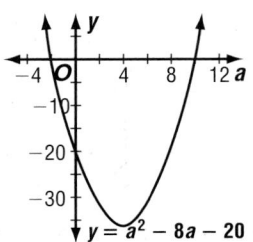

15. $-7, -5$

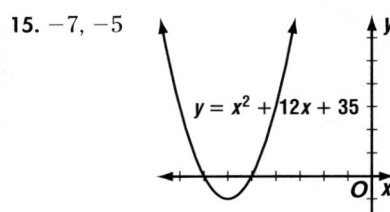

17. $-4, 8$ **19.** $-4, 4$ **21.** $-25, -4$ **23.** $-\frac{4}{5}, 0, \frac{9}{5}$

25. $-\frac{1}{2} \pm \frac{\sqrt{205}}{10}$ **27.** $3 \pm \sqrt{5}$ **29.** $2 \pm \sqrt{11}$ **31.** $-1 \pm i\sqrt{6}$ **33.** $\frac{5}{2} \pm \frac{i\sqrt{11}}{2}$ **35.** $-\frac{7}{6} \pm \frac{\sqrt{73}}{6}$ **37.** $x^2 - x - 42 = 0$ **39.** $2x^2 + 21x + 52 = 0$ **41.** $8x^2 - 22x - 105 = 0$
43. $(-2, 3)$, $x = -2$, down

45. $f(x) = 5\left(x - \frac{7}{2}\right)^2 - \frac{13}{4}$; $\left(\frac{7}{2}, -\frac{13}{4}\right)$, $x = \frac{7}{2}$, up

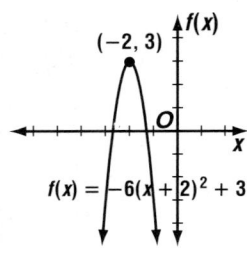
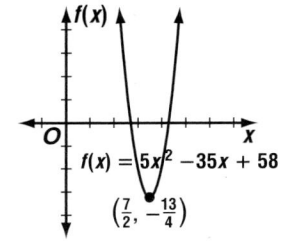

47. $f(x) = -\frac{1}{3}(x - 12)^2 + 48$; $(12, 48)$, $x = 12$, down   **49.**

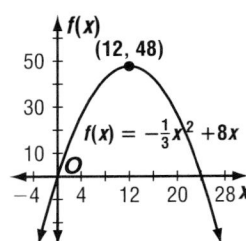
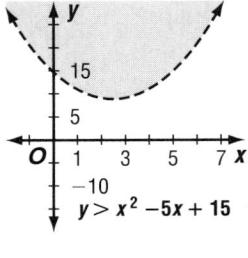

51.   **53.**

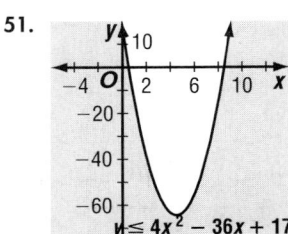
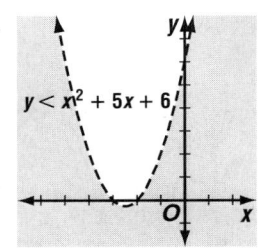

55. $\bar{x} = 157.5$, $\sigma = 23.98$ **57.** $\bar{x} = 315$, $\sigma = 18.47$
59a. 68%   **59b.** 2%   **59c.** 4 students   **59d.** 0.5%
59e. 20 students   **61.** 10 seconds

CHAPTER 7   ANALYZING CONIC SECTIONS

Pages 411–414   Lesson 7–1
5. $\sqrt{2.61}$ units   **7.** $\left(\frac{5}{2}, \frac{9}{4}\right)$ **9.** $AB = 2\sqrt{5}$, $BC = 2\sqrt{10}$, $AC = 2\sqrt{5}$ **11.** 13 units   **13.** $3\sqrt{17}$ units   **15.** 16 units
17. $\sqrt{65}$ units   **19.** $(12, 5)$ **21.** $(-4, -2)$ **23.** $(0.075, -1)$ **25.** -1 or -13 **27.** about 12.8 or -6.8 **29.** $(-1, 1)$
31. $(-4.3, 2.8)$ **33a.** $y = -\frac{2}{5}x - 2$ **33b.** $\sqrt{29}$ units

33c. $\left(-\frac{5}{2}, -1\right)$ **35.** $\sqrt{65} + 2\sqrt{2} + \sqrt{122} + \sqrt{277}$ units

37a. $2\sqrt{58}$ and $6\sqrt{2}$ units   **37b.** $(9, 5)$ **39a.** 5 units
39b. about 18.97 miles
39c. : ClrHome
    : Disp "X1 ="
    : Input A
    : Disp "Y1 ="
    : Input B
    : Disp "X2 ="
    : Input C
    : Disp "Y2 ="
    : Input D
    : (A + C)/2 → X
    : (B + D)/2 → Y
    : Disp "X VALUE OF MIDPOINT IS"
    : Disp X
    : Disp "Y VALUE OF MIDPOINT IS"
    : Disp Y

39d. $(3.5, -1.5)$ **41a.** about 300 miles   **41b.** about 1.6 hours   **43a.** Sample answer: Draw several lines across the U.S. One should go from the northeast corner to the southwest corner, another should go from the southeast corner to the northwest corner, another should go across the middle of the U.S. from east to west, etc. Find the centers of these lines. Find a point that represents all of the points.   **43d.** Sample answer: Cut out Alaska and Hawaii and place them next to continental U.S. Follow the procedure described in Exercise 43a.   **43f.** Sample answer: Because certain sections of the U.S. are more populated, the geographical and population centers differ. **45.** $t - 5 + \dfrac{12}{t + 2}$ **47.** $(11, 2, 24)$ **49.** 15 inches by 28 inches **51.** 21

Pages 419–422   Lesson 7–2
5. $y = 2(x - 3)^2 - 12$ **7.** $x = 3\left(y + \dfrac{5}{6}\right)^2 - 11\dfrac{1}{12}$

9. $(3, -4)$; $\left(3, -3\dfrac{3}{4}\right)$; $x = 3$;   **11.** $\left(1\dfrac{1}{3}, \dfrac{2}{3}\right)$; $\left(1\dfrac{1}{3}, \dfrac{3}{4}\right)$; $x = 1\dfrac{1}{3}$;
$y = -4\dfrac{1}{4}$; up; 1 unit $y = \dfrac{7}{12}$; up; $\dfrac{1}{3}$ unit

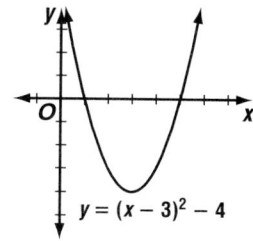
$y = (x - 3)^2 - 4$

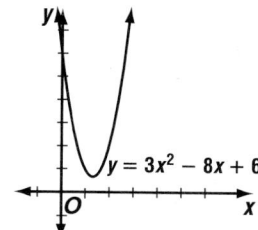
$y = 3x^2 - 8x + 6$

13. $y = -2x^2 + 3$ **15.** $x = -\dfrac{1}{8}(y + 1)^2 + 5$

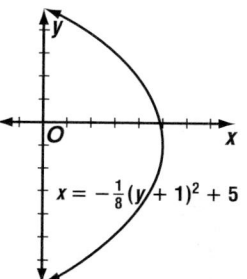
$x = -\dfrac{1}{8}(y + 1)^2 + 5$

19. $(2, -3)$; $(3, -3)$; $y = -3$; $x = 1$; right; 4 units

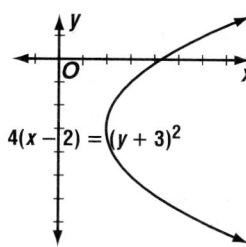
$4(x - 2) = (y + 3)^2$

23. $\left(1\dfrac{1}{4}, -6\dfrac{7}{8}\right)$; $\left(1\dfrac{1}{4}, -7\right)$; $x = 1\dfrac{1}{4}$; $y = -6\dfrac{3}{4}$; down; $\dfrac{1}{2}$ unit

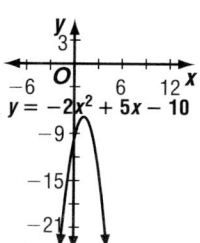
$y = -2x^2 + 5x - 10$

27. $(123, -18)$; $\left(122\dfrac{1}{4}, -18\right)$; $y = -18$; $x = 123\dfrac{3}{4}$; left; 3 units

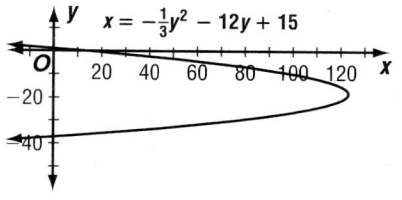
$x = -\dfrac{1}{3}y^2 - 12y + 15$

31. $x = 2(y - 2)^2 + 1$

33. $y = \dfrac{1}{8}(x + 3)^2 - 4$

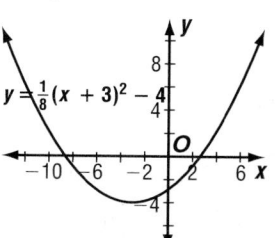
$y = \dfrac{1}{8}(x + 3)^2 - 4$

35. $y = -\dfrac{1}{18}(x - 4)^2 + \dfrac{3}{2}$

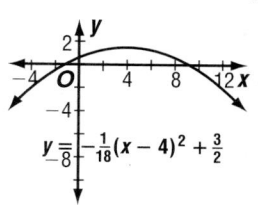
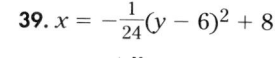
$y = -\dfrac{1}{18}(x - 4)^2 + \dfrac{3}{2}$

37. $x = \dfrac{1}{12}(y - 0)^2 + 1$

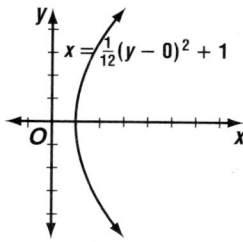
$x = \dfrac{1}{12}(y - 0)^2 + 1$

39. $x = -\dfrac{1}{24}(y - 6)^2 + 8$

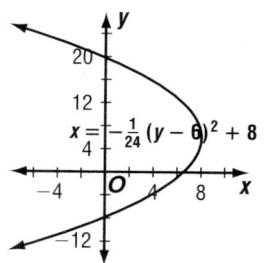
$x = -\dfrac{1}{24}(y - 6)^2 + 8$

41. $y = \dfrac{1}{16}(x - 1)^2 + 7$

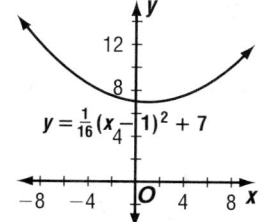
$y = \dfrac{1}{16}(x - 1)^2 + 7$

43. $x = \dfrac{1}{4}(y - 3)^2 + 4$

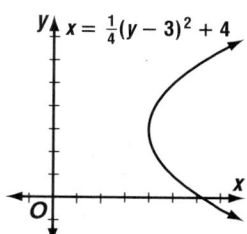
$x = \dfrac{1}{4}(y - 3)^2 + 4$

45a. 4 and 12   **45b.** $4 < x < 12$ **45c.** $x < 4$ or $x > 12$
47. 3 units   **49.** $y = -\dfrac{1}{26,200}x^2 + 6550$ **51.** $2\dfrac{5}{32}$ inches
53. $-7 \pm \sqrt{61}$ **55.** $201{,}600$; 2.016×10^5 **57.** 5
59. $d \geq -105$

Pages 426–429   Lesson 7–3
7. $(x + 12)^2 + y^2 = 23$ **9.** $(4, 1)$; 3 units

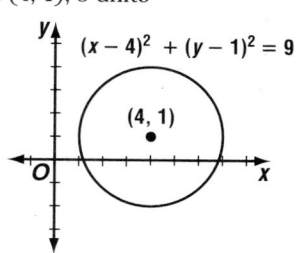
$(x - 4)^2 + (y - 1)^2 = 9$
$(4, 1)$

11. $(4, 0)$; $\dfrac{4}{5}$ units

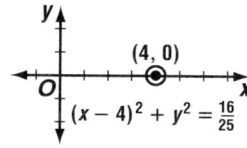
$(4, 0)$
$(x - 4)^2 + y^2 = \dfrac{16}{25}$

13. $(-4, 3)$; 5 units

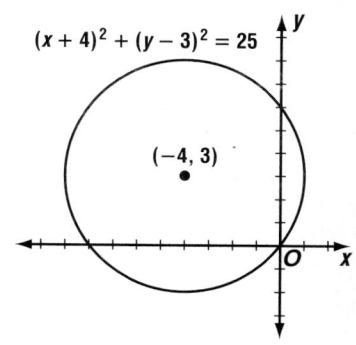
$(x + 4)^2 + (y - 3)^2 = 25$
$(-4, 3)$

15. $(x - 3)^2 + (y + 1)^2 = 9$ **17.** $(x + 1)^2 + (y + 5)^2 = 4$
19. $(x + 8)^2 + (y - 7)^2 = \frac{1}{4}$
21. $(x - 0.5)^2 + (y - 0.7)^2 = 182.25$
23. $(0, -2)$; 2 units   **27.** $(3, 0)$; 4 units

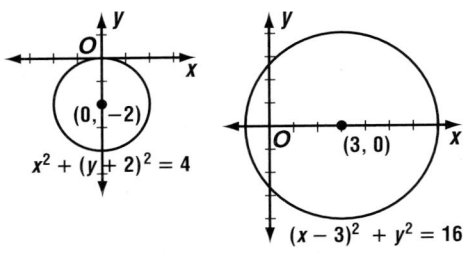

31. $(-7, -3)$, $2\sqrt{2}$ units   **35.** $\left(0, -\frac{9}{2}\right)$, $\sqrt{19}$ units

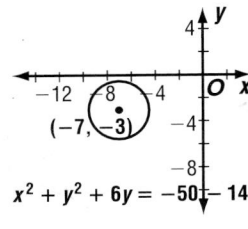

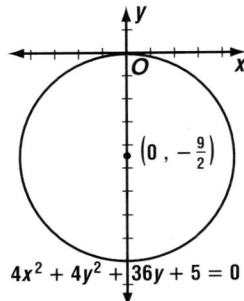

39. $(-1, -2)$, $\sqrt{14}$ units

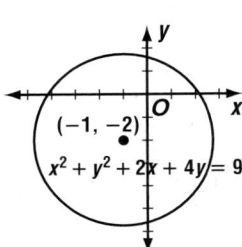

41. $(x - 2)^2 + (y + 1)^2 = 4$ **43.** $(x + 3)^2 + (y + 2)^2 = 1$
45. $(x + \sqrt{13})^2 + (y - 42)^2 = 1777$ **47.** $(x + 8)^2 +$
$(y + 7)^2 = 64$ **49.** $(x + 3)^2 + (y + 6)^2 = 9$

51. $y = 0$

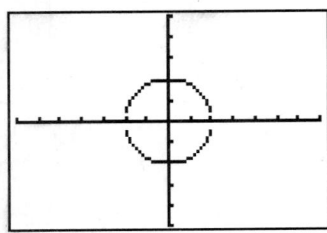

53. $y = \sqrt{4 - x^2}$, $y = -\sqrt{4 - x^2}$

55. circles with a radius of 8 and centers on the graph of
$x = 3$ **57a.** $x^2 + y^2 = 841,000,000$

57b.

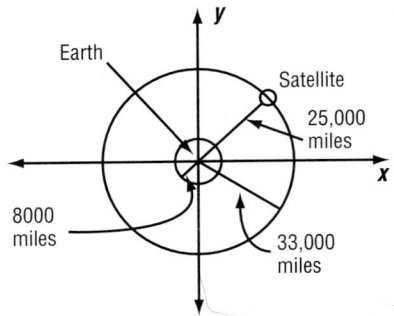

59. $x = \frac{1}{6}y^2$ **61.** $\frac{12 + 3i}{17}$ **63.** $4, -1$ **65.** $16, 36$ **67.** 120

Page 430   Lesson 7–4A
1.

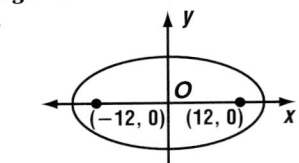

3.

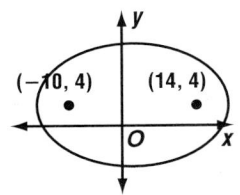

Pages 436–439   Lesson 7–4
7. $(4, -6)$, horizontal   **9.** $\frac{(x - 1)^2}{30} + \frac{(y + 1)^2}{5} = 1$
11. $(1, -2)$; $(-3, -2)$,   **13.** $(4, -2)$; $(4 \pm 2\sqrt{6}, -2)$;
$(5, -2)$; $4\sqrt{5}$; 4   10; 2

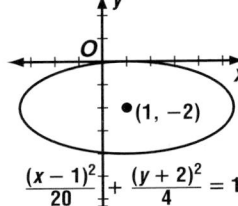

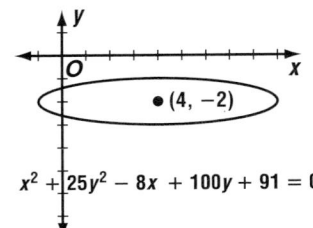

15. $\frac{x^2}{36} + \frac{y^2}{100} = 1$ **17.** $\frac{(x - 5)^2}{64} + \frac{(y - 4)^2}{9} = 1$
19. $(0, 0)$; $(0, \pm\sqrt{5})$;   **23.** $(0, 0)$; $(\pm 3\sqrt{5}, 0)$;
$2\sqrt{10}$; $2\sqrt{5}$ 18; 12

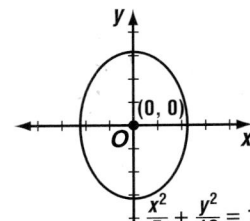

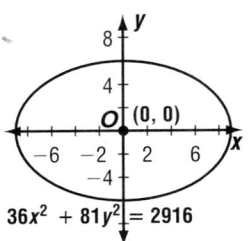

27. $(-3, 1)$; $(-3, 5)$,   **31.** $(1, -2)$; $(1 \pm \sqrt{7}, -2)$;
$(-3, -3)$; $4\sqrt{6}$; $4\sqrt{2}$ 8; 6

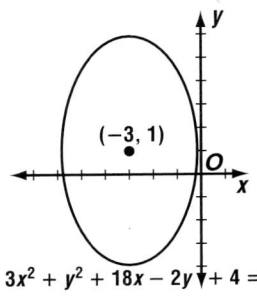

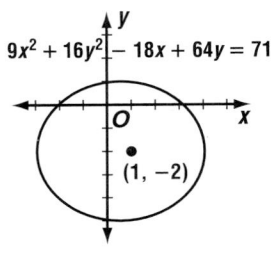

33. $\frac{(x-1)^2}{81} + \frac{(y-2)^2}{56} = 1$ **35.** $\frac{x^2}{169} + \frac{y^2}{25} = 1$ **37.** $\frac{(x-2)^2}{4} + \frac{(y-4)^2}{64} = 1$ **39.** As k increases, the length of the major axis increases, but the minor axis length stays the same. **41.** $20\sqrt{3}$ cm or about 34.6 cm **43.** about $\frac{x^2}{2.02 \times 10^{16}} + \frac{y^2}{2.00 \times 10^{16}} = 1$

45a.

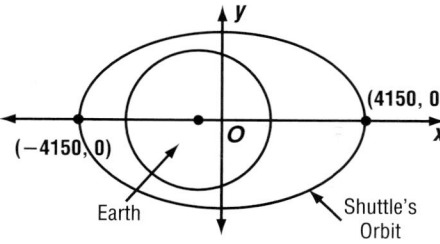

45b. $\frac{x^2}{17,222,500} + \frac{y^2}{17,200,000} = 1$

47.

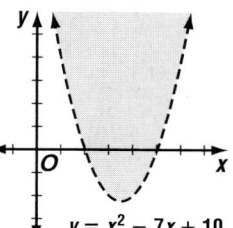

$y = x^2 - 7x + 10$

49. 6.25×10^{-11}

51. $(4.27, -5.11)$ **53.** 19°F; 16°F; about 18.5°F

Page 439   Self Test
1. 13 units **3.** $\left(\frac{3}{2}, 6\right)$

5. $(0, 0)$; $\left(\frac{3}{2}, 0\right)$; $y = 0$; $x = -\frac{3}{2}$; right; 6 **7.** $(0, 4)$, 7 units

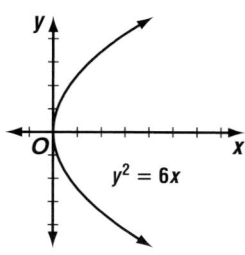

$y^2 = 6x$

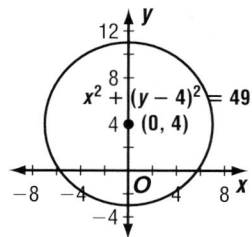

$x^2 + (y-4)^2 = 49$

9. $\frac{(x-1)^2}{5} + \frac{(y-2.5)^2}{10} = 1$

Pages 445–447   Lesson 7-5
5. hyperbola **7.** hyperbola **9.** $\frac{(y+3)^2}{24} - \frac{(x-1)^2}{8} = 1$

11. $(\pm 6, 0)$; $(\pm\sqrt{37}, 0)$; $\pm\frac{1}{6}$ **13.** $(4 \pm 2\sqrt{5}, -2)$; $(4 \pm 3\sqrt{5}, -2)$; $\pm\frac{\sqrt{5}}{2}$

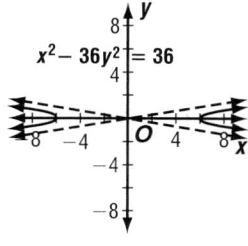

$x^2 - 36y^2 = 36$

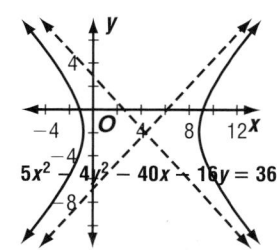

$5x^2 - 4y^2 - 40x - 16y = 36$

15. $\frac{x^2}{1} - \frac{y^2}{16} = 1$ **17.** $\frac{(y+5.5)^2}{6.25} - \frac{x^2}{6} = 1$

19. $(\pm 9, 0)$; $(\pm\sqrt{130}, 0)$; $\pm\frac{7}{9}$ **23.** $(\pm\sqrt{2}, 0)$; $(\pm\sqrt{3}, 0)$; $\pm\frac{\sqrt{2}}{2}$

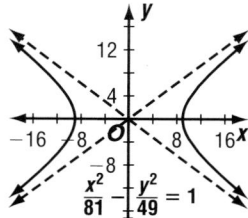

$\frac{x^2}{81} - \frac{y^2}{49} = 1$

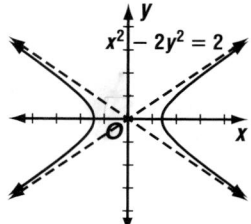

$x^2 - 2y^2 = 2$

27. $(-2, 0)$, $(-2, 8)$; $(-2, -1)$, $(-2, 9)$; $\pm\frac{4}{3}$ **31.** $(-5, 2)$, $(-1, 2)$; $(-3 \pm \sqrt{5}, 2)$; $\pm\frac{1}{2}$

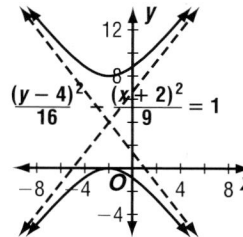

$\frac{(y-4)^2}{16} - \frac{(x+2)^2}{9} = 1$

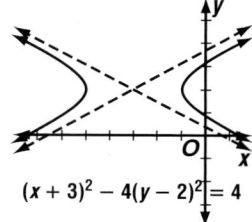

$(x+3)^2 - 4(y-2)^2 = 4$

35. $\frac{(x-2)^2}{49} - \frac{(y+3)^2}{4} = 1$ **37.** $\frac{x^2}{25} - \frac{y^2}{36} = 1$

39.

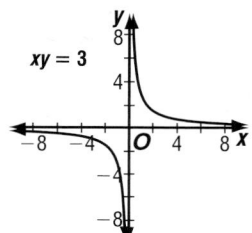

$xy = 3$

41.

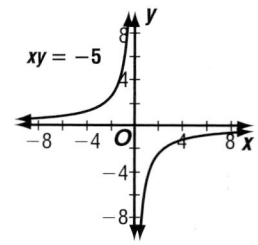

$xy = -5$

43. One branch is always in the first quadrant and the other is always in the third quadrant. As the value of c increases, the vertices move away from the origin.
45. The graph becomes the x- and y-axes.

47.

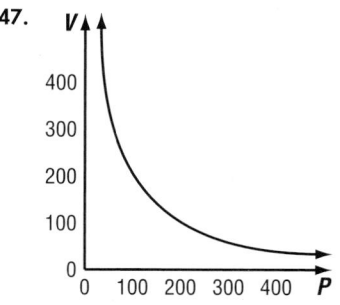

49. $\frac{(x-1)^2}{25} + \frac{(y-4)^2}{9} = 1$ **51.** $-7, \frac{3}{2}$

53.

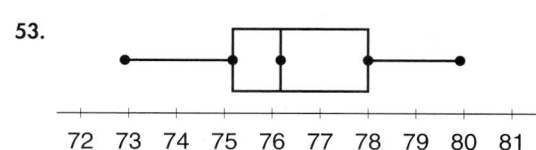

55. -12 **57.** $2x + 17y$

Page 449   Lesson 7–6A

1. parabola

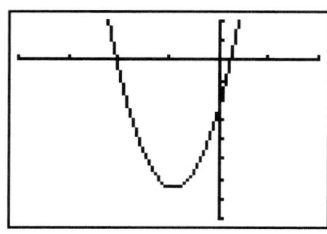

3. ellipse

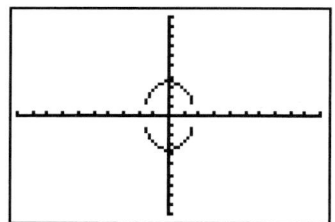

5. ellipse

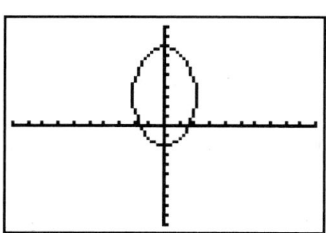

7. ellipse

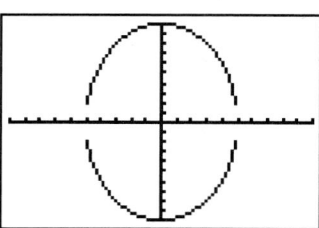

9. parabola

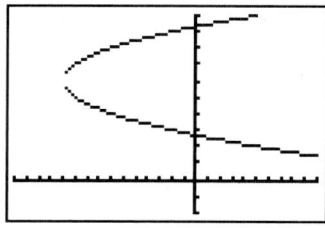

Pages 453–455   Lesson 7–6

5. circle   **7.** hyperbola

9. $y = \left(x + \frac{3}{2}\right)^2 - \frac{5}{4}$; parabola

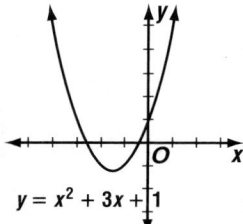

11. $\left(x - \frac{1}{2}\right)^2 + y^2 = \left(\frac{3}{2}\right)^2$; circle

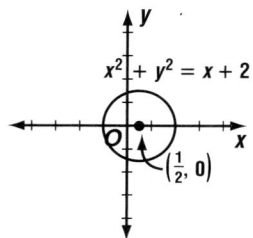

13. $y = -\left(x - \frac{1}{2}\right)^2 + \frac{9}{4}$

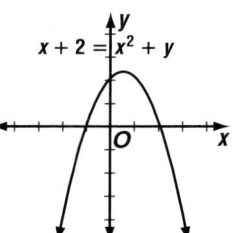

15. $y = \frac{1}{8}x^2$; parabola

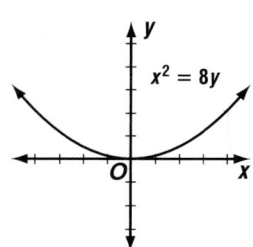

19. $y = (x - 2)^2 - 4$; parabola

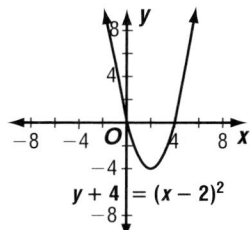

23. $y = -(x + 4)^2 - 7$; parabola

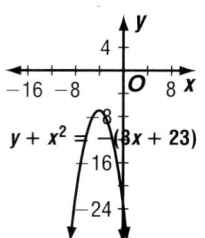

27. $\dfrac{(x - 3)^2}{25} + \dfrac{(y - 1)^2}{9} = 1$; ellipse

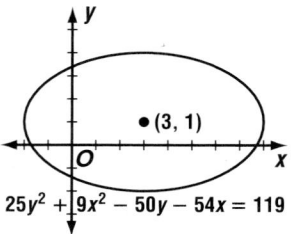

31. $\dfrac{(x - 2)^2}{5} + \dfrac{(y + 1)^2}{6} = 1$; hyperbola

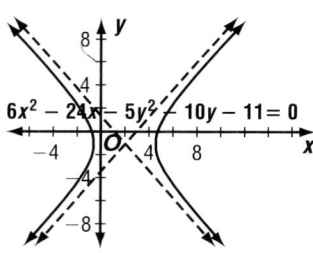

33. isolated point

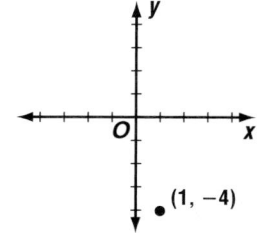

35a. ellipse   **35b.** hyperbola   **35c.** circle   **35d.** parabola
37. $y = \pm\frac{1}{8}x^2$ **41.** quadratic, $4x^2$; linear, $-8x$; constant, -2 **43.** -27 **45.**

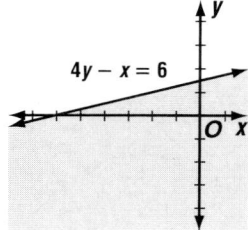

Page 457 Lesson 7–6B

1. The points are equidistant from the focus and the directrix. **3.** Each branch of the hyperbola becomes more narrow, and the vertices become farther apart. Each branch of the hyperbola becomes wider, and the vertices become closer.

Page 459 Lesson 7–7A

1. $(\pm 1.68, 3.63)$ **3.** $(0, -1)$, $(\pm 1.36, 0.85)$
5. $(\pm 2.98, \pm 0.93)$ **7.** $(0, 5)$, $(4, 3)$

Pages 464–467 Lesson 7–7

7. ellipse, hyperbola; $\left(\frac{\pm\sqrt{39}}{2}, \frac{\pm\sqrt{3}}{2}\right)$; **9.** parabola, hyperbola; no solution **11.** $(\pm 1, 5)$, $(\pm 1, \pm 5)$

13. 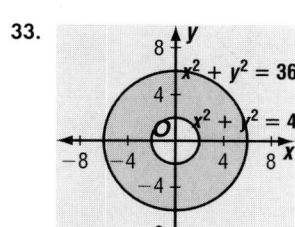 **15.** $(\pm 3\sqrt{3}, 6)$ **17.** no solution

19. $(\pm 8, 0)$ **21.** $(\pm 4, \pm 3)$ **23.** $(5, \pm 2)$, $(-1, \pm 4)$
25. $\left(\sqrt{5}, \sqrt{5}\right), \left(-\sqrt{5}, -\sqrt{5}\right)$ **27.** $\left(-1 + \sqrt{17}, 1 + \sqrt{17}\right),$ $\left(-1 - \sqrt{17}, 1 - \sqrt{17}\right)$ **29.** no solution
31. $\left(\pm\sqrt{6}, \pm\sqrt{5}\right)$

33. **35.**
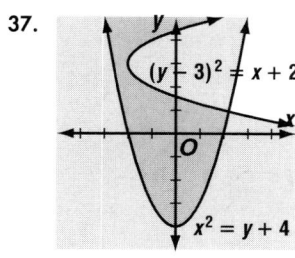

37. **39.** $y = -x - 1,$ $x = (y - 1)^2 - 4$

41. $y = -2x^2$, $x^2 + y^2 = 5$ **43.** $x^2 + y^2 \geq 25$, $x^2 + y^2 \leq 100$
45. impossible **47.** $x^2 + y^2 = 36$, $\frac{(x + 2)^2}{16} - \frac{y^2}{4} = 1$
49. $x^2 + y^2 = 81$, $\frac{x^2}{4} + \frac{y^2}{100} = 1$ **51a.** $(-2, -2)$, $\left(0, \sqrt{2}\right)$
51b. $(0, 1)$ **53.** $(40, 30)$ **55.** 1.3 **57.** $5b - 4 + 7a$
59. $(2, 3)$

61a.

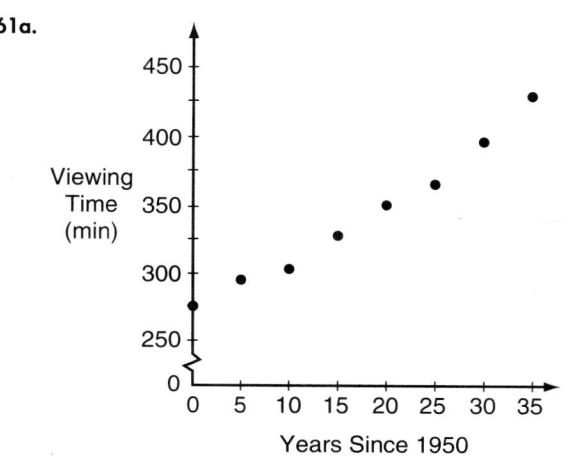

Sample answer: $y = 4.4x - 8305$ **61b.** Sample answer:
8 h 15 min **63.** $-\frac{19}{2}$

Page 469 Chapter 7 Highlights

1. true **3.** true **5.** true **7.** false; A parabola is the set of all points that are the same distance from a given point called the focus and a given line called the directrix. **9.** false; The conjugate axis is the line segment perpendicular to the transverse axis. **11.** false; A hyperbola is the set of all points in a plane such that the absolute value of the difference of the distances from any point on the hyperbola to two given points is constant. **13.** false; The midpoint formula is given by the following: $\left(\frac{x_1 + x_2}{2}, \frac{y_1 + y_2}{2}\right)$. **15.** true

Pages 470–472 Chapter 7 Study Guide and Assessment

17. $\sqrt{290}$ **19.** $\sqrt{61}$ **21.** $\left(\frac{5}{2}, 4\right)$ **23.** $(4a, -4b)$
25. $(1, 1)$; $(1, 4)$; $x = 1$; **27.** $(4, -2)$; $(4, -4)$;
$y = -2$; up; 12 units $\qquad$ $x = 4$; $y = 0$; down; 8 units

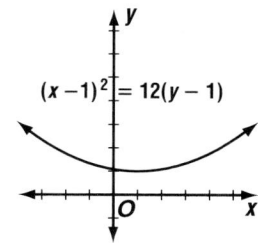 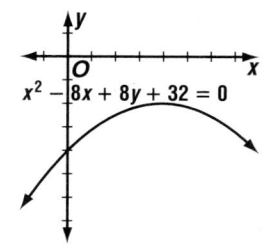

29. $(0, 0)$; 13 units **31.** $(3, -8)$; 15 units

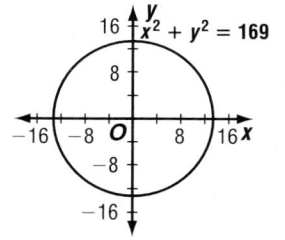 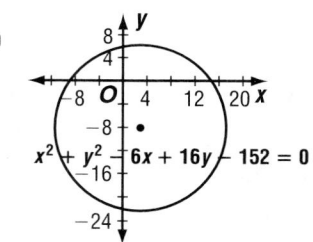

33. $(0, 0); (0, \pm\sqrt{33}); 14; 8$ **35.** $(0, 0); (\pm\sqrt{39}, 0); 16; 10$

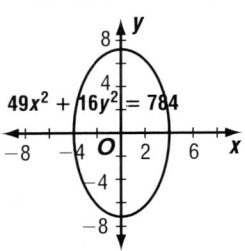

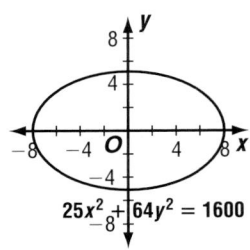

37. $(0, \pm2); (0, \pm\sqrt{13}); \pm\frac{2}{3}$ **39.** $(0, \pm4); (0, \pm5); \pm\frac{4}{3}$

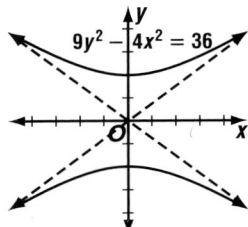

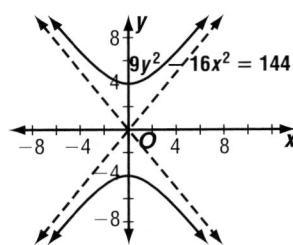

41. ellipse **43.** parabola **45.** circle **47.** $(4, 1)$
49. 75 miles **51.** $(x - 9)^2 + (y - 23)^2 = 2025$

CHAPTER 8 EXPLORING POLYNOMIAL FUNCTIONS

Pages 482–484   Lesson 8-1
7. 3 **9.** 5 **11.** quintic, 5, 5 **13.** quartic, 4, 4 **15.** a
17. c **19.** 12, −12 **21.** $2x + 2h − 3$ **23.** 15.625 units
25. odd, 2 **27.** 21, −4 **29.** 37, −13 **31.** −28, 37 **33.** $x + h + 2$ **35.** $5x^2 + 10xh + 5h^2$ **37.** $3x^2 + 6xh + 3h^2 + 7$
39. $4x^2 + 20$ **41.** $x^3 + \frac{x^2}{4} − 8$ **43.** $f(x) = x^4 − 9x^3 + 27x^2 − 31x + 12$

45a.

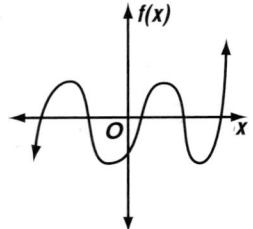

45b.

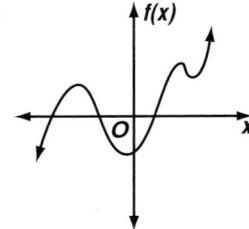

45c.

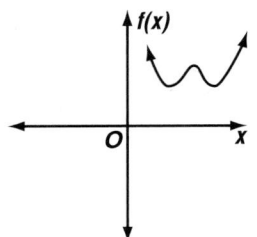

47. $3a^2 − 2a + 10$

49. $4x + 30$ **51.** $2x^2 + 18x + 60$ **53.** There is no real
number x that can make the equation $0 = x^4 + x^2 + 1$
true. **55.** 109 lumens **57.** Approximately $(13.6, 21)$ or
$(13.6, −21)$; that is, the epicenter could have been
13.6 miles east and 21 miles north or south of the first
station. **59.** −20, 20 **61.** 1.6×10^6; 1.7×10^6
63. $\begin{bmatrix} -6 & -12 \\ 9 & 7 \end{bmatrix}$ **65.** $(4, −28)$

Pages 489–490   Lesson 8-2
7. $(x^4 − 16) = (x^3 + 2x^2 + 4x + 8)(x − 2) + 0$; yes
9. −4, −1 **11.** $x − 3, x − 1$ **13.** $x − 1; x^2 + 2x + 3$
15. $x + 1, x + 3$ **17.** $(x^3 − 2x^2 − 5x + 6) = (x^2 + x − 2)(x − 3) + 0$; yes **19.** $(x^3 + 27) = (x^2 − 3x + 9)$
$(x + 3) + 0$; yes **21.** $(6x^3 + 9x^2 − 6x + 2) = (6x^2 − 3x)$
$(x + 2) + 2$; no **23.** $(4x^4 − 2x^2 + x + 1) = (4x^3 + 4x^2 + 2x + 3)(x − 1) + 4$; no. **25.** −2, 16 **27.** −23, −2 **29.** 62,
11 **31.** $x − 1, x + 2$ **33.** $2x + 1, x − 4$ **35.** $x − 2, x^2 + 2x + 4$ **37.** $2x − 3, 2x + 3, 4x^2 + 9$ **39.** $(x + 4)(x + 1)^3$
41. 8 **43.** 1, 4 **45a.** 7.5, 8, 7.5 **45b.** 0; The elevator is
stopping or is stopped. **47.** $65,892 **49.** yes
51. $2x^2 − 3x − 2$ **53.** $(1, 2)$

Page 492   Lesson 8–3A
1. $[−10, 10]$ by $[−5, 25]$, 3 **3.** $[−15, 10]$ by $[−175, 75]$, 2

5. −0.41, 2.41

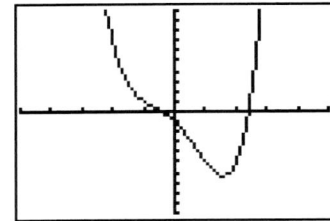

7. −0.78

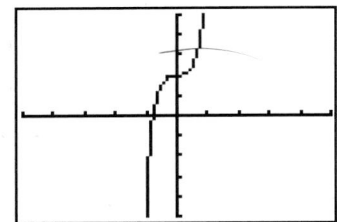

9. −2.38, −0.45, 0.11

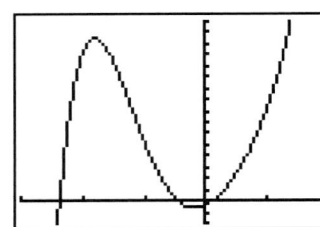

Pages 497–499   Lesson 8-3
5. −0.8
7.

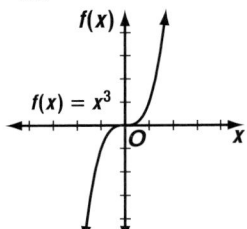

9.

11. odd; 1 min, 1 max   **13.** even; 0 min, 1 max   **15.** −1.3
17. 1.4   **19.** −2.6, 1.1   **21.** −0.3, 1.4, 4.3

23.
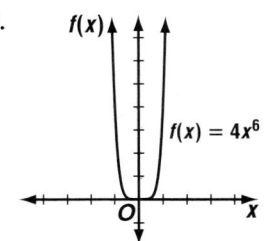
$f(x) = 4x^6$

25.
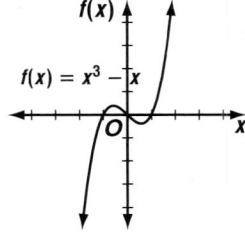
$f(x) = x^3 - x$

27.
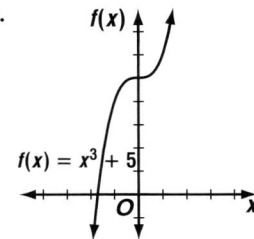
$f(x) = x^3 + 5$

29.
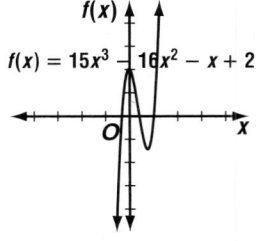
$f(x) = 15x^3 - 16x^2 - x + 2$

31.
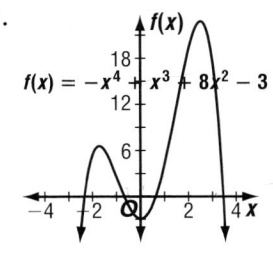
$f(x) = -x^4 + x^3 + 8x^2 - 3$

33. −3.6, −1.6, −0.7, 0.6, 1.3
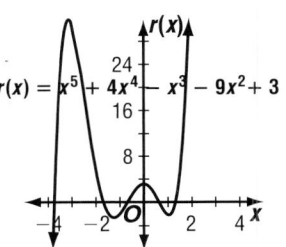
$r(x) = x^5 + 4x^4 - x^3 - 9x^2 + 3$

35. −0.7, 1.0, 2.7

37. −1.9, −0.1

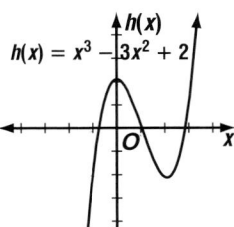
$h(x) = x^3 - 3x^2 + 2$

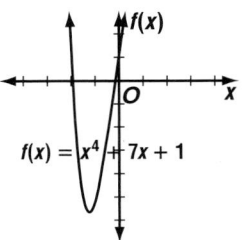
$f(x) = x^4 + 7x + 1$

39a.
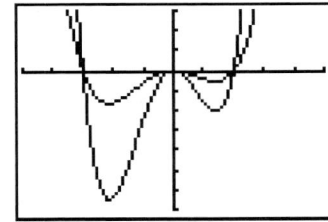

39b. The second graph is a vertical stretch of the first.
41. Sample answer: The ends of an even-degree function both point up or down and the ends of an odd-degree function point in opposite directions.   **43.** radius = 4.2 m, height = 21.2 m   **45.** 12 cm³   **47.** (3, −11), $x = 3$
49. $\pm i\sqrt{6}$

51.
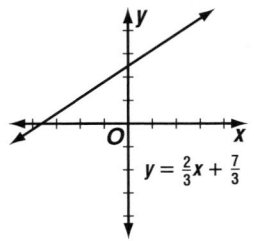
$y = \frac{2}{3}x + \frac{7}{3}$

Page 501   Lesson 8–3B
1. Sample answer: [0, 15], Xscl=3; [0, 11,000], Yscl=1000

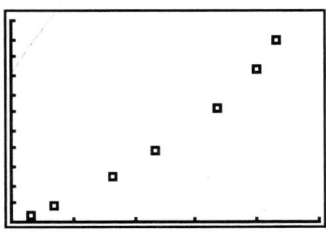

Pages 506–508   Lesson 8–4
7. $f(-x) = 6x^4 + 3x^3 + 5x^2 + x + 2$ **9.** 2 or 0; 1; 0 or 2
11. 4, $1 + i$, $1 - i$ **13.** $5i$, $-5i$, −7   **15.** $f(x) = x^3 - 2x^2 - 19x + 20$ **17.** $\ell = 8$ in., $w = 5$ in., $h = 3$ in.   **19.** 1; 3 or 1; 2 or 0   **21.** 1; 1; 2   **23.** 3 or 1; 1; 10 or 12   **25.** 5, 3, or 1; 5, 3, or 1; 0, 2, 4, 6, or 8   **27.** 2, $1 + i$, $1 - i$
29. $2i$, $-2i$, $\frac{i}{2}$, $-\frac{i}{2}$ **31.** $\frac{1}{2}$, $4 + 5i$, $4 - 5i$
33. $3 - 2i$, $3 + 2i$, −1, 1   **35.** $y = x^3 - 2x^2 - 5x + 6$
37. $y = x^4 + 7x^2 - 144$ **39.** $y = x^5 - x^4 + 13x^3 - 13x^2 + 36x - 36$ **41.** 1

43a. −22.3, −4.2, 0, 3.2
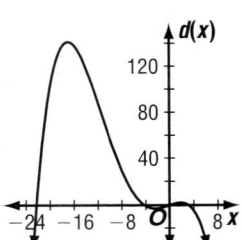
43b. 3.2

45.
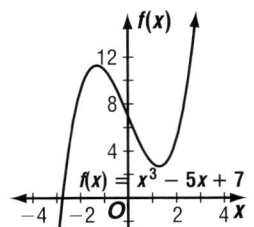
$f(x) = x^3 - 5x + 7$

47. $x^2 + 2x - 15$

49.
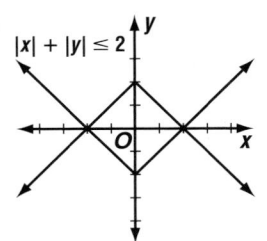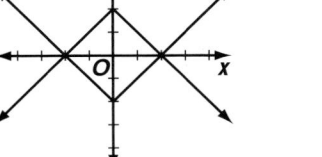
$|x| + |y| \le 2$

51. $\frac{19}{11}$

Page 508   Self Test
1. $4a^6 - 3a^4 + 2a^2 - 5$ **3.** $x + 6$, $x - 2$

5.

$g(x) = x^5 - 5$

7. 2 or 0; 1; 2 or 0   **9.** even, 4

5. 4, staircase out

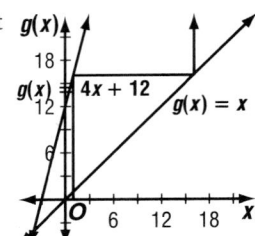

$g(x) = 4x + 12$ $g(x) = x$

7. -2, spiral out   **9.** $\frac{1}{4}$, staircase in

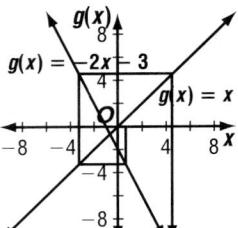

$g(x) = -2x - 3$ $g(x) = x$

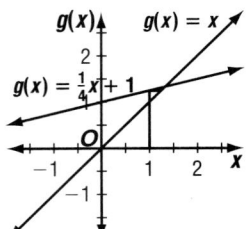

$g(x) = x$ $g(x) = \frac{1}{4}x + 1$

Pages 512–514   Lesson 8-5
7. $\pm1, \pm2, \pm3, \pm6$ **9.** $-2, -4, 7$ **11.** $0, 3$ **13.** 3,
$\frac{2}{3}, -\frac{2}{3}, \frac{-3 \pm \sqrt{13}}{2}$ **15.** 12 cm by 8 cm by 4 cm   **17.** $\pm1, \pm2,$
$\pm5, \pm10$ **19.** $\pm1, \pm\frac{1}{3}, \pm3$ **21.** $\pm1, \pm\frac{1}{3}, \pm\frac{1}{9}, \pm3, \pm9, \pm27$
23. $-6, -5, 10$ **25.** $3, 3, -\frac{1}{2}$ **27.** $-2, -4$ **29.** $2, -2, 3,$
-3 **31.** $-7, 1, 3$ **33.** $-\frac{1}{2}, \frac{1}{3}, \frac{1}{2}, \frac{3}{4}$ **35.** $-2, \frac{4}{3}, \frac{-3 \pm i}{2}$
37. $-1, -2, 5, i, -i$ **39a.** \$59   **39b.** They are the same.
41a. 16 regions   **41b.** 8 points

43. $x = \frac{1}{7}y^2$ **45.** $T = \frac{2\pi\sqrt{mrF_c}}{F_c}$ **47.** $(3, -1)$ **49.** 4 feet

Pages 517–519   Lesson 8-6
5. $x(2x^2 + 7x - 8)$ **7.** $-y(y^4 - y^2 + 100)$
9. $(x - 9)(x^2 + 9x + 81)$ **11.** $(a^4)^2 + 10(a^4) - 16 = 0$
13. $0, -4, -3$ **15.** $3.8, -7, 0$ **17.** $(x^4)^2 + 10(x^4) +$
$13.2 = 0$ **19.** $84(n^2)^2 - 62(n^2) = 0$ **21.** impossible
23. $-2, 0, 5$ **25.** $11, \frac{-11 \pm 11i\sqrt{3}}{2}$ **27.** 400   **29.** 125, 64
31. $0, 1.4, -1.4$ **33.** $3.2, -4.7, 0$ **35.** $x^3 + x^2 - 6x = 0$
37. $a = 0.885; b = 3.185$ **39a.** 317.29 miles
39b. 119 pounds   **41.** 5 or -1 **43.** 4 seconds
45. $(-6, -8)$

Pages 523–525   Lesson 8-7
9. 5, 5   **11.** $8x - 4, 8x - 1$ **13.** 9   **15.** $x^2 + 2$
17. 6, 12, 24, 48   **19.** $-3, -3$ **21.** 16, 22   **23.** 9, 12
25. $x + 11, x + 11$ **27.** $x^2 - 2, x^2 - 4x + 4$ **29.** $x^3 + 1,$
$x^3 + 3x^2 + 3x + 1$ **31.** 7   **33.** 8   **35.** -9 **37.** $4x^2$
39. $16x^2 - 32x + 16$ **41.** 6, 18, 54, 162   **43.** 0.2, 0.4, 1.2,
4.8   **45.** 25, 104, 425, 1716   **47.** $f \circ g$ does not exist; $g \circ f =$
$\{(3, 6), (4, 4), (6, 6), (7, 8)\}$ **49.** 244   **51a.** $r(x) = x - 5,$
$p(x) = x - 0.25x$ **51b.** \$23.50; taking the discount first
51c. \$24.75; taking the rebate first   **53.** $(x^3)^2 + 3(x^3) -$
$10 = 0$ **55.** 11.09'; 0.097'
57. $\begin{bmatrix} 10 & 17 & -4 & -5 \\ -28 & 3 & 48 & -9 \\ -13 & 1 & 22 & -4 \end{bmatrix}$ **59.** b   **61.** $9a + 10b$

Page 527   Lesson 8-7B
1. 1, 5, 25   **3.** 4.6, 1.16, 2.54

11. Sample answer: Functions whose slopes are positive form staircase paths, functions whose slopes are negative form spiral paths.

Pages 532–534   Lesson 8-8
7. $\{(2, 3), (2, 4)\}$, no
9. $y^{-1} = \frac{x}{7}$ **11.** $f^{-1}(x) = x + 6$

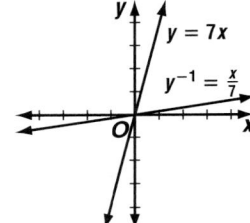

$y = 7x$ $y^{-1} = \frac{x}{7}$

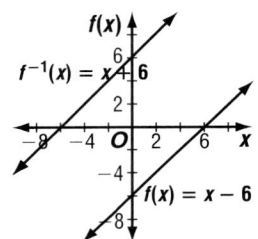

$f^{-1}(x) = x + 6$ $f(x) = x - 6$

13. no   **15.** $\{(4, 2), (1, -3), (8, 2)\}$, yes
17. $\{(3, 1), (-1, 1), (-3, 1), (1, 1)\}$, yes
19. $x = 6$ **23.** $f^{-1}(x) = -x$

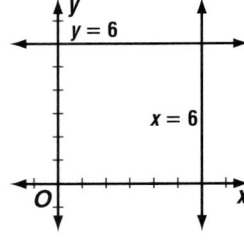

$y = 6$ $x = 6$

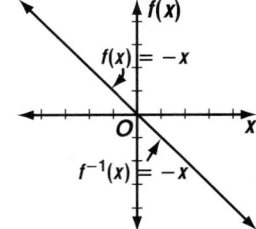

$f(x) = -x$ $f^{-1}(x) = -x$

27. $f^{-1}(x) = \frac{3x + 1}{2}$ **29.** $y^{-1} = \pm\sqrt{x} + 4$

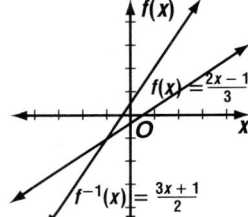

$f(x) = \frac{2x - 1}{3}$ $f^{-1}(x) = \frac{3x + 1}{2}$

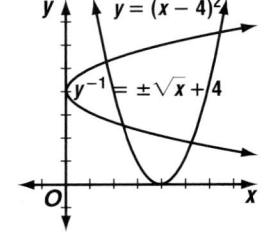

$y = (x - 4)^2$ $y^{-1} = \pm\sqrt{x} + 4$

31. yes **33.** no
35. yes **37.** no

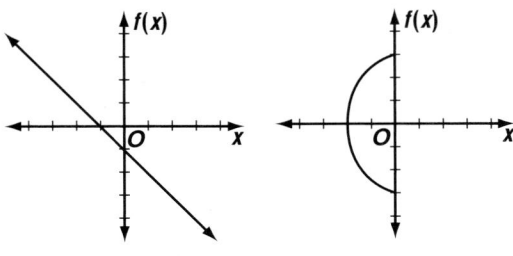

39. Sample answer: $f(x) = x$ and $f^{-1}(x) = x$ or $f(x) = -x$ and $f^{-1}(x) = -x$. **41.** 12 **43.** 288°K **45.** 300 ft, 2.5 seconds **47.** $\begin{bmatrix} 5 & -3 \\ 7 & 3 \end{bmatrix}$ **49.** D = $\{-11, 0, 1, 3, 9, 12\}$, R = $\{-6, -4, -3, 0, 1, 7, 8\}$; no

Page 537–538 Lesson 8–8B

5.
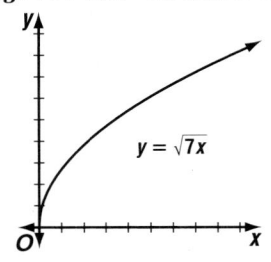
D: $x \geq 0$; R: $y \geq 0$

9.
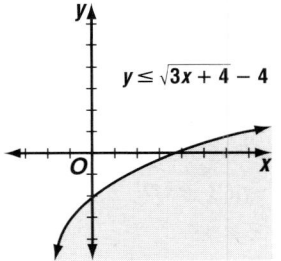
$y \leq \sqrt{3x + 4} - 4$

13.
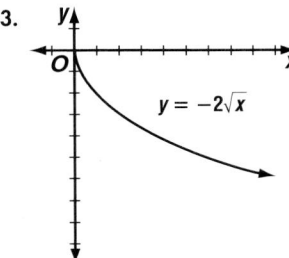
D: $x \geq 0$; R: $y \leq 0$

17.
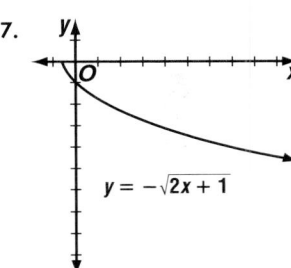
$y = -\sqrt{2x + 1}$
D: $x \geq -\frac{1}{2}$; R: $y \leq 0$

21.
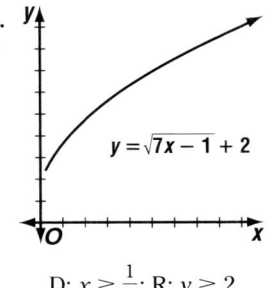
$y = \sqrt{7x - 1} + 2$
D: $x \geq \frac{1}{7}$; R: $y \geq 2$

25.
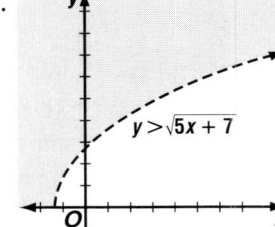
$y > \sqrt{5x + 7}$

29a.

If:	$h > 0$	$h < 0$	$k > 0$	$k < 0$
$a > 0$	h is minimum value in domain.	$-h$ is minimum value in domain.	k is minimum value in range.	
$a < 0$			k is maximum value in range.	

29b. $y = -2\sqrt{x}$ is a little wider than $y = \sqrt{x}$, and falls downward to the right. $y = \sqrt{x - 4}$ begins at $(4, 0)$. $y = \sqrt{x + 3}$ begins at $(0, 3)$. $y = 3\sqrt{x - 1} + 5$ is wider than $y = \sqrt{x}$, and begins at $(1, 5)$.

Page 539 Chapter 8 Highlights
1. h **3.** a **5.** g **7.** f

Pages 540–542 Chapter 8 Study Guide and Assessment
9. $-6, x + h - 2$ **11.** $-21, 6x + 6h + 3$ **13.** $20, x^2 + 2xh + h^2 - x - h$ **15.** $4, -1$ **17.** $20, -20$ **19.** $x + 2, x + 2$ **21.** $x - 1, x - 2$

25. -2.0 **27.** $-1.9, 0.3, 1.4$

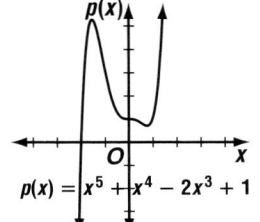
$p(x) = x^5 + x^4 - 2x^3 + 1$

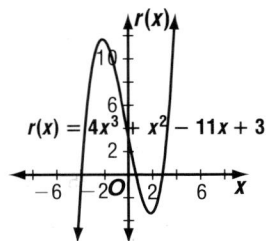
$r(x) = 4x^3 + x^2 - 11x + 3$

29. 3 or 1; 1; 2 or 0 **31.** 0; 0 or 2; 4 or 2 **33.** 2 or 0; 2 or 0; 4, 2, or 0 **35.** $-1, -1$ **37.** $-3, 5, \frac{1}{2}$ **39.** $\frac{5}{3}, -3, 0$ **41.** $4, -2 \pm 2i\sqrt{3}$ **43.** $2, -2$ **45.** $6x + 1, 6x + 7$ **47.** $-2x^2 - 1, 4x^2 - 4x + 2$ **49.** $x^3 - 2, x^3 - 6x^2 + 12x - 8$ **53.** $g^{-1}(x) = 3x - 6$ **55.** $y^{-1} = \pm\sqrt{x}$

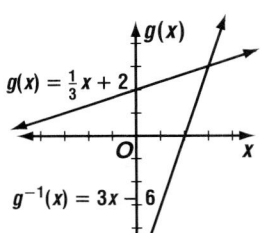
$g(x) = \frac{1}{3}x + 2$
$g^{-1}(x) = 3x - 6$

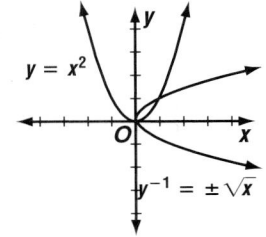
$y = x^2$
$y^{-1} = \pm\sqrt{x}$

57a. $A = 1000(1 + r)^6 + 1000(1 + r)^5 + 1000(1 + r)^4 + 1200(1 + r)^3 + 1200(1 + r)^2 + 2000(1 + r)$ **57b.** $8916.76
59. $7.50

CHAPTER 9 EXPLORING RATIONAL EXPRESSIONS

Page 549 Lesson 9–1A
1. $x = 4$
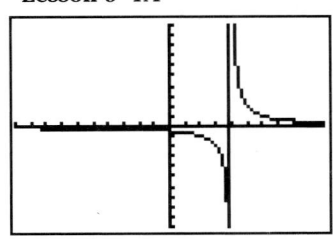

5. $x = 4$
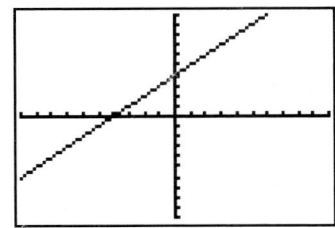

9. $y = 4$

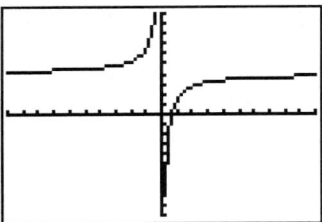

Pages 553–555 Lesson 9–1

5. $x = 1, x = -3, y = 0$ **7.**

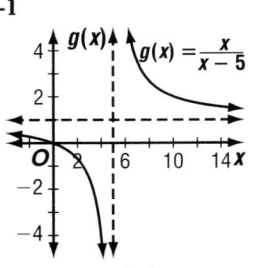

9. 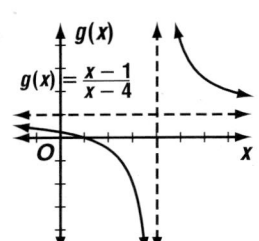 **11.**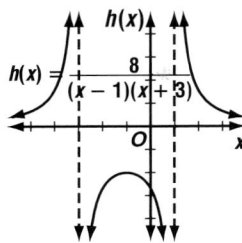

13. $x = 4, y = 1$ **15.** $x = 1, x = -5, y = 0$
17. $x = 3, y = 1$

19. 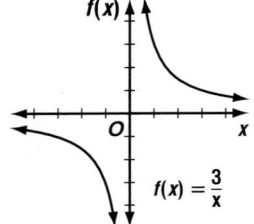 **21.**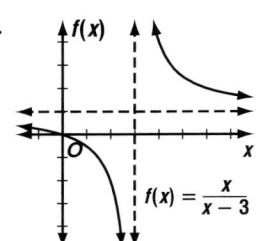

23. 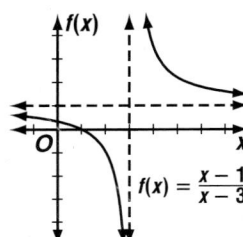 **25.**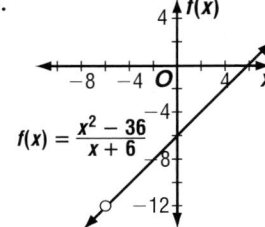

27. 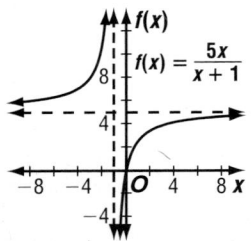 **29.**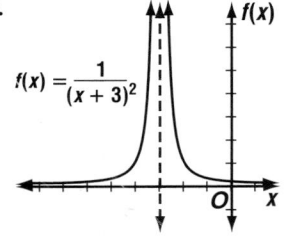

31.

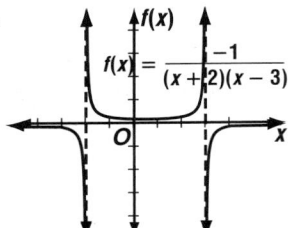

33.

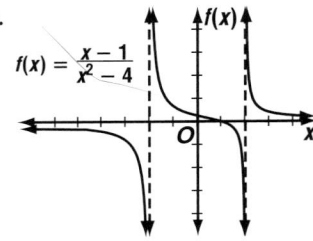

35a. 100 mg **35b.** The graph has a vertical asymptote at $y = -12$ and a horizontal asymptote at $C = 1$.

37. no **39.** $\dfrac{y^2}{34} - \dfrac{x^2}{6} = 1$

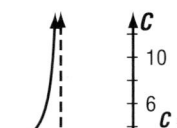 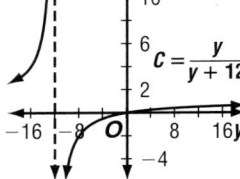

41. 0; 1 R, Q; 5 **43.** $x{:}4; y{:}-6$

Pages 559–561 Lesson 9–2

5. inverse, -5 **7.** direct, $\dfrac{1}{3}$ **9.** $y = 5x, 60$ **11a.** $I = \dfrac{k}{d^2}$

11b.

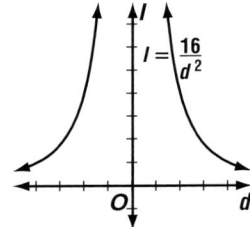

11c. The sound will be heard $\dfrac{1}{4}$ as intensely.

13. direct, $-\dfrac{1}{7}$ **15.** inverse, 1 **17.** direct, $-\dfrac{4}{3}$
19. $rt = -54, 4.91$ **21.** $xy = 50, 1.25$ **23.** 38 meters
25. $\dfrac{22}{5}$ **27.** 4 **29a.** directly **29b.** 2π **31a.** $P = 0.43d$
31b. 25.8 psi **31c.** about 150 ft

31d.

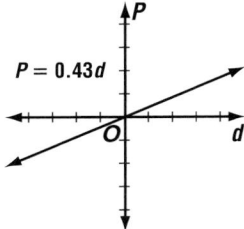

33a. like one third of a month

33b. like 5 weeks **33c.** like 1 hour **35.** 1500 BTU/h
37. x, x **39.** $\dfrac{(x+2)^2}{16} + \dfrac{(y-3)^2}{36} = 1$ **41.** 175 passengers

Pages 565–568   Lesson 9–3

5. $6x, \frac{5y}{2x}$ 7. $c+5, \frac{1}{2}$ 9. $\frac{3a^2}{2bc}$ 11. 3   13. $\frac{4}{3}$ 15. $\frac{2y(y-2)}{3(y+2)}$

17. $\frac{9x}{4y^4}$ 19. $\frac{5}{x+1}$ 21. y 23. $\frac{3c}{20b}$ 25. $\frac{xz}{8y}$ 27. $\frac{10x}{3y^2}$ 29. $\frac{b^3}{xy^2}$

31. $\frac{4}{5xyz^2}$ 33. $\frac{6}{5(x-1)}$ 35. $\frac{2x^2}{3(x+1)^2}$ 37. $\frac{5(x-3)}{2(x+1)}$

39. $\frac{3(m+n)}{m^2+n^2}$ 41. $\frac{4}{3}$ 43. $4x^2+6x+2$ square feet

45. $21\frac{1}{3}$ m³   47. $x^2+2xh+h^2-\frac{1}{2}x+\frac{1}{2}h$ 49a. 8.8

49b. 5.9   49c. 6.7   49d. Fitright Shoes   51. -8

Page 568   Self Test

1. $x=4, y=0$ 3.   5. 22

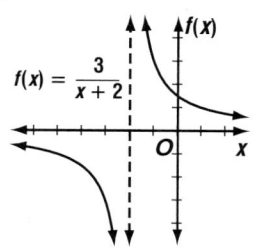

$f(x)=\frac{3}{x+2}$

7. $\$2560$ 9. $\frac{7}{2}$

Pages 573–575   Lesson 9–4

5. $70x^2y^2$ 7. $\frac{6+8b}{ab}$ 9. $\frac{2x+3}{x+1}$ 11. $\frac{5x+16}{(x+2)^2}$ 13. $\frac{q+p}{pq}$

15. $4w-12$ 17. $x^2(x-y)(x+y)$ 19. $2(x-5)(x+3)$

21. $\frac{6x+7}{x+2}$ 23. $\frac{y+3}{y-4}$ 25. $\frac{y^2-6y+10}{y-3}$ 27. $\frac{x(x-9)}{(x+3)(x-3)}$

29. $\frac{3m-10}{(m-5)(m+4)}$ 31. $\frac{-4x^2-5x-2}{(x+1)^2}$ 33. $\frac{-12x+21xy-4y}{6x^2y}$

35. $\frac{4x^2-2x-14}{x^2-4}$ 37. $\frac{2x^2+x-4}{(x-1)(x-2)}$ 39. $\frac{1}{x+1}$

41. Sample answer: 2, 4; LCM = 4, GCF = 2; 4 × 2 = 8

43a.

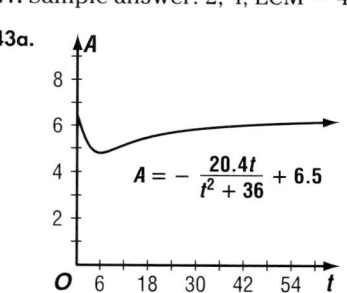

43b. 5.8

$A=-\frac{20.4t}{t^2+36}+6.5$

43c. It quickly drops below normal and then slowly rises back to normal.   43d. after 6 minutes   45. $-\frac{1}{2}, \frac{3}{2}, \frac{7}{2}$

47. $y=-\frac{1}{6}(x-11)^2+\frac{1}{2}$ 49. $-1, 3$

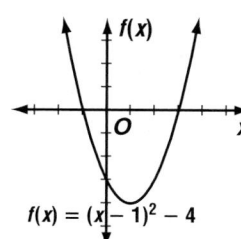

$f(x)=(x-1)^2-4$

51. $x=24y+18$

Pages 581–583   Lesson 9–5

7. $(m-4)(m-2); 4, 2; 6$ 9. $\frac{21}{5}$ 11. $x>5$ 13. 4; none; $\frac{1}{2}$

15. $m^2; 0; \frac{3}{2}$ 17. $(a-6)(a-2); 6, 2; 6.94, 1.73$ 19. 2

21. $y\le-6$ 23. $-12, 1$ 25. $-6, \frac{3}{2}$ 27. -1 29. $\varnothing$

31. $\frac{b}{bc+1}$ 33a. $\frac{x+11}{x+20}\ge0.70$ 33b. $x\ge10$

33c. Selena needs at least 10 consecutive free throws.

35. 56.67 mph   37. 32   39. 0; 1; 2   41. $\sqrt{4.58}$ units

43. about 7 in. × 7 in. × 7 in.

Page 585   Chapter 9   Highlights

1. false, point discontinuity   3. true   5. false, asymptote

7. false, $x=-2$

Pages 586–588   Chapter 9   Study Guide and Assessment

9. $x=2; y=0$ 11. $x=0; y=0$

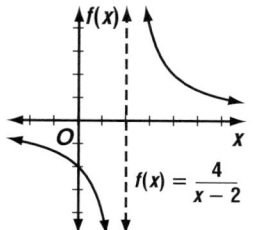

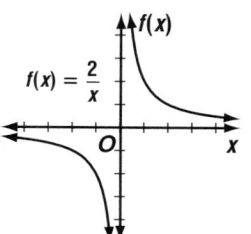

$f(x)=\frac{2}{x}$

$f(x)=\frac{4}{x-2}$

13. $x=3, x=-1; y=0$

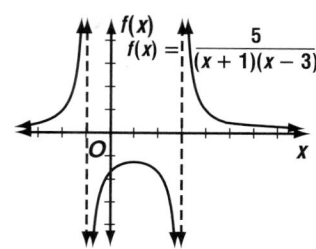

$f(x)=\frac{5}{(x+1)(x-3)}$

15. $xy=22.5; -37.5$ 17. $y=\frac{9}{14}x; 98$

19. $y=\frac{25}{8}xz; 0.192$ 21. $y-2$ 23. $\frac{a}{5x^2}$ 25. $\frac{x-2}{x+2}$

27. $\frac{2}{n-3}$ 29. $\frac{28a-27b}{12ab}$ 31. $\frac{7}{5(x+1)}$ 33. $\frac{18}{y-2}$

35. $\frac{3(3m^2-14m+27)}{(m+3)(m-3)^2}$ 37. 31   39. 0   41a. $I=\frac{6}{R}$ 41b. 6

43. 42 lb/in²

CHAPTER 10   EXPLORING EXPONENTIAL AND LOGARITHMIC FUNCTIONS

Page 595   Lesson 10–1A

1.

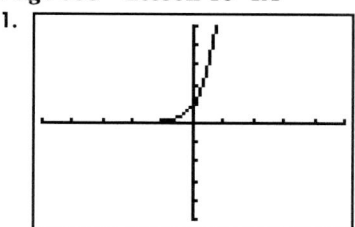

3.

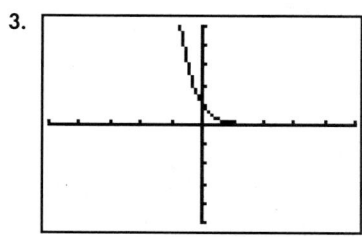

5.

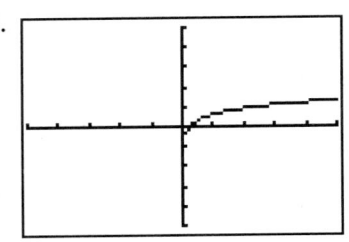

7. 1.50   **9.** 8.06   **11.** −0.77, 2, 4   **13.** 1.70   **15.** 1.73

Pages 600–602   Lesson 10–1
7. 4.7   **9.** $5^{4\sqrt{2}}$ **11.** $3^{2\sqrt{5}}$ **13.** $n \le -2$ **15.** −2   **17.** 16 cells
19. 0.7   **21.** 0.5   **23.** 0.5   **25.** $4^{3\sqrt{2}}$ **27.** y^6 **29.** $2^{5\sqrt{7}}$
31. a^{10} **33.** 4   **35.** −1   **37.** $\frac{1}{9}$ **39.** $\frac{3}{5}$ **41.** 6   **43.** $n < -3$
45. 12   **47.** −3, 5   **49.**

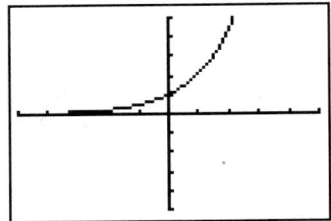

51.

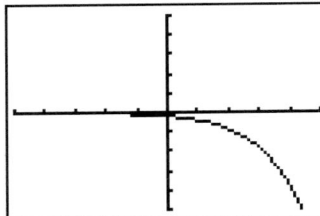

53a. 2, 4, 8, 16   **53b.** $y = 2^x$ **53c.** $y = 0.003(2)^x$
53d. about 3,221,225.47 in.   **55a.** about 14.7 psi
55b. about 14.0 psi   **55c.** about 11.4 psi
55d. about 16.0 psi
55e.

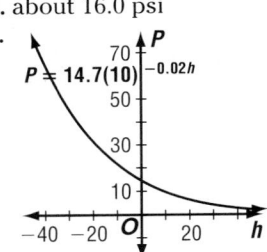

They represent the atmospheric pressure for places below sea level.

57a. $d = 1.30h^{\frac{3}{2}}$ **57b.** about 997 cm   **59.** 0; 2 or 0; 2 or 4

61. 36 mi   **63.** $\frac{3+4i}{25}$ **65.** $\begin{bmatrix} -15 & 6 \\ 19 & 43 \\ -6 & -24 \end{bmatrix}$ **67.** no

---

Page 604   Lesson 10–1B
1.

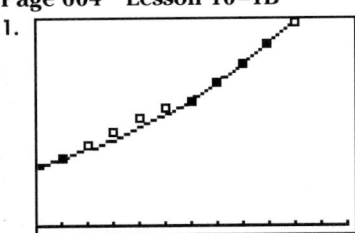

3. about 90.088   **5.** Louisiana Purchase

Pages 608–610   Lesson 10–2
7. $\log_2 32 = 5$ **9.** $5^3 = 125$ **11.** $10^{-3} = 0.001$ **13.** $\frac{1}{2}$
15. $\frac{1}{49}$ **17.** −3   **19.** ±8   **21.** 10^3 or 1000 times stronger
23. 2   **25.** −4   **27.** $\frac{1}{2}$ **29.** 243   **31.** 32   **33.** 3   **35.** 125
37. 1000   **39.** 2   **41.** $x < \frac{1}{2}$ **43.** no solution   **45.** 10   **47.** 1
49.

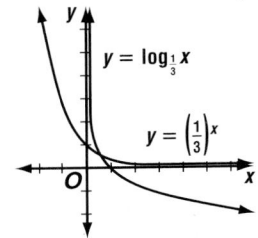

51. In the graphs of $y = \log_{10} x$ and $y = \log_5 x$, the value of y increases as the value of x increases. In the graph of $y = \log_{\frac{1}{3}} x$ and $y = \log_{\frac{1}{2}} x$, the value of y decreases as the value of x increases. The value of y in the graph of $y = \log_5 x$ increases more as x increases than in the graph of $y = \log_{10} x$. The graphs of $y = \log_{\frac{1}{3}} x$ and $y = \log_{\frac{1}{2}} x$ are reflections of $y = \log_{10} x$ over the x-axis. All of the graphs cross the x-axis at 1.

53. $\log_4 16 \overset{?}{=} 2 \log_4 4$ **55.** $\log_{10} [\log_3 (\log_4 64)] \overset{?}{=} 0$
$\qquad 2 \overset{?}{=} 2(1) \qquad\qquad\qquad \log_{10} [\log_3 (3)] \overset{?}{=} 0$
$\qquad 2 = 2 \qquad\qquad\qquad\qquad\quad \log_{10} [1] \overset{?}{=} 0$
$\qquad\qquad\qquad\qquad\qquad\qquad\qquad\quad 0 = 0$

57. about 1.26 times greater   **59.** $11^{4\sqrt{5}}$ **61.** $f \circ g$ does not exist; $g \circ f = \{(2, 5), (-1, -1), (3, 2)\}$ **63.** $x^2 - 36$
65. about 3.54 s   **67.**   **69.** $x < -8$

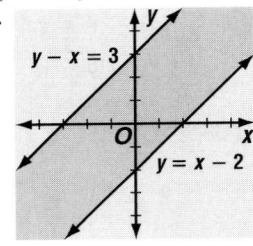

Pages 614–616   Lesson 10–3
5. $2 \log_4 x + \log_4 y$ **7.** $\log_5 a + \log_5 c - \log_5 b$ **9.** 1.222
11. −0.415   **13.** 4   **15.** 6   **17.** 1.262   **19.** 1.1402
21. −0.3690   **23.** 4.2620   **25.** −0.2288   **27.** 7   **29.** 14
31. 4   **33.** 4   **35.** 12   **37.** 6   **39.** $\frac{x^4}{2}$ **41.** $\frac{1}{2}(n + 1)$

43. $\log_b \frac{x}{y^3}$ **45a.** $\text{pH} = 6.1 + \log_{10} B - \log_{10} C$ **45b.** a very weak base   **45c.** 7.197   **47.** 5   **49.** $2a^3 + 5a^2 + 8a + 5$ **51.** no   **53.** $x(x + 7)(x - 5)$

Pages 619–621   Lesson 10–4
5. 2   **7.** 1.1367; 0.1367; 1   **9.** 2.8662   **11.** about 1.58   **13.** 3
15. 0.8129   **17.** 0.8129   **19.** 2.9544; 0.9544; 2   **21.** 0.8046;
0.8046; 0   **23.** −3.1549; 0.8451; −4   **25.** 141.0912
27. 0.0153   **29.** 0.0014   **31a.** about 316,227,766 times
31b. about 19,952,623 times   **31c.** about 93.7%   **33.** 72

35. $y = 3(x - 4)^2 - 1$ **37.** $\begin{bmatrix} 28 \\ -16 \\ 3 \end{bmatrix}$

Page 621   Self Test
1. −2   **3.** 4   **5.** $\frac{1}{2}$ **7.** 8   **9.** 2.7786; 0.7786; 2

Pages 623–625   Lesson 10–5
5. −3.1011   **7.** 1.2961   **9.** 0.2066   **11.** 2.0732   **13.** 4.0483

15. 1.0000   **17.** −5.2983   **19.** −4.6052   **21.** 1.0000
23. 0.8940   **25.** 2.7319   **27.** 4.7115
29.

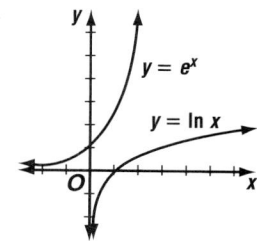

31. The graphs of $y = \ln x$ and $y = e^x$ are reflections of each other along the line $x = y$.   **33a.** about 2.7167
33b. about 2.7183   **33c.** the second one
33d. about 0.06%   **35a.** about 1,587,209,679 drives
35b.

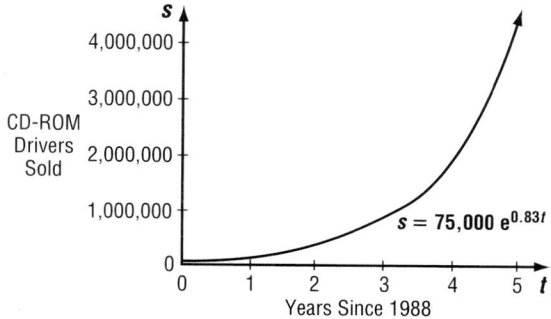

Sample answer: Both graphs show the sales increasing more and more each year.   **37a.** 1.8882   **37b.** −2.2518
37c. 3603.2965   **37d.** 0.0006   **39.** $(x + 1)(x - 1)$
$(x^3 + x^2 + x + 1)$ **41.** 10 in. by 12 in.
43. vertices: (0, 0), (0, 3), (2, 0), (1.5, 1.5), max: $f(0, 3) = 12$, min: $f(0, 0) = 0$

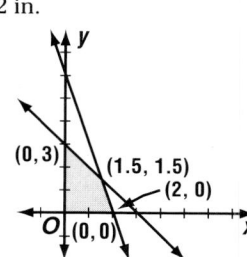

Pages 628–630   Lesson 10–6
5. 1.833   **7.** 1.159   **9.** 3.483   **11.** −7.638   **13.** 1.723
15. 4.395   **17.** 1.295   **19.** 1.732   **21.** 2.191   **23.** −0.412
25. 2.558   **27.** 2.548   **29.** 4.836   **31.** 0.779   **33.** ±1.915
35. 0   **37.** 11.567   **39.** 8.849   **41.** 0.587   **43.** 0.205
45. 2.125   **47.** about 23 years   **49.** 0.0023   **51.** ±2, ±1
53. $\{x \mid x > - 2$ or $x < -9\}$ **55.** $x = 2, y = 3$
57. $3x + 2y = 24$

Pages 633–636   Lesson 10–7
3a. about 92.3 megahertz   **3b.** about 9.4 cm from the left
side   **5a.** about −0.00043   **5b.** $y = ne^{-0.00043t}$ **5c.** about
2.33 grams   **5d.** in about 3224 years   **5e.** Never; the
amount left will always be half of the amount that existed
1620 years ago.   **7.** about 8.1 days   **9a.** about $y =$
$546,488e^{0.0070t}$ **9b.** about 674,190 people   **9c.** Sample
answer: If the economy of a city is doing well, more
people than expected may move into the area. If the
economy is not doing well, people may lose their jobs
and move out of the area.   **11.** No; the bone is only about
19,000 years old and dinosaurs died out 63,000,000 years
ago.   **13a.** about $t = 1.0208n^{0.7776}$ **13b.** about 2.4 min;
about 3.6 min   **15a.** 11 payments   **15b.** 16 payments

15c. 48 payments   **15d.** 60 payments   **15e.** 81 payments
15f. 157 payments   **17.** about 2.45   **19a.** about 2.8 s
19b. about 4.7 s   **19c.** about 6.2 s   **19d.** No, the formula is
not a direct variation.   **21.** 128 cm^2

Page 637   Chapter 10   Highlights
1. h   **3.** b   **5.** a   **7.** g

**Pages 638–640   Chapter 10   Study Guide and
Assessment**
9. x^{10} **11.** $8^{3\sqrt{3}}$ **13.** $-\frac{7}{4}$ **15.** −2   **17.** $\log_5 \frac{1}{25} = -2$
19. $\log_4 8 = \frac{3}{2}$ **21.** $8^{\frac{1}{3}} = 2$ **23.** $6^0 = 1$ **25.** −3   **27.** 2

29. 2   **31.** 4   **33.** 3   **35.** no solution   **37.** 1.7712
39. 1.2547   **41.** 3   **43.** 6   **45.** 1.6680   **47.** −2.3298
49. 562.3413   **51.** 0.2710   **53.** 0.8329   **55.** 3.9120
57. 7.2102   **59.** 0.6132   **61.** 1.683   **63.** 1.090   **65.** 2.131
67. 5.7286   **69.** ±2.2452   **71.** −1.8928   **73.** 3.2838
75. about 12.8 psi   **77.** 5.05 days

**CHAPTER 11   INVESTIGATING SEQUENCES
AND SERIES**

Page 647   Lesson 11–1A
1. 117   **3.** 12.112   **5.** 48

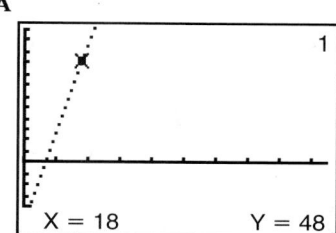

7. −126

9. 144.49,
143.32, 142.15

Pages 652–655   Lesson 11–1
5. 24, 28, 32, 36   **7.** 5, 8, 11, 14, 17   **9.** −112   **11.** 15
13a. 56, 68, 80   **13b.**   **15.** $88

17. 10, 3, −4, −11   **19.** 4.2, 8.2, 12.2, 16.2   **21.** 41,
46, 51, 56, 61   **23.** $\frac{5}{8}$, 1, $\frac{11}{8}$, $\frac{14}{8}$, $\frac{17}{8}$ **25.** −175   **27.** $-12\frac{1}{12}$
29. 340   **31.** 19   **33.** 27   **35.** 173

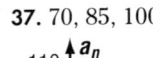

37. 70, 85, 100

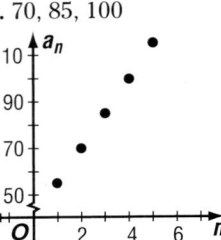

39. −8, −2, 1, 7

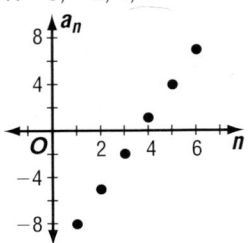

41. 3 **43.** 5 **45.** $a_n = 9n - 2$ **47a.** 35 **47b.** $a_n = 4n + 3$

49. 3, 347 **51.** 46

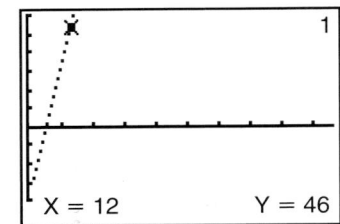

53. 17

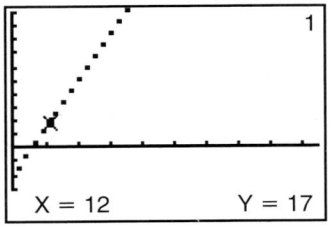

55. 50

57. 37.5 in. **59a.** The bottom right box is divided into 4 parts.

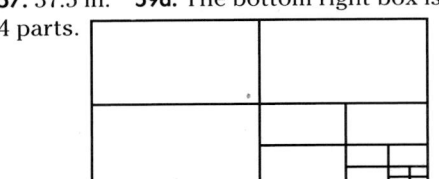

59b. 9 **59c.** 19, 29 **59d.** 499 **59e.** $10n - 1$ **61.** about 1.3863 **63.** $\dfrac{y^3 - w^2 y - y + w^2}{w^2 - y^2}$ **65.** parabola **67.** (9, 0); $x = 9$; downward **69.** $\dfrac{1}{7}\begin{bmatrix} -4 & -5 \\ -1 & -3 \end{bmatrix}$ **71.** all reals

Pages 660–661 Lesson 11-2
5. 6, 1.4 11.6, 5 **7.** 30, −18, −78, 7 **9.** 230 **11.** 552 **13.** 260 **15a.** 1,001,000 **15b.** 166,833 **17.** 663 **19.** 2646 **21.** −140 **23.** 182 **25.** 225 **27.** 119 **29.** 735 **31.** −204 **33.** 92 + 97 + 102 + 107 + 112; 510 **35.** 17, 26, 35 **37.** −12, −9, −6 **39.** $\displaystyle\sum_{n=1}^{12} \frac{1}{5}n; \frac{78}{5}$ **41.** 3649 **43.** 12 days **45.** 780 ft **47.** −2.8824 **49.** even; 2 **51.** $f(x) = x^2 + 4x - 2; x^2; 4x; -2$

Pages 667–669 Lesson 11-3
7. yes, −5 **9.** yes, $\frac{2}{3}$ **11.** 2, −4 **13.** 56 **15.** $\frac{15}{64}$ **17.** 635 **19.** 192, 256 **21.** 48, 32 **23.** $-1, \frac{1}{4}$ **25.** 243, 81, 27, 9, 3 **27.** 729 **29.** 243 **31.** 1 **33.** 78,125 **35.** −8748

37. ±18, 36, ±72

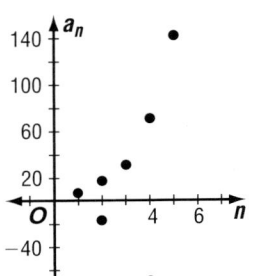

39. 16, 8, 4, 2

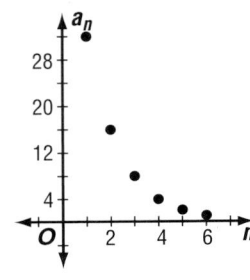

41. 25 **43.** $-\frac{4}{3}$ **45.** $a_n = 2^n$ **47.** $a_n = \frac{4}{25}(-5)^{n-1}$ **49.** 2, 4, 8, 16, 32 **51.** 243, 81, 27, 9, 3, 1, $\frac{1}{3}$, $\frac{1}{9}$

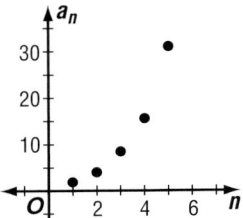

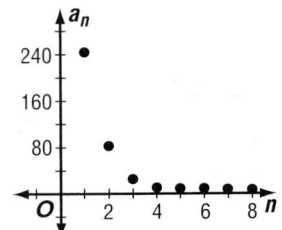

53. Sample answer: In a geometric sequence, each subsequent term can be found by multiplying the preceding term by r. So the nth term of a sequence a_n would equal $a_{n-1}r$. Each term of the sequence can be expressed in terms of the first term. The second term is $a_1 \cdot r$, the third term is $a_1 \cdot r \cdot r$ or $a_1 r^2$, the fourth term is $a_1 \cdot r \cdot r \cdot r$ or $a_1 r^3$, and so on. Notice each term is a product of a_1 and $n-1$ factors of r. So the nth term would be $a_1 r^{n-1}$. Since $a_{n-1}r$ and $a_1 r^{n-1}$ both describe the nth term, they are equivalent. **55.** 262.144 lb **57a.** 307 mg/L **57b.** 29 tiers **57c.** at most 47 tiers **59.** 9 **61.** 64 **63.** $x = \frac{1}{18}(y+7)^2 + \frac{5}{2}$ **65.** $13 - 13i$ **67.** O'Hare, 31.2 million; Heathrow, 24 million

Pages 673–675 Lesson 11-4
5. 6, −3, −162, 4 **7.** 4, 3, 324, 5 **9.** 732 **11.** 81,915 **13.** $147\frac{7}{9}$ **15.** $81 + 27 + 9 + 3 + 1 + \frac{1}{3} + \frac{1}{9}; 121\frac{4}{9}$ **17.** 1441 **19.** 1,328,600 **21.** 244 **23.** 1111 **25.** 300 **27.** 7.875 **29.** 1040.984 **31.** 144 **33.** 2101 **35.** 5 + 10 + 20 + 40 + 80 + 160 + 320 + 640 + 1280; 2555 **37.** $2 - 6 + 18 - 54 + 162 - 486; -364$ **39.** 5 **41.** 243 **43.** $\displaystyle\sum_{n=1}^{12} 243\left(\frac{2}{3}\right)^{n-1}, 723.38$ **45.** −1,048,575 **47a.** $10.23, $10,485.75 **47b.** The amount you have to put away each day exceeds the funds available to anyone. **49.** $\frac{1}{24}, \frac{1}{4}$ **51.** 3 **53.** $\left(\frac{3}{2}, -4\right)$; about 6.2 units **55.** 11

Page 675 Self Test
1. 46 **3.** 816 **5.** 30, 36, 42

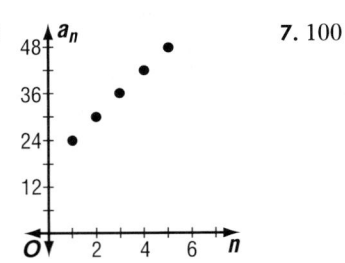

7. 100

9. ± 6, 12, ± 24

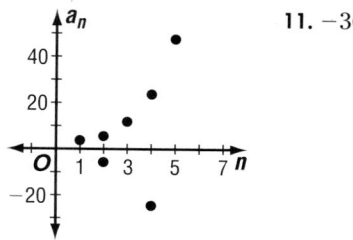

11. -364

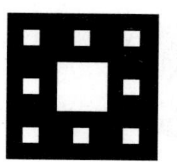 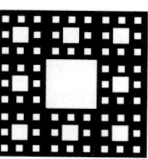

Stage 3   Stage 4

9b. $a_1 = 1$, $a_{n+1} = 8a_n$ **9c.** 32,768   **9d.** $a_1 = 81$, $a_{n+1} = \frac{8}{9} a_n$ **9e.** The area approaches zero.   **11.** Sample answer: Small changes produced unexpected results. **15.** -2.8824 **17.** $5x + y = 26$, $2x - 3y = 41$

Pages 680–682   Lesson 11–5
7. 36, $\frac{2}{3}$; 108   **9.** 6, $-\frac{2}{3}$; $\frac{18}{5}$ **11.** 16, $\frac{3}{2}$; none   **13.** $\frac{5}{9}$
15. $\frac{175}{999}$ **17.** 14   **19.** 45   **21.** 45   **23.** 3   **25.** $\frac{54}{5}$ **27.** 5
29. 144   **31.** -15 **33.** $\frac{1}{9}$ **35.** $\frac{82}{99}$ **37.** $\frac{41}{90}$ **39.** 1
41. $\sum\limits_{n=1}^{\infty} 3(9)^{n-1}$, does not exist   **43.** 27, 18, 12   **45.** 24, $16\frac{1}{2}$, $11\frac{11}{32}$, $7\frac{409}{512}$ **47.** Yes, it is possible to have infinite arithmetic series, but a sum does not exist since the sum increases with each term, or decreases with each term if d is negative.   **49.** 50 ft   **51.** $\sum\limits_{n=1}^{6} (-3)^{n-1}$
53. $\frac{3y^2 + 10y + 5}{2(y-5)(y+3)}$ **55a.** 340 items   **55b.** 475 items
55c. 495 items   **55d.** 170 items   **55e.** 237.5 items
57. $(4, 12, -3)$ **59.** $y = \frac{3}{4}x - 4$

Pages 699–701   Lesson 11–8
5. 40,320   **7.** 66   **9.** $t^6 + 12t^5 + 60t^4 + 160t^3 + 240t^2 + 192t + 64$ **11.** $56a^5b^3$ **13.** $(8x^3 - 36x^2 + 54x - 27)$ cm^3
15. 120   **17.** 72   **19.** 32,760   **21.** 210   **23.** $r^7 + 7r^6s + 21r^5s^2 + 35r^4s^3 + 35r^3s^4 + 21r^2s^5 + 7rs^6 + s^7$
25. $m^5 - 5m^4a + 10m^3a^2 - 10m^2a^3 + 5ma^4 - a^5$
27. $16b^4 - 32b^3x + 24b^2x^2 - 8bx^3 + x^4$ **29.** $81x^4 + 216x^3y + 216x^2y^2 + 96xy^3 + 16y^4$ **31.** $243 + 135m + 30m^2 + \frac{10m^3}{3} + \frac{5m^4}{27} + \frac{m^5}{243}$ **33.** $924x^6y^6$ **35.** $145{,}152x^6y^3$
37. 0.111477   **39.** $(k + 3)!$ **41.** $\frac{7}{n-3}$ **43a.** eighth
43b. 22nd power   **43c.** $319{,}770a^{14}b^8$ **45.** 1, 2, 4, 8, 16; 512
47. about 13.6%   **49.** $\frac{3}{2}$ **51.** 0.2393   **53.** 42 lb/in^2
55. circle   **57.** $y = \frac{2}{3}x + \frac{10}{3}$

Page 703   Chapter 11   Highlights
1. k   **3.** l   **5.** d   **7.** b, e   **9.** c

Pages 704–706   Chapter 11   Study Guide and Assessment
11. 38   **13.** -11 **15.** 97   **17.** 12   **19.** 10   **21.** -3, 1, 5
23. 9, 3, 0, -6 **25.** 2322   **27.** 1155   **29.** 7, 10, 13, 16, 19, 22, 25, 28, 31, 34, 37, 40; 282   **31.** $\frac{64}{3}$ **33.** 56   **35.** ± 6, 12, ± 24
37. 4, 2, 1, $\frac{1}{2}$ **39.** ± 14, ± 686, -4802 **41.** $\frac{21}{8}$ **43.** $\frac{11}{16}$
45. 625   **47.** -61 **49.** 72   **51.** $-\frac{16}{13}$ **53.** 1, 1, 1   **55.** -5, 19, 355   **57.** $x^3 + 3x^2y + 3xy^2 + y^3$ **59.** $243r^5 + 405r^4s + 270r^3s^2 + 90r^2s^3 + 15rs^4 + s^5$ **61.** $-13{,}107{,}200x^9$ **63.** 43

CHAPTER 12   INVESTIGATING DISCRETE MATHEMATICS AND PROBABILITY

Pages 686–687   Lesson 11–6
5. 1, 2, 5, 14, 41, 122   **7.** 3, 11, 123   **9.** -7.2, -22.4, -52.8, -113.6, -235.2 **11.** 9, 14, 24, 44, 84, 164   **13.** 13, 18, 23, 28, 33, 38   **15.** $1, \frac{1}{2}, \frac{1}{3}, \frac{1}{4}, \frac{1}{5}, \frac{1}{6}$ **17.** 16, 142, 1276
19. -7, -16, -43 **21.** -1, -1, -1 **23.** 19, 71, 279, 1111, 4439   **25.** -3, -17, -73, -297, -1193 **27.** 67   **29.** 0.33, 0.333, 0.3333, ...; The values approach $\frac{1}{3}$.
31a. 1, 3, 6, 10, 15   **31b.** $a_{n+1} = a_n + (n + 1)$ **31c.** 3240
33. 27   **35.** $\frac{(y - 0.5)^2}{1} - \frac{(x - 3)^2}{11.25} = 1$

37a.
Numbers of Job-Related Injuries
```
x   x xxx x x xxx                        x
+---+---+---+---+---+---+---+
10  15  20  25  30  35  40
```

37b.
Stem	Leaf
1	0 3 5 6 7 9
2	1 3 4 5
3	9 9

$2 | 1 = 21$

37c. 20; 39; 21.75

Pages 692–694   Lesson 11–7
5a. $a_1 = 1$, $a_{n+1} = 3a_n$ **5b.** 81

7.

Pages 715–717   Lesson 12–1
7.

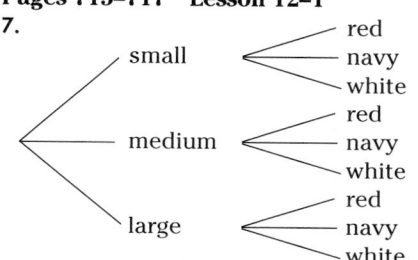

9 possibilities

9. dependent   **11.** 16 choices

13.

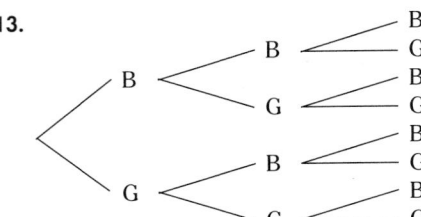

8 possibilities

15.

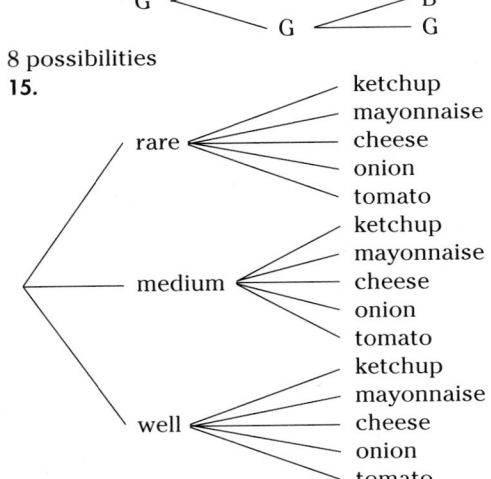

15 possibilities   **17.** dependent, if a person can hold only one office   **19.** dependent   **21.** 120 routes
23. 3125 passwords   **25.** 240 ways   **27a.** 10,140,000
27b. 1,872,000   **29.** 210 combinations   **31.** 96 meals
33. 27,600 ways   **35.** 11.26

37.

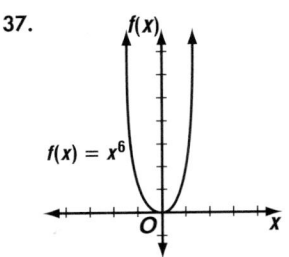

$f(x) = x^6$

39. $\dfrac{4\sqrt{3} + 4n\sqrt{6} + \sqrt{3n} + n\sqrt{6n}}{16 - n}$

Pages 723–725   Lesson 12–2
7. 24   **9.** 3   **11.** linear, not reflection; 362,880   **13a.** Order is important since each place winner receives a different cash prize and awards.   **13b.** $\dfrac{52}{\text{winner}}, \dfrac{51}{\text{1st}}, \dfrac{50}{\text{2nd}}, \dfrac{49}{\text{3rd}}$
13c. 6,497,400 ways   **15.** 5040   **17.** 3360   **19.** 12   **21.** 120
23. 2520   **25.** 453,600   **27.** circular, not reflection; 24
29. circular, not reflection; 5040   **31.** circular, reflection; 60   **33.** 72   **35.** 8   **37.** 42   **39.** 720 ways   **41.** 5040 ways
43. 1440 ways   **45.** 255 combinations   **47.** 13 ways
49. 12 hours   **51.** 2 or 0; 1; 2 or 4   **53.** $\begin{bmatrix} 15 & -8 & -10 \\ -7 & 23 & 16 \end{bmatrix}$
55. $p \le 400$

Pages 729–731   Lesson 12–3
7. combination   **9.** 6   **11.** 168   **13.** 840 bouquets
15. permutation   **17.** permutation   **19.** combination
21. permutation   **23.** 10   **25.** 70   **27.** 792   **29.** 27,720
31. 2100   **33.** 56 triangles   **35.** 12   **37.** 14
39. $C(n, r) = \dfrac{n!}{(n-r)!r!}$ or $\dfrac{n!}{(n-r)!} \cdot \dfrac{1}{r!}$; $P(n, r) = \dfrac{n!}{(n-r)!}$

By substituting, $C(n, r) = P(n, r) \cdot \dfrac{1}{r!}$ or $\dfrac{P(n, r)}{r!}$.   **41a.** 56

41b. 252   **41c.** 1680   **41d.** 2520   **43a.** 6   **43b.** 0   **43c.** 126
43d. 2808   **43e.** 810   **45.** 75,287,520 senatorial committees   **47.** 120   **49a.** $y = 100(1.08)^n$ **49b.** $342.59

51. 1.20×10^{25} molecules   **53.** $m = \dfrac{3}{4}$; $b = \dfrac{5}{2}$

Pages 735–738   Lesson 12–4
7. 3:1   **9.** $\dfrac{6}{11}$ **11.** 1:4   **13.** $\dfrac{1}{7}$ **15.** $\dfrac{4}{7}$ **17.** 1:1   **19.** 11:1
21. 1:4   **23.** $\dfrac{6}{7}$ **25.** $\dfrac{5}{11}$ **27.** $\dfrac{1}{9}$ **29a.** $\dfrac{1}{4}$ **29b.** $\dfrac{2}{5}$
29c. 80 jelly beans   **31.** $\dfrac{21}{55}$, 21:34   **33.** $\dfrac{14}{575}$ **35.** $\dfrac{7}{115}$
37. $\dfrac{24}{115}$ **39.** $\dfrac{40\pi}{225}$ or about 56%   **41.** $\dfrac{\pi - 1}{\pi}$ **43.** $\dfrac{1}{8}$ **45a.** $\dfrac{1}{10}$
45b. 1   **45c.** $\dfrac{21}{50}$ **45d.** 0   **47.** $\dfrac{3}{16}$ **49.** 420   **51.** $\dfrac{1}{9}, \dfrac{1}{3}, 1$
53. $f(m) = 4m^2 - 20m + 25$ **55.** $(0, 0), (1, 2), (-3, 1)$
57. $-9 \le x \le 24$

Page 738   Self Test
1a. tree diagram   **1b.** 24   **3.** 14,400 ways   **5.** permutation
7. 1320   **9.** 1716 teams

Pages 742–745   Lesson 12–5
7. dependent, $\dfrac{21}{220}$ **9a.** independent, $\dfrac{8}{243}$
9b. dependent, $\dfrac{1}{21}$ **11.** $\dfrac{1}{4}$ **13.** dependent, $\dfrac{3}{28}$
15. independent, $\dfrac{81}{2401}$ **17.** $\dfrac{10}{21}$ **19a.** (2, 6), (6, 2), (4, 4), (3, 5), (5, 3)   **19b.** (4, 4) is one of 5 possible outcomes when rolling a sum of 8.   **21.** $\dfrac{2}{15}$ **23.** $\dfrac{1}{5}$ **25.** $\dfrac{1}{36}$ **27.** $\dfrac{1}{36}$
29. $\dfrac{1}{6}$ **31a.** $\dfrac{7}{170}$ **31b.** $\dfrac{168}{4913}$ **33.** $\dfrac{1}{635,013,559,600}$
35. $\dfrac{1}{158,753,389,900}$ **37a.** $\dfrac{1}{12}$ **37b.** $\dfrac{1}{6}$ **37c.** 6 outcomes;
GR, GB, BR, BG, RG, RB   **39a.** $\dfrac{1}{12} \cdot \dfrac{1}{11} \cdot \dfrac{1}{10}$ or $\dfrac{1}{1320}$
39b. $\left(\dfrac{1}{12}\right)^5$ or $\dfrac{1}{248,832}$ **41a.** $\dfrac{1}{204}$ **41b.** $\dfrac{9}{340}$ **41c.** $\dfrac{6}{119}$ **43.** 0, 1
45. 276 milligrams   **47.** $-\dfrac{7}{8}$

Pages 748–751   Lesson 12–6
5a. mutually exclusive, $\dfrac{2}{13}$ **5b.** inclusive, $\dfrac{4}{13}$ **7.** $\dfrac{8}{26}$
9. mutually exclusive, $\dfrac{5}{9}$ **11.** inclusive, $\dfrac{1}{2}$ **13.** $\dfrac{5}{14}$
15. $\dfrac{25}{42}$ **17.** $\dfrac{55}{221}$ **19.** $\dfrac{188}{663}$ **21.** $\dfrac{35}{143}$ **23.** $\dfrac{3}{143}$ **25.** $\dfrac{1}{780}$
27. $\dfrac{11}{156}$ **29.** $P(A), P(B), P(C), P(A \text{ and } B), P(B \text{ and } C),$
$P(A \text{ and } C), P(A \text{ and } B \text{ and } C); P(A \text{ or } B \text{ or } C) = P(A) + P(B) + P(C) - P(A \text{ and } B) - P(B \text{ and } C) - P(A \text{ and } C) + P(A \text{ and } B \text{ and } C)$ **31.** 96%
33a.

	First Serve	Second Serve	Point
75%	in 80%	yes	
		20%	no
	90% in	35%	yes
25%	out	65%	no
	10% out		

33b. 67.9%   **33c.** 88.4%   **35a.** $\dfrac{1}{204}$ **35b.** $\dfrac{1}{119}$ **35c.** $\dfrac{3}{340}$
37. $\dfrac{1 - 2x}{3x + 1}$ **39a.** $\dfrac{x^2}{2304} + \dfrac{y^2}{529} = 1$ **39b.** The desk is at one focus point; about 84 feet.

Pages 756–757   Lesson 12–7
9a. binomial, $\frac{1}{28,561}$ **9b.** not binomial   **11.** $\frac{3024}{15,625} \approx 0.194$

13. $\frac{3}{8} = 0.375$ **15.** $\frac{3125}{7776} \approx 0.402$ **17.** $\frac{625}{648} \approx 0.965$

19. $\frac{1}{64} \approx 0.016$ **21.** $\frac{45}{512} \approx 0.088$ **23.** $\frac{319}{512} \approx 0.623$
25. ≈ 0.309 **27.** ≈ 0.031 **29.** ≈ 0.0000168 **33.** about 0.353
35. 2.4330; 2; 0.5440   **37.** $\sqrt{37}$ units   **39.** $\begin{bmatrix} -2 & 9 & 22 \\ 20 & 12 & -1 \end{bmatrix}$

Pages 760–762   Lesson 12–8
3. yes   **5.** 4%; There is a 95% probability that the value of
p in the population is between $68 - 4$ or 64% and $68 + 4$
or 72% and that 64% to 72% of high school students are
involved in extracurricular activities.   **7a.** There is a 95%
probability that the value of p in the population is
between $77 - 5$ or 72% and $77 + 5$ or 82% and that 72%
to 82% of Americans age 12 or older listen to the radio
every day.   **7b.** 283 people   **9.** yes   **11.** no   **13.** yes
15. 2%; There is a 95% probability that the value of p in
the population is between $33 - 2$ or 31% and $33 + 2$ or
35% and that 33% to 35% of adults believe that skim milk
is a good calcium source.   **19a.** There is a 95% probability
that the value of p in the population is between $90 - 2$ or
88% and $90 + 2$ or 92% and that 88% to 92% of people
registered to vote actually vote on election day.

19b. 900 people   **23.** 72   **25.** $\{x \mid -\sqrt{6} \le x \le \sqrt{6}\}$

Page 763   Chapter 12   Highlights
1. e   **3.** b   **5.** d   **7.** i   **9.** f   **11.** h

**Pages 764–766   Chapter 12   Study Guide and
Assessment**
13. 3,628,800   **15.** 5040 patterns   **17.** 19,958,400
19. 604,800   **21.** 462 ways   **23.** 28 pizzas   **25.** $\frac{5}{42}$ **27.** $\frac{1}{7}$

29. mutually exclusive, $\frac{1}{2}$ **31.** mutually exclusive, $\frac{1}{2}$ **33.** $\frac{1}{32}$

35a. $\frac{1}{2,176,782,336}$ **35b.** $\frac{585,937,500}{2,176,782,336}$ **35c.** $\frac{14,437,500}{2,176,782,336}$

37. 460 mothers   **39.** 1260 varieties

CHAPTER 13   EXPLORING TRIGONOMETRIC FUNCTIONS

Pages 777–779   Lesson 13–1
7. $\sin \theta \approx 0.8944$, $\cos \theta \approx 0.4472$, $\csc \theta \approx 1.1180$,
$\sec \theta \approx 2.2361$, $\cot \theta = 0.5$ **9.** $\tan x° = \frac{33}{15}$, $x \approx 66°$
11. $8°$ **13.** $b = 5$, $A \approx 67°$, $B \approx 23°$ **15.** about 300 feet
17. $\sin \theta \approx 0.9682$, $\tan \theta \approx 3.8730$, $\csc \theta \approx 1.0328$,
$\sec \theta = 4$, $\cot \theta \approx 0.2582$ **19.** $\sin \theta = 0.4$, $\cos \theta \approx 0.9165$,
$\tan \theta \approx 0.4364$, $\sec \theta \approx 1.0911$, $\cot \theta \approx 2.2913$
21. $\cos \theta \approx 0.8660$, $\tan \theta \approx 0.5774$, $\sec \theta \approx 1.1547$,
$\csc \theta = 2$, $\cot \theta \approx 1.7321$ **23.** $\sin 54° = \frac{17.8}{x}$, $x \approx 22.0$

25. $\tan x° = \frac{15}{21}$, $x \approx 36°$ **27.** $\sin x° = \frac{18}{33}$, $x \approx 33°$

29. $58°$ **31.** $89°$ **33.** $30°$ **35.** $a \approx 3.9$, $b \approx 13.5$, $B = 74°$
37. $b \approx 1.3$, $c \approx 4.1$, $A = 72°$ **39.** $B = 45°$, $a = 7$, $b = 7$
41. $a \approx 19.1°$, $c = 22$, $A = 60°$ **43.** $b \approx 13.7$, $c \approx 15.4$,
$B = 63°$ **45.** $b \approx 14.1$, $A \approx 19.5°$, $B \approx 70.5°$ **47.** 4.05 cm
49. Sample answer: The legs of a right triangle are never
greater than the hypotenuse, but one leg may be greater
than the other leg.   **51.** 99 feet   **53.** $58.3°$ **55.** no   **57.** $\frac{5}{6}$
59. 100   **61.** x^4 **63.** about 2.99×10^1 or 2.99

Pages 783–785   Lesson 13–2
7. a   **9.** g   **11.** $-\frac{2\pi}{3}$ **13.** $120°$ **15.** $286.48°$ **17.** Sample

answer: $\frac{9\pi}{4}$, $-\frac{7\pi}{4}$ **19.** $-60°$, $-\frac{\pi}{3}$ **21.** π **23.** $\frac{20\pi}{3}$

25. $-\frac{40\pi}{9}$ **27.** 3π **29.** $180°$ **31.** $-480°$ **33.** $900°$
35. $401.07°$ **37.** $-85.94°$ **39.** Sample answer: $670°$, $-50°$
41. Sample answer: $\frac{\pi}{4}$, $-\frac{7\pi}{4}$ **43.** Sample answer: $360°$,
$-360°$ **45.** Sample answer: $180°$, $-180°$
47. $a^2 + (-b)^2 = a^2 + b^2 = 1$, $b^2 + a^2 = a^2 + b^2 = 1$,
$b^2 + (-a)^2 = a^2 + b^2 = 1$ **49.** $\frac{4\pi}{3}$ radians/second
51. 129 meters   **53.** 35   **55.** no solution
57. $x^2 - 10x + 24 = 0$ **59.** additive identity

Pages 790–791   Lesson 13–3
7. 0   **9.** $\sin \theta = -\frac{8\sqrt{65}}{65}$, $\cos \theta = \frac{\sqrt{65}}{65}$, $\tan \theta = -8$
11. $-\frac{1}{2}$ **13.** $-\frac{\sqrt{2}}{2}$ **15.** $\sin \theta = \frac{2\sqrt{5}}{5}$, $\cos \theta = \frac{\sqrt{5}}{5}$,
$\cot \theta = \frac{1}{2}$, $\csc \theta = \frac{\sqrt{5}}{2}$, $\sec \theta = \sqrt{5}$ **17.** $-$ **19.** $-$
21. $+$ **23.** undefined   **25.** $-$ **27.** $\sin \theta = 0$, $\cos \theta = -1$,
$\tan \theta = 0$ **29.** $\sin \theta = -\frac{3\sqrt{34}}{34}$, $\cos \theta = \frac{5\sqrt{34}}{34}$, $\tan \theta = -\frac{3}{5}$
31. $\sin \theta = \frac{\sqrt{2}}{2}$, $\cos \theta = \frac{\sqrt{2}}{2}$, $\tan \theta = 1$ **33.** $\frac{1}{2}$ **35.** $-\frac{\sqrt{3}}{2}$
37. $\frac{1}{2}$ **39.** 1   **41.** $\sin \theta = \frac{\sqrt{3}}{2}$, $\tan \theta = -\sqrt{3}$, $\csc \theta = \frac{2\sqrt{3}}{3}$,
$\sec \theta = -2$, $\cot \theta = -\frac{\sqrt{3}}{3}$ **43.** $\sin \theta = -\frac{3\sqrt{10}}{10}$,
$\cos \theta = -\frac{\sqrt{10}}{10}$, $\cot \theta = \frac{1}{3}$, $\csc \theta = -\frac{\sqrt{10}}{3}$,
$\sec \theta = -\sqrt{10}$ **45.** $\sin \theta = \frac{\sqrt{26}}{26}$, $\cos \theta = -\frac{5\sqrt{26}}{26}$,
$\tan \theta = -\frac{1}{5}$, $\csc \theta = \sqrt{26}$, $\sec \theta = -\frac{\sqrt{26}}{5}$
47. $\sin \theta = \pm\frac{\sqrt{5}}{3}$ **49.** $\sin \theta = \pm 1$, $\tan \theta =$ undefined
51. II   **53.** $\dfrac{\cos \theta}{\sin \theta} = \dfrac{\frac{x}{r}}{\frac{y}{r}} = \dfrac{x}{r} \cdot \dfrac{r}{y} = \dfrac{x}{y} = \cot \theta$ **55.** 8 feet
57. $a \approx 10.8$, $A = 31°$, $B = 59°$ **59.** $\frac{8}{9}$ **61.** 100 times
63. $x^3 - 14x^2 + 68x - 120$ **65.** 8

Pages 796–798 Lesson 13–4
7. $\frac{1}{2} bc \sin A$, 135.9   **9.** $B = 80°$, $a \approx 13.1$, $c \approx 17.6$
11. two; $B \approx 51.6°$, $C \approx 92.4°$, $c \approx 10.2$; $B \approx 128.4°$, $C \approx$
$15.6°$, $c \approx 2.7$ **13.** one; $A \approx 44.8°$, $B \approx 37.2°$, $b \approx 55.0$
15. $\frac{1}{2}ab \sin C$, 65.4   **17.** $\frac{1}{2} ab \sin C$, 229.5

19. $\frac{1}{2} ac \sin B$, 37.4   **21.** $C = 74°$, $b \approx 8.9$, $c \approx 10.2$

23. $B \approx 35.3°$, $A \approx 34.7°$, $a \approx 7.9$ **25.** $A = 67°$, $b \approx 47.2$,
$c \approx 30.1$ **27.** $B = 80°$, $a \approx 4.3$, $b \approx 8.4$ **29.** $B \approx 31.5°$,
$C \approx 45.5°$, $c \approx 10.3$ **31.** no solution   **33.** no solution
35. no solution   **37.** no solution   **39.** two solutions;
$B \approx 53.5°$, $C \approx 86.5°$, $c \approx 12.4$; $B \approx 126.5°$, $C \approx 13.5°$,
$c \approx 2.9$ **41.** There is no solution.   **43.** 46.8 miles to
Norfolk   **45.** 4.9 and 10.4 miles   **47.** $\sin 180° = 0$,
$\cos 180° = -1$, $\tan 180° = 0$, $\csc 180°$ undefined,
$\sec 180° = -1$, $\cot 180°$ undefined   **49.** 30,875   **51.** 1
53. $(2x + 3)(x - 7)$

Page 798 Self Test

1. $B = 41°$, $c \approx 9.3$, $b \approx 6.1$ **3.** $\frac{5\pi}{6}$ **5.** $270°$ **7.** $114.6°$
9. 82.1 watts

Pages 802–804 Lesson 13–5
7. cosines; $c \approx 11.5$, $B \approx 81.6°$, $A \approx 56.4°$ **9.** sines; $b \approx$
21.0, $C \approx 44.3°$, $B \approx 101.7°$ **11.** $35.7°$ **13.** cosines; $c \approx$
6.5, $A \approx 76.1°$, $B \approx 68.9°$ **15.** cosines; $A \approx 46.6°$, $B \approx$
73.8°, $C \approx 59.6°$ **17.** cosines; $A = 90°$, $B \approx 67.4°$,
$C \approx 22.6°$ **19.** cosines; $b \approx 17.9$, $A \approx 54.7°$, $C \approx 78.3°$
21. sines; $C = 81°$, $a \approx 9.1$, $b \approx 12.2$ **23.** sines; $b \approx 31.7$,
$C \approx 77.0°$, $c \approx 31.6$ **25.** cosines; $A \approx 30.0°$, $B \approx 69.2°$, $C \approx$
80.8° **27.** cosines; $a \approx 2.4$, $B \approx 78.0°$, $C \approx 74.0°$
29. cosines; $A \approx 14.6°$, $B \approx 130.8°$, $C \approx 34.6°$ **31.** 26.4 cm
33a. 4.08 **33b.** 14.15 **33c.** 9.12 **35.** 1434 feet; 86,756.06
square feet **37.** about 561 miles **39.** $-\frac{\pi}{4}$
41.

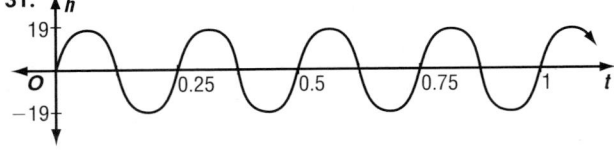

43. $8, -12$ **45a.** Sample answer: $y = 249x + 5000$
45b. Sample answer based on equation in Exercise 45a:
28,157

Pages 808–810 Lesson 13–6
5. $\sin \theta = \frac{4}{5}$, $\cos \theta = -\frac{3}{5}$, $\tan \theta = -\frac{4}{3}$ **7.** 9 **9.** $-\frac{\sqrt{3}}{2}$

11. $\sin \theta = \frac{\sqrt{2}}{2}$, $\cos \theta = \frac{\sqrt{2}}{2}$, $\tan \theta = 1$ **13.** $\sin \theta = 0$,
$\cos \theta = 1$, $\tan \theta = 0$ **15.** $720°$ **17.** $-\frac{\sqrt{3}}{2}$ **19.** 0 **21.** -1

23. 1 **25.** $-\frac{\sqrt{3}}{3}$ **27.** 1
31.

33. about 228 miles **35.** No; the figure is not self-similar.
37. $-3, 3, -2, 2$ **39.** $(2, -2)$

Pages 813–814 Lesson 13–7
5. $y = \text{Arctan } -3$ **7.** $45°$ **9.** $\frac{4}{5}$ **11.** $-45°$ **13.** $90°$
15. $b = \text{Arccos } a$ **17.** $\alpha = \text{Arctan } \beta$ **19.** $\text{Arccos } y = 45°$
21. $90°$ **23.** $30°$ **25.** $30°$ **27.** $120°$ **29.** $0°$ **31.** $\frac{\sqrt{11}}{5}$
33. $\frac{1}{2}$ **35.** 3 **37.** $60°$ **39.** $\frac{1}{2}$ **41.** $\frac{\sqrt{2}}{2}$ **43.** $\frac{\sqrt{3}}{2}$ **47.** $60°$
49. 61.1 meters **51.** 0.2392 **53.** 0, 7
55.

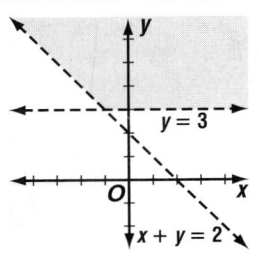

Page 815 Chapter 13 Highlights
1. false, coterminal **3.** true **5.** true **7.** true **9.** false,
$\tan \theta$

**Pages 816–818 Chapter 13 Study Guide and
Assessment**
11. $\sin \theta \approx 0.9911$, $\tan \theta \approx 7.4330$, $\csc \theta \approx 1.0090$,
$\sec \theta = 7.5$, $\cot \theta \approx 0.1345$ **13.** $\cos \theta \approx 0.6614$,
$\tan \theta \approx 1.1339$, $\csc \theta \approx 1.3333$, $\sec \theta \approx 1.5119$,
$\cot \theta \approx 0.8819$ **15.** $15°$ **17.** $75°$ **19.** $b \approx 14.4$, $A \approx 26°$,
$B \approx 64°$ **21.** $A = 45°$, $a \approx 8.5$, $b \approx 8.5$ **23.** $A \approx 41°$,
$b \approx 10.4$, $c \approx 13.7$ **25.** $\frac{17\pi}{12}$ **27.** $\frac{13\pi}{36}$ **29.** $315°$ **31.** $-720°$
33. $-\frac{\sqrt{3}}{2}$ **35.** $-\frac{\sqrt{2}}{2}$ **37.** $\frac{2\sqrt{3}}{3}$ **39.** $\sin \theta = \frac{5\sqrt{29}}{29}$
41. $B = 67°$, $C = 63°$, $c = 11.7$ **43.** $C = 105°$, $a = 28.3$,
$c = 38.6$ **45.** no solution **47.** one solution; $B = 38°$,
$A = 76°$, $a = 15.9$ **49.** $c \approx 6.4$, $A \approx 35°$, $B \approx 80°$
51. $c \approx 4.5$, $A \approx 59°$, $B \approx 81°$ **53.** $A \approx 45°$, $B \approx 58°$,
$C \approx 77°$ **55.** $\frac{1}{2}$ **57.** $-\frac{\sqrt{2}}{2}$ **59.** $-\sqrt{3}$ **61.** $30°$ **63.** 0
65. $\frac{15}{8}$ **67.** $55.75°$

**CHAPTER 14 USING TRIGONOMETRIC GRAPHS
AND IDENTITIES**

Page 824 Lesson 14–1A
1.

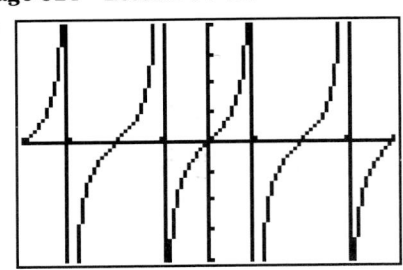

3.

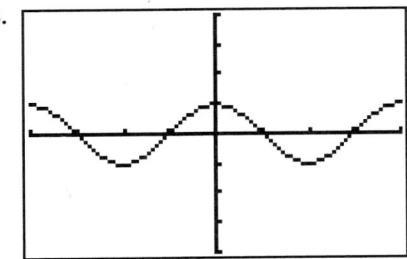

5.

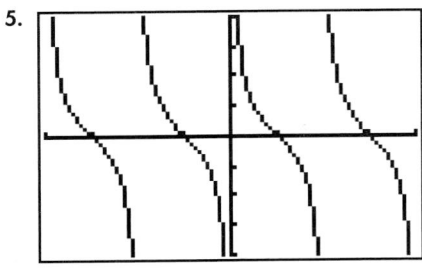

7.

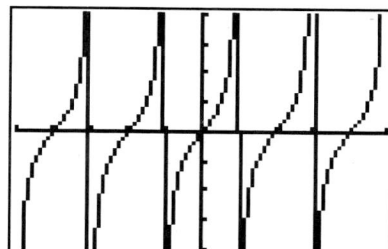

9.

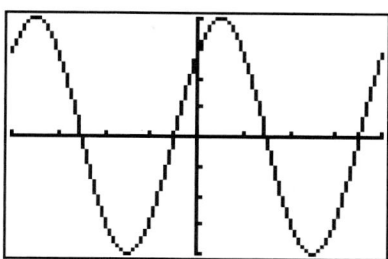

11.

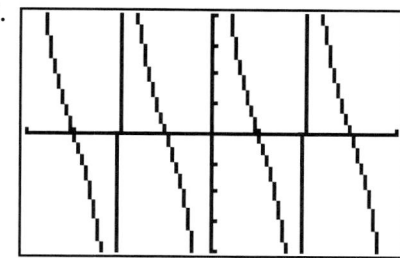

13.

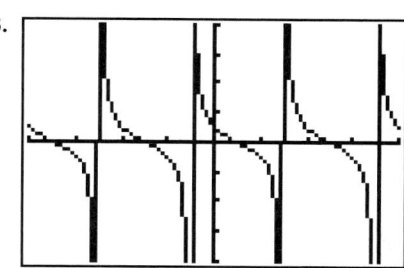

Pages 832–834 Lesson 14–1

7. a **9.** b **11.** 6, 540°

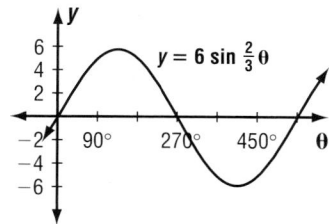

$y = 6 \sin \frac{2}{3} \theta$

13. none, 120°

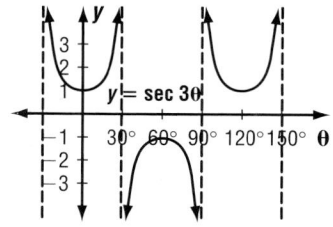

$y = \sec 3\theta$

15. none, 36°

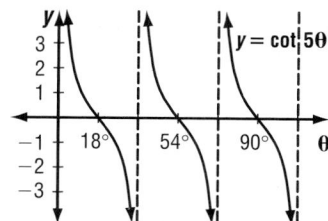

$y = \cot 5\theta$

17. c **19.**

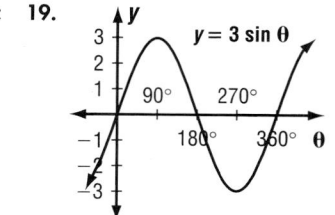

$y = 3 \sin \theta$

21.

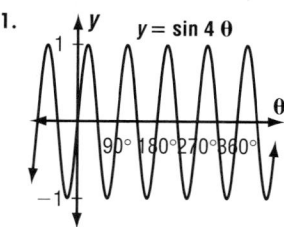

$y = \sin 4\theta$

23.

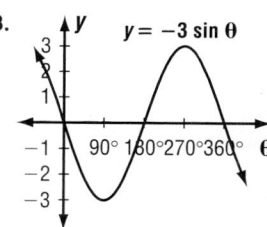

$y = -3 \sin \theta$

25.

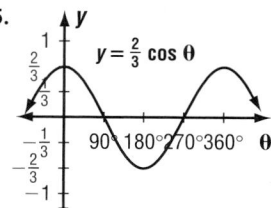

$y = \frac{2}{3} \cos \theta$

27.

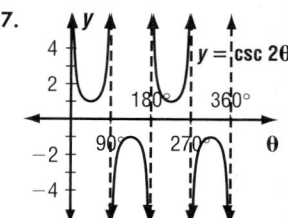

$y = \csc 2\theta$

29.

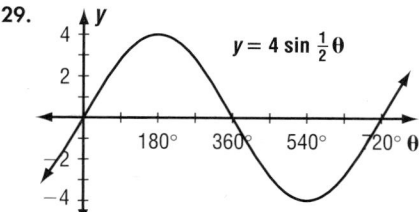

$y = 4 \sin \frac{1}{2}\theta$

31.

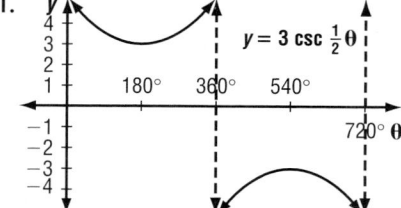

$y = 3 \csc \frac{1}{2}\theta$

33.

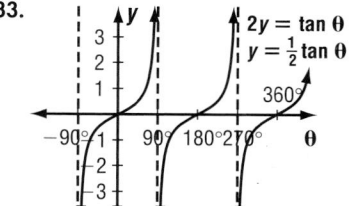

$2y = \tan \theta$
$y = \frac{1}{2} \tan \theta$

35.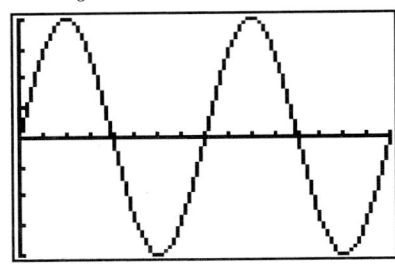

$\frac{3}{4}y = \frac{2}{3}\sin\frac{3}{5}\theta$

$y = \frac{8}{9}\sin\frac{3}{5}\theta$

37. $y = 0.6\cos\frac{1}{2}\theta$ **39.** $y = \frac{1}{3}\cos 4\theta$ **41.** $y = 6.7\sin 3\theta$

43. $y = \cos 2\theta$ **45a.**

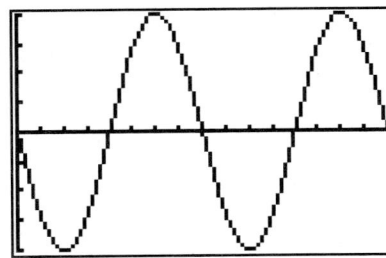

45b.

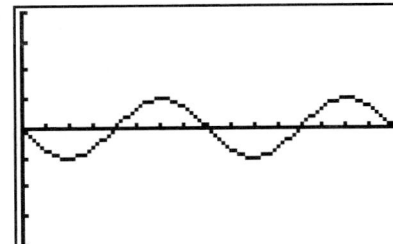

45c.

45d. The greater the absolute value of a, the greater the amplitude. If a is negative, then the curve is the same as would be obtained by reflecting the curve $y = -a\sin\theta$ about the x-axis.

47a. 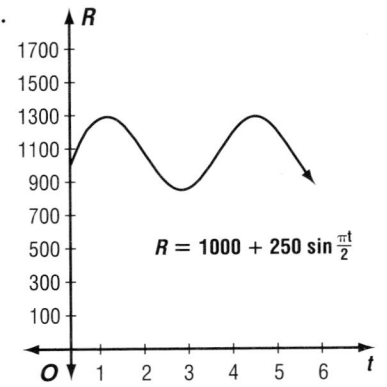   **47b.** 1000

$R = 1000 + 250\sin\frac{\pi t}{2}$

47c. 1250; January 1, 1991   **47d.** 750; January 1, 1993

49. $\frac{8}{17}$ **51.** 2.375, 4.75   **53.** $\pm\frac{\sqrt{6}}{2}$ **55.** $16^{\frac{1}{3}}a^{\frac{5}{3}}b^{\frac{7}{3}}$

57. Sample answer: $a(b + c) = ab + ac$

Pages 839–840   Lesson 14–2

5. $\frac{3}{5}$ **7.** $\frac{3\sqrt{5}}{5}$ **9.** $\sin\theta\cos\theta$ **11.** $\cot^2 x$ **13.** $\frac{5}{4}$ **15.** $-\frac{4}{5}$

17. $-\frac{5}{4}$ **19.** $-\frac{\sqrt{17}}{4}$ **21.** $-\frac{12}{13}$ **23.** $\frac{3}{5}$ **25.** 1   **27.** $\tan^2\theta$

29. $\sec\beta$ **31.** 1   **33.** $\csc\beta$ **35.** 1

37. $1 + \tan^2\theta = 1 + \frac{\sin^2\theta}{\cos^2\theta}$

$= \frac{\cos^2\theta}{\cos^2\theta} + \frac{\sin^2\theta}{\cos^2\theta}$

$= \frac{\cos^2\theta + \sin^2\theta}{\cos^2\theta}$

$= \frac{1}{\cos^2\theta}$

$= \sec^2\theta$

39. $\sec\alpha - \cos\alpha = \frac{1}{\cos\alpha} - \cos\alpha$

$= \frac{1}{\cos\alpha} - \frac{\cos^2\alpha}{\cos\alpha}$

$= \frac{1 - \cos^2\alpha}{\cos\alpha}$

$= \frac{\sin^2\alpha}{\cos\alpha}$

$= \sin\alpha \cdot \frac{\sin\alpha}{\cos\alpha}$

$= \sin\alpha \cdot \tan\alpha$

41. $\frac{9}{16}$ **43.** about $14.5°$ **45.**

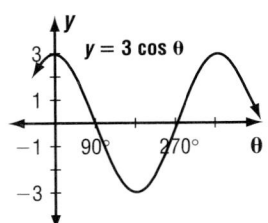

$y = 3\cos\theta$

47. 792   **49.** 18, 23, 28   **51.** 3, -7

Page 841   Lesson 14–3A

1. yes   **3.** yes   **5.** yes   **7.** no   **9.** yes   **11.** yes

Pages 845–846   Lesson 14–3

3. $\sin\theta\sec\theta\cot\theta \overset{?}{=} 1$

$\sin\theta \cdot \frac{1}{\cos\theta} \cdot \frac{\cos\theta}{\sin\theta} \overset{?}{=} 1$

$1 = 1$

5. $\csc y\sec y \overset{?}{=} \cot y + \tan y$

$\csc y\sec y \overset{?}{=} \frac{\cos y}{\sin y} + \frac{\sin y}{\cos y}$

$\csc y\sec y \overset{?}{=} \frac{\sin^2 y + \cos^2 y}{\sin y\cos y}$

$\csc y\sec y \overset{?}{=} \frac{1}{\sin y\cos y}$

$\csc y\sec y \overset{?}{=} \csc y\sec y$

7. $\frac{\frac{1}{\cos\beta} + \frac{1}{\sin\beta}}{1 + \frac{\sin\beta}{\cos\beta}} \overset{?}{=} \csc\beta$

$\frac{\frac{\sin\beta + \cos\beta}{\sin\beta\cos\beta}}{\frac{\sin\beta + \cos\beta}{\cos\beta}} \overset{?}{=} \csc\beta$

$\frac{\sin\beta + \cos\beta}{\sin\beta\cos\beta} \cdot \frac{\cos\beta}{\sin\beta + \cos\beta} \overset{?}{=} \csc\beta$

$\frac{\cos\beta}{\sin\beta\cos\beta} \overset{?}{=} \csc\beta$

$\frac{1}{\sin\beta} \overset{?}{=} \csc\beta$

$\csc\beta = \csc\beta$

9. about 20 cm **11.** $\cos^2 \theta + \sin^2 \theta \stackrel{?}{=} 1$

$$1 = 1$$

13.
$$\dfrac{\dfrac{1}{\cos \alpha}}{\sin \alpha} - \dfrac{\sin \alpha}{\cos \alpha} \stackrel{?}{=} \cot \alpha$$

$$\dfrac{1}{\sin \alpha \cos \alpha} - \dfrac{\sin^2 \alpha}{\sin \alpha \cos \alpha} \stackrel{?}{=} \cot \alpha$$

$$\dfrac{1 - \sin^2 \alpha}{\sin \alpha \cos \alpha} \stackrel{?}{=} \cot \alpha$$

$$\dfrac{\cos^2 \alpha}{\sin \alpha \cos \alpha} \stackrel{?}{=} \cot \alpha$$

$$\dfrac{\cos \alpha}{\sin \alpha} \stackrel{?}{=} \cot \alpha$$

$$\cot \alpha = \cot \alpha$$

15.
$$\dfrac{\sin \theta}{\sec \theta} \stackrel{?}{=} \dfrac{1}{\dfrac{\sin \theta}{\cos \theta} + \dfrac{\cos \theta}{\sin \theta}}$$

$$\dfrac{\sin \theta}{\sec \theta} \stackrel{?}{=} \dfrac{1}{\dfrac{\sin^2 \theta + \cos^2 \theta}{\sin \theta \cos \theta}}$$

$$\dfrac{\sin \theta}{\sec \theta} \stackrel{?}{=} \dfrac{\sin \theta \cos \theta}{\sin^2 \theta + \cos^2 \theta}$$

$$\dfrac{\sin \theta}{\sec \theta} \stackrel{?}{=} \dfrac{\sin \theta \cos \theta}{1}$$

$$\dfrac{\sin \theta}{\sec \theta} = \dfrac{\sin \theta}{\sec \theta}$$

17.
$$\dfrac{\sec \theta + 1}{\tan \theta} \stackrel{?}{=} \dfrac{\tan \theta}{\sec \theta - 1} \cdot \dfrac{\sec \theta + 1}{\sec \theta + 1}$$

$$\dfrac{\sec \theta + 1}{\tan \theta} \stackrel{?}{=} \dfrac{\tan \theta \cdot (\sec \theta + 1)}{\sec^2 \theta - 1}$$

$$\dfrac{\sec \theta + 1}{\tan \theta} \stackrel{?}{=} \dfrac{\tan \theta \cdot (\sec \theta + 1)}{\tan^2 \theta}$$

$$\dfrac{\sec \theta + 1}{\tan \theta} = \dfrac{\sec \theta + 1}{\tan \theta}$$

19. $\cos^2 x + \tan^2 x \cos^2 x \stackrel{?}{=} 1$

$$\cos^2 x + \dfrac{\sin^2 x}{\cos^2 x} \cdot \cos^2 x \stackrel{?}{=} 1$$

$$\cos^2 x + \sin^2 x \stackrel{?}{=} 1$$

$$1 = 1$$

21.
$$\dfrac{1 + \tan^2 \theta}{\csc^2 \theta} \stackrel{?}{=} \tan^2 \theta$$

$$\dfrac{\sec^2 \theta}{\csc^2 \theta} \stackrel{?}{=} \tan^2 \theta$$

$$\dfrac{\dfrac{1}{\cos^2 \theta}}{\dfrac{1}{\sin^2 \theta}} \stackrel{?}{=} \tan^2 \theta$$

$$\dfrac{1}{\cos^2 \theta} \cdot \sin^2 \theta \stackrel{?}{=} \tan^2 \theta$$

$$\tan^2 \theta = \tan^2 \theta$$

23.
$$\dfrac{\cos y}{1 + \sin y} \cdot \dfrac{1 - \sin y}{1 - \sin y} + \dfrac{\cos y}{1 - \sin y} \cdot \dfrac{1 + \sin y}{1 + \sin y} \stackrel{?}{=} 2 \sec y$$

$$\dfrac{\cos y(1 - \sin y) + \cos y(1 + \sin y)}{(1 + \sin y)(1 - \sin y)} \stackrel{?}{=} 2 \sec y$$

$$\dfrac{\cos y - \sin y \cos y + \cos y + \sin y \cos y}{1 - \sin^2 y} \stackrel{?}{=} 2 \sec y$$

$$\dfrac{2 \cos y}{\cos^2 y} \stackrel{?}{=} 2 \sec y$$

$$\dfrac{2}{\cos y} \stackrel{?}{=} 2 \sec y$$

$$2 \sec y = 2 \sec y$$

25. $\cot x(\cot x + \tan x) \stackrel{?}{=} \csc^2 x$

$$\cot^2 x + \cot x \tan x \stackrel{?}{=} \csc^2 x$$

$$\csc^2 x - 1 + \dfrac{\sin x}{\cos x} \cdot \dfrac{\cos x}{\sin x} \stackrel{?}{=} \csc^2 x$$

$$\csc^2 x - 1 + 1 \stackrel{?}{=} \csc^2 x$$

$$\csc^2 x = \csc^2 x$$

27.
$$\dfrac{1 + \tan \alpha}{1 + \cot \alpha} \stackrel{?}{=} \dfrac{\sin \alpha}{\cos \alpha}$$

$$\dfrac{1 + \dfrac{\sin \alpha}{\cos \alpha}}{1 + \dfrac{\cos \alpha}{\sin \alpha}} \stackrel{?}{=} \dfrac{\sin \alpha}{\cos \alpha}$$

$$\dfrac{\dfrac{\sin \alpha + \cos \alpha}{\cos \alpha}}{\dfrac{\sin \alpha + \cos \alpha}{\sin \alpha}} \stackrel{?}{=} \dfrac{\sin \alpha}{\cos \alpha}$$

$$\dfrac{\sin \alpha + \cos \alpha}{\cos \alpha} \cdot \dfrac{\sin \alpha}{\sin \alpha + \cos \alpha} \stackrel{?}{=} \dfrac{\sin \alpha}{\cos \alpha}$$

$$\dfrac{\sin \alpha}{\cos \alpha} = \dfrac{\sin \alpha}{\cos \alpha}$$

29. $1 + \sec^2 x \sin^2 x \stackrel{?}{=} \sec^2 x$

$$1 + \dfrac{1}{\cos^2 x} \cdot \sin^2 x \stackrel{?}{=} \sec^2 x$$

$$1 + \tan^2 x \stackrel{?}{=} \sec^2 x$$

$$\sec^2 x = \sec^2 x$$

31. \$85.80 **33.** $\cot \alpha$ **35.** $\dfrac{1}{36}$ **37.** $x^7 y^7$

Page 846 Self Test
1. $1; \dfrac{\pi}{2}$ or $90°$ **3.** none; 5π or $900°$

5.

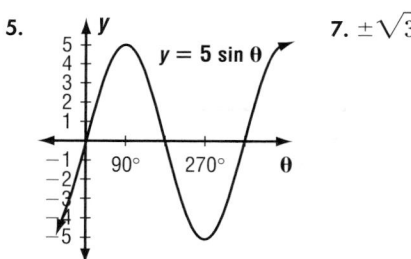

7. $\pm\sqrt{3}$

9. $(1 - \sin^2 \theta)(1 + \tan^2 \theta) \stackrel{?}{=} 1$

$$\cos^2 \theta\left(1 + \dfrac{\sin^2 \theta}{\cos^2 \theta}\right) \stackrel{?}{=} 1$$

$$\cos^2 \theta + \sin^2 \theta \stackrel{?}{=} 1$$

$$1 = 1$$

Pages 850–852 Lesson 14–4
3. $\dfrac{\sqrt{6} - \sqrt{2}}{4}$ **5.** $\dfrac{\sqrt{2} - \sqrt{6}}{4}$

7. $\sin(270° - \theta) \stackrel{?}{=} \sin 270° \cos \theta - \cos 270° \sin \theta$

$$\stackrel{?}{=} -1 \cos \theta - 0$$

$$\stackrel{?}{=} -\cos \theta$$

9. $\sin(x + 30°) + \cos(x + 60°)$

$$\stackrel{?}{=} \sin x \cos 30° + \cos x \sin 30° + \cos x \cos 60° - \sin x \sin 60°$$

$$\stackrel{?}{=} \dfrac{\sqrt{3}}{2} \sin x + \dfrac{1}{2} \cos x + \dfrac{1}{2} \cos x - \dfrac{\sqrt{3}}{2} \sin x$$

$$\stackrel{?}{=} \dfrac{1}{2} \cos x + \dfrac{1}{2} \cos x$$

$$= \cos x$$

11. $\dfrac{-\sqrt{2}-\sqrt{6}}{4}$ **13.** $\dfrac{-\sqrt{6}-\sqrt{2}}{4}$ **15.** $\dfrac{\sqrt{6}+\sqrt{2}}{4}$ **17.** $\dfrac{1}{2}$

19. $\dfrac{\sqrt{3}}{2}$

21. $\sin(90° + \theta) \stackrel{?}{=} \sin 90° \cos\theta + \cos 90° \sin\theta$

$\qquad \stackrel{?}{=} 1 \cdot \cos\theta + 0$

$\qquad = \cos\theta$

23. $\sin(90° - \theta) \stackrel{?}{=} \sin 90° \cos\theta - \cos 90° \sin\theta$

$\qquad = 1 \cdot \cos\theta - 0 \cdot \sin\theta$

$\qquad = \cos\theta$

25. $\sin(x + y)\sin(x - y)$

$\qquad \stackrel{?}{=} (\sin x \cos y + \cos x \sin y)(\sin x \cos y - \cos x \sin y)$

$\qquad \stackrel{?}{=} \sin^2 x \cos^2 y - \cos^2 x \sin^2 y$

$\qquad \stackrel{?}{=} \sin^2 x(1 - \sin^2 y) - (1 - \sin^2 x)\sin^2 y$

$\qquad \stackrel{?}{=} \sin^2 x - \sin^2 x \sin^2 y - \sin^2 y + \sin^2 x \sin^2 y$

$\qquad = \sin^2 x - \sin^2 y$

27. $-2 - \sqrt{3}$ **29.** $\sqrt{3}$ **31.** $90°$ **33.** $90°$

35. $\tan(\alpha + \beta) = \dfrac{\sin(\alpha + \beta)}{\cos(\alpha + \beta)}$

$\qquad = \dfrac{\sin\alpha\cos\beta + \cos\alpha\sin\beta}{\cos\alpha\cos\beta - \sin\alpha\sin\beta}$

$\qquad = \dfrac{\dfrac{\sin\alpha\cos\beta}{\cos\alpha\cos\beta} + \dfrac{\cos\alpha\sin\beta}{\cos\alpha\cos\beta}}{\dfrac{\cos\alpha\cos\beta}{\cos\alpha\cos\beta} - \dfrac{\sin\alpha\sin\beta}{\cos\alpha\cos\beta}}$

$\qquad = \dfrac{\tan\alpha + \tan\beta}{1 - \tan\alpha\tan\beta}$

37a. 0.5549 E **37b.** 0.3584 E **37c.** 0.9171 E

39. $\sin\theta\sec\theta\cot\theta \stackrel{?}{=} 1$

$\quad \sin\theta \cdot \dfrac{1}{\cos\theta} \cdot \dfrac{\cos\theta}{\sin\theta} \stackrel{?}{=} 1$

$\qquad\qquad\qquad 1 \stackrel{?}{=} 1$

41. $\dfrac{175}{221}$ **43.**

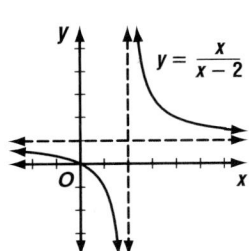

$y = \dfrac{x}{x - 2}$

45. 15 colonial, 45 ranch

Pages 857–859 Lesson 14–5

5. $-\dfrac{4\sqrt{6}}{25}, \dfrac{23}{25}, \dfrac{\sqrt{10}}{5}, -\dfrac{\sqrt{15}}{5}$ **7.** $\dfrac{\sqrt{2 - \sqrt{2}}}{2}$

9. $\dfrac{1}{\sin x \cos x} - \dfrac{\cos x}{\sin x} \stackrel{?}{=} \tan x$

$\qquad \dfrac{1 - \cos^2 x}{\sin x \cos x} \stackrel{?}{=} \tan x$

$\qquad \dfrac{\sin^2 x}{\sin x \cos x} \stackrel{?}{=} \tan x$

$\qquad\qquad \tan x = \tan x$

11. $-\dfrac{24}{25}, -\dfrac{7}{25}, \dfrac{2\sqrt{5}}{5}, \dfrac{\sqrt{5}}{5}$ **13.** $\dfrac{4\sqrt{2}}{9}, -\dfrac{7}{9}, \dfrac{\sqrt{6}}{3}, -\dfrac{\sqrt{3}}{3}$

15. $\dfrac{24}{25}, \dfrac{7}{25}, \dfrac{3\sqrt{10}}{10}, -\dfrac{\sqrt{10}}{10}$ **17.** $-\dfrac{4\sqrt{2}}{9}, -\dfrac{7}{9}, \dfrac{\sqrt{6}}{3}, -\dfrac{\sqrt{3}}{3}$

19. $\dfrac{\sqrt{2 + \sqrt{3}}}{2}$ **21.** $\dfrac{\sqrt{2 - \sqrt{2}}}{2}$

---

23. $\cos^2 2x + 4\sin^2 x \cos^2 x \stackrel{?}{=} 1$

$\qquad \cos^2 2x + \sin^2 2x \stackrel{?}{=} 1$

$\qquad\qquad\qquad\qquad 1 = 1$

25. $\sin 2x \stackrel{?}{=} 2\cot x \sin^2 x$

$\quad 2\sin x \cos x \stackrel{?}{=} 2\dfrac{\cos x}{\sin x} \cdot \sin^2 x$

$\quad 2\sin x \cos x = 2\sin x \cos x$

27. $2\cos^2 \dfrac{x}{2} \stackrel{?}{=} 1 + \cos x$

$\quad 2\left(\pm\sqrt{\dfrac{1 + \cos x}{2}}\right)^2 \stackrel{?}{=} 1 + \cos x$

$\quad 2\left(\dfrac{1 + \cos x}{2}\right) \stackrel{?}{=} 1 + \cos x$

$\qquad 1 + \cos x = 1 + \cos x$

29. $\pm\dfrac{\sqrt{18 - 3\sqrt{18 - 6\sqrt{5}}}}{6}$ **31.** $61.3°$

33. $b = 11.0,\ c = 12.2,\ C = 78$ **35.** 1.814 **37.** $h = \dfrac{3V}{\pi r^2}$

Page 860 Lesson 14–6A

1. $11.5°, 168.5°$ **3.** $0°, 60°, 180°, 300°$ **5.** $-170°, -51.6°,$
$185.2°, 308.4°$

Pages 864–867 Lesson 14–6

5. $15°, 75°, 195°, 255°$ **7.** $0°, 180°$ **9.** $\dfrac{\pi}{2} + k\pi, \dfrac{2\pi}{3} + 2k\pi,$

$\dfrac{4\pi}{3} + 2k\pi$ **11.** $\dfrac{\pi}{3} + 2k\pi, \dfrac{5\pi}{3} + 2k\pi$ **13.** $60°, 120°, 240°,$

$300°$ **15.** $90°, 270°$ **17.** $30°, 150°$ **19.** $\dfrac{7\pi}{6}, \dfrac{11\pi}{6}$ **21.** $\dfrac{4\pi}{3}, \dfrac{5\pi}{3}$

23. $\dfrac{\pi}{2}, \dfrac{7\pi}{6}, \dfrac{11\pi}{6}$ **25.** $45° + k \cdot 180°$ **27.** $0° + k \cdot 180°$

29. $120° + k \cdot 360°, 240° + k \cdot 360°$ **31.** $0° + k \cdot 180°,$

$90° + k \cdot 360°$ **33.** $\dfrac{7\pi}{6} + 2k\pi, \dfrac{11\pi}{6} + 2k\pi$

35. $\pi + 2k\pi, \dfrac{\pi}{3} + 2k\pi, \dfrac{5\pi}{3} + 2k\pi$ **37.** $\dfrac{2\pi}{3} + 2k\pi, \dfrac{4\pi}{3} + 2k\pi$

39. $0 + k\pi, \dfrac{\pi}{6} + 2k\pi, \dfrac{5\pi}{6} + 2k\pi$

41a. two times; $90°$ and $270°$ **41b.** Exercise 15; yes
41c. They are the same. **41d.** If the equation is an
identity then the two curves are identical. Otherwise, the
curves intersect only at certain points. **43.** All reals
except $0 + \dfrac{k\pi}{2}$, where k is any integer **45.** $0°$ **47.** $\dfrac{2}{3}$

49. 24.576% **51.** $\dfrac{1}{2}, \dfrac{1}{4}$ **53.** $29x \le 609;\ x \le 21$

Page 869 Chapter 14 Highlights
1. g **3.** b **5.** e **7.** i **9.** a

**Pages 870–872 Chapter 14 Study Guide and
Assessment**

11.

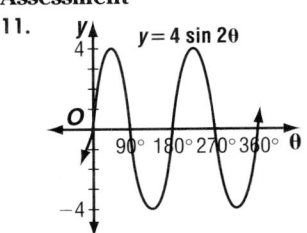

$y = 4\sin 2\theta$

13.

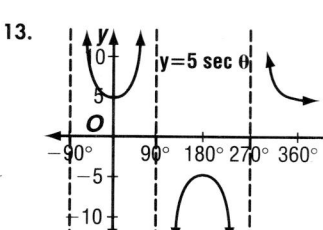

15.

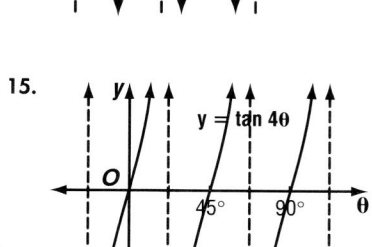

17. $\dfrac{2\sqrt{3}}{3}$

43. 0°   **45.** 15°, 75°, 195°, 255°   **47.** 30°, 210°   **49.** 65.3 miles

19. cot θ   **21.** csc α

23. $\dfrac{\sin \theta}{1 - \cos \theta} \overset{?}{=} \dfrac{1}{\sin \theta} + \dfrac{\cos \theta}{\sin \theta}$

$\dfrac{\sin \theta}{1 - \cos \theta} \overset{?}{=} \dfrac{1 + \cos \theta}{\sin \theta} \cdot \dfrac{1 - \cos \theta}{1 - \cos \theta}$

$\dfrac{\sin \theta}{1 - \cos \theta} \overset{?}{=} \dfrac{1 - \cos^2 \theta}{\sin \theta(1 - \cos \theta)}$

$\dfrac{\sin \theta}{1 - \cos \theta} \overset{?}{=} \dfrac{\sin^2 \theta}{\sin \theta(1 - \cos \theta)}$

$\dfrac{\sin \theta}{1 - \cos \theta} = \dfrac{\sin \theta}{1 - \cos \theta}$

25. $\sec x (\sec x - \cos x) \overset{?}{=} \tan^2 x$

$\sec^2 x - \sec x \cos x \overset{?}{=} \tan^2 x$

$\sec^2 x - 1 \overset{?}{=} \tan^2 x$

$\tan^2 x = \tan^2 x$

27. $\dfrac{\csc \theta + 1}{\cot \theta} \overset{?}{=} \dfrac{\cot \theta}{\csc \theta - 1} \cdot \dfrac{\csc \theta + 1}{\csc \theta + 1}$

$\dfrac{\csc \theta + 1}{\cot \theta} \overset{?}{=} \dfrac{\cot \theta (\csc \theta + 1)}{\csc^2 \theta}$

$\dfrac{\csc \theta + 1}{\cot \theta} \overset{?}{=} \dfrac{\cot \theta (\csc \theta + 1)}{\cot^2 \theta}$

$\dfrac{\csc \theta + 1}{\cot \theta} = \dfrac{\csc \theta + 1}{\cot \theta}$

29. $\dfrac{\sqrt{6} - \sqrt{2}}{4}$ **31.** $\dfrac{\sqrt{2} + \sqrt{6}}{-4}$ **33.** $-\dfrac{\sqrt{3}}{2}$

35. $\cos (180° - \alpha) = \cos 180° \cos \alpha + \sin \alpha \sin 180°$

$= -1 \cdot \cos \alpha + \sin \alpha \cdot 0$

$= -\cos \alpha$

37. $\dfrac{120}{169}, \dfrac{119}{169}, \dfrac{5\sqrt{26}}{26}, -\dfrac{\sqrt{26}}{26}$ **39.** $-\dfrac{120}{169}, \dfrac{119}{169}, \dfrac{\sqrt{26}}{26}, -\dfrac{5\sqrt{26}}{26}$

41. $\dfrac{1}{2} \sin 2x \overset{?}{=} \dfrac{\tan x}{1 + \tan^2 x}$

$\dfrac{1}{2} \sin 2x \overset{?}{=} \dfrac{\frac{\sin x}{\cos x}}{\sec^2 x}$

$\dfrac{1}{2} \sin 2x \overset{?}{=} \dfrac{\sin x}{\cos x} \cdot \cos^2 x$

$\dfrac{1}{2} \sin 2x \overset{?}{=} \dfrac{2 \sin x \cos x}{2}$

$\dfrac{1}{2} \sin 2x = \dfrac{1}{2} \sin 2x$

Cover (l)Kunio Owaki/The Stock Market, (r)Matt Meadows, (bkgd)Pete Saloutos/The Stock Market; **iii** (t)Ron Batzdorff/Universal City Studios/Shooting Star, (b)National Portrait Gallery, Smithsonian Institution/Art Resource, NY; **viii** (l)Chris Slagerman, (r)Michael Springer/Gamma Liaison; **ix** (t)John Youger, (b)Jonathon Daniel/Allsport USA; **x** (tl)Steve Lissau, (tr)Walt Disney Studios/Shooting Star, (b)UPI/Corbis-Bettmann; **xi** (t)H. Armstrong Roberts, (b)Ed Bock/The Stock Market; **xii** (t)Wayne Shields/Valan Photos, (b)Shattil/Rozinski; **xiii** (t)Arthur Tilley/FPG, (b)NASA; **xiv** (t)Giraudon/Art Resource, NY, (c)CNRI/Science Photo Library/Photo Researchers, (b)Frank Abbeloos/Gamma Liaison; **xix** (t)Tony Quinn Photography, (bl)Lee Balgeman, (br)Charles Krebs/The Stock Market; **xv** (t)S. Kermani/Gamma Liaison, (b)Deborah Davis/PhotoEdit; **xvi** (t)Buddy Mays/FPG, (b)Photofest; **xvii** (t)Bruce Byers/FPG, (c)First Image, (b)Bill Miles/The Stock Market; **4** (l)John Elk III/Stock Boston, (b)Culver Pictures **5** (tl)J.D. Cuban/Allsport USA, (tr)Chris Cole/Allsport USA, (c)H.P. Merten/The Stock Market, (bl)Don Mason/The Stock Market, (br)UPI/Corbis-Bettmann; **7** Chris Slagerman; **8** Morton & White; **9** (l)Corbis-Bettmann, (r)Frederic Stein/FPG; **10** Per Eide/The Image Bank; **12** Jose L. Pelaez/The Stock Market; **15** Morton & White; **16** Doug Martin; **17** Aaron Haupt; **18** (t)Rob Tringali Jr./Sportschrome, (b)AP/Wide World Photos; **19** Aaron Haupt; **20** Halebian/Gamma Liaison; **21** (t)Richard Hutchings/Photo Researchers, (b)Aaron Haupt; **22** (t)Craig Tuttle/The Stock Market, (b)Al Behrman/AP/Wide World Photos; **24** Corbis-Bettmann; **25** (t)Jim Patrico/David R. Frazier Photolibrary, (b)Jim Steinberg/Photo Researchers; **26** (t)David Pollack/The Stock Market, (b)Steven Gottlieb/FPG; **27** (l)Joseph Martin/Corbis-Bettmann, (r)Corbis-Bettmann; **31** Richard Pasley/Stock Boston; **32** Franco D'Ambrosio/Everett Collection; **33** (t)Michael Schneps/The Image Bank, (b)Don Smith/Sports Photo Master; **34** Rick Weber; **37** (t)David Lawrence/The Stock Market, (b)Aaron Haupt; **38** Bob Daemmrich; **39** Howard M. Paul/Emergency! Stock; **41** Suzanne Murphy-Larronde/FPG; **42** Peter Steiner/The Stock Market; **43** (t)Bob Daemmrich/Stock Boston, (b)Michael A. Keller Studios Ltd./The Stock Market; **46** Gary Bistram/The Image Bank; **48** (l)Doug Martin, (r)Thierry Cariou/The Stock Market; **49** Morton & White; **51** (t)Universal City Studios/Photofest, (b)Lucasfilm Ltd./Paramount Pictures, courtesy Kobal Collection; **52** Michael Furman/The Stock Market; **53** Marvy!/The Stock Market; **54** Jose L. Pelaez/The Stock Market; **58** Donna McLaughlin/The Stock Market; **59** Frank Moscati/The Stock Market; **60** (l)Geoff Butler, (r)UPI/Corbis-Bettmann; **61** Animals Animals/Zig Leszczynski; **62** (t)Stephen Simpson/FPG, (b)Animals Animals/Breck P. Kent; **63** (tl)Jerry Jacka Photography, (tr)Dick Dietrich/FPG, (cl)FPG, (cr)Archive Photos, (b)First Image; **64** (t)Ron Thomas/FPG, (b)courtesy University of AZ/George Kew; **65** Lou Florence/David R. Frazier Photolibrary; **66** Robert A. Jureit/The Stock Market; **70** First Image; **71** Donald C. Johnson/The Stock Market; **75** Geoff Butler; **77** (l)Carl Roessler/FPG, (r)Frank S. Balthis; **78** Richard Foreman/Everett Collection; **80** (l)Corbis-Bettmann, (r)David Madison; **81** Corbis-Bettmann; **86** (t)Ulf Sjostedt/FPG, (b)FPG; **88** Rick Weber; **90** (t)Charles E. Zirkle, (b)KS Studio; **91** Morton & White; **92** Marcus Brooke/FPG; **93** John Youger; **94** Steve Lissau; **95** NASA; **97** (t)Michael Paris/Shooting Star, (b)Everett Collection; **99** (t)Telegraph Colour Library/FPG, (b)Joseph DiCello; **100** Tom Bean/The Stock Market; **101** Arthur Tilley/FPG; **104** Morton & White; **106** Gerald French/FPG; **107** (l)Doug Martin, (r)Morton & White; **110** Geoff Butler; **112** Kenji Kerins; **113** Morton & White; **114** Vladimir Pcholkin/FPG; **118** Geoff Butler; **119** (t)Etienne De Malglaive/Liaison International, (b)Animals Animals/Raymond A. Mendez; **122** (t)ATC Productions/The Stock Market, (c)Corbis-Bettmann, (b)Ancient Art & Architecture Collection; **123** (t)courtesy William Morrow & Co., Inc., (c)Roger Tully/Tony Stone Images, (bl)Paul L. Ruben Archives, (br) Steve Allen/The Gamma Liaison Liaison Network; **126** (t)Aaron Haupt, (b)Tom Tracy/Tony Stone Images; **128** Mark D. Phillips/Photo Researchers; **132** David Cannon/Allsport USA; **133, 135** Aaron Haupt; **139** Glencoe photo; **140** Richard Hutchings; **145** John G. Davenport/Shooting Star; **146** courtesy Peter Shor; **148** NASA; **150** Ralph Morse, Life Magazine/Time Inc.; **151** Doris DeWitt/Tony Stone Images; **153** Mitch Kezar/Tony Stone Images; **154** Willard Clay/Tony Stone Images; **157** Corbis-Bettmann; **158** (l)Kenji Kerins, (r)Stephen Wade/Allsport USA; **160** Aaron Haupt; **161** Geoff Butler; **162** Ken Levine/Allsport USA; **163** Alan Carey; **165** Geoff Butler; **168** First Image; **170** Larry Hamill; **171** Jonathon Daniel/Allsport USA; **174** (t)Glencoe photo, (b)William J. Weber; **179, 180** First Image; **181** (t)Rick Weber, (b)H. Abernathy/H. Armstrong Roberts; **182** (t)Fred Ward/Black Star, (c)Geoff Butler, (b)Mak-I Photo Design; **183** (t)courtesy Judy Hehs, (c)David R.

Frazier Photolibrary, (bl)Alan Becker/The Image Bank, (br)Dennis Brack/Black Star; **186** (t)First Image, (b)courtesy Marietta College; **187** (t)Cheryl Blair, (bl)Steve Lissau, (br)Tim Courlas; **188** Ken Frick; **191** Geoff Butler; **192** courtesy Gloria P. Mondragon; **194** (t)Robert A. Jureit/The Stock Market, (b)Andrew Holbrooke/The Stock Market; **195** Mark E. Gibson; **197** Aaron Haupt; **198** Matt Meadows; **199** Geoff Butler; **200** (t)David M. Dennis, (b)Kenji Kerins; **203** Mark E. Gibson; **204** King/Zefa/H. Armstrong Roberts; **205** John M. Roberts/The Stock Market; **208** Trent Steffler/David R. Frazier Photolibrary; **210** George Holton/Photo Researchers; **211** Mark E. Gibson; **212** (t)Ron Lowery/The Stock Market, (b)National Archives; **216** courtesy Rebecca Marier; **218** Michael Simpson/FPG; **219, 222** Mark Steinmetz/Amanita Pictures; **225** Matt Meadows; **226** Aaron Haupt; **228** R. Kord/H. Armstrong Roberts; **231** Geoff Butler; **235** Matt Meadows; **236** Doug Pensinger/Allsport USA; **237** Stewart/Allsport USA; **239** (tl)H. Armstrong Roberts, (tr)courtesy Angela Wimberly, (b)Manfred Gottschalk/Tom Stack & Associates; **241** H. Armstrong Roberts; **243** (t)Jim Stratford/Black Star, (b)Steve Lissau; **244** Aaron Haupt; **252** (t)Geoff Butler, (cl)J. Snyder/The Stock Market, (cr)Scala/Art Resource, NY, (b)Culver Pictures; **253** (tl)courtesy David Bray, (tr)Aaron Haupt, (cl)Carol Halebian/Gamma Liaison, (cr)courtesy Josephine Kong Chan, (bl)Corbis-Bettmann, (br)Larry Downing/Sygma; **258** Aaron Haupt; **260** UPI/Corbis-Bettmann; **261** Mark E. Gibson; **265** Aaron Haupt; **266** David L. Perry; **267** Chisholm Rich & Associates/The Stock Market; **270** Ed Bock/The Stock Market; **272** Everett Collection; **280** Viviane Moos/The Stock Market; **281** John Paul/FSP/Gamma Liaison; **284** Tony Freeman/PhotoEdit; **286** Reuters/Corbis-Bettmann; **288** Tony Freeman/PhotoEdit; **291** Rick Weber; **294** (l)Yann Guichaouva/Allsport USA, (r)Rick Weber; **295** (t)UPI/Corbis-Bettmann, (b)Geoff Butler; **296** National Portrait Gallery, Smithsonian Institution/Art Resource, NY; **297** Everett Collection; **301** (l)Barry Seidman/The Stock Market, (r)Wernher Krutein/Gamma Liaison; **303** Larry Hamill; **305** Alan Oddie/PhotoEdit; **308** Comnet/Westlight; **309** Kaku Kurita/Gamma Liaison; **313** UPI/Corbis-Bettmann; **316** (t)Corbis-Bettmann, (b)Mehau Kulyk/Science Photo Library/Photo Researchers; **317** Tibor Bognar/The Stock Market; **318, 321** Digital Art/Westlight; **322** Corbis-Bettmann; **327** (l)Doug Martin, (r)Glencoe photo; **328** David Ball/The Stock Market; **329** (t)Peter Beck/The Stock Market, (b)David Young-Wolff/PhotoEdit; **330** (t)David Young-Wolff/PhotoEdit, (b)Morton & White; **331** (tl)Jesse Frohman, (tr)Massimo Mastrorillo/The Stock Market, (c)reprinted with permission of Texas Instruments, (bl)Presse Sports, (br)M. Jamet/Sygma; **334** (t)Ron Batzdorff/Universal City Studios/Shooting Star, (b)Raven/Explorer/Photo Researchers; **337** Doug Martin; **341** (t)Simon Bruty/Allsport USA, (b)Richard Martin/Allsport USA; **342** ©1995 Sotheby's, Inc.; **345** J. Barry O'Rourke/The Stock Market; **346** (l)Lee Snider, (r)Rick Weber; **349** Mark E. Gibson; **350** Aaron Haupt; **352** Vandystadt/Allsport USA; **353** Wes Thompson/The Stock Market; **355** Mark E. Gibson; **358** courtesy Bryan Starr; **359** Warren Morgan/Westlight; **361** Dave Lawrence/The Stock Market; **364** Kennon Cooke/Valan Photos; **367** Maggie Steber; **368** J. Pat Carter/Gamma Liaison; **370** Buddy Mays/FPG; **371** Bill Ross/Westlight; **374** The Kobal Collection; **377** First Image; **378** E.H. Wallop/The Stock Market; **382** Bob Mullenix; **383** T. Kitchin/Tom Stack & Associates; **384** J. Barry O'Rourke/The Stock Market; **386** Tim Defrisco/Allsport USA; **387** Wayne Shiels/Valan Photos; **388** C.W. Norris/Valan Photos; **389** Geoff Butler; **390** John Eastcott & Yva Momatiuk/Valan Photos; **391** Aaron Haupt; **392** (l)Mario Elie Dunks/Allsport USA, (r)Al Bello/Allsport USA; **393** Myrleen Ferguson/PhotoEdit; **394** Richard Dunoff/The Stock Market; **395** Chuck Savage/The Stock Market; **396** (l)Aaron Haupt, (r)Bob Daemmrich; **397** Joe Towers/The Stock Market; **398** Doug Martin; **403** Rick Weber; **406** (t)Jon Bradley/Tony Stone Images, (c)Glencoe photo, (b)Corbis-Bettmann; **407** (tl)National Geographic Society Image Collection, (tr)Everett Collection, (cl)Morton & White, (cr)National Portrait Gallery, Washington D.C./Art Resource, NY, (b)Corbis-Bettmann; **408** Paul Vandevelder/Gamma Liaison; **412** Mark E. Gibson; **413** James Blank/FPG; **414** Bob Newland/courtesy Belle Fourche Chamber of Commerce; **415** David R. Frazier Photolibrary; **418** (l)Hank Morgan/Science Source/Photo Researchers, (r)Arthur Tilley/FPG; **419** Morton & White; **421** (t)C. Moore/Westlight, (b)Tom Tracy/FPG; **423** (t)Robert Landau/Westlight, (b)Alex Quesada/Matrix; **424** (l)Lara Jo Regan/SABA, (r)Ron Watts/Westlight; **427** Shattil/Rozinski; **428** Ted Rice; **429** Mary Lou Uttermohlen; **431** Peter Pearson/Tony Stone Images; **434** NASA/Science Source/Photo Researchers; **438** Jeff Kaufman/FPG; **440** R. Stockton/H. Armstrong Roberts; **447** David W. Hamilton/The Image Bank; **449** First Image;

450 Herman Kokojan/Black Star; **451** Dallas & John Heaton/Westlight; **452** A. Tovy/H. Armstrong Roberts; **454** (t)David Austen/FPG, (b)First Image; **455** Doug Martin; **460, 463** Geoff Butler; **466** (l)Photofest, (r)Gamma Liaison; **467** Francois Gohier/Photo Researchers; **468** Steven Gottlieb/FPG; **472** David Lissy/FPG; **473** (l)Doug Martin, (r)Harald Sund/The Image Bank; **476** (t)Aaron Haupt, (cl)First Image, (cr)Earl Young/FPG, (b)Jerry Cooke/Photo Researchers; **477** (tl)Andy Freeberg/Time Magazine, (tr)Rick Weber, (c)Bridgeman/Art Resource, NY, (b)courtesy The National Inventors Hall of Fame; **478** Bill Watterson/Universal Press Syndicate; **479** Wolfpro; **482** Dallas & John Heaton/Westlight; **484** (t)E.R. Degginger/Photo Researchers, (c)Prisma/Westlight, (b)First Image; **485** David Stoecklein/F-Stock; **488** George Haling/Photo Researchers; **490** Vic Bider/PhotoEdit; **493** Morton & White; **495** Alfred Gescheidt/The Image Bank; **497** Glencoe photo; **498** Dick Luria/Science Source/Photo Researchers; **500** Vicki Silbert/PhotoEdit; **501** Ben Simmons/The Stock Market; **502** (t)Bill Bachman/PhotoEdit, (b)First Image; **507** Geoff Butler; **509** (t)Jose Fuste Rage/The Stock Market, (b)AP/Wide World Photos; **514** (t)Geoff Butler, (b)Mak-I Photo Design; **516** J. Messerschmidt/The Stock Market; **519** (t)NASA/The Stock Market, (b)Thomas Del Brase/The Stock Market; **520** CNRI/Science Photo Library/Photo Researchers; **525** (t)Kenji Kerins, (bl)Morton & White, (br)NASA; **528** David Ball/The Stock Market; **532** Tim Courlas; **534** Aaron Haupt; **535** Comstock, Inc.; **546** (t)David Young-Wolff/PhotoEdit, (b)Culver Pictures; **547** (tl)James A. Sugar, (tr)Jeff Greenberg/PhotoEdit, (cl)courtesy Motorcycle Heritage Museum, (cr)Roland Hiltscher Photography, (b)NASA; **550** Corbis-Bettmann; **552** Mark Steinmetz/Amanita Pictures; **554** Doug Martin; **556** David R. Frazier Photolibrary; **557** Bob Jones/Gamma Liaison; **559** Bonnie Kamin/PhotoEdit; **560** Steven M. Barnett/Gamma Liaison: **561** David R. Frazier Photolibrary; **567** (t)Aaron Haupt, (b)Jon Feingersh/The Stock Market; **569** First Image; **570** KS Studio; **572** from *Men of Mark: Eminent, Progressive and Rising*, Geo. M. Rewell & Co. 1887; **573** Frank Abbeloos/Gamma Liaison; **574** Morton & White; **576** (t)Tom McCarthy/PhotoEdit, (b)Mike Powell/Allsport USA; **579** (t)Raphael Gaillarde/Gamma Liaison, (b)David Noble/FPG; **580, 581, 582** David R. Frazier Photolibrary; **583** Robert Brenner/PhotoEdit; **584** Doug Martin; **588** Steve Grohe/The Stock Market; **589** Morton & White; **590** Dave Nagel/Gamma Liaison; **591** First Image; **592** (t)Rick Reinhard/Impact Visuals, (c)Eric Sanford/F-Stock, (b)Fujifotos/The Image Works; **593** (t)Lori Grinker/Contact Press Images, (cl)NE State Historical Society, (cr)David Young-Wolff/PhotoEdit, (bl br)First Image; **596** Warner Bros./Everett Collection; **598** (t)David R. Frazier Photolibrary, (b)Stephen Green-Armytage/The Stock Market; **600** Oliver Meckes/Ottawa/Photo Researchers; **601** Harry Engels/Photo Researchers; **602** Dave Brown/The Stock Market; **603** First Image; **604** Geopress/H. Armstrong Roberts; **605** S. Kermani/Gamma Liaison; **606** Corbis-Bettmann; **608** First Image; **609** Alon Reininger/Contact Press/The Stock Market; **610** Corbis-Bettmann; **611** Morton & White; **612** David R. Frazier Photolibrary; **614** Morton & White; **615** (l)Fisk University Library, (r)David R. Frazier Photolibrary; **617** (t)William Stevens/Gamma Liaison, (bl)David R. Frazier Photolibrary, (br)Jeff Zaruba/The Stock Market; **619** Fotex/H. Kuehn/Shooting Star; **620** (t)David Nunuk/Science Photo Library/Photo Researchers, (b)Aaron Haupt; **622** First Image; **624** (t)Rick Weber, (b)Kevin Galvin/The Stock Market; **625, 626** First Image; **627** David R. Frazier Photolibrary; **631** (t)David R. Frazier Photolibrary, (c)Mark E. Gibson/The Stock Market, (b)FPG; **632** Robin Prange/The Stock Market; **633** David R. Frazier Photolibrary; **634** (l)Mike Penney/David R. Frazier Photolibrary, (r)First Image; **635** Rick Weber; **636** (l)Joe Mahoney/AP/Wide World Photos, (r)Lee Balterman/FPG International; **641** FPG; **644** (t)First Image, (c)Telegraph Colour Library/FPG, (b)Erich Lessing/Art Resource, NY; **645** (tl)Lou Coopey, (tr)First Image, (cl)First Image, (cr)Evan Agostini/Gamma Liaison, (bl)Painting by Betsy Graves/Corbis-Bettmann, (b)Aaron Haupt; **648** Morton & White; **649** Aaron Haupt; **650** (l)courtesy WSBC Radio, (r)Aaron Haupt; **652** John Kelly/The Image Bank; **654** Kenji Kerins; **656** Richard Hutchings/PhotoEdit; **658** Charles O'Rear/Westlight; **661** Aaron Haupt; **662** The Walt Disney Company/Everett Collection; **667** Morton & White from *Rumpelstiltskin* by Jacob & Wilhelm Grimm, illustrated by Bernadette Watts, published by North-South Books Inc., New York. ©1993 by NordSud Verlag, Gossau Zurich, Switzerland. **668** (l)Kunio Owaki/The Stock Market, (r)Ed Bock/The Stock Market; **670** Lawrence Migdale; **671** Chris Jones/The Stock Market; **673** Bob Daemmrich; **675** Don Mason/The Stock Market; **677** Morton & White; **678** Larry Hamill; **680** Nick Nicholson/The Image Bank; **681** (t)Globus Brothers/The Stock Market, (b)Kenji Kerins; **683** Grant Faint/The Image Bank; **685** Jon Feingersh/The Stock Market; **686** Morton & White; **687** Robert W. Ginn/PhotoEdit; **688** (l)Marc Romanelli/The Image Bank, (c)Animals Animals/Ed Wolff, (r)First Image; **690** Zefa-Sauer/The Stock Market; **692** Craig Hammell/The Stock Market; **693**

(t)Geospace/Science Photo Library/Photo Researchers, (b)Cordon Art B.V.; **694** Deborah Davis/PhotoEdit; **695** Gianni Cigolini/The Image Bank; **698** Aaron Haupt; **700** David W. Hamilton/The Image Bank; **701** Thomas Zimmerman BFF/FPG; **702** Blaine Harrington III/The Stock Market; **707** William Weber; **710** (t)Rod Joslin, (b)The Science Museum/Science & Society Picture Library; **711** (tl)courtesy Jim McHugh, (tr)Scott Cunningham, (bl)Corbis-Bettmann, (br)James D. Wilson/Gamma Liaison; **712** FPG; **713** (l)Nino Mascardi/The Image Bank, (r)First Image; **714** PEANUTS reprinted by permission of United Feature Syndicate, Inc.; **715** (l)Bob Daemmrich, (r)StudiOhio; **716** Latent Image; **718** (t)Bruce Byers/FPG, (bl)McBroom/Tristar Pictures/Shooting Star, (br)Walt Disney Studios/Shooting Star; **719** Geoff Butler; **722** Aaron Haupt; **723** (t)Morton & White, (b)Geoff Butler; **724** Geoff Butler; **726** First Image; **727** Geoff Butler; **728** Aaron Haupt; **729** (t)William D. Popejoy, (b)James N. Westwater; **730** Geoff Butler; **731** David R. Frazier Photolibrary; **732** Ron Watts/Westlight; **733** Aaron Haupt; **734** First Image; **735** John Gajda/FPG; **736** First Image; **738** Geoff Butler; **739** Aaron Haupt; **740** Geoff Butler; **741** Gail Meese/Meese Photo Research; **742** (t)Geoff Butler, (b)Aaron Haupt; **743** (t)Geoff Butler, (b)Frederick McKinney/FPG; **744** courtesy Denis Stone; **746** THE BORN LOSER reprinted by permission of Newspaper Enterprise Association, Inc.; **747** Ken Frick; **748** Morton & White; **749** Bob Daemmrich; **750** Otto Greule/Allsport USA; **751** Billy Barnes/FPG; **752** AP/Wide World Photos; **754** (t)Kunio Owaki/The Stock Market, **756** Greg Davis/The Stock Market; **757** Canadian Museum of Civilization; **760, 761** Aaron Haupt; **762** courtesy Gary Locke; **767** Mak-I Photo Design; **768** courtesy Kennywood; **769** (l)courtesy Six Flags Amusement Park, (r)Cedar Point Photo by Dan Feicht; **770** (t)Mary Lou Uttermohlen, (c)J. Ciganovic/FPG, (b)Mercury Archives/The Image Bank; **771** (tl)courtesy Anna Maria Padilla, (tr)Ted Horowitz/The Stock Market, (bl)Corbis-Bettmann, (br)Everett Collection; **772** (t)Photofest, (b)Morton & White; **776** Kunio Owaki/The Stock Market; **777** Cralle/The Image Bank; **779** Kenneth Redding/The Image Bank; **780** Frank L. Psaute; **781** Terje Rakke/The Image Bank; **784** Morton & White; **786** Geoffrey Clifford/The Stock Market; **789** Francisco Hidalgo/The Image Bank; **790** Mik Miles/The Stock Market; **791** Jim Erickson/The Stock Market; **792** M.L. Sinibaldi/The Stock Market; **793** J.B. Diedrich/The Stock Market; **797** James Blank/The Stock Market; **798** David Ball/The Stock Market; **799** (l)Universal City Studios/Photofest, (r)Universal City Studios/The Kobal Collection; **801** (t)The Natural History Museum, London, (b)Tom Bean/The Stock Market; **802** Alan Carey; **804** (t)Kenji Kerins, (b)Anne van der Vaeren/The Image Bank; **805** Morton & White; **808** Bob Daemmrich; **810** Stephen Marks/The Stock Market; **811** Tom Craig/FPG; **814** KS Studio; **819** Joseph DiChello; **822** (t)Veryl Oakland/Retna, (bl)Alan Hipwell/FPG, (br)First Image; **823** (tl)Mark Sennet/Onyx, (tr)Morton & White, (cl)Tretiakov Gallery, Moscow/A. Burkatousky/SuperStock, (cr)John Harrington/Black Star, (b)courtesy NE Historical Society **826** Mercury Archives/The Image Bank; **829** (l)Morton & White, (r)Peter Aprahamian/Science Photo Library/Photo Researchers; **832** Kenji Kerins; **833** (t)Ralph A. Clevenger/Westlight, (b)Alvin E. Staffan; **835** Gerard Vandystadt/Allsport USA; **838** Dallas & John Heaton/Westlight; **840** (t)Pete Saloutos/The Stock Market, (b)Charles Krebs/The Stock Market; **842** Aaron Haupt; **845** First Image; **846** Tony Quinn Photography; **847** Paul McCormick/The Image Bank; **849** H. Richard Johnston/FPG; **851** (l)David Noble/FPG, (r)Paul McCormick/The Image Bank; **852** (t)Larry Lee/Westlight, (b)Kenji Kerins; **853** Corbis-Bettmann; **856** Hank Delespin/The Image Bank; **858** (l)UPI/Corbis-Bettmann, (r)Rick Gayle/The Stock Market; **861** Aaron Haupt; **863** Steve Proehl/The Image Bank; **864** Telegraph Colour Library/FPG; **865** Zefa-Kalt/The Stock Market; **866** Aaron Haupt; **867** Morton & White; **868** Lee Balgeman; **875** (tl)Kennon Cooke/Valan Photos, (tr)David Lawrence/The Stock Market, (bl)Zefa-Kalt/The Stock Market, (br)Larry Hamill.

A

Absolute value, 37–39
  equations, 36–42
  functions, 104–105, 111
  inequalities, 49–54, 149
  parent graph, 105

Addition
  of complex numbers, 312, 314
  of matrices, 194–198
  of probabilities, 746–751
  of radicals, 290, 293
  of rational expressions, 569–574
  of roots of quadratic equations, 359–364

Addition properties
  of equality, 28
  for inequalities, 43

Additive inverse property, 14–15

Agnesi, Maria Gaetana, 550

Algebra, fundamental theorem of, 505

Algebraic expressions, 8–9, 15, 16, 27, 31
  rational, 562–575

Algebraic methods
  for solving systems of equations, 133–140
  for solving systems of quadratic equations, 461, 462

Alternative Assessment, 59, 119, 179, 249, 327, 403, 473, 543, 589, 641, 707, 767, 819

Ambiguous case, 795

Amplitude, 828–832

Angles
  coterminal, 783
  of depression, 776
  double–angle formulas, 853–854, 856–859
  of elevation, 776
  formulas for sum and differences of, 847–852
  general, trigonometric functions of, 786–791
  half-angle formulas, 855–859
  of incidence, 786, 789
  of incline, 835, 838, 840
  initial side of, 780
  measure of, 780–785
  measuring, 528
  quadrantal, 787
  of quadrilaterals, 265
  of refraction, 786, 789
  standard position, 780, 787
  step, 799, 801, 802
  supplementary, 137
  terminal side of, 780
  of triangles, 230, 793–795
  vertices of, 131, 144

Angular velocity, 784

Antilogarithms, 618, 622

Applications. *See* Applications Index

Approximating zeros, 497

Arccosine function, 811

Arcsine function, 812

Arctangent function, 812

Area, 327
  of parallelogram, 264, 265, 560, 562, 564, 567
  of polygon, 210
  of rectangle, 10, 18, 262, 265, 293, 345, 379, 562, 565, 642, 643, 682
  of square, 279, 340, 346, 358
  surface, 12, 308, 495
  of trapezoid, 11, 18, 146, 558, 636
  of triangle, 11, 12, 207–208, 209, 265, 268, 336, 560, 566, 792

Area diagrams, 739, 742

Arithmetic means, 650

Arithmetic sequences, 646–655, 659, 660
  common difference, 649
  defined, 649
  formula for nth term of, 649–651

Arithmetic series, 656–661

Aryabhata, 531

Assessment
  Alternative, 59, 119, 179, 249, 327, 403, 473, 543, 589, 641, 707, 767, 819
  Chapter Tests, 912–925
  College Entrance Exams, 120–121, 250–251, 404–405, 544–545, 642–643, 708–709, 820–821
  Math Journal, 23, 52, 76, 84, 91, 129, 150, 162, 190, 196, 229, 270, 284, 300, 357, 362, 381, 395, 426, 453, 464, 482, 490, 508, 526, 572, 581, 614, 619, 666, 680, 715, 729, 735, 748, 777, 808, 832, 864
  Portfolio, 59, 119, 179, 249, 322, 327, 403, 468, 473, 543, 584, 589, 641, 702, 707, 767, 819
  Self Evaluation, 59, 119, 179, 249, 327, 403, 473, 543, 589, 641, 707, 767, 819
  Self Tests, 34, 87, 152, 211, 287, 358, 439, 510, 568, 621, 675, 738, 846
  Study Guide and Assessment, 56–59, 116–119, 176–179, 246–249, 324–327, 400–403, 470–473, 540–543, 586–589, 638–641, 704–707, 764–767, 816–819

Associative property, 14, 311

Asymptotes, 441, 550–551

Augmented matrices, 226–234

Average(s), income, 15–16

Average error, 273

Axis
  conjugate, 441
  of ellipse, 432
  of hyperbola, 441
  major, 432
  minor, 432
  of symmetry, 335, 368, 369, 416
  transverse, 441
  x-, 64
  y-, 64

B

Back-to-back stem-and-leaf plot, 20–21, 71

Bar graphs, 19, 403

Base formula, change of, 628

Bell curve, 393

Best-fit line, 95, 97, 682

Binomial(s)
  defined, 261
  multiplying, 262–263, 264, 276

Binomial experiments, 752–757

Binomial theorem, 695–701

Boundary, 110–111, 378

Box-and-whisker plots, 235–244, 260, 447

Braces {}, 8

Brackets [], 8

Break-even point, 128

C

Calculators
  exponents, 301
  trigonometric functions, 774–775, 788, 803, 812–813
  See also Graphing calculators; Scientific calculators

Career Choices
  aircraft mechanic, 408
  architect, 812
  astronaut, 335
  astronomer, 847
  banking, 228
  computer scientist, 663
  electrical engineer, 321
  electrician, 570
  financial planner, 103
  industrial designer, 495
  meteorologist, 746
  occupations in demand, 122
  physician, 596
  statistician, 19
  stockbroker, 170

Cartesian coordinate plane, 64

Celsius temperature scale, 10, 523, 525, 536, 537

Center
  of circle, 410–411, 423
  of ellipse, 432
  of hyperbola, 441

Central tendency, measures of, 19, 21–26, 33, 34, 42, 48, 71, 87, 94, 193, 266, 439, 687

Change of base formula, 628

Chaos theory, 690–694

Chapter Projects
  AIDS research, 593